Sp	span (Definition 2.4.3)
σ	standard deviation (Definition 4.2.7). Also denotes permutation, e.g.
$\sum$	sum (Section 0.1)
sup	supremum; least upper bound (Definitions 0.4.1, 1.6.4)
$\mathrm{Supp}(f)$	support of a function (Definition 4.1.2)
τ	(tau) torsion (Definition 3.8.11)
$T_{\mathbf{a}}C$	tangent space to curve (Definition 3.1.8)
$T_{\mathbf{x}}X$	tangent space to manifold (Definition 3.2.6)
tr	trace of a matrix (Definition 4.8.20)
$\vec{\mathbf{v}} \cdot \vec{\mathbf{w}}$	dot product of two vectors, (Definition 1.4.1)
$\vec{\mathbf{v}} \times \vec{\mathbf{w}}$	cross product of two vectors (Definition 1.4.17)
$|\vec{\mathbf{v}}|$	length of vector $\vec{\mathbf{v}}$ (Definition 1.4.2)

Notation particular to this book

$B_r(\mathbf{x})$	ball of radius r around $\mathbf{x}$ (Definition 1.5.2)
$\mathcal{D}_N(\mathbb{R}^n)$	dyadic paving (Definition 4.1.6)
$[\mathbf{Df}(\mathbf{a})]$	derivative of $\mathbf{f}$ at $\mathbf{a}$ (Definition 1.7.11)
$\mathbb{D}$	set of finite decimals. (Definition 0.4.4)
$\Phi_{\vec{F}}$	flux form (Definition 6.4.2)
$\Phi_{\{\underline{\mathbf{v}}\}}$	concrete to abstract function (Definition 2.6.12)
$\mathcal{I}_n^k$	set of multi-exponents (Definition 3.3.6)
I-integrable	improperly integrable (Definition 4.11.2)
$[\mathbf{Jf}(\mathbf{a})]$	Jacobian matrix (Definition 1.7.8)
$L(f)$	lower integral (Definition 4.1.8)
$m_A(f)$	infimum of $f(\mathbf{x})$ for $\mathbf{x} \in A$ (Definition 4.1.3)
$M_A(f)$	supremum of $f(\mathbf{x})$ for $\mathbf{x} \in A$ (Definition 4.1.3)
$\mathrm{Mat}\,(n, m)$	space of $n \times m$ matrices
$|d^n\mathbf{x}|$	integrand for multiple integral (Section 4.1)
$\mathrm{osc}_A(f)$	oscillation of f over A (Definition 4.1.4)
$P_{f,\mathbf{a}}^k$	Taylor polynomial (Definition 3.3.15)
$P_{\mathbf{x}}(\vec{\mathbf{v}}_1, \ldots \vec{\mathbf{v}}_k)$	k-parallelogram (Definition 5.1.1)
$P_{\mathbf{x}}^o(\vec{\mathbf{v}}_1, \ldots \vec{\mathbf{v}}_k)$	oriented k-parallelogram (Definition 6.3.1)
$\overline{\mathbb{R}}$	$\mathbb{R} \cup \{\pm\infty\}$: union of $\mathbb{R}$ and $\pm\infty$ (Section 4.11)
ρ_f	density form (Definition 6.4.3)
U^{ok}, V^{ok}	(Theorem 5.2.8)
$U(f)$	upper integral (Definition 4.1.8)
$\vec{\mathbf{v}}$	column vector (Definition 1.1.2)
$\underline{\mathbf{v}}$	element of abstract vector space (Section 2.6)
vol_n	n-dimensional volume (Definition 4.1.13)
$W_{\vec{F}}$	work form (Definition 6.4.1)
$\dot{x}, \dot{y}$	used in denoting tangent space (Figure 3.1.6)

Vector Calculus, Linear Algebra,
And Differential Forms

A Unified Approach

Vector Calculus, Linear Algebra, And Differential Forms

A Unified Approach

John Hamal Hubbard
Cornell University

Barbara Burke Hubbard

PRENTICE HALL
Upper Saddle River, New Jersey 07458

Library of Congress Cataloging–in–Publication Data

Hubbard, John Hamal
    Vector calculus, linear algebra, and differential forms: a
unified approach   /   John H. Hubbard and Barbara B. Hubbard.

      p.   cm.
   Includes index.
   ISBN 0–13–657446–7
   1. Calculus. 2. Algebra, linear.   I. Hubbard, Barbara Burke.
   II. Title.
QA303.H79   1999
515-dc21   98–35966
                                    CIP

Editoral director, Tim Bozik
Editor-in-chief, Jerome Grant
Acquisition editor, George Lobell
Executive managing editor, Kathleen Schiaparelli
Managing editor, Linda Mihatov Behrens
Editorial assistants, Gale Epps, Nancy Bauer
Assistant VP production/manufacturing, David W. Riccardi
Manufacturing manager, Trudy Pisciotti
Manufacturing buyer, Alan Fischer
Creative director, Paula Maylahn
Art director, Jayne Conte
Marketing manager, Melody Marcus
Cover design, Kiwi Design

© 1999 by Prentice-Hall, Inc.
Simon & Schuster / A Viacom Company
Upper Saddle River, New Jersey 07458

Printed in the United States of America
10 9 8 7 6 5 4 3

ISBN 0–13–657446–7

Prentice-Hall International (UK) Limited, *London*
Prentice-Hall of Australia Pty. Limited, *Sidney*
Prentice-Hall Canada, Inc., *Toronto*
Prentice-Hall Hispanoamericana, S.A., *Mexico*
Prentice-Hall of India Private Limited, *New Delhi*
Prentice-Hall of Japan, Inc., *Tokyo*
Simon & Schuster Asia Pte. Ltd., *Singapore*
Editora Prentice-Hall do Brasil, Ltda., *Rio de Janeiro*

A Adrien et Régine Douady, pour l'inspiration d'une vie entière

Contents

Preface

> *... The numerical interpretation ... is however necessary. ... So long as it is not obtained, the solutions may be said to remain incomplete and useless, and the truth which it is proposed to discover is no less hidden in the formulae of analysis than it was in the physical problem itself.*
> –Joseph Fourier, *The Analytic Theory of Heat*

This book covers most of the standard topics in multivariate calculus, and a substantial part of a standard first course in linear algebra. The teacher may find the organization rather less standard.

There are three guiding principles which led to our organizing the material as we did. One is that at this level linear algebra should be more a convenient setting and language for multivariate calculus than a subject in its own right. We begin most chapters with a treatment of a topic in linear algebra and then show how the methods apply to corresponding nonlinear problems. In each chapter, enough linear algebra is developed to provide the tools we need in teaching multivariate calculus (in fact, somewhat more: the spectral theorem for symmetric matrices is proved in Section 3.7). We discuss abstract vector spaces in Section 2.6, but the emphasis is on $\mathbb{R}^n$, as we believe that most students find it easiest to move from the concrete to the abstract.

Another guiding principle is that one should emphasize computationally effective algorithms, and prove theorems by showing that those algorithms really work: to marry theory and applications by using practical algorithms as theoretical tools. We feel this better reflects the way this mathematics is used today, in both applied and in pure mathematics. Moreover, it can be done with no loss of rigor.

For linear equations, row reduction (the practical algorithm) is the central tool from which everything else follows, and we use row reduction to prove all the standard results about dimension and rank. For nonlinear equations, the cornerstone is Newton's method, the best and most widely used method for solving nonlinear equations. We use Newton's method both as a computational tool and as the basis for proving the inverse and implicit function theorem, rather than basing those proofs on Picard iteration, which converges too slowly to be of practical interest.

In keeping with our emphasis on computations, we include a section on numerical methods of integration, and we encourage the use of computers both to reduce tedious calculations (row reduction in particular) and as an aid in visualizing curves and surfaces. We have also included a section on probability and integrals, as this seems to us too important a use of integration to be ignored.

A third principle is that differential forms are the right way to approach the various forms of Stokes's theorem. We say this with some trepidation, especially after some of our most distinguished colleagues told us they had never really understood what differential forms were about. We believe that differential forms can be taught to freshmen and sophomores, if forms are presented geometrically, as integrands that take an oriented piece of a curve, surface, or manifold, and return a number. We are aware that students taking courses in other fields need to master the language of vector calculus, and we devote three sections of Chapter 6 to integrating the standard vector calculus into the language of forms.

The great conceptual simplifications gained by doing electromagnetism in the language of forms is a central motivation for using forms, and we will apply the language of forms to electromagnetism in a subsequent volume.

Although most long proofs have been put in Appendix A, we made an exception for the material in Section 1.6. These theorems in topology are often not taught, but we feel we would be doing the beginning student a disservice not to include them, particularly the mean value theorem and the theorems concerning convergent subsequences in compact sets and the existence of minima and maxima of functions. In our experience, students do not find this material particularly hard, and systematically avoiding it leaves them with an uneasy feeling that the foundations of the subject are shaky.

Jean Dieudonné, for many years a leader of Bourbaki, is the very personification of rigor in mathematics. In his book *Infinitesimal Calculus*, he put the harder proofs in small type, saying " ... a beginner will do well to accept plausible results without taxing his mind with subtle proofs ...."

Following this philosophy, we have put many of the more difficult proofs in the appendix, and feel that for a first course, these proofs should be omitted. Students should learn how to drive before they learn how to take the car apart.

Different ways to use the book

This book can be used either as a textbook in multivariate calculus or as an accessible textbook for a course in analysis.

We see calculus as analogous to learning how to drive, while analysis is analogous to learning how and why a car works. To use this book to "learn how to drive," the proofs in Appendix A should be omitted. To use it to "learn how a car works," the emphasis should be on those proofs. For most students, this will be best attempted when they already have some familiarity with the material in the main text.

Students who have studied first year calculus only

(1) For a one-semester course taken by students who have studied neither linear algebra nor multivariate calculus, we suggest covering only the first four

chapters, omitting the sections marked "optional," which, in the analogy of learning to drive rather than learning how a car is built, correspond rather to learning how to drive on ice. (These sections include the part of Section 2.8 concerning a stronger version of the Kantorovitch theorem, and Section 4.4 on measure 0). Other topics that can be omitted in a first course include the proof of the fundamental theorem of algebra in Section 1.6, the discussion of criteria for differentiability in Section 1.9, Section 3.2 on manifolds, and Section 3.8 on the geometry of curves and surfaces. (In our experience, beginning students do have trouble with the proof of the fundamental theorem of algebra, while manifolds do not pose much of a problem.)

(2) The entire book could also be used for a full year's course. This could be done at different levels of difficulty, depending on the students' sophistication and the pace of the class. Some students may need to review the material in Sections 0.3 and 0.5; others may be able to include some of the proofs in the appendix, such as those of the central limit theorem and the Kantorovitch theorem.

(3) With a year at one's disposal (and excluding the proofs in the appendix), one could also cover more than the present material, and a second volume is planned, covering

> differential equations;
> abstract vector spaces, inner product spaces, and Fourier series;
> eigenvalues, eigenvectors, and differential equations;
> applications of differential forms;
> electromagnetism.

We favor this third approach; in particular, we feel that the last two topics above are of central importance. Indeed, we feel that three semesters would not be too much to devote to linear algebra, multivariate calculus, differential forms, differential equations, and an introduction to Fourier series and partial differential equations. This is more or less what the engineering and physics departments expect students to learn in second year calculus, although we feel this is unrealistic.

Students who have studied some linear algebra or multivariate calculus

The book can also be used for students who have some exposure to either linear algebra or multivariate calculus, but who are not ready for a course in analysis. We used an earlier version of this text with students who had taken a course in linear algebra, and feel they gained a great deal from seeing how linear algebra and multivariate calculus mesh. Such students could be expected to cover Chapters 1–6, possibly omitting some of the optional material discussed above. For a less fast-paced course, the book could also be covered in an entire year, possibly including some proofs from the appendix.

We view Chapter 0 primarily as a resource for students, rather than as part of the material to be covered in class. An exception is Section 0.4, which might well be covered in a class on analysis.

Mathematical notation is not always uniform. For example, $|A|$ can mean the length of a matrix A (the meaning in this book) or it can mean the determinant of A. Different notations for partial derivatives also exist. This should not pose a problem for readers who begin at the beginning and end at the end, but for those who are using only selected chapters, it could be confusing. Notations used in the book are listed on the front inside cover, along with an indication of where they are first introduced.

Students ready for a course in analysis

If the book is used as a text for an analysis course, then in one semester one could hope to cover all six chapters and some or most of the proofs in Appendix A. This could be done at varying levels of difficulty; students might be expected to follow the proofs, for example, or they might be expected to understand them well enough to construct similar proofs. Several exercises in Appendix A and in Section 0.4 are of this nature.

Numbering of theorems, examples, and equations

Theorems, lemmas, propositions, corollaries, and examples share the same numbering system. For example, Proposition 2.3.8 is not the eighth proposition of Section 2.3; it is the eighth numbered item of that section. We often refer back to theorems, examples, and so on, and hope this numbering will make them easier to find.

Figures are numbered independently; Figure 3.2.3 is the third figure of Section 3.2. All displayed equations are numbered, with the numbers given at right; Equation 4.2.3 is the third equation of Section 4.2. When an equation is displayed a second time, it keeps its original number, but the number is in parentheses.

We use the symbol $\triangle$ to mark the end of an example or remark, and the symbol $\square$ to mark the end of a proof.

Exercises

Exercises are given at the end of each chapter, grouped by section. They range from very easy exercises intended to make the student familiar with vocabulary, to quite difficult exercises. The hardest exercises are marked with a star (or, in rare cases, two stars). On occasion, figures and equations are numbered in the exercises. In this case, they are given the number of the exercise to which they pertain.

In addition, there are occasional "mini-exercises" incorporated in the text, with answers given in footnotes. These are straightforward questions containing no tricks or subtleties, and are intended to let the student test his or her understanding (or be reassured that he or she has understood). We hope that even the student who finds them too easy will answer them; working with pen and paper helps vocabulary and techniques sink in.

Web page

Errata will be posted on the web page

http://www.math.cornell.edu/~hubbard/vectorcalculus.html

The three programs given in Appendix B are also available there. We plan to expand the web page, making the programs available on more platforms, and adding new programs and examples of their uses.

Readers are encouraged to write the authors at jhh8@cornell.edu to signal errors, or to suggest new exercises, which will then be shared with other readers via the web page.

Acknowledgments

Many people contributed to this book. We would in particular like to express our gratitude to Robert Terrell of Cornell University, for his many invaluable suggestions, including ideas for examples and exercises, and advice on notation; Adrien Douady of the University of Paris at Orsay, whose insights shaped our presentation of integrals in Chapter 4; and Régine Douady of the University of Paris-VII, who inspired us to distinguish between points and vectors. We would also like to thank Allen Back, Harriet Hubbard, Peter Papadopol, Birgit Speh, and Vladimir Veselov, for their many contributions.

Cornell undergraduates in Math 221, 223, and 224 showed exemplary patience in dealing with the inevitable shortcomings of an unfinished text in photocopied form. They pointed out numerous errors, and they made the course a pleasure to teach. We would like to thank in particular Allegra Angus, Daniel Bauer, Vadim Grinshpun, Michael Henderson, Tomohiko Ishigami, Victor Kam, Paul Kautz, Kevin Knox, Mikhail Kobyakov, Jie Li, Surachate Limkumnerd, Mon-Jed Liu, Karl Papadantonakis, Marc Ratkovic, Susan Rushmer, Samuel Scarano, Warren Scott, Timothy Slate, and Chan-ho Suh. Another Cornell student, Jon Rosenberger, produced the *Newton* program in Appendix B.1. Karl Papadantonakis helped produce the picture used on the cover.

For insights concerning the history of linear algebra, we are indebted to the essay by J.-L. Dorier in *L'Enseignement de l'algèbre linéaire en question*. Other books that were influential include *Infinitesimal Calculus* by Jean Dieudonné, *Advanced Calculus* by Lynn Loomis and Shlomo Sternberg, *Calculus on Manifolds* by Michael Spivak, and *Mathematical Methods of Classical Mechanics* by Vladimir Arnold.

Ben Salzberg of Blue Sky Research saved us from despair when a new computer refused to print files in Textures. Barbara Beeton of American Mathematical Society's Technical Support gave prompt and helpful answers to technical questions.

We would also like to thank George Lobell at Prentice Hall, who encouraged us to write this book; Nicholas Romanelli for his editing and advice, and Gale Epps, as well as the mathematicians who served as reviewers for Prentice-Hall and made many helpful suggestions and criticisms: Robert Boyer, Drexel University; Ashwani Kapila, Rensselaer Polytechnic Institute; Krystyna Kuperberg, Auburn University; Ralph Oberste-Vorth, University of South Florida; and Ernest Stitzinger, North Carolina State University.

Many people contributed to making this a better book, by alerting us to errors in the first two printings. We are grateful to Adam Barth, Nils Barth, Ryan Budney, Walter Chang, Robin Chapman, Gregory Clinton, David Easley, Dion Harmon, Matt Holland, Margo Levine, Anselm Levskaya, Colm Mulcahy, Peter Papadopol, Robert Piche, Daniel Alexander Ramras, Oswald Riemenschneider, Dierk Schleicher, George Sclavos, Scott Selikoff, John Shaw, Ted Shifrin, Mike Stevens, Chan-Ho Suh, Robert Terrell, and Hans van den Berg. We apologize to anyone who has been inadvertently omitted. We are of course responsible for any remaining errors.

We are indebted to our son, Alexander, for his suggestions, for writing numerous solutions, and for writing a program to help with the indexing. We thank our oldest daughter, Eleanor, for the goat figure of Section 3.8, and for earning part of her first-year college tuition by reading through the text and pointing out both places where it was not clear and passages she liked—the first invaluable for the text, the second for our morale. With her sisters, Judith and Diana, she also put food on the table when we were too busy to do so. Finally, we thank Diana for discovering errors in the table of contents.

Third printing

In this third printing we have corrected, as far as possible, the errors of which we were aware at the time corrections were prepared for the printers. We have not changed any numbers for equations, theorems, exercises, etc., so that students in the same class will be able to work with different printings. When we found that exercises were placed in the wrong section, we did not move them, but rather placed notes in the margin explaining where they should be.

Page breaks are virtually the same as in the previous printings, with two exceptions. In the first printing, exercises for Chapter 4 begin on page 449; in this and in the second printing, they begin on page 450, because we added material on the Fourier and Laplace transforms to Section 4.11. In the third printing, page numbers for many sections of the appendix have been changed, because we have completed the proof Kantorovitch's theorem with a discussion of uniqueness. This additional material is also available on the web page.

John H. Hubbard
Barbara Burke Hubbard

Ithaca, N.Y.
jhh8@cornell.edu

John H. Hubbard is a professor of mathematics at Cornell University and the author of several books on differential equations. His research mainly concerns complex analysis, differential equations, and dynamical systems. He believes that mathematics research and teaching are activities that enrich each other and should not be separated.

Barbara Burke Hubbard is the author of *The World According to Wavelets*, which was awarded the prix d'Alembert by the French Mathematical Society in 1996.

0
Preliminaries

0.0 INTRODUCTION

This chapter is intended as a resource, providing some background for those who may need it. In Section 0.1 we share some guidelines that in our experience make reading mathematics easier, and discuss a few specific issues like sum notation. Section 0.2 analyzes the rather tricky business of negating mathematical statements. (To a mathematician, the statement "All seven-legged alligators are orange with blue spots" is an obviously true statement, not an obviously meaningless one.) Section 0.3 reviews set theory notation; Section 0.4 discusses the real numbers; Section 0.5 discusses countable and uncountable sets and Russell's paradox; and Section 0.6 discusses complex numbers.

0.1 READING MATHEMATICS

The most efficient logical order for a subject is usually different from the best psychological order in which to learn it. Much mathematical writing is based too closely on the logical order of deduction in a subject, with too many definitions without, or before, the examples which motivate them, and too many answers before, or without, the questions they address.—
William Thurston

We recommend not spending much time on Chapter 0. In particular, if you are studying multivariate calculus for the first time you should definitely skip certain parts of Section 0.4 (Definition 0.4.4 and Proposition 0.4.6). However, Section 0.4 contains a discussion of sequences and series which you may wish to consult when we come to Section 1.5 about convergence and limits, if you find you don't remember the convergence criteria for sequences and series from first year calculus.

Reading mathematics is different from other reading. We think the following guidelines can make it easier. First, keep in mind that there are two parts to understanding a theorem: understanding the statement, and understanding the proof. *The first is more important than the second.*

What if you don't understand the statement? If there's a symbol in the formula you don't understand, perhaps a δ, look to see whether the next line continues, "where δ is such-and-such." In other words, read the whole sentence before you decide you can't understand it. In this book we have tried to define all terms before giving formulas, but we may not have succeeded everywhere.

If you're still having trouble, *skip ahead to examples.* This may contradict what you have been told—that mathematics is sequential, and that you must understand each sentence before going on to the next. In reality, although mathematical writing is necessarily sequential, mathematical understanding is not: you (and the experts) never understand perfectly up to some point and

The Greek Alphabet

Greek letters that look like Roman letters are not used as mathematical symbols; for example, A is capital a, not capital α. The letter χ is pronounced "kye," to rhyme with "sky"; φ, ψ and ξ may rhyme with either "sky" or "tea."

α	A	alpha
β	B	beta
γ	Γ	gamma
δ	Δ	delta
ϵ	E	epsilon
ζ	Z	zeta
η	H	eta
θ	Θ	theta
ι	I	iota
κ	K	kappa
λ	Λ	lambda
μ	M	mu
ν	N	nu
ξ	Ξ	xi
o	O	omicron
π	Π	pi
ρ	P	rho
σ	Σ	sigma
τ	T	tau
υ	Υ	upsilon
φ, ϕ	Φ	phi
χ	X	chi
ψ	Ψ	psi
ω	Ω	omega

In Equation 0.1.3, the symbol $\sum_{k=1}^{n}$ says that the sum will have n terms. Since the expression being summed is $a_{i,k}b_{k,j}$, each of those n terms will have the form ab.

not at all beyond. The "beyond," where understanding is only partial, is an essential part of the motivation and the conceptual background of the "here and now." You may often (perhaps usually) find that when you return to something you left half-understood, it will have become clear in the light of the further things you have studied, even though the further things are themselves obscure.

Many students are very uncomfortable in this state of partial understanding, like a beginning rock climber who wants to be in stable equilibrium at all times. To learn effectively one must be willing to leave the cocoon of equilibrium. So *if you don't understand something perfectly, go on ahead and then circle back.*

In particular, an example will often be easier to follow than a general statement; you can then go back and reconstitute the meaning of the statement in light of the example. Even if you still have trouble with the general statement, you will be ahead of the game if you understand the examples. We feel so strongly about this that we have sometimes flouted mathematical tradition and given examples before the proper definition.

Read with pencil and paper in hand, making up little examples for yourself as you go on.

Some of the difficulty in reading mathematics is notational. A pianist who has to stop and think whether a given note on the staff is A or F will not be able to sight-read a Bach prelude or Schubert sonata. The temptation, when faced with a long, involved equation, may be to give up. You need to take the time to identify the "notes."

Learn the names of Greek letters—not just the obvious ones like alpha, beta, and pi, but the more obscure psi, xi, tau, omega. The authors know a mathematician who calls all Greek letters "xi," (ξ) except for omega (ω), which he calls "w." This leads to confusion. Learn not just to recognize these letters, but how to pronounce them. Even if you are not reading mathematics out loud, it is hard to think about formulas if $\xi, \psi, \tau, \omega, \varphi$ are all "squiggles" to you.

Sum and product notation

Sum notation can be confusing at first; we are accustomed to reading in one dimension, from left to right, but something like

$$\sum_{k=1}^{n} a_{i,k}b_{k,j} \qquad 0.1.1$$

requires what we might call two-dimensional (or even three-dimensional) thinking. It may help at first to translate a sum into a linear expression:

$$\sum_{i=0}^{\infty} 2^i = 2^0 + 2^1 + 2^2 \ldots \qquad 0.1.2$$

or

$$c_{i,j} = \sum_{k=1}^{n} a_{i,k}b_{k,j} = a_{i,1}b_{1,j} + a_{i,2}b_{2,j} + \cdots + a_{i,n}b_{n,j}. \qquad 0.1.3$$

Two $\sum$ placed side by side do not denote the product of two sums; one sum is used to talk about one index, the other about another. The same thing could be written with one $\sum$, with information about both indices underneath. For example,

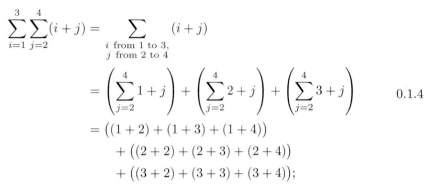

$$\sum_{i=1}^{3}\sum_{j=2}^{4}(i+j) = \sum_{\substack{i \text{ from } 1 \text{ to } 3, \\ j \text{ from } 2 \text{ to } 4}}(i+j)$$

$$= \left(\sum_{j=2}^{4}1+j\right) + \left(\sum_{j=2}^{4}2+j\right) + \left(\sum_{j=2}^{4}3+j\right) \qquad 0.1.4$$

$$= \bigl((1+2)+(1+3)+(1+4)\bigr)$$
$$+ \bigl((2+2)+(2+3)+(2+4)\bigr)$$
$$+ \bigl((3+2)+(3+3)+(3+4)\bigr);$$

this double sum is illustrated in Figure 0.1.1.

Rules for product notation are analogous to those for sum notation:

$$\prod_{i=1}^{n}a_i = a_1 \cdot a_2 \cdots a_n; \quad \text{for example,} \quad \prod_{i=1}^{n}i = n!.$$

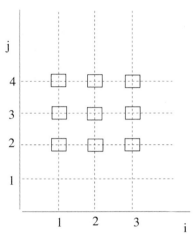

FIGURE 0.1.1.

In the double sum of Equation 0.1.4, each sum has three terms, so the double sum has nine terms.

Proofs

We said earlier that it is more important to understand a mathematical statement than to understand its proof. We have put some of the harder proofs in the appendix; these can safely be skipped by a student studying multivariate calculus for the first time. We urge you, however, to read the proofs in the main text. By reading many proofs you will learn what a proof is, so that (for one thing) you will know when you have proved something and when you have not.

When Jacobi complained that Gauss's proofs appeared unmotivated, Gauss is said to have answered, You build the building and remove the scaffolding. Our sympathy is with Jacobi's reply: he likened Gauss to the fox who erases his tracks in the sand with his tail.

In addition, a good proof doesn't just convince you that something is true; it tells you why it is true. You presumably don't lie awake at night worrying about the truth of the statements in this or any other math textbook. (This is known as "proof by eminent authority"; you assume the authors know what they are talking about.) But reading the proofs will help you understand the material.

If you get discouraged, keep in mind that the content of this book represents a cleaned-up version of many false starts. For example, John Hubbard started by trying to prove Fubini's theorem in the form presented in Equation 4.5.1. When he failed, he realized (something he had known and forgotten) that the statement was in fact false. He then went through a stack of scrap paper before coming up with a correct proof. *Other statements in the book represent the efforts of some of the world's best mathematicians over many years.*

0.2 How to negate mathematical statements

Even professional mathematicians have to be careful not to get confused when negating a complicated mathematical statement. The rules to follow are:

(1) The opposite of

$$[\text{For all } x, P(x) \text{ is true}]$$
$$\text{is} \quad [\text{There exists } x \text{ for which } P(x) \text{ is not true}].$$

0.2.1

Above, P stands for "property." Symbolically the same sentence is written:

$$\text{The opposite of} \quad \forall\, x, P(x) \quad \text{is} \quad \exists x|\ \text{not } P(x). \qquad 0.2.2$$

Instead of using the bar | to mean "such that" we could write the last line $(\exists x)\big(\text{not } P(x)\big)$. Sometimes (not in this book) the symbols $\sim$ and $\neg$ are used to mean "not."

(2) The opposite of

$$[\text{There exists } x \text{ for which } R(x) \text{ is true}]$$
$$\text{is} \quad [\text{For all } x, R(x) \text{ is not true}].$$

0.2.3

Symbolically the same sentence is written:

$$\text{The opposite of} \quad (\exists x)\big(P(x)\big) \quad \text{is} \quad (\forall\, x) \text{ not } P(x). \qquad 0.2.4$$

These rules may seem reasonable and simple. Clearly the opposite of the (false) statement, "All rational numbers equal 1," is the statement, "There exists a rational number that does not equal 1."

However, by the same rules, the statement, "All seven-legged alligators are orange with blue spots" is true, since if it were false, then there would exist a seven-legged alligator that is not orange with blue spots. The statement, "All seven-legged alligators are black with white stripes" is equally true.

Statements that to the ordinary mortal are false or meaningless are thus accepted as true by mathematicians; if you object, the mathematician will retort, "find me a counter-example."

In addition, mathematical statements are rarely as simple as "All rational numbers equal 1." Often there are many quantifiers and even the experts have to watch out. At a lecture attended by one of the authors, it was not clear to the audience in what order the lecturer was taking the quantifiers; when he was forced to write down a precise statement, he discovered that he didn't know what he meant and the lecture fell apart.

Here is an example where the order of quantifiers really counts: in the definitions of continuity and uniform continuity. A function f is *continuous* if for all x, and for all $\epsilon > 0$, there exists $\delta > 0$ such that for all y, if $|x - y| < \delta$, then $|f(x) - f(y)| < \epsilon$. That is, f is continuous if

$$(\forall x)(\forall \epsilon > 0)(\exists \delta > 0)(\forall y)\,\big(|x - y| < \delta \implies |f(x) - f(y)| < \epsilon\big). \qquad 0.2.5$$

A function f is *uniformly continuous* if for all $\epsilon > 0$, there exists $\delta > 0$ for all x and all y such that if $|x - y| < \delta$, then $|f(x) - f(y)| < \epsilon$. That is, f is uniformly continuous if

$$(\forall \epsilon > 0)(\exists \delta > 0)(\forall x)(\forall y) \left(|x - y| < \delta \implies |f(x) - f(y)| < \epsilon \right). \qquad 0.2.6$$

For the continuous function, we can choose *different* δ for different x; for the uniformly continuous function, we start with ϵ and have to find a *single* δ that works for all x.

For example, the function $f(x) = x^2$ is continuous but not uniformly continuous: as you choose bigger and bigger x, you will need a smaller δ if you want the statement $|x - y| < \delta$ to imply $|f(x) - f(y)| < \epsilon$, because the function keeps climbing more and more steeply. But $\sin x$ is uniformly continuous; you can find one δ that works for all x and all y.

0.3 SET THEORY

At the level at which we are working, set theory is a language, with a vocabulary consisting of seven words. In the late 1960's and early 1970's, under the sway of the "New Math," they were a standard part of the elementary school curriculum, and set theory was taught as a subject in its own right. This was a resounding failure, possibly because many teachers, understandably not knowing why set theory was being taught at all, made mountains out of molehills. As a result the schools (elementary, middle, high) have often gone to the opposite extreme, and some have dropped the subject altogether.

The seven vocabulary words are

There is nothing new about the concept of "set" denoted by $\{a|p(a)\}$. Euclid spoke of geometric *loci*, a locus being the set of points defined by some property. (The Latin word *locus* means "place"; its plural is *loci*.)

$\in$	"is an element of"
$\{a \mid p(a)\}$	"the set of a such that $p(a)$ is true"
$\subset$	"is a subset of" (or equals, when $A \subset A$)
$\cap$	"intersect": $A \cap B$ is the set of elements of both A and B.
$\cup$	"union": $A \cup B$ is the set of elements of either A or B or both.
$\times$	"cross": $A \times B$ is the set of pairs (a, b) with $a \in A$ and $b \in B$.
$-$	"complement": $A - B$ is the set of elements in A that are not in B.

One set has a standard name: the empty set ϕ, which has no elements. There are also sets of numbers that have standard names; they are written in *black-board bold*, a font we use only for these sets. Throughout this book and most other mathematics books (with the exception of $\mathbb{N}$, as noted in the margin below), they have exactly the same meaning:

$\mathbb{N}$ is for "natural," $\mathbb{Z}$ is for "Zahl," the German for number, $\mathbb{Q}$ is for "quotient," $\mathbb{R}$ is for "real," and $\mathbb{C}$ is for "complex." Mathematical notation is not quite standard: some authors do not include 0 in $\mathbb{N}$.

When writing with chalk on a black board, it's hard to distinguish between normal letters and bold letters.   Black-board bold font is characterized by double lines, as in $\mathbb{N}$ and $\mathbb{R}$.

$\mathbb{N}$ "the natural numbers" $\{0, 1, 2, \dots\}$
$\mathbb{Z}$ "the integers," i.e., signed whole numbers $\{\dots, -1, 0, 1, \dots\}$
$\mathbb{Q}$ "the rational numbers" p/q, with $p, q \in \mathbb{Z}$, $q \neq 0$
$\mathbb{R}$ "the real numbers," which we will think of as infinite decimals
$\mathbb{C}$ "the complex numbers" $\{a + ib \mid a, b \in \mathbb{R}\}$

Often we use slight variants of the notation above: $\{3, 5, 7\}$ is the set consisting of $3, 5$, and 7, and more generally, the set consisting of some list of elements is denoted by that list, enclosed in curly brackets, as in

$$\{n \mid n \in \mathbb{N} \text{ and } n \text{ is even}\} = \{0, 2, 4, \dots\}, \qquad 0.3.1$$

where again the vertical line $\mid$ means "such that."

The symbols are sometimes used backwards; for example, $A \supset B$ means $B \subset A$, as you probably guessed. Expressions are sometimes condensed:

$$\{x \in \mathbb{R} \mid x \text{ is a square}\} \quad \text{means} \quad \{x \mid x \in \mathbb{R} \text{ and } x \text{ is a square}\}, \qquad 0.3.2$$

i.e., the set of non-negative real numbers.

A slightly more elaborate variation is *indexed unions and intersections*: if S_α is a collection of sets indexed by $\alpha \in A$, then

$$\bigcap_{\alpha \in A} S_\alpha \text{ denotes the intersection of all the } S_\alpha, \text{ and}$$

Although it may seem a bit pedantic, you should notice that

$$\bigcup_{n \in \mathbb{Z}} l_n \quad \text{and} \quad \{l_n \mid n \in \mathbb{Z}\}$$

are not the same thing: the first is a subset of the plane; an element of it is a point on one of the lines. The second is a set of lines, not a set of points. This is similar to one of the molehills which became mountains in the new-math days: telling the difference between ϕ and $\{\phi\}$, the set whose only element is the empty set.

$$\bigcup_{\alpha \in A} S_\alpha \quad \text{denotes their union.}$$

For instance, if $l_n \subset \mathbb{R}^2$ is the line of equation $y = n$, then $\bigcup_{n \in \mathbb{Z}} l_n$ is the set of points in the plane whose y-coordinate is an integer.

We will use exponents to denote multiple products of sets; $A \times A \times \dots \times A$ with n terms is denoted A^n: the set of n-tuples of elements of A.

If this is all there is to set theory, what is the fuss about? For one thing, historically, mathematicians apparently did not think in terms of sets, and the introduction of set theory was part of a revolution at the end of the 19th century that included topology and measure theory. We explore another reason in Section 0.5, concerning infinite sets and Russell's paradox.

0.4 REAL NUMBERS

Showing that all such constructions lead to the same numbers is a fastidious exercise, which we will not pursue.

All of calculus, and to a lesser extent linear algebra, is about *real numbers*. In this introduction, we will present real numbers, and establish some of their most useful properties. Our approach privileges the writing of numbers in base 10; as such it is a bit unnatural, but we hope you will like our real numbers being exactly the numbers you are used to. Also, addition and multiplication will be defined in terms of finite decimals.

There are more elegant approaches to defining real numbers, (Dedekind cuts, for instance (see, for example, Michael Spivak, *Calculus*, second edition, 1980, pp. 554–572), or Cauchy sequences of rational numbers; one could also mirror the present approach, writing numbers in any base, for instance 2. Since this section is partially motivated by the treatment of floating-point numbers on computers, base 2 would seem very natural.

Numbers and their ordering

By definition, the set of real numbers is the set of infinite decimals: expressions like $2.95765392045756\ldots$, preceded by a plus or a minus sign (in practice the plus sign is usually omitted). The number that you usually think of as 3 is the infinite decimal $3.0000\ldots$, ending in all zeroes.

The following identification is absolutely vital: a number ending in all 9's is equal to the "rounded up" number ending in all 0's:

$$0.34999999\cdots = 0.350000\ldots. \qquad\qquad 0.4.1$$

Also, $+.0000\cdots = -.0000\ldots$. Other than these exceptions, there is only one way of writing a real number.

Numbers that start with a $+$ sign, except $+0.000\ldots$, are positive; those that start with a $-$ sign, except $-0.00\ldots$, are negative. If x is a real number, then $-x$ has the same string of digits, but with the opposite sign in front. For $k \geq 0$, we will denote by $[x]_k$ the truncated finite decimal consisting of all the digits of x before the decimal, and exactly k digits after the decimal. To avoid ambiguity, if x is a real number with two decimal expressions, $[x]_k$ will be the finite decimal built from the infinite decimal ending in 0's; for the number in Equation 0.4.1, $[x]_3 = 0.350$.

Given any two different numbers x and y, one is always bigger than the other. This is defined as follows: if x is positive and y is non-positive, then $x > y$. If both are positive, then in their decimal expansions there is a first digit in which they differ; whichever has the larger digit in that position is larger. If both are negative, then $x > y$ if $-y > -x$.

The least upper bound property

The least upper bound property of the reals is often taken as an axiom; indeed, it characterizes the real numbers, and it sits at the foundation of every theorem in calculus. However, at least with the description above of the reals, it is a theorem, not an axiom.

The least upper bound $\sup X$ is sometimes denoted l.u.b. X; the notation $\max X$ is also sometimes used, but suggests to some people that $\max X \in X$.

Definition 0.4.1 (Upper bound; least upper bound). A number a is an upper bound for a subset $X \subset \mathbb{R}$ if for every $x \in X$ we have $x \leq a$. A least upper bound is an upper bound b such that for any other upper bound a, we have $b \leq a$. The least upper bound is denoted sup.

Theorem 0.4.2. *Every non-empty subset $X \subset \mathbb{R}$ that has an upper bound has a least upper bound* $\sup X$.

Proof. We will construct successive decimals of $\sup X$. Let us suppose that $x \in X$ is an element (which we know exists, since $X \neq \emptyset$) and that a is an upper bound. We will assume that $x > 0$ (the case $x \leq 0$ is slightly different). If $x = a$, we are done: the least upper bound is a.

Recall that $[a]_j$ denotes the finite decimal consisting of all the digits of a before the decimal, and j digits after the decimal.

We use the symbol □ to mark the end of a proof, and the symbol △ to denote the end of an example or a remark.

Because you learned to add, subtract, divide, and multiply in elementary school, the algorithms used may seem obvious. But understanding how computers simulate real numbers is not nearly as routine as you might imagine. A real number involves an infinite amount of information, and computers cannot handle such things: they compute with finite decimals. This inevitably involves rounding off, and writing arithmetic subroutines that minimize round-off errors is a whole art in itself. In particular, computer addition and multiplication are not commutative or associative. Anyone who really wants to understand numerical problems has to take a serious interest in "computer arithmetic."

If $x \neq a$, there is a first j such that the jth digit of x is smaller than the jth digit of a. Consider all the numbers in $[x, a]$ that can be written using only j digits after the decimal, then all zeroes. This is a finite non-empty set; in fact it has at most 10 elements, and $[a]_j$ is one of them. Let b_j be the largest which is not an upper bound. Now consider the set of numbers in $[b_j, a]$ that have only $j + 1$ digits after the decimal point, then all zeroes. Again this is a finite non-empty set, so you can choose the largest which is not an upper bound; call it b_{j+1}. It should be clear that b_{j+1} is obtained by adding one digit to b_j. Keep going this way, defining numbers $b_{j+2}, b_{j+3}, \ldots$, each time adding one digit to the previous number. We can let b be the number whose kth decimal digit is the same as that of b_k; we claim that $b = \sup X$.

Indeed, if there exists $y \in X$ with $y > b$, then there is a first digit k of y which differs from the kth digit of b, and then b_k was not the largest number with k digits which is not an upper bound, since using the kth digit of y would give a bigger one. So b is an upper bound.

Now suppose that $b' < b$ is also an upper bound. Again there is a first digit k of b which is different from that of b'. This contradicts the fact that b_k was not an upper bound, since then $b_k > b'$.   □

Arithmetic of real numbers

The next task is to make arithmetic work for the reals: defining addition, multiplication, subtraction, and division, and to show that the usual rules of arithmetic hold. This is harder than one might think: addition and multiplication always start at the right, and for reals there is no right.

The underlying idea is to show that if you take two reals, truncate (cut) them further and further to the right and add them (or multiply them, or subtract them, etc.) and look only at the digits to the left of any fixed position, the digits we see will not be affected by where the truncation takes place, once it is well beyond where we are looking. The problem with this is that it isn't quite true.

Example 0.4.3 (Addition). Consider adding the following two numbers:

$$.222222\ldots222\ldots$$
$$.777777\ldots778\ldots$$

The sum of the truncated numbers will be $.9999\ldots9$ if we truncate before the position of the 8, and $1.0000\ldots0$ if we truncate after the 8. So there cannot be any rule which says: the 100th digit will stay the same if you truncate after the Nth digit, however large N is. The carry can come from arbitrarily far to the right.

If you insist on defining everything in terms of digits, it can be done but is quite involved: even showing that addition is associative involves at least six different cases, and although none is hard, keeping straight what you are doing is quite delicate. Exercise 0.4.1 should give you enough of a taste of this approach. Proposition 0.4.6 allows a general treatment; the development is quite abstract, and you should definitely not think you need to understand this in order to proceed.

Let us denote by $\mathbb{D}$ the set of finite decimals.

$\mathbb{D}$ stands for "finite decimal."

Definition 0.4.4 (Finite decimal continuity). A mapping $f : \mathbb{D}^n \to \mathbb{D}$ will be called finite decimal continuous ($\mathbb{D}$-continuous) if for all integers N and k, there exists an integer l such that if $(x_1, \dots, x_n)$ and $(y_1, \dots, y_n)$ are two elements of $\mathbb{D}^n$ with all $|x_i|, |y_i| < N$, and if $|x_i - y_i| < 10^{-l}$ for all $i = 1, \dots, n$, then

$$|f(x_1, \dots, x_n) - f(y_1, \dots, y_n)| < 10^{-k}. \qquad 0.4.2$$

We use A for addition, M for multiplication, and S for subtraction; the function *Assoc* is needed to prove associativity of addition.

Since we don't yet have a notion of subtraction in $\mathbb{R}$, we can't write $|x - y| < \epsilon$, much less $\sum (x_i - y_i)^2 < \epsilon^2$, which involves addition and multiplication besides. Our definition of k-close uses only subtraction of finite decimals.

For definitions of sup and inf, see Definitions 1.6.4 and 1.6.6.

The notion of k-close is the correct way of saying that two numbers agree to k digits after the decimal point. It takes into account the convention by which a number ending in all 9's is equal to the rounded up number ending in all 0's: the numbers .9998 and 1.0001 are 3-close.

We define points in $\mathbb{R}^n$ in Section 1.1; sup and inf (Equation 0.4.3) are defined in Section 1.6 (Definitions 1.6.4 and 1.6.5).

The functions $\widetilde{A}$ and $\widetilde{M}$ satisfy the conditions of Proposition 0.4.6; thus they apply to the real numbers, while A and M without tildes apply to finite decimals.

Exercise 0.4.3 asks you to show that the functions $A(x, y) = x + y$, $M(x, y) = xy$, $S(x, y) = x - y$, $Assoc(x, y) = (x + y) + z$ are $\mathbb{D}$-continuous, and that $1/x$ is not.

To see why Definition 0.4.4 is the right definition, we need to define what it means for two points $\mathbf{x}, \mathbf{y} \in \mathbb{R}^n$ to be close.

Definition 0.4.5 (k-close). Two points $\mathbf{x}, \mathbf{y} \in \mathbb{R}^n$ are k-close if for each $i = 1, \dots, n$, then $\big|[x_i]_k - [y_i]_k\big| \leq 10^{-k}$.

Notice that if two numbers are k-close for all k, then they are equal (see Exercise 0.4.2).

If $f : \mathbb{D}^n \to \mathbb{D}$ is $\mathbb{D}$-continuous, then define $\widetilde{f} : \mathbb{R}^n \to \mathbb{R}$ by the formula

$$\widetilde{f}(\mathbf{x}) = \sup_k \inf_{l \geq k} f([x_1]_l, \dots, [x_n]_l). \qquad 0.4.3$$

Proposition 0.4.6. *The function $\widetilde{f} : \mathbb{R}^n \to \mathbb{R}$ is the unique function that coincides with f on $\mathbb{D}^n$ and which satisfies the continuity condition that for all $k \in \mathbb{N}$, for all $N \in \mathbb{N}$, there exists $l \in \mathbb{N}$ such that when $\mathbf{x}, \mathbf{y} \in \mathbb{R}^n$ are l-close and all coordinates x_i of $\mathbf{x}$ satisfy $|x_i| < N$, then $\widetilde{f}(\mathbf{x})$ and $\widetilde{f}(\mathbf{y})$ are k-close.*

The proof of Proposition 0.4.6 is the object of Exercise 0.4.4.

With this proposition, setting up arithmetic for the reals is plain sailing.

Consider the $\mathbb{D}$-continuous functions $A(x, y) = x + y$ and $M(x, y) = xy$; then we define addition of reals by setting

$$x + y = \widetilde{A}(x, y) \quad \text{and} \quad xy = \widetilde{M}(x, y). \qquad 0.4.4$$

It isn't harder to show that the basic laws of arithmetic hold:

$$x + y = y + x \qquad \textit{Addition is commutative.}$$
$$(x + y) + z = x + (y + z) \quad \textit{Addition is associative.}$$
$$x + (-x) = 0 \qquad \textit{Existence of additive inverse.}$$
$$xy = yx \qquad \textit{Multiplication is commutative.}$$
$$(xy)z = x(yz) \qquad \textit{Multiplication is associative.}$$
$$x(y + z) = xy + xz \qquad \textit{Multiplication is distributive over addition.}$$

These are all proved the same way: let us prove the last. Consider the function $\mathbb{D}^3 \to \mathbb{D}$ given by

$$F(x, y, z) = M\big(x, A(y, z)\big) - A\big(M(x, y), M(x, z)\big). \qquad 0.4.5$$

We leave it to you to check that F is $\mathbb{D}$-continuous, and that

$$\widetilde{F}(x, y, z) = \widetilde{M}\big(x, \widetilde{A}(y, z)\big) - \widetilde{A}\big(\widetilde{M}(x, y), \widetilde{M}(x, z)\big). \qquad 0.4.6$$

But F is identically 0 on $\mathbb{D}^3$, and the identically 0 function on $\mathbb{R}^3$ is a function which coincides with 0 on $\mathbb{D}^3$ and satisfies the continuity condition of Proposition 0.4.6, so $\widetilde{F}$ vanishes identically by the uniqueness part of Proposition 0.4.6. That is what was to be proved.

It is one of the basic irritants of elementary school math that division is not defined in the world of finite decimals.

This sets up almost all of arithmetic; the missing piece is division. Exercise 0.4.5 asks you to define division in the reals.

Sequences and series

A sequence is an infinite list (of numbers or vectors or matrices ...).

Definition 0.4.7 (Convergent sequence). A sequence a_n of real numbers is said to converge to the limit a if for all $\epsilon > 0$, there exists N such that for all $n > N$, we have $|a - a_n| < \epsilon$.

All of calculus is based on this definition, and the closely related definition of limits of functions.

If a series converges, then the same list of numbers viewed as a sequence must converge to 0. The converse is not true. For example, the harmonic series

$$1 + \frac{1}{2} + \frac{1}{3} + \dots$$

does not converge, although the terms tend to 0.

Many important sequences appear as partial sums of series. A *series* is a sequence where the terms are to be added. If $a_1, a_2, \dots$ is a series of numbers, then the associated sequence of partial sums is the sequence $s_1, s_2, \dots$, where

$$s_N = \sum_{n=1}^{N} a_n. \qquad 0.4.7$$

For example, if $a_1 = 1, a_2 = 2, a_3 = 3$, and so on, then $s_4 = 1 + 2 + 3 + 4$.

Definition 0.4.8 (Convergent series). If the sequence of partial sums of a series has a limit S, we say that the series converges, and its limit is

$$\sum_{n=1}^{\infty} a_n = S. \qquad\qquad 0.4.8$$

Example 0.4.9 (Geometric series). If $|r| < 1$, then

$$\sum_{n=0}^{\infty} ar^n = \frac{a}{1-r}. \qquad\qquad 0.4.9$$

Example of geometric series:

$$2.020202 =$$

$$2 + 2(.01) + 2(.01)^2 + \dots$$

$$= \frac{2}{1 - (.01)}$$

$$= \frac{200}{99}.$$

Indeed, the following subtraction shows that $S_n(1 - r) = a - ar^{n+1}$:

$$S_n = a + ar + ar^2 + ar^3 + \dots + ar^n$$

$$S_n r = \qquad ar + ar^2 + ar^3 + \dots + ar^n + ar^{n+1}$$

$$\overline{} \qquad 0.4.10$$

$$S_n(1 - r) = a \qquad\qquad\qquad\qquad\quad - ar^{n+1}$$

But $\lim_{n \to \infty} ar^{n+1} = 0$ when $|r| < 1$, so we can forget about the $-ar^{n+1}$: as $n \to \infty$, we have $S_n \to a/(1-r)$. $\triangle$

Proving convergence

The weakness of the definition of a convergent sequence is that it involves the limit value. At first, it is hard to see how you will ever be able to prove that a sequence has a limit if you don't know the limit ahead of time.

The first result along these lines is the following theorem.

It is hard to overstate the importance of this problem: proving that a limit exists without knowing ahead of time what it is. It was a watershed in the history of mathematics, and remains a critical dividing point between first year calculus and multivariate calculus, and more generally, between elementary mathematics and advanced mathematics.

It is of course also true that A non-increasing sequence converges if and only if it is bounded.

Theorem 0.4.10. *A non-decreasing sequence a_n converges if and only if it is bounded.*

Proof. If a sequence a_n converges, it is clearly bounded. If it is bounded, it has a least upper bound A. We claim that A is the limit. This means that for any $\epsilon > 0$, there exists N such that if $n > N$, then $|a_n - A| < \epsilon$. Choose $\epsilon > 0$; if $A - a_n > \epsilon$ for all n, then $A - \epsilon$ is an upper bound for the sequence, contradicting the definition of A. So there is a first N with $A - a_N < \epsilon$, and it will do, since when $n > N$, we must have $A - a_n < A - a_N < \epsilon$. $\square$

Theorem 0.4.10 has the following consequence:

Theorem 0.4.11. *If a_n is a series such that the series of absolute values*

$$\sum_{n=1}^{\infty} |a_n| \quad \text{converges, then so does the series} \quad \sum_{n=1}^{\infty} a_n.$$

Proof. The series $\sum_{n=1}^{\infty} a_n + |a_n|$ is a series of non-negative numbers, and so the partial sums $b_m = \sum_{n=1}^{m}(a_n + |a_n|)$ are non-decreasing. They are also bounded:

$$b_m = \sum_{n=1}^{m}(a_n + |a_n|) \leq \sum_{n=1}^{m} 2|a_n| = 2\sum_{n=1}^{m}|a_n| \leq 2\sum_{n=1}^{\infty}|a_n|. \qquad 0.4.11$$

So (by Theorem 0.4.10) the b_m form a convergent sequence, and finally

$$\sum_{n=1}^{\infty} a_n = \sum_{n=1}^{\infty}\left(a_n + |a_n|\right) + \left(-\sum_{n=1}^{\infty}|a_n|\right) \qquad 0.4.12$$

represents the series $\sum_{n=1}^{\infty} a_n$ as the sum of two numbers, each one the sum of a convergent series. $\square$

The intermediate value theorem

The intermediate value theorem appears to be obviously true, and is often useful. Moreover, it follows easily from Theorem 0.4.2 and the definition of continuity.

One unsuccessful 19th century definition of continuity stated that a function f is continuous if it satisfies the intermediate value theorem: if, for all $a < b$, f takes on all values between $f(a)$ and $f(b)$ at some $c \in [a, b]$. You are asked in Exercise 0.4.7 to show that this does not coincide with the usual definition (and presumably not with anyone's intuition of what continuity should mean).

Theorem 0.4.12 (Intermediate value theorem). *If $f : [a, b] \to \mathbb{R}$ is a continuous function such that $f(a) \leq 0$ and $f(b) \geq 0$, then there exists $c \in [a, b]$ such that $f(c) = 0$.*

Proof. Let X be the set of $x \in [a, b]$ such that $f(x) \leq 0$. Note that X is non-empty (a is in it) and it has an upper bound, namely b, so that it has a least upper bound, which we call c. We claim $f(c) = 0$.

Since f is continuous, for any $\epsilon > 0$, there exists $\delta > 0$ such that when $|x - c| < \delta$, then $|f(x) - f(c)| < \epsilon$. Therefore, if $f(c) > 0$, we can set $\epsilon = f(c)$, and there exists $\delta > 0$ such that if $|x - c| < \delta$, then $|f(x) - f(c)| < f(c)$. In particular, we see that if $x > c - \delta/2$, $f(x) > 0$, so $c - \delta/2$ is also an upper bound for X, which is a contradiction.

If $f(c) < 0$, a similar argument shows that there exists $\delta > 0$ such that $f(c + \delta/2) < 0$, contradicting the assumption that c is an upper bound for X. The only choice left is $f(c) = 0$. $\square$

0.5 INFINITE SETS AND RUSSELL'S PARADOX

One reason set theory is accorded so much importance is that Georg Cantor (1845–1918) discovered that two infinite sets need not have the same "number" of elements; there isn't just one infinity. You might think this is just obvious, for instance because there are more whole numbers than even whole numbers. But with the definition Cantor gave, two sets A and B have the same number of elements (the same *cardinality*) if you can set up a one-to-one correspondence between them. For instance

$$\begin{array}{ccccccc} 0 & 1 & 2 & 3 & 4 & 5 & 6 & \ldots \\ 0 & 2 & 4 & 6 & 8 & 10 & 12 & \ldots \end{array}$$

0.5.1

establishes a one to one correspondence between the natural numbers and the even natural numbers. More generally, any set whose elements you can list has the same cardinality as $\mathbb{N}$. But Cantor discovered that $\mathbb{R}$ does not have the same cardinality as $\mathbb{N}$: it has a *bigger infinity of elements*. Indeed, imagine making any infinite list of real numbers, say between 0 and 1, so that written as decimals, your list might look like

$$.\mathbf{1}5436278645342982376349065236734754 8757\ldots$$
$$.9\mathbf{8}735462194375659867356294065734932 7658\ldots$$
$$.22\mathbf{9}5735219035643554230354655233900 80742\ldots$$
$$.104\mathbf{7}5201874626765320936572368907656 5787\ldots$$
$$.0263\mathbf{2}8560082356835654432879897652377 327\ldots$$

0.5.2

$$\ldots$$

Now consider the decimal formed by the elements of the diagonal digits (in bold above) $.18972\ldots$, and modify it (almost any way you want) so that every digit is changed, for instance according to the rule "change 7's to 5's and change anything that is not a 7 to a 7": in this case, your number becomes $.77757\ldots$. Clearly this last number does not appear in your list: it is not the nth element of the list, because it doesn't have the same nth decimal.

Sets that can be put in one-to-one correspondence with the integers are called *countable*, other infinite sets are called *uncountable*; the set $\mathbb{R}$ of real numbers is uncountable.

All sorts of questions naturally arise from this proof: are there other infinities besides those of $\mathbb{N}$ and $\mathbb{R}$? (There are: Cantor showed that there are infinitely many of them.) Are there infinite subsets of $\mathbb{R}$ that cannot be put into one to one correspondence with either $\mathbb{R}$ or $\mathbb{Z}$? This statement is called the *continuum hypothesis*, and has been shown to be unsolvable: it is consistent with the other axioms of set theory to assume it is true (Gödel, 1938) or false (Cohen, 1965). This means that if there is a contradiction in set theory assuming the continuum hypothesis, then there is a contradiction without assuming it, and if there is a contradiction in set theory assuming that the continuum hypothesis is false, then again there is a contradiction without assuming it is false.

Russell's paradox

In 1902, Bertrand Russell (1872–1970) wrote the logician Gottlob Frege a letter containing the following argument: Consider the set X of all sets that do not

This argument simply flabbergasted the mathematical world; after thousands of years of philosophical speculation about the infinite, Cantor found a fundamental notion that had been completely overlooked.

It would seem likely that $\mathbb{R}$ and $\mathbb{R}^2$ have different infinities of elements, but that is not the case (see Exercise 0.4.6).

contain themselves. If $X \in X$, then X does contain itself, so $X \notin X$. But if $X \notin X$, then X is a set which does not contain itself, so $X \in X$.

"Your discovery of the contradiction caused me the greatest surprise and, I would almost say, consternation," Frege replied, "since it has shaken the basis on which I intended to build arithmetic ... your discovery is very remarkable and will perhaps result in a great advance in logic, unwelcome as it may seem at first glance." Russell's paradox was (and remains) extremely perplexing. The "solution," such as it is, is to say that the naive idea that any property defines a set is untenable, and that sets must be built up, allowing you to take subsets, unions, products, ... of sets already defined; moreover, to make the theory interesting, you must assume the existence of an infinite set. Set theory (still an active subject of research) consists of describing exactly the allowed construction procedures, and seeing what consequences can be derived.

This paradox has a long history, in various guises: the Greeks knew it as the paradox of the barber, who lived on the island of Milos, and decided to shave all the men of the island who did not shave themselves. Does the barber shave himself?

These letters by Russell and Frege are published in *From Frege to Gödel: A Source Book in Mathematical Logic, 1879–1931*, by Jean van Heijenoort (who in his youth was bodyguard to Leon Trotsky).

0.6 COMPLEX NUMBERS

Complex numbers are written $a + bi$, where a and b are real numbers, and addition and multiplication are defined in Equations 0.6.1 and 0.6.2. It follows from those rules that $i = \sqrt{-1}$.

The complex number $a + ib$ is often plotted as the point $\begin{pmatrix} a \\ b \end{pmatrix} \in \mathbb{R}^2$. The real number a is called the *real part* of $a + ib$, denoted $\mathrm{Re}\,(a + ib)$, and the real number b is called the *imaginary part*, denoted $\mathrm{Im}\,(a + ib)$. The reals $\mathbb{R}$ can be considered as a subset of the complex numbers $\mathbb{C}$, by identifying $a \in \mathbb{R}$ with $a + i0 \in \mathbb{C}$; such complex numbers are called "real," as you might imagine. Real numbers are systematically identified with the real complex numbers, and $a + i0$ is usually denoted simply a.

Numbers of the form $0 + ib$ are called *purely imaginary*. What complex numbers, if any, are both real and purely imaginary?[1] If we plot $a + ib$ as the point $\begin{pmatrix} a \\ b \end{pmatrix} \in \mathbb{R}^2$, what do the purely real numbers correspond to? The purely imaginary numbers?[2]

Complex numbers (long considered "impossible" numbers) were first used in 16th century Italy, as a crutch that made it possible to find *real* roots of *real* cubic polynomials. But they turned out to have immense significance in many fields of mathematics, leading John Stillwell to write in his *Mathematics and Its History* that "this resolution of the paradox of $\sqrt{-1}$ was so powerful, unexpected and beautiful that only the word 'miracle' seems adequate to describe it."

Arithmetic in $\mathbb{C}$

Complex numbers are added in the obvious way:

$$(a_1 + ib_1) + (a_2 + ib_2) = (a_1 + a_2) + i(b_1 + b_2). \qquad 0.6.1$$

Thus the identification with $\mathbb{R}^2$ preserves the operation of addition.

---

[1] The only complex number which is both real and purely imaginary is $0 = 0 + 0i$.

[2] The purely real numbers are all found on the x-axis, the purely imaginary numbers on the y-axis.

Equation 0.6.2 is not the only definition of multiplication one can imagine. For instance, we could define $(a_1 + ib_1) * (a_2 + ib_2) = (a_1 a_2) + i(b_1 b_2)$. But in that case, there would be lots of elements by which one could not divide, since the product of any purely real number and any purely imaginary number would be 0:

$$(a_1 + i0) * (0 + ib_2) = 0.$$

If the product of any two nonzero numbers α and β is 0: $\alpha\beta = 0$, then division by either is impossible; if we try to divide by α, we arrive at the contradiction $\beta = 0$:

$$\beta = \beta\frac{\alpha}{\alpha} = \frac{\beta\alpha}{\alpha} = \frac{0}{\alpha} = 0.$$

These four properties, concerning addition, don't depend on the special nature of complex numbers; we can similarly define addition for n-tuples of real numbers, and these rules will still be true.

The multiplication in these five properties is of course the special multiplication of complex numbers, defined in Equation 0.6.2. Multiplication can only be defined for *pairs* of real numbers. If we were to define a new kind of number as the 3-tuple (a, b, c) there would be no way to multiply two such 3-tuples that satisfies these five requirements.

There is a way to define multiplication for 4-tuples that satisfies all but commutativity, called *Hamilton's quaternions*.

What makes $\mathbb{C}$ interesting is that complex numbers can also be multiplied:

$$(a_1 + ib_1)(a_2 + ib_2) = (a_1 a_2 - b_1 b_2) + i(a_1 b_2 + a_2 b_1). \qquad 0.6.2$$

This formula consists of multiplying $a_1 + ib_1$ and $a_2 + ib_2$ (treating i like the variable x of a polynomial) to find

$$(a_1 + ib_1)(a_2 + ib_2) = a_1 a_2 + i(a_1 b_2 + a_2 b_1) + i^2(b_1 b_2) \qquad 0.6.3$$

and then setting $i^2 = -1$.

Example 0.6.1 (Multiplying complex numbers).

(a) $(2 + i)(1 - 3i) = (2 + 3) + i(1 - 6) = 5 - 5i$ (b) $(1 + i)^2 = 2i.$ $\triangle$ 0.6.4

Addition and multiplication of reals viewed as complex numbers coincides with ordinary addition and multiplication:

$$(a + i0) + (b + i0) = (a + b) + i0 \qquad (a + i0)(b + i0) = (ab) + i0. \qquad 0.6.5$$

Exercise 0.6.1 asks you to check the following nine rules, for $z_1, z_2 \in \mathbb{C}$:

(1) $(z_1 + z_2) + z_3 = z_1 + (z_2 + z_3)$ Addition is associative.

(2) $z_1 + z_2 = z_2 + z_1$ Addition is commutative.

(3) $z + 0 = z$ 0 (i.e., the complex number $0 + 0i$) is an additive identity.

(4) $(a + ib) + (-a - ib) = 0$ $(-a - ib)$ is the additive inverse of $a + ib$.

(5) $(z_1 z_2) z_3 = z_1 (z_2 z_3)$ Multiplication is associative.

(6) $z_1 z_2 = z_2 z_1$ Multiplication is commutative.

(7) $1z = z$ 1 (i.e., the complex number $1 + 0i$) is a multiplicative identity.

(8) $(a + ib)\left(\frac{a}{a^2+b^2} - i\frac{b}{a^2+b^2}\right) = 1$ If $z \neq 0$, then z has a multiplicative inverse.

(9) $z_1(z_2 + z_3) = z_1 z_2 + z_1 z_3$ Multiplication is distributive over addition.

The complex conjugate

Definition 0.6.2 (Complex conjugate). The complex conjugate of the complex number $z = a + ib$ is the number $\overline{z} = a - ib$.

Complex conjugation preserves all of arithmetic:

$$\overline{z + w} = \overline{z} + \overline{w} \quad \text{and} \quad \overline{zw} = \overline{z}\,\overline{w}. \qquad 0.6.6$$

The real numbers are the complex numbers z that are equal to their complex conjugates: $\bar{z} = z$, and the purely imaginary complex numbers are those that are the opposites of their complex conjugates: $\bar{z} = -z$.

There is a very useful way of writing the length of a complex number in terms of complex conjugates: If $z = a + ib$, then $z\bar{z} = a^2 + b^2$. The number

$$|z| = \sqrt{a^2 + b^2} = \sqrt{z\bar{z}} \qquad 0.6.7$$

is called the *absolute value* (or the *modulus*) of z. Clearly, $|a+ib|$ is the distance from the origin to $\begin{pmatrix} a \\ b \end{pmatrix}$.

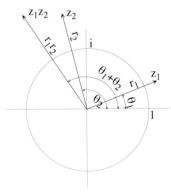

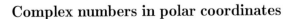

FIGURE 0.6.1.

When multiplying two complex numbers, the absolute values are multiplied and the arguments (polar angles) are added.

Complex numbers in polar coordinates

Let $z = a + ib \neq 0$ be a complex number. Then the point $\begin{pmatrix} a \\ b \end{pmatrix}$ can be represented in polar coordinates as $\begin{pmatrix} r\cos\theta \\ r\sin\theta \end{pmatrix}$, where

$$r = \sqrt{a^2 + b^2} = |z|, \qquad 0.6.8$$

and θ is an angle such that

$$\cos\theta = \frac{a}{r} \quad \text{and} \quad \sin\theta = \frac{b}{r}, \qquad 0.6.9$$

so that

$$z = r(\cos\theta + i\sin\theta). \qquad 0.6.10$$

The polar angle θ, called the *argument* of z, is determined by Equation 0.6.9 up to addition of a multiple of 2π.

The marvelous thing about this polar representation is that it gives a geometric representation of multiplication, as shown in Figure 0.6.1.

Proposition 0.6.3 (Geometrical representation of multiplication of complex numbers). *The modulus of the product $z_1 z_2$ is the product of the moduli $|z_1|\,|z_2|$.*

The polar angle of the product is the sum of the polar angles θ_1, θ_2:

$$\Big(r_1(\cos\theta_1 + i\sin\theta_1)\Big)\Big(r_2(\cos\theta_2 + i\sin\theta_2)\Big) = r_1 r_2 \Big(\cos(\theta_1 + \theta_2) + i\sin(\theta_1 + \theta_2)\Big).$$

Proof. Multiply out, and apply the addition rules of trigonometry:

$$\cos(\theta_1 + \theta_2) = \cos\theta_1\cos\theta_2 - \sin\theta_1\sin\theta_2$$
$$\sin(\theta_1 + \theta_2) = \sin\theta_1\cos\theta_2 + \cos\theta_1\sin\theta_2. \quad \square \qquad 0.6.11$$

The following formula, known as *de Moivre's formula*, follows immediately:

Corollary 0.6.4 (De Moivre's formula). *If* $z = r(\cos\theta + i\sin\theta)$, *then*

$$z^n = r^n(\cos n\theta + i\sin n\theta). \qquad 0.6.12$$

De Moivre's formula itself has a very important consequence, showing that in the process of adding a square root of -1 to the real numbers, we have actually added all the roots of complex numbers one might hope for.

Proposition 0.6.5. *Every complex number* $z = r(\cos\theta + i\sin\theta)$ *with* $r \neq 0$ *has* n *distinct complex* n*th roots, which are the numbers*

$$r^{1/n}\left(\cos\frac{\theta + 2k\pi}{n} + i\sin\frac{\theta + 2k\pi}{n}\right), \quad k = 0, \ldots, n-1. \qquad 0.6.13.$$

Note that $r^{1/n}$ stands for the positive real nth root of the positive number r. Figure 0.6.2 illustrates Proposition 0.6.5 for $n = 5$.

Proof. All that needs to be checked is that

(1) $\left(r^{1/n}\right)^n = r$, which is true by definition;

(2)

$$\cos n\frac{\theta + 2k\pi}{n} = \cos\theta \quad\text{and}\quad \sin n\frac{\theta + 2k\pi}{n} = \sin\theta, \qquad 0.6.14$$

which is true since $n\frac{\theta+2k\pi}{n} = \theta + 2k\pi$, and sin and cos are periodic with period 2π; and

(3) The numbers in Equation 0.6.13 are distinct, which is true since the polar angles do not differ by an integer multiple of 2π; they differ by multiples $2\pi k/n$, with $0 < k < n$. $\square$

A great deal more is true: *all polynomial equations with complex coefficients have all the roots one might hope for.* This is the content of the fundamental theorem of algebra, Theorem 1.6.10, proved by d'Alembert in 1746 and by Gauss around 1799. This milestone of mathematics followed by some 200 years the first introduction of complex numbers, about 1550, by several Italian mathematicians who were trying to solve cubic equations. Their work represented the rebirth of mathematics in Europe after a long sleep, of over 15 centuries.

Historical background: solving the cubic equation

We will show that a cubic equation can be solved using formulas analogous to the formula

$$\frac{-b \pm \sqrt{b^2 - 4ac}}{2a} \qquad 0.6.15$$

for the quadratic equation $ax^2 + bx + c = 0$.

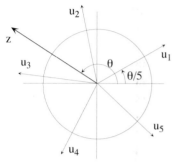

FIGURE 0.6.2.

The fifth roots of z form a regular pentagon, with one vertex at polar angle $\theta/5$, and the others rotated from that one by multiples of $2\pi/5$.

Immense psychological difficulties had to be overcome before complex numbers were accepted as an integral part of mathematics; when Gauss came up with his proof of the fundamental theorem of algebra, complex numbers were still not sufficiently respectable that he could use them in his statement of the theorem (although the proof depends on them).

Let us start with two examples; the explanation of the tricks will follow.

Example 0.6.6 (Solving a cubic equation).   Let us solve the equation $x^3 + x + 1 = 0$. First substitute $x = u - 1/3u$, to get

$$\left(u - \frac{1}{3u}\right)^3 + \left(u - \frac{1}{3u}\right) + 1 = u^3 - u + \frac{1}{3u} - \frac{1}{27u^3} + u - \frac{1}{3u} + 1 = 0. \quad 0.6.16$$

After simplification and multiplication by u^3 this becomes

$$u^6 + u^3 - \frac{1}{27} = 0. \qquad\qquad 0.6.17$$

This is a quadratic equation for u^3, which can be solved by formula 0.6.15, to yield

$$u^3 = \frac{1}{2}\left(-1 \pm \sqrt{\frac{31}{27}}\right) \approx 0.0358\ldots, -1.0358\ldots . \qquad 0.6.18$$

Both of these numbers have real cube roots: approximately $u_1 \approx 0.3295$ and $u_2 \approx -1.0118$.

This allows us to find $x = u - 1/3u$:

$$x = u_1 - \frac{1}{3u_1} = u_2 - \frac{1}{3u_2} \approx -0.6823. \qquad \triangle \quad 0.6.19$$

Example 0.6.7. Let us solve the equation $x^3 - 3x + 1 = 0$. As we will explain below, the right substitution to make in this case is $x = u + 1/u$, which leads to

$$\left(u + \frac{1}{u}\right)^3 - 3\left(u + \frac{1}{u}\right) + 1 = 0. \qquad\qquad 0.6.20$$

Here we see something bizarre: in Example 0.6.6, the polynomial has only one real root and we can find it using only real numbers, but in Example 0.6.7 there are three real roots, and we can't find any of them using only real numbers. We will see below that it is always true that when Cardano's formula is used, then if a real polynomial has one real root, we can always find it using only real numbers, but if it has three real roots, we never can find any of them using real numbers.

After multiplying out, canceling and multiplying by u^3, this gives the quadratic equation

$$u^6 + u^3 + 1 = 0 \quad \text{with solutions} \quad v_{1,2} = \frac{-1 \pm i\sqrt{3}}{2} = \cos\frac{2\pi}{3} \pm i\sin\frac{2\pi}{3}. \quad 0.6.21$$

The cube roots of v_1 (with positive imaginary part) are

$$\cos\frac{2\pi}{9} + i\sin\frac{2\pi}{9}, \ \cos\frac{8\pi}{9} + i\sin\frac{8\pi}{9}, \ \cos\frac{14\pi}{9} + i\sin\frac{14\pi}{9}. \qquad 0.6.22$$

In all three cases, we have $1/u = \bar{u}$, so that $u + 1/u = 2\,\mathrm{Re}\,u$, leading to the three roots

$$x_1 = 2\cos\frac{2\pi}{9} \approx 1.532088, \quad x_2 = 2\cos\frac{8\pi}{9} \approx -1.879385,$$

$$x_3 = 2\cos\frac{14\pi}{9} \approx 0.347296. \quad \triangle$$

$$0.6.23$$

Derivation of Cardano's formulas

The substitutions $x = u - 1/3u$ in Example 0.6.6 and $x = u + 1/u$ in Example 0.6.7 were special cases.

Eliminating the term in x^2 means changing the roots so that their sum is 0: If the roots of a cubic polynomial are a_1, a_2, and a_3, then we can write the polynomial as

$$p = (x - a_1)(x - a_2)(x - a_3)$$
$$= x^3 - (a_1 + a_2 + a_3)x^2$$
$$+ (a_1a_2 + a_1a_3 + a_2a_3)x$$
$$- a_1a_2a_3.$$

Thus eliminating the term in x^2 means that $a_1 + a_2 + a_3 = 0$. We will use this to prove Proposition 0.6.9.

If we start with the equation $x^3 + ax^2 + bx + c = 0$, we can eliminate the term in x^2 by setting $x = y - a/3$: the equation becomes

$$y^3 + py + q = 0, \quad \text{where } p = b - \frac{a^2}{3} \text{ and } q = c - \frac{ab}{3} + \frac{2a^3}{27}. \qquad 0.6.24$$

Now set $y = u - \frac{p}{3u}$; the equation $y^3 + py + q = 0$ then becomes

$$u^6 + qu^3 - \frac{p^3}{27} = 0, \qquad 0.6.25$$

which is a quadratic equation for u^3.

Let v_1 and v_2 be the two solutions of the quadratic equation $v^2 + qv - \frac{p^3}{27}$, and let $u_{i,1}, u_{i,2}, u_{i,3}$ be the three cubic roots of v_i for $i = 1, 2$. We now have apparently six roots for the equation $x^3 + px + q = 0$: the numbers

$$y_{i,j} = u_{i,j} - \frac{p}{3u_{i,j}}, \quad i = 1, 2; \ j = 1, 2, 3. \qquad 0.6.26$$

Exercise 0.6.2 asks you to show that $-p/(3u_{1,j})$ is a cubic root of v_2, and that we can renumber the cube roots of v_2 so that $-p/(3u_{1,j}) = u_{2,j}$. If that is done, we find that $y_{1,j} = y_{2,j}$ for $j = 1, 2, 3$; this explains why the apparently six roots are really only three.

The discriminant of the cubic

Definition 0.6.8 (Discriminant of cubic equation). The number $\Delta = 27q^2 + 4p^3$ is called the *discriminant* of the cubic equation $x^3 + px + q$.

Proposition 0.6.9. *The discriminant Δ vanishes exactly when $x^3 + px + q = 0$ has a double root.*

Proof. If there is a double root, then the roots are necessarily $\{a, a, -2a\}$ for some number a, since the sum of the roots is 0. Multiply out

$$(x - a)^2(x + 2a) = x^3 - 3a^2x + 2a^3, \quad \text{so } p = -3a^2 \text{ and } q = 2a^3,$$

and indeed $4p^3 + 27q^2 = -4 \cdot 27a^6 + 4 \cdot 27a^6 = 0$.

Now we need to show that if the discriminant is 0, the polynomial has a double root. Suppose $\Delta = 0$, and call α the square root of $-p/3$ such that $2\alpha^3 = q$; such a square root exists since $4\alpha^6 = 4(-p/3)^3 = -4p^3/27 = q^2$. Now multiply out

$$(x - \alpha)^2(x + 2\alpha) = x^3 + x(-4\alpha^2 + \alpha^2) + 2\alpha^3 = x^3 + px + q,$$

and we see that α is a double root of our cubic polynomial. $\quad \square$

Cardano's formula for real polynomials

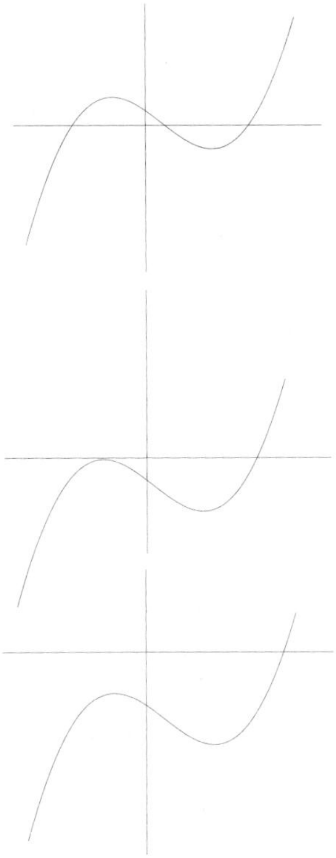

Suppose p, q are real. Figure 0.6.3 should explain why equations with double roots are the boundary between equations with one real root and equations with three real roots.

> **Proposition 0.6.10 (Number of real roots of a polynomial).** *The real cubic polynomial $x^3 + px + q$ has three real roots if the discriminant $27q^2 + 4p^3 < 0$, and one real root if $27q^2 + 4p^3 > 0$.*

Proof. If the polynomial has three real roots, then it has a positive maximum at $-\sqrt{-p/3}$, and a negative minimum at $\sqrt{-p/3}$. In particular, p must be negative. Thus we must have

$$\left(\left(\sqrt{-\frac{p}{3}} \right)^3 + p \left(\sqrt{-\frac{p}{3}} \right) + q \right) \left(\left(-\sqrt{-\frac{p}{3}} \right)^3 - p \left(\sqrt{-\frac{p}{3}} \right) + q \right) < 0. \quad 0.6.27$$

After a bit of computation, this becomes the result we want:

$$q^2 + \frac{4p^3}{27} < 0. \quad \square \qquad\qquad 0.6.28$$

FIGURE 0.6.3.

The graphs of three cubic polynomials. The polynomial at the top has three roots. As it is varied, the two roots to the left coalesce to give a double root, as shown by the middle figure. If the polynomial is varied a bit further, the double root vanishes (actually becoming a pair of complex conjugate roots).

Thus indeed, if a real cubic polynomial has three real roots, and you want to find them by Cardano's formula, you must use complex numbers, even though both the problem and the result involve only reals. Faced with this dilemma, the Italians of the 16th century, and their successors until about 1800, held their noses and computed with complex numbers. The name "imaginary" they used for such numbers expresses what they thought of them.

Several cubics are proposed in the exercises, as well as an alternative to Cardano's formula which applies to cubics with three real roots (Exercise 0.6.6), and a sketch of how to deal with quartic equations (Exercise 0.6.7) .

0.7 Exercises

**Exercises for Section 0.4:
Real Numbers**

0.4.1 (a) Let x and y be two positive reals. Show that $x + y$ is well defined by showing that for any k, the digit in the kth position of $[x]_N + [y]_N$ is the same for all sufficiently large N. Note that N cannot depend just on k, but must depend also on x and y.

Stars (*) denote difficult exercises. Two stars indicate a particularly challenging exercise.

Many of the exercises for Chapter 0 are quite theoretical, and too difficult for students taking multivariate calculus for the first time. They are intended for use when the book is being used for a first analysis class. Exceptions include Exercises 0.5.1 and part (a) of 0.5.2.

*(b) Now drop the hypothesis that the numbers are positive, and try to define addition. You will find that this is quite a bit harder than part (a).

*(c) Show that addition is commutative. Again, this is a lot easier when the numbers are positive.

**(d) Show that addition is associative, i.e., $x + (y + z) = (x + y) + z$. This is much harder, and requires separate consideration of the cases where each of x, y and z is positive and negative.

0.4.2 Show that if two numbers are k-close for all k, then they are equal.

*0.4.3 Show that the functions $A(x, y) = x + y$, $M(x, y) = xy$, $S(x, y) = x - y$, $(x + y) + z$ are $\mathbb{D}$-continuous, and that $1/x$ is not. Notice that for A and S, the l of Definition 0.4.4 does not depend on N, but that for M, l does depend on N.

**0.4.4 Prove Proposition 0.4.6. This can be broken into the following steps.

(a) Show that $\sup_k \inf_{l \geq k} f([x_1]_l, \ldots, [x_n]_l)$ is well defined, i.e., that the sets of numbers involved are bounded. Looking at the function S from Exercise 0.4.3, explain why both the sup and the inf are there.

(b) Show that the function $\widetilde{f}$ has the required continuity properties.

(c) Show the uniqueness.

*0.4.5 Define division of reals, using the following steps.

(a) Show that the algorithm of long division of a positive finite decimal a by a positive finite decimal b defines a repeating decimal a/b, and that $b(a/b) = a$.

(b) Show that the function $\text{inv}(x)$ defined for $x > 0$ by the formula

$$\text{inv}(x) = \inf_k 1/[x]_k$$

satisfies $x \, \text{inv}(x) = 1$ for all $x > 0$.

(c) Define the inverse for any $x \neq 0$, and show that $x \, \text{inv}(x) = 1$ for all $x \neq 0$.

digit position	0	1
even	left	right
odd	right	left

Table 0.4.6

The *image* of a mapping is discussed in Section 2.5.

**0.4.6 In this exercise we will construct a continuous mapping $\gamma : [0, 1] \to \mathbb{R}^2$, the image of which is a (full) triangle T; such a mapping is called a *Peano curve*. We will write our numbers in $[0, 1]$ in base 2, so such a number might be something like $.0011101000011\ldots$, and we will use Table 0.4.6.

Take a right triangle T. We will associate to a string $\underline{s} = s_1, s_2, \ldots$ of digits 0 and 1 a sequence of points $\mathbf{x}_0, \mathbf{x}_1, \mathbf{x}_2, \ldots$ of T by starting at the right angle $\mathbf{x}_0(\underline{s})$, dropping the perpendicular to the opposite side, landing at $\mathbf{x}_1(\underline{s})$, and deciding to turn left or right according to the digit s_1, as interpreted by the bottom line of the table, since this digit is the first digit (and therefore in an odd position): on 0 turn right and on 1 turn left.

Now drop the perpendicular to the opposite side, landing at $\mathbf{x}_2(\underline{s})$, and turn right or left according to the digit s_2, as interpreted by the top line of the table, etc.

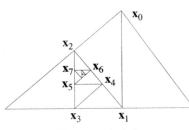

FIGURE 0.4.6.

This sequence corresponds to the string of digits

00100010010....

This construction is illustrated in Figure 0.4.6.

(a) Show that for any string of digits $(\underline{s})$, the sequence $\mathbf{x}_n(\underline{s})$ converges.

(b) Suppose a number $t \in [0,1]$ can be written in base 2 in two different ways (one ending in 0's and the other in 1's), and call $(\underline{s})$, $(\underline{s}')$ the two strings of digits. Show that

$$\lim_{n \to \infty} \mathbf{x}_n(\underline{s}) = \lim_{n \to \infty} \mathbf{x}_n(\underline{s}').$$

Hint: Construct the sequences associated to .1000... and .0111....
This allows us to define $\gamma(t) = \lim_{n \to \infty} \mathbf{x}_n(\underline{s})$.

(c) Show that γ is continuous.

(d) Show that every point in T is in the image of γ. What is the maximum number of distinct numbers $t_1, \dots, t_k$ such that $\gamma(t_1) = \cdots = \gamma(t_k)$? Hint: Choose a point in T, and draw a path of the sort above which leads to it.

0.4.7   (a) Show that the function

$$f(x) = \begin{cases} \sin \frac{1}{x} & \text{if } x \neq 0 \\ 0 & \text{if } x = 0 \end{cases} \quad \text{is not continuous.}$$

(b) Show that f satisfies the conclusion of the intermediate value theorem: if $f(x_1) = a_1$ and $f(x_2) = a_2$, then for any number a between a_1 and a_2, there exists a number x between x_1 and x_2 such that $f(x) = a$.

Exercises for Section 0.5

**Infinite Sets
and Russell's Paradox**

0.5.1   (a) Show that the set of rational numbers is countable, i.e., that you can list all rational numbers.

(b) Show that the set $\mathbb{D}$ of finite decimals is countable.

0.5.2   (a) Show that the open interval $(-1, 1)$ has the same infinity of points as the reals. Hint: Consider the function $g(x) = \tan(\pi x/2)$.

*(b) Show that the closed interval $[-1, 1]$ has the same infinity of points as the reals. For some reason, this is much trickier than (a). Hint: Choose two sequences, (1) $a_0 = 1, a_1, a_2, \dots$; and (2) $b_0 = -1, b_1, b_2, \dots$ and consider the map

$$g(x) = x \quad \text{if } x \text{ is not in either sequence.}$$
$$g(a_n) = a_{n+1}.$$
$$g(b_n) = b_{n+1}.$$

*(c) Show that the points of the circle

$$\left\{ \begin{pmatrix} x \\ y \end{pmatrix} \in \mathbb{R}^2 \mid x^2 + y^2 = 1 \right\}$$

have the same infinity of elements as $\mathbb{R}$. Hint: Again, try to choose an appropriate sequence.

*(d) Show that $\mathbb{R}^2$ has the same infinity of elements as $\mathbb{R}$.

*0.5.3 Is it possible to make a list of the rationals in $[0, 1]$, written as decimals, so that the entries on the diagonal also give a rational number?

*0.5.4 An *algebraic number* is a root of a polynomial equation with integer coefficients; for instance, the rational number p/q is algebraic, since it is a solution of $qx - p = 0$, and so is $\sqrt{2}$, since it is a root of $x^2 - 2 = 0$. A number that is not algebraic is called *transcendental*. It isn't obvious that there are any transcendental numbers; the following exercise gives a (highly unsatisfactory) proof for their existence.

Exercise 0.5.4, part (b): This proof, due to Cantor, proves that transcendental numbers exist without exhibiting a single one. Many contemporaries of Cantor were scandalized, largely for this reason.

(a) Show that the set of all algebraic numbers is countable. Hint: List the finite collection of all roots of linear polynomials with coefficients with absolute value ≤ 1. Then list the roots of all quadratic equations with coefficients ≤ 2 (which will include the linear equations, for instance $0x^2 + 2x - 1 = 0$), then all roots of cubic equation with coefficients ≤ 3, etc.

(b) Derive from part (a) that there exist transcendental numbers, in fact uncountably many of them.

Exercise 0.5.5 is the one-dimensional case of the celebrated *Brouwer fixed point theorem*, to be discussed in a subsequent volume. In dimension one it is an easy consequence of the intermediate value theorem, but in higher dimensions (even two) it is quite a delicate result.

0.5.5 Show that if $f : [a, b] \to [a, b]$ is continuous, there exists $c \in [a, b]$ with $f(c) = c$.

0.5.6 Show that if $p(x)$ is a polynomial of odd degree with real coefficients, then there is a real number c such that $f(c) = 0$.

Exercises for Section 0.6: Complex Numbers

0.6.1 Verify the nine rules for addition and multiplication of complex numbers. Statements (5) and (9) are the only ones that are not immediate.

For Exercise 0.6.2, see the subsection on the derivation of Cardano's formulas (Equation 0.6.26 in particular).

0.6.2 Show that $-p/(3u_{1,j})$ is a cubic root of v_2, and that we can renumber the cube roots of v_2 so that $-p/(3u_{1,j}) = u_{2,j}$.

0.6.3 (a) Find all the cubic roots of 1.

(b) Find all the 4th roots of 1.

*(c) Find all the 5th roots of 1. Use your formula to construct a regular pentagon using ruler and compass.

(d) Find all the 6th roots of 1.

0.6.4 Show that the following cubics have exactly one real root, and find it.
(a) $x^3 - 18x + 35 = 0$
(b) $x^3 + 3x^2 + x + 2 = 0$

0.6.5 Show that the polynomial $x^3 - 7x + 6$ has three real roots, and find them.

0.6.6   There is a way of finding the roots of real cubics with three real roots, using only real numbers and a bit of trigonometry.

(a) Prove the formula $4\cos^3\theta - 3\cos\theta - \cos 3\theta = 0$.

In Exercise 0.6.6, part (a), use de Moivre's formula:

$$\cos n\theta + i\sin n\theta = (\cos\theta + i\sin\theta)^n.$$

(b) Set $y = ax$ in the equation $x^3 + px + q = 0$, and show that there is a value of a for which the equation becomes $4y^3 - 3y - q_1 = 0$; find the value of a and of q_1.

(c) Show that there exists an angle θ such that $\cos(3\theta) = q_1$ precisely when $27q^2 + 4p^3 < 0$, i.e., precisely when the original polynomial has three real roots.

(d) Find a formula (involving arccos) for all three roots of a real cubic polynomial with three real roots.

Exercise 0.6.7 uses results from Section 3.1.

*0.6.7   In this exercise, we will find formulas for the solution of 4th degree polynomials, known as *quartics*. Let $w^4 + aw^3 + bw^2 + cw + d$ be a quartic polynomial.

(a) Show that if we set $w = x - a/4$, the quartic equation becomes

$$x^4 + px^2 + qx + r = 0,$$

and compute p, q and r in terms of a, b, c, d.

(b) Now set $y = x^2 + p/2$, and show that solving the quartic is equivalent to finding the intersections of the parabolas Γ_1 and Γ_2 of equation

$$x^2 - y + p/2 = 0 \quad \text{and} \quad y^2 + qx + r - \frac{p^2}{4} = 0$$

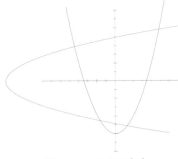

FIGURE 0.6.7(A).

The two parabolas of Equation 0.7.1; note that their axes are respectively the y-axis and the x-axis.

respectively, pictured in Figure 0.6.7 (A).

The parabolas Γ_1 and Γ_2 intersect (usually) in four points, and the curves of equation

$$f_m\begin{pmatrix}x\\y\end{pmatrix} = x^2 - y + p/2 + m\left(y^2 + qx + r - \frac{p^2}{4}\right) = 0 \qquad 0.7.1$$

are exactly the curves given by quadratic equations which pass through those four points; some of these curves are shown in Figure 0.6.7 (C).

(c) What can you say about the curve given by Equation 0.7.1 when $m = 1$? When m is negative? When m is positive?

(d) The assertion in (b) is not quite correct: there is one curve that passes through those four points, and which is given by a quadratic equation, that is missing from the family given by Equation 0.7.1. Find it.

(e) The next step is the really clever part of the solution. Among these curves, there are three, shown in Figure 0.6.7(B), that consist of a pair of lines, i.e., each such "degenerate" curve consists of a pair of diagonals of the quadrilateral formed by the intersection points of the parabolas. Since there are three of these, we may hope that the corresponding values of m are solutions of a cubic equation, and this is indeed the case. Using the fact that a pair of lines is not a smooth curve near the point where they intersect, show that the numbers

m for which the equation $f_m = 0$ defines a pair of lines, and the coordinates x, y of the point where they intersect, are the solutions of the system of three equations in three unknowns,

$$y^2 + qx + r - \frac{p^2}{4} + m(x^2 - y - p/2) = 0$$

$$2y - m = 0$$

$$q + 2mx = 0.$$

(f) Expressing x and y in terms of m using the last two equations, show that m satisfies the equation

$$m^3 - 2pm^2 + (p^2 - 4r)m + q^2 = 0$$

for m; this equation is called the *resolvent cubic* of the original quartic equation.

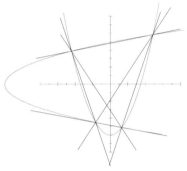

FIGURE 0.6.7(B).

The three pairs of lines that go through the intersections of the two parabolas.

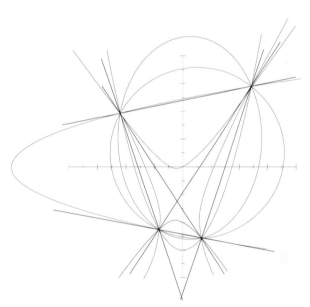

FIGURE 0.6.7 (C). The curves $f_m \begin{pmatrix} x \\ y \end{pmatrix} = x^2 - y + p/2 + m \left(y^2 + qx + r - \frac{p^2}{4} \right) = 0$ for seven different values of m.

Let m_1, m_2 and m_3 be the roots of the equation, and let $\begin{pmatrix} x_1 \\ y_1 \end{pmatrix}$, $\begin{pmatrix} x_2 \\ y_2 \end{pmatrix}$ and $\begin{pmatrix} x_3 \\ y_3 \end{pmatrix}$ be the corresponding points of intersection of the diagonals. This doesn't quite give the equations of the lines forming the two diagonals. The next part gives a way of finding them.

(g) Let $\begin{pmatrix} x_1 \\ y_1 \end{pmatrix}$ be one of the points of intersection, as above, and consider the line l_k through the point $\begin{pmatrix} x_1 \\ y_1 \end{pmatrix}$ with slope k, of equation

$$y - y_1 = k(x - x_1).$$

Show that the values of k for which l_k is a diagonal are also the values for which the restrictions of the two quadratic functions $y^2 + qx + r - \frac{p^2}{4}$ and $x^2 - y - p/2$ to l_k are proportional. Show that this gives the equations

$$\frac{1}{k^2} = \frac{-k}{2k(-kx_1 + y_1) + q} = \frac{kx_1 - y_1 + p/2}{(kx_1 - y_1)^2 - p^2/4 + r},$$

which can be reduced to the single quadratic equation

$$k^2(x_1^2 - y_1 + a/2) = y_1^2 + bx_1 - a^2/4 + c.$$

Now the full solution is at hand: compute (m_1, x_1, y_1) and (m_2, x_2, y_2); you can ignore the third root of the resolvent cubic or use it to check your answers. Then for each of these compute the slopes $k_{i,1}$ and $k_{i,2} = -k_{i,1}$ from the equation above. You now have four lines, two through A and two through B. Intersect them in pairs to find the four intersections of the parabolas.

(h) Solve the quartic equations

$$x^4 - 4x^2 + x + 1 = 0 \quad \text{and} \quad x^4 + 4x^3 + x - 1 = 0.$$

1

Vectors, Matrices, and Derivatives

It is sometimes said that the great discovery of the nineteenth century was that the equations of nature were linear, and the great discovery of the twentieth century is that they are not.—Tom Körner, *Fourier Analysis*

1.0 INTRODUCTION

In this chapter, we introduce the principal actors of linear algebra and multivariable calculus.

By and large, first year calculus deals with functions f that associate *one* number $f(x)$ to *one* number x. In most realistic situations, this is inadequate: the description of most systems depends on many functions of many variables.

In physics, a gas might be described by pressure and temperature as a function of position and time, two functions of four variables. In biology, one might be interested in numbers of sharks and sardines as functions of position and time; a famous study of sharks and sardines in the Adriatic, described in *The Mathematics of the Struggle for Life* by Vito Volterra, founded the subject of mathematical ecology.

In micro-economics, a company might be interested in production as a function of input, where that function has as many coordinates as the number of products the company makes, each depending on as many inputs as the company uses. Even thinking of the variables needed to describe a macro-economic model is daunting (although economists and the government base many decisions on such models). The examples are endless and found in every branch of science and social science.

Mathematically, all such things are represented by functions $\mathbf{f}$ that take n numbers and return m numbers; such functions are denoted $\mathbf{f} : \mathbb{R}^n \to \mathbb{R}^m$. In that generality, there isn't much to say; we must impose restrictions on the functions we will consider before any theory can be elaborated.

The strongest requirement one can make is that $\mathbf{f}$ should be *linear*; roughly speaking, a function is linear if when you double the input, you double the output. Such linear functions are fairly easy to describe completely, and a thorough understanding of their behavior is the foundation for everything else.

The first four sections of this chapter are devoted to laying the foundations of linear algebra. We will introduce the main actors, vectors and matrices, relate them to the notion of function (which we will call *transformation*), and develop the geometrical language (length of a vector, length of a matrix, ...) that we will need in multi-variable calculus. In Section 1.5 we will discuss sequences, subsequences, limits, and convergence. In Section 1.6 we will expand on that discussion, developing the topology needed for a rigorous treatment of calculus.

Most functions are not linear, but very often they are well approximated by linear functions, at least for some values of the variables. For instance, as long as there are few hares, their number may well double every year, but as soon as they become numerous, they will compete with each other, and their rate of increase (or decrease) will become more complex. In the last three sections of this chapter we will begin exploring how to approximate a nonlinear function by a linear function—specifically, by its higher-dimensional derivative.

1.1 Introducing the Actors: Vectors

Much of linear algebra and multivariate calculus takes place within $\mathbb{R}^n$. This is the space of ordered lists of n real numbers.

The notion that one can think about and manipulate higher dimensional spaces by considering a point in n-dimensional space as a list of its n "coordinates" did not always appear as obvious to mathematicians as it does today. In 1846, the English mathematician Arthur Cayley pointed out that a point with four coordinates can be interpreted geometrically without recourse to "any metaphysical notion concerning the possibility of four-dimensional space."

You are probably used to thinking of a point in the plane in terms of its two coordinates: the familiar Cartesian plane with its x, y axes is $\mathbb{R}^2$. Similarly, a point in space (after choosing axes) is specified by its three coordinates: Cartesian space is $\mathbb{R}^3$. Analogously, a point in $\mathbb{R}^n$ is specified by its n coordinates; it is a list of n real numbers. Such ordered lists occur everywhere, from grades on a transcript to prices on the stock exchange.

Seen this way, higher dimensions are no more complicated than $\mathbb{R}^2$ and $\mathbb{R}^3$; the lists of coordinates just get longer. But it is not obvious how to think about such spaces geometrically. Even the experts understand such objects only by educated analogy to objects in $\mathbb{R}^2$ or $\mathbb{R}^3$; the authors cannot "visualize $\mathbb{R}^4$" and we believe that no one really can. The object of linear algebra is at least in part to extend to higher dimensions the geometric language and intuition we have concerning the plane and space, familiar to us all from everyday experience. It will enable us to speak for instance of the "space of solutions" of a particular system of equations as being a four-dimensional subspace of $\mathbb{R}^7$.

Example 1.1.1 (The stock market). The following data is from the Ithaca Journal, Dec. 14, 1996.

"Vol" denotes the number of shares traded, "High" and "Low," the highest and lowest price paid per share, "Close," the price when trading stopped at the end of the day, and "Chg," the difference between the closing price and the closing price on the previous day.

LOCAL NYSE STOCKS

	Vol	High	Low	Close	Chg
Airgas	193	$24\,^1/_2$	$23\,^1/_8$	$23\,^5/_8$	$-^3/_8$
AT&T	36606	$39\,^1/_4$	$38\,^3/_8$	39	$^3/_8$
Borg Warner	74	$38\,^3/_8$	38	38	$-^3/_8$
Corning	4575	$44\,^3/_4$	43	$44\,^1/_4$	$^1/_2$
Dow Jones	1606	$33\,^1/_4$	$32\,^1/_2$	$33\,^1/_4$	$^1/_8$
Eastman Kodak	7774	$80\,^5/_8$	$79\,^1/_4$	$79\,^3/_8$	$-^3/_4$
Emerson Elec.	3335	$97\,^3/_8$	$95\,^5/_8$	$95\,^5/_8$	$-1\,^1/_8$
Federal Express	5828	$42\,^1/_2$	41	$41\,^5/_8$	$1\,^1/_2$

Each of these lists of eight numbers is an element of $\mathbb{R}^8$; if we were listing the full New York Stock Exchange, they would be elements of $\mathbb{R}^{3356}$.

We can think of this table as five columns, each an element of $\mathbb{R}^8$:

$$\vec{\textbf{Vol}} = \begin{bmatrix} 193 \\ 36606 \\ 74 \\ 4575 \\ 1606 \\ 7774 \\ 3335 \\ 5828 \end{bmatrix} \qquad \textbf{High} = \begin{pmatrix} 24\,{}^1/_2 \\ 39\,{}^1/_4 \\ 38\,{}^3/_8 \\ 44\,{}^3/_4 \\ 33\,{}^1/_4 \\ 80\,{}^5/_8 \\ 97\,{}^3/_8 \\ 42\,{}^1/_2 \end{pmatrix} \qquad \textbf{Low} = \begin{pmatrix} 23\,{}^1/_8 \\ 38\,{}^3/_8 \\ 38 \\ 43 \\ 32\,{}^1/_2 \\ 79\,{}^1/_4 \\ 95\,{}^5/_8 \\ 41 \end{pmatrix}$$

$$\textbf{Close} = \begin{pmatrix} 23\,{}^5/_8 \\ 39 \\ 38 \\ 44\,{}^1/_4 \\ 33\,{}^1/_4 \\ 79\,{}^3/_8 \\ 95\,{}^5/_8 \\ 41\,{}^5/_8 \end{pmatrix} \qquad \vec{\textbf{Chg}} = \begin{bmatrix} -{}^3/_8 \\ {}^3/_8 \\ -{}^3/_8 \\ {}^1/_2 \\ {}^1/_8 \\ -{}^3/_4 \\ -1\,{}^1/_8 \\ 1\,{}^1/_2 \end{bmatrix} \qquad \triangle$$

The Swiss mathematician Leonhard Euler (1707-1783) touched on all aspects of the mathematics and physics of his time. He wrote textbooks on algebra, trigonometry, and infinitesimal calculus; all texts in these fields are in some sense rewrites of Euler's. He set the notation we use from high school on: sin, cos, and tan for the trigonometric functions, $f(x)$ to indicate a function of the variable x are all due to him. Euler's complete works fill 85 large volumes—more than the number of mystery novels published by Agatha Christie; some were written after he became completely blind in 1771. Euler spent much of his professional life in St. Petersburg. He and his wife had thirteen children, five of whom survived to adulthood.

Note that we write elements of $\mathbb{R}^n$ as *columns*, not rows. The reason for preferring columns will become clear later: we want the order of terms in matrix multiplication to be consistent with the notation $f(x)$, where the function is placed before the variable—notation established by the famous mathematician Euler. Note also that we use parentheses for "positional" data and brackets for "incremental" data; the distinction is discussed below.

Points and vectors: positional data versus incremental data

An element of $\mathbb{R}^n$ is simply an ordered list of n numbers, but such a list can be interpreted in two ways: as a *point* representing a position or as a *vector* representing a displacement or increment.

Definition 1.1.2 (Point, vector, and coordinates). The element of $\mathbb{R}^n$ with coordinates $x_1, x_2, \cdots, x_n$ can be interpreted in two ways: as the point

$$\mathbf{x} = \begin{pmatrix} x_1 \\ \vdots \\ x_n \end{pmatrix}, \text{ or as the vector } \vec{\mathbf{x}} = \begin{bmatrix} x_1 \\ \vdots \\ x_n \end{bmatrix}, \text{ which represents an increment.}$$

Example 1.1.3 (An element of $\mathbb{R}^2$ as a point and as a vector). The element of $\mathbb{R}^2$ with coordinates $x = 2, y = 3$ can be interpreted as the point $\begin{pmatrix} 2 \\ 3 \end{pmatrix}$ in the plane, as shown in Figure 1.1.1. But it can also be interpreted as the instructions "start anywhere and go two units right and three units up," rather like instructions for a treasure hunt: "take two giant steps to the east, and three to the north"; this is shown in Figure 1.1.2. Here we are interested in the displacement: if we start at *any* point and travel $\begin{bmatrix} 2 \\ 3 \end{bmatrix}$, how far will we have gone, in what direction? When we interpret an element of $\mathbb{R}^n$ as a position, we call it a *point*; when we interpret it as a displacement, or increment, we call it a *vector*.   △

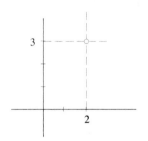

FIGURE 1.1.1.

The point $\begin{pmatrix} 2 \\ 3 \end{pmatrix}$.

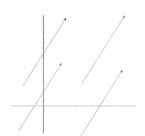

FIGURE 1.1.2.

All the arrows represent the same vector, $\begin{bmatrix} 2 \\ 3 \end{bmatrix}$.

As shown in Figure 1.1.2, in the plane (and in three-dimensional space) a vector can be depicted as an arrow pointing in the direction of the displacement. The amount of displacement is the length of the arrow. This does not extend well to higher dimensions. How are we to picture the "arrow" in $\mathbb{R}^{3356}$ representing the change in prices on the stock market? How long is it, and in what "direction" does it point? We will show how to compute these magnitudes and directions for vectors in $\mathbb{R}^n$ in Section 1.4.

Example 1.1.4 (A point as a state of a system). It is easy to think of a point in $\mathbb{R}^2$ or $\mathbb{R}^3$ as a position; in higher dimensions, it can be more helpful to think of a point as a "state" of a system. If 3356 stocks are listed on the New York Stock Exchange, the list of closing prices for those stocks is an element of $\mathbb{R}^{3356}$, and every element of $\mathbb{R}^{3356}$ is one theoretically possible state of the stock market. This corresponds to thinking of an element of $\mathbb{R}^{3356}$ as a point.

The list telling how much each stock gained or lost compared with the previous day is also an element of $\mathbb{R}^{3356}$, but this corresponds to thinking of the element as a vector, with direction and magnitude: did the price of each stock go up or down? How much?   △

Remark. In physics textbooks and some first year calculus books, vectors are often said to represent quantities (velocity, forces) that have both "magnitude" and "direction," while other quantities (length, mass, volume, temperature) have only "magnitude" and are represented by numbers (scalars). We think this focuses on the wrong distinction, suggesting that some quantities are always represented by vectors while others never are, and that it takes more information to specify a quantity with direction than one without.

The volume of a balloon is a single number, but so is the vector expressing the difference in volume between an inflated balloon and one that has popped. The first is a number in $\mathbb{R}$, while the second is a vector in $\mathbb{R}$. The height of a child is a single number, but so is the vector expressing how much he has grown since his last birthday. A temperature can be a "magnitude," as in "It got down to -20 last night," but it can also have "magnitude and direction," as in "It is 10 degrees colder today than yesterday." Nor can "static" information always be expressed by a single number: the state of the Stock Market at a given instant requires as many numbers as there are stocks listed—as does the vector describing the change in the Stock Market from one day to the next. △

Points can't be added; vectors can

As a rule, it doesn't make sense to add points together, any more than it makes sense to "add" the positions "Boston" and "New York" or the temperatures 50 degrees Fahrenheit and 70 degrees Fahrenheit. (If you opened a door between two rooms at those temperatures, the result would not be two rooms at 120

We will not consistently use different notation for the point zero and the zero vector, although philosophically the two are quite different. The zero vector, i.e., the "zero increment," has a universal meaning, the same regardless of the frame of reference. The point zero is arbitrary, just as "zero degrees" is arbitrary, and has a different meaning in the Centigrade system and in Fahrenheit.

Sometimes, often at a key point in the proof of a hard theorem, we will suddenly start thinking of points as vectors, or vice versa; this happens in the proof of Kantorovitch's theorem in Appendix A.2, for example.

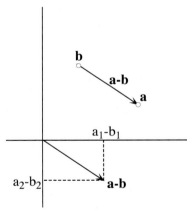

FIGURE 1.1.3.

The difference $\mathbf{a} \overset{\longrightarrow}{-} \mathbf{b}$ between point $\mathbf{a}$ and point $\mathbf{b}$ is the vector joining them. The difference can be computed by subtracting the coordinates of $\mathbf{b}$ from those of $\mathbf{a}$.

degrees!) But it does make sense to measure the difference between points (i.e., to subtract them): you can talk about the distance between Boston and New York, or about the difference in temperature between two rooms. The result of subtracting one point from another is thus a vector specifying the increment you need to add to get from one point to another.

You can also add increments (vectors) together, giving another increment. For instance the vectors "advance five meters east then take two giant steps south" and "take three giant steps north and go seven meters west" can be added, to get "advance 2 meters west and one giant step north."

Similarly, in the NYSE table in Example 1.1.1, adding the *Close* columns on two successive days does not produce a meaningful answer. But adding the *Chg* columns for each day of a week produces a perfectly meaningful increment: the change in the market over that week. It is also meaningful to add increments to points (giving a point): adding a *Chg* column to the previous day's *Close* column produces the current day's *Close*—the new state of the system.

To help distinguish these two kinds of elements of $\mathbb{R}^n$, we will denote them differently: points will be denoted by boldface lower case letters, and vectors will be lower case boldface letters with arrows above them. Thus $\mathbf{x}$ is a point in $\mathbb{R}^2$, while $\vec{\mathbf{x}}$ is a vector in $\mathbb{R}^2$. We do not distinguish between entries of points and entries of vectors; they are all written in plain type, with subscripts. However, when we write elements of $\mathbb{R}^n$ as columns, we will use parentheses for a point $\mathbf{x}$ and square brackets for a vector $\vec{\mathbf{x}}$: in $\mathbb{R}^2$, $\mathbf{x} = \begin{pmatrix} x_1 \\ x_2 \end{pmatrix}$ and $\vec{\mathbf{x}} = \begin{bmatrix} x_1 \\ x_2 \end{bmatrix}$.

Remark. An element of $\mathbb{R}^n$ is an element of $\mathbb{R}^n$—i.e., an ordered list of numbers—whether it is interpreted as a point or as a vector. But we have very different images of points and vectors, and we hope that sharing them with you explicitly will help you build a sound intuition. In linear algebra, you should just think of elements of $\mathbb{R}^n$ as vectors. However, differential calculus is all about increments to points. It is because the increments are vectors that linear algebra is a prerequisite for multivariate calculus: it provides the right language and tools for discussing these increments.

Subtraction and addition of vectors and points

The difference between point $\mathbf{a}$ and point $\mathbf{b}$ is the vector $\overset{\longrightarrow}{\mathbf{a} - \mathbf{b}}$, as shown in Figure 1.1.3.

Vectors are added by adding the corresponding coordinates:

$$\underbrace{\begin{bmatrix} v_1 \\ v_2 \\ \vdots \\ v_n \end{bmatrix}}_{\vec{\mathbf{v}}} + \underbrace{\begin{bmatrix} w_1 \\ w_2 \\ \vdots \\ w_n \end{bmatrix}}_{\vec{\mathbf{w}}} = \underbrace{\begin{bmatrix} v_1 + w_1 \\ v_2 + w_2 \\ \vdots \\ v_n + w_n \end{bmatrix}}_{\vec{\mathbf{v}} + \vec{\mathbf{w}}} ; \qquad 1.1.1$$

the result is a vector. Similarly, vectors are subtracted by subtracting the corresponding coordinates to get a new vector. A point and a vector are added by adding the corresponding coordinates; the result is a point.

In the plane, the sum $\vec{v} + \vec{w}$ is the diagonal of the parallelogram of which two adjacent sides are $\vec{v}$ and $\vec{w}$, as shown in Figure 1.1.4 (left). We can also add vectors by placing the beginning of one vector at the end of the other, as shown in Figure 1.1.4 (right).

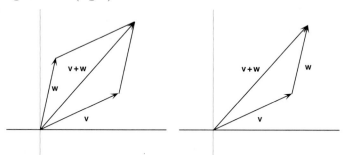

FIGURE 1.1.4. In the plane, the sum $\vec{v} + \vec{w}$ is the diagonal of the parallelogram at left. We can also add them by putting them head to tail.

If we were working with complex vector spaces, our scalars would be complex numbers; in number theory, scalars might be the rational numbers; in coding theory, they might be elements of a finite field. (You may have run into such things under the name of "clock arithmetic.") We use the word "scalar" rather than "real number" because most theorems in linear algebra are just as true for complex vector spaces or rational vector spaces as for real ones, and we don't want to restrict the validity of the statements unnecessarily.

Multiplying vectors by scalars

Multiplication of a vector by a *scalar* is straightforward:

$$a \begin{bmatrix} x_1 \\ \vdots \\ x_n \end{bmatrix} = \begin{bmatrix} ax_1 \\ \vdots \\ ax_n \end{bmatrix}; \quad \text{for example,} \quad \sqrt{3} \begin{bmatrix} 3 \\ -1 \\ 2 \end{bmatrix} = \begin{bmatrix} 3\sqrt{3} \\ -1\sqrt{3} \\ 2\sqrt{3} \end{bmatrix}. \qquad 1.1.2$$

In this book, our vectors will be lists of real numbers, so that our scalars—the kinds of numbers we are allowed to multiply vectors or matrices by—are real numbers.

The symbol $\in$ means "element of." Out loud, one says "in." The expression "$\vec{x}, \vec{y} \in V$" means "$\vec{x} \in V$ and $\vec{y} \in V$." If you are unfamiliar with the notation of set theory, see the discussion in Section 0.3.

Subspaces of $\mathbb{R}^n$

A subspace of $\mathbb{R}^n$ is a subset of $\mathbb{R}^n$ that is closed under addition and multiplication by scalars.[1] (This $\mathbb{R}^n$ should be thought of as made up of vectors, not points.)

---

[1]In Section 2.6 we will discuss abstract vector spaces. These are sets in which one can add and multiply by scalars, and where these operations satisfy rules (ten of them) that make them clones of $\mathbb{R}^n$. Subspaces of $\mathbb{R}^n$ will be our main examples of vector spaces.

Definition 1.1.5 (Subspace of $\mathbb{R}^n$). A non-empty subset $V \subset \mathbb{R}^n$ is called a subspace if it is closed under addition and closed under multiplication by scalars; i.e., V is a subspace if when

$$\vec{\mathbf{x}}, \vec{\mathbf{y}} \in V, \text{ and } a \in \mathbb{R}, \qquad \text{then} \qquad \vec{\mathbf{x}} + \vec{\mathbf{y}} \in V \text{ and } a\vec{\mathbf{x}} \in V.$$

To be closed under multiplication a subspace must contain the zero vector, so that

$$0 \cdot \vec{\mathbf{v}} = \vec{\mathbf{0}}.$$

For example, a straight line through the origin is a subspace of $\mathbb{R}^2$ and of $\mathbb{R}^3$. A plane through the origin is a subspace of $\mathbb{R}^3$. The set consisting of just the zero vector $\{\vec{\mathbf{0}}\}$ is a subspace of any $\mathbb{R}^n$, and $\mathbb{R}^n$ is a subspace of itself. These last two, $\{\vec{\mathbf{0}}\}$ and $\mathbb{R}^n$, are considered *trivial subspaces*.

Intuitively, it is clear that a line that is a subspace has dimension 1, and a plane that is a subspace has dimension 2. Being precise about what this means requires some "machinery" (mainly the notions of linear independence and span), introduced in Section 2.4.

The standard basis vectors

The notation for the standard basis vectors is ambiguous; at right we have three different vectors, all denoted $\vec{\mathbf{e}}_1$. The subscript tells us which entry is 1 but does not say how many entries the vector has—i.e., whether it is a vector in $\mathbb{R}^2$, $\mathbb{R}^3$ or what.

We will meet one particular family of vectors in $\mathbb{R}^n$ often: the *standard basis vectors*. In $\mathbb{R}^2$ there are two standard basis vectors, $\vec{\mathbf{e}}_1$ and $\vec{\mathbf{e}}_2$; in $\mathbb{R}^3$, there are three:

$$\text{in } \mathbb{R}^2 : \vec{\mathbf{e}}_1 = \begin{bmatrix} 1 \\ 0 \end{bmatrix}, \vec{\mathbf{e}}_2 = \begin{bmatrix} 0 \\ 1 \end{bmatrix}; \quad \text{in } \mathbb{R}^3 : \vec{\mathbf{e}}_1 = \begin{bmatrix} 1 \\ 0 \\ 0 \end{bmatrix}, \vec{\mathbf{e}}_2 = \begin{bmatrix} 0 \\ 1 \\ 0 \end{bmatrix}, \vec{\mathbf{e}}_3 = \begin{bmatrix} 0 \\ 0 \\ 1 \end{bmatrix}.$$

Similarly, in $\mathbb{R}^5$ there are five standard basis vectors:

$$\vec{\mathbf{e}}_1 = \begin{bmatrix} 1 \\ 0 \\ 0 \\ 0 \\ 0 \end{bmatrix}, \quad \vec{\mathbf{e}}_2 = \begin{bmatrix} 0 \\ 1 \\ 0 \\ 0 \\ 0 \end{bmatrix}, \quad \dots, \quad \vec{\mathbf{e}}_5 = \begin{bmatrix} 0 \\ 0 \\ 0 \\ 0 \\ 1 \end{bmatrix}.$$

The standard basis vectors in $\mathbb{R}^2$ and $\mathbb{R}^3$ are often denoted $\vec{i}, \vec{j}$, and $\vec{k}$:

$$\vec{i} = \vec{\mathbf{e}}_1 = \begin{bmatrix} 1 \\ 0 \end{bmatrix} \text{ or } \begin{bmatrix} 1 \\ 0 \\ 0 \end{bmatrix};$$

$$\vec{j} = \vec{\mathbf{e}}_2 = \begin{bmatrix} 0 \\ 1 \end{bmatrix} \text{ or } \begin{bmatrix} 0 \\ 1 \\ 0 \end{bmatrix};$$

$$\vec{k} = \vec{\mathbf{e}}_3 = \begin{bmatrix} 0 \\ 0 \\ 1 \end{bmatrix}.$$

We do not use this notation but mention it in case you encounter it elsewhere.

Definition 1.1.6 (Standard basis vectors). The standard basis vectors in $\mathbb{R}^n$ are the vectors $\vec{\mathbf{e}}_j$ with n entries, the jth entry 1 and the others zero.

Geometrically, there is a close connection between the standard basis vectors in $\mathbb{R}^2$ and a choice of axes in the Euclidean plane. When in school you drew an x-axis and y-axis on a piece of paper and marked off units so that you could plot a point, you were identifying the plane with $\mathbb{R}^2$: each point on the plane corresponded to a pair of real numbers—its coordinates with respect to those axes. A set of axes providing such an identification must have an origin, and each axis must have a direction (so you know what is positive and what is

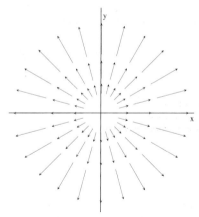

FIGURE 1.1.5.

The point marked with a circle is the point $\begin{pmatrix} 2 \\ 3 \end{pmatrix}$ in this non-orthogonal coordinate system.

negative) and it must have units (so you know, for example, where $x = 3$ or $y = 2$ is).

Such axes need not be at right angles, and the units on one axis need not be the same as those on the other, as shown in Figure 1.1.5. However, the identification is more useful if we choose the axes at right angles (orthogonal) and the units equal; the plane with such axes, generally labeled x and y, is known as the Cartesian plane. We can think that $\vec{e}_1$ measures one unit along the x-axis, going to the right, and $\vec{e}_2$ measures one unit along the y-axis, going "up."

Vector fields

Virtually all of physics deals with fields. The electric and magnetic fields of electromagnetism, the gravitational and other force fields of mechanics, the velocity fields of fluid flow, the wave function of quantum mechanics, are all "fields." Fields are also used in other subjects, epidemiology and population studies, for instance.

By "field" we mean data that varies from point to point. Some fields, like temperature or pressure distribution, are scalar fields: they associate a number to every point. Some fields, like the Newtonian gravitation field, are best modeled by vector fields, which associate a vector to every point. Others, like the electromagnetic field and charge distributions, are best modeled by *form fields*, discussed in Chapter 6. Still others, like the Einstein field of general relativity (a field of pseudo inner products), are none of the above.

Definition 1.1.7 (Vector field). A vector field on $\mathbb{R}^n$ is a function whose input is a point in $\mathbb{R}^n$ and whose output is a vector (also in $\mathbb{R}^n$) emanating from that point.

We will distinguish between functions and vector fields by putting arrows on vector fields, as in $\vec{F}$ in Example 1.1.8.

FIGURE 1.1.6.

A vector field associates a vector to each point. Here we show the radial vector field

$$\vec{F} \begin{pmatrix} x \\ y \end{pmatrix} = \begin{bmatrix} x \\ y \end{bmatrix}.$$

Vector fields generally are easier to depict when one scales the vectors down, as we have done above and in Figure 1.1.7.

Example 1.1.8 (Vector fields in $\mathbb{R}^2$). The identity *function* in $\mathbb{R}^2$

$$f \begin{pmatrix} x \\ y \end{pmatrix} = \begin{pmatrix} x \\ y \end{pmatrix} \qquad 1.1.3$$

takes a point in $\mathbb{R}^2$ and returns the same point. But the *vector field*

$$\vec{F} \begin{pmatrix} x \\ y \end{pmatrix} = \begin{bmatrix} x \\ y \end{bmatrix} \qquad 1.1.4$$

takes a point in $\mathbb{R}^2$ and assigns to it the vector corresponding to that point, as shown in Figure 1.1.6. To the point with coordinates $(1, 1)$ it assigns the vector $\begin{bmatrix} 1 \\ 1 \end{bmatrix}$; to the point with coordinates $(4, 2)$ it assigns the vector $\begin{bmatrix} 4 \\ 2 \end{bmatrix}$.

Actually, a vector field simply associates to each point a vector; how you imagine that vector is up to you. But it is always helpful to imagine each vector anchored at, or emanating from, the corresponding point.

Similarly, the vector field $\vec{F}\begin{pmatrix} x \\ y \end{pmatrix} = \begin{bmatrix} xy - 2 \\ x - y \end{bmatrix}$, shown in Figure 1.1.7, takes a point in $\mathbb{R}^2$ and assigns to it the vector $\begin{bmatrix} xy - 2 \\ x - y \end{bmatrix}$. $\triangle$

Vector fields are often used to describe the flow of fluids or gases: the vector assigned to each point gives the velocity and direction of the flow. For flows that don't change over time (steady-state flows), such a vector field gives a complete description. In more realistic cases where the flow is constantly changing, the vector field gives a snapshot of the flow at a given instant. Vector fields are also used to describe force fields such as electric fields or gravitational fields.

1.2 INTRODUCING THE ACTORS: MATRICES

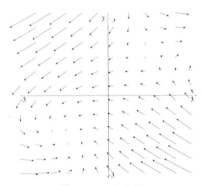

FIGURE 1.1.7.
The vector field
$$\vec{F}\begin{pmatrix} x \\ y \end{pmatrix} = \begin{bmatrix} xy - 2 \\ x - y \end{bmatrix}.$$

Probably no other area of mathematics has been applied in such numerous and diverse contexts as the theory of matrices. In mechanics, electromagnetics, statistics, economics, operations research, the social sciences, and so on, the list of applications seems endless. By and large this is due to the utility of matrix structure and methodology in conceptualizing sometimes complicated relationships and in the orderly processing of otherwise tedious algebraic calculations and numerical manipulations.—James Cochran, *Applied Mathematics: Principles, Techniques, and Applications*

The other central actor in linear algebra is the *matrix*.

Definition 1.2.1 (Matrix). An $m \times n$ matrix is a rectangular array of entries, m high and n wide.

We use capital letters to denote matrices. Usually our matrices will be arrays of numbers, real or complex, but matrices can be arrays of polynomials, or of more general functions; a matrix can even be an array of other matrices. A vector $\vec{v} \in \mathbb{R}^m$ is an $m \times 1$ matrix.

Addition of matrices, and multiplication of a matrix by a scalar, work in the obvious way:

Example 1.2.2 (Addition of matrices and multiplication by a scalar).

When a matrix is described, height is given first, then width: an $m \times n$ matrix is m high and n wide. After struggling for years to remember which goes first, one of the authors hit on a mnemonic: first take the elevator, then walk down the hall.

$$\begin{bmatrix} 1 & 0 \\ 2 & -1 \\ 4 & 2 \end{bmatrix} + \begin{bmatrix} 0 & -3 \\ 1 & -2 \\ 3 & 1 \end{bmatrix} = \begin{bmatrix} 1 & -3 \\ 3 & -3 \\ 7 & 3 \end{bmatrix} \quad \text{and} \quad 2\begin{bmatrix} 1 & 4 \\ -2 & 3 \end{bmatrix} = \begin{bmatrix} 2 & 8 \\ -4 & 6 \end{bmatrix} \quad \triangle$$

So far, it's not clear that matrices gain us anything. Why put numbers (or other entries) into a rectangular array? What do we gain by talking about the

2×2 matrix $\begin{bmatrix} a & b \\ c & d \end{bmatrix}$ rather than the point $\begin{bmatrix} a \\ b \\ c \\ d \end{bmatrix} \in \mathbb{R}^4$? The answer is that the matrix format allows another operation to be performed: *matrix multiplication.* We will see in Section 1.3 that every linear transformation corresponds to multiplication by a matrix. This is one reason matrix multiplication is a natural and important operation; other important applications of matrix multiplication are found in probability theory and graph theory.

How would you add the matrices

$$\begin{bmatrix} 1 & 2 & 5 \\ 0 & 2 & 3 \end{bmatrix} \quad \text{and} \quad \begin{bmatrix} 1 & 2 \\ 0 & 2 \end{bmatrix} \ ?$$

You can't: matrices can be added only if they have the same height and same width.

Matrices were introduced by Arthur Cayley, a lawyer who became a mathematician, in *A Memoir on the Theory of Matrices,* published in 1858. He denoted the multiplication of a 3×3 matrix by the vector $\begin{bmatrix} x \\ y \\ z \end{bmatrix}$ using the format

$$\begin{array}{c} (\ a, \quad b, \quad c\ )(x, y, z) \\ \begin{vmatrix} a', & b', & c' \\ a'', & b'', & c'' \end{vmatrix}. \end{array}$$

" ... when Werner Heisenberg discovered 'matrix' mechanics in 1925, he didn't know what a matrix was (Max Born had to tell him), and neither Heisenberg nor Born knew what to make of the appearance of matrices in the context of the atom."—Manfred R. Schroeder, "Number Theory and the Real World," *Mathematical Intelligencer,* Vol. 7, No. 4

Matrix multiplication is best learned by example. The simplest way to multiply A times B is to write B above and to the right of A. Then the product AB fits in the space to the right of A and below B, the i, jth entry of AB being the intersection of the ith row of A and the jth column of B, as shown in Example 1.2.3. Note that for AB to exist, *the width of A must equal the height of B.* The resulting matrix then has the height of A and the width of B.

Example 1.2.3 (Matrix multiplication). The first entry of the product AB is obtained by multiplying, one by one, the entries of the first *row* of A by those of the first *column* of B, and adding these products together: in Equation 1.2.1, $(2 \times 1) + (-1 \times 3) = -1$. The second entry is obtained by multiplying the first row of A by the second column of B: $(2 \times 4) + (-1 \times 0) = 8$. After multiplying the first row of A by all the columns of B, the process is repeated with the second row of A: $(3 \times 1) + (2 \times 3) = 9$, and so on.

$$[A][B] = [AB] \qquad \begin{bmatrix} & B & \\ \hline A & & AB \end{bmatrix} \qquad \underbrace{\begin{bmatrix} 2 & -1 \\ 3 & 2 \end{bmatrix}}_{A} \underbrace{\begin{bmatrix} 1 & 4 & -2 \\ 3 & 0 & 2 \\ \hline -1 & 8 & -6 \\ 9 & 12 & -2 \end{bmatrix}}_{AB} \triangle$$

$$1.2.1$$

Given the matrices

$$A = \begin{bmatrix} 1 & 0 \\ 2 & 3 \end{bmatrix} \quad B = \begin{bmatrix} 0 & 1 \\ 0 & 1 \end{bmatrix} \quad C = \begin{bmatrix} 1 & -1 & 1 \\ 1 & 0 & -1 \end{bmatrix} \quad D = \begin{bmatrix} 1 & 0 \\ 2 & 2 \\ 1 & 1 \end{bmatrix},$$

what are the products AB, AC and CD? Check your answers below.[2] Now compute BA. What do you notice? What if you try to compute CA?[3]

---

[2] $AB = \begin{bmatrix} 0 & 1 \\ 0 & 5 \end{bmatrix}$; $\quad AC = \begin{bmatrix} 1 & -1 & 1 \\ 5 & -2 & -1 \end{bmatrix}$; $\quad CD = \begin{bmatrix} 0 & -1 \\ 0 & -1 \end{bmatrix}$.

[3] Matrix multiplication is *not* commutative; $BA = \begin{bmatrix} 2 & 3 \\ 2 & 3 \end{bmatrix}$, which is *not* equal to $AB = \begin{bmatrix} 0 & 1 \\ 0 & 5 \end{bmatrix}$. Although the product AC exists, you cannot compute CA.

Below we state the formal definition of the process we've just described. If the indices bother you, do refer to Figure 1.2.1.

Definition 1.2.4 says nothing new, but it provides some practice moving between the concrete (multiplying two particular matrices) and the symbolic (expressing this operation so that it applies to any two matrices of appropriate dimensions, even if the entries are complex numbers or even functions, rather than real numbers.) In linear algebra one is constantly moving from one form of representation (one "language") to another. For example, as we have seen, a point in $\mathbb{R}^n$ can be considered as a single entity, $\mathbf{b}$, or as the ordered list of its coordinates; matrix A can be thought of as a single entity or as a rectangular array of its entries.

In Example 1.2.3, A is a 2×2 matrix and B is a 2×3 matrix, so that $n = 2$, $m = 2$ and $p = 3$; the product C is then a 2×3 matrix. If we set $i = 2$ and $j = 3$, we see that the entry $c_{2,3}$ of the matrix C is

$$c_{2,3} = a_{2,1}b_{1,3} + a_{2,2}b_{2,3}$$
$$= (3 \cdot -2) + (2 \cdot 2)$$
$$= -6 + 4 = -2.$$

Using the format for matrix multiplication shown in Example 1.2.3, the i,jth entry is the entry at the intersection of the ith row and jth column.

Definition 1.2.4 (Matrix multiplication). If A is an $m \times n$ matrix whose (i,j)th entry is $a_{i,j}$, and B is an $n \times p$ matrix whose (i,j)th entry is $b_{i,j}$, then $C = AB$ is the $m \times p$ matrix with entries

$$c_{i,j} = \sum_{k=1}^{n} a_{i,k}b_{k,j}$$

$$= a_{i,1}b_{1,j} + a_{i,2}b_{2,j} + \cdots + a_{i,n}b_{n,j}.$$

1.2.2

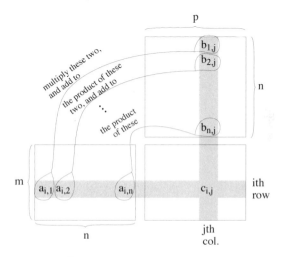

FIGURE 1.2.1. The entry $c_{i,j}$ of the matrix $C = AB$ is the sum of the products of the entries of the $a_{i,k}$ of the matrix A and the corresponding entry $b_{k,j}$ of the matrix B. The entries $a_{i,k}$ are all in the ith row of A; the first index i is constant, and the second index k varies. The entries $b_{k,j}$ are all in the jth column of B; the first index k varies, and the second index j is constant. Since the width of A equals the height of B, the entries of A and those of B can be paired up exactly.

Remark. Often people write a problem in matrix multiplication in a row: $[A][B] = [AB]$. The format shown in Example 1.2.3 avoids confusion: the product of the ith row of A and the jth column of B lies at the intersection of that row and column. It also avoids recopying matrices when doing repeated multiplications, for example A times B times C times D:

$$\begin{bmatrix} A \end{bmatrix} \begin{bmatrix} B \end{bmatrix} \begin{bmatrix} (AB) \end{bmatrix} \begin{bmatrix} C \end{bmatrix} \begin{bmatrix} (AB)C \end{bmatrix} \begin{bmatrix} D \end{bmatrix} \begin{bmatrix} (ABC)D \end{bmatrix} \quad \triangle \qquad 1.2.3$$

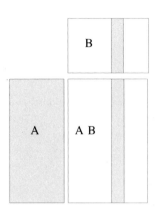

FIGURE 1.2.2.

The ith column of the product AB depends on all the entries of A but only the ith column of B.

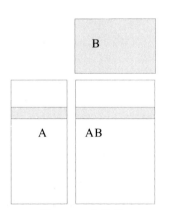

FIGURE 1.2.3.

The jth row of the product AB depends on all the entries of B but only the jth row of A.

Multiplying a matrix by a standard basis vector

Observe that multiplying a matrix A by the standard basis vector $\vec{e}_i$ selects out the ith column of A, as shown in the following example. We will use this fact often.

Example 1.2.5 (The ith column of A is $A\vec{e}_i$). Below, we show that the second column of A is $A\vec{e}_2$:

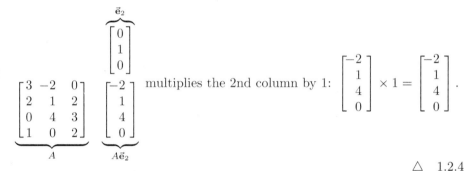

$$\underbrace{\begin{bmatrix} 3 & -2 & 0 \\ 2 & 1 & 2 \\ 0 & 4 & 3 \\ 1 & 0 & 2 \end{bmatrix}}_{A} \overbrace{\begin{bmatrix} 0 \\ 1 \\ 0 \end{bmatrix}}^{\vec{e}_2} \underbrace{\begin{bmatrix} -2 \\ 1 \\ 4 \\ 0 \end{bmatrix}}_{A\vec{e}_2} \quad \text{multiplies the 2nd column by 1:} \quad \begin{bmatrix} -2 \\ 1 \\ 4 \\ 0 \end{bmatrix} \times 1 = \begin{bmatrix} -2 \\ 1 \\ 4 \\ 0 \end{bmatrix}.$$

$$\triangle \quad 1.2.4$$

Similarly, the ith column of AB is $A\vec{b}_i$, where $\vec{b}_i$ is the ith column of B, as shown in Example 1.2.6 and represented in Figure 1.2.2. The jth *row* of AB is the product of the jth row of A and the matrix B, as shown in Example 1.2.7 and Figure 1.2.3.

Example 1.2.6. The second column of the product AB is the same as the product of the second column of A and the matrix B:

$$\underbrace{\begin{bmatrix} 2 & -1 \\ 3 & 2 \end{bmatrix}}_{A} \underbrace{\begin{bmatrix} -1 & 8 & -6 \\ 9 & 12 & -2 \end{bmatrix}}_{AB} \overbrace{\begin{bmatrix} 1 & 4 & -2 \\ 3 & 0 & 2 \end{bmatrix}}^{B} \quad \underbrace{\begin{bmatrix} 2 & -1 \\ 3 & 2 \end{bmatrix}}_{A} \underbrace{\begin{bmatrix} 8 \\ 12 \end{bmatrix}}_{A\vec{b}_2} \overbrace{\begin{bmatrix} 4 \\ 0 \end{bmatrix}}^{\vec{b}_2} \qquad 1.2.5$$

Example 1.2.7. The second row of the product AB is the same as the product of the second row of A and the matrix B:

In his 1858 article on matrices, Cayley stated that matrix multiplication is associative but gave no proof. The impression one gets is that he played around with matrices (mostly 2×2 and 3×3) to get some feeling for how they behave, without worrying about rigor. Concerning another matrix result (the Cayley-Hamilton theorem) he verifies it for 3×3 matrices, adding *I have not thought it necessary to undertake the labour of a formal proof of the theorem in the general case of a matrix of any degree.*

$$
\underbrace{\begin{bmatrix} 2 & -1 \\ 3 & 2 \end{bmatrix}}_{A} \underbrace{\begin{bmatrix} \overbrace{1 \quad 4 \quad -2}^{B} \\ 3 \quad 0 \quad 2 \\ -1 \quad 8 \quad -6 \\ 9 \quad 12 \quad -2 \end{bmatrix}}_{AB} \qquad \begin{bmatrix} 3 & 2 \end{bmatrix} \begin{bmatrix} \overbrace{1 \quad 4 \quad -2}^{B} \\ 3 \quad 0 \quad 2 \\ 9 \quad 12 \quad -2 \end{bmatrix} \qquad 1.2.6
$$

Matrix multiplication is associative

When multiplying the matrices $A, B,$ and C, we could set up the repeated multiplication as we did in Equation 1.2.3, which corresponds to the product $(AB)C$. We can use another format to get the product $A(BC)$:

$$
\begin{array}{cc} & \begin{bmatrix} B \end{bmatrix} \begin{bmatrix} C \end{bmatrix} \\ \begin{bmatrix} A \end{bmatrix} & \begin{bmatrix} AB \end{bmatrix} \begin{bmatrix} (AB)C \end{bmatrix} \end{array} \quad \text{or} \quad \begin{array}{cc} & \begin{bmatrix} C \end{bmatrix} \\ \begin{bmatrix} B \end{bmatrix} & \begin{bmatrix} (BC) \end{bmatrix} \\ \begin{bmatrix} A \end{bmatrix} & \begin{bmatrix} A(BC) \end{bmatrix} \end{array}. \qquad 1.2.7
$$

Is $(AB)C$ the same as $A(BC)$? In Section 1.3 we give a conceptual reason why they are; here we give a computational proof.

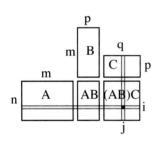

FIGURE 1.2.4.

This way of writing the matrices corresponds to calculating $(AB)C$.

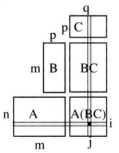

FIGURE 1.2.5.

This way of writing the matrices corresponds to calculating $A(BC)$.

Proposition 1.2.8 (Matrix multiplication is associative). *If A is an $n \times m$ matrix, B is an $m \times p$ matrix and C is a $p \times q$ matrix, so that $(AB)C$ and $A(BC)$ are both defined, then they are equal:*

$$
(AB)C = A(BC). \qquad 1.2.8
$$

Proof. Figures 1.2.4 and 1.2.5 show that the i, jth entry of both $A(BC)$ and $(AB)C$ depend only on the ith row of A and the jth column of C (but on all the entries of B), and that without loss of generality we can assume that A is a line matrix and that C is a column matrix, i.e., that $n = q = 1$, so that both $(AB)C$ and $A(BC)$ are numbers. The proof is now an application of associativity of multiplication of numbers:

$$
(AB)C = \sum_{l=1}^{p} \underbrace{\left(\sum_{k=1}^{m} a_k b_{k,l} \right)}_{l\text{th entry of } AB} c_l
$$

$$
= \sum_{l=1}^{p} \sum_{k=1}^{m} a_k b_{k,l} c_l = \sum_{k=1}^{m} a_k \underbrace{\left(\sum_{l=1}^{p} b_{k,l} c_l \right)}_{k\text{th entry of } BC} = A(BC). \qquad \Box
$$

$$1.2.9$$

Non-commutativity of matrix multiplication

Exercise 1.2.2 provides practice on matrix multiplication. At the end of this section, Example 1.2.22, involving graphs, shows a setting where matrix multiplication is a natural and powerful tool.

As we saw earlier, matrix multiplication is *most definitely not commutative.* It may well be possible to multiply A by B but not B by A. Even if both matrices have the same number of rows and columns, AB will usually not equal BA, as shown in Example 1.2.9.

Example 1.2.9 (Matrix multiplication is not commutative). If you multiply the matrix $\begin{bmatrix} 0 & 1 \\ 1 & 1 \end{bmatrix}$ by the matrix $\begin{bmatrix} 0 & 1 \\ 1 & 0 \end{bmatrix}$, the answer you get will depend on which one you put first:

$$\begin{bmatrix} 0 & 1 \\ 1 & 1 \end{bmatrix} \overbrace{\begin{bmatrix} 0 & 1 \\ 1 & 0 \end{bmatrix} \begin{bmatrix} 1 & 0 \\ 1 & 1 \end{bmatrix}} \quad \text{is not equal to} \quad \begin{bmatrix} 0 & 1 \\ 1 & 0 \end{bmatrix} \overbrace{\begin{bmatrix} 0 & 1 \\ 1 & 1 \end{bmatrix} \begin{bmatrix} 1 & 1 \\ 0 & 1 \end{bmatrix}} \qquad \triangle \quad 1.2.10$$

The identity matrix

The identity matrix I plays the same role in matrix multiplication as the number 1 does in multiplication of numbers: $IA = A = AI$.

The main diagonal is also called the *diagonal*. The diagonal from bottom left to top right is the *anti-diagonal*.

Definition 1.2.10 (Identity matrix). The identity matrix I_n is the $n \times n$-matrix with 1's along the main diagonal (the diagonal from top left to bottom right) and 0's elsewhere.

Multiplication by the identity matrix I does not change the matrix being multiplied.

For example,

$$I_2 = \begin{bmatrix} 1 & 0 \\ 0 & 1 \end{bmatrix} \quad \text{and} \quad I_4 = \begin{bmatrix} 1 & 0 & 0 & 0 \\ 0 & 1 & 0 & 0 \\ 0 & 0 & 1 & 0 \\ 0 & 0 & 0 & 1 \end{bmatrix}. \qquad 1.2.11$$

If A is an $n \times m$-matrix, then

$$IA = AI = A, \qquad \text{or, more precisely,} \qquad I_n A = A I_m = A, \qquad 1.2.12$$

since if $n \neq m$ one must change the size of the identity matrix to match the size of A. When the context is clear, we will omit the index.

The columns of the identity matrix I_n are of course the standard basis vectors $\vec{e}_1, \dots, \vec{e}_n$:

$$I_4 = \begin{bmatrix} 1 & 0 & 0 & 0 \\ 0 & 1 & 0 & 0 \\ 0 & 0 & 1 & 0 \\ 0 & 0 & 0 & 1 \end{bmatrix}.$$
$$ \ \vec{e}_1 \ \ \vec{e}_2 \ \ \vec{e}_3 \ \ \vec{e}_4$$

Matrix inverses

The inverse A^{-1} of a matrix A plays the same role in matrix multiplication as the inverse $1/a$ does for the number a. We will see in Section 2.3 that we can use the inverse of a matrix to solve systems of linear equations.

The only number that does not have an inverse is 0, but many matrices do not have inverses. In addition, the non-commutativity of matrix multiplication makes the definition more complicated.

Definition 1.2.11 (Left and right inverses of matrices). Let A be a matrix. If there is another matrix B such that

$$BA = I,$$

then B is called a left inverse of A. If there is another matrix C such that

$$AC = I,$$

then C is called a right inverse of A.

It is possible for a nonzero matrix to have neither a right nor a left inverse.

We will see in Section 2.3 that only square matrices can have a two-sided inverse, i.e., an inverse. Furthermore, if a square matrix has a left inverse then that left inverse is necessarily also a right inverse; similarly, if it has a right inverse, that right inverse is necessarily a left inverse.

It is possible for a non-square matrix to have lots of left inverses and no right inverse, or lots of right inverses and no left inverse, as explored in Exercise 1.2.20.

Example 1.2.12 (A matrix with neither right nor left inverse). The matrix $\begin{bmatrix} 1 & 0 \\ 0 & 0 \end{bmatrix}$ does not have a right or a left inverse. To see this, assume it has a right inverse. Then there exists a matrix $\begin{bmatrix} a & b \\ c & d \end{bmatrix}$ such that

$$\begin{bmatrix} 1 & 0 \\ 0 & 0 \end{bmatrix} \begin{bmatrix} a & b \\ c & d \end{bmatrix} = \begin{bmatrix} 1 & 0 \\ 0 & 1 \end{bmatrix}. \qquad 1.2.13$$

But that product is $\begin{bmatrix} a & b \\ 0 & 0 \end{bmatrix}$, i.e., in the bottom right-hand corner, $0 = 1$. A similar computation shows that there is no left inverse. $\triangle$

Definition 1.2.13 (Invertible matrix). An invertible matrix is a matrix that has both a left inverse and a right inverse.

Associativity of matrix multiplication gives the following result:

While we can write the inverse of a number x either as x^{-1} or as $1/x$, giving $x\,x^{-1} = x\,(1/x) = 1$, the inverse of a matrix A is only written A^{-1}. We *cannot divide by a matrix.* If for two matrices A and B you were to write A/B, it would be unclear whether this meant

$$B^{-1}A \quad \text{or} \quad AB^{-1}.$$

Proposition and Definition 1.2.14. *If a matrix A has both a left and a right inverse, then it has only one left inverse and one right inverse, and they are identical; such a matrix is called the inverse of A and is denoted A^{-1}.*

Proof. If a matrix A has a right inverse B, then $AB = I$. If it has a left inverse C, then $CA = I$. So

$$C(AB) = CI = C \quad \text{and} \quad (CA)B = IB = B, \quad \text{so} \quad C = B. \quad \square \qquad 1.2.14$$

We discuss how to find inverses of matrices in Section 2.3. A formula exists for 2×2 matrices: the inverse of

$$A = \begin{bmatrix} a & b \\ c & d \end{bmatrix} \quad \text{is} \quad A^{-1} = \frac{1}{ad - bc} \begin{bmatrix} d & -b \\ -c & a \end{bmatrix}, \qquad 1.2.15$$

as Exercise 1.2.12 asks you to confirm by matrix multiplication of AA^{-1} and $A^{-1}A$. (Exercise 1.4.12 discusses the formula for the inverse of a 3×3 matrix.)

Notice that a 2×2 matrix is invertible if $ad - bc \neq 0$. The converse is also true: if $ad - bc = 0$, the matrix is not invertible, as you are asked to show in Exercise 1.2.13.

Associativity of matrix multiplication is also used to prove that the inverse of the product of two invertible matrices is the product of their inverses, in reverse order:

We are indebted to Robert Terrell for the mnemonic, "socks on, shoes on; shoes off, socks off." To undo a process, you undo first the last thing you did.

Proposition 1.2.15 (The inverse of the product of matrices). *If A and B are invertible matrices, then AB is invertible, and the inverse is given by the formula*

$$(AB)^{-1} = B^{-1}A^{-1}. \qquad \text{1.2.16}$$

Proof. The computation

$$(AB)(B^{-1}A^{-1}) = A(BB^{-1})A^{-1} = AA^{-1} = I \qquad \text{1.2.17}$$

If $\vec{v} = \begin{bmatrix} 1 \\ 0 \\ 1 \end{bmatrix}$, then its transpose is

$$\vec{v}^\top = [1\ 0\ 1].$$

and a similar one for $(B^{-1}A^{-1})(AB)$ prove the result. $\square$

Where was associativity used in the proof? Check your answer below.[4]

The transpose

The *transpose* is an operation on matrices that will be useful when we come to the dot product, and in many other places.

Do not confuse a matrix with its transpose, and in particular, *never write a vector horizontally.* If you write a vector written horizontally you have actually written its transpose; confusion between a vector (or matrix) and its transpose leads to endless difficulties with the order in which things should be multiplied, as you can see from Theorem 1.2.17.

Definition 1.2.16 (Transpose). The transpose $A^\top$ of a matrix A is formed by interchanging all the rows and columns of A, reading the rows from left to right, and columns from top to bottom.

For example, if $A = \begin{bmatrix} 1 & 4 & -2 \\ 3 & 0 & 2 \end{bmatrix}$, then $A^\top = \begin{bmatrix} 1 & 3 \\ 4 & 0 \\ -2 & 2 \end{bmatrix}$.

The transpose of a single row of a matrix is a vector; we will use this in Section 1.4.

---

[4]Associativity is used for the first two equalities below:

$$\underbrace{(AB)}_{(AB)}\underbrace{(B^{-1}A^{-1})}_{C} = \underbrace{A}_{A}\ \underbrace{(\overbrace{B}^{D}\ \overbrace{(B^{-1}A^{-1})}^{(EF)})}_{(BC)} = A\Big(\overbrace{(BB^{-1})}^{(DE)}\ \overbrace{A^{-1}}^{F}\Big) = A(IA^{-1}) = I.$$

The proof of Theorem 1.2.17 is straightforward and is left as Exercise 1.2.14.

Theorem 1.2.17 (The transpose of a product). *The transpose of a product is the product of the transposes in reverse order:*

$$(AB)^\top = B^\top A^\top. \hspace{2cm} 1.2.18$$

$$\begin{bmatrix} 1 & 1 & 0 \\ 1 & 0 & 3 \\ 0 & 3 & 0 \end{bmatrix}$$

A symmetric matrix

Some special kinds of matrices

Definition 1.2.18 (Symmetric matrix). A symmetric matrix is equal to its transpose. An anti-symmetric matrix is equal to minus its transpose.

$$\begin{bmatrix} 0 & 1 & 2 \\ -1 & 0 & 3 \\ -2 & -3 & 0 \end{bmatrix}$$

An anti-symmetric matrix

Definition 1.2.19 (Triangular matrix). An upper triangular matrix is a square matrix with nonzero entries only on or above the main diagonal. A lower triangular matrix is a square matrix with nonzero entries only on or below the main diagonal.

$$\begin{bmatrix} 1 & 1 & 0 & 3 \\ \mathbf{0} & 2 & 0 & 0 \\ \mathbf{0} & \mathbf{0} & 1 & 0 \\ \mathbf{0} & \mathbf{0} & \mathbf{0} & 0 \end{bmatrix}$$

An upper triangular matrix

Definition 1.2.20 (Diagonal matrix). A diagonal matrix is a square matrix with nonzero entries (if any) only on the main diagonal.

What happens if you square the diagonal matrix $\begin{bmatrix} a & 0 \\ 0 & a \end{bmatrix}$? If you cube it?[5]

$$\begin{bmatrix} \mathbf{2} & 0 & 0 & 0 \\ 0 & \mathbf{2} & 0 & 0 \\ 0 & 0 & \mathbf{1} & 0 \\ 0 & 0 & 0 & \mathbf{1} \end{bmatrix}$$

A diagonal matrix

Exercise 1.2.10 asks you to show that if A and B are upper triangular $n \times n$ matrices, then so is AB.

Applications of matrix multiplication: probabilities and graphs

While from the perspective of this book matrices are most important because they represent linear transformations, discussed in the next section, there are other important applications of matrix multiplication. Two good examples are probability theory and graph theory.

Example 1.2.21 (Matrices and probabilities). Suppose you have three reference books on a shelf: a thesaurus, a French dictionary, and an English dictionary. Each time you consult one of these books, you put it back on the shelf at the far left. When you need a reference, we denote the probability that it will be the thesaurus P_1, the French dictionary P_2 and the English dictionary P_3. There are six possible arrangements on the shelf: 1 2 3 (thesaurus, French dictionary, English dictionary), 1 3 2, and so on.

---

[5] $\begin{bmatrix} a & 0 \\ 0 & a \end{bmatrix}^2 = \begin{bmatrix} a^2 & 0 \\ 0 & a^2 \end{bmatrix}; \begin{bmatrix} a & 0 \\ 0 & a \end{bmatrix}^3 = \begin{bmatrix} a^3 & 0 \\ 0 & a^3 \end{bmatrix}.$

For example, the move from $(2\,1\,3)$ to $(3\,2\,1)$ has probability P_3 (associated with the English dictionary), since if you start with the order $(2\,1\,3)$ (French dictionary, thesaurus, English dictionary), consult the English dictionary, and put it back to the far left, you will then have the order $(3\,2\,1)$. So the entry at the 3rd row, 6th column is P_3. The move from $(2\,1\,3)$ to $(3\,1\,2)$ has probability 0, since moving the English dictionary from third to first position won't change the position of the other books. So the entry at the 3rd row, 5th column is 0.

A situation like this one, where each outcome depends only on the one just before it, it called a *Markov chain.*

Sometimes easy access isn't the goal. In Zola's novel *Au Bonheur des Dames,* the epic story of the growth of the first big department store in Paris, the hero has an inspiration: he places his merchandise in the most inconvenient arrangement possible, forcing his customers to pass through parts of the store where they otherwise wouldn't set foot, and which are mined with temptations for impulse shopping.

We can then write the following 6×6 transition matrix, indicating the probability of going from one arrangement to another:

	$(1,2,3)$	$(1,3,2)$	$(2,1,3)$	$(2,3,1)$	$(3,1,2)$	$(3,2,1)$
$(1,2,3)$	P_1	0	P_2	0	P_3	0
$(1,3,2)$	O	P_1	P_2	0	P_3	0
$(2,1,3)$	P_1	0	P_2	0	0	P_3
$(2,3,1)$	P_1	0	0	P_2	0	P_3
$(3,1,2)$	0	P_1	0	P_2	P_3	0
$(3,2,1)$	0	P_1	0	P_2	0	P_3

Now say you start with the fourth arrangement, $(2,3,1)$. Multiplying the line matrix $(0,0,0,1,0,0)$ (probability 1 for the fourth choice, 0 for the others) by the transition matrix T gives the probabilities $P_1, 0, 0, P_2, 0, P_3$. This is of course just the 4th row of the matrix. The interesting point here is to explore the long-term probabilities. At the second step, we would multiply the line matrix $P_1, 0, 0, P_2, 0, P_3$ by T; at the third we would multiply that product by T, ... If we know actual values for P_1, P_2, and P_3 we can compute the probabilities for the various configurations after a great many iterations. If we don't know the probabilities, we can use this system to deduce them from the configuration of the bookshelf after different numbers of iterations.

This kind of approach is useful in determining efficient storage. How should a lumber yard store different sizes and types of woods, so as little time as possible is lost digging out a particular plank from under others? For computers, what applications should be easier to access than others? Based on the way you use your computer, how should its operating system store data most efficiently?   △

Example 1.2.22 is important for many applications. It introduces no new theory and can be skipped if time is at a premium, but it provides an entertaining setting for practice at matrix multiplication, while showing some of its power.

Example 1.2.22 (Matrices and graphs). We are going to take walks on the edges of a unit cube; if in going from a vertex V_i to another vertex V_k we walk along n edges, we will say that our walk is of length n. For example, in Figure 1.2.6, if we go from vertex V_1 to V_6, passing by V_4 and V_5, the total length of our walk is 3. We will stipulate that each segment of the walk has to take us from one vertex to a different vertex; the shortest possible walk from a vertex to itself is of length 2.

How many walks of length n are there that go from a vertex to itself, or, more generally, from a given vertex to a second vertex? As we will see in Proposition 1.2.23, we answer that question by raising to the nth power the *adjacency matrix* of the graph. The adjacency matrix for our cube is the 8×8 matrix whose rows and columns are labeled by the vertices $V_1, \ldots, V_8$, and such that the i, jth entry is 1 if there is an edge joining V_i to V_j, and 0 if not, as shown in Figure 1.2.6. For example, the entry 4, 1 is 1 (underlined in the matrix) because there is an edge joining V_4 to V_1; the entry 4, 6 is 0 (also underlined) because there is no edge joining V_4 to V_6.

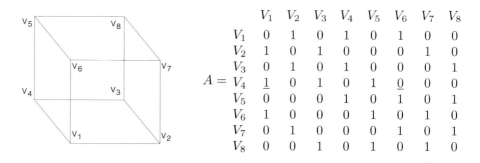

$$A = \begin{array}{c|cccccccc} & V_1 & V_2 & V_3 & V_4 & V_5 & V_6 & V_7 & V_8 \\ V_1 & 0 & 1 & 0 & 1 & 0 & 1 & 0 & 0 \\ V_2 & 1 & 0 & 1 & 0 & 0 & 0 & 1 & 0 \\ V_3 & 0 & 1 & 0 & 1 & 0 & 0 & 0 & 1 \\ V_4 & \underline{1} & 0 & 1 & 0 & 1 & \underline{0} & 0 & 0 \\ V_5 & 0 & 0 & 0 & 1 & 0 & 1 & 0 & 1 \\ V_6 & 1 & 0 & 0 & 0 & 1 & 0 & 1 & 0 \\ V_7 & 0 & 1 & 0 & 0 & 0 & 1 & 0 & 1 \\ V_8 & 0 & 0 & 1 & 0 & 1 & 0 & 1 & 0 \end{array}$$

FIGURE 1.2.6. Left: The graph of a cube. Right: Its adjacency matrix A. If two vertices V_i and V_j are joined by a single edge, the (i, j)th and (j, i)th entries of the matrix are 1; otherwise they are 0.

The reason this matrix is important is the following.

You may appreciate this result more if you try to make a rough estimate of the number of walks of length 4 from a vertex to itself. The authors did and discovered later that they had missed quite a few possible walks.

Proposition 1.2.23. *For any graph formed of vertices connected by edges, the number of possible walks of length n from vertex V_i to vertex V_j is given by the i, jth entry of the matrix A^n formed by taking the nth power of the graph's adjacency matrix A.*

For example, there are 20 different walks of length 4 from V_5 to V_7 (or vice versa), but no walks of length 4 from V_4 to V_3 because

As you would expect, all the 1's in the adjacency matrix A have turned into 0's in A^4; if two vertices are connected by a single edge, then when n is even there will be no walks of length n between them.

Of course we used a computer to compute this matrix. For all but simple problems involving matrix multiplication, use MATLAB or an equivalent.

$$A^4 = \begin{bmatrix} 21 & 0 & 20 & 0 & 20 & 0 & 20 & 0 \\ 0 & 21 & 0 & 20 & 0 & 20 & 0 & 20 \\ 20 & 0 & 21 & 0 & 20 & 0 & 20 & 0 \\ 0 & 20 & 0 & 21 & 0 & 20 & 0 & 20 \\ 20 & 0 & 20 & 0 & 21 & 0 & 20 & 0 \\ 0 & 20 & 0 & 20 & 0 & 21 & 0 & 20 \\ 20 & 0 & 20 & 0 & 20 & 0 & 21 & 0 \\ 0 & 20 & 0 & 20 & 0 & 20 & 0 & 21 \end{bmatrix}.$$

Proof. This will be proved by induction, in the context of the graph above; the general case is the same. Let B_n be the 8×8 matrix whose i, jth entry is the number of walks from V_i to V_j of length n, for a graph with eight vertices;

we must prove $B_n = A^n$. First notice that $B_1 = A^1 = A$: the entry $A_{i,j}$ of the matrix A is exactly the number of walks of length 1 from v_i to v_j.

Next, suppose it is true for n, and let us see that it is true for $n + 1$. A walk of length $n+1$ from V_i to V_j must be at some vertex V_k at time n. The number of such walks is the sum, over all such V_k, of the number of ways of getting from V_i to V_k in n steps, times the number of ways of getting from V_k to V_j in one step (which will be 1 if V_k is next to V_j, and 0 otherwise). In symbols, this is

$$\underbrace{(B_{n+1})_{i,j}}_{\substack{\text{No. of ways} \\ i \text{ to } j \text{ in } n+1 \text{ steps}}} = \sum_{k=1}^{8} \underbrace{}_{\substack{\text{for all} \\ \text{vertices } k}} \underbrace{(B_n)_{i,k}}_{\substack{\text{No. ways } i \text{ to} \\ k \text{ in } n \text{ steps}}} \underbrace{(B_1)_{k,j}}_{\substack{\text{No. ways } k \text{ to} \\ j \text{ in 1 step}}}$$

1.2.19

$$= \sum_{k=1}^{8} \underbrace{(A^n)_{i,k}}_{\substack{\text{inductive} \\ \text{hypothesis}}} \underbrace{A_{k,j}}_{\substack{\text{def.} \\ \text{of } A}} \underbrace{=}_{\substack{\text{Def.} \\ 1.2.4}} (A^{n+1})_{i,j},$$

which is precisely the definition of A^{n+1}. $\square$

Like the transition matrices of probability theory, matrices representing the length of walks from one vertex of a graph to another have important applications for computers and multiprocessing.

Above, what do we mean by A^n? If you look at the proof, you will see that what we used was

$$A^n = \underbrace{\Big(\big(\dots (A)A \big) A \Big) A}_{n \text{ factors}}.$$

1.2.20

Matrix multiplication is associative, so you can also put the parentheses any way you want; for example,

$$A^n = \Big(A \big(A(A) \dots \big) \Big).$$

1.2.21

In this case, we can see that it is true, and simultaneously make the associativity less abstract: with the definition above, $B_n B_m = B_{n+m}$. Indeed, a walk of length $n + m$ from V_i to V_j is a walk of length n from V_i to some V_k, followed by a walk of length m from V_k to V_j. In formulas, this gives

Exercise 1.2.15 asks you to construct the adjacency matrix for a triangle and for a square. We can also make a matrix that allows for one-way streets (one-way edges), as Exercise 1.2.18 asks you to show.

$$(B_{n+m})_{i,j} = \sum_{k=1}^{8} (B_n)_{i,k} (B_m)_{k,j}.$$

1.2.22

1.3 WHAT THE ACTORS DO: A MATRIX AS A TRANSFORMATION

In Section 2.2 we will see how matrices are used to solve systems of linear equations, but first let us consider a different view of matrices. In this view, multiplication of a matrix by a vector is seen as a *linear transformation*, a special kind of mapping. This is the central notion of linear algebra, which

The words *mapping* (or *map*) and *function* are synonyms, generally used in different contexts. A *function* normally takes a point and gives a number. *Mapping* is a more recent word; it was first used in topology and geometry and has spread to all parts of mathematics. In higher dimensions, we tend to use the word *mapping* rather than *function*, but there is nothing wrong with calling a mapping from $\mathbb{R}^5 \to \mathbb{R}^5$ a function.

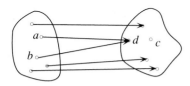

FIGURE 1.3.1.

A mapping: every point on the left goes to only one point on the right.

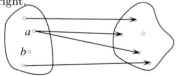

FIGURE 1.3.2.

Not a mapping: not well defined at a, not defined at b.

The domain of our mathematical "final grade function" is $\mathbb{R}^n$; its range is $\mathbb{R}$. In practice this function has a "socially acceptable" domain of the realistic grade vectors (no negative numbers, for example) and also a "socially acceptable" range, the set of possible final grades. Often a mathematical function modeling a real system has domain and range considerably larger than the realistic values.

allows us to put matrices in context and to see them as something other than "pushing numbers around."

Mappings

A mapping associates elements of one set to elements of another. In common speech, we deal with mappings all the time. Like the character in Molière's play *Le Bourgeois Gentilhomme*, who discovered that he had been speaking prose all his life without knowing it, we use mappings from early childhood, typically with the word "of" or its equivalent: "the price of a book" goes from books to money; "the capital of a country" goes from countries to cities.

This is not an analogy intended to ease you into the subject. "The father of" *is* a mapping, not "sort of like" a mapping. We could write it with symbols: $f(x) = y$ where $x = $ a person and $y = $ that person's father: $f(\text{John Jr.}) = $ John. (Of course in English it would be more natural to say, "John Jr.'s father" rather than "the father of John Jr." A school of algebraists exists that uses this notation: they write $(x)f$ rather than $f(x)$.)

The difference between expressions like "the father of" in everyday speech and mathematical mappings is that in mathematics one must be explicit about things that are taken for granted in speech.

Rigorous mathematical terminology requires specifying three things about a mapping:

(1) the set of departure (the *domain*),
(2) the set of arrival (the *range*),
(3) a rule going from one to the other.

If the domain of a mapping M is the real numbers $\mathbb{R}$ and its range is the rational numbers $\mathbb{Q}$, we denote it $M : \mathbb{R} \to \mathbb{Q}$, which we read "$M$ from $\mathbb{R}$ to $\mathbb{Q}$." Such a mapping takes a real number as input and gives a rational number as output.

What about a mapping $T : \mathbb{R}^n \to \mathbb{R}^m$? Its input is a vector with n entries; its output is a vector with m entries: for example, the mapping from $\mathbb{R}^n$ to $\mathbb{R}$ that takes n grades on homework, tests, and the final exam and gives you a final grade in a course.

The rule for the "final grade" mapping above consists of giving weights to homework, tests, and the final exam. But the rule for a mapping does not have to be something that can be stated in a neat mathematical formula. For example, the mapping $M : \mathbb{R} \to \mathbb{R}$ that changes every digit 3 and turns it into a 5 is a valid mapping. When you invent a mapping you enjoy the rights of an absolute dictator; you don't have to justify your mapping by saying that "look, if you square a number x, then multiply it by the cosine of 2π, subtract 7 and then raise the whole thing to the power 3/2, and finally do such-and-such, then if x contains a 3, that 3 will turn into a 5, and everything else will remain unchanged." There isn't any such sequence of operations that will "carry out" your mapping for you, and you don't need one.[6]

A mapping going "from" $\mathbb{R}^n$ "to" $\mathbb{R}^m$ is said to be *defined on* its domain $\mathbb{R}^n$. A mapping in the mathematical sense must be *well defined*: it must be defined at every point of the domain, and for each, must return a unique element of the range. A mapping takes you, unambiguously, from one element of the set of departure to one element of the set of arrival, as shown in Figures 1.3.1 and 1.3.2. (This does not mean that you can go unambiguously (or at all) in the reverse direction; in Figure 1.3.1, going backwards from the point d in the range will take you to either a or b in the domain, and there is no path from c in the range to any point in the domain.)

Not all expressions "the this of the that" are true mappings in this sense. "The daughter of," as a mapping from people to girls and women, is not everywhere defined, because not everyone has a daughter; it is not well defined because some people have more than one daughter. It is not a mapping. But "the number of daughters of," as a mapping from women to numbers, is everywhere defined and well defined, at a particular time. And "the father of," as a mapping from people to men, *is* everywhere defined, and well defined; every person has a father, and only one. (We speak here of biological fathers.)

Note that in correct mathematical usage, "the father of" as a mapping from people to people is not the same mapping as "the father of" as a mapping from people to men. A mapping includes a domain, a range, and a rule going from the first to the second.

Remark. We use the word "range" to mean the space of arrival, or "target space"; some authors use "range" to mean those elements of the arrival space that are actually reached. In that usage, the range of the squaring function $F : \mathbb{R} \to \mathbb{R}$ given by $F(x) = x^2$ is the non-negative real numbers, while in our usage the range is $\mathbb{R}$. We will see in Section 2.5 that what these authors call the range, we call the image. Some authors who use "range" to denote those elements of the arrival space that are actually reached denote the space of arrival by "codomain." Others either have no word for the space of arrival, or use the word "range" interchangeably to mean both space of arrival and image. $\triangle$

---

[6]Here's another "pathological" but perfectly valid mapping: the mapping $M : \mathbb{R} \to \mathbb{R}$ that takes every number in the interval $[0, 1]$ that can be written in base 3 without using 1's, changes every 2 to a 1, and then considers the result as a number in base 2. If the number has a 1, it changes all the digits after the first 1 into 0's and considers the result as a number in base 2. Cantor proposed this mapping to point out the need for greater precision in a number of theorems, in particular the fundamental theorem of calculus. At the time it was viewed as pathological but it turns out to be important for understanding Newton's method for complex cubic polynomials. Mappings just like it occur everywhere in complex dynamics—a surprising discovery of the early 1980's.

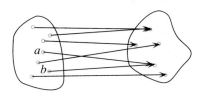

FIGURE 1.3.3.

An onto mapping, not 1–1, *a* and *b* go to the same point.

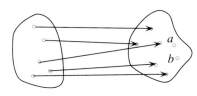

FIGURE 1.3.4.

A mapping: 1–1, not onto, no points go to *a* or to *b*.

"Onto" is a way to talk about the *existence* of solutions: a mapping T is onto if there is a solution to the equation $T(x) = b$, for every b in the set of arrival (the range of T). "One to one" is a way to talk about the *uniqueness* of solutions: T is one to one if for every b there is at most one solution to the equation $T(x) = b$.

Existence and uniqueness of solutions

Given a mapping T, is there a solution to the equation $T(x) = b$, for every b in the range (set of arrival)? If so, the mapping is said to be *onto*, or *surjective*. "Onto" is thus a way to talk about the *existence of solutions*. The mapping "the father of" as a mapping from people to men is not onto, because not all men are fathers. There is no solution to the equation "The father of x is Mr. Childless." An onto mapping is shown in Figure 1.3.3.

A second question of interest concerns uniqueness of solutions. Is there at most *one* solution to the equation $T(x) = b$, for every b in the set of arrival, or might there be many? If there is at most one solution to the equation $T(x) = b$, the mapping T is said to be *one to one*, or *injective*. The mapping "the father of" is not one to one. There are, in fact, four solutions to the equation "The father of x is John Hubbard." But the mapping "the twin sibling of," as a mapping from twins to twins, is one to one: the equation "the twin sibling of $x = y$" has a unique solution for each y. "One to one" is thus a way to talk about the *uniqueness of solutions*. A one to one mapping is shown in Figure 1.3.4.

A mapping T that is both *onto* and *one to one* (also called *bijective*) has an inverse mapping T^{-1} that undoes it. Because T is onto, T^{-1} is everywhere defined; because T is one to one, T^{-1} is well defined. So T^{-1} qualifies as a mapping. To summarize:

Definition 1.3.1 (Onto). A mapping is onto (or *surjective*) if every element of the set of arrival corresponds to at least one element of the set of departure.

Definition 1.3.2 (One to one). A mapping is one to one (or *injective*) if every element of the set of arrival corresponds to at most one element of the set of departure.

Definition 1.3.3 (Bijective). A mapping is bijective if it is both onto and one to one. A bijective mapping is invertible.

Example 1.3.4 (One to one and onto). The mapping "the Social Security number of" as a mapping from Americans to numbers is not onto because there exist numbers that aren't Social Security numbers. But it is one to one: no two Americans have the same Social Security number.

The mapping $f(x) = x^2$ from real numbers to real positive numbers is onto because every real positive number has a real square root, but it is not one to one because every real positive number has both a positive and a negative square root. $\triangle$

Composition of mappings

Often one wishes to apply, consecutively, more than one mapping. This is known as *composition*.

A composition is written from left to right but computed from right to left: you apply the mapping g to the argument x and then apply the mapping f to the result. Exercise 1.3.12 provides some practice.

Definition 1.3.5 (Composition). The composition $f \circ g$ of two mapping, f and g, is

$$(f \circ g)(x) = f\big(g(x)\big). \qquad \qquad 1.3.1$$

Example 1.3.6 (Composition of "the father of" and "the mother of"). Consider the following two mappings from the set of persons to the set of persons (alive or dead): F, "the father of," and M, "the mother of." Composing these gives:

$$F \circ M \quad \text{(the father of the mother of = maternal grandfather of)}$$
$$M \circ F \quad \text{(the mother of the father of = paternal grandmother of).}$$

It is clear in this case that composition is associative:

$$F \circ (F \circ M) = (F \circ F) \circ M. \qquad \qquad 1.3.2$$

When computers do compositions it is not quite true that composition is associative. One way of doing the calculation may be more computationally effective than another; because of round-off errors, the computer may even come up with different answers, depending on where the parentheses are placed.

The father of David's maternal grandfather is the same person as the paternal grandfather of David's mother. Of course it is not commutative: the "father of the mother" is not the "mother of the father.") $\triangle$

Example 1.3.7 (Composition of two functions). If $f(x) = x - 1$, and $g(x) = x^2$, then

$$(f \circ g)(x) = f\big(g(x)\big) = x^2 - 1. \quad \triangle \qquad \qquad 1.3.3$$

Although composition is associative, in many settings,

$$\Big((f \circ g) \circ h\Big) \quad \text{and} \quad \Big(f \circ (g \circ h)\Big)$$

correspond to different ways of thinking. Already, the "father of the maternal grandfather" and "the paternal grandfather of the mother" are two ways of thinking of the same person; the author of a biography might use the first term when focusing on the relationship between the subject's grandfather and that grandfather's father, and use the other when focusing on the relationship between the subject's mother and her grandfather.

Proposition 1.3.8 (Composition is associative). *Composition is associative:*

$$f \circ g \circ h = (f \circ g) \circ h = f \circ (g \circ h). \qquad \qquad 1.3.4$$

Proof. This is simply the computation

$$((f \circ g) \circ h)(x) = (f \circ g)\big(h(x)\big) = f\big(g(h(x))\big) \quad \text{whereas}$$
$$\big(f \circ (g \circ h)\big)(x) = f\big((g \circ h)(x)\big) = f\big(g(h(x))\big). \quad \square \qquad 1.3.5$$

You may find this "proof" devoid of content. Composition of mappings is part of our basic thought processes: you use a composition any time you speak of "the this of the that of the other."

Matrices and transformations

A special class of mappings consists of those mappings that are encoded by matrices. By "encoded" we mean that multiplication by a matrix is the rule that turns an input vector into an output vector: just as $f(x) = y$ takes a number x and gives y, $A\vec{v} = \vec{w}$ takes a vector $\vec{v}$ and gives a vector $\vec{w}$.

Such mappings, called *linear transformations*, are of central importance in linear algebra (and every place else in mathematics). Throughout mathematics, the constructs of central interest are the mappings that preserve whatever structure is at hand. In linear algebra, "preserve structure" means that you can first add, then map, or first map, then add, and get the same answer; similarly, first multiplying by a scalar and then mapping gives the same result as first mapping and then multiplying by a scalar. One of the great discoveries at the end of the 19th century was that the natural way to do mathematics is to look at sets with structure, such as $\mathbb{R}^n$, with addition and multiplication by scalars, and to consider the mappings that preserve that structure.

We give a mathematical definition of linear transformations in Definition 1.3.11, but first let's see an example.

The words *transformation* and *mapping* are synonyms, so we could call the matrix A of Figure 1.3.5 a mapping. But in linear algebra the word *transformation* is more common. In fact, the matrix A is a *linear transformation*, but we haven't formally defined that term yet.

Example 1.3.9 (Frozen dinners). In a food processing plant making three types of frozen dinners, one might associate the number of dinners of various sorts produced to the total ingredients needed (beef, chicken, noodles, cream, salt, ...). As shown in Figure 1.3.5, this mapping is given by multiplication (on the left) by the matrix A, which gives the amount of each ingredient needed for each dinner: A tells how to go from $\vec{b}$, which tells how many dinners of each kind are produced, to the product $\vec{c}$, which tells the total ingredients needed. For example, 21 pounds of beef are needed, because $(.25 \times 60) + (.20 \times 30) + (0 \times 40) = 21$. For chicken, $(0 \times 60) + (0 \times 30) + (.45 \times 40) = 18$.

Mathematicians usually denote a linear transformation by its associated matrix; rather than saying that the "dinners to shopping list" transformation is the multiplication $A\vec{b} = \vec{c}$, they would call this transformation A.

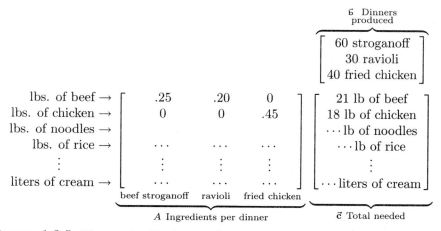

FIGURE 1.3.5. The matrix A is the transformation associating the number of dinners of various sorts produced to the total ingredients needed. △

Example 1.3.10 (Frozen foods: composition). For the food plant of Example 1.3.9, one might make a matrix D, 1 high and n wide (n being the total number of ingredients), that would list the price of each ingredient, per pound or liter. The product DA would then tell the cost of each ingredient in each dinner, since A tells how much of each ingredient is in each dinner. The product $(DA)\vec{\mathbf{b}}$ would give the total cost of the ingredients for all $\vec{\mathbf{b}}$ dinners. We could also compose these transformations in a different order, first figuring how much of each ingredient we need for all $\vec{\mathbf{b}}$ dinners—the product $A\vec{\mathbf{b}}$. Then, using D, we could figure the total cost: $D(A\vec{\mathbf{b}})$. Clearly, $(DA)\vec{\mathbf{b}} = D(A\vec{\mathbf{b}})$, although the two correspond to slightly different perspectives. $\triangle$

Notice that matrix multiplication emphatically does *not* allow for feedback. For instance, it does not allow for the possibility that if you buy more you will get a discount for quantity, or that if you buy even more you might create scarcity and drive prices up. *This is a key feature of linearity,* and is the fundamental weakness of all models that linearize mappings and interactions.

Real-life matrices

We kept Example 1.3.9 simple, but you can easily see how this works in a more realistic situation. In real life—modeling the economy, designing buildings, modeling airflow over the wing of an airplane—vectors of input data contain tens of thousands of entries, or more, and the matrix giving the transformation has millions of entries.

We hope you can begin to see that a matrix might be a very useful way of mapping from $\mathbb{R}^n$ to $\mathbb{R}^m$. To go from $\mathbb{R}^3$, where vectors all have three entries,

$$\vec{\mathbf{v}} = \begin{bmatrix} v_1 \\ v_2 \\ v_3 \end{bmatrix}, \text{ to } \mathbb{R}^4, \text{ where vectors have four entries, } \vec{\mathbf{w}} = \begin{bmatrix} w_1 \\ w_2 \\ w_3 \\ w_4 \end{bmatrix}, \text{ you would}$$

multiply $\vec{\mathbf{v}}$ on the left by a 4×3 matrix:

$$\begin{bmatrix} \cdots & \cdots & \cdots \\ \cdots & \cdots & \cdots \\ \cdots & \cdots & \cdots \\ \cdots & \cdots & \cdots \end{bmatrix} \begin{bmatrix} v_1 \\ v_2 \\ v_3 \\ w_1 \\ w_2 \\ w_3 \\ w_4 \end{bmatrix}. \tag{1.3.6}$$

One can imagine doing the same thing when the n and m of $\mathbb{R}^n$ and $\mathbb{R}^m$ are arbitrarily large. One can somewhat less easily imagine extending the same idea to infinite-dimensional spaces, but making sense of the notion of multiplication of infinite matrices gets into some deep water, beyond the scope of this book. Our matrices are finite: rectangular arrays, m high and n wide.

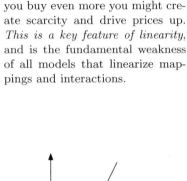

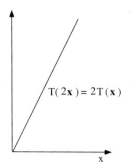

FIGURE 1.3.6.
For *any* linear transformation T,
$$T(\alpha\mathbf{x}) = \alpha T(\mathbf{x}).$$

Linearity

The assumption that a transformation is linear is the main simplifying assumption that scientists and social scientists (especially economists) make to understand their models of the world. Roughly speaking, *linearity* means that if you double the input, you double the output; triple the input, triple the output . . . .

In Example 1.3.9, the transformation A is linear: each frozen beef stroganoff dinner will require the same amount of beef, whether one is making one dinner or 10,000. We treated the price function D in Example 1.3.10 as linear, but in real life it is cheaper per pound to buy 10,000 pounds of beef than one. Many, perhaps most, real-life problems are nonlinear. It is always easier to treat them as if they were linear; knowing when it is safe to do so is a central issue of applied mathematics.

The Italian mathematician Salvatore Pincherle, one of the early pioneers of linear algebra, called a linear transformation a *distributive transformation* (*operazioni distributive*), a name that is perhaps more suggestive of the formulas than is "linear."

Definition 1.3.11 (Linear transformation). A linear transformation $T : \mathbb{R}^n \to \mathbb{R}^m$ is a mapping such that for all scalars a and all $\vec{v}, \vec{w} \in \mathbb{R}^n$,

$$T(\vec{v} + \vec{w}) = T(\vec{v}) + T(\vec{w}) \quad \text{and} \quad T(a\vec{v}) = aT(\vec{v}). \qquad 1.3.7$$

The two formulas can be combined into one (where b is also a scalar):

$$T(a\vec{v} + b\vec{w}) = aT(\vec{v}) + bT(\vec{w}). \qquad 1.3.8$$

Example 1.3.12 (Linearity at the checkout counter). Suppose you need to buy three gallons of cider and six packages of doughnuts for a Halloween party. The transformation T is performed by the scanner at the checkout counter, reading the UPC code to determine the price. Equation 1.3.7 is nothing but the obvious statement that if you do your shopping all at once, it will cost you exactly the same amount as it will if you go through the checkout line nine times, once for each item:

$$T(3\,\text{gal.cider} + 6\,\text{pkg. doughnuts}) = 3\big(T(1\,\text{gal.cider})\big) + 6\big(T(1\,\text{pkg. doughnuts})\big),$$

unless the supermarket introduces nonlinearities such as "buy two, get one free."

Every linear transformation is given by a matrix. The matrix can be found by seeing how the transformation acts on the standard basis vectors

$$\vec{e}_1 = \begin{bmatrix} 1 \\ 0 \\ \vdots \\ 0 \end{bmatrix}, \quad \dots \quad ,\vec{e}_n = \begin{bmatrix} 0 \\ \vdots \\ 0 \\ 1 \end{bmatrix}.$$

Example 1.3.13 (A matrix gives a linear transformation). Let A be an $m \times n$ matrix. Then A defines a linear transformation $T : \mathbb{R}^n \to \mathbb{R}^m$ by matrix multiplication:

$$T(\vec{v}) = A\vec{v}. \qquad 1.3.9$$

Such mappings are indeed linear, because $A(\vec{v} + \vec{w}) = A\vec{v} + A\vec{w}$ and $A(c\vec{v}) = cA\vec{v}$, as you are asked to check in Exercise 1.3.14. $\triangle$

The crucial result of Theorem 1.3.14 below is that *every linear transformation* $\mathbb{R}^n \to \mathbb{R}^m$ is given by a matrix, which one can construct by seeing how the transformation acts on the standard basis vectors. This is rather remarkable. A priori the notion of a transformation from $\mathbb{R}^n$ to $\mathbb{R}^m$ is quite vague and abstract; one might not think that merely by imposing the condition of linearity one could say something so precise about this shapeless set of mappings as saying that each is given by a matrix.

To find the matrix for a linear transformation, ask: what is the result of applying that transformation to the standard basis vectors? The ith column of the matrix for a linear transformation T is $T(\vec{e}_i)$; to get the ith column of the matrix, just ask: what does the transformation do to $\vec{e}_i$?

Theorem 1.3.14 (Linear transformations given by matrices). *Every linear transformation $T : \mathbb{R}^n \to \mathbb{R}^m$ is given by multiplication by the $m \times n$ matrix $[T]$, the ith column of which is $T(\vec{e}_i)$.*

Putting the columns together, this gives $T(\vec{v}) = [T]\vec{v}$. This means that Example 1.3.13 is "the general" linear transformation in $\mathbb{R}^n$.

Proof. Start with a linear transformation $T : \mathbb{R}^n \to \mathbb{R}^m$, and manufacture the matrix $[T]$ according to the rule given immediately above: the ith column of $[T]$ is $T(\vec{e}_i)$. We may write any vector $\vec{v} \in \mathbb{R}^n$ in terms of the standard basis vectors:

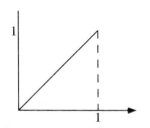

FIGURE 1.3.7.

The orthogonal projection of the point $\begin{pmatrix} 1 \\ 1 \end{pmatrix}$ onto the x-axis is the point $\begin{pmatrix} 1 \\ 0 \end{pmatrix}$. "Projection" means we draw a line from the point to the x-axis. "Orthogonal" means we make that line perpendicular to the x-axis.

$$
\vec{v} = \begin{bmatrix} v_1 \\ v_2 \\ \vdots \\ \vdots \\ v_n \end{bmatrix} = v_1 \underbrace{\begin{bmatrix} 1 \\ 0 \\ \vdots \\ \vdots \\ 0 \end{bmatrix}}_{\vec{e}_1} + v_2 \underbrace{\begin{bmatrix} 0 \\ 1 \\ 0 \\ \vdots \\ 0 \end{bmatrix}}_{\vec{e}_2} + \cdots + v_n \underbrace{\begin{bmatrix} 0 \\ \vdots \\ \vdots \\ 0 \\ 1 \end{bmatrix}}_{\vec{e}_n}. \qquad 1.3.10
$$

We can write this more succinctly:

$$
\vec{v} = v_1\vec{e}_1 + v_2\vec{e}_2 + \cdots + v_n\vec{e}_n, \quad \text{or, with sum notation,} \quad \vec{v} = \sum_{i=1}^{n} v_i\vec{e}_i. \quad 1.3.11
$$

Then by linearity,

$$
T(\vec{v}) = T \sum_{i=1}^{n} v_i\vec{e}_i = \sum_{i=1}^{n} v_i T(\vec{e}_i), \qquad 1.3.12
$$

which is precisely the column vector $[T]\vec{v}$. $\square$

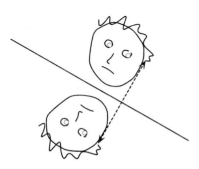

FIGURE 1.3.8.

Every point on one face is reflected to the corresponding point of the other.

If this isn't apparent, try translating it out of sum notation:

$$
T(\vec{v}) = \sum_{i=1}^{n} v_i T(\vec{e}_i) = v_1 \underbrace{\begin{bmatrix} \\ \\ \end{bmatrix}}_{\substack{\text{1st col.} \\ \text{of } [T]}}^{T(\vec{e}_1)} + v_2 \underbrace{\begin{bmatrix} \\ \\ \end{bmatrix}}_{\substack{\text{2nd col.} \\ \text{of } [T]}}^{T(\vec{e}_2)} + \cdots + v_n \underbrace{\begin{bmatrix} \\ \\ \end{bmatrix}}_{\substack{n\text{th col.} \\ \text{of } [T]}}^{T(\vec{e}_n)}
$$

$$
= \begin{bmatrix} & & \\ & T & \\ & & \end{bmatrix} \begin{bmatrix} v_1 \\ v_2 \\ \vdots \\ v_n \end{bmatrix} = [T]\vec{v} \qquad 1.3.13
$$

Example 1.3.15 (Finding the matrix of a linear transformation). What is the matrix for the transformation that takes any point in $\mathbb{R}^2$ and gives its orthogonal (perpendicular) projection on the x-axis, as illustrated in Figure 1.3.7? You should assume this transformation is linear. Check your answer in the footnote below.[7]

What is the orthogonal projection on the line of equation $x = y$ of the point $\begin{pmatrix} 3 \\ -1 \end{pmatrix}$? Again, assume this is a linear transformation, and check below.[8]

Example 1.3.16 (Reflection with respect to a line through the origin). Let us show that the transformation that reflects a point through a line through the origin is linear. This is the transformation that takes a point on one side of the line and moves it perpendicular to the line, crosses it, and continues the same distance away from the line, as shown in Figure 1.3.8.

We will first assume that the transformation T is linear, and thus given by a matrix whose ith column is $T(\vec{e}_i)$. Again, *all we have to do* is figure out what the T does to $\vec{e}_1$ and $\vec{e}_2$. We can then apply that transformation to any point we like, by multiplying it by the matrix. There's no need to do an elaborate computation for each point.

To obtain the first column of our matrix we thus consider where $\vec{e}_1$ is mapped to. Suppose that our line makes an angle θ (theta) with the x-axis, as shown in Figure 1.3.9. Then $\vec{e}_1$ is mapped to $\begin{bmatrix} \cos 2\theta \\ \sin 2\theta \end{bmatrix}$. To get the second column, we

---

[7]The matrix is $\begin{bmatrix} 1 & 0 \\ 0 & 0 \end{bmatrix}$, which you will note is consistent with Figure 1.3.7, since $\begin{bmatrix} 1 & 0 \\ 0 & 0 \end{bmatrix} \begin{bmatrix} 1 \\ 1 \end{bmatrix} = \begin{bmatrix} 1 \\ 0 \end{bmatrix}$. If you had trouble with this question, you are making life too hard for yourself. The power of Theorem 1.3.14 is that *you don't need to look for the transformation itself to construct its matrix.* Just ask: what is the result of applying that transformation to the standard basis vectors? The ith column of the matrix for a linear transformation T is $T(\vec{e}_i)$. So to get the first column of the matrix, ask, what does the transformation do to $\vec{e}_1$? Since $\vec{e}_1$ lies on the x-axis, it is projected onto itself. The first column of the transformation matrix is thus $\vec{e}_1 = \begin{bmatrix} 1 \\ 0 \end{bmatrix}$. The second standard basis vector, $\vec{e}_2$, lies on the y-axis and is projected onto the origin, so the second column of the matrix is $\begin{bmatrix} 0 \\ 0 \end{bmatrix}$.

[8]The matrix for this linear transformation is $\begin{bmatrix} 1/2 & 1/2 \\ 1/2 & 1/2 \end{bmatrix}$, since the perpendicular line from $\begin{bmatrix} 1 \\ 0 \end{bmatrix}$ to the line of equation $x = y$ intersects that line at $\begin{pmatrix} 1/2 \\ 1/2 \end{pmatrix}$, as does the perpendicular line from $\begin{bmatrix} 0 \\ 1 \end{bmatrix}$. To determine the orthogonal projection of the point $\begin{pmatrix} 3 \\ -1 \end{pmatrix}$, we multiply $\begin{bmatrix} 1/2 & 1/2 \\ 1/2 & 1/2 \end{bmatrix} \begin{bmatrix} 3 \\ -1 \end{bmatrix} = \begin{bmatrix} 1 \\ 1 \end{bmatrix}$. Note that we have to consider the point $\begin{pmatrix} 3 \\ -1 \end{pmatrix}$ as a vector in order to carry out the multiplication; we can't multiply a matrix and a point.

see that $\vec{e}_2$ is mapped to

$$\begin{bmatrix} \cos{(2\theta - 90°)} \\ \sin{(2\theta - 90°)} \end{bmatrix} = \begin{bmatrix} \sin 2\theta \\ -\cos 2\theta \end{bmatrix}. \qquad 1.3.14$$

So the "reflection" matrix is

$$\begin{bmatrix} \cos 2\theta & \sin 2\theta \\ \sin 2\theta & -\cos 2\theta \end{bmatrix}. \qquad 1.3.15$$

For example, we can compute that the point with coordinates $x = 2, y = 1$ reflects to the point $\begin{bmatrix} 2\cos 2\theta + \sin 2\theta \\ 2\sin 2\theta - \cos 2\theta \end{bmatrix}$, since

$$\begin{bmatrix} \cos 2\theta & \sin 2\theta \\ \sin 2\theta & -\cos 2\theta \end{bmatrix} \begin{bmatrix} 2 \\ 1 \end{bmatrix} = \begin{bmatrix} 2\cos 2\theta + \sin 2\theta \\ 2\sin 2\theta - \cos 2\theta \end{bmatrix}. \qquad 1.3.16$$

The transformation is indeed linear because given two vectors $\vec{v}$ and $\vec{w}$, we have $T(\vec{v} + \vec{w}) = T(\vec{v}) + T(\vec{w})$, as shown in Figure 1.3.10. It is also apparent from the figure that $T(c\vec{v}) = cT(\vec{v})$. $\triangle$

Example 1.3.17 (Rotation by an angle θ). The matrix giving the transformation R ("rotation by θ around the origin") is

$$[R(\vec{e}_1), R(\vec{e}_2)] = \begin{bmatrix} \cos\theta & -\sin\theta \\ \sin\theta & \cos\theta \end{bmatrix}.$$

The transformation is linear, as shown in Figure 1.3.11: rather than thinking of rotating just the vectors $\vec{v}$ and $\vec{w}$, we can rotate the whole parallelogram $P(\vec{v}, \vec{w})$ that they span. Then $R(P(\vec{v}, \vec{w}))$ is the parallelogram spanned by $R(\vec{v}), R(\vec{w})$, and in particular the diagonal of $R(P(\vec{v}, \vec{w}))$ is $R(\vec{v} + \vec{w})$. $\triangle$

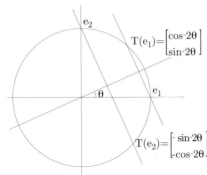

FIGURE 1.3.9.

The reflection maps

$$\vec{e}_1 = \begin{bmatrix} 1 \\ 0 \end{bmatrix} \quad \text{to} \quad \begin{bmatrix} \cos 2\theta \\ \sin 2\theta \end{bmatrix},$$

and

$$\vec{e}_2 = \begin{bmatrix} 0 \\ 1 \end{bmatrix} \quad \text{to} \quad \begin{bmatrix} \sin 2\theta \\ -\cos 2\theta \end{bmatrix}.$$

Exercise 1.3.15 asks you to use composition of the transformation in Example 1.3.17 to derive the fundamental theorems of trigonometry.

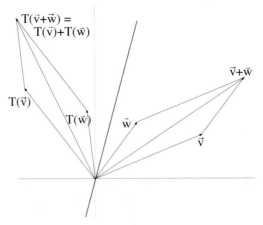

FIGURE 1.3.10. Reflection is linear: the sum of the reflections is the reflection of the sum.

Exercise 1.3.13 asks you to find the matrix for the transformation $\mathbb{R}^3 \to \mathbb{R}^3$ that rotates by 30° around the y-axis.

Now we will see that composition corresponds to matrix multiplication.

Theorem 1.3.18 (Composition corresponds to matrix multiplication). *Suppose $S : \mathbb{R}^n \to \mathbb{R}^m$ and $T : \mathbb{R}^m \to \mathbb{R}^l$ are linear transformations given by the matrices $[S]$ and $[T]$ respectively. Then the matrix of the composition $T \circ S$ equals the product $[T][S]$ of the matrices of S and T:*

$$[T \circ S] = [T][S]. \qquad 1.3.17$$

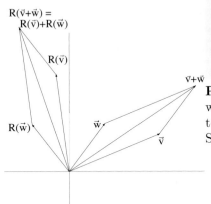

FIGURE 1.3.11.

Rotation is linear: the sum of the rotations is the rotation of the sum.

Many mathematicians would say that Theorem 1.3.18 justifies the definition of matrix multiplication. This may seem odd to the novice, who probably feels that composition of linear mappings is more baroque than matrix multiplication.

Exercise 1.3.16 asks you to confirm by matrix multiplication that reflecting a point across the line, and then back again, lands you back at the original point.

Proof. This is a statement about matrix multiplication and cannot be proved without explicit reference to how matrices are multiplied. Our only references to the multiplication algorithm will be the following facts, both discussed in Section 1.1.

(1) $A\vec{e}_i$ is the ith column of A (as illustrated by Example 1.2.5);
(2) the ith column of AB is $A\vec{b}_i$, where $\vec{b}_i$ is the ith column of B (as illustrated by Example 1.2.6).

Now to prove the theorem; to make it unambiguous when we are applying a transformation to a variable and when we are multiplying matrices, we will write matrix multiplication with a star $*$.

The composition $(T \circ S)$ is itself a linear transformation and thus can be given by a matrix, which we will call $[T \circ S]$, accounting for the first equality below. The definition of composition gives the second equality. Next we replace S by its matrix $[S]$, and finally we replace T by its matrix:

$$[T \circ S] * \vec{e}_i = (T \circ S)(\vec{e}_i) = T\big(S(\vec{e}_i)\big) = T([S] * \vec{e}_i) = [T] * \big([S] * \vec{e}_i\big). \quad 1.3.18$$

So the first term in this sequence, $[T \circ S] * \vec{e}_i$, which is the ith column of $[T \circ S]$ by fact (1), is equal to

$$[T] \ * \ \text{the } i\text{th column of } [S], \qquad 1.3.19$$

which is the ith column of $[T] * [S]$ by fact (2).

Each column of $[T \circ S]$ is equal to the corresponding column of $[T] * [S]$, so the two matrices are equal. $\square$

We gave a computational proof of the associativity of matrix multiplication in Proposition 1.2.8; this associativity is also an immediate consequence of Theorem 1.3.18.

Corollary 1.3.19. *Matrix multiplication is associative: if A, B, C are matrices such that the matrix multiplication $(AB)C$ is allowed, then so is $A(BC)$, and they are equal.*

Proof. Composition of mappings is associative. $\square$

1.4 GEOMETRY OF $\mathbb{R}^n$

> ... *To acquire the feeling for calculus that is indispensable even in the most abstract speculations, one must have learned to distinguish that which is "big" from that which is "little," that which is "preponderant" and that which is "negligible."*—Jean Dieudonné, *Calcul infinitésimal*

Whereas algebra is all about equalities, calculus is about inequalities: about things being arbitrarily small or large, about some terms being dominant or negligible compared to others. Rather than saying that things are exactly true, we need to be able to say that they are almost true, so they "become true in the limit."

For example, $(5 + h)^3 = 125 + 75h + \dots$, so if $h = .01$, we could use the approximation

$$(5.01)^3 \approx 125 + (75 \cdot .01) = 125.75. \qquad 1.4.1$$

The issue then is to quantify the error.

Such notions cannot be discussed in the language about $\mathbb{R}^n$ that has been developed so far: we need lengths of vectors to say that vectors are small, or that points are close to each other. We will also need lengths of matrices to say that linear transformations are "close" to each other. Having a notion of distance between transformations will be crucial in proving that under appropriate circumstances Newton's method converges to a solution (Section 2.7).

In this section we introduce these notions. The formulas are all more or less immediate generalizations of the Pythagorean theorem and the cosine law, but they acquire a whole new meaning in higher dimensions (and more yet in infinitely many dimensions).

The dot product

The dot product in $\mathbb{R}^n$ is the basic construct that gives rise to all the geometric notions of lengths and angles.

Definition 1.4.1 (Dot product). The dot product $\vec{x} \cdot \vec{y}$ of two vectors $\vec{x}, \vec{y} \in \mathbb{R}^n$ is:

The dot product is also known as the *standard inner product*.

$$\vec{x} \cdot \vec{y} = \begin{bmatrix} x_1 \\ x_2 \\ \vdots \\ x_n \end{bmatrix} \cdot \begin{bmatrix} y_1 \\ y_2 \\ \vdots \\ y_n \end{bmatrix} = x_1 y_1 + x_2 y_2 + \dots + x_n y_n. \qquad 1.4.2$$

For example,

$$\begin{bmatrix} 1 \\ 2 \\ 3 \end{bmatrix} \cdot \begin{bmatrix} 1 \\ 0 \\ 1 \end{bmatrix} = (1 \times 1) + (2 \times 0) + (3 \times 1) = 4.$$

The dot product is obviously commutative:

$$\vec{\mathbf{x}} \cdot \vec{\mathbf{y}} = \vec{\mathbf{y}} \cdot \vec{\mathbf{x}}, \qquad\qquad 1.4.3$$

and it is not much harder to check that it is distributive, i.e., that

$$\vec{\mathbf{x}} \cdot (\vec{\mathbf{y}}_1 + \vec{\mathbf{y}}_2) = (\vec{\mathbf{x}} \cdot \vec{\mathbf{y}}_1) + (\vec{\mathbf{x}} \cdot \vec{\mathbf{y}}_2), \text{ and}$$

$$(\vec{\mathbf{x}}_1 + \vec{\mathbf{x}}_2) \cdot \vec{\mathbf{y}} = (\vec{\mathbf{x}}_1 \cdot \vec{\mathbf{y}}) + (\vec{\mathbf{x}}_2 \cdot \vec{\mathbf{y}}). \qquad 1.4.4$$

The dot product of two vectors can be written as the matrix product of the transpose of one vector by the other: $\vec{\mathbf{x}} \cdot \vec{\mathbf{y}} = \vec{\mathbf{x}}^\top \vec{\mathbf{y}} = \vec{\mathbf{y}}^\top \vec{\mathbf{x}}$.

$$\underbrace{\begin{bmatrix} x_1 \\ x_2 \\ \vdots \\ x_n \end{bmatrix} \cdot \begin{bmatrix} y_1 \\ y_2 \\ \vdots \\ y_n \end{bmatrix}}_{\vec{\mathbf{x}} \cdot \vec{\mathbf{y}}} \text{ is the same as } \underbrace{\begin{bmatrix} x_1 & x_2 & \cdots & x_n \end{bmatrix}}_{\text{transpose } \vec{\mathbf{x}}^\top} \underbrace{\begin{bmatrix} y_1 \\ y_2 \\ \vdots \\ y_n \end{bmatrix}}_{} \underbrace{\begin{bmatrix} x_1 y_1 + x_2 y_2 + \cdots + x_n y_n \end{bmatrix}}_{\vec{\mathbf{x}}^\top \vec{\mathbf{y}}}.$$

$$1.4.5$$

Conversely, the i, jth entry of the matrix product AB is the dot product of the jth column of B and the transpose of the ith row of A. For example, the entry $1, 2$ of AB below is 5, which is the dot product of the transpose of the first row of A and the second column of B:

$$\underset{A}{\underbrace{\begin{bmatrix} 1 & 2 \\ 3 & 4 \end{bmatrix}}} \overset{\overset{B}{\begin{bmatrix} 1 & 3 \\ 1 & 1 \end{bmatrix}}}{\underset{AB}{\underbrace{\begin{bmatrix} 3 & 5 \\ 7 & 13 \end{bmatrix}}}}; \qquad 5 = \underset{\substack{\text{transpose,} \\ \text{1st row of } A}}{\underbrace{\begin{bmatrix} 1 \\ 2 \end{bmatrix}}} \cdot \underset{\substack{\text{2nd col.} \\ \text{of } B}}{\underbrace{\begin{bmatrix} 3 \\ 1 \end{bmatrix}}}. \qquad 1.4.6$$

What we call the length of a vector is often called the *Euclidean norm*.

Some texts use double lines to denote the length of a vector: $\|\vec{\mathbf{v}}\|$ rather than $|\vec{\mathbf{v}}|$. We reserve double lines to denote the norm of a matrix, defined in Section 2.8. Please do not confuse the length of a vector with the absolute value of a number. In one dimension, the two are the same; the "length" of the one-entry vector $\vec{\mathbf{v}} = [-2]$ is $\sqrt{2^2} = 2$.

Definition 1.4.2 (Length of a vector). The length $|\vec{\mathbf{x}}|$ of a vector $\vec{\mathbf{x}}$ is

$$|\vec{\mathbf{x}}| = \sqrt{\vec{\mathbf{x}} \cdot \vec{\mathbf{x}}} = \sqrt{x_1^2 + x_2^2 + \cdots + x_n^2}. \qquad 1.4.7$$

What is the length $|\vec{\mathbf{v}}|$ of $\vec{\mathbf{v}} = \begin{bmatrix} 1 \\ 1 \\ 1 \end{bmatrix}$?[9]

Length and dot product: geometric interpretation in $\mathbb{R}^2$ and $\mathbb{R}^3$

In the plane and in space, the length of a vector has a geometric interpretation: $|\vec{\mathbf{x}}|$ is then the ordinary distance between 0 and $\vec{\mathbf{x}}$. As indicated by Figure 1.4.1,

---

[9]Its length is $|\vec{\mathbf{v}}| = \sqrt{1^2 + 1^2 + 1^2} = \sqrt{3}$.

this is exactly what the Pythagorean theorem says in the case of the plane; in space, this it is still true, since OAB is still a right triangle.

Definition 1.4.2 is a version of the Pythagorean theorem: in two dimensions, the vector $\vec{\mathbf{x}} = \begin{bmatrix} x_1 \\ x_2 \end{bmatrix}$ is the hypotenuse of a right triangle of which the other two sides have lengths x_1 and x_2:

$$\mathbf{x}^2 = {x_1}^2 + {x_2}^2.$$

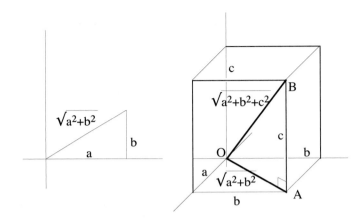

FIGURE 1.4.1. In the plane, the length of the vector with coordinates (a, b) is the ordinary distance between 0 and the point $\begin{pmatrix} a \\ b \end{pmatrix}$. In space, the length of the vector with coordinates (a, b, c) is the ordinary distance between 0 and the point with coordinates (a, b, c).

The dot product also has an interpretation in ordinary geometry:

Proposition 1.4.3 (Geometric interpretation of the dot product).
If $\vec{\mathbf{x}}, \vec{\mathbf{y}}$ are two vectors in $\mathbb{R}^2$ or $\mathbb{R}^3$, then

$$\vec{\mathbf{x}} \cdot \vec{\mathbf{y}} = |\vec{\mathbf{x}}|\,|\vec{\mathbf{y}}|\cos\alpha, \qquad \text{1.4.8}$$

where α is the angle between $\vec{\mathbf{x}}$ and $\vec{\mathbf{y}}$.

Remark. Proposition 1.4.3 says that the dot product is independent of the coordinate system we use. You can rotate a pair of vectors in the plane, or in space, without changing the dot product, as long as you don't change the angle between them. $\triangle$

Proof. This is an application of the *cosine law* from trigonometry, which says that if you know all of a triangle's sides, or two sides and the angle between them, or two angles and a side, then you can determine the others. Let a triangle have sides of length a, b, c, and let γ be the angle opposite the side with length c. Then

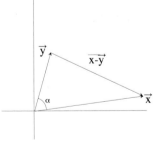

FIGURE 1.4.2.
The cosine law gives
$|\vec{\mathbf{x}} - \vec{\mathbf{y}}|^2 = |\vec{\mathbf{x}}|^2 + |\vec{\mathbf{y}}|^2 - 2|\vec{\mathbf{x}}||\vec{\mathbf{y}}|\cos\alpha.$

$$c^2 = a^2 + b^2 - 2ab\cos\gamma. \qquad \text{Cosine Law} \quad \text{1.4.9}$$

Consider the triangle formed by the three vectors $\vec{\mathbf{x}}, \vec{\mathbf{y}}$ and $\vec{\mathbf{x}} - \vec{\mathbf{y}}$, and let α be the angle between $\vec{\mathbf{x}}$ and $\vec{\mathbf{y}}$, as shown in Figure 1.4.2.

Applying the cosine law, we find

$$|\vec{\mathbf{x}} - \vec{\mathbf{y}}|^2 = |\vec{\mathbf{x}}|^2 + |\vec{\mathbf{y}}|^2 - 2|\vec{\mathbf{x}}||\vec{\mathbf{y}}| \cos\alpha. \qquad 1.4.10$$

But we can also write (remembering that the dot product is distributive):

$$\begin{aligned}
|\vec{\mathbf{x}} - \vec{\mathbf{y}}|^2 &= (\vec{\mathbf{x}} - \vec{\mathbf{y}}) \cdot (\vec{\mathbf{x}} - \vec{\mathbf{y}}) = \big((\vec{\mathbf{x}} - \vec{\mathbf{y}}) \cdot \vec{\mathbf{x}}\big) - \big((\vec{\mathbf{x}} - \vec{\mathbf{y}}) \cdot \vec{\mathbf{y}}\big) \\
&= (\vec{\mathbf{x}} \cdot \vec{\mathbf{x}}) - (\vec{\mathbf{y}} \cdot \vec{\mathbf{x}}) - (\vec{\mathbf{x}} \cdot \vec{\mathbf{y}}) + (\vec{\mathbf{y}} \cdot \vec{\mathbf{y}}) \qquad 1.4.11 \\
&= (\vec{\mathbf{x}} \cdot \vec{\mathbf{x}}) + (\vec{\mathbf{y}} \cdot \vec{\mathbf{y}}) - 2\vec{\mathbf{x}} \cdot \vec{\mathbf{y}} = |\vec{\mathbf{x}}|^2 + |\vec{\mathbf{y}}|^2 - 2\vec{\mathbf{x}} \cdot \vec{\mathbf{y}}.
\end{aligned}$$

This leads to

$$\vec{\mathbf{x}} \cdot \vec{\mathbf{y}} = |\vec{\mathbf{x}}||\vec{\mathbf{y}}| \cos\alpha, \qquad (1.4.8)$$

which is the formula we want. $\square$

If you don't see how we got the numerator in Equation 1.4.12, note that the dot product of a standard basis vector $\vec{\mathbf{e}}_i$ and any vector $\vec{\mathbf{v}}$ is the ith entry of $\vec{\mathbf{v}}$. For example, in $\mathbb{R}^3$,

$$\vec{\mathbf{v}} \cdot \vec{\mathbf{e}}_2 = \begin{bmatrix} v_1 \\ v_2 \\ v_3 \end{bmatrix} \cdot \begin{bmatrix} 0 \\ 1 \\ 0 \end{bmatrix}$$
$$= 0 + v_2 + 0 = v_2.$$

Example 1.4.4 (Finding an angle). What is the angle between the diagonal of a cube and any side? Let us assume our cube is the unit cube $0 \leq x, y, z \leq 1$, so that the standard basis vectors $\vec{\mathbf{e}}_1, \vec{\mathbf{e}}_2, \vec{\mathbf{e}}_3$ are sides, and the vector $\vec{\mathbf{d}} = \begin{bmatrix} 1 \\ 1 \\ 1 \end{bmatrix}$

is a diagonal. The length of the diagonal is $|\vec{\mathbf{d}}| = \sqrt{3}$, so the required angle α satisfies

$$\cos\alpha = \frac{\vec{\mathbf{d}} \cdot \vec{\mathbf{e}}_i}{|\vec{\mathbf{d}}||\vec{\mathbf{e}}_i|} = \frac{1}{\sqrt{3}}. \qquad 1.4.12$$

Thus $\alpha = \arccos \sqrt{3}/3 \approx 54.7°$. $\triangle$

Corollary 1.4.5 restates Proposition 1.4.3 in terms of projections; it is illustrated by Figure 1.4.3.

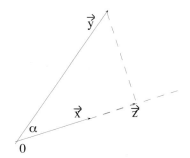

FIGURE 1.4.3.
The projection of $\vec{\mathbf{y}}$ onto the line spanned by $\vec{\mathbf{x}}$ is $\vec{\mathbf{z}}$. This gives

$$\vec{\mathbf{x}} \cdot \vec{\mathbf{y}} = |\vec{\mathbf{x}}||\vec{\mathbf{y}}| \cos\alpha$$
$$= |\vec{\mathbf{x}}||\vec{\mathbf{y}}| \frac{|\vec{\mathbf{z}}|}{|\vec{\mathbf{y}}|} = |\vec{\mathbf{x}}||\vec{\mathbf{z}}|.$$

Corollary 1.4.5 (The dot product in terms of projections). *If $\vec{\mathbf{x}}$ and $\vec{\mathbf{y}}$ are two vectors in $\mathbb{R}^2$ or $\mathbb{R}^3$, then $\vec{\mathbf{x}} \cdot \vec{\mathbf{y}}$ is the product of $|\vec{\mathbf{x}}|$ and the signed length of the projection of $\vec{\mathbf{y}}$ onto the line spanned by $\vec{\mathbf{x}}$. The signed length of the projection is positive if it points in the direction of $\vec{\mathbf{x}}$; it is negative if it points in the opposite direction.*

Defining angles between vectors in $\mathbb{R}^n$

We want to use Equation 1.4.8 backwards, to provide a definition of angles in $\mathbb{R}^n$, where we can't invoke elementary geometry when $n > 3$. Thus, we want to define

$$\alpha = \arccos \frac{\vec{\mathbf{v}} \cdot \vec{\mathbf{w}}}{|\vec{\mathbf{v}}||\vec{\mathbf{w}}|}, \quad \text{i.e., define } \alpha \text{ so that} \quad \cos\alpha = \frac{\vec{\mathbf{v}} \cdot \vec{\mathbf{w}}}{|\vec{\mathbf{v}}||\vec{\mathbf{w}}|}. \qquad 1.4.13$$

But there's a problem: how do we know that

$$-1 \leq \frac{\vec{\mathbf{v}} \cdot \vec{\mathbf{w}}}{|\vec{\mathbf{v}}||\vec{\mathbf{w}}|} \leq 1, \qquad\qquad 1.4.14$$

so that the arccosine exists? *Schwarz's inequality* provides the answer.[10] It is an absolutely fundamental result regarding dot products.

Theorem 1.4.6 (Schwarz's inequality). *For any two vectors* $\vec{\mathbf{v}}$ *and* $\vec{\mathbf{w}}$,

$$|\vec{\mathbf{v}} \cdot \vec{\mathbf{w}}| \leq |\vec{\mathbf{v}}|\,|\vec{\mathbf{w}}|. \qquad\qquad 1.4.15$$

The two sides are equal if and only if $\vec{\mathbf{v}}$ *or* $\vec{\mathbf{w}}$ *is a multiple of the other by a scalar.*

Proof. Consider the function $|\vec{\mathbf{v}} + t\vec{\mathbf{w}}|^2$ as a function of t. It is a second degree polynomial of the form $at^2 + bt + c$; in fact,

$$|t\vec{\mathbf{w}} + \vec{\mathbf{v}}|^2 = |\vec{\mathbf{w}}|^2 t^2 + 2(\vec{\mathbf{v}} \cdot \vec{\mathbf{w}})t + |\vec{\mathbf{v}}|^2. \qquad\qquad 1.4.16$$

All its values are ≥ 0, since it is the left-hand term squared; therefore, the graph of the polynomial must not cross the t-axis. But remember the quadratic formula you learned in high school: for an equation of the form $at^2 + bt + c = 0$,

$$t = \frac{-b \pm \sqrt{b^2 - 4ac}}{2a}. \qquad\qquad 1.4.17$$

If the *discriminant* (the quantity $b^2 - 4ac$ under the square root sign) is positive, the equation will have two distinct solutions, and its graph will cross the t-axis twice, as shown in the left-most graph in Figure 1.4.4.

Substituting $|\vec{\mathbf{w}}|^2$ for a, $2\vec{\mathbf{v}} \cdot \vec{\mathbf{w}}$ for b and $|\vec{\mathbf{v}}|^2$ for c, we see that the discriminant of Equation 1.4.16 is

$$4(\vec{\mathbf{v}} \cdot \vec{\mathbf{w}})^2 - 4|\vec{\mathbf{v}}|^2|\vec{\mathbf{w}}|^2. \qquad\qquad 1.4.18$$

All the values of Equation 1.4.16 are ≥ 0, so its discriminant can't be positive:

$$4(\vec{\mathbf{v}} \cdot \vec{\mathbf{w}})^2 - 4|\vec{\mathbf{v}}|^2|\vec{\mathbf{w}}|^2 \leq 0, \quad \text{and therefore} \quad |\vec{\mathbf{v}} \cdot \vec{\mathbf{w}}| \leq |\vec{\mathbf{v}}||\vec{\mathbf{w}}|,$$

which is what we wanted to show.

The second part of Schwarz's inequality, that $|\vec{\mathbf{v}} \cdot \vec{\mathbf{w}}| = |\vec{\mathbf{v}}|\,|\vec{\mathbf{w}}|$ if and only if $\vec{\mathbf{v}}$ or $\vec{\mathbf{w}}$ is a multiple of the other by a scalar, has two directions. If $\vec{\mathbf{w}}$ is a multiple of $\vec{\mathbf{v}}$, say $\vec{\mathbf{w}} = t\vec{\mathbf{v}}$, then

$$|\vec{\mathbf{v}} \cdot \vec{\mathbf{w}}| = |t||\vec{\mathbf{v}}|^2 = (|\vec{\mathbf{v}}|)(|t||\vec{\mathbf{v}}|) = |\vec{\mathbf{v}}||\vec{\mathbf{w}}|, \qquad\qquad 1.4.19$$

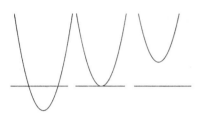

FIGURE 1.4.4.

Left to right: a positive discriminant gives two roots; a zero discriminant gives one root; a negative discriminant gives no roots.

---

[10]A more abstract form of Schwarz's inequality concerns inner products of vectors in possibly infinite-dimensional vector spaces, not just the standard dot product in $\mathbb{R}^n$. The general case is no more difficult to prove: the definition of an abstract inner product is precisely what is required to make this proof work.

The proof of Schwarz's inequality is clever; you can follow it line by line, like any proof which is written out in detail, but you won't find it by simply following your nose! There is considerable contention for the credit: Cauchy and Bunyakovski are often considered the inventors, particularly in France and in Russia.

We see that the dot product of two vectors is positive if the angle between them is less than $\pi/2$, and negative if it is bigger than $\pi/2$.

We prefer the word *orthogonal* to its synonym *perpendicular* for etymological reasons. Orthogonal comes from the Greek for "right angle," while perpendicular comes from the Latin for "plumb line," which suggests a vertical line. The word *normal* is also used, both as a noun and as an adjective, to express a right angle.

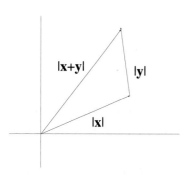

FIGURE 1.4.5.
The triangle inequality:
$|\vec{x} + \vec{y}| < |\vec{x}| + |\vec{y}|$.

so that Schwarz's inequality is satisfied as an equality.

Conversely, if $|\vec{v} \cdot \vec{w}| = |\vec{v}||\vec{w}|$, then the discriminant in Equation 1.4.18 is zero, so the polynomial has a single root t_0:

$$|\vec{v} + t_0\vec{w}|^2 = 0, \quad \text{i.e.,} \quad \vec{v} = -t_0\vec{w} \qquad 1.4.20$$

and $\vec{v}$ is a multiple of $\vec{w}$. $\square$

Schwarz's inequality allows us to define the angle between two vectors, since we are now assured that

$$-1 \;\leq\; \frac{\vec{a} \cdot \vec{b}}{|\vec{a}||\vec{b}|} \;\leq\; 1, \qquad (1.4.14)$$

Definition 1.4.7 (The angle between two vectors). The angle between two vectors $\vec{v}$ and $\vec{w}$ in $\mathbb{R}^n$ is that angle α satisfying $0 \leq \alpha \leq \pi$ such that

$$\cos\alpha = \frac{\vec{v} \cdot \vec{w}}{|\vec{v}|\,|\vec{w}|}. \qquad 1.4.21$$

Corollary 1.4.8. *Two vectors are orthogonal if their dot product is zero.*

Schwarz's inequality also gives us the *triangle inequality*: when traveling from London to Paris, it is shorter to go across the English Channel than by way of Moscow.

Theorem 1.4.9 (The triangle inequality). *For any vectors $\vec{x}$ and $\vec{y}$ in $\mathbb{R}^n$,*

$$|\vec{x} + \vec{y}| \leq |\vec{x}| + |\vec{y}|. \qquad 1.4.22$$

Proof. This inequality is proved by the following computation:

$$|\vec{x}+\vec{y}|^2 \;=\; |\vec{x}|^2+2\vec{x}\cdot\vec{y}+|\vec{y}|^2 \;\underset{\text{Schwarz}}{\leq}\; |\vec{x}|^2+2|\vec{x}||\vec{y}|+|\vec{y}|^2 \;=\; \big(|\vec{x}|+|\vec{y}|\big)^2, \;1.4.23$$

so that $|\vec{x} + \vec{y}| \leq |\vec{x}| + |\vec{y}|$. $\square$

This is called the triangle inequality because it can be interpreted (in the case of strict inequality, not $\leq$) as the statement that the length of one side of a triangle is less than the sum of the lengths of the other two sides. If a triangle has vertices $\mathbf{0}$, $\vec{x}$ and $\vec{x}+\vec{y}$, then the lengths of the sides are $|\vec{x}|$, $|\vec{x}+\vec{y}-\vec{x}| = |\vec{y}|$ and $|\vec{x} + \vec{y}|$, as shown in Figure 1.4.5.

Measuring matrices

The dot product gives a way to measure the length of vectors. We will also need a way to measure the "length" of a matrix (not to be confused with either its height or its width). There is an obvious way to do this: consider an $m \times n$ matrix as a point in $\mathbb{R}^{nm}$, and use the ordinary dot product.

In some texts, $|A|$ denotes the determinant of the matrix A. We use $\det A$ to denote the determinant.

The length $|A|$ is also called the *Frobenius norm* (and the *Schur norm* and *Hilbert-Schmidt norm*). We find it simpler to call it the length, generalizing the notion of length of a vector. Indeed, the length of an $n \times 1$ matrix is identical to the length of the vector in $\mathbb{R}^n$ with the same entries.

You shouldn't take the word "length" too literally; it's just a name for one way to measure matrices. (A more sophisticated measure, considerably harder to compute, is discussed in Section 2.8.)

Definition 1.4.10 (The length of a matrix). If A is an $n \times m$ matrix, its length $|A|$ is the square root of the sum of the squares of all its entries:

$$|A|^2 = \sum_{i=1}^{n} \sum_{j=1}^{m} a_{i,j}^2. \qquad 1.4.24$$

For example, the length $|A|$ of the matrix $A = \begin{bmatrix} 1 & 2 \\ 0 & 1 \end{bmatrix}$ is $\sqrt{6}$, since $1 + 4 + 0 + 1 = 6$. What is the length of the matrix $B = \begin{bmatrix} 1 & 2 & 1 \\ 1 & 0 & 3 \end{bmatrix}$?[11]

If you find double sum notation confusing, Equation 1.4.24 can be rewritten as a single sum:

$$|A|^2 = \sum_{\substack{i=1,\dots,n \\ j=1,\dots,m}} a_{i,j}^2 : \quad \text{we sum all } a_{i,j}^2 \text{ for } i \text{ from 1 to } n \text{ and } j \text{ from 1 to } m.$$

As in the case of the length of a vector, do not confuse the length $|A|$ of a matrix with the absolute value of a number. (But the length of the 1×1 matrix consisting of the single entry $[n]$ is indeed the absolute value of n.)

Length and matrix multiplication

We said earlier that the point of writing the entries of $\mathbb{R}^{mn}$ as matrices is to allow matrix multiplication, yet it isn't clear that this notion of length, in which a matrix is considered simply as a list of numbers, is in any way related to matrix multiplication. The following proposition says that it is.

Thinking of an $m \times n$ matrix as a point in $\mathbb{R}^{nm}$, we can see that two matrices A and B (and therefore, the corresponding linear transformations) are close if the length of their difference is small; i.e., if $|A - B|$ is small.

Proposition 1.4.11. (a) If A is an $n \times m$ matrix, and $\vec{\mathbf{b}}$ is a vector in $\mathbb{R}^m$, then

$$|A\vec{\mathbf{b}}| \le |A||\vec{\mathbf{b}}|. \qquad 1.4.25$$

(b) If A is an $n \times m$ matrix, and B is a $m \times k$ matrix, then

$$|AB| \le |A|\,|B|. \qquad 1.4.26$$

---

[11] $|B| = 4$, since $1 + 4 + 0 + 1 + 1 + 9 = 16$.

Proposition 1.4.11 will soon become an old friend; it is a very useful tool in a number of proofs.

Of course, part (a) is the special case of part (b) (where $k = 1$), but the intuitive content is sufficiently different that we state the two parts separately. In any case, the proof of the second part follows from the first.

Remark 1.4.12. It follows from Proposition 1.4.11 that a linear transformation is continuous. Saying that a linear transformation A is continuous means that for every ϵ and every $\mathbf{x} \in \mathbb{R}^n$, there exists a δ such that if $|\vec{x} - \vec{y}| < \delta$, then $|A\vec{x} - A\vec{y}| < \epsilon$. By Proposition 1.4.11,

$$|A\vec{x} - A\vec{y}| = |A(\vec{x} - \vec{y})| \le |A||\vec{x} - \vec{y}|.$$

So, set

$$\delta = \frac{\epsilon}{|A|}.$$

Then if we have $|\vec{x} - \vec{y}| < \delta$,

$$|\vec{x} - \vec{y}| < \frac{\epsilon}{|A|}$$

and

$$|A\vec{x} - A\vec{y}| < \frac{|A|\epsilon}{|A|} = \epsilon.$$

We have actually proved more: the δ we found did not depend on $\mathbf{x}$; this means that a linear transformation $\mathbb{R}^n \to \mathbb{R}^n$ is always *uniformly continuous*. The definition of uniform continuity was given in Equation 0.2.6. $\triangle$

Proof. First note that if the matrix A consists of a single row, i.e., if $A = \vec{a}^\top$ is the transpose of a vector $\vec{a}$, the assertion of the theorem is exactly Schwarz's inequality:

$$\underbrace{|A\vec{b}|}_{|\vec{a}^\top \vec{b}|} = |\vec{a} \cdot \vec{b}| \le |\vec{a}| |\vec{b}| = \underbrace{|A||\vec{b}|}_{|\vec{a}^\top||\vec{b}|}. \qquad 1.4.27$$

The idea of the proof is to consider that the rows of A are the transposes of vectors $\vec{a}_1, \ldots \vec{a}_n$, as shown in Figure 1.4.6, and to apply the argument above to each row separately. Remember that since the ith row of A is $\vec{a}_i^\top$, the ith entry $(A\vec{b})_i$ of $A\vec{b}$ is precisely the dot product $\vec{a}_i \cdot \mathbf{b}$. (This accounts for the equal sign marked (1) in Equation 1.4.28.)

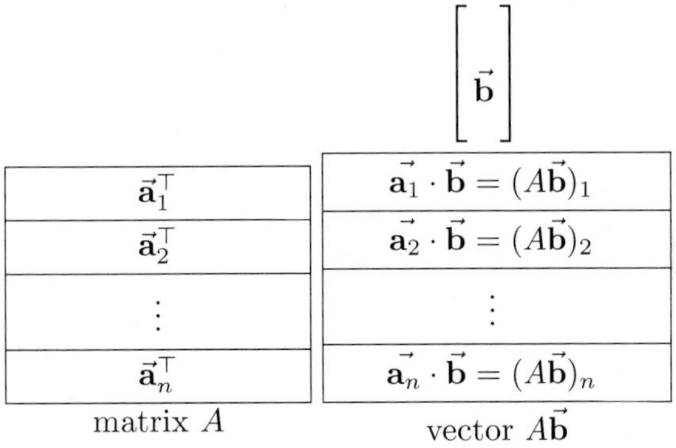

FIGURE 1.4.6. Think of the rows of A as the *transposes* of the vectors $\vec{a}_1, \vec{a}_2, \ldots, \vec{a}_n$. Then the product $\vec{a}_i^\top \vec{b}$ is the same as the dot product $\vec{a}_i \cdot \mathbf{b}$. Note that $A\vec{b}$ is a vector, not a matrix.

This leads to

$$|A\vec{b}|^2 \quad = \quad \sum_{i=1}^{n} (A\vec{b})_i^2 \quad \underset{(1)}{=} \quad \sum_{i=1}^{n} (\vec{a}_i \cdot \mathbf{b})^2. \qquad 1.4.28$$

Now use Schwarz's inequality (2); factor out $|\vec{b}|^2$ (step 3), and consider (step 4) the length squared of A to be the sum of the squares of the lengths of $\vec{a}_i$. (Of course, $|\vec{a}_i|^2 = |\vec{a}_i^\top|^2$). Thus,

$$\sum_{i=1}^{n} (\vec{a}_i \cdot \mathbf{b})^2 \quad \underset{(2)}{\le} \quad \sum_{i=1}^{n} |\vec{a}_i|^2 |\vec{b}|^2 \quad \underset{(3)}{=} \quad \left(\sum_{i=1}^{n} |\vec{a}_i|^2 \right) |\vec{b}|^2 \quad \underset{(4)}{=} \quad |A|^2 |\vec{b}|^2. \qquad 1.4.29$$

This gives us the result we wanted:

$$|A\vec{b}|^2 \le |A|^2 |\vec{b}|^2.$$

For the second, we decompose the matrix B into its columns and proceed as above. Let $\vec{\mathbf{b}}_1, \ldots, \vec{\mathbf{b}}_k$ be the columns of B. Then

$$|AB|^2 = \sum_{j=1}^{k} |A\vec{\mathbf{b}}_j|^2 \leq \sum_{j=1}^{k} |A|^2 |\vec{\mathbf{b}}_j|^2 = |A|^2 \sum_{j=1}^{k} |\vec{\mathbf{b}}_j|^2 = |A|^2 |B|^2, \qquad 1.4.30$$

which proves the second part. $\square$

When solving big systems of linear questions was in any case out of the question, determinants were a reasonable approach to the theory of linear equations. With the advent of computers they lost importance, as systems of linear equations can be solved far more effectively with row reduction (to be discussed in Sections 2.1 and 2.2). However, determinants have an interesting geometric interpretation; in Chapters 4, 5, and especially 6, we use determinants constantly.

Determinants in $\mathbb{R}^2$

The *determinant* is a function of square matrices: it takes a square matrix as input and gives a number as output.

Definition 1.4.13 (Determinant in $\mathbb{R}^2$). The determinant of a 2×2 matrix $\begin{bmatrix} a_1 & b_1 \\ a_2 & b_2 \end{bmatrix}$ is

$$\det \begin{bmatrix} a_1 & b_1 \\ a_2 & b_2 \end{bmatrix} = a_1 b_2 - a_2 b_1. \qquad 1.4.31$$

Recall that the formula for the inverse of a 2×2 matrix $A = \begin{bmatrix} a & b \\ c & d \end{bmatrix}$ is

$$A^{-1} = \frac{1}{ad - bc} \begin{bmatrix} d & -b \\ -c & a \end{bmatrix}.$$

So a 2×2 matrix A is invertible if and only if $\det A \neq 0$.

The determinant is an interesting number to associate to a matrix because if we think of the determinant as a function of the vectors $\vec{\mathbf{a}}$ and $\vec{\mathbf{b}}$ in $\mathbb{R}^2$, then it has a geometric interpretation, illustrated by Figure 1.4.7:

Proposition 1.4.14 (Geometric interpretation of the determinant in $\mathbb{R}^2$). *(a) The area of the parallelogram spanned by the vectors*

$$\vec{\mathbf{a}} = \begin{bmatrix} a_1 \\ a_2 \end{bmatrix} \quad and \quad \vec{\mathbf{b}} = \begin{bmatrix} b_1 \\ b_2 \end{bmatrix}$$

is $|\det[\vec{\mathbf{a}}, \vec{\mathbf{b}}]|$

(b) The determinant $\det[\vec{\mathbf{a}}, \vec{\mathbf{b}}]$ is positive if and only if $\vec{\mathbf{b}}$ lies counterclockwise from $\vec{\mathbf{a}}$; it is negative if and only if $\vec{\mathbf{b}}$ lies clockwise from $\vec{\mathbf{a}}$.

In this section we limit our discussion to determinants of 2×2 and 3×3 matrices; we discuss determinants in higher dimensions in Section 4.8.

Proof. (a) The area of the parallelogram is its height times its base. Its base is $|\vec{\mathbf{b}}| = \sqrt{b_1^2 + b_2^2}$. If θ is the angle between $\vec{\mathbf{a}}$ and $\vec{\mathbf{b}}$, its height h is

$$h = \sin \theta |\vec{\mathbf{a}}| = \sin \theta \sqrt{a_1^2 + a_2^2}. \qquad 1.4.32$$

We can compute $\cos \theta$ by using Equation 1.4.8:

$$\cos \theta = \frac{\vec{\mathbf{a}} \cdot \vec{\mathbf{b}}}{|\vec{\mathbf{a}}||\vec{\mathbf{b}}|} = \frac{a_1 b_1 + a_2 b_2}{\sqrt{a_1^2 + a_2^2} \sqrt{b_1^2 + b_2^2}}. \qquad 1.4.33$$

So we get $\sin \theta$ as follows:

$$\sin \theta = \sqrt{1 - \cos^2 \theta} = \sqrt{\frac{(a_1^2 + a_2^2)(b_1^2 + b_2^2) - (a_1 b_1 + a_2 b_2)^2}{(a_1^2 + a_2^2)(b_1^2 + b_2^2)}}$$

$$= \sqrt{\frac{a_1^2 b_1^2 + a_1^2 b_2^2 + a_2^2 b_1^2 + a_2^2 b_2^2 - a_1^2 b_1^2 - 2a_1 b_1 a_2 b_2 - a_2^2 b_2^2}{(a_1^2 + a_2^2)(b_1^2 + b_2^2)}} \qquad 1.4.34$$

$$= \sqrt{\frac{(a_1 b_2 - a_2 b_1)^2}{(a_1^2 + a_2^2)(b_1^2 + b_2^2)}}.$$

Using this value for $\sin \theta$ in the equation for the area of a parallelogram gives

$$\text{Area} = \underbrace{|\vec{b}|}_{\text{base}} \underbrace{|\vec{a}| \sin \theta}_{\text{height}}$$

$$= \underbrace{\sqrt{b_1^2 + b_2^2}}_{\text{base}} \underbrace{\sqrt{a_1^2 + a_2^2} \sqrt{\frac{(a_1 b_2 - a_2 b_1)^2}{(a_1^2 + a_2^2)(b_1^2 + b_2^2)}}}_{\text{height}} = \underbrace{|a_1 b_2 - a_2 b_1|}_{\text{determinant}}. \qquad 1.4.35$$

(b) The vector $\vec{c}$ obtained by rotating $\vec{a}$ counterclockwise by $\pi/2$ is $\vec{c} = \begin{bmatrix} -a_2 \\ a_1 \end{bmatrix}$, and we see that $\vec{c} \cdot \vec{b} = \det[\vec{a}, \vec{b}]$:

$$\begin{bmatrix} -a_2 \\ a_1 \end{bmatrix} \cdot \begin{bmatrix} b_1 \\ b_2 \end{bmatrix} = -a_2 b_1 + a_1 b_2 = \det \begin{bmatrix} a_1 & b_1 \\ a_2 & b_2 \end{bmatrix}. \qquad 1.4.36$$

Since (Proposition 1.4.3) the dot product of two vectors is positive if the angle between them is less than $\pi/2$, the determinant is positive if the angle between $\vec{b}$ and $\vec{c}$ is less than $\pi/2$. So $\vec{b}$ lies counterclockwise from $\vec{a}$, as shown in Figure 1.4.8. $\quad \square$

Exercise 1.4.6 gives a more geometric proof of Proposition 1.4.14.

Determinants in $\mathbb{R}^3$

Definition 1.4.15 (Determinant in $\mathbb{R}^3$). The determinant of a 3×3 matrix is

$$\det \begin{bmatrix} a_1 & b_1 & c_1 \\ a_2 & b_2 & c_2 \\ a_3 & b_3 & c_3 \end{bmatrix} = a_1 \det \begin{bmatrix} b_2 & c_2 \\ b_3 & c_3 \end{bmatrix} - a_2 \det \begin{bmatrix} b_1 & c_1 \\ b_3 & c_3 \end{bmatrix} + a_3 \det \begin{bmatrix} b_1 & c_1 \\ b_2 & c_2 \end{bmatrix}$$

$$= a_1(b_2 c_3 - b_3 c_2) - a_2(b_1 c_3 - b_3 c_1) + a_3(b_1 c_2 - b_2 c_1).$$

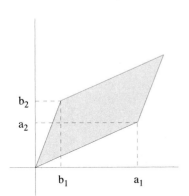

FIGURE 1.4.7.

The area of the parallelogram spanned by $\vec{a}$ and $\vec{b}$ is $|\det[\vec{a}, \vec{b}]|$.

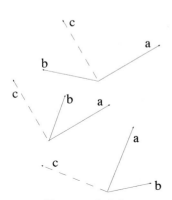

FIGURE 1.4.8.

In the two cases at the top, the angle between $\vec{b}$ and $\vec{c}$ is less than $\pi/2$, so $\det(\vec{a}, \vec{b}) > 0$; this corresponds to $\vec{b}$ being counterclockwise from $\vec{a}$. At the bottom, the angle between $\vec{b}$ and $\vec{c}$ is more than $\pi/2$; in this case, $\vec{b}$ is clockwise from $\vec{a}$, and $\det(\vec{a}, \vec{b})$ is negative.

Exercise 1.4.12 shows that a 3×3 matrix is invertible if its determinant is not 0.

For larger matrices, the formulas rapidly get out of hand; we will see in Section 4.8 that such determinants can be computed much more reasonably by row (or column) reduction.

Each entry of the first column of the original matrix serves as the coefficient for the determinant of a 2×2 matrix; the first and third (a_1 and a_3) are positive, the middle one is negative. To remember which 2×2 matrix goes with which coefficient, cross out the row and column the coefficient is in; what is left is the matrix you want. To get the 2×2 matrix for the coefficient a_2:

$$\begin{bmatrix} a_1 & b_1 & c_1 \\ a_2 & b_2 & c_2 \\ a_3 & b_3 & c_3 \end{bmatrix} = \begin{bmatrix} b_1 & c_1 \\ b_3 & c_3 \end{bmatrix}. \tag{1.4.37}$$

The determinant can also be computed using the entries of the first row, rather than of the first column, as coefficients.

Example 1.4.16 (Determinant of a 3×3 matrix).

$$\det \begin{bmatrix} 3 & 1 & -2 \\ 1 & 2 & 4 \\ 2 & 0 & 1 \end{bmatrix} = 3 \det \begin{bmatrix} 2 & 4 \\ 0 & 1 \end{bmatrix} - 1 \det \begin{bmatrix} 1 & -2 \\ 0 & 1 \end{bmatrix} + 2 \det \begin{bmatrix} 1 & -2 \\ 2 & 4 \end{bmatrix} \tag{1.4.38}$$

$$= 3\,(2 - 0) - (1 + 0) + 2\,(4 + 4) = 21. \quad \triangle$$

The cross product of two vectors

Although the determinant is a *number*, as is the dot product, the *cross product* is a vector:

Definition 1.4.17 (Cross product in $\mathbb{R}^3$). The cross product $\vec{\mathbf{a}} \times \vec{\mathbf{b}}$ in $\mathbb{R}^3$ is

$$\begin{bmatrix} a_1 \\ a_2 \\ a_3 \end{bmatrix} \times \begin{bmatrix} b_1 \\ b_2 \\ b_3 \end{bmatrix} = \begin{bmatrix} \det \begin{bmatrix} a_2 & b_2 \\ a_3 & b_3 \end{bmatrix} \\ -\det \begin{bmatrix} a_1 & b_1 \\ a_3 & b_3 \end{bmatrix} \\ \det \begin{bmatrix} a_1 & b_1 \\ a_2 & b_2 \end{bmatrix} \end{bmatrix} = \begin{bmatrix} a_2 b_3 - a_3 b_2 \\ -a_1 b_3 + a_3 b_1 \\ a_1 b_2 - a_2 b_1 \end{bmatrix}. \tag{1.4.39}$$

The cross product exists only in $\mathbb{R}^3$ (and to some extent in $\mathbb{R}^7$).

Think of your vectors as a 3×2 matrix; first cover up the first row and take the determinant of what's left. That gives the first entry of the cross product. Then cover up the second row and take *minus* the determinant of what's left, giving the second entry of the cross product. The third entry is obtained by covering up the third row and taking the determinant of what's left.

Example 1.4.18 (Cross product of two vectors in $\mathbb{R}^3$).

$$\begin{bmatrix} 3 \\ 0 \\ 1 \end{bmatrix} \times \begin{bmatrix} 2 \\ 1 \\ 4 \end{bmatrix} = \begin{bmatrix} \det \begin{bmatrix} 0 & 1 \\ 1 & 4 \end{bmatrix} \\ -\det \begin{bmatrix} 3 & 2 \\ 1 & 4 \end{bmatrix} \\ \det \begin{bmatrix} 3 & 2 \\ 0 & 1 \end{bmatrix} \end{bmatrix} = \begin{bmatrix} -1 \\ -10 \\ 3 \end{bmatrix}. \quad \triangle \qquad 1.4.40$$

Like the determinant, the cross product has a geometric interpretation.

The right-hand rule: if you put the thumb of your right hand on $\vec{\mathbf{a}}$ and your index finger on $\vec{\mathbf{b}}$, while bending your third finger, then your third finger points in the direction of $\vec{\mathbf{a}} \times \vec{\mathbf{b}}$. (Alternatively, curl the fingers of your right hand from $\vec{\mathbf{a}}$ to $\vec{\mathbf{b}}$; then your right thumb will point in the direction of $\vec{\mathbf{a}} \times \vec{\mathbf{b}}$.)

Proposition 1.4.19 (Geometric interpretation of the cross product). *The cross product $\vec{\mathbf{a}} \times \vec{\mathbf{b}}$ is the vector satisfying three properties:*

(1) *It is orthogonal to the plane spanned by $\vec{\mathbf{a}}$ and $\vec{\mathbf{b}}$; i.e.,*

$$\vec{\mathbf{a}} \cdot (\vec{\mathbf{a}} \times \vec{\mathbf{b}}) = 0 \quad and \quad \vec{\mathbf{b}} \cdot (\vec{\mathbf{a}} \times \vec{\mathbf{b}}) = 0. \qquad 1.4.41$$

(2) *Its length $|\vec{\mathbf{a}} \times \vec{\mathbf{b}}|$ is the area of the parallelogram spanned by $\vec{\mathbf{a}}$ and $\vec{\mathbf{b}}$;*

(3) *The three vectors $\vec{\mathbf{a}}, \vec{\mathbf{b}},$ and $\vec{\mathbf{a}} \times \vec{\mathbf{b}}$ satisfy the right-hand rule.*

Proof. (1) It is not difficult to check that the cross product $\vec{\mathbf{a}} \times \vec{\mathbf{b}}$ is orthogonal to both $\vec{\mathbf{a}}$ and $\vec{\mathbf{b}}$: we check that the dot product in each case is zero (Corollary 1.4.8). Thus $\vec{\mathbf{a}} \times \vec{\mathbf{b}}$ is orthogonal to $\vec{\mathbf{a}}$ because

$$\vec{\mathbf{a}} \cdot (\vec{\mathbf{a}} \times \vec{\mathbf{b}}) = \begin{bmatrix} a_1 \\ a_2 \\ a_3 \end{bmatrix} \cdot \overbrace{\begin{bmatrix} a_2 b_3 - a_3 b_2 \\ -a_1 b_3 + a_3 b_1 \\ a_1 b_2 - a_2 b_1 \end{bmatrix}}^{\text{Definition 1.4.17 of } \vec{\mathbf{a}} \times \vec{\mathbf{b}}} \qquad 1.4.42$$

$$= a_1 a_2 b_3 - a_1 a_3 b_2 - a_1 a_2 b_3 + a_2 a_3 b_1 + a_1 a_3 b_2 - a_2 a_3 b_1 = 0.$$

(2) The area of the parallelogram spanned by $\vec{\mathbf{a}}$ and $\vec{\mathbf{b}}$ is $|\vec{\mathbf{a}}| \cdot |\vec{\mathbf{b}}| \sin\theta$, where θ is the angle between $\vec{\mathbf{a}}$ and $\vec{\mathbf{b}}$. We know (Equation 1.4.8) that

$$\cos\theta = \frac{\vec{\mathbf{a}} \cdot \vec{\mathbf{b}}}{|\vec{\mathbf{a}}||\vec{\mathbf{b}}|} = \frac{a_1 b_1 + a_2 b_2 + a_3 b_3}{\sqrt{a_1^2 + a_2^2 + a_3^2}\sqrt{b_1^2 + b_2^2 + b_3^2}}, \qquad 1.4.43$$

so we have

$$\sin\theta = \sqrt{1 - \cos^2\theta} = \sqrt{1 - \frac{(a_1 b_1 + a_2 b_2 + a_3 b_3)^2}{(a_1^2 + a_2^2 + a_3^2)(b_1^2 + b_2^2 + b_3^2)}}$$

$$= \sqrt{\frac{(a_1^2 + a_2^2 + a_3^2)(b_1^2 + b_2^2 + b_3^2) - (a_1 b_1 + a_2 b_2 + a_3 b_3)^2}{(a_1^2 + a_2^2 + a_3^2)(b_1^2 + b_2^2 + b_3^2)}}, \qquad 1.4.44$$

so that

$$|\vec{\mathbf{a}}||\vec{\mathbf{b}}| \sin\theta = \sqrt{(a_1^2 + a_2^2 + a_3^2)(b_1^2 + b_2^2 + b_3^2) - (a_1b_1 + a_2b_2 + a_3b_3)^2}. \quad 1.4.45$$

The last equality in Equation 1.4.47 comes of course from Definition 1.4.17.

You may object that the middle term of the square root looks different than the middle entry of the cross product as given in Definition 1.4.17, but since we are squaring it,

$$(-a_1b_3 + a_3b_1)^2 = (a_1b_3 - a_3b_1)^2.$$

Carrying out the multiplication results in a formula for the area that looks worse than it is: a long string of terms too big to fit on this page under one square root sign. That's a good excuse for omitting it here. But if you do the computations you'll see that after cancellations we have for the right-hand side:

$$\sqrt{\underbrace{a_1^2b_2^2 + a_2^2b_1^2 - 2a_1b_1a_2b_2}_{(a_1b_2 - a_2b_1)^2} + \underbrace{a_1^2b_3^2 + a_3^2b_1^2 - 2a_1b_1a_3b_3}_{(a_1b_3 - a_3b_1)^2} + \underbrace{a_2^2b_3^2 + a_3^2b_2^2 - 2a_2b_2a_3b_3}_{(a_2b_3 - a_3b_2)^2}},$$

$$1.4.46$$

which conveniently gives us

$$\text{Area} = |\vec{\mathbf{a}}||\vec{\mathbf{b}}| \sin\theta = \sqrt{(a_1b_2 - a_2b_1)^2 + (a_1b_3 - a_3b_1)^2 + (a_2b_3 - a_3b_2)^2}$$

$$= |\vec{\mathbf{a}} \times \vec{\mathbf{b}}|.$$

$$1.4.47$$

(3) So far, then, we have seen that the cross product $\vec{\mathbf{a}} \times \vec{\mathbf{b}}$ is orthogonal to $\vec{\mathbf{a}}$ and $\vec{\mathbf{b}}$, and that its length is the area of the parallelogram spanned by $\vec{\mathbf{a}}$ and $\vec{\mathbf{b}}$. What about the right-hand rule? Equation 1.4.39 for the cross product cannot actually specify that the three vectors obey the right-hand rule, because your right hand is not an object of mathematics.

What we can show is that if one of your hands fits $\vec{\mathbf{e}}_1, \vec{\mathbf{e}}_2, \vec{\mathbf{e}}_3$, then it will also fit $\vec{\mathbf{a}}, \vec{\mathbf{b}}, \vec{\mathbf{a}} \times \vec{\mathbf{b}}$. Suppose $\vec{\mathbf{a}}$ and $\vec{\mathbf{b}}$ are not collinear. You have one hand that fits $\vec{\mathbf{a}}, \vec{\mathbf{b}}, \vec{\mathbf{a}} \times \vec{\mathbf{b}}$; i.e., you can put the thumb in the direction of $\vec{\mathbf{a}}$, your index finger in the direction of $\vec{\mathbf{b}}$ and the middle finger in the direction of $\vec{\mathbf{a}} \times \vec{\mathbf{b}}$ without bending your knuckles backwards. You can move $\vec{\mathbf{a}}$ to point in the same direction as $\vec{\mathbf{e}}_1$, for instance, by rotating all of space (in particular $\vec{\mathbf{b}}$, $\vec{\mathbf{a}} \times \vec{\mathbf{b}}$ and your hand) around the line perpendicular to the plane containing $\vec{\mathbf{a}}$ and $\vec{\mathbf{e}}_1$. Now rotate all of space (in particular $\vec{\mathbf{a}} \times \vec{\mathbf{b}}$ and your hand) around the x-axis, until $\vec{\mathbf{b}}$ is in the (x, y)-plane, with the y-coordinate positive. These movements simply rotated your hand, so it still fits the vectors.

Now we see that our vectors have become

$$\vec{\mathbf{a}} = \begin{bmatrix} a \\ 0 \\ 0 \end{bmatrix} \quad \text{and} \quad \vec{\mathbf{b}} = \begin{bmatrix} b_1 \\ b_2 \\ 0 \end{bmatrix}, \quad \text{so} \quad \vec{\mathbf{a}} \times \vec{\mathbf{b}} = \begin{bmatrix} 0 \\ 0 \\ ab_2 \end{bmatrix}. \quad 1.4.48$$

Thus, your thumb is in the direction of the positive x-axis, your index finger is horizontal, pointing into the part of the (x, y)-plane where y is positive, and since both a and b_2 are positive, your middle finger points straight up. So the same hand will fit the vectors as will fit the standard basis vectors: the right hand if you draw them the standard way (x-axis coming out of the paper straight at you, y-axis pointing to the right, z-axis up.)   △

Geometric interpretation of the determinant in $\mathbb{R}^3$

The determinant of three vectors $\vec{a}, \vec{b}$ and $\vec{c}$ can also be thought of as the dot product of one vector with the cross product of the other two, $\vec{a} \cdot (\vec{b} \times \vec{c})$:

$$\begin{bmatrix} a_1 \\ a_2 \\ a_3 \end{bmatrix} \cdot \begin{bmatrix} \det \begin{bmatrix} b_2 & c_2 \\ b_3 & c_3 \end{bmatrix} \\ -\det \begin{bmatrix} b_1 & c_1 \\ b_3 & c_3 \end{bmatrix} \\ \det \begin{bmatrix} b_1 & c_1 \\ b_2 & c_2 \end{bmatrix} \end{bmatrix} = a_1 \det \begin{bmatrix} b_2 & c_2 \\ b_3 & c_3 \end{bmatrix} - a_2 \det \begin{bmatrix} b_2 & c_2 \\ b_3 & c_3 \end{bmatrix} + a_3 \det \begin{bmatrix} b_1 & c_1 \\ b_2 & c_2 \end{bmatrix}.$$

1.4.49

As such it has a geometric interpretation:

The word *parallelepiped* seems to have fallen into disuse; we've met students who got a 5 on the Calculus BC exam who don't know what the term means. It is simply a possibly slanted box: a box with six faces, each of which is a parallelogram; opposite faces are equal.

The determinant is 0 if the three vectors are co-planar.

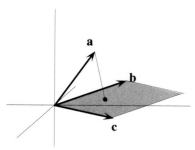

FIGURE 1.4.9.

The volume of the parallelepiped spanned by the vectors $\vec{a}, \vec{b}, \vec{c}$ is $|\det[\vec{a}, \vec{b}, \vec{c}]|$.

Proposition 1.4.20 (Geometric interpretation of the determinant in $\mathbb{R}^3$). (a) *The absolute value of the determinant of three vectors $\vec{a}, \vec{b}, \vec{c}$ forming a 3×3 matrix gives the volume of the parallelepiped they span.*

(b) *The determinant is positive if the vectors satisfy the right-hand rule, and negative otherwise.*

Proof. (a) The volume is height times the area of the base, the base shown in Figure 1.4.9 as the parallelogram spanned by $\vec{b}$ and $\vec{c}$. That area is given by the length of the cross product, $|\vec{b} \times \vec{c}|$. The height h is the projection of $\vec{a}$ onto a line orthogonal to the base. Let's choose the line spanned by the cross product $\vec{b} \times \vec{c}$—that is, the line in the same direction as that vector. Then $h = |\vec{a}| \cos \theta$, where θ is the angle between $\vec{a}$ and $\vec{b} \times \vec{c}$, and we have

$$\text{Volume of parallelepiped } = \underbrace{|\vec{b} \times \vec{c}|}_{\text{base}} \underbrace{|\vec{a}| \cos \theta}_{\text{height}} = \underbrace{|\vec{a} \cdot (\vec{b} \times \vec{c})|}_{\text{determinant}}. \qquad 1.4.50$$

(b) The determinant is positive if $\cos \theta > 0$ (i.e., if the angle between $\vec{a}$ and $\vec{b} \times \vec{c}$ is less than $\pi/2$). Put your right hand to fit $\vec{b} \times \vec{c}, \vec{b}, \vec{c}$; since $\vec{b} \times \vec{c}$ is perpendicular to the plane spanned by $\vec{b}$ and $\vec{c}$, you can move your thumb in any direction by any angle less than $\pi/2$, in particular, in the direction of $\vec{a}$. (This requires a mathematically correct, very supple thumb.) $\triangle$

Remark. The correspondence between algebra and geometry is a constant theme of mathematics. Figure 1.4.10 summarizes the relationships discussed in this section. $\triangle$

Correspondence of Algebra and Geometry		
Operation	Algebra	Geometry
dot product	$\vec{\mathbf{v}} \cdot \vec{\mathbf{w}} = \sum v_j w_i$	$\vec{\mathbf{v}} \cdot \vec{\mathbf{w}} = |\vec{\mathbf{v}}||\vec{\mathbf{w}}| \cos\theta$
determinant of 2×2 matrix	$\det \begin{bmatrix} a_1 & b_1 \\ a_2 & b_2 \end{bmatrix} = a_1 b_2 - a_2 b_1$	$\left| \det \begin{bmatrix} a_1 & b_1 \\ a_2 & b_2 \end{bmatrix} \right| = $ Area of parallelogram
cross product	$\begin{bmatrix} a_1 \\ a_2 \\ a_3 \end{bmatrix} \times \begin{bmatrix} b_1 \\ b_2 \\ b_3 \end{bmatrix} = \begin{bmatrix} a_2 b_3 - b_2 a_3 \\ b_1 a_3 - a_1 b_3 \\ a_1 b_2 - a_2 b_1 \end{bmatrix}$	$(\vec{\mathbf{a}} \times \vec{\mathbf{b}}) \perp \vec{\mathbf{a}}, \quad (\vec{\mathbf{a}} \times \vec{\mathbf{b}}) \perp \vec{\mathbf{b}}$ Length = area of parallelogram Right-hand rule
determinant of 3×3 matrix	$\det \begin{bmatrix} a_1 & b_1 & c_1 \\ a_2 & b_2 & c_2 \\ a_3 & b_3 & c_3 \end{bmatrix} = \vec{\mathbf{a}} \cdot (\vec{\mathbf{b}} \times \vec{\mathbf{c}})$	$\left| \det[\vec{\mathbf{a}}, \vec{\mathbf{b}}, \vec{\mathbf{c}}] \right| = $ Volume of parallelepiped

FIGURE 1.4.10. Mathematical "objects" often have two interpretations: algebraic and geometric.

1.5 CONVERGENCE AND LIMITS

In this section, we collect the relevant definitions of limits and continuity. Integrals, derivatives, series, approximations: calculus is all about convergence and limits. It could easily be argued that these notions are the hardest and deepest of all of mathematics. They give students a lot of trouble, and historically, mathematicians struggled to come up with correct definitions for two hundred years. Fortunately, these notions do not become more difficult in several variables than they are in one variable.

More students have foundered on these definitions than on anything else in calculus: the combination of Greek letters, precise order of quantifiers, and inequalities is a hefty obstacle. Working through a few examples will help you understand what the definitions mean, but a proper appreciation can probably only come from use; we hope you have already started on this path in one-variable calculus.

The inventors of calculus in the 17th century did not have rigorous definitions of limits and continuity; these were achieved only in the 1870s. Rigor is ultimately necessary in mathematics, but it does not always come first, as Archimedes acknowledged about his own work, in a manuscript discovered in 1906. In it Archimedes reveals that his deepest results were found using dubious infinitary arguments, and only later proved rigorously, because *"it is of course easier to supply the proof when we have previously acquired some knowledge of the questions by the method, than it is to find it without any previous knowledge."* (We found this story in John Stillwell's *Mathematics and Its History*.)

Open and closed sets

In mathematics we often need to speak of an *open set U*; whenever we want to approach points of a set U from every side, U must be open.

Think of a set or subset as your property, surrounded by a fence. The set is open (Figure 1.5.1) if the entire fence belongs to your neighbor. As long as you stay on your property, you can get closer and closer to the fence, but you can

FIGURE 1.5.1.

An open set includes none of the fence; no matter how close a point in the open set is to the fence, you can always surround it with a ball of other points in the open set.

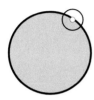

FIGURE 1.5.2.

A closed set includes its fence.

Note that $|\mathbf{x} - \mathbf{y}|$ must be *less than* r for the ball to be open; it *cannot* be $= r$.

The symbol $\subset$ used in Definition 1.5.3 means "subset of." If you are not familiar with the symbols used in set theory, you may wish to read the discussion of set theoretic notation in Section 0.3.

never reach it. No matter how close you are to your neighbor's property, there is always an epsilon-thin buffer zone of your property between you and it—just as no matter how close a non zero point on the real number line is to 0, you can always find points that are closer.

The set is closed (Figure 1.5.2) if you own the fence. Now, if you sit on your fence, there is nothing between you and your neighbor's property. If you move even an epsilon further, you will be trespassing.

What if some of the fence belongs to you and some belongs to your neighbors? Then the set is neither open nor closed.

Remark 1.5.1. Even very good students often don't see the point of specifying that a set is open. But it is absolutely essential, for example in computing derivatives. If a function f is defined on a set that is not open, and thus contains at least one point x that is part of the fence, then talking of the derivative of f at x is meaningless. To compute $f'(x)$ we need to compute

$$f'(x) = \lim_{h \to 0} \frac{1}{h}\big(f(x+h) - f(x)\big), \qquad 1.5.1$$

but $f(x+h)$ won't necessarily exist for h arbitrarily small, since $x+h$ may be outside the fence and thus not in the domain of f. This situation gets much worse in $\mathbb{R}^n$.[12]   △

In order to define open and closed sets in proper mathematical language, we first need to define an open ball. Imagine a balloon of radius r, centered around a point $\mathbf{x}$. The open ball of radius r around $\mathbf{x}$ consists of all points $\mathbf{y}$ inside the balloon, but *not* the skin of the balloon itself: whatever $\mathbf{y}$ you choose, the distance between $\mathbf{x}$ and $\mathbf{y}$ is always less than the radius r.

Definition 1.5.2 (Open ball). For any $\mathbf{x} \in \mathbb{R}^n$ and any $r > 0$, the open ball of radius r around $\mathbf{x}$ is the subset

$$B_r(\mathbf{x}) = \{\mathbf{y} \in \mathbb{R}^n \text{ such that } |\mathbf{x} - \mathbf{y}| < r\}. \qquad 1.5.2$$

We use a subscript to indicate the radius of a ball B; the argument gives the center of the ball: a ball of radius 2 centered at the point $\mathbf{y}$ would be written $B_2(\mathbf{y})$.

A subset is open if every point in it is contained in an open ball that itself is contained in the subset:

Definition 1.5.3 (Open set of $\mathbb{R}^n$). A subset $U \subset \mathbb{R}^n$ is open in $\mathbb{R}^n$ if for every point $\mathbf{x} \in U$, there exists $r > 0$ such that $B_r(\mathbf{x}) \subset U$.

---

[12]It is possible to make sense of the notion of derivatives in closed sets, but these results, due to the great American mathematician Hassler Whitney, are extremely difficult, well beyond the scope of this book.

However close a point in the open subset U is to the "fence" of the set, by choosing r small enough, you can surround it with an open ball in $\mathbb{R}^n$ that is entirely in the open set, not touching the fence.

A set that is not open is not necessarily closed: an open set owns *none* of its fence. A closed set owns *all* of its fence:

Note that parentheses denote an open set: (a, b), while brackets denote a closed set: $[a, b]$. Sometimes, especially in France, backwards brackets are used to denote an open set: $]a, b[= (a, b)$.

> **Definition 1.5.4 (Closed set of $\mathbb{R}^n$).** A closed set of $\mathbb{R}^n$, $C \subset \mathbb{R}^n$, is a set whose complement $\mathbb{R}^n - C$ is open.

Example 1.5.5 (Open sets).

(1) If $a < b$, then the interval

$$(a, b) = \{x \in \mathbb{R} \mid a < x < b\} \qquad\qquad 1.5.3$$

is open. Indeed, if $x \in (a, b)$, set $r = \min\{x - a, b - x\}$. Both these numbers are strictly positive, since $a < r < b$, and so is their minimum. Then the ball $\{y \mid y - x < r\}$ is a subset of (a, b).

(2) The infinite intervals (a, ∞), $(-\infty, b)$ are also open, but the intervals

$$(a, b] = \{x \in \mathbb{R} \mid a < x \leq b\} \quad \text{and} \quad [a, b] = \{x \in \mathbb{R} \mid a \leq x \leq b\} \qquad 1.5.4$$

are not.

The use of the word *domain* in Example 1.5.6 is not really mathematically correct: a function is the triple of

(1) a set X: the domain;
(2) a set Y: the range;
(3) a rule f that associates an element $f(x) \in Y$ to each element $x \in X$.

Strictly speaking, the formula $1/(y - x^2)$ isn't a function until we have specified the domain and the range, and nobody says that the domain must be the complement of the parabola of equation $y = x^2$: it could be any subset of this set. Mathematicians usually disregard this, and think of a formula as defining a function, whose domain is the *natural domain* of the formula, i.e., those arguments for which the formula is defined.

(3) The rectangle

$$(a, b) \times (c, d) = \left\{ \begin{pmatrix} x \\ y \end{pmatrix} \in \mathbb{R}^2 \mid a < x < b \, , \; c < y < d \right\} \qquad 1.5.5$$

is also open. $\quad \triangle$

Natural domains of functions

We will often be interested in whether the domain of definition of a function— what we will call its *natural domain*—is open or closed, or neither.

Example 1.5.6 (Checking whether the domain of a function is open or closed). The natural domain of the function $1/(y - x^2)$ is the subset of $\mathbb{R}^2$ where the denominator is not 0, i.e., the natural domain is the complement of the parabola P of equation $y = x^2$. This is more or less obviously an open set, as suggested by Figure 1.5.3.

We can see it rigorously as follows. Suppose $\begin{pmatrix} a \\ b \end{pmatrix} \notin P$, so that $|b - a^2| = C > 0$, for some constant C. Then if

$$|u|, |v| < \min\left\{ 1, \frac{C}{3}, \frac{C}{6|a|} \right\} = r, \qquad\qquad 1.5.6$$

we have

$$|(b+v) - (a+u)^2| = |b - a^2 + v - 2au - u^2| \geq C - (|v| + 2|a||u| + |u|^2)$$

$$> C - \left(\frac{C}{3} + \frac{C}{3} + \frac{C}{3}\right) = 0. \qquad 1.5.7$$

Therefore, $\begin{pmatrix} a+u \\ b+v \end{pmatrix}$ is not on the parabola. This means that we can draw a square of side length $2r$ around the point $\begin{pmatrix} a \\ b \end{pmatrix}$ and know that any point in that open square will not be on the parabola. (We used that since $|u| < 1$, we have $|u|^2 < |u|$.)

If we had defined an open set in terms of squares around points rather than balls around points, we would now be finished: we would have shown that the complement of the parabola P is open. But to be complete we now need to point out the obvious fact that there is an open ball that fits in that open square. We do this by saying that if $\left| \begin{pmatrix} a+u \\ b+v \end{pmatrix} - \begin{pmatrix} a \\ b \end{pmatrix} \right| < r$ (i.e., if $\begin{pmatrix} a+u \\ b+v \end{pmatrix}$ is in the circle of radius r around $\begin{pmatrix} a \\ b \end{pmatrix}$) then $|u|, |v| < r$ (i.e., it is also in the square of side length $2r$ around $\begin{pmatrix} a \\ b \end{pmatrix}$). Therefore the complement of the parabola is open.[13]
$\triangle$

This seems like a lot of work to prove something that was obvious to begin with. However, now we can actually compute the radius of an open disk around any point off the parabola. For the point $\begin{pmatrix} 2 \\ 3 \end{pmatrix}$, what is the radius of such a disk? Check your answer below.[14] The answer you get will not be sharp: there are points between that disk and the parabola. Exercise 1.5.6 asks you to find a sharper result; Exercise 1.5.7 asks you to find the exact result. $\triangle$

Example 1.5.7 (Natural domain). What is the natural domain of the function

$$f\begin{pmatrix} x \\ y \end{pmatrix} = \sqrt{\frac{x}{y}}, \qquad 1.5.8$$

i.e., those arguments for which the formula is defined? If the argument of the square root is non-negative, the square root can be evaluated, so the first and the third quadrants are in the natural domain. The x-axis is not (since $y = 0$ there), but the y-axis with the origin removed is in the natural domain, since

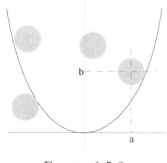

FIGURE 1.5.3.

It seems obvious that given a point off the parabola P, you can draw a disk around the point that avoids the parabola. Actually finding a formula for the radius of such a disk is more tedious than you might expect.

---

[13]How did we make up this proof? We fiddled, starting at the end and seeing what r should be in order for the computation to come out. Note that if $a = 0$, then $C/(6|a|$ is infinite, but this does not affect the choice of r since we are choosing a minimum.

[14]For $x = 2, y = 3$ we have $C = |y - x^2| = |3 - 4| = 1$, so $r = \min\left\{1, \frac{1}{3}, \frac{1}{6|2|}\right\} = 1/12$. The open disk of radius $1/12$ around $\begin{pmatrix} 2 \\ 3 \end{pmatrix}$ does not intersect the parabola.

x/y is zero there. So the natural domain is the region drawn in Figure 1.5.4.
$\triangle$

Several similar examples are suggested in Exercise 1.5.8.

FIGURE 1.5.4.

The natural domain of the function

$$f\begin{pmatrix} x \\ y \end{pmatrix} = \sqrt{\dfrac{x}{y}}.$$

is neither open nor closed.

Infinite decimals are actually limits of convergent sequences. If $a_0 = 3, a_1 = 3.1, a_2 = 3.14, \ldots, a_n = \pi$ to n decimal places, how large does M have to be so that if $n \geq M$, then $|a_n - \pi| < 10^{-3}$? The answer is $M = 3$: $\pi - 3.141 = .0005926\ldots$. The same argument holds for any real number.

Convergence

Unless we state explicitly that a sequence is finite, sequences will be infinite. A sequence of points $\mathbf{a}_1, \mathbf{a}_2 \ldots$ *converges* to $\mathbf{a}$ if, by choosing a point $\mathbf{a}_M$ far enough out in the sequence, you can make the distance between all subsequent points in the sequence and $\mathbf{a}$ as small as you like:

Definition 1.5.8 (Convergent sequence). A sequence of points $\mathbf{a}_1, \mathbf{a}_2 \ldots$ in $\mathbb{R}^n$ converges to $\mathbf{a} \in \mathbb{R}^n$ if for all $\epsilon > 0$ there exists M such that when $m > M$, then $|\mathbf{a}_m - \mathbf{a}| < \epsilon$. We then call $\mathbf{a}$ the limit of the sequence.

Exactly the same definition applies to a sequence of vectors: just replace $\mathbf{a}$ in Definition 1.5.8 by $\vec{\mathbf{a}}$, and substitute the word "vector" for "point."

Convergence in $\mathbb{R}^n$ is just n separate convergences in $\mathbb{R}$:

Proposition 1.5.9 (Convergent sequence in $\mathbb{R}^n$). *A sequence $(\mathbf{a}_m) = \mathbf{a}_1, \mathbf{a}_2, \ldots$ with $\mathbf{a}_i \in \mathbb{R}^n$ converges to $\mathbf{a}$ if and only if each coordinate converges; i.e., if for all j with $1 \leq j \leq n$ the coordinate $(\mathbf{a}_m)_j$ converges to a_j, the jth coordinate of the limit $\mathbf{a}$.*

The proof is a good setting for understanding how the $\epsilon - M$ game is played (where M is the M of Definition 1.5.8). You should imagine that your opponent gives you an epsilon and challenges you to find an M that works, i.e., an M such that when $m > M$, then $|(\mathbf{a}_m)_j - a_j| < \epsilon$. You get extra points for style for finding a small M, but it is not necessary in order to win the game.

Proof. Let us first see the easy direction: the statement that $\mathbf{a}_m = \begin{bmatrix} (a_m)_1 \\ \vdots \\ (a_m)_n \end{bmatrix}$ converges to $\mathbf{a}$ implies that for each $j = 1, \ldots, n$, the sequence of numbers $(a_m)_j$ converges. The challenger hands you an epsilon. Fortunately you have a teammate who knows how to play the game for the sequence $\mathbf{a}_m$, and you hand her the epsilon you just got. She promptly hands you back an M with the guarantee that when $m > M$, then $|\mathbf{a}_m - \mathbf{a}| < \epsilon$ (since the sequence $\mathbf{a}_m$ is convergent). The length of the vector $\mathbf{a}_m - \mathbf{a}$ is

$$|\mathbf{a}_m - \mathbf{a}| = \sqrt{\big((a_m)_1 - a_1\big)^2 + \cdots + \big((a_m)_n - a_n\big)^2},$$

This is typical of all proofs involving convergence and limits: you are given an ϵ and challenged to come up with a δ (or M or whatever) such that a certain quantity is less than ϵ.

Your "challenger" can give you any $\epsilon > 0$ he likes; statements concerning limits and continuity are of the form "for *all* epsilon, there exists ... ."

so you give that M to your challenger, with the argument that

$$|(a_m)_j - a_j| \leq |\mathbf{a}_m - \mathbf{a}| < \epsilon. \tag{1.5.9}$$

He promptly concedes defeat.

Now let us try the opposite direction: the convergence of the coordinate sequences $(a_m)_j$ implies the convergence of the sequence $\mathbf{a}_m$. Again the challenger hands you an $\epsilon > 0$. This time you have n teammates, each of whom knows how the play the game for a single convergent coordinate sequence $(a_m)_j$. After a bit of thought and scribbling on a piece of paper, you pass along $\epsilon/\sqrt{n}$ to each of them. They dutifully return to you cards containing numbers $M_1, \ldots, M_n$, with the guarantee that

$$|(a_m)_j - a_j| < \frac{\epsilon}{\sqrt{n}} \quad \text{when } m > M_j. \tag{1.5.10}$$

You sort through the cards and choose the one with the largest number,

$$M = \max\{M_1, \ldots, M_n\}, \tag{1.5.11}$$

which you pass on to the challenger with the following message:

if $m > M$, then $m > M_j$ for each $j = 1 \cdots = n$, so $|(a_m)_j - a_j| < \epsilon/\sqrt{n}$, so

$$|\mathbf{a}_m - \mathbf{a}| = \sqrt{\left((a_m)_1 - a_1\right)^2 + \cdots + \left((a_m)_n - a_n\right)^2}$$
$$< \sqrt{\left(\frac{\epsilon}{\sqrt{n}}\right)^2 + \cdots + \left(\frac{\epsilon}{\sqrt{n}}\right)^2} = \sqrt{\frac{n\epsilon^2}{n}} = \epsilon. \quad \square \tag{1.5.12}$$

The scribbling you did was to figure out that handing $\epsilon/\sqrt{n}$ to your teammates would work. What if you can't figure out how to "slice up" ϵ so that the final answer will be precisely ϵ? In that case, just work directly with ϵ and see where it takes you. If you use ϵ instead of $\epsilon/\sqrt{n}$ in Equations 1.5.10 and 1.5.12, you will end up with

$$|\mathbf{a}_m - \mathbf{a}| < \epsilon\sqrt{n}. \tag{1.5.13}$$

You can then see that to land on the exact answer, you should have chosen $\epsilon/\sqrt{n}$.

In fact, the answer in Equation 1.5.13 is good enough and you don't really need to go back and fiddle. Intuitively, "less than epsilon" for any $\epsilon > 0$ and "less than some quantity that goes to 0 when epsilon goes to 0" achieve the same goal: showing that you can make some quantity arbitrarily small. The following theorem states this precisely; you are asked to prove it in Exercise 1.5.12.

Theorem 1.5.10 (Elegance is not required). *Let $u(\epsilon)$, with $\epsilon > 0$, be a function such that $u(\epsilon) \to 0$ as $\epsilon \to 0$. Then the following two statements are equivalent:*

(1) *For all $\epsilon > 0$, there exists a $\delta > 0$ such that when $|\mathbf{x} - \mathbf{x}_0| < \delta$, then $|f(\mathbf{x}) - f(\mathbf{x}_0)| < u(\epsilon)$.*

(2) *For all $\epsilon > 0$, there exists a $\delta > 0$ such that when $|\mathbf{x} - \mathbf{x}_0| < \delta$, then $|f(\mathbf{x}) - f(\mathbf{x}_0)| < \epsilon$.*

In practice, the first statement is the one mathematicians use most often.

The following result is of great importance, saying that the notion of limit is well defined: if the limit is something, then it isn't something else. It could be reduced to the one-dimensional case as above, but we will use it as an opportunity to play the ϵ, M game in more sober fashion.

Proposition 1.5.11. *If the sequence of points $\mathbf{a}_1, \mathbf{a}_2 \ldots$ in $\mathbb{R}^n$ converges to $\mathbf{a}$ and to $\mathbf{b}$, then $\mathbf{a} = \mathbf{b}$.*

Proof. Suppose $\mathbf{a} \neq \mathbf{b}$, and set $\epsilon_0 = (|\mathbf{a} - \mathbf{b}|)/4$; our assumption $\mathbf{a} \neq \mathbf{b}$ implies that $\epsilon_0 > 0$. Thus, by the definition of the limit, there exists M_1 such that $|\mathbf{a}_n - \mathbf{a}| < \epsilon_0$ when $n > M_1$, and M_2 such that $|\mathbf{a}_n - \mathbf{b}| < \epsilon_0$ when $n > M_2$. Set $M = \max\{M_1, M_2\}$. If $n > M$, then by the triangle inequality (Theorem 1.4.9),

$$|\mathbf{a} - \mathbf{b}| = |(\mathbf{a} - \mathbf{a}_n) + (\mathbf{a}_n - \mathbf{b})| \leq \underbrace{|\mathbf{a} - \mathbf{a}_n|}_{<\epsilon_0} + \underbrace{|\mathbf{a}_n - \mathbf{b}|}_{<\epsilon_0} < 2\epsilon_0 = |\mathbf{a} - \mathbf{b}|/2. \quad \text{1.5.14}$$

This is a contradiction, so $\mathbf{a} = \mathbf{b}$. $\square$

Theorem 1.5.13 states rules concerning limits. First, we need to define a *bounded set*.

Definition 1.5.12 (Bounded set). A subset $X \subset \mathbb{R}^n$ is bounded if it is contained in a ball in $\mathbb{R}^n$ centered at the origin:

$$X \subset B_R(0) \quad \text{for some } R < \infty. \quad \text{1.5.15}$$

Recall that B_R denotes a ball of radius R; the ball $B_R(0)$ is a ball of radius R centered at the origin.

The ball containing the bounded set can be very big, but its radius must be finite.

Theorem 1.5.13 (Limits). *Let $\mathbf{a}_m, \mathbf{b}_m$ be two sequences of points in $\mathbb{R}^n$, and c_m be a sequence of numbers. Then*

(a) If $\mathbf{a}_m$ and $\mathbf{b}_m$ both converge, then so does $\mathbf{a}_m + \mathbf{b}_m$, and

$$\lim_{m \to \infty} (\mathbf{a}_m + \mathbf{b}_m) = \lim_{m \to \infty} \mathbf{a}_m + \lim_{m \to \infty} \mathbf{b}_m.$$

Illustration for part (d): Let

$$c_m = 1/m \quad \text{and} \quad \mathbf{a}_m = \begin{pmatrix} 2 \\ m \end{pmatrix}.$$

Then c_m converges to 0, but

$$\lim_{m \to \infty} (c_m \mathbf{a}_m) \neq 0.$$

Why is the limit not 0 as in part (d)? Because $\mathbf{a}_m$ is not bounded.

(b) If $\mathbf{a}_m$ and c_m both converge, then so does $c_m \mathbf{a}_m$, and

$$\lim_{m \to \infty} (c_m \mathbf{a}_m) = \left(\lim_{m \to \infty} c_m \right) \left(\lim_{m \to \infty} \mathbf{a}_m \right).$$

(c) If $\mathbf{a}_m$ and $\mathbf{b}_m$ both converge, then so does $\mathbf{a}_m \cdot \mathbf{b}_m$, and

$$\lim_{m \to \infty} (\mathbf{a}_m \cdot \mathbf{b}_m) = \left(\lim_{m \to \infty} \mathbf{a}_m \right) \cdot \left(\lim_{m \to \infty} \mathbf{b}_m \right).$$

(d) If $\mathbf{a}_m$ is bounded and c_m converges to 0, then

$$\lim_{m \to \infty} (c_m \mathbf{a}_m) = \mathbf{0}.$$

We will not prove Theorem 1.5.13, since Proposition 1.5.9 reduces it to the one-dimensional case; the proof is left as Exercise 1.5.16.

There is an intimate relationship between limits of sequences and closed sets: closed sets are "closed under limits."

Exercise 1.5.13 asks you to prove the converse: if every convergent sequence in a set $C \subset \mathbb{R}^n$ converges to a point in C, then C is closed.

Proposition 1.5.14. *If $\mathbf{x}_1, \mathbf{x}_2, \ldots$ is a convergent sequence in a closed set $C \subset \mathbb{R}^n$, converging to a point $\mathbf{x}_0 \in \mathbb{R}^n$, then $\mathbf{x}_0 \in C$.*

Intuitively, this is not hard to see: a convergent sequence in a closed set can't approach a point outside the set without leaving the set. (But a sequence in a set that is not closed can converge to a point of the fence that is not in the set.)

Proof. Indeed, if $\mathbf{x}_0 \notin C$, then $\mathbf{x}_0 \in (\mathbb{R}^n - C)$, which is open, so there exists $r > 0$ such that $B_r(\mathbf{x}_0) \subset (\mathbb{R}^n - C)$. Then for all m we have $|\mathbf{x}_m - \mathbf{x}_0| \geq r$. On the other hand, by the definition of convergence, we must have that for any $\epsilon > 0$ we have $|\mathbf{x}_m - \mathbf{x}_0| < \epsilon$ for m sufficiently large. Taking $\epsilon = r/2$, we see that this is a contradiction. $\square$

Subsequences

Subsequences are a useful tool, as we will see in Section 1.6. They are not particularly difficult, but they require somewhat complicated indices, which are scary on the page and tedious to type.

Definition 1.5.15 (Subsequence). A subsequence of a sequence $a_1, a_2, \ldots$ is a sequence formed by taking first some element of the original sequence, then another element further on, and yet another, yet further along $\ldots$. It is denoted $a_{i(1)}, a_{i(2)}, \ldots$, where $i(k) > i(j)$ when $k > j$.

You might take all the even terms, or all the odd terms, or all those whose index is a prime, etc. Of course, any sequence is a subsequence of itself. The index i is the function that associates to the position in the subsequence the position of the same entry in the original sequence. For example, if the original sequence is

Sometimes the subsequence
$$a_{i(1)}, a_{i(2)}, \ldots$$
is denoted $a_{i_1}, a_{i_2}, \ldots$.

$$\underbrace{\frac{1}{1}}_{a_1}, \underbrace{\frac{1}{2}}_{a_2}, \underbrace{\frac{1}{3}}_{a_3}, \underbrace{\frac{1}{4}}_{a_4}, \underbrace{\frac{1}{5}}_{a_5}, \underbrace{\frac{1}{6}}_{a_6} \ldots \quad \text{and the subsequence is} \quad \underbrace{\frac{1}{2}}_{a_{i(1)}}, \underbrace{\frac{1}{4}}_{a_{i(2)}}, \underbrace{\frac{1}{6}}_{a_{i(3)}} \ldots$$

we see that $i(1) = 2$, since $1/2$ is the second entry of the original sequence. Similarly, $i(2) = 4, i(3) = 6, \ldots$. (In specific cases, figuring out what $i(1), i(2)$, etc., correspond to can be a major challenge.)

The proof of Proposition 1.5.16 is left as Exercise 1.5.17, largely to provide practice with the language.

Proposition 1.5.16. *If a sequence* $\mathbf{a}_k$ *converges to* $\mathbf{a}$*, then any subsequence of* $\mathbf{a}_k$ *converges to the same limit.*

Limits of functions

Limits like $\lim_{\mathbf{x} \to \mathbf{x}_0} f(\mathbf{x})$ can only be defined if you can approach $\mathbf{x}_0$ by points where f can be evaluated. The notion of *closure* of a set is designed to make this precise.

The closure of A is thus A plus its fence. If A is closed, then $\overline{A} = A$.

Definition 1.5.17 (Closure). If $A \subset \mathbb{R}^n$ is a subset, the closure of A, denoted $\overline{A}$, is the set of all limits of sequences in A that converge in $\mathbb{R}^n$.

The *boundary* of a subset $A \subset \mathbb{R}^n$ is those points for which every neighborhood intersects both A and the complement of A. It is also the intersection of the closure of A and the closure of the complement of A.

For example, if $A = (0, 1)$ then $\overline{A} = [0, 1]$; the point 0 is the limit of the sequence $1/n$, which is a sequence in A and converges to a point in $\mathbb{R}$.

When $\mathbf{x}_0$ is in the closure of the domain of f, we can define the limit of a function, $\lim_{\mathbf{x} \to \mathbf{x}_0} f(\mathbf{x})$. Of course, this includes the case when $\mathbf{x}_0$ is in the domain of f, but the really interesting case is when it is in the boundary of the domain.

Example 1.5.18. (a) If $A = (0, 1)$ then $\overline{A} = [0, 1]$, so that 0 and 1 are in $\overline{A}$. Thus, it makes sense to talk about

$$\lim_{x \to 0} (1 + x)^{1/x} \qquad\qquad 1.5.16$$

because although you cannot evaluate the function at 0, the natural domain of the function contains 0 in its closure.

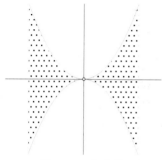

FIGURE 1.5.5.

The region in example 1.5.18, (c). You can approach the origin from this region, but only in rather special ways.

Definition 1.5.19 is not standard in the United States but is quite common in France. The standard version substitutes $0 < |\mathbf{x} - \mathbf{x}_0| < \delta$ for our $|\mathbf{x} - \mathbf{x}_0| < \delta$. The definition we have adopted makes little difference in applications, but has the advantage that allowing for the case where $\mathbf{x} = \mathbf{x}_0$ makes limits better behaved under composition. With the standard version, Theorem 1.5.22 is not true.

A mapping $\mathbf{f} : \mathbb{R}^n \to \mathbb{R}^m$ is an "$\mathbb{R}^m$-valued" mapping; its argument is in $\mathbb{R}^n$ and its values are in $\mathbb{R}^m$. Often such mappings are called "vector-valued" mappings (or functions), but usually we are thinking of its values as points rather than vectors. Note that we denote an $\mathbb{R}^m$-valued mapping whose values are points in $\mathbb{R}^m$ with a boldface letter without arrow: $\mathbf{f}$. Sometimes we do want to think of the values of a mapping $\mathbb{R}^n \to \mathbb{R}^n$ as vectors: when we are thinking of vector fields. We denote a vector field with an arrow: $\vec{F}$ or $\vec{\mathbf{f}}$.

(b) The point $\begin{pmatrix} 0 \\ 0 \end{pmatrix}$ is in the closure of

$$U = \left\{ \begin{pmatrix} x \\ y \end{pmatrix} \in \mathbb{R}^2 \mid 0 < x^2 + y^2 < 1 \right\} \text{ (the unit disk with the origin removed)}$$

(c) The point $\begin{pmatrix} 0 \\ 0 \end{pmatrix}$ is also in the closure of U (the region between two parabolas touching at the origin, shown in Figure 1.5.5):

$$U = \left\{ \begin{pmatrix} x \\ y \end{pmatrix} \in \mathbb{R}^2 \mid |y| < x^2 \right\} \quad \square$$

Definition 1.5.19 (Limit of a function). Let U be a subset of $\mathbb{R}^n$. A function $f : U \to \mathbb{R}$ has the limit a at $\mathbf{x}_0$:

$$\lim_{\mathbf{x} \to \mathbf{x}_0} f(\mathbf{x}) = a \qquad\qquad 1.5.17$$

if $\mathbf{x}_0 \in \overline{U}$ and if for all $\epsilon > 0$, there exists $\delta > 0$ such that when $|\mathbf{x} - \mathbf{x}_0| < \delta$, and $\mathbf{x} \in U$, then $|f(\mathbf{x}) - a| < \epsilon$.

In other words, $f(\mathbf{x})$ can be made arbitrarily close to a by choosing $\mathbf{x} \in U$ sufficiently close to $\mathbf{x}_0$.

Since we are not requiring that $\mathbf{x}_0 \in U$, $f(\mathbf{x}_0)$ is not necessarily defined, but if it is defined, then for the limit to exist we must have

$$\lim_{\mathbf{x} \to \mathbf{x}_0} f(\mathbf{x}) = f(\mathbf{x}_0). \qquad\qquad 1.5.18$$

Limits of mappings with values in $\mathbb{R}^m$

As is the case for sequences (Proposition 1.5.9), it is the same thing to claim that an $\mathbb{R}^m$-valued mapping $\mathbf{f} : \mathbb{R}^n \to \mathbb{R}^m$ has a limit, and that its components have limits, as shown in Proposition 1.5.20. Such a mapping is sometimes written in terms of the "sub-functions" (coordinate functions) that define each new coordinate. For example, the mapping $\mathbf{f} : \mathbb{R}^2 \to \mathbb{R}^3$,

$$\mathbf{f} \begin{pmatrix} x \\ y \end{pmatrix} = \begin{pmatrix} xy \\ x^2 y \\ x - y \end{pmatrix} \quad \text{can be written} \quad \mathbf{f} = \begin{pmatrix} f_1 \\ f_2 \\ f_3 \end{pmatrix}, \qquad 1.5.19$$

where $f_1(\mathbf{x}) = xy$, $f_2(\mathbf{x}) = x^2 y$, and $f_3(\mathbf{x}) = x - y$.

Proposition 1.5.20. *Let* $\mathbf{f}(\mathbf{x}) = \begin{pmatrix} f_1(\mathbf{x}) \\ \vdots \\ f_m(\mathbf{x}) \end{pmatrix}$ *be a function defined on a domain* $U \subset \mathbb{R}^n$, *and let* $\mathbf{x}_0 \in \mathbb{R}^n$ *be a point in* $\overline{U}$. *Then* $\lim_{\mathbf{x} \to \mathbf{x}_0} \mathbf{f} = \mathbf{a}$ *exists if and only if each of* $\lim_{\mathbf{x} \to \mathbf{x}_0} f_i = a_i$ *exists, and*

$$\lim_{\mathbf{x} \to \mathbf{x}_0} \mathbf{f} = \begin{pmatrix} \lim_{\mathbf{x} \to \mathbf{x}_0} f_1 \\ \vdots \\ \lim_{\mathbf{x} \to \mathbf{x}_0} f_m \end{pmatrix}; \quad i.e., \quad \mathbf{a} = \begin{pmatrix} a_1 \\ \vdots \\ a_m \end{pmatrix}. \qquad 1.5.20$$

Recall (Definition 1.5.17) that $\overline{U}$ denotes the closure of U: the subset of $\mathbb{R}^n$ made up of the set of all limits of sequences in U which converge in $\mathbb{R}^n$.

Proof. Let's go through the picturesque description again. The proof has an "if" part and an "only if" part.

For the "if" part, the challenger hands you an $\epsilon > 0$. You pass it on to a teammate who returns a δ with the guarantee that when $|\mathbf{x} - \mathbf{x}_0| < \delta$, and $\mathbf{f}(\mathbf{x})$ is defined, then $|\mathbf{f}(\mathbf{x}) - \mathbf{a}| < \epsilon$. You pass on the same δ, and a_i, to the challenger, with the explanation:

If you gave $\epsilon/\sqrt{m}$ to your teammates, as in the proof of Proposition 1.5.9, you would end up with

$$|\mathbf{f}(\mathbf{x}) - \mathbf{a}| < \epsilon,$$

rather than $|\mathbf{f}(\mathbf{x}) - \mathbf{a}| < \epsilon\sqrt{m}$. In some sense this is more "elegant." But Theorem 1.5.10 says that it is mathematically *just as good* to arrive at less than or equal to epsilon times some fixed number or, more generally, anything that goes to 0 when ϵ goes to 0.

$$|f_i(\mathbf{x}) - a_i| \le |\mathbf{f}(\mathbf{x}) - \mathbf{a}| < \epsilon. \qquad 1.5.21$$

For the "only if" part, the challenger hands you an $\epsilon > 0$. You pass this ϵ to your teammates, who know how to deal with the coordinate functions. They hand you back $\delta_1, \ldots, \delta_m$. You look through these, and select the smallest one, which you call δ, and pass on to the challenger, with the message

"If $|\mathbf{x} - \mathbf{x}_0| < \delta$, then $|\mathbf{x} - \mathbf{x}_0| < \delta_i$, so that $|f_i(\mathbf{x}) - a_i| < \epsilon$, so that

$$|f(\mathbf{x}) - \mathbf{a}| = \sqrt{\big(f_1(\mathbf{x}) - a_1\big)^2 + \cdots + \big(f_m(\mathbf{x}) - a_m\big)^2} < \underbrace{\sqrt{\epsilon^2 + \cdots + \epsilon^2}}_{m \text{ terms}} = \epsilon\sqrt{m},$$

$$1.5.22$$

which goes to 0 as ϵ goes to 0. You win!

Theorem 1.5.21 (Limits of functions). *Let* $\mathbf{f}$ *and* $\mathbf{g}$ *be functions from* $U \to \mathbb{R}^m$, *and* h *a function from* $U \to \mathbb{R}$.

(a) If $\lim_{\mathbf{x} \to \mathbf{x}_0} \mathbf{f}(\mathbf{x})$ *and* $\lim_{\mathbf{x} \to \mathbf{x}_0} \mathbf{g}(\mathbf{x})$ *exist, then* $\lim_{\mathbf{x} \to \mathbf{x}_0}(\mathbf{f} + \mathbf{g})(\mathbf{x})$ *exists, and*

$$\lim_{\mathbf{x} \to \mathbf{x}_0} \mathbf{f}(\mathbf{x}) + \lim_{\mathbf{x} \to \mathbf{x}_0} \mathbf{g}(\mathbf{x}) = \lim_{\mathbf{x} \to \mathbf{x}_0}(\mathbf{f} + \mathbf{g})(\mathbf{x}). \qquad 1.5.23$$

(b) If $\lim_{\mathbf{x} \to \mathbf{x}_0} \mathbf{f}(\mathbf{x})$ *and* $\lim_{\mathbf{x} \to \mathbf{x}_0} h(\mathbf{x})$ *exist, then* $\lim_{\mathbf{x} \to x_0} h\mathbf{f}(\mathbf{x})$ *exists, and*

$$\lim_{\mathbf{x} \to \mathbf{x}_0} h(\mathbf{x}) \lim_{\mathbf{x} \to \mathbf{x}_0} \mathbf{f}(\mathbf{x}) = \lim_{\mathbf{x} \to \mathbf{x}_0} h\mathbf{f}(\mathbf{x}). \qquad 1.5.24$$

(c) If $\lim_{\mathbf{x}\to\mathbf{x}_0} \mathbf{f}(\mathbf{x})$ exists, and $\lim_{\mathbf{x}\to\mathbf{x}_0} h(\mathbf{x})$ exists and is different from 0, then $\lim_{\mathbf{x}\to x_0}\left(\frac{f}{h}\right)(\mathbf{x})$ exists, and

$$\frac{\lim_{\mathbf{x}\to\mathbf{x}_0}\mathbf{f}(\mathbf{x})}{\lim_{\mathbf{x}\to\mathbf{x}_0}h(\mathbf{x})} = \lim_{\mathbf{x}\to\mathbf{x}_0}\left(\frac{f}{h}\right)(\mathbf{x}) \qquad 1.5.25$$

(d) If $\lim_{\mathbf{x}\to\mathbf{x}_0}\mathbf{f}(\mathbf{x})$ and $\lim_{\mathbf{x}\to\mathbf{x}_0}\mathbf{g}(\mathbf{x})$ exist, then so does $\lim_{\mathbf{x}\to\mathbf{x}_0}(\mathbf{f}\cdot\mathbf{g})$, and

$$\lim_{\mathbf{x}\to\mathbf{x}_0}\mathbf{f}(\mathbf{x})\cdot\lim_{\mathbf{x}\to\mathbf{x}_0}\mathbf{g}(\mathbf{x}) = \lim_{\mathbf{x}\to\mathbf{x}_0}(\mathbf{f}\cdot\mathbf{g})(\mathbf{x}). \qquad 1.5.26$$

(e) If $\mathbf{f}$ is bounded and $\lim_{\mathbf{x}\to\mathbf{x}_0}h(\mathbf{x}) = 0$, then

$$\lim_{\mathbf{x}\to\mathbf{x}_0}(h\mathbf{f})(\mathbf{x}) = \mathbf{0}. \qquad 1.5.27$$

(f) If $\lim_{\mathbf{x}\to\mathbf{x}_0}\mathbf{f}(\mathbf{x}) = 0$ and $h(\mathbf{x})$ is bounded, then

$$\lim_{\mathbf{x}\to\mathbf{x}_0}(h\mathbf{f})(\mathbf{x}) = \mathbf{0}. \qquad 1.5.28$$

We could substitute

$$\frac{\epsilon}{2(|\mathbf{g}(\mathbf{x}_0)| + \epsilon)} \quad \text{for } \epsilon$$

in Equation 1.5.29, and

$$\frac{\epsilon}{2(|\mathbf{f}(\mathbf{x}_0)| + \epsilon)} \quad \text{for } \epsilon$$

in Equation 1.5.30. This would give

$$|\mathbf{f}(\mathbf{x})\cdot\mathbf{g}(\mathbf{x}) - \mathbf{f}(\mathbf{x}_0)\cdot\mathbf{g}(\mathbf{x}_0)|$$
$$\leq \frac{\epsilon|\mathbf{g}(\mathbf{x})|}{2(|\mathbf{g}(\mathbf{x}_0)| + \epsilon)} + \frac{\epsilon|\mathbf{f}(\mathbf{x}_0)|}{2(|\mathbf{f}(\mathbf{x}_0)| + \epsilon)}$$
$$\leq \epsilon.$$

Again, if you want to land exactly on epsilon, fine, but mathematically it is completely unnecessary.

Proof. The proofs of all these statements are very similar; we will do only (d), which is the hardest. Choose ϵ (think of the challenger giving it to you). Then

(1) Find a δ_1 such that when $|\mathbf{x} - \mathbf{x}_0| < \delta_1$, then

$$|\mathbf{g}(\mathbf{x}) - \mathbf{g}(\mathbf{x}_0)| < \epsilon. \qquad 1.5.29$$

(2) Next find a δ_2 such that when $|\mathbf{x} - \mathbf{x}_0| < \delta_2$, then

$$|\mathbf{f}(\mathbf{x}) - \mathbf{f}(\mathbf{x}_0)| < \epsilon. \qquad 1.5.30$$

Now set δ to be the smallest of δ_1 and δ_2, and consider the sequence of inequalities

$$|\mathbf{f}(\mathbf{x})\cdot\mathbf{g}(\mathbf{x}) - \mathbf{f}(\mathbf{x}_0)\cdot\mathbf{g}(\mathbf{x}_0)|$$
$$= |\mathbf{f}(\mathbf{x})\cdot\mathbf{g}(\mathbf{x}) \underbrace{-\mathbf{f}(\mathbf{x}_0)\cdot\mathbf{g}(\mathbf{x}) + \mathbf{f}(\mathbf{x}_0)\cdot\mathbf{g}(\mathbf{x})}_{=0} -\mathbf{f}(\mathbf{x}_0)\cdot\mathbf{g}(\mathbf{x}_0)|$$
$$\leq |\mathbf{f}(\mathbf{x})\cdot\mathbf{g}(\mathbf{x}) - \mathbf{f}(\mathbf{x}_0)\cdot\mathbf{g}(\mathbf{x})| + |\mathbf{f}(\mathbf{x}_0)\cdot\mathbf{g}(\mathbf{x}) - \mathbf{f}(\mathbf{x}_0)\cdot\mathbf{g}(\mathbf{x}_0)| \qquad 1.5.31$$
$$= |(\mathbf{f}(\mathbf{x}) - \mathbf{f}(\mathbf{x}_0))\cdot\mathbf{g}(\mathbf{x})| + |\mathbf{f}(\mathbf{x}_0)\cdot(\mathbf{g}(\mathbf{x}) - \mathbf{g}(\mathbf{x}_0))|$$
$$\leq |(\mathbf{f}(\mathbf{x}) - \mathbf{f}(\mathbf{x}_0))||\mathbf{g}(\mathbf{x})| + |\mathbf{f}(\mathbf{x}_0)||(\mathbf{g}(\mathbf{x}) - \mathbf{g}(\mathbf{x}_0))|$$
$$\leq \epsilon|\mathbf{g}(\mathbf{x})| + \epsilon|\mathbf{f}(\mathbf{x}_0)| = \epsilon(|\mathbf{g}(\mathbf{x})| + |\mathbf{f}(\mathbf{x}_0)|).$$

Since $\mathbf{g}(\mathbf{x})$ is a function, not a point, we might worry that it could get big faster than ϵ gets small. But we know that when $|\mathbf{x} - \mathbf{x}_0| < \delta$, then $|\mathbf{g}(\mathbf{x}) - \mathbf{g}(\mathbf{x}_0)| < \epsilon$, which gives

$$|\mathbf{g}(\mathbf{x})| < \epsilon + |\mathbf{g}(\mathbf{x}_0)|. \qquad 1.5.32$$

So continuing Equation 1.5.31, we get

$$\epsilon\big(|\mathbf{g}(\mathbf{x})| + |\mathbf{f}(\mathbf{x}_0)|\big) < \epsilon\big(\epsilon + |\mathbf{g}(\mathbf{x}_0)|\big) + \epsilon|\mathbf{f}(\mathbf{x}_0)|, \qquad 1.5.33$$

which goes to 0 as ϵ goes to 0. $\square$

Limits also behave well with respect to compositions.

Theorem 1.5.22 (Limit of a composition). *If $U \subset \mathbb{R}^n$, $V \subset \mathbb{R}^m$ are subsets, and $\mathbf{f} : U \to V$ and $\mathbf{g} : V \to \mathbb{R}^k$ are mappings, so that $\mathbf{g} \circ \mathbf{f}$ is defined, and if $\mathbf{y}_0 \overset{\text{def}}{=} \lim_{\mathbf{x} \to \mathbf{x}_0} \mathbf{f}(\mathbf{x})$, and $\lim_{\mathbf{y} \to \mathbf{y}_0} \mathbf{g}(\mathbf{y})$ both exist, then $\lim_{\mathbf{x} \to \mathbf{x}_0} \mathbf{g} \circ \mathbf{f}(\mathbf{x})$, exists, and*

$$\lim_{\mathbf{x} \to \mathbf{x}_0} (\mathbf{g} \circ \mathbf{f})(\mathbf{x}) = \lim_{\mathbf{y} \to \mathbf{y}_0} \mathbf{g}(\mathbf{y}). \qquad 1.5.34$$

There is no natural condition that will guarantee that

$$\mathbf{f}(\mathbf{x}) \neq \mathbf{f}(\mathbf{x}_0);$$

if we had required $\mathbf{x} \neq \mathbf{x}_0$ in our definition of limit, this argument would not work.

Proof. For all $\epsilon > 0$ there exists δ_1 such that if $|\mathbf{y} - \mathbf{y}_0| < \delta_1$, then $|\mathbf{g}(\mathbf{y}) - \mathbf{g}(\mathbf{y}_0)| < \epsilon$. Next, there exists δ such that if $|\mathbf{x} - \mathbf{x}_0| < \delta$, then $|\mathbf{f}(\mathbf{x}) - \mathbf{f}(\mathbf{x}_0)| < \delta_1$. Hence

$$|\mathbf{g}\big(\mathbf{f}(\mathbf{x})\big) - \mathbf{g}\big(\mathbf{f}(\mathbf{x}_0)\big)| < \epsilon \quad \text{when} \quad |\mathbf{x} - \mathbf{x}_0| < \delta. \quad \square \qquad 1.5.35$$

Theorems 1.5.21 and 1.5.22 show that if you have a function $f : \mathbb{R}^n \to \mathbb{R}$ given by a formula involving addition, multiplication, division, and composition of continuous functions, and which is defined at a point $\mathbf{x}_0$, then $\lim_{\mathbf{x} \to \mathbf{x}_0} f(\mathbf{x})$ exists, and is equal to $f(\mathbf{x}_0)$.

Example 1.5.23 (Limit of a function). We have

$$\lim_{\binom{x}{y} \to \binom{3}{-1}} x^2 \sin(xy) = 3^2 \sin(-3) \sim -1.27\dots. \qquad 1.5.36$$

In fact, the function $x^2 \sin(xy)$ has limits at all points of the plane, and the limit is always precisely the value of the function at the point. Indeed, xy is the product of two continuous functions, as is x^2, and sine is continuous at every point, so $\sin(xy)$ is continuous everywhere; hence also $x^2 \sin(xy)$. $\triangle$

In Example 1.5.23 we just have multiplication and sines, which are pretty straightforward. But whenever there is a division we need to worry: are we dividing by 0? We also need to worry whenever we see tan: what happens if the argument of tan is $\pi/2 + k\pi$? Similarly, $\ln, \cot, \sec, \csc$ all introduce complications.

In one dimension, these problems are often addressed using l'Hôpital's rule (although Taylor expansions often work better).

Much of the subtlety of limits in higher dimensions is that there are lots of different ways of approaching a point, and different approaches may yield different limits, in which case the limit may not exist. The following example illustrates some of the difficulties.

Example 1.5.24 (A case where different approaches give different limits). Consider the function

$$f\begin{pmatrix} x \\ y \end{pmatrix} = \begin{cases} \dfrac{|y|e^{-\frac{|y|}{x^2}}}{x^2} & \text{if } x \neq 0 \\ 0 & \text{if } x = 0, \end{cases} \qquad 1.5.37$$

shown in Figure 1.5.6. Does $\lim_{\begin{pmatrix} x \\ y \end{pmatrix} \to \begin{pmatrix} 0 \\ 0 \end{pmatrix}} f\begin{pmatrix} x \\ y \end{pmatrix}$ exist?

A first idea is to approach the origin along straight lines. Set $y = mx$. When $m = 0$, the limit is obviously 0, and when $m \neq 0$, the limit becomes

$$\lim_{x \to 0} \left| \frac{m}{x} \right| e^{-\left|\frac{m}{x}\right|}; \qquad 1.5.38$$

this limit exists and is always 0, for all values of m. Indeed,

$$\lim_{t \to 0} \frac{1}{t} e^{-1/t} = \lim_{s \to \infty} \frac{s}{e^s} = 0. \qquad 1.5.39$$

So however you approach the origin along straight lines, the limit always exists, and is always 0. But if you set $y = kx^2$ and let $x \to 0$, approaching 0 along a parabola, you find something quite different:

$$\lim_{x \to 0} |k|e^{-|k|} = |k|e^{-|k|}, \qquad 1.5.40$$

which is some number that varies between 0 and $1/e$ (see Exercise 1.5.18). Thus if you approach the origin in different ways, the limits may be different. $\triangle$

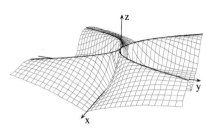

FIGURE 1.5.6.

The function of Example 1.5.24 is continuous except at the origin. Its value is $1/e$ along the "crest line" $y = \pm x^2$, but vanishes on both axes, forming a very deep canyon along the x-axis. If you approach the origin along any straight line $y = mx$ with $m \neq 0$, the path will get to the broad valley along the y-axis before it reaches the origin, so along any such path the limit of f exists and is 0.

Continuous functions

Continuity is *the* fundamental notion of topology, and it arises throughout calculus also. It took mathematicians 200 years to arrive at a correct definition. (Historically, we have our presentation out of order: it was the search for a usable definition of continuity that led to the correct definition of limits.)

Definition 1.5.25 (Continuous function). Let $X \subset \mathbb{R}^n$. Then a mapping $\mathbf{f} : X \to \mathbb{R}^m$ is continuous at $\mathbf{x}_0 \in X$ if

$$\lim_{\mathbf{x} \to \mathbf{x}_0} \mathbf{f}(\mathbf{x}) = \mathbf{f}(\mathbf{x}_0); \qquad 1.5.41$$

$\mathbf{f}$ is continuous on X if it is continuous at every point of X.

The following criterion shows that it is enough to consider $\mathbf{f}$ on sequences converging to $\mathbf{x}_0$.

A map $\mathbf{f}$ is continuous at $\mathbf{x}_0$ if you can make the difference between $\mathbf{f}(\mathbf{x})$ and $\mathbf{f}(\mathbf{x}_0)$ arbitrarily small by choosing $\mathbf{x}$ sufficiently close to $\mathbf{x}_0$. Note that $|\mathbf{f}(\mathbf{x}) - \mathbf{f}(\mathbf{x}_0)|$ must be small for *all* $\mathbf{x}$ "sufficiently close" to $\mathbf{x}_0$. It is not enough to find a δ such that for one particular value of $\mathbf{x}$ the statement is true. However, the "sufficiently close" (i.e., the choice of δ) can be *different* for different values of $\mathbf{x}$. (If a single δ works for all $\mathbf{x}$, then the mapping is *uniformly continuous*, as discussed in Section 0.2.)

We started by trying to write this in one simple sentence, and found it was impossible to do so and avoid mistakes. If definitions of continuity sound stilted, it is because any attempt to stray from the "for all this, there exists that..." inevitably leads to ambiguity.

You are asked to prove Theorem 1.5.27 in Exercise 1.5.19.

Note that with the definition of limit we have given, it would be the same to say that a function $\mathbf{f} : U \to \mathbb{R}^m$ is continuous at $\mathbf{x}_0 \in U$ if and only if $\lim_{\mathbf{x} \to \mathbf{x}_0} \mathbf{f}(\mathbf{x})$ exists.

Proposition 1.5.26 (Criterion for continuity). *The map* $\mathbf{f} : X \to \mathbb{R}^m$ *is continuous at* $\mathbf{x}_0$ *if and only if for every sequence* $\mathbf{x}_i \in X$ *converging to* $\mathbf{x}_0$,

$$\lim_{i \to \infty} \mathbf{f}(\mathbf{x}_i) = \mathbf{f}(\mathbf{x}_0).$$

Proof. Suppose the ϵ, δ condition is satisfied, and let $\mathbf{x}_i$, $i = 1, 2, \ldots$ be a sequence in X that converges to $\mathbf{x}_0 \in X$. We must show that the sequence $\mathbf{f}(\mathbf{x}_i)$, $i = 1, 2, \ldots$ converges to $\mathbf{f}(\mathbf{x}_0)$, i.e., that for any $\epsilon > 0$, there exists N such that when $n > N$ we have $|\mathbf{f}(\mathbf{x}) - \mathbf{f}(\mathbf{x}_0)| < \epsilon$. To find this N, first find the δ such that $|\mathbf{x} - \mathbf{x}_0| < \delta$ implies that $|\mathbf{f}(\mathbf{x}) - \mathbf{f}(\mathbf{x}_0)| < \epsilon$. Next apply the definition of a convergence sequence to the sequence $\mathbf{x}_i$: there exists N such that if $n > N$, then $|\mathbf{x}_n - \mathbf{x}_0| < \delta$. Clearly this N works.

For the converse, remember how to negate sequences of quantifiers (Section 0.2). Suppose the ϵ, δ condition is not satisfied; then there exists $\epsilon_0 > 0$ such that for all δ, there exists $\mathbf{x} \in X$ such that $|\mathbf{x} - \mathbf{x}_0| < \delta$ but $|\mathbf{f}(\mathbf{x}) - \mathbf{f}(\mathbf{x}_0)| \geq \epsilon_0$. Let $\delta_n = 1/n$, and let $\mathbf{x}_n \in X$ be such a point; i.e.,

$$|\mathbf{x}_n - \mathbf{x}_0| < \frac{1}{n} \quad \text{and} \quad |\mathbf{f}(\mathbf{x}_n) - \mathbf{f}(\mathbf{x}_0)| \geq \epsilon_0. \qquad 1.5.42$$

The first part shows that the sequence $\mathbf{x}_n$ converges to $\mathbf{x}_0$, and the second part shows that $\mathbf{f}(\mathbf{x}_n)$ does not converge to $\mathbf{f}(\mathbf{x}_0)$. $\quad \square$

The following theorem is a reformulation of Theorem 1.5.21.

Theorem 1.5.27 (Combining continuous mappings). *Let U be a subset of $\mathbb{R}^n$, $\mathbf{f}$ and $\mathbf{g}$ mappings $U \to \mathbb{R}^m$, and h a function $U \to \mathbb{R}$.*

(a) If $\mathbf{f}$ and $\mathbf{g}$ are continuous at $\mathbf{x}_0$, then so is $\mathbf{f} + \mathbf{g}$.

(b) If $\mathbf{f}$ and h are continuous at $\mathbf{x}_0$, then so is $h\mathbf{f}$.

(c) If $\mathbf{f}$ and h are continuous at $\mathbf{x}_0$, and $h(\mathbf{x}_0) \neq 0$, then so is $\frac{\mathbf{f}}{h}$.

(d) If $\mathbf{f}$ and $\mathbf{g}$ are continuous at $\mathbf{x}_0$, then so is $\mathbf{f} \cdot \mathbf{g}$

(e) If h is continuous at $\mathbf{x}_0$, with $h(\mathbf{x}_0) = 0$, and $\mathbf{f}$ is bounded in a neighborhood of $\mathbf{x}_0$, then $h\mathbf{f}$ is continuous at $\mathbf{x}_0$ (even if $\mathbf{f}$ is not defined at $\mathbf{x}_0$).

We can now write down a fairly large collection of continuous functions on $\mathbb{R}^n$: polynomials and rational functions.

A *monomial* function on $\mathbb{R}^n$ is an expression of the form $x_1^{k_1} \ldots x_n^{k_n}$ with integer exponents $k_1, \ldots, k_n \geq 0$. For instance, $x^2 y z^5$ is a monomial on $\mathbb{R}^3$, and $x_1 x_2 x_4^2$ is a monomial on $\mathbb{R}^4$ (or perhaps $\mathbb{R}^n$ with $n > 4$). A polynomial function is a finite sum of monomials with real coefficients, like $x^2 y + 3yz$. A *rational function* is a ratio of two polynomials, like $\frac{x+y}{xy+z^2}$.

Corollary 1.5.28 (Continuity of polynomials and rational functions).

(a) Any polynomial function $\mathbb{R}^n \to \mathbb{R}$ is continuous on all of $\mathbb{R}^n$.

(b) Any rational function is continuous on the subset of $\mathbb{R}^n$ where the denominator does not vanish.

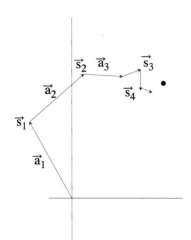

FIGURE 1.5.7.

A convergent series of vectors. The kth partial sum is gotten by putting the first k vectors nose to tail.

Series of vectors

As is the case for numbers (Section 0.4), many of the most interesting sequences arise as partial sums of series.

Definition 1.5.29 (Convergent series of vectors). A series $\sum_{i=1}^{\infty} \vec{\mathbf{a}}_i$ is convergent if the sequence of partial sums

$$\vec{\mathbf{s}}_n = \sum_{i=1}^{n} \vec{\mathbf{a}}_i \qquad 1.5.43$$

is a convergent sequence of vectors. In that case the infinite sum is

$$\sum_{i=1}^{\infty} \vec{\mathbf{a}}_i = \lim_{n \to \infty} \vec{\mathbf{s}}_n. \qquad 1.5.44$$

Absolute convergence means that the absolute values converge.

Proposition 1.5.30 (Absolute convergence implies convergence).

$$\text{If} \quad \sum_{i=1}^{\infty} |\vec{\mathbf{a}}_i| \quad \text{converges, then} \quad \sum_{i=1}^{\infty} \vec{\mathbf{a}}_i \quad \text{converges.}$$

Proposition 1.5.30 is very important; we use it in particular to prove that Newton's method's converges.

Proof. Proposition 1.5.9 says that it is enough to check this component by component; in one variable, it is a standard statement of elementary calculus (Theorem 0.4.11). $\square$

Geometric series of matrices

When he introduced matrices, Cayley remarked that square matrices "comport themselves as single quantities." In many ways, one can think of a square matrix as a generalized number; many constructions possible with numbers are also possible with matrices. Here we will see that a standard result about the sum of a geometric series applies to matrices as well; we will need this result when we discuss Newton's method for solving nonlinear equations, in Section 2.7. In the exercises we will explore other series of matrices.

Definitions 1.5.8 and 1.5.29 apply just as well to matrices as to vectors, since when distances are concerned, if we denote by Mat (n, m) the set of $n \times m$

matrices, then Mat (n, m) is the same as $\mathbb{R}^{nm}$. In particular, Proposition 1.5.9 applies: a series

$$\sum_{k=1}^{\infty} A_k \qquad 1.5.45$$

of $n \times m$ matrices converges if for each position (i, j) of the matrix, the series of the entries $(A_n)_{(i,j)}$ converges.

Recall (Example 0.4.9) that the geometric series $S = a + ar + ar^2 + \cdots$ converges if $|r| < 1$, and that the sum is $a/(1 - r)$. We want to generalize this to matrices:

Example: Let
$$A = \begin{bmatrix} 0 & 1/4 \\ 0 & 0 \end{bmatrix}. \quad \text{Then}$$

$A^2 = 0$ (surprise!), so that the infinite series of Equation 1.5.48 becomes a finite sum:

$$(I - A)^{-1} = I + A,$$

and

$$\begin{bmatrix} 1 & -1/4 \\ 0 & 1 \end{bmatrix}^{-1} = \begin{bmatrix} 1 & 1/4 \\ 0 & 1 \end{bmatrix}.$$

Proposition 1.5.31. *Let A be a square matrix. If $|A| < 1$, the series*

$$S = I + A + A^2 + \cdots \qquad 1.5.48$$

converges to $(I - A)^{-1}$.

Proof. We use the same trick used in the scalar case of Example 0.4.9. Denote by S_k the sum of the first k terms of the series, and subtract from S_k the product $S_k A$, to get $S_l(I - A) = I - A^{k+1}$:

$$\begin{aligned} S_k &= I + A + A^2 + A^3 + \cdots + A^k \\ S_k A &= \underline{\quad A + A^2 + A^3 + \cdots + A^k + A^{k+1}} \\ S_k(I - A) &= I \qquad\qquad\qquad\qquad\qquad - A^{k+1}. \end{aligned} \qquad 1.5.49$$

We know (Proposition 1.4.11 b) that

$$|A^{k+1}| \le |A|^k |A| = |A|^{k+1}, \qquad 1.5.50$$

so $\lim_{k \to \infty} A^{k+1} = 0$ when $|A| < 1$, which gives us

$$S(I - A) = \lim_{k \to \infty} S_k(I - A) = \lim_{k \to \infty} \left(I - A^{k+1} \right) = I - \underbrace{\lim_{k \to \infty} A^{k+1}}_{0} = I. \qquad 1.5.51$$

Since $S(I - A) = I$, S is a left inverse of $(I - A)$. If in Equation 1.5.49 we had written AS_k instead of $S_k A$, the same computation would have given us $(I - A)S = I$, showing that S is a right inverse. So by Proposition 1.2.14, S is *the* inverse of $(1 - A)$. $\square$

We will see in Section 2.3 that if a square matrix has either a right or a left inverse, that inverse is necessarily a true inverse; checking both directions is not actually necessary.

Corollary 1.5.32. *If $|A| < 1$, then $(I - A)$ is invertible.*

Corollary 1.5.33. *The set of invertible $n \times n$ matrices is open.*

Proof. Suppose B is invertible, and $|H| < 1/|B^{-1}|$. Then $|-B^{-1}H| < 1$, so $I + B^{-1}H$ is invertible (by Corollary 1.5.32), and

$$(I + B^{-1}H)^{-1}B^{-1} = \left(B(I + B^{-1}H) \right)^{-1} = (B + H)^{-1}. \qquad 1.5.52$$

Thus if $|H| < 1/|B^{-1}|$, the matrix $B + H$ is invertible, giving an explicit neighborhood of B made up of invertible matrices. $\square$

1.6 Four Big Theorems

When they were discovered, the examples of Peano and Cantor were thought of as aberrations. "I turn with terror and horror from this lamentable scourge of continuous functions with no derivatives . . . ," wrote Charles Hermite in 1893. Six years later, the French mathematician Henri Poincaré lamented the rise of "a rabble of functions . . . whose only job, it seems, is to look as little as possible like decent and useful functions."

"What will the poor student think?" Poincaré worried. "He will think that mathematical science is just an arbitrary heap of useless subtleties; either he will turn from it in aversion, or he will treat it like an amusing game."

Ironically, although Poincaré wrote that these functions, "specially invented only to show up the arguments of our fathers," would never have any other use, he was ultimately responsible for showing that seemingly "pathological" functions are essential in describing nature, leading to such fields as chaos and fractals.

Definition 1.6.1 is amazingly important, invading whole chapters of mathematics; it is the basic "finiteness criterion" for spaces. Something like half of mathematics consists of showing that some space is compact.

In this section we describe a number of results, most only about 100 years old or so. They are not especially hard, and were mainly discovered after various mathematicians (Peano, Weierstrass, Cantor) found that many statements earlier thought to be obvious were in fact false.

For example, the statement *a curve in the plane has area 0* may seem obvious. Yet it is possible to construct a continuous curve that completely fills up a triangle, visiting every point at least once! The discovery of this kind of thing forced mathematicians to rethink their definitions and statements, putting calculus on a rigorous basis.

These results are usually avoided in first and second year calculus. Two key statements typically glossed over are the *mean value theorem* and the *integrability of continuous functions*. These are used—indeed, they are absolutely central—but often they are not proved.[15] In fact they are not so hard to prove when one knows a bit of topology: notions like open and closed sets, and maxima and minima of functions, for example.

In Section 1.5 we introduced some basic notions of topology. Now we will use them to prove Theorem 1.6.2, a remarkable non-constructive result that will enable us to prove the existence of a convergent subsequence without knowing where it is. We will use this theorem in crucial ways to prove the mean value theorem and the fundamental theorem of algebra (this section), to prove the spectral theorem for symmetric matrices (Theorem 3.7.12) and to see what functions can be integrated (Section 4.3).

In Definition 1.6.1 below, recall that a subset $X \subset \mathbb{R}^n$ is bounded if it is contained in a ball centered at the origin (Definition 1.5.12).

Definition 1.6.1 (Compact set).   A subset $C \subset \mathbb{R}^n$ is compact if it is closed and bounded.

The following theorem is as important as the definition, if not more so.

Theorem 1.6.2 (Convergent subsequence in a compact set).   *If a compact set $C \subset \mathbb{R}^n$ contains a sequence $\mathbf{x}_1, \mathbf{x}_2, \ldots$, then that sequence has a convergent subsequence $\mathbf{x}_{i(1)}, \mathbf{x}_{i(2)}, \ldots$ whose limit is in C.*

Note that Theorem 1.6.2 says nothing about what the convergent subsequence converges to; it just says that a convergent subsequence exists.

---

[15]One exception is Michael Spivak's *Calculus*.

Proof.   The set C is contained in a box $-10^N \leq x_i < 10^N$ for some N. Decompose this box into boxes of side 1 in the obvious way. Then at least one of these boxes, which we'll call B_0, must contain infinitely many terms of the sequence, since the sequence is infinite and we have a finite number of boxes. Choose some term $\mathbf{x}_{i(0)}$ in B_0, and cut up B_0 into 10^n boxes of side $1/10$ (in the plane, 100 boxes; in $\mathbb{R}^3$, 1,000 boxes). At least one of these smaller boxes must contain infinitely many terms of the sequence. Call this box B_1, choose $\mathbf{x}_{i(1)} \in B_1$ with $i(1) > i(0)$. Now keep going: cut up B_1 into 10^n boxes of side $1/10^2$; again, one of these boxes must contain infinitely many terms of the sequence; call one such box B_2 and choose an element $\mathbf{x}_{i(2)} \in B_2$ with $i(2) > i(1) \ldots$

Think of the first box B_0 as giving the coordinates, up to the decimal point, of all the points in B_0. (Because it is hard to illustrate many levels for a decimal system, Figure 1.6.1 illustrates the process for a binary system.) The next box, B_1, gives the first digit after the decimal point.[16] Suppose, for example, that B_0 has vertices $(1,2)$, $(2,2)$, $(1,3)$ and $(2,3)$; i.e., the point $(1,2)$ has coordinates $x = 1, y = 2$, and so on. Suppose further that B_1 is the small square at the top right-hand corner. Then *all* the points in B_1 have coordinates $(x = 1.9 \ldots, y = 2.9 \ldots)$. When you divide B_1 into 10^2 smaller boxes, the choice of B_2 will determine the next digit; if B_2 is at the bottom right-hand corner, then all points in B_2 will have coordinates $(x = 1.99 \ldots, y = 2.90 \ldots)$, and so on.

Of course you don't actually know what the coordinates of your points are, because you don't know that B_1 is the small square at the top right-hand corner, or that B_2 is at the bottom right-hand corner. All you know is that there exists a first box B_0 of side 1 that contains infinitely many terms of the original sequence, a second box $B_1 \subset B_0$ of side $1/10$ that also contains infinitely many terms of the original sequence, and so on.

Construct in this way a sequence of nested boxes

$$B_0 \supset B_1 \supset B_2 \supset \ldots \qquad \qquad 1.6.1$$

with B_m of side 10^{-m}, and each containing infinitely many terms of the sequence; further choose $\mathbf{x}_{i(m)} \in B_m$ with $i(m+1) > i(m)$.

Clearly the sequence $\mathbf{x}_{i(m)}$ converges; in fact the mth term beyond the decimal point never changes after the mth choice. The limit is in C since C is closed. □

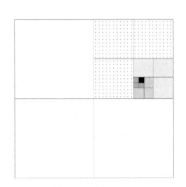

FIGURE 1.6.1.

If the large box contains an infinite sequence, then one of the four quadrants must contain a convergent subsequence. If that quadrant is divided into four smaller boxes, one of those small boxes must contain a convergent subsequence, and so on.

Several more properties of compact sets are stated and proved in Appendix A.17.

You may think "what's the big deal?" To see the troubling implications of the proof, consider Example 1.6.3.

Example 1.6.3 (Convergent subsequence).   Consider the sequence

$$x_m = \sin 10^m. \qquad \qquad 1.6.2$$

---

[16]To ensure that all points in the same box have the same decimal expansion, we should say that our boxes are all open on the top and on the right.

This is certainly a sequence in the compact set $C = [-1, 1]$, so it contains a convergent subsequence. But how do you find it? The first step of the construction above is to divide the interval $[-1, 1]$ into three sub-intervals (our "boxes"), writing

$$[-1, 1] = [-1, 0) \cup [0, 1) \cup \{1\}. \qquad 1.6.3$$

Remember (Section 0.3) that $\cup$ means "union": $A \cup B$ is the set of elements of either A or B or both.

Now how do we choose which of the three "boxes" above should be the first box B_0? We know that x_m will never be in the box $\{1\}$, since $\sin\theta = 1$ if and only if $\theta = \pi/2 + 2k\pi$ for some integer k and (since π is irrational) 10^m cannot be $\pi/2 + 2k\pi$. But how do we choose between $[-1, 0)$ and $[0, 1)$? If we want to choose $[0, 1)$, we must be sure that we have infinitely many positive x_m. So, when is $x_m = \sin 10^m$ positive?

Since $\sin\theta$ is positive for $0 < \theta < \pi$, then x_m is positive when the fractional part of $10^m/(2\pi)$ is greater than 0 and less than 1/2. (By "fractional part" we mean the part after the decimal; for example $5/3 = 1 + 2/3 = 1.666\ldots$; the fractional part is $.666\ldots$.) If you don't see this, consider that (as shown in Figure 1.6.2) $\sin 2\pi\alpha$ depends only on the fractional part of α:

$$\sin 2\pi\alpha \begin{cases} = 0 & \text{if } \alpha \text{ is an integer or half-integer} \\ > 0 & \text{if the fractional part of } \alpha \text{ is } < 1/2 \qquad 1.6.4 \\ < 0 & \text{if the fractional part of } \alpha \text{ is } > 1/2 \end{cases}$$

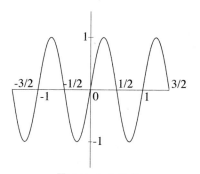

FIGURE 1.6.2.

Graph of $\sin 2\pi x$. If the fractional part of a number x is between 0 and 1/2, then $\sin 2\pi x > 0$; if it is between 1/2 and 1, then $\sin 2\pi x < 0$.

If instead of writing $x_m = \sin 10^m$ we write

$$x_m = \sin 2\pi \frac{10^m}{2\pi}, \quad \text{i.e.} \quad \alpha = \frac{10^m}{2\pi}, \qquad 1.6.5$$

we see, as stated above, that x_m is positive when the fractional part of $10^m/(2\pi)$ is less than 1/2.

So if a convergent subsequence of x_m is contained in the box $[0, 1)$, an infinite number of $10^m/(2\pi)$ must have a fractional part that is less than 1/2. This will ensure that we have infinitely many $x_m = \sin 10^m$ in the box $[0, 1)$.

For any single x_m, it is enough to know that the first digit of the fractional part of $10^m/(2\pi)$ is 0, 1, 2, 3 or 4: knowing the first digit after the decimal point tells you whether the fractional part is less than or greater than 1/2. Since multiplying by 10^m just moves the decimal point to the right by m, knowing whether the fractional part of *every* $10^m/(2\pi)$ starts this way is really a question about the decimal expansion of $\frac{1}{2\pi}$: do the digits 0,1,2,3 or 4 appear infinitely many times in the decimal expansion of

$$\frac{1}{2\pi} = .1591549\ldots? \qquad 1.6.6$$

Note that we are not saying that all the $10^m/(2\pi)$ must have the decimal point followed by 0,1,2,3, or 4! Clearly they don't. We are not interested in all the x_m; we just want to know that we can find a subsequence of x_m that converges to something inside the box $[0, 1)$. For example, x_1 is not in the box $[0, 1)$, since $10 \times .1591549\ldots = 1.591549\ldots$; the fractional part starts with 5. Nor is x_2, since $10^2 \times .1591549\ldots = 15.91549\ldots$; the fractional part starts with 9. But x_3 is in the box $[0, 1)$, since the fractional part of $10^3 \times .1591549\ldots = 159.1549\ldots$ starts with a 1.

Everyone believes that the digits 0,1,2,3 or 4 appear infinitely many times in the decimal expansion of $\frac{1}{2\pi}$: it is widely believed that π is a *normal number*, i.e., where every digit appears roughly 1/10th of the time, every pair of digits appears roughly 1/100th of the time, etc. The first 4 billion digits of π have been computed and appear to bear out this conjecture. Still, no one knows how to prove it; as far as we know it is conceivable that all the digits after the 10 billionth are 6's, 7's and 8's.

Thus, even choosing the first "box" B_0 requires some god-like ability to "see" this whole infinite sequence, when there is simply no obvious way to do it. $\triangle$

> The point is that although the sequence $x_m = \sin 10^m$ is a sequence in a compact set, and therefore (by Theorem 1.6.2) contains a convergent subsequence, we can't begin to "locate" that subsequence. We can't even say whether it is in $[-1, 0)$ or $[0, 1)$.

Theorem 1.6.2 is *non-constructive*: it proves that something exists but gives not the slightest hint of how to find it. Many mathematicians of the end of the 19th century were deeply disturbed by this type of proof; even today, a school of mathematicians called the intuitionists reject this sort of thinking. They demand that in order for a number to be determined, one give a rule which allows the computation of the successive decimals. Intuitionists are pretty scarce these days: we have never met one. But we have a certain sympathy with their views, and much prefer proofs that involve effectively computable algorithms, at least implicitly.

Continuous functions on compact sets

We can now explore some of the consequences of Theorem 1.6.2.

> Recall that *compact* means closed and bounded.

One is that a continuous function defined on a compact subset has both a maximum and a minimum. Recall from first year calculus and from Section 0.4 the difference between a *least upper bound* and a *maximum* (similarly, the difference between a *greatest lower bound* and a *minimum*).

> Although the use of the words *least upper bound* and *sup* is completely standard, some people use *maximum* as another synonym for least upper bound, not a least upper bound that is achieved, as we have defined it. Similarly, some people use the words *greatest lower bound* and *minimum* interchangeably; we do not.

Definition 1.6.4 (Least upper bound). A number x is the *least upper bound* of a function f defined on a set C if x is the smallest number such that $f(a) \leq x$ for all $a \in C$. It is also called *supremum*, abbreviated *sup*.

Definition 1.6.5 (Maximum). A number x is the *maximum* of a function f defined on a set C if it is the least upper bound of f and there exists $b \in C$ such that $f(b) = x$.

For example, on the open set $(0, 1)$ the least upper bound of $f(x) = x^2$ is 1, and f has no maximum. On the closed set $[0, 1]$, 1 is both the least upper bound and the maximum of f.

On the open set $(0, 1)$ the greatest lower bound of $f(x) = x^2$ is 0, and f has no minimum. On the closed set $[0, 1]$, 0 is both the greatest lower bound and the minimum of f.

Recall that "compact" means closed and bounded.

Definition 1.6.6 (Greatest lower bound, minimum). A number y is the *greatest lower bound* of a function f defined on a set C if y is the largest number such that $f(a) \geq y$ for all $a \in C$. The word *infimum*, abbreviated *inf*, is a synonym for greatest lower bound. The number y is the *minimum* of f if there exists $b \in C$ such that $f(b) = y$.

Theorem 1.6.7 (Existence of minima and maxima). *Let $C \subset \mathbb{R}^n$ be a compact subset, and $f : C \to \mathbb{R}$ be a continuous function. Then there exists a point $\mathbf{a} \in C$ such that $f(\mathbf{a}) \geq f(\mathbf{x})$ for all $\mathbf{x} \in C$, and a point $\mathbf{b} \in C$ such that $f(\mathbf{b}) \leq f(\mathbf{x})$ for all $\mathbf{x} \in C$.*

Here are some examples to show that the conditions in the theorem are necessary. Consider the function

$$f(x) = \begin{cases} 0 & \text{when } x = 0 \\ \frac{1}{x} & \text{otherwise,} \end{cases} \qquad 1.6.7$$

defined on the compact set $[0, 1]$. As $x \to 0$, we see that $f(x)$ blows up to infinity; the function does not have a maximum (it is not bounded). This function is not continuous, so Theorem 1.6.7 does not apply to it.

The function $f(x) = 1/x$, defined on $(0, 1]$, is continuous but it has no maximum either; this time the problem is that $(0, 1]$ is not closed, hence not compact. And the function $f(x) = x$, defined on all of $\mathbb{R}$, is not bounded either; this time the problem is that $\mathbb{R}$ is not bounded, hence not compact. Exercise 1.6.1 asks you to show that if $A \subset \mathbb{R}^n$ is any non-compact subset, then there always is a continuous unbounded function on A.

Proof. We will prove the statement for the maximum. The proof is by contradiction. Assume f is unbounded. Then for any integer N, no matter how large, there exists a point $\mathbf{x}_N \in C$ such that $|f(\mathbf{x}_N)| > N$. By Theorem 1.6.2, the sequence $\mathbf{x}_N$ must contain a convergent subsequence $\mathbf{x}_{N(j)}$, which converges to some point $\mathbf{b} \in C$. Since f is continuous at $\mathbf{b}$, then for any ϵ, there exists a $\delta > 0$ such that when $|\mathbf{x} - \mathbf{b}| < \delta$, then $|f(\mathbf{x}) - f(\mathbf{b})| < \epsilon$; i.e., $|f(\mathbf{x})| < |f(\mathbf{b})| + \epsilon$.

Since the $\mathbf{x}_{N(j)}$ converge to $\mathbf{b}$, we will have $|\mathbf{x}_{N(j)} - \mathbf{b}| < \delta$ for j sufficiently large. But as soon as $N(j) > |f(\mathbf{b})| + \epsilon$, we have

$$|f(\mathbf{x}_{N(j)})| > N(j) > |f(\mathbf{b})| + \epsilon, \qquad 1.6.8$$

a contradiction.

Therefore, the set of values of f is bounded, which means that f has a least upper bound M. What we now want to show is that f has a maximum: that there exists a point $\mathbf{a} \in C$ such that $f(\mathbf{a}) = M$.

There is a sequence $\mathbf{x}_i$ such that

$$\lim_{i \to \infty} f(\mathbf{x}_i) = M. \qquad\qquad 1.6.9$$

We can again extract a convergent subsequence $\mathbf{x}_{i(m)}$ that converges to some point $\mathbf{a} \in C$. Then, since $\mathbf{a} = \lim_{m \to \infty} \mathbf{x}_{i(m)}$,

$$f(\mathbf{a}) = \lim_{m \to \infty} f(\mathbf{x}_{i(m)}) = M. \qquad\qquad 1.6.10$$

The proof for the minimum works the same way. $\square$

We will have several occasions to use Theorem 1.6.7. First, we need the following proposition, which you no doubt proved in first year calculus.

Proposition 1.6.8. *If a function g defined and differentiable on an open interval in $\mathbb{R}$ has a maximum (respectively a minimum) at c, then its derivative at c is 0.*

Notice that the mean value theorem does not require that the derivative be continuous. But it does require that the derivative take on a value at every point (that's what being differentiable means). Thus (surprise!) a car cannot jump from going 59 mph to going 61 mph, without ever passing through 60 mph. We will see in Example 1.9.4 a differentiable function with a discontinuous derivative. (See also Exercise 0.4.7.) This kind of oscillating discontinuity is the only kind of discontinuity a derivative can have; this is more or less what the mean value theorem says. More precisely, the derivative of a differentiable function satisfies the intermediate value property.

The special case of the mean value theorem where

$$f(a) = f(b) = 0$$

is called *Rolle's theorem.*

Proof. We will prove it only for the maximum. If g has a maximum at c, then $g(c) - g(c + h) \geq 0$, so

$$\frac{g(c) - g(c+h)}{h} \begin{cases} \geq 0 & \text{if } h > 0 \\ \leq 0 & \text{if } h < 0; \end{cases} \quad \text{i.e.,} \quad \lim_{h \to 0} \frac{g(c) - g(c+h)}{h} \qquad 1.6.11$$

is simultaneously ≤ 0 and ≥ 0, so it is 0. $\square$

An essential application of Theorem 1.6.7 and Proposition 1.6.8 is the *mean value theorem*, without which practically nothing in differential calculus can be proved. The mean value theorem says that you can't drive 60 miles in an hour without going exactly 60 mph at one instant at least: the average change in f over the interval (a, b) is the derivative of f at some point $c \in (a, b)$.

Theorem 1.6.9 (Mean value theorem). *If $f : [a, b] \to \mathbb{R}$ is continuous, and f is differentiable on (a, b), then there exists $c \in (a, b)$ such that*

$$f'(c) = \frac{f(b) - f(a)}{b - a}. \qquad\qquad 1.6.12$$

Note that f is defined on the closed and bounded interval $[a, b]$, but we must specify the open interval (a, b) when we talk about where f is differentiable.[17] If we think that f measures position as a function of time, then the right-hand side of Equation 1.6.12 measures average speed over the time interval $b - a$.

---

[17]One could have a left-hand and right-hand derivative at the endpoints, but we are not assuming that such one-sided derivatives exist.

Proof. Think of f as representing distance traveled (by a car or, as in Figure 1.6.3, by a hare). The distance the hare travels in the time interval $b - a$ is $f(b) - f(a)$, so its average speed is

$$m = \frac{f(b) - f(a)}{b - a}. \qquad \text{1.6.13}$$

The function g represents the steady progress of a tortoise starting at $f(a)$ and constantly maintaining that average speed (alternatively, a car set on cruise control):

$$g(x) = f(a) + m(x - a).$$

The function h measures the distance between f and g:

$$h(x) = f(x) - g(x) = f(x) - \big(f(a) + m(x - a)\big). \qquad \text{1.6.14}$$

It is a continuous function on $[a, b]$, and $h(a) = h(b) = 0$. (The hare and the tortoise start together and finish in a dead heat.)

If h is 0 everywhere, then $f(x) = g(x) = f(a) + m(x - a)$ has derivative m everywhere, so the theorem is true.

If h is not 0 everywhere, then it must take on positive values or negative values somewhere, so it must have a positive maximum or a negative minimum, or both. Let c be a point where it has such an extremum; then $c \in (a, b)$, so h is differentiable at c, and by Proposition 1.6.8, $h'(c) = 0$.

This gives $0 = h'(c) = f'(c) - m$. (In Equation 1.6.14, x appears only twice; the $f(x)$ contributes $f'(c)$ and the $-mx$ contributes $-m$.) $\quad\square$

The fundamental theorem of algebra

The fundamental theorem of algebra is one of the most important results of all mathematics, with a history going back to the Greeks and Babylonians. It was not proved satisfactorily until about 1830. The theorem asserts that every polynomial has roots.

> **Theorem 1.6.10 (Fundamental theorem of algebra).** *Let*
>
> $$p(z) = z^k + a_{k-1}z^{k-1} + \cdots + a_0 \qquad \text{1.6.15}$$
>
> *be a polynomial of degree $k > 0$ with complex coefficients. Then p has a root: there exists a complex number z_0 such that $p(z_0) = 0$.*

When $k = 1$, this is clear: the unique root is $z_0 = -a_0$.

When $k = 2$, the famous *quadratic formula* tells you that the roots are

$$\frac{-a_1 \pm \sqrt{a_1^2 - 4a_0}}{2}. \qquad \text{1.6.16}$$

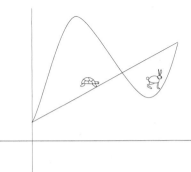

FIGURE 1.6.3.

A race between hare and tortoise ends in a dead heat. The function f represents the progress of the hare, starting at time a and ending at time b. He speeds ahead, overshoots the mark, and returns. Slow-and-steady tortoise is represented by $g(x) = f(a) + m(x - a)$.

Even if the coefficients a_n are real, the fundamental theorem of algebra does not guarantee that the polynomial has any real roots; the roots may be complex.

(Recall above that the coefficient of z^2 is 1.) This was known to the Greeks and Babylonians.

The cases $k = 3$ and $k = 4$ were solved in the 16th century by Cardano and others; their solutions are presented in Section 0.6.

For the next two centuries, an intense search failed to find anything analogous for equations of higher degree. Finally, around 1830, two young mathematicians with tragic personal histories, the Norwegian Hans Erik Abel and the Frenchman Evariste Galois, proved that no analogous formulas exist in degrees 5 and higher. Again, these discoveries opened new fields in mathematics.

Several mathematicians (Laplace, d'Alembert, Gauss) had earlier come to suspect that the fundamental theorem was true, and tried their hands at proving it. In the absence of topological tools, their proofs were necessarily short on rigor, and the criticism each heaped on his competitors does not reflect well on any of them. Although the first correct proof is usually attributed to Gauss (1799), we will present a modern version of d'Alembert's argument (1746).

Unlike the quadratic formula and Cardano's formulas, our proof *does not provide a recipe to find a root.* (Indeed, as we mentioned above, Abel and Galois proved that no recipes analogous to Equation 1.6.16 exist.) This is a serious problem: one very often needs to solve polynomials, and to this day there is no really satisfactory way to do it; the picture on the cover of this text is an attempt to solve a polynomial of degree 256. There is an enormous literature on the subject.

Proof of 1.6.10. We want to show that there exists a number z such that $p(z) = 0$. The strategy of the proof is first to establish that $|p(z)|$ has a minimum, and next to establish that its minimum is in fact 0. To establish that $|p(z)|$ has a minimum, we will show that there is a disk around the origin such that every z outside the disk gives a value $|p(z)|$ that is greater than $|p(0)|$. The disk we create is closed and bounded, and $|p(z)|$ is a continuous function, so by Theorem 1.6.7 there is a point z_0 inside the disk such that $|p(z_0)|$ is the minimum of the function on the disk. It is also the minimum of the function everywhere, by the preceding argument. Finally—and this will be the main part of the argument—we will show that $p(z_0) = 0$.

We shall create our disk in a rather crude fashion; the radius of the disk we establish will be greater than we really need. First, $|p(z)|$ can be at least as small as $|a_0|$, since when $z = 0$, Equation 1.6.15 gives $p(0) = a_0$. So we want to show that for $|z|$ big enough, $|p(z)| > |a_0|$. The "big enough" will be the radius of our disk; we will then know that the minimum inside the disk is the global minimum for the function.

It it is clear that for $|z|$ large, $|z^k|$ is much larger. What we have to ascertain is that when $|z|$ is very big, $|p(z)| > |a_0|$: the size of the other terms,

$$|a_{k-1}z^{k-1} + \cdots + a_1 z + a_0|, \qquad \text{1.6.17}$$

will not compensate enough to make $|p(z)| < |a_0|$.

Niels Henrik Abel, born in 1802, assumed responsibility for a younger brother and sister after the death of their alcoholic father in 1820. For years he struggled against poverty and illness, trying to obtain a position that would allow him to marry his fiancée; he died from tuberculosis at the age of 26, without learning that he had been appointed professor in Berlin.

Evariste Galois, born in 1811, twice failed to win admittance to Ecole Polytechnique in Paris, the second time shortly after his father's suicide. In 1831 he was imprisoned for making an implied threat against the king at a republican banquet; he was acquitted and released about a month later. He was 20 years old when he died from wounds received in a duel.

At the time Gauss gave his proof of Theorem 1.6.10, complex numbers were not sufficiently respectable that they could be mentioned in a rigorous paper; Gauss stated his theorem in terms of real polynomials. For a discussion of complex numbers, see Section 0.6.

The absolute value of a complex number $z = x + iy$ is

$$|z| = \sqrt{x^2 + y^2}.$$

The notation

$$\sup\{|a_{k-1}|,\ldots,|a_0|\}$$

means the largest of

$$|a_{k-1}|,\ldots,|a_0|.$$

First, choose the largest of the coefficients $|a_{k-1}|,\ldots,|a_0|$ and call it A:

$$A = \sup\{|a_{k-1}|,\ldots,|a_0|\}. \qquad 1.6.18$$

Then if $|z| = R$, and $R > 1$, we have

$$|a_{k-1}z^{k-1} + \cdots + a_1z + a_0| \le AR^{k-1} + \cdots + AR + A$$
$$< AR^{k-1} + \cdots + AR^{k-1} + AR^{k-1} = kAR^{k-1}.$$
$$1.6.19$$

To get from the first to the second line of Equation 1.6.19 we multiplied all the terms on the right-hand side, except the first, by $R^1, R^2 \ldots$ up to R^{k-1} in order to get an R^{k-1} in all k terms, giving kAR^{k-1} in all. (We don't need to make this term so very big; we're being extravagant in order to get a relatively simple expression for the sum. This is not a case where one has be delicate with inequalities.)

The triangle inequality can also be stated as

$$|\vec{\mathbf{v}}| - |\vec{\mathbf{w}}| \le |\vec{\mathbf{v}} + \vec{\mathbf{w}}|,$$

since

$$|\vec{\mathbf{v}}| = |\vec{\mathbf{v}} + \vec{\mathbf{w}} - \vec{\mathbf{w}}|$$
$$\le |\vec{\mathbf{v}} + \vec{\mathbf{w}}| + |-\vec{\mathbf{w}}|$$
$$= |\vec{\mathbf{v}} + \vec{\mathbf{w}}| + |\vec{\mathbf{w}}|.$$

Now, when $|z| = R$, we have

$$|p(z)| = |\underbrace{z^k}_{R^k} + \underbrace{a_{k-1}z^{k-1} + \cdots + a_0}_{\text{abs. value} < kAR^{k-1}}|, \qquad 1.6.20$$

so using the triangle inequality,

$$|p(z)| \ge |z^k| - |a_{k-1}z^{k-1} + \cdots + a_1z + a_0|$$
$$> R^k - kAR^{k-1} = R^{k-1}(R - kA). \qquad 1.6.21$$

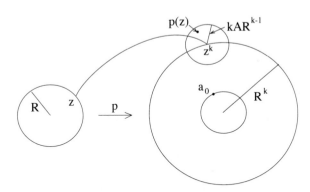

FIGURE 1.6.4. Any z outside the disk of radius R will give $|p(z)| > |a_0|$.

Of course $R^{k-1}|R - kA| > |a_0|$ when $R = \max\{kA + |a_0|, 1\}$. So now we know that any z chosen outside the disk of radius R will give $|p(z)| > |a_0|$, as shown in Figure 1.6.4. If the function has a minimum, that minimum has to be *inside* the disk. Moreover, we know by Theorem 1.6.7 that it *does* have a minimum inside the disk. We will denote by z_0 a point inside the disk at which the function achieves its minimum.

The big question is: is z_0 a root of the polynomial? Is it true that $|p(z_0)| = 0$? Earlier we used the fact that for $|z|$ large, $|z^k|$ is very large. Now we will use the fact that for $|z|$ small, $|z^k|$ is very small. We will also take into account that we are dealing with complex numbers. The preceding argument works just as well with real numbers, but now we will need the fact that when a complex number is written in terms of its length r and polar angle θ, taking its power has the following effect:

Equation 1.6.22 is de Moivre's formula, which we stated earlier as Corollary 0.6.4.

$$\left(r(\cos\theta + i\sin\theta)\right)^k = r^k(\cos k\theta + i\sin k\theta). \qquad 1.6.22$$

As you choose different values of θ then $z = r(\cos\theta + i\sin\theta)$ travels on a circle of radius r. If you raise that number to the kth power, then it travels around a much smaller circle (for r small), going much faster—k times around for every one time around the original circle.

The formulas in this last part of the proof may be hard to follow, so first we will outline what we are going to do. We are going to argue by contradiction, saying that $p(z_0) \neq 0$, and seeing that we land on an impossibility. We will then see that $p(z_0)$ is *not* the minimum, because there exists a point z such that $|p(z)| < |p(z_0)|$. Since we have already proved that $|p(z_0)|$ *is* the minimum, our assumption that $p(z_0) \neq 0$ is false.

We start with a change of variables; it will be easier to consider numbers in a circle around z_0 if we treat z_0 as the origin. So set $z = z_0 + u$, and consider the function

You might object, what happens to the middle terms, for example, the $2a_2 z_0 u$ in $a_2(z_0+u)^2 = a_2 z_0^2 + 2a_2 z_0 u + a_2 u^2$? But that is a term in u with coefficient $a_2 2z_0$, so the coefficient $a_2 2z_0$ just gets added to b_1, the coefficient of u.

$$p(z) = z^k + a_{k-1}z^{k-1} + \cdots + a_0 = (z_0 + u)^k + a_{k-1}(z_0 + u)^{k-1} + \cdots + a_0$$
$$= u^k + b_{k-1}u^{k-1} + \cdots + b_0 = q(u),$$
$$1.6.23$$

where

$$b_0 = z_0^k + a_{k-1}z_0^{k-1} + \cdots + a_0 = p(z_0). \qquad 1.6.24$$

This is a polynomial of degree k in u. We have grouped together all the terms that don't contain u and called them b_0.

Now, looking at our function $q(u)$ of Equation 1.6.23, we choose the term with the smallest power $j > 0$ that has a nonzero coefficient. (For example, if we had $q(u) = u^4 + 2u^2 + 3u + 10$, that term, which we call $b_j u^j$, would be $3u$; if we had $q(u) = u^5 + 2u^4 + 5u^3 + 1$, that term would be $5u^3$.) We rewrite our function as follows

$$q(u) = b_0 + b_j u^j + \underbrace{(b_{j+1}u^{j+1} + \cdots + u^k)}_{\text{abs.val. smaller than } |b_j u^j| \text{ for small } u}. \qquad 1.6.25$$

Exercise 1.6.2 asks you to justify that $|(b_{j+1}u^{j+1} + \cdots + u^k)| < |b_j u^j|$ for small u. The construction is illustrated in Figure 1.6.5.

Because there may be lots of little terms $b_{j+1}u^{j+1}+\cdots+u^k$, you might imagine that the first dog holds a shorter leash for a smaller dog who is running around him, that smaller dog holding a yet shorter leash for a yet smaller dog who is running around him . . . .

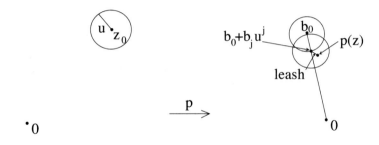

FIGURE 1.6.5. The point $p(z_0) = b_0$ (the flagpole) is the closest that $p(z)$ ever comes to the origin, for all z. The assumption that the flagpole is different from the origin ($b_0 \neq 0$) leads to a contradiction: if $|u|$ is small, then as $z = z_0 + u$ takes a walk around z_0 (shown at left), $p(z)$ (the dog) goes around the flagpole and will at some point be closer to the origin than is the flagpole itself (shown at right).

Exercise 1.6.6 asks you to prove that every polynomial over the complex numbers can be factored into linear factors, and that every polynomial over the real numbers can be factored into linear factors and quadratic factors with complex roots.

Recall that ρ is pronounced "rho."

Now consider our number u written in terms of length ρ and polar angle θ:

$$u = \rho(\cos\theta + i\sin\theta). \qquad 1.6.26$$

The numbers $z = z_0 + u$ then turn in a circle of radius ρ around z_0 as we change the angle θ. What about the numbers $p(z)$? If we were to forget about the small terms grouped in parentheses on the right-hand side of Equation 1.6.25, we would say that these points travel in a circle of radius ρ^j (smaller than ρ for $\rho < 1$) around the point $b_0 = p(z_0)$. We would then see that, as shown in Figure 1.6.5, that if $\rho^j < |b_0|$, some of these points are *between* b_0 and 0; i.e., they are smaller than b_0. If we ignore the small terms, this would mean that there exists a number z such that $|p(z)| < |p(z_0)|$, contradicting the fact, which we have proved that $|p(z_0)|$ is the minimum of the function.

Of course we can't quite ignore the small terms, but we can show that they don't affect our conclusion. Think of b_0 as a flagpole and $b_0 + b_j u^j$, with $|u| = \rho$ as a man walking on a circle of radius $|b_j|\rho^j$ around that flagpole. He is walking a dog that is running circles around him, restrained by a leash of radius *less than* $|b_j|\rho^j$, for ρ sufficiently small. The leash represents the small terms. So when the man is between 0 and the flagpole, the dog, which represents the point $p(z)$), is closer to 0 than is the flagpole. That is, $|p(z)|$ is *less than* $|b_0| = |p(z_0)|$. This is impossible, because we proved that $|p(z_0)|$ is the minimum of our function. Therefore, our assumption that $p(z_0) \neq 0$ is false. $\square$

The proof of the fundamental theorem of algebra illustrates the kind of thing we meant when we said, in the beginning of Section 1.4, that calculus is about "some terms being dominant or negligible compared to other terms."

1.7 Differential Calculus: Replacing nonlinear Transformations by Linear Transformations

> Born: *I should like to put to Herr Einstein a question, namely, how quickly the action of gravitation is propagated in your theory . . . .*
>
> Einstein: *It is extremely simple to write down the equations for the case when the perturbations that one introduces in the field are infinitely small. . . . The perturbations then propagate with the same velocity as light.*
>
> Born: *But for great perturbations things are surely very complicated?*
>
> Einstein: *Yes, it is a mathematically complicated problem. It is especially difficult to find exact solutions of the equations, as the equations are nonlinear.*—Discussion after lecture by Einstein in 1913

As mentioned in Section 1.3, in real life (and in pure mathematics as well) a great many problems of interest are not linear; one must consider the effects of feedback. A pendulum is an obvious example: if you push it so that it moves away from you, eventually it will swing back. Second-order effects in other problems may be less obvious. If one company cuts costs by firing workers, it will probably increase profits; if all its competitors do the same, no one company will gain a competitive advantage; if enough workers lose jobs, who will buy the company's products? Modeling the economy is notoriously difficult, but second-order effects also complicate behavior of mechanical systems.

The object of differential calculus is to study nonlinear mappings by replacing them with linear transformations; we replace nonlinear equations with linear equations, curved surfaces by their tangent planes, and so on.

The object of differential calculus is to study nonlinear mappings by replacing them with linear transformations. Of course, this linearization is useful only if you understand linear objects reasonably well. Also, this replacement is only more or less justified. Locally, near the point of tangency, a curved surface may be very similar to its tangent plane, but further away it isn't. The hardest part of differential calculus is determining when replacing a nonlinear object by a linear one is justified.

In Section 1.3 we studied linear transformations in $\mathbb{R}^n$. Now we will see what this study contributes to the study of nonlinear transformations, more commonly called mappings.

This isn't actually a reasonable description: nonlinear is much too broad a class to consider. Dividing mappings into linear and nonlinear is like dividing people into left-handed cello players and everyone else. We will study a limited subset of nonlinear mappings: those that are, in a sense we will study with care, "well approximated by linear transformations."

Derivatives and linear approximation in one dimension

In one dimension, the derivative is the main tool used to linearize a function. Recall from one variable calculus the definition of the derivative:

The derivative of a function $f : \mathbb{R} \to \mathbb{R}$, evaluated at a, is

$$f'(a) = \lim_{h \to 0} \frac{1}{h}\big(f(a+h) - f(a)\big). \qquad 1.7.1$$

Although it sounds less friendly, we really should say:

Limiting the domain as we do in Definition 1.7.1 is necessary, because many interesting functions are not defined on all of $\mathbb{R}$, but they are defined on an appropriate open subset $U \subset \mathbb{R}$. Such functions as $\ln x$, $\tan x$ and $1/x$ are not defined on all of $\mathbb{R}$; for example, $1/x$ is not defined at 0. So if we used Equation 1.7.1 as our definition, $\tan x$ or $\ln x$ or $1/x$ would not be differentiable.

Definition 1.7.1 (Derivative). Let U be an open subset of $\mathbb{R}$, and $f : U \to \mathbb{R}$ a function. Then f is differentiable at $a \in U$ with derivative $f'(a)$ if the limit

$$f'(a) = \lim_{h \to 0} \frac{1}{h}\big(f(a+h) - f(a)\big) \quad \text{exists.} \qquad 1.7.2$$

Students often find talk about open sets $U \subset \mathbb{R}$ and domains of definition pointless; what does it mean when we talk about a function $f : U \to \mathbb{R}$? This is the same as saying $f : \mathbb{R} \to \mathbb{R}$, except that $f(x)$ is only defined if x is in U.

Example 1.7.2 (Derivative of a function from $\mathbb{R} \to \mathbb{R}$). If $f(x) = x^2$, then $f'(x) = 2x$. This is proved by writing

$$f'(x) = \lim_{h \to 0} \frac{1}{h}\big((x+h)^2 - x^2\big) = \lim_{h \to 0} \frac{1}{h}(2xh + h^2) = 2x + \lim_{h \to 0} h = 2x. \quad 1.7.3$$

We discussed in Remark 1.5.1 why it is necessary to specify an *open* set when talking about derivatives.

Exercises 1.7.1, 1.7.2, 1.7.3, and 1.7.4 provide some review of tangents and derivatives.

The derivative $2x$ of the function $f(x) = x^2$ is the slope of the line tangent to f at x; one also says that $2x$ is the *slope of the graph of f at x*. In higher dimensions, this idea of the slope of the tangent to a function still holds, although already in two dimensions, picturing a plane tangent to a surface is considerably more difficult than picturing a line tangent to a curve. $\triangle$

Partial derivatives

One kind of derivative of a function of several variables works just like a derivative of a function of one variable: take the derivative *with respect to one variable*, treating all the others as constants.

Definition 1.7.3 (Partial derivative). Let U be an open subset of $\mathbb{R}^n$ and $f : U \to \mathbb{R}$ a function. The partial derivative of f with respect to the ith variable, and evaluated at $\mathbf{a}$, is the limit

$$D_i f(\mathbf{a}) = \lim_{h \to 0} \frac{1}{h}\left(f\begin{pmatrix} a_1 \\ \vdots \\ a_i + h \\ \vdots \\ a_n \end{pmatrix} - f\begin{pmatrix} a_1 \\ \vdots \\ a_i \\ \vdots \\ a_n \end{pmatrix} \right), \qquad 1.7.4$$

if the limit exists, of course.

The partial derivative $D_1 f$ measures change in the direction of the vector $\vec{e}_1$; the partial derivative $D_2 f$ measures change in the direction of the vector $\vec{e}_2$; and so on.

We can rewrite Equation 1.7.4, using standard basis vectors:

$$D_i f(\mathbf{a}) = \lim_{h \to 0} \frac{f(\mathbf{a} + h\vec{e}_i) - f(\mathbf{a})}{h}, \qquad 1.7.5$$

since all the entries of $\vec{e}_i$ are 0 except for the ith entry, which is 1, so that

$$\mathbf{a} + h\vec{e}_i = \begin{pmatrix} a_1 \\ \vdots \\ a_i + h_i \\ \vdots \\ a_n \end{pmatrix}. \qquad 1.7.6$$

The partial derivative $D_i f(\mathbf{a})$ answers the question, how fast does the function change when you vary the ith variable, keeping the other variables constant? It is computed exactly the same way as derivatives are computed in first year calculus. To take the partial derivative with respect to the first variable of the function $f\begin{pmatrix} x \\ y \end{pmatrix} = xy$, one considers y to be a constant and computes $D_1 f = y$.

What is $D_1 f$ if $f\begin{pmatrix} x \\ y \end{pmatrix} = x^3 + x^2 y + y^2$? What is $D_2 f$? Check your answers below.[18]

Different notations for the partial derivative exist:

$$D_1 f = \frac{\partial f}{\partial x_1}, \quad D_2 f = \frac{\partial f}{\partial x_2}$$
$$\dots \; D_i f = \frac{\partial f}{\partial x_i}.$$

A notation often used in partial differential equations is

$$f_{x_i} = D_i f.$$

Remark. There are at least four commonly used notations for partial derivatives, the most common being

$$\frac{\partial f}{\partial x_1}, \frac{\partial f}{\partial x_2}, \dots, \frac{\partial f}{\partial x_i} \qquad 1.7.7$$

for the partial derivative with respect to the first, second, $\dots$, ith variable. We prefer the notation $D_i f$, because it focuses on the important information: with respect to *which variable* the partial derivative is being taken. (In problems in economics, for example, where there may be no logical order to the variables, one might assign letters rather than numbers: $D_w f$ for the "wages" variable, $D_p f$ for the "prime rate," etc.) It is also simpler to write and looks better in matrices. But we will occasionally use the other notation in examples and exercises, so that you will be familiar with it. $\triangle$

Pitfalls of partial derivatives

One eminent French mathematician, Adrien Douady, complains that the notation for the partial derivative omits the most important information: which variables are being kept constant.

---

[18] $D_1 f = 3x^2 + 2xy$ and $D_2 f = x^2 + 2y$.

For instance, consider the very real question: does increasing the minimum wage increase or decrease the number of minimum wage jobs? This is a question about the sign of

$$D_{\text{minimum wage}}\, f,$$

where x is the economy and $f(x)$ = number of minimum wage jobs.

All notations for the partial derivative omit crucial information: which variables are being kept constant. In modeling real phenomena, it can be difficult even to know what all the variables are. But if you don't, your partial derivatives may be meaningless.

But this partial derivative is meaningless until you state what is being held constant, and it isn't at all easy to see what this means. Is public investment to be held constant, or the discount rate, or is the discount rate to be adjusted to keep total unemployment constant, as appears to be the present policy? There are many other variables to consider, who knows how many. You can see here why economists disagree about the sign of this partial derivative: it is hard if not impossible to say what the partial derivative is, never mind evaluating it.

Similarly, if you are studying pressure of a gas as a function of temperature, it makes a big difference whether the volume of gas is kept constant or whether the gas is allowed to expand, for instance because it fills a balloon.

Partial derivatives of vector-valued functions

Note that the partial derivative of a vector-valued function is a vector.

We use the standard expression, "vector-valued function," but note that the values of such a function could be points rather than vectors; the difference in Equation 1.7.8 would still be a vector.

The definition of a partial derivative makes just as good sense for a vector-valued function (a function from $\mathbb{R}^n$ to $\mathbb{R}^m$). In such a case, we evaluate the limit for each component of $\mathbf{f}$, defining

$$\overrightarrow{D_i}\mathbf{f}(\mathbf{a}) = \lim_{h \to 0} \frac{1}{h}\left(\mathbf{f}\begin{pmatrix} a_1 \\ \vdots \\ a_i + h \\ \vdots \\ a_n \end{pmatrix} - \mathbf{f}\begin{pmatrix} a_1 \\ \vdots \\ a_i \\ \vdots \\ a_n \end{pmatrix} \right) = \begin{bmatrix} D_i f_1(\mathbf{a}) \\ \vdots \\ D_i f_m(\mathbf{a}). \end{bmatrix} \qquad 1.7.8$$

Example 1.7.4. Let $\mathbf{f}: \mathbb{R}^2 \to \mathbb{R}^3$ be given by

$$f_1\begin{pmatrix} x \\ y \end{pmatrix} = xy$$

$$f_2\begin{pmatrix} x \\ y \end{pmatrix} = \sin(x+y), \quad \text{written more simply} \quad \mathbf{f}\begin{pmatrix} x \\ y \end{pmatrix} = \begin{pmatrix} xy \\ \sin(x+y) \\ x^2 - y^2 \end{pmatrix}.$$

$$f_3\begin{pmatrix} x \\ y \end{pmatrix} = x^2 - y^2$$

$$1.7.9$$

We give two versions of Equation 1.7.10 to illustrate the two notations and to emphasize the fact that although we used x and y to define the function, we can evaluate it at variables that look different.

The partial derivative of $\mathbf{f}$ with respect to the first variable is

$$\overrightarrow{D_1}\mathbf{f}\begin{pmatrix} x \\ y \end{pmatrix} = \begin{bmatrix} y \\ \cos(x+y) \\ 2x \end{bmatrix} \quad \text{or} \quad \frac{\partial \mathbf{f}}{\partial x_1}\begin{pmatrix} a \\ b \end{pmatrix} = \begin{bmatrix} b \\ \cos(a+b) \\ 2a \end{bmatrix}. \qquad 1.7.10$$

What is the partial derivative with respect to the second variable?[19] What are the partial derivatives at $\begin{pmatrix} a \\ b \end{pmatrix}$ of the function

$$\mathbf{f}\begin{pmatrix} x \\ y \end{pmatrix} = \begin{pmatrix} x^2 y \\ \cos y \end{pmatrix} ?$$

How would you rewrite the answer, using the notation of Equation 1.7.7?[20]

Directional derivatives

The partial derivative

$$\lim_{h \to 0} \frac{\mathbf{f}(\mathbf{a} + h\vec{e}_i) - \mathbf{f}(\mathbf{a})}{h} \qquad 1.7.11$$

measures the rate at which $\mathbf{f}$ varies as the variable $\mathbf{x}$ moves from $\mathbf{a}$ in the direction of the standard basis vector $\vec{e}_i$. It is natural to want to know how $\mathbf{f}$ varies when the variable moves in any direction $\vec{v}$:

Partial derivatives measure the rate at which $\mathbf{f}$ varies as the variable moves in the direction of the standard basis vectors. Directional derivatives measure the rate at which $\mathbf{f}$ varies when the variable moves in any direction.

Some authors consider that only vectors $\vec{v}$ of length 1 can be used in the definition of directional derivatives. We feel this is an undesirable restriction, as it loses the essential linear character of the directional derivative as a function of $\vec{v}$.

Definition 1.7.5 (Directional derivative). The directional derivative of $\mathbf{f}$ at $\mathbf{a}$ in the direction $\vec{v}$,

$$\lim_{h \to 0} \frac{\mathbf{f}(\mathbf{a} + h\vec{v}) - \mathbf{f}(\mathbf{a})}{h}, \qquad 1.7.12$$

measures the rate at which $\mathbf{f}$ varies when $\mathbf{x}$ moves from $\mathbf{a}$ in the direction $\vec{v}$.

Example 1.7.6 (Computing a directional derivative). Let us compute the derivative in the direction $\vec{v} = \begin{bmatrix} 1 \\ 2 \\ 1 \end{bmatrix}$ of the function $f\begin{pmatrix} x \\ y \\ z \end{pmatrix} = xy \sin z$, evaluated at the point $\mathbf{a} = \begin{pmatrix} 1 \\ 1 \\ \pi/2 \end{pmatrix}$. We have $h\vec{v} = h \begin{bmatrix} 1 \\ 2 \\ 1 \end{bmatrix} = \begin{bmatrix} h \\ 2h \\ h \end{bmatrix}$, so Equation 1.7.12 becomes

$$\lim_{h \to 0} \frac{1}{h} \overbrace{(1+h)(1+2h)\sin(\frac{\pi}{2}+h)}^{\mathbf{f}(\mathbf{a}+h\vec{v})} - \overbrace{(1 \cdot 1 \cdot \sin \frac{\pi}{2})}^{\mathbf{f}(\mathbf{a})}. \qquad 1.7.13$$

---

[19] $\overrightarrow{D_2}\mathbf{f}\begin{pmatrix} x \\ y \end{pmatrix} = \begin{bmatrix} x \\ \cos(x+y) \\ -2y \end{bmatrix}.$

[20] $\overrightarrow{D_1}\mathbf{f}\begin{pmatrix} a \\ b \end{pmatrix} = \begin{bmatrix} 2ab \\ 0 \end{bmatrix}; \quad \overrightarrow{D_2}\mathbf{f}\begin{pmatrix} a \\ b \end{pmatrix} = \begin{bmatrix} a^2 \\ -\sin b \end{bmatrix}.$

$\frac{\partial \mathbf{f}}{\partial x_1}\begin{pmatrix} a \\ b \end{pmatrix} = \begin{bmatrix} 2ab \\ 0 \end{bmatrix}; \quad \frac{\partial \mathbf{f}}{\partial x_2}\begin{pmatrix} a \\ b \end{pmatrix} = \begin{bmatrix} a^2 \\ -\sin b \end{bmatrix}.$

Using the formula $\sin(a+b) = \sin a \cos b + \cos a \sin b$, this becomes

$$\lim_{h \to 0} \frac{1}{h}\left(\left((1+h)(1+2h)\left(\sin\frac{\pi}{2}\overbrace{\cos h}^{=1} + \overbrace{\cos\frac{\pi}{2}}^{=0}\sin h\right)\right) - \sin\frac{\pi}{2}\right)$$

$$= \lim_{h \to 0} \frac{1}{h}\left(\left((1+3h+2h^2)(\cos h)\right) - 1\right) \qquad\qquad 1.7.14$$

$$= \lim_{h \to 0}\frac{1}{h}(\cos h - 1) + \lim_{h \to 0}\frac{1}{h}3h\cos h + \lim_{h \to 0}\frac{1}{h}2h^2\cos h = 0 + 3 + 0 = 3. \quad \triangle$$

The derivative in several variables

Often we will want to see how a system changes when *all* the variables are allowed to vary; we want to compute the whole derivative of the function. We will see that this derivative consists of a matrix, called the *Jacobian matrix*, whose entries are the partial derivatives of the function. We will also see that if a function is differentiable, we can extrapolate all its directional derivatives from the Jacobian matrix.

Definition 1.7.1 from first year calculus defines the derivative as the limit

$$\frac{\text{change in } f}{\text{change in } x}, \quad \text{i.e.,} \quad \frac{f(a+h) - f(a)}{h}, \qquad\qquad 1.7.15$$

as h (the increment to the variable x) approaches 0. This does not generalize well to higher dimensions. When f is a function of several variables, then an increment to the variable will be a vector, and we can't divide by vectors.

It is tempting just to divide by $|h|$, the length of h:

$$f'(\mathbf{a}) = \lim_{\vec{\mathbf{h}} \to 0} \frac{1}{|\vec{\mathbf{h}}|}\left(f(\mathbf{a}+\vec{\mathbf{h}}) - f(\mathbf{a})\right). \qquad\qquad 1.7.16$$

This would allow us to rewrite Definition 1.7.1 in higher dimensions, since we can divide by the length of a vector, which is a number. But this wouldn't work even in dimension 1, because the limit changes sign when h approaches 0 from the left and from the right. In higher dimensions it's much worse. All the different directions from which $\vec{\mathbf{h}}$ could approach 0 give different limits. By dividing by $|\vec{\mathbf{h}}|$ in Equation 1.7.16 we are canceling the magnitude but not the direction.

We will rewrite it in a form that does generalize well. This definition will emphasize the idea that a function f is differentiable at a point a if the increment Δf to the function is well approximated by a *linear function of the increment h to the variable*. This linear function is $f'(a)h$.

When we call $f'(a)h$ a linear function, we mean the linear function that takes the variable h and *multiplies* it by $f'(a)$—i.e., the function $h \mapsto f'(a)\, h$ (to be read, "h maps to $f'(a)\, h$"). Usually the derivative of f at a is *not* a linear function of a. If $f(x) = \sin x$ or $f(x) = x^3$, or just about anything except $f(x) = x^2$, then $f'(a)$ is not a linear function of a. But $h \mapsto f'(a)\, h$ *is* a linear function of h. For example, $h \mapsto (\sin x)h$ is a linear function of h, since $(\sin x)(h_1 + h_2) = (\sin x)h_1 + (\sin x)h_2$.

Note the difference between $\mapsto$ ("maps to") and $\to$ ("to"). The first has a "pusher."

Definition 1.7.7 (Alternate definition of the derivative). A function f is differentiable at a, with derivative m, if and only if

$$\lim_{h \to 0} \frac{1}{h}\left(\overbrace{(f(a+h) - f(a))}^{\Delta f} - \overbrace{(mh)}^{\substack{\text{linear function} \\ \text{of } \Delta x}} \right) = 0. \qquad 1.7.17$$

The letter Δ, named "delta," denotes "change in"; Δf is the change in the function; $\Delta x = h$ is the change in the variable x. The function mh that multiplies h by the derivative m is thus a linear function of the change in x.

We are taking the limit as $h \to 0$, so h is small, and dividing by it makes things big; the numerator—the difference between the increment to the function and the approximation of that increment—must be *very small* when h is near 0 for the limit to be zero anyway (see Exercise 1.7.11).

The following computation shows that Definition 1.7.7 is just a way of restating Definition 1.7.1:

$$\lim_{h \to 0} \frac{1}{h}\left((f(a+h) - f(a)) - [f'(a)]h \right) = \lim_{h \to 0} \left(\overbrace{\frac{f(a+h) - f(a)}{h}}^{f'(a) \text{ by Equation 1.7.2}} - \frac{f'(a)h}{h} \right)$$

$$= f'(a) - f'(a) = 0. \qquad 1.7.18$$

Moreover, the linear function $h \mapsto f'(a)h$ is the *only* linear function satisfying Equation 1.7.17. Indeed, any linear function of one variable can be written $h \mapsto mh$, and

$$0 = \lim_{h \to 0} \frac{1}{h}\left((f(a+h) - f(a)) - mh \right) = \lim_{h \to 0} \left(\frac{f(a+h) - f(a)}{h} - \frac{mh}{h} \right) = f'(a) - m,$$

$$1.7.19$$

so $f'(a) = m$.

The derivative in several variables: the Jacobian matrix

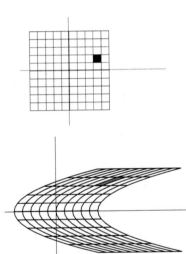

FIGURE 1.7.1.

The mapping

$$\mathbf{f}\begin{pmatrix} x \\ y \end{pmatrix} = \begin{pmatrix} x^2 + y \\ x \end{pmatrix}$$

takes the shaded square in the square at top to the shaded area at bottom.

The point of rewriting the definition of the derivative is that with Definition 1.7.7, we can divide by $|h|$ rather than h; $m = f'(a)$ is also the unique number such that

$$\lim_{h \to 0} \frac{1}{|h|}\left(\overbrace{(f(a+h) - f(a))}^{\Delta f} - \overbrace{(f'(a)h)}^{\text{linear function of } h} \right) = 0. \qquad 1.7.20$$

It doesn't matter if the limit changes sign, since the limit is 0; a number close to 0 is close to 0 whether it is positive or negative.

Therefore we can generalize Equation 1.7.20 to mappings in higher dimensions, like the one in Figure 1.7.1. As in the case of functions of one variable, the key to understanding derivatives of functions of several variables is to think that the increment to the function (the output) is *approximately a linear function* of the increment to the variable (the input): i.e., that the increment

$$\Delta \mathbf{f} = \mathbf{f}(\mathbf{a} + \vec{\mathbf{h}}) - \mathbf{f}(\mathbf{a}) \qquad 1.7.21$$

is approximately a linear function of the increment $\vec{\mathbf{h}}$.

In the one-dimensional case, Δf is well approximated by the linear function $h \mapsto f'(a)h$. We saw in Section 1.3 that every linear transformation is given by a matrix; the linear transformation $h \mapsto f'(a)h$ is given by multiplication by the 1×1 matrix $[f'(a)]$.

For a mapping from $\mathbb{R}^n \to \mathbb{R}^m$, the role of this 1×1 matrix is played by a $m \times n$ matrix composed of the partial derivatives of the mapping at $\mathbf{a}$. This matrix is called the *Jacobian matrix* of the mapping $\mathbf{f}$; we denote it $[\mathbf{Jf}(\mathbf{a})]$:

> **Definition 1.7.8 (Jacobian matrix).** The Jacobian matrix of a function $\mathbf{f} : \mathbb{R}^n \to \mathbb{R}^m$ is the $m \times n$ matrix composed of the partial derivatives of $\mathbf{f}$ evaluated at $\mathbf{a}$:
>
> $$[\mathbf{Jf}(\mathbf{a})] = \begin{bmatrix} D_1 f_1(\mathbf{a}) & \dots & D_n f_1(\mathbf{a}) \\ \vdots & & \vdots \\ D_1 f_m(\mathbf{a}) & \dots & D_n f_m(\mathbf{a}) \end{bmatrix}. \qquad 1.7.22$$

Example 1.7.9. The Jacobian matrix of the function in Example 1.7.4 is

$$\left[\mathbf{Jf}\begin{pmatrix} x \\ y \end{pmatrix}\right] = \begin{bmatrix} y & x \\ \cos(x+y) & \cos(x+y) \\ 2x & -2y \end{bmatrix}. \qquad 1.7.23$$

Note that in the Jacobian matrix we write the components of $\mathbf{f}$ from top to bottom, and the variables from left to right. The first column gives the partial derivatives with respect to the first variable; the second column gives the partial derivatives with respect to the second variable, and so on.

The first column of the Jacobian matrix gives $\overrightarrow{D_1\mathbf{f}}$, the partial derivative with respect to the first variable, x; the second column gives $\overrightarrow{D_2\mathbf{f}}$, the partial derivative with respect to the second variable, y. $\triangle$

What is the Jacobian matrix of the function $\mathbf{f}\begin{pmatrix} x \\ y \end{pmatrix} = \begin{pmatrix} x^3 y \\ 2x^2 y^2 \\ xy \end{pmatrix}$? Check your answer below.[21]

---

[21] $\left[\mathbf{Jf}\begin{pmatrix} x \\ y \end{pmatrix}\right] = \begin{bmatrix} 3x^2 y & x^3 \\ 4xy^2 & 4x^2 y \\ y & x \end{bmatrix}$. The first column is $\overrightarrow{D_1\mathbf{f}}$ (the partial derivatives with respect to the first variable); the second is $\overrightarrow{D_2\mathbf{f}}$. The first row gives the partial derivatives for $f\begin{pmatrix} x \\ y \end{pmatrix} = x^3 y$; the second row gives the partial derivatives for $f\begin{pmatrix} x \\ y \end{pmatrix} = 2x^2 y^2$, and the third gives the partial derivatives for $f\begin{pmatrix} x \\ y \end{pmatrix} = xy$.

When is the Jacobian matrix of a function its derivative?

We would like to say that if the Jacobian matrix exists, then it is the derivative of $\mathbf{f}$. That is, we would like to say that the increment $\mathbf{f}(\mathbf{a} + \vec{\mathbf{h}}) - \mathbf{f}(\mathbf{a})$ is approximately $[\mathbf{Jf}(\mathbf{a})]\vec{\mathbf{h}}$, in the sense that

$$\lim_{\vec{\mathbf{h}} \to 0} \frac{1}{|\vec{\mathbf{h}}|} \left((\mathbf{f}(\mathbf{a} + \vec{\mathbf{h}}) - \mathbf{f}(\mathbf{a})) - [\mathbf{J}f(\mathbf{a})]\vec{\mathbf{h}} \right) = 0. \qquad 1.7.24$$

The fact that the derivative of a function in several variables is represented by the Jacobian matrix is the reason why linear algebra is a prerequisite to multivariable calculus.

This is the higher-dimensional analog of Equation 1.7.17, which we proved in one dimension. Usually it is true in higher dimensions: you can calculate the derivative of a function with several variables by computing its partial derivatives, using techniques you already know, and putting them in a matrix.

We will examine the issue of such pathological functions in Section 1.9.

Unfortunately, it isn't always true: it is possible for all partial derivatives of a function $\mathbf{f}$ to exist, and yet for $\mathbf{f}$ not to be differentiable! The best we can do without extra hypotheses is the following statement.

Theorem 1.7.10 (The Jacobian matrix and the derivative). *If there is any linear transformation L such that*

$$\lim_{\vec{\mathbf{h}} \to 0} \frac{1}{|\vec{\mathbf{h}}|} \left((\mathbf{f}(\mathbf{a} + \vec{\mathbf{h}}) - \mathbf{f}(\mathbf{a})) - (L(\vec{\mathbf{h}})) \right) = 0, \qquad 1.7.25$$

then all partial derivatives of $\mathbf{f}$ at $\mathbf{a}$ exist, and the matrix representing L is $[\mathbf{Jf}(\mathbf{a})]$. In particular, such a linear transformation is unique.

Definition 1.7.11 (Derivative). If the linear transformation of Theorem 1.7.10 exists, $\mathbf{f}$ is differentiable at $\mathbf{a}$, and the linear transformation represented by $[\mathbf{Jf}(\mathbf{a})]$ is its derivative $[\mathbf{Df}(\mathbf{a})]$: the derivative of $\mathbf{f}$ at $\mathbf{a}$.

Remark. It is essential to remember that the derivative $[\mathbf{Df}(\mathbf{a})]$ is a *matrix* (in the case of a function $f : \mathbb{R} \to \mathbb{R}$, a 1×1 matrix, i.e., a number). It is convenient to write $[\mathbf{Df}(\mathbf{a})]$ rather than writing the Jacobian matrix in full:

$$[\mathbf{Df}(\mathbf{a})] = [\mathbf{Jf}(\mathbf{a})] = \begin{bmatrix} D_1 f_1(\mathbf{a}) & \dots & D_n f_1(\mathbf{a}) \\ \vdots & & \vdots \\ D_1 f_m(\mathbf{a}) & \dots & D_n f_m(\mathbf{a}) \end{bmatrix}. \qquad 1.7.26$$

But when you see $[\mathbf{Df}(\mathbf{a})]$, you should always be aware of its dimensions. Given a function $\mathbf{f} : \mathbb{R}^3 \to \mathbb{R}^2$, what are the dimensions of its derivative at $\mathbf{a}$, $[\mathbf{Df}(\mathbf{a})]$? Check your answer below.[22] $\triangle$

---

[22]Since $\mathbf{f} : \mathbb{R}^3 \to \mathbb{R}^2$ takes a point in $\mathbb{R}^3$ and gives a point in $\mathbb{R}^2$, similarly, $[\mathbf{Df}(\mathbf{a})]$ takes a vector in $\mathbb{R}^3$ and gives a vector in $\mathbb{R}^2$. Therefore $[\mathbf{Df}(\mathbf{a})]$ is a 2×3 matrix.

We will prove Theorem 1.7.10 after some further discussion of directional derivatives, and a couple of extended examples of the Jacobian matrix as derivative.

Extrapolating directional derivatives from partial derivatives

If a function is differentiable, we can extrapolate all its directional derivatives from its partial derivatives—i.e., from its derivative:

In Equation 1.7.27 for the directional derivative, we use the increment vector $h\vec{v}$ rather than $\vec{h}$ because we are measuring the derivative only in the direction of a particular vector $\vec{v}$.

Proposition 1.7.12 (Computing directional derivatives from the derivative). If $\mathbf{f}$ is differentiable at $\mathbf{a}$, then all directional derivatives of $\mathbf{f}$ at $\mathbf{a}$ exist; the directional derivative in the direction $\vec{v}$ is given by the formula

$$\lim_{h \to 0} \frac{\mathbf{f}(\mathbf{a} + h\vec{v}) - \mathbf{f}(\mathbf{a})}{h} = [\mathbf{Df}(\mathbf{a})]\vec{v}. \qquad 1.7.27$$

Example: You are standing at the origin on a hill with height

$$f\begin{pmatrix} x \\ y \end{pmatrix} = 3x + 8y.$$

When you step in direction $\vec{v} = \begin{bmatrix} 1 \\ 2 \end{bmatrix}$, your rate of ascent is

$$\left[Df\begin{pmatrix} 0 \\ 0 \end{pmatrix}\right]\vec{v} = [3, \ 8]\begin{bmatrix} 1 \\ 2 \end{bmatrix} = 19.$$

Example 1.7.13 (Computing a directional derivative from the Jacobian matrix). Let us use Proposition 1.7.12 to compute the directional derivative of Example 1.7.6. The partial derivatives of $f\begin{pmatrix} x \\ y \\ z \end{pmatrix} = xy\sin z$ are

$D_1 f = y\sin z, D_2 f = x\sin z$ and $D_3 f = xy\cos z$, so its derivative evaluated at the point $\begin{pmatrix} 1 \\ 1 \\ \pi/2 \end{pmatrix}$ is the one-row matrix $[1, \ 1, \ 0]$. (The commas may be misleading but omitting them might lead to confusion with multiplication.) Multiplying this by the vector $\vec{v} = \begin{bmatrix} 1 \\ 2 \\ 1 \end{bmatrix}$ does indeed give the answer 3, which is what we got before. $\triangle$

Proof of Proposition 1.7.12. The expression

$$\mathbf{r}(\vec{h}) = \big(\mathbf{f}(\mathbf{a} + \vec{h}) - \mathbf{f}(\mathbf{a})\big) - [\mathbf{Df}(\mathbf{a})]\vec{h} \qquad 1.7.28$$

defines the "remainder" $\mathbf{r}(\vec{h})$—the difference between the increment to the function and its linear approximation—as a function of the increment $\vec{h}$. The hypothesis that $\mathbf{f}$ is differentiable at $\mathbf{a}$ says that

$$\lim_{\vec{h} \to \vec{0}} \frac{\mathbf{r}(\vec{h})}{|\vec{h}|} = 0. \qquad 1.7.29$$

Substituting $h\vec{v}$ for $\vec{h}$ in Equation 1.7.28, we find

$$\mathbf{r}(h\vec{v}) = \mathbf{f}(\mathbf{a} + h\vec{v}) - \mathbf{f}(\mathbf{a}) - [\mathbf{Df}(\mathbf{a})]h\vec{v}, \qquad 1.7.30$$

and dividing by h gives

$$|\vec{\mathbf{v}}| \frac{\mathbf{r}(h\vec{\mathbf{v}})}{h|\vec{\mathbf{v}}|} = \frac{\mathbf{f}(\mathbf{a} + h\vec{\mathbf{v}}) - \mathbf{f}(\mathbf{a})}{h} - [\mathbf{Df}(\mathbf{a})]\vec{\mathbf{v}}, \qquad 1.7.31$$

To follow Equation 1.7.32, recall that for any linear transformation T, we have

$$T(\alpha\vec{\mathbf{v}}) = \alpha T(\vec{\mathbf{v}}).$$

The derivative $[\mathbf{Df}(\mathbf{a})]$ gives a linear transformation, so

$$[\mathbf{Df}(\mathbf{a})]h\vec{\mathbf{v}} = h[\mathbf{Df}(\mathbf{a})]\vec{\mathbf{v}}.$$

where we have used the linearity of the derivative to write

$$\frac{[\mathbf{Df}(\mathbf{a})]h\vec{\mathbf{v}}}{h} = \frac{h[\mathbf{Df}(\mathbf{a})]\vec{\mathbf{v}}}{h} = [\mathbf{Df}(\mathbf{a})]\vec{\mathbf{v}}. \qquad 1.7.32$$

The term

$$\frac{\mathbf{r}(h\vec{\mathbf{v}})}{h|\vec{\mathbf{v}}|} \qquad 1.7.33$$

on the left side of Equation 1.7.31 has limit 0 as $h \to 0$ by Equation 1.7.29, so

Once we know the partial derivatives of $\mathbf{f}$, which measure rate of change in the direction of the standard basis vectors, we can compute the derivatives in *any* direction.

This should not come as a surprise. As we saw in Example 1.3.15, the matrix for any linear transformation is formed by seeing what the transformation does to the standard basis vectors: The ith column of the matrix $[T]$ is $T(\vec{\mathbf{e}}_i)$,. One can then see what T does to any vector $\vec{\mathbf{v}}$ by multiplying $[T]\vec{\mathbf{v}}$. The Jacobian matrix is the matrix for the "rate of change" transformation, formed by seeing what that transformation does to the standard basis vectors.

$$\lim_{h \to 0} \frac{\mathbf{f}(\mathbf{a} + h\vec{\mathbf{v}}) - \mathbf{f}(\mathbf{a})}{h} - [\mathbf{Df}(\mathbf{a})]\vec{\mathbf{v}} = 0. \quad \square \qquad 1.7.34$$

Example 1.7.14 (The Jacobian matrix of a function $\mathbf{f} : \mathbb{R}^2 \to \mathbb{R}^2$). Let's see, for a fairly simple nonlinear mapping from $\mathbb{R}^2$ to $\mathbb{R}^2$, that the Jacobian matrix does indeed provide the desired approximation of the change in the mapping. The Jacobian matrix of the mapping

$$\mathbf{f}\begin{pmatrix} x \\ y \end{pmatrix} = \begin{pmatrix} xy \\ x^2 - y^2 \end{pmatrix} \quad \text{is} \quad \left[\mathbf{Jf}\begin{pmatrix} x \\ y \end{pmatrix}\right] = \begin{bmatrix} y & x \\ 2x & -2y \end{bmatrix}, \qquad 1.7.35$$

since the partial derivative of xy with regard to x is y, the partial derivative of xy with regard to y is x, and so on. Our increment vector will be $\begin{bmatrix} h \\ k \end{bmatrix}$.

Plugging this vector, and the Jacobian matrix of Equation 1.7.35, into Equation 1.7.24, we get

$$\lim_{\begin{bmatrix} h \\ k \end{bmatrix} \to \begin{bmatrix} 0 \\ 0 \end{bmatrix}} \frac{1}{\sqrt{h^2 + k^2}} \left(\left(\underbrace{\mathbf{f}\begin{pmatrix} a+h \\ b+k \end{pmatrix}}_{\mathbf{f}(\mathbf{a}+\vec{\mathbf{h}})} - \underbrace{\mathbf{f}\begin{pmatrix} a \\ b \end{pmatrix}}_{\mathbf{f}(\mathbf{a})} \right) - \left(\underbrace{\begin{bmatrix} b & a \\ 2a & -2b \end{bmatrix}}_{\text{Jacobian matrix}} \underbrace{\begin{bmatrix} h \\ k \end{bmatrix}}_{\vec{\mathbf{h}}} \right) \right) \stackrel{?}{=} 0.$$

$$1.7.36$$

The $\sqrt{h^2 + k^2}$ at left is $|\vec{\mathbf{h}}|$, the length of the increment vector (as defined in Equation 1.4.7). Evaluating $\mathbf{f}$ at $\begin{pmatrix} a+h \\ b+k \end{pmatrix}$ and at $\begin{pmatrix} a \\ b \end{pmatrix}$, we have

$$\lim_{\begin{bmatrix} h \\ k \end{bmatrix} \to \begin{bmatrix} 0 \\ 0 \end{bmatrix}} \frac{1}{\sqrt{h^2 + k^2}} \left(\left(\begin{pmatrix} (a+h)(b+k) \\ (a+h)^2 - (b+k)^2 \end{pmatrix} - \begin{pmatrix} ab \\ a^2 - b^2 \end{pmatrix} \right) - \begin{bmatrix} bh + ak \\ 2ah - 2bk \end{bmatrix} \right) \stackrel{?}{=} 0.$$

$$1.7.37$$

After some computations the left-hand side becomes

$$\lim_{\begin{bmatrix}h\\k\end{bmatrix}\to\begin{bmatrix}0\\0\end{bmatrix}} \frac{1}{\sqrt{h^2+k^2}}\left[\begin{matrix}ab+ak+bh+\underline{hk}-ab-bh-ak\\a^2+2ah+\underline{h^2}-b^2-2bk-\underline{k^2}-a^2+b^2-2ah+2bk\end{matrix}\right],$$

1.7.38

which looks forbidding. But all the terms of the vector cancel out except those that are underlined, giving us

$$\lim_{\begin{bmatrix}h\\k\end{bmatrix}\to\begin{bmatrix}0\\0\end{bmatrix}} \frac{1}{\sqrt{h^2+k^2}}\left[\begin{matrix}hk\\h^2-k^2\end{matrix}\right] \overset{?}{=} \begin{bmatrix}0\\0\end{bmatrix}.$$

1.7.39

For example, at $\begin{pmatrix}a\\b\end{pmatrix}=\begin{pmatrix}1\\1\end{pmatrix}$, the function

$$\mathbf{f}\begin{pmatrix}x\\y\end{pmatrix}=\begin{pmatrix}xy\\x^2-y^2\end{pmatrix}$$

of Equation 1.7.35 gives $f\begin{pmatrix}1\\1\end{pmatrix}=\begin{pmatrix}1\\0\end{pmatrix}$, and we are asking whether

$$f\begin{pmatrix}1\\1\end{pmatrix}+\begin{bmatrix}1&1\\2&-2\end{bmatrix}\begin{bmatrix}h\\k\end{bmatrix}$$
$$=\begin{pmatrix}1+h+k\\2h-2k\end{pmatrix}$$

is a good approximation to

$$f\begin{pmatrix}1+h\\1+k\end{pmatrix}$$
$$=\begin{pmatrix}1+h+k+hk\\2h-2k+h^2-k^2\end{pmatrix}.$$

(That is, we are asking whether the difference is smaller than linear. In this case, it clearly is: the first entry differs by the quadratic term hk, the second entry by the quadratic term h^2-k^2.)

Indeed, the hypotenuse of a triangle is longer than either of its other sides, $0\le|h|\le\sqrt{h^2+k^2}$ and $0\le|k|\le\sqrt{h^2+k^2}$, so

$$0\le\left|\frac{hk}{\sqrt{h^2+k^2}}\right|=\left|\frac{h}{\sqrt{h^2+k^2}}\right||k|\le|k|,$$

1.7.40

and we have

$$0\le\overbrace{\lim_{\begin{bmatrix}h\\k\end{bmatrix}\to\begin{bmatrix}0\\0\end{bmatrix}}\left|\frac{hk}{\sqrt{h^2+k^2}}\right|}^{\substack{\text{squeezed between}\\ \text{0 and 0}}}\le\lim_{\begin{bmatrix}h\\k\end{bmatrix}\to\begin{bmatrix}0\\0\end{bmatrix}}|k|=0.$$

1.7.41

Similarly,

$$0\le\left|\frac{h^2-k^2}{\sqrt{h^2+k^2}}\right|\le|h|\left|\frac{h}{\sqrt{h^2+k^2}}\right|+|k|\left|\frac{k}{\sqrt{h^2+k^2}}\right|\le|h|+|k|,$$

1.7.42

so

$$0\le\overbrace{\lim_{\begin{bmatrix}h\\k\end{bmatrix}\to\begin{bmatrix}0\\0\end{bmatrix}}\left|\frac{h^2-k^2}{\sqrt{h^2+k^2}}\right|}^{\substack{\text{squeezed between}\\ \text{0 and 0}}}\le\lim_{\begin{bmatrix}h\\k\end{bmatrix}\to\begin{bmatrix}0\\0\end{bmatrix}}(|h|+|k|)=0+0=0.$$

1.7.43

When we speak of $\mathrm{Mat}\,(n,n)$ as the space of $n \times n$ matrices we mean that we can identify an $n \times n$ matrix with an element of $\mathbb{R}^{n^2}$. In Section 2.6 we will see that $\mathrm{Mat}\,(n,n)$ is an example of an abstract vector space, and we will be more precise about what it means to identify such a space with an appropriate $\mathbb{R}^N$.

If you wonder how we found the result of Equation 1.7.45, look at the comment accompanying Equation 1.7.48.

We could express the derivative of the function $f(x) = x^2$ as $f'(x) : h \mapsto 2xh$.

Example 1.7.15 (The derivative of a matrix squared). In most serious calculus texts, the first example of a derivative is that if $f(x) = x^2$, then $f'(x) = 2x$, as shown in Equation 1.7.3. Let us compute the same thing when a matrix, not a number, is being squared. This could be written as a function $\mathbb{R}^{n^2} \to \mathbb{R}^{n^2}$, and you are asked to spell this out for $n = 2$ and $n = 3$ in Exercise 1.7.15. But the expression that you get is very unwieldy as soon as $n > 2$, as you will see if you try to solve the exercise. This is one time when a linear transformation is easier to deal with than the corresponding matrix. It is much easier to denote by $\mathrm{Mat}\,(n,n)$ the space of $n \times n$ matrices, and to consider the mapping $S : \mathrm{Mat}\,(n,n) \to \mathrm{Mat}\,(n,n)$ given by

$$S(A) = A^2. \qquad 1.7.44$$

(The S stands for "square.")

In this case we can compute the derivative without computing the Jacobian matrix. We shall see that S is differentiable and that its derivative $[\mathbf{D}S(A)]$ is the linear transformation that maps H to $AH + HA$:

$$[\mathbf{D}S(A)]H = AH + HA, \text{ also written } [\mathbf{D}S(A)] : H \mapsto AH + HA. \qquad 1.7.45$$

Since the increment is a matrix, we denote it H. Note that if matrix multiplication were commutative, we could denote this derivative $2AH$ or $2HA$—very much like the derivative $f' = 2x$ for the function $f(x) = x^2$.

To make sense of Equation 1.7.45, a first thing to realize is that the map

$$[\mathbf{D}S(A)] : \mathrm{Mat}\,(n,n) \to \mathrm{Mat}\,(n,n)\,, \quad H \mapsto AH + HA \qquad 1.7.46$$

is a linear transformation. Exercise 2.6.4 asks you to check this, along with some extensions.

Now, how do we prove Equation 1.7.45?
Well, the assertion is that

$$\lim_{H \to 0} \frac{1}{|H|} \Big| \big(S(A + H) - S(A)\big) \underbrace{-\ \ (AH + HA)}_{\substack{\text{linear function of} \\ \text{increment to variable}}} \Big| = 0. \qquad 1.7.47$$

$$\underbrace{}_{\substack{\text{increment} \\ \text{to mapping}}}$$

Equation 1.7.48 shows that

$$AH + HA$$

is exactly the linear terms in H of the increment to the function, so that subtracting them leaves only higher degree terms; i.e., $AH+HA$ is the derivative.

Since $S(A) = A^2$,

$$|S(A + H) - S(A) - (AH + HA)| = |(A + H)^2 - A^2 - AH - HA|$$

$$= | A^2 + AH + HA + H^2 - A^2 - AH - HA |$$

$$= |H^2|. \qquad 1.7.48$$

So the object is to show that

$$\lim_{H \to 0} \frac{|H^2|}{|H|} = 0. \qquad 1.7.49$$

Since $|H^2| \leq |H|^2$ (by Proposition 1.4.11), this is true.   △

Exercise 1.7.18 asks you to prove that the derivative $AH + HA$ is the "same" as the Jacobian matrix computed with partial derivatives, for 2×2 matrices. Much of the difficulty is in understanding S as a mapping from $\mathbb{R}^4 \to \mathbb{R}^4$.

Here is another example of the same kind of thing. Recall that if $f(x) = 1/x$, then $f'(a) = -1/a^2$. Proposition 1.7.16 generalizes this to matrices.

> **Exercise 1.7.20** asks you to compute the Jacobian matrix and verify Proposition 1.7.16 in the case of 2×2 matrices. It should be clear from the exercise that using this approach even for 3×3 matrices would be extremely unpleasant.

Proposition 1.7.16. *The set of invertible matrices is open in* $\mathrm{Mat}\,(n, n)$, *and if* $f(A) = A^{-1}$, *then* f *is differentiable, and*

$$[\mathbf{D}f(A)]H = -A^{-1}HA^{-1}. \qquad 1.7.50$$

Note the interesting way in which this reduces to $f'(a)h = -h/a^2$ in one dimension.

Proof. (Optional) We proved that the set of invertible matrices is open in Corollary 1.5.33. Now we need to show that

$$\lim_{H \to 0} \frac{1}{|H|}\left(\underbrace{(A+H)^{-1} - A^{-1}}_{\text{increment to mapping}} - \underbrace{-A^{-1}HA^{-1}}_{\text{linear function of } H} \right) = 0. \qquad 1.7.51$$

Our strategy (as in the proof of Corollary 1.5.33) will be to use Proposition 1.5.31, which says that if B is a square matrix such that $|B| < 1$, then the series $I + B + B^2 + \cdots$ converges to $(I - B)^{-1}$. (We restated the proposition here changing the A's to B's to avoid confusion with our current A's.) We also use Proposition 1.2.15 concerning the inverse of a product of invertible matrices. This gives the following computation:

$$
\begin{aligned}
(A+H)^{-1} = \big(A(I + A^{-1}H)\big)^{-1} &\overset{\text{Prop.\,1.2.15}}{=} (I + A^{-1}H)^{-1}A^{-1} \\
&= \underbrace{\big(I - (-A^{-1}H)\big)^{-1}}_{\text{sum of series in line below}} A^{-1}
\end{aligned}
$$

> Since $H \to 0$ in Equation 1.7.51, we can assume that $|A^{-1}H| < 1$, so treating $(I + A^{-1}H)^{-1}$ as the sum of the series is justified.

$$= \underbrace{\Big(I + (-A^{-1}H) + (-A^{-1}H)^2 + \cdots\Big)}_{\text{series } I + B + B^2 + \ldots,\ \text{where } B = -A^{-1}H} A^{-1} \qquad 1.7.52$$

(Now we consider the first term, second terms, and remaining terms:)

$$= \underbrace{A^{-1}}_{\text{1st}} \underbrace{- A^{-1}HA^{-1}}_{\text{2nd}} + \underbrace{\big((-A^{-1}H)^2 + (-A^{-1}H)^3 + \ldots\big)A^{-1}}_{\text{others}}$$

It may not be immediately obvious why we did the computations above. The point is that subtracting $(A^{-1} - A^{-1}HA^{-1})$ from both sides of Equation 1.7.52

gives us, on the left, the quantity that really interests us: the difference between the increment of the function and its approximation by the linear function of the increment to the variable:

Switching from matrices to lengths of matrices in Equation 1.7.54 has several important consequences. First, it allows us, via Proposition 1.4.11, to establish an inequality. Next, it explains why we could multiply the $|A^{-1}|^2$ and $|A^{-1}|$ to get $|A^{-1}|^3$: matrix multiplication isn't commutative, but multiplication of matrix lengths is, since the length of a matrix is a number. Finally, it explains why the sum of the series is a fraction rather than a matrix inverse.

$$\underbrace{\left((A+H)^{-1}-A^{-1}\right)}_{\text{increment to function}} - \underbrace{\left(-A^{-1}HA^{-1}\right)}_{\substack{\text{linear function of}\\\text{increment to variable}}}$$

$$= \left((-A^{-1}H)^2 + (-A^{-1}H)^3 + \cdots\right)A^{-1} \qquad 1.7.53$$

$$= (A^{-1}H)^2\left(1 + (-A^{-1}H) + (-A^{-1}H)^2 + \cdots\right)A^{-1}.$$

Now applying Proposition 1.4.11 to the right-hand side gives us

$$\left|(A+H)^{-1}-A^{-1}+A^{-1}HA^{-1}\right|$$

$$\leq |A^{-1}H|^2\left|1 + (-A^{-1}H) + (-A^{-1}H)^2 + \cdots\right||A^{-1}|,$$

and the triangle inequality gives

$$\qquad\qquad\qquad\qquad\qquad\qquad\overbrace{\qquad\qquad\qquad\qquad\qquad}^{\text{convergent geometric series}} \qquad 1.7.54$$

$$\leq |A^{-1}H|^2|A^{-1}|\left(1 + |-A^{-1}H| + |-A^{-1}H|^2 + \cdots\right)$$

$$\leq |H|^2\,|A^{-1}|^3\,\frac{1}{1-|A^{-1}H|}.$$

Recall (Exercise 1.5.3) that the triangle inequality applies to convergent infinite sums.

Now suppose H so small that $|A^{-1}H| < 1/2$, so that

$$\frac{1}{1-|A^{-1}H|} \leq 2. \qquad 1.7.55$$

We see that

$$\lim_{H\to 0}\frac{1}{|H|}\left|(A+H)^{-1}-A^{-1}+A^{-1}HA^{-1}\right| \leq \lim_{H\to 0} 2|H||A^{-1}|^3 = 0.$$

$$\square \quad 1.7.56$$

Proving Theorem 1.7.10 about the Jacobian matrix

Now it's time that we proved Theorem 1.7.10. We restate it here:

Theorem 1.7.10 (The Jacobian matrix as derivative). *If there is any linear transformation L such that*

$$\lim_{\vec{\mathbf{h}}\to\mathbf{0}}\frac{\left(\mathbf{f}(\mathbf{a}+\vec{\mathbf{h}})-\mathbf{f}(\mathbf{a})\right)-L(\vec{\mathbf{h}})}{|\vec{\mathbf{h}}|} = 0, \qquad (1.7.25)$$

then all partial derivatives of $\mathbf{f}$ at $\mathbf{a}$ exist, and the matrix representing L is $[\mathbf{Jf}(\mathbf{a})]$.

Proof. We know (Theorem 1.3.14) that the linear transformation L is represented by the matrix whose ith column is $L\vec{e}_i$, so we need to show that

$$L\vec{e}_i = \overrightarrow{D_i}\mathbf{f}, \qquad 1.7.57$$

$\overrightarrow{D_i}\mathbf{f}$ being by definition the ith column of the Jacobian matrix $[\mathbf{Jf(a)}]$.

Equation 1.7.25 is true for any vector $\vec{\mathbf{h}}$, including $t\vec{e}_i$, where t is a number:

$$\lim_{t\vec{e}_i \to \mathbf{0}} \frac{\big(\mathbf{f}(\mathbf{a} + t\vec{e}_i) - f(\mathbf{a})\big) - L(t\vec{e}_i)}{|t\vec{e}_i|} = 0. \qquad 1.7.58$$

We want to get rid of the absolute value signs in the denominator. Since $|t\vec{e}_i| = |t||\vec{e}_i|$ (remember t is a number) and $|\vec{e}_i| = 1$, we can replace $|t\vec{e}_i|$ by $|t|$. The limit in Equation 1.7.58 is 0 for t small, whether t is positive or negative, so we can replace $|t|$ by t:

$$\lim_{t\vec{e}_i \to \mathbf{0}} \frac{\mathbf{f}(\mathbf{a} + t\vec{e}_i) - f(\mathbf{a}) - L(t\vec{e}_i)}{t} = 0. \qquad 1.7.59$$

Using the linearity of the derivative, we see that

$$L(t\vec{e}_i) = tL(\vec{e}_i), \qquad 1.7.60$$

so we can rewrite Equation 1.7.59 as

$$\lim_{t\vec{e}_i \to \mathbf{0}} \frac{\mathbf{f}(\mathbf{a} + t\vec{e}_i) - \mathbf{f}(\mathbf{a})}{t} - L(\vec{e}_i) = 0. \qquad 1.7.61$$

The first term is precisely Definition 1.7.5 of the partial derivative. So $L(\vec{e}_i) = \overrightarrow{D_i}\mathbf{f(a)}$: the ith column of the matrix corresponding to the linear transformation L is indeed $\overrightarrow{D_i}\mathbf{f}$. In other words, the matrix corresponding to L is the Jacobian matrix. $\square$

This proof proves that if there is a derivative, it is unique, since a linear transformation has just one matrix.

1.8 Rules for Computing Derivatives

In Theorem 1.8.1, we considered writing $\mathbf{f}$ and $\mathbf{g}$ in as $\vec{f}$ and $\vec{g}$, since some of the computations only make sense for vectors: for example, the dot product $(\mathbf{f} \cdot \mathbf{g})(x)$. We did not do so partly to avoid heavy notation, but also because these rules often are applied in computations where one is not thinking in terms of points or vectors. In practice, you can go ahead and compute without worrying about the distinction.

In this section we state rules for computing derivatives. Some are grouped in Theorem 1.8.1 below; the chain rule is discussed separately, in Theorem 1.8.2. These rules allow you to differentiate any function that is given by a formula.

Theorem 1.8.1 (Rules for computing derivatives). *Let $U \subset \mathbb{R}^n$ be an open set.*

(1) If $\mathbf{f} : U \to \mathbb{R}^m$ is a constant function, then $\mathbf{f}$ is differentiable, and its derivative is $[\mathbf{0}]$ (i.e., it is the zero linear transformation $\mathbb{R}^n \to \mathbb{R}^m$, represented by the $m \times n$ matrix filled with zeroes)

(2) If $\mathbf{f} : \mathbb{R}^n \to \mathbb{R}^m$ is linear, then it is differentiable everywhere, and its derivative at all points $\mathbf{a}$ is $\mathbf{f}$:

$$[\mathbf{Df(a)}]\vec{\mathbf{v}} = \mathbf{f}(\vec{\mathbf{v}}). \qquad 1.8.1$$

Note that the terms on the right of Equation 1.8.4 belong to the indicated spaces, and therefore the whole expression makes sense; it is the sum of two vectors in $\mathbb{R}^m$, each of which is the product of a vector in $\mathbb{R}^m$ and a number. Note that $[\mathbf{D}f(\mathbf{a})]\vec{v}$ is the product of a line matrix and a vector, hence it is a number.

The expression $\mathbf{f}, \mathbf{g} : U \to \mathbb{R}^m$ in (4) and (6) is shorthand for $\mathbf{f} : U \to \mathbb{R}^m$ and $\mathbf{g} : U \to \mathbb{R}^m$. We discussed in Remark 1.5.1 the importance of limiting the domain to an open subset.

(5) Example of $f\mathbf{g}$: if

$$f(x) = x^2 \text{ and } \mathbf{g}(x) = \begin{pmatrix} \sin x \\ \cos x \end{pmatrix},$$

then $f\mathbf{g}(x) = \begin{pmatrix} x^2 \sin x \\ x^2 \cos x \end{pmatrix}.$

(6) Example of $\mathbf{f} \cdot \mathbf{g}$: if

$$\mathbf{f}(x) = \begin{pmatrix} x \\ x^2 \end{pmatrix}, \mathbf{g}(x) = \begin{pmatrix} \sin x \\ \cos x \end{pmatrix},$$

then their dot product is

$$(\mathbf{f} \cdot \mathbf{g})(x) = x \sin x + x^2 \cos x.$$

(3)   If $f_1, \ldots, f_m : U \to \mathbb{R}$ are m scalar-valued functions differentiable at $\mathbf{a}$, then the vector-valued mapping

$$\mathbf{f} = \begin{pmatrix} f_1 \\ \vdots \\ f_m \end{pmatrix} : U \to \mathbb{R}^m \quad \text{is differentiable at } \mathbf{a}, \text{ with derivative}$$

$$[\mathbf{D}\mathbf{f}(\mathbf{a})]\vec{v} = \begin{bmatrix} [\mathbf{D}f_1(\mathbf{a})]\vec{v} \\ \vdots \\ [\mathbf{D}f_m(\mathbf{a})]\vec{v} \end{bmatrix}. \qquad 1.8.2$$

(4)   If $\mathbf{f}, \mathbf{g} : U \to \mathbb{R}^m$ are differentiable at $\mathbf{a}$, then so is $\mathbf{f} + \mathbf{g}$, and

$$[\mathbf{D}(\mathbf{f} + \mathbf{g})(\mathbf{a})] = [\mathbf{D}\mathbf{f}(\mathbf{a})] + [\mathbf{D}\mathbf{g}(\mathbf{a})]. \qquad 1.8.3$$

(5)   If $f : U \to \mathbb{R}$ and $\mathbf{g} : U \to \mathbb{R}^m$ are differentiable at $\mathbf{a}$, then so is $f\mathbf{g}$, and the derivative is given by

$$[\mathbf{D}(f\mathbf{g})(\mathbf{a})]\vec{v} = \underbrace{f(\mathbf{a})}_{\mathbb{R}}\underbrace{[\mathbf{D}\mathbf{g}(\mathbf{a})]\vec{v}}_{\mathbb{R}^m} + \underbrace{([\mathbf{D}f(\mathbf{a})]\vec{v})}_{\mathbb{R}}\underbrace{\mathbf{g}(\mathbf{a})}_{\mathbb{R}^m}. \qquad 1.8.4$$

(6)   If $\mathbf{f}, \mathbf{g} : U \to \mathbb{R}^m$ are both differentiable at $\mathbf{a}$, then so is the dot product $\mathbf{f} \cdot \mathbf{g} : U \to \mathbb{R}$, and (as in one dimension)

$$[\mathbf{D}(\mathbf{f} \cdot \mathbf{g})(\mathbf{a})]\vec{v} = \underbrace{[\mathbf{D}\mathbf{f}(\mathbf{a})]\vec{v}}_{\mathbb{R}^m} \cdot \underbrace{\mathbf{g}(\mathbf{a})}_{\mathbb{R}^m} + \underbrace{\mathbf{f}(\mathbf{a})}_{\mathbb{R}^m} \cdot \underbrace{[\mathbf{D}\mathbf{g}(\mathbf{a})]\vec{v}}_{\mathbb{R}^m}. \qquad 1.8.5$$

As shown in the proof below, the rules are either immediate, or they are straightforward applications of the corresponding one-dimensional statements. However, we hesitate to call them (or any other proof) easy; when we are struggling to learn a new piece on the piano, we do not enjoy seeing that it has been labeled an "easy piece for beginners."

Proof of 1.8.1. (1) If $\mathbf{f}$ is a constant function, then $\mathbf{f}(\mathbf{a} + \vec{h}) = \mathbf{f}(\mathbf{a})$, so the derivative $[\mathbf{D}\mathbf{f}(\mathbf{a})]$ is the zero linear transformation:

$$\lim_{\vec{h} \to 0} \frac{1}{|\vec{h}|} \Big(\mathbf{f}(\mathbf{a} + \vec{h}) - \mathbf{f}(\mathbf{a}) - \underbrace{\vec{0}\,\vec{h}}_{([\mathbf{D}\mathbf{f}(\mathbf{a})]\vec{h}} \Big) = \lim_{h \to 0} \frac{1}{|\vec{h}|}\vec{0} = \vec{0}. \qquad 1.8.6$$

(2) Suppose $\mathbf{f}(\mathbf{a})$ is linear. Then $[\mathbf{D}\mathbf{f}(\mathbf{a})] = \mathbf{f}$:

$$\lim_{\vec{h} \to 0} \frac{1}{|\vec{h}|}\Big(\mathbf{f}(\mathbf{a} + \vec{h}) - \mathbf{f}(\mathbf{a}) - \mathbf{f}(\vec{h}) \Big) = 0, \qquad 1.8.7$$

since $\mathbf{f}(\mathbf{a} + \vec{h}) = \mathbf{f}(\mathbf{a}) + \mathbf{f}(\vec{h})$.

(3) Just write everything out:

$$\lim_{\vec{\mathbf{h}}\to 0}\frac{1}{|\vec{\mathbf{h}}|}\left(\begin{pmatrix} f_1(\mathbf{a}+\vec{\mathbf{h}}) \\ \vdots \\ f_m(\mathbf{a}+\vec{\mathbf{h}}) \end{pmatrix} - \begin{pmatrix} f_1(\mathbf{a}) \\ \vdots \\ f_m(\mathbf{a}) \end{pmatrix} - \begin{bmatrix} [\mathbf{D}f_1(\mathbf{a})]\vec{\mathbf{h}} \\ \vdots \\ [\mathbf{D}f_m(\mathbf{a})]\vec{\mathbf{h}} \end{bmatrix}\right)$$

$$= \begin{bmatrix} \lim_{\vec{\mathbf{h}}\to 0}\frac{1}{|\vec{\mathbf{h}}|}\big(f_1(\mathbf{a}+\vec{\mathbf{h}})-f_1(\mathbf{a})-[\mathbf{D}f_1(\mathbf{a})]\vec{\mathbf{h}}\big) \\ \vdots \\ \lim_{\vec{\mathbf{h}}\to 0}\frac{1}{|\vec{\mathbf{h}}|}\big(f_m(\mathbf{a}+\vec{\mathbf{h}})-f_m(\mathbf{a})-[\mathbf{D}f_m(\mathbf{a})]\vec{\mathbf{h}}\big) \end{bmatrix} = \mathbf{0}. \qquad 1.8.8$$

According to a contemporary, the French mathematician Laplace (1749–1827) used the formula *il est aisé à voir* ("it's easy to see") whenever he himself couldn't remember the details of how his reasoning went, but was sure his conclusions were correct. "I never come across one of Laplace's '*Thus it plainly appears*' without feeling sure that I have hours of hard work before me to fill up the chasm and find out and show *how* it plainly appears," wrote the nineteenth century mathematician N. Bowditch.

(4) Functions are added point by point, so we can separate out $\mathbf{f}$ and $\mathbf{g}$:

$$(\mathbf{f}+\mathbf{g})(\mathbf{a}+\vec{\mathbf{h}})-(\mathbf{f}+\mathbf{g})(\mathbf{a})-\big([\mathbf{D}f(\mathbf{a})]+[\mathbf{D}g(\mathbf{a})]\big)\vec{\mathbf{h}} \qquad 1.8.9$$
$$= \big(\mathbf{f}(\mathbf{a}+\vec{\mathbf{h}})-\mathbf{f}(\mathbf{a})-[\mathbf{D}f(\mathbf{a})]\mathbf{h}\big)+\big(\mathbf{g}(\mathbf{a}+\vec{\mathbf{h}})-\mathbf{g}(\mathbf{a})-[\mathbf{D}g(\mathbf{a})]\vec{\mathbf{h}}\big).$$

Now divide by $|\vec{\mathbf{h}}|$, and take the limit as $|\vec{\mathbf{h}}|\to 0$. The right-hand side gives $0+0=0$, so the left-hand side does too.

(5) By part (3), we may assume that $m=1$, i.e., that $\mathbf{g}=g$ is scalar valued. Then

$$[\mathbf{D}fg(\mathbf{a})]\vec{\mathbf{h}} = \overbrace{[(D_1fg)(\mathbf{a}),\ldots,(D_nfg)(\mathbf{a})]}^{\text{Jacobian matrix of }fg}\vec{\mathbf{h}} \qquad 1.8.10$$

$$= \underbrace{[f(\mathbf{a})(D_1g)(\mathbf{a})+(D_1f)(\mathbf{a})g(\mathbf{a}),\ldots,f(\mathbf{a})(D_ng)(\mathbf{a})+(D_nf)(\mathbf{a})g(\mathbf{a})]}_{\text{in one variable, }(fg)'=fg'+f'g}\vec{\mathbf{h}}$$

$$= f(\mathbf{a})\underbrace{[(D_1g)(\mathbf{a}),\ldots,(D_ng)(\mathbf{a})]}_{\text{Jacobian matrix of }g}\vec{\mathbf{h}}+\underbrace{[(D_1f)(\mathbf{a}),\ldots,(D_nf)(\mathbf{a})]}_{\text{Jacobian matrix of }f}g(\mathbf{a})\vec{\mathbf{h}}$$

$$= f(\mathbf{a})\big([\mathbf{D}g(\mathbf{a})]\vec{\mathbf{h}}\big)+\big([\mathbf{D}f(\mathbf{a})]\vec{\mathbf{h}}\big)g(\mathbf{a}).$$

(6) Again, write everything out:

$$[\mathbf{D}(\mathbf{f}\cdot\mathbf{g})(\mathbf{a})]\vec{\mathbf{h}} \stackrel{\substack{\text{def. of}\\ \text{dot prod.}}}{=} \left[\mathbf{D}\left(\sum f_ig_i\right)(\mathbf{a})\right]\vec{\mathbf{h}} \stackrel{(4)}{=} \sum_{i=1}^{n}[\mathbf{D}(f_ig_i)(\mathbf{a})]\vec{\mathbf{h}}$$

$$\stackrel{(5)}{=} \sum_{i=1}^{n}\big([\mathbf{D}f_i(\mathbf{a})]\vec{\mathbf{h}}\big)g_i(\mathbf{a})+f_i(\mathbf{a})\big([\mathbf{D}g_i(\mathbf{a})]\vec{\mathbf{h}}\big) \qquad 1.8.11$$

$$\stackrel{\substack{\text{def. of}\\ \text{dot prod.}}}{=} \big([\mathbf{D}\mathbf{f}(\mathbf{a})]\vec{\mathbf{h}}\big)\cdot\mathbf{g}(\mathbf{a})\quad+\quad \mathbf{f}(\mathbf{a})\cdot\big([\mathbf{D}\mathbf{g}(\mathbf{a})]\vec{\mathbf{h}}\big).$$

The second equality uses rule (4) above: $\mathbf{f}\cdot\mathbf{g}$ is the sum of the f_ig_i, so the derivative of the sum is the sum of the derivatives. The third equality uses rule (5). A more conceptual proof of (5) and (6) is sketched in Exercise 1.8.1. $\square$

The chain rule

One rule for differentiation is so fundamental that it deserves a subsection of its own: the *chain rule*, which states that *the derivative of a composition is the composition of the derivatives*, as shown in Figure 1.8.1.

Some physicists claim that the chain rule is the most important theorem in all of all mathematics.

The chain rule is proved in Appendix A.1.

Theorem 1.8.2 (Chain rule).   *Let $U \subset \mathbb{R}^n, V \subset \mathbb{R}^m$ be open sets, let $\mathbf{g} : U \to V$ and $\mathbf{f} : V \to \mathbb{R}^p$ be mappings and let $\mathbf{a}$ be a point of U. If $\mathbf{g}$ is differentiable at $\mathbf{a}$ and $\mathbf{f}$ is differentiable at $\mathbf{g(a)}$, then the composition $\mathbf{f} \circ \mathbf{g}$ is differentiable at $\mathbf{a}$, and its derivative is given by*

$$[\mathbf{D(f \circ g)(a)}] = [\mathbf{Df(g(a))}] \circ [\mathbf{Dg(a)}]. \qquad 1.8.12$$

In practice, when we use the chain rule, most often these linear transformations will be represented by their matrices, and we will compute the right-hand side of Equation 1.8.12 by multiplying the matrices together:

$$[\mathbf{D(f \circ g)(a)}] = [\mathbf{Df(g(a))}][\mathbf{Dg(a)}]. \qquad 1.8.13$$

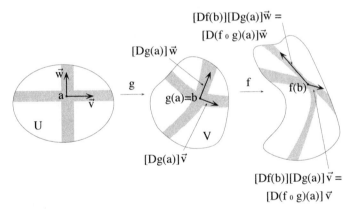

FIGURE 1.8.1. The function $\mathbf{g}$ maps a point $\mathbf{a} \in U$ to a point $\mathbf{g(a)} \in V$. The function $\mathbf{f}$ maps the point $\mathbf{g(a)} = \mathbf{b}$ to the point $\mathbf{f(b)}$. The derivative of $\mathbf{g}$ maps the vector $\vec{v}$ to $[\mathbf{Dg(a)}](\vec{v})$. The derivative of $\mathbf{f} \circ \mathbf{g}$ maps $\vec{v}$ to $[\mathbf{Df(b)}][\mathbf{Dg(a)}]\vec{v}$, i.e., to $[\mathbf{Df(g(a))}][\mathbf{Dg(a)}]\vec{v}$.

Remark.   One motivation for discussing matrices, matrix multiplication, linear transformations and the relation of composition of linear transformations to matrix multiplication at the beginning of this chapter was to have these tools available now. In coordinates, and using matrix multiplication, the chain rule states that

$$\overrightarrow{D_j}(\mathbf{f} \circ \mathbf{g})_i(\mathbf{a}) = \sum_{k=1}^{m} D_k f_i(\mathbf{g(a)}) D_j g_k(\mathbf{a}). \qquad 1.8.14$$

We will need this form of the chain rule often, but as a statement, it is a disaster: it makes a fundamental and transparent statement into a messy formula, the proof of which seems to be a computational miracle. $\triangle$

Example 1.8.3 (The derivative of a composition). Suppose $\mathbf{g} : \mathbb{R} \to \mathbb{R}^3$ and $f : \mathbb{R}^3 \to \mathbb{R}$ are the functions

In Example 1.8.3, $\mathbb{R}^3$ plays the role of V in Theorem 1.8.2.

$$f\begin{pmatrix} x \\ y \\ z \end{pmatrix} = x^2 + y^2 + z^2; \qquad \mathbf{g}(t) = \begin{pmatrix} t \\ t^2 \\ t^3 \end{pmatrix}. \qquad 1.8.15$$

The derivatives (Jacobian matrices) of these functions are computed by computing separately the partial derivatives, giving, for f,

You can see why the range of $\mathbf{g}$ and the domain of f must be the same (i.e., V in the theorem; $\mathbb{R}^3$ in this example): the width of $[\mathbf{D}f(\mathbf{g}(t))]$ must equal the height of $[\mathbf{D}\mathbf{g}(t)]$ for the multiplication to be possible.

$$\left[\mathbf{D}f\begin{pmatrix} x \\ y \\ z \end{pmatrix} \right] = [2x, 2y, 2z]. \qquad 1.8.16$$

(The derivative of f is a one-row matrix.) The derivative of f evaluated at $\mathbf{g}(t)$ is thus $[2t, 2t^2, 2t^3]$. The derivative of $\mathbf{g}$ at t is

$$[\mathbf{D}\mathbf{g}(t)] = \begin{bmatrix} 1 \\ 2t \\ 3t^2 \end{bmatrix}. \qquad 1.8.17$$

So the derivative at t of the composition $f \circ \mathbf{g}$ is

$$[\mathbf{D}(f \circ \mathbf{g})(t)] = [\mathbf{D}f(\mathbf{g}(t))] \circ [\mathbf{D}\mathbf{g}(t)] = \underbrace{[2t, 2t^2, 2t^3]}_{[\mathbf{D}f(\mathbf{g}(t))]} \underbrace{\begin{bmatrix} 1 \\ 2t \\ 3t^2 \end{bmatrix}}_{[\mathbf{D}\mathbf{g}(t)]} = 2t + 4t^3 + 6t^5.$$

$$\triangle \quad 1.8.18$$

Equation 1.7.45 says that the derivative of the "squaring function" f is

$$[\mathbf{D}f(A)]H = AH + HA.$$

In the second line of Equation 1.8.19, $g(A) = A^{-1}$ plays the role of A above, and $-A^{-1}HA^{-1}$ plays the role of H.

Notice the interesting way this result is related to the one-variable computation: if $f(x) = x^{-2}$, then $f'(x) = -2x^{-3}$. Notice also how much easier this computation is, using the chain rule, than the proof of Proposition 1.7.16, without the chain rule.

Example 1.8.4 (Composition of linear transformations). Here is a case where it is easier to think of the derivative as a linear transformation than as a matrix, and of the chain rule as speaking of a composition of linear transformations rather than a product of matrices. If A and H are $n \times n$ matrices, and $f(A) = A^2, g(A) = A^{-1}$, then $(f \circ g)(A) = A^{-2}$. To compute the derivative of $f \circ g$ we use the chain rule in the first line, Proposition 1.7.16 in the second and Equation 1.7.45 in the third:

$$\begin{aligned} [\mathbf{D}f \circ g(A)]H &= [\mathbf{D}f(g(A))][\mathbf{D}g(A)]H \\ &= [\mathbf{D}f(g(A))](-A^{-1}HA^{-1}) \\ &= A^{-1}(-A^{-1}HA^{-1}) + (-A^{-1}HA^{-1})A^{-1} \\ &= -\left(A^{-2}HA^{-1} + A^{-1}HA^{-2} \right). \end{aligned} \qquad 1.8.19$$

Exercise 1.8.7 asks you to compute the derivatives of the maps $A \mapsto A^{-3}$ and $A \mapsto A^{-n}$.

1.9 THE MEAN VALUE THEOREM AND CRITERIA FOR DIFFERENTIABILITY

> *I turn with terror and horror from this lamentable scourge of continuous functions with no derivatives.*—Charles Hermite, in a letter to Thomas Stieltjes, 1893

In this section we discuss two applications of the mean value theorem (Theorem 1.6.9). The first extends that theorem to functions of several variables, and the second gives a criterion for when a function is differentiable.

The mean value theorem for functions of several variables

The derivative measures the difference of the values of functions at different points. For functions of one variable, the mean value theorem says that if $f : [a, b] \to \mathbb{R}$ is continuous, and f is differentiable on (a, b), then there exists $c \in (a, b)$ such that

$$f(b) - f(a) = f'(c)(b - a). \qquad 1.9.1$$

The analogous statement in several variables is the following.

Theorem 1.9.1 (Mean value theorem for functions of several variables). *Let $U \subset \mathbb{R}^n$ be open, $f : U \to \mathbb{R}$ be differentiable, and the segment $[\mathbf{a}, \mathbf{b}]$ joining $\mathbf{a}$ to $\mathbf{b}$ be contained in U. Then there exists $\mathbf{c} \in [\mathbf{a}, \mathbf{b}]$ such that*

$$f(\mathbf{b}) - f(\mathbf{a}) = [\mathbf{D}f(\mathbf{c})](\mathbf{b} - \mathbf{a}). \qquad 1.9.2$$

Corollary 1.9.2. *If f is a function as defined in Theorem 1.9.1, then*

$$|f(\mathbf{b}) - f(\mathbf{a})| \le \left(\sup_{\mathbf{c} \in [\mathbf{a}, \mathbf{b}]} \Big| [\mathbf{D}f(\mathbf{c})] \Big| \right) |\mathbf{b} - \mathbf{a}|. \qquad 1.9.3$$

Proof of Corollary 1.9.2. This follows immediately from Theorem 1.9.1 and Proposition 1.4.11.   □

Proof of Theorem 1.9.1. Note that as t varies from 0 to 1, the point $(1-t)\mathbf{a}+t\mathbf{b}$ moves from $\mathbf{a}$ to $\mathbf{b}$. Consider the mapping $g(t) = f((1-t)\mathbf{a}+t\mathbf{b})$. By the chain rule, g is differentiable, and by the one-variable mean value theorem, there exists t_0 such that

$$g(1) - g(0) = g'(t_0)(1 - 0) = g'(t_0). \qquad 1.9.4$$

Set $\mathbf{c} = (1 - t_0)\mathbf{a} + t_0\mathbf{b}$. By Proposition 1.7.12, we can express $g'(t_0)$ in terms of the derivative of f:

$$g'(t_0) = \lim_{s \to 0} \frac{g(t_0 + s) - g(t_0)}{s} = \lim_{s \to 0} \frac{f(\mathbf{c} + s(\mathbf{b} - \mathbf{a})) - f(\mathbf{c})}{s} = [\mathbf{D}f(\mathbf{c})](\mathbf{b} - \mathbf{a}).$$

$$1.9.5$$

So Equation 1.9.4 reads

$$f(\mathbf{b}) - f(\mathbf{a}) = [\mathbf{D}f(\mathbf{c})](\mathbf{b} - \mathbf{a}). \quad \square \qquad 1.9.6$$

Criterion for differentiability

Most often, the Jacobian matrix of a function is its derivative. But as we mentioned in Section 1.7, this isn't always true. It is perfectly possible for all partial derivatives of f to exist, and even all directional derivatives, and yet for f not to be differentiable! In such a case the Jacobian matrix exists but does not represent the derivative. This happens even for the innocent-looking function

$$f\begin{pmatrix} x \\ y \end{pmatrix} = \frac{x^2 y}{x^2 + y^2}$$

shown in Figure 1.9.1. Actually, we should write this function

$$f\begin{pmatrix} x \\ y \end{pmatrix} = \begin{cases} \dfrac{x^2 y}{x^2 + y^2} & \text{if } \begin{pmatrix} x \\ y \end{pmatrix} \neq \begin{pmatrix} 0 \\ 0 \end{pmatrix} \\ 0 & \text{if } \begin{pmatrix} x \\ y \end{pmatrix} = \begin{pmatrix} 0 \\ 0 \end{pmatrix}, \end{cases} \qquad 1.9.7$$

You have probably learned to be suspicious of functions that are defined by different formulas for different values of the variable. In this case, the value at $\begin{pmatrix} 0 \\ 0 \end{pmatrix}$ is really natural, in the sense that as $\begin{pmatrix} x \\ y \end{pmatrix}$ approaches $\begin{pmatrix} 0 \\ 0 \end{pmatrix}$, the function f approaches 0. This is not one of those functions whose value takes a sudden jump; indeed, f is continuous everywhere. Away from the origin, this is obvious; f is then defined by an algebraic formula, and we can compute both its partial derivatives at any point $\begin{pmatrix} x \\ y \end{pmatrix} \neq \begin{pmatrix} 0 \\ 0 \end{pmatrix}$.

That f is continuous at the origin requires a little checking, as follows. If $x^2 + y^2 = r^2$, then $|x| \leq r$ and $|y| \leq r$ so $|x^2 y| \leq r^3$. Therefore,

$$\left| f\begin{pmatrix} x \\ y \end{pmatrix} \right| \leq \frac{r^3}{r^2} = r, \quad \text{and} \quad \lim_{\begin{pmatrix} x \\ y \end{pmatrix} \to \begin{pmatrix} 0 \\ 0 \end{pmatrix}} f\begin{pmatrix} x \\ y \end{pmatrix} = 0. \qquad 1.9.8$$

So f is continuous at the origin. Moreover, f vanishes identically on both axes, so both partial derivatives of f vanish at the origin.

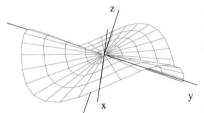

FIGURE 1.9.1.

The graph of f is made up of straight lines through the origin, so if you leave the origin in any direction, the directional derivative in that direction certainly exists. Both axes are among the lines making up the graph, so the directional derivatives in those directions are 0. But clearly there is no tangent plane to the graph at the origin.

"Vanish" means to equal 0. "Identically" means "at every point."

So far, f looks perfectly civilized: it is continuous, and both partial derivatives exist everywhere. But consider the derivative in the direction of the vector $\begin{bmatrix} 1 \\ 1 \end{bmatrix}$, i.e., the directional derivative

$$\lim_{t \to 0} \frac{f\left(\begin{pmatrix} 0 \\ 0 \end{pmatrix} + t \begin{bmatrix} 1 \\ 1 \end{bmatrix}\right) - f\begin{pmatrix} 0 \\ 0 \end{pmatrix}}{t} = \lim_{t \to 0} \frac{t^3}{2t^3} = \frac{1}{2}. \qquad 1.9.9$$

This is *not* what we get when we compute the same directional derivative by multiplying the Jacobian matrix of f by the vector $\begin{bmatrix} 1 \\ 1 \end{bmatrix}$, as in the right-hand side of Equation 1.7.27:

$$\underbrace{\left[D_1 f\begin{pmatrix} 0 \\ 0 \end{pmatrix}, D_2 f\begin{pmatrix} 0 \\ 0 \end{pmatrix} \right]}_{\text{Jacobian matrix } [\mathbf{J}f(\mathbf{0})]} \begin{bmatrix} 1 \\ 1 \end{bmatrix} = [0,0] \begin{bmatrix} 1 \\ 1 \end{bmatrix} = 0. \qquad 1.9.10$$

If we change our function, replacing the x^2y in the numerator of the algebraic formula by xy, then the resulting function, which we'll call g, will *not* be continuous at the origin. If $x = y$, then $g = 1/2$ no matter how close $\begin{pmatrix} x \\ y \end{pmatrix}$ gets to the origin: we then have

$$g\begin{pmatrix} x \\ x \end{pmatrix} = \frac{x^2}{2x^2} = \frac{1}{2}.$$

Thus, by Proposition 1.7.12, f is not differentiable.

In fact, things can get worse. The function we just discussed is continuous, but it is possible for all directional derivatives of a function to exist, and yet for the function not to be continuous, or even bounded in a neighborhood of 0, as we saw in Example 1.5.24; Exercise 1.9.2 provides another example.

Continuously differentiable functions

The lesson so far is that knowing a function's partial derivatives or directional derivatives does not tell you either that the function is differentiable or that it is continuous. Even in one variable, derivatives alone reveal much less than you might expect; we will see in Example 1.9.4 that a function $f : \mathbb{R} \to \mathbb{R}$ can have a positive derivative at x although it does not increase in a neighborhood of x!

Of course we don't claim that derivatives are worthless. The problem in these pathological cases is that the function is not *continuously differentiable*: its derivative is not continuous. As long as a function is continuously differentiable, things behave nicely.

Example 1.9.3 (A function that has partial derivatives but is not differentiable). Let us go back to the function of Equation 1.9.7, which we just saw is not differentiable:

$$f\begin{pmatrix} x \\ y \end{pmatrix} = \begin{cases} \dfrac{x^2y}{x^2 + y^2} & \text{if } \begin{pmatrix} x \\ y \end{pmatrix} \neq \begin{pmatrix} 0 \\ 0 \end{pmatrix} \\ 0 & \text{if } \begin{pmatrix} x \\ y \end{pmatrix} = \begin{pmatrix} 0 \\ 0 \end{pmatrix}, \end{cases}$$

and reconsider its partial derivatives. We find that both partials are 0 at the origin, and that away from the origin—i.e., if $\begin{pmatrix} x \\ y \end{pmatrix} \neq \begin{pmatrix} 0 \\ 0 \end{pmatrix}$—then

$$D_1 f \begin{pmatrix} x \\ y \end{pmatrix} = \frac{(x^2 + y^2)(2xy) - x^2 y(2x)}{(x^2 + y^2)^2} = \frac{2xy^3}{(x^2 + y^2)^2}$$

$$D_2 f \begin{pmatrix} x \\ y \end{pmatrix} = \frac{(x^2 + y^2)(x^2) - x^2 y(2y)}{(x^2 + y^2)^2} = \frac{x^4 - x^2 y^2}{(x^2 + y^2)^2}.$$

1.9.11

These partial derivatives are not continuous at the origin, as you will see if you approach the origin from any direction other than one of the axes. For example, if you compute the first partial derivative at the point $\begin{pmatrix} t \\ t \end{pmatrix}$ of the diagonal, you find the limit

$$\lim_{t \to 0} D_1 f \begin{pmatrix} t \\ t \end{pmatrix} = \frac{2t^4}{(2t^2)^2} = \frac{1}{2},$$

1.9.12

which is not the value of

$$D_1 f \begin{pmatrix} 0 \\ 0 \end{pmatrix} = 0. \quad \triangle$$

1.9.13

Example 1.9.4 (A differentiable yet pathological function in one variable). Consider the function

$$f(x) = \frac{x}{2} + x^2 \sin \frac{1}{x},$$

1.9.14

a variant of which is shown in Figure 1.9.2. To be precise, one should add $f(0) = 0$, since $\sin 1/x$ is not defined there, but this was the only reasonable value, since

$$\lim_{x \to 0} x^2 \sin \frac{1}{x} = 0.$$

1.9.15

Moreover, we will see that the function f is differentiable at the origin, with derivative

$$f'(0) = \frac{1}{2}.$$

1.9.16

This is one case where you *must* use the definition of the derivative as a limit; you *cannot* use the rules for computing derivatives blindly. In fact, let's try. We find

$$f'(x) = \frac{1}{2} + 2x \sin \frac{1}{x} + x^2 \left(\cos \frac{1}{x} \right) \left(-\frac{1}{x^2} \right) = \frac{1}{2} + 2x \sin \frac{1}{x} - \cos \frac{1}{x}.$$

1.9.17

This formula is certainly correct for $x \neq 0$, but $f'(x)$ doesn't have a limit when $x \to 0$. Indeed,

$$\lim_{x \to 0} \frac{1}{2} + 2x \sin \frac{1}{x} = \frac{1}{2}$$

1.9.18

FIGURE 1.9.2.

Graph of the function $f(x) = \frac{x}{2} + 6x^2 \sin \frac{1}{x}$. The derivative of f does not have a limit at the origin, but the curve still has slope $1/2$ there.

does exist, but $\cos 1/x$ oscillates infinitely many times between -1 and 1. So f' will oscillate from a value near $-1/2$ to a value near $3/2$. *This does not mean that f isn't differentiable at 0.* We can compute the derivative at 0 using the definition of the derivative:

$$f'(0) = \lim_{h \to 0} \frac{1}{h} \left(\frac{0+h}{2} + (0+h)^2 \sin \frac{1}{0+h} \right) = \lim_{h \to 0} \frac{1}{h} \left(\frac{h}{2} + h^2 \sin \frac{1}{h} \right)$$

$$= \frac{1}{2} + \lim_{h \to 0} h \sin \frac{1}{h} = \frac{1}{2},$$

1.9.19

since (by Theorem 1.5.21, part (f)) $\lim_{h \to 0} h \sin \frac{1}{h}$ exists, and indeed vanishes.

Finally, we can see that although the derivative *at* 0 is positive, the function is not increasing *in any neighborhood* of 0, since in any interval arbitrarily close to 0 the derivative takes negative values; as we saw above, it oscillates from a value near $-1/2$ to a value near $3/2$. $\triangle$

The moral of the story is: only study *continuously differentiable* functions.

This is *very bad.* Our whole point is that the function should behave like its best linear approximation, and in this case it emphatically doesn't. We could easily make up examples in several variables where the same occurs: where the function is differentiable, so that the Jacobian matrix represents the derivative, but where that derivative doesn't tell you much.

Determining whether a function is continuously differentiable

Fortunately, you can do a great deal of mathematics without ever dealing with such pathological functions. Moreover, there is a nice criterion that allows us to check whether a function in several variables is continuously differentiable:

Theorem 1.9.5 (Criterion for differentiability). *If U is an open subset of $\mathbb{R}^n$, and $\mathbf{f} : U \to \mathbb{R}^m$ is a mapping such that all partial derivatives of $\mathbf{f}$ exist and are continuous on U, then $\mathbf{f}$ is differentiable on U, and its derivative is given by its Jacobian matrix.*

A function that is continuously differentiable—i.e., whose derivative is continuous—is known as a C^1 function.

Definition 1.9.6 (Continuously differentiable function). A function is continuously differentiable on $U \subset \mathbb{R}^n$ if all its partial derivatives exist and are continuous on U.

If you come across a function that is not continuously differentiable (and you may find such functions particularly interesting) you should be aware that none of the usual tools of calculus can be relied upon. Each such function is an outlaw, obeying none of the standard theorems.

Most often, when checking that a function is differentiable, the criterion of Theorem 1.9.5 is the tool used. Note that the last part, " ... and its derivative is given by its Jacobian matrix," is obvious; if a function is differentiable, Theorem 1.7.10 tells us that its derivative is given by its Jacobian matrix. So the point is to prove that the function is differentiable. Since we are told that the partial derivatives of $\mathbf{f}$ are continuous, if we prove that $\mathbf{f}$ is differentiable, we will have proved that it is continuously differentiable.

In Equation 1.9.20 we use the interval $(a, a + h)$, rather than (a, b), making the statement

$$f'(c) = \frac{f(a + h) - f(a)}{h},$$

or

$$hf'(c) = f(a + h) - f(a).$$

It will become clearer in Chapter 2 why we emphasize the dimensions of the derivative $[\mathbf{Df(a)}]$. The object of differential calculus is to study nonlinear mappings by studying their linear approximations, using the derivative. We will want to have at our disposal the techniques of linear algebra. Many will involve knowing the dimensions of a matrix.

(What are the dimensions of the derivative of the function $\mathbf{f}$ described in Theorem 1.9.5? Check your answer below.[23])

Proof. This is an application of Theorem 1.6.9, the mean value theorem. What we need to show is that

$$\lim_{\vec{\mathbf{h}} \to \mathbf{0}} \frac{1}{|\vec{\mathbf{h}}|} \mathbf{f}(\mathbf{a} + \vec{\mathbf{h}}) - \mathbf{f}(\mathbf{a}) - [\mathbf{Jf(a)}]\vec{\mathbf{h}} = 0. \qquad 1.9.20$$

First, note (Theorem 1.8.1, part (3)), that it is enough to prove it when $m = 1$ (i.e., $f : U \to \mathbb{R}$).

Next write

$$f(\mathbf{a} + \vec{\mathbf{h}}) - f(\mathbf{a}) = f \begin{pmatrix} a_1 + h_1 \\ a_2 + h_2 \\ \vdots \\ a_n + h_n \end{pmatrix} - f \begin{pmatrix} a_1 \\ a_2 \\ \vdots \\ a_n \end{pmatrix} \qquad 1.9.21$$

in expanded form, subtracting and adding inner terms:

$$f(\mathbf{a} + \vec{\mathbf{h}}) - f(\mathbf{a}) =$$

$$f \begin{pmatrix} a_1 + h_1 \\ a_2 + h_2 \\ \vdots \\ a_n + h_n \end{pmatrix} - f \underbrace{\begin{pmatrix} a_1 \\ a_2 + h_2 \\ \vdots \\ a_n + h_n \end{pmatrix}}_{\text{subtracted}}$$

$$+ f \underbrace{\begin{pmatrix} a_1 \\ a_2 + h_2 \\ a_3 + h_3 \\ \vdots \\ a_n + h_n \end{pmatrix}}_{\text{added}} - f \begin{pmatrix} a_1 \\ a_2 \\ a_3 + h_3 \\ \vdots \\ a_n + h_n \end{pmatrix} \qquad 1.9.22$$

$$+ \dots$$

$$+ f \begin{pmatrix} a_1 \\ a_2 \\ \vdots \\ a_{n-1} \\ a_n + h_n \end{pmatrix} - f \begin{pmatrix} a_1 \\ a_2 \\ \vdots \\ a_{n-1} \\ a_n \end{pmatrix}.$$

---

[23]The function $\mathbf{f}$ goes from a subset of $\mathbb{R}^n$ to $\mathbb{R}^m$, so its derivative takes a vector in $\mathbb{R}^n$ and gives a vector in $\mathbb{R}^m$. Therefore it is an $m \times n$ matrix, m tall and n wide.

By the mean value theorem, the ith term above is

$$
f\begin{pmatrix} a_1 \\ a_2 \\ \vdots \\ a_i + h_i \\ a_{i+1} + h_{i+1} \\ \vdots \\ a_n + h_n \end{pmatrix} - f\begin{pmatrix} a_1 \\ a_2 \\ \vdots \\ a_i \\ a_{i+1} + h_{i+1} \\ \vdots \\ a_n + h_n \end{pmatrix} = h_i D_i f\begin{pmatrix} a_1 \\ a_2 \\ \vdots \\ b_i \\ a_{i+1} + h_{i+1} \\ \vdots \\ a_n + h_n \end{pmatrix} \qquad 1.9.23
$$

$$\underbrace{}_{i\text{th term}}$$

for some $b_i \in [a_i,\, a_i + h_i]$: there is some point b_i in the interval $[a_i,\, a_i + h_i]$ such that the partial derivative $D_i f$ at b_i gives the average change of the function f over that interval, when all variables except the ith are kept constant.

Since f has n variables, we need to find such a point for every i from 1 to n. We will call these points $\mathbf{c}_i$:

$$
\mathbf{c}_i = \begin{pmatrix} a_1 \\ a_2 \\ \vdots \\ b_i \\ a_{i+1} + h_{i+1} \\ \vdots \\ a_n + h_n \end{pmatrix} ; \qquad \text{this gives} \qquad f(\mathbf{a} + \vec{\mathbf{h}}) - f(\mathbf{a}) = \sum_{i=1}^{n} h_i D_i f(\mathbf{c}_i).
$$

Thus we find that

$$
\underbrace{f(\mathbf{a} + \vec{\mathbf{h}}) - f(\mathbf{a})}_{=\sum_{i=1}^{n} D_i f(\mathbf{c}_i) h_i} - \sum_{i=1}^{n} D_i f(\mathbf{a}) h_i = \sum h_i \big(D_i f(\mathbf{c}_i) - D_i f(\mathbf{a}) \big). \qquad 1.9.24
$$

So far we haven't used the hypothesis that the partial derivatives $D_i f$ are continuous. Now we do. Since $D_i f$ is continuous, and since $\mathbf{c}_i$ tends to $\mathbf{a}$ as $\vec{\mathbf{h}} \to 0$, we see that the theorem is true:

The inequality in the second line of Equation 1.9.25 comes from the fact that $|h_i|/|\vec{\mathbf{h}}| \le 1$.

$$
\lim_{\vec{\mathbf{h}} \to 0} \frac{\left| f(\mathbf{a} + \vec{\mathbf{h}}) - f(\mathbf{a}) - \overbrace{[\mathbf{J}f(\mathbf{a})]\vec{\mathbf{h}}}^{\sum_{i=1}^{n} D_i f(\mathbf{a}) h_i} \right|}{|\vec{\mathbf{h}}|} = \lim_{\vec{\mathbf{h}} \to 0} \sum_{i=1}^{n} \frac{|h_i|}{|\vec{\mathbf{h}}|} |D_i f(\mathbf{c}_i) - D_i f(\mathbf{a})|
$$

$$
\le \lim_{\vec{\mathbf{h}} \to 0} \sum_{i=1}^{n} |D_i f(\mathbf{c}_i) - D_i f(\mathbf{a})| \;=\; 0.
$$

$$\square \qquad 1.9.25$$

Example 1.9.7. Here we work out the above computation when f is a function on $\mathbb{R}^2$:

$$f\begin{pmatrix} a_1 + h_1 \\ a_2 + h_2 \end{pmatrix} - f\begin{pmatrix} a_1 \\ a_2 \end{pmatrix}$$

$$= f\begin{pmatrix} a_1 + h_1 \\ a_2 + h_2 \end{pmatrix} \overbrace{- f\begin{pmatrix} a_1 \\ a_2 + h_2 \end{pmatrix} + f\begin{pmatrix} a_1 \\ a_2 + h_2 \end{pmatrix}}^{0} - f\begin{pmatrix} a_1 \\ a_2 \end{pmatrix}$$

$$= h_1 D_1 f\begin{bmatrix} b_1 \\ a_2 + h_2 \end{bmatrix} + h_2 D_2 f\begin{bmatrix} a_1 \\ b_2 \end{bmatrix}$$

$$= h_1 D_1 f(\mathbf{c}_1) + h_2 D_2 f(\mathbf{c}_2). \quad \triangle \qquad\qquad 1.9.26$$

1.10 Exercises for Chapter One

Exercises for Section 1.1:
Vectors

1.1.1   Compute the following vectors by coordinates and sketch what you did:

(a) $\begin{bmatrix} 1 \\ 3 \end{bmatrix} + \begin{bmatrix} 2 \\ 1 \end{bmatrix}$ (b) $2\begin{bmatrix} 2 \\ 4 \end{bmatrix}$ (c) $\begin{bmatrix} 1 \\ 3 \end{bmatrix} - \begin{bmatrix} 2 \\ 1 \end{bmatrix}$ (d) $\begin{bmatrix} 3 \\ 2 \end{bmatrix} + \vec{e}_1$

1.1.2   Compute the following vectors:

(a) $\begin{bmatrix} 3 \\ \pi \\ 1 \end{bmatrix} + \begin{bmatrix} 1 \\ -1 \\ \sqrt{2} \end{bmatrix}$ (b) $\begin{bmatrix} 1 \\ 4 \\ c \\ 2 \end{bmatrix} + \vec{e}_2$ (c) $\begin{bmatrix} 1 \\ 4 \\ c \\ 2 \end{bmatrix} - \vec{e}_4$

1.1.3   Name the two trivial subspaces of $\mathbb{R}^n$.

1.1.4   Which of the following lines are subspaces of $\mathbb{R}^2$ (or $\mathbb{R}^n$)? For any that is not, why not?

(a) $y = -2x - 5$ (b) $y = 2x + 1$ (c) $y = \dfrac{5x}{2}$.

1.1.5   Sketch the following vector fields:

(a) $\vec{F}\begin{pmatrix} x \\ y \end{pmatrix} = \begin{bmatrix} 0 \\ 1 \end{bmatrix}$ (b) $\vec{F}\begin{pmatrix} x \\ y \end{pmatrix} = \begin{bmatrix} x \\ 0 \end{bmatrix}$ (c) $\vec{F}\begin{pmatrix} x \\ y \end{pmatrix} = \begin{bmatrix} x \\ y \end{bmatrix}$

(d) $\vec{F}\begin{pmatrix} x \\ y \end{pmatrix} = \begin{bmatrix} x \\ -y \end{bmatrix}$ (e) $\vec{F}\begin{pmatrix} x \\ y \end{pmatrix} = \begin{bmatrix} y \\ x \end{bmatrix}$ (f) $\vec{F}\begin{pmatrix} x \\ y \end{pmatrix} = \begin{bmatrix} -y \\ x \end{bmatrix}$

(g) $\vec{F}\begin{pmatrix} x \\ y \end{pmatrix} = \begin{bmatrix} y \\ x - y \end{bmatrix}$ (h) $\vec{F}\begin{pmatrix} x \\ y \end{pmatrix} = \begin{bmatrix} x - y \\ x + y \end{bmatrix}$ (i) $\vec{F}\begin{pmatrix} x \\ y \end{pmatrix} = \begin{bmatrix} x^2 - y - 1 \\ x - y \end{bmatrix}$

(j) $\vec{F}\begin{pmatrix} x \\ y \\ z \end{pmatrix} = \begin{bmatrix} 0 \\ 0 \\ x^2 + y^2 \end{bmatrix}$ (k) $\vec{F}\begin{pmatrix} x \\ y \\ z \end{pmatrix} = \begin{bmatrix} y \\ -x \\ -z \end{bmatrix}$ (l) $\vec{F}\begin{pmatrix} x \\ y \\ z \end{pmatrix} = \begin{bmatrix} x - y \\ x + y \\ -z \end{bmatrix}$

1.1.6   Suppose that in a circular pipe of radius a, water is flowing in the direction of the pipe, with speed $a^2 - r^2$, where r is the distance to the axis of the pipe.

(a) Write the vector field describing the flow if the pipe is in the direction of the z-axis.

(b) Write the vector field describing the flow if the axis of the pipe is the unit circle in the (x, y)-plane.

Exercises for Section 1.2:

Matrices

1.2.1   (a) What are the dimensions of the following matrices?

$$
\text{(a)} \begin{bmatrix} a & b & c \\ d & e & f \end{bmatrix} \quad \text{(b)} \begin{bmatrix} 4 & 1 \\ 0 & 2 \end{bmatrix} \quad \text{(c)} \begin{bmatrix} \pi & 1 \\ 0 & 1 \\ 1 & 0 \end{bmatrix};
$$

$$
\text{(d)} \begin{bmatrix} 1 & 0 & 0 & 1 \\ 0 & 1 & 0 & 1 \\ 1 & 0 & 1 & 0 \end{bmatrix} \quad \text{(e)} \begin{bmatrix} 1 & 0 & 0 \\ 0 & 1 & 0 \\ 0 & 0 & 1 \end{bmatrix}.
$$

(b) Which of the above matrices can be multiplied together?

In Exercise 1.2.2, remember to use the format:

$$
\begin{bmatrix} 1 & 2 & 3 \\ 4 & 5 & 6 \end{bmatrix} \begin{bmatrix} 7 & 8 \\ 9 & 0 \\ 1 & 2 \end{bmatrix} \begin{bmatrix} .. & .. \\ .. & .. \end{bmatrix}
$$

1.2.2   Perform the following matrix multiplications when it is possible.

$$
\text{(a)} \begin{bmatrix} 1 & 2 & 3 \\ 4 & 5 & 6 \end{bmatrix} \begin{bmatrix} 7 & 8 \\ 9 & 0 \\ 1 & 2 \end{bmatrix}; \quad \text{(b)} \begin{bmatrix} 1 & 2 \\ 0 & 3 \end{bmatrix} \begin{bmatrix} 1 & 4 \\ -1 & 3 \\ -2 & 2 \end{bmatrix}
$$

$$
\text{(c)} \begin{bmatrix} 1 & -1 & 1 \\ -1 & 0 & 2 \\ -1 & 1 & 1 \end{bmatrix} \begin{bmatrix} 0 & 1 & -1 \\ -1 & 1 & 2 \\ 2 & 0 & -2 \end{bmatrix}; \quad \text{(d)} \begin{bmatrix} 7 & 1 \\ -1 & 0 \\ 2 & 3 \end{bmatrix} \begin{bmatrix} 5 \\ -4 \end{bmatrix}
$$

$$
\text{(e)} \begin{bmatrix} 1 & 2 \\ 0 & 3 \end{bmatrix} \begin{bmatrix} 1 & 4 \\ -1 & 3 \end{bmatrix} \begin{bmatrix} 0 & 1 \\ -1 & 3 \end{bmatrix}; \quad \text{(f)} \begin{bmatrix} 0 & 2 & 1 \\ 1 & 3 & 2 \end{bmatrix} \begin{bmatrix} 0 & 1 \\ 3 & 5 \end{bmatrix}
$$

1.2.3   Compute the following without doing any arithmetic.

$$
\text{(a)} \begin{bmatrix} 7 & 2 & \sqrt{3} & 4 \\ 6 & 8 & a^2 & 2 \\ 3 & \sqrt{5} & a & 7 \end{bmatrix} \begin{bmatrix} 0 \\ 1 \\ 0 \\ 0 \end{bmatrix} \quad \text{(b)} \begin{bmatrix} 6a & 2 & 3a^2 \\ 4 & 2\sqrt{a} & 2 \\ 5 & 12 & 3 \end{bmatrix} \vec{e}_2 \quad \text{(c)} \begin{bmatrix} 2 & 1 & 8 & 6 \\ 3 & 2 & \sqrt{3} & 4 \end{bmatrix} \vec{e}_3
$$

$$
A = \begin{bmatrix} 1 & 2 & 0 \\ 3 & 1 & -1 \end{bmatrix}
$$

$$
B = \begin{bmatrix} 2 & 5 & 1 \\ 1 & 4 & 2 \\ 1 & 3 & 3 \end{bmatrix}
$$

Matrices for Exercise 1.2.4

1.2.4   Given the matrices A and B in the margin at left,

(a) Compute the third column of AB *without* computing the entire matrix AB.

(b) Compute the second row of AB, again *without* computing the entire matrix AB.

1.2.5   For what values of a do the matrices

$$
A = \begin{bmatrix} 1 & 1 \\ 1 & 0 \end{bmatrix} \quad \text{and} \quad B = \begin{bmatrix} 1 & 0 \\ a & 1 \end{bmatrix} \quad \text{satisfy } AB = BA?
$$

1.2.6   For what values of a and b do the matrices

$$A = \begin{bmatrix} 1 & a \\ a & 0 \end{bmatrix} \quad \text{and} \quad B = \begin{bmatrix} 1 & 0 \\ b & 1 \end{bmatrix} \quad \text{satisfy } AB = BA?$$

1.2.7   From the matrices below, find those that are transposes of each other.

$$(a) \begin{bmatrix} 1 & 2 & 3 \\ x & 0 & \sqrt{3} \\ 1 & x^2 & 2 \end{bmatrix} \quad (b) \begin{bmatrix} 1 & x & 1 \\ 2 & 0 & \sqrt{3} \\ 3 & x^2 & 2 \end{bmatrix} \quad (c) \begin{bmatrix} 1 & x^2 & 2 \\ x & 0 & \sqrt{3} \\ 1 & 2 & 3 \end{bmatrix}$$

$$(d) \begin{bmatrix} 3 & \sqrt{3} & 2 \\ 2 & 0 & x^2 \\ 1 & x & 1 \end{bmatrix} \quad (e) \begin{bmatrix} 1 & x & 1 \\ x^2 & 0 & 2 \\ 2 & \sqrt{3} & 3 \end{bmatrix} \quad (f) \begin{bmatrix} 1 & 2 & 3 \\ x & 0 & x^2 \\ 1 & \sqrt{3} & 2 \end{bmatrix}$$

1.2.8   Given the two matrices $A = \begin{bmatrix} 1 & 0 \\ 1 & 0 \end{bmatrix}$ and $B = \begin{bmatrix} 1 & 0 & 1 \\ 2 & 1 & 0 \end{bmatrix}$:

(a) What are their transposes?

(b) Without computing AB what is $(AB)^\top$?

(c) Confirm your result by computing AB.

(d) What happens if you do part (b) using the *incorrect* formula $(AB)^\top = A^\top B^\top$?

$$A = \begin{bmatrix} 1 & 0 \\ 1 & 0 \end{bmatrix}$$

$$B = \begin{bmatrix} 1 & 0 & 1 \\ 1 & 0 & 1 \end{bmatrix}$$

$$C = \begin{bmatrix} 1 & 1 & 0 \\ 1 & 0 & 1 \\ 1 & 1 & 0 \end{bmatrix}$$

Matrices for Exercise 1.2.9

1.2.9   Given the matrices A, B, and C at left, which of the following expressions make no sense?

(a) AB (b) BA (c) $A + B$ (d) AC

(e) BC (f) CB (g) $\frac{A+B}{C}$ (h) $B^\top A$ (i) $B^\top C$

1.2.10   Show that if A and B are upper-triangular $n \times n$ matrices, then so is AB.

1.2.11   (a) What is the inverse of the matrix $A = \begin{bmatrix} a & b \\ 0 & a \end{bmatrix}$ for $a \neq 0$?

Exercise 1.2.11 (b) should be with the exercises to Section 1.4.

Recall that Mat (n, m) denotes the set of $n \times m$ matrices.

(b) If we identify Mat $(2, 2)$ with $\mathbb{R}^4$ in the standard way, what is the angle between A and A^{-1}? Under what condition are A and A^{-1} orthogonal?

1.2.12   Confirm by matrix multiplication that the inverse of

$$A = \begin{bmatrix} a & b \\ c & d \end{bmatrix} \quad \text{is} \quad A^{-1} = \frac{1}{ad - bc} \begin{bmatrix} d & -b \\ -c & a \end{bmatrix}.$$

1.2.13   Prove that a matrix $\begin{bmatrix} a & b \\ c & d \end{bmatrix}$ is not invertible if $ad - bc = 0$.

1.2.14   Prove Theorem 1.2.17: that the transpose of a product is the product of the transposes in reverse order:

$$(AB)^\top = B^\top A^\top.$$

1.2.15   Recall from Proposition 1.2.23, and the discussion preceding it, what the adjacency graph of a matrix is.

(a) Compute the adjacency matrix A_T for a triangle and A_S for a square.

(b) For each of these, compute the powers up to 5, and explain the meaning of the diagonal entries.

(c) For the triangle, observe that the diagonal terms differ by 1 from the off-diagonal terms. Can you prove that this will be true for all powers of A_T?

(d) For the square, you should observe that half the terms are 0 for even powers, and the other half are 0 for odd powers. Can you prove that this will be true for all powers of A_S?

"Stars" indicate difficult exercises.

*(e) Show that you can color the vertices of a connected graph (one on which you can walk from any vertex to any other) in two colors, such that no two adjacent vertices have the same color, if and only if for all sufficiently high powers n of the adjacency matrix, those entries that are 0 for A^n are nonzero for A^{n+1}, and those that are nonzero for A^n are zero for A^{n+1}.

1.2.16   (a) For the adjacency matrix A corresponding to the cube (shown in Figure 1.2.6), compute A^2, A^3 and A^4. Check directly that $(A^2)(A^2) = (A^3)A$.

(b) The diagonal entries of A^4 should all be 21; count the number of walks of length 4 from a vertex to itself directly.

(c) For this same matrix A, some entries of A^n are always 0 when n is even, and others (the diagonal entries for instance) are always 0 when n is odd. Can you explain why? Think of coloring the vertices of the cube in two colors, so that each edge connects vertices of opposite colors.

(d) Is this phenomenon true for A_T, A_S? Explain why, or why not.

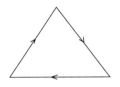

1.2.17   Suppose we redefined a walk on the cube to allow stops: in one time unit you may either go to an adjacent vertex, or stay where you are.

(a) Find a matrix B such that $B_{i,j}^n$ counts the walks of length n from V_i to V_j.

(b) Compute B^2, B^3 and explain the diagonal entries of B^3.

1.2.18   Suppose all the edges of a graph are oriented by an arrow on them. We allow multiple edges joining vertices, so that there might be many (a superhighway) joining two vertices, or two going in opposite directions (a 2-way street). Define the oriented adjacency matrix to be the square matrix with both rows and columns labeled by the vertices, where the (i, j) entry is m if there are m oriented edges leading from vertex i to vertex j.

What are the oriented adjacency matrices of the graphs at left?

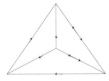

Graphs for Exercise 1.2.18

1.2.19   An oriented walk of length n on an oriented graph consists of a sequence of vertices $V_0, V_1, \ldots, V_n$ such that V_i, V_{i+1} are, respectively, the beginning and the end of an oriented edge.

(a) Show that if A is the oriented adjacency matrix of an oriented graph, then the (i, j) entry of A^n is the number of oriented walks of length n going from vertex i to vertex j.

(b) What does it mean for the oriented adjacency matrix of an oriented graph to be upper triangular? lower triangular? diagonal?

"Stars" indicate difficult exercises.

1.2.20   (a) Show that $\begin{bmatrix} a & 1 & 0 \\ b & 0 & 1 \end{bmatrix}$ is a left inverse of $\begin{bmatrix} 0 & 0 \\ 1 & 0 \\ 0 & 1 \end{bmatrix}$.

(b) Show that the matrix $\begin{bmatrix} 0 & 0 \\ 1 & 0 \\ 0 & 1 \end{bmatrix}$ has no right inverse.

(c) Find a matrix that has infinitely many right inverses. (Try transposing.)

1.2.21   Show that

$$\begin{bmatrix} 1 & a & b \\ 0 & 1 & c \\ 0 & 0 & 1 \end{bmatrix} \text{ has an inverse of the form } \begin{bmatrix} 1 & x & y \\ 0 & 1 & z \\ 0 & 0 & 1 \end{bmatrix},$$

and find it.

*1.2.22   What 2×2 matrices A satisfy

$$\text{(a) } A^2 = 0, \qquad \text{(b) } A^2 = I, \qquad \text{(c) } A^2 = -I \, ?$$

Exercises for Section 1.3:
A Matrix as a Transformation

1.3.1   Are the following true functions? That is, are they both everywhere defined and well defined?

(a) "The aunt of," from people to people.

(b) $f(x) = \frac{1}{x}$, from real numbers to real numbers.

(c) "The capital of," from countries to cities (careful—at least two countries, the Netherlands and Bolivia, have two capitals.)

1.3.2   (a) Give one example of a linear transformation $T : \mathbb{R}^4 \to \mathbb{R}^2$.

(b) What is the matrix of the linear transformation $S_1 : \mathbb{R}^3 \to \mathbb{R}^3$ corresponding to reflection in the plane of equation $x = y$? What is the matrix corresponding to reflection $S_2 : \mathbb{R}^3 \to \mathbb{R}^3$ in the plane $y = z$? What is the matrix of $S_1 \circ S_2$?

1.3.3   Of the functions in Exercise 1.3.1, which are onto? One to one?

1.3.4   (a) Make up a non-mathematical transformation that is bijective (both onto and one to one).   (b) Make up a mathematical transformation that is bijective.

1.3.5   (a) Make up a non-mathematical transformation that is onto but not one to one. (b) Make up a mathematical transformation that is onto but not one to one.

1.3.6   (a) Make up a non-mathematical transformation that is one to one but not onto. (b) Make up a mathematical transformation that is one to one but not onto.

1.3.7   The transformation $f(x) = x^2$ from real numbers to real positive numbers is onto but not one to one.

(a) Can you make it 1-1 by changing its domain? By changing its range?

(b) Can you make it not onto by changing its domain? By changing its range?

1.3.8   Which of the following are characterized by linearity? Justify your answer.

(a) The increase in height of a child from birth to age 18.

(b) "You get what you pay for."

(c) The value of a bank account at 5 percent interest, compounded daily, as a function of time.

(d) "Two can live as cheaply as one."

(e) "Cheaper by the dozen"

In Exercise 1.3.9, remember that the height of a matrix is given first: a 3×2 matrix is 3 tall and 2 wide.

1.3.9   For each of the following linear transformations, what must be the dimensions of the corresponding matrix?
  (a) $T : \mathbb{R}^2 \to \mathbb{R}^3$ (b) $T : \mathbb{R}^3 \to \mathbb{R}^3$
  (c) $T : \mathbb{R}^4 \to \mathbb{R}^2$ (d) $T : \mathbb{R}^4 \to \mathbb{R}$

(a) $A = \begin{bmatrix} 1 & 3 & 0 & 1 \\ 0 & 3 & 1 & 5 \\ 1 & 2 & 0 & 1 \end{bmatrix}$.

1.3.10   For the matrices at left, what is the domain and range of the corresponding transformation?

(b) $B = \begin{bmatrix} a_1 & b_1 \\ a_2 & b_2 \\ a_3 & b_3 \\ a_4 & b_4 \\ a_5 & b_5 \end{bmatrix}$

(c) $C = \begin{bmatrix} \pi & 1 & 0 & \sqrt{2} \\ 0 & -1 & 2 & 1 \end{bmatrix}$

$D = \begin{bmatrix} 1 & 0 & -2 & 5 \end{bmatrix}$.

Matrices for Exercise 1.3.10

1.3.11   For a class of 150 students, grades on a mid-term exam, 10 homework assignments, and the final were entered in matrix form, each row corresponding to a student, the first column corresponding to the grade on the mid-term, the next 10 columns corresponding to grades on the homeworks and the last column corresponding to the grade on the final. The final counts for 50 percent, the mid-term counts for 25 percent, and each homework for 2.5 percent of the final grade. What is the transformation $T : \mathbb{R}^{12} \to \mathbb{R}$ that assigns to each student his or her final grade?

1.3.12   Perform the composition $f \circ g \circ h$ for the following functions and values of x.

(a) $f(x) = x^2 - 1$, $g(x) = 3x$, $h(x) = -x + 2$, for $x = 2$.

(b) $f(x) = x^2$, $g(x) = x - 3$, $h(x) = x - 3$, for $x = 1$.

1.3.13   Find the matrix for the transformation from $\mathbb{R}^3 \to \mathbb{R}^3$ that rotates by $30°$ around the y-axis.

1.3.14   Show that the mapping from $\mathbb{R}^n$ to $\mathbb{R}^m$ described by the product $A\vec{v}$ is indeed linear.

1.3.15   Use composition of transformations to derive from the transformation in Example 1.3.17 the fundamental theorems of trigonometry:

$$\cos(\theta_1 + \theta_2) = \cos\theta_1\cos\theta_2 - \sin\theta_1\sin\theta_2$$

$$\sin(\theta_1 + \theta_2) = \sin\theta_1\cos\theta_2 + \cos\theta_1\sin\theta_2.$$

In Exercise 1.3.18 note that the symbol $\mapsto$ (to be read, "maps to") is different from the symbol $\to$ (to be read "to"). While $\to$ describes the relationship between the domain and range of a mapping, as in $T: \mathbb{R}^2 \to \mathbb{R}$, the symbol $\mapsto$ describes what a mapping does to a particular input. One could write $f(x) = x^2$ as $f: x \mapsto x^2$.

1.3.16   Confirm (Example 1.3.16) by matrix multiplication that reflecting a point across the line, and then back again, lands you back at the original point.

1.3.17   If A and B are $n \times n$ matrices, their *Jordan product* is

$$\frac{AB + BA}{2}.$$

Show that this product is commutative but not associative.

1.3.18   Consider $\mathbb{R}^2$ as identified to $\mathbb{C}$ by identifying $\begin{pmatrix} a \\ b \end{pmatrix}$ to $z = a + ib$.

Show that the following mappings $\mathbb{C} \to \mathbb{C}$ are linear transformations, and give their matrices:

(a) $\mathrm{Re} : z \mapsto \mathrm{Re}\,(z)$ (the real part of z);

(b) $\Im : z \mapsto \Im(z)$ (the imaginary part of z);

(c) $c : z \mapsto \bar{z}$ (the complex conjugate of z, i.e., $\bar{z} = a - ib$ if $z = a + ib$);

(d) $m_w : z \mapsto wz$, where $w = u + iv$ is a fixed complex number.

(a) $\begin{bmatrix} 1 \\ 2 \end{bmatrix}$ (b) $\begin{bmatrix} \sqrt{2} \\ \sqrt{7} \end{bmatrix}$

(c) $\begin{bmatrix} 1 \\ -1 \\ 1 \end{bmatrix}$ (d) $\begin{bmatrix} 1 \\ -2 \\ 2 \end{bmatrix}$

Vectors for Exercise 1.4.2

1.3.19   Show that the set of complex numbers $\{z \mid \mathrm{Re}\,(wz) = 0\}$ with fixed $w \in \mathbb{C}$ is a subspace of $\mathbb{R}^2 = \mathbb{C}$. Describe this subspace.

Exercises for Section 1.4:

Geometry in $\mathbb{R}^n$

1.4.1   If $\vec{v}$ and $\vec{w}$ are vectors, and A is a matrix, which of the following are numbers? Which are vectors?

$$\vec{v} \times \vec{w}; \quad \vec{v} \cdot \vec{w}; \quad |\vec{v}|; \quad |A|; \quad \det A; \quad A\vec{v}.$$

1.4.2   What are the lengths of the vectors in the margin?

1.4.3   (a) What is the angle between the vectors (a) and (b) in Exercise 1.4.2?

(b) What is the angle between the vectors (c) and (d) in Exercise 1.4.2?

1.4.4   Calculate the angles between the following pairs of vectors:

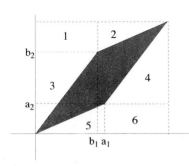

FIGURE 1.4.6.

(When figures and equations are numbered in the exercises, they are given the number of the exercise to which they pertain.)

(a) $\begin{bmatrix} 3 & 1 \\ 0 & 2 \end{bmatrix}$ (b) $\begin{bmatrix} 1 & 0 & 2 \\ 2 & 4 & 1 \\ 0 & 1 & 3 \end{bmatrix}$

(c) $\begin{bmatrix} -2 & 5 & 3 \\ -1 & 3 & 4 \\ -2 & 3 & 7 \end{bmatrix}$

(d) $\begin{bmatrix} 1 & 2 & -6 \\ 0 & 1 & -3 \\ 1 & 0 & -2 \end{bmatrix}$

Matrices for Exercise 1.4.9

(a) $\begin{bmatrix} 1 & 2 & 3 \\ -1 & 1 & 1 \\ 2 & 2 & 2 \end{bmatrix}$

(b) $\begin{bmatrix} a & b & c \\ 0 & d & e \\ 0 & 0 & f \end{bmatrix}$

(c) $\begin{bmatrix} a & b & 0 \\ c & d & 0 \\ e & f & g \end{bmatrix}$

Matrices for Exercise 1.4.11

(a) $\begin{bmatrix} 1 \\ 0 \\ 0 \end{bmatrix}$, $\begin{bmatrix} 1 \\ 1 \\ 1 \end{bmatrix}$ (b) $\begin{bmatrix} 1 \\ 0 \\ -1 \\ 0 \end{bmatrix}$, $\begin{bmatrix} 1 \\ 1 \\ 1 \\ 1 \end{bmatrix}$.

(c) $\lim_{n\to\infty}$ (angle between $\begin{bmatrix} 1 \\ 0 \\ 0 \\ 0 \\ \vdots \end{bmatrix}$, $\begin{bmatrix} 1 \\ 1 \\ 1 \\ 1 \\ \vdots \end{bmatrix}$ as vectors in $\mathbb{R}^n$).

1.4.5   Let P be the parallelepiped $0 \le x \le a$, $0 \le y \le b$, $0 \le z \le c$.

(a) What angle does a diagonal make with the sides? What relation is there between the length of a side and the corresponding angle?

(b) What are the angles between the diagonal and the faces of the parallelepiped? What relation is there between the area of a face and the corresponding angle?

1.4.6   (a) Prove Proposition 1.4.14 in the case where the coordinates **a** and **b** are positive, by subtracting pieces 1–6 from $(a_1 + b_1)(a_2 + b_2)$, as suggested by Figure 1.4.6.

(b) Repeat for the case where b_1 is negative.

1.4.7   (a) Find the equation of the line in the plane through the origin and perpendicular to the vector $\begin{bmatrix} 2 \\ -1 \end{bmatrix}$.

(b) Find the equation of the line in the plane through the point $\begin{pmatrix} 2 \\ 3 \end{pmatrix}$ and perpendicular to the vector $\begin{bmatrix} 2 \\ -4 \end{bmatrix}$.

1.4.8   (a) What is the length of $\vec{v}_n = \vec{e}_1 + \cdots + \vec{e}_n \in \mathbb{R}^n$?

(b) What is the angle α_n between $\vec{v}_n$ and $\vec{e}_1$? What is $\lim_{n\to\infty} \alpha_n$?

1.4.9   Compute the determinants of the matrices at left.

1.4.10   Compute the determinants of the matrices

(a) $\begin{bmatrix} 2 & -1 \\ 1 & 0 \end{bmatrix}$ (b) $\begin{bmatrix} 1 & 1 \\ 1 & 1 \end{bmatrix}$ (c) $\begin{bmatrix} a & b \\ 0 & d \end{bmatrix}$

1.4.11   Compute the determinants of the matrices in the margin at left.

1.4.12   Confirm the following formula for the inverse of a 3×3 matrix:

$$\begin{bmatrix} a_1 & b_1 & c_1 \\ a_2 & b_2 & c_2 \\ a_3 & b_3 & c_3 \end{bmatrix}^{-1} = \frac{1}{\det A} \begin{bmatrix} b_2c_3 - b_3c_2 & b_3c_1 - b_1c_3 & b_1c_2 - b_2c_1 \\ a_3c_2 - a_2c_3 & a_1c_3 - a_3c_1 & a_2c_1 - a_1c_2 \\ a_2b_3 - a_3b_2 & a_3b_1 - a_1b_3 & a_1b_2 - a_2b_1 \end{bmatrix}.$$

1.4.13   (a) What is the area of the parallelogram with vertices at

$$\begin{pmatrix} 0 \\ 0 \end{pmatrix}, \begin{pmatrix} 1 \\ 2 \end{pmatrix}, \begin{pmatrix} 5 \\ 1 \end{pmatrix}, \begin{pmatrix} 6 \\ 3 \end{pmatrix}?$$

(b) What is the area of the parallelogram with vertices at

$$\begin{pmatrix} 0 \\ 0 \end{pmatrix}, \begin{pmatrix} 1 \\ 2 \end{pmatrix}, \begin{pmatrix} 5 \\ -1 \end{pmatrix}, \begin{pmatrix} 6 \\ 1 \end{pmatrix}?$$

1.4.14   Compute the following cross products:

(a) $\begin{bmatrix} 2x \\ -y \\ 3z \end{bmatrix} \times \begin{bmatrix} x \\ 2y \\ 0 \end{bmatrix}$ (b) $\begin{bmatrix} 1 \\ 2 \\ 5 \end{bmatrix} \times \begin{bmatrix} 2 \\ 0 \\ 3 \end{bmatrix}$ (c) $\begin{bmatrix} 2 \\ -1 \\ -6 \end{bmatrix} \times \begin{bmatrix} 3 \\ 0 \\ 2 \end{bmatrix}$

1.4.15   Show that the cross product of two vectors pointing in the same direction is zero.

1.4.16   Given the vectors $\vec{u} = \begin{bmatrix} 1 \\ 2 \\ 1 \end{bmatrix}$, $\vec{v} = \begin{bmatrix} 2 \\ 0 \\ 1 \end{bmatrix}$, $\vec{w} = \begin{bmatrix} 1 \\ 0 \\ -1 \end{bmatrix}$:

(a) Compute $\vec{u} \times (\vec{v} \times \vec{w})$ and $(\vec{u} \times \vec{v}) \times \vec{w}$.

(b) Confirm that $\vec{v} \cdot (\vec{v} \times \vec{w}) = 0$. What is the geometrical relationship of $\vec{v}$ and $\vec{v} \times \vec{w}$?

1.4.17   Given two vectors, $\vec{v}$ and $\vec{w}$, show that $(\vec{v} \times \vec{w}) = -(\vec{w} \times \vec{v})$.

1.4.18   Let A be a 3×3 matrix with columns $\vec{a}, \vec{b}, \vec{c}$, and let Q_A be the 3×3 matrix with rows

$$(\vec{b} \times \vec{c})^\top, (\vec{c} \times \vec{a})^\top, (\vec{a} \times \vec{b})^\top.$$

(a) Compute Q_A when

$$A = \begin{bmatrix} 1 & 2 & 0 \\ 0 & -1 & 1 \\ 1 & 1 & -1 \end{bmatrix}.$$

In part (c) of Exercise 1.4.18 think of the geometric definition of the cross product, and the definition of the determinant of a 3×3 matrix in terms of cross products.

(b) What is the product $Q_A A$ in the case of (a) above?

(c) What is $Q_A A$ for any 3×3 matrix A?

(d) Can you relate this problem to Exercise 1.4.12?

1.4.19   (a) What is the length of

$$\vec{w}_n = \vec{e}_1 + 2\vec{e}_2 + \cdots + n\vec{e}_n = \sum_{i=1}^{n} i\vec{e}_i\,?$$

(b) What is the angle $\alpha_{n,k}$ between $\vec{w}_n$ and $\vec{e}_k$?

*(c) What are the limits

$$\lim_{n \to \infty} \alpha_{n,k} \quad , \quad \lim_{n \to \infty} \alpha_{n,n} \quad , \quad \lim_{n \to \infty} \alpha_{n,[n/2]}$$

where $[n/2]$ stands for the largest integer not greater than $n/2$?

1.4.20   For the two matrices and the vector

$$A = \begin{bmatrix} 1 & 0 \\ 1 & 2 \end{bmatrix}, \quad B = \begin{bmatrix} 2 & 0 \\ 0 & 1 \end{bmatrix}, \quad \vec{c} = \begin{bmatrix} 1 \\ 3 \end{bmatrix},$$

(a) compute $|A|, |B|, |\vec{c}|$;

(b) confirm that: $|AB| \leq |A||B|$, $\quad |A\vec{c}| \leq |A||\vec{c}|$, and $|B\vec{c}| \leq |B||\vec{c}|$.

1.4.21   Use direct computation to prove Schwarz's inequality (Theorem 1.4.6) in $\mathbb{R}^2$ for the standard inner product (dot product); i.e., show that for any numbers x_1, x_2, y_1, y_2, we have

$$|x_1 y_1 + x_2 y_2| \leq \sqrt{x_1^2 + x_2^2}\, \sqrt{y_1^2 + y_2^2}.$$

Exercises for Section 1.5:
Convergence and Limits

$$B = \begin{bmatrix} 1 & \epsilon & \epsilon \\ 0 & 1 & \epsilon \\ 0 & 0 & 1 \end{bmatrix},$$

Matrix for Exercise 1.5.1

$$B = \begin{bmatrix} 1 & -\epsilon \\ +\epsilon & 1 \end{bmatrix},$$

Matrix for Exercise 1.5.2

1.5.1   Find the inverse of the matrix B at left, by finding the matrix A such that $B = I - A$ and computing the value of the series $S = I + A + A^2 + A^3 + \dots$. This is easier than you might fear!

1.5.2   Following the procedure in Exercise 1.5.1, compute the inverse of the matrix B at left, where $|\epsilon| < 1$, using a geometric series.

1.5.3   Suppose $\sum_{i=1}^{\infty} \mathbf{x}_i$ is a convergent series in $\mathbb{R}^n$. Show that the triangle inequality applies:

$$\left| \sum_{i=1}^{\infty} \mathbf{x}_i \right| \leq \sum_{i=1}^{\infty} |\mathbf{x}_i|.$$

1.5.4   Let A be a square $n \times n$ matrix, and define

$$e^A = \sum_{k=0}^{\infty} \frac{1}{k!} A^k = I + A + \frac{1}{2} A^2 + \frac{1}{3!} A^3 + \dots.$$

(a) Show that the series converges for all A, and find a bound for $|e^A|$ in terms of $|A|$ and n.

(b) Compute explicitly e^A for the following values of A:

$$(1) \begin{bmatrix} a & 0 \\ 0 & b \end{bmatrix}, \quad (2) \begin{bmatrix} 0 & a \\ 0 & 0 \end{bmatrix}, \quad (3) \begin{bmatrix} 0 & a \\ -a & 0 \end{bmatrix}.$$

For the third above, you might look up the power series for $\sin x$ and $\cos x$.

(c) Prove the following, or find counterexamples:

(1) Do you think that $e^{A+B} = e^A e^B$ for all A and B? What if $AB = BA$?

(2) Do you think that $e^{2A} = \left(e^A\right)^2$ for all A?

1.5.5   For each of the following subsets X of $\mathbb{R}$ and $\mathbb{R}^2$, state whether it is open or closed (or both or neither), and prove it.

(a) $\{x \in \mathbb{R} \mid 0 < x \leq 1\}$

(b) $\left\{\begin{pmatrix} x \\ y \end{pmatrix} \in \mathbb{R}^2 \mid 1 < x^2 + y^2 < 2\right\}$

(c) $\left\{\begin{pmatrix} x \\ y \end{pmatrix} \in \mathbb{R}^2 \mid xy \neq 0\right\}$

(d) $\left\{\begin{pmatrix} x \\ y \end{pmatrix} \in \mathbb{R}^2 \mid y = 0\right\}$

*(e) $\{\mathbb{Q} \subset \mathbb{R}\}$ (the rational numbers)

1.5.6   (a) Show that the expression

$$\left| \begin{pmatrix} x \\ x^2 \end{pmatrix} - \begin{pmatrix} 2 \\ 3 \end{pmatrix} \right|^2$$

is a polynomial $p(x)$ of degree 4, and compute it.

(b) Use a computer to plot it; observe that it has two minima and a maximum. Evaluate approximately the absolute minimum: you should find something like .0663333 ... .

(c) What does this say about the radius of the largest disk centered at $\begin{pmatrix} 2 \\ 3 \end{pmatrix}$ which does not intersect the parabola of equation $y = x^2$. Is the number $1/12$ found in Example 1.5.6 sharp?

(d) Can you explain the meaning of the other local maxima and minima?

1.5.7   Find a formula for the radius r of the largest disk centered at $\begin{pmatrix} 2 \\ 3 \end{pmatrix}$ that doesn't intersect the parabola of equation $y = x^2$, using the following steps:

(a) Find the distance squared $\left| \begin{pmatrix} 2 \\ 3 \end{pmatrix} - \begin{pmatrix} x \\ x^2 \end{pmatrix} \right|^2$ as a 4th degree polynomial in x.

(b) Find the zeroes of the derivative by the method of Exercise 0.6.6.

(c) Find r.

1.5.8   For each of the following formulas, find its natural domain, and show whether it is open, closed or neither.

(a) $\sin \frac{1}{xy}$ (b) $\ln \sqrt{x^2 - y}$ (c) $\ln \ln x$

(d) $\arcsin \frac{3}{x^2 + y^2}$ (e) $\sqrt{e^{\cos xy}}$ (f) $\frac{1}{xyz}$

1.5.9   What subset of $\mathbb{R}$ is the natural domain of the function

$$(1 + x)^{1/x} \quad \text{of Example 1.5.18?}$$

1.5.10   This exercise has been deleted because it referred to material not in the published book.

1.5.11   Show that if X is a subset of $\mathbb{R}^n$, then $\overline{X}$ is closed.

1.5.12   Suppose that $\varphi : (0, \infty) \to (0, \infty)$ is a function such that $\lim_{\epsilon \to 0} \varphi(\epsilon) = 0$.

(a) Let $\mathbf{a}_1, \mathbf{a}_2, \ldots$ be a sequence in $\mathbb{R}^n$. Show that this sequence converges to $\mathbf{a}$ if and only if for any $\epsilon > 0$, there exists N such that for $n > N$, we have $|\mathbf{a}_n - \mathbf{a}| \leq \varphi(\epsilon)$.

(b) Find an analogous statement for limits of functions.

1.5.13   Prove the converse of Proposition 1.5.14: i.e., prove that if every convergent sequence in a set $C \subset \mathbb{R}^n$ converges to a point in C, then C is closed.

1.5.14   State whether the following limits exist, and prove it.

(a) $\lim\limits_{\begin{pmatrix} x \\ y \end{pmatrix} \to \begin{pmatrix} 1 \\ 2 \end{pmatrix}} \dfrac{x^2}{x + y}$

(b) $\lim\limits_{\begin{pmatrix} x \\ y \end{pmatrix} \to \begin{pmatrix} 0 \\ 0 \end{pmatrix}} \dfrac{\sqrt{|x|}\, y}{x^2 + y^2}$

(c) $\lim\limits_{\begin{pmatrix} x \\ y \end{pmatrix} \to \begin{pmatrix} 0 \\ 0 \end{pmatrix}} \dfrac{\sqrt{|xy|}}{\sqrt{x^2 + y^2}}$

(d) $\lim\limits_{\begin{pmatrix} x \\ y \end{pmatrix} \to \begin{pmatrix} 1 \\ 2 \end{pmatrix}} x^2 + y^3 - 3$

(e) $\lim\limits_{\begin{pmatrix} x \\ y \end{pmatrix} \to \begin{pmatrix} 0 \\ 0 \end{pmatrix}} \dfrac{x + y}{x^2 - y^2}$

(f) $\lim\limits_{\begin{pmatrix} x \\ y \end{pmatrix} \to \begin{pmatrix} 0 \\ 0 \end{pmatrix}} \dfrac{(x^2 + y^2)^2}{x + y}$

* (g) $\lim\limits_{\begin{pmatrix} x \\ y \end{pmatrix} \to \begin{pmatrix} 0 \\ 0 \end{pmatrix}} (x^2 + y^2)(\ln|xy|)$, defined when $xy \neq 0$.

(h) $\lim\limits_{\begin{pmatrix} x \\ y \end{pmatrix} \to \begin{pmatrix} 0 \\ 0 \end{pmatrix}} (x^2 + y^2) \ln(x^2 + y^2)$

1.5.15   (a) Let $D^* \subset \mathbb{R}^2$ be the region $0 < x^2 + y^2 < 1$, and let $f : D^* \to \mathbb{R}$ be a function. What does the following assertion mean?

$$\lim\limits_{\begin{pmatrix} x \\ y \end{pmatrix} \to \begin{pmatrix} 0 \\ 0 \end{pmatrix}} f\begin{pmatrix} x \\ y \end{pmatrix} = a.$$

(b) For the two functions below, defined on $\mathbb{R}^2 - \left\{ \begin{pmatrix} 0 \\ 0 \end{pmatrix} \right\}$, either show that the limit exists at $\begin{pmatrix} 0 \\ 0 \end{pmatrix}$ and find it, or show that it does not exist:

$$f\begin{pmatrix} x \\ y \end{pmatrix} = \dfrac{\sin(x + y)}{\sqrt{x^2 + y^2}} \qquad g\begin{pmatrix} x \\ y \end{pmatrix} = (|x| + |y|) \ln(x^2 + y^4)$$

1.5.16   Prove Theorem 1.5.13.

1.5.17   Prove Proposition 1.5.16: If a sequence $\vec{a}_k$ converges to $\vec{a}$, then any subsequence converges to the same limit.

1.5.18   (a) Show that the function $f(x) = |x|e^{-|x|}$ has an absolute maximum at some $x > 0$.

(b) What is the maximum of the function?

(c) Show that the image of f is $[0, 1/e]$.

1.5.19   Prove Theorem 1.5.27.

1.5.20   For what numbers θ does the sequence of matrices A_m (shown at left) converge? When does it have a convergent subsequence?

$$A_m = \begin{bmatrix} \cos m\theta & \sin m\theta \\ -\sin m\theta & \cos m\theta \end{bmatrix}$$

Sequence for Exercise 1.5.20

1.5.21   (a) Let $\operatorname{Mat}(n, m)$ denote the space of $n \times m$ matrices, which we will identify with $\mathbb{R}^{nm}$. For what numbers $a \in \mathbb{R}$ does the sequence of matrices $A^n \in \operatorname{Mat}(2, 2)$ converge as $n \to \infty$, when $A = \begin{bmatrix} a & a \\ a & a \end{bmatrix}$? What is the limit?

(b) What about 3×3 matrices filled with a's, or $n \times n$ matrices?

1.5.22   Let $U \subset \operatorname{Mat}(2, 2)$ be the set of matrices A such that $I - A$ is invertible.

(a) Show that U is open, and find a sequence in U that converges to I.

(b) Consider the mapping $f : U \to \operatorname{Mat}(2, 2)$ given by

$$f(A) = (A^2 - I)(A - I)^{-1}.$$

Does $\lim_{A \to I} f(A)$ exist? If so, what is the limit?

*(c) Let $B = \begin{bmatrix} 1 & 0 \\ 0 & -1 \end{bmatrix}$, and let $V \subset \operatorname{Mat}(2, 2)$ be the set of matrices A such that $A - B$ is invertible. Again, show that V is open, and that B can be approximated by elements of V.

*(d) Consider the mapping $g : V \to \operatorname{Mat}(2, 2)$ given by

$$g(A) = (A^2 - B^2)(A - B)^{-1}.$$

Does $\lim_{A \to B} g(A)$ exist? If so, what is the limit?

1.5.23   (a) Show that the matrix $A = \begin{bmatrix} 2 & 2 \\ 0 & 1 \end{bmatrix}$ represents a continuous mapping $\mathbb{R}^2 \to \mathbb{R}^2$.

*(b) Find an explicit δ in terms of ϵ.

(c) Now show that the mapping

$$\begin{bmatrix} x \\ y \end{bmatrix} \mapsto \begin{bmatrix} a & b \\ c & d \end{bmatrix} \begin{bmatrix} x \\ y \end{bmatrix} \quad \text{is continuous for any } a, b, d, c.$$

1.5.24   Let $\mathbf{a}_n = \begin{bmatrix} 3.14\ldots \\ 2.78\ldots \end{bmatrix}$; i.e., the two entries are π and e, to n places (for example, 3.14 is π to two places). How large does n have to be so that $\left| \mathbf{a}_n - \begin{bmatrix} \pi \\ e \end{bmatrix} \right| < 10^{-3}$? How large does n have to be so that $\left| \mathbf{a}_n - \begin{bmatrix} \pi \\ e \end{bmatrix} \right| < 10^{-4}$?

1.5.25   For the following functions, can you choose a value for f at $\begin{pmatrix} 0 \\ 0 \end{pmatrix}$ to make the function continuous?

(a) $f\begin{pmatrix} x \\ y \end{pmatrix} = \dfrac{1}{x^2 + y^2 + 1}$ (b) $f\begin{pmatrix} x \\ y \end{pmatrix} = \sqrt{1 - x^2 - y^2}$

(c) $f\begin{pmatrix} x \\ y \end{pmatrix} = (x^2 + y^2) \ln|x + y|$ (d) $f\begin{pmatrix} x \\ y \end{pmatrix} = (x^2 + y^2) \ln(x^2 + 2y^2)$

(e) $f\begin{pmatrix} x \\ y \end{pmatrix} = \dfrac{\sqrt{x^2 + y^2}}{|x| + |y|^{1/3}}$

Exercises for Section 1.6:
Four Big Theorems

1.6.1   Let $A \subset \mathbb{R}^n$ be a subset that is not compact. Show that there exists a continuous unbounded function on A.

Hint: If A is not bounded, then consider $f(\mathbf{x}) = |\mathbf{x}|$. If A is not closed, then consider $f(\mathbf{x}) = 1/|\mathbf{x} - \mathbf{a}|$ for an appropriate $\mathbf{a}$.

1.6.2   In the proof of the fundamental theorem of algebra (Theorem 1.6.10), justify the statement (Equation 1.6.25) that

$$|(b_{j+1}u^{j+1} + \cdots + u^k)| < |b_j u^j| \qquad \text{for small } u.$$

1.6.3   Set $z = x + iy$, where $x, y \in \mathbb{R}$. Show that the polynomial

$$p(z) = 1 + x^2 y^2$$

has no roots. Why doesn't this contradict Theorem 1.6.10?

1.6.4   Find, with justification, a number R such that there is a root of $p(z) = z^5 + 4z^3 + 3iz - 3$ in the disk $|z| \le R$. (You may use that a minimum of $|p|$ is a root of p.)

1.6.5   Consider the polynomial

$$p(z) = z^6 + 4z^4 + z + 2 = z^6 + q(z).$$

(a) Find R such that $|z^6| > |q(z)|$ when $|z| > R$.

(b) Find a number R_1 such that you are sure that the minimum of $|p(z)|$ occurs for $|z| < R_1$.

1.6.6   Prove that:

(a) Every polynomial over the complex numbers can be factored into linear factors.

(b) Every polynomial over the real numbers can be factored into real linear factors and real quadratic factors with complex roots.

1.6.7   Find a number R for which you can prove that the polynomial

$$p(z) = z^{10} + 2z^9 + 3z^8 + \cdots + 10z + 11$$

has a root for $|z| < R$. Explain your reasoning.

Exercises for Section 1.7:
Differential Calculus

1.7.1   Find the equation of the line tangent to the graph of $f(x)$ at $\begin{pmatrix} a \\ f(a) \end{pmatrix}$ for the following functions:

(a) $f(x) = \sin x$, $a = 0$ (b) $f(x) = \cos x$, $a = \pi/3$

(c) $f(x) = \cos x$, $a = 0$ (d) $f(x) = 1/x$, $a = 1/2$.

1.7.2   For what a is the tangent to the graph of $f(x) = e^{-x}$ at $\begin{pmatrix} a \\ e^{-a} \end{pmatrix}$ a line of the form $y = mx$?

1.7.3   Example 1.7.2 may lead you to expect that if f is differentiable at a, then $f(a+h) - f(a) - f'(a)h$ has something to do with h^2. It is not true that once you get rid of the linear term you always have a term that includes h^2. Try computing the derivative of

(a) $f(x) = |x|^{3/2}$ at 0   (b) $f(x) = x \ln|x|$ at 0   (c) $f(x) = x/\ln|x|$, also at 0

1.7.4   Find $f'(x)$ for the following functions f.

(a) $f(x) = \sin^3(x^2 + \cos x)$ (b) $f(x) = \cos^2((x + \sin x)^2)$

(c) $f(x) = (\cos x)^4 \sin x$ (d) $f(x) = (x + \sin^4 x)^3$

(e) $f(x) = \dfrac{\sin x^2 \sin^3 x}{2 + \sin x}$ (f) $f(x) = \sin\left(\dfrac{x^3}{\sin x^2}\right)$

1.7.5   Using Definition 1.7.1, show that $\sqrt{x^2}$ and $\sqrt[3]{x^2}$ are not differentiable at 0, but that $\sqrt{x^4}$ is.

1.7.6   What are the partial derivatives $D_1 f$ and $D_2 f$ of the following functions, at the points $\begin{pmatrix} 2 \\ 1 \end{pmatrix}$ and $\begin{pmatrix} 1 \\ -2 \end{pmatrix}$:

(a) $f\begin{pmatrix} x \\ y \end{pmatrix} = \sqrt{x^2 + y}$;   (b) $f\begin{pmatrix} x \\ y \end{pmatrix} = x^2 y + y^4$,

(c) $f\begin{pmatrix} x \\ y \end{pmatrix} = \cos xy + y \cos y$;   (d) $f\begin{pmatrix} x \\ y \end{pmatrix} = \dfrac{xy^2}{\sqrt{x + y^2}}$.

You really must learn the notation for partial derivatives used in Exercise 1.7.7, as it is used practically everywhere, but we much prefer $D_1 \vec{f}$, etc.

1.7.7   Calculate the partial derivatives

$$\frac{\partial \mathbf{f}}{\partial x}, \quad \text{and} \quad \frac{\partial \mathbf{f}}{\partial y} \quad \text{for the vector-valued functions:}$$

(a) $\mathbf{f}\begin{pmatrix} x \\ y \end{pmatrix} = \begin{pmatrix} \cos x \\ x^2 y + y^2 \\ \sin(x^2 - y) \end{pmatrix}$ and   (b) $\mathbf{f}\begin{pmatrix} x \\ y \end{pmatrix} = \begin{pmatrix} \sqrt{x^2 + y^2} \\ xy \\ \sin^2 xy \end{pmatrix}$.

1.7.8   Write the answers to Exercise 1.7.7 in the form of the Jacobian matrix.

1.7.9   (a) Given a vector-valued function

$$\mathbf{f}\begin{pmatrix} x \\ y \end{pmatrix} = \begin{pmatrix} f_1 \\ f_2 \end{pmatrix}, \text{ with Jacobian matrix } \begin{bmatrix} 2x\cos(x^2+y) & \cos(x^2+y) \\ ye^{xy} & xe^{xy} \end{bmatrix},$$

what is D_1 of the function f_1? D_2 of the function f_1? D_2 of f_2?

(b) What are the dimensions of the Jacobian matrix of a vector-valued function

$$\mathbf{f}\begin{pmatrix} x \\ y \end{pmatrix} = \begin{pmatrix} f_1 \\ f_2 \\ f_3 \end{pmatrix} ?$$

1.7.10   What is the derivative of the function $f : \mathbb{R}^n \to \mathbb{R}^n$ given by the formula $f(\mathbf{x}) = |\mathbf{x}|^2 \mathbf{x}$?

1.7.11   Show that if $f(x) = |x|$, then for any number m,

$$\lim_{h \to 0} \big(f(0+h) - f(0) - mh\big) = 0,$$

but that

$$\lim_{h \to 0} \frac{1}{h}\big(f(0+h) - f(0) - mh\big) = 0$$

never exists: there is no number m such that mh is a "good approximation" to $f(h) - f(0)$ in the sense of Definition 1.7.7.

1.7.12   (a) Show that the mapping

$$\text{Mat}\,(n,n) \to \text{Mat}\,(n,n), \quad A \mapsto A^3$$

is differentiable, and compute its derivative.

(b) Compute the derivative of the mapping

$$\text{Mat}\,(n,n) \to \text{Mat}\,(n,n), \quad A \mapsto A^k \quad \text{for any integer } k \geq 1.$$

1.7.13   (a) Define what it means for a mapping $F : \text{Mat}\,(n,m) \to \text{Mat}\,(k,l)$ to be differentiable at a point $A \in \text{Mat}\,(n,m)$.

(b) Consider the function $F : \text{Mat}\,(n,m) \to \text{Mat}\,(n,n)$ given by

$$F(A) = AA^\top.$$

Show that F is differentiable, and compute the derivative $[\mathbf{D}F(A)]$.

1.7.14   Compute the derivative of the mapping $A \mapsto AA^\top$.

1.7.15   Let $A = \begin{bmatrix} a & b \\ c & d \end{bmatrix}$ and $A^2 = \begin{bmatrix} a_1 & b_1 \\ c_1 & d_1 \end{bmatrix}$.

$$S \begin{pmatrix} a \\ b \\ c \\ d \end{pmatrix} = \begin{pmatrix} a_1 \\ b_1 \\ c_1 \\ d_1 \end{pmatrix}.$$

The function of Exercise 1.7.15

Exercise 1.7.16, part (a) should be with the exercises for Section 1.8.

(a) Write the formula for the function $S : \mathbb{R}^4 \to \mathbb{R}^4$ defined at left.

(b) Find the Jacobian matrix of S.

(c) Check that your answer agrees with Example 1.7.15.

(d) (For the courageous): Do the same for 3×3 matrices.

1.7.16   Which of the following functions are differentiable at $\begin{pmatrix} 0 \\ 0 \end{pmatrix}$?

(a) $f \begin{pmatrix} x \\ y \end{pmatrix} = \sin(e^{xy})$

(b) $f \begin{pmatrix} x \\ y \end{pmatrix} = \begin{cases} \dfrac{\sin(xy)}{x^2 + y^2} & \text{if } \begin{pmatrix} x \\ y \end{pmatrix} \neq \begin{pmatrix} 0 \\ 0 \end{pmatrix} \\ 0 & \text{if } \begin{pmatrix} x \\ y \end{pmatrix} = \begin{pmatrix} 0 \\ 0 \end{pmatrix} \end{cases}$

(c) $f \begin{pmatrix} x \\ y \end{pmatrix} = |x + y|$

(d) $f \begin{pmatrix} x \\ y \end{pmatrix} = \begin{cases} \dfrac{x^2 y}{x^2 + y^2} & \text{if } \begin{pmatrix} x \\ y \end{pmatrix} \neq \begin{pmatrix} 0 \\ 0 \end{pmatrix} \\ 0 & \text{if } \begin{pmatrix} x \\ y \end{pmatrix} = \begin{pmatrix} 0 \\ 0 \end{pmatrix} \end{cases}$

1.7.17   Find the Jacobian matrices of the following mappings:

(a) $f \begin{pmatrix} x \\ y \end{pmatrix} = \sin(xy)$

b) $f \begin{pmatrix} x \\ y \end{pmatrix} = e^{(x^2 + y^3)}$

(c) $\mathbf{f} \begin{pmatrix} x \\ y \end{pmatrix} = \begin{pmatrix} xy \\ x + y \end{pmatrix}$

d) $\mathbf{f} \begin{pmatrix} r \\ \theta \end{pmatrix} = \begin{pmatrix} r \cos \theta \\ r \sin \theta \end{pmatrix}$

1.7.18   In Example 1.7.15, prove that the derivative $AH + HA$ is the "same" as the Jacobian matrix computed with partial derivatives.

1.7.19   (a) Let $U \subset \mathbb{R}^n$ be open and $\mathbf{f} : U \to \mathbb{R}^m$ be a mapping. When is $\mathbf{f}$ differentiable at $\mathbf{a} \in U$? What is its derivative?

(b) Is the mapping $\mathbf{f} : \mathbb{R}^n \to \mathbb{R}^n$ given by

$$\mathbf{f}(\vec{x}) = |\vec{x}|\vec{x}$$

differentiable at the origin? If so, what is its derivative?

Hint for 1.7.20 (a): Think of

$$A = \begin{bmatrix} a & b \\ c & d \end{bmatrix} \text{ as the element } \begin{pmatrix} a \\ b \\ c \\ d \end{pmatrix}$$

of $\mathbb{R}^4$. Use the formula for computing the inverse of a 2×2 matrix (Equation 1.2.15).

1.7.20   (a) Compute the derivative (Jacobian matrix) for the function $\mathbf{f}(A) = A^{-1}$ described in Proposition 1.7.16, when A is a 2×2 matrix.

(b) Show that your result agrees with the result of Proposition 1.7.16.

1.7.21   Considering the determinant as a function only of 2×2 matrices, i.e., $\det : \text{Mat}\,(2,2) \mapsto \mathbb{R}$, show that

$$[\mathbf{D} \det(I)]H = h_{1,1} + h_{2,2},$$

where I of course is the identity and H is the increment matrix

$$H = \begin{bmatrix} h_{1,1} & h_{1,2} \\ h_{2,1} & h_{2,2} \end{bmatrix}.$$

1.8.1   (a) Prove Leibnitz's rule (part (5) of Theorem 1.8.1) directly when $f : U \to \mathbb{R}$ and $\mathbf{g} : U \to \mathbb{R}^m$ are differentiable at $\mathbf{a}$, by writing

$$\lim_{\vec{\mathbf{h}} \to 0} \frac{\left| f(\mathbf{a} + \vec{\mathbf{h}})\mathbf{g}(\mathbf{a} + \vec{\mathbf{h}}) - f(\mathbf{a})\mathbf{g}(\mathbf{a}) - f(\mathbf{a}) \left([\mathbf{Dg}(\mathbf{a})]\vec{\mathbf{h}} \right) - \left([\mathbf{D}f(\mathbf{a})]\vec{\mathbf{h}} \right) \mathbf{g}(\mathbf{a}) \right|}{|\vec{\mathbf{h}}|} = 0,$$

and then developing the term under the limit:

$$\left(f(\mathbf{a} + \vec{\mathbf{h}}) - f(\mathbf{a}) \right) \frac{\mathbf{g}(\mathbf{a} + \vec{\mathbf{h}}) - \mathbf{g}(\mathbf{a})}{|\vec{\mathbf{h}}|} + f(\mathbf{a}) \left(\frac{\mathbf{g}(\mathbf{a} + \vec{\mathbf{h}}) - \mathbf{g}(\mathbf{a}) - [\mathbf{Dg}(\mathbf{a})]\vec{\mathbf{h}}}{|\vec{\mathbf{h}}|} \right)$$

$$+ \left(\frac{f(\mathbf{a} + \vec{\mathbf{h}}) - f(\mathbf{a}) - [\mathbf{D}f(\mathbf{a})]\vec{\mathbf{h}}}{|\vec{\mathbf{h}}|} \right) \mathbf{g}(\mathbf{a}).$$

(b) Prove the rule for differentiating dot products (part (6) of Theorem 1.8.1) by a similar decomposition.

(c) Show by a similar argument that if $\mathbf{f}, \mathbf{g} : U \to \mathbb{R}^3$ are both differentiable at $\mathbf{a}$, then so is the cross product $\mathbf{f} \times \mathbf{g} : U \to \mathbb{R}^3$. Find the formula for this derivative.

1.8.2   (a) What is the derivative of the function

Hint for Exercise 1.8.2: think of the composition of

$$t \mapsto \begin{pmatrix} t \\ t^2 \end{pmatrix} \quad \text{and}$$

$$\begin{pmatrix} x \\ y \end{pmatrix} \mapsto \int_x^y \frac{ds}{s + \sin s},$$

both of which you should know how to differentiate.

$$f(t) = \int_t^{t^2} \frac{ds}{s + \sin s}, \quad \text{defined for } t > 1?$$

(b) When is f increasing or decreasing?

1.8.3   Consider the function

$$f \begin{pmatrix} x_1 \\ \vdots \\ x_n \end{pmatrix} = \sum_{i=1}^{n-1} x_i x_{i+1} \quad \text{and the curve } \gamma : \mathbb{R} \to \mathbb{R}^n \text{ given by} \quad \gamma(t) = \begin{pmatrix} t \\ t^2 \\ \vdots \\ t^n \end{pmatrix}.$$

What is the derivative of the function $t \mapsto f\big(\gamma(t)\big)$?

1.8.4   True or false? Justify your answer. If $\mathbf{f} : \mathbb{R}^2 \to \mathbb{R}^2$ is a differentiable function with

$$\mathbf{f} \begin{pmatrix} 0 \\ 0 \end{pmatrix} = \begin{pmatrix} 1 \\ 1 \end{pmatrix} \quad \text{and} \quad \left[\mathbf{Df} \begin{pmatrix} 0 \\ 0 \end{pmatrix} \right] = \begin{bmatrix} 1 & 1 \\ 1 & 1 \end{bmatrix},$$

there is no differentiable mapping $\mathbf{g} : \mathbb{R}^2 \to \mathbb{R}^2$ with

$$\mathbf{g} \begin{pmatrix} 1 \\ 1 \end{pmatrix} = \begin{pmatrix} 0 \\ 0 \end{pmatrix} \quad \text{and} \quad \mathbf{f} \circ \mathbf{g} \begin{pmatrix} x \\ y \end{pmatrix} = \begin{pmatrix} y \\ x \end{pmatrix}.$$

1.8.5   Let $\varphi : \mathbb{R} \to \mathbb{R}$ be any differentiable function. Show that the function

$$f \begin{pmatrix} x \\ y \end{pmatrix} = y\varphi(x^2 - y^2)$$

satisfies the equation

$$\frac{1}{x}D_1 f\left(\begin{array}{c}x\\y\end{array}\right) + \frac{1}{y}D_2 f\left(\begin{array}{c}x\\y\end{array}\right) = \frac{1}{y^2}f\left(\begin{array}{c}x\\y\end{array}\right).$$

1.8.6   (a) Show that if a function $\mathbf{f} : \mathbb{R}^2 \to \mathbb{R}^2$ can be written $\varphi(x^2 + y^2)$ for some function $\varphi : \mathbb{R} \to \mathbb{R}$, then it satisfies

$$x\overrightarrow{D_2}\mathbf{f} - y\overrightarrow{D_1}\mathbf{f} = 0.$$

Hint for part (b): What is the "partial derivative of f with respect to the polar angle θ"?

*(b) Show the converse: every function satisfying $x\overrightarrow{D_2}\mathbf{f} - y\overrightarrow{D_1}\mathbf{f} = 0$ can be written $\varphi(x^2 + y^2)$ for some function $\varphi : \mathbb{R} \to \mathbb{R}$.

Exercise 1.8.7: It's a lot easier to think of this as the composition of $A \mapsto A^3$ and $A \mapsto A^{-1}$ and to apply the chain rule, than to compute the derivative directly.

1.8.7   Referring to Example 1.8.4: (a) Compute the derivative of the map $A \mapsto A^{-3}$;

(b) Compute the derivative of the map $A \mapsto A^{-n}$.

1.8.8   If $f\left(\begin{array}{c}x\\y\end{array}\right) = \varphi\left(\dfrac{x+y}{x-y}\right)$ for some differentiable function $\varphi : \mathbb{R} \to \mathbb{R}$, show that

$$xD_1 f + yD_2 f = 0.$$

1.8.9   True or false? Explain your answers. (a) If $\mathbf{f} : \mathbb{R}^2 \to \mathbb{R}^2$ is differentiable, and $[D\mathbf{f}(\mathbf{0})]$ is not invertible, then there is no differentiable function $\mathbf{g} : \mathbb{R}^2 \to \mathbb{R}^2$ such that $\mathbf{g} \circ \mathbf{f}(\mathbf{x}) = \mathbf{x}$.

(b) Differentiable functions have continuous partial derivatives.

Exercises for Section 1.9:

Criteria for Differentiability

1.9.1   Show that the function

$$f(x) = \begin{cases} \frac{x}{2} + x^2 \sin\frac{1}{x} & \text{if } x \neq 0 \\ 0 & \text{if } x = 0 \end{cases}$$

is differentiable at 0, with derivative $f'(0) = 1/2$.

Exercise 1.9.2:   Remember, sometimes you have to use the definition of the derivative, rather than the rules for computing derivatives.

1.9.2   (a) Show that for

$$f\left(\begin{array}{c}x\\y\end{array}\right) = \begin{cases} \dfrac{3x^2 y - y^3}{x^2 + y^2} & \text{if } \left(\begin{array}{c}x\\y\end{array}\right) \neq \left(\begin{array}{c}0\\0\end{array}\right) \\ 0 & \text{if } \left(\begin{array}{c}x\\y\end{array}\right) = \left(\begin{array}{c}0\\0\end{array}\right), \end{cases}$$

all directional derivatives exist, but that f is not differentiable at the origin.

**(b) Show that there exists a function which has directional derivatives everywhere and isn't continuous, or even bounded. (Hint: Consider Example 1.5.24.)

$$f\begin{pmatrix} x \\ y \end{pmatrix} = \begin{cases} \dfrac{xy}{x^2+y^2} & \text{if } \begin{pmatrix} x \\ y \end{pmatrix} \neq \begin{pmatrix} 0 \\ 0 \end{pmatrix} \\ 0 & \text{if } \begin{pmatrix} x \\ y \end{pmatrix} = \begin{pmatrix} 0 \\ 0 \end{pmatrix} \end{cases}$$

Function for Exercise 1.9.3

1.9.3   Consider the function defined on $\mathbb{R}^2$ given by the formula at left.

(a) Show that both partial derivatives exist everywhere.

(b) Where is f differentiable?

1.9.4   Consider the function $f : \mathbb{R}^2 \to \mathbb{R}$ given by

$$f\begin{pmatrix} x \\ y \end{pmatrix} = \begin{cases} \dfrac{\sin(x^2 y^2)}{x^2+y^2} & \text{if } \begin{pmatrix} x \\ y \end{pmatrix} \neq \begin{pmatrix} 0 \\ 0 \end{pmatrix} \\ 0 & \text{if } \begin{pmatrix} x \\ y \end{pmatrix} = \begin{pmatrix} 0 \\ 0 \end{pmatrix}. \end{cases}$$

(a) What does it mean to say that f is differentiable at $\begin{pmatrix} 0 \\ 0 \end{pmatrix}$?

(b) Show that both partial derivatives $D_1 f \begin{pmatrix} 0 \\ 0 \end{pmatrix}$ and $D_2 f \begin{pmatrix} 0 \\ 0 \end{pmatrix}$ exist, and compute them.

Hint for Exercise 1.9.4, part (c): You may find the following fact useful: $|\sin x| \leq |x|$ for all $x \in \mathbb{R}$.

(c) Is f differentiable at $\begin{pmatrix} 0 \\ 0 \end{pmatrix}$?

1.9.5   Consider the function defined on $\mathbb{R}^3$ defined by the formulas

$$f\begin{pmatrix} x \\ y \\ z \end{pmatrix} = \begin{cases} \dfrac{xyz}{x^4+y^4+z^4} & \text{if } \begin{pmatrix} x \\ y \\ z \end{pmatrix} \neq \begin{pmatrix} 0 \\ 0 \\ 0 \end{pmatrix} \\ 0 & \text{if } \begin{pmatrix} x \\ y \\ z \end{pmatrix} = \begin{pmatrix} 0 \\ 0 \\ 0 \end{pmatrix}. \end{cases}$$

(a) Show that all partial derivatives exist everywhere.

(b) Where is f differentiable?

2

Solving Equations

In 1985, John Hubbard was asked to testify before the Committee on Science and Technology of the U.S. House of Representatives. He was preceded by a chemist from DuPont who spoke of modeling molecules, and by an official from the geophysics institute of California, who spoke of exploring for oil and attempting to predict tsunamis.

When it was his turn, he explained that when chemists model molecules, they are solving Schrödinger's equation, that exploring for oil requires solving the Gelfand-Levitan equation, and that predicting tsunamis means solving the Navier-Stokes equation. Astounded, the chairman of the committee interrupted him and turned to the previous speakers. "Is that true, what Professor Hubbard says?" he demanded. "Is it true that what you do is solve <u>equations</u>?"

2.0 INTRODUCTION

In every subject, language is intimately related to understanding.

"It is impossible to dissociate language from science or science from language, because every natural science always involves three things: the sequence of phenomena on which the science is based; the abstract concepts which call these phenomena to mind; and the words in which the concepts are expressed. To call forth a concept a word is needed; to portray a phenomenon, a concept is needed. All three mirror one and the same reality."—Antoine Lavoisier, 1789.

"Professor Hubbard, you always underestimate the difficulty of vocabulary."—Helen Chigirinskaya, Cornell University, 1997.

All readers of this book will have solved systems of simultaneous *linear* equations. Such problems arise throughout mathematics and its applications, so a thorough understanding of the problem is essential.

What most students encounter in high school is systems of n equations in n unknowns, where n might be general or might be restricted to $n = 2$ and $n = 3$. Such a system usually has a unique solution, but sometimes something goes wrong: some equations are "consequences of others," and have infinitely many solutions; other systems of equations are "incompatible," and have no solutions. This chapter is largely concerned with making these notions systematic.

A language has evolved to deal with these concepts, using the words "linear transformation," "linear combination," "linear independence," "kernel," "span," "basis," and "dimension." These words may sound unfriendly, but they correspond to notions which are unavoidable and actually quite transparent if thought of in terms of linear equations. They are needed to answer questions like: "how many equations are consequences of the others?"

The relationship of these words to linear equations goes further. Theorems in linear algebra can be proved with abstract induction proofs, but students generally prefer the following method, which we discuss in this chapter:

Reduce the statement to a statement about linear equations, row reduce the resulting matrix, and see whether the statement becomes obvious.

If so, the statement is true; otherwise it is likely to be false.

Solving nonlinear equations is much harder. In the days before computers, finding solutions was virtually impossible; even in the good cases, where mathematicians could prove that solutions existed, they were usually not concerned with whether their proof could be turned into a practical algorithm to find the solutions in question. The advent of computers has made such an abstract approach unreasonable. Knowing that a system of equations has solutions is no longer enough; we want a practical algorithm that will enable us to solve them. The algorithm most often used is *Newton's method*. In Section 2.7 we will show Newton's method in action, and state Kantorovitch's theorem, which guarantees that under appropriate circumstances Newton's method converges to a solution; in Section 2.8 we discuss the superconvergence of Newton's method and state a stronger version of Kantorovitch's theorem, using the *norm* of a matrix rather than its length.

In Section 2.9 we will base the implicit and inverse function theorems on Newton's method. This gives more precise statements than the standard approach, and we do not believe that it is harder.

2.1 THE MAIN ALGORITHM: ROW REDUCTION

Suppose we want to solve the system of linear equations

$$
\begin{aligned}
2x + y + 3z &= 1 \\
x - y \quad\;\; &= 1 \\
2x \quad\;\; + z &= 1.
\end{aligned}
\qquad 2.1.1
$$

We could add together the first and second equations to get $3x + 3z = 2$. Substituting $(2 - 3z)/3$ for x in the third equation will give $z = 1/3$, hence $x = 1/3$; putting this value for x into the second equation then gives $y = -2/3$.

In this section we will show how to make this approach systematic, using *row reduction*. The big advantage of row reduction is that it requires no cleverness, as we will see in Theorem 2.1.8. It gives a recipe so simple that the dumbest computer can follow it.

The first step is to write the system of Equation 2.1.1 in matrix form. We can write the coefficients as one matrix, the unknowns as a vector and the constants on the right as another vector:

$$
\underbrace{\begin{bmatrix} 2 & 1 & 3 \\ 1 & -1 & 0 \\ 2 & 0 & 1 \end{bmatrix}}_{\text{coefficient matrix } (A)}
\qquad
\underbrace{\begin{bmatrix} x \\ y \\ z \end{bmatrix}}_{\text{vector of unknowns } (\vec{\mathbf{x}})}
\qquad
\underbrace{\begin{bmatrix} 1 \\ 1 \\ 1 \end{bmatrix}}_{\text{constants } (\vec{\mathbf{b}})} .
$$

Our system of equations can thus be written as the matrix multiplication $A\vec{\mathbf{x}} = \vec{\mathbf{b}}$:

The matrix A uses position to impart information, as do Arabic numbers; in both cases, 0 plays a crucial role as place holder. In the number 4 084, the two 4's have very different meanings, as do the 1's in the matrix: the 1 in the first column is the coefficient of x, the 1's in the second column are the coefficients of y, and that in the third column is the coefficient of z.

Using position to impart information allows for concision; in Roman numerals, 4 084 is

MMMMLXXXIIII.

(To some extent, we use position when writing Roman numerals, as in IV = 4 and VI = 6, but the Romans themselves were quite happy writing their numbers in any order, MMXXM for 3 020, for example.)

The ith column of the matrix A corresponds to the ith unknown.

The first subscript in a pair of subscripts refers to vertical position, and the second to horizontal position: $a_{1,n}$ is the coefficient for the top row, nth column: *first take the elevator, then walk down the hall.*

The matrix $[A, \vec{b}]$ is shorthand for the equation $A\vec{x} = \vec{b}$.

$$\overbrace{\underbrace{\begin{bmatrix} 2 & 1 & 3 \\ 1 & -1 & 0 \\ 2 & 0 & 1 \end{bmatrix}}_{A} \overbrace{\begin{bmatrix} x \\ y \\ z \end{bmatrix}}^{\vec{x}} \underbrace{\begin{bmatrix} 1 \\ 1 \\ 1 \end{bmatrix}}_{\vec{b}}} \qquad 2.1.2$$

We now use a shorthand notation, omitting the vector $\vec{x}$, and writing A and $\vec{b}$ as a single matrix, with $\vec{b}$ the last column of the new matrix:

$$\begin{bmatrix} 2 & 1 & 3 & 1 \\ 1 & -1 & 0 & 1 \\ 2 & 0 & 1 & 1 \end{bmatrix}. \qquad 2.1.3$$

$$\underbrace{}_{A} \underbrace{}_{\vec{b}}$$

More generally, we see that a system of equations

$$\begin{matrix} a_{1,1}x_1 & +\cdots+ & a_{1,n}x_n & = b_1 \\ \vdots & \cdots & \vdots & \vdots \\ \vdots & \cdots & \vdots & \vdots \\ a_{m,1}x_1 & +\cdots+ & a_{m,n}x_n & = b_m \end{matrix} \qquad 2.1.4$$

is the same as $A\vec{x} = \vec{b}$:

$$\underbrace{\begin{bmatrix} a_{1,1} & \cdots & a_{1,n} \\ \vdots & & \vdots \\ a_{m,1} & \cdots & a_{m,n} \end{bmatrix}}_{A} \underbrace{\begin{bmatrix} x_1 \\ \vdots \\ x_n \end{bmatrix}}_{\vec{x}} = \underbrace{\begin{bmatrix} b_1 \\ \vdots \\ b_m \end{bmatrix}}_{\vec{b}}, \quad \text{i.e.,} \quad \underbrace{\begin{bmatrix} a_{1,1} & \cdots & a_{1,n} & b_1 \\ \vdots & \cdots & \vdots & \vdots \\ a_{m,1} & \cdots & a_{m,n} & b_m \end{bmatrix}}_{[A,\vec{b}]}. \qquad 2.1.5$$

We denote by $[A, \vec{b}]$, with a comma, the matrix obtained by putting side-by-side the matrix A of coefficients and the vector $\vec{b}$, as in the right-hand side of Equation 2.1.5. The comma is intended to avoid confusion with multiplication; we are not multiplying A and $\vec{b}$.

How would you write in matrix form the system of equations

$$x + 3z = 2$$
$$2x + y + z = 0$$
$$2y + z = 1?$$

Check your answer below.[1]

Row operations

We can solve a system of linear equations by *row reducing* the corresponding matrix, using *row operations*.

Definition 2.1.1 (Row operations). A row operation on a matrix is one of three operations:

    (1) Multiplying a row by a nonzero number;

    (2) Adding a multiple of a row onto another row;

    (3) Exchanging two rows.

Exercise 2.1.3 asks you to show that the third operation is not necessary; one can exchange rows using operations (1) and (2).

There are two good reasons why row operations are important. The first is that they require only arithmetic: addition, subtraction, multiplication, and division. This is what computers do well; in some sense it is all they can do. And they spend a lot of time doing it: row operations are fundamental to most other mathematical algorithms.

Remark 2.1.2. We could just as well talk about *column operations*, substituting the word *column* for the word *row* in Definition 2.1.1. We will use column operations in Section 4.8.

The other reason is that they will enable us to solve systems of linear equations:

Theorem 2.1.3. *If the matrix $[A, \vec{b}]$ representing a system of linear equations $A\vec{x} = \vec{b}$ can be turned into $[A', \vec{b}']$ by a sequence of row operations, then the set of solutions of $A\vec{x} = \vec{b}$ and set of solutions of $A'\vec{x} = \vec{b}'$ coincide.*

Proof. Row operations consist of multiplying one equation by a nonzero number, adding a multiple of one equation to another and exchanging two equations. Any solution of $A\vec{x} = \vec{b}$ is thus a solution of $A'\vec{x} = \vec{b}'$. In the other direction, any row operation can be undone by another row operation (Exercise 2.1.4), so any solution $A'\vec{x} = \vec{b}'$ is also a solution of $A\vec{x} = \vec{b}$. $\square$

Theorem 2.1.3 suggests that we solve $A\vec{x} = \vec{b}$ by using row operations to bring the system of equations to the most convenient form. In Example 2.1.4 we apply this technique to Equation 2.1.1. For now, don't worry about how the row reduction was achieved; this will be discussed soon, in the proof of Theorem 2.1.8. Concentrate instead on what the row reduced matrix tells us about solutions to the system of equations.

$$
{}_1\begin{bmatrix} 1 & 0 & 3 & 2 \\ 2 & 1 & 1 & 0 \\ 0 & 2 & 1 & 1 \end{bmatrix} \; ; \text{ i.e., } \begin{bmatrix} 1 & 0 & 3 \\ 2 & 1 & 1 \\ 0 & 2 & 1 \end{bmatrix} \begin{bmatrix} x \\ y \\ z \end{bmatrix} = \begin{bmatrix} 2 \\ 0 \\ 1 \end{bmatrix}.
$$

Example 2.1.4 (Solving a system of equations with row operations).
To solve

$$2x + y + 3z = 1$$
$$x - y = 1 \qquad \qquad 2.1.6$$
$$2x + z = 1,$$

we can use row operations to bring the matrix

$$\begin{bmatrix} 2 & 1 & 3 & 1 \\ 1 & -1 & 0 & 1 \\ 2 & 0 & 1 & 1 \end{bmatrix} \text{ to the form } \begin{bmatrix} 1 & 0 & 0 & 1/3 \\ 0 & 1 & 0 & -2/3. \\ 0 & 0 & 1 & 1/3 \end{bmatrix} \qquad 2.1.7$$
$$\underbrace{}_{\widetilde{A}} \underbrace{}_{\widetilde{\mathbf{b}}}$$

We said not to worry about how we did the row reduction in Equation 2.1.7. But if you do worry, here are the steps: To get (1), divide Row 1 by 2, and add $-1/2$ Row 1 to Row 2, and subtract Row 1 from Row 3. To get from (1) to (2), multiply Row 2 by $-2/3$, and then add that result to Row 3. From (2) to (3), subtract half of Row 2 from Row 1. For (4), subtract Row 3 from Row 1. For (5), subtract Row 3 from Row 2.

(To distinguish the new A and $\vec{\mathbf{b}}$ from the old, we put a "tilde" on top: $\widetilde{A}, \widetilde{\mathbf{b}}$.) In this case, the solution can just be read off the matrix. If we put the unknowns back in the matrix, we get

$$\begin{bmatrix} x & 0 & 0 & 1/3 \\ 0 & y & 0 & -2/3 \\ 0 & 0 & z & 1/3 \end{bmatrix} \text{ or } \begin{array}{r} x = 1/3 \\ y = -2/3 \\ z = 1/3 \end{array} \triangle \qquad 2.1.8$$

$$(1) \begin{bmatrix} 1 & 1/2 & 3/2 & 1/2 \\ 0 & -3/2 & -3/2 & 1/2 \\ 0 & -1 & -2 & 0 \end{bmatrix}$$

$$(2) \begin{bmatrix} 1 & 1/2 & 3/2 & 1/2 \\ 0 & 1 & 1 & -1/3 \\ 0 & 0 & -1 & -1/3 \end{bmatrix}$$

$$(3) \begin{bmatrix} 1 & 0 & 1 & 2/3 \\ 0 & 1 & 1 & -1/3 \\ 0 & 0 & 1 & 1/3 \end{bmatrix}$$

$$(4) \begin{bmatrix} 1 & 0 & 0 & 1/3 \\ 0 & 1 & 1 & -1/3 \\ 0 & 0 & 1 & 1/3 \end{bmatrix}$$

$$(5) \begin{bmatrix} 1 & 0 & 0 & 1/3 \\ 0 & 1 & 0 & -2/3 \\ 0 & 0 & 1 & 1/3 \end{bmatrix}.$$

Echelon form is generally considered best for solving systems of linear equations. (But it is not best for all purposes. See Exercise 2.1.9.)

Echelon form

Of course some systems of linear equations may have no solutions, and others may have infinitely many. But if a system has solutions, they can be found by an appropriate sequence of row operations, called *row reduction*, bringing the matrix to *echelon form*, as in the second matrix of Equation 2.1.7.

Definition 2.1.5 (Echelon form). A matrix is in echelon form if:

(1) In every row, the first nonzero entry is 1, called a pivotal 1.

(2) The pivotal 1 of a lower row is always to the right of the pivotal 1 of a higher row;

(3) In every column that contains a pivotal 1, all other entries are 0.

(4) Any rows consisting entirely of 0's are at the bottom.

Clearly, the identity matrix is in echelon form.

Example 2.1.6 (Matrices in echelon form). The following matrices are in echelon form; the pivotal 1's are underlined:

$$\begin{bmatrix} \underline{1} & 0 & 0 & 3 \\ 0 & \underline{1} & 0 & -2 \\ 0 & 0 & \underline{1} & 1 \end{bmatrix}, \quad \begin{bmatrix} \underline{1} & 1 & 0 & 0 \\ 0 & 0 & \underline{1} & 0 \\ 0 & 0 & 0 & \underline{1} \end{bmatrix}, \quad \begin{bmatrix} 0 & \underline{1} & 3 & 0 & 0 & 3 & 0 & -4 \\ 0 & 0 & 0 & \underline{1} & -2 & 1 & 0 & 1 \\ 0 & 0 & 0 & 0 & 0 & 0 & \underline{1} & 2 \end{bmatrix}.$$

Row reduction to echelon form is really a systematic form of elimination of variables. The goal is to arrive, if possible, at a situation where each row of the row-reduced matrix corresponds to just one variable. Then, as in Equation 2.1.8, the solution can be just be read off the matrix.

Essentially every result in the first six sections of this chapter is an elaboration of Theorem 2.1.8.

In MATLAB, the command rref ("row reduce echelon form") brings a matrix to echelon form.

Once you've gotten the hang of row reduction you'll see that it is perfectly simple (although we find it astonishingly easy to make mistakes). There's no need to look for tricks; you just trudge through the calculations.

Computers use algorithms that are somewhat faster than the one we have outlined. Exercise 2.1.9 explores the computational cost of solving a system of n equations in n unknowns. Partial row reduction with back-substitution, defined in the exercise, is roughly a third cheaper than full row reduction. You may want to take shortcuts too; for example, if the first row of your matrix starts with a 3, and the third row starts with a 1, you might want to make the third row the first one, rather than dividing through by 3.

Example 2.1.7 Matrices not in echelon form). The following matrices are *not* in echelon form. Can you say why not?[2]

$$\begin{bmatrix} 1 & 0 & 0 & 2 \\ 0 & 0 & 1 & -1 \\ 0 & 1 & 0 & 1 \end{bmatrix}, \begin{bmatrix} 1 & 1 & 0 & 1 \\ 0 & 0 & 2 & 0 \\ 0 & 0 & 0 & 1 \end{bmatrix}, \begin{bmatrix} 0 & 0 & 0 \\ 1 & 0 & 0 \\ 0 & 1 & 0 \end{bmatrix}, \begin{bmatrix} 0 & 1 & 0 & 3 & 0 & -3 \\ 0 & 0 & -1 & 1 & 1 & 1 \\ 0 & 0 & 0 & 0 & 1 & 2 \end{bmatrix}.$$

Exercise 2.1.5 asks you to bring them to echelon form.

How to row reduce a matrix

The following result and its proof are absolutely fundamental:

Theorem 2.1.8. *Given any matrix A, there exists a unique matrix $\widetilde{A}$ in echelon form that can be obtained from A by row operations.*

Proof. The proof of this theorem is more important than the result: it is an explicit algorithm for computing $\widetilde{A}$. Called *row reduction* or *Gaussian elimination* (or several other names), it is the main tool for solving linear equations.

Row reduction: the algorithm. To bring a matrix to echelon form:

(1) Look down the first column until you find a nonzero entry, called a *pivot*. If there is none, look down the second column, etc.

(2) Put the row containing the pivot in the first row position, and then divide it by the pivot to make its first entry a pivotal 1, as defined above.

(3) Add appropriate multiples of this row onto the other rows to cancel the entries in the first column of each of the other rows.

Now look down the next column over, (and then the next column if necessary, etc.) starting beneath the row you just worked with, and look for a nonzero entry (the next pivot). As above, exchange its row with the second row, divide through, etc.

This proves existence of a matrix in echelon form that can be obtained from a given matrix. Uniqueness is more subtle and will have to wait; it uses the notion of linear independence, and is proved in Exercise 2.4.10.   □

Example 2.1.9 (Row reduction). Here we row reduce a matrix. The R's refer in each case to the rows of the immediately preceding matrix. For example, the second row of the second matrix is labeled $R_1 + R_2$, because that row is obtained by adding the first and second rows of the preceding matrix.

---

[2]The first matrix violates rule (2); the second violates rules (1) and (3); the third violates rule (4), and the fourth violates rules (1) and (3).

$$\begin{bmatrix} 1 & 2 & 3 & 1 \\ -1 & 1 & 0 & 2 \\ 1 & 0 & 1 & 2 \end{bmatrix} \rightarrow \begin{matrix} R_1 + R_2 \\ R_3 - R_1 \end{matrix} \begin{bmatrix} \underline{1} & 2 & 3 & 1 \\ 0 & 3 & 3 & 3 \\ 0 & -2 & -2 & 1 \end{bmatrix} \rightarrow R_2/3 \begin{bmatrix} \underline{1} & 2 & 3 & 1 \\ 0 & \underline{1} & 1 & 1 \\ 0 & -2 & -2 & 1 \end{bmatrix}$$

$$\rightarrow \begin{matrix} R_1 - 2R_2 \\ R_3 + 2R_2 \end{matrix} \begin{bmatrix} \underline{1} & 0 & 1 & -1 \\ 0 & \underline{1} & 1 & 1 \\ 0 & 0 & 0 & 3 \end{bmatrix} \rightarrow R_3/3 \begin{bmatrix} \underline{1} & 0 & 1 & -1 \\ 0 & \underline{1} & 1 & 1 \\ 0 & 0 & 0 & \underline{1} \end{bmatrix} \rightarrow \begin{matrix} R_1 + R_3 \\ R_2 - R_3 \end{matrix} \begin{bmatrix} \underline{1} & 0 & 1 & 0 \\ 0 & \underline{1} & 1 & 0 \\ 0 & 0 & 0 & \underline{1} \end{bmatrix}.$$

Note that in the fourth matrix we were unable to find a nonzero entry in the third column, third row, so we had to look in the next column over, where there is a 3.   △

Just as you should know how to add and multiply, you should know how to row reduce, but the goal is not to compete with a computer, or even a scientific calculator; that's a losing proposition.

Exercise 2.1.7 provides practice in row reducing matrices. It should serve also to convince you that it is indeed possible to bring any matrix to echelon form.

When computers row reduce: avoiding loss of precision

Matrices generated by computer operations often have entries that are really zero but are made nonzero by round-off error: for example, a number may be subtracted from a number that in theory is the same, but in practice is off by, say, 10^{-50}, because it has been rounded off. Such an entry is a poor choice for a pivot, because you will need to divide its row through by it, and the row will then contain very large entries. When you then add multiples of that row onto another row, you will be committing the *basic sin of computation*: adding numbers of very different sizes, which leads to loss of precision. So, what do you do? You skip over that almost-zero entry and choose another pivot. There is, in fact, no reason to choose the first nonzero entry in a given column; in practice, when computers row reduce matrices, they always choose the largest.

This is not a small issue. Computers spend most of their time solving linear equations by row reduction. Keeping loss of precision due to round-off errors from getting out of hand is critical. Entire professional journals are devoted to this topic; at a university like Cornell perhaps half a dozen mathematicians and computer scientists spend their lives trying to understand it.

Example 2.1.10 (Thresholding to minimize round-off errors). If you are computing to 10 significant digits, then $1 + 10^{-10} = 1.0000000001 = 1$. So consider the system of equations

$$\begin{aligned} 10^{-10}x + 2y &= 1 \\ x + y &= 1, \end{aligned} \qquad 2.1.9$$

the solution of which is

$$x = \frac{1}{2 - 10^{-10}}, \quad y = \frac{1 - 10^{-10}}{2 - 10^{-10}}. \qquad 2.1.10$$

If you are computing to 10 significant digits, this is $x = y = .5$. If you actually use 10^{-10} as a pivot, the row reduction, to 10 significant digits, goes as follows:

$$\begin{bmatrix} 10^{-10} & 2 & 1 \\ 1 & 1 & 1 \end{bmatrix} \rightarrow \begin{bmatrix} \underline{1} & 2 \cdot 10^{10} & 10^{10} \\ 1 & 1 & 1 \end{bmatrix} \rightarrow \begin{bmatrix} \underline{1} & 2 \cdot 10^{10} & 10^{10} \\ 0 & -2 \cdot 10^{10} & -10^{10} \end{bmatrix}$$

$$\rightarrow \begin{bmatrix} \underline{1} & 0 & 0 \\ 0 & \underline{1} & .5 \end{bmatrix}. \qquad\qquad 2.1.11$$

The "solution" shown by the last matrix reads $x = 0$, which is badly wrong: x is supposed to be .5. Now do the row reduction treating 10^{-10} as zero; what do you get? If you have trouble, check the answer in the footnote.[3] △

Exercise 2.1.8 asks you to analyze precisely where the troublesome errors occurred. All computations have been carried out to 10 significant digits only.

2.2 Solving Equations Using Row Reduction

Recall (Equations 2.1.4 and 2.1.5) that $A\vec{\mathbf{x}} = \vec{\mathbf{b}}$ represents a system of equations, the matrix A giving the coefficients, the vector $\vec{\mathbf{x}}$ giving the unknowns (for example, for a system with three unknowns,

$$\vec{\mathbf{x}} = \begin{bmatrix} x \\ y \\ z \end{bmatrix}$$

), and the vector $\vec{\mathbf{b}}$ contains the solutions. The matrix $[A, \vec{\mathbf{b}}]$ is shorthand for $A\vec{\mathbf{x}} = \vec{\mathbf{b}}$.

In this section we will see, in Theorem 2.2.4, what a row-reduced matrix representing a system of linear equations tells us about its solutions. To solve the system of linear equations $A\vec{\mathbf{x}} = \vec{\mathbf{b}}$, form the matrix $[A, \vec{\mathbf{b}}]$ and row reduce it to echelon form, giving $[\widetilde{A}, \widetilde{\vec{\mathbf{b}}}]$. If the system has a unique solution, it can then be read off the matrix, as in Example 2.1.4. If it does not, the matrix will tell you whether there is no solution, or infinitely many solutions. Although the theorem is practically obvious, it is the backbone of the entire part of linear algebra that deals with linear equations, dimension, bases, rank, and so forth.

Remark. In Theorem 2.1.8 we used the symbol tilde to denote the echelon form of a matrix: $\widetilde{A}$ is the echelon form of A, obtained by row reduction. Here, $[\widetilde{A}, \widetilde{\vec{\mathbf{b}}}]$ represents the echelon form of the entire "augmented" matrix $[A, \vec{\mathbf{b}}]$: i.e., it is $\widetilde{[A, \vec{\mathbf{b}}]}$. We use two tildes rather than one wide one because we need to talk about $\widetilde{\vec{\mathbf{b}}}$ independently of $\widetilde{A}$.   △

In the matrix $[A, \vec{\mathbf{b}}]$, the columns of A correspond in the obvious way to the unknowns x_i of the system $A\vec{\mathbf{x}} = \vec{\mathbf{b}}$: the ith column corresponds to the ith unknown. In Theorem 2.2.4 we will want to distinguish between those unknowns corresponding to *pivotal columns* and those corresponding to *non-pivotal columns*.

Definition 2.2.1 (Pivotal column). A pivotal column of A is a column of A such that the corresponding column of $\widetilde{A}$ contains a pivotal 1.

---

[3]Remember to put the second row in the first row position, as we do in the third step below:

$$\begin{bmatrix} 10^{-10} & 2 & 1 \\ 1 & 1 & 1 \end{bmatrix} \rightarrow \begin{bmatrix} 0 & 2 & 1 \\ 1 & 1 & 1 \end{bmatrix} \rightarrow \begin{bmatrix} \underline{1} & 1 & 1 \\ 0 & 2 & 1 \end{bmatrix} \rightarrow \begin{bmatrix} \underline{1} & 1 & 1 \\ 0 & 1 & .5 \end{bmatrix} \rightarrow \begin{bmatrix} \underline{1} & 0 & .5 \\ 0 & \underline{1} & .5 \end{bmatrix}.$$

A non-pivotal column is a column of A such that the corresponding column of $\tilde{A}$ does not contain a pivotal 1.

The terms "pivotal" and "non-pivotal" do not describe some intrinsic quality of a particular unknown. If a system of equations has both pivotal and non-pivotal unknowns, which are pivotal and which are not may depend on the order in which you order the unknowns, as illustrated by Exercise 2.2.1.

Definition 2.2.2 (Pivotal unknown). A pivotal unknown (or pivotal variable) of a system of linear equations $A\vec{x} = \vec{b}$ is an unknown corresponding to a pivotal column of A: x_i is a pivotal unknown if the ith column of $\tilde{A}$ contains a pivotal 1. A non-pivotal unknown corresponds to a non-pivotal column of A: x_j is a non-pivotal unknown if the jth column of $\tilde{A}$ does not contain a pivotal 1.

Example 2.2.3 (Pivotal and non-pivotal unknowns). The matrix

The row reduction in Example 2.2.3 is unusually simple in the sense that it involves no fractions; this is the exception rather than the rule. Don't be alarmed if your calculations look a lot messier.

$$[A, \vec{b}] = \begin{bmatrix} 2 & 1 & 3 & 1 \\ 1 & -1 & 0 & 1 \\ 1 & 1 & 2 & 1 \end{bmatrix}$$

corresponding to the system of equations

$$
\begin{aligned}
2x + y + 3z &= 1 \\
x - y &= 1 \quad \text{row reduces to} \\
x + y + 2z &= 1
\end{aligned}
\qquad
\underbrace{\begin{bmatrix} \underline{1} & 0 & 1 & 0 \\ 0 & \underline{1} & 1 & 0 \\ 0 & 0 & 0 & \underline{1} \end{bmatrix}}_{[\tilde{A},\tilde{\vec{b}}]},
$$

In Example 2.2.3 the non-pivotal unknown z corresponds to the third entry of $\vec{x}$; the system of equations

$$
\begin{aligned}
2x + y + 3z &= 1 \\
x - y &= 1 \\
x + y + 2z &= 1
\end{aligned}
$$

corresponds to the multiplication

$$\begin{bmatrix} 2 & 1 & 3 \\ 1 & -1 & 0 \\ 1 & 1 & 2 \end{bmatrix} \begin{bmatrix} x \\ y \\ z \end{bmatrix} \begin{bmatrix} 1 \\ 1 \\ 1 \end{bmatrix}.$$

so x and y are pivotal unknowns, and z is a non-pivotal unknown. $\triangle$

Here is what Theorems 2.1.3 and 2.1.8 do for us:

Theorem 2.2.4 (Solutions to linear equations). *Represent the system $A\mathbf{x} = \vec{b}$, involving m linear equations in n unknowns, by the $m \times (n+1)$ matrix $[A, \vec{b}]$, which row reduces to $[\tilde{A}, \tilde{\vec{b}}]$. Then*

(1) *If the row-reduced vector $\tilde{\vec{b}}$ contains a pivotal 1, the system has no solutions.*

(2) *If $\tilde{\vec{b}}$ does not contain a pivotal 1, then:*

    (a) *if there are no non-pivotal unknowns (i.e., each column of $\tilde{A}$ contains a pivotal 1), the system has a unique solution;*

    (b) *if at least one unknown is non-pivotal, there are infinitely many solutions; you can choose freely the values of the non-pivotal unknowns, and these values will determine the values of the pivotal unknowns.*

One case is so important that we isolate it as a separate theorem, even though it is a special case of part (2a).

> **Theorem 2.2.5.** *A system $A\vec{\mathbf{x}} = \vec{\mathbf{b}}$ has a unique solution for every $\vec{\mathbf{b}}$ if and only if A row reduces to the identity. (For this to occur, there must be as many equations as unknowns, i.e., A must be square.)*

The nonlinear versions of these two theorems are the inverse function theorem and the implicit function theorem, discussed in Section 2.9. In the nonlinear case, we define the pivotal and non-pivotal unknowns as being those of the linearized problems; as in the linear case, the pivotal unknowns are *implicit functions* of the non-pivotal unknowns. But those implicit functions will be defined only in a small region, and which variables are pivotal and which are not depends on where we compute our linearization.

We will prove Theorem 2.2.4 after looking at some examples. Let us consider the case where the results are most intuitive, where $n = m$. The case where the system of equations has a unique solution is illustrated by Example 2.1.4. The other two—no solution and infinitely many solutions—are illustrated below.

Example 2.2.6 (A system with no solutions). Let us solve

$$
\begin{aligned}
2x + y + 3z &= 1 \\
x - y &= 1 \\
x + y + 2z &= 1.
\end{aligned} \tag{2.2.1}
$$

The matrix

$$
\begin{bmatrix} 2 & 1 & 3 & 1 \\ 1 & -1 & 0 & 1 \\ 1 & 1 & 2 & 1 \end{bmatrix} \quad \text{row reduces to} \quad \begin{bmatrix} \underline{1} & 0 & 1 & 0 \\ 0 & \underline{1} & 1 & 0 \\ 0 & 0 & 0 & \underline{1} \end{bmatrix}. \tag{2.2.2}
$$

Note, as illustrated by Equation 2.2.2, that if $\widetilde{\vec{\mathbf{b}}}$ (i.e., the last column in the row-reduced matrix $[\widetilde{A}, \widetilde{\vec{\mathbf{b}}}]$) contains a pivotal 1, then necessarily all the entries to the left of the pivotal 1 are zero, by definition.

so the equations are incompatible and there are no solutions; the last row tells us that $0 = 1$. $\triangle$

Example 2.2.7 (A system with infinitely many solutions). Let us solve

$$
\begin{aligned}
2x + y + 3z &= 1 \\
x - y \phantom{{}+ 2z} &= 1 \\
x + y + 2z &= 1/3.
\end{aligned} \tag{2.2.3}
$$

The matrix

$$
\begin{bmatrix} 2 & 1 & 3 & 1 \\ 1 & -1 & 0 & 1 \\ 1 & 1 & 2 & 1/3 \end{bmatrix} \quad \text{row reduces to} \quad \begin{bmatrix} \underline{1} & 0 & 1 & 2/3 \\ 0 & \underline{1} & 1 & -1/3 \\ 0 & 0 & 0 & 0 \end{bmatrix}. \tag{2.2.4}
$$

The first row of the matrix says that $x + z = 2/3$; the second that $y + z = -1/3$. You can choose z arbitrarily, giving the solutions

In this case, the solutions form a family that depends on the single non-pivotal variable, z; $\widetilde{A}$ has one column that does not contain a pivotal 1.

$$
\begin{bmatrix} 2/3 - z \\ -1/3 - z \\ z \end{bmatrix}; \tag{2.2.5}
$$

there are as many solutions as there are possible values of z—an infinite number. In this system of equations, the third equation provides no new information; it is a consequence of the first two. If we denote the three equations R_1, R_2 and R_3 respectively, then $R_3 = 1/3 \, (2R_1 - R_2)$:

$$\begin{array}{rl} 2R_1 & 4x + 2y + 6z = 2 \\ -R_2 & -x + y = -1 \\ \hline 2R_1 - R_2 = 3R_3 & 3x + 3y + 6z = 1. \quad \triangle \end{array}$$

In the examples we have seen so far, $\vec{\mathbf{b}}$ was a vector with numbers as entries. What if its entries are symbolic? Depending on the values of the symbols, different cases of Theorem 2.2.4 may apply.

Example 2.2.8 (Equations with symbolic coefficients). Suppose we want to know what solutions, if any, exist for the system of equations

$$\begin{aligned} x_1 + x_2 &= a_1 \\ x_2 + x_3 &= a_2 \\ x_3 + x_4 &= a_3 \\ x_4 + x_1 &= a_4. \end{aligned} \qquad 2.2.6$$

Row operations bring the matrix

$$\begin{bmatrix} 1 & 1 & 0 & 0 & a_1 \\ 0 & 1 & 1 & 0 & a_2 \\ 0 & 0 & 1 & 1 & a_3 \\ 1 & 0 & 0 & 1 & a_4 \end{bmatrix} \quad \text{to} \quad \begin{bmatrix} 1 & 0 & 0 & 1 & a_1 + a_3 - a_2 \\ 0 & 1 & 0 & -1 & a_2 - a_3 \\ 0 & 0 & 1 & 1 & a_3 \\ 0 & 0 & 0 & 0 & a_2 + a_4 - a_1 - a_3 \end{bmatrix}, \qquad 2.2.7$$

so a first thing to notice is that there are no solutions if $a_2 + a_4 - a_1 - a_3 \neq 0$: we are then in case (1) of Theorem 2.2.4. Solutions exist only if $a_2 + a_4 - a_1 - a_3 = 0$. If that condition is met, we are in case (2b) of Theorem 2.2.4: there is no pivotal 1 in the last column, so the system has infinitely many solutions, depending on the value of the single non-pivotal variable, x_4, corresponding to the fourth column. $\triangle$

Proof of Theorem 2.2.4.   Case (1). *If the row-reduced vector $\widetilde{\mathbf{b}}$ contains a pivotal 1, the system has no solutions.*

Proof: The set of solutions of $A\mathbf{x} = \vec{\mathbf{b}}$ is the same as that of $\widetilde{A}\vec{\mathbf{x}} = \widetilde{\mathbf{b}}$ by Theorem 2.1.3. If $\widetilde{b}_j$ is a pivotal 1, then the jth equation of $\widetilde{A}\vec{\mathbf{x}} = \widetilde{\mathbf{b}}$ reads $0 = 1$ (as illustrated by the matrix in Figure 2.2.1), so the system is inconsistent.

Case (2a). *If $\widetilde{\mathbf{b}}$ does not contain a pivotal 1, and each column of $\widetilde{A}$ contains a pivotal 1, the system has a unique solution.*

Proof: This occurs only if there are at least as many equations as unknowns (there may be more, as shown in Figure 2.2.2). If each column of $\widetilde{A}$ contains a pivotal 1, and $\widetilde{\mathbf{b}}$ has no pivotal 1, then for each variable x_i there is a unique solution $x_i = \widetilde{b}_i$; all other entries in the ith row will be 0, by the rules of row reduction. If there are more equations than unknowns, the extra equations do not make the system incompatible, since by the rules of row reduction,

---

If we had arranged the columns differently, a different variable would be non-pivotal; the four variables here play completely symmetrical roles.

$$[\widetilde{A}, \widetilde{\mathbf{b}}] = \begin{bmatrix} 1 & 0 & 1 & 0 \\ 0 & \underline{1} & 1 & 0 \\ 0 & 0 & 0 & \underline{1} \end{bmatrix}$$
$$\phantom{[\widetilde{A}, \widetilde{\mathbf{b}}] = \begin{bmatrix} 1 & 0 & 1 & \end{bmatrix}} \widetilde{\mathbf{b}}$$

FIGURE 2.2.1.

Case 1: No solution. The row-reduced column $\widetilde{\mathbf{b}}$ contains a pivotal 1; the third line reads $0 = 1$. (The left-hand side of that line must contain all 0's; if the third entry were not 0, it would be a pivotal 1, and then $\widetilde{\mathbf{b}}$ would contain no pivotal 1.)

$$[\widetilde{A}, \widetilde{\mathbf{b}}] = \begin{bmatrix} \underline{1} & 0 & 0 & \widetilde{b}_1 \\ 0 & \underline{1} & 0 & \widetilde{b}_2 \\ 0 & 0 & \underline{1} & \widetilde{b}_3 \\ 0 & 0 & 0 & 0 \end{bmatrix}$$

FIGURE 2.2.2.

Case 2a: Unique solution. Each column of $\widetilde{A}$ contains a pivotal 1, giving

$$x_1 = \widetilde{b}_1; \quad x_2 = \widetilde{b}_2; \quad x_3 = \widetilde{b}_3.$$

the corresponding rows will contain all 0's, giving the correct if uninformative equation $0 = 0$.

Case (2b) *If $\widetilde{\mathbf{b}}$ does not contain a pivotal 1, and at least one column of $\widetilde{A}$ contains no pivotal 1, there are infinitely many solutions: you can choose freely the values of the non-pivotal unknowns, and these values will determine the values of the pivotal unknowns.*

Proof: A pivotal 1 in the ith column corresponds to the pivotal variable x_i. The row containing this pivotal 1 (which is often the ith row but may not be, as shown in Figure 2.2.3, matrix $\widetilde{B}$) contains no other pivotal 1's: all other non-zero entries in that row correspond to non-pivotal unknowns. (For example, in the row-reduced matrix $\widetilde{A}$ of Figure 2.2.3, the -1 in the first row, and the 2 in the second row, both correspond to the non-pivotal variable x_3.)

Thus if there is a pivotal 1 in the jth row, corresponding to the pivotal unknown x_i, then x_i equals $\widetilde{b}_j$ minus the sum of the products of the non-pivotal unknowns x_k and their (row-reduced) coefficients in the jth row:

$$x_i = \widetilde{b}_j - \underbrace{\sum \widetilde{a}_{j,k} x_k}_{\substack{\text{sum of products of the} \\ \text{non-pivotal unknowns in} \\ j\text{th row and their coefficients}}} \qquad \square \qquad 2.2.8$$

For the matrix $\widetilde{A}$ of Figure 2.2.3 we get

$$x_1 = \widetilde{b}_1 + x_3 \quad \text{and} \quad x_2 = \widetilde{b}_2 - 2x_3;$$

we can make x_3 equal anything we like; our choice will determine the values of the pivotal variables x_1 and x_2. What are the equations for the pivotal variables of matrix $\widetilde{B}$ in Figure 2.2.3?[4]

$$[\widetilde{A}, \widetilde{\mathbf{b}}] = \begin{bmatrix} \underline{1} & 0 & -1 & \widetilde{b}_1 \\ 0 & \underline{1} & 2 & \widetilde{b}_2 \\ 0 & 0 & 0 & 0 \end{bmatrix}$$

$$[\widetilde{B}, \widetilde{\mathbf{b}}] = \begin{bmatrix} \underline{1} & 0 & 3 & 0 & 2 & \widetilde{b}_1 \\ 0 & \underline{1} & 1 & 0 & 0 & \widetilde{b}_2 \\ 0 & 0 & 0 & \underline{1} & 1 & \widetilde{b}_3 \end{bmatrix}$$

FIGURE 2.2.3.

Case 2b: Infinitely many solutions (one for each value of non-pivotal variables).

How many equations in how many unknowns?

In most cases, the outcomes given by Theorem 2.2.4 can be predicted by considering how many equations you have for how many unknowns. If you have n equations for n unknowns, most often there will be a unique solution. In terms of row reduction, A will be square, and most often row reduction will result in every row of A having a pivotal 1; i.e., $\widetilde{A}$ will be the identity. This is not always the case, however, as we saw in Examples 2.2.6 and 2.2.7.

---

[4]The pivotal variables x_1, x_2 and x_4 depend on our choice of values for the non-pivotal variables x_3 and x_5:

$$x_1 = \widetilde{b}_1 - 3x_3 - 2x_5$$
$$x_2 = \widetilde{b}_2 - x_3$$
$$x_4 = \widetilde{b}_3 - x_5.$$

If you have more equations than unknowns, as in Exercise 2.1.7(b), you would expect there to be no solutions; only in very special cases can $n - 1$ unknowns satisfy n equations. In terms of row reduction, in this case A will have more rows than columns, and at least one row of $\widetilde{A}$ will not have a pivotal 1. A row of $\widetilde{A}$ without a pivotal 1 will consist of 0's; if the adjacent entry of $\widetilde{\mathbf{b}}$ is nonzero (as is likely), then the solution will have no solutions.

If you have fewer equations than unknowns, as in Exercise 2.2.2(e), you would expect infinitely many solutions. In terms of row reduction, A will have fewer rows than columns, so at least one column of $\widetilde{A}$ will contain no pivotal 1: there will be at least one non-pivotal unknown. In most cases, $\widetilde{\mathbf{b}}$ will not contain a pivotal 1. (If it does, then that pivotal 1 is preceded by a row of 0's.)

Geometric interpretation of solutions

These examples have a geometric interpretation. The top graph in Figure 2.2.4 shows the case where two equations in two unknowns have a unique solution.

As you surely know, two equations in two unknowns,

$$
\begin{aligned}
a_1x + b_1y &= c_1 \\
a_2x + b_2y &= c_2,
\end{aligned}
\qquad 2.2.9
$$

are incompatible if and only if the lines ℓ_1 and ℓ_2 in $\mathbb{R}^2$ with equations $a_1x + b_1y = c_1$ and $a_2x + b_2y = c_2$ are parallel (middle graph, Figure 2.2.4). The equations have infinitely many solutions if and only if $\ell_1 = \ell_2$ (bottom graph, Figure 2.2.4).

When you have three equations in three unknowns, each equation describes a plane in $\mathbb{R}^3$. The top graph of Figure 2.2.5 shows three planes meeting in a single point, the case where three equations in three unknowns have a unique solution.

There are two ways for the equations in $\mathbb{R}^3$ to be incompatible, which means that the planes do not intersect. One way is that two of the planes are parallel, but this is not the only, or even the usual way: they will also be incompatible if no two are parallel, but the line of intersection of any two is parallel to the third, as shown by the middle graph of Figure 2.2.5. This latter possibility occurs in Example 2.2.6.

There are also two ways for equations in $\mathbb{R}^3$ to have infinitely many solutions. The three planes may coincide, but again this is not necessary or usual. The equations will also have infinitely many solutions if the planes intersect in a common line, as shown by the bottom graph of Figure 2.2.5. (This second possibility occurs in Example 2.2.7.)

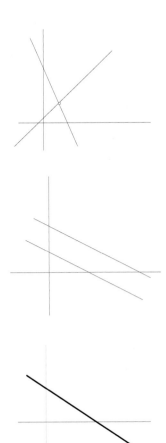

FIGURE 2.2.4.

Top: Two lines meet in a single point, representing the unique solution to two equations in two unknowns. Middle: A case where two equations in two unknowns have no solution. Bottom: Two lines are colinear, representing a case where two equations in two unknowns have infinitely many solutions.

Solving several systems of linear equations with one matrix

Theorem 2.2.4 has an additional spinoff. If you want to solve several systems of n linear equations in n unknowns that have the same matrix of coefficients, you can deal with them all at once, using row reduction. This will be useful when we compute inverses of matrices in Section 2.3.

> **Corollary 2.2.9 (Solving several systems of equations simultaneously).** *Several systems of n linear equations in n unknowns, with the same coefficients (e.g. $A\vec{x} = \vec{b}_1, \ldots,   A\vec{x} = \vec{b}_k$) can be solved at once with row reduction. Form the matrix*
>
> $$[A, \vec{b}_1, \ldots, \vec{b}_k] \quad \text{and row reduce it to get} \quad [\widetilde{A}, \widetilde{\vec{b}}_1, \ldots, \widetilde{\vec{b}}_k].$$
>
> *If $\widetilde{A}$ is the identity, then $\widetilde{\vec{b}}_i$ is the solution to the ith equation $A\vec{x} = \vec{b}_i$.*

If A row reduces to the identity, the row reduction is completed by the time one has dealt with the last row of A. The row operations needed to turn A into $\widetilde{A}$ affect each $\vec{b}_i$, but the $\vec{b}_i$ do not affect each other.

Example 2.2.10 (Several systems of equations solved simultaneously). Suppose we want to solve the three systems

$$
\begin{array}{lll}
& 2x + y + 3z = 1 & \qquad 2x + y + 3z = 2 & \qquad 2x + y + 3z = 0 \\
(1) & x - y + z = 1 & \quad(2)\quad x - y + z = 0 & \quad(3)\quad x - y + z = 1 \\
& x + y + 2z = 1. & \qquad x + y + 2z = 1. & \qquad x + y + 2z = 1.
\end{array}
$$

We form the matrix

$$
\begin{bmatrix}
2 & 1 & 3 & 1 & 2 & 0 \\
1 & -1 & 1 & 1 & 0 & 1 \\
1 & 1 & 2 & 1 & 1 & 1
\end{bmatrix},
\quad \text{which row reduces to} \quad
\begin{bmatrix}
1 & 0 & 0 & -2 & 2 & -5 \\
0 & 1 & 0 & -1 & 1 & -2 \\
0 & 0 & 1 & 2 & -1 & 4
\end{bmatrix}.
$$

$$
\underbrace{}_{A} \quad \underbrace{}_{\vec{b}_1\ \vec{b}_2\ \vec{b}_3} \qquad\qquad \underbrace{}_{I} \quad \underbrace{}_{\widetilde{\vec{b}}_1\ \widetilde{\vec{b}}_2\ \widetilde{\vec{b}}_3}
$$

The solution to the first system of equations is $\begin{bmatrix} -2 \\ -1 \\ 2 \end{bmatrix}$, i.e. $\begin{bmatrix} x \\ y \\ z \end{bmatrix} = \begin{bmatrix} -2 \\ -1 \\ 2 \end{bmatrix}$; the solution to the second is $\begin{bmatrix} 2 \\ 1 \\ -1 \end{bmatrix}$; the solution to the third is $\begin{bmatrix} -5 \\ -2 \\ 4 \end{bmatrix}$. $\triangle$

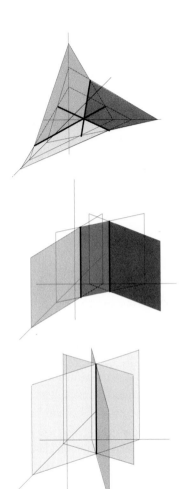

FIGURE 2.2.5.
Top: Three equations in three unknowns meet in a single point, representing the unique solution to three equations in three unknowns. Middle: Three equations in three unknowns have no solution. Bottom: Three equations in three unknowns have infinitely many solutions.

2.3 MATRIX INVERSES AND ELEMENTARY MATRICES

In this section we will see that matrix inverses give another way to solve equations. We will also introduce the modern view of row reduction: that a row operation is equivalent to multiplying a matrix by an *elementary matrix*.

Solving equations with matrix inverses

Recall from Section 1.2 that the inverse of a matrix A is another matrix A^{-1} such that $AA^{-1} = A^{-1}A = I$, the identity. In that section we discussed two results involving inverses, Propositions 1.2.14 and 1.2.15. The first says that if a matrix has both a left and a right inverse, then those inverses are identical. The second says that the product of two invertible matrices is invertible, and that the inverse of the product is the product of the inverses, in reverse order.

Inverses give another way to solve equations. If a matrix A has an inverse A^{-1}, then for any $\vec{\mathbf{b}}$ the equation $A\vec{\mathbf{x}} = \vec{\mathbf{b}}$ has a unique solution, namely $\vec{\mathbf{x}} = A^{-1}\vec{\mathbf{b}}$.

Only square matrices can have inverses: Exercise 2.3.1 asks you to (1) derive this from Theorem 2.2.4, and (2) show an example where $AB = I$, but $BA \neq I$. Such a B would be only a "one-sided inverse" for A, not a real inverse; a "one-sided inverse" can give uniqueness *or* existence of solutions to $A\vec{\mathbf{x}} = \vec{\mathbf{b}}$, but not both.

One can verify that $A^{-1}\vec{\mathbf{b}}$ is a solution by plugging it into the equation $A\vec{\mathbf{x}} = \vec{\mathbf{b}}$:

$$A(A^{-1}\vec{\mathbf{b}}) = (AA^{-1})\vec{\mathbf{b}} = I\vec{\mathbf{b}} = \vec{\mathbf{b}}. \qquad 2.3.1$$

This makes use of the associativity of matrix multiplication.

The following computation proves uniqueness:

$$A\vec{\mathbf{x}} = \vec{\mathbf{b}}, \quad \text{so} \quad A^{-1}A\vec{\mathbf{x}} = A^{-1}\vec{\mathbf{b}}; \quad \text{since} \quad A^{-1}A\vec{\mathbf{x}} = \vec{\mathbf{x}}, \quad \text{we have} \quad \vec{\mathbf{x}} = A^{-1}\vec{\mathbf{b}}.$$
$$2.3.2$$

Again we use the associativity of matrix multiplication. Note that in Equation 2.3.1 the inverse of A is on the right; in Equation 2.3.2 it is on the left.

The above argument, plus Theorem 2.2.5, proves the following proposition.

Proposition 2.3.1. *A matrix A is invertible if and only if it row reduces to the identity.*

In particular, to be invertible a matrix must be square.

Computing matrix inverses

To construct the matrix $[A|I]$ of Theorem 2.3.2, you put A to the left of the corresponding identity matrix. By "corresponding" we mean that if A is $n \times n$, then the identity matrix I must be $n \times n$.

Computing matrix inverses is rarely a good way to solve linear equations, but it is nevertheless a very important construction. Equation 1.2.15 shows how to compute the inverse of a 2×2 matrix. Analogous formulas exist for larger matrices, but they rapidly get out of hand. The effective way to compute matrix inverses for larger matrices is by row reduction:

Theorem 2.3.2 (Computing a matrix inverse). *If A is a $n \times n$ matrix, and you construct the $n \times 2n$ augmented matrix $[A|I]$ and row reduce it, then either:*

(1) *The first n columns row reduce to the identity, in which case the last n columns of the row-reduced matrix are the inverse of A, or*

(2) *The first n columns do not row reduce to the identity, in which case A does not have an inverse.*

Example 2.3.3 (Computing a matrix inverse).

$$A = \begin{bmatrix} 2 & 1 & 3 \\ 1 & -1 & 1 \\ 1 & 1 & 2 \end{bmatrix} \quad \text{has inverse} \quad A^{-1} = \begin{bmatrix} 3 & -1 & -4 \\ 1 & -1 & -1 \\ -2 & 1 & 3 \end{bmatrix}, \qquad 2.3.3$$

because

$$\begin{bmatrix} 2 & 1 & 3 & 1 & 0 & 0 \\ 1 & -1 & 1 & 0 & 1 & 0 \\ 1 & 1 & 2 & 0 & 0 & 1 \end{bmatrix} \quad \text{row reduces to} \quad \begin{bmatrix} \underline{1} & 0 & 0 & 3 & -1 & -4 \\ 0 & \underline{1} & 0 & 1 & -1 & -1 \\ 0 & 0 & \underline{1} & -2 & 1 & 3 \end{bmatrix}. \quad 2.3.4$$

Exercise 2.3.3 asks you to confirm that you can use this inverse matrix to solve the system of Example 2.2.10.   △

Example 2.3.4 (A matrix with no inverse). Consider the matrix of Examples 2.2.6 and 2.2.7, for two systems of linear equations, neither of which has a unique solution:

$$A = \begin{bmatrix} 2 & 1 & 3 \\ 1 & -1 & 0 \\ 1 & 1 & 2 \end{bmatrix}. \qquad 2.3.5$$

This matrix has *no* inverse A^{-1} because

We haven't row reduced the matrix to echelon form; as soon as we see that the first three columns are not the identity matrix, there's no point in continuing; we already know that A has no inverse.

$$\begin{bmatrix} 2 & 1 & 3 & 1 & 0 & 0 \\ 1 & -1 & 0 & 0 & 1 & 0 \\ 1 & 1 & 2 & 0 & 0 & 1 \end{bmatrix} \quad \text{row reduces to} \quad \begin{bmatrix} \underline{1} & 0 & 1 & 1 & 0 & -1 \\ 0 & \underline{1} & 1 & -1 & 0 & 2 \\ 0 & 0 & 0 & -2 & 1 & 3 \end{bmatrix}.$$
△ 2.3.6

Proof of Theorem 2.3.2. Suppose $[A|I]$ row reduces to $[I|B]$. Since A row reduces to the identity, the ith column of B is the solution $\vec{x}_i$ to the equation $A\vec{x}_i = \vec{e}_i$.

This uses Corollary 2.2.9. In Example 2.2.10 illustrating that corollary, AB row reduces to $I\widetilde{B}$, so the ith column of $\widetilde{B}$ (i.e., $\widetilde{\mathbf{b}}_i$) is the solution to the equation $A\vec{x}_i = \vec{b}_i$. We repeat the row reduction of that example here:

$$\begin{bmatrix} 2 & 1 & 3 & 1 & 2 & 0 \\ 1 & -1 & 1 & 1 & 0 & 1 \\ 1 & 1 & 2 & 1 & 1 & 1 \end{bmatrix} \quad \text{row reduces to} \quad \begin{bmatrix} 1 & 0 & 0 & -2 & 2 & -5 \\ 0 & 1 & 0 & -1 & 1 & -2 \\ 0 & 0 & 1 & 2 & -1 & 4 \end{bmatrix},$$
$$\quad A \qquad\quad \mathbf{b}_1\ \mathbf{b}_2\ \mathbf{b}_3 \qquad\qquad\qquad\quad I \qquad\quad \widetilde{\mathbf{b}}_1\ \widetilde{\mathbf{b}}_2\ \widetilde{\mathbf{b}}_3$$

so $A\widetilde{\mathbf{b}}_i = \vec{b}_i$.

Similarly, when AI row reduces to IB, the ith column of B (i.e., $\vec{b}_i$) is the solution to the equation $A\vec{x}_i = \vec{e}_i$:

$$\begin{bmatrix} 2 & 1 & 3 & 1 & 0 & 0 \\ 1 & -1 & 1 & 0 & 1 & 0 \\ 1 & 1 & 2 & 0 & 0 & 1 \end{bmatrix} \quad \text{row reduces to} \quad \begin{bmatrix} 1 & 0 & 0 & 3 & -1 & -4 \\ 0 & 1 & 0 & 1 & -1 & -1 \\ 0 & 0 & 1 & -2 & 1 & 3 \end{bmatrix}, \quad 2.3.7$$
$$\quad A \qquad\qquad \vec{e}_1\ \vec{e}_2\ \vec{e}_3 \qquad\qquad\qquad\quad I \qquad\quad \vec{b}_1\ \vec{b}_2\ \vec{b}_3$$

$$\begin{bmatrix} 1 & 0 & \ldots & 0 & \ldots & 0 \\ 0 & 1 & \ldots & 0 & \ldots & 0 \\ \vdots & \vdots & \ddots & \vdots & & \vdots \\ 0 & 0 & \ldots & x & \ldots & 0 \\ \vdots & \vdots & & \vdots & \ddots & \vdots \\ 0 & 0 & \ldots & 0 & \ldots & 1 \end{bmatrix} \begin{matrix} \\ \\ \\ i \\ \\ \\ \end{matrix}$$

Type 1: $E_1(i,x)$

$$\begin{bmatrix} 1 & 0 & 0 & 0 \\ 0 & 1 & 0 & 0 \\ 0 & 0 & 2 & 0 \\ 0 & 0 & 0 & 1 \end{bmatrix}$$

Example type 1: $E_1(3,2)$

Recall (Figure 1.2.3) that the ith row of $E_1 A$ depends on all the entries of A but only the ith row of E_1.

$$\begin{bmatrix} 1 & \ldots & 0 & \ldots & 0 & \ldots & 0 \\ \vdots & \ddots & \vdots & & \vdots & & \vdots \\ 0 & \ldots & 1 & \ldots & x & \ldots & 0 \\ \vdots & & \vdots & \ddots & \vdots & & \vdots \\ 0 & \ldots & 0 & \ldots & 1 & \ldots & 0 \\ \vdots & & \vdots & & \vdots & \ddots & \vdots \\ 0 & \ldots & 0 & \ldots & 0 & \ldots & 1 \end{bmatrix} \begin{matrix} \\ \\ i \\ \\ j \\ \\ \end{matrix}$$

Type 2: $E_2(i,j,x)$

$$\begin{bmatrix} 1 & 0 & -3 \\ 0 & 1 & 0 \\ 0 & 0 & 1 \end{bmatrix}$$

Example type 2: $E_2(1,3,-3)$

so $A\vec{\mathbf{b}}_i = \vec{\mathbf{e}}_i$. So we have:

$$A \underbrace{[\vec{\mathbf{b}}_1, \vec{\mathbf{b}}_2, \ldots, \vec{\mathbf{b}}_n]}_{B} = \underbrace{[\vec{\mathbf{e}}_1, \vec{\mathbf{e}}_2, \ldots, \vec{\mathbf{e}}_n]}_{I}. \qquad 2.3.8$$

This tells us that B is a right inverse of A: that $AB = I$.

We already know by Proposition 2.3.1 that if A row reduces to the identity it is invertible, so by Proposition 1.2.14, B is also a left inverse, hence the inverse of A. (At the end of this section we give a slightly different proof in terms of elementary matrices.) $\square$

Elementary matrices

After introducing matrix multiplication in Chapter 1, we may appear to have dropped it. We haven't really. The modern view of row reduction is that any row operation can be performed on a matrix by multiplying A *on the left* by an elementary matrix. Elementary matrices will simplify a number of arguments further on in the book.

There are three types of elementary matrices, *all square*, corresponding to the three kinds of row operations. They are defined in terms of the main diagonal, from top left to bottom right. We refer to them as "type 1," "type 2,"and "type 3," but there is no standard numbering; we have listed them in the same order that we listed the corresponding row operations in Definition 2.1.1.

Definition 2.3.5 (Elementary matrices).

(1) The type 1 elementary matrix $E_1(i,x)$ is the square matrix where every entry on the main diagonal is 1 except for the (i,i)th entry, which is $x \neq 0$, and in which all other entries are zero.

(2) The type 2 elementary matrix $E_2(i,j,x)$, for $i \neq j$, is the matrix where all the entries on the main diagonal are 1, and all other entries are 0 except for the (i,j)th, which is x. (Remember that the first index, i, refers to which row, and the second, j, refers to which column. While the (i,j)th entry is x, the (j,i)th entry is 0.)

(3) The type 3 elementary matrix $E_3(i,j)$, $i \neq j$, is the matrix where the entries i,j and j,i are 1, as are all entries on the main diagonal except i,i and j,j, which are 0. All the others are 0.

• Multiplying A on the left by E_1 multiplies the ith row of A by x: $E_1 A$ is identical to A except that every entry of the ith row has been multiplied by x.

• Multiplying A on the left by E_2 adds (x times the jth row) to the ith row.

• The matrix $E_3 A$ is the matrix A with the ith and the jth rows exchanged.

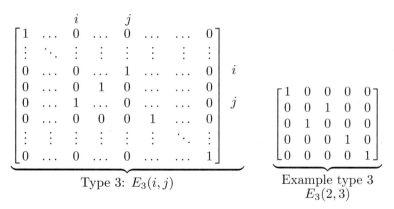

Type 3: $E_3(i,j)$

Example type 3
$E_3(2,3)$

The type 3 elementary matrix $E_3(i,j)$ is shown at right. It is the matrix where the entries i,j and j,i are 1, as are all entries on the main diagonal except i,i and j,j, which are 0. All the others are 0.

Example 2.3.6 (Multiplication by an elementary matrix). We can multiply by 2 the third row of the matrix A, by multiplying it on the left by the type 1 elementary matrix $E_1(3,2)$ shown at left.   △

Exercise 2.3.8 asks you to confirm that multiplying a matrix A by the other types of elementary matrices is equivalent to performing the corresponding row operation. Exercise 2.1.3 asked you to show that it is possible to exchange rows using only the first two row operations. Exercise 2.3.14 asks you to show this in terms of elementary matrices. Exercise 2.3.12 asks you to check that column operations can be achieved by multiplication *on the right* by an elementary matrix of types 1,2, and 3 respectively.

$$A$$
$$\begin{bmatrix} 1 & 0 \\ 0 & 2 \\ 2 & 1 \\ 1 & 1 \end{bmatrix}$$

$$\underbrace{\begin{bmatrix} 1 & 0 & 0 & 0 \\ 0 & 1 & 0 & 0 \\ 0 & 0 & 2 & 0 \\ 0 & 0 & 0 & 1 \end{bmatrix}}_{\text{elementary matrix}} \begin{bmatrix} 1 & 0 \\ 0 & 2 \\ 4 & 2 \\ 1 & 1 \end{bmatrix}$$

Multiplying A by this elementary matrix multiplies the third row of A by 2.

Elementary matrices are invertible

One very important property of elementary matrices is that they are invertible, and that their inverses are also elementary matrices. This is another way of saying that any row operation can be undone by another elementary operation. It follows from Proposition 1.2.15 that any product of elementary matrices is also invertible.

Proposition 2.3.7 (Elementary matrices invertible). *Any elementary matrix is invertible. More precisely,*

(1) $\left(E_1(i,x)\right)^{-1} = E_1(i,\frac{1}{x})$: *the inverse is formed by replacing the x in the (i,i)th position by $1/x$. This undoes multiplication of the ith row by x.*

(2) $\left(E_2(i,j,x)\right)^{-1} = E_2(i,j,-x)$: *the inverse is formed by replacing the x in the (i,j)th position by $-x$. This subtracts x times the jth row from the ith row.*

(3) $\left(E_3(i,j)\right)^{-1} = E_3(i,j)$: *multiplication by the inverse exchanges rows i and j a second time, undoing the first change.*

The proof is left as Exercise 2.3.13.

Proving Theorem 2.3.2 with elementary matrices

Of course in ordinary arithmetic you can't conclude from $4 \times 6 = 3 \times 8$ that $4 = 3$ and $6 = 8$, but in matrix multiplication if

$$[E][A|I] = [I|B], \quad \text{then}$$

$$[E][A] = I \quad \text{and} \quad [E][I] = B,$$

since the multiplication of each column of A and of I by $[E]$ occurs independently of all other columns:

$$\begin{array}{cc} & [A|I] \\ [E] & [EA|EI]. \end{array}$$

Equation 2.3.10 shows that when row reducing a matrix of the form $[A|I]$, the right-hand side of that augmented matrix serves to keep track of the row operations needed to reduce the matrix to echelon form; at the end of the procedure, I has been row reduced to $E_k \ldots E_1 = B$, which is precisely (in elementary matrix form) the series of row operations used.

We can now give a slightly different proof of Theorem 2.3.2 using elementary matrices.

(1) Suppose that $[A|I]$ row reduces to $[I|B]$. This can be expressed as multiplication on the left by elementary matrices:

$$E_k \ldots E_1 [A|I] = [I|B]. \qquad 2.3.9$$

The left and right halves of Equation 2.3.9 give

$$E_k \ldots E_1 A = I \quad \text{and} \quad E_k \ldots E_1 I = B. \qquad 2.3.10$$

Thus B is a product of elementary matrices, which are invertible, so (by Proposition 1.2.15) B is invertible: $B^{-1} = E_1^{-1} \ldots E_k^{-1}$. Moreover, substituting the right equation of 2.3.10 into the left equation gives $BA = I$, so B is a left inverse of A. We don't need to check that it is also a right inverse, but doing so is straightforward: multiplying $BA = I$ by B^{-1} on the left and B on the right gives

$$I = B^{-1}IB = B^{-1}(BA)B = (B^{-1}B)AB = AB. \qquad 2.3.11$$

So B is also a right inverse of A.

(2) If row reducing $[A|I]$ row reduces to $[A'|A'']$, where A' is *not* the identity, then (by Theorem 2.2.5), the equation $A\vec{x}_i = \vec{e}_i$ either has no solution or has infinitely many solutions for each $i = 1, \ldots, n$. In either case, A is noninvertible. $\square$

2.4 Linear Combinations, Span, and Linear Independence

In 1750, questioning the general assumption that every system of n linear equations in n unknowns has a unique solution, the great mathematician Leonhard Euler pointed out the case of the two equations $3x - 2y = 5$ and $4y = 6x - 10$. "We will see that it is not possible to determine the two unknowns x and y," he wrote, "since when one is eliminated, the other disappears by itself, and we are left with an identity from which we can determine nothing. The reason for this accident is immediately obvious, since the second equation can be changed to $6x - 4y = 10$, which, being just the double of the first, is in no way different from it."

Euler concluded by noting that when claiming that n equations are sufficient to determine n unknowns, "one must add the restriction that all the equations be different from each other, and that none of them is included in the others." Euler's "descriptive and qualitative approach" represented the beginning of a

More generally, these ideas apply in all linear settings, such as function spaces and integral and differential equations. Any time the notion of linear combination makes sense one can talk about linear independence, span, kernels, and so forth.

Example 2.4.1 (Linear combination). The vector $\begin{bmatrix} 3 \\ 4 \end{bmatrix}$ is a *linear combination* of the standard basis vectors $\vec{\mathbf{e}}_1$ and $\vec{\mathbf{e}}_2$, since

$$\begin{bmatrix} 3 \\ 4 \end{bmatrix} = 3 \begin{bmatrix} 1 \\ 0 \end{bmatrix} + 4 \begin{bmatrix} 0 \\ 1 \end{bmatrix}.$$

But the vector $\begin{bmatrix} 3 \\ 4 \\ 1 \end{bmatrix}$ is not a linear combination of the vectors

$$\vec{\mathbf{e}}_1 = \begin{bmatrix} 1 \\ 0 \\ 0 \end{bmatrix} \quad \text{and} \quad \vec{\mathbf{e}}_2 = \begin{bmatrix} 0 \\ 1 \\ 0 \end{bmatrix}.$$

new way of thinking.[5] At the time, mathematicians were interested in solving individual systems of equations, not in analyzing them. Even Euler began his argument by pointing out that attempts to solve the system fail; only then did he explain this failure by the obvious fact that $3x - 2y = 5$ and $4y = 6x - 10$ are really the same equation.

Today, linear algebra provides a systematic approach to both analyzing and solving systems of linear equations, which was completely unknown in Euler's time. We have already seen something of its power. Row reduction to echelon form puts a system of linear equations in a form where it is easy to analyze. Theorem 2.2.4 then tells us how to read that matrix, to find out whether the system has no solution, infinitely many solutions, or a unique solution (and, in the latter case, what it is).

Now we will introduce vocabulary that describes concepts implicit in what what we have done so far. The notions *linear combinations*, *span*, and *linear independence* give a precise way to answer the questions, given a collection of linear equations, how many genuinely different equations do we have? How many can be derived from the others?

Definition 2.4.2 (Linear combinations). If $\vec{\mathbf{v}}_1, \ldots, \vec{\mathbf{v}}_k$ is a collection of vectors in $\mathbb{R}^n$, then a linear combination of the $\vec{\mathbf{v}}_i$ is a vector $\vec{\mathbf{w}}$ of the form

$$\vec{\mathbf{w}} = \sum_{i=1}^{k} a_i \vec{\mathbf{v}}_i. \qquad 2.4.1$$

for any scalars a_i.

In other words, the vector $\vec{\mathbf{w}}$ is the sum of the vectors $\vec{\mathbf{v}}_1, \ldots, \vec{\mathbf{v}}_k$, each multiplied by a coefficient.

Span is a way of talking about the existence of solutions to linear equations.

Definition 2.4.3 (Span). The span of $\vec{\mathbf{v}}_1, \ldots, \vec{\mathbf{v}}_k$ is the set of linear combinations $a_1 \vec{\mathbf{v}}_1 + \cdots + a_k \vec{\mathbf{v}}_k$. It is denoted $\mathrm{Sp}\,(\vec{\mathbf{v}}_1, \ldots, \vec{\mathbf{v}}_k)$.

The word *span* is also used as a verb. For instance, the standard basis vectors $\vec{\mathbf{e}}_1$ and $\vec{\mathbf{e}}_2$ *span* $\mathbb{R}^2$ but not $\mathbb{R}^3$. They span the plane, because any vector in the plane is a linear combination $a_1 \vec{\mathbf{e}}_1 + a_2 \vec{\mathbf{e}}_2$.

Geometrically, this means that any point in the x, y plane can be written in terms of its x and y coordinates. The vectors $\vec{\mathbf{u}}$ and $\vec{\mathbf{v}}$ shown in Figure 2.4.1 also span the plane.

---

[5]Jean-Luc Dorier, ed., *L'Enseignement de l'algèbre linéaire en question*, La Pensée Sauvage, Editions, 1997. Euler's description, which we have roughly translated, is from "Sur une Contradiction Apparente dans la Doctrine des Lignes Courbes," *Mémoires de l'Académie des Sciences de Berlin* 4 (1750).

You are asked to show in Exercise 2.4.1 that $\mathrm{Sp}\,(\vec{v}_1,\dots,\vec{v}_k)$ is a subspace of $\mathbb{R}^n$ and is the smallest subspace containing $\vec{v}_1,\dots,\vec{v}_k$.

Examples 2.4.4 (Span: two easy cases). In simple cases it is possible to see immediately whether a given vector is in the span of a set of vectors.

(1) Is the vector $\vec{u} = \begin{bmatrix} 2 \\ 1 \\ 1 \end{bmatrix}$ in the span of $\vec{w} = \begin{bmatrix} 2 \\ 0 \\ 1 \end{bmatrix}$? Clearly not; no multiple of 0 will give the 1 in the second position of $\vec{u}$.

(2) Given the vectors

$$\vec{v}_1 = \begin{bmatrix} 1 \\ 0 \\ 0 \\ -1 \end{bmatrix}, \quad \vec{v}_2 = \begin{bmatrix} -2 \\ -1 \\ 1 \\ 0 \end{bmatrix}, \quad \vec{v}_3 = \begin{bmatrix} 1 \\ 1 \\ -1 \\ 1 \end{bmatrix}, \quad \vec{v}_4 = \begin{bmatrix} 0 \\ 0 \\ 1 \\ 0 \end{bmatrix}, \qquad 2.4.2$$

is $\vec{v}_4$ in the span of $\{\vec{v}_1, \vec{v}_2, \vec{v}_3\}$? Check your answer below.[6]

Example 2.4.5 (Row reducing to check span). When it's not immediately obvious whether a vector is in the span of other vectors, row reduction gives the answer. Given the vectors

$$\vec{w}_1 = \begin{bmatrix} 2 \\ 1 \\ 1 \end{bmatrix}, \quad \vec{w}_2 = \begin{bmatrix} 1 \\ -1 \\ 1 \end{bmatrix}, \quad \vec{w}_3 = \begin{bmatrix} 3 \\ 0 \\ 2 \end{bmatrix}, \quad \vec{v} = \begin{bmatrix} 3 \\ 3 \\ 1 \end{bmatrix}, \qquad 2.4.3$$

is $\vec{v}$ in the span of the other three? Here the answer is not apparent, so we can take a more systematic approach. If $\vec{v}$ is in the span of $\{\vec{w}_1, \vec{w}_2, \vec{w}_3\}$, then $x_1\vec{w}_1 + x_2\vec{w}_2 + x_3\vec{w}_3 = \vec{v}$; i.e., (writing $\vec{w}_1, \vec{w}_2$ and $\vec{w}_3$ in terms of their entries) there is a solution to the set of equations

$$\begin{aligned} 2x_1 + x_2 + 3x_3 &= 3 \\ x_1 - x_2 \qquad\;\; &= 3 \qquad\qquad 2.4.4 \\ x_1 + x_2 + 2x_3 &= 1. \end{aligned}$$

Theorem 2.2.4 tells us how to solve this system; we make a matrix and row reduce it:

$$\begin{bmatrix} 2 & 1 & 3 & 3 \\ 1 & -1 & 0 & 3 \\ 1 & 1 & 2 & 1 \end{bmatrix} \quad \text{row reduces to} \quad \begin{bmatrix} 1 & 0 & 1 & 2 \\ 0 & 1 & 1 & -1 \\ 0 & 0 & 0 & 0 \end{bmatrix}. \qquad 2.4.5$$

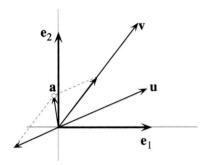

FIGURE 2.4.1.

The vectors $\vec{u}$ and $\vec{v}$ span the plane: any vector, such as $\vec{a}$, can be expressed as the sum of components in the directions $\vec{u}$ and $\vec{v}$, i.e., multiples of $\vec{u}$ and $\vec{v}$.

Like the word *onto*, the word *span* is a way to talk about the existence of solutions.

We used MATLAB to row reduce the matrix in Equation 2.4.5, as we don't enjoy row reduction and tend to make mistakes.

---

[6]No. It is impossible to write $\vec{v}_4$ as a linear combination of the other three vectors. Since the second and third entries of $\vec{v}_1$ are 0, if $\vec{v}_4$ were in the span of $\{\vec{v}_1, \vec{v}_2, \vec{v}_3\}$, its second and third entries would depend only on $\vec{v}_2$ and $\vec{v}_3$. To achieve the 0 of the second position, we must give equal weights to $\vec{v}_2$ and $\vec{v}_3$, but then we would also have a 0 in the third position, whereas we need a 1.

Since the last column of the row reduced matrix contains no pivotal 1, there is a solution: $\vec{v}$ is in the span of the other three vectors. But the solution is not unique: $\widetilde{A}$ has a column with no pivotal 1, so there are infinitely many ways to express $\vec{v}$ as a linear combination of $\{\vec{w}_1, \vec{w}_2, \vec{w}_3\}$. For example,

$$\vec{v} = 2\vec{w}_1 - \vec{w}_2 = \mathbf{w}_1 - 2\vec{w}_2 + \vec{w}_3. \quad \triangle$$

Is the vector $\vec{v} = \begin{bmatrix} 0 \\ 1 \\ 1 \end{bmatrix}$ in the span of the vectors $\begin{bmatrix} 1 \\ 0 \\ 1 \end{bmatrix}$, $\begin{bmatrix} 2 \\ 1 \\ 1 \end{bmatrix}$, and $\begin{bmatrix} 1 \\ 3 \\ 0 \end{bmatrix}$? Is

$\vec{w} = \begin{bmatrix} 0 \\ 1 \\ 1 \end{bmatrix}$ in the span of $\begin{bmatrix} 2 \\ 2 \\ 0 \end{bmatrix}$, $\begin{bmatrix} -2 \\ -1 \\ 2 \end{bmatrix}$, and $\begin{bmatrix} 1 \\ 1 \\ 0 \end{bmatrix}$? Check your answers below.[7]

The vectors $\vec{e}_1$ and $\vec{e}_2$ are linearly independent. There is only one way to write $\begin{bmatrix} 3 \\ 4 \end{bmatrix}$ in terms of $\vec{e}_1$ and $\vec{e}_2$:

$$3\begin{bmatrix} 1 \\ 0 \end{bmatrix} + 4\begin{bmatrix} 0 \\ 1 \end{bmatrix} = \begin{bmatrix} 3 \\ 4 \end{bmatrix}.$$

But if we give ourselves a third vector, say $\vec{v} = \begin{bmatrix} 3 \\ 2 \end{bmatrix}$, then we can also write

$$\begin{bmatrix} 3 \\ 4 \end{bmatrix} = \begin{bmatrix} 3 \\ 2 \end{bmatrix} + 2\begin{bmatrix} 0 \\ 1 \end{bmatrix}.$$

The three vectors $\vec{e}_1, \vec{e}_2$ and $\vec{v}$ are not linearly independent.

Linear independence

Linear independence is a way to talk about *uniqueness* of solutions to linear equations.

Definition 2.4.6 (Linear independence). The vectors $\vec{v}_1, \dots, \vec{v}_k$ are linearly independent if there is at most one way of writing a vector $\vec{w}$ as a linear combination of $\vec{v}_1, \dots, \vec{v}_k$, i.e., if

$$\vec{w} = \sum_{i=1}^{k} x_i \vec{v}_i = \sum_{i=1}^{k} y_i \vec{v}_i \text{ implies } x_1 = y_1, \ x_2 = y_2, \dots, x_k = y_k. \quad 2.4.6$$

(Note that the unknowns in Definition 2.4.6 are the coefficients x_i.) In other words, if $\vec{v}_1, \dots, \vec{v}_k$ are linearly independent, then if the system of equations

$$\vec{w} = x_1 \vec{v}_1 + x_2 \vec{v}_2 + \cdots + x_k \vec{v}_k \quad 2.4.7$$

has a solution, the solution is unique.

_____

[7]Yes, $\vec{v}$ is in the span of the others: $\vec{v} = 3\begin{bmatrix} 1 \\ 0 \\ 1 \end{bmatrix} - 2\begin{bmatrix} 2 \\ 1 \\ 1 \end{bmatrix} + \begin{bmatrix} 1 \\ 3 \\ 0 \end{bmatrix}$, since the matrix

$\begin{bmatrix} 1 & 2 & 1 & 0 \\ 0 & 1 & 3 & 1 \\ 1 & 1 & 0 & 1 \end{bmatrix}$ row reduces to $\begin{bmatrix} 1 & 0 & 0 & 3 \\ 0 & 1 & 0 & -2 \\ 0 & 0 & 1 & 1 \end{bmatrix}$. No, $\vec{w}$ is not in the span of the

others (as you might have suspected, since $\begin{bmatrix} 2 \\ 2 \\ 0 \end{bmatrix}$ is a multiple of $\begin{bmatrix} 1 \\ 1 \\ 0 \end{bmatrix}$). If we row

reduce the appropriate matrix we get $\begin{bmatrix} 1 & 0 & 1/2 & 0 \\ 0 & 1 & 0 & 0 \\ 0 & 0 & 0 & 1 \end{bmatrix}$: the system of equations has

no solution.

Example 2.4.7 (Linearly independent vectors). Are the vectors

$$\vec{w}_1 = \begin{bmatrix} 1 \\ 2 \\ 3 \end{bmatrix}, \quad \vec{w}_2 = \begin{bmatrix} -2 \\ 1 \\ 2 \end{bmatrix}, \quad \vec{w}_3 = \begin{bmatrix} -1 \\ 1 \\ -1 \end{bmatrix},$$

linearly independent? Theorem 2.2.5 says that a system $A\mathbf{x} = \vec{b}$ of n equations in n unknowns has a unique solution for every $\vec{b}$ if and only if A row reduces to the identity. The matrix

$$\begin{bmatrix} 1 & -2 & -1 \\ 2 & 1 & 1 \\ 3 & 2 & -1 \end{bmatrix} \quad \text{row reduces to} \quad \begin{bmatrix} 1 & 0 & 0 \\ 0 & 1 & 0 \\ 0 & 0 & 1 \end{bmatrix},$$
$$\vec{w}_1 \quad \vec{w}_2 \quad \vec{w}_3$$

so the vectors are linearly independent: there is only one way to write a vector $\vec{v}_i$ in $\mathbb{R}^3$ as the linear combination $a_i\vec{w}_1 + b_1\vec{w}_2 + c_1\vec{w}_3$. These three vectors also *span* $\mathbb{R}^3$, since we know from Theorem 2.2.5 that any vector in $\mathbb{R}^3$ can be written as a linear combination of them. $\triangle$

Like the term *one to one*, the term *linear independence* is a way to talk about uniqueness of solutions.

The pivotal columns of a matrix are linearly independent.

Non-pivotal is another way of saying *linearly dependent*.

Example 2.4.8 (Linearly dependent vectors). If we make the collection of vectors in Example 2.4.7 *linearly dependent,* by adding a vector that is a linear combination of some of them, say $\vec{w}_4 = 2\vec{w}_2 + \vec{w}_3$:

$$\vec{w}_1 = \begin{bmatrix} 1 \\ 2 \\ 3 \end{bmatrix}, \ \vec{w}_2 = \begin{bmatrix} -2 \\ 1 \\ 2 \end{bmatrix}, \ \vec{w}_3 = \begin{bmatrix} -1 \\ 1 \\ -1 \end{bmatrix}, \ \vec{w}_4 = \begin{bmatrix} -5 \\ 3 \\ 3 \end{bmatrix}, \quad 2.4.8$$

and use them to express some arbitrary vector[8] in $\mathbb{R}^3$, say $\vec{v} = \begin{bmatrix} -7 \\ -2 \\ 1 \end{bmatrix}$, we get

$$\begin{bmatrix} 1 & -2 & -1 & -5 & -7 \\ 2 & 1 & 1 & 3 & -2 \\ 3 & 2 & -1 & 3 & 1 \end{bmatrix} \quad \text{which row reduces to} \quad \begin{bmatrix} 1 & 0 & 0 & 0 & -2 \\ 0 & 1 & 0 & 2 & 3 \\ 0 & 0 & 1 & 1 & -1 \end{bmatrix}.$$
$$2.4.9$$

Since the fourth column is non-pivotal and the last column has no pivotal 1, the system has infinitely many solutions: there are infinitely many ways to write $\vec{v}$ as a linear combination of the vectors $\vec{w}_1, \vec{w}_2, \vec{w}_3, \vec{w}_4$. The vector $\vec{v}$ is in the span of those vectors, but they are not linearly independent. $\triangle$

Saying that $\vec{v}$ is in the span of the vectors $\vec{w}_1, \vec{w}_2, \vec{w}_3, \vec{w}_4$ means that the system of equations has a solution; since the four vectors are not linearly independent, the solution is not unique.

In the case of three linearly independent vectors in $\mathbb{R}^3$, there is a unique solution for every $\mathbf{b} \in \mathbb{R}^3$, but uniqueness is irrelevant to the question of whether the vectors span $\mathbb{R}^3$; span is concerned only with existence of solutions.

It is clear from Theorem 2.2.5 that three linearly independent vectors in $\mathbb{R}^3$ span $\mathbb{R}^3$: three linearly independent vectors in $\mathbb{R}^3$ row reduce to the identity,

---

[8]Actually, not quite arbitrary. The first choice was $\begin{bmatrix} 1 \\ 1 \\ 1 \end{bmatrix}$, but that resulted in messy fractions, so we looked for a vector that gave a neater answer.

which means that (considering the three vectors as the matrix A) there will be a unique solution for $A\vec{x} = \vec{b}$ for every $\vec{b} \in \mathbb{R}^3$.

But linear independence does not guarantee the existence of a solution, as we see below.

Calling a single vector linearly independent may seem bizarre; the word *independence* seems to imply that there is something to be independent from. But one can easily verify that the case of one vector is simply Definition 2.4.6 with $k = 1$; excluding that case from the definition would create all sorts of difficulties.

Example 2.4.9. Let us modify the vectors of Example 2.4.7 to get

$$\vec{u}_1 = \begin{bmatrix} 1 \\ 2 \\ 3 \\ 0 \end{bmatrix}, \quad \vec{u}_2 = \begin{bmatrix} -2 \\ 1 \\ 2 \\ -1 \end{bmatrix}, \quad \vec{u}_3 = \begin{bmatrix} -1 \\ 1 \\ -1 \\ 1 \end{bmatrix}, \quad \vec{v} = \begin{bmatrix} -7 \\ -2 \\ 1 \\ 1 \end{bmatrix}. \qquad 2.4.10$$

The vectors $\vec{u}_1, \vec{u}_2, \vec{u}_3 \in \mathbb{R}^4$ are linearly independent, but $\vec{v}$ is not in their span: the matrix

$$\begin{bmatrix} 1 & -2 & -1 & -7 \\ 2 & 1 & 1 & -2 \\ 3 & 2 & -1 & 1 \\ 0 & -1 & 1 & 1 \end{bmatrix} \quad \text{row reduces to} \quad \begin{bmatrix} 1 & 0 & 0 & 0 \\ 0 & 1 & 0 & 0 \\ 0 & 0 & 1 & 0 \\ 0 & 0 & 0 & 1 \end{bmatrix}. \qquad 2.4.11$$

The pivotal 1 in the last column tells us that the system of equations has no solution, as you would expect: it is rare that four equations in three unknowns will be compatible. $\triangle$

Linear independence is not restricted to vectors in $\mathbb{R}^n$: it also applies to matrices (and more generally, to elements of arbitrary vector spaces). The matrices A, B and C are linearly independent if the only solution to

$$\alpha_1 A + \alpha_2 B + \alpha_3 C = 0 \quad \text{is}$$
$$\alpha_1 = \alpha_2 = \alpha_3 = 0.$$

Example 2.4.10 (Geometrical interpretation of linear independence).

(1) One vector is linearly independent if it isn't the zero vector.

(2) Two vectors are linearly independent if they do not both lie on the same line.

(3) Three vectors are linearly independent if they do not all lie in the same plane.

These are not separate definitions; they are all examples of Definition 2.4.6.

Alternative definitions of linear independence

Many books define linear independence as follows, then prove the equivalence of our definition:

Definition 2.4.11 (Alternative definition of linear independence). A set of k vectors $\vec{v}_1, \dots, \vec{v}_k$ is linearly independent if and only if the *only* solution to

$$a_1 \vec{v}_1 + a_2 \vec{v}_2 + \cdots + a_k \vec{v}_k = \vec{0} \quad \text{is} \quad a_1 = a_2 = \cdots = a_k = 0. \qquad 2.4.12$$

In one direction, we know that $0\vec{v}_1 + 0\vec{v}_2 + \cdots + 0\vec{v}_k = \vec{0}$, so if the coefficients do not all equal 0, there will be two ways of writing the zero vector as a linear combination, which contradicts Definition 2.4.6. In the other direction, if

$$b_1\vec{v}_1 + \cdots + b_k\vec{v}_k \quad \text{and} \quad c_1\vec{v}_1 + \cdots + c_k\vec{v}_k \qquad 2.4.13$$

are two ways of writing a vector $\vec{u}$ as a linear combination of $\vec{v}_1, \ldots, \vec{v}_k$, then

$$(b_1 - c_1)\vec{v}_1 + \cdots + (b_k - c_k)\vec{v}_k = \vec{0}. \qquad 2.4.14$$

But if the only solution to that equation is $b_1 - c_1 = 0, \ldots, b_k - c_k = 0$, then there is only one way of writing $\vec{u}$ as a linear combination of the $\vec{v}_i$.

Yet another equivalent statement (as you are asked to show in Exercise 2.4.2) is to say that $\vec{v}_i$ are linearly independent if none of the $\vec{v}_i$ is a linear combination of the others.

How many vectors that span $\mathbb{R}^n$ can be linearly independent?

The following theorem is basic to the entire theory:

> More generally, more than n vectors in $\mathbb{R}^n$ are never linearly independent, and fewer than n vectors never span.

Theorem 2.4.12. *In $\mathbb{R}^n$, $n + 1$ vectors are never linearly independent, and $n - 1$ vectors never span.*

Proof. First, we will show that in $\mathbb{R}^4$, five vectors are never linearly independent; the general case is exactly the same, and a bit messier to write.

If we express a vector $\vec{w} \in \mathbb{R}^4$ using the five vectors

$$\vec{v}_1, \vec{v}_2, \vec{v}_3, \vec{v}_4, \vec{u}, \qquad 2.4.15$$

> The matrix of Equation 2.4.16 could have fewer than four pivotal columns, and if it has four, they need not be the first four. All that matters for the proof is to know that at least one of the first five columns must be non-pivotal.

at least one column is non-pivotal, since there can be at most four pivotal 1's:

$$\begin{bmatrix} 1 & 0 & 0 & 0 & \tilde{u}_1 & \tilde{w}_1 \\ 0 & 1 & 0 & 0 & \tilde{u}_2 & \tilde{w}_2 \\ 0 & 0 & 1 & 0 & \tilde{u}_3 & \tilde{w}_3 \\ 0 & 0 & 0 & 1 & \tilde{u}_4 & \tilde{w}_4 \end{bmatrix}. \qquad 2.4.16$$

(The tilde indicates the row reduced entry: row reduction turns u_1 into $\tilde{u}_1$.) So there are infinitely many solutions: infinitely many ways that $\vec{w}$ can be expressed as a linear combination of the vectors $\vec{v}_1, \vec{v}_2, \vec{v}_3, \vec{v}_4, \vec{u}$.

> The matrix of Equation 2.4.18,
>
> $$\big[\underbrace{\vec{v}_1, \vec{v}_2, \vec{v}_3, \vec{v}_4,}_{A} \underbrace{\vec{w}}_{\vec{b}} \big],$$
>
> corresponds to the equation
>
> $$\begin{bmatrix} v_{1,1} & v_{1,2} & v_{1,3} & v_{1,4} \\ v_{2,1} & v_{2,2} & v_{2,3} & v_{2,4} \\ v_{3,1} & v_{3,2} & v_{3,3} & v_{3,4} \\ v_{4,1} & v_{4,2} & v_{4,3} & v_{4,4} \\ v_{5,1} & v_{5,2} & v_{5,3} & v_{5,4} \end{bmatrix} \begin{bmatrix} a_1 \\ a_2 \\ a_3 \\ a_4 \end{bmatrix} = \begin{bmatrix} w_1 \\ w_2 \\ w_3 \\ w_4 \\ w_5 \end{bmatrix}.$$

Next, we need to prove that $n - 1$ vectors never span $\mathbb{R}^n$. We will show that four vectors never span $\mathbb{R}^5$; the general case is the same.

Saying that four vectors do not span $\mathbb{R}^5$ means that there exists a vector $\vec{w} \in \mathbb{R}^5$ such that the equation

$$a_1\vec{v}_1 + a_2\vec{v}_2 + a_3\vec{v}_3 + a_4\vec{v}_4 = \vec{w} \qquad 2.4.17$$

has no solutions. Indeed, if we row reduce the matrix

$$\underbrace{[\vec{v}_1, \vec{v}_2, \vec{v}_3, \vec{v}_4,}_{A} \underbrace{\vec{w}]}_{\vec{b}} \qquad 2.4.18$$

we end up with at least one row with at least four zeroes: any row of $\widetilde{A}$ must either contain a pivotal 1 or be all 0's, and we have five rows but at most four pivotal 1's:

$$\begin{bmatrix} 1 & 0 & 0 & 0 & \widetilde{w}_1 \\ 0 & 1 & 0 & 0 & \widetilde{w}_2 \\ 0 & 0 & 1 & 0 & \widetilde{w}_3 \\ 0 & 0 & 0 & 1 & \widetilde{w}_4 \\ 0 & 0 & 0 & 0 & \widetilde{w}_5 \end{bmatrix}. \qquad 2.4.19$$

Set $\widetilde{w}_5 = 1$, and set the other entries of $\widetilde{\mathbf{w}}$ to 0; then $\widetilde{w}_5$ is a pivotal 1, and the system has no solutions; $\vec{\mathbf{w}}$ is outside the span of our four vectors. (See the note in the margin if you don't see why we were allowed to set $\widetilde{w}_5 = 1$.) $\square$

What gives us the right to set $\widetilde{w}_5$ to 1 and set the other entries of $\widetilde{\mathbf{w}}$ to 0? To prove that four vectors never span $\mathbb{R}^5$, we need to find just *one* vector $\vec{\mathbf{w}}$ for which the equations are incompatible. Since any row operation can be undone, we can assign any values we like to our $\widetilde{\mathbf{w}}$, and then bring the echelon matrix back to where it started, $[\vec{\mathbf{v}}_1, \vec{\mathbf{v}}_2, \vec{\mathbf{v}}_3, \vec{\mathbf{v}}_4, \vec{\mathbf{w}}]$. The vector $\vec{\mathbf{w}}$ that we get by starting with

$$\widetilde{\mathbf{w}} = \begin{bmatrix} 0 \\ 0 \\ 0 \\ 0 \\ 1 \end{bmatrix}$$

and undoing the row operations is a vector that makes the system incompatible.

We can look at the same thing in terms of multiplication by elementary matrices; here we will treat the general case of $n-1$ vectors in $\mathbb{R}^n$. Suppose that the row reduction of $[\vec{\mathbf{v}}_1, \dots, \vec{\mathbf{v}}_{n-1}]$ is achieved by multiplying on the left by the product of elementary matrices $E = E_k \dots E_1$, so that

$$E([\vec{\mathbf{v}}_1, \dots, \vec{\mathbf{v}}_{n-1}]) = \widetilde{V} \qquad 2.4.20$$

is in echelon form; hence its bottom row is all zeroes.

Thus, to show that our $n-1$ vectors do not span $\mathbb{R}^n$, we want the last column of the augmented, row-reduced matrix to be

$$\widetilde{\mathbf{w}} = \begin{bmatrix} 0 \\ 0 \\ \vdots \\ 1 \end{bmatrix} = \vec{\mathbf{e}}_n; \qquad 2.4.21$$

we will then have a system of equations with no solution. We can achieve that by taking $\vec{\mathbf{w}} = E^{-1} \vec{\mathbf{e}}_n$: the system of linear equations $a_1 \vec{\mathbf{v}}_1 + \cdots + a_{n-1} \vec{\mathbf{v}}_{n-1} = E^{-1} \vec{\mathbf{e}}_n$ has no solutions.

A set of vectors as a basis

Choosing a *basis* for a subspace of $\mathbb{R}^n$, or for $\mathbb{R}^n$ itself, is like choosing axes (with units marked) in the plane or in space. This allows us to pass from non-coordinate geometry (synthetic geometry) to coordinate geometry (analytic geometry). Bases provide a "frame of reference" for vectors in a subspace.

Definition 2.4.13 (Basis). Let $V \subset \mathbb{R}^n$ be a subspace. An ordered set of vectors $\vec{\mathbf{v}}_1, \dots, \vec{\mathbf{v}}_k \in V$ is called a basis of V if it satisfies one of the three equivalent conditions.

(1) The set is a *maximal linearly independent set*: it is independent, and if you add one more vector, the set will no longer be linearly independent.

The direction of basis vectors gives the direction of the axes; the length of the basis vectors provides units for those axes.

Recall (Definition 1.1.5) that a subspace of $\mathbb{R}^n$ is a subset of $\mathbb{R}^n$ that is closed under addition and closed under multiplication by scalars. Requiring that the vectors be ordered is just a convenience.

$$\vec{e}_1 = \begin{bmatrix} 1 \\ 0 \\ 0 \\ \vdots \\ 0 \end{bmatrix}, \ldots, \vec{e}_n = \begin{bmatrix} 0 \\ 0 \\ 0 \\ \vdots \\ 1 \end{bmatrix}$$

The standard basis vectors

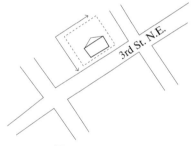

FIGURE 2.4.2.

The standard basis would not be convenient when surveying this yard. Use a basis suited to the job.

(2) The set is a *minimal spanning set*: it spans V, and if you drop one vector, it will no longer span V.

(3) The set is a *linearly independent set* spanning V.

Before proving that these conditions are indeed equivalent, let's see some examples.

Example 2.4.14 (Standard basis). The fundamental example of a basis is the *standard basis* of $\mathbb{R}^n$; our vectors are already lists of numbers, written with respect to the "standard basis" of standard basis vectors, $\vec{e}_1, \ldots, \vec{e}_n$.

Clearly every vector in $\mathbb{R}^n$ is in the span of $\vec{e}_1, \ldots, \vec{e}_n$:

$$\begin{bmatrix} a_1 \\ \vdots \\ \vdots \\ a_n \end{bmatrix} = a_1 \vec{e}_1 + \cdots + a_n \vec{e}_n; \qquad 2.4.22$$

and it is equally clear that $\vec{e}_1, \ldots, \vec{e}_n$ are linearly independent (Exercise 2.4.3).

Example 2.4.15 (Basis formed of n vectors in $\mathbb{R}^n$). The standard basis is not the only one. For instance,

$$\begin{bmatrix} 1 \\ 1 \end{bmatrix}, \begin{bmatrix} 1 \\ -1 \end{bmatrix} \text{ form a basis in } \mathbb{R}^2, \text{ as do } \begin{bmatrix} 2 \\ 0 \end{bmatrix}, \begin{bmatrix} 0.5 \\ -3 \end{bmatrix} \left(\text{but not } \begin{bmatrix} 2 \\ 0 \end{bmatrix}, \begin{bmatrix} 0.5 \\ 0 \end{bmatrix}. \right)$$
$\triangle$

In general, if you choose at random n vectors in $\mathbb{R}^n$, they will form a basis. In $\mathbb{R}^2$, the odds are completely against picking two vectors on the same line; in $\mathbb{R}^3$ the odds are completely against picking three vectors in the same plane.

You might think that the standard basis should be enough. But there are times when a problem becomes much more straightforward in a different basis. The best examples of this are beyond the scope of this chapter (eigenvectors, orthogonal polynomials), but a simple case is illustrated by Figure 2.4.2. (Think also of decimals and fractions. It is a great deal simpler to write $1/7$ than $0.142857142857\ldots$, yet at other times computing with decimals is easier.)

In addition, for a subspace $V \subset \mathbb{R}^n$ it is usually inefficient to describe vectors using all n numbers:

Example 2.4.16 (Using two basis vectors in a subspace of $\mathbb{R}^3$). In the subspace $V \subset \mathbb{R}^3$ of equation $x + y + z = 0$, rather than writing a vector by giving its three entries, we could write them using only two coefficients, a and b, and the vectors $\vec{w}_1 = \begin{bmatrix} 1 \\ -1 \\ 0 \end{bmatrix}$ and $\vec{w}_2 = \begin{bmatrix} 1 \\ 0 \\ -1 \end{bmatrix}$. For instance,

$$\vec{v} = a\vec{w}_1 + b\vec{w}_2. \qquad 2.4.23$$

What other two vectors might you choose as a basis for V?[9]

Orthonormal bases

One student asked, "When $n > 3$, how can vectors be orthogonal, or is that some weird math thing you just can't visualize?" Precisely! Two vectors are orthogonal if their dot product is 0. In $\mathbb{R}^2$ and $\mathbb{R}^3$, this corresponds to the vectors being perpendicular to each other. The geometrical relation in higher dimensions is analogous to that in $\mathbb{R}^2$ and $\mathbb{R}^3$, but you shouldn't expect to be able to visualize four (or 17, or 98) vectors all perpendicular to each other.

For a long time, the impossibility of visualizing higher dimensions hobbled mathematicians. In 1827, August Moebius wrote that if two flat figures can be moved in space so that they coincide at every point, they are "equal and similar." To speak of equal and similar objects in three dimensions, he continued, one would have to be able to move them in four-dimensional space, to make them coincide. "But since such space cannot be imagined, coincidence in this case is impossible."

When doing geometry, it is almost always best to work with an *orthonormal basis*. Below, recall that two vectors are orthogonal if their dot product is zero (Corollary 1.4.80).

> **Definition 2.4.17 (Orthonormal basis).** A basis $\vec{v}_1, \ldots, \vec{v}_k$ of a subspace $V \subset \mathbb{R}^n$ is orthonormal if each vector in the basis is orthogonal to every other vector in the basis, and if all basis vectors have length 1:
> $$\vec{v}_i \cdot \vec{v}_j = 0 \quad \text{for } i \neq j \quad \text{and} \quad |\vec{v}_i| = 1 \quad \text{for all } i \leq k.$$

The standard basis is of course orthonormal.

One reason orthonormal bases are interesting is that the length squared of a vector is the sum of the squares of its coordinates, with respect to any orthonormal basis. If $\vec{v}_1, \ldots, \vec{v}_k$ and $\vec{w}_1, \ldots, \vec{w}_k$ are two orthonormal bases, and

$$a_1 \vec{v}_1 + \cdots + a_k \vec{v}_k = b_1 \vec{w}_1 + \cdots + b_k \vec{w}_k, \quad \text{then} \quad a_1^2 + \cdots + a_k^2 = b_1^2 + \cdots + b_k^2.$$

The proof is left as Exercise 2.4.4.

If all vectors in the basis are orthogonal to each other, but they don't all have length 1, then the basis is *orthogonal*. Is either of the two bases of Example 2.4.15 orthogonal? orthonormal?[10]

> **Proposition 2.4.18.** *An orthogonal set of nonzero vectors $\vec{v}_1, \ldots, \vec{v}_k$ is linearly independent.*

Proof. Suppose $a_1 \vec{v}_1 + \cdots + a_k \vec{v}_k = \vec{0}$. Take the dot product of both sides with $\vec{v}_i$:

$$(a_1 \vec{v}_1 + \cdots + a_k \vec{v}_k) \cdot \vec{v}_i = \vec{0} \cdot \vec{v}_i = 0. \qquad 2.4.24$$

---

[9]The vectors $\begin{bmatrix} -1 \\ 0 \\ 1 \end{bmatrix}, \begin{bmatrix} 0 \\ 1 \\ -1 \end{bmatrix}$ are a basis for V, as are $\begin{bmatrix} -1/2 \\ -1/2 \\ 1 \end{bmatrix}, \begin{bmatrix} 1 \\ -1 \\ 0 \end{bmatrix}$; the vectors just need to be linearly independent and the sum of the entries of each vector must be 0 (satisfying $x + y + z = 0$). Part of the "structure" of the subspace V is thus built into the basis vectors.

[10]The first is orthogonal, since $\begin{bmatrix} 1 \\ 1 \end{bmatrix} \cdot \begin{bmatrix} 1 \\ -1 \end{bmatrix} = 1 - 1 = 0$; the second is not, since $\begin{bmatrix} 2 \\ 0 \end{bmatrix} \cdot \begin{bmatrix} 0.5 \\ -3 \end{bmatrix} = 1 + 0 = 1$. Neither is orthonormal; the vectors of the first basis each have length $\sqrt{2}$, and those of the second have lengths 2 and $\sqrt{9.25}$.

So

The surprising thing about Proposition 2.4.18 is that it allows us to assert that a set of vectors is linearly independent looking only at pairs of vectors. It is of course not true that if you have a set of vectors and every pair is linearly independent, the whole set is linearly independent; consider, for instance, the three vectors $\begin{bmatrix} 1 \\ 0 \end{bmatrix}$, $\begin{bmatrix} 0 \\ 1 \end{bmatrix}$ and $\begin{bmatrix} 1 \\ 1 \end{bmatrix}$.

By "nontrivial," we mean a solution *other than*

$$a_1 = a_2 = \cdots = a_n = b = 0.$$

$$a_1(\vec{v}_1 \cdot \vec{v}_i) + \cdots + a_i(\vec{v}_i \cdot \vec{v}_i) + \cdots + a_k(\vec{v}_k \cdot \vec{v}_i) = 0. \qquad 2.4.25$$

Since the $\vec{v}_j$ form an orthogonal set, all the dot products on the left are zero except for the ith, so $a_i(\vec{v}_i \cdot \vec{v}_i) = 0$. Since the vectors are assumed to be nonzero, this says that $a_i = 0$. $\square$

Equivalence of the three conditions for a basis

We need to show that the three conditions for a basis given in Definition 2.4.13 are indeed equivalent.

We will show that (1) implies (2): that if a set of vectors is a maximal linearly independent set, it is a minimal spanning set. Let $V \subset \mathbb{R}^n$ be a subspace. If an ordered set of vectors $\vec{v}_1, \ldots, \vec{v}_k \in V$ is a maximal linearly independent set, then for any other vector $\vec{w} \in V$, the set $\{\vec{v}_1, \ldots, \vec{v}_k, \vec{w}\}$ is linearly *dependent*, and (by Definition 2.4.11) there exists a nontrivial relation

$$a_1\vec{v}_1 + \cdots + a_k\vec{v}_k + b\vec{w} = 0. \qquad 2.4.26$$

The coefficient b is not zero, because if it were, the relation would then involve only the $\vec{v}$'s, which are linearly independent by hypothesis. Therefore we can divide through by b, expressing $\vec{w}$ as a linear combination of the $\vec{v}$'s:

$$\frac{a_1}{b}\vec{v}_1 + \cdots + \frac{a_k}{b}\vec{v}_k = -\vec{w}. \qquad 2.4.27$$

Since $\vec{w} \in V$ can be any vector in V, we see that the $\vec{v}$'s do span V.

Moreover, $\vec{v}_1, \ldots, \vec{v}_k$ is a *minimal* spanning set: if one of the $\vec{v}_i$'s is omitted, the set no longer spans, since the omitted $\vec{v}_i$ is linearly independent of the others and hence cannot be in the span of the others.

This shows that (1) implies (2); the other implications are similar and left as Exercise 2.4.7.

Now we can restate Theorem 2.4.12:

Corollary 2.4.19. *Every basis of $\mathbb{R}^n$ has exactly n elements.*

Indeed a set of vectors in $\mathbb{R}^n$ never spans $\mathbb{R}^n$ if it has fewer than n elements, and it is never linearly independent if it has more than n elements (see Theorem 2.4.12).

The notion of the dimension of a subspace will allow us to talk about such things as the size of the space of solutions to a set of equations, or the number of genuinely different equations.

Proposition and Definition 2.4.20 (Dimension). *Every subspace $E \subset \mathbb{R}^n$ has a basis, and any two bases of a subspace E have the same number of elements, called the dimension of E.*

Proof.   First we construct a basis of E. If $E = \{\vec{0}\}$, the empty set is a basis of E. Otherwise, choose a sequence of vectors $\vec{v}_1, \vec{v}_2, \dots$ in E as follows: choose $\vec{v}_1 \neq \vec{0}$, then $\vec{v}_2 \notin \mathrm{Sp}\,(\vec{v}_1)$, then $\vec{v}_3 \notin \mathrm{Sp}\,(\vec{v}_1, \vec{v}_2)$, etc. Vectors chosen this way are clearly linearly independent. Therefore we can choose at most n such vectors, and for some $m \leq n$, $\vec{v}_1, \dots, \vec{v}_k$ will span E. (If they don't span, we can choose another.) Since these vectors are linearly independent, they form a basis of E.

Now to see that any two bases have the same number of elements. Suppose $\vec{v}_1, \dots, \vec{v}_k$ and $\vec{w}_1, \dots, \vec{w}_p$ are two bases of E. Then there exists an $k \times p$ matrix A with entries $a_{i,j}$ such that

We can express any $\vec{w}_j$ as a linear combination of the $\vec{v}_i$ because the $\vec{v}_i$ span E.

$$\vec{w}_j = \sum_{i=1}^{k} a_{i,j} \vec{v}_i. \qquad 2.4.28$$

i.e., that $\vec{w}_j$ can be expressed as a linear combination of the $\vec{v}_i$. There also exists a $p \times k$ matrix B with entries $b_{l,i}$ such that

It might seem more natural to express the $\vec{w}$'s as linear combinations of the $\vec{v}$'s in the following way:

$$\vec{v}_i = \sum_{l=1}^{p} b_{l,i} \vec{w}_l. \qquad 2.4.29$$

$$\vec{w}_1 = a_{1,1} \vec{v}_1 + \cdots + a_{1,k} \vec{v}_k$$

Substituting the value for $\vec{v}_i$ of Equation 2.4.29 into Equation 2.4.28 gives

$$\vdots \qquad \qquad \vdots$$

$$\vec{w}_p = a_{p,1}\vec{v}_1 + \cdots + a_{p,k}\vec{v}_k.$$

$$\vec{w}_j = \sum_{i=1}^{k} a_{i,j} \sum_{l=1}^{p} b_{l,i} \vec{w}_l = \sum_{l=1}^{p} \underbrace{\left(\sum_{i=1}^{k} b_{l,i} a_{i,j} \right)}_{l,j\text{th entry of } BA} \vec{w}_l. \qquad 2.4.30$$

The a's then form the transpose of matrix A we have written. We use A because it is the change of basis matrix, which we will see again in Section 2.6 (Theorem 2.6.16).

This expresses $\mathbf{w}_j$ as a linear combination of the $\vec{w}$'s, but since the $\vec{w}$'s are linearly independent, Equation 2.4.30 must read

The sum in Equation 2.4.28 is not a matrix multiplication. For one thing it is the sum of products of numbers with *vectors*; for another, the indices are in the wrong order.

$$\vec{w}_j = 0\vec{w}_1 + 0\vec{w}_2 + \cdots + 1\vec{w}_j + \cdots + 0\vec{w}_p. \qquad 2.4.31$$

So $(\sum_i b_{l,i} a_{i,j})$ is 0, unless $l = j$, in which case it is 1. In other words, $BA = I$.

The same argument, exchanging the roles of the $\vec{v}$'s and the $\vec{w}$'s, shows that $AB = I$. Thus A is invertible, hence square, and $k = p$. $\square$

Corollary 2.4.21. *The only n-dimensional subspace of $\mathbb{R}^n$ is $\mathbb{R}^n$ itself.*

Remark.   We said earlier that the terms linear combinations, span, and linear independence give a precise way to answer the questions, given a collection of linear equations, how many genuinely different equations do we have. We have seen that row reduction provides a systematic way to determine how many of the columns of a matrix are linearly independent. But the equations correspond to the rows of a matrix, not to its columns. In the next section we will see that the number of linearly independent equations in the system $A\vec{x} = \vec{b}$ is the same as the number of linearly independent columns in A.

2.5 Kernels and Images

The *kernel* and the *image* of a linear transformation are important but rather abstract concepts. They are best understood in terms of linear equations. Kernels are related to uniqueness of solutions of linear equations, whereas images are related to their existence.

Definition 2.5.1 (Kernel). The kernel of a linear transformation T, denoted $\ker T$, is the set of vectors $\vec{x}$ such that $T(\vec{x}) = \mathbf{0}$. When T is represented by a matrix $[T]$, the kernel is the set of solutions to the system of linear equations $[T]\vec{x} = \mathbf{0}$.

The kernel is sometimes called the "null-space."

The vector $\begin{bmatrix} -2 \\ -1 \\ 3 \end{bmatrix}$ is in the kernel of $\begin{bmatrix} 1 & 1 & 1 \\ 2 & -1 & 1 \end{bmatrix}$, because

$$\begin{bmatrix} 1 & 1 & 1 \\ 2 & -1 & 1 \end{bmatrix} \begin{bmatrix} -2 \\ -1 \\ 3 \end{bmatrix} = \begin{bmatrix} 0 \\ 0 \end{bmatrix}.$$

Kernels are a way to talk about uniqueness of solutions of linear equations.

Proposition 2.5.2 (Unique solution equivalent to kernel 0). *The system of linear equations $T(\vec{x}) = \vec{b}$ has at most one solution for every $\vec{b}$ if and only if $\ker T = \{\vec{0}\}$ (that is, if the only vector in the kernel is the zero vector).*

Proof. If the kernel of T is not 0, then there is more than one solution to $T(\vec{x}) = 0$ (the other one being of course $\mathbf{x} = 0$).

In the other direction, if there exists a $\vec{b}$ for which $T(\vec{x}) = \vec{b}$ has more than one solution, i.e.,

$$T(\vec{x}_1) = T(\vec{x}_2) = \vec{b} \quad \text{and} \quad \vec{x}_1 \neq \vec{x}_2, \quad \text{then}$$
$$T(\vec{x}_1 - \vec{x}_2) = T(\vec{x}_1) - T(\vec{x}_2) = \vec{b} - \vec{b} = 0. \tag{2.5.1}$$

So $(\vec{x}_1 - \vec{x}_2)$ is a nonzero element of the kernel. $\square$

It is the same to say that $\vec{b}$ is in the image of A and to say that $\vec{b}$ is in the span of the columns of A. We can rewrite $A\vec{x} = \vec{b}$ as

$$\vec{a}_1 x_1 + \vec{a}_2 x_2 + \cdots + \vec{a}_n x_n = \vec{b}.$$

If $A\vec{x} = \vec{b}$, then $\vec{b}$ is in the span of $\vec{a}_1, \ldots, \vec{a}_n$, since it can be written as a linear combination of those vectors.

The image of a transformation is a way to talk about existence of solutions.

Definition 2.5.3 (Image). The image of T, denoted $\operatorname{Img} T$, is the set of vectors $\vec{b}$ for which there exists a solution of $T(\vec{x}) = \vec{b}$.

For example, $\begin{bmatrix} 7 \\ 2 \end{bmatrix}$ is in the image of $\begin{bmatrix} 1 & 3 \\ 2 & 0 \end{bmatrix}$, since

$$\begin{bmatrix} 1 & 3 \\ 2 & 0 \end{bmatrix} \begin{bmatrix} 1 \\ 2 \end{bmatrix} = \begin{bmatrix} 7 \\ 2 \end{bmatrix}. \tag{2.5.2}$$

Remark. The word *image* is not restricted to linear algebra; for example, the image of $f(x) = x^2$ is the set of positive reals. $\triangle$

Given the matrix and vectors below, which if any vectors are in the kernel of A? Check your answer below.[11]

$$A = \begin{bmatrix} 1 & 0 & 1 & 1 \\ 2 & 1 & 1 & 3 \\ 1 & 0 & 2 & 2 \end{bmatrix}, \quad \vec{v}_1 = \begin{bmatrix} 0 \\ 2 \\ 1 \\ -1 \end{bmatrix}, \quad \vec{v}_2 = \begin{bmatrix} 1 \\ 0 \\ 1 \\ 0 \end{bmatrix}, \quad \vec{v}_3 = \begin{bmatrix} 0 \\ 4 \\ 2 \\ -2 \end{bmatrix}.$$

The image and kernel of a linear transformation provide a third language for talking about existence and uniqueness of solutions to linear equations, as summarized in Figure 2.5.1. It is important to master all three and understand their equivalence. We may think of the first language as computational: does a system of linear equations have a solution? Is it unique? The second language, that of span and linear independence of vectors, is more algebraic.

The third language, that of image and kernel, is more geometric, concerning subspaces. The kernel is a subspace of the domain (set of departure) of a linear transformation; the image is a subspace of its *range* (set of arrival).

> The image is sometimes called the "range"; this usage is a source of confusion since many authors, including ourselves, use "range" to mean the entire set of arrival: the range of a transformation $T : \mathbb{R}^n \to \mathbb{R}^m$ is $\mathbb{R}^m$.
>
> The image of T is sometimes denoted $\operatorname{im} T$, but we will stick to $\operatorname{Img} T$ to avoid confusion with "the imaginary part," which is also denoted im. For any complex matrix, both "image" and "imaginary part" make sense.

Algorithms	Algebra	Geometry
Row reduction	Inverses of matrices Solving linear equations	Subspaces
Existence of solutions Uniqueness of solutions	Span Linear independence	Images Kernels

FIGURE 2.5.1. Three languages for discussing solutions to linear equations: algorithms, algebra, geometry

> The following statements are equivalent:
>
> (1) the kernel of A is 0;
>
> (2) the only solution to the equation $A\vec{x} = \vec{0}$ is $\vec{x} = \mathbf{0}$;
>
> (3) the columns $\vec{a}_i$ making up A are linearly independent;
>
> (4) the transformation given by A is one to one;
>
> (5) the transformation given by A is injective;
>
> (6) if the equation $A\vec{x} = \vec{b}$ has a solution, it is unique.

This may be clearer if we write our definitions more precisely, specifying the domain and range of our transformation. Let $T : \mathbb{R}^n \to \mathbb{R}^m$ be a linear transformation given by the $m \times n$ matrix $[T]$. Then:

(1) The kernel of T is the set of all vectors $\vec{v} \in \mathbb{R}^n$ such that $[T]\vec{v} = 0$. (Note that the vectors in the kernel are in $\mathbb{R}^n$, the domain of the transformation.)

(2) The image of T is the set of vectors $\vec{w} \in \mathbb{R}^m$ such that there is a vector $\vec{v} \in \mathbb{R}^n$ with $[T]\vec{v} = \vec{w}$. (Note that the vectors in the image are in $\mathbb{R}^m$, the range of T.)

---

[11] The vectors $\vec{v}_1$ and $\vec{v}_3$ are in the kernel of A, since $A\vec{v}_1 = 0$ and $A\vec{v}_3 = 0$. But $\vec{v}_2$ is not, since $A\vec{v}_2 = \begin{bmatrix} 2 \\ 3 \\ 3 \end{bmatrix}$. The vector $\begin{bmatrix} 2 \\ 3 \\ 3 \end{bmatrix}$ is in the image of A.

Proposition 2.5.4 means that the kernel and the image are closed under addition and under multiplication by scalars; if you add two elements of the kernel you get an element of the kernel, and so on.

Thus by definition, the kernel of a transformation is a subset of its domain, and the image is a subset of its range. In fact, they are also *subspaces* of the domain and range respectively.

Proposition 2.5.4. *If $T : \mathbb{R}^n \to \mathbb{R}^m$ is a linear transformation given by the $m \times n$ matrix A, then the kernel of A is a subspace of $\mathbb{R}^n$, and the image of A is a subspace of $\mathbb{R}^m$.*

The proof is left as Exercise 2.5.1.

Given the vectors and the matrix T below:

$$T = \begin{bmatrix} 2 & -1 & 3 & 2 & 1 \\ 1 & 0 & 1 & 3 & 0 \\ 2 & -1 & 1 & 0 & 1 \end{bmatrix}, \ \vec{w}_1 = \begin{bmatrix} 1 \\ 2 \\ 3 \end{bmatrix}, \ \vec{w}_2 = \begin{bmatrix} 0 \\ 1 \\ 1 \\ 2 \end{bmatrix}, \ \vec{w}_3 = \begin{bmatrix} 1 \\ 0 \\ 1 \end{bmatrix}, \ \vec{w}_4 = \begin{bmatrix} 2 \\ 1 \\ 2 \\ 0 \\ 0 \end{bmatrix},$$

which vectors have the right height to be in the kernel of T? To be in its image? Can you find an element of its kernel? Check your answer below.[12]

Finding bases for the image and kernel

Suppose A is the matrix of T. If we row reduce A to echelon form $\widetilde{A}$, we can find a basis for the image, using the following theorem. Recall (Definition 2.2.1) that a *pivotal column* of A is one whose corresponding column in $\widetilde{A}$ contains a pivotal 1.

Theorem 2.5.5 (A basis for the image). *The pivotal columns of A form a basis for* Img A.

We will prove this theorem, and the analogous theorem for the kernel, after giving some examples.

Example 2.5.6 (Finding a basis for the image). Consider the matrix A below, which describes a linear transformation from $\mathbb{R}^5$ to $\mathbb{R}^4$:

---

[12]The matrix T represents a transformation from $\mathbb{R}^5$ to $\mathbb{R}^3$; it takes a vector in $\mathbb{R}^5$ and gives a vector in $\mathbb{R}^3$. Therefore $\vec{w}_4$ has the right height to be in the kernel (although it isn't), and $\vec{w}_1$ and $\vec{w}_3$ have the right height to be in its image. Since the sum of the second and fifth columns of T is 0, one element of the kernel is $\begin{bmatrix} 0 \\ 1 \\ 0 \\ 0 \\ 1 \end{bmatrix}$.

$$A = \begin{bmatrix} 1 & 0 & 1 & 3 & 0 \\ 0 & 1 & 1 & 2 & 0 \\ 1 & 1 & 2 & 5 & 1 \\ 0 & 0 & 0 & 0 & 0 \end{bmatrix}, \text{ which row reduces to } \widetilde{A} = \begin{bmatrix} \underline{1} & 0 & 1 & 3 & 0 \\ 0 & \underline{1} & 1 & 2 & 0 \\ 0 & 0 & 0 & 0 & \underline{1} \\ 0 & 0 & 0 & 0 & 0 \end{bmatrix}.$$

The vectors of Equation 2.5.3 are *not* the only basis for the image.

Note that while the pivotal columns of the original matrix A form a basis for the image, it is not necessarily the case that the columns of the row reduced matrix $\widetilde{A}$ containing pivotal 1's form such a basis. For example, the matrix $\begin{bmatrix} 1 & 2 \\ 1 & 2 \end{bmatrix}$ row reduces to $\begin{bmatrix} 1 & 2 \\ 0 & 0 \end{bmatrix}$. The vector $\begin{bmatrix} 1 \\ 1 \end{bmatrix}$ forms a basis for the image, but $\begin{bmatrix} 1 \\ 0 \end{bmatrix}$ does not.

The pivotal 1's of the row-reduced matrix $\widetilde{A}$ are in columns 1, 2 and 5, so columns 1, 2 and 5 of the original matrix A are a basis for the image:

$$\begin{bmatrix} 1 \\ 0 \\ 1 \\ 0 \end{bmatrix}, \begin{bmatrix} 0 \\ 1 \\ 1 \\ 0 \end{bmatrix}, \text{ and } \begin{bmatrix} 0 \\ 0 \\ 1 \\ 0 \end{bmatrix}. \qquad 2.5.3$$

For example, the $\vec{w}$ below, which is in the image of A, can be expressed uniquely as a linear combination of the image basis vectors:

$$\vec{w} = 2\vec{a}_1 + \vec{a}_2 - \vec{a}_3 + 2\vec{a}_4 - 3\vec{a}_5 = \begin{bmatrix} 7 \\ 4 \\ 8 \\ 0 \end{bmatrix} = 7\begin{bmatrix} 1 \\ 0 \\ 1 \\ 0 \end{bmatrix} + 4\begin{bmatrix} 0 \\ 1 \\ 1 \\ 0 \end{bmatrix} - 3\begin{bmatrix} 0 \\ 0 \\ 1 \\ 0 \end{bmatrix}. \qquad 2.5.4$$

Note that each vector in the basis for the image has four entries, as it must, since the image is a subspace of $\mathbb{R}^4$. (The image is not of course $\mathbb{R}^4$ itself; a basis for $\mathbb{R}^4$ must have four elements.)   △

A basis for the kernel

Finding a basis for the kernel is more complicated; you may find it helpful to refer to Example 2.5.8 to understand the statement of Theorem 2.5.7.

A basis for the kernel is of course a set of vectors such that any vector in the kernel (any vector $\vec{w}$ satisfying $A\vec{w} = \vec{0}$) can be expressed as a linear combination of those basis vectors. The basis vectors must themselves be in the kernel, and they must be linearly independent.

Theorem 2.2.4 says that if a system of linear equations has a solution, then it has a unique solution for any value you choose of the non-pivotal unknowns. Clearly $A\vec{w} = \vec{0}$ has a solution, namely $\vec{w} = \vec{0}$. So the tactic is to choose the values of the non-pivotal unknowns in a convenient way. We take our inspiration from the standard basis vectors, which each have one entry equal to 1, and the others 0. We construct one vector for each non-pivotal column, by setting the entry corresponding to that non-pivotal unknown to be 1, and the entries corresponding to the other non-pivotal unknowns to be 0. (The entries corresponding to the pivotal unknowns will be whatever they have to be to satisfy the equation $A\vec{v}_i = \vec{0}$.)

In Example 2.5.6, the matrix A has two non-pivotal columns, so $p = 2$; those two columns are the third and fourth columns of A, so $k_1 = 3$ and $k_2 = 4$.

An equation $A\vec{x} = \vec{0}$ (i.e., $A\vec{x} = \vec{b}$ where $\vec{b} = \vec{0}$) is called *homogeneous*.

These two vectors are clearly linearly independent; no "linear combination" of $\vec{v}_1$ could produce the 1 in the fourth entry of $\vec{v}_2$, and no "linear combination" of $\vec{v}_2$ could produce the 1 in the third entry of $\vec{v}_1$. Basis vectors found using the technique given in Theorem 2.5.7 will always be linearly independent, since for each entry corresponding to a non-pivotal unknown, one basis vector will have 1 and all the others will have 0.

Note that each vector in the basis for the kernel has five entries, as it must, since the domain of the transformation is $\mathbb{R}^5$.

Theorem 2.5.7 (A basis for the kernel). *Let p be the number of non-pivotal columns of A, and $k_1, \ldots, k_p$ be their positions. For each non-pivotal column form the vector $\vec{v}_i$ satisfying $A\vec{v}_i = \mathbf{0}$, and such that its k_ith entry is 1, and its k_jth entries are all 0, for $j \neq i$. The vectors $\vec{v}_1, \ldots, \vec{v}_p$ form a basis of $\ker A$.*

We prove Theorem 2.5.7 below. First, an example.

Example 2.5.8 (Finding a basis for the kernel). The third and fourth columns of A in Example 2.5.6 above are non-pivotal, so the system has a unique solution for any values we choose of the third and fourth unknowns. In particular, there is a unique vector $\vec{v}_1$ whose third entry is 1 and fourth entry is 0, such that $A\vec{v}_1 = 0$. There is another, $\vec{v}_2$, whose fourth entry is 1 and third entry is 0, such that $A\vec{v}_2 = 0$:

$$\vec{v}_1 = \begin{bmatrix} - \\ - \\ 1 \\ 0 \\ - \end{bmatrix}, \quad \vec{v}_2 = \begin{bmatrix} - \\ - \\ 0 \\ 1 \\ - \end{bmatrix}. \qquad 2.5.5$$

Now we need to fill in the blanks, finding the first, second, and fifth entries of these vectors, which correspond to the pivotal unknowns. We read these values from the first three rows of $[\widetilde{A}, \widetilde{\mathbf{0}}]$ (remembering that a solution for $\widetilde{A}\mathbf{x} = \widetilde{\mathbf{0}}$ is also a solution for $A\mathbf{x} = \mathbf{0}$):

$$[\widetilde{A}, \widetilde{\mathbf{0}}] = \begin{bmatrix} 1 & 0 & 1 & 3 & 0 & 0 \\ 0 & 1 & 1 & 2 & 0 & 0 \\ 0 & 0 & 0 & 0 & 1 & 0 \\ 0 & 0 & 0 & 0 & 0 & 0 \end{bmatrix}, \quad \text{i.e.,} \quad \begin{aligned} x_1 + x_3 + 3x_4 &= 0 \\ x_2 + x_3 + 2x_4 &= 0 \\ x_5 &= 0, \end{aligned} \qquad 2.5.6$$

which gives

$$\begin{aligned} x_1 &= -x_3 - 3x_4 \\ x_2 &= -x_3 - 2x_4 \\ x_5 &= 0. \end{aligned} \qquad 2.5.7$$

So for $\vec{v}_1$, where $x_3 = 1$ and $x_4 = 0$, the first entry is $x_1 = -1$, the second is -1 and the fifth is 0; the corresponding entries for $\vec{v}_2$ are -3, -2 and 0:

$$\vec{v}_1 = \begin{bmatrix} -1 \\ -1 \\ 1 \\ 0 \\ 0 \end{bmatrix}; \quad \vec{v}_2 = \begin{bmatrix} -3 \\ -2 \\ 0 \\ 1 \\ 0 \end{bmatrix}. \qquad 2.5.8$$

These two vectors form a basis of the kernel of A. For example, the vector

$$\vec{v} = \begin{bmatrix} 0 \\ -1 \\ 3 \\ -1 \\ 0 \end{bmatrix} \text{ is in the kernel of } A, \text{ since } A\vec{v} = 0, \text{ so it should be possible to}$$

express $\vec{v}$ as a linear combination of the vectors of the basis for the kernel. Indeed it is: $\vec{v} = 3\vec{v}_1 - \vec{v}_2$. $\triangle$

Now find a basis for the image and kernel of the following matrix:

$$\begin{bmatrix} 2 & 1 & 3 & 1 \\ 1 & -1 & 0 & 1 \\ 1 & 1 & 2 & 1 \end{bmatrix}, \quad \text{which row reduces to} \quad \begin{bmatrix} \underline{1} & 0 & 1 & 0 \\ 0 & \underline{1} & 1 & 0 \\ 0 & 0 & 0 & \underline{1} \end{bmatrix}, \qquad 2.5.9$$

checking your answer below.[13]

Proof of Theorem 2.5.5 (A basis for the image). Let $A = [\vec{a}_1 \dots \vec{a}_m]$. To prove that the pivotal columns of A form a basis for the image of A we need to prove: (1) that the pivotal columns of A are in the image, (2) that they are linearly independent and (3) that they span the image.

(1) The pivotal columns of A (in fact, all columns of A) are in the image, since $A\vec{e}_i = \vec{a}_i$.

(2) The vectors are linearly independent, since when all non-pivotal entries of $\vec{x}$ are 0, the only solution of $A\vec{x} = \mathbf{0}$ is $\vec{x} = \mathbf{0}$. (If the pivotal unknowns are also 0, i.e., if $\vec{x} = 0$, then clearly $A\vec{x} = \mathbf{0}$. This is the only such solution, because the system has a *unique* solution for each value we choose of the non-pivotal unknowns.)

(3) They span the image, since each non-pivotal vector $\vec{v}_k$ is a linear combination of the preceding pivotal ones (Equation 2.2.8). $\square$

Proof of Theorem 2.5.7 (A basis for the kernel). Similarly, to prove that the vectors $\vec{v}_i = \vec{v}_1, \dots, \vec{v}_p$ form a basis for the kernel of A, we must show

---

[13]The vectors $\begin{bmatrix} 2 \\ 1 \\ 1 \end{bmatrix}$, $\begin{bmatrix} 1 \\ -1 \\ 1 \end{bmatrix}$, and $\begin{bmatrix} 1 \\ 1 \\ 1 \end{bmatrix}$ form a basis for the image; the vector $\begin{bmatrix} -1 \\ -1 \\ 1 \\ 0 \end{bmatrix}$

is a basis for the kernel. The row-reduced matrix $[\widetilde{A}, \widetilde{\mathbf{0}}]$ is

$$\begin{bmatrix} \underline{1} & 0 & 1 & 0 & 0 \\ 0 & \underline{1} & 1 & 0 & 0 \\ 0 & 0 & 0 & \underline{1} & 0 \end{bmatrix}, \quad \text{i.e.,} \quad \begin{aligned} x_1 + x_3 &= 0 \\ x_2 + x_3 &= 0 \\ x_4 &= 0. \end{aligned}$$

The third column of the original matrix is non-pivotal, so for the vector of the basis of the kernel we set $x_3 = 1$, which gives $x_1 = -1, x_2 = -1$.

that they are in the kernel, that they are linearly independent, and that they span the kernel.

For a transformation $T : \mathbb{R}^n \to \mathbb{R}^m$ the following statements are equivalent:

(1) the columns of $[T]$ span $\mathbb{R}^m$;

(2) the image of T is $\mathbb{R}^m$;

(3) the transformation T is onto;

(4) the transformation T is surjective;

(5) the rank of T is m;

(6) the dimension of $\mathrm{Img}(T)$ is m.

(7) the row reduced matrix $\widetilde{T}$ has no row containing all zeroes.

(8) the row reduced matrix $\widetilde{T}$ has a pivotal 1 in every row.

For a transformation from $\mathbb{R}^n$ to $\mathbb{R}^n$, if $\ker(T) = 0$, then the image is all of $\mathbb{R}^n$.

Recall (Definition 2.4.20) that the *dimension* of a subspace of $\mathbb{R}^n$ is the number of basis vectors of the subspace. It is denoted dim.

The dimension formula says there is a conservation law concerning the kernel and the image: saying something about uniqueness says something about existence.

The rank of a matrix is the most important number to associate to it.

(1) By definition, $A\vec{v}_i = 0$, so $\vec{v}_i \in \ker A$.

(2) As pointed out in Example 2.5.8, the $\vec{v}_i$ are linearly independent, since exactly one has a nonzero number in each position corresponding to non-pivotal unknown.

(3) Saying that the $\vec{v}_i$ span the kernel means that any $\vec{x}$ such that $A\vec{x} = \mathbf{0}$ can be written as a linear combination of the $\vec{v}_i$. Indeed, suppose that $A\vec{x} = \mathbf{0}$. We can construct a vector $\vec{w} = x_{k_1}\vec{v}_1 + x_{k_2}\vec{v}_2 + \cdots + x_{k_p}\vec{v}_p$ that has the same entry x_{k_i} in the non-pivotal column k_i as does $\vec{x}$. Since $A\vec{v}_i = \mathbf{0}$, we have $A\vec{w} = \mathbf{0}$. But for each value of the non-pivotal variables, there is a unique vector $\vec{x}$ such that $A\vec{x} = 0$. Therefore $\vec{x} = \vec{w}$. $\square$

Uniqueness and existence: the dimension formula

Much of the power of linear algebra comes from the following theorem, known as the *dimension formula*.

Theorem 2.5.9 (Dimension formula). *Let* $T : \mathbb{R}^n \to \mathbb{R}^m$ *be a linear transformation. Then*

$$\dim(\ker T) + \dim(\mathrm{Img}\, T) = n, \quad \textit{the dimension of the domain.} \qquad 2.5.10$$

Definition 2.5.10 (Rank and Nullity). The dimension of the image of a linear transformation is called its rank, and the dimension of its kernel is called its nullity.

Thus the dimension formula says that for any linear transformation, the rank plus the nullity is equal to the dimension of the domain.

Proof. Suppose T is given by the matrix A. Then, by Theorems 2.5.5 and 2.5.7 above, the image has one basis vector for each pivotal column of A, and the kernel has one basis vector for each non-pivotal column, so in all we find

$$\dim(\ker T) + \dim(\mathrm{Img}\, T) = \text{number of columns of } A = n. \quad \square$$

Given a transformation T represented by a 3×4 matrix $[T]$ with rank 2, what is the domain and its range of the transformation? What is the dimension of its kernel? Is it onto? Check your answers below.[14]

---

[14]The domain of T is $\mathbb{R}^4$ and its range is $\mathbb{R}^3$. The dimension of its kernel is 2, since the dimension of the kernel and that of the image equal the dimension of the domain. The transformation is not onto, since a basis for $\mathbb{R}^3$ must have three basis elements.

The most important case of the dimension formula is when the domain and range have the same dimension. In this case, one can deduce existence of solutions from uniqueness, and vice versa. Most often, the first approach is most useful; it is often easier to prove that $T(\vec{x}) = 0$ has a unique solution than it is to construct a solution of $T(\vec{x}) = \vec{b}$. It is quite remarkable that knowing that $T(\vec{x}) = 0$ has a unique solution guarantees existence of solutions for all $T(\vec{x}) = \vec{b}$. This is, of course, an elaboration of Theorem 2.2.4. But that theorem depends on knowing a matrix. Corollary 2.5.11 can be applied when there is no matrix to write down, as we will see in Example 2.5.14, and in exercises mentioned at left.

The *power of linear algebra* comes from Corollary 2.5.11. See Example 2.5.14, and Exercises 2.5.10, 2.5.16 and 2.5.17. These exercises deduce major mathematical results from this corollary.

Corollary 2.5.11 (Deducing existence from uniqueness). *If $T : \mathbb{R}^n \to \mathbb{R}^n$ is a linear transformation, then the equation $T(\vec{x}) = \vec{b}$ has a solution for any $\vec{b} \in \mathbb{R}^n$ if and only if the only solution to the equation $T(\vec{x}) = \mathbf{0}$ is $\vec{x} = \mathbf{0}$, (i.e., if the kernel is zero).*

Since Corollary 2.5.11 is an "if and only if" statement, it can also be used to deduce uniqueness from existence; in practice this is not quite so useful.

Proof. Saying that $T(\vec{x}) = \vec{b}$ has a solution for any $\vec{b} \in \mathbb{R}^n$ means that $\mathbb{R}^n$ is the image of T, so $\dim \operatorname{Img} T = n$, which is equivalent to $\dim \ker(T) = 0$. $\quad\square$

The following result is really quite surprising: it says that the number of linearly independent columns and the number of linearly independent rows of a matrix are equal.

Proposition 2.5.12. *Let A be a matrix. Then the span of the columns of A and the span of the rows of A have the same dimension.*

One way to understand this result is to think of *constraints on the kernel of A*. Think of A as the $m \times n$ matrix made up of its rows:

$$A = \begin{bmatrix} -- A_1 -- \\ -- A_2 -- \\ \vdots \\ -- A_m -- \end{bmatrix}. \qquad\qquad 2.5.11$$

Then the kernel of A is made up of the vectors $\vec{x}$ satisfying the linear constraints $A_1\vec{x} = 0, \ldots, A_m\vec{x} = 0$. Think of adding in these constraints one at a time. Before any constraints are present, the kernel is all of $\mathbb{R}^n$. Each time you add one constraint, you cut down the dimension of the kernel by 1. But this is only true if the new constraint is genuinely new, not a consequence of the previous ones, i.e., if A_i is linearly independent from $A_1, \ldots, A_{i-1}$.

Let us call the number of linearly independent rows A_i the *row rank* of A. The argument above leads to the formula

$$\dim \ker A = n - \operatorname{row\ rank}(A).$$

The dimension formula says exactly that

$$\dim \ker A = n - \operatorname{rank}(A),$$

so the rank of A and the row rank of A should be equal.

The argument above isn't quite rigorous: it used the intuitively plausible but unjustified "Each time you add one constraint, you cut down the dimension of the kernel by 1." This is true and not hard to prove, but the following argument is shorter (and interesting too).

Proof. Given a matrix, we will call the span of the columns the *column space* of A and the span of the rows the *row space* of A. Indeed, the rows of $\widetilde{A}$ are linear combinations of the rows of A, and vice versa since row operations are reversible. In particular, the row space of A and of $\widetilde{A}$ coincide, where A row-reduces to $\widetilde{A}$.

The rows of $\widetilde{A}$ that contain pivotal 1's are a basis of the row space of $\widetilde{A}$: the other rows are zero so they definitely don't contribute to the row space, and the pivotal rows of A are linearly independent, since all the other entries in a column containing a pivotal 1 are 0. So the dimension of the row space of A is the number of pivotal 1's of $\widetilde{A}$, which we have seen is the dimension of the column space of A. $\square$

Corollary 2.5.13. *A matrix A and its transpose $A^\top$ have the same rank.*

Remark. Proposition 2.5.12 gives us the statement we wanted in Section 2.4: the number of linearly independent equations in a system of linear equations $A\vec{x} = \vec{b}$ is the number of pivotal columns of A. Basing linear algebra on row reduction can be seen as a return to Euler's way of thinking. It is, as Euler said, immediately apparent why you can't determine x and y from the two equations $3x - 2y = 5$ and $4y = 6x - 10$. (In the original, "La raison de cet accident saute d'abord aux yeux": *the reason for this accident leaps to the eyes*). In that case, it is obvious that the second equation is twice the first. When the linear dependence of a system of linear equations no longer *leaps to the eyes*, row reduction provides a way to make it obvious.

Unfortunately for the history of mathematics, in the same year 1750 that Euler wrote his analysis, Gabriel Cramer published a treatment of linear equations based on determinants, which rapidly took hold, and the more qualitative approach begun by Euler was forgotten. As Jean-Luc Dorier writes in his essay on the history of linear algebra,[15]

> *... even if determinants proved themselves a valuable tool for studying linear equations, it must be admitted that they introduced a certain complexity, linked to the technical skill their use requires. This fact had the undeniable effect of masking certain intuitive aspects of the nature of linear equations ...*   △

We defined linear combinations in terms of linear combinations of vectors, but (as we will see in Section 2.6) the same definition can apply to linear combinations of other objects, such as matrices, functions. In this proof we are applying it to row matrices.

Gauss is a notable exception; when he needed to solve linear equations, he used row reduction. In fact, row reduction is also called *Gaussian elimination.*

The rise of the computer, with emphasis on computationally effective schemes, has refocused attention on row reduction as a way to solve linear equations.

[15]J.-L. Dorier, ed., *L'Enseignement de l'algèbre linéaire en question*, La Pensée Sauvage, Editions, 1997.

Now let us see an example of the power of Corollary 2.5.11.

Example 2.5.14 (Partial fractions). Let

$$p\,(x) = (x - a_1)^{n_1} \cdots (x - a_k)^{n_k} \qquad 2.5.12$$

be a polynomial of degree $n = n_1 + \cdots + n_k$, with the a_i distinct; for example,

$$x^2 - 1 = (x+1)(x-1), \text{ with } a_1 = -1, a_2 = 1; n_1 = n_2 = 1, \text{ so that } n = 2$$

$$x^3 - 2x^2 + x = x(x-1)^2, \text{ with } a_1 = 0, a_2 = 1; n_1 = 1, n_2 = 2, \text{ so that } n = 3.$$

The claim of partial fractions is the following:

Proposition 2.5.15 (Partial fractions). *For any such polynomial p of degree n, and any polynomial q of degree $< n$, the rational function q/p can be written uniquely as a sum of simpler terms, called partial fractions:*

$$\frac{q\,(x)}{p\,(x)} = \frac{q_1(x)}{(x - a_1)^{n_1}} + \cdots + \frac{q_k(x)}{(x - a_k)^{n_k}}, \qquad 2.5.13$$

with each q_i a polynomial of degree $< n_i$.

For example, when $q(x) = 2x + 3$ and $p(x) = x^2 - 1$, Proposition 2.5.15 says that there exist polynomials q_1 and q_2 of degree less than 1 (i.e., numbers, which we will call A_0 and B_0, the subscript indicating that they are coefficients of the term of degree 0) such that

$$\frac{2x + 3}{x^2 - 1} = \frac{A_0}{x + 1} + \frac{B_0}{x - 1}. \qquad 2.5.14$$

If $q(x) = x^3 - 1$ and $p(x) = (x+1)^2(x-1)^2$, then the proposition says that there exist two polynomials of degree 1, $q_1 = A_1 x + A_0$ and $q_2 = B_1 x + B_0$, such that

$$\frac{x^3 - 1}{(x+1)^2(x-1)^2} = \frac{A_1 x + A_0}{(x+1)^2} + \frac{B_1 x + B_0}{(x-1)^2}. \qquad 2.5.15$$

In simple cases, it's clear how to find these terms. In the first case above, to find the numerators A_0 and B_0, we multiply out to get a common denominator:

$$\frac{2x + 3}{x^2 - 1} = \frac{A_0}{x+1} + \frac{B_0}{x-1} = \frac{A_0(x-1) + B_0(x+1)}{x^2 - 1} = \frac{(A_0 + B_0)x + (B_0 - A_0)}{x^2 - 1},$$

so that we get two linear equations in two unknowns:

$$\begin{matrix} -A_0 + B_0 = 3 \\ A_0 + B_0 = 2, \end{matrix} \quad \text{i.e., the constants} \quad B_0 = \frac{5}{2}, \; A_0 = -\frac{1}{2}. \qquad 2.5.16$$

We can think of the system of linear equations on the left-hand side of Equation 2.5.16 as the matrix multiplication

$$\begin{bmatrix} -1 & 1 \\ 1 & 1 \end{bmatrix} \begin{bmatrix} A_0 \\ B_0 \end{bmatrix} = \begin{bmatrix} 3 \\ 2 \end{bmatrix}. \qquad 2.5.17$$

In Equation 2.5.12, note that we are requiring that the a_i be distinct. For example, although

$$p(x) = x(x-1)(x-1)(x-1)$$

can be written

$$p(x) = x(x-1)(x-1)^2,$$

this is not an allowable decomposition for use in Proposition 2.5.15, while

$$p(x) = x(x-1)^3 \text{ is.}$$

Note that by the fundamental theorem of algebra (Theorem 1.6.10), every polynomial can be written as a product of powers of degree 1 polynomials (Equation 2.5.12). Of course finding the a_i means finding the roots of the polynomial, which may be very difficult.

What is the analogous matrix multiplication for Equation 2.5.15?[16]

What about the general case? If we put the right-hand side of Equation 2.5.13 on a common denominator we see that $q(x)/p(x)$ is equal to

$$\frac{q_1(x)(x-a_2)^{n_2}\dots(x-a_k)^{n_k}+q_2(x)(x-a_1)^{n_1}(x-a_3)^{n_3}\dots(x-a_k)^{n_k}+\cdots+q_k(x)(x-a_1)^{n_1}\dots(x-a_{k-1})^{n_{k-1}}}{(x-a_1)^{n_1}(x-a_2)^{n_2}\dots(x-a_k)^{n_k}}.$$

2.5.18

As we did in our simpler cases, we could write this as a system of linear equations for the coefficients of the q_i and solve by row reduction. But except in the simplest cases, computing the matrix would be a big job. Worse, how do we know that the system of equations we get *has* solutions? We might worry about investing a lot of work only to discover that the equations were incompatible.

Proposition 2.5.15 assures us that there will always be a solution, and Corollary 2.5.11 provides the key.

Proof of Proposition 2.5.15 (Partial fractions). Note that the matrix we would get following the above procedure would necessarily be an $n \times n$ matrix. This matrix gives a linear transformation that has as its input a vector whose entries are the coefficients of $q_1, \dots, q_k$. There are n such coefficients in all. (Each polynomial q_i has n_i coefficients, for terms of degree 0 through $(n_i - 1)$, and the sum of the n_i equals n.) It has as its output a vector giving the n coefficients of q (since q is of degree $< n$, it has n coefficients, $0, \dots, n - 1$.)

Thus the matrix can be thought of a linear transformation $T : \mathbb{R}^n \to \mathbb{R}^n$, and by Corollary 2.5.11, Proposition 2.5.15 is true *if and only if the only solution of $T(q_1, \dots, q_k) = \mathbf{0}$ is $q_1 = \cdots = q_k = 0$.* This will follow from Lemma 2.5.16:

Corollary 2.5.11: if $T : \mathbb{R}^n \to \mathbb{R}^n$ is a linear transformation, the equation $T(\vec{x}) = \vec{b}$ has a solution for any $\vec{b} \in \mathbb{R}^n$ if and only if the only solution to the equation $T(\vec{x}) = \mathbf{0}$ is $\vec{x} = \mathbf{0}$.

We are thinking of the transformation T both as the matrix that takes the coefficients of the q_i and returns the coefficients of q, and as the linear function that takes $q_i, \dots, q_k$ and returns the polynomial q.

---

[16]Multiplying out, we get

$$\frac{x^3(A_1 + B_1) + x^2(-2A_1 + A_0 + 2B_1 + B_0) + x(A_1 - 2A_0 + B_1 + 2B_0) + A_0 + B_0}{(x+1)^2(x-1)^2},$$

so

$$A_0 + B_0 = -1 \quad \text{(coefficient of term of degree 0)}$$
$$A_1 - 2A_0 + B_1 + 2B_0 = 0 \quad \text{(coefficient of term of degree 1)}$$
$$-2A_1 + A_0 + 2B_1 + B_0 = 0 \quad \text{(coefficient of term of degree 2)}$$
$$A_1 + B_1 = 1 \quad \text{(coefficient of term of degree 3);}$$

i.e.,

$$\begin{bmatrix} 0 & 1 & 0 & 1 \\ 1 & -2 & 1 & 2 \\ -2 & 1 & 2 & 1 \\ 1 & 0 & 1 & 0 \end{bmatrix} \begin{bmatrix} A_1 \\ A_0 \\ B_1 \\ B_0 \end{bmatrix} = \begin{bmatrix} -1 \\ 0 \\ 0 \\ 1 \end{bmatrix}.$$

Lemma 2.5.16. *If $q_i \neq 0$ is a polynomial of degree $< n_i$, then*

$$\lim_{x \to a_i} \left| \frac{q_i(x)}{(x - a_i)^{n_i}} \right| = \infty. \qquad 2.5.19$$

That is, if $q_i \neq 0$, then $q_i(x)/(x - a_i)^{n_i}$ blows up to infinity.

Proof. It is clear that for values of x very close to a_i, the denominator $(x - a_i)^{n_i}$ will get very small; if all goes well the entire term will then get very big. But we have to work a bit to make sure both that the numerator does not get small equally fast, and that the other terms of the polynomial don't compensate.

Let us make the change of variables $u = x - a_i$, so that

$$q_i(x) = q_i(u + a_i), \quad \text{which we will denote} \quad \widetilde{q}_i(u). \qquad 2.5.20$$

The numerator in Equation 2.5.21 is of degree $< n_i$, while the denominator is of degree n_i.

Then we have

$$\lim_{u \to 0} \left| \frac{\widetilde{q}_i(u)}{u^{n_i}} \right| = \infty \quad \text{if } q_i \neq 0. \qquad 2.5.21$$

Indeed, if $q_i \neq 0$, then $\widetilde{q}_i \neq 0$, and there exists a number $m < n_i$ such that

$$\widetilde{q}_i(u) = a_m u^m + \cdots + a_{n_i - 1} u^{n_i - 1} \qquad 2.5.22$$

with $a_m \neq 0$. (This a_m is the first nonzero coefficient; as $u \to 0$, the term $a_m u^m$ is bigger than all the other terms.) Dividing by u^{n_i} we can write

$$\frac{\widetilde{q}_i(u)}{u^{n_i}} = \frac{1}{u^{n_1 - m}} (a_m + \dots), \qquad 2.5.23$$

where the dots $\dots$ represent terms containing u to a positive power, since $m < n_i$. In particular,

$$\text{as} \quad u \to 0, \quad \left| \frac{1}{u^{n_i - m}} \right| \to \infty \quad \text{and} \quad (a_m + \dots) \to a_m. \qquad 2.5.24$$

We see that as $x \to a_i$, the term $q_i(x)/(x - a_i)^{n_i}$ blows up to infinity: the denominator gets smaller and smaller while the numerator tends to $a_m \neq 0$.

This ends the proof of Lemma 2.5.16.

Proof of Proposition 2.5.15, continued. Suppose $q_i \neq 0$. For all the other terms $q_j, j \neq i$, the rational functions

$$\frac{q_j(x)}{(x - a_j)^{n_j}} \qquad 2.5.25$$

have the finite limits $q_j(a_i)/(a_i - a_j)^{n_j}$ as $x \to a_i$, and therefore the sum

$$\frac{q(x)}{p(x)} = \frac{q_1(x)}{(x - a_1)^{n_1}} + \cdots + \frac{q_k(x)}{(x - a_k)^{n_k}}, \qquad 2.5.26$$

has infinite limit as $x \to a_i$ and q cannot vanish identically. So $T(q_1, \dots, q_k) \neq \mathbf{0}$ if some $q_i \neq 0$, and we can conclude—without having to compute any matrices

This example really put linear algebra to work. Even after translating the problem into linear algebra, via the linear transformation T, the answer was not clear; only after using the dimension formula is the result apparent. The dimension formula (or rather, Corollary 2.5.11, the dimension formula applied to transformations from $\mathbb{R}^n$ to $\mathbb{R}^n$) tells us that if $T : \mathbb{R}^n \to \mathbb{R}^n$ is one to one (solutions are unique), then it is onto (solutions exist).

Still, all of this is nothing more than the intuitively obvious statement that either n equations in n unknowns are independent, the good case, or everything goes wrong at once–the transformation is not one to one and therefore not onto.

or solve any systems of equations—that Proposition 2.5.15 is correct: for any polynomial p of degree n, and any polynomial q of degree $< n$, the rational function q/p can be written uniquely as a sum of partial fractions. $\square$

2.6 An Introduction to Abstract Vector Spaces

In this section we give a very brief introduction to *abstract vector spaces*, introducing vocabulary that will be useful later in the book, particularly in Chapter 6 on forms.

We have already used vector spaces that are not $\mathbb{R}^n$: in Section 1.4 we considered an $m \times n$ matrix as a point in $\mathbb{R}^{nm}$, and in Example 1.7.15 we spoke of the "space" P_k of polynomials of degree at most k. In each case we "identified" the space with $\mathbb{R}^N$ for some appropriate N: we identified the space of $m \times n$ matrices with $\mathbb{R}^{nm}$, and we identified P_k with $\mathbb{R}^{k+1}$. But just what "identifying" means is not quite clear, and difficulties with such identifications become more and more cumbersome.

As we will see in a moment, a vector space is a set in which elements can be added and multiplied by numbers. We need to decide what numbers we are using, and for our purposes there are only two interesting choices: real or complex numbers. Mainly to keep the psychological load lighter, we will restrict our discussion to real numbers, and consider only *real vector spaces*, to be called simply "vector spaces" from now on. (Virtually everything to be discussed would work just as well for complex numbers.)

We will denote a vector in an abstract vector space by an underlined bold letter, to distinguish it from a vector in $\mathbb{R}^n$: $\underline{\mathbf{v}} \in V$ as opposed to $\vec{\mathbf{v}} \in \mathbb{R}^n$.

A vector space is anything that satisfies the following rules.

Definition 2.6.1 (Vector space). A vector space is a set V of vectors such that two vectors can be added to form another vector, and a vector can be multiplied by a scalar in $\mathbb{R}$ to form another vector. This addition and multiplication must satisfy the following eight rules:

(1) *Additive identity.*   There exists a vector $\mathbf{0} \in V$ such that for any $\underline{\mathbf{v}} \in V$, $\mathbf{0} + \underline{\mathbf{v}} = \underline{\mathbf{v}}.$

(2) *Additive inverse.*   For any $\underline{\mathbf{v}} \in V$, there exists a vector $-\underline{\mathbf{v}} \in V$ such that $\underline{\mathbf{v}} + (-\underline{\mathbf{v}}) = \mathbf{0}.$

(3) *Commutative law for addition.*   For all $\underline{\mathbf{v}}, \underline{\mathbf{w}} \in V$,
$$\underline{\mathbf{v}} + \underline{\mathbf{w}} = \underline{\mathbf{w}} + \underline{\mathbf{v}}.$$

(4) *Associative law for addition.*   For all $\underline{\mathbf{v}}_1, \underline{\mathbf{v}}_2, \underline{\mathbf{v}}_3 \in V$,
$$\underline{\mathbf{v}}_1 + (\underline{\mathbf{v}}_2 + \underline{\mathbf{v}}_3) = (\underline{\mathbf{v}}_1 + \underline{\mathbf{v}}_2) + \underline{\mathbf{v}}_3.$$

You may think of these eight rules as the "essence of $\mathbb{R}^n$," abstracting from the vector space $\mathbb{R}^n$ all its most important properties, *except its distinguished standard basis.* This allows us to work with other vector spaces, whose elements are not naturally defined in terms of lists of numbers.

(5) *Multiplicative identity.*   For all $\underline{\mathbf{v}} \in V$ we have $1\underline{\mathbf{v}} = \underline{\mathbf{v}}.$

(6) *Associative law for multiplication.*   For all $\alpha, \beta \in \mathbb{R}$ and all $\underline{\mathbf{v}} \in V$, $\alpha(\beta\underline{\mathbf{v}}) = (\alpha\beta)\underline{\mathbf{v}}.$

(7) *Distributive law for scalar addition.*   For all scalars $\alpha, \beta \in \mathbb{R}$ and all $\underline{\mathbf{v}} \in V$, $(\alpha + \beta)\underline{\mathbf{v}} = \alpha\underline{\mathbf{v}} + \beta\underline{\mathbf{v}}.$

(8) *Distributive law for vector addition.*   For all scalars $\alpha \in \mathbb{R}$ and $\underline{\mathbf{v}}, \underline{\mathbf{w}} \in V$, we have $\alpha(\underline{\mathbf{v}} + \underline{\mathbf{w}}) = \alpha\underline{\mathbf{v}} + \alpha\underline{\mathbf{w}}.$

The primordial example of a vector space is of course $\mathbb{R}^n$ itself. More generally, a subset of $\mathbb{R}^n$ (endowed with the same addition and multiplication by scalars as $\mathbb{R}^n$ itself) is a vector space in its own right if and only if it is a subspace of $\mathbb{R}^n$ (Definition 1.1.5).

Other examples that are fairly easy to understand are the space $\mathrm{Mat}\,(n, m)$ of $n \times m$ matrices, with addition and multiplication defined in Section 1.2, and the space P_k of polynomials of degree at most k. In fact, these are easy to "identify with $\mathbb{R}^n$."

But other vector spaces have a different flavor: they are somehow much too big.

> Definition 1.1.5: A non-empty subset $V \subset \mathbb{R}^n$ is a subspace if it is closed under addition and closed under multiplication by scalars.

> Note that in Example 2.6.2 our assumption that addition is well defined in $\mathcal{C}[0,1]$ uses the fact that the sum of two continuous functions is continuous. Similarly, multiplication by scalars is well defined, because the product of a continuous function by a constant is continuous.

Example 2.6.2 (An infinite-dimensional vector space). Consider the space $\mathcal{C}(0,1)$ of continuous real-valued functions $f(x)$ defined for $0 < x < 1$. The "vectors" of this space are functions $f : (0,1) \to \mathbb{R}$, with addition defined as usual by $(f+g)(x) = f(x) + g(x)$ and multiplication by $(\alpha f)(x) = \alpha\, f(x)$. $\triangle$

Exercise 2.6.1 asks you to show that this space satisfies all eight requirements for a vector space.

The vector space $\mathcal{C}(0,1)$ cannot be identified with $\mathbb{R}^n$; there is no linear transformation from any $\mathbb{R}^n$ to this space that is onto, as we will see in detail in Example 2.6.20. But it has subspaces that can be identified with appropriate $\mathbb{R}^n$'s, as seen in Example 2.6.3, and also subspaces that cannot.

Example 2.6.3 (A finite-dimensional subspace of $\mathcal{C}(0,1)$). Consider the space of twice differentiable functions $f : \mathbb{R} \to \mathbb{R}$ such that $D^2 f = 0$ (i.e., functions of one variable whose second derivatives are 0; we could also write this $f'' = 0$). This is a subspace of the vector space of Example 2.6.2, and is a vector space itself. But since a function has a vanishing second derivative if and only if it is a polynomial of degree at most 1, we see that this space is the set of functions

> In some sense, this space "is" $\mathbb{R}^2$, by identifying $f_{a,b}$ with $\begin{bmatrix} a \\ b \end{bmatrix} \in \mathbb{R}^2$; this was not obvious from the definition.

$$f_{a,b}(x) = a + bx. \qquad\qquad 2.6.1$$

Precisely two numbers are needed to specify each element of this vector space; we could choose as our basis 1 and x.

On the other hand, the subspace $\mathcal{C}^1(0,1) \subset \mathcal{C}(0,1)$ of once continuously differentiable functions on $(0,1)$ also cannot be identified with any $\mathbb{R}^n$; the elements are more restricted than those of $\mathcal{C}(0,1)$, but not enough so that an element can be specified by finitely many numbers.

Linear transformations

In Sections 1.2 and 1.3 we investigated linear transformations $\mathbb{R}^n \to \mathbb{R}^m$. Now we wish to define linear transformations from one (abstract) vector space to another.

Equation 2.6.2 is a shorter way of writing both

$$T(\underline{\mathbf{v}}_1 + \underline{\mathbf{v}}_2) = T(\underline{\mathbf{v}}_1) + T(\underline{\mathbf{v}}_2)$$

and

$$T(\alpha\underline{\mathbf{v}}_1) = \alpha T(\underline{\mathbf{v}}_1).$$

In order to write a linear transformation from one abstract vector space to another as a matrix, you have to choose bases: one in the domain, one in the range. As long as you are in finite-dimensional vector spaces, you can do this. In infinite-dimensional vector spaces, bases usually do not exist.

In Example 2.6.6, $\mathcal{C}[0,1]$ is the space of continuous real-valued functions $f(x)$ defined for $0 \leq x \leq 1$.

The function g in Example 2.6.6 is very much like a matrix, and the formula for T_g looks a lot like $\sum g_{i,j} f_j$. This is the kind of thing we meant above when we referred to "analogs" of matrices; it is as much like a matrix as you can hope to get in this particular infinite dimensional setting. But it is not true that all transformations from $\mathcal{C}[0,1]$ to $\mathcal{C}[0,1]$ are of this sort; even the identity cannot be written in the form T_g.

Definition 2.6.4 (Linear transformation). If V and W are vector spaces, a linear transformation $T : V \to W$ is a mapping satisfying

$$T(\alpha\underline{\mathbf{v}}_1 + \beta\underline{\mathbf{v}}_2) = \alpha T(\underline{\mathbf{v}}_1) + \beta T(\underline{\mathbf{v}}_2) \qquad 2.6.2$$

for all scalars $\alpha, \beta \in \mathbb{R}$ and all $\underline{\mathbf{v}}_1, \underline{\mathbf{v}}_2 \in V$.

In Section 1.3 we saw (Theorem 1.3.14) that every linear transformation $T : \mathbb{R}^m \to \mathbb{R}^n$ is given by the $n \times m$ matrix whose ith column is $T(\vec{\mathbf{e}}_i)$. This provides a complete understanding of linear transformations from $\mathbb{R}^m$ to $\mathbb{R}^n$.

In the setting of more abstract vector spaces, linear transformations don't have this wonderful concreteness. In finite-dimensional vector spaces, it is still possible to understand a linear transformation as a matrix but you have to work at it; in particular, you must choose a basis for the domain and a basis for the range. (For infinite-dimensional vector spaces, bases usually do not exist, and matrices and their analogs are usually not available.)

Even when it is possible to write a linear transformation as a matrix, it may not be the easiest way to deal with things, as shown in Example 2.6.5.

Example 2.6.5 (A linear transformation difficult to write as a matrix). If $A \in \text{Mat}(n, n)$, then the transformation $\text{Mat}(n, n) \to \text{Mat}(n, n)$ given by $H \mapsto AH + HA$ is a linear transformation, which we encountered in Example 1.7.15 as the derivative of the mapping $S : A \mapsto A^2$:

$$[DS(A)]H = AH + HA. \qquad 2.6.3$$

Even in the case $n = 3$ it would be difficult, although possible, to write this transformation as a 9×9 matrix; the language of abstract linear transformations is more appropriate. $\triangle$

Example 2.6.6 (Showing that a transformation is linear). Let us show that if $g \begin{pmatrix} x \\ y \end{pmatrix}$ is a continuous function on $[0,1] \times [0,1]$, then the transformation $T_g : \mathcal{C}[0,1] \to \mathcal{C}[0,1]$ given by

$$\left(T_g(f)\right)(x) = \int_0^1 g \begin{pmatrix} x \\ y \end{pmatrix} f(y) \, dy \qquad 2.6.4$$

is a linear transformation. For example, if $g \begin{pmatrix} x \\ y \end{pmatrix} = |x - y|$, then we would have the linear transformation $\left(T_g(f)\right)(x) = \int_0^1 |x - y| f(y) \, dy$.

To show that Equation 2.6.4 is a linear transformation, we first show that

$$T_g(f_1 + f_2) = T_g(f_1) + T_g(f_2), \qquad 2.6.5$$

which we do as follows:

$$\big(T_g(f_1 + f_2)\big)(x) = \int_0^1 g\left(\begin{matrix} x \\ y \end{matrix}\right)(f_1 + f_2)(y)\,dy = \int_0^1 g\left(\begin{matrix} x \\ y \end{matrix}\right) \overbrace{\big(f_1(y) + f_2(y)\big)}^{\substack{\text{definition of addition} \\ \text{in vector space}}}\,dy$$

$$= \int_0^1 \left(g\left(\begin{matrix} x \\ y \end{matrix}\right) f_1(y) + g\left(\begin{matrix} x \\ y \end{matrix}\right) f_2(y)\right)\,dy$$

$$= \int_0^1 g\left(\begin{matrix} x \\ y \end{matrix}\right) f_1(y)\,dy + \int_0^1 g\left(\begin{matrix} x \\ y \end{matrix}\right) f_2(y)\,dy$$

$$= \big(T_g(f_1)\big)(x) + \big(T_g(f_2)\big)(x) = \big(T_g(f_1) + T_g(f_2)\big)(x). \qquad 2.6.6$$

Next we show that $T_g(\alpha f)(x) = \alpha T_g(f)(x)$:

$$T_g(\alpha f)(x) = \int_0^1 g\left(\begin{matrix} x \\ y \end{matrix}\right)(\alpha f)(y)\,dy = \alpha \int_0^1 g\left(\begin{matrix} x \\ y \end{matrix}\right) f(y)\,dy = \alpha T_g(f)(x). \quad 2.6.7$$

We denote by $\mathcal{C}^2$ the space of C^2 functions: functions that are twice continuously differentiable.

The linear transformation in Example 2.6.7 is a special kind of linear transformation, called a *linear differential operator*. Solving a differential equation is same as looking for the kernel of such a linear transformation. The coefficients could be any functions of x, so it's an example of an important class.

Example 2.6.7 (A linear differential operator). The transformation $T : \mathcal{C}^2(\mathbb{R}) \to \mathcal{C}(\mathbb{R})$ given by the formula

$$\big(T(f)\big)(x) = (x^2 + 1)f''(x) - xf'(x) + 2f(x) \qquad 2.6.8$$

is a linear transformation, as Exercise 2.6.2 asks you to show.

Linear independence, span and bases

In Section 2.4 we discussed linear independence, span, and bases for $\mathbb{R}^n$ and subspaces of $\mathbb{R}^n$. Extending these notions to arbitrary real vector spaces requires somewhat more work. However, we will be able to tap into what we have already done.

Let V be a vector space and let $\{\underline{\mathbf{v}}\} = \underline{\mathbf{v}}_1, \ldots, \underline{\mathbf{v}}_m$ be a finite collection of vectors in V.

Definition 2.6.8 (Linear combination). A linear combination of the vectors $\underline{\mathbf{v}}_1, \ldots, \underline{\mathbf{v}}_m$ is a vector $\underline{\mathbf{v}}$ of the form

$$\underline{\mathbf{v}} = \sum_{i=1}^{m} a_i\,\underline{\mathbf{v}}_i, \quad \text{with} \quad a_1, \ldots, a_m \in \mathbb{R}. \qquad 2.6.9$$

Definition 2.6.9 (Span). The collection of vectors $\{\underline{\mathbf{v}}\} = \underline{\mathbf{v}}_1, \ldots, \underline{\mathbf{v}}_m$ spans V if and only if all vectors of V are linear combinations of $\underline{\mathbf{v}}_1, \ldots, \underline{\mathbf{v}}_m$.

Definition 2.6.10 (Linear independence). The vectors $\underline{\mathbf{v}}_1, \ldots, \underline{\mathbf{v}}_m$ are linearly independent if and only if any one (hence all) of the following three equivalent conditions are met:

(1) There is only one way of writing a given linear combination; i.e., if

$$\sum_{i=1}^{m} a_i \underline{\mathbf{v}}_i = \sum_{i=1}^{m} b_i \underline{\mathbf{v}}_i \quad \text{implies} \quad a_1 = b_1,\ a_2 = b_2, \ldots,\ a_m = b_m. \qquad 2.6.10$$

(2) The only solution to

$$a_1\underline{\mathbf{v}}_1 + a_2\underline{\mathbf{v}}_2 + \cdots + a_m \underline{\mathbf{v}}_m = 0 \quad \text{is} \quad a_1 = a_2 = \cdots = a_m = 0. \qquad 2.6.11$$

(3) None of the $\underline{\mathbf{v}}_i$ is a linear combination of the others.

Definition 2.6.11 (Basis). A set of vectors $\underline{\mathbf{v}}_1, \ldots, \underline{\mathbf{v}}_m \in V$ is a basis of V if and only if it is linearly independent and spans V.

The following definition is central. It enables us to move from the concrete world of $\mathbb{R}^n$ to the abstract world of a vector space V.

Definition 2.6.12 ("Concrete to abstract" function $\Phi_{\{\underline{\mathbf{v}}\}}$). Let V be a vector space, and let $\{\underline{\mathbf{v}}\} = \underline{\mathbf{v}}_1, \ldots, \underline{\mathbf{v}}_m$ be a finite collection of vectors in V. The "concrete to abstract" function $\Phi_{\{\underline{\mathbf{v}}\}}$ is the linear transformation $\Phi_{\{\underline{\mathbf{v}}\}} : \mathbb{R}^m \to V$ given the formula

The *concrete to abstract* function $\Phi_{\{\underline{\mathbf{v}}\}}$ ("Phi v") takes a column vector $\vec{a} \in \mathbb{R}^m$ and gives an abstract vector $\sum_{i=1}^{n} a_i \underline{\mathbf{v}}_i$.

$$\Phi_{\{\underline{\mathbf{v}}\}}(\vec{a}) = \Phi_{\{\underline{\mathbf{v}}\}}\left(\begin{bmatrix} a_1 \\ \vdots \\ a_m \end{bmatrix}\right) = a_1\underline{\mathbf{v}}_1 + \cdots + a_m\underline{\mathbf{v}}_m. \qquad 2.6.12$$

Example 2.6.13 (Concrete to abstract function). Let P_2 be the space of polynomials of degree at most 2, and consider its basis $\underline{\mathbf{v}}_1 = 1$, $\underline{\mathbf{v}}_2 = x$, $\underline{\mathbf{v}}_3 = x^2$. Then $\Phi_{\{\underline{\mathbf{v}}\}} \begin{bmatrix} a_1 \\ a_2 \\ a_3 \end{bmatrix} = a_1 + a_2 x + a_3 x^2$ identifies P_2 with $\mathbb{R}^3$. $\quad \triangle$

Example 2.6.14 (To interpret a column vector, the basis matters). If $V = \mathbb{R}^2$ and $\vec{e}$ is the standard basis, then

$$\Phi_{\{\vec{e}\}}\left(\begin{bmatrix} a \\ b \end{bmatrix}\right) = \begin{bmatrix} a \\ b \end{bmatrix}, \quad \text{since} \quad \Phi_{\{\vec{e}\}}\left(\begin{bmatrix} a \\ b \end{bmatrix}\right) = a\vec{e}_1 + b\vec{e}_2 = \begin{bmatrix} a \\ b \end{bmatrix}. \qquad 2.6.13$$

(If $V = \mathbb{R}^n$, and $\{\vec{e}\}$ is the standard basis, then $\Phi_{\{\vec{e}\}}$ is always the identity.)

Choosing a basis is analogous to choosing a language. A language gives names to an object or an idea; a basis gives a name to a vector living in an abstract vector space. A vector has many embodiments, just as the words *book, livre, Buch* ... all mean the same thing, in different languages.

In Example 2.6.14, the function $\Phi_{\{\underline{v}\}}$ is given by the matrix $\begin{bmatrix} 1 & 1 \\ 1 & -1 \end{bmatrix}$; in this case, both the domain and the range are $\mathbb{R}^2$.

Exercise 2.4.9 asked you to prove Proposition 2.6.15 when V is a subspace of $\mathbb{R}^m$.

Why study abstract vector spaces? Why not just stick to $\mathbb{R}^n$? One reason is that $\mathbb{R}^n$ comes with the standard basis, which may not be the best basis for the problem at hand. Another is that when you prove something about $\mathbb{R}^n$, you then need to check that your proof was "basis independent" before you can extend it to an arbitrary vector space.

If instead we used the basis $\vec{v}_1 = \begin{bmatrix} 1 \\ 1 \end{bmatrix}, \vec{v}_2 = \begin{bmatrix} 1 \\ -1 \end{bmatrix}$, then

$$\Phi_{\{\underline{v}\}}\left(\begin{bmatrix} a \\ b \end{bmatrix} \right) = a\vec{v}_1 + b\vec{v}_2 = \begin{bmatrix} a+b \\ a-b \end{bmatrix};$$ 2.6.14

$\begin{bmatrix} a \\ b \end{bmatrix}$ in the new basis equals $\begin{bmatrix} a+b \\ a-b \end{bmatrix}$ in the standard basis. $\triangle$

Proposition 2.6.15 says that if $\{\underline{v}\}$ is a basis of V, then the linear transformation $\Phi_{\{\underline{v}\}} : \mathbb{R}^m \to V$ allows us to identify $\mathbb{R}^n$ with V, and replace questions about V with questions about the coefficients in $\mathbb{R}^n$; any vector space with a basis is "just like" $\mathbb{R}^n$. A look at the proof should convince you that this is just a change of language, without mathematical content.

Proposition 2.6.15 (Linear independence, span, basis). *If V is a vector space, and $\{\underline{v}\} = \underline{v}_1, \ldots, \underline{v}_n$ are vectors in V, then:*

(1) *The set $\{\underline{v}\}$ is linearly independent if and only if $\Phi_{\{\underline{v}\}}$ is one to one.*

(2) *The set $\{\underline{v}\}$ spans V if and only if $\Phi_{\{\underline{v}\}}$ is onto.*

(3) *The set $\{\underline{v}\}$ is a basis of V if and only if $\Phi_{\{\underline{v}\}}$ is one to one and onto (i.e., invertible).*

When $\{\underline{v}\}$ is a basis, then $\Phi_{\{\underline{v}\}}^{-1}$ is the "abstract to concrete" transformation. It takes an element in V and gives the ordered list of its coordinates, with regard to the basis $\{\underline{v}\}$. While $\Phi_{\{\underline{v}\}}$ synthesizes, $\Phi_{\{\underline{v}\}}^{-1}$ decomposes: taking the function of Example 2.6.13, we have

$$\Phi_{\{\underline{v}\}}\left(\begin{bmatrix} a_1 \\ a_2 \\ a_3 \end{bmatrix} \right) = a_1 + a_2 x + a_3 x^2$$

2.6.15

$$\Phi_{\{\underline{v}\}}^{-1}(a_1 + a_2 x + a_3 x^2) = \begin{bmatrix} a_1 \\ a_2 \\ a_3 \end{bmatrix}.$$

Proof. (1) Definition 2.6.10 says that $\underline{v}_1, \ldots, \underline{v}_n$ are linearly independent if

$$\sum_{i=1}^{m} a_i \underline{v}_i = \sum_{i=1}^{m} b_i \underline{v}_i \quad \text{implies} \quad a_1 = b_1, \ a_2 = b_2, \ldots, \ a_m = b_m.$$ 2.6.16

That is exactly saying that $\Phi_{\{\underline{v}\}}(\vec{a}) = \Phi_{\{\underline{v}\}}(\vec{b})$ if and only if $\vec{a} = \vec{b}$, i.e., that $\Phi_{\{\underline{v}\}}$ is one to one.

The use of $\Phi_{\{\underline{v}\}}$ and its inverse to identify an abstract vector space with $\mathbb{R}^n$ is very effective but is generally considered ugly; working directly in the world of abstract vector spaces is seen as more aesthetically pleasing. We have some sympathy with this view.

(2) Definition 2.6.9 says that $\{\underline{v}\} = \underline{v}_1, \ldots, \underline{v}_m$ span V if and only if all vectors of V are linear combinations of $\underline{v}_1, \ldots, \underline{v}_m$; i.e., any vector $\underline{v} \in V$ can be written

$$\underline{v} = a_1\underline{v}_1 + \cdots + a_n\underline{v}_n = \Phi_{\{\underline{v}\}}(\vec{a}). \qquad 2.6.17$$

In other words, $\Phi_{\{\underline{v}\}}$ is onto.

(3) Putting these together, $\underline{v}_1, \ldots, \underline{v}_n$ is a basis if and only if it is linearly independent and spans V, i.e., if $\Phi_{\{\underline{v}\}}$ is one to one and onto. $\square$

The dimension of a vector space

The most important result about bases is the following statement.

Exercise 2.6.3 asks you to show that in a vector space of dimension n, more than n vectors are never linearly independent, and fewer than n vectors never span.

> **Theorem 2.6.16.** *Any two bases of a vector space have the same number of elements.*

The number of elements in a basis of a vector space V is called the *dimension* of V, denoted dim:

> **Definition 2.6.17 (Dimension of a vector space).** The dimension of a vector space is the number of elements of a basis of that space.

How do we know that $\Phi^{-1}_{\{\underline{w}\}}$ and $\Phi^{-1}_{\{\underline{v}\}}$ exist? By Proposition 2.6.15, the fact that $\{\underline{v}\}$ and $\{\underline{w}\}$ are bases means that $\Phi_{\{\underline{v}\}}$ and $\Phi_{\{\underline{w}\}}$ are invertible.

It is often easier to understand a composition if one writes it in diagram form, as in

$$\mathbb{R}^k \underset{\Phi_{\{\underline{v}\}}}{\to} V \underset{\Phi^{-1}_{\{\underline{w}\}}}{\to} \mathbb{R}^p,$$

in Equation 2.6.18. When writing this diagram, one reverses the order, following the order in which the computations are done.

Equation 2.6.18 is the general case of Equation 2.4.30, where we showed that any two bases of a subspace of $\mathbb{R}^n$ have the same number of elements.

Proof of Theorem 2.6.16. Let $\{\underline{v}\}$ and $\{\underline{w}\}$ be two bases of a vector space V: $\{\underline{v}\}$ the set of k vectors $\underline{v}_1, \ldots, \underline{v}_k$, so that $\Phi_{\{\underline{v}\}}$ is a linear transformation from $\mathbb{R}^k$ to V, and $\{\underline{w}\}$ the set of p vectors $\underline{w}_1, \ldots, \underline{w}_p$, so that $\Phi_{\{\underline{w}\}}$ is a linear transformation $\mathbb{R}^p$ to V. Then the linear transformation

$$\underbrace{\Phi^{-1}_{\{\underline{w}\}} \circ \Phi_{\{\underline{v}\}}}_{\text{change of basis matrix}} : \mathbb{R}^k \to \mathbb{R}^p \quad (\text{i.e.} \quad \mathbb{R}^k \underset{\Phi_{\{\underline{v}\}}}{\to} V \underset{\Phi^{-1}_{\{\underline{w}\}}}{\to} \mathbb{R}^p), \qquad 2.6.18$$

is invertible. (Indeed, we can undo the transformation, using $\Phi^{-1}_{\{\underline{v}\}} \circ \Phi_{\{\underline{w}\}}$.) But it is given by an $p \times k$ matrix (since it takes us from $\mathbb{R}^k$ to $\mathbb{R}^p$), and we know that a matrix can be invertible only if it is square. Thus $k = p$. $\square$

Remark. There is something a bit miraculous about this proof; we are able to prove an important result about abstract vector spaces, using a matrix that seemed to drop out of the sky. Without the material developed earlier in this chapter, this result would be quite difficult to prove. The realization that the dimension of a vector space needed to be well defined was a turning point in the development of linear algebra. Dedekind's proof of this theorem in 1893 was a variant of row reduction. $\triangle$

With our definition (Definition 2.6.11), a basis is necessarily finite, but we could have allowed infinite bases. We stick to finite bases because in infinite-dimensional vector spaces, bases tend to be useless. The interesting notion for infinite-dimensional vector spaces is not expressing an element of the space as a linear combination of a finite number of basis elements, but expressing it as a linear combination that uses *infinitely* many basis vectors, i.e., as an infinite series $\sum_{i=0}^{\infty} a_i \underline{\mathbf{v}}_i$ (for example, power series or Fourier series). This introduces questions of convergence, which are interesting indeed, but a bit foreign to the spirit of linear algebra.

It is quite surprising that there *is* a one to one and onto map from $\mathbb{R}$ to $\mathcal{C}[0,1]$; the infinities of elements they have are not different infinities. But this map is not linear. Actually, it is already surprising that the infinities of points in $\mathbb{R}$ and in $\mathbb{R}^2$ are equal; this is illustrated by the existence of Peano curves, described in Exercise 0.4.6. Analogs of Peano curves can be constructed in $\mathcal{C}[0,1]$.

Example 2.6.18 (Change of basis). Let us see that the matrix A in the proof of Proposition and Definition 2.4.20 (Equation 2.4.28) is indeed the change of basis matrix

$$\Phi_{\{\underline{\mathbf{v}}\}}^{-1} \circ \Phi_{\{\underline{\mathbf{w}}\}} : \mathbb{R}^p \to \mathbb{R}^k, \qquad 2.6.19$$

expressing the new vectors (the $\vec{\mathbf{w}}$'s) in terms of the old (the $\vec{\mathbf{v}}$'s.)

Like any linear transformation $\mathbb{R}^p \to \mathbb{R}^k$, the transformation $\Phi_{\{\underline{\mathbf{v}}\}}^{-1} \circ \Phi_{\{\underline{\mathbf{w}}\}}$ has a $k \times p$ matrix A whose jth column $\begin{bmatrix} a_{1,j} \\ \vdots \\ a_{k,j} \end{bmatrix}$ is $A(\vec{\mathbf{e}}_j)$. This means

$$\begin{bmatrix} a_{1,j} \\ \vdots \\ a_{k,j} \end{bmatrix} = A(\vec{\mathbf{e}}_j) = \Phi_{\{\underline{\mathbf{v}}\}}^{-1} \circ \Phi_{\{\underline{\mathbf{w}}\}}(\vec{\mathbf{e}}_j) = \Phi_{\{\underline{\mathbf{v}}\}}^{-1}(\vec{\mathbf{w}}_j), \qquad 2.6.20$$

or, multiplying the first and last term above by $\Phi_{\{\underline{\mathbf{v}}\}}$,

$$\Phi_{\{\underline{\mathbf{v}}\}} \begin{bmatrix} a_{1,j} \\ \vdots \\ a_{k,j} \end{bmatrix} = a_{1,j}\vec{\mathbf{v}}_1 + \cdots + a_{k,j}\vec{\mathbf{v}}_k = \vec{\mathbf{w}}_j. \quad \triangle \qquad 2.6.21$$

Example 2.6.19 (Dimension of vector spaces). The space $\mathrm{Mat}\,(n,m)$ is a vector space of dimension nm. The space P_k of polynomials of degree at most k is a vector space of dimension $k+1$. $\triangle$

Earlier we talked a bit loosely of "finite-dimensional" and "infinite-dimensional" vector spaces. Now we can be precise: a vector space is *finite dimensional* if it has a finite basis, and it is infinite dimensional if it does not.

Example 2.6.20 (An infinite-dimensional vector space). The vector space $\mathcal{C}[0,1]$ of continuous functions on $[0,1]$, which we saw in Example 2.6.2, is infinite dimensional. Intuitively it is not hard to see that there are too many such functions to be expressed with any finite number of basis vectors. We can pin it down as follows.

Assume functions $f_1, \ldots, f_n$ are a basis, and pick $n+1$ distinct points $0 = x_1 < x_2 \cdots < x_{n+1} = 1$ in $[0,1]$. Then given any values $c_1, \ldots, c_{n+1}$, there certainly exists a continuous function $f(x)$ with $f(x_i) = c_i$, for instance, the piecewise linear one whose graph consists of the line segments joining up the points $\begin{pmatrix} x_i \\ c_i \end{pmatrix}$.

If we can write $f = \sum_{k=1}^{n} a_k f_k$, then evaluating at the x_i, we get

$$f(x_i) = c_i = \sum_{k=1}^{n} a_k f_k(x_i), \qquad i = 1, \ldots, n+1. \qquad 2.6.22$$

This, for given c_i's is a system of $n+1$ equations for the n unknowns $a_1, \ldots, a_n$; we know by Theorem 2.2.4 that for appropriate c_i's the equations will be incompatible. Therefore there are functions that are not linear combinations of $f_1, \ldots, f_n$, so $f_1, \ldots, f_n$ do not span $\mathcal{C}[0,1]$.

2.7 Newton's Method

When John Hubbard was teaching first year calculus in France in 1976, he wanted to include some numerical content in the curriculum. Those were the early days of programmable calculators; computers for undergraduates did not then exist. Newton's method to solve cubic polynomials just about fit into the 50 steps of program and eight memory registers available, so he used that as his main example. Writing the program was already a problem, but then came the question of the place to start: what should the initial guess be?

At the time he assumed that even though he didn't know where to start, the experts surely did; after all, Newton's method was in practical use all over the place. It took some time to discover that no one knew anything about the global behavior of Newton's method. A natural thing to do was to color each point of the complex plane according to what root (if any) starting at that point led to. (But this was before the time of color screens and color printers: what he actually did was to print some character at every point of some grid: x's and 0's, for example.)

The resulting printouts were the first pictures of fractals arising from complex dynamical systems, with its archetype the Mandelbrot set.

Recall that the derivative $[\mathbf{D}\vec{\mathbf{f}}(\mathbf{a}_0)]$ is a *matrix*, the Jacobian matrix, whose entries are the partial derivatives of $\vec{\mathbf{f}}$ at $\mathbf{a}_0$. The increment to the variable, $\mathbf{x} - \mathbf{a}_0$, is a *vector*.

We put an arrow over $\mathbf{f}$ to indicate that elements of the range of $\vec{\mathbf{f}}$ are vectors; $\mathbf{a}_0$ is a point and $\vec{\mathbf{f}}(\mathbf{a}_0)$ is a vector. In this way $\vec{\mathbf{f}}$ is like a vector field, taking a point and giving a vector. But whereas a vector field $\vec{F}$ takes a point in one space and turns it into a vector in the same space, the domain and range of $\vec{\mathbf{f}}$ can be different spaces, with different units. The only requirement is that there must be as many equations as unknowns: the dimensions of the two spaces must be equal. Newton's method has such wide applicability that being more precise is impossible.

Theorem 2.2.4 gives a quite complete understanding of linear equations. In practice, one often wants to solve nonlinear equations. This is a genuinely hard problem, and when confronted with such equations, the usual response is: apply Newton's method and hope for the best.

Let $\vec{\mathbf{f}}$ be a differentiable function from $\mathbb{R}^n$ (or from an open subset of $\mathbb{R}^n$) to $\mathbb{R}^n$. Newton's method consists of starting with some guess $\mathbf{a}_0$ for a solution of $\vec{\mathbf{f}}(\mathbf{x}) = \vec{\mathbf{0}}$. Then linearize the equation at $\mathbf{a}_0$: replace the increment to the function, $\vec{\mathbf{f}}(\mathbf{x}) - \vec{\mathbf{f}}(\mathbf{a}_0)$, by a linear function of the increment, $[\mathbf{D}\vec{\mathbf{f}}(\mathbf{a}_0)](\mathbf{x} - \mathbf{a}_0)$. Now solve the corresponding *linear equation*:

$$\vec{\mathbf{f}}(\mathbf{a}_0) + [\mathbf{D}\vec{\mathbf{f}}(\mathbf{a}_0)](\mathbf{x} - \mathbf{a}_0) = \vec{\mathbf{0}}. \qquad 2.7.1$$

This is a system of n linear equations in n unknowns. We can rewrite it

$$\underbrace{[\mathbf{D}\vec{\mathbf{f}}(\mathbf{a}_0)]}_{A}\underbrace{(\mathbf{x} - \mathbf{a}_0)}_{\vec{x}} = \underbrace{-\vec{\mathbf{f}}(\mathbf{a}_0)}_{\vec{b}}. \qquad 2.7.2$$

Remember that if a matrix A has an inverse A^{-1}, then for any $\vec{b}$ the equation $A\vec{x} = \vec{b}$ has the unique solution $A^{-1}\vec{b}$, as discussed in Section 2.3. So if $[\mathbf{D}\vec{f}(\mathbf{a}_0)]$ is invertible, which will usually be the case, then

$$\mathbf{x} = \mathbf{a}_0 - [\mathbf{D}\vec{f}(\mathbf{a}_0)]^{-1}\vec{f}(\mathbf{a}_0);\qquad 2.7.3$$

Call this solution $\mathbf{a}_1$, use it as your new "guess," and solve

$$[\mathbf{D}\vec{f}(\mathbf{a}_1)](\mathbf{x} - \mathbf{a}_1) = -\vec{f}(\mathbf{a}_1),\qquad 2.7.4$$

calling the solution $\mathbf{a}_2$, and so on. The hope is that $\mathbf{a}_1$ is a better approximation to a root than $\mathbf{a}_0$, and that the sequence $\mathbf{a}_0, \mathbf{a}_1, \ldots$ converges to a root of the equation. This hope is sometimes justified on theoretical grounds, and actually works much more often than any theory explains.

Note that Newton's method requires inverting a matrix, which is a lot harder than inverting a number; this is why Newton's method is so much harder in higher dimensions than in one dimension.

In practice, rather than find the inverse of $[\mathbf{D}\vec{f}(\mathbf{a}_0)]$, one solves Equation 2.7.1 by row reduction, or better, by partial row reduction and back substitution, discussed in Exercise 2.1.9. When applying Newton's method, the vast majority of the computational time is spent doing row operations.

Example 2.7.1 (Finding a square root). How do calculators compute the square root of a positive number b? They apply Newton's method to the equation $f(x) = x^2 - b = 0$. In this case, this means the following: choose a_0 and plug it into Equation 2.7.2. Our equation is in one variable, so we can replace $[\mathbf{D}\vec{f}(\mathbf{a}_0)]$ by $f'(a_0) = 2a_0$, as shown in Equation 2.7.5.

This method is sometimes introduced in middle school, under the name *divide and average*.

$$a_1 = \underbrace{a_0 - \frac{1}{2a_0}(a_0^2 - b)}_{\text{Newton's method}} = \underbrace{\frac{1}{2}\left(a_0 + \frac{b}{a_0}\right)}_{\text{divide and average}}.\qquad 2.7.5$$

(Exercise 2.7.3 asks you to find the corresponding formula for nth roots.)

How do you come by your initial guess $\mathbf{a}_0$? You might have a good reason to think that nearby there is a solution, for instance because $|\vec{f}(\mathbf{a}_0)|$ is small; we will see many examples of this later: in good cases you can then prove that the scheme works. Or it might be wishful thinking: you know roughly what solution you want. Or you might pull your guess out of thin air, and start with a collection of initial guesses $\mathbf{a}_0$, hoping that you will be lucky and that at least one will converge. In some cases, this is just a hope.

The motivation for *divide and average* is the following: let a be a first guess at $\sqrt{b}$. If your guess is too big, i.e., if $a > \sqrt{b}$, then b/a will be too small, and the average of the two will be better than the original guess. This seemingly naive explanation is quite solid and can easily be turned into a proof that Newton's method works in this case.

Suppose first that $a_0 > \sqrt{b}$; then we want to show that $\sqrt{b} < a_1 < a_0$. Since $a_1 = \frac{1}{2}(a_0 + b/a_0)$, this comes down to showing

$$b < \underbrace{\left(\frac{1}{2}\left(a_0 + \frac{b}{a_0}\right)\right)^2}_{a_1} < a_0^2,\qquad 2.7.6$$

or, if you develop, $4b < a_0^2 + 2b + \frac{b^2}{a_0^2} < 4a_0^2$. To see the left-hand inequality, subtract $4b$ from each side:

$$a_0^2 + 2b + \frac{b^2}{a_0^2} - 4b = a_0^2 - 2b + \frac{b^2}{a_0^2} = \left(a_0 - \frac{b}{a_0}\right)^2 > 0.\qquad 2.7.7$$

The right-hand inequality follows immediately from $b < a_0^2$, hence $b^2/a_0^2 < a_0^2$:

$$a_0^2 + \underbrace{2b}_{<2a_0^2} + \underbrace{\frac{b^2}{a_0^2}}_{<a_0^2} < 4a_0^2. \qquad 2.7.8$$

Two theorems from first year calculus.

(1) If a decreasing sequence is bounded below, it converges (see Theorem 0.4.10).

(2) If a_n is a convergent sequence, and f is a continuous function in a neighborhood of the limit of the a_n, then

$$\lim f(a_n) = f(\lim a_n).$$

Recall from first year calculus (or from Theorem 0.4.10) that if a decreasing sequence is bounded below, it converges. Hence the a_i converge. The limit a must satisfy

$$a = \lim_{i \to \infty} a_{i+1} = \lim_{i \to \infty} \frac{1}{2}\left(a_i + \frac{b}{a_i}\right) = \frac{1}{2}\left(a + \frac{b}{a}\right) \quad , \text{ i.e., } \quad a = \sqrt{b}. \qquad 2.7.9$$

What if you choose $0 < a_0 < \sqrt{b}$? In this case as well, $a_1 > \sqrt{b}$:

$$4a_0^2 < 4b < \underbrace{a_0^2 + 2b + \frac{b^2}{a_0^2}}_{4a_1^2}. \qquad 2.7.10$$

We get the right-hand inequality using the same argument used in Equation 2.7.7: $2b < a_0^2 + \frac{b^2}{a_0^2}$, since subtracting $2b$ from both sides gives $0 < \left(a_0 - \frac{b}{a_0}\right)^2$. Then the same argument as above applies to show that $a_2 < a_1$.

This "divide and average" method can be interpreted geometrically in terms of Newton's method: Each time we calculate a_{n+1} from a_n we are calculating the intersection with the x-axis of the line tangent to the parabola $y = x^2 - b$ at $\left(\begin{smallmatrix} a_n \\ a_n^2 - b \end{smallmatrix}\right)$, as shown in Figure 2.7.1.

There aren't many cases where Newton's method is really well understood far away from the roots; Example 2.7.2 shows one of the problems that can arise, and there are many others.

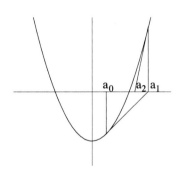

FIGURE 2.7.1.

Newton's method: we start with a_0, drawing the tangent to the parabola $y = x^2 - b$ at the point with x-coordinate a_0. The point where that tangent intersects the x-axis is a_1. Now we draw the tangent to the parabola at the point with x-coordinate a_1. That tangent intersects the x-axis at a_2 .... Each time we calculate a_{n+1} from a_n we are calculating the intersection with the x-axis of the line tangent to the parabola at the point with x-coordinate a_n.

Example 2.7.2 (A case where Newton's method doesn't work). Let's apply Newton's method to the equation

$$x^3 - x + \frac{\sqrt{2}}{2} = 0, \qquad 2.7.11$$

starting at $x = 0$ (i.e., our "guess" a_0 is 0). The derivative is $f'(x) = 3x^2 - 1$, so $f'(0) = -1$, and $f(0) = \sqrt{2}/2$, giving

$$a_1 = a_0 - \frac{1}{f'(a_0)}f(a_0) = a_0 - \frac{a_0^3 - a_0 + \frac{\sqrt{2}}{2}}{3a_0^2 - 1} = 0 + \frac{\frac{\sqrt{2}}{2}}{1} = \frac{\sqrt{2}}{2}. \qquad 2.7.12$$

Since $a_1 = \sqrt{2}/2$, we have $f'(a_1) = 1/2$, and

$$a_2 = \frac{\sqrt{2}}{2} - 2\left(\frac{\sqrt{2}}{4} - \frac{\sqrt{2}}{2} + \frac{\sqrt{2}}{2}\right) = 0. \qquad 2.7.13$$

We're back to where we started, at $a_0 = 0$. If we continue, we'll bounce back and forth between $\frac{\sqrt{2}}{2}$ and 0, never converging to any root:

$$0 \to \frac{\sqrt{2}}{2} \to 0\dots \qquad 2.7.14$$

Don't be too discouraged by this example. Most of the time Newton's method does work. It is the best method available for solving nonlinear equations.

Now let's try starting at some small $\epsilon > 0$. We have $f'(\epsilon) = 3\epsilon^2 - 1$, and $f(\epsilon) = \epsilon^3 - \epsilon + \sqrt{2}/2$, giving us

$$a_1 = \epsilon - \frac{1}{3\epsilon^2 - 1}\left(\epsilon^3 - \epsilon + \frac{\sqrt{2}}{2}\right) = \epsilon + \left(\epsilon^3 - \epsilon + \frac{\sqrt{2}}{2}\right)\frac{1}{1 - 3\epsilon^2}. \qquad 2.7.15$$

Now we can treat

$$\frac{1}{1 - 3\epsilon^2} \text{ as the sum of the geometric series } (1 + 3\epsilon^2 + 9\epsilon^4 + \dots).$$

This uses Equation 0.4.9 the sum of a geometric series: If $|r| < 1$, then

$$\sum_{n=0}^{\infty} ar^n = \frac{a}{1 - r}.$$

We can substitute $3\epsilon^2$ for r in that equation because ϵ^2 is small.

This gives us

$$a_1 = \epsilon + \left(\epsilon^3 - \epsilon + \frac{\sqrt{2}}{2}\right)(1 + 3\epsilon^2 + 9\epsilon^4 + \dots). \qquad 2.7.16$$

Now we just ignore terms that are smaller than ϵ^2, getting

$$a_1 = \epsilon + \left(\frac{\sqrt{2}}{2} - \epsilon\right)(1 + 3\epsilon^2) + \text{ remainder}$$

$$= \frac{\sqrt{2}}{2} + \frac{3\sqrt{2}\epsilon^2}{2} + \text{ remainder}. \qquad 2.7.17$$

Remember that we said in the introduction to Section 1.4, that calculus is about " ... about some terms being dominant or negligible compared to other terms." Just because a computation is there doesn't mean we have to do it to the bitter end.

Ignoring the remainder, and repeating the process, we get

$$a_2 \approx \frac{\sqrt{2}}{2} + \frac{3\sqrt{2}\epsilon^2}{2} - \frac{(\frac{\sqrt{2}}{2} + \frac{3\sqrt{2}}{2}\epsilon^2)^3 - (\frac{\sqrt{2}}{2} + \frac{3\sqrt{2}}{2}\epsilon^2) + \frac{\sqrt{2}}{2}}{3(\frac{\sqrt{2}}{2} + \frac{3\sqrt{2}}{2}\epsilon^2)^2 - 1}. \qquad 2.7.18$$

This looks unpleasant; let's throw out all the terms with ϵ^2. We get

$$a_2 \approx \frac{\sqrt{2}}{2} - \frac{(\frac{\sqrt{2}}{2})^3 - \frac{\sqrt{2}}{2} + \frac{\sqrt{2}}{2}}{3(\frac{\sqrt{2}}{2})^2 - 1} = \frac{\sqrt{2}}{2} - \frac{\frac{1}{2}(\frac{\sqrt{2}}{2})}{\frac{1}{2}} = \frac{\sqrt{2}}{2} - \frac{\sqrt{2}}{2} = 0, \text{ so that}$$

$$a_2 = 0 + c\epsilon^2, \text{ where } c \text{ is a constant.} \qquad 2.7.19$$

If we continue, we'll bounce between a region around $\frac{\sqrt{2}}{2}$ and a region around 0, getting closer and closer to these points each time.

We started at $0 + \epsilon$ and we've been sent back to $0 + c\epsilon^2$!

We're not getting anywhere; does that mean there are no roots? Not at all.[17] Let's try once more, with $a_0 = -1$. We have

$$a_1 = a_0 - \frac{a_0^3 - a_0 + \frac{\sqrt{2}}{2}}{3a_0^2 - 1} = \frac{2a_0^3 - \frac{\sqrt{2}}{2}}{3a_0^2 - 1}. \qquad 2.7.20$$

---

[17]Of course not. All odd-degree polynomials have real roots by the intermediate value theorem, Theorem 0.4.12

A computer or programmable calculator can be programmed to keep iterating this formula. It's slightly more tedious with a simple scientific calculator; with the one the authors have at hand, we enter "1 +/− Min" to put −1 in the memory ("MR") and then:

$$(2 \times MR \times MR \times MR - 2\sqrt{}\operatorname{div} 2)\operatorname{div}(3 \times MR \times MR - 1).$$

We get $a_1 = -1.35355\ldots$; entering that in memory by pushing on the "Min" (or "memory in") key we repeat the process to get:

$$a_2 = -1.26032\ldots \qquad a_4 = -1.25107\ldots$$
$$a_3 = -1.25116\ldots \qquad a_5 = -1.25107\ldots. \qquad\qquad 2.7.21$$

It's then simple to confirm that a_5 is indeed a root, to the limits of precision of the calculator or computer.   △

Does Newton's method depend on starting with a lucky guess? Luck sometimes enters into it; with a fast computer one can afford to try out several guesses and see if one converges. But, you may ask, how do we really *know* that solutions are converging? Checking by plugging in a root into the equation isn't entirely convincing, because of round-off errors. We shall see that we can say something more precise. Kantorovitch's theorem *guarantees* that under appropriate circumstances Newton's method converges. Even stating the theorem is difficult. But the effort will pay off.

Lipschitz conditions

Imagine an airplane beginning its approach to its destination, its altitude represented by f. If it loses altitude *gradually*, the derivative f' allows one to approximate the function very well; if you know how high the airplane is at the moment t, and what its derivative is at t, you can get a good idea of how high the airplane will be at the moment $t + h$:

$$f(t + h) \approx f(t) + f'(t)h. \qquad\qquad 2.7.22$$

But if the airplane suddenly loses power and starts plummeting to earth, the derivative changes abruptly: the derivative of f at t will no longer be a reliable gauge of the airplane's altitude a few seconds later.

The natural way to limit how fast the derivative can change is to bound the second derivative; you probably ran into this when studying Taylor's theorem with remainder. In one variable this is a good idea. If you put an appropriate limit to f'' at t, then the airplane will not suddenly change altitude. Bounding the second derivative of an airplane's altitude function is indeed a pilot's primary goal, except in rare emergencies.

To guarantee that Newton's method starting at a certain point will converge to a root, we will need an explicit bound on how good an approximation

$$[\mathbf{D}\vec{\mathbf{f}}(\mathbf{x}_0)]\vec{\mathbf{h}} \quad \text{is to} \quad \vec{\mathbf{f}}(\mathbf{x}_0 + \vec{\mathbf{h}}) - \vec{\mathbf{f}}(\mathbf{x}_0). \qquad\qquad 2.7.23$$

As we said earlier, the reason Newton's method has become so important is that people no longer have to carry out the computations by hand.

Any statement that guarantees that you can find solutions to nonlinear equations in any generality at all is bound to be tremendously important. In addition to the immediate applicability of Newton's method to solutions of all sorts of nonlinear equations, it gives a practical algorithm for finding implicit and inverse functions. Kantorovitch's theorem then gives a proof that these algorithms actually work.

As in the case of the airplane, to do this we will need some assumption on how fast the derivative of $\vec{\mathbf{f}}$ changes.

But in several variables there are lots of second derivatives, so bounding the second derivative doesn't work so well. We will adopt a different approach: demanding that the derivative of $\vec{\mathbf{f}}$ satisfy a *Lipschitz condition*.

In Definition 2.7.3, U can be a subset of $\mathbb{R}^n$; the domain and the range of $\mathbf{f}$ do not need to have the same dimension. But when we use this definition in the Kantorovitch theorem, those dimensions will have to be the same.

> **Definition 2.7.3 (Lipschitz condition).** Let $\mathbf{f} : U \to \mathbb{R}^m$ be a differentiable mapping. The derivative $[\mathbf{Df}(\mathbf{x})]$ satisfies a Lipschitz condition on a subset $V \subset U$ with Lipschitz ratio M if for all $\mathbf{x}, \mathbf{y} \in V$
>
> $$\underbrace{\big| [\mathbf{Df}(\mathbf{x})] - [\mathbf{Df}(\mathbf{y})] \big|}_{\substack{\text{distance} \\ \text{between deriv.}}} \leq M \underbrace{|\mathbf{x} - \mathbf{y}|}_{\substack{\text{distance} \\ \text{between points}}} . \qquad 2.7.24$$

Note that a function whose derivative satisfies a Lipschitz condition is certainly *continuously differentiable*. Having the derivative Lipschitz is a requirement that the derivative is especially nicely continuous (it is actually close to demanding that the function be twice continuously differentiable).

A Lipschitz ratio tells us something about how fast the derivative of a function changes.

It is often called a *Lipschitz constant*. But M is not a true constant; it depends on the problem at hand; in addition, a mapping will almost always have different M at different points or on different regions. When there is a single Lipschitz ratio that works on all of $\mathbb{R}^n$, we will call it a *global Lipschitz ratio*.

Example 2.7.4 is misleading: there is usually no Lipschitz ratio valid on the entire space.

Example 2.7.4 (Lipschitz ratio: a simple case). Consider the mapping $\mathbf{f} : \mathbb{R}^2 \to \mathbb{R}^2$

$$\mathbf{f}\begin{pmatrix} x_1 \\ x_2 \end{pmatrix} = \begin{pmatrix} x_1 - x_2^2 \\ x_1^2 + x_2 \end{pmatrix} \text{ with derivative } \left[\mathbf{Df}\begin{pmatrix} x_1 \\ x_2 \end{pmatrix}\right] = \begin{bmatrix} 1 & -2x_2 \\ 2x_1 & 1 \end{bmatrix}. \qquad 2.7.25$$

Given two points $\mathbf{x}$ and $\mathbf{y}$,

$$\left[\mathbf{Df}\begin{pmatrix} x_1 \\ x_2 \end{pmatrix}\right] - \left[\mathbf{Df}\begin{pmatrix} y_1 \\ y_2 \end{pmatrix}\right] = \begin{bmatrix} 0 & -2(x_2 - y_2) \\ 2(x_1 - y_1) & 0 \end{bmatrix}. \qquad 2.7.26$$

Calculating the length of the matrix above gives

$$\left| \begin{bmatrix} 0 & -2(x_2 - y_2) \\ 2(x_1 - y_1) & 0 \end{bmatrix} \right| = 2\sqrt{(x_1 - y_1)^2 + (x_2 - y_2)^2} = 2 \left| \begin{matrix} x_1 - y_1 \\ x_2 - y_2 \end{matrix} \right|,$$

so

$$\big|[\mathbf{Df}(\mathbf{x})] - [\mathbf{Df}(\mathbf{y})]\big| = 2|\mathbf{x} - \mathbf{y}|; \qquad 2.7.27$$

in this case $M = 2$ is a Lipschitz ratio for $[\mathbf{Df}]$.

Example 2.7.5 (Lipschitz ratio: a more complicated case). Consider the mapping $\mathbf{f} : \mathbb{R}^2 \to \mathbb{R}^2$ given by

$$\mathbf{f}\begin{pmatrix} x_1 \\ x_2 \end{pmatrix} = \begin{pmatrix} x_1 - x_2^3 \\ x_1^3 + x_2 \end{pmatrix}, \text{ with derivative } \left[\mathbf{Df}\begin{pmatrix} x_1 \\ x_2 \end{pmatrix}\right] = \begin{bmatrix} 1 & -3x_2^2 \\ 3x_1^2 & 1 \end{bmatrix}. \qquad 2.7.28$$

Given two points $\mathbf{x}$ and $\mathbf{y}$ we have

$$\left[\mathbf{Df}\begin{pmatrix} x_1 \\ x_2 \end{pmatrix}\right] - \left[\mathbf{Df}\begin{pmatrix} y_1 \\ y_2 \end{pmatrix}\right] = \begin{bmatrix} 0 & -3(x_2^2 - y_2^2) \\ 3(x_1^2 - y_1^2) & 0 \end{bmatrix}, \qquad 2.7.29$$

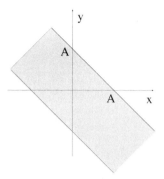

FIGURE 2.7.2.

Equations 2.7.31 and 2.7.32 say that when $\begin{pmatrix} x_1 \\ y_1 \end{pmatrix}$ and $\begin{pmatrix} x_2 \\ y_2 \end{pmatrix}$ are in the shaded region above, then $3A$ is a Lipschitz ratio for $[\mathbf{Df}(\mathbf{x})]$. It is not immediately obvious how to translate this statement into a condition on the points

$$\mathbf{x} = \begin{pmatrix} x_1 \\ x_2 \end{pmatrix} \quad \text{and} \quad \mathbf{y} = \begin{pmatrix} y_1 \\ y_2 \end{pmatrix}.$$

By $\sup\{(x_1 + y_1)^2, (x_2 + y_2)^2\}$ we mean "the greater of $(x_1 + y_1)^2$ and $(x_2 + y_2)^2$." In this computation, we are using the fact that for any two numbers a and b, we always have

$$(a + b)^2 \le 2(a^2 + b^2),$$

since $0 \le (a - b)^2$.

and taking the length gives

$$\left| \left[\mathbf{Df}\begin{pmatrix} x_1 \\ x_2 \end{pmatrix} \right] - \left[\mathbf{Df}\begin{pmatrix} y_1 \\ y_2 \end{pmatrix} \right] \right|$$
$$= 3\sqrt{(x_1 - y_1)^2(x_1 + y_1)^2 + (x_2 - y_2)^2(x_2 + y_2)^2}. \tag{2.7.30}$$

Therefore, when

$$(x_1 + y_1)^2 \le A^2 \quad \text{and} \quad (x_2 + y_2)^2 \le A^2, \tag{2.7.31}$$

as shown in Figure 2.7.2, we have

$$\left| \left[\mathbf{Df}\begin{pmatrix} x_1 \\ x_2 \end{pmatrix} \right] - \left[\mathbf{Df}\begin{pmatrix} y_1 \\ y_2 \end{pmatrix} \right] \right| \le 3A \left| \begin{pmatrix} x_1 \\ x_2 \end{pmatrix} - \begin{pmatrix} y_1 \\ y_2 \end{pmatrix} \right|, \tag{2.7.32}$$

i.e., $3A$ is a Lipschitz ratio for $[\mathbf{Df}(\mathbf{x})]$.

When is the condition of Equation 2.7.31 satisfied? It isn't really clear that we need to ask this question: why can't we just say that it is satisfied when it is satisfied; in what sense can we be more explicit? There isn't anything wrong with this view, but the requirement in Equation 2.7.31 describes some more or less unimaginable region in $\mathbb{R}^4$. (Keep in mind that Equation 2.7.32 concerns points $\mathbf{x}$ with coordinates x_1, x_2 and $\mathbf{y}$ with coordinates y_1, y_2, *not* the points of Figure 2.7.2, which have coordinates x_1, y_1 and x_2, y_2 respectively.) Moreover, in many settings, what we really want is a ball of radius R such that when two points are in the ball, the Lipschitz condition is satisfied:

$$|[\mathbf{Df}(\mathbf{x})] - [\mathbf{Df}(\mathbf{y})]| \le 3A|\mathbf{x} - \mathbf{y}| \quad \text{when} \quad |\mathbf{x}| \le R \quad \text{and} \quad |\mathbf{y}| \le R. \tag{2.7.33}$$

If we require that $|\mathbf{x}|^2 = x_1^2 + x_2^2 \le A^2/4$ and $|\mathbf{y}|^2 = y_1^2 + y_2^2 \le A^2/4$, then

$$\sup\{(x_1 + y_1)^2, (x_2 + y_2)^2\} \le 2(x_1^2 + y_1^2 + x_2^2 + y_2^2) = 2(|\mathbf{x}|^2 + |\mathbf{y}|^2) \le A^2. \tag{2.7.34}$$

Thus we can assert that if

$$|\mathbf{x}|, |\mathbf{y}| \le \frac{A}{2}, \quad \text{then} \quad \left| [\mathbf{Df}(\mathbf{x})] - [\mathbf{Df}(\mathbf{y})] \right| \le 3A|\mathbf{x} - \mathbf{y}|. \quad \triangle \tag{2.7.35}$$

Computing Lipschitz ratios using higher partial derivatives

Most students can probably follow the computation in Example 2.7.5 line by line, but even well above average students will probably feel that the tricks used are way beyond anything they can be expected to come up with on their own. Finding ratios M as we did above is a delicate art, and finding M's that are as small as possible is harder yet. The manipulation of inequalities is a hard skill to acquire, and no one seems to know how to teach it very well. Fortunately, there is a systematic way to compute Lipschitz ratios, using higher partial derivatives.

Higher partial derivatives are essential throughout mathematics and in science. Mathematical physics is essentially the theory of partial differential equations. Electromagnetism is based on Maxwell's equations, general relativity

Originally we introduced higher partial derivatives in a separate section in Chapter 3. They are so important in all scientific applications of mathematics that it seems mildly scandalous to slip them in here, just to solve a computational problem. But in our experience students have such trouble computing Lipschitz ratios—each problem seeming to demand a new trick—that we feel it worthwhile to give a "recipe."

Different notations for partial derivatives exist:

$$D_j(D_i f)(\mathbf{a}) = \frac{\partial^2 f}{\partial x_j \partial x_i}(\mathbf{a})$$
$$= f_{x_i x_j}(\mathbf{a}).$$

As usual, we specify the point $\mathbf{a}$ at which the derivative is evaluated.

In Example 2.7.7 we evaluated the partial derivatives of f at both $\begin{pmatrix} a \\ b \\ c \end{pmatrix}$ and at $\begin{pmatrix} x \\ y \\ z \end{pmatrix}$ to emphasize the fact that although we used x, y and z to define f, we can evaluate it on variables that look different.

A function is C^2 if it is twice continuously differentiable, i.e., if its second partial derivatives exist and are continuous.

on Einstein's equation, fluid dynamics on the Navier-Stokes equation, quantum mechanics on Schrödinger's equation. Understanding partial differential equations is a prerequisite for any serious study of these phenomena. Here, however, we will use them as a computational tool.

Definition 2.7.6 (Second partial derivative). Let $U \subset \mathbb{R}^n$ be open, and $f : U \to \mathbb{R}$ be a differentiable function. If the function $D_i f$ is itself differentiable, then its partial derivative with respect to the jth variable,

$$D_j(D_i f),$$

is called a second partial derivative of f.

Example 2.7.7 (Second partial derivative). Let f be the function

$$f\begin{pmatrix} x \\ y \\ z \end{pmatrix} = 2x + xy^3 + 2yz^2. \quad \text{Then} \quad D_2(D_1 f)\begin{pmatrix} a \\ b \\ c \end{pmatrix} = D_2 \underbrace{(2 + b^3)}_{D_1 f} = 3b^2.$$

Similarly, $D_3(D_2 f)\begin{pmatrix} x \\ y \\ z \end{pmatrix} = D_3 \underbrace{(3xy^2 + 2z^2)}_{D_2 f} = 4z.$ △

We can denote $D_1(D_1 f)$ by $D_1^2 f$, $D_2(D_2 f)$ by $D_2^2 f$, ... For the function $f\begin{pmatrix} x \\ y \end{pmatrix} = xy^2 + \sin x$, what are $D_1^2 f, D_2^2 f, D_1(D_2 f)$, and $D_2(D_1 f)$?[18]

Proposition 2.7.8 says that the derivative of $\mathbf{f}$ is Lipschitz if $\mathbf{f}$ is of class C^2.

Proposition 2.7.8 (Derivative of a C^2 mapping is Lipschitz). Let $U \subset \mathbb{R}^n$ be open, and $\mathbf{f} : U \to \mathbb{R}^n$ be a C^2 mapping. If $|D_k D_j f_i(\mathbf{x})| \leq c_{i,j,k}$ for all triples of indices $1 \leq i, j, k \leq n$, then

$$|[\mathbf{Df}(\mathbf{u})] - [\mathbf{Df}(\mathbf{v})]| \leq \left(\sum_{1 \leq i,j,k \leq n} (c_{i,j,k})^2 \right)^{1/2} |\mathbf{u} - \mathbf{v}|. \qquad 2.7.36$$

Proof. Each of the $D_j f_i$ is a scalar-valued function, and Corollary 1.9.2 tells us that

---

[18] $D_1 f = y^2 + \cos x$ and $D_2 f = 2xy$, so

$$D_1^2 f = D_1(y^2 + \cos x) = -\sin x, \; D_2^2 f = D_2(2xy) = 2x,$$
$$D_1(D_2 f) = D_1(2xy) = 2y, \quad \text{and} \quad D_2(D_1 f) = D_2(y^2 + \cos x) = 2y.$$

Equation 2.7.37 uses the fact that for any function g (in our case, $D_j f_i$),

$$\left| g(\mathbf{a} + \vec{\mathbf{h}}) - g(\mathbf{a}) \right|$$

$$\leq \left(\sup_{[0,1]} \left| [\mathbf{D}g(\mathbf{a} + t\vec{\mathbf{h}})] \right| \right) |\vec{\mathbf{h}}|;$$

remember that

$$\left| [\mathbf{D}g(\mathbf{a} + t\vec{\mathbf{h}})] \right|$$

$$= \sqrt{\sum_{k=1}^{n} \left(D_k g(\mathbf{a} + t\vec{\mathbf{h}}) \right)^2}.$$

$$|D_j f_i(\mathbf{a} + \vec{\mathbf{h}}) - D_j f_i(\mathbf{a})| \leq \left(\sum_{k=1}^{n} (c_{i,j,k})^2 \right)^{1/2} |\vec{\mathbf{h}}|. \qquad 2.7.37$$

By definition,

$$|[\mathbf{Df}(\mathbf{a} + \vec{\mathbf{h}})] - [\mathbf{Df}(\mathbf{a})]| = \left(\sum_{i,j=1}^{n} \left(D_j f_i(\mathbf{a} + \vec{\mathbf{h}}) - D_j f_i(\mathbf{a}) \right)^2 \right)^{1/2}. \qquad 2.7.38$$

So $\quad |[\mathbf{Df}(\mathbf{a} + \vec{\mathbf{h}})] - [\mathbf{Df}(\mathbf{a})]| \leq \left(\sum_{i,j=1}^{n} \left(\left(\sum_{k=1}^{n} (c_{i,j,k})^2 \right)^{1/2} |\vec{\mathbf{h}}| \right)^2 \right)^{1/2}$

$$2.7.39$$

$$= \left(\sum_{1 \leq i,j,k \leq n} (c_{i,j,k})^2 \right)^{1/2} |\vec{\mathbf{h}}|.$$

The proposition follows by setting $\mathbf{u} = \mathbf{a} + \vec{\mathbf{h}}$, and $\mathbf{v} = \mathbf{a}$. $\quad \square$

Example 2.7.9 (Redoing Example 2.7.5 the easy way). Let's see how much easier it is to find a Lipschitz ratio in Example 2.7.5 using higher partial derivatives. First we compute the first and second derivatives, for $\mathbf{f}_1 = x_1 - x_2^3$ and $\mathbf{f}_2 = x_1^3 + x_2$:

$$D_1\mathbf{f} = 1; \quad D_2\mathbf{f}_1 = -3x_2^2; \quad D_1\mathbf{f}_2 = 3x_1^2; \quad D_2\mathbf{f}_2 = 1. \qquad 2.7.40$$

In Equation 2.7.41 we use the fact that crossed partials of $\mathbf{f}$ are equal (Theorem 3.3.9).

This gives

$$D_1 D_1 \mathbf{f}_1 = 0; \quad D_1 D_2 \mathbf{f}_1 = D_2 D_1 \mathbf{f}_1 = 0; \quad D_2 D_2 \mathbf{f}_1 = -6x_2$$
$$D_1 D_1 \mathbf{f}_2 = 6x_1; \quad D_1 D_2 \mathbf{f}_2 = D_2 D_1 \mathbf{f}_2 = 0; \quad D_2 D_2 \mathbf{f}_2 = 0. \qquad 2.7.41$$

So if $|\mathbf{x}|, |\mathbf{y}| \leq \frac{A}{2}$, we can take

Earlier we got $3A$, a better result. A blunderbuss method guaranteed to work in all cases is unlikely to give results as good as techniques adapted to the problem at hand. But the higher partial derivative method gives results that are often good enough.

$$c_{2,2,1} = c_{1,1,2} = 3A \quad \text{with all others 0, so} \quad \sqrt{c_{2,2,1}^2 + c_{1,1,2}^2} = 3A\sqrt{2}.$$

Thus we have

$$|[\mathbf{Df}(\mathbf{x})] - [\mathbf{Df}(\mathbf{y})]| \leq 3\sqrt{2}A|\mathbf{x} - \mathbf{y}|. \quad \triangle \qquad 2.7.42$$

Using higher partial derivatives, recompute the Lipschitz ratio of Example 2.7.4. Do you get the same answer we did?[19]

---

[19]The higher partial derivative method gives $2\sqrt{2}$.

Example 2.7.10 (Finding a Lipschitz ratio using second derivatives: a second example). Let us find a Lipschitz ratio for the derivative of $\mathbf{F}\begin{pmatrix} x \\ y \end{pmatrix} = \begin{pmatrix} \sin(x+y) \\ \cos(xy) \end{pmatrix}$, for $|x| < 2, |y| < 2$. We compute

$$D_1 D_1 F_1 = D_2 D_2 F_1 = D_2 D_1 F_1 = D_1 D_2 F_1 = -\sin(x+y),$$

$$D_1 D_1 F_2 = -y^2 \cos(xy), \quad D_2 D_1 F_2 = D_1 D_2 F_2 = -\big(\sin(xy) + yx\cos(xy)\big),$$

$$D_2 D_2 F_2 = -x^2 \cos xy. \qquad\qquad 2.7.43$$

Since $|\sin|$ and $|\cos|$ are bounded by 1, if we set $|x| < 2, |y| < 2$, this gives

$$|D_1 D_1 F_1| = |D_1 D_2 F_1| = |D_2 D_1 F_1| = |D_2 D_2 F_1| \le 1$$
$$|D_1 D_1 F_2|, |D_2 D_2 F_2| \le 4, \quad |D_2 D_1 F_2| = |D_1 D_2 F_2| \le 5. \qquad 2.7.44$$

So for $|x| < 2$, $|y| < 2$, we have a Lipschitz ratio

$$M \le \sqrt{4 + 16 + 16 + 25 + 25} = \sqrt{86} < 9.3;$$

i.e.,

$$\Big|[\mathbf{Df}(\mathbf{u})] - [\mathbf{Df}(\mathbf{v})]\Big| \le 9.3\,|\mathbf{u} - \mathbf{v}|. \quad \triangle \qquad\qquad 2.7.45$$

By fiddling with the trigonometry, one can get the $\sqrt{86}$ down to $\sqrt{78} \approx 8.8$, but the advantage of Proposition 2.7.8 is that it gives a systematic way to compute Lipschitz ratios; you don't have to worry about being clever.

Kantorovitch's theorem

Now we are ready to tackle Kantorovitch's theorem. It says that if the product of three quantities is $\le 1/2$, then the equation $\vec{\mathbf{f}}(\mathbf{x}) = \vec{\mathbf{0}}$ has a unique root in a neighborhood U_0, and if you start with initial guess $\mathbf{a}_0$ *in that neighborhood*, Newton's method will converge to that root.

The Kantorovitch theorem says that if certain conditions are met, the equation

$$\vec{\mathbf{f}}(\mathbf{x}) = \vec{\mathbf{0}}$$

has a unique root *in a neighborhood* U_0. In our airplane analogy, where is the neighborhood mentioned? It is implicit in the Lipschitz condition: the derivative is Lipschitz with Lipschitz ratio M in the neighborhood U_0.

The basic idea is simple. The first of the three quantities that must be small is the value of the function at $\mathbf{a}_0$. If you are in an airplane flying close to the ground, you are more likely to crash (find a root) than if you are several kilometers up. The second quantity is the square of the inverse of the derivative of the function at $\mathbf{a}_0$. In one dimension, we can think that the derivative must be big.[20]

If your plane is approaching the ground steeply, it is much more likely to crash than if it is flying almost parallel to the ground.

The third quantity is the Lipschitz ratio M, measuring the change in the derivative (i.e., acceleration). If at the last minute the pilot pulls the plane out of a nose dive, some passengers or flight attendants may be thrown to the floor as the derivative changes sharply, but a crash will be avoided.

---

[20]Why the theorem stipulates the *square* of the inverse of the derivative is more subtle. We think of it this way: the theorem should remain true if one changes the scale. Since the "numerator" $\vec{\mathbf{f}}(\mathbf{a}_0)M$ in Equation 2.7.48 contains two terms, scaling up will change it by the scale factor squared. So the "denominator" $|[\mathbf{D\vec{f}}(\mathbf{a}_0)]^{-1}|^2$ must also contain a square.

But it is not each quantity individually that must be small: the product must be small. If the airplane starts its nose dive too close to the ground, even a sudden change in derivative may not save it. If it starts its nose dive from an altitude of several kilometers, it will still crash if it falls straight down. And if it loses altitude progressively, rather than plummeting to earth, it will still crash (or at least land) if the derivative never changes.

Note that the domain and the range of the mapping $\vec{\mathbf{f}}$ have the *same dimension*. In other words, setting $\vec{\mathbf{f}}(\mathbf{x}) = \vec{\mathbf{0}}$, we get the same number of equations as unknowns. This is a reasonable requirement. If we had fewer equations than unknowns we wouldn't expect them to specify a unique solution, and if we had more equations than unknowns it would be unlikely that there will be any solutions at all.

In addition, if $n \neq m$, then $[\mathbf{D}\vec{\mathbf{f}}(\mathbf{a}_0)]$ would not be a square matrix, so it would not be invertible.

The Kantorovitch theorem is proved in Appendix A.2.

Theorem 2.7.11 (Kantorovitch's theorem). *Let $\mathbf{a}_0$ be a point in $\mathbb{R}^n$, U an open neighborhood of $\mathbf{a}_0$ in $\mathbb{R}^n$ and $\vec{\mathbf{f}} : U \to \mathbb{R}^n$ a differentiable mapping, with its derivative $[\mathbf{D}\vec{\mathbf{f}}(\mathbf{a}_0)]$ invertible. Define*

$$\vec{\mathbf{h}}_0 = -[\mathbf{D}\vec{\mathbf{f}}(\mathbf{a}_0)]^{-1}\vec{\mathbf{f}}(\mathbf{a}_0) \quad , \quad \mathbf{a}_1 = \mathbf{a}_0 + \vec{\mathbf{h}}_0 \quad , \quad U_0 = \left\{ \mathbf{x} \mid |\mathbf{x} - \mathbf{a}_1| \leq |\vec{\mathbf{h}}_0| \right\}. \tag{2.7.46}$$

If $U_0 \subset U$ and the derivative $[\mathbf{D}\vec{\mathbf{f}}(\mathbf{x})]$ satisfies the Lipschitz condition

$$\left| [\mathbf{D}\vec{\mathbf{f}}(\mathbf{u}_1)] - [\mathbf{D}\vec{\mathbf{f}}(\mathbf{u}_2)] \right| \leq M|\mathbf{u}_1 - \mathbf{u}_2| \quad \text{for all points } \mathbf{u}_1, \mathbf{u}_2 \in U_0, \tag{2.7.47}$$

and if the inequality

$$\left|\vec{\mathbf{f}}(\mathbf{a}_0)\right| \left| [\mathbf{D}\vec{\mathbf{f}}(\mathbf{a}_0)]^{-1} \right|^2 M \leq \frac{1}{2} \tag{2.7.48}$$

is satisfied, the equation $\vec{\mathbf{f}}(\mathbf{x}) = \vec{\mathbf{0}}$ has a unique solution in U_0, and Newton's method with initial guess $\mathbf{a}_0$ converges to it.

If Inequality 2.7.48 is satisfied, then at each iteration we create a new ball inside the previous ball, and with at most half the radius of the previous: U_1 is in U_0, U_2 is in $U_1, \ldots$, as shown to the right of Figure 2.7.3. In particular, the Lipschitz condition that is valid for U_0 is valid for all subsequent balls. As the radius of the balls goes to zero, the sequence $\mathbf{a}_0, \mathbf{a}_1, \ldots$ converges to $\mathbf{a}$, which we will see is a root.

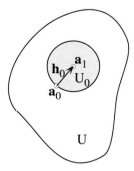
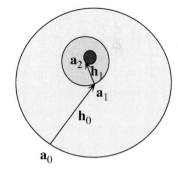

FIGURE 2.7.3. Equation 2.7.46 defines the neighborhood U_0 for which Newton's method is guaranteed to work when the inequality of Equation 2.7.48 is satisfied. Left: the neighborhood U_0 is the ball of radius $|\vec{\mathbf{h}}_0| = |\mathbf{a}_1 - \mathbf{a}_0|$ around $\mathbf{a}_1$, so $\mathbf{a}_0$ is on the border of U_0. Right: a blow-up of U_0, showing the neighborhood U_1.

The right units.   Note that the right-hand side of Equation 2.7.48 is the unitless number $1/2$:

$$\left|\vec{\mathbf{f}}(\mathbf{a}_0)\right|\left|[\mathbf{D}\vec{\mathbf{f}}(\mathbf{a}_0)]^{-1}\right|^2 M \leq \frac{1}{2} \qquad 2.7.49$$

All the units of the left-hand side cancel. This is fortunate, because there is no reason to think that the domain U will have the same units as the range; although both spaces have the same dimension, they can be very different. For example, the units of U might be temperature and the units of the range might be volume, with $\vec{\mathbf{f}}$ measuring volume as a function of temperature. (In this one-dimensional case, $\vec{\mathbf{f}}$ would be f.)

Let us see that the units on the left-hand side of Equation 2.7.48 cancel. We'll denote by u the units of the domain, U, and by r the units of the range, $\mathbb{R}^n$. The term $|\vec{\mathbf{f}}(\mathbf{a}_0)|$ has units r. A derivative has units range/domain (typically, distance divided by time), so the inverse of the derivative has units domain/range $= u/r$, and the term $|[\mathbf{D}\vec{\mathbf{f}}(\mathbf{a}_0)]^{-1}|^2$ has units u^2/r^2. The Lipschitz ratio M is the distance between derivatives divided by a distance in the domain, so its units are r/u divided by u. This gives the following units:

$$r \times \frac{u^2}{r^2} \times \frac{r}{u^2}; \qquad 2.7.50$$

both the r's and the u's cancel out.

Equation 2.7.48:

$$\left|\vec{\mathbf{f}}(\mathbf{a}_0)\right|\left|[\mathbf{D}\vec{\mathbf{f}}(\mathbf{a}_0)]^{-1}\right|^2 M \leq \frac{1}{2}$$

A good way to check whether some equation makes sense is to make sure that both sides have the same units. In physics this is essential.

Example 2.7.12 (Using Newton's method).   Suppose we want to solve the two equations

$$\begin{array}{l} \cos(x - y) = y \\ \sin(x + y) = x, \end{array} \quad \text{i.e.} \quad \vec{F}\begin{pmatrix} x \\ y \end{pmatrix} = \begin{bmatrix} \cos(x-y) - y \\ \sin(x+y) - x \end{bmatrix} = \begin{bmatrix} 0 \\ 0 \end{bmatrix}. \qquad 2.7.51$$

We just happen to notice that the equation is close to being satisfied at $\begin{pmatrix} 1 \\ 1 \end{pmatrix}$:

$$\cos(1 - 1) - 1 = 0 \quad \text{and} \quad \sin(1 + 1) - 1 = -.0907\ldots . \qquad 2.7.52$$

We "happened to notice" that $\sin 2 - 1 = -.0907$ with the help of a calculator. Finding an initial condition for Newton's method is always the delicate part.

Let us check that starting Newton's method at $\mathbf{a}_0 = \begin{pmatrix} 1 \\ 1 \end{pmatrix}$ works. To do this we must see that the inequality of Equation 2.7.48 is satisfied. We just saw that $|\vec{F}(\mathbf{a}_0)| \sim .0907 < .1$. The derivative at $\mathbf{a}_0$ isn't much worse:

The Kantorovitch theorem does *not* say that the system of equations has a unique solution; it may have many. But it has one unique solution in the neighborhood U_0, and if you start with the initial guess $\mathbf{a}_0$, Newton's method will find it for you.

$$[\mathbf{D}\vec{F}(\mathbf{a}_0)] = \begin{bmatrix} 0 & -1 \\ \cos 2 - 1 & \cos 2 \end{bmatrix}, \text{ so } [\mathbf{D}\vec{F}(\mathbf{a}_0)]^{-1} = \frac{1}{\cos 2 - 1}\begin{bmatrix} \cos 2 & 1 \\ 1 - \cos 2 & 0 \end{bmatrix}, \qquad 2.7.53$$

Recall that the inverse of

$$A = \begin{bmatrix} a & b \\ c & d \end{bmatrix} \quad \text{is}$$

$$A^{-1} = \frac{1}{ad - bc}\begin{bmatrix} d & -b \\ -c & a \end{bmatrix},$$

and

$$\left|[\mathbf{D}\vec{F}(\mathbf{a}_0)]^{-1}\right|^2 = \frac{1}{(\cos 2 - 1)^2}\left((\cos 2)^2 + 1 + (1 - \cos 2)^2\right) \sim 1.1727 < 2,$$

as you will see if you put it in your calculator.

Rather than compute the Lipschitz ratio for the derivative using higher partial derivatives, we will do it directly, taking advantage of the helpful formulas $|\sin a - \sin b| \le |a-b|$, $|\cos a - \cos b| \le |a-b|$.[21] These make the computation manageable:

$$
\left| [\mathbf{D}\vec{F}(\begin{smallmatrix} x_1 \\ y_1 \end{smallmatrix})] - [\mathbf{D}\vec{F}(\begin{smallmatrix} x_2 \\ y_2 \end{smallmatrix})] \right| = \left| \begin{bmatrix} -\sin(x_1 - y_1) + \sin(x_2 - y_2) & \sin(x_1 - y_1) - \sin(x_2 - y_2) \\ \cos(x_1 + y_1) - \cos(x_2 - y_2) & \cos(x_1 + y_1) - \cos(x_2 - y_2) \end{bmatrix} \right|
$$

$$
\le \left| \begin{bmatrix} |-(x_1 - y_1) + (x_2 - y_2)| & |(x_1 - y_1) - (x_2 - y_2)| \\ |(x_1 + y_1) - (x_2 - y_2)| & |(x_1 + y_1) - (x_2 - y_2)| \end{bmatrix} \right|
$$

$$
= \sqrt{4\left((x_1 - x_2)^2 + (y_1 - y_2)^2\right)} = 2\left| \begin{pmatrix} x_1 \\ y_1 \end{pmatrix} - \begin{pmatrix} x_2 \\ y_2 \end{pmatrix} \right|. \qquad 2.7.54
$$

Thus $M = 2$ is a Lipschitz constant for $[\mathbf{D}\vec{F}]$. Putting these together, we see that

$$
|\vec{F}(\mathbf{a}_0)| \left| [\mathbf{D}\vec{F}(\mathbf{a}_0)]^{-1} \right|^2 M \le .1 \cdot 2 \cdot 2 = .4 < .5 \qquad 2.7.55
$$

The Kantorovitch theorem does *not* say that if Inequality 2.7.48 is not satisfied, the equation has no solutions; it does not even say that if the inequality is not satisfied, there are no solutions in the neighborhood U_0. In Section 2.8 we will see that if we use a different way to measure $[\mathbf{D}\vec{f}(\mathbf{a}_0)]$, which is harder to compute, then inequality 2.7.48 is easier to satisfy. That version of Kantorovitch's theorem thus guarantees convergence for some equations about which this somewhat weaker version of the theorem is silent.

The MATLAB program "Newton.m" is found in Appendix B.1.

so the equation has a solution, and Newton's method starting at $\begin{pmatrix} 1 \\ 1 \end{pmatrix}$ will converge to it. Moreover,

$$
\vec{h}_0 = \frac{1}{\cos 2 - 1} \begin{bmatrix} \cos 2 & 1 \\ 1 - \cos 2 & 0 \end{bmatrix} \begin{bmatrix} 0 \\ \sin 2 - 1 \end{bmatrix} = \begin{bmatrix} \frac{\sin 2 - 1}{1 - \cos 2} \\ 0 \end{bmatrix} \sim \begin{bmatrix} -.064 \\ 0 \end{bmatrix}, \qquad 2.7.56
$$

so Kantorovitch's theorem guarantees that the solution is within .064 of $\begin{pmatrix} .936 \\ 1 \end{pmatrix}$. The computer says that the solution is actually $\begin{pmatrix} .935 \\ .998 \end{pmatrix}$, correct to three decimal places. $\triangle$

Example 2.7.13 (Newton's method, using a computer). Now we will use the MATLAB program to solve the equations

$$
x^2 - y + \sin(x - y) = 2 \quad \text{and} \quad y^2 - x = 3, \qquad 2.7.57
$$

starting at $\begin{pmatrix} 2 \\ 2 \end{pmatrix}$ and at $\begin{pmatrix} -2 \\ 2 \end{pmatrix}$.

The equation we are solving is

$$
\vec{F}\begin{pmatrix} x \\ y \end{pmatrix} = \begin{bmatrix} x^2 - y + \sin(x - y) - 2 \\ y^2 - x - 3 \end{bmatrix} = \begin{bmatrix} 0 \\ 0 \end{bmatrix}. \qquad 2.7.58
$$

---

[21]By the mean value theorem, there exists a c between a and b such that

$$
|\sin a - \sin b| = |\underbrace{\cos}_{\sin' c} c||a - b|;
$$

since $|\cos c| \le 1$, we have $|\sin a - \sin b| \le |a - b|$.

Starting at $\begin{pmatrix} 2 \\ 2 \end{pmatrix}$, the MATLAB Newton program gives the following values:

$$\mathbf{x}_0 = \begin{pmatrix} 2 \\ 2 \end{pmatrix}, \quad \mathbf{x}_1 = \begin{pmatrix} 2.\overline{1} \\ 2.\overline{27} \end{pmatrix}, \quad \mathbf{x}_2 = \begin{pmatrix} 2.10131373055664 \\ 2.25868946388913 \end{pmatrix},$$

$$\mathbf{x}_3 = \begin{pmatrix} 2.10125829441818 \\ 2.25859653392414 \end{pmatrix}, \quad \mathbf{x}_4 = \begin{pmatrix} 2.10125829294805 \\ 2.25859653168689 \end{pmatrix}, \qquad 2.7.59$$

In fact, it superconverges: the number of correct decimals roughly doubles at each iteration; we see 1, then 3, then 8, then 14 correct decimals. We will discuss superconvergence in detail in Section 2.8.

and the first 14 decimals don't change after that. Newton's method certainly does appear to converge.

But are the conditions of Kantorovitch's theorem satisfied? The MATLAB program prints out a "condition number," *cond*, at each iteration, which is $|\vec{F}(\mathbf{x}_i)| \, \left| [\mathbf{D}\vec{F}(\mathbf{x}_i)]^{-1} \right|^2$. Kantorovitch's Theorem says that Newton's method will converge if $cond \cdot M \leq 1/2$, where M is a Lipschitz constant for $[\mathbf{D}\vec{F}]$ on U_i.

We first computed this Lipschitz constant without higher partial derivatives, and found it quite tricky. It's considerably easier with higher partial derivatives:

$$D_1 D_1 f_1 = 2 - \sin(x - y); \quad D_1 D_2 f_1 = \sin(x - y); \quad D_2 D_2 f_1 = -\sin(x - y)$$

$$D_1 D_1 f_2 = 0; \quad D_1 D_2 f_2 = 0; \quad D_2 D_2 f_2 = 2, \qquad 2.7.60$$

so

$$\sum_{i,j} \left(D_i D_j f_k \right)^2 = \left(2 - \sin(x - y) \right)^2 + 2\left(\sin(x - y) \right)^2 + \left(\sin(x - y) \right)^2 + 4;$$

$$\text{since } -1 \leq \sin \leq 1, \quad \left| \sum_{i,j} \left(D_i D_j f_k \right)^2 \right| \leq 9 + 2 + 1 + 4 = 16. \qquad 2.7.61$$

Thus $M = 4$ is a Lipschitz constant for F on all of $\mathbb{R}^2$.

Let us see what we get when $cond \cdot M \leq 1/2$. At the first iteration, *cond* $= 0.1419753$ (the exact value is $\sqrt{46}/18$), and $4 \times 0.1419753 > .5$. So Kantorovitch's theorem does not assert convergence, but it isn't far off. At the next iteration, we find *cond* $= 0.00874714275069$, and this works with a lot to spare.

What happens if we start at $\begin{pmatrix} -2 \\ 2 \end{pmatrix}$? The computer gives

$$\mathbf{x}_0 = \begin{pmatrix} -2 \\ 2 \end{pmatrix}, \quad \mathbf{x}_1 = \begin{pmatrix} -1.78554433070248 \\ 1.30361391732438 \end{pmatrix}, \quad \mathbf{x}_2 = \begin{pmatrix} -1.82221637692367 \\ 1.10354485721642 \end{pmatrix},$$

$$\mathbf{x}_3 = \begin{pmatrix} -1.82152790765992 \\ 1.08572086062422 \end{pmatrix}, \quad \mathbf{x}_4 = \begin{pmatrix} -1.82151878937233 \\ 1.08557875385529 \end{pmatrix},$$

$$\mathbf{x}_5 = \begin{pmatrix} -1.82151878872556 \\ 1.08557874485200 \end{pmatrix}, \quad \dots,$$

and again the numbers do not change if we iterate the process further. It certainly converges fast. The condition numbers are

$$0.3337, \ 0.1036, \ 0.01045, \ \dots. \qquad 2.7.62$$

The computation we had made for the Lipschitz constant of the derivative is still valid, so we see that the condition of Kantorovitch's theorem fails rather badly at the first step (and indeed, the first step is rather large), but succeeds (just barely) at the second. $\triangle$

Remark. Although both the domain and the range of Newton's method are n-dimensional, you should think of them as different spaces. As we mentioned, in many practical applications they have different units. It is further a good idea to think of the domain as made up of points, and the range as made up of vectors. Thus $\vec{\mathbf{f}}(\mathbf{a}_i)$ is a vector, and $\vec{\mathbf{h}}_i = -[\mathbf{D}\vec{\mathbf{f}}(\mathbf{a}_i)]^{-1}\vec{\mathbf{f}}(\mathbf{a}_i)$ is an increment in the domain, i.e., a vector. The next point $\mathbf{a}_{i+1} = \mathbf{a}_i + \vec{\mathbf{h}}_i$ is really a point: the sum of a point and an increment.

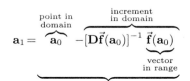

$$\mathbf{a}_1 = \overbrace{\mathbf{a}_0}^{\substack{\text{point in} \\ \text{domain}}} \overbrace{-[\mathbf{D}\vec{\mathbf{f}}(\mathbf{a}_0)]^{-1}\underbrace{\vec{\mathbf{f}}(\mathbf{a}_0)}_{\substack{\text{vector} \\ \text{in range}}}}^{\substack{\text{increment} \\ \text{in domain}}}$$

point minus vector equals point

Remark. You may not find Newton's method entirely satisfactory; what if you don't know an initial "seed" $\mathbf{a}_0$? *Newton's method is guaranteed to work only when you know something to start out.* If you don't, you have to guess and hope for the best. Actually, this isn't quite true. In the nineteenth century, Cayley showed that for any quadratic equation, Newton's method essentially *always* works. But quadratic equations form the only case where Newton's method does not exhibit chaotic behavior.[22]

2.8 SUPERCONVERGENCE

Kantorovitch's theorem is in some sense optimal: you cannot do better than the given inequalities unless you strengthen the hypotheses.

Example 2.8.1 (Slow convergence). Consider solving $f(x) = (x-1)^2 = 0$ by Newton's method, starting at $a_0 = 0$. Exercise 2.8.1 asks you to show that the best Lipschitz ratio for f' is 2, so the product

$$|f(a_0)|\,\left|\left(f'(a_0)\right)^{-1}\right|^2 M = 1 \cdot \left(-\frac{1}{2}\right)^2 \cdot 2 = \frac{1}{2}, \qquad 2.8.1$$

and Theorem 2.7.11 guarantees that Newton's method will work, and will converge to the unique root $a = 1$. The exercise further asks you to check that $h_n = 1/2^{n+1}$ so $a_n = 1 - 1/2^{n+1}$, exactly the rate of convergence advertised. $\triangle$

Example 2.8.1 is both true and squarely misleading. If at each step Newton's method only halved the distance between guess and root, a number of simpler algorithms (bisection, for example) would work just as well.

---

[22]For a precise description of how Newton's method works for quadratic equations, and for a description of how things can go wrong in other cases, see J. Hubbard and B. West, *Differential Equations, A Dynamical Systems Approach, Part I*, Texts in Applied Mathematics No. 5, Springer-Verlag, N.Y., 1991, pp. 227–235.

Newton's method is the favorite scheme for solving equations because usually it converges *much, much faster* than in Example 2.8.1. If, instead of allowing the product in Inequality 2.7.48 to be $\leq 1/2$, we insist that it be strictly less than $1/2$:

$$|\vec{\mathbf{f}}(\mathbf{a}_0)|\,|[\mathbf{D}\vec{\mathbf{f}}(\mathbf{a}_0)]^{-1}|^2 M = k < \frac{1}{2}, \qquad 2.8.2$$

then Newton's method *superconverges*.

How soon Newton's method starts superconverging depends on the problem at hand. But once it starts, it is so fast that within four more steps you will have computed your answer to as many digits as a computer can handle. In practice, when Newton's method works at all, it starts superconverging soon.

What do we mean when we say that a sequence $a_0, a_1, \ldots$ superconverges? Our definition is the following:

As a rule of thumb, if Newton's method hasn't converged to a root in seven or eight steps, you've chosen a poor initial condition.

The $1/2$ in $x_0 = 1/2$ is unrelated to the $1/2$ of Equation 2.8.2. If we were to define superconvergence using digits in base 10, then the same sequence would superconverge starting at $x_3 \leq 1/10$. For it to start superconverging at x_0, we would have to have $x_0 \leq 1/10$.

Definition 2.8.2 (Superconvergence). Set $x_i = |a_{i+1} - a_i|$; i.e., x_i represents the difference between two successive entries of the sequence. We will say that the sequence $a_0, a_1, \ldots$ superconverges if, when the x_i are written in base 2, then each number x_i starts with $2^i - 1 \approx 2^i$ zeroes.

For example, the sequence $x_{n+1} = x_n^2$, starting with $x_0 = 1/2$ (written .1 in base 2), superconverges to zero, as shown in the left-hand side of Figure 2.8.1. By comparison, the right-hand side of Figure 2.8.1 shows the convergence achieved in Example 2.8.1, again starting with $x_0 = 1/2$.

$x_0 = .1$	$x_0 = .1$
$x_1 = .01$	$x_1 = .01$
$x_2 = .0001$	$x_2 = .001$
$x_3 = .00000001$	$x_3 = .0001$
$x_4 = .0000000000000001.$	$x_4 = .00001.$

FIGURE 2.8.1. Left: superconvergence. Right: the convergence guaranteed by Kantorovitch's theorem. In both cases, numbers are written in base 2: $.1 = 1/2, .01 = 1/4, .001 = 1/8, \ldots$.

We will see that what goes wrong for Example 2.8.1 is that at the root $a = 1$, $f'(a) = 0$, so the derivative of f is not invertible at the limit point: $1/f'(1)$ does not exist. Whenever the derivative is invertible at the limit point, we do have superconvergence. This occurs as soon as Equation 2.8.2 is satisfied: as soon as the product in the Kantorovitch inequality is strictly less than $1/2$.

Theorem 2.8.3 (Newton's method superconverges). *Let the conditions of the Kantorovitch theorem 2.7.11 be satisfied, but with the stronger assumption that*

$$\left|\vec{\mathbf{f}}(\mathbf{a}_0)\right| \left|[\mathbf{D}\vec{\mathbf{f}}(\mathbf{a}_0)]^{-1}\right|^2 M = k < \frac{1}{2}. \tag{2.8.2}$$

$$Set \quad c = \frac{1-k}{1-2k}\left|[\mathbf{D}\vec{\mathbf{f}}(\mathbf{a}_0)]^{-1}\right|\frac{M}{2}. \tag{2.8.3}$$

$$If \quad |\vec{\mathbf{h}}_n| \leq \frac{1}{2c}, \quad then \quad |\vec{\mathbf{h}}_{n+m}| \leq \frac{1}{c}\cdot\left(\frac{1}{2}\right)^{2^m}. \tag{2.8.4}$$

Equation 2.8.4 means superconvergence. Since $\vec{\mathbf{h}}_n = |\mathbf{a}_{n+1} - \mathbf{a}_n|$, starting at step n and using Newton's method for m iterations causes the distance between $\mathbf{a}_n$ and $\mathbf{a}_{n+m}$ to shrink to practically nothing before our eyes. For example, if $m = 10$:

Even if k is almost $1/2$, so that c is large, the factor $(1/2)^{2^m}$ will soon predominate.

$$|\vec{\mathbf{h}}_{n+m}| \leq \frac{1}{c}\cdot\left(\frac{1}{2}\right)^{1024}. \tag{2.8.5}$$

The proof requires the following lemma, proved in Appendix A.3.

Lemma 2.8.4. *If the conditions of Theorem 2.8.3 are satisfied, then for all i,*

$$|\vec{\mathbf{h}}_{i+1}| \leq c|\vec{\mathbf{h}}_i|^2. \tag{2.8.6}$$

Proof of Theorem 2.8.3. Let $x_i = c|\vec{\mathbf{h}}_i|$. Then

$$x_{i+1} = c|\vec{\mathbf{h}}_{i+1}| \leq c^2|\vec{\mathbf{h}}_i|^2 = x_i^2. \tag{2.8.7}$$

Our assumption that $|\vec{\mathbf{h}}_n| \leq \frac{1}{2c}$ tells us that $x_n \leq 1/2$. So

$$x_{n+1} \leq x_n^2 \leq \frac{1}{4} = \left(\frac{1}{2}\right)^{2^1},$$

$$x_{n+2} \leq (x_{n+1})^2 \leq x_n^4 \leq \frac{1}{16} = \left(\frac{1}{2}\right)^{2^2}, \tag{2.8.8}$$

$$\vdots$$

$$x_{n+m} \leq x_n^{2^m} \leq \left(\frac{1}{2}\right)^{2^m}.$$

Since $|\vec{\mathbf{h}}_n| \leq \frac{1}{2c}$, we have the result we want, Equation 2.8.4:

$$if \quad |\vec{\mathbf{h}}_n| \leq \frac{1}{2c}, \quad then \quad |\vec{\mathbf{h}}_{n+m}| \leq \frac{1}{c}\cdot\left(\frac{1}{2}\right)^{2^m}. \quad \square \tag{2.8.9}$$

Kantorovitch's theorem: a stronger version (optional)

We have seen that Newton's method converges much faster than guaranteed by Kantorovitch's theorem. In this subsection we show that it is possible to state Kantorovitch's theorem in such a way that it will apply to a larger class of functions. We do this by using a different way to measure linear mappings: the *norm* $\|A\|$ of a matrix A.

Definition 2.8.5 (The norm of a matrix). The norm $\|A\|$ of a matrix A is

$$\|A\| = \sup |A\vec{x}|, \text{ when } |\vec{x}| = 1. \qquad 2.8.10$$

This means that $\|A\|$ is the maximum amount by which multiplication by A will stretch a vector.

Multiplication by the matrix A of Example 2.8.6 can *at most* double the length of a vector; it does not always do so; the product $A\vec{b}$, where $A = \begin{bmatrix} 2 & 0 \\ 0 & 1 \end{bmatrix}$ and $\vec{b} = \begin{bmatrix} 0 \\ 1 \end{bmatrix}$, is $\begin{bmatrix} 0 \\ 1 \end{bmatrix}$, with length 1.

Example 2.8.6 (Norm of a matrix). Take

$$A = \begin{bmatrix} 2 & 0 \\ 0 & 1 \end{bmatrix} \text{ and } \vec{x} = \begin{bmatrix} x \\ y \end{bmatrix}, \text{ so that } A\mathbf{x} = \begin{bmatrix} 2x \\ y \end{bmatrix}.$$

Since by definition $|\vec{x}| = \sqrt{x^2 + y^2} = 1$, we have

$$\|A\| = \sup_{|\vec{x}|=1} |A\vec{x}| = \underbrace{\sup \sqrt{4x^2 + y^2} = 2}_{\text{setting } x=1, y=0}. \quad \triangle \qquad 2.8.11$$

In Example 2.8.6, note that $\|A\| = 2$, while $|A| = \sqrt{5}$. It is always true that

$$\|A\| \le |A|; \qquad 2.8.12$$

this follows from Proposition 1.4.11, as you are asked to show in Exercise 2.8.2.

There are many equations for which convergence is guaranteed if one uses the norm, but not if one uses the length.

This is why using the norm $\|A\|$ rather than the length $|A|$ makes Kantorovitch's theorem stronger: the theorem applies equally as well when we use the norm rather than the length to measure the derivative $[\mathbf{Df}(\mathbf{x})]$ and its inverse, and the key inequality of that theorem, Equation 2.7.48, is easier to satisfy using the norm.

Theorem 2.8.7 (Kantorovitch's Theorem: a stronger version). *Kantorovitch's theorem 2.7.11 still holds if you replace all lengths of matrices by norms of matrices.*

Proof. In the proof of Theorem 2.7.11 we only used the triangle inequality and Proposition 1.4.11, and these hold for the norm $\|A\|$ of a matrix A as well as for its length $|A|$, as Exercises 2.8.3 and 2.8.4 ask you to show. $\square$

Unfortunately, the norm is usually *much* harder to compute than the length. In Equation 2.8.11 above, it is not difficult to see that 2 is the largest value of $\sqrt{4x^2 + y^2}$ compatible with the requirement that $\sqrt{x^2 + y^2} = 1$, obtained by setting $x = 1$ and $y = 0$. Computing the norm is not often that easy.

Example 2.8.8 (Norm is harder to compute). The length of the matrix $A = \begin{bmatrix} 1 & 1 \\ 0 & 1 \end{bmatrix}$ is $\sqrt{1^2 + 1^2 + 1^2} = \sqrt{3}$, or about 1.732. The norm is $\frac{1+\sqrt{5}}{2}$, or about 1.618; arriving at that figure takes some work, as follows. A vector $\begin{bmatrix} x \\ y \end{bmatrix}$ with length 1 can be written $\begin{bmatrix} \cos t \\ \sin t \end{bmatrix}$, and the product of A and that vector is $\begin{bmatrix} \cos t + \sin t \\ \sin t \end{bmatrix}$, so the object is to find

$$\sup \sqrt{(\cos t + \sin t)^2 + \sin^2 t} \ . \tag{2.8.13}$$

At its maximum and minimum, the derivative of a function is 0, so we need to see where the derivative of $(\cos t + \sin t)^2 + \sin^2 t$ vanishes. That derivative is $2 \cos 2t + \sin 2t$, which vanishes for $2t = \arctan(-2)$. We have two possible angles to look for, t_1 and t_2, as shown in Figure 2.8.2; they can be computed with a calculator or with a bit of trigonometry, and we can choose the one that gives the biggest value for Equation 2.8.13. Since the entries of the matrix A are all positive, we choose t_1, in the first quadrant, as being the best bet.

By similar triangles, we find that

$$\cos 2t_1 = -\frac{1}{\sqrt{5}} \quad \text{and} \quad \sin 2t_1 = \frac{2}{\sqrt{5}}. \tag{2.8.14}$$

Using the formula $\cos 2t_1 = 2\cos^2 t_1 - 1 = 1 - 2\sin^2 t$, we find that

$$\cos t_1 = \sqrt{\frac{1}{2}\left(1 - \frac{1}{\sqrt{5}}\right)}, \quad \text{and} \quad \sin t_1 = \sqrt{\frac{1}{2}\left(1 + \frac{1}{\sqrt{5}}\right)}, \tag{2.8.15}$$

which, after some computation, gives

$$\left|\begin{bmatrix} \cos t_1 + \sin t_1 \\ \sin t_1 \end{bmatrix}\right|^2 = \frac{3 + \sqrt{5}}{2}, \tag{2.8.16}$$

and finally $\quad \|A\| = \left\| \begin{bmatrix} 1 & 1 \\ 0 & 1 \end{bmatrix} \right\| = \sqrt{\dfrac{3 + \sqrt{5}}{2}}.$ $\qquad\qquad$ 2.8.17

Remark. We could have used the following formula for computing the norm of a 2×2 matrix from its length and its determinant:

$$\|A\| = \sqrt{\frac{|A|^2 + \sqrt{|A|^4 - 4(\det A)^2}}{2}}. \tag{2.8.18}$$

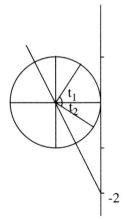

FIGURE 2.8.2.

The diagram of the trigonometric circle for Example 2.8.8.

$\left((\cos t + \sin t)^2 + \sin^2 t\right)'$

$= 2(\cos t + \sin t)(\cos t - \sin t)$

$\quad + 2\sin t \cos t$

$= 2(\cos^2 t - \sin^2 t) + 2\sin t \cos t$

$= 2\cos 2t + \sin 2t.$

In higher dimensions things are much worse. It was to avoid this kind of complication that we used the length rather than the norm when we proved Kantorovitch's theorem in Section 2.7.   △

In some cases, however, the norm is easier to use than the length, as in the following example. In particular, norms of multiples of the identity matrix are easy to compute: such a norm is just the absolute value of the multiple.

Example 2.8.9 (Using the norm in Newton's method). Suppose we want to find a 2×2 matrix A such that $A^2 = \begin{bmatrix} 8 & 1 \\ -1 & 10 \end{bmatrix}$. So we define $F :$ $\mathrm{Mat}\,(2,2) \to \mathrm{Mat}\,(2,2)$ by

"Mat" of course stands for "matrix"; $\mathrm{Mat}\,(2,2)$ is the space of 2×2 matrices.

$$F(A) = A^2 - \begin{bmatrix} 8 & 1 \\ -1 & 10 \end{bmatrix}, \qquad 2.8.19$$

and try to solve it by Newton's method. First we choose an initial point A_0. A logical place to start would seem to be the matrix

You might think that something like $\begin{bmatrix} 3 & 1 \\ -1 & 3 \end{bmatrix}$ would be even better, but squaring that gives $\begin{bmatrix} 8 & 6 \\ -6 & 8 \end{bmatrix}$. In addition, starting with a diagonal matrix makes our computations easier.

$$A_0 = \begin{bmatrix} 3 & 0 \\ 0 & 3 \end{bmatrix}, \quad \text{so that} \quad A_0^2 = \begin{bmatrix} 9 & 0 \\ 0 & 9 \end{bmatrix}. \qquad 2.8.20$$

We want to see whether the Kantorovitch inequality 2.7.48 is satisfied, i.e., that

$$|F(A_0)| \cdot M \|[\mathbf{D}F(A_0)]^{-1}\|^2 \le \frac{1}{2}. \qquad 2.8.21$$

First, compute the derivative:

You may recognize the $AB + BA$ in Equation 2.8.22 from Equation 1.7.45, Example 1.7.15.

$$[\mathbf{D}F(A)]B = AB + BA. \qquad 2.8.22$$

The following computation shows that $A \mapsto [\mathbf{D}F(A)]$ is Lipschitz with respect to the norm, with Lipschitz ratio 2 on all of $\mathrm{Mat}\,(2,2)$:

$$\|[\mathbf{D}F(A_1)] - [\mathbf{D}F(A_2)]\| = \sup_{|B|=1} \left| \Big([\mathbf{D}F(A_1)] - [\mathbf{D}F(A_2)]\Big)B \right|$$

$$= \sup_{|B|=1} |A_1 B + B A_1 - A_2 B - B A_2| = \sup_{|B|=1} |(A_1 - A_2)B + B(A_1 - A_2)|$$

$$\le \sup_{|B|=1} |(A_1 - A_2)B| + |B(A_1 - A_2)| \le \sup_{|B|=1} |A_1 - A_2||B| + |B||A_1 - A_2|$$

$$\le \sup_{|B|=1} 2|B||A_1 - A_2| = 2|A_1 - A_2|. \qquad 2.8.23$$

Now we insert A_0 into Equation 2.8.19, getting

$$F(A_0) = \begin{bmatrix} 9 & 0 \\ 0 & 9 \end{bmatrix} - \begin{bmatrix} 8 & 1 \\ -1 & 10 \end{bmatrix} = \begin{bmatrix} 1 & -1 \\ 1 & -1 \end{bmatrix}, \qquad 2.8.24$$

so that $|F(A_0)| = \sqrt{4} = 2.$

Now we need to compute $\|[\mathbf{D}F(A_0)]^{-1}\|^2$. Using Equation 2.8.22 and the fact that A_0 is three times the identity, we get

$$[\mathbf{D}F(A_0)]B = A_0B + BA_0 = 3B + 3B = 6B. \qquad 2.8.25$$

So we have

$$[\mathbf{D}F(A_0)]^{-1}B = \frac{B}{6},$$

$$\|[\mathbf{D}F(A_0)]^{-1}\| = \sup_{|B|=1} |B/6| = \sup_{|B|=1} \frac{|B|}{6} = 1/6, \qquad 2.8.26$$

$$\|[\mathbf{D}F(A_0)]^{-1}\|^2 = \frac{1}{36}.$$

The left-hand side of Equation 2.8.21 is $2 \cdot 2 \cdot 1/36 = 1/9$, and we see that the inequality is satisfied with room to spare: if we start at $\begin{bmatrix} 3 & 0 \\ 0 & 3 \end{bmatrix}$ and use Newton's method, we can compute the square root of $\begin{bmatrix} 8 & 1 \\ -1 & 10 \end{bmatrix}$.

Trying to solve Equation 2.8.19 without Newton's method would be unpleasant. In a draft of this book we proposed a different example, finding a 2×2 matrix A such that

$$A^2 + A = \begin{bmatrix} 1 & 1 \\ 1 & 1 \end{bmatrix}.$$

A friend pointed out that this problem can be solved explicitly (and more easily) without Newton's method, as Exercise 2.8.5 asks you to do.

2.9 THE INVERSE AND IMPLICIT FUNCTION THEOREMS

In Section 2.2 we completely analyzed systems of linear equations. Given a system of *nonlinear* equations, what solutions do we have? What variables depend on others? Our tools for answering these questions are the implicit function theorem and its special case, the inverse function theorem. These two theorems are the backbone of differential calculus, just as their linear analogs, Theorem 2.2.4 and its special case, Theorem 2.2.5, are the backbone of linear algebra. We will start with inverse functions, and then move to the more general case.

The inverse and implicit function theorems are a lot harder than the corresponding linear theorems, but most of the hard work is contained in the proof of Kantorovitch's theorem concerning the convergence of Newton's method.

"Implicit" means "implied." The statement $2x - 8 = 0$ implies that $x = 4$; it does not say it explicitly (directly).

Inverse functions in one dimension

An inverse function is a function that "undoes" the original function. If $f(x) = 2x$, clearly there is a function $g(f(x)) = x$, mainly, $g(y) = y/2$. Usually finding an inverse isn't so straightforward. But the basic condition for a continuous function in one variable to have an inverse is simple: the function must be *monotone*.

Definition 2.9.1 (Monotone function). A function is monotone if its graph always goes up or always goes down: if $x < y$ always implies $f(x) < f(y)$, the function is *monotone increasing*; if $x < y$ always implies $f(x) > f(y)$, the function is *monotone decreasing*.

If a function f that expresses x in terms of y is monotone, then it has an inverse function g expressing y in terms of x. In addition you can find $g(y)$ by a series of guesses that converge to the solution, and knowing the derivative of f tells you how to compute the derivative of g.

More precisely:

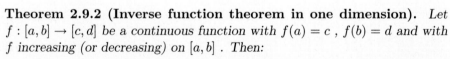

FIGURE 2.9.1.

The function $f(x) = 2x + \sin x$ is monotone increasing; it has an inverse function $g(2x + \sin x) = x$, but finding it requires solving the equation $2x + \sin x = y$, with x the unknown and y known. This can be done, but it requires an approximation technique; you can't find a formula for the solution using algebra, trigonometry, or even more advanced techniques.

Part (c) justifies the use of implicit differentiation: such statements as

$$\arcsin'(x) = \frac{1}{\sqrt{1 - x^2}}.$$

Theorem 2.9.2 (Inverse function theorem in one dimension). *Let* $f : [a, b] \to [c, d]$ *be a continuous function with* $f(a) = c$, $f(b) = d$ *and with* f *increasing (or decreasing) on* $[a, b]$. *Then:*

(a) There exists a unique continuous function $g : [c, d] \to [a, b]$ *such that*

$$f\big(g(y)\big) = y, \text{ for all } y \in [c, d], \quad \text{and} \quad \qquad 2.9.1$$

$$g\big(f(x)\big) = x, \text{ for all } x \in [a, b]. \qquad \qquad 2.9.2$$

(b) You can find $g(y)$ *by solving the equation* $y - f(x) = 0$ *for* x *by bisection (described below).*

(c) If f *is differentiable at* $x \in (a, b)$, *and* $f'(x) \neq 0$, *then* g *is differentiable at* $f(x)$, *and its derivative satisfies* $g'\big(f(x)\big) = 1/f'(x)$.

You are asked to prove Theorem 2.9.2 in Exercise 2.9.1.

Example 2.9.3 (An inverse function in one dimension). Take $f(x) = 2x + \sin x$, shown in Figure 2.9.1, and choose $[a, b] = [-k\pi, k\pi]$ for some positive integer k . Then

$$f(a) = f(-k\pi) = -2k\pi + \underbrace{\sin(-k\pi)}_{= 0} \text{ and } f(b) = f(k\pi) = 2k\pi + \underbrace{\sin(k\pi)}_{= 0}; \quad 2.9.3$$

i.e., $f(a) = 2a$ and $f(b) = 2b$, and since $f'(x) = 2 + \cos x$, which is ≥ 1, we see that f is strictly increasing. Thus Theorem 2.9.2 says that $y = 2x + \sin x$ expresses x implicitly as a function of y for $y \in [-2k\pi, 2k\pi]$: there is a function $g : [-2k\pi, 2k\pi] \to [-k\pi, k\pi]$ such that $g\big(f(x)\big) = g(2x + \sin x) = x$.

But if you take a hardnosed attitude and say, "Okay, so what is $g(1)$?", you will see that this question is not so easy to answer. The equation $1 = 2x + \sin x$, is not a particularly hard equation to "solve," but you can't find a formula for the solution using algebra, trigonometry or even more advanced techniques. Instead you must apply some approximation technique. $\triangle$

In several variables the approximation technique we will use is Newton's method; in one dimension, we can use *bisection*. Suppose you want to solve $f(x) = y$, and you know a and b such that $f(a) < y$ and $f(b) > y$. First try the x in the middle of $[a, b]$, computing $f(\frac{a+b}{2})$. If the answer is too small, try the midpoint of the right half-interval; if the answer is too big, try the midpoint of the left half-interval. Next choose the midpoint of the quarter-interval to the

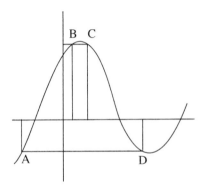

FIGURE 2.9.2.

The function graphed above is not monotone and has no global inverse; the same value of y gives both $x = B$ and $x = C$. Similarly, the same value of y gives $x = A$ and $x = D$. But it has many local inverses; the arc AB and the arc CD both represent x as a function of y.

The inverse function theorem is really the "local" inverse function theorem, carefully specifying the domain and the image of the inverse function.

As in Section 1.7, we are using the derivative to linearize a nonlinear problem.

Fortunately, proving this constructivist version of the inverse function theorem is no harder than proving the standard version

right (if your answer was too small) or to the left (if your answer was too big). The sequence of x_n chosen this way will converge to $g(y)$.

Note, as shown in Figure 2.9.2, that if a function is not monotone, we cannot expect to find a global inverse function, but there will usually be monotone stretches of the function for which local inverse functions exist.

Inverse functions in higher dimensions

In one dimension, monotonicity of a function is a sufficient (and necessary) criterion for an inverse function to exist, and bisection can be used to solve the equations. The point of the inverse function theorem is to show that inverse functions exist in higher dimensions, even though monotonicity and bisection do not generalize. In higher dimensions, we can't speak of a mapping always increasing or always decreasing. The requirement of monotonicity is replaced by the requirement that *the derivative of the mapping be invertible*. Bisection is replaced by Newton's method. The theorem is a great deal harder in higher dimensions, and you should not expect to breeze through it.

The inverse function theorem deals with the case where we have as many equations as unknowns: $\mathbf{f}$ maps U to W, where U and W are both subsets of $\mathbb{R}^n$. By definition, $\mathbf{f}$ is invertible if the equation $\mathbf{f}(\mathbf{x}) = \mathbf{y}$ has a unique solution $\mathbf{x} \in U$ for every $\mathbf{y} \in W$.

But generally we must be satisfied with asking, if $\mathbf{f}(\mathbf{x}_0) = \mathbf{y}_0$, *in what neighborhood* of $\mathbf{y}_0$ does there exist a local inverse? The name "inverse function theorem" is somewhat misleading. We said in Definition 1.3.3 that a transformation has an inverse if it is both onto and one to one. Such an inverse is global. Very often a mapping will not have a global inverse but it will have a *local* inverse (or several local inverses): there will be a neighborhood $V \subset W$ of $\mathbf{y}_0$ and a mapping $\mathbf{g} : V \to U$ such that $(\mathbf{f} \circ \mathbf{g})(\mathbf{y}) = \mathbf{y}$ for all $\mathbf{y} \in V$.

The statement of Theorem 2.9.4 is involved. The key message to retain is:

If the derivative is invertible, the mapping is locally invertible.

More precisely:

If the derivative of a mapping $\mathbf{f}$ is invertible at some point $\mathbf{x}_0$, the mapping is locally invertible in some neighborhood of the point $\mathbf{f}(\mathbf{x}_0)$

All the rest is spelling out just what we mean by "locally" and "neighborhood." The standard statement of the inverse function theorem doesn't spell that out; it guarantees the existence of an inverse, in the abstract: the theorem is shorter, but also less useful. If you ever want to use Newton's method to compute an inverse function, you'll need to know in what neighborhood such a function exists.[23]

We saw an example of local vs. global inverses in Figure 2.9.2; another example is $f(x) = x^2$. First, any inverse function of f can only be defined on the image of f, the positive real numbers. Second, there are two such "inverses," $g_1(y) = +\sqrt{y}$ and $g_2(y) = -\sqrt{y}$, and they both satisfy $f\big(g(y)\big) = y$, but they do not satisfy

$$g\big(f(x)\big) = x.$$

However, g_1 is an inverse if the domain of f is restricted to $x \geq 0$, and g_2 is an inverse if the domain of f is restricted to $x \leq 0$.

Theorem 2.9.4 (The inverse function theorem).   *Let $W \subset \mathbb{R}^m$ be an open neighborhood of $\mathbf{x}_0$, and $\mathbf{f} : W \to \mathbb{R}^m$ be a continuously differentiable function. Set $\mathbf{y}_0 = \mathbf{f}(\mathbf{x}_0)$, and suppose that the derivative $L = [\mathbf{Df}(\mathbf{x}_0)]$ is invertible.*

Let $R > 0$ be a number satisfying the following hypotheses:

(1) *The ball W_0 of radius $2R|L^{-1}|$ and centered at $\mathbf{x}_0$ is contained in W.*

(2) *In W_0, the derivative satisfies the Lipschitz condition*

$$|[\mathbf{Df}(\mathbf{u})] - [\mathbf{Df}(\mathbf{v})]| \leq \overbrace{\frac{1}{2R|L^{-1}|^2}}^{\text{Lipschitz ratio}} |\mathbf{u} - \mathbf{v}|. \qquad 2.9.4$$

There then exists a unique continuously differentiable mapping $\mathbf{g}$ from the ball of radius R centered at $\mathbf{y}_0$ (which we will denote V) to the ball W_0:

$$\mathbf{g} : V \to W_0, \qquad \text{such that} \qquad 2.9.5$$

$$\mathbf{f}\big(\mathbf{g}(\mathbf{y})\big) = \mathbf{y} \quad \text{and} \quad [\mathbf{Dg}(\mathbf{y})] = [\mathbf{Df}(\mathbf{g}(\mathbf{y}))]^{-1}. \qquad 2.9.6$$

The statement "suppose that the derivative $L = [\mathbf{Df}(\mathbf{x}_0)]$ is invertible" is the *key condition* of the theorem.

Moreover, the image of $\mathbf{g}$ contains the ball of radius R_1 around $\mathbf{x}_0$, where

$$R_1 = 2R|L^{-1}|^2 \left(\sqrt{|L|^2 + \frac{1}{|L^{-1}|^2}} - |L| \right). \qquad 2.9.7$$

We could write $\mathbf{f}\Big(\mathbf{g}(\mathbf{y})\Big) = \mathbf{y}$ as the composition

$$(\mathbf{f} \circ \mathbf{g})(\mathbf{y}) = \mathbf{y}.$$

On first reading, *skip* the last sentence concerning the little ball with radius R_1, centered at $\mathbf{x}_0$. It is a minor point, and we will discuss it later. Do notice that we have two main balls, W_0 centered at $\mathbf{x}_0$ and V centered at $\mathbf{y}_0 = \mathbf{f}(\mathbf{x}_0)$, as shown in Figure 2.9.3.

The ball V gives a lower bound for the domain of $\mathbf{g}$; the actual domain may be bigger.

The theorem tells us that if certain conditions are satisfied, then $\mathbf{f}$ has a local inverse function $\mathbf{g}$. The function $\mathbf{f}$ maps every point in the lumpy-shaped region $\mathbf{g}(V)$ to a point in V, and the inverse function $\mathbf{g}$ will undo that mapping, sending every point in V to a point in $\mathbf{g}(V)$.

Note that not every point $\mathbf{f}(\mathbf{x})$ is in the domain of $\mathbf{g}$; as shown in Figure 2.9.3, $\mathbf{f}$ maps some points in W to points outside of V. For this reason we had to write $\mathbf{f}\big(\mathbf{g}(\mathbf{y})\big) = \mathbf{y}$ in Equation 2.9.6, rather than $\mathbf{g}\big(\mathbf{f}(\mathbf{x})\big) = \mathbf{x}$. In addition, the function $\mathbf{f}$ may map more than one point to the same point in V, but only one can come from W_0 (and any point from W_0 must come from the subset $\mathbf{g}(V)$). But $\mathbf{g}$ maps a point in V to only one point. (Indeed, if $\mathbf{g}$ mapped the same point in V to more than one point, then $\mathbf{g}$ would not be a well-defined

---

[23]But once your exams are over, you can safely forget the details of how to compute that neighborhood, as long as you remember (1) if the derivative is invertible, the mapping is locally invertible, and (2) that you can look up statements that spell out what "locally" means.

mapping, as discussed in Section 1.3.) Moreover, that point is in W_0. This may appear obvious from Figure 2.9.3; after all, $\mathbf{g}(V)$ is the image of $\mathbf{g}$ and we can see that $\mathbf{g}(V)$ is in W_0. But the picture is illustrating what we have to prove, not what is given; the punch line of the theorem is precisely that " ... then there exists a unique continuously differentiable mapping from ... V to the ball W_0."

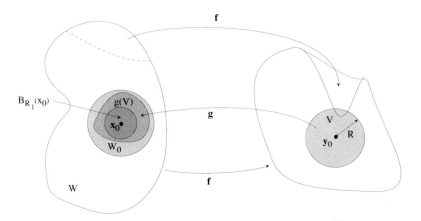

FIGURE 2.9.3. The function $\mathbf{f} : W \to \mathbb{R}^m$ maps every point in $\mathbf{g}(V)$ to a point in V; in particular, it sends $\mathbf{x}_0$ to $\mathbf{y}_0$. Its inverse function $\mathbf{g} : V \to W_0$ sends every point in V to a point in $\mathbf{g}(V)$. Note that $\mathbf{f}$ can well map other points outside W_0 into V.

Do you still remember the main point of the theorem? Where is the mapping

$$\mathbf{f}\begin{pmatrix} x \\ y \end{pmatrix} = \begin{pmatrix} xy \\ x^2 - y^2 \end{pmatrix}$$

guaranteed by the inverse function theorem to be locally invertible?[24]

---

[24]You're not being asked to spell out how big a neighborhood "locally" refers to, so you can forget about R, V, etc. Remember, *if the derivative of a mapping is invertible, the mapping is locally invertible.* The derivative is

$$\left[\mathbf{Df}\begin{pmatrix} x \\ y \end{pmatrix} \right] = \begin{bmatrix} y & x \\ 2x & -2y \end{bmatrix}.$$

The formula for the inverse of a 2×2 matrix

$$A = \begin{bmatrix} a & b \\ c & d \end{bmatrix} \quad \text{is} \quad A^{-1} = \frac{1}{ad - bc}\begin{bmatrix} d & -b \\ -c & a \end{bmatrix},$$

and here $ad - bc = -2(x^2 + y^2)$, which is 0 only if $x = 0$ and $y = 0$. The function is locally invertible near every point except $\mathbf{f}\begin{pmatrix} x \\ y \end{pmatrix} = \begin{pmatrix} 0 \\ 0 \end{pmatrix}$. To determine whether a larger matrix is invertible, use Theorem 2.3.2. Exercise 1.4.12 shows that a 3×3 matrices is invertible if its determinant is not 0.

In emphasizing the "main point" we don't mean to suggest that the details are unimportant. They are crucial if you want to compute an inverse function, since they provide an effective algorithm for computing the inverse: Newton's method. This requires knowing a lower bound for the natural domain of the inverse: where it is defined. To come to terms with the details, it may help to imagine different quantities as being big or little, and see how that affects the statement. First, in an ideal situation, would we want R to be big or little? We'd like it to be big, because then V will be big (remember R is the radius of V) and that will mean that the inverse function $\mathbf{g}$ is defined in a bigger neighborhood. What might keep R from being big? First, look at condition (1) of the theorem. We need W_0 to be in W, the domain of $\mathbf{f}$. Since the radius of W_0 is $2R|L^{-1}|$, if R is too big, W_0 may no longer fit in W.

That constraint is pretty clear. Condition (2) of the theorem is more delicate. Suppose that on W the derivative $[\mathbf{Df}(\mathbf{x})]$ is locally Lipschitz. It will then be Lipschitz on each $W_0 \subset W$, but with a best Lipschitz constant M_R which starts out at some probably non-zero value when W_0 is just a point (i.e., when $R = 0$), and gets bigger and bigger as R increases (it's harder to satisfy a Lipschitz ratio over a large area than a small one). On the other hand, the quantity $1/(2R|L^{-1}|^2)$ starts at infinity when $R = 0$, and decreases as R increases (see Figure 2.9.4). So Inequality 2.9.4 will be satisfied when R is small; but usually the graphs of M_R and $1/(2R|L^{-1}|^2)$ will cross for some R_0, and the inverse function theorem does not guarantee the existence of an inverse in any V with radius larger than R_0.

The conditions imposed on R may look complicated; do we need to worry that maybe no suitable R exists? The answer is no. If $\mathbf{f}$ is differentiable, and the derivative is Lipschitz (with *any* Lipschitz ratio) in some neighborhood of $\mathbf{x}_0$, then the function M_R exists, so the hypotheses on R will be satisfied as soon as $R < R_0$. Thus a differentiable map with Lipschitz derivative has a local inverse near any point where the derivative is invertible: if L^{-1} exists, we can find an R that works.

Do we really have to check that the derivative of a function is Lipschitz? *The answer is no*: as we will see in Corollary 2.7.8, *if the second partial derivatives of f are continuous, then the derivative is automatically Lipschitz in some neighborhood of $\mathbf{x}_0$.* Often this is enough.

Remark.   The standard statement of the inverse function theorem, which guarantees the existence of an inverse function in the abstract, doesn't require the derivative to be Lipschitz, just continuous.[25]   Because we want a lower

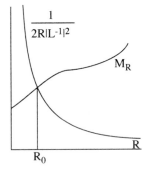

FIGURE 2.9.4.

The graph of the "best Lipschitz constant" M_R for $[\mathbf{Df}]$ on the ball of radius $2R|L^{-1}|$ increases with R, and the graph of the function

$$\frac{1}{2R|L^{-1}|^2}$$

decreases. The inverse function theorem only guarantees an inverse on a neighborhood V of radius R when

$$\frac{1}{2R|L^{-1}|^2} > M_R.$$

The main difficulty in applying these principles is that M_R is usually very hard to compute, and $|L^{-1}|$, although usually easier, may be unpleasant too.

---

[25]Requiring that the derivative be continuous *is* necessary, as you can see by looking at Example 1.9.3, in which we described a function whose partial derivatives are not continuous at the origin; see Exercise 2.9.2

bound for R (to know how big V is), we must impose some condition about how the derivative is continuous. We chose the Lipschitz condition because we want to use Newton's method to compute the inverse function, and Kantorovitch's theorem requires the Lipschitz condition.

Proof of the inverse function theorem

We show below that if the conditions of the inverse function theorem are satisfied, then Kantorovitch's theorem applies, and Newton's method can be used to find the inverse function.

Given $\mathbf{y} \in V$, we want to find $\mathbf{x}$ such that $\mathbf{f}(\mathbf{x}) = \mathbf{y}$. Since we wish to use Newton's method, we will restate the problem: Define

$$\mathbf{f_y}(\mathbf{x}) \stackrel{\text{def}}{=} \mathbf{f}(\mathbf{x}) - \mathbf{y} = 0. \qquad 2.9.8$$

We wish to solve the equation $\mathbf{f_y}(\mathbf{x}) = 0$ for $\mathbf{y} \in V$, using Newton's method with initial point $\mathbf{x}_0$.

We will use the notation of Theorem 2.7.11, but since the problem depends on $\mathbf{y}$, we will write $\vec{\mathbf{h}}_0(\mathbf{y}), U_0(\mathbf{y})$, etc. Note that

$$[\mathbf{Df_y}(\mathbf{x}_0)] = [\mathbf{Df}(\mathbf{x}_0)] = L, \quad \text{and} \quad \mathbf{f_y}(\mathbf{x}_0) = \underbrace{\mathbf{f}(\mathbf{x}_0)}_{=\mathbf{y}_0} - \mathbf{y} = \mathbf{y}_0 - \mathbf{y}, \qquad 2.9.9$$

so that

$$\vec{\mathbf{h}}_0(\mathbf{y}) = -\underbrace{[\mathbf{Df_y}(\mathbf{x}_0)]}_{L}^{-1} \mathbf{f_y}(\mathbf{x}_0) = -L^{-1}(\mathbf{y}_0 - \mathbf{y}). \qquad 2.9.10$$

This implies that $|\vec{\mathbf{h}}_0(\mathbf{y})| \leq |L^{-1}|R$, since $\mathbf{y}_0$ is the center of V, $\mathbf{y}$ is in V, and the radius of V is R, giving $|\mathbf{y}_0 - \mathbf{y}| \leq R$. Now we compute $\mathbf{x}_1 = \mathbf{x}_0 + \vec{\mathbf{h}}_0(\mathbf{y})$ (as in Equation 2.7.46, where $\mathbf{a}_1 = \mathbf{a}_0 + \vec{\mathbf{h}}_0$). Since $|\vec{\mathbf{h}}_0(\mathbf{y})|$ is at most half the radius of W_0 (i.e., half $2R|L^{-1}|$), we see that $U_0(\mathbf{y})$ (the ball of radius $|\vec{\mathbf{h}}_0(\mathbf{y})|$ centered at $\mathbf{x}_1$) is contained in W_0, as suggested by Figure 2.9.5.

Now we see that the Kantorovitch inequality (Equation 2.7.48) is satisfied:

$$\underbrace{|\mathbf{f_y}(\mathbf{x}_0)|}_{|\mathbf{y}_0 - \mathbf{y}| \leq R} \underbrace{|[\mathbf{Df}(\mathbf{x}_0)]^{-1}|^2}_{|L^{-1}|^2} M \leq R|L^{-1}|^2 \underbrace{\frac{1}{2R|L^{-1}|^2}}_{M} = \frac{1}{2}. \qquad 2.9.11$$

Thus Newton's method applied to the equation $\mathbf{f_y}(\mathbf{x}) = 0$ starting at $\mathbf{x}_0$ converges; denote the limit by $\mathbf{g}(\mathbf{y})$. Certainly on V, $\mathbf{f} \circ \mathbf{g}$ is the identity: as we have just shown, $\mathbf{f}(\mathbf{g}(\mathbf{y})) = \mathbf{y}$. $\square$

We now have our inverse function $\mathbf{g}$. A complete proof requires showing that $\mathbf{g}$ is continuously differentiable. This is done in Appendix A.4.

Example 2.9.5 (Where is f invertible?). Where is the function

---

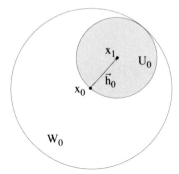

FIGURE 2.9.5.

Newton's method applied to the equation $f_{\mathbf{y}}(\mathbf{x}) = 0$ starting at $\mathbf{x}_0$ converges to a root in U_0. Kantorovitch's theorem tells us this is the unique root in U_0; the inverse function theorem tells us that it is the unique root in all of W_0.

We get the first equality in Equation 2.9.10 by plugging in appropriate values to the definition of $\vec{\mathbf{h}}_0$ given in the statement of Kantorovitch's theorem (Equation 2.7.46):

$$\vec{\mathbf{h}}_0 = -[\mathbf{Df}(\mathbf{a}_0)]^{-1}\mathbf{f}(\mathbf{a}_0).$$

Recall that in Equation 2.9.10 we write $\vec{\mathbf{h}}_0(\mathbf{y})$ rather than $\vec{\mathbf{h}}_0$ because our problem depends on $\mathbf{y}$: we are solving $\mathbf{f_y} = 0$.

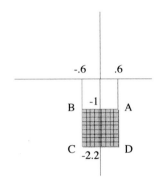

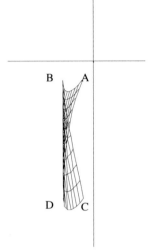

$$\mathbf{f}\begin{pmatrix} x \\ y \end{pmatrix} = \begin{pmatrix} \sin(x+y) \\ x^2 - y^2 \end{pmatrix} \qquad 2.9.12$$

locally invertible? The derivative is

$$\left[\mathbf{Df}\begin{pmatrix} x \\ y \end{pmatrix}\right] = \begin{bmatrix} \cos(x+y) & \cos(x+y) \\ 2x & -2y \end{bmatrix}, \qquad 2.9.13$$

which is invertible if $-2y\cos(x+y) - 2x\cos(x+y) \neq 0$. (Remember the formula for the inverse of a 2×2 matrix.[26]) So $\mathbf{f}$ is locally invertible at all points $\mathbf{f}\begin{pmatrix} x_0 \\ y_0 \end{pmatrix}$ that satisfy $-y \neq x$ and $\cos(x+y) \neq 0$ (i.e., $x+y \neq \pi/2 + k\pi$).   △

Remark. We strongly recommend using a computer to understand the mapping $\mathbf{f} : \mathbb{R}^2 \to \mathbb{R}^2$ of Example 2.9.5 and, more generally, any mapping from $\mathbb{R}^2$ to $\mathbb{R}^2$. (One thing we can say without a computer's help is that the first coordinate of every point in the image of $\mathbf{f}$ cannot be bigger than 1 or less than -1, since the sine function oscillates between -1 and 1. So if we graph the image using x, y coordinates, it will be contained in a band between $x = -1$ and $x = 1$.) Figures 2.9.6 and 2.9.7 show just two examples of regions of the domain of $\mathbf{f}$ and the corresponding region of the image. Figure 2.9.6 shows a region of the image that is folded over; in that region the function has no inverse.   △

FIGURE 2.9.6.

Top: The square $-.6 < x < .6, -2.2 < y < 1$. Bottom: Its image under the mapping $\mathbf{f}$ of Example 2.9.5. Note that the square is folded over itself along the line $x + y = -\pi/2$ (the line from B to D); $\mathbf{f}$ is not invertible in the neighborhood of the square.

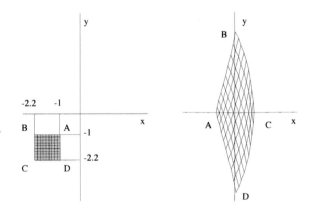

FIGURE 2.9.7. The function $\mathbf{f}$ of Example 2.9.5 maps the region at left to the region at right. In this region, $\mathbf{f}$ is invertible.

Example 2.9.6. Let C_1 be the circle of radius 3 centered at the origin in $\mathbb{R}^2$, and C_2 be the circle of radius 1 centered at $\begin{pmatrix} 10 \\ 0 \end{pmatrix}$. What is the loci of centers of line segments drawn from a point of C_1 to a point of C_2?

$$26 \begin{bmatrix} a & b \\ c & d \end{bmatrix}^{-1} = \frac{1}{ad - bc} \begin{bmatrix} d & -b \\ -c & a \end{bmatrix}.$$

The center of the segment joining

$$\begin{pmatrix} 3\cos\theta \\ 3\sin\theta \end{pmatrix} \in C_1 \qquad \text{to} \qquad \begin{pmatrix} \cos\varphi + 10 \\ \sin\varphi \end{pmatrix} \in C_2 \qquad \text{2.9.14}$$

is the point

$$F\begin{pmatrix} \theta \\ \varphi \end{pmatrix} = \frac{1}{2}\begin{pmatrix} 3\cos\theta + \cos\varphi + 10 \\ 3\sin\theta + \sin\varphi \end{pmatrix}. \qquad \text{2.9.15}$$

We want to find the image of F. A point $\begin{pmatrix} \theta \\ \varphi \end{pmatrix}$ where $\left[\mathbf{D}F\begin{pmatrix} \theta \\ \varphi \end{pmatrix}\right]$ is invertible will certainly be in the *interior* of the image (since points in the neighborhood of that point are also in the image), so the candidates to be in the *boundary* of the image are those points $F\begin{pmatrix} \theta \\ \varphi \end{pmatrix}$ where $\left[\mathbf{D}F\begin{pmatrix} \theta \\ \varphi \end{pmatrix}\right]$ is *not* invertible. Since

$$\det\left[\mathbf{D}F\begin{pmatrix} \theta \\ \varphi \end{pmatrix}\right] = \frac{1}{4}\det\begin{bmatrix} -3\sin\theta & -\sin\varphi \\ 3\cos\theta & \cos\varphi \end{bmatrix} \qquad \text{2.9.16}$$

$$= \frac{-3}{4}(\sin\theta\cos\varphi - \cos\theta\sin\varphi) = \frac{-3}{4}\sin(\theta - \varphi),$$

It follows from the formula for the inverse of a 2×2 matrix that the matrix is not invertible if its determinant is 0; Exercise 1.4.12 shows that the same is true of 3×3 matrices. Theorem 4.8.6 generalizes this to $n \times n$ matrices.

which vanishes when $\theta = \varphi$ and when $\theta = \varphi + \pi$, we see that the candidates for the boundary of the image are the points

$$F\begin{pmatrix} \theta \\ \theta \end{pmatrix} = \begin{pmatrix} 2\cos\theta + 5 \\ 2\sin\theta \end{pmatrix} \quad \text{and} \quad F\begin{pmatrix} \theta \\ \theta - \pi \end{pmatrix} = \begin{pmatrix} \cos\theta + 5 \\ \sin\theta \end{pmatrix}, \qquad \text{2.9.17}$$

Does Example 2.9.6 seem artificial? It's not. Problems like this come up all the time in robotics; the question of knowing where a robot arm can reach is a question just like this.

i.e., the circles of radius 2 and 1 centered at $\mathbf{p} = \begin{pmatrix} 5 \\ 0 \end{pmatrix}$. The only regions whose boundaries are subsets of these sets are the whole disk of radius 2 and the annular region between the two circles. We claim that the image of F is the annular region, since the symmetric of C_2 with respect to $\mathbf{p}$ is the circle of radius 1 centered at the origin, which does not intersect C_1, so $\mathbf{p}$ is not in the image of F. $\triangle$

We say "guaranteed to exist" because the actual domain of the inverse function may be larger than the ball V.

Example 2.9.7 (Quantifying "locally"). Now let's return to the function $\mathbf{f}$ of Example 2.9.5; let's choose a point $\mathbf{x}_0$ where the derivative is invertible and see in how big a neighborhood of $\mathbf{f}(\mathbf{x}_0)$ an inverse function is guaranteed to exist. We know from Example 2.9.5 that the derivative is invertible at $\mathbf{x}_0 = \begin{pmatrix} 0 \\ \pi \end{pmatrix}$. This gives $L = \left[\mathbf{Df}\begin{pmatrix} 0 \\ \pi \end{pmatrix}\right] = \begin{bmatrix} -1 & -1 \\ 0 & -2\pi \end{bmatrix}$, so

$$L^{-1} = \frac{1}{2\pi}\begin{bmatrix} -2\pi & 1 \\ 0 & -1 \end{bmatrix}, \text{ and } |L^{-1}|^2 = \frac{4\pi^2 + 2}{4\pi^2}. \qquad \text{2.9.18}$$

Next we need to compute the Lipschitz ratio M (Equation 2.9.23). We have

$$\left|[\mathbf{Df(u)}] - [\mathbf{Df(v)}]\right|$$

$$= \left|\left[\begin{matrix} \cos(u_1+u_2) - \cos(v_1+v_2) & \cos(u_1+u_2) - \cos(v_1+v_2) \\ 2(u_1-v_1) & 2(v_2-v_1) \end{matrix}\right]\right|$$

For the first inequality of Equation 2.9.19, remember that

$$|\cos a - \cos b| \le |a-b|,$$

and set $a = u_1+u_2$ and $b = v_1+v_2$.

$$\le \left|\left[\begin{matrix} u_1+u_2-v_1-v_2 & u_1+u_2-v_1-v_2 \\ 2(u_1-v_1) & 2(v_2-u_2) \end{matrix}\right]\right|$$

$$= \sqrt{2\big((u_1-v_1)+(u_2-v_2)\big)^2 + 4\big((u_1-v_1)^2 + (u_2-v_2)^2\big)}$$

In going from the first square root to the second, we use

$$(a+b)^2 \le 2(a^2+b^2),$$

setting $a = u_1-v_1$ and $b = u_2-v_2$.

$$= \sqrt{4\big((u_1-v_1)^2+(u_2-v_2)^2\big) + 4\big((u_1-v_1)^2+(u_2-v_2)^2\big)}$$

$$= \sqrt{8}\,|\mathbf{u}-\mathbf{v}|. \qquad\qquad 2.9.19$$

Our Lipschitz ratio M is thus $\sqrt{8} = 2\sqrt{2}$, allowing us to compute R:

Since the domain W of $\mathbf{f}$ is $\mathbb{R}^2$, the value of R in Equation 2.9.20 clearly satisfies the requirement that the ball W_0 with radius $2R|L^{-1}|$ be in W.

$$\frac{1}{2R|L^{-1}|^2} = 2\sqrt{2}, \quad\text{so}\quad R = \frac{4\pi^2}{4\sqrt{2}\,(4\pi^2+2)} \approx 0.16825. \qquad 2.9.20$$

The minimum domain V of our inverse function is a ball with radius ≈ 0.17.

What does this say about actually computing an inverse? For example, since

$$\mathbf{f}\begin{pmatrix} 0 \\ \pi \end{pmatrix} = \begin{pmatrix} 0 \\ -\pi^2 \end{pmatrix}, \quad\text{and}\quad \begin{pmatrix} 0.1 \\ -10 \end{pmatrix}\ \text{is within .17 of}\ \begin{pmatrix} 0 \\ -\pi^2 \end{pmatrix},$$

then the inverse function theorem tells us that by using Newton's method we can solve for $\mathbf{x}$ the equation $\mathbf{f(x)} = \begin{pmatrix} 0.1 \\ -10 \end{pmatrix}$. $\triangle$

The implicit function theorem

We have seen that the inverse function theorem deals with the case where we have n equations in n unknowns. Forgetting the detail, it says that if $U \subset \mathbb{R}^n$ is open, $\mathbf{f} : U \to \mathbb{R}^n$ is differentiable, $\mathbf{f(x_0)} = \mathbf{y_0}$ and $[\mathbf{Df(x_0)}]$ is invertible, then there exists a neighborhood V of $\mathbf{y_0}$ and an inverse function $\mathbf{g} : V \to \mathbb{R}^n$ with $\mathbf{g(y_0)} = \mathbf{x_0}$, and $\mathbf{f} \circ \mathbf{g(y)} = \mathbf{y}$. Near $\begin{pmatrix} \mathbf{x_0} \\ \mathbf{y_0} \end{pmatrix}$, the equation $\mathbf{f(x)} = \mathbf{y}$ (or equivalently, $\mathbf{f(x)} - \mathbf{y} = 0$) expresses $\mathbf{x}$ implicitly as a function of $\mathbf{y}$.

Stated this way, there is no reason why the dimensions of the variables $\mathbf{x}$ and $\mathbf{y}$ should be the same.

Example 2.9.8 (Three variables, one equation). The equation $x^2 + y^2 + z^2 - 1 = 0$ expresses z as an implicit function of $\begin{pmatrix} x \\ y \end{pmatrix}$ near $\begin{pmatrix} 0 \\ 0 \\ 1 \end{pmatrix}$. This implicit function can be made explicit: $z = \sqrt{1 - x^2 - y^2}$; you can solve for z as a function of x and y. $\triangle$

More generally, if we have n equations in $n + m$ variables, we can think of m variables as "known," leaving n equations in the n "unknown" variables, and try to solve them. If a solution exists, then we will have expressed the n unknown variables in terms of the m known variables. In this case, the original equation expresses the n unknown variables implicitly in terms of the others.

If all we want to know is that an implicit function exists on some unspecified neighborhood, then we can streamline the statement of the implicit function theorem; the important question to ask is, *"is the derivative onto?"*

Recall that C^1 means continuously differentiable: differentiable with continuous derivative. We saw (Theorem 1.9.5) that this is equivalent to requiring that all the partial derivatives be continuous. As in the case of the inverse function theorem, Theorem 2.9.9 would *not* be true if we did not require $[\mathbf{DF}(\mathbf{x})]$ to be continuous with respect to $\mathbf{x}$. Exercise 2.9.2 shows what goes wrong in that case. But such functions are pathological; in practice you are unlikely to run into any.

Theorem 2.9.9 is true as stated, but the proof we give requires that the derivative be Lipschitz.

Theorem 2.9.9 (Stripped-down version of the implicit function theorem). *Let U be an open subset of $\mathbb{R}^{n+m}$. Let $\mathbf{F} : U \to \mathbb{R}^n$ be a C^1 mapping such that $\mathbf{F}(\mathbf{c}) = \mathbf{0}$, and such that its derivative, the linear transformation $[\mathbf{DF}(\mathbf{c})]$, is onto. Then the system of linear equations $[\mathbf{DF}(\mathbf{c})](\mathbf{x}) = \mathbf{0}$ has n pivotal variables and m non-pivotal variables, and there exists a neighborhood of $\mathbf{c}$ for which $\mathbf{F} = \mathbf{0}$ implicitly defines the n pivotal variables in terms of the m non-pivotal variables.*

The implicit function theorem thus says that *locally, the mapping behaves like its derivative—i.e., like its linearization.* Since $\mathbf{F}$ goes from a subset of $\mathbb{R}^{n+m}$ to $\mathbb{R}^n$, its derivative goes from $\mathbb{R}^{n+m}$ to $\mathbb{R}^n$. The derivative $[\mathbf{DF}(\mathbf{c})]$ being onto means that it spans $\mathbb{R}^n$. Therefore $[\mathbf{DF}(\mathbf{c})]$ has n pivotal columns and m non-pivotal columns. We are then in the case (2b) of Theorem 2.2.4; we can choose freely the values of the m non-pivotal variables; those values will determine the values of the n pivotal variables. The theorem says that locally, what is true of the derivative of $\mathbf{F}$ is true of $\mathbf{F}$.

The full statement of the implicit function theorem

In Sections 3.1 and 3.2, we will see that the stripped-down version of the implicit function theorem is enough to tell us when an equation defines a smooth curve, surface or higher dimensional analog. But in these days of computations, we often need to compute implicit functions; for those, having a precise bound on the domain is essential. For this we need the full statement.

Note that in the long version of the theorem, we replace the condition that the derivative be continuous by a more demanding condition, requiring that the derivative be Lipschitz. Both conditions are ways of ensuring that the derivative not change too quickly. In exchange for the more demanding hypothesis, we get an explicit domain for the implicit function.

The theorem is long and involved, so we'll give some commentary.

The assumption that we are trying to express the first n variables in terms of the last m is a convenience; in practice the question of what to express in terms of what will depend on the context.

First line, through the line immediately following Equation 2.9.21: Not only is $[\mathbf{DF}(\mathbf{c})]$ is onto, but also the first n columns of $[\mathbf{DF}(\mathbf{c})]$ are pivotal. (Since $\mathbf{F}$ goes from a subset of $\mathbb{R}^{n+m}$ to $\mathbb{R}^n$, so does $[\mathbf{DF}(\mathbf{c})]$. Since the matrix of Equation 2.9.21, formed by the first n columns of that matrix, is invertible, the first n columns of $[\mathbf{DF}(\mathbf{c})]$ are linearly independent, i.e., pivotal, and $[\mathbf{DF}(\mathbf{c})]$ is onto.)

The next sentence: We need the matrix L to be invertible because we will use its inverse in the Lipschitz condition.

Definition of W_0: Here we get precise about neighborhoods.

We represent by $\mathbf{a}$ the first n coordinates of $\mathbf{c}$ and by $\mathbf{b}$ the last m coordinates. For example, if $n = 2$ and $m = 1$, the point $\mathbf{c} \in \mathbb{R}^3$ might be $\begin{pmatrix} 1 \\ 0 \\ 2 \end{pmatrix}$, with $\mathbf{a} = \begin{pmatrix} 1 \\ 0 \end{pmatrix} \in \mathbb{R}^2$, and $\mathbf{b} = 2 \in \mathbb{R}$.

Equation 2.9.23: This Lipschitz condition replaces the requirement in the stripped-down version that the derivative be continuous.

Equation 2.9.24: Here we define the implicit function $\mathbf{g}$.

Theorem 2.9.10 (The implicit function theorem). *Let W be an open neighborhood of $\mathbf{c} = \begin{pmatrix} \mathbf{a} \\ \mathbf{b} \end{pmatrix} \in \mathbb{R}^{n+m}$, and $\mathbf{F} : W \to \mathbb{R}^n$ be differentiable, with $\mathbf{F}(\mathbf{c}) = \mathbf{0}$. Suppose that the $n \times n$ matrix*

$$[D_1\mathbf{F}(\mathbf{c}), \dots, D_n\mathbf{F}(\mathbf{c})], \qquad 2.9.21$$

representing the first n columns of the derivative of $\mathbf{F}$, is invertible.

Then the following matrix, which we denote L, is invertible also:

If it isn't clear why L is invertible, see Exercise 2.3.6.

The $\mathbf{0}$ stands for the $m \times n$ zero matrix; I_m is the $m \times m$ identity matrix. So L is $(n+m) \times (n+m)$. If it weren't square, it would not be invertible.

$$L = \begin{bmatrix} [D_1\mathbf{F}(\mathbf{c}), \dots, D_n\mathbf{F}(\mathbf{c})] & [D_{n+1}\mathbf{F}(\mathbf{c}), \dots, D_m\mathbf{F}(\mathbf{c})] \\ \mathbf{0} & I_m \end{bmatrix}. \qquad 2.9.22$$

Let $W_0 = B_{2R|L^{-1}|}(\mathbf{c}) \subset \mathbb{R}^{n+m}$ be the ball of radius $2R|L^{-1}|$ centered at $\mathbf{c}$. Suppose that $R > 0$ satisfies the following hypotheses:

(1) It is small enough so that $W_0 \subset W$.

(2) In W_0, the derivative satisfies the Lipschitz condition

$$\left| [\mathbf{DF}(\mathbf{u})] - [\mathbf{DF}(\mathbf{v})] \right| \leq \frac{1}{2R|L^{-1}|^2} |\mathbf{u} - \mathbf{v}|. \qquad 2.9.23$$

Then there exists a unique continuously differentiable mapping

Equation 2.9.25, which tells us how to compute the derivative of an implicit function, is important; we will use it often. *What would we do with an implicit function if we didn't know how to differentiate it?*

$$\mathbf{g} : B_R(\mathbf{b}) \to B_{2R|L^{-1}|}(\mathbf{a}) \quad \text{such that} \quad \mathbf{F}\begin{pmatrix} \mathbf{g}(\mathbf{y}) \\ \mathbf{y} \end{pmatrix} = \mathbf{0} \quad \text{for all } \mathbf{y} \in B_R(\mathbf{b}),$$
$$2.9.24$$

and the derivative of the implicit function $\mathbf{g}$ at $\mathbf{b}$ is

$$[\mathbf{Dg}(\mathbf{b})] = -\underbrace{[D_1\mathbf{F}(\mathbf{c}), \dots, D_n\mathbf{F}(\mathbf{c})]}_{\substack{\text{partial deriv. for the} \\ n \text{ pivotal variables}}}^{-1} \underbrace{[D_{n+1}\mathbf{F}(\mathbf{c}), \dots, D_{n+m}\mathbf{F}(\mathbf{c})]}_{\substack{\text{partial deriv. for the} \\ m \text{ non-pivotal variables}}}. \qquad 2.9.25$$

Summary. We assume that we have an n-dimensional (unknown) variable $\mathbf{x}$, an m-dimensional (known) variable $\mathbf{y}$, an equation $\mathbf{F} : \mathbb{R}^{n+m} \to \mathbb{R}^n$, and a point $\begin{pmatrix} \mathbf{a} \\ \mathbf{b} \end{pmatrix}$ such that $\mathbf{F} \begin{pmatrix} \mathbf{a} \\ \mathbf{b} \end{pmatrix} = \mathbf{0}$. We ask whether the equation $\mathbf{F} \begin{pmatrix} \mathbf{x} \\ \mathbf{y} \end{pmatrix} = \mathbf{0}$ expresses $\mathbf{x}$ implicitly in terms of $\mathbf{y}$ near $\begin{pmatrix} \mathbf{a} \\ \mathbf{b} \end{pmatrix}$. The implicit function theorem asserts that this is true if the linearized equation

$$\left[\mathbf{DF} \begin{pmatrix} \mathbf{a} \\ \mathbf{b} \end{pmatrix} \right] \begin{bmatrix} \mathbf{u} \\ \mathbf{v} \end{bmatrix} = \mathbf{0} \qquad 2.9.26$$

Since the range of $\mathbf{F}$ is $\mathbb{R}^n$, saying that $[\mathbf{DF}(\mathbf{c})]$ is onto is the same as saying that it has rank n. Many authors state the implicit function theorem in terms of the rank.

expresses $\mathbf{u}$ implicitly in terms of $\mathbf{v}$, which we know is true if the first n columns of $\left[\mathbf{DF} \begin{pmatrix} \mathbf{a} \\ \mathbf{b} \end{pmatrix} \right]$ are linearly independent. $\triangle$

The theorem is proved in Appendix A.5.

The inverse function theorem is the special case of the implicit function theorem where we have $2n$ variables: the unknown n-dimensional variable $\mathbf{x}$ and the known n-dimensional variable $\mathbf{y}$, and where our original equation is $\mathbf{f}(\mathbf{x}) - \mathbf{y} = \mathbf{0}$; it is the case where we can separate out the $\mathbf{y}$ from $\mathbf{F} \begin{pmatrix} \mathbf{x} \\ \mathbf{y} \end{pmatrix}$.

There is a sneaky way of making the implicit function theorem be a special case of the inverse function theorem; we use this in our proof.

Example 2.9.11 (The unit circle and the implicit function theorem). The unit circle is the set of points $\mathbf{c} = \begin{pmatrix} x \\ y \end{pmatrix}$ such that $F(\mathbf{c}) = 0$ when F is the function $F \begin{pmatrix} x \\ y \end{pmatrix} = x^2 + y^2 - 1$. The function is differentiable, with derivative

$$DF \begin{pmatrix} a \\ b \end{pmatrix} = [2a, \, 2b]. \qquad 2.9.27$$

In this case, the matrix of Equation 2.9.21 is the 1×1 matrix $[2a]$, so requiring it to be invertible simply means requiring $a \neq 0$.

Therefore, if $a \neq 0$, the stripped-down version guarantees that in some neighborhood of $\begin{pmatrix} a \\ b \end{pmatrix}$, the equation $x^2 + y^2 - 1 = 0$ implicitly expresses x as a function of y. (Similarly, if $b \neq 0$, then in some neighborhood of $\begin{pmatrix} a \\ b \end{pmatrix}$ the equation $x^2 + y^2 - 1 = 0$ expresses implicitly y as a function of x.)

Let's see what the strong version of the implicit function theorem says about the domain of this implicit function.

The matrix L of Equation 2.9.22 is

Equation 2.9.28: In the lower right-hand corner of L we have the number 1, not the identity matrix I; our function F goes from $\mathbb{R}^2$ to $\mathbb{R}$, so $n = m = 1$, and the 1×1 identity matrix is the number 1.

$$L = \begin{bmatrix} 2a & 2b \\ 0 & 1 \end{bmatrix}, \quad \text{and} \quad L^{-1} = \frac{1}{2a} \begin{bmatrix} 1 & -2b \\ 0 & 2a \end{bmatrix}. \qquad 2.9.28$$

So we have

$$|L^{-1}| = \frac{1}{2|a|} \sqrt{1 + 4a^2 + 4b^2} = \frac{\sqrt{5}}{2|a|}. \qquad 2.9.29$$

The derivative of F is Lipschitz with Lipschitz ratio 2:

$$\left| \left[\mathbf{D}f \begin{pmatrix} u_1 \\ u_2 \end{pmatrix} \right] - \left[\mathbf{D}F \begin{pmatrix} v_1 \\ v_2 \end{pmatrix} \right] \right| = |[2u_1 - 2v_1, \, 2u_2 - 2v_2]|$$
$$= 2|[u_1 - v_1, \, u_2 - v_2]| \leq 2|\mathbf{u} - \mathbf{v}|, \qquad 2.9.30$$

so (by Equation 2.9.23) we can satisfy condition (2) by choosing an R such that

Equation 2.9.31: Note the way the radius R of the interval around b shrinks, without ever disappearing, as $a \to 0$. At the point $\begin{pmatrix} 0 \\ \pm 1 \end{pmatrix}$, the equation

$$x^2 + y^2 - 1 = 0$$

does not express x in terms of y, but it does express x in terms of y when a is arbitrarily close to 0.

Of course there are two possible x's. One will be found by starting Newton's method at a, the other by starting at $-a$.

In Equation 2.9.33 we write $1/D_1 F$ rather than $(D_1 F)^{-1}$ because $D_1 F$ is a 1×1 matrix, i.e., a number.

$$2 = \frac{1}{2R|L^{-1}|^2}; \quad \text{i.e.,} \quad R = \frac{1}{4|L^{-1}|^2} = \frac{a^2}{5}. \qquad 2.9.31$$

We then see that W_0 is the ball of radius

$$2R|L^{-1}| = \frac{2a^2}{5} \frac{\sqrt 5}{2|a|} = \frac{|a|}{\sqrt 5}; \qquad 2.9.32$$

since W is all of $\mathbb{R}^2$, condition (1) is satisfied.

Therefore, for all $\begin{pmatrix} a \\ b \end{pmatrix}$ when $a \neq 0$, the equation $x^2 + y^2 - 1 = 0$ expresses x (in the interval of radius $|a|/\sqrt 5$ around a) as a function of y (in the interval of radius $a^2/5$ around b).

Of course we don't need the implicit function theorem to understand the unit circle; we already knew that we could write $x = \pm\sqrt{1 - y^2}$. But let's pretend we don't, and go further. The implicit function theorem says that if we know that a point $\begin{pmatrix} a \\ b \end{pmatrix}$ is a root of the equation $x^2 + y^2 - 1 = 0$, then for any y within $a^2/5$ of b, we can find the corresponding x by starting with the guess $x_0 = a$ and applying Newton's method, iterating

$$x_{n+1} = x_n - \frac{F\begin{pmatrix} x_n \\ y \end{pmatrix}}{D_1 F \begin{pmatrix} x_n \\ y \end{pmatrix}} = x_n - \frac{x_n^2 + y^2 - 1}{2x_n}. \quad \triangle \qquad 2.9.33$$

Example 2.9.12 (An implicit function in several variables). In what neighborhood of $\begin{pmatrix} 0 \\ 0 \end{pmatrix}$ do the equations

$$\begin{aligned} x^2 - y &= a \\ y^2 - z &= b \\ z^2 - x &= 0 \end{aligned} \qquad 2.9.34$$

determine $\begin{pmatrix} x \\ y \\ z \end{pmatrix}$ as an implicit function $\mathbf{g}\begin{pmatrix} a \\ b \end{pmatrix}$, with $\mathbf{g}\begin{pmatrix} 0 \\ 0 \end{pmatrix} = \begin{pmatrix} 0 \\ 0 \\ 0 \end{pmatrix}$? Here, $n = 3$, $m = 2$; the relevant function is $\mathbf{F} : \mathbb{R}^5 \to \mathbb{R}^3$, given by

$$\mathbf{F}\begin{pmatrix} x \\ y \\ z \\ a \\ b \end{pmatrix} = \begin{pmatrix} x^2 - y - a \\ y^2 - z - b \\ z^2 - x \end{pmatrix}; \qquad 2.9.35$$

the derivative of $\mathbf{F}$ is $\left[\mathbf{Df}\begin{pmatrix} x \\ y \\ z \\ a \\ b \end{pmatrix} \right] = \begin{bmatrix} 2x & -1 & 0 & -1 & 0 \\ 0 & 2y & -1 & 0 & -1 \\ -1 & 0 & 2z & 0 & 0 \end{bmatrix}, \qquad 2.9.36$

and $M = 2$ is a global Lipschitz constant for this derivative:

$$
\left\| \begin{bmatrix} 2x_1 & -1 & 0 & -1 & 0 \\ 0 & 2y_1 & -1 & 0 & -1 \\ -1 & 0 & 2z_1 & 0 & 0 \end{bmatrix} - \begin{bmatrix} 2x_2 & -1 & 0 & -1 & 0 \\ 0 & 2y_2 & -1 & 0 & -1 \\ -1 & 0 & 2z_2 & 0 & 0 \end{bmatrix} \right\|
$$

$$
= 2\sqrt{(x_1-x_2)^2 + (y_1-y_2)^2 + (z_1-z_2)^2} \le 2 \left| \begin{pmatrix} x_1 \\ y_1 \\ z_1 \\ a_1 \\ b_1 \end{pmatrix} - \begin{pmatrix} x_2 \\ y_2 \\ z_2 \\ a_2 \\ b_2 \end{pmatrix} \right|. \qquad 2.9.37
$$

Setting $x = y = z = 0$ and adding the appropriate two bottom lines, we find that

$$
L = \left[\begin{bmatrix} 0 & -1 & 0 \\ 0 & 0 & -1 \\ -1 & 0 & 0 \\ 0 & 0 & 0 \\ 0 & 0 & 0 \end{bmatrix} \begin{bmatrix} -1 & 0 \\ 0 & -1 \\ 0 & 0 \\ 1 & 0 \\ 0 & 1 \end{bmatrix} \right], \quad L^{-1} = \begin{bmatrix} 0 & 0 & -1 & 0 & 0 \\ -1 & 0 & 0 & -1 & 0 \\ 0 & -1 & 0 & 0 & -1 \\ 0 & 0 & 0 & 1 & 0 \\ 0 & 0 & 0 & 0 & 1 \end{bmatrix}. \qquad 2.9.38
$$

Since the function $\mathbf{F}$ is defined on all of $\mathbb{R}^5$, the first restriction on R is vacuous. The second restriction requires that

$$
\frac{1}{2R|L^{-1}|^2} \ge 2, \quad \text{i.e.,} \quad R \le \frac{1}{28}. \qquad 2.9.39
$$

The $\begin{pmatrix} a \\ b \end{pmatrix}$ of this discussion is the $\mathbf{y}$ of Equation 2.9.24, and the origin here is the $\mathbf{b}$ of that equation.

Thus we can be sure that for any $\begin{pmatrix} a \\ b \end{pmatrix}$ in the ball of radius $1/28$ around the origin (i.e., satisfying $\sqrt{a^2 + b^2} \le 1/28$), there will be a unique solution to Equation 2.9.34 with

$$
\left| \begin{pmatrix} x \\ y \\ z \end{pmatrix} \right| \le \frac{2\sqrt{7}}{28} = \frac{1}{2\sqrt{7}}. \qquad 2.9.40
$$

2.10 EXERCISES FOR CHAPTER TWO

Exercises for Section 2.1:
Row Reduction

2.1.1   (a) Write the following system of linear equations as the multiplication of a matrix by a vector, using the format of Exercise 1.2.2.

$$
3x + y - 4z = 0
$$
$$
2y + z = 4
$$
$$
x - 3y = 1.
$$

(b) Write the same system as a single matrix, using the shorthand notation discussed in Section 2.1.

(c) Write the following system of equations as a single matrix:

$$x_1 - 7x_2 + 2x_3 = 1$$
$$x_1 - 3x_2 = 2$$
$$2x_1 - 2x_2 = -1.$$

2.1.2   Write each of the following systems of equations as a single matrix:

$$
\begin{aligned}
3y - z &= 0 \\
\text{(a)} \quad -2x + y + 2z &= 0\,; \\
x - 5z &= 0
\end{aligned}
\qquad
\begin{aligned}
2x_1 + 3x_2 - x_3 &= 1 \\
\text{(b)} \quad -2x_2 + x_3 &= 2 \\
x_1 - 2x_3 &= -1.
\end{aligned}
$$

2.1.3   Show that the row operation that consists of exchanging two rows is not necessary; one can exchange rows using the other two row operations: (1) multiplying a row by a nonzero number, and (2) adding a multiple of a row onto another row.

2.1.4   Show that any row operation can be undone by another row operation. Note the importance of the word "nonzero" in the algorithm for row reduction.

2.1.5   For each of the four matrices in Example 2.1.7, find (and label) row operations that will bring them to echelon form.

2.1.6   Show that if A is square, and $\widetilde{A}$ is what you get after row reducing A to echelon form, then either $\widetilde{A}$ is the identity, or the last row is a row of zeroes.

2.1.7   Bring the following matrices to echelon form, using row operations.

(a) $\begin{bmatrix} 1 & 2 & 3 \\ 4 & 5 & 6 \end{bmatrix}$ (b) $\begin{bmatrix} 1 & -1 & 1 \\ -1 & 0 & 2 \\ -1 & 1 & 1 \end{bmatrix}$ (c) $\begin{bmatrix} 1 & 2 & 3 & 5 \\ 2 & 3 & 0 & -1 \\ 0 & 1 & 2 & 3 \end{bmatrix}$

(d) $\begin{bmatrix} 1 & 3 & -1 & 4 \\ 1 & 2 & 1 & 2 \\ 3 & 7 & 1 & 9 \end{bmatrix}$ (e) $\begin{bmatrix} 1 & 1 & 1 & 1 \\ 2 & -3 & 3 & 3 \\ 1 & -4 & 2 & 2 \end{bmatrix}$

2.1.8   For Example 2.1.10, analyze precisely where the troublesome errors occur.

In Exercise 2.1.9 we use the following rules: a single addition, multiplication, or division has unit cost; administration (i.e., relabeling entries when switching rows, and comparisons) is free.

2.1.9   In this exercise, we will estimate how expensive it is to solve a system $A\vec{x} = \vec{b}$ of n equations in n unknowns, assuming that there is a unique solution, i.e., that A row reduces to the identity. In particular, we will see that *partial row reduction* and *back substitution* (to be defined below) is roughly a third cheaper than full row reduction.

In the first part, we will show that the number of operations required to row reduce the augmented matrix $[A|\vec{b}]$ is

$$R(n) = n^3 + n^2/2 - n/2.$$

(a) Compute $R(1), R(2)$, and show that this formula is correct when $n = 1$ and 2.

Hint: There will be $n - k + 1$ divisions, $(n - 1)(n - k + 1)$ multiplications and $(n-1)(n-k+1)$ additions.

(b) Suppose that columns $1, \ldots, k - 1$ each contain a pivotal 1, and that all other entries in those columns are 0. Show that you will require another $(2n - 1)(n - k + 1)$ operations for the same to be true of k.

(c) Show that

$$\sum_{k=1}^{n}(2n - 1)(n - k + 1) = n^3 + \frac{n^2}{2} - \frac{n}{2}.$$

$$\begin{bmatrix} 1 & * & * & \cdots & * & \widetilde{b}_1 \\ 0 & 1 & * & \cdots & * & \widetilde{b}_2 \\ \vdots & \vdots & \vdots & \vdots & \vdots & \vdots \\ 0 & 0 & 0 & \cdots & 1 & \widetilde{b}_n \end{bmatrix}$$

Now we will consider an alternative approach, in which we will do all the steps of row reduction, except that we do not make the entries above pivotal 1's be 0. We end up with a matrix of the form at left, where $*$ stands for terms which are whatever they are, usually nonzero. Putting the variables back in, when $n = 3$, our system of equations might be

$$x + 2y - z = 2$$
$$y - 3z = -1$$
$$z = 5, \quad \text{which can be solved by } back \; substitution \text{ as follows:}$$

$$z = 5, \quad y = -1 + 3z = 14, \quad x = 2 - 2y + z = 2 - 28 + 5 = -21.$$

We will show that partial row reduction and back substitution takes

$$Q(n) = \frac{2}{3}n^3 + \frac{3}{2}n^2 - \frac{1}{6}n - 1 \quad \text{operations.}$$

(d) Compute $Q(1), Q(2), Q(3)$. Show that $Q(n) < R(n)$ when $n \geq 3$.

(e) Following the same steps as in part (b), show that the number of operations needed to go from the $(k - 1)$th step to the kth step of partial row reduction is $(n - k + 1)(2n - 2k + 1)$.

(f) Show that

$$\sum_{k=1}^{n}(n - k + 1)(2n - 2k + 1) = \frac{2}{3}n^3 + \frac{1}{2}n^2 - \frac{1}{6}n.$$

(g) Show that the number of operations required by back substitution is $n^2 - 1$.

(h) Compute $Q(n)$.

Exercises for Section 2.2:

Solving Equations with Row Reduction

2.2.1   Rewrite the system of equations in Example 2.2.3 so that y is the first variable, z the second. Now what are the pivotal unknowns?

2.2.2   Predict whether each of the following systems of equations will have a unique solution, no solution, or infinitely many solutions. Solve, using row operations. If your results do not confirm your predictions, can you suggest an explanation for the discrepancy?

(a) $\begin{aligned} 2x + 13y - 3z &= -7 \\ x + y &= 1 \\ x + 7z &= 22 \end{aligned}$ (b) $\begin{aligned} x - 2y - 12z &= 12 \\ 2x + 2y + 2z &= 4 \\ 2x + 3y + 4z &= 3 \end{aligned}$ (c) $\begin{aligned} x + y + z &= 5 \\ x - y - z &= 4 \\ 2x + 6y + 6z &= 12 \end{aligned}$

(d)
$$\begin{aligned} x + 3y + z &= 4 \\ -x - y + z &= -1 \\ 2x + 4y &= 0 \end{aligned}$$

(e) $\begin{aligned} x + 2y + z - 4w + v &= 0 \\ x + 2y - z + 2w - v &= 0 \\ 2x + 4y + z - 5w + v &= 0 \\ x + 2y + 3z - 10w + 2v &= 0 \end{aligned}$

2.2.3   Confirm the solution for Exercise 2.2.2 (e), without using row reduction.

2.2.4   Compose a system of $(n - 1)$ equations in n unknowns, in which $\widetilde{\mathbf{b}}$ contains a pivotal 1.

2.2.5   On how many parameters does the family of solutions for Exercise 2.2.2 (e) depend?

2.2.6   Symbolically row reduce the system of linear equations

$$x + y + 2z = 1$$
$$x - y + az = b$$
$$2x - bz = 0.$$

(a) For what values of a, b does the system have a unique solution? Infinitely many solutions? No solutions?

(b) Which of the possibilities above correspond to open subsets of the (a, b)-plane? Closed subsets? Neither?

For example, for $k = 2$ we are asking about the system of equations

$$\begin{bmatrix} 1 & -1 \\ -2 & 2 \\ 0 & 2 \\ 2 & -6 \end{bmatrix} \begin{bmatrix} x_1 \\ x_2 \end{bmatrix} = \begin{bmatrix} 3 \\ -6 \\ 5 \\ -4 \end{bmatrix}.$$

2.2.7   (a) Row reduce the matrix

$$A = \begin{bmatrix} 1 & -1 & 3 & 0 & -2 \\ -2 & 2 & -6 & 0 & 4 \\ 0 & 2 & 5 & -1 & 0 \\ 2 & -6 & -4 & 2 & -4 \end{bmatrix}.$$

(b) Let $\vec{\mathbf{v}}_k, \ k = 1, \dots, 5$ be the columns of A. What can you say about the systems of equations

$$[\vec{\mathbf{v}}_1, \dots, \vec{\mathbf{v}}_m] \begin{bmatrix} x_1 \\ \vdots \\ x_m \end{bmatrix} = \vec{\mathbf{v}}_{m+1}$$

for $m = 1, 2, 3, 4$.

*2.2.8   Given the system of equations

$$x_1 - x_2 - x_3 - 3x_4 + x_5 = 1$$
$$x_1 + x_2 - 5x_3 - x_4 + 7x_5 = 2$$
$$-x_1 + 2x_2 + 2x_3 + 2x_4 + x_5 = 0$$
$$-2x_1 + 5x_2 - 4x_3 + 9x_4 + 7x_5 = \beta,$$

for what values of β does the system have solutions? When solutions exist, give values of the pivotal variables in terms of the non-pivotal variables.

Exercises for Section 2.3:
Inverses and
Elementary Matrices

$$A = \begin{bmatrix} 2 & 1 & 3 & a \\ 1 & -1 & 1 & b \\ 1 & 1 & 2 & c \end{bmatrix}$$

$$B = \begin{bmatrix} 2 & 1 & 3 \\ 1 & -1 & 1 \\ 1 & 1 & 2 \end{bmatrix}$$

Matrices for Exercise 2.3.2

$$C = \begin{bmatrix} 1 & -2 & 4 \\ 0 & 5 & -5 \\ 3 & a & b \end{bmatrix}$$

Matrix for Exercise 2.3.5

$$\begin{bmatrix} 0 & 0 & 0 \\ 0 & 0 & 0 \end{bmatrix}$$

Example of a "0 matrix."

$$A = \begin{bmatrix} 1 & -6 & 3 \\ 2 & -7 & 3 \\ 4 & -12 & 5 \end{bmatrix}$$

Matrix for Exercise 2.3.7

2.3.1   (a) Derive from Theorem 2.2.4 the fact that only square matrices have inverses.

(b) Construct an example where $AB = I$, but $BA \neq I$.

2.3.2   (a) Row reduce symbolically the matrix A at left.

(b) Compute the inverse of the matrix B at left.

(c) What is the relation between the answers in parts (a) and (b)?

2.3.3   Use A^{-1} to solve the system of Example 2.2.10.

2.3.4   Find the inverse, or show it does not exist, for each of the following matrices:

(a) $\begin{bmatrix} 1 & -5 \\ 9 & 9 \end{bmatrix}$;   (b) $\begin{bmatrix} 1 & 3 \\ 3 & 9 \end{bmatrix}$;   (c) $\begin{bmatrix} 1 & 2 & 3 \\ 2 & 3 & 0 \\ 0 & 1 & 2 \end{bmatrix}$;   (d) $\begin{bmatrix} 1 & 2 \\ 0 & 3 \\ 1 & 0 \end{bmatrix}$

(e) $\begin{bmatrix} 3 & 2 & -1 \\ 0 & 1 & 1 \\ 8 & 3 & 9 \end{bmatrix}$;   (f) $\begin{bmatrix} 1 & 0 & 1 \\ 2 & 1 & -1 \\ 1 & 1 & -1 \end{bmatrix}$;   (g) $\begin{bmatrix} 1 & 1 & 1 & 1 \\ 1 & 2 & 3 & 4 \\ 1 & 3 & 6 & 10 \\ 1 & 4 & 10 & 20 \end{bmatrix}$.

2.3.5   (a) For what values of a and b is the matrix C at left invertible?

(b) For those values, compute the inverse.

2.3.6   (a) Show that if A is an invertible $n \times n$ matrix, B is an invertible $m \times m$ matrix, and C is any $n \times m$ matrix, then the $(n+m) \times (n+m)$ matrix

$$\begin{bmatrix} A & C \\ 0 & B \end{bmatrix},$$ where 0 stands for the $m \times n$ 0 matrix, is invertible.

(b) Find a formula for the inverse.

2.3.7   For the matrix A at left, (a) compute the matrix product AA;

(b) use the result in (a) to solve the system of equations

$$x - 6y + 3z = 5$$
$$2x - 7y + 3z = 7$$
$$4x - 12y + 5z = 11.$$

In both cases, remember that the elementary matrix goes on the left of the matrix to be multiplied.

2.3.8   (a) Confirm that multiplying a matrix by a type 2 elementary matrix as described in Definition 2.3.5 is equivalent to adding rows or multiples of rows.

(b) Confirm that multiplying a matrix by a type 3 elementary matrix is equivalent to switching rows.

2.3.9   (a) Predict the effect of multiplying the matrix $\begin{bmatrix} 1 & 0 & -1 \\ 2 & 1 & 1 \\ 0 & 1 & 2 \end{bmatrix}$ by each of the elementary matrices, with the elementary matrix on the left.

$$(1) \begin{bmatrix} 1 & 0 & 0 \\ 0 & 3 & 0 \\ 0 & 0 & 1 \end{bmatrix} \quad (2) \begin{bmatrix} 1 & 0 & 0 \\ 0 & 0 & 1 \\ 0 & 1 & 0 \end{bmatrix} \quad (3) \begin{bmatrix} 1 & 0 & 0 \\ 0 & 1 & 0 \\ 2 & 0 & 1 \end{bmatrix}.$$

(b) Confirm your answer by carrying out the multiplication.

(c) Redo part (a) and part (b) placing the elementary matrix on the right.

$$\begin{bmatrix} 1 & 2 & 0 & 1 \\ 1 & 1 & 3 & 3 \\ 0 & 1 & 0 & 1 \\ 2 & 1 & 1 & 3 \end{bmatrix}$$

The matrix A of Exercise 2.3.10

2.3.10   When A is the matrix at left, multiplication by what elementary matrix corresponds to:

(a) Exchanging the first and second rows of A?

(b) Multiplying the fourth row of A by 3?

(c) Adding 2 times the third row of A to the first row of A?

2.3.11   (a) Predict the effect of multiplying the matrix B at left by each of the matrices. (The matrices below will be on the left.)

$$\begin{bmatrix} 1 & 3 & -2 \\ 0 & 2 & 3 \\ 1 & 0 & 4 \end{bmatrix}$$

The matrix B of Exercise 2.3.11

$$(1) \begin{bmatrix} 1 & 0 & -3 \\ 0 & 1 & 0 \\ 0 & 0 & 1 \end{bmatrix} \quad (2) \begin{bmatrix} 1 & 0 & 0 \\ 0 & 2 & 0 \\ 0 & 0 & 1 \end{bmatrix} \quad (3) \begin{bmatrix} 1 & 0 & 0 \\ 0 & 0 & 1 \\ 0 & 1 & 0 \end{bmatrix}.$$

(b) Verify your prediction by carrying out the multiplication.

2.3.12   Show that column operations (Definition 2.1.11) can be achieved by multiplication on the right by an elementary matrix of type 1, 2, and 3 respectively.

2.3.13   Prove Proposition 2.3.7.

2.3.14   Show that it is possible to switch rows using multiplication by only the first two types of elementary matrices, as described in Definition 2.3.5.

2.3.15   Row reduce the matrices in Exercise 2.1.7, using elementary matrices.

Exercises for Section 2.4:

Linear Independence

2.4.1   Show that $\operatorname{Sp}(\vec{v}_1, \ldots, \vec{v}_k)$ is a subspace of $\mathbb{R}^n$ and is the smallest subspace containing $\vec{v}_1, \ldots, \vec{v}_k$.

2.4.2   Show that the following two statements are equivalent to saying that a set of vectors $\vec{v}_1, \ldots, \vec{v}_k$ is linearly independent:

(a) The only way to write the zero vector $\vec{0}$ as a linear combination of the $\vec{v}_i$ is to use only zero coefficients.

(b) None of the $\vec{v}_i$ is a linear combination of the others.

2.4.3   Show that the standard basis vectors $\vec{e}_1, \ldots, \vec{e}_k$ are linearly independent.

2.4.4   (a) For vectors in $\mathbb{R}^2$, prove that the length squared of a vector is the sum of the squares of its coordinates, with respect to any orthonormal basis.

(b) Prove the same thing for vectors in $\mathbb{R}^3$.

(c) Repeat for $\mathbb{R}^n$: i.e., show that if $\vec{v}_1, \ldots, \vec{v}_n$ and $\vec{w}_1, \ldots, \vec{w}_n$ are two orthonormal bases, and

$$a_1\vec{v}_1 + \cdots + a_n\vec{v}_n = b_1\vec{w}_1 + \cdots + b_n\vec{w}_n, \quad \text{then} \quad a_1^2 + \cdots + a_n^2 = b_1^2 + \cdots + b_n^2.$$

2.4.5   Consider the following vectors: $\begin{bmatrix} 1 \\ 1 \\ 0 \end{bmatrix}$, $\begin{bmatrix} 1 \\ 2 \\ 1 \end{bmatrix}$, and $\begin{bmatrix} 0 \\ 1 \\ \alpha \end{bmatrix}$.

(a) For what values of α are these three vectors linearly dependent?

(b) Show that for each such α the three vectors lie in the same plane, and give an equation of the plane.

2.4.6   (a) Let $\vec{v}_1, \ldots, \vec{v}_k$ be vectors in $\mathbb{R}^n$. What does it mean to say that they are linearly independent? That they span $\mathbb{R}^n$? That they form a basis of $\mathbb{R}^n$?

Recall that Mat (n, m) denotes the set of $n \times m$ matrices.

(b) Let $A = \begin{bmatrix} 1 & 2 \\ 2 & 1 \end{bmatrix}$. Are the elements I, A, A^2, A^3 linearly independent in Mat $(2, 2)$? What is the dimension of the subspace $V \subset$ Mat $(2, 2)$ that they span?

(c) Show that the set W of matrices $B \in$ Mat $(2, 2)$ that satisfy $AB = BA$ is a subspace of Mat $(2, 2)$. What is its dimension?

(d) Show that $V \subset W$. Are they equal?

2.4.7   Finish the proof that the three conditions in Definition 2.4.13 are equivalent: show that (2) implies (3) and (3) implies (1).

2.4.8   Let $\vec{v}_1 = \begin{bmatrix} 1 \\ 1 \end{bmatrix}$ and $\vec{v}_2 = \begin{bmatrix} 1 \\ 3 \end{bmatrix}$. Let x and y be the coordinates with respect to the standard basis $\{\vec{e}_1, \vec{e}_2\}$ and let u and v be the coordinates with respect to $\{\vec{v}_1, \vec{v}_2\}$. Write the equations to translate from (x, y) to (u, v) and back. Use these equations to write the vector $\begin{bmatrix} 3 \\ -5 \end{bmatrix}$ in terms of $\vec{v}_1$ and $\vec{v}_2$.

2.4.9   Let $\vec{v}_1, \dots, \vec{v}_n$ be vectors in $\mathbb{R}^m$, and let $P_{\{\vec{v}\}} \colon \mathbb{R}^n \to \mathbb{R}^m$ be given by

$$P_{\{\vec{v}\}} \begin{pmatrix} a_1 \\ \vdots \\ a_n \end{pmatrix} = \sum a_i \vec{v}_i.$$

(a) Show that $\vec{v}_1, \dots, \vec{v}_n$ are linearly independent if and only if the map $P_{\{\mathbf{v}\}}$ is one to one.

(b) Show that $\vec{v}_1, \dots, \vec{v}_n$ span $\mathbb{R}^m$ if and only if $P_{\{\mathbf{v}\}}$ is onto.

(c) Show that $\vec{v}_1, \dots, \vec{v}_n$ is a basis of $\mathbb{R}^m$ if and only if $P_{\{\mathbf{v}\}}$ is one to one and onto.

2.4.10   The object of this exercise is to show that a matrix A has a unique row echelon form $\widetilde{A}$: i.e., that all sequences of row operations that turn A into a matrix in row echelon form produce the same matrix, $\widetilde{A}$. This is the harder part of Theorem 2.1.8.

We will do this by saying explicitly what this matrix is. Let A be an $n \times m$ matrix with columns $\vec{a}_1, \dots, \vec{a}_m$. Make the matrix $\widetilde{A} = [\widetilde{\vec{a}}_1, \dots, \widetilde{\vec{a}}_m]$ as follows:

Let $i_1 < \cdots < i_k$ be the indices of the columns that are not linear combinations of the earlier columns; we will refer to these as the *marked columns*.

Set $\widetilde{\vec{a}}_{i_j} = \vec{e}_j$; this defines the marked columns of $\widetilde{A}$.

If $\vec{a}_l$ is a linear combination of the earlier columns, let $j(l)$ be the largest marked index such that $j(l) < l$, and write

$$\vec{a}_l = \sum_{j=1}^{j(l)} \alpha_j \vec{a}_{i_j}, \qquad \text{setting} \qquad \widetilde{\vec{a}}_l = \begin{bmatrix} \alpha_1 \\ \vdots \\ \alpha_{j(l)} \\ 0 \\ \vdots \\ 0 \end{bmatrix}. \qquad 2.4.10$$

Hint for Exercise 2.4.10, part (b): Work by induction on the number m of columns. First check that it is true if $m = 1$. Next, suppose it is true for $m - 1$, and view an $n \times m$ matrix as an augmented matrix, designed to solve n equations in $m - 1$ unknowns.

After row reduction there is a pivotal 1 in the last column exactly if $\vec{a}_m$ is not in the span of $\vec{a}_1, \dots, \vec{a}_{m-1}$, and otherwise the entries of the last column satisfy Equation 2.4.10.

(When figures and equations are numbered in the exercises, they are given the number of the exercise to which they pertain.)

This defines the unmarked columns of $\widetilde{A}$.

(a) Show that $\widetilde{A}$ is in row echelon form.

(b) Show that if you row reduce A, you get $\widetilde{A}$.

2.4.11   Let $\vec{v}_1, \dots, \vec{v}_k$ be vectors in $\mathbb{R}^n$, and set $V = [\vec{v}_1, \dots, \vec{v}_k]$.

(a) Show that the set $\vec{v}_1, \dots, \vec{v}_k$ is orthogonal if and only if $V^\top V$ is diagonal.

(b) Show that the set $\vec{v}_1, \dots, \vec{v}_k$ is orthonormal if and only if $V^\top V = I_k$.

Exercise 2.4.12 says that any linearly independent set can be extended to form a basis. In French treatments of linear algebra, this is called the *theorem of the incomplete basis*; it plus induction can be used to prove all the theorems of linear algebra in Chapter 2.

2.4.12   (a) Let V be a finite-dimensional vector space, and $\vec{v}_1, \dots, \vec{v}_k \in V$ be linearly independent vectors. Show that there exist $\vec{v}_{k+1}, \dots, \vec{v}_n$ such that $\vec{v}_1, \dots, \vec{v}_n \in V$ is a basis of V.

(b) Let V be a finite-dimensional vector space, and $\vec{v}_1, \dots, \vec{v}_k \in V$ be a set of vectors that spans V. Show that there exists a subset $i_1, i_2, \dots, i_m$ of $\{1, 2, \dots, k\}$ such that $\vec{v}_{i_1}, \dots, \vec{v}_{i_m}$ is a basis of V.

Exercises for Section 2.5:
Kernels and Images

(a) $\begin{bmatrix} 1 & 1 & 3 \\ 2 & 2 & 6 \end{bmatrix}$

(b) $\begin{bmatrix} 1 & 2 & 3 \\ -1 & 1 & 1 \\ -1 & 4 & 5 \end{bmatrix}$

(c) $\begin{bmatrix} 1 & 1 & 1 \\ 1 & 2 & 3 \\ 2 & 3 & 4 \end{bmatrix}$

Matrices for Exercise 2.5.2.

2.5.1   Prove that if $T : \mathbb{R}^n \to \mathbb{R}^m$ is a linear transformation, then the kernel of T is a subspace of $\mathbb{R}^n$, and the image of T is a subspace of $\mathbb{R}^m$.

2.5.2   For each of the matrices at left, find a basis for the kernel and a basis for the image, using Theorems 2.5.5 and 2.5.7.

2.5.3   True or false? (Justify your answer). Let $f : \mathbb{R}^m \to \mathbb{R}^k$ and $g : \mathbb{R}^n \to \mathbb{R}^m$ be linear transformations. Then

$$f \circ g = 0 \quad \text{implies} \quad \text{Img } g = \ker f.$$

2.5.4   Let P_2 be the space of polynomials of degree ≤ 2, identified with $\mathbb{R}^3$ by identifying $a + bx + cx^2$ to $\begin{pmatrix} a \\ b \\ c \end{pmatrix}$.

(a) Write the matrix of the linear transformation $T : P_2 \to P_2$ given by

$$(T(p))(x) = xp'(x) + x^2 p''(x).$$

(b) Find a basis for the image and the kernel of T.

2.5.5   (a) Let P_k be the space of polynomials of degree $\le k$. Suppose $T : P_k \to \mathbb{R}^{k+1}$ is a linear transformation. What relation is there between the dimension of the image of T and the dimension of the kernel of T?

$A = \begin{bmatrix} 1 & b \\ a & 2 \end{bmatrix}$

$B = \begin{bmatrix} 1 & 2 & a \\ a & b & a \\ b & b & a \end{bmatrix}$

Matrices for Exercise 2.5.6.

(b) Consider the mapping $T_k : P_k \to \mathbb{R}^{k+1}$ given by $T_k(p) = \begin{bmatrix} p(0) \\ p(1) \\ \vdots \\ p(k) \end{bmatrix}$. What is the matrix of T_2, where P_2 is identified to $\mathbb{R}^3$ by identifying $a + bx + cx^2$ to $\begin{pmatrix} a \\ b \\ c \end{pmatrix}$?

(c) What is the kernel of T_k?

(d) Show that there exist numbers $c_0, \ldots, c_k$ such that

$$\int_0^n p(t)\,dt = \sum_{i=0}^k c_i p(i) \quad \text{for all polynomials } p \in P_k.$$

$A = \begin{bmatrix} 1 & 1 & 3 & 6 & 2 \\ 2 & -1 & 0 & 4 & 1 \\ 4 & 1 & 6 & 16 & 5 \end{bmatrix}$

$B = \begin{bmatrix} 2 & 1 & 3 & 6 & 2 \\ 2 & -1 & 0 & 4 & 1 \end{bmatrix}$

Matrices for Exercise 2.5.7

2.5.6   Make a sketch, in the (a, b)-plane, of the sets where the kernels of the matrices at left have kernels of dimension $0, 1, 2, \ldots$. Indicate on the same sketch the dimensions of the images.

2.5.7   For the matrices A and B at left, find a basis for the image and the kernel, and verify that the dimension formula is true.

2.5.8   Let P be the space of polynomials of degree at most 2 in the two variables x, y, which we will identify to $\mathbb{R}^6$ by identifying

$$a_1 + a_2 x + a_3 y + a_4 x^2 + a_5 xy + a_6 y^2 \quad \text{with} \quad \begin{bmatrix} a_1 \\ \vdots \\ a_6 \end{bmatrix}.$$

(a) What are the matrices of the linear transformations $S, T : P \to P$ given by

$$S(p)\begin{pmatrix} x \\ y \end{pmatrix} = x D_1 p \begin{pmatrix} x \\ y \end{pmatrix} \quad \text{and} \quad T(p)\begin{pmatrix} x \\ y \end{pmatrix} = y D_2 p \begin{pmatrix} x \\ y \end{pmatrix}?$$

(b) What are the kernel and the image of of the linear transformation

$$p \mapsto 2p - S(p) - T(p)?$$

2.5.9   Let $a_1, \ldots, a_k, b_1, \ldots, b_k$ be any $2k$ numbers. Show that there exists a unique polynomial p of degree at most $2k - 1$ such that

$$p(i) = a_i, \quad p'(i) = b_i$$

for all integers i with $1 \leq i \leq k$. In other words, show that the values of p and p' at $1, \ldots, k$ determine p. Hint: you should use the fact that a polynomial p of degree d such that $p(i) = p'(i) = 0$ can be written $p(x) = (x - i)^2 q(x)$ for some polynomial q of degree $d - 2$.

2.5.10   Decompose the following into partial fractions, as requested, being explicit in each case about the system of linear equations involved and showing that its matrix is invertible:

(a) Write

$$\frac{x + x^2}{(x + 1)(x + 2)(x + 3)} \quad \text{as} \quad \frac{A}{x + 1} + \frac{B}{x + 2} + \frac{C}{x + 3}.$$

(b) Write

$$\frac{x + x^3}{(x + 1)^2 (x - 1)^3} \quad \text{as} \quad \frac{Ax + B}{(x + 1)^2} + \frac{Cx^2 + Dx + F}{(x - 1)^3}.$$

2.5.11   (a) For what value of a can you *not* write

$$\frac{x - 1}{(x + 1)(x^2 + ax + 5)} = \frac{A_0}{x + 1} + \frac{B_1 x + B_0}{x^2 + ax + 5}?$$

(b) Why does this not contradict Proposition 2.5.15?

2.5.12   (a) Let $f(x) = x + Ax^2 + Bx^3$. Find a polynomial $g(x) = x + \alpha x^2 + \beta x^3$ such that $g(f(x)) - x$ is a polynomial starting with terms of degree 4.

(b) Show that if

$$f(x) = x + \sum_{i=2}^{k} a_i x^i \quad \text{is a polynomial, then there exists a unique polynomial}$$

$$g(x) = x + \sum_{i=2}^{k} b_i x^i \quad \text{with } g \circ f(x) = x + x^{k+1} p(x) \text{ for some polynomial } p.$$

2.5.13 A square $n \times n$ matrix P such that $P^2 = P$ is called a *projector*.

(a) Show that P is a projector if and only if $I - P$ is a projector. Show that if P is invertible, then P is the identity.

(b) Let $V_1 = \text{Img } P$ and $V_2 = \ker P$. Show that any vector $\vec{v} \in \mathbb{R}^n$ can be written uniquely $\vec{v} = \vec{v}_1 + \vec{v}_2$ with $\vec{v}_1 \in V_1$ and $\vec{v}_2 \in V_2$. Hint: $\vec{v} = P(\vec{v}) + (\vec{v} - P(\vec{v}))$.

(c) Show that there exists a basis $\vec{v}_1, \ldots, \vec{v}_n$ of $\mathbb{R}^n$ and a number $k \leq n$ such that $P(\vec{v}_1) = \vec{v}_1, \ldots, \quad P(\vec{v}_k) = \vec{v}_k, P(\vec{v}_{k+1}) = 0, \ldots, \quad P(\vec{v}_n) = 0$.

(d) Show that, if P_1 and P_2 are projectors such that $P_1 P_2 = 0$, then $Q = P_1 + P_2 - (P_2 P_1)$ is a projector, $\ker Q = \ker P_1 \cap \ker P_2$, and the image of Q is the space spanned by the image of P_1 and the image of P_2.

The polynomial p which Exercise 2.5.16 constructs is called the *Lagrange interpolation polynomial*, which "interpolates" between the assigned values.

Hint for Exercise 2.5.16: Consider the map from the space of P_n of polynomials of degree n to $\mathbb{R}^{n+1}$ given by

$$p \mapsto \begin{bmatrix} p(x_0) \\ \vdots \\ p(x_n) \end{bmatrix}.$$

You need to show that this map is onto; by Corollary 2.5.11 it is enough to show that its kernel is $\{0\}$.

2.5.14 Show that if A and B are $n \times n$ matrices, and AB is invertible, then A and B are invertible.

*2.5.15 Let $T_1, T_2 : \mathbb{R}^n \to \mathbb{R}^n$ be linear transformations.

(a) Show that there exists $S : \mathbb{R}^n \to \mathbb{R}^n$ such that $T_1 = S \circ T_2$ if and only if $\ker T_2 \subset \ker T_1$.

(b) Show that there exists $S : \mathbb{R}^n \to \mathbb{R}^n$ such that $T_1 = T_2 \circ S$ if and only if $\text{Img } T_1 \subset \text{Img } T_2$.

*2.5.16 (a) Find a polynomial $p(x) = a + bx + cx^2$ of degree 2 such that

$$p(0) = 1, \quad p(1) = 4, \quad \text{and} \quad p(3) = -2.$$

(b) Show that if $x_0, \ldots, x_n$ are $n + 1$ distinct points in $\mathbb{R}$, and $a_0, \ldots, a_n$ are any numbers, there exists a unique polynomial of degree n such that $p(x_i) = a_i$ for each $i = 0, \ldots, n$.

(c) Let the x_i and a_i be as above, and let $b_0, \ldots, b_n$ be some further set of numbers. Find a number k such that there exists a unique polynomial of degree k with

$$p(x_i) = a_i \quad \text{and} \quad p'(x_i) = b_i \quad \text{for all } i = 0, \ldots, n.$$

*2.5.17 This exercise gives a proof of *Bezout's theorem*. Let p_1 and p_2 be polynomials of degree k_1 and k_2 respectively, and consider the mapping

$$T : (q_1, q_2) \to p_1 q_1 + p_2 q_2,$$

where q_1 and q_2 are polynomials of degrees $k_2 - 1$ and $k_1 - 1$ respectively, so that $p_1q_1 + p_2q_2$ is of degree$\leq k_1 + k_2 - 1$.

Note that the space of such (q_1, q_2) is of dimension $k_1 + k_2$, and the space of polynomials of degree $k_1 + k_2 - 1$ is also of dimension $k_1 + k_2$.

(a) Show that ker $T = \{0\}$ if and only if p_1 and p_2 are *relatively prime* (have no common factors).

(b) Use Corollary 2.5.11 to show that if p_1, p_2 are relatively prime, then there exist unique q_1 and q_2 as above such that

$$p_1q_1 + p_2q_2 = 1. \qquad\qquad \text{(Bezout's theorem)}$$

Exercises for Section 2.6:

Abstract Vector Spaces

2.6.1   Show that the space $\mathcal{C}(0, 1)$ of continuous real-valued functions $f(x)$ defined for $0 < x < 1$ (Example 2.6.2) satisfies all eight requirements for a vector space.

2.6.2   Show that the transformation $T : \mathcal{C}^2(\mathbb{R}) \to \mathcal{C}(\mathbb{R})$ given by the formula

$$\big(T(f)\big)(x) = (x^2 + 1)f''(x) - xf'(x) + 2f(x)$$

of Example 2.6.7 is a linear transformation.

2.6.3   Show that in a vector space of dimension n, more than n vectors are never linearly independent, and fewer than n vectors never span.

2.6.4   Denote by $\mathcal{L}\big(\mathrm{Mat}\,(n, n), \mathrm{Mat}\,(n, n)\big)$ the space of linear transformations from $\mathrm{Mat}\,(n, n)$ to $\mathrm{Mat}\,(n, n)$.

(a) Show that $\mathcal{L}\big(\mathrm{Mat}\,(n, n), \mathrm{Mat}\,(n, n)\big)$ is a vector space, and that it is finite dimensional. What is its dimension?

(b) Prove that for any $A \in \mathrm{Mat}\,(n, n)$, the transformations

$$L_A,\, R_A : \mathrm{Mat}\,(n, n) \to \mathrm{Mat}\,(n, n) \qquad \text{given by}$$
$$L_A(B) = AB, \quad R_A(B) = BA$$

are linear transformations.

(c) What is the dimension of the subspace of transformations of the form L_A, R_A?

(d) Show that there are linear transformations $T : \mathrm{Mat}\,(2, 2) \to \mathrm{Mat}\,(2, 2)$ that cannot be written as $L_A + R_B$. Can you find an explicit one?

2.6.5   (a) Let V be a vector space. When is a subset $W \subset V$ a subspace of V?

Note: To show that a space is not a vector space, you will need to show that it is not $\{0\}$.

(b) Let V be the vector space of C^1 functions on $(0, 1)$. Which of the following are subspaces of V:

(i) $\{f \in V \mid f(x) = f'(x) + 1\}$;

(ii) $\{f \in V \mid f(x) = xf'(x)\}$;

(iii) $\{f \in V \mid f(x) = (f'(x))^2\}$.

2.6.6 Let $V, W \subset \mathbb{R}^n$ be two subspaces.

(a) Show that $V \cap W$ is a subspace of $\mathbb{R}^n$.

(b) Show that if $V \cup W$ is a subspace of $\mathbb{R}^n$, then either $V \subset W$ or $W \subset V$.

2.6.7 Let P_2 be the space of polynomials of degree at most two, identified to $\mathbb{R}^3$ via the coefficients; i.e.,

$$p(x) = a + bx + cx^2 \in P_2 \quad \text{is identified to} \quad \begin{pmatrix} a \\ b \\ c \end{pmatrix}.$$

Consider the mapping $T : P_2 \to P_2$ given by

$$T(p)(x) = (x^2 + 1)p''(x) - xp'(x) + 2p(x).$$

(a) Verify that T is linear, i.e., that $T(ap_1 + bp_2) = aT(p_1) + bT(p_2)$.

(b) Choose the basis of P_2 consisting of the polynomials $p_1(x) = 1, p_2(x) = x, p_3(x) = x^2$. Denote $\Phi_{\{p\}} : \mathbb{R}^3 \to P_2$ the corresponding concrete-to-abstract linear transformation. Show that the matrix of

$$\Phi_{\{p\}}^{-1} \circ T \circ \Phi_{\{p\}} \quad \text{is} \quad \begin{bmatrix} 2 & 0 & 2 \\ 0 & 1 & 0 \\ 0 & 0 & 2 \end{bmatrix}.$$

(c) Using the basis $1, x, x^2, \ldots, x^n$, compute the matrices of the same differential operator T, viewed as an operator from P_3 to P_3, from P_4 to P_4, $\ldots$, P_n to P_n (polynomials of degree at most $3, 4, \ldots, n$).

2.6.8 Suppose we use the same operator $T : P_2 \to P_2$ as in Exercise 2.6.7, but choose instead to work with the basis

$$q_1(x) = x^2, \quad q_2(x) = x^2 + x, \quad q_3(x) = x^2 + x + 1.$$

Now what is the matrix $\Phi_{\{q\}}^{-1} \circ T \circ \Phi_{\{q\}}$?

Exercises for Section 2.7:
Newton's Method

2.7.1 (a) What happens if you compute $\sqrt{b}$ by Newton's method, i.e., by setting

$$a_{n+1} = \frac{1}{2}\left(a_n + \frac{b}{a_n}\right), \quad \text{starting with } a_0 < 0?$$

(b) What happens if you compute $\sqrt[3]{b}$ by Newton's method, with $b > 0$, starting with $a_0 < 0$?

2.7.2 Show (a) that the function $|x|$ is Lipschitz with Lipschitz ratio 1 and (b) that the function $\sqrt{|x|}$ is not Lipschitz.

2.7.3 (a) Find the formula $a_{n+1} = g(a_n)$ to compute the kth root of a number by Newton's method.

(b) Interpret this formula as a weighted average.

2.7.4   (a) Compute by hand the number $9^{1/3}$ to six decimals, using Newton's method, starting at $a_0 = 2$.

(b) Find the relevant quantities h_0, a_1, M of Kantorovitch's theorem in this case.

(c) Prove that Newton's method does converge. (You are allowed to use Kantorovitch's theorem, of course.)

2.7.5   (a) Find a global Lipschitz ratio for the derivative of the mapping $F : \mathbb{R}^2 \to \mathbb{R}^2$ given by

$$F \begin{pmatrix} x \\ y \end{pmatrix} = \begin{pmatrix} x^2 - y - 12 \\ y^2 - x - 11 \end{pmatrix}.$$

(b) Do one step of Newton's method to solve $F \begin{pmatrix} x \\ y \end{pmatrix} = \begin{pmatrix} 0 \\ 0 \end{pmatrix}$, starting at $\begin{pmatrix} 4 \\ 4 \end{pmatrix}$.

(c) Find a disk which you are sure contains a root.

2.7.6   (a) Find a global Lipschitz ratio for the derivative of the mapping $F : \mathbb{R}^2 \to \mathbb{R}^2$ given by

In Exercise 2.7.7 we advocate using a program like MATLAB (Newton.m), but it is not too cumbersome for a calculator.

$$F \begin{pmatrix} x \\ y \end{pmatrix} = \begin{pmatrix} \sin(x - y) + y^2 \\ \cos(x + y) - x \end{pmatrix}.$$

(b) Do one step of Newton's method to solve

$$F \begin{pmatrix} x \\ y \end{pmatrix} - \begin{pmatrix} .5 \\ 0 \end{pmatrix} = \begin{pmatrix} 0 \\ 0 \end{pmatrix} \quad \text{starting at } \begin{pmatrix} 0 \\ 0 \end{pmatrix}.$$

(c) Can you be sure that Newton's method converges?

2.7.7   Consider the system of equations

$$\cos x + y = 1.1$$
$$x + \cos(x + y) = .9$$

(a) Carry out four steps of Newton's method, starting at $\begin{pmatrix} 0 \\ 0 \end{pmatrix}$. How many decimals change between the third and the fourth step?

(b) Are the conditions of Kantorovitch's theorem satisfied at the first step? At the second step?

2.7.8   Using Newton's method, solve the equation

$$A^3 = \begin{bmatrix} 9 & 0 & 1 \\ 0 & 7 & 0 \\ 0 & 2 & 8 \end{bmatrix}.$$

For Exercise 2.7.8, note that $[2\,I]^3 = [8\,I]$, i.e.,

$$\begin{bmatrix} 2 & 0 & 0 \\ 0 & 2 & 0 \\ 0 & 0 & 2 \end{bmatrix}^3 = \begin{bmatrix} 8 & 0 & 0 \\ 0 & 8 & 0 \\ 0 & 0 & 8 \end{bmatrix}.$$

2.7.9   Use the MATLAB program Newton.m (or the equivalent) to solve the systems of equations:

(a)
$$\begin{aligned} x^2 - y + \sin(x - y) &= 2 \\ y^2 - x &= 3 \end{aligned} \quad \text{starting at } \begin{pmatrix} 2 \\ 2 \end{pmatrix}, \begin{pmatrix} -2 \\ 2 \end{pmatrix}$$

(b)
$$\begin{aligned} x^3 - y + \sin(x - y) &= 5 \\ y^2 - x &= 3 \end{aligned} \quad \text{starting at } \begin{pmatrix} 2 \\ 2 \end{pmatrix}, \begin{pmatrix} -2 \\ 2 \end{pmatrix}.$$

"Superconvergence" is defined in Section 2.8. "Does Newton's method appear to superconverge" means "does the number of correct decimals appear to double at every step."

Does Newton's method appear to superconverge?

In all cases, determine the numbers that appear in Kantorovitch's theorem, and check whether the theorem guarantees convergence.

2.7.10   Find a number $\epsilon > 0$ such that the set of equations

$$x + y^2 = a$$
$$y + z^2 = b \quad \text{has a unique solution near 0 when } |a|, |b|, |c| < \epsilon.$$
$$z + x^2 = c$$

2.7.11   Do one step of Newton's method to solve the system of equations

$$\begin{aligned} x + \cos y - 1.1 &= 0 \\ x^2 - \sin y + .1 &= 0 \end{aligned} \quad \text{starting at } \mathbf{a_0} = \begin{pmatrix} 0 \\ 0 \end{pmatrix}.$$

2.7.12   (a) Write one step of Newton's method to solve $x^5 - x - 6 = 0$, starting at $x_0 = 2$.

(b) Prove that this Newton's method converges.

Hint for Exercise 2.7.14 b: This is a bit harder than for Newton's method. Consider the intervals bounded by a_n and b/a_n^{k-1}, and show that they are nested.

A drawing is recommended for part (c), as computing cube roots is considerably harder than computing square roots.

2.7.13   Does a 2×2 matrix of the form $I + \epsilon B$ have a square root A near

$$\begin{bmatrix} 1 & 0 \\ 0 & -1 \end{bmatrix}?$$

*2.7.14   (a) Prove that if you compute $\sqrt[k]{b}$ by Newton's method, as in Exercise 2.7.3, choosing $a_0 > 0$ and $b > 0$, then the sequence a_n converges to the positive nth root.

(b) Show that this would not be true if you simply applied a divide and average algorithm:

$$a_{n+1} = \frac{1}{2}\left(a_n + \frac{b}{a_n^{k-1}} \right).$$

(c) Use Newton's method and "divide and average" (and a calculator or computer, of course) to compute $\sqrt[3]{2}$, starting at $a_0 = 2$. What can you say about the speeds of convergence?

2.8.1   Show (Example 2.8.1) that when solving $f(x) = (x-1)^2 = 0$ by Newton's method, starting at $a_0 = 0$, the best Lipschitz ratio for f' is 2, so

$$|f(a_0)|\,|(f'(a_0))^{-1}|^2 M = 1 \cdot \left(-\frac{1}{2}\right)^2 \cdot 2 = \frac{1}{2}$$

and Theorem 2.7.11 guarantees that Newton's method will work, and will converge to the unique root $a = 1$. Check that $h_n = 1/2^{n+1}$ so $a_n = 1 - 1/2^n$, on the nose the rate of convergence advertised.

2.8.2   (a) Prove (Equation 2.8.12) that the norm of a matrix is at most its length: $\|A\| \leq |A|$.

**(b) When are they equal?

2.8.3   Prove that Proposition 1.4.11 is true for the norm $\|A\|$ of a matrix A as well as for its length $|A|$: i.e., prove:

(a) If A is an $n \times m$ matrix, and $\vec{\mathbf{b}}$ is a vector in $\mathbb{R}^m$, then

$$\|A\vec{\mathbf{b}}\| \leq \|A\|\,|\vec{\mathbf{b}}|.$$

(b) If A is an $n \times m$ matrix, and B is a $m \times k$ matrix, then

$$\|AB\| \leq \|A\|\,\|B\|.$$

2.8.4   Prove that the triangle inequality (Theorem 1.4.9) holds for the norm $\|A\|$ of a matrix A, i.e., that for any matrices A and B in $\mathbb{R}^n$,

$$\|A + B\| \leq \|A\| + \|B\|.$$

2.8.5   (a) Find a 2×2 matrix A such that

Hint for Exercise 2.8.5: Try a matrix all of whose entries are equal.

$$A^2 + A = \begin{bmatrix} 1 & 1 \\ 1 & 1 \end{bmatrix}.$$

(b) Show that Newton's method converges when it is used to solve the above equation, starting at the identity.

2.8.6   For what matrices C can you be sure that the equation $A^2 + A = C$ in $\mathrm{Mat}\,(2,2)$ has a solution that can be found starting at 0? At I?

2.8.7   There are other plausible ways to measure matrices other than the length and the norm; for example, we could declare the size $|A|$ of a matrix A to be the largest absolute value of an entry. In this case, $|A + B| \leq |A| + |B|$, but the statement $|A\vec{\mathbf{x}}| \leq |A||\vec{\mathbf{x}}|$ (where $|\vec{\mathbf{x}}|$ is the ordinary length of a vector) is false. Find an ϵ so that it is false for

$$A = \begin{bmatrix} 1 & 1 & 1+\epsilon \\ 0 & 0 & 0 \\ 0 & 0 & 0 \end{bmatrix}, \quad \text{and} \quad \vec{\mathbf{x}} = \begin{bmatrix} 1 \\ 1 \\ 0 \end{bmatrix}.$$

Starred exercises are difficult; exercises with two stars are particularly challenging.

**2.8.8   If $A = \begin{bmatrix} a & b \\ c & d \end{bmatrix}$ is a 2×2 real matrix, show that

$$\|A\| = \left(\frac{|A|^2 + \sqrt{|A|^4 - 4D}}{2} \right)^{1/2}, \quad \text{where } D = ad - bc = \det A.$$

Exercises for Section 2.9:

Inverse and Implicit Function Theorems

Exercise 2.9.1 should be in Appendix A.4.

2.9.1   Prove Theorem 2.9.2 (the inverse function theorem in 1 dimension).

2.9.2   Consider the function

$$f(x) = \begin{cases} \frac{x}{2} + x^2 \sin \frac{1}{x} & \text{if } x \neq 0, \\ 0 & \text{if } x = 0, \end{cases}$$

discussed in Example 1.9.4. (a) Show that f is differentiable at 0 and that the derivative is $1/2$.

(b) Show that f does not have an inverse on any neighborhood of 0.

(c) Why doesn't this contradict the inverse function theorem, Theorem 2.9.2?

2.9.3   (a) See by direct calculation where the equation $y^2 + y + 3x + 1 = 0$ defines y implicitly as a function of x.

(b) Check that your answer agrees with the answer given by the implicit function theorem.

2.9.4   Consider the mapping $\mathbf{f} : \mathbb{R}^2 - \begin{pmatrix} 0 \\ 0 \end{pmatrix} \to \mathbb{R}^2$ given by

$$\mathbf{f} \begin{pmatrix} x \\ y \end{pmatrix} = \begin{pmatrix} \frac{(x^2 - y^2)}{(x^2 + y^2)} \\ \frac{xy}{(x^2 + y^2)} \end{pmatrix}.$$

Does $\mathbf{f}$ have a local inverse at every point of $\mathbb{R}^2$?

2.9.5   Let $y(x)$ be defined implicitly by

$$x^2 + y^3 + e^y = 0.$$

Compute $y'(x)$ in terms of x and y.

2.9.6   (a) True or false? The equation $\sin(xyz) = z$ expresses x implicitly as a differentiable function of y and z near the point

$$\begin{pmatrix} x \\ y \\ z \end{pmatrix} = \begin{pmatrix} \pi/2 \\ 1 \\ 1 \end{pmatrix}.$$

(b) True or false? The equation $\sin(xyz) = z$ expresses z implicitly as a differentiable function of x and y near the same point.

2.9.7   Does the system of equations

$$x + y + \sin(xy) = a$$
$$\sin(x^2 + y) = 2a$$

have a solution for sufficiently small a?

2.9.8   Consider the mapping $S : \text{Mat}\,(2,2) \to \text{Mat}\,(2,2)$ given by $S(A) = A^2$. Observe that $S(-I) = I$. Does there exist an inverse mapping g, i.e., a mapping such that $S(g(A)) = A$, defined in a neighborhood of I, such that $g(I) = -I$?

2.9.9   True or false? (Explain your answer.) There exists $r > 0$ and a differentiable map

$$g : B_r \left(\begin{bmatrix} -3 & 0 \\ 0 & -3 \end{bmatrix} \right) \to \text{Mat}\,(2,2) \text{ such that } g \left(\begin{bmatrix} -3 & 0 \\ 0 & -3 \end{bmatrix} \right) = \begin{bmatrix} 1 & 2 \\ -2 & -1 \end{bmatrix}$$

and $(g(A))^2 = A$ for all $A \in B_r \left(\begin{bmatrix} -3 & 0 \\ 0 & -3 \end{bmatrix} \right)$.

2.9.10   True or false? If $f : \mathbb{R}^3 \to \mathbb{R}$ is continuously differentiable, and

$$D_2 f \begin{pmatrix} a \\ b \\ c \end{pmatrix} \neq 0 \quad , \quad D_3 f \begin{pmatrix} a \\ b \\ c \end{pmatrix} \neq 0, \quad \text{then there exists}$$

a function h of $\begin{pmatrix} y \\ z \end{pmatrix}$, defined near $\begin{pmatrix} b \\ c \end{pmatrix}$, such that $f \begin{pmatrix} h \begin{pmatrix} y \\ z \end{pmatrix} \\ y \\ z \end{pmatrix} = 0$.

2.9.11   (a) Show that the mapping

$$F \begin{pmatrix} x \\ y \end{pmatrix} = \begin{pmatrix} e^x + e^y \\ e^x + e^{-y} \end{pmatrix} \quad \text{is locally invertible at every point } \begin{pmatrix} x \\ y \end{pmatrix} \in \mathbb{R}^2.$$

(b) If $F(\mathbf{a}) = \mathbf{b}$, what is the derivative of F^{-1} at $\mathbf{b}$?

2.9.12   True or false: There exists a neighborhood $U \subset \text{Mat}\,(2,2)$ of $\begin{bmatrix} 5 & 0 \\ 0 & 5 \end{bmatrix}$ and a C^1 mapping $F : U \to \text{Mat}\,(2,2)$ with

(1) $F \left(\begin{bmatrix} 5 & 0 \\ 0 & 5 \end{bmatrix} \right) = \begin{bmatrix} 1 & 2 \\ 2 & -1 \end{bmatrix}$, and

(2) $(F(A))^2 = A$.

You may use the fact that if $S : \text{Mat}\,(2,2) \to \text{Mat}\,(2,2)$ denotes the squaring map $S(A) = A^2$, then $[\mathbf{D}S(A)]B = AB + BA$.

3

Higher Partial Derivatives, Quadratic Forms, and Manifolds

Thomson [Lord Kelvin] had predicted the problems of the first [transatlantic] cable by mathematics. On the basis of the same mathematics he now promised the company a rate of eight or even 12 words a minute. Half a million pounds was being staked on the correctness of a partial differential equation.—T.W. Körner, *Fourier Analysis*

3.0 INTRODUCTION

When a computer calculates sines, it is not looking up the answer in some mammoth table of sines; stored in the computer is a polynomial that very well approximates $\sin x$ for x in some particular range. Specifically, it uses the formula

$$\sin x = x + a_3 x^3 + a_5 x^5 + a_7 x^7$$
$$+ a_9 x^9 + a_{11} x^{11} + \epsilon(x),$$

where the coefficients are

$$a_3 = -.1666666664$$
$$a_5 = .0083333315$$
$$a_7 = -.0001984090$$
$$a_9 = .0000027526$$
$$a_{11} = -.0000000239.$$

When $|x| \leq \pi/2$, the error is guaranteed to be less than 2×10^{-9}, good enough for a calculator that computes to eight significant digits.

This chapter is something of a grab bag. The various themes are related, but the relationship is not immediately apparent. We begin with two sections on geometry. In Section 3.1 we use the implicit function theorem to define just what we mean by a smooth curve and a smooth surface. Section 3.2 extends these definitions to more general k-dimensional "surfaces" in $\mathbb{R}^n$, called *manifolds*: surfaces in space (possibly, higher-dimensional space) that locally are graphs of differentiable mappings.

We switch gears in Section 3.3, where we use higher partial derivatives to construct the Taylor polynomial of a function in several variables. We saw in Section 1.7 how to approximate a nonlinear function by its derivative; here we will see that, as in one dimension, we can make higher-degree approximations using a function's Taylor polynomial. This is a useful fact, since polynomials, unlike sines, cosines, exponentials, square roots, logarithms, ... can actually be computed using arithmetic. Computing Taylor polynomials by calculating higher partial derivatives can be quite unpleasant; in Section 3.4 we give some rules for computing them by combining the Taylor polynomials of simpler functions.

In Section 3.5 we take a brief detour, introducing quadratic forms, and seeing how to classify them according to their "signature." In Section 3.6 we see that if we consider the second degree terms of a function's Taylor polynomial as a quadratic form, the signature of that form usually tells us whether at a particular point the function is a minimum, a maximum or some kind of *saddle*. In Section 3.7 we look at extrema of a function f when f is restricted to some manifold $M \subset \mathbb{R}^n$.

Finally, in Section 3.8 we give a brief introduction to the vast and important subject of the geometry of curves and surfaces. To define curves and surfaces in the beginning of the chapter, we did not need the higher-degree approximations provided by Taylor polynomials. To discuss the geometry of curves and surfaces, we do need Taylor polynomials: the *curvature* of a curve or surface depends on the quadratic terms of the functions defining it.

3.1 CURVES AND SURFACES

Everyone knows what a curve is, until he has studied enough mathematics to become confused through the countless number of possible exceptions— F. Klein

As familiar as these objects are, the mathematical definitions of smooth curves and smooth surfaces exclude some objects that we ordinarily think of as smooth: a figure eight, for example. Nor are these familiar objects simple: already, the theory of soap bubbles is a difficult topic, with a complicated partial differential equation controlling the shape of the film.

We are all familiar with smooth curves and surfaces. Curves are idealizations of things like telephone wires or a tangled garden hose. Beautiful surfaces are produced when you blow soap bubbles, especially big ones that wobble and slowly vibrate as they drift through the air, almost but not quite spherical. More prosaic surfaces can be imagined as an infinitely thin inflated inner tube (forget the valve), or for that matter the surface of any smooth object.

In this section we will see how to define these objects mathematically, and how to tell whether the locus defined by an equation or set of equations is a smooth curve or smooth surface. We will cover the same material three times, once for curves in the plane (also known as *plane curves*), once for surfaces in space and once for curves in space. The entire material will be repeated once more in Section 3.2 for more general k-dimensional "surfaces" in $\mathbb{R}^n$.

Smooth curves in the plane

When is a subset $X \subset \mathbb{R}^2$ a smooth curve? There are many possible answers, but today there seems to be a consensus that the objects defined below are the right curves to study. Our form of the definition, which depends on the chosen coordinates, might not achieve the same consensus: with this definition, it isn't obvious that if you rotate a smooth curve it is still smooth. (We will see in Theorem 3.2.8 that it is.)

Recall that the graph $\Gamma(f)$ of a function $f : \mathbb{R}^n \to \mathbb{R}$:

$$\Gamma(f) \subset \mathbb{R}^{n+1}$$

is the set of pairs $(\mathbf{x}, y) \in \mathbb{R}^n \times \mathbb{R}$ such that $f(\mathbf{x}) = y$.

Remember from the discussion of set theory notation that $I \times J$ is the set of pairs (x, y) with $x \in I$ and $y \in J$: e.g., the shaded rectangle of Figure 3.1.1.

Definition 3.1.1 looks more elaborate than it is. It says that a subset $X \subset \mathbb{R}^2$ is a smooth curve *if X is locally the graph of a differentiable function*, either of x in terms of y or of y in terms of x; the detail below simply spells out what the word "locally" means. Actually, this is the definition of a "C^1 curve"; as discussed in the remark following the definition, for our purposes here we will consider C^1 curves to be "smooth."

Definition 3.1.1 (Smooth curve in the plane).   A subset $X \subset \mathbb{R}^2$ is a C^1 curve if for every point $\begin{pmatrix} a \\ b \end{pmatrix} \in X$, there exist open neighborhoods I of a and J of b, and either a C^1 mapping $f : I \to J$ or a C^1 mapping $g : J \to I$ (or both) such that $X \cap (I \times J)$ is the graph of f or of g.

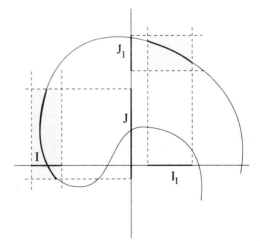

Note that we do not require that the *same* differentiable mapping work for every point: we can switch horses in mid-stream, and often we will need to, as in Figure 3.1.1.

FIGURE 3.1.1. Above, I and I_1 are intervals on the x-axis, while J and J_1 are intervals on the y-axis. The darkened part of the curve in the shaded rectangle $I \times J$ is the graph of a function expressing $x \in I$ as a function of $y \in J$, and the darkened part of the curve in $I_1 \times J_1$ is the graph of a function expressing $y \in J_1$ as a function of $x \in I_1$. Note that the curve in $I_1 \times J_1$ can also be thought of as the graph of a function expressing $x \in I_1$ as a function of $y \in J_1$. But we cannot think of the darkened part of the curve in $I \times J$ as the graph of a function expressing $y \in J$ as a function of $x \in I$; there are values of x that would give two different values of y, so such a "function" is not well defined.

A function is C^2 ("twice continuously differentiable") if its first and second partial derivatives exist and are continuous. It is C^3 if its first, second, and third partial derivatives exist and are continuous.

Some authors use "smooth" to mean "infinitely many times differentiable"; for our purposes, this is overkill.

Exercise 3.1.4 asks you to show that every straight line in the plane is a smooth curve.

Remark 3.1.2 (Fuzzy definition of "smooth"). For the purposes of this section, "smooth" means "of class C^1." We don't want to give a precise definition of smooth; its meaning depends on context and means "as many times differentiable as is relevant to the problem at hand." In this and the next section, only the first derivatives matter, but later, in Section 3.7 on constrained extrema, the curves, surfaces, etc. will need to be twice continuously differentiable (of class C^2), and the curves of Section 3.8 will need to be three times continuously differentiable (of class C^3). In the section about Taylor polynomials, it will really matter exactly how many derivatives exist, and there we won't use the word smooth at all. When objects are labeled smooth, we will compute derivatives without worrying about whether the derivatives exist.

Example 3.1.3 (Graph of any smooth function). The graph of any smooth function is a smooth curve: for example, the curve of equation $y = x^2$, which is the graph of y as a function of x, or the curve of equation $x = y^2$, which is the graph of x as a function of y.

For the first, for every point $\begin{pmatrix} x \\ y \end{pmatrix}$ with $y = x^2$, we can take $I = \mathbb{R}$, $J = \mathbb{R}$ and $f(x) = x^2$. $\triangle$

Example 3.1.4 (Unit circle). A more representative example is the unit circle of equation $x^2 + y^2 = 1$, which we denote S. Here we need the graphs of four functions to cover the entire circle: the unit circle is only locally the graph of a function. For the upper half of the circle, made up of points $\begin{pmatrix} x \\ y \end{pmatrix}$ with $y > 0$, we can take

> Think of $I = (-1, 1)$ as an interval on the x-axis, and $J = (0, 2)$ as an interval on the y-axis.

$$I = (-1, 1), \quad J = (0, 2) \quad \text{and } f : I \to J \text{ given by } f(x) = \sqrt{1 - x^2}. \qquad 3.1.1$$

We could also take $J = (0, \infty)$, or $J = (0, 1.2)$, but $J = (0, 1)$ will not do, as then J will not contain 1, so the point $\begin{pmatrix} 0 \\ 1 \end{pmatrix}$, which is in the circle, will not be in the graph. Remember that I and J are open.

> Note that for the upper half circle we could not have taken $J = \mathbb{R}$. Of course, f does map $(-1, 1) \to \mathbb{R}$, but the intersection
>
> $$S \cap ((-1, 1) \times \mathbb{R})$$
>
> (where $\mathbb{R}$ is the y-axis) is the whole circle with the two points
>
> $$\begin{pmatrix} 1 \\ 0 \end{pmatrix} \text{ and } \begin{pmatrix} -1 \\ 0 \end{pmatrix}$$
>
> removed, and not just the graph of f, which is just the top half of the circle.

Near the point $\begin{pmatrix} 1 \\ 0 \end{pmatrix}$, S is not the graph of any function f expressing y as a function of x, but it is the graph of a function g expressing x as a function of y, for example, the function $g : (-1, 1) \to (0, 2)$ given by $x = \sqrt{1 - y^2}$. (In this case, $J = (-1, 1)$ and $I = (0, 2)$.) Similarly, near the point $\begin{pmatrix} -1 \\ 0 \end{pmatrix}$, S is the graph of the function $g : (-1, 1) \to (-2, 0)$ given by $x = -\sqrt{1 - y^2}$.

For the lower half of the circle, when $y < 0$, we can choose $I = (-1, 1)$, $J = (0, -12)$, and the function $f : I \to J$ given by $f(x) = -\sqrt{1 - x^2}$. $\triangle$

Above, we expressed all but two points of the unit circle as the graph of functions of y in terms of x; we divided the circle into top and bottom. When we analyzed the unit circle in Example 2.9.11 we divided the circle into right-hand and left-hand sides, expressing all but two (different) points as the graph of functions expressing x in terms of y. In both cases we use the same four functions and we can use the same choices of I and J.

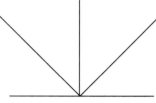

FIGURE 3.1.2.

The graph of $f(x) = |x|$ is not a smooth curve.

Example 3.1.5 (Graphs that are not smooth curves). The graph of the function $f : \mathbb{R} \to \mathbb{R}$, $f(x) = |x|$, shown in Figure 3.1.2, is not a smooth curve; it is the graph of the function f expressing y as a function of x, of course, but f is not differentiable. Nor is it the graph of a function g expressing x as a function of y, since in a neighborhood of $\begin{pmatrix} 0 \\ 0 \end{pmatrix}$ the same value of y gives two values of x.

The set $X \subset \mathbb{R}^2$ of equation $xy = 0$ (i.e., the union of the two axes) is also not a smooth curve; in any neighborhood of $\begin{pmatrix} 0 \\ 0 \end{pmatrix}$, there are infinitely many y's corresponding to $x = 0$, and infinitely many x's corresponding to $y = 0$, so it isn't a graph of a function either way.

In contrast, the graph of the function $f(x) = x^{1/3}$, shown in Figure 3.1.3, is a smooth curve; f is not differentiable at the origin, but the curve is the graph of the function $x = y^3$, which is differentiable.

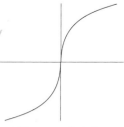

FIGURE 3.1.3.

The graph of $f(x) = x^{1/3}$ is a smooth curve: although f is not differentiable at the origin, the function $g(y) = y^3$ is.

Example 3.1.6 (A smooth curve can be disconnected). The union X of the x and y axes, shown on the left in Figure 3.1.4, is not a smooth curve, but $X - \left\{ \begin{pmatrix} 0 \\ 0 \end{pmatrix} \right\}$ *is* a smooth curve—even though it consists of four distinct pieces.

Tangent lines and tangent space

Definition 3.1.7 (Tangent line to a smooth plane curve). The tangent line to a smooth plane curve C at a point $\begin{pmatrix} a \\ f(a) \end{pmatrix}$ is the line of equation $y - f(a) = f'(a)(x - a)$. The tangent line to C at a point $\begin{pmatrix} g(b) \\ b \end{pmatrix}$ is the line of equation $x - g(b) = g'(b)(y - b)$.

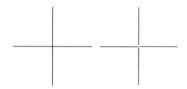

FIGURE 3.1.4.

Left: The graph of the x and y axes is not a smooth curve. Right: The graph of the axes minus the origin *is* a smooth curve.

You should recognize this as saying that the slope of the graph of f is given by f'.

At a point where the curve is neither vertical nor horizontal, it can be thought of locally as either a graph of x as a function of y or as a graph of y as a function of x. Will this give us two different tangent lines? No. If we have a point

$$\begin{pmatrix} a \\ b \end{pmatrix} = \begin{pmatrix} a \\ f(a) \end{pmatrix} = \begin{pmatrix} g(b) \\ b \end{pmatrix} \in C, \qquad 3.1.2$$

where C is a graph of $f : I \to J$ and $g : J \to I$, then $g \circ f(x) = x$ (i.e., $g(f(x)) = x$). In particular, $g'(b)f'(a) = 1$ by the chain rule, so the line of equation $y - f(a) = f'(a)(x - a)$ is also the line of equation $x - g(b) = g'(b)(y - b)$, and our definition of the tangent line is consistent.[1]

Very often the interesting thing to consider is not the tangent line but the tangent *vectors* at a point. Imagine that the curve is a hill down which you are skiing or sledding. At any particular moment, you would be interested in the slope of the tangent *line* to the curve: how steep is the hill? But you would also be interested in how fast you are going. Mathematically, we would represent your speed at a point **a** by a velocity vector lying on the tangent line to the curve at **a**. The arrow of the velocity vector would indicate what direction you are skiing, and its length would say how fast. If you are going very fast, the velocity vector will be long; if you have come to a halt while trying to get up nerve to proceed, the velocity vector will be the zero vector.

The *tangent space* to a smooth curve at **a** is the collection of vectors of all possible lengths, anchored at **a** and lying on the tangent line, as shown at the middle of Figure 3.1.5.

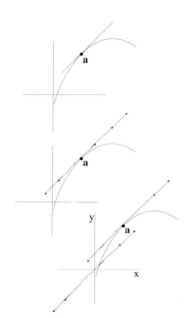

FIGURE 3.1.5.

Top: The tangent line. Middle: the tangent space. Bottom: The tangent space at the tangent point and translated to the origin.

Definition 3.1.8 (Tangent space to a smooth curve). The tangent space to C at **a**, denoted $T_{\mathbf{a}}C$, is the set of vectors tangent to C at **a**: i.e., vectors from the point of tangency to a point of the tangent line.

---

[1]Since $g'(b)f'(a) = 1$, we have $f'(a) = 1/g'(b)$, so $y - f(a) = f'(a)(x - a)$ can be written

$$y - b = \frac{x - a}{g'(b)} = \frac{x - g(b)}{g'(b)}; \quad \text{i.e.,} \quad x - g(b) = g'(b)(y - b).$$

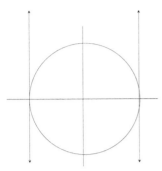

FIGURE 3.1.6.

The unit circle with tangent spaces at $\begin{pmatrix} 1 \\ 0 \end{pmatrix}$ and at $\begin{pmatrix} -1 \\ 0 \end{pmatrix}$. The two tangent spaces are the same; they consist of vectors such that the increment in the x direction is 0. They can be denoted $\dot{x} = 0$, where $\dot{x}$ denotes the first entry of the vector $\begin{bmatrix} \dot{x} \\ \dot{y} \end{bmatrix}$; it is not a coordinate of a point in the tangent line.

The tangent space will be essential in the discussion of constrained extrema, in Section 3.7, and in the discussion of orientation, in Section 6.5.

Note that a function of the form $F \begin{pmatrix} x \\ y \end{pmatrix} = c$ is of a different species than the functions f and g used to define a smooth curve; it is a function of two variables, while f and g are functions of one variable. If f is a function of one variable, its graph is the smooth curve of equation $f(x) - y = 0$. Then the curve is also given by the equation $F \begin{pmatrix} x \\ y \end{pmatrix} = 0$, where

$$F \begin{pmatrix} x \\ y \end{pmatrix} = f(x) - y.$$

The vectors making up the tangent space represent increments to the point $\mathbf{a}$; they include the zero vector representing a zero increment. The tangent space can be freely translated, as shown at the bottom of Figure 3.1.5: an increment has meaning independent of its location in the plane, or in space. Often we make use of such translations when describing a tangent space by an equation. In Figure 3.1.6, the tangent space to the circle at the point where $x = 1$ is the *same* as the tangent space to the circle where $x = -1$; this tangent space consists of vectors with no increment in the x direction. (But the equation for the tangent *line* at the point where $x = 1$ is $x = 1$, and the equation for the tangent line at the point where $x = -1$ is $x = -1$; the tangent line is made of points, not vectors, and points have a definite location.) To distinguish the tangent space from the line $x = 0$, we will say that the equation for the tangent space in Figure 3.1.6 is $\dot{x} = 0$. (This use of a dot above a variable is consistent with the use of dots by physicists to denote increments.)

Level sets as smooth curves

Graphs of smooth functions are the "obvious" examples of smooth curves. Very often, the locus (set of points) we are asked to consider is not the graph of any function we can write down explicitly. We can still determine whether such a locus is a smooth curve.

Suppose a locus is defined by an equation of the form $F \begin{pmatrix} x \\ y \end{pmatrix} = c$, such as $x^2 - .2x^4 - y^2 = -2$. One way to imagine this locus is to think of cutting the graph of $F \begin{pmatrix} x \\ y \end{pmatrix} = x^2 - .2x^4 - y^2$ by the plane $z = -2$. The intersection of the graph and the plane is called a *level curve*; three such intersections, for different values of z, are shown in Figure 3.1.7. How can we tell whether such a level set is a smooth curve? We will see that the implicit function theorem is the right tool to handle this question.

Theorem 3.1.9 (Equations for a smooth curve in $\mathbb{R}^2$). *(a) If U is open in $\mathbb{R}^2$, $F : U \to \mathbb{R}$ is a differentiable function with Lipschitz derivative, and $X_c = \{\mathbf{x} \in U \mid F(\mathbf{x}) = c\}$, then X_c is a smooth curve in $\mathbb{R}^2$ if $[\mathbf{D}F(\mathbf{a})]$ is onto for all $\mathbf{a} \in X_c$; i.e., if*

$$\left[DF \begin{pmatrix} a \\ b \end{pmatrix} \right] \neq 0 \quad \text{for all} \quad \mathbf{a} = \begin{pmatrix} a \\ b \end{pmatrix} \in X_c. \qquad 3.1.3$$

(b) If Equation 3.1.3 is satisfied, then the tangent space to X_c at $\mathbf{a}$ is $\ker[\mathbf{D}F(\mathbf{a})]$:

$$T_\mathbf{a} X_c = \ker[\mathbf{D}F(\mathbf{a})].$$

The condition that $[\mathbf{D}F(\mathbf{a})]$ be onto is the crucial condition of the implicit function theorem.

Because $[\mathbf{D}F(\mathbf{a})]$ is a 1×2 matrix (a transformation from $\mathbb{R}^2$ to $\mathbb{R}$), the following statements mean the same thing:

for all $\mathbf{a} = \begin{pmatrix} a \\ b \end{pmatrix} \in X_c$,

(1) $[\mathbf{D}F(\mathbf{a})]$ is onto.
(2) $[\mathbf{D}F(\mathbf{a})] \neq 0$.
(3) At least one of $D_1F(\mathbf{a})$ or $D_2F(\mathbf{a})$ is not 0.

Note that

$[\mathbf{D}F(\mathbf{a})] = [D_1F(\mathbf{a}), D_2F(\mathbf{a})]$; saying that $[\mathbf{D}F(\mathbf{a})]$ is onto is saying that any real number can be expressed as a linear combination $D_1F(\mathbf{a})\alpha + D_2F(\mathbf{a})\beta$ for some $\begin{bmatrix} \alpha \\ \beta \end{bmatrix} \in \mathbb{R}^2$.

Part (b) of Theorem 3.1.9 relates the algebraic notion of $\ker[\mathbf{D}F(\mathbf{a})]$ to the geometrical notion of a tangent space

Saying that $\ker[\mathbf{D}F(\mathbf{a})]$ is the tangent space to X_c at $\mathbf{a}$ says that every vector $\vec{v}$ tangent to X_c at $\mathbf{a}$ satisfies the equation

$$[\mathbf{D}F(\mathbf{a})]\vec{v} = 0.$$

This puzzled one student, who argued that for this equation to be true, either $[\mathbf{D}F(\mathbf{a})]$ or $\vec{v}$ must be $\mathbf{0}$, yet Equation 3.1.3 says that $[\mathbf{D}F(\mathbf{a})] \neq 0$. This is forgetting that $[\mathbf{D}F(\mathbf{a})]$ is a matrix. For example: if $[\mathbf{D}F(\mathbf{a})]$ is the line matrix $[2, -2]$, then $[2, -2]\begin{bmatrix} 1 \\ 1 \end{bmatrix} = 0$.

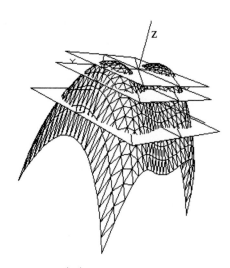

FIGURE 3.1.7. The surface $F\begin{pmatrix} x \\ y \end{pmatrix} = x^2 - .2x^4 - y^2$ sliced horizontally by setting z equal to three different constants. The intersection of the surface and the plane $z = c$ used to slice it is known as a *level set*. (This intersection is of course the same as the locus of equation $F\begin{pmatrix} x \\ y \end{pmatrix} = c$.) The three level sets shown above are smooth curves. If we were to "slice" the surface at a maximum of F, we would get a point, not a smooth curve. If we were to slice it at a *saddle point* (also a point where the derivative of F is 0), we would get a figure eight, not a smooth curve.

Example 3.1.10 (Finding the tangent space). We have no idea what the locus X_c defined by $x^9 + 2x^3 + y + y^5 = c$ looks like, but the derivative of the function $F\begin{pmatrix} x \\ y \end{pmatrix} = x^9 + 2x^3 + y + y^5$ is

$$\left[\mathbf{D}F\begin{pmatrix} x \\ y \end{pmatrix}\right] = [\underbrace{9x^8 + 6x^2}_{D_1F}, \underbrace{1 + 5y^4}_{D_2F}], \qquad 3.1.4$$

which is never 0, so X_c is a smooth curve for all c. At the point $\begin{pmatrix} 1 \\ 1 \end{pmatrix} \in X_5$, the derivative $\left[\mathbf{D}F\begin{pmatrix} x \\ y \end{pmatrix}\right]$ is $[15, 6]$, so the equation of the tangent space to X_5 at that point is $15\dot{x} + 6\dot{y} = 0$. $\triangle$

Proof of Theorem 3.1.9. (a) Choose $\mathbf{a} = \begin{pmatrix} a \\ b \end{pmatrix} \in X$. The hypothesis $[\mathbf{D}F(\mathbf{a})] \neq 0$ implies that at least one of $D_1F\begin{pmatrix} a \\ b \end{pmatrix}$ or $D_2F\begin{pmatrix} a \\ b \end{pmatrix}$ is not 0; let us suppose $D_2F\begin{pmatrix} a \\ b \end{pmatrix} \neq 0$ (i.e., the second variable, y, is the pivotal variable, which will be expressed as a function of the non-pivotal variable x).

This is what is needed in order to apply the short version of the implicit function theorem (Theorem 2.9.9): $F\left(\begin{array}{c}x\\y\end{array}\right) = 0$ then expresses y implicitly as a function of x in a neighborhood of a.

More precisely, there exists a neighborhood U of a in $\mathbb{R}$, a neighborhood V of b, and a continuously differentiable mapping $f : U \to V$ such that $F\left(\begin{array}{c}x\\f(x)\end{array}\right) = 0$ for all $x \in U$. The implicit function theorem also guarantees that we can choose U and V so that when x is chosen in U, then $f(x)$ is the only $y \in V$ such that $F\left(\begin{array}{c}x\\y\end{array}\right) = 0$. In other words, $X \cap (U \times V)$ is exactly the graph of f, which is our definition of a curve.

(b) Now we need to prove that the tangent space $T_{\mathbf{a}}X_c$ is $\ker[\mathbf{D}F(\mathbf{a})]$. For this we need the formula for the derivative of the implicit function, in Theorem 2.9.10 (the long version of the implicit function theorem). Let us suppose that $D_2F(\mathbf{a}) \neq 0$, so that, as above, the curve has the equation $y = f(x)$ near $\mathbf{a} = \left(\begin{array}{c}a\\b\end{array}\right)$, and its tangent space has equation $\dot{y} = f'(a)\dot{x}$.

Note that the derivative of the implicit function, in this case f', is evaluated at a, not at $\mathbf{a} = \left(\begin{array}{c}a\\b\end{array}\right)$.

The implicit function theorem (Equation 2.9.25) says that the derivative of the implicit function f is

$$f'(a) = [\mathbf{D}f(a)] = -D_2F(\mathbf{a})^{-1}D_1F(\mathbf{a}). \qquad 3.1.5$$

Substituting this value for $f'(a)$ in the equation $\dot{y} = f'(a)\dot{x}$, we get

$$\dot{y} = -D_2F(\mathbf{a})^{-1}D_1F(\mathbf{a})\dot{x}. \qquad 3.1.6$$

Multiplying through by $D_2F(\mathbf{a})$ gives $D_2F(\mathbf{a})\dot{y} = -D_1F(\mathbf{a})\dot{x}$, so

$$0 = D_1F(\mathbf{a})\dot{x} + D_2F(\mathbf{a})\dot{y} = \underbrace{[D_1F(\mathbf{a}),\ D_2F(\mathbf{a})]}_{[\mathbf{D}F(\mathbf{a})]}\left[\begin{array}{c}\dot{x}\\\dot{y}\end{array}\right]. \quad \square \qquad 3.1.7$$

If you know a curve as a graph, this procedure will give you the tangent space as a graph. If you know it as an equation, it will give you an equation for the tangent space. If you know it by a parametrization, it will give you a parametrization for the tangent space.

The same rule applies to surfaces and higher-dimensional manifolds.

Remark. Part (b) is one instance of the golden rule: to find the tangent space to a curve, *do unto the increment* $\left[\begin{array}{c}\dot{x}\\\dot{y}\end{array}\right]$ *with the derivative whatever you did to points with the function to get your curve.* For instance:

• If the curve is the graph of f, i.e, has equation $y = f(x)$, the tangent space at $\left(\begin{array}{c}a\\f(a)\end{array}\right)$ is the graph of $f'(a)$, i.e. has equation $\dot{y} = f'(a)\dot{x}$.

• If the curve has equation $F\left(\begin{array}{c}x\\y\end{array}\right) = 0$, then the tangent space at $\left(\begin{array}{c}x_0\\y_0\end{array}\right)$ has equation $\left[\mathbf{D}F\left(\begin{array}{c}x_0\\y_0\end{array}\right)\right]\left[\begin{array}{c}\dot{x}\\\dot{y}\end{array}\right] = 0$.

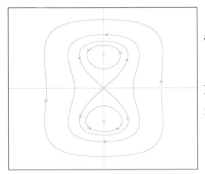

Figure 3.1.8.

The locus of equation

$$x^4 + y^4 + x^2 - y^2 = -1/4$$

consists of the two points on the y-axis where $y = \pm 1/\sqrt{2}$; it is not a smooth curve. Nor is the figure eight, which is the locus of equation $x^4 + y^4 + x^2 - y^2 = 0$. The other curves are smooth curves. The arrows on the lines are an artifact of the drawing program.

Why? The result of "do unto the increment ... " will be the best linear approximation to the locus defined by "whatever you did to points ... ."   △

Example 3.1.11 (When is a level set a smooth curve?). Consider the function $F\begin{pmatrix} x \\ y \end{pmatrix} = x^4 + y^4 + x^2 - y^2$. We have

$$\left[\mathbf{D}F\begin{pmatrix} x \\ y \end{pmatrix} \right] = [\underbrace{4x^3 + 2x}_{D_1 F}, \underbrace{4y^3 - 2y}_{D_2 F}] = \left[2x(2x^2 + 1),\ 2y(2y^2 - 1) \right]. \qquad 3.1.8$$

There are no real solutions to $2x^2 + 1 = 0$; the only places where both partials vanish are

$$\begin{pmatrix} 0 \\ 0 \end{pmatrix}, \ \begin{pmatrix} 0 \\ \pm 1/\sqrt{2} \end{pmatrix}, \qquad 3.1.9$$

where F takes on the value 0 and $-1/4$. Thus for any number $c \neq 0$ and $c \neq -1/4$, the locus of equation $c = x^4 + y^4 + x^2 - y^2$ is a smooth curve.

Some examples are plotted in Figure 3.1.8. Indeed, the locus of equation $x^4 + y^4 + x^2 - y^2 = -1/4$ consists of precisely two points, and is nothing you would want to call a curve, while the locus of equation $x^4 + y^4 + x^2 - y^2 = 0$ is a figure eight, and near the origin looks like two intersecting lines; to make it a smooth curve we would have to take out the point where the lines intersect. The others really are things one would want to call smooth curves.

Smooth surfaces in $\mathbb{R}^3$

"Smooth curve" means something different in mathematics and in common speech: a figure eight is not a smooth curve, while the four separate straight lines of Example 3.1.6 form a smooth curve. In addition, by our definition the empty set (which arises in Example 3.1.11 if $c < -1/4$) is also a smooth curve! Allowing the empty set to be a smooth curve makes a number of statements simpler.

Our definition of a smooth surface in $\mathbb{R}^3$ is a clone of the definition of a curve.

Definition 3.1.12 (Smooth surface). A subset $S \subset \mathbb{R}^3$ is a smooth surface if for every point $\mathbf{a} = \begin{pmatrix} a \\ b \\ c \end{pmatrix} \in S$, there are neighborhoods I of a, J of b and K of c, and either a differentiable mapping

- $f : I \times J \ \rightarrow K$, i.e., z as a function of (x, y) or
- $g : I \times K \ \rightarrow J$, i.e., y as a function of (x, z) or
- $h : J \times K \ \rightarrow I$, i.e., x as a function of (y, z),

such that $S \cap (I \times J \times K)$ is the graph of $f, g,$ or h.

We will see in Proposition 3.2.8 that the choice of coordinates doesn't matter; if you rotate a smooth surface in any way, it is still a smooth surface.

Definition 3.1.13 (Tangent plane to a smooth surface). The tangent plane to a smooth surface S at $\begin{pmatrix} a \\ b \\ c \end{pmatrix}$ is the plane of the equations

$$z - c = \left[\mathbf{D}f\begin{pmatrix} a \\ b \end{pmatrix} \right] \begin{bmatrix} x - a \\ y - b \end{bmatrix} = D_1 f\begin{pmatrix} a \\ b \end{pmatrix}(x - a) + D_2 f\begin{pmatrix} a \\ b \end{pmatrix}(y - b)$$

$$y - b = \left[\mathbf{D}g\begin{pmatrix} a \\ c \end{pmatrix} \right] \begin{bmatrix} x - a \\ z - c \end{bmatrix} = D_1 g\begin{pmatrix} a \\ c \end{pmatrix}(x - a) + D_2 g\begin{pmatrix} a \\ c \end{pmatrix}(z - c) \quad 3.1.10$$

$$x - a = \left[\mathbf{D}h\begin{pmatrix} b \\ c \end{pmatrix} \right] \begin{bmatrix} y - b \\ z - c \end{bmatrix} = D_1 h\begin{pmatrix} b \\ c \end{pmatrix}(y - b) + D_2 h\begin{pmatrix} b \\ c \end{pmatrix}(z - c)$$

in the three cases above.

If at a point $\mathbf{x}_0$ the surface is simultaneously the graph of z as a function of x and y, y as a function of x and z, and x as a function of y and z, then the corresponding equations for the tangent planes to the surface at $\mathbf{x}_0$ denote the same plane, as you are asked to show in Exercise 3.1.9.

As in the case of curves, we will distinguish between the tangent plane, given above, and the tangent *space*.

Definition 3.1.14 (Tangent space to a smooth surface). The tangent space to a smooth surface S at $\mathbf{a}$ is the plane composed of the vectors tangent to the surface at $\mathbf{a}$, i.e., vectors going from the point of tangency $\mathbf{a}$ to a point of the tangent plane. It is denoted $T_{\mathbf{a}}S$.

As before, $\dot{x}$ denotes an increment in the x direction, $\dot{y}$ an increment in the y direction, and so on. When the tangent space is anchored at $\mathbf{a}$, the vector $\begin{bmatrix} \dot{x} \\ \dot{y} \\ \dot{z} \end{bmatrix}$ is an increment from the point $\begin{pmatrix} a \\ b \\ c \end{pmatrix}$.

The equation for the tangent space to a surface is:

$$\dot{z} = \left[\mathbf{D}f\begin{pmatrix} a \\ b \end{pmatrix} \right] \begin{bmatrix} \dot{x} \\ \dot{y} \end{bmatrix} = D_1 f\begin{pmatrix} a \\ b \end{pmatrix}\dot{x} + D_2 f\begin{pmatrix} a \\ b \end{pmatrix}\dot{y}$$

$$\dot{y} = \left[\mathbf{D}g\begin{pmatrix} a \\ c \end{pmatrix} \right] \begin{bmatrix} \dot{x} \\ \dot{z} \end{bmatrix} = D_1 g\begin{pmatrix} a \\ c \end{pmatrix}\dot{x} + D_2 g\begin{pmatrix} a \\ c \end{pmatrix}\dot{z} \quad 3.1.11$$

$$\dot{x} = \left[\mathbf{D}h\begin{pmatrix} b \\ c \end{pmatrix} \right] \begin{bmatrix} \dot{y} \\ \dot{z} \end{bmatrix} = D_1 h\begin{pmatrix} b \\ c \end{pmatrix}\dot{y} + D_2 h\begin{pmatrix} b \\ c \end{pmatrix}\dot{z}.$$

Example 3.1.15 (Sphere in $\mathbb{R}^3$). Consider the unit sphere: the set

$$S^2 = \left\{ \begin{pmatrix} x \\ y \\ z \end{pmatrix} \text{ such that } x^2 + y^2 + z^2 = 1 \right\}. \quad 3.1.12$$

This is a smooth surface. Let

$$D_{x,y} = \left\{ \begin{pmatrix} x \\ y \\ z \end{pmatrix} \text{ such that } x^2 + y^2 < 1, \ z = 0 \right\} \quad 3.1.13$$

be the unit disk in the (x, y)-plane and $\mathbb{R}_z^+$ the positive part of the z-axis. Then

$$S^2 \cap (D_{x,y} \times \mathbb{R}_z^+) \qquad\qquad 3.1.14$$

Many students find it very hard to call the sphere of equation

$$x^2 + y^2 + z^2 = 1$$

two-dimensional. But when we say that Chicago is "at" x latitude and y longitude, we are treating the surface of the earth as two-dimensional.

is the graph of the function $D_{x,y} \to \mathbb{R}_z^+$ given by $\sqrt{1 - x^2 - y^2}$.

This shows that S^2 is a surface near every point where $z > 0$, and considering $-\sqrt{1 - x^2 - y^2}$ should convince you that S^2 is also a smooth surface near any point where $z < 0$.

In the case where $z = 0$, we can consider

(1) $D_{x,z}$ and $D_{y,z}$;

(2) the half-axes $\mathbb{R}_x^+$ and $\mathbb{R}_y^+$; and

(3) the mappings $\pm\sqrt{1 - x^2 - z^2}$ and $\pm\sqrt{1 - y^2 - z^2}$,

as Exercise 3.1.5 asks you to do. $\triangle$

Most often, surfaces are defined by an equation like $x^2 + y^2 + z^2 = 1$, which is probably familiar, or $\sin(x + yz) = 0$, which is surely not. That the first is a surface won't surprise anyone, but what about the second? Again, the implicit function theorem comes to the rescue, showing how to determine whether a given locus is a smooth surface.

In Theorem 3.1.16 we could say "if $[\mathbf{D}F(\mathbf{a})]$ is onto, then X is a smooth surface." Since $\mathbf{F}$ goes from $U \subset \mathbb{R}^3$ to $\mathbb{R}$, the derivative $[\mathbf{D}F(\mathbf{a})]$ is a row matrix with three entries, $D_1\mathbf{F}, D_2\mathbf{F}$, and $D_3\mathbf{F}$. The only way it can fail to be onto is if all three entries are 0.

Theorem 3.1.16 (Smooth surface in $\mathbb{R}^3$). *(a) Let U be an open subset of $\mathbb{R}^3$, $F : U \to \mathbb{R}$ a differentiable function with Lipschitz derivative and*

$$X = \left\{ \begin{pmatrix} x \\ y \\ z \end{pmatrix} \in \mathbb{R}^3 \ \bigg| \ F(\mathbf{x}) = 0 \right\}. \qquad\qquad 3.1.15$$

If at every $\mathbf{a} \in X$ we have $[\mathbf{D}F(\mathbf{a})] \neq 0$, then X is a smooth surface.

(b) The tangent space $T_\mathbf{a}X$ to the smooth surface is $\ker[\mathbf{D}F(\mathbf{a})]$.

You should be impressed by Example 3.1.17. The implicit function theorem is hard to prove, but the work pays off. Without having any idea what the set defined by Equation 3.1.16 might look like, we were able to determine, with hardly any effort, that it is a smooth surface. Figuring out what the surface looks like— or even whether the set is empty— is another matter. Exercise 3.1.15 outlines what it looks like in this case, but usually this kind of thing can be quite hard indeed.

Example 3.1.17 (Smooth surface in $\mathbb{R}^3$). Consider the set X defined by the equation

$$F \begin{pmatrix} x \\ y \\ z \end{pmatrix} = \sin(x + yz) = 0. \qquad\qquad 3.1.16$$

The derivative is

$$\left[\mathbf{D}F \begin{pmatrix} a \\ b \\ c \end{pmatrix}\right] = [\underbrace{\cos(a + bc)}_{D_1 F}, \underbrace{c\cos(a + bc)}_{D_2 F}, \underbrace{b\cos(a + bc)}_{D_3 F}]. \qquad 3.1.17$$

On X, by definition, $\sin(a + bc) = 0$, so $\cos(a + bc) \neq 0$, so X is a smooth surface. $\triangle$

Proof of Theorem 3.1.16.   Again, this is an application of the implicit function theorem. If for instance $D_1F(\mathbf{a}) \neq 0$ at some point $\mathbf{a} \in X$, then the condition $F(\mathbf{x}) = 0$ locally expresses x as a function h of y and z (see Definition 3.1.12). This proves (a).

For part (b), recall Definition 3.1.11, which says that in this case the tangent space $T_{\mathbf{a}}X$ has equation

$$\dot{x} = \left[\mathbf{D}h\begin{pmatrix} b \\ c \end{pmatrix}\right]\begin{bmatrix} \dot{y} \\ \dot{z} \end{bmatrix}. \tag{3.1.18}$$

But the implicit function theorem says that

$$\left[\mathbf{D}h\begin{pmatrix} b \\ c \end{pmatrix}\right] = -\left[D_1F(\mathbf{a})\right]^{-1}\left[D_2F(\mathbf{a}), D_3F(\mathbf{a})\right]. \tag{3.1.19}$$

(Can you explain how Equation 3.1.19 follows from the implicit function theorem? Check your answer below.[2])

Substituting this value for $\left[\mathbf{D}h\begin{pmatrix} b \\ c \end{pmatrix}\right]$ in Equation 3.1.18 gives

$$\dot{x} = -\left[D_1F(\mathbf{a})\right]^{-1}\left[D_2F(\mathbf{a}), D_3F(\mathbf{a})\right]\begin{bmatrix} \dot{y} \\ \dot{z} \end{bmatrix}, \tag{3.1.20}$$

and multiplying through by $D_1F(\mathbf{a})$, we get

$$[D_1F(\mathbf{a})]\dot{x} = -\overbrace{[D_1F(\mathbf{a})][D_1F(\mathbf{a})]^{-1}}^{I}[D_2F(\mathbf{a}), D_3F(\mathbf{a})]\begin{bmatrix} \dot{y} \\ \dot{z} \end{bmatrix}, \text{ so} \tag{3.1.21}$$

$$[D_2F(\mathbf{a}), D_3F(\mathbf{a})]\begin{bmatrix} \dot{y} \\ \dot{z} \end{bmatrix} + [D_1F(\mathbf{a})]\dot{x} = 0; \text{ i.e.,}$$

$$\underbrace{[D_1F(\mathbf{a}), D_2F(\mathbf{a}), D_3F(\mathbf{a})]}_{[\mathbf{D}F(\mathbf{a})]}\begin{bmatrix} \dot{x} \\ \dot{y} \\ \dot{z} \end{bmatrix} = 0, \quad \text{or} \quad [\mathbf{D}F(\mathbf{a})]\begin{bmatrix} \dot{x} \\ \dot{y} \\ \dot{z} \end{bmatrix} = 0 \tag{3.1.22}$$

So the tangent space is the kernel of $[\mathbf{D}F(\mathbf{a})]$. $\square$

---

[2]Recall Equation 2.9.25 for the derivative of the implicit function:

$$[\mathbf{D}\mathbf{g}(\mathbf{b})] = -\underbrace{[\vec{D_1}F(\mathbf{c}), \dots, \vec{D_n}F(\mathbf{c})]}_{\substack{\text{partial deriv. for} \\ \text{pivotal variables}}}^{-1}\underbrace{[\vec{D}_{n+1}F(\mathbf{c}), \dots, \vec{D}_{n+m}F(\mathbf{c})]}_{\substack{\text{partial deriv. for} \\ \text{non-pivotal variables}}}.$$

Our assumption was that at some point $\mathbf{a} \in X$ the equation $F(\mathbf{x}) = 0$ locally expresses x as a function of y and z. In Equation 3.1.19 $D_1F(\mathbf{a})$ is the partial derivative with respect to the pivotal variable, while $D_2F(\mathbf{a})$ and $D_3F(\mathbf{a})$ are the partial derivatives with respect to the non-pivotal variables.

Smooth curves in $\mathbb{R}^3$

A subset $X \subset \mathbb{R}^3$ is a smooth curve if it is locally the graph of either

- y and z as functions of x or
- x and z as functions of y or
- x and y as functions of z.

Let us spell out the meaning of "locally."

Definition 3.1.18 (Smooth curve in $\mathbb{R}^3$). A subset $X \subset \mathbb{R}^3$ is a smooth curve if for every $\mathbf{a} = \begin{pmatrix} a \\ b \\ c \end{pmatrix} \in X$, there exist neighborhoods I of a, J of b and K of c, and a differentiable mapping

- $\mathbf{f} : I \to J \times K$, i.e., y, z as a function of x or
- $\mathbf{g} : J \to I \times K$, i.e., x, z as a function of y or
- $\mathbf{k} : K \to I \times J$, i.e., x, y as a function of z,

such that $X \cap (I \times J \times K)$ is the graph of $\mathbf{f}, \mathbf{g}$, or $\mathbf{k}$ respectively.

For smooth curves in $\mathbb{R}^2$ or smooth surfaces in $\mathbb{R}^3$, we always had *one* variable expressed as a function of the other variable or variables. Now we have *two* variables expressed as a function of the other variable.

This means that curves in space have two degrees of freedom, as opposed to one for curves in the plane and surfaces in space; they have more freedom to wiggle and get tangled. A sheet can get a little tangled in a washing machine, but if you put a ball of string in the washing machine you will have a fantastic mess. Think too of tangled hair. That is the natural state of curves in $\mathbb{R}^3$.

Note that our functions $\mathbf{f}, \mathbf{g}$, and $\mathbf{k}$ are bold. The function $\mathbf{f}$, for example, is

$$\mathbf{f}(x) = \begin{bmatrix} f_1(x) \\ f_2(x) \end{bmatrix} = \begin{pmatrix} y \\ z \end{pmatrix}.$$

If y and z are functions of x, then the tangent line to X at $\begin{pmatrix} a \\ b \\ c \end{pmatrix}$ is the line intersection of the two planes

$$y - b = f_1'(a)(x - a) \quad \text{and} \quad z - c = f_2'(a)(x - a). \qquad 3.1.23$$

What are the equations if x and z are functions of y? If x and y are functions of z? Check your answers below.[3]

The tangent space is the subspace given by the same equations, where the increment $x - a$ is written $\dot{x}$ and similarly $y - b = \dot{y}$, and $z - c = \dot{z}$. What are the relevant equations?[4]

---

[3]If x and z are functions of y, the tangent line is the intersection of the planes

$$x - a = g_1'(b)(y - b) \quad \text{and} \quad z - c = g_2'(b)(y - b).$$

If x and y are functions of z, it is the intersection of the planes

$$x - a = k_1'(c)(z - c) \quad \text{and} \quad y - b = k_2'(c)(z - c).$$

[4]The tangent space can be written as $\begin{bmatrix} \dot{y} \\ \dot{z} \end{bmatrix} = \begin{bmatrix} f_1'(a)(\dot{x}) \\ f_2'(a)(\dot{x}) \end{bmatrix}$ or

$$\begin{bmatrix} \dot{x} \\ \dot{z} \end{bmatrix} = \begin{bmatrix} g_1'(b)(\dot{y}) \\ g_2'(b)(\dot{y}) \end{bmatrix} \quad \text{or} \quad \begin{bmatrix} \dot{x} \\ \dot{y} \end{bmatrix} = \begin{bmatrix} k_1'(c)(\dot{z}) \\ k_2'(c)(\dot{z}) \end{bmatrix}.$$

Since the range of $[\mathbf{DF}(\mathbf{a})]$ is $\mathbb{R}^2$, saying that it has rank 2 is the same as saying that it is onto; both are ways of saying that its columns span $\mathbb{R}^2$.

Proposition 3.1.19 says that another natural way to think of a smooth curve in $\mathbb{R}^3$ is as the intersection of two surfaces. If the surfaces S_1 and S_2 are given by equations $f_1(\mathbf{x}) = 0$ and $f_2(\mathbf{x}) = 0$, then $C = S_1 \cap S_2$ is given by the equation $\mathbf{F}(\mathbf{x}) = 0$, where $\mathbf{F}(\mathbf{x}) = \begin{pmatrix} f_1(\mathbf{x}) \\ f_2(\mathbf{x}) \end{pmatrix}$ is a mapping from $\mathbb{R}^3 \to \mathbb{R}^2$.

Below we speak of the derivative having rank 2 instead of the derivative being onto; as the margin note explains, in this case the two mean the same thing.

Proposition 3.1.19 (Smooth curves in $\mathbb{R}^3$).   (a) Let $U \subset \mathbb{R}^3$ be open, $\mathbf{F} : U \to \mathbb{R}^2$ be differentiable with Lipschitz derivative, and let C be the set of equation $\mathbf{F}(\mathbf{x}) = 0$. If $[\mathbf{DF}(\mathbf{a})]$ has rank 2 for every $\mathbf{a} \in C$, then C is a smooth curve in $\mathbb{R}^3$.

(b) The tangent vector space to X at $\mathbf{a}$ is $\ker[\mathbf{DF}(\mathbf{a})]$.

In Equation 3.1.24, the partial derivatives on the right-hand side are evaluated at $\mathbf{a} = \begin{pmatrix} a \\ b \\ c \end{pmatrix}$. The derivative of the implicit function $\mathbf{k}$ is evaluated at c; it is a function of one variable, z, and is not defined at $\mathbf{a}$.

Proof.   Once more, this is the implicit function theorem. Let $\mathbf{a}$ be a point of C. Since $[\mathbf{DF}(\mathbf{a})]$ is a 2×3 matrix with rank 2, it has two columns that are linearly independent. By changing the names of the variables, we may assume that they are the first two. Then the implicit function theorem asserts that near $\mathbf{a}$, x and y are expressed implicitly as functions of z by the relation $\mathbf{F}(\mathbf{x}) = 0$.

The implicit function theorem further tells us (Equation 2.9.25) that the derivative of the implicit function $\mathbf{k}$ is

$$[\mathbf{Dk}(c)] = -[\vec{D_1}\mathbf{F}(\mathbf{a}), \; \vec{D_2}\mathbf{F}(\mathbf{a})]^{-1}[\vec{D_3}\mathbf{F}(\mathbf{a})]. \qquad 3.1.24$$

$$\underbrace{\phantom{[\vec{D_1}\mathbf{F}(\mathbf{a}), \; \vec{D_2}\mathbf{F}(\mathbf{a})]^{-1}}}_{\substack{\text{partial deriv. for} \\ \text{pivotal variables}}} \quad \underbrace{\phantom{[\vec{D_3}\mathbf{F}(\mathbf{a})]}}_{\substack{\text{for non-} \\ \text{pivotal} \\ \text{variable}}}$$

Here $[\mathbf{DF}(\mathbf{a})]$ is a 2×3 matrix, so the partial derivatives are vectors, not numbers; because they are vectors we write them with arrows, as in $\vec{D_1}\mathbf{F}(\mathbf{a})$.

We saw (footnote 4) that the tangent space is the subspace of equation

$$\begin{bmatrix} \dot{x} \\ \dot{y} \end{bmatrix} = \begin{bmatrix} k_1'(c)\dot{z} \\ k_2'(c)\dot{z} \end{bmatrix} = [\mathbf{Dk}(c)]\dot{z}, \qquad 3.1.25$$

Once again, we distinguish between the tangent line and the tangent space, which is the set of vectors from the point of tangency to a point of the tangent line.

where once more $\dot{x}, \dot{y}$ and $\dot{z}$ are increments to x, y and z. Inserting the value of $[\mathbf{Dk}(c)]$ from Equation 3.1.24 and multiplying through by $[\vec{D_1}\mathbf{F}(\mathbf{a}), \vec{D_2}\mathbf{F}(\mathbf{a})]$ gives

$$-\overbrace{[\vec{D_1}\mathbf{F}(\mathbf{a}), \vec{D_2}\mathbf{F}(\mathbf{a})] \, [\vec{D_1}\mathbf{F}(\mathbf{a}), \vec{D_2}\mathbf{F}(\mathbf{a})]^{-1}}^{I} [\vec{D_3}\mathbf{F}(\mathbf{a})]\dot{z} = [\vec{D_1}\mathbf{F}(\mathbf{a}), \vec{D_2}\mathbf{F}(\mathbf{a})]\begin{bmatrix} \dot{x} \\ \dot{y} \end{bmatrix},$$

This should look familiar; we did the same thing in Equations 3.1.20–3.1.22.

$$\text{so} \quad 0 = \underbrace{[\vec{D_1}\mathbf{F}(\mathbf{a}), \; \vec{D_2}\mathbf{F}(\mathbf{a}), \; \vec{D_3}\mathbf{F}(\mathbf{a})]}_{[\mathbf{DF}(\mathbf{a})]}\begin{bmatrix} \dot{x} \\ \dot{y} \\ \dot{z} \end{bmatrix}; \quad \text{i.e.,} \quad [\mathbf{DF}(\mathbf{a})]\begin{bmatrix} \dot{x} \\ \dot{y} \\ \dot{z} \end{bmatrix} = 0. \quad \square$$

$$3.1.26$$

Parametrizations of curves and surfaces

We can think of curves and surfaces as being defined by equations, but there is another way to think of them (and of the higher-dimensional analogs we will encounter in Section 3.2): *parametrizations*. Actually, *local* parametrizations have been built into our definitions of curves and surfaces. Locally, as we have defined them, smooth curves and surfaces come provided both with equations and parametrizations. The graph of $f\begin{pmatrix} x \\ y \end{pmatrix}$ is both the locus of equation $z = f\begin{pmatrix} x \\ y \end{pmatrix}$ (expressing z as a function of x and y) and the image of the *parametrization*

$$\begin{pmatrix} x \\ y \end{pmatrix} \mapsto \begin{pmatrix} x \\ y \\ f\begin{pmatrix} x \\ y \end{pmatrix} \end{pmatrix}. \qquad 3.1.27$$

In Equation 3.1.27 we *parametrize* the surface by the variables x and y. But another part of the surface may be the graph of a function expressing x as a function of y and z; we would then be locally parametrizing the surface by the variables y and z.

Out loud, Equation 3.1.27 is

"x, y maps to $x, y, f\begin{pmatrix} x \\ y \end{pmatrix}$".

How would you interpret Example 3.1.4 (the unit circle) in terms of local parametrizations?[5]

Global parametrizations really represent a different way of thinking.

The first thing to know about parametrizations is that *practically any mapping is a "parametrization" of something.*

The second thing to know about parametrizations is that trying to find a global parametrization for a curve or surface that you know by equations (or even worse, by a picture on a computer monitor) is *very hard*, and often impossible. There is no general rule for solving such problems.

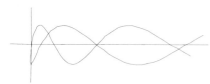

FIGURE 3.1.9.

A curve in the plane, known by the parametrization

$$t \mapsto \begin{pmatrix} t^2 - \sin t \\ 6 \sin t \cos t \end{pmatrix}.$$

By the first statement we mean that if you fill in the blanks of $t \mapsto \begin{pmatrix} - \\ - \end{pmatrix}$, where $-$ represents a function of t (t^3, $\sin t$, whatever) and ask a computer to plot it, it will draw you something that looks like a curve in the plane. If you happen to choose $t \mapsto \begin{pmatrix} \cos t \\ \sin t \end{pmatrix}$, it will draw you a circle; $t \mapsto \begin{pmatrix} \cos t \\ \sin t \end{pmatrix}$ *parametrizes* the circle. If you choose $t \mapsto \begin{pmatrix} t^2 - \sin t \\ 6 \sin t \cos t \end{pmatrix}$, you will get the curve shown in Figure 3.1.9.

---

[5]In Example 3.1.4, where the unit circle $x^2 + y^2 = 1$ is composed of points $\begin{pmatrix} x \\ y \end{pmatrix}$, we parametrized the top and bottom of the unit circle ($y > 0$ and $y < 0$) by x: we expressed the pivotal variable y as a function of the non-pivotal variable x, using the functions $y = f(x) = \sqrt{1 - x^2}$ and $y = f(x) = -\sqrt{1 - x^2}$. In the neighborhood of the points $\begin{pmatrix} 1 \\ 0 \end{pmatrix}$ and $\begin{pmatrix} -1 \\ 0 \end{pmatrix}$ we parametrized the circle by y: we expressed the pivotal variable x as a function of the non-pivotal variable y, using the functions $x = f(y) = \sqrt{1 - y^2}$ and $x = f(y) = -\sqrt{1 - y^2}$.

FIGURE 3.1.10.

A curve in space, known by the parametrization $t \mapsto \begin{pmatrix} \cos t \\ \sin t \\ at \end{pmatrix}$.

If you choose three functions of t, the computer will draw something that looks like a curve in space; if you happen to choose $t \mapsto \begin{pmatrix} \cos t \\ \sin t \\ at \end{pmatrix}$, you'll get the helix shown in Figure 3.1.10.

If you fill in the blanks of $\begin{pmatrix} u \\ v \end{pmatrix} \mapsto \begin{pmatrix} - \\ - \\ - \end{pmatrix}$, where $-$ represents a function of u and v (for example, $\sin^2 u \cos v$, for some such thing) the computer will draw you a surface in $\mathbb{R}^3$. The most famous parametrization of surfaces parametrizes the unit sphere in $\mathbb{R}^3$ by latitude u and longitude v:

$$\begin{pmatrix} u \\ v \end{pmatrix} \mapsto \begin{pmatrix} \cos u \cos v \\ \cos u \sin v \\ \sin u \end{pmatrix}. \qquad 3.1.28$$

But virtually whatever you type in, the computer will draw you something. For example, if you type in $\begin{pmatrix} u \\ v \end{pmatrix} \mapsto \begin{pmatrix} u^3 \cos v \\ u^2 + v^2 \\ v^2 \cos u \end{pmatrix}$, you will get the surface shown in Figure 3.1.11.

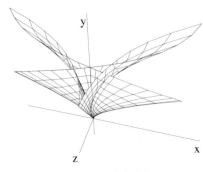

FIGURE 3.1.11.

How does the computer do it? It plugs some numbers into the formulas to find points of the curve or surface, and then it connects up the dots. Finding points on a curve or surface that you know by a parametrization is easy.

But the curves or surfaces we get by such "parametrizations" are not necessarily *smooth* curves or surfaces. If you typed random parametrizations into a computer (as we hope you did), you will have noticed that often what you get is not a smooth curve or surface; the curve or surface may intersect itself, as shown in Figures 3.1.9 and 3.1.11. If we want to define parametrizations of smooth curves and surfaces, we must be more demanding.

In Definition 3.1.20 we could write "$[\mathbf{D}\gamma(t)]$ is one to one" instead of "$\vec{\gamma}'(t) \neq 0$"; $\vec{\gamma}'(t)$ and $[\mathbf{D}\gamma(t)]$ are the same $n \times 1$ column matrix, and the linear transformation given by the matrix $[\mathbf{D}\gamma(t)]$ is one to one exactly when

$$\vec{\gamma}'(t) = [\mathbf{D}\gamma(t)] \neq \vec{0}.$$

Recall that γ is pronounced "gamma."

We could replace "one to one and onto" by "bijective."

Definition 3.1.20 (Parametrization of a curve). A parametrization of a smooth curve $C \subset \mathbb{R}^n$ is a mapping $\gamma : I \to C$ satisfying the following conditions:

(1) I is an open interval of $\mathbb{R}$.
(2) γ is C^1, one to one, and onto
(3) $\vec{\gamma}'(t) \neq 0$ for every $t \in I$.

Think of I as an interval of time; if you are traveling along the curve, the parametrization tells you where you are on the curve at a given time, as shown in Figure 3.1.12.

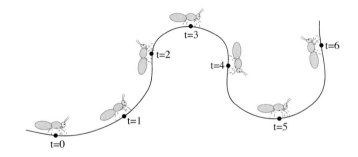

FIGURE 3.1.12. We imagine a parametrized curve as an ant taking a walk in the plane or in space. The parametrization tells where the ant is at any particular time.

In Definition 3.1.21, $[\mathbf{D}\gamma(\mathbf{u})]$ is a 3×2 matrix. Saying that it is one to one is the same as saying that the two partial derivatives $\vec{D}_1\gamma, \vec{D}_2\gamma$ are linearly independent. Recall that the kernel of a linear transformation represented by a matrix is 0 if and only if its columns are linearly independent; it takes two linearly independent vectors to span a plane, in this case the tangent plane.

In the case of the parametrization of a curve (Definition 3.1.20), the requirement that $\vec{\gamma}(t) \neq 0$ could also be stated in these terms: for one vector, being linearly independent means not being 0.

Definition 3.1.21 (Parametrization of a surface). A parametrization of a surface $S \subset \mathbb{R}^3$ is a smooth mapping $\gamma : U \to S$ such that

(1) $U \subset \mathbb{R}^2$ is open.
(2) γ is one to one and onto.
(3) $[\mathbf{D}\gamma(\mathbf{u})]$ is one to one for every $\mathbf{u} \in U$.

The parametrization

$$t \mapsto \begin{pmatrix} \cos t \\ \sin t \end{pmatrix},$$

which parametrizes the circle, is of course not one to one, but its restriction to $(0, 2\pi)$ is; unfortunately, this restriction misses the point $\begin{pmatrix} 1 \\ 0 \end{pmatrix}$.

It is generally far easier to get a picture of a curve or surface if you know it by a parametrization than if you know it by equations. In the case of the curve whose parametrization is given in Equation 3.1.29, it will take a computer milliseconds to compute the coordinates of enough points to give you a good picture of the curve.

It is rare to find a mapping γ that meets the criteria for a parametrization given by Definitions 3.1.20 and 3.1.21, and which parametrizes the entire curve or surface. A circle is not like an open interval: if you bend a strip of tubing into a circle, the two endpoints become a single point. A cylinder is not like an open subspace of the plane: if you roll up a piece of paper into a cylinder, two edges become a single line. Neither parametrization is one to one.

The sphere is similar. The parametrization by latitude and longitude (Equation 3.1.28) satisfies our definition only if we remove the curve going from the North Pole to the South Pole through Greenwich (for example).

Example 3.1.22 (Parametrizations vs. equations). If you know a curve by a global parametrization, it is easy to find points of the curve, but difficult to check whether a given point is on the curve. The opposite is true if you know the curve by an equation: then it may well be difficult to find points of the curve, but checking whether a point is on the curve is straightforward. For example, given the parametrization

$$\gamma : t \mapsto \begin{pmatrix} \cos^3 t - 3\sin t \cos t \\ t^2 - t^5 \end{pmatrix}, \qquad 3.1.29$$

you can find a point by substituting some value of t, like $t = 0$ or $t = 1$. But checking whether some particular point $\begin{pmatrix} a \\ b \end{pmatrix}$ is on the curve would be very difficult. That would require showing that the set of nonlinear equations

$$a = \cos^3 t - 3\sin t \cos t$$
$$b = t^2 - t^5 \qquad\qquad 3.1.30$$

has a solution.

Now suppose you are given the equation

$$y + \sin xy + \cos(x + y) = 0, \qquad\qquad 3.1.31$$

which defines a different curve. It's not clear how you would go about finding a point of the curve. But you could check whether a given point is on the curve simply by inserting the values for x and y in the equation.[6]   △

Remark. It is not true that if $\gamma : I \to C$ is a smooth mapping satisfying $\gamma'(t) \neq 0$ for every t, then C is necessarily a smooth curve. Nor is it true that if $\gamma : U \to S$ is a smooth mapping such that $[\mathbf{D}\gamma(\mathbf{u})]$ is one to one, then necessarily S is a smooth surface. This is true only locally: if I and U are small enough, then the image of the corresponding γ will be a smooth curve or smooth surface. A sketch of how to prove this is given in Exercise 3.1.20.   △

3.2 MANIFOLDS

A mathematician trying to picture a manifold is rather like a blindfolded person who has never met or seen a picture of an elephant seeking to identify one by patting first an ear, then the trunk or a leg.

In Section 3.1 we explored smooth curves and surfaces. We saw that a subset $X \subset \mathbb{R}^2$ is a smooth curve if X is locally the graph of a differentiable function, either of x in terms of y or of y in terms of x. We saw that $S \subset \mathbb{R}^3$ is a smooth surface if it is locally the graph of a differentiable function of one coordinate in terms of the other two. Often, we saw, a patchwork of graphs of function is required to express a curve or a surface.

This generalizes nicely to higher dimensions. You may not be able to visualize a five-dimensional manifold (we can't either), but you should be able to guess how we will determine whether some five-dimensional subset of $\mathbb{R}^n$ is a manifold: given a subset of $\mathbb{R}^n$ defined by equations, we use the implicit function theorem

---

[6]You might think, why not use Newton's method to find a point of the curve given by Equation 3.1.31? But Newton's method requires that you know a point of the curve to start out. What we could do is wonder whether the curve crosses the y-axis. That means setting $x = 0$, which gives $y + \cos y = 0$. This certainly has a solution by the intermediate value theorem: $y + \cos y$ is positive when $y > 1$, and negative when $y < -1$. So you might think that using Newton's method starting at $y = 0$ should converge to a root. In fact, the inequality of Kantorovitch's theorem (Equation 2.7.48) is not satisfied, so that convergence isn't guaranteed. But starting at $y = -\pi/4$ is guaranteed to work: this gives

$$\frac{M|f(y_0)|}{\left(f'(y_0)\right)^2} \leq 0.027 < \frac{1}{2}.$$

to determine whether every point of the subset has a neighborhood in which the subset is the graph of a function of several variables in terms of the others. If so, the set is a smooth manifold: manifolds are loci which are *locally the graphs of functions expressing some of the standard coordinate functions in terms of others*. Again, it is rare that a manifold is the graph of a single function.

Example 3.2.1 (Linked rods). Linkages of rods are everywhere, in mechanics (consider a railway bridge or the Eiffel tower), in biology (the skeleton), in robotics, in chemistry. One of the simplest examples is formed of four rigid rods, with assigned lengths $l_1, l_2, l_3, l_4 > 0$, connected by universal joints that can achieve any position, to form a quadrilateral, as shown in Figure 3.2.1.

In order to guarantee that our sets are not empty, we will require that each rod be shorter than the sum of the other three.

What is the set X_2 of positions the linkage can achieve if the points are restricted to a plane? Or the set X_3 of positions the linkage can achieve if the points are allowed to move in space? These sets are easy to describe by equations. For X_2 we have

$$X_2 = \text{the set of all } (\mathbf{x}_1, \mathbf{x}_2, \mathbf{x}_3, \mathbf{x}_4) \in (\mathbb{R}^2)^4 \quad \text{such that} \qquad 3.2.1$$

$$|\mathbf{x}_1 - \mathbf{x}_2| = l_1, \quad |\mathbf{x}_2 - \mathbf{x}_3| = l_2, \quad |\mathbf{x}_3 - \mathbf{x}_4| = l_3, \quad |\mathbf{x}_4 - \mathbf{x}_1| = l_4.$$

Thus X_2 is a subset of $\mathbb{R}^8$. Another way of saying this is that X_2 is the subset defined by the equation $\mathbf{f}(\mathbf{x}) = 0$, where $\mathbf{f} : (\mathbb{R}^2)^4 \to \mathbb{R}^4$ is the mapping

$$\mathbf{f}\left(\underbrace{\begin{pmatrix} x_1 \\ y_1 \end{pmatrix}}_{\mathbf{x}_1}, \underbrace{\begin{pmatrix} x_2 \\ y_2 \end{pmatrix}}_{\mathbf{x}_2}, \underbrace{\begin{pmatrix} x_3 \\ y_3 \end{pmatrix}}_{\mathbf{x}_3}, \underbrace{\begin{pmatrix} x_4 \\ y_4 \end{pmatrix}}_{\mathbf{x}_4}\right) = \begin{bmatrix} (x_2 - x_1)^2 + (y_2 - y_1)^2 - l_1^2 \\ (x_3 - x_2)^2 + (y_3 - y_2)^2 - l_2^2 \\ (x_4 - x_3)^2 + (y_4 - y_3)^2 - l_3^2 \\ (x_1 - x_4)^2 + (y_1 - y_4)^2 - l_4^2 \end{bmatrix}. \qquad 3.2.2$$

Similarly, the set X_3 of positions in space is also described by Equation 3.2.1, if we take $\mathbf{x}_i \in \mathbb{R}^3$; X_3 is a subset of $\mathbb{R}^{12}$. (Of course, to make equations corresponding to Equation 3.2.2 we would have to add a third entry to the $\mathbf{x}_i$, and instead of writing $(x_2 - x_1)^2 + (y_2 - y_1)^2 - l_1^2$ we would need to write $(x_2 - x_1)^2 + (y_2 - y_1)^2 + (z_2 - z_1)^2 - l_1^2$.)

Can we express some of the $\mathbf{x}_i$ as functions of the others? You should feel, on physical grounds, that if the linkage is sitting on the floor, you can move two opposite connectors any way you like, and that the linkage will follow in a unique way. This is not quite to say that $\mathbf{x}_2$ and $\mathbf{x}_4$ are a function of $\mathbf{x}_1$ and $\mathbf{x}_3$ (or that $\mathbf{x}_1$ and $\mathbf{x}_3$ are a function of $\mathbf{x}_2$ and $\mathbf{x}_4$). This isn't true, as is suggested by Figure 3.2.2.

In fact, usually knowing $\mathbf{x}_1$ and $\mathbf{x}_3$ determines either no positions of the linkage (if the $\mathbf{x}_1$ and $\mathbf{x}_3$ are farther apart than $l_1 + l_2$ or $l_3 + l_4$) or exactly four (if a few other conditions are met; see Exercise 3.2.3). But $\mathbf{x}_2$ and $\mathbf{x}_4$ are *locally* functions of $\mathbf{x}_1, \mathbf{x}_3$. It is true that for a given $\mathbf{x}_1$ and $\mathbf{x}_3$, four positions

Making some kind of global sense of such a patchwork of graphs of functions can be quite challenging indeed, especially in higher dimensions. It is a subject full of open questions, some fully as interesting and demanding as, for example, Fermat's last theorem, whose solution after more than three centuries aroused such passionate interest. Of particular interest are four-dimensional manifolds (4-manifolds), in part because of applications in representing spacetime.

This description is remarkably concise and remarkably uninformative. It isn't even clear how many dimensions X_2 and X_3 have; this is typical when you know a set by equations.

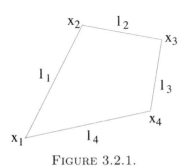

FIGURE 3.2.1.

One possible position of four linked rods, of lengths l_1, l_2, l_3, and l_4, restricted to a plane.

are possible in all, but if you move $\mathbf{x}_1$ and $\mathbf{x}_3$ a *small* amount from a given position, only one position of $\mathbf{x}_2$ and $\mathbf{x}_4$ is near the old position of $\mathbf{x}_2$ and $\mathbf{x}_4$. Locally, knowing $\mathbf{x}_1$ and $\mathbf{x}_3$ uniquely determines $\mathbf{x}_2$ and $\mathbf{x}_4$.

If you object that you cannot visualize what this manifold looks like, you have our sympathy; neither can we. Precisely for this reason, it gives a good idea of the kind of problem that comes up: you have a collection of equations defining some set but you have no idea what the set looks like. For example, as of this writing we don't know precisely when X_2 is connected—that is, whether we can move continuously from any point in X_2 to any other point in X_2. (A manifold can be disconnected, as we saw already in the case of smooth curves, in Example 3.1.6.) It would take a bit of thought to figure out for what lengths of bars X_2 is, or isn't, connected.

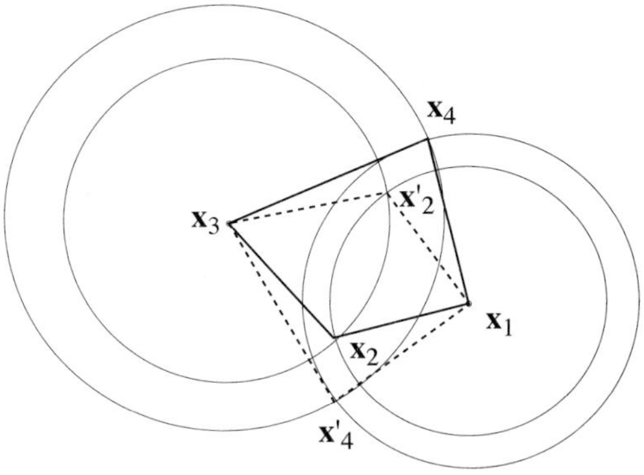

FIGURE 3.2.2. Two of the possible positions of a linkage with the same $\mathbf{x}_1$ and $\mathbf{x}_3$ are shown in solid and dotted lines. The other two are $\mathbf{x}_1, \mathbf{x}_2, \mathbf{x}_3, \mathbf{x}_4'$ and $\mathbf{x}_1, \mathbf{x}_2', \mathbf{x}_3, \mathbf{x}_4$.

Even this isn't always true: if any three are aligned, or if one rod is folded back against another, as shown in Figure 3.2.3, then the endpoints cannot be used as parameters (as the variables that determine the values of the other variables). For example, if $\mathbf{x}_1, \mathbf{x}_2$ and $\mathbf{x}_3$ are aligned, then you cannot move $\mathbf{x}_1$ and $\mathbf{x}_3$ arbitrarily, as the rods cannot be stretched. But it is still true that the position is a locally a function of $\mathbf{x}_2$ and $\mathbf{x}_4$.

There are many other possibilities: for instance, we could choose $\mathbf{x}_2$ and $\mathbf{x}_4$ as the variables that locally determine $\mathbf{x}_1$ and $\mathbf{x}_3$, again making X_2 locally a graph. Or we could use the coordinates of $\mathbf{x}_1$ (two numbers), the polar angle of the first rod with the horizontal line passing through $\mathbf{x}_1$ (one number), and the angle between the first and the second (one number): four numbers in all, the same number we get using the coordinates of $\mathbf{x}_1$ and $\mathbf{x}_3$.[7] We said above that usually knowing $\mathbf{x}_1$ and $\mathbf{x}_3$ determines either no positions of the linkage or exactly four positions. Exercise 3.2.4 asks you to determine how many positions are possible using $\mathbf{x}_1$ and the two angles above—again, except in a few cases. Exercise 3.2.5 asks you to describe X_2 and X_3 when $l_1 = l_2 + l_3 + l_4$.   △

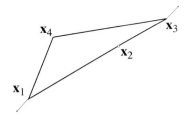

FIGURE 3.2.3.

If three vertices are aligned, the end-vertices cannot move freely: for instance, they can't moved in the directions of the arrows without stretching the rods.

A manifold: locally the graph of a function

The set X_2 of Example 3.2.1 is a four-dimensional manifold in $\mathbb{R}^8$; locally, it is the graph of a function expressing four variables (two coordinates each for two points) in terms of four other variables (the coordinates of the other two points

---

[7]Such a system is said to have four *degrees of freedom*.

A k-dimensional manifold M embedded in $\mathbb{R}^n$, denoted $M \subset \mathbb{R}^n$, is sometimes called a *submanifold* of $\mathbb{R}^n$. Strictly speaking, it should not be referred to simply as a "manifold," which can mean an abstract manifold, not embedded in any space. The manifolds in this book are all submanifolds of $\mathbb{R}^n$.

Definition 3.2.2 is not friendly. Unfortunately, it is difficult to be precise about what it means to be "locally the graph of a function" without getting involved. But we have seen examples of just what this means in the case of 1-manifolds (curves) and 2-manifolds (surfaces), in Section 3.1.

A k-manifold in $\mathbb{R}^n$ is locally the graph of a mapping expressing $n-k$ variables in terms of the other k variables.

If $U \subset \mathbb{R}^n$ is open, then U is a manifold. This corresponds to the case where $E_1 = \mathbb{R}^n$, $E_2 = \{\mathbf{0}\}$.

Figure 3.2.4 reinterprets Figure 3.1.1 (illustrating a smooth curve) in the language of Definition 3.2.2.

or some other choice). It doesn't have to be the same function everywhere. In most neighborhoods, X_2 is the graph of a function of $\mathbf{x}_1$ and $\mathbf{x}_3$, but we saw that this is not true when $\mathbf{x}_1, \mathbf{x}_2$ and $\mathbf{x}_3$ are aligned; near such points, X_2 is the graph of a function expressing $\mathbf{x}_1$ and $\mathbf{x}_3$ in terms of $\mathbf{x}_2$ and $\mathbf{x}_4$.[8]

Now it's time to define a manifold more precisely.

Definition 3.2.2 (Manifold). A subset $M \subset \mathbb{R}^n$ is a k-dimensional manifold embedded in $\mathbb{R}^n$ if it is locally the graph of a C^1 mapping expressing $n - k$ variables as functions of the other k variables. More precisely, for every $\mathbf{x} \in M$, we can find

(1) k standard basis vectors $\vec{\mathbf{e}}_{i_1}, \ldots, \vec{\mathbf{e}}_{i_k}$ corresponding to the k variables that, near $\mathbf{x}$, will determine the values of the other variables. Denote by E_1 the span of these, and by E_2 the span of the remaining $n - k$ standard basis vectors; let $\mathbf{x}_1$ be the projection of $\mathbf{x}$ onto E_1, and $\mathbf{x}_2$ its projection onto E_2;

(2) a neighborhood U of $\mathbf{x}$ in $\mathbb{R}^n$;

(3) a neighborhood U_1 of $\mathbf{x}_1$ in E_1;

(4) a mapping $\mathbf{f} : U_1 \to E_2$;

such that $M \cap U$ is the graph of $\mathbf{f}$.

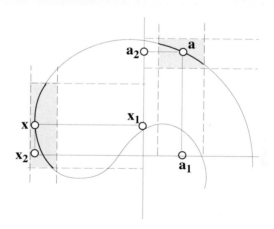

FIGURE 3.2.4. In the neighborhood of $\mathbf{x}$, the curve is the graph of a function expressing x in terms of y. The point $\mathbf{x}_1$ is the projection of $\mathbf{x}$ onto E_1 (i.e., the y-axis); the point $\mathbf{x}_2$ is its projection onto E_2 (i.e., the x-axis). In the neighborhood of $\mathbf{a}$, we can consider the curve the graph of a function expressing y in terms of x. For this point, E_1 is the x-axis, and E_2 is the y-axis.

---

[8] For some lengths, X_2 is no longer a manifold in a neighborhood of some positions: if all four lengths are equal, then X_2 is not a manifold near the position where it is folded flat.

A curve in $\mathbb{R}^2$ is a 1-manifold in $\mathbb{R}^2$; a surface in $\mathbb{R}^3$ is a 2-manifold in $\mathbb{R}^3$; a curve in $\mathbb{R}^3$ is a 1-manifold in $\mathbb{R}^3$.

Recall that for both curves in $\mathbb{R}^2$ and surfaces in $\mathbb{R}^3$, we had $n-k = 1$ variable expressed as a function of the other k variables. For curves in $\mathbb{R}^3$, there are $n - k = 2$ variables expressed as a function of one variable; in Example 3.2.1 we saw that for X_2, we had four variables expressed as a function of four other variables: X_2 is a 4-manifold in $\mathbb{R}^8$.

Of course, once manifolds get a bit more complicated it is impossible to draw them or even visualize them. So it's not obvious how to use Definition 3.2.2 to see whether a set is a manifold. Fortunately, Theorem 3.2.3 will give us a more useful criterion.

Manifolds known by equations

Since $\mathbf{f} : U \to \mathbb{R}^{n-k}$, saying that $[\mathbf{Df}(\mathbf{x})]$ is onto is the same as saying that it has $n - k$ linearly independent columns, which is the same as saying that those $n - k$ columns span $\mathbb{R}^{n-k}$: the equation

$$[\mathbf{Df}(\mathbf{x})]\vec{\mathbf{v}} = \vec{\mathbf{b}}$$

has a solution for every $\vec{\mathbf{b}} \in \mathbb{R}^{n-k}$. (This is the crucial hypothesis of the stripped-down version of the implicit function theorem, Theorem 2.9.9.)

How do we know that our linkage spaces X_2 and X_3 of Example 3.2.1 are manifolds? Our argument used some sort of intuition about how the linkage would move if we moved various points on it, and although we could prove this using a bit of trigonometry, we want to see directly that it is a manifold from Equation 3.2.1. This is a matter of saying that $\mathbf{f}(\mathbf{x}) = \mathbf{0}$ expresses some variables implicitly as functions of others, and this is exactly what the implicit function theorem is for.

Theorem 3.2.3 (Knowing a manifold by equations). *Let $U \subset \mathbb{R}^n$ be an open subset, and $\mathbf{f} : U \to \mathbb{R}^{n-k}$ be a differentiable mapping with Lipschitz derivative (for instance a C^2 mapping). Let $M \subset U$ be the set of solutions to the equation $\mathbf{f}(\mathbf{x}) = \mathbf{0}$.*

If $[\mathbf{Df}(\mathbf{x})]$ is onto, then M is a k-dimensional manifold embedded in $\mathbb{R}^n$.

In the proof of Theorem 3.2.3 we would prefer to write

$$\mathbf{f} \begin{pmatrix} \mathbf{u} \\ \mathbf{g}(\mathbf{u}) \end{pmatrix} = 0 \quad \text{rather than}$$

$$\mathbf{f}\Big(\mathbf{u} + \mathbf{g}(\mathbf{u})\Big) = \mathbf{0},$$

but that's not quite right because E_1 may not be spanned by the first k basis vectors. We have $\mathbf{u} \in E_1$ and $\mathbf{g}(\mathbf{u}) \in E_2$; since both E_1 and E_2 are subspaces of $\mathbb{R}^n$, it makes sense to add them, and $\mathbf{u} + \mathbf{g}(\mathbf{u})$ is a point of the graph of $\mathbf{u}$. This is a fiddly point; if you find it easier to think of $\mathbf{f} \begin{pmatrix} \mathbf{u} \\ \mathbf{g}(\mathbf{u}) \end{pmatrix}$, go ahead; just pretend that E_1 is spanned by $\vec{\mathbf{e}}_1, \ldots, \vec{\mathbf{e}}_k$, and E_2 by $\vec{\mathbf{e}}_{k+1}, \ldots, \vec{\mathbf{e}}_n$.

This theorem is a generalization of part (a) of Theorems 3.1.9 (for curves) and 3.1.16 (for surfaces). Note that we cannot say—as we did for surfaces in Theorem 3.1.16—that M is a k-manifold if $[\mathbf{Df}(\mathbf{x})] \neq 0$. Here $[\mathbf{Df}(\mathbf{x})]$ is a matrix $n - k$ high and n wide; it could be nonzero and still fail to be onto. Note also that k, the dimension of M, is $n - (n - k)$, i.e., the dimension of the domain of $\mathbf{f}$ minus the dimension of its range.

Proof.   This is very close to the statement of the implicit function theorem, Theorem 2.9.10. Choose $n - k$ of the basis vectors $\vec{\mathbf{e}}_i$ such that the corresponding columns of $[\mathbf{Df}(\mathbf{x})]$ are linearly independent (corresponding to pivotal variables). Denote by E_2 the subspace of $\mathbb{R}^n$ spanned by these vectors, and by E_1 the subspace spanned by the remaining k standard basis vectors. Clearly $\dim E_2 = n - k$ and $\dim E_1 = k$.

Let $\mathbf{x}_1$ be the projection of $\mathbf{x}$ onto E_1, and $\mathbf{x}_2$ be its projection onto E_2. The implicit function theorem then says that there exists a ball U_1 around $\mathbf{x}_1$, a ball U_2 around $\mathbf{x}_2$, and a differentiable mapping $\mathbf{g} : U_1 \to U_2$ such that $\mathbf{f}\big(\mathbf{u} + \mathbf{g}(\mathbf{u})\big) = \mathbf{0}$, so that the graph of $\mathbf{g}$ is a subset of M. Moreover, if U is the set of points with E_1-coordinates in U_1 and E_2-coordinates in U_2, then the implicit function theorem guarantees that the graph of $\mathbf{g}$ is $M \cap U$. This proves the theorem. $\square$

Example 3.2.4 (Using Theorem 3.2.3 to check that the linkage space X_2 is a manifold). In Example 3.2.1, X_2 is given by the equation

$$
\mathbf{f}\begin{pmatrix} x_1 \\ y_1 \\ x_2 \\ y_2 \\ x_3 \\ y_3 \\ x_4 \\ y_4 \end{pmatrix} = \begin{bmatrix} (x_2 - x_1)^2 + (y_2 - y_1)^2 - l_1^2 \\ (x_3 - x_2)^2 + (y_3 - y_2)^2 - l_2^2 \\ (x_4 - x_3)^2 + (y_4 - y_3)^2 - l_3^2 \\ (x_1 - x_4)^2 + (y_1 - y_4)^2 - l_4^2 \end{bmatrix} = 0. \qquad 3.2.3
$$

Each partial derivative at right is a vector with four entries: e.g.,

$$
\vec{D_1}\mathbf{f}(\mathbf{x}) = \begin{bmatrix} D_1 f_1(\mathbf{x}) \\ D_1 f_2(\mathbf{x}) \\ D_1 f_3(\mathbf{x}) \\ D_1 f_4(\mathbf{x}) \end{bmatrix}
$$

and so on.

The derivative is composed of the eight partial derivatives (in the second line we label the partial derivatives explicitly by the names of the variables):

$$
[\mathbf{Df}(\mathbf{x})] = [\vec{D_1}\mathbf{f}(\mathbf{x}), \vec{D_2}\mathbf{f}(\mathbf{x}), \vec{D_3}\mathbf{f}(\mathbf{x}), \vec{D_4}\mathbf{f}(\mathbf{x}), \vec{D_5}\mathbf{f}(\mathbf{x}), \vec{D_6}\mathbf{f}(\mathbf{x}), \vec{D_7}\mathbf{f}(\mathbf{x}), \vec{D_8}\mathbf{f}(\mathbf{x})]
$$

$$
= [\vec{D_{x_1}}\mathbf{f}(\mathbf{x}), \vec{D_{y_1}}\mathbf{f}(\mathbf{x}), \vec{D_{x_2}}\mathbf{f}(\mathbf{x}), \vec{D_{y_2}}\mathbf{f}(\mathbf{x}), \vec{D_{x_3}}\mathbf{f}(\mathbf{x}), \vec{D_{y_3}}\mathbf{f}(\mathbf{x}), \vec{D_{x_4}}\mathbf{f}(\mathbf{x}), \vec{D_{y_4}}\mathbf{f}(\mathbf{x})].
$$

Unfortunately we had to put the matrix on two lines to make it fit. The second line contains the last four columns of the matrix.

Computing the partial derivatives gives

$$
[\mathbf{Df}(\mathbf{x})] = \begin{bmatrix} 2(x_1 - x_2) & 2(y_1 - y_2) & -2(x_1 - x_2) & -2(y_1 - y_2) \\ 0 & 0 & 2(x_2 - x_3) & 2(y_2 - y_3) \\ 0 & 0 & 0 & 0 \\ -2(x_4 - x_1) & -2(y_4 - y_1) & 0 & 0 \end{bmatrix} \qquad 3.2.4
$$

$$
\begin{bmatrix} 0 & 0 & 0 & 0 \\ -2(x_2 - x_3) & -2(y_2 - y_3) & 0 & 0 \\ (x_3 - x_4) & 2(y_3 - y_4) & -2(x_3 - x_4) & -2(y_3 - y_4) \\ 0 & 0 & 2(x_4 - x_1) & 2(y_4 - y_1) \end{bmatrix}.
$$

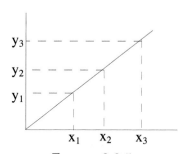

FIGURE 3.2.5.

If the points

$$
\begin{pmatrix} x_1 \\ y_1 \end{pmatrix}, \begin{pmatrix} x_2 \\ y_2 \end{pmatrix}, \begin{pmatrix} x_3 \\ y_3 \end{pmatrix}
$$

are aligned, then the first two columns of Equation 3.2.5 cannot be linearly independent: $y_1 - y_2$ is necessarily a multiple of $x_1 - x_2$, and $y_2 - y_3$ is a multiple of $x_2 - x_3$.

Since $\mathbf{f}$ is a mapping from $\mathbb{R}^8$ to $\mathbb{R}^4$, so that E_2 has dimension $n - k = 4$, four standard basis vectors can be used to span E_2 if the four corresponding column vectors are linearly independent. For instance, here you can never use the first four, or the last four, because in both cases there is a row of zeroes. How about the third, fourth, seventh, and eighth, i.e., the points $\mathbf{x}_2 = \begin{pmatrix} x_2 \\ y_2 \end{pmatrix}, \mathbf{x}_4 = \begin{pmatrix} x_4 \\ y_4 \end{pmatrix}$? These work as long as the corresponding columns of the matrix

$$
\begin{bmatrix} -2(x_1 - x_2) & -2(y_1 - y_2) & 0 & 0 \\ 2(x_2 - x_3) & 2(y_2 - y_3) & 0 & 0 \\ 0 & 0 & -2(x_3 - x_4) & -2(y_3 - y_4) \\ 0 & 0 & 2(x_4 - x_1) & 2(y_4 - y_1) \end{bmatrix} \qquad 3.2.5
$$

$$
\quad D_{x_2}\mathbf{f}(\mathbf{x}) \qquad D_{y_2}\mathbf{f}(\mathbf{x}) \qquad D_{x_4}\mathbf{f}(\mathbf{x}) \qquad D_{y_4}\mathbf{f}(\mathbf{x})
$$

are linearly independent. The first two columns are linearly independent precisely when $\mathbf{x}_1, \mathbf{x}_2$, and $\mathbf{x}_3$ are not aligned as they are in Figure 3.2.5, and the last two are linearly independent when $\mathbf{x}_3, \mathbf{x}_4$, and $\mathbf{x}_1$ are not aligned. The same argument holds for the first, second, fifth, and sixth columns, corresponding to $\mathbf{x}_1$ and $\mathbf{x}_3$. Thus you can use the positions of opposite points to locally parametrize X_2, as long as the other two points are aligned with neither of the two opposite points. The points are never all four in line, unless either one length is the sum of the other three, or $l_1 + l_2 = l_3 + l_4$, or $l_2 + l_3 = l_4 + l_1$. In all other cases, X_2 is a manifold, and even in these last two cases, it is a manifold except perhaps at the positions where all four rods are aligned.

Equations versus parametrizations

As in the case of curves and surfaces, there are two different ways of knowing a manifold: equations and parametrizations. Usually we start with a set of equations. Technically, such a set of equations gives us a complete description of the manifold. In practice (as we saw in Example 3.1.22 and Equation 3.2.2) such a description is not satisfying; the information is not in a form that can be understood as a global picture of the manifold. Ideally, we also want to know the manifold by a global parametrization; indeed, we would like to be able to move freely between these two representations. This duality repeats a theme of linear algebra, as suggested by Figure 3.2.6.

	Algorithms	Algebra	Geometry
Linear Algebra	Row reduction	Inverses of matrices Solving linear equations	Subspaces Kernels and images
Differential Calculus	Newton's method	Inverse function theorem Implicit function theorem	Manifolds Defining manifolds by equations and parametrizations

FIGURE 3.2.6. Correspondences: algorithms, algebra, geometry

William Thurston, arguably the best geometer of the 20th century, says that the right way to know a k-dimensional manifold embedded in n-dimensional space is neither by equations nor by parametrizations but from the inside: imagine yourself inside the manifold, walking in the dark, aiming a flashlight first at one spot, then another. If you point the flashlight straight ahead, will you see anything? Will anything be reflected back? Or will you see the light to your side? ...

Mappings that meet these criteria, and which parametrize the entire manifold, are rare. Choosing even a local parametrization that is well adapted to the problem at hand is a difficult and important skill, and exceedingly difficult to teach.

The definition of a parametrization of a manifold is simply a generalization of our definitions of a parametrization of a curve and of a surface:

Definition 3.2.5 (Parametrization of a manifold). A parametrization of a k-dimensional manifold $M \subset \mathbb{R}^n$ is a mapping $\gamma : U \to M$ satisfying the following conditions:

(1) U is an open subset of $\mathbb{R}^k$.
(2) γ is C^1, one to one, and onto;
(3) $[\mathbf{D}\gamma(\mathbf{u})]$ is one to one for every $\mathbf{u} \in U$.

The tangent space to a manifold

In this sense, a manifold is a surface in space (possibly, higher-dimensional space) that looks flat if you look closely at a small region.

As mentioned in Section 3.1, the tangent space will be essential in the discussion of constrained extrema, in Section 3.7, and in the discussion of orientation, in Section 6.5.

The essence of a k-dimensional differentiable manifold is that it is well approximated, near every point, by a k-dimensional subspace of $\mathbb{R}^n$. Everyone has an intuition of what this means: a curve is approximated by its tangent line at a point, a surface by its tangent plane.

Just as in the cases of curves and surfaces, we want to distinguish the *tangent vector space* $T_{\mathbf{x}}M$ to a manifold M at a point $\mathbf{x} \in M$ from the tangent line (or plane, ...) to the manifold at $\mathbf{x}$. The tangent space $T_{\mathbf{x}}M$ is the set of vectors tangent to M at $\mathbf{x}$.

Definition 3.2.6 (Tangent space of a manifold). Let $M \subset \mathbb{R}^n$ be a k-dimensional manifold and let $\mathbf{x} \in M$, so that

- k standard basis vectors span E_1;
- the remaining $n - k$ standard basis vectors span E_2;
- $U_1 \subset E_1$, $U \subset \mathbb{R}^n$ are open sets, and
- $\mathbf{g} : U_1 \to E_2$ is a C^1 mapping,

such that $\mathbf{x} \in U$ and $M \cap U$ is the graph of $\mathbf{g}$. Then the tangent vector space to the manifold at $\mathbf{x}$, denoted $T_{\mathbf{x}}M$, is the graph of $[\mathbf{Dg}(\mathbf{x})]$: the linear approximation to the graph is the graph of the linear approximation.

If we know a manifold by the equation $\mathbf{f} = 0$, then the tangent space to the manifold is the kernel of the derivative of $\mathbf{f}$.

Part (b) of Theorems 3.1.9 (for curves) and 3.1.16 (for surfaces) are special cases of Theorem 3.2.7.

Theorem 3.2.7 (Tangent space to a manifold). *If $\mathbf{f} = 0$ describes a manifold, under the same conditions as in Theorem 3.2.3, then the tangent space $T_{\mathbf{x}}M$ is the kernel of $[\mathbf{Df}(\mathbf{x})]$.*

Proof. Let $\mathbf{g}$ be the function of which M is locally the graph, as discussed in the proof of Theorem 3.2.3. The implicit function theorem gives not only the existence of $\mathbf{g}$ but also its derivative (Equation 2.9.25): the matrix

$$[\mathbf{Dg}(\mathbf{x}_1)] = -\underbrace{[D_{j_1}\mathbf{f}(\mathbf{x}), \ \dots, \ D_{j_{n-k}}\mathbf{f}(\mathbf{x})]}_{\substack{\text{partial deriv. for} \\ \text{pivotal variables}}}^{-1} \underbrace{[D_{i_1}\mathbf{f}(\mathbf{x}), \ \dots, \ D_{i_k}\mathbf{f}(\mathbf{x})]}_{\substack{\text{partial deriv. for} \\ \text{non-pivotal variables}}} \qquad 3.2.6$$

where $D_{j_1}, \dots, D_{j_{n-k}}$ are the partial derivatives with respect to the $n-k$ pivotal variables, and $D_{i_1}, \dots, D_{i_k}$ are the partial derivatives with respect to the k non-pivotal variables.

By definition, the tangent space to M at $\mathbf{x}$ is the graph of the derivative of $\mathbf{g}$. Thus the tangent space is the space of equation

$$\vec{\mathbf{w}} = -[D_{j_1}\mathbf{f}(\mathbf{x}), \ \dots, \ D_{j_{n-k}}\mathbf{f}(\mathbf{x})]^{-1}[D_{i_1}\mathbf{f}(\mathbf{x}), \ \dots, \ D_{i_k}\mathbf{f}(\mathbf{x})]\vec{\mathbf{v}}, \qquad 3.2.7$$

One thing needs checking: if the same manifold can be represented as a graph in two different ways, then the tangent spaces should be the same. This should be clear from Theorem 3.2.7. Indeed, if an equation $\mathbf{f}(\mathbf{x})$ expresses some variables in terms of others in several different ways, then in all cases, the tangent space is the kernel of the derivative of $\mathbf{f}$ and does not depend on the choice of pivotal variables.

In Theorem 3.2.8, T^{-1} is not an inverse mapping; indeed, since T goes from $\mathbb{R}^n$ to $\mathbb{R}^m$, such an inverse mapping does not exist when $n \neq m$. By $T^{-1}(M)$ we denote the inverse image: the set of points $\mathbf{x} \in \mathbb{R}^n$ such that $T(\mathbf{x})$ is in M.

A graph is automatically given by an equation. For instance, the graph of $f : \mathbb{R} \to \mathbb{R}$ is the curve of equation $y - f(x) = 0$.

where $\vec{\mathbf{v}}$ is a variable in E_1, and $\vec{\mathbf{w}}$ is a variable in E_2. This can be rewritten

$$[D_{j_1}\mathbf{f}(\mathbf{x}), \ \ldots, \ D_{j_{n-k}}\mathbf{f}(\mathbf{x})]\vec{\mathbf{w}} + [D_{i_1}\mathbf{f}(\mathbf{x}), \ \ldots, \ D_{i_k}\mathbf{f}(\mathbf{x})]\vec{\mathbf{v}} = 0, \qquad 3.2.8$$

which is simply saying $[\mathbf{Df}(\vec{\mathbf{v}} + \vec{\mathbf{w}})] = 0$. $\square$

Manifolds are independent of coordinates

We defined smooth curves, surfaces, and higher-dimensional manifolds in terms of coordinate systems, but these objects are independent of coordinates; it doesn't matter if you translate a curve in the plane, or rotate a surface in space. In fact Theorem 3.2.8 says a great deal more.

Theorem 3.2.8. *Let* $T : \mathbb{R}^n \to \mathbb{R}^m$ *be a linear transformation that is onto. If* $M \subset \mathbb{R}^m$ *is a smooth* k-*dimensional manifold, then* $T^{-1}(M)$ *is a smooth manifold, of dimension* $k + n - m$.

Proof.   Choose a point $\mathbf{a} \in T^{-1}(M)$, and set $\mathbf{b} = T(\mathbf{a})$. Using the notation of Definition 3.2.2, there exists a neighborhood U of $\mathbf{b}$ such that the subset $M \cap U$ is defined by the equation $F(\mathbf{x}) = \mathbf{0}$, where $F : U \to E_2$ is given by

$$F\left(\begin{matrix}\mathbf{x}_1 \\ \mathbf{x}_2\end{matrix}\right) = f(\mathbf{x}_1) - \mathbf{x}_2 = \mathbf{0}. \qquad 3.2.9$$

Moreover, $[\mathbf{D}F(\mathbf{b})]$ is certainly onto, since the columns corresponding to the variables in E_2 make up the identity matrix.

The set $T^{-1}(M \cap U) = T^{-1}M \cap T^{-1}(U)$ is defined by the equation $F \circ T(\mathbf{y}) = 0$. Moreover,

$$[\mathbf{D}F \circ T(\mathbf{a})] = [\mathbf{D}F(T(\mathbf{a}))] \circ [\mathbf{D}T(\mathbf{a})] = [\mathbf{D}F(\mathbf{b})] \circ T \qquad 3.2.10$$

is also onto, since it is a composition of two mappings which are both onto. So $T^{-1}M$ is a manifold by Theorem 3.2.3.

For the dimension of the smooth manifold $T^{-1}(M)$, we use Theorem 3.2.3 to say that it is n (the dimension of the domain of $F \circ T$) minus $m - k$ (the dimension of the range of $F \circ T$), i.e., $n - m + k$. $\square$

Corollary 3.2.9 follows immediately from Theorem 3.2.8, as applied to T^{-1}:

$$T(M) = (T^{-1})^{-1}(M).$$

Corollary 3.2.9 (Manifolds are independent of coordinates).   *If* $T : \mathbb{R}^m \to \mathbb{R}^m$ *is an invertible linear transformation, and* $M \subset \mathbb{R}^m$ *is a* k-*dimensional manifold, then* $T(M)$ *is also a* k-*dimensional manifold.*

Corollary 3.2.9 says in particular that if you rotate a manifold the result is still a manifold, and our definition, which appeared to be tied to the coordinate system, is in fact coordinate-independent.

3.3 Taylor Polynomials in Several Variables

In Sections 3.1 and 3.2 we used first-degree approximations (derivatives) to discuss curves, surfaces, and higher-dimensional manifolds. Now we will discuss higher-degree approximations, using Taylor polynomials.

Approximation of functions by polynomials is a central issue in calculus in one and several variables. It is also of great importance in such fields as interpolation and curve fitting, computer graphics, and computer aided design; when a computer graphs a function, most often it is approximating it with cubic piecewise polynomial functions. In Section 3.8 we will apply these notions to the geometry of curves and surfaces. (The geometry of manifolds is quite a bit harder.)

Almost the only functions that can be computed are polynomials, or rather piecewise polynomial functions, also known as *splines*: functions formed by stringing together bits of different polynomials. Splines can be computed, since you can put *if* statements in the program that computes your function, allowing you to compute different polynomials for different values of the variables. (Approximation by rational functions, which involves division, is also important in practical applications.)

Taylor's theorem in one variable

In one variable, you learned that at a point x near a, a function is well approximated by its Taylor polynomial at a. Below, recall that $f^{(n)}$ denotes the nth derivative of f.

One proof, sketched in Exercise 3.3.8, consists of using l'Hôpital's rule k times. The theorem is also a special case of Taylor's theorem in several variables.

Theorem 3.3.1 (Taylor's theorem without remainder, one variable).
If $U \subset \mathbb{R}$ is an open subset and $f : U \to \mathbb{R}$ is k times continuously differentiable on U, then the polynomial

$$\underbrace{p_{f,a}^k(a+h)}_{\text{Taylor polynomial}} = f(a) + f'(a)h + \frac{1}{2!}f''(a)h^2 + \cdots + \frac{1}{k!}f^{(k)}(a)h^k \qquad 3.3.1$$

is the best approximation to f at a in the sense that it is the unique polynomial of degree $\leq k$ such that

$$\lim_{h \to 0} \frac{f(a+h) - p_{f,a}^k(a+h)}{h^k} = 0. \qquad 3.3.2$$

We will see that there is a polynomial in n variables that in the same sense best approximates functions of n variables.

Multi-exponent notation for polynomials in higher dimensions

First we must introduce some notation. In one variable, it is easy to write the "general polynomial" of degree k as

$$a_0 + a_1 x + a_2 x^2 + \cdots + a_k x^k = \sum_{i=0}^{k} a_i x^i. \qquad 3.3.3$$

For example,

$$3 + 2x - x^2 + 4x^4 = 3x^0 + 2x^1 - 1\,x^2 + 0\,x^3 + 4x^4 \qquad 3.3.4$$

Polynomials in several variables really are a lot more complicated than in one variable: even the first questions involving factoring, division, etc. lead rapidly to difficult problems in algebraic geometry.

can be written as

$$\sum_{i=0}^{4} a_i x^i, \quad \text{where} \quad a_0 = 3, \ a_1 = 2, \ a_2 = -1, \ a_3 = 0, \ a_4 = 4. \qquad 3.3.5$$

But it isn't obvious how to find a "general notation" for expressions like

$$1 + x + yz + x^2 + xyz + y^2 z - x^2 y^2. \qquad 3.3.6$$

One effective if cumbersome notation uses *multi-exponents*. A multi-exponent is a way of denoting one term of an expression like Equation 3.3.6.

Definition 3.3.2 (Multi-exponent). A multi-exponent I is an ordered finite sequence of non-negative whole numbers, which definitely may include 0:

$$I = (i_1, \ldots, i_n). \qquad 3.3.7$$

Example 3.3.3 (Multi-exponents). In the following polynomial with $n = 3$ variables:

$$1 + x + yz + x^2 + xyz + y^2 z - x^2 y^2, \qquad (3.3.6)$$

each multi-exponent I can be used to describe one term:

$$\begin{aligned}
1 &= x^0 y^0 z^0 \quad \text{corresponds to} \quad I = (0,0,0) \\
x &= x^1 y^0 z^0 \quad \text{corresponds to} \quad I = (1,0,0) \qquad 3.3.8 \\
yz &= x^0 y^1 z^1 \quad \text{corresponds to} \quad I = (0,1,1). \quad \triangle
\end{aligned}$$

What multi-exponents describe the terms $x^2, xyz, y^2 z$, and $x^2 y^2$?[9]

The set of multi-exponents with n entries is denoted $\mathcal{I}_n$:

$$\mathcal{I}_n = \{(i_1, \ldots, i_n)\} \qquad 3.3.9$$

The set $\mathcal{I}_3$ includes the seven multi-exponents of Equation 3.3.8, but many others as well, for example $I = (0,1,0)$, which corresponds to the term y, and $I = (2,2,2)$, which corresponds to the term $x^2 y^2 z^2$. (In the case of the polynomial of Equation 3.3.6, these terms have coefficient 0.)

We can group together elements of $\mathcal{I}_n$ according to their *degree*:

---
[9]

$$\begin{aligned}
x^2 &= x^2 y^0 z^0 \quad \text{corresponds to} \quad I = (2,0,0). \\
xyz &= x^1 y^1 z^1 \quad \text{corresponds to} \quad I = (1,1,1). \\
y^2 z &= x^0 y^2 z^1 \quad \text{corresponds to} \quad I = (0,2,1). \\
x^2 y^2 &= x^2 y^2 z^0 \quad \text{corresponds to} \quad I = (2,2,0).
\end{aligned}$$

Definition 3.3.4 (Degree of a multi-exponent). For any multi-exponent $I \in \mathcal{I}_n$, the total degree of I is $\deg I = i_1 + \cdots + i_n$.

The degree of xyz is 3, since $1 + 1 + 1 = 3$; the degree of $y^2 z$ is also 3.

Recall that $0! = 1$, not 0.

Definition 3.3.5 (I!). For any multi-exponent $I \in \mathcal{I}_n$,

$$I! = i_1! \cdot \ldots \cdot i_n!. \qquad 3.3.10$$

For example, if $I = (2, 0, 3)$, then $I! = 2!0!3! = 12$.

For example, the set $\mathcal{I}_3^2$ of multi-exponents with three entries and total degree 2 consists of $(0, 1, 1)$, $(1, 1, 0)$, $(1, 0, 1)$, $(2, 0, 0)$, $(0, 2, 0)$, and $(0, 0, 2)$.

Definition 3.3.6 ($\mathcal{I}_n^k$). We denote by $\mathcal{I}_n^k$ the set of multi-exponents with n entries and of total degree k.

What are the elements of the set $\mathcal{I}_2^3$? Of $\mathcal{I}_3^3$? Check your answers below.[10]

Using multi-exponents, we can break up a polynomial into a sum of *monomials* (as we already did in Equation 3.3.8).

The monomial $x_2^2 x_4^3$ is of degree 5; it can be written

$$\mathbf{x}^I = \mathbf{x}^{(0,2,0,3)}.$$

Definition 3.3.7 (Monomial). For any $I \in \mathcal{I}_n$, the function

$$\mathbf{x}^I = x_1^{i_1} \ldots x_n^{i_n} \quad \text{on } \mathbb{R}^n \text{ will be called a monomial of degree } \deg I.$$

In Equation 3.3.12, m is just a placeholder indicating the degree. To write a polynomial with n variables, first we consider the single multi-exponent I of degree $m = 0$, and determine its coefficient. Next we consider the set $\mathcal{I}_n^1$ (multi-exponents of degree $m = 1$) and for each we determine its coefficient. Then we consider the set $\mathcal{I}_n^2$ (multi-exponents of degree $m = 2$), and so on. Note that we could use the multi-exponent notation without grouping by degree, expressing a polynomial as

$$\sum_{I \in \mathcal{I}_n} a_I \mathbf{x}^I.$$

But it is often useful to group together terms of a polynomial by degree: constant term, linear terms, quadratic terms, cubic terms, etc.

Here i_1 gives the power of x_1, while i_2 gives the power of x_2, and so on. If $I = (2, 3, 1)$, then $\mathbf{x}^I$ is a monomial of degree 6:

$$\mathbf{x}^I = \mathbf{x}^{(2,3,1)} = x_1^2 x_2^3 x_3^1. \qquad 3.3.11$$

We can now write the general polynomial of degree k as a sum of monomials, each with its own coefficient a_I:

$$\sum_{m=0}^{k} \sum_{I \in \mathcal{I}_n^m} a_I \mathbf{x}^I. \qquad 3.3.12$$

Example 3.3.8 (Multi-exponent notation). To apply this notation to the polynomial

$$2 + x_1 - x_2 x_3 + 4 x_1 x_2 x_3 + 2 x_1^2 x_2^2, \qquad 3.3.13$$

we break it up into the terms:

$$2 = 2 x_1^0 x_2^0 x_3^0 \qquad I = (0, 0, 0), \text{ degree 0, with coefficient 2}$$
$$x_1 = 1 x_1^1 x_2^0 x_3^0 \qquad I = (1, 0, 0), \text{ degree 1, with coefficient 1}$$
$$-x_2 x_3 = -1 x_1^0 x_2^1 x_3^1 \qquad I = (0, 1, 1), \text{ degree 2, with coefficient } -1$$
$$4 x_1 x_2 x_3 = 4 x_1^1 x_2^1 x_3^1 \qquad I = (1, 1, 1), \text{ degree 3, with coefficient 4}$$
$$2 x_1^2 x_2^2 = 2 x_1^2 x_2^2 x_3^0 \qquad I = (2, 2, 0), \text{ degree 4, with coefficient 2}.$$

---

[10] $\mathcal{I}_2^3 = \{(1, 2), (2, 1), (0, 3), (3, 0);$ $\mathcal{I}_3^3 = \{(1, 1, 1), (2, 1, 0), (2, 0, 1), (1, 2, 0),$ $(1, 0, 2), (0, 2, 1), (0, 1, 2), (3, 0, 0), (0, 3, 0), (0, 0, 3).$

Thus we can write the polynomial as

$$\sum_{m=0}^{4} \sum_{I \in \mathcal{I}_3^m} a_I \mathbf{x}^I, \quad \text{where} \qquad\qquad 3.3.14$$

We write $I \in \mathcal{I}_3^m$ under the second sum in Equation 3.3.14 because the multi-exponents I that we are summing are sequences of three numbers, x_1, x_2 and x_3, and have total degree m.

$$a_{(0,0,0)} = 2, \qquad a_{(1,0,0)} = 1, \qquad a_{(0,1,1)} = -1,$$
$$a_{(1,1,1)} = 4, \qquad\quad a_{(2,2,0)} = 2, \qquad\qquad\qquad 3.3.15$$

and all other $a_I = 0$, for $I \in \mathcal{I}_3^m$, with $m \leq 4$. (There are 30 such terms.) $\triangle$

What is the polynomial

Exercise 3.3.6 provides more practice with multi-exponent notation.

$$\sum_{m=0}^{3} \sum_{I \in \mathcal{I}_2^m} a_I \mathbf{x}^I, \qquad\qquad 3.3.16$$

where $a_{(0,0)} = 3$, $a_{(1,0)} = -1$, $a_{(1,2)} = 3$, $a_{(2,1)} = 2$, and all the other coefficients a_I are 0? Check your answer below.[11]

Multi-exponent notation and equality of crossed partial derivatives

Multi-exponent notation also provides a concise way to describe the higher partial derivatives in Taylor polynomials in higher dimensions. Recall (Definition 2.7.6) that if the function $D_i f$ is differentiable, then its partial derivative with respect to the jth variable, $D_j(D_i f)$, exists[12]; it is is called a *second partial derivative* of f.

Recall that different notations for partial derivatives exist:

$$D_j(D_i f)(\mathbf{a}) = \frac{\partial^2 f}{\partial x_j \partial x_i}(\mathbf{a}).$$

To apply multi-exponent notation to higher partial derivatives, let

$$D_I f = D_1^{i_1} D_2^{i_2} \ldots D_n^{i_n} f. \qquad\qquad 3.3.17$$

For example, for a function f in three variables,

$$D_1\Big(D_1\big(D_2(D_2 f)\big)\Big) = D_1^2(D_2^2 f) \quad \text{can be written} \quad D_1^2\big(D_2^2(D_3^0 f)\big), \qquad 3.3.18$$

Of course $D_I f$ is only defined if all partials up to order deg I exist, and it is also a good idea to assume that they are all continuous, so that the order in which the partials are calculated doesn't matter (Theorem 3.3.9).

which can be written $D_{(2,2,0)} f$, i.e., $D_I f$, where $I = (i_1, i_2, i_3) = (2, 2, 0)$.

What is $D_{(1,0,2)} f$, written in our standard notation for higher partial derivatives? What is $D_{(0,1,1)} f$? Check your answer below.[13]

---

[11]It is $3 - x_1 + 3x_1 x_2^2 + 2x_1^2 x_2$.

[12]This assumes, of course, that $f : U \to \mathbb{R}$ is a differentiable function, and $U \subset \mathbb{R}^n$ is open.

[13]The first is $D_1(D_3^2 f)$, which can also be written $D_1\big(D_3(D_3 f)\big)$. The second is $D_2(D_3 f)$.

Recall, however, that a multi-exponent I is an *ordered* finite sequence of non-negative whole numbers. Using multi-exponent notation, how can we distinguish between $D_1(D_3 f)$ and $D_3(D_1 f)$? Both are written $D_{(1,0,1)}$. Similarly, $D_{1,1}$ could denote $D_1(D_2 f)$ or $D_2(D_1 f)$. Is this a problem?

No. If you compute the second partials $D_1(D_3 f)$ and $D_3(D_1 f)$ of the function $x^2 + xy^3 + xz$, you will see that they are equal:

$$D_1(D_3 f)\begin{pmatrix} x \\ y \\ z \end{pmatrix} = D_3(D_1 f)\begin{pmatrix} x \\ y \\ z \end{pmatrix} = 1. \qquad 3.3.19$$

We will see when we define Taylor polynomials in higher dimensions (Definition 3.3.15) that a major benefit of multi-exponent notation is that it takes advantage of the equality of crossed partials, writing them only once; for instance, $D_1(D_2 f)$ and $D_2(D_1 f)$ are written $D_{(1,1)}$.

Similarly, $D_1(D_2 f) = D_2(D_1 f)$, and $D_2(D_3 f) = D_3(D_2 f)$.

Normally, crossed partials are equal. They can fail to be equal only if the second partials are not continuous; you are asked in Exercise 3.3.1 to verify that this is the case in Example 3.3.11. (Of course the second partials do not exist unless the first partials exist and are continuous, in fact, differentiable.)

Theorem 3.3.9 (Crossed partials equal). *Let $f : U \to \mathbb{R}$ be a function such that all second partial derivatives exist and are continuous. Then for every pair of variables x_i, x_j, the crossed partials are equal:*

$$D_j(D_i f)(\mathbf{a}) \;=\; D_i(D_j f)(\mathbf{a}). \qquad 3.3.20$$

With the techniques now available to us, Theorem 3.3.9 is a surprisingly difficult result, proved in Appendix A.6. In Exercise 4.5.11 we give a very simple proof that uses Fubini's theorem.

Corollary 3.3.10. *If $f : U \to \mathbb{R}$ is a function all of whose partial derivatives up to order k are continuous, then the partial derivatives of order up to k do not depend on the order in which they are computed.*

For example, $D_i\big(D_j(D_k f)\big)(\mathbf{a}) = D_k\big(D_j(D_i f)\big)(\mathbf{a})$, and so on.

The requirement that the partial derivatives be continuous is essential, as shown by Example 3.3.11

Example 3.3.11 (A case where crossed partials aren't equal). Consider the function

Don't take this example too seriously. The function f here is pathological; such things do not show up unless you go looking for them. You should think that crossed partials are equal.

$$f\begin{pmatrix} x \\ y \end{pmatrix} = \begin{cases} xy\,\dfrac{x^2 - y^2}{x^2 + y^2} & \text{if } \begin{pmatrix} x \\ y \end{pmatrix} \neq \begin{pmatrix} 0 \\ 0 \end{pmatrix} \\[2ex] 0 & \text{if } \begin{pmatrix} x \\ y \end{pmatrix} = \begin{pmatrix} 0 \\ 0 \end{pmatrix} \end{cases}. \qquad 3.3.21$$

Then $\quad D_1 f\begin{pmatrix} x \\ y \end{pmatrix} = \dfrac{4x^2 y^3 + x^4 y - y^5}{(x^2 + y^2)^2} \quad$ and $\quad D_2 f\begin{pmatrix} x \\ y \end{pmatrix} = \dfrac{x^5 - 4x^3 y^2 - xy^4}{(x^2 + y^2)^2}$

$$3.3.22$$

when $\begin{pmatrix} x \\ y \end{pmatrix} \neq \begin{pmatrix} 0 \\ 0 \end{pmatrix}$, and both partials vanish at the origin. So

$$D_1 f \begin{pmatrix} 0 \\ y \end{pmatrix} = \left\{ \begin{array}{ll} -y & \text{if } y \neq 0 \\ 0 & \text{if } y = 0 \end{array} \right. = -y \quad \text{and} \quad D_2 f \begin{pmatrix} x \\ 0 \end{pmatrix} = \left\{ \begin{array}{ll} x & \text{if } x \neq 0 \\ 0 & \text{if } x = 0 \end{array} \right. = x,$$

giving $\quad D_2(D_1 f) \begin{pmatrix} 0 \\ y \end{pmatrix} = D_2(-y) = -1 \quad \text{and} \quad D_1(D_2 f) \begin{pmatrix} x \\ 0 \end{pmatrix} = D_1(x) = 1,$

the first for any value of y and the second for any value of x; at the origin, the crossed partials $D_2(D_1 f)$ and $D_1(D_2 f)$ are not equal. $\triangle$

The coefficients of polynomials as derivatives

We can express the coefficients of a polynomial in one variable in terms of the derivatives of the polynomial at 0. If p is a polynomial of degree k with coefficients $a_0 \ldots a_k$, i.e.,

For example, take the polynomial $x + 2x^2 + 3x^3$ (i.e., $a_1 = 1, a_2 = 2, a_3 = 3$.) Then

$f'(x) = 1 + 4x + 9x^2$, so

$f'(0) = 1$;   indeed, $1! \, a_1 = 1$.

$f''(x) = 4 + 18x$,   so

$f''(0) = 4$;   indeed, $2! \, a_2 = 4$

$f^{(3)}(x) = 18$;

indeed, $3! \, a_3 = 6 \cdot 3 = 18$.

Evaluating the derivatives at 0 gets rid of terms that come from higher-degree terms. For example, in $f''(x) = 4 + 18x$, the $18x$ comes from the original $3x^3$.

$$p(x) = a_0 + a_1 x + a_2 x^2 + \cdots + a_k x^k, \qquad 3.3.23$$

then, denoting by $p^{(i)}$ the ith derivative of p, we have

$$i! \, a_i = p^{(i)}(0); \qquad \text{i.e.,} \qquad a_i = \frac{1}{i!} p^{(i)}(0). \qquad 3.3.24$$

Evaluating the ith derivative of a polynomial at 0 isolates the coefficient of x^i: the ith derivative of lower terms vanishes (is 0), and the ith derivative of higher degree terms contains positive powers of x, and vanishes when evaluated at 0.

We will want to translate this to the case of several variables. You may wonder why. Our goal is to approximate differentiable functions by polynomials. We will see in Proposition 3.3.19 that if, at a point $\mathbf{a}$, all derivatives up to order k of a function vanish, then the function is small in a neighborhood of that point (small in a sense that depends on k). If we can manufacture a polynomial with the same derivatives up to order k as the function we want to approximate, then the function representing the *difference* between the function being approximated and the polynomial approximating it will have vanishing derivatives up to order k: hence it will be small.

So, how does Equation 3.3.23 translate to the case of several variables?

As in one variable, the coefficients of a polynomial in several variables can be expressed in terms of the partial derivatives of the polynomial at 0.

In Proposition 3.3.12 we use J to denote the multi-exponents we sum over to express a polynomial, and I to denote a particular multi-exponent.

Proposition 3.3.12 (Coefficients expressed in terms of partial derivatives at 0). *Let p be the polynomial*

$$p(\mathbf{x}) = \sum_{m=0}^{k} \sum_{J \in \mathcal{I}_n^m} a_J \mathbf{x}^J. \qquad 3.3.25$$

Then for any particular $I \in \mathcal{I}$, we have $I! \, a_I = D_I p(0)$.

For example, if $f(x) = x^2$, then $f''(x) = 2$.

Proof. First, note that it is sufficient to show that

$$D_I \mathbf{x}^I(0) = I! \quad \text{and} \quad D_I \mathbf{x}^J(0) = 0 \text{ for all } J \neq I. \qquad 3.3.26$$

We can see that this is enough by writing:

$$D_I p(0) = D_I \overbrace{\left(\sum_{m=0}^{k} \sum_{J \in \mathcal{I}_n^m} a_J \mathbf{x}^J \right)}^{\substack{p \text{ written in} \\ \text{multi-exponent form}}} (0) \qquad 3.3.27$$

$$= \sum_{m=0}^{k} \sum_{J \in \mathcal{I}_n^m} a_J D_I \mathbf{x}^J(0);$$

If you find it hard to focus on this proof written in multi-exponent notation, look at Example 3.3.13.

if we prove the statements in Equation 3.3.26, then all the terms $a_J D_I \mathbf{x}^J(0)$ for $J \neq I$ drop out, leaving $D_I p(0) = I! \, a_I$.

To prove that $D_I \mathbf{x}^I(0) = I!$, write

$$D_I \mathbf{x}^I = D_1^{i_1} \dots D_n^{i_n} x_1^{i_1} \cdot \dots \cdot x_n^{i_n} = D_1^{i_1} x_1^{i_1} \cdot \dots \cdot D_n^{i_n} x_n^{i_n}$$

$$= i_1! \cdot \dots \cdot i_n! = I!. \qquad 3.3.28$$

To prove $D_I \mathbf{x}^J(0) = 0$ for all $J \neq I$, write similarly

$$D_I \mathbf{x}^J = D_1^{i_1} \dots D_n^{i_n} x_1^{j_1} \cdot \dots \cdot x_n^{j_n} = D_1^{i_1} x_1^{j_1} \cdot \dots \cdot D_n^{i_n} x_n^{j_n}. \qquad 3.3.29$$

At least one j_m must be different from i_m, either bigger or smaller. If it is smaller, then we see a higher derivative than the power, and the derivative is 0. If it is bigger, then there is a positive power of x_m left over after the derivative, and evaluated at 0, we get 0 again. $\square$

Example 3.3.13 (Coefficients of a polynomial in terms of its partial derivatives at 0).

What is $D_1^2 D_2^3 p$, where $p = 3x_1^2 x_2^3$? We have $D_2 p = 9x_2^2 x_1^2$, $D_2^2 p = 18x_2 x_1^2$, and so on, ending with $D_1 D_1 D_2 D_2 D_2 p = 36$.

Multi-exponent notation takes some getting used to; Example 3.3.13 translates multi-exponent notation into more standard (and less concise) notation.

In multi-exponent notation, $p = 3x_1^2 x_2^3$ is written $3\mathbf{x}^{(2,3)}$, i.e., $a_I \mathbf{x}^I$, where $I = (2,3)$ and $a_{(2,3)} = 3$. The higher partial derivative $D_1^2 D_2^3 p$ is written $D_{(2,3)} p$. By definition (Equation 3.3.10), when $I = (2,3)$, $I! = 2!3! = 12$.

Proposition 3.3.12 says

$$a_I = \frac{1}{I!} D_I p(0); \quad \text{here, } \frac{1}{I!} D_{(2,3)} p(0) = \frac{36}{12} = 3, \text{ which is indeed } a_{(2,3)}.$$

What if the multi-exponent I for the higher partial derivatives is not the same as the multi-exponent J for $\mathbf{x}$? As mentioned in the proof of Proposition 3.3.12, the result is 0. For example, if we take $D_1^2 D_2^2$ of the polynomial $p = 3x_1^2 x_2^3$, so that $I = (2,2)$ and $J = (2,3)$, we get $36x_2$; evaluated at $p = 0$, this becomes 0. If $I > J$, the result is also 0; for example, what is $D_I p(0)$ when $I = (2,3)$, $p = a_J \mathbf{x}^J$, $a_J = 3$, and $J = (2,2)$?[14] $\triangle$

---

[14] This corresponds to $D_1^2 D_2^3 (3x_1^2 x_2^2)$; already, $D_2^3 (3x_1^2 x_2^2) = 0$.

Taylor polynomials in higher dimensions

Although the polynomial in Equation 3.3.30 is called the Taylor polynomial of f at $\mathbf{a}$, it is evaluated at $\mathbf{a} + \vec{\mathbf{h}}$, and its value there depends on $\vec{\mathbf{h}}$, the increment to $\mathbf{a}$.

In Equation 3.3.30, remember that I is a multi-exponent; if you want to write the polynomial out in particular cases, it can get complicated, especially if k or n is big.

Now we are ready to define Taylor polynomials in higher dimensions, and to see in what sense they can be used to approximate functions in n variables.

Definition 3.3.14 (C^k function). A C^k function on $U \subset \mathbb{R}^n$ is a function that is k-times continuously differentiable—i.e., all of its partial derivatives up to order k exist and are continuous on U.

Example 3.3.16 illustrates notation; it has no mathematical content.

The first term—the term of degree m=0—corresponds to the 0th derivative, i.e., the function f itself.

Remember (Definition 3.3.7) that $\mathbf{x}^I = x_1^{i_1} \ldots x_n^{i_n}$; similarly, $\vec{\mathbf{h}}^I = h_1^{i_1} \ldots h_n^{i_n}$. For instance, if $I = (1,1)$ we have

$$\vec{\mathbf{h}}^I = \vec{\mathbf{h}}^{(1,1)} = h_1 h_2;$$

if $I = (2,0,3)$ we have

$$\vec{\mathbf{h}}^I = \vec{\mathbf{h}}^{(2,0,3)} = h_1^2 h_3^3.$$

Since the crossed partials of f are equal,

$$D_{(1,1)} f(\mathbf{a}) h_1 h_2 =$$
$$\frac{1}{2} D_1 D_2 f(\mathbf{a}) h_1 h_2$$
$$+ \frac{1}{2} D_2 D_1 f(\mathbf{a}) h_1 h_2.$$

The term $1/I!$ in the formula for the Taylor polynomial gives appropriate weights to the various terms to take into account the existence of crossed partials.

This is the big advantage of multi-exponent notation, which is increasingly useful as n gets big.

Definition 3.3.15 (Taylor polynomial in higher dimensions). Let $U \subset \mathbb{R}^n$ be an open subset and $f : U \to \mathbb{R}$ be a C^k function. Then the polynomial of degree k,

$$P_{f,\mathbf{a}}^k(\mathbf{a} + \vec{\mathbf{h}}) = \sum_{m=0}^{k} \sum_{I \in \mathcal{I}_n^m} \frac{1}{I!} D_I f(\mathbf{a}) \vec{\mathbf{h}}^I, \qquad 3.3.30$$

is called the Taylor polynomial of degree k of f at $\mathbf{a}$.

Example 3.3.16 (Multi-exponent notation for a Taylor polynomial of a function in two variables). Suppose f is a function in two variables. The formula for the Taylor polynomial of degree 2 of f at $\mathbf{a}$ is then

$$P_{f,\mathbf{a}}^2(\mathbf{a} + \vec{\mathbf{h}}) = \sum_{m=0}^{2} \sum_{I \in \mathcal{I}_2^m} \frac{1}{I!} D_I f(\mathbf{a}) \vec{\mathbf{h}}^I \qquad 3.3.31$$

$$= \underbrace{\frac{1}{0!0!} D_{(0,0)} f(\mathbf{a}) h_1^0 h_2^0}_{f(\mathbf{a})} + \underbrace{\frac{1}{1!0!} D_{(1,0)} f(\mathbf{a}) h_1^1 h_2^0 + \frac{1}{0!1!} D_{(0,1)} f(\mathbf{a}) h_1^0 h_2^1}_{\text{terms of degree 1: first derivatives}}$$

$$+ \underbrace{\frac{1}{2!0!} D_{(2,0)} f(\mathbf{a}) h_1^2 h_2^0 + \frac{1}{1!1!} D_{(1,1)} f(\mathbf{a}) h_1 h_2 + \frac{1}{0!2!} D_{(0,2)} f(\mathbf{a}) h_1^0 h_2^2,}_{\text{terms of degree 2: second derivatives}}$$

which we can write more simply as

$$P_{f,\mathbf{a}}^2(\mathbf{a} + \vec{\mathbf{h}}) = f(\mathbf{a}) + D_{(1,0)} f(\mathbf{a}) h_1 + D_{(0,1)} f(\mathbf{a}) h_2$$
$$+ \frac{1}{2} D_{(2,0)} f(\mathbf{a}) h_1^2 + D_{(1,1)} f(\mathbf{a}) h_1 h_2 + \frac{1}{2} D_{(0,2)} f(\mathbf{a}) h_2^2. \quad \triangle \qquad 3.3.32$$

Remember that $D_{(1,0)} f$ corresponds to the partial derivative with respect to the first variable, $D_1 f$, while $D_{(0,1)} f$ corresponds to the partial derivative with respect to the second variable, $D_2 f$. Similarly, $D_{(1,1)} f$ corresponds to $D_1 D_2 f = D_2 D_1 f$, and $D_{(2,0)} f$ corresponds to $D_1 D_1 f$. $\quad \triangle$

What are the terms of degree 2 (second derivatives) of the Taylor polynomial at $\mathbf{a}$, of degree 2, of a function with three variables?[15]

Example 3.3.17 (Computing a Taylor polynomial). What is the Taylor polynomial of degree 2 of the function $f\begin{pmatrix} x \\ y \end{pmatrix} = \sin(x + y^2)$, at $\begin{pmatrix} 0 \\ 0 \end{pmatrix}$? The first term, of degree 0, is $f\begin{pmatrix} 0 \\ 0 \end{pmatrix} = \sin 0 = 0$. For the terms of degree 1 we have

$$D_{(1,0)}f\begin{pmatrix} x \\ y \end{pmatrix} = \cos(x + y^2) \quad \text{and} \quad D_{(0,1)}f\begin{pmatrix} x \\ y \end{pmatrix} = 2y\cos(x + y^2), \qquad 3.3.33$$

so $D_{(1,0)}f\begin{pmatrix} 0 \\ 0 \end{pmatrix} = 1$ and $D_{(0,1)}f\begin{pmatrix} 0 \\ 0 \end{pmatrix} = 0$. For the terms of degree 2, we have

$$D_{(2,0)}f\begin{pmatrix} x \\ y \end{pmatrix} = -\sin(x + y^2)$$

$$D_{(1,1)}f\begin{pmatrix} x \\ y \end{pmatrix} = -2y\sin(x + y^2) \qquad 3.3.34$$

$$D_{(0,2)}f\begin{pmatrix} x \\ y \end{pmatrix} = 2\cos(x + y^2) - 4y^2\sin(x + y^2);$$

In Example 3.4.5 we will see how to reduce this computation to two lines, using rules we will give for computing Taylor polynomials.

evaluated at $\begin{pmatrix} 0 \\ 0 \end{pmatrix}$, these give 0, 0, and 2 respectively. So the Taylor polynomial of degree 2 is

$$P^2_{f,\begin{pmatrix} 0 \\ 0 \end{pmatrix}}\left(\begin{bmatrix} h_1 \\ h_2 \end{bmatrix}\right) = 0 + h_1 + 0 + 0 + 0 + \frac{2}{2}h_2^2. \qquad 3.3.35$$

What would we have to add to make this the Taylor polynomial of degree 3 of f at $\begin{pmatrix} 0 \\ 0 \end{pmatrix}$? The third partial derivatives are

$$D_{(3,0)}f\begin{pmatrix} x \\ y \end{pmatrix} = D_1 D_1^2 f\begin{pmatrix} x \\ y \end{pmatrix} = D_1\left(-\sin(x + y^2)\right) = -\cos(x + y^2)$$

$$D_{(0,3)}f\begin{pmatrix} x \\ y \end{pmatrix} = D_2 D_2^2 f\begin{pmatrix} x \\ y \end{pmatrix} = D_2\left(2\cos(x + y^2) - 4y^2\sin(x + y^2)\right)$$

$$= 4y\sin(x + y^2) - 8y\sin(x + y^2) - 8y^3\cos(x + y^2)$$

---

[15]The third term of $\quad P^2_{f,\mathbf{a}}(\mathbf{a} + \vec{\mathbf{h}}) = \sum_{m=0}^{2} \sum_{I \in \mathcal{I}_3^m} \frac{1}{I!} D_I f(\mathbf{a})\vec{\mathbf{h}}^I$

is $\quad \overbrace{D_{(1,1,0)}f(\mathbf{a})h_1 h_2}^{D_1 D_2} + \overbrace{D_{(1,0,1)}f(\mathbf{a})h_1 h_3}^{D_1 D_3} + \overbrace{D_{(0,1,1)}f(\mathbf{a})h_2 h_3}^{D_2 D_3}$

$\qquad + \underbrace{\frac{1}{2}D_{(2,0,0)}f(\mathbf{a})h_1^2}_{D_1^2} + \underbrace{\frac{1}{2}D_{(0,2,0)}f(\mathbf{a})h_2^2}_{D_2^2} + \underbrace{\frac{1}{2}D_{(0,0,2)}f(\mathbf{a})h_3^2}_{D_3^2}.$

$$D_{(2,1)}f\begin{pmatrix}x\\y\end{pmatrix} = D_1(D_1D_2)f\begin{pmatrix}x\\y\end{pmatrix} = D_1\big(-2y\sin(x+y^2)\big) = -2y\cos(x+y^2)$$

$$D_{(1,2)}f\begin{pmatrix}x\\y\end{pmatrix} = D_1D_2^2f\begin{pmatrix}x\\y\end{pmatrix} = D_1\big(2\cos(x+y^2) - 4y^2\sin(x+y^2)\big)$$

$$= -2\sin(x+y^2) - 4y^2\cos(x+y^2). \qquad\qquad 3.3.36$$

At $\begin{pmatrix}0\\0\end{pmatrix}$ all are 0 except $D_{(3,0)}$, which is -1. So the term of degree 3 is $(-\frac{1}{3}!)h_1^3 = -\frac{1}{6}h_1^3$, and the Taylor polynomial of degree 3 of f at $\begin{pmatrix}0\\0\end{pmatrix}$ is

$$P^3_{f,\begin{pmatrix}0\\0\end{pmatrix}}\left(\begin{bmatrix}h_1\\h_2\end{bmatrix}\right) = h_1 + h_2^2 - \frac{1}{6}h_1^3. \qquad\qquad \triangle \quad 3.3.37$$

Taylor's theorem with remainder is discussed in Appendix A.9.

Taylor's theorem without remainder in higher dimensions

Theorem 3.3.18 (Taylor's theorem without remainder in higher dimensions). *(a) The polynomial $P^k_{f,a}(\mathbf{a}+\vec{\mathbf{h}})$ is the unique polynomial of total degree k that has the same partial derivatives up to order k at $\mathbf{a}$ as f.*

(b) The polynomial $P^k_{f,\mathbf{a}}(\mathbf{a}+\vec{\mathbf{h}})$ is the unique polynomial of degree $\leq k$ that best approximates f when $\vec{\mathbf{h}} \to 0$, in the sense that it is the unique polynomial of degree $\leq k$ such that

Note that since we are dividing by a high power of $|\vec{\mathbf{h}}|$, the limit being 0 means that the numerator is very small.

$$\lim_{\vec{\mathbf{h}}\to 0} \frac{f(\mathbf{a}+\vec{\mathbf{h}}) - P^k_{f,\mathbf{a}}(\mathbf{a}+\vec{\mathbf{h}})}{|\vec{\mathbf{h}}|^k} = 0. \qquad\qquad 3.3.38$$

To prove Theorem 3.3.18 we need the following proposition, which says that if all the partial derivatives of f up to some order k equal 0 at a point $\mathbf{a}$, then the function is small in a neighborhood of $\mathbf{a}$.

Proposition 3.3.19 is proved in Appendix A.7.

We must require that the partial derivatives be continuous; if the aren't, the statement isn't true even when $k = 1$, as you will see if you go back to Equation 1.9.9, where f is the function of Example 1.9.3, a function whose partial derivatives are not continuous.

Proposition 3.3.19 (Size of a function with many vanishing partial derivatives). *Let U be an open subset of $\mathbb{R}^n$ and $f : U \to \mathbb{R}$ be a C^k function. If at $\mathbf{a} \in U$ all partial derivatives up to order k vanish (including the 0th partial derivative, i.e., $f(\mathbf{a})$), then*

$$\lim_{\vec{\mathbf{h}}\to 0} \frac{f(\mathbf{a}+\vec{\mathbf{h}})}{|\vec{\mathbf{h}}|^k} = 0. \qquad\qquad 3.3.39$$

Proof of Theorem 3.3.18. Part (a) follows from Proposition 3.3.12. Consider the polynomial $Q_{f,\mathbf{a}}^k$ that, evaluated at $\vec{\mathbf{h}}$, gives the same result as the Taylor polynomial $P_{f,\mathbf{a}}^k$ evaluated at $\mathbf{a} + \vec{\mathbf{h}}$:

$$P_{f,\mathbf{a}}^k(\mathbf{a} + \vec{\mathbf{h}}) = Q_{f,\mathbf{a}}^k(\vec{\mathbf{h}}) = \sum_{m=0}^{k} \sum_{J \in \mathcal{I}_n^m} \frac{1}{J!} D_J f(\mathbf{a}) \vec{\mathbf{h}}^J. \qquad 3.3.40$$

Now consider the Ith derivative of that polynomial, at $\mathbf{0}$:

The expression in Equation 3.3.41 is $D_I p(\mathbf{0})$, where $p = Q_{f,\mathbf{a}}^k$.

$$D_I \underbrace{\left(\overbrace{\sum_{m=0}^{k} \sum_{J \in \mathcal{I}_n^m} \underbrace{\frac{1}{J!} D_J f(\mathbf{a})}_{\text{coefficient of } p} \vec{\mathbf{h}}^J}^{p = Q_{f,\mathbf{a}}^k \vec{\mathbf{h}}^J} . \right)(\mathbf{0}) = \frac{1}{I!} D_I f(\mathbf{a}) \vec{\mathbf{h}}^I(\mathbf{0}).}_{D_I p(\mathbf{0})} \qquad 3.3.41$$

We get the equality of Equation 3.3.41 by the same argument as in the proof of Proposition 3.3.12: all partial derivatives where $I \neq J$ vanish.

Proposition 3.3.12 says that for a polynomial p, we have $I! a_I = D_I p(0)$, where the a_I are the coefficients. This gives

$$\underbrace{I! \overbrace{\frac{1}{I!} \big(D_I f(\mathbf{a}) \big)}^{\substack{I\text{th coeff. of} \\ Q_{f,\mathbf{a}}^k}}}_{I! a_I} = \underbrace{D_I Q_{f,\mathbf{a}}^k(\mathbf{0})}_{D_I p(\mathbf{0})}; \quad \text{i.e., } D_I f(\mathbf{a}) = D_I Q_{f,\mathbf{a}}^k(\mathbf{0}). \qquad 3.3.42$$

Now, when $\vec{\mathbf{h}} = \mathbf{0}$, then $P_{f,\mathbf{a}}^k(\mathbf{a} + \vec{\mathbf{h}})$ becomes $P_{f,\mathbf{a}}^k(\mathbf{a})$, so

$$D_I Q_{f,\mathbf{a}}^k(\mathbf{0}) = D_I P_{f,\mathbf{a}}^k(\mathbf{a}), \quad \text{so} \quad D_I P_{f,\mathbf{a}}^k(\mathbf{a}) = D_I f(\mathbf{a}); \qquad 3.3.43$$

the partial derivatives of $P_{f,\mathbf{a}}^k$, up to order k, are the same as the partial derivatives of f, up to order k. Therefore all the partials of order at most k of the difference $f(\mathbf{a} + \vec{\mathbf{h}}) - P_{f,\mathbf{a}}^k(\mathbf{a} + \vec{\mathbf{h}})$ vanish.

Part (b) then follows from Proposition 3.3.19. To lighten the notation, denote by $g(\mathbf{a} + \vec{\mathbf{h}})$ the difference between $f(\mathbf{a} + \vec{\mathbf{h}})$ and the Taylor polynomial of f at $\mathbf{a}$. Since all the partials of g up to order k vanish, Proposition 3.3.19 says that

$$\lim_{\vec{\mathbf{h}} \to 0} \frac{g(\mathbf{a} + \vec{\mathbf{h}})}{|\vec{\mathbf{h}}|^k} = 0. \quad \square \qquad 3.3.44$$

3.4 Rules for Computing Taylor Polynomials

Computing Taylor polynomials is very much like computing derivatives; in fact, when the degree is 1, they are essentially the same. Just as we have rules for differentiating sums, products, compositions, etc., there are rules for computing Taylor polynomials of functions obtained by combining simpler functions. Since computing partial derivatives rapidly becomes unpleasant, we strongly recommend making use of these rules.

"Since the computation of successive derivatives is always *painful*, we recommend (when it is possible) considering the function as being obtained from simpler functions by elementary operations (sum, product, power, etc.). ... Taylor polynomials are most often a *theoretical, not a practical, tool.*"—Jean Dieudonné, *Calcul Infinitésimal*

To write down the Taylor polynomials of some standard functions, we will use notation invented by Landau to express the idea that one is computing "up to terms of degree k": the notation o, or "little o." While in the equations of Proposition 3.4.2 the "little o" term may look like a remainder, such terms do not give a precise, computable remainder. Little o provides a way to bound one function by another function, in an *unspecified* neighborhood of the point at which you are computing the Taylor polynomial.

Definition 3.4.1 (Little o). Little o, denoted o, means "smaller than," in the following sense: if $h(x) > 0$ in some neighborhood of 0, then $f \in o(h)$ if for all $\epsilon > 0$, there exists $\delta > 0$ such that if $|x| < \delta$, then

$$|f(x)| \leq \epsilon h(x).   \qquad 3.4.1$$

Alternatively, we can say that $f \in o(h)$ if

$$\lim_{x \to 0} \frac{f(x)}{h(x)} = 0;   \qquad 3.4.2$$

in some unspecified neighborhood, $|f|$ is smaller than h; as $x \to 0$, $|f(x)|$ becomes infinitely smaller than $h(x)$.

A famous example of an asymptotic development is the *prime number theorem*, which states that if $\pi(x)$ represents the number of prime numbers smaller than x, then, for x near ∞,

$$\pi(x) = \frac{x}{\ln x} + o\left(\frac{x}{\ln x}\right).$$

(Here π has nothing to do with $\pi \approx 3.1415$.) This was proved independently in 1898 by Hadamard and de la Vallé-Poussin, after being conjectured a century earlier by Gauss.

Anyone who proves the stronger statement,

$$\pi(x) = \int_1^x \frac{1}{\ln u}\, du + o\left(|x|^{\frac{1}{2}+\epsilon}\right),$$

for all $\epsilon > 0$ will have proved the *Riemann hypothesis*, one of the two most famous outstanding problems of mathematics, the other being the *Poincaré conjecture*.

Very often Taylor polynomials written in terms of bounds with little o are good enough. But in settings where you want to know the error for some particular x, something stronger is required: Taylor's theorem with remainder, discussed in Appendix A.9.

Remark. In the setting of functions that can be approximated by Taylor polynomials, the only functions $h(x)$ of interest are the functions $|x|^k$ for $k \geq 0$. In other settings, it is interesting to compare nastier functions (not of class C^k) to a broader class of functions, for instance, one might be interested in bounding functions by functions $h(x)$ such as $\sqrt{|x|}$ or $|x| \ln |x| \ldots$. (An example of what we mean by "nastier functions" is Equation 5.3.10.) The art of making such comparisons is called the theory of *asymptotic developments*. But any place that a function is C^k it has to look like an positive integer power of x. $\triangle$

In Proposition 3.4.2 we list the functions whose Taylor polynomials we expect you to know from first year calculus. We will write them only near 0, but by translation they can be written anywhere. Note that in the equations of Proposition 3.4.2, the Taylor polynomial is the expression on the right-hand side *excluding* the little o term, which indicates how good an approximation the Taylor polynomial is to the corresponding function, without giving any precision.

Proposition 3.4.2 (Taylor polynomials of some standard functions).
The following formulas give the Taylor polynomials at 0 of the corresponding functions:

$$e^x = 1 + x + \frac{x^2}{2!} + \cdots + \frac{x^n}{n!} + o(|x|^n) \qquad\qquad 3.4.3$$

$$\sin(x) = x - \frac{x^3}{3!} + \frac{x^5}{5!} - \cdots + (-1)^n \frac{x^{2n+1}}{(2n+1)!} + o(|x|^{2n+1}) \qquad 3.4.4$$

$$\cos(x) = 1 - \frac{x^2}{2!} + \frac{x^4}{4!} - \cdots + (-1)^n \frac{x^{2n}}{(2n)!} + o(|x|^{2n}) \qquad 3.4.5$$

$$\ln(1+x) = x - \frac{x^2}{2} + \cdots + (-1)^{n+1}\frac{x^n}{n} + o(|x|^n) \qquad\qquad 3.4.6$$

$$(1+x)^m = 1 + mx + \frac{m(m-1)}{2!}x^2 + \frac{m(m-1)(m-2)}{3!}x^3 + \cdots$$

$$+ \frac{m(m-1)\ldots(m-(n-1))}{n!}x^n + o(|x|^n). \qquad 3.4.7$$

Equation 3.4.7 is the *binomial formula*.

The proof is left as Exercise 3.4.1. Note that the Taylor polynomial for sine contains only odd terms, with alternating signs, while the Taylor polynomial for cosine contains only even terms, again with alternating signs. All odd functions (functions f such that $f(-x) = -f(x)$) have Taylor polynomials with only odd terms, and all even functions (functions f such that $f(-x) = f(x)$) have Taylor polynomials with only even terms. Note also that in the Taylor polynomial of $\ln(1+x)$, there are no factorials in the denominators.

Now let us see how to combine these Taylor polynomials.

Propositions 3.4.3 and 3.4.4 are proved in Appendix A.8. We state them for scalar-valued functions, largely because we only defined Taylor polynomials for scalar-valued functions. However, they are true for vector-valued functions, at least whenever the latter make sense. For instance, the product should be replaced by a dot product (or the product of a scalar with a vector-valued function). When composing functions, of course we can consider only compositions where the range of one function is the domain of the other. The proofs of all these variants are practically identical to the proofs for Propositions 3.4.3 and 3.4.4.

Proposition 3.4.3 (Sums and products of Taylor polynomials). *Let $U \subset \mathbb{R}^n$ be open, and $f, g : U \to \mathbb{R}$ be C^k functions. Then $f + g$ and fg are also of class C^k, and their Taylor polynomials are computed as follows.*

(a) The Taylor polynomial of the sum is the sum of the Taylor polynomials:

$$P_{f+g,\mathbf{a}}^k(\mathbf{a}+\vec{\mathbf{h}}) = P_{f,\mathbf{a}}^k(\mathbf{a}+\vec{\mathbf{h}}) + P_{g,\mathbf{a}}^k(\mathbf{a}+\vec{\mathbf{h}}). \qquad 3.4.8$$

(b) The Taylor polynomial of the product fg is obtained by taking the product

$$P_{f,\mathbf{a}}^k(\mathbf{a}+\vec{\mathbf{h}}) \cdot P_{g,\mathbf{a}}^k(\mathbf{a}+\vec{\mathbf{h}}) \qquad 3.4.9$$

and discarding the terms of degree $> k$.

Please notice that the composition of two polynomials is a polynomial.

Why does the composition in Proposition 3.4.4 make sense? $P_{f,b}^k(b+u)$ is a good approximation to $f(b+u)$ only when $|u|$ is small. But our requirement that $g(\mathbf{a}) = b$ guarantees precisely that $P_{g,\mathbf{a}}^k(\mathbf{a}+\vec{\mathbf{h}}) = b + \text{something small}$ when $\vec{\mathbf{h}}$ is small. So it is reasonable to substitute that "something small" for the increment u when evaluating the polynomial $P_{f,b}^k(b+u)$.

Proposition 3.4.4 (Chain rule for Taylor polynomials). *Let $U \subset \mathbb{R}^n$ and $V \subset \mathbb{R}$ be open, and $g : U \to V$ and $f : V \to \mathbb{R}$ be of class C^k. Then $f \circ g : U \to \mathbb{R}$ is of class C^k, and if $g(\mathbf{a}) = b$, then the Taylor polynomial $P_{f\circ g,\mathbf{a}}^k(\mathbf{a}+\vec{\mathbf{h}})$ is obtained by considering the polynomial*

$$P_{f,b}^k\big(P_{g,\mathbf{a}}^k(\mathbf{a}+\vec{\mathbf{h}})\big)$$

and discarding the terms of degree $> k$.

Example 3.4.5 (Computing a Taylor polynomial: an easy example). Let's use these rules to compute the Taylor polynomial of degree 3 of the function $f\begin{pmatrix} x \\ y \end{pmatrix} = \sin(x + y^2)$, at $\begin{pmatrix} 0 \\ 0 \end{pmatrix}$, which we already saw in Example 3.3.17. According to Proposition 3.4.4, we simply substitute $x + y^2$ for u in $\sin u = u - u^3/6 + o(u^3)$, and omit all the terms of degree > 3:

$$\sin(x + y^2) = (x + y^2) - \frac{(x + y^2)^3}{6} + o\Big((x^2 + y^2)^{3/2}\Big)$$

$$= \underbrace{x + y^2 - \frac{x^3}{6}}_{\text{Taylor polynomial}} + \underbrace{o\Big((x^2 + y^2)^{3/2}\Big)}_{\text{error term}}.$$

3.4.10

Presto: half a page becomes two lines.

Example 3.4.6 (Computing a Taylor polynomial: a harder example). Let $U \subset \mathbb{R}$ be open, and $f : U \to \mathbb{R}$ be of class C^2. Let $V \subset U \times U$ be the subset of $\mathbb{R}^2$ where $f(x) + f(y) \neq 0$. Compute the Taylor polynomial of degree 2 of the function $F : V \to \mathbb{R}$, at a point $\begin{pmatrix} a \\ b \end{pmatrix} \in V$.

$$F\begin{pmatrix} x \\ y \end{pmatrix} = \frac{1}{f(x) + f(y)}.$$

3.4.11

Whenever you are trying to compute the Taylor polynomial of a quotient, a good tactic is to factor out the constant terms (here, $f(a) + f(b)$), and apply Equation 3.4.7 to what remains.

Choose $\begin{pmatrix} a \\ b \end{pmatrix} \in V$, and set $\begin{pmatrix} x \\ y \end{pmatrix} = \begin{pmatrix} a + u \\ b + v \end{pmatrix}$. Then

$$F\begin{pmatrix} a + u \\ b + v \end{pmatrix} = \frac{1}{\big(f(a) + f'(a)u + f''(a)u^2/2 + o(u^2)\big) + \big(f(b) + f'(b)v + f''(b)v^2/2 + o(v^2)\big)}$$

$$= \left(\underbrace{\frac{1}{f(a) + f(b)}}_{\text{a constant}} \overbrace{\frac{1}{1 + \dfrac{f'(a)u + f''(a)u^2/2 + f'(b)v + f''(b)v^2/2}{f(a) + f(b)}}}^{(1+x)^{-1}, \text{ where } x \text{ is the fraction in the denominator}} \right) + o(u^2 + v^2).$$

3.4.12

The equation

$$(1 + x)^{-1} = 1 - x + x^2 - \dots$$

is a special case of Equation 3.4.7, where $m = -1$. We already saw this case in Example 0.4.9, where we had

$$\sum_{n=0}^{\infty} ar^n = \frac{a}{1-r}.$$

The point of this is that the second factor is something of the form $(1+x)^{-1} = 1 - x + x^2 - \dots$, leading to

$$F\begin{pmatrix} a + u \\ b + v \end{pmatrix}$$

$$= \frac{1}{f(a) + f(b)} \left(1 - \frac{f'(a)u + f''(a)u^2/2 + f'(b)v + f''(b)v^2/2}{f(a) + f(b)} \right.$$

$$\left. + \left(\frac{f'(a)u + f''(a)u^2/2 + f'(b)v + f''(b)v^2/2}{f(a) + f(b)} \right)^2 + \dots \right). \qquad 3.4.13$$

In this expression, we discard the terms of degree > 2, to find

$$P^2_{F,\begin{pmatrix} a \\ b \end{pmatrix}} \begin{pmatrix} a + u \\ b + v \end{pmatrix} = \frac{1}{f(a) + f(b)} - \frac{f'(a)u + f'(b)v}{(f(a) + f(b))^2} - \frac{f''(a)u^2 + f''(b)v^2}{2(f(a) + f(b))^2} + \frac{(f'(a)u + f'(b)v)^2}{(f(a) + f(b))^3}.$$

$$3.4.14$$

Taylor polynomials of implicit functions

Among the functions whose Taylor polynomials we are particularly interested in are those furnished by the inverse and implicit function theorems. Although these functions are only known via some limit process like Newton's method, their Taylor polynomials can be computed algebraically.

Assume we are in the setting of the implicit function theorem (Theorem 2.9.10), where $\mathbf{F}$ is a function from a neighborhood of $\mathbf{c} = \begin{pmatrix} \mathbf{a} \\ \mathbf{b} \end{pmatrix} \in \mathbb{R}^{n+m}$ to $\mathbb{R}^n$, and $\mathbf{g}$ is an implicit function such that $\mathbf{F}\begin{pmatrix} \mathbf{g(y)} \\ \mathbf{y} \end{pmatrix} = \mathbf{0}$ for all $\mathbf{y}$ in some neighborhood of $\mathbf{b}$.

It follows from Theorem 3.4.7 that if you write the Taylor polynomial of the implicit function with undetermined coefficients, insert it into the equation specifying the implicit function, and identify like terms, you will be able to determine the coefficients.

We do not include a proof of Theorem 3.4.7 in this book. The difficult part is proving that the implicit function is k times differentiable.

Theorem 3.4.7 (Taylor polynomials of implicit functions). *If $\mathbf{F}$ is of class C^k for some $k \geq 1$, then the implicit function $\mathbf{g}$ is also of class C^k, and its Taylor polynomial of degree k, $P^k_{\mathbf{g},\mathbf{b}} : \mathbb{R}^m \to \mathbb{R}^n$, satisfies*

$$P^k_{F,\begin{pmatrix} \mathbf{a} \\ \mathbf{b} \end{pmatrix}} \begin{pmatrix} P^k_{\mathbf{g},\mathbf{b}}(\mathbf{b} + \mathbf{u}) \\ \mathbf{b} + \mathbf{u} \end{pmatrix} \in o(|\mathbf{u}|^k). \qquad 3.4.15$$

It is the unique polynomial map of degree at most k that does so.

Example 3.4.8 (Taylor polynomial of an implicit function. The equation $F\begin{pmatrix} x \\ y \\ z \end{pmatrix} = x^2 + y^3 + xyz^3 - 3 = 0$ determines z as a function of x and y in a neighborhood of $\begin{pmatrix} 1 \\ 1 \\ 1 \end{pmatrix}$, since $D_3 F\begin{pmatrix} 1 \\ 1 \\ 1 \end{pmatrix} = 3 \neq 0$. Let us compute the Taylor polynomial of this implicit function g to degree 2. We will set

$$g \begin{pmatrix} x \\ y \end{pmatrix} = g \begin{pmatrix} 1+u \\ 1+v \end{pmatrix} = 1 + a_1 u + a_2 v + \frac{a_{1,1}}{2} u^2 + a_{1,2} uv + \frac{a_{2,2}}{2} v^2 + o(u^2 + v^2).$$

3.4.16

Inserting this expression for z into $x^2 + y^3 + xyz^3 - 3 = 0$ leads to

$$(1+u)^2 + (1+v)^3 + \left((1+u)(1+v) \left(1 + a_1 u + a_2 v + \frac{a_{1,1}}{2} u^2 + a_{1,2} uv + \frac{a_{2,2}}{2} v^2 \right)^3 \right) - 3 \in o(u^2 + v^2).$$

Now it is a matter of multiplying out and identifying like terms. We get:

Constant terms: $3 - 3 = 0$.

Linear terms:

$$2u + 3v + u + v + 3a_1 u + 3a_2 v = 0, \quad \text{i.e.,} \quad a_1 = -1, \ a_2 = -\frac{4}{3}.$$

Quadratic terms:

The linear terms could have been derived from Equation 2.9.25, which gives in this case

$$\left[\mathbf{Dg} \begin{pmatrix} 1 \\ 1 \end{pmatrix} \right] = -[3]^{-1}[3, 4]$$

$$= -[1/3][3, 4] = [-1, -4/3].$$

$$u^2 \left(1 + 3a_1 + 3a_1^2 + \frac{3}{2} a_{1,1} \right) + v^2 \left(3 + 3a_2 + 3a_2^2 + \frac{3}{2} a_{2,2} \right) + uv(1 + 3a_1 + 3a_2 + 6a_1 a_2 + 3a_{1,2}).$$

Identifying the coefficients to 0, and using $a_1 = -1$ and $a_2 = -4/3$ now gives

$$a_{1,1} = -2/3, \ a_{2,2} = -26/9, \ a_{1,2} = 10/9.$$

3.4.17

Finally, this gives the Taylor polynomial of g:

$$g \begin{pmatrix} x \\ y \end{pmatrix} = 1 - (x-1) - \frac{4}{3}(y-1) - \frac{1}{3}(x-1)^2 - \frac{13}{9}(y-1)^2 + \frac{10}{9}(x-1)(y-1) + o\left((x-1)^2 + (y-1)^2 \right).$$

3.4.18

3.5 QUADRATIC FORMS

A quadratic form is a polynomial all of whose terms are of degree 2. For instance, $x^2 + y^2$ and xy are quadratic forms in two variables, as is $4x^2 + xy - y^2$. The polynomial xz is also a quadratic form (probably in three variables). But xyz is not a quadratic form; it is a cubic form in three variables.

Exercises 3.5.1 and 3.5.2 give a more intrinsic definition of a quadratic form on an abstract vector space.

Definition 3.5.1 (Quadratic form). A quadratic form $Q : \mathbb{R}^n \to \mathbb{R}$ is a polynomial in the variables $x_1, \ldots, x_n$, all of whose terms are of degree 2.

Although we will spend much of this section working on quadratic forms that look like $x^2 + y^2$ or $4x^2 + xy - y^2$, the following is a more realistic example. Most often, the quadratic forms one encounters in practice are integrals of functions, often functions in higher dimensions.

Example 3.5.2 (An integral as a quadratic form). The integral

$$Q(p) = \int_0^1 \left(p(t) \right)^2 dt,$$

3.5.1

where p is the polynomial $p(t) = a_0 + a_1 t + a_2 t^2$, is a quadratic form, as we can confirm by computing the integral:

$$Q(p) = \int_0^1 (a_0 + a_1 t + a_2 t^2)^2 \, dt$$

$$= \int_0^1 (a_0^2 + a_1^2 t^2 + a_2^2 t^4 + 2 a_0 a_1 t + 2 a_0 a_2 t^2 + 2 a_1 a_2 t^3) \, dt$$

$$= \left[a_0^2 t \right]_0^1 + \left[\frac{a_1^2 \, t^3}{3} \right]_0^1 + \left[\frac{a_2^2 \, t^5}{5} \right]_0^1 + \left[\frac{2 a_0 a_1 t^2}{2} \right]_0^1 + \left[\frac{2 a_0 a_2 t^3}{3} \right]_0^1 + \left[\frac{2 a_1 a_2 t^4}{4} \right]_0^1$$

$$= a_0^2 + \frac{a_1^2}{3} + \frac{a_2^2}{5} + a_0 a_1 + \frac{2 a_0 a_2}{3} + \frac{a_1 a_2}{2}. \qquad\qquad 3.5.2$$

The quadratic form of Example 3.5.2 is absolutely fundamental in physics. The energy of an electromagnetic field is the integral of the square of the field, so if p is the electromagnetic field, the quadratic form $Q(p)$ gives the amount of energy between 0 and 1.

Above, p is a quadratic polynomial, but $Q(p)$ is a quadratic form if p is a polynomial of *any* degree, not just quadratic. This is obvious if p is linear: if $a_2 = 0$, Equation 3.5.2 becomes $Q(p) = a_0^2 + a_1^2/3 + a_0 a_1$. Exercise 3.5.3 asks you to show that Q is a quadratic form if p is a cubic polynomial. $\triangle$

In various guises, quadratic forms have been an important part of mathematics since the ancient Greeks. The quadratic formula, always the centerpiece of high school math, is one aspect.

A famous theorem due to Fermat (*Fermat's little theorem*) asserts that a prime number $p \neq 2$ can be written as a sum of two squares if and only if the remainder after dividing p by 4 is 1. The proof of this and a world of analogous results (due to Fermat, Euler, Lagrange, Legendre, Gauss, Dirichlet, Kronecker, ...) led to algebraic number theory and the development of abstract algebra.

A much deeper problem is the question: what whole numbers a can be written in the form $x^2 + y^2$? Of course any number a can be written $\sqrt{a}^2 + 0^2$, but suppose you impose that x and y be whole numbers. For instance, $2^2 + 1^2 = 5$, so that 5 can be written as a sum of two squares, but 3 and 7 cannot.

The classification of quadratic forms over the integers is thus a deep and difficult problem, though now reasonably well understood. But the classification over the reals, where we are allowed to extract square roots of positive numbers, is relatively easy. We will be discussing quadratic forms over the reals. In particular, we will be interested in classifying such quadratic forms by associating to each quadratic form two integers, together called its *signature*.

In contrast, no one knows anything about cubic forms. This has ramifications for the understanding of manifolds. The abstract, algebraic view of a four-dimensional manifold is that it is a quadratic form over the integers; because integral quadratic forms are so well understood, a great deal of progress has been made in understanding 4-manifolds. But even the foremost researchers don't know how to approach six-dimensional manifolds; that would require knowing something about cubic forms.

In Section 3.6 we will see that quadratic forms can be used to analyze the behavior of a function at a critical point: the signature of a quadratic form will enable us to determine whether the critical point is a maximum, a minimum or some flavor of *saddle*, where the function goes up in some directions and down in others, as in a mountain pass.

Quadratic forms as sums of squares

Essentially everything there is to say about real quadratic forms is summed up by Theorem 3.5.3, which says that a quadratic form can be represented as a *sum of squares of linearly independent linear functions* of the variables.

We know that $m \leq n$, since there can't be more than n linearly independent linear functions on $\mathbb{R}^n$ (Exercise 2.6.3).

The term "sum of squares" is traditional; it would perhaps be more accurate to call it a *combination* of squares, since some squares may be subtracted rather than added.

Theorem 3.5.3 (Quadratic forms as sums of squares). *(a) For any quadratic form $Q(\vec{x})$ on $\mathbb{R}^n$, there exist $m = k+l$ linearly independent linear functions $\alpha_1(\vec{x}), \ldots, \alpha_m(\vec{x})$ such that*

$$Q(\vec{x}) = \big(\alpha_1(\vec{x})\big)^2 + \cdots + \big(\alpha_k(\vec{x})\big)^2 - \big(\alpha_{k+1}(\vec{x})\big)^2 - \cdots - \big(\alpha_{k+l}(\vec{x})\big)^2. \quad 3.5.3$$

(b) The number k of plus signs and the number l of minus signs in a decomposition like that of Equation 3.5.3 depends only on Q and not on the specific linear functions chosen.

Definition 3.5.4 (Signature). The signature of a quadratic form is the pair of integers (k, l).

Of course more than one quadratic form can have the same signature. The quadratic forms in Examples 3.5.6 and 3.5.7 below both have signature $(2, 1)$.

The signature of a quadratic form is sometimes called its *type*.

The word suggests, correctly, that the signature remains unchanged regardless of how the quadratic form is decomposed into a sum of linearly independent linear functions; it suggests, incorrectly, that the signature identifies a quadratic form.

Before giving a proof, or even a precise definition of the terms involved, we want to give some examples of the main technique used in the proof; a careful look at these examples should make the proof almost redundant.

Completing squares to prove the quadratic formula

The proof is provided by an algorithm for finding the linearly independent functions α_i: "completing squares." This technique is used in high school to prove the quadratic formula.

Indeed, to solve $ax^2 + bx + c = 0$, write

The key point is that $ax^2 + Bx$ can be rewritten

$$ax^2 + Bx + \left(\frac{B}{2\sqrt{a}}\right)^2 - \left(\frac{B}{2\sqrt{a}}\right)^2$$

$$= \left(\sqrt{a}x + \frac{B}{2\sqrt{a}}\right)^2 - \left(\frac{B}{2\sqrt{a}}\right)^2.$$

(We have written a lower case and B upper case because in our applications, a will be a number, but B will be a linear function.)

$$ax^2 + bx + c = ax^2 + bx + \left(\frac{b}{2\sqrt{a}}\right)^2 - \left(\frac{b}{2\sqrt{a}}\right)^2 + c = 0, \qquad 3.5.4$$

which gives

$$\left(\sqrt{a}x + \frac{b}{2\sqrt{a}}\right)^2 = \frac{b^2}{4a} - c. \qquad 3.5.5$$

Taking square roots gives

$$\sqrt{a}x + \frac{b}{2\sqrt{a}} = \sqrt{\frac{b^2 - 4ac}{4a}}, \quad \text{leading to the famous formula} \qquad 3.5.6$$

$$x = \frac{-b \pm \sqrt{b^2 - 4ac}}{2a}. \qquad 3.5.7$$

Example 3.5.5 (Quadratic form as a sum of squares).

$$x^2 + xy = x^2 + xy + \frac{1}{4}y^2 - \frac{1}{4}y^2 = \left(x + \frac{y}{2}\right)^2 - \left(\frac{y}{2}\right)^2. \qquad 3.5.8$$

In this case, the linear functions are

$$\alpha_1\begin{pmatrix} x \\ y \end{pmatrix} = x + \frac{y}{2} \quad \text{and} \quad \alpha_2\begin{pmatrix} x \\ y \end{pmatrix} = \frac{y}{2}. \qquad 3.5.9$$

Clearly the functions

$$\alpha_1\begin{pmatrix} x \\ y \end{pmatrix} = x + \frac{y}{2}, \quad \alpha_2\begin{pmatrix} x \\ y \end{pmatrix} = \frac{y}{2}$$

are linearly independent: no multiple of $y/2$ can give $x + y/2$. If we like, we can be systematic and write these functions as rows of a matrix:

$$\begin{bmatrix} 1 & 1/2 \\ 0 & 1/2 \end{bmatrix}.$$

It is not necessary to row reduce this matrix to see that the rows are linearly independent.

Express the quadratic form $x^2 + xy - y^2$ as a sum of squares, checking your answer below.[16]

Example 3.5.6 (Completing squares: a more complicated example). Consider the quadratic form

$$Q(\vec{\mathbf{x}}) = x^2 + 2xy - 4xz + 2yz - 4z^2. \qquad 3.5.10$$

We take all the terms in which x appears, which gives us $x^2 + (2y - 4z)x$; we see that $B = 2y - 4z$ will allow us to complete the square; adding and subtracting $(y - 2z)^2$ yields

$$\begin{aligned} Q(\vec{\mathbf{x}}) &= (x + y - 2z)^2 - (y^2 - 4yz + 4z^2) + 2yz - 4z^2 \\ &= (x + y - 2z)^2 - y^2 + 6yz - 8z^2. \end{aligned} \qquad 3.5.11$$

Collecting all remaining terms in which y appears and completing the square gives:

$$Q(\vec{\mathbf{x}}) = (x + y - 2z)^2 - (y - 3z)^2 + (z)^2. \qquad 3.5.12$$

In this case, the linear functions are

This decomposition of $Q(\vec{\mathbf{x}})$ is not the only possible one. Exercise 3.5.7 asks you to derive two alternative decompositions.

$$\alpha_1\begin{pmatrix} x \\ y \\ z \end{pmatrix} = x + y - 2z, \quad \alpha_2\begin{pmatrix} x \\ y \\ z \end{pmatrix} = y - 3z, \quad \text{and} \quad \alpha_2\begin{pmatrix} x \\ y \\ z \end{pmatrix} = z. \quad 3.5.13$$

If we write each function as the row of a matrix and row reduce:

$$\begin{bmatrix} 1 & 1 & -2 \\ 0 & 1 & -3 \\ 0 & 0 & 1 \end{bmatrix} \quad \text{row reduces to} \quad \begin{bmatrix} 1 & 0 & 0 \\ 0 & 1 & 0 \\ 0 & 0 & 1 \end{bmatrix}, \qquad 3.5.14$$

we see that the functions are linearly independent. $\triangle$

The algorithm for completing squares should be pretty clear: as long as the square of some coordinate function actually figures in the expression, every

---

16

$$x^2 + xy - y^2 = x^2 + xy + \frac{y^2}{4} - \frac{y^2}{4} - y^2 = \left(x + \frac{y}{2}\right)^2 - \left(\frac{\sqrt{5}y}{2}\right)^2$$

appearance of that variable can be incorporated into a perfect square; by subtracting off that perfect square, you are left with a quadratic form in precisely one fewer variable. (The "precisely one fewer variable" guarantees linear independence.) This works when there is at least one square, but what should you do with something like the following?

Example 3.5.7 (Quadratic form with no squares). Consider the quadratic form

$$Q(\vec{x}) = xy - xz + yz. \qquad 3.5.15$$

There wasn't anything magical about the choice of u, as Exercise 3.5.8 asks you to show; almost anything would have done.

One possibility is to introduce the new variable $u = x - y$, so that we can trade x for $u + y$, getting

$$(u+y)y - (u+y)z + yz = y^2 + uy - uz$$
$$= \left(y + \frac{u}{2}\right)^2 - \frac{u^2}{4} - uz - z^2 + z^2$$
$$= \left(y + \frac{u}{2}\right)^2 - \left(\frac{u}{2} + z\right)^2 + z^2 \qquad 3.5.16$$
$$= \left(\frac{x}{2} + \frac{y}{2}\right)^2 - \left(\frac{x}{2} - \frac{y}{2} + z\right)^2 + z^2.$$

Again, to check that the functions

$$\alpha_1 \begin{pmatrix} x \\ y \\ z \end{pmatrix} = \frac{x}{2} + \frac{y}{2}, \quad \alpha_2 \begin{pmatrix} x \\ y \\ z \end{pmatrix} = \frac{x}{2} - \frac{y}{2} + z, \quad \alpha_3 \begin{pmatrix} x \\ y \\ z \end{pmatrix} = z \qquad 3.5.17$$

There is another meaning one can imagine for linear independence, which applies to any functions $\alpha_1, \ldots, \alpha_m$, not necessarily linear: one can interpret the equation

$$c_1\alpha_1 + \cdots + c_m\alpha_m = 0$$

as meaning that $(c_1\alpha_1 + \cdots + c_m\alpha_m)$ is the zero function: i.e., that

$$(c_1\alpha_1 + \cdots + c_m\alpha_m)(\vec{x}) = 0$$

for any $\vec{x} \in \mathbb{R}^n$, and say that $\alpha_1, \ldots, \alpha_m$ are linearly independent if $c_1 = \cdots = c_m = 0$. In fact, these two meanings coincide: for a matrix to represent the linear transformation 0, it must be the 0 matrix (of whatever size is relevant, here $1 \times n$).

are linearly independent, we can write them as rows of a matrix:

$$\begin{bmatrix} 1/2 & 1/2 & 0 \\ 1/2 & -1/2 & 1 \\ 0 & 0 & 1 \end{bmatrix} \quad \text{row reduces to} \quad \begin{bmatrix} 1 & 0 & 0 \\ 0 & 1 & 0 \\ 0 & 0 & 1 \end{bmatrix}. \quad \triangle \qquad 3.5.18$$

Theorem 3.5.3 says that a quadratic form can be expressed as a sum of linearly independent functions of its variables, but it does not say that whenever a quadratic form is expressed as a sum of squares, those squares are necessarily linearly independent.

Example 3.5.8 (Squares that are not linearly independent). We can write

$$2x^2 + 2y^2 + 2xy = x^2 + y^2 + (x+y)^2 \qquad 3.5.19$$

or

$$2x^2 + 2y^2 + 2xy = \left(\sqrt{2}x + \frac{y}{\sqrt{2}}\right)^2 + \left(\sqrt{\frac{3}{2}}\,y\right)^2. \qquad 3.5.20$$

Only the second decomposition reflects Theorem 3.5.3. In the first, the linear functions x, y and $x + y$ are not linearly independent, since $x + y$ is a linear combination of x and y.

Proof of Theorem 3.5.3

All the essential ideas for the proof of Theorem 3.5.3, part (a) are contained in the examples; a formal proof is in Appendix A.10.

Before proving part (b), which says that the signature (k, l) of a quadratic form does not depend on the specific linear functions chosen for its decomposition, we need to introduce some new vocabulary.

Definition 3.5.9 is equivalent to saying that a quadratic form is positive definite if its signature is $(n, 0)$ and negative definite if its signature is $(0, n)$, as Exercise 3.5.14 asks you to show.

The fact that the quadratic form of Example 3.5.10 is negative definite means that the *Laplacian* in one dimension (i.e., the transformation that takes p to p'') is negative. This has important ramifications; for example, it leads to stable equilibria in elasticity.

When we write $Q(p)$ we mean that Q is a function of the coefficients of p. For example, if $p = x^2 + 2x + 1$, then $Q(p) = Q \begin{pmatrix} 1 \\ 2 \\ 1 \end{pmatrix}$.

Definition 3.5.9 (Positive and negative definite). A quadratic form $Q(\vec{x})$ is positive definite if and only if $Q(\vec{x}) > 0$ when $\vec{x} \neq 0$. It is negative definite if and only if $Q(\vec{x}) < 0$ when $\vec{x} \neq 0$.

The fundamental example of a positive definite quadratic form is $Q(\vec{x}) = |\vec{x}|^2$. The quadratic form of Example 3.5.2,

$$Q(p) = \int_0^1 \big(p(t)\big)^2 \, dt, \quad \text{is also positive definite.} \tag{3.5.1}$$

Here is an important example of a negative definite quadratic form.

Example 3.5.10 (Negative definite quadratic form). Let P_k be the space of polynomials of degree $\leq k$, and $V_{a,b} \subset P_k$ the space of polynomials p that vanish at a and b for some $a < b$. Consider the quadratic form $Q : V_{a,b} \to \mathbb{R}$ given by

$$Q(p) = \int_a^b p(t) p''(t) \, dt. \tag{3.5.21}$$

Using integration by parts,

$$Q(p) = \int_a^b p(t) p''(t) \, dt = \underbrace{p(b) p'(b) - p(a) p'(a)}_{= 0 \text{ by def.}} - \int_a^b \big(p'(t)\big)^2 \, dt \; < \; 0. \tag{3.5.22}$$

Since $p \in V_{a,b}$, $p(a) = p(b) = 0$ by definition; the integral is negative unless $p' = 0$ (i.e., unless p is constant); the only constant in $V_{a,b}$ is 0. $\triangle$

Proof of Theorem 3.5.3 (b)

Now to prove that the signature (k, l) of a quadratic form Q on $\mathbb{R}^n$ depends only on Q and not on how Q is decomposed. This follows from Proposition 3.5.11.

Recall that when a quadratic form is written as a "sum" of squares of linearly independent functions, k is the number of squares preceded by a plus sign, and l is the number squares preceded by a minus sign.

Proposition 3.5.11. *The number k is the largest dimension of a subspace of $\mathbb{R}^n$ on which Q is positive definite and the number l is the largest dimension of a subspace on which Q is negative definite.*

Proof. First let us show that Q cannot be positive definite on any subspace of dimension $> k$. Suppose

$$Q(\vec{\mathbf{x}}) = \big(\underbrace{\alpha_1(\vec{\mathbf{x}})^2 + \cdots + \alpha_k(\vec{\mathbf{x}})^2}_{k \text{ terms}} \big) - \big(\underbrace{\alpha_{k+1}(\vec{\mathbf{x}})^2 + \cdots + \alpha_{k+l}(\vec{\mathbf{x}})^2}_{l \text{ terms}} \big) \qquad 3.5.23$$

is a decomposition of Q into squares of linearly independent linear functions, and that $W \subset \mathbb{R}^n$ is a subspace of dimension $k_1 > k$. Consider the linear transformation $W \to \mathbb{R}^k$ given by

$$\vec{\mathbf{w}} \mapsto \begin{bmatrix} \alpha_1(\vec{\mathbf{w}}) \\ \vdots \\ \alpha_k(\vec{\mathbf{w}}) \end{bmatrix}. \qquad 3.5.24$$

"Non-trivial" kernel means the kernel is not 0.

Since the domain has dimension k_1, which is greater than the dimension k of the range, this mapping has a non-trivial kernel. Let $\vec{\mathbf{w}} \neq 0$ be an element of this kernel. Then, since the terms $\alpha_1(\vec{\mathbf{w}})^2 + \cdots + \alpha_k(\vec{\mathbf{w}})^2$ vanish, we have

$$Q(\vec{\mathbf{w}}) = -\big(\alpha_{k+1}(\vec{\mathbf{w}})^2 + \cdots + \alpha_{k+l}(\vec{\mathbf{w}})^2 \big) \leq 0. \qquad 3.5.25$$

So Q cannot be positive definite on any subspace of dimension $> k$.

Now we need to exhibit a subspace of dimension k on which Q is positive definite. So far we have $k + l$ linearly independent linear functions $\alpha_1, \dots, \alpha_{k+l}$. Add to this set linear functions $\alpha_{k+l+1}, \dots, \alpha_n$ such that $\alpha_1, \dots, \alpha_n$ form a maximal family of linearly independent linear functions, i.e., a basis of the space of $1 \times n$ row matrices (see Exercise 2.4.12).

Consider the linear transformation $T : \mathbb{R}^n \to \mathbb{R}^{n-k}$

$$T : \vec{\mathbf{x}} \mapsto \begin{bmatrix} \alpha_{k+1}(\vec{\mathbf{x}}) \\ \vdots \\ \alpha_n(\vec{\mathbf{x}}) \end{bmatrix}. \qquad 3.5.26$$

The rows of the matrix corresponding to T are thus the linearly independent row matrices $\alpha_{k+1}, \dots, \alpha_n$; like Q, they are defined on $\mathbb{R}^n$ so the matrix T is n wide. It is $n - k$ tall.

Let us see that $\ker T$ has dimension k, and is thus a subspace of dimension k on which Q is positive definite. The rank of T is equal to the number of its linearly independent rows (Theorem 2.5.13), i.e., $\dim \operatorname{Img} T = n - k$, so by the dimension formula,

$$\dim \ker T + \underbrace{\dim \operatorname{Img} T}_{n-k} = n, \quad \text{i.e.,} \quad \dim \ker T = k. \qquad 3.5.27$$

For any $\vec{v} \in \ker T$, the terms $\alpha_{k+1}(\vec{v}), \dots, \alpha_{k+l}(\vec{v})$ of $Q(\vec{v})$ vanish, so

$$Q(\vec{v}) = \alpha_1(\vec{v})^2 + \cdots + \alpha_k(\vec{v})^2 \geq 0. \qquad 3.5.28$$

If $Q(\vec{v}) = 0$, this means that every term is zero, so

$$\alpha_1(\vec{v}) = \cdots = \alpha_n(\vec{v}) = 0, \qquad 3.5.29$$

which implies that $\vec{v} = 0$. So we see that if $\vec{v} \neq 0$, Q is strictly positive.

The argument for l is identical. $\quad \square$

Proof of Theorem 3.5.3(b). Since the proof of Proposition 3.5.11 says nothing about any particular choice of decomposition, we see that k and l depend only on the quadratic form, not on the particular linearly independent functions we use to represent it as a sum of squares. $\quad \square$

Classification of quadratic forms

The quadratic form of Example 3.5.5 has rank 2; the quadratic form of Example 3.5.6 has rank 3.

Definition 3.5.12 (Rank of a quadratic form). The rank of a quadratic form on $\mathbb{R}^n$ is the number of linearly independent squares that appear when the quadratic form is represented as a sum of linearly independent squares.

It follows from Exercise 3.5.14 that only nondegenerate forms can be positive definite or negative definite.

Definition 3.5.13 (Degenerate and nondegenerate quadratic forms). A quadratic form on $\mathbb{R}^n$ with rank m is nondegenerate if $m = n$. It is degenerate if $m < n$.

The examples we have seen so far in this section are all nondegenerate; a degenerate one is shown in Example 3.5.15.

The following proposition is important; we will use it to prove Theorem 3.6.6 about using quadratic forms to classify critical points of functions.

Proposition 3.5.14. *If $Q : \mathbb{R}^n \to \mathbb{R}$ is a positive definite quadratic form, then there exists a constant $C > 0$ such that*

$$Q(\vec{x}) \geq C|\vec{x}|^2 \qquad 3.5.30$$

for all $\vec{x} \in \mathbb{R}^n$.

Proof. Since Q has rank n, we can write

$$Q(\vec{x}) = (\alpha_1(\vec{x}))^2 + \cdots + (\alpha_n(\vec{x}))^2 \qquad 3.5.31$$

Another proof (shorter and less constructive) is sketched in Exercise 3.5.15.

Of course Proposition 3.5.14 applies equally well to negative definite quadratic forms; just use $-C$.

as a sum of squares of n linearly independent functions. The linear transformation $T : \mathbb{R}^n \to \mathbb{R}^n$ whose rows are the α_i is invertible.

Since Q is positive definite, all the squares in Equation 3.5.31 are preceded by plus signs, and we can consider $Q(\vec{x})$ as the length squared of the vector $T\vec{x}$. Thus we have

$$Q(\vec{x}) = |T\vec{x}|^2 \geq \frac{|\vec{x}|^2}{|T^{-1}|^2}, \qquad 3.5.32$$

so you can take $C = 1/|T^{-1}|^2$. (For the inequality in Equation 3.5.32, recall that $|\vec{x}| = |T^{-1}T\vec{x}| \leq |T^{-1}||T\vec{x}|$.) □

Example 3.5.15 (Degenerate quadratic form). The quadratic form

$$Q(p) = \int_0^1 \left(p'(t)\right)^2 dt \qquad 3.5.33$$

on the space P_k of polynomials of degree at most k is a degenerate quadratic form, because Q vanishes on the constant polynomials.   △

3.6 Classifying Critical Points of Functions

In this section we see what the quadratic terms of a function's Taylor polynomial tell us about the function's behavior. The quadratic terms of a function's Taylor polynomial constitute a quadratic form. If that quadratic form is nondegenerate (which is usually the case), its signature tells us whether a *critical point* (a point where the first derivative vanishes) is a minimum of the function, a maximum, or a saddle (illustrated by Figure 3.6.1).

Finding maxima and minima

A standard application of one-variable calculus is to find the maximum or minimum of a function by finding the places where the derivative vanishes, according to the following theorem, which elaborates on Proposition 1.6.8.

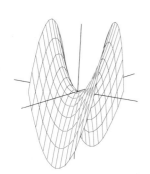

FIGURE 3.6.1.

The graph of $x^2 - y^2$, a typical saddle.

By "strict maximum" we mean $f(x_0) > f(x)$, not $f(x_0) \geq f(x)$; by "strict minimum" we mean $f(x_0) < f(x)$, not $f(x_0) \leq f(x)$.

Theorem 3.6.1 (Maximum and minimum of function of one variable). *(a) Let $U \subset \mathbb{R}$ be an open interval and $f : U \to \mathbb{R}$ be a differentiable function. If $x_0 \in U$ is a maximum or a minimum of f, then $f'(x_0) = 0$.*

(b) If f is twice differentiable, and if $f'(x_0) = 0$ and $f''(x_0) < 0$, then x_0 is a strict local maximum of f, i.e., there exists a neighborhood $V \subset U$ of x_0 such that $f(x_0) > f(x)$ for all $x \in V - \{x_0\}$.

(c) If f is twice differentiable, and if $f'(x_0) = 0$ and $f''(x_0) > 0$, then x_0 is a strict local minimum of f, i.e., there exists a neighborhood $V \subset U$ of x_0 such that $f(x_0) < f(x)$ for all $x \in V - \{x_0\}$.

The plural of extremum is extrema.

Part (a) of Theorem 3.6.1 generalizes in the most obvious way. So as not to privilege maxima over minima, we define an *extremum* of a function to be either a maximum or a minimum.

Note that the equation

$$[\mathbf{D}f(\mathbf{x})] = 0$$

is really n equations in n variables, just the kind of thing Newton's method is suited to. Indeed, one important use of Newton's method is finding maxima and minima of functions.

Theorem 3.6.2 (Derivative zero at an extremum). *Let $U \subset \mathbb{R}^n$ be an open subset and $f : U \to \mathbb{R}$ be a differentiable function. If $\mathbf{x}_0 \in U$ is an extremum of f, then $[\mathbf{D}f(\mathbf{x}_0)] = 0$.*

Proof. The derivative is given by the Jacobian matrix, so it is enough to show that if $\mathbf{x}_0$ is an extremum of f, then $D_i f(\mathbf{x}_0) = 0$ for all $i = 1, \ldots, n$. But $D_i f(\mathbf{x}_0) = g'(0)$, where g is the function of one variable $g(t) = f(\mathbf{x}_0 + t\vec{\mathbf{e}}_i)$, and our hypothesis also implies that g has an extremum at $t = 0$, so $g'(0) = 0$ by Theorem 3.6.1. $\square$

It is not true that every point at which the derivative vanishes is an extremum. When we find such a point (called a *critical point*), we will have to work harder to determine whether it is indeed a maximum or minimum.

In Definition 3.6.3, saying that the derivative vanishes means that all the partial derivatives vanish. Finding a place where all partial derivatives vanish means solving n equations in n unknowns. Usually there is no better approach than applying Newton's method, and finding critical points is an important application of Newton's method.

Definition 3.6.3 (Critical point). Let $U \subset \mathbb{R}^n$ be open, and $f : U \to \mathbb{R}$ be a differentiable function. A critical point of f is a point where the derivative vanishes.

Example 3.6.4 (Finding critical points). What are the critical points of the function

$$f\begin{pmatrix} x \\ y \end{pmatrix} = x + x^2 + xy + y^3?$$ 3.6.1

The partial derivatives are

$$D_1 f\begin{pmatrix} x \\ y \end{pmatrix} = 1 + 2x + y, \qquad D_2 f\begin{pmatrix} x \\ y \end{pmatrix} = x + 3y^2.$$ 3.6.2

In this case we don't need Newton's method, since the system can be solved explicitly: substitute $x = -3y^2$ from the second equation into the first, to find

$$1 + y - 6y^2 = 0; \quad \text{i.e.,} \quad y = \frac{1 \pm \sqrt{1 + 24}}{12} = \frac{1}{2} \text{ or } -\frac{1}{3}.$$ 3.6.3

Substituting this into $x = -(1 + y)/2$ (or into $x = -3y^2$) gives two critical points:

$$\mathbf{a}_1 = \begin{pmatrix} -3/4 \\ 1/2 \end{pmatrix} \quad \text{and} \quad \mathbf{a}_2 = \begin{pmatrix} -1/3 \\ -1/3 \end{pmatrix}. \quad \triangle$$ 3.6.4

Remark 3.6.5 (Maxima on closed sets). Just as in the case of one variable, a major problem in using Theorem 3.6.2 is the hypothesis that U is open. Often we want to find an extremum of a function on a closed set, for instance the maximum of x^2 on $[0, 2]$. The maximum, which is 4, occurs when $x = 2$,

In Section 3.7, we will see that sometimes we can analyze the behavior of the function restricted to the boundary, and use the variant of critical point theory developed there.

which is not a point where the derivative of x^2 vanishes. Especially when we have used Theorem 1.6.7 to assert that a maximum exists in a compact subset, we need to check that this maximum occurs in the interior of the region under consideration, not on the boundary, before we can say that it is a critical point. $\triangle$

The second derivative criterion

Is either of the critical points given by Equation 3.6.4 an extremum? In one variable, we would answer this question by looking at the sign of the second derivative. The right generalization of "the second derivative" to higher dimensions is "the quadratic form given by the quadratic terms of the Taylor polynomial." It seems reasonable to hope that since (like every sufficiently differentiable function) the function is well approximated near these points by its Taylor polynomial, the function should behave like its Taylor polynomial.

In evaluating the second derivative, remember that $D_1^2 f(\mathbf{a})$ means the second partial derivative $D_1 D_1 f$, evaluated at $\mathbf{a}$. It does not mean D_1^2 times $f(\mathbf{a})$. In this case we have $D_1^2 f = 2$ and $D_1 D_2 f = 1$; these are constants, so where we evaluate the derivative doesn't matter. But $D_2^2(f) = 6y$; evaluated at $\mathbf{a}$ this gives 3.

Let us apply this to the function in Example 3.6.4, $f(\mathbf{x}) = x + x^2 + xy + y^3$. Evaluating its Taylor polynomial at $\mathbf{a}_1 = \begin{pmatrix} -3/4 \\ 1/2 \end{pmatrix}$, we get

$$P_{f,\mathbf{a}_1}^2(\mathbf{a}_1 + \vec{\mathbf{h}}) = \underbrace{-\frac{7}{16}}_{f(\mathbf{a})} + \underbrace{\frac{1}{2}2h_1^2 + h_1 h_2 + \frac{1}{2}3h_2^2}_{\text{second derivative}}. \qquad 3.6.5$$

The second derivative is a positive definite quadratic form:

$$h_1^2 + h_1 h_2 + \frac{3}{2}h_2^2 = \left(h_1 + \frac{h_2}{2} \right)^2 + \frac{5}{4}h_2^2, \quad \text{with signature } (2,0). \qquad 3.6.6$$

What happens at the critical point $\mathbf{a}_2 = \begin{pmatrix} -1/3 \\ -1/3 \end{pmatrix}$? Check your answer below.[17]

How should we interpret these results? If we believe that the function behaves near a critical point like its second degree Taylor polynomial, then the critical point $\mathbf{a}_1$ is a minimum; as the increment vector $\vec{\mathbf{h}} \to \mathbf{0}$, the quadratic form goes to 0 as well, and as $\vec{\mathbf{h}}$ gets bigger (i.e., we move further from $\mathbf{a}_1$), the quadratic form gets bigger. Similarly, if at a critical point the second derivative is a negative definite quadratic form, we would expect it to be a maximum. But what about a critical point like $\mathbf{a}_2$, where the second derivative is a quadratic form with signature $(1,1)$?

You may recall that even in one variable, a critical point is not necessarily an extremum: if the second derivative vanishes also, there are other possibilities

---

17

$$P_{f,\mathbf{a}_2}^2(\mathbf{a}_2 + \vec{\mathbf{h}}) = -\frac{4}{27} + \frac{1}{2}2h_1^2 + h_1 h_2 + \frac{1}{2}(-2)h_2^2, \quad \text{with quadratic form}$$

$$h_1^2 + h_1 h_2 - h_2^2 = \left(h_1 + \frac{h_2}{2} \right)^2 - \frac{5}{4}h_2^2, \quad \text{which has signature } (1,1).$$

(the point of inflection of $f(x) = x^3$, for instance). However, such points are exceptional: zeroes of the first and second derivative do not usually coincide. Ordinarily, for functions of one variable, critical points are extrema.

This is not the case in higher dimensions. The right generalization of "the second derivative of f does not vanish" is "the quadratic terms of the Taylor polynomial are a non-degenerate quadratic form." A critical point at which this happens is called a *non-degenerate critical point*. This is the ordinary course of events (degeneracy requires coincidences). But a non-degenerate critical point need not be an extremum. Even in two variables, there are three signatures of non-degenerate quadratic forms: (2,0), (1,1) and (0,2). The first and third correspond to extrema, but signature (1,1) corresponds to a *saddle point*.

> The quadratic form in Equation 3.6.7 is the second degree term of the Taylor polynomial of f at $\mathbf{a}$.

> We state the theorem as we do, rather than saying simply that the quadratic form is not positive definite, or that it is not negative definite, because if a quadratic form on $\mathbb{R}^n$ is degenerate (i.e., $k + l < n$), then if its signature is $(k, 0)$, it is positive, but not positive definite, and the signature does not tell you that there is a local minimum. Similarly, if the signature is $(0, k)$, it does not tell you that there is a local maximum.

> We will say that a critical point has signature (k, l) if the corresponding quadratic form has signature (k, l). For example, $x^2 + y^2 - z^2$ has a saddle of signature $(2, 1)$ at the origin.

> The origin is a saddle for the function $x^2 - y^2$.

The following theorems confirm that the above idea really works.

Theorem 3.6.6 (Quadratic forms and extrema). *Let $U \subset \mathbb{R}^n$ be an open set, $f : U \to \mathbb{R}$ be twice continuously differentiable (i.e., of class C^2), and let $\mathbf{a} \in U$ be a critical point of f, i.e., $[\mathbf{D}f(\mathbf{a})] = 0$.*

(a) If the quadratic form

$$Q(\vec{\mathbf{h}}) = \sum_{I \in \mathcal{I}_n^2} \frac{1}{I!} \left(D_I f(\mathbf{a}) \right) \vec{\mathbf{h}}^I \qquad 3.6.7$$

is positive definite (i.e., has signature $(n, 0)$), then $\mathbf{a}$ is a strict local minimum of f. If the signature of the quadratic form is (k, l) with $l > 0$, then the critical point is not a local minimum.

(b) If the quadratic form is negative definite, (i.e., has signature $(0, n)$), then $\mathbf{a}$ is a strict local maximum of f. If the signature of the quadratic form is (k, l) with $k > 0$, then the critical point is not a maximum.

Definition 3.6.7 (Saddle). If the quadratic form has signature (k, l) with $k > 0$ and $l > 0$, then the critical point is a saddle.

Theorem 9.8.5 (Behavior of functions near saddle points). *Let $U \subset \mathbb{R}^n$ be an open set, and let $f : U \to \mathbb{R}$ be a C^2 function. If f has a saddle at $\mathbf{a} \in U$, then in every neighborhood of $\mathbf{a}$ there are points $\mathbf{b}$ with $f(\mathbf{b}) > f(\mathbf{a})$, and points $\mathbf{c}$ with $f(\mathbf{c}) < f(\mathbf{a})$.*

Proof of 3.6.6 (Quadratic forms and extrema). We will treat case (a) only; case (b) can be derived from it by considering $-f$ rather than f.

We can write

$$f(\mathbf{a} + \vec{\mathbf{h}}) = f(\mathbf{a}) + Q(\vec{\mathbf{h}}) + r(\vec{\mathbf{h}}), \qquad 3.6.8$$

where the remainder $r(\vec{\mathbf{h}})$ satisfies

$$\lim_{\vec{\mathbf{h}} \to 0} \frac{r(\vec{\mathbf{h}})}{|\vec{\mathbf{h}}|^2} = 0. \qquad 3.6.9$$

Thus if Q is positive definite,

Equation 3.6.10 uses Proposition 3.5.14.

$$\frac{f(\mathbf{a} + \vec{\mathbf{h}}) - f(\mathbf{a})}{|\vec{\mathbf{h}}|^2} = \frac{Q(\vec{\mathbf{h}})}{|\vec{\mathbf{h}}|^2} + \frac{r(\vec{\mathbf{h}})}{|\vec{\mathbf{h}}|^2} \geq C + \frac{r(\vec{\mathbf{h}})}{|\vec{\mathbf{h}}|^2}, \qquad 3.6.10$$

The constant C depends on Q, not on the vector on which Q is evaluated, so $Q(\vec{\mathbf{h}}) \geq C|\vec{\mathbf{h}}|^2$; i.e.,

$$\frac{Q(\vec{\mathbf{h}})}{|\vec{\mathbf{h}}|^2} \geq \frac{C|\vec{\mathbf{h}}|^2}{|\vec{\mathbf{h}}|^2} = C.$$

where C is the constant of Proposition 3.5.14—the constant $C > 0$ such that $Q(\vec{\mathbf{x}}) \geq C|\mathbf{x}|^2$ for all $\vec{\mathbf{x}} \in \mathbb{R}^n$, when Q is positive definite.

The right-hand side is positive for $\vec{\mathbf{h}}$ sufficiently small (see Equation 3.6.9), so the left-hand side is also, i.e., $f(\mathbf{a} + \vec{\mathbf{h}}) > f(\mathbf{a})$ for $\vec{\mathbf{h}}$ sufficiently small; i.e., $\mathbf{a}$ is a strict local minimum of f.

If Q has signature (k, l) with $l > 0$, then there is a subspace $V \subset \mathbb{R}^n$ of dimension l on which Q is negative definite. Suppose that Q is given by the quadratic terms of the Taylor polynomial of f at a critical point $\mathbf{a}$ of f. Then the same argument as above shows that if $\mathbf{h} \in V$ and $|\mathbf{x}|$ is sufficiently small, then the increment $f(\mathbf{a} + \vec{\mathbf{h}}) - f(\mathbf{a})$ will be negative, certainly preventing $\mathbf{a}$ from being a minimum of f. $\square$

Proof of 9.8.5 (Behavior of functions near saddle points). Write

$$f(\mathbf{a} + \vec{\mathbf{h}}) = f(\mathbf{a}) + Q(\vec{\mathbf{h}}) + r(\vec{\mathbf{h}}) \quad \text{and} \quad \lim_{\vec{\mathbf{h}} \to 0} \frac{r(\vec{\mathbf{h}})}{|\vec{\mathbf{h}}|^2} = 0.$$

as in Equations 3.6.8 and 3.6.9.

By Proposition 3.5.11 there exist subspaces V and W of $\mathbb{R}^n$ such that Q is positive definite on V and negative definite on W.

If $\vec{\mathbf{h}} \in V$, and $t > 0$, there exists $C > 0$ such that

Since Q is a quadratic form, $Q(t\vec{\mathbf{h}}) = t^2 Q(\vec{\mathbf{h}})$.

$$\frac{f(\mathbf{a} + t\vec{\mathbf{h}}) - f(\mathbf{a})}{t^2} = \frac{t^2 Q(\vec{\mathbf{h}}) + r(t\vec{\mathbf{h}})}{t^2} \geq C|\vec{\mathbf{h}}|^2 + \frac{r(t\vec{\mathbf{h}})}{t^2}, \qquad 3.6.11$$

and since

$$\lim_{t \to 0} \frac{r(t\vec{\mathbf{h}})}{t^2} = 0, \qquad 3.6.12$$

A similar argument about W shows that there are also points $\mathbf{c}$ where $f(\mathbf{c}) < f(\mathbf{a})$. Exercise 3.6.3 asks you to spell out this argument.

it follows that $f(\mathbf{a} + t\vec{\mathbf{h}}) > f(\mathbf{a})$ for $t > 0$ sufficiently small. $\square$

Degenerate critical points

When $f(\mathbf{x})$ has a critical point at $\mathbf{a}$ such that the quadratic terms of the Taylor polynomial of f at $\mathbf{a}$ are a nondegenerate quadratic form, the function near $\mathbf{a}$ behaves just like that quadratic form. We have just proved this when the quadratic form is positive or negative definite, and the only thing preventing us from proving it for any signature of a nondegenerate form is an accurate definition of "behave just like its quadratic terms."[18]

But if the quadratic form is degenerate, there are many possibilities; we will not attempt to classify them (it is a big job), but simply give some examples.

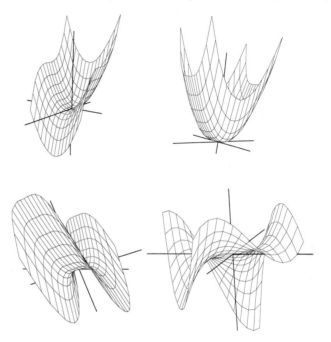

FIGURE 3.6.2. The upper left-hand figure is the surface of equation $z = x^2 + y^3$, and the upper right-hand figure is the surface of equation $z = x^2 + y^4$. The lower left-hand figure is the surface of equation $z = x^2 - y^4$. Although the three graphs look very different, all three functions have the same degenerate quadratic form for the Taylor polynomial of degree 2. The lower right-hand figure shows the monkey saddle; it is the graph of $z = x^3 - 2xy^2$, whose quadratic form is 0.

Example 3.6.9 (Degenerate critical points). The three functions $x^2 + y^3$, $x^2 + y^4$, and $x^2 - y^4$ all have the same degenerate quadratic form for the Taylor polynomial of degree 2: x^2. But they behave very differently, as shown in Figure 3.6.2 (upper left, upper right and lower left). The second one has a minimum, the other two do not.   △

Example 3.6.10 (Monkey saddle). The function $f\begin{pmatrix} x \\ y \end{pmatrix} = x^3 - 2xy^2$ has a critical point that goes up in three directions and down in three also (two for the legs, one for the tail). Its graph is shown in Figure 3.6.2, lower right.   △

---

[18]A precise statement is called the Morse lemma; it can be found (Lemma 2.2) on p. 6 of J. Milnor, *Morse Theory*, Princeton University Press, 1963.

3.7 CONSTRAINED CRITICAL POINTS AND LAGRANGE MULTIPLIERS

The shortest path between two points is a straight line. But what is the shortest path if you are restricted to paths that lie on a sphere (for example, because you are flying from New York to Paris)? This example is intuitively clear but actually quite difficult to address. In this section we will look at problems in the same spirit, but easier. We will be interested in extrema of a function f when f is restricted to some manifold $X \subset \mathbb{R}^n$.

In the case of the set $X \subset \mathbb{R}^8$ describing the position of a link of four rods in the plane (Example 3.2.1) we might imagine that the origin is attracting, and that each vertex $\mathbf{x}_i$ has a "potential" $|\mathbf{x}_i|^2$, perhaps realized by rubber bands connecting the origin to the joints. Then what is the equilibrium position, where the link realizes the minimum of the potential energy? Of course, all four vertices try to be at the origin, but they can't. Where will they go?

In this section we provide tools to answer this sort of question.

Yet another example occurs in the optional subsection of Section 2.8: the norm of a matrix A is

$$\sup_{|\vec{\mathbf{x}}|=1} |A\vec{\mathbf{x}}|.$$

What is $\sup |A\vec{\mathbf{x}}|$ when we require that $\vec{\mathbf{x}}$ have length 1?

Finding constrained critical points using derivatives

A characterization of extrema in terms of derivatives should say that in some sense the derivative vanishes at an extremum. But when we take a function defined on $\mathbb{R}^n$ and consider its *restriction* to a manifold of $\mathbb{R}^n$, we cannot assert that an extremum of the restricted function is a point at which the derivative of the function vanishes. The derivative of the function may vanish at points not in the manifold (the shortest "unrestricted" path from New York to Paris would require tunneling under the Atlantic Ocean). In addition, only very seldom will a constrained maximum be an unconstrained maximum (the tallest child in kindergarten is unlikely to be the tallest child in the entire elementary school). So only very seldom will the derivative of the function vanish at a critical point of the restricted function.

Recall (Definition 3.2.6) that $T_{\mathbf{a}}X$ is the tangent space to a manifold X at $\mathbf{a}$.

Another way to state Equation 3.7.1 is to say that if $\vec{\mathbf{x}} \in T_{\mathbf{a}}X$, then $[\mathbf{D}\varphi(\mathbf{a})]\vec{\mathbf{x}} = 0$.

Geometrically, Theorem 3.7.1 means that a critical point of φ restricted to X is a point $\mathbf{a}$ such that the tangent space to the constraint, $T_{\mathbf{a}}X$, is a subspace of $\ker[\mathbf{D}\varphi(\mathbf{a})]$, the tangent space to a *level set* of φ.

What we can say is that at an extremum of the function restricted to a manifold, the derivative of the function is 0 *when evaluated on any tangent vector to the manifold*: i.e., the derivative vanishes on the tangent space to the manifold.

Theorem 3.7.1. *If $X \subset \mathbb{R}^n$ is a manifold, $U \subset \mathbb{R}^n$ is open, $\varphi : U \to \mathbb{R}$ is a C^1 function and $\mathbf{a} \in X \cap U$ is a local extremum of φ restricted to X, then*

$$T_{\mathbf{a}}X \subset \ker[\mathbf{D}\varphi(\mathbf{a})]. \qquad 3.7.1$$

Definition 3.7.2 (Constrained critical point). A point $\mathbf{a}$ such that $T_{\mathbf{a}}X \subset \ker[\mathbf{D}\varphi(\mathbf{a})]$ is called a critical point of φ constrained to X.

A level set of a function φ is those points such that $\varphi = c$, where c is some constant. We used level sets in Section 3.1.

Example 3.7.3 (Constrained critical point: a simple example). Suppose we wish to maximize the function $\varphi \begin{pmatrix} x \\ y \end{pmatrix} = xy$ on the first quadrant of the circle $x^2 + y^2 = 1$. As shown in Figure 3.7.1, some level sets of that function do not intersect the circle, and some intersect it in two points, but one, $xy = 1/2$, intersects it at the point $\mathbf{a} = \begin{pmatrix} 1/\sqrt{2} \\ 1/\sqrt{2} \end{pmatrix}$. That point is the critical point of φ constrained to the circle. The tangent space at $\mathbf{a}$ to the constraint (i.e., to the circle) consists of the vectors $\begin{bmatrix} \dot{x} \\ \dot{y} \end{bmatrix}$ where $\dot{x} = -\dot{y}$. This tangent space is a subspace of the tangent space to the level set $xy = 1/2$. In fact, the two are the same.

Example 3.7.4 (Constrained critical point in higher dimensions). Suppose we wish to find the minimum of the function $\varphi \begin{pmatrix} x \\ y \\ z \end{pmatrix} = x^2 + y^2 + z^2$, when it is constrained to the ellipse (denoted X) that is the intersection of the cylinder $x^2 + y^2 = 1$, and the plane of equation $x = z$, shown in Figure 3.7.2.

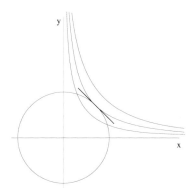

FIGURE 3.7.1.
The unit circle and several level curves of the function xy. The level curve $xy = 1/2$, which realizes the maximum of xy restricted to the circle, is tangent to the circle at the point $\begin{pmatrix} 1/\sqrt{2} \\ 1/\sqrt{2} \end{pmatrix}$, where the maximum is realized.

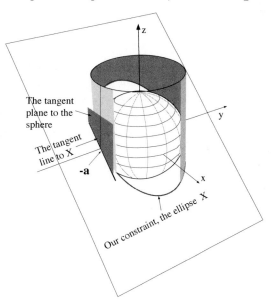

Since φ measures the square of the distance from the origin, we wish to find the points on the ellipse X that are closest to the origin.

FIGURE 3.7.2. At the point $-\mathbf{a} = \begin{pmatrix} 0 \\ -1 \\ 0 \end{pmatrix}$, the distance to the origin has a minimum on the ellipse; at this point, the tangent space to the ellipse is a subspace of the tangent space to the sphere.

The points on the ellipse nearest the origin are $\mathbf{a} = \begin{pmatrix} 0 \\ 1 \\ 0 \end{pmatrix}$ and $-\mathbf{a} = \begin{pmatrix} 0 \\ -1 \\ 0 \end{pmatrix}$;

they are the minima of φ constrained to X. In this case

$$\ker[\mathbf{D}\varphi(\mathbf{a})] = \ker[0\,,2\,,0] \quad \text{and} \quad \ker[\mathbf{D}\varphi(-\mathbf{a})] = \ker[0\,,-2\,,0],$$

In keeping with the notation introduced in Section 3.1, $\dot{y} = 0$ indicates the plane where there is zero increment in the direction of the y-axis; thus $\dot{y} = 0$ denotes the plane tangent to the sphere (and to the cylinder) at $\mathbf{a} = \begin{pmatrix} 0 \\ -1 \\ 0 \end{pmatrix}$; it also denotes the plane tangent to the sphere (and the cylinder) at $\mathbf{a} = \begin{pmatrix} 0 \\ 1 \\ 0 \end{pmatrix}$. It is the plane $y = 0$ translated from the origin.

i.e., at these critical points, $\ker[\mathbf{D}\varphi]$ is the space $y = 0$. The tangent space to the ellipse at the points $\mathbf{a}$ and $-\mathbf{a}$ is the intersection of the planes of equation $\dot{y} = 0$ (the tangent space to the cylinder) and $x = z$ (which is both the plane and the tangent space to the plane). Certainly this is a subspace of

$$\ker\left[\mathbf{D}\varphi\begin{pmatrix} 0 \\ 1 \\ 0 \end{pmatrix}\right]. \quad \triangle$$

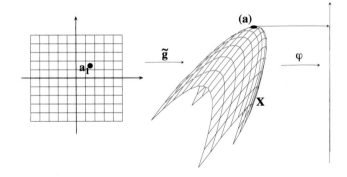

FIGURE 3.7.3. The composition $\varphi \circ \tilde{\mathbf{g}}$: the parametrization $\tilde{\mathbf{g}}$ takes a point in $\mathbb{R}^2$ to the constraint manifold X; φ takes it to $\mathbb{R}$. An extremum of the composition, at $\mathbf{a}_1$, corresponds to an extremum of φ restricted to X, at $\mathbf{a}$; the constraint is incorporated into the parametrization.

The proof of Theorem 3.7.1 is easier to understand if you think in terms of parametrizations. Suppose we want to find a maximum of a function $\varphi\begin{pmatrix} x \\ y \end{pmatrix}$ on the unit circle in $\mathbb{R}^2$. One approach is to parametrize the circle by $t \mapsto \begin{pmatrix} \cos t \\ \sin t \end{pmatrix}$, and look for the *unconstrained* maximum of the new function of one variable, $\varphi_1(t) = \varphi\begin{pmatrix} \cos t \\ \sin t \end{pmatrix}$. We did just this in Example 2.8.8. In this way, the restriction is incorporated in the parametrization.

Proof of Theorem 3.7.1.   Since X is a manifold, near $\mathbf{a}$, X is the graph of a map $\mathbf{g}$ from some subset U_1 of the space E_1 spanned by k standard basis vectors, to the space E_2 spanned by the other $n - k$ standard basis vectors. Call $\mathbf{a}_1$ and $\mathbf{a}_2$ the projections of $\mathbf{a}$ onto E_1 and E_2 respectively.

Then the mapping $\tilde{\mathbf{g}}(\mathbf{x}_1) = \mathbf{x}_1 + \mathbf{g}(\mathbf{x}_1)$ is a parametrization of X near $\mathbf{a}$, and X, which is locally the *graph* of $\mathbf{g}$, is locally the *image* of $\tilde{\mathbf{g}}$. Similarly, $T_\mathbf{a}X$, which is locally the graph of $[\mathbf{Dg}(\mathbf{a}_1)]$, is also locally the image of $[\mathbf{D\tilde{g}}(\mathbf{a}_1)]$.

Then saying that φ on X has a local extremum at $\mathbf{a}$ is the same as saying that the composition $\varphi \circ \tilde{\mathbf{g}}$ has an (unconstrained) extremum at $\mathbf{a}_1$, as sketched in Figure 3.7.3. Thus $[\mathbf{D}(\varphi \circ \tilde{\mathbf{g}})(\mathbf{a}_1)] = 0$. This means exactly that $[\mathbf{D}\varphi(\mathbf{a})]$ vanishes on the image of $[\mathbf{D\tilde{g}}(\mathbf{a}_1)]$, which is the tangent space $T_\mathbf{a}X$. $\square$

This proof provides a straightforward approach to finding constrained critical points, *provided* you know the "constraint manifold" by a parametrization.

Example 3.7.5 (Finding constrained critical points using a parametrization). Say we want to find local critical points of the function

$$\varphi\begin{pmatrix} x \\ y \\ z \end{pmatrix} = x+y+z, \text{ on the surface parametrized by } \mathbf{g}: \begin{pmatrix} u \\ v \end{pmatrix} \mapsto \begin{pmatrix} \sin uv + u \\ u + v \\ uv \end{pmatrix},$$

Exercise 3.7.1 asks you to show that $\mathbf{g}$ really is a parametrization.

shown in Figure 3.7.4 (left). Instead of looking for constrained critical points of φ, we will look for (ordinary) critical points of $\varphi \circ \mathbf{g}$. We have

Multiplying the first equation by u and the second by v gives

$$uv \cos uv + 2u + uv = 0,$$
$$uv \cos uv + v + uv = 0;$$

subtracting the second equation from the first gives $2u = v$.

$$\varphi \circ \mathbf{g} = \sin uv + u + (u+v) + uv, \quad \text{so} \quad \begin{aligned} D_1(\varphi \circ \mathbf{g}) &= v \cos uv + 2 + v \\ D_2(\varphi \circ \mathbf{g}) &= u \cos uv + 1 + u; \end{aligned}$$

setting these both to 0 and solving them gives $2u - v = 0$. In the parameter space, the critical points lie on this line, so the actual constrained critical points lie on the image of that line by the parametrization. Plugging $v = 2u$ into $D_1(\varphi \circ \mathbf{g})$ gives

We could have substituted $v = 2u$ into $D_2(\varphi \circ \mathbf{g})$ instead.

$$u \cos 2u^2 + u + 1 = 0, \qquad\qquad 3.7.2$$

whose graph is shown in Figure 3.7.4 (right).

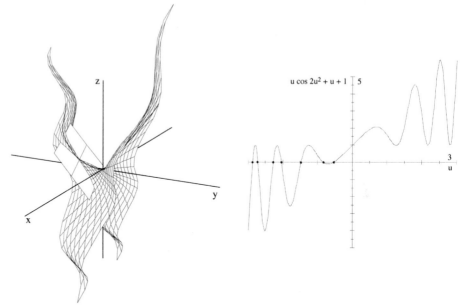

FIGURE 3.7.4.
Left: The surface X parametrized by $\begin{pmatrix} u \\ v \end{pmatrix} \mapsto \begin{pmatrix} \sin uv + u \\ u + v \\ uv \end{pmatrix}$. The critical point where the white tangent plane is tangent to the surface corresponds to $u = -1.48$.
Right: The graph of $u \cos 2u^2 + u + 1 = 0$. The roots of that equation, marked with black dots, give values of the first coordinates of critical points of $\varphi(\mathbf{x}) = x+y+z$ restricted to X.

This function has infinitely many zeroes, each one the first coordinate of a critical point; the seven visible in Figure 3.7.4 are approximately $u = -2.878$, -2.722, -2.28, -2.048, $-1.48 - .822$, $-.548$. The image of the line $v = 2u$ is represented as a dark curve on the surface in Figure 3.7.4, together with the tangent plane at the point corresponding to $u = -1.48$.

Solving that equation (not necessarily easy, of course) will give us the values of the first coordinate (and because $2u = v$, of the second coordinate) of the points that are critical points of φ constrained to X.

Notice that the same computation works if instead of $\mathbf{g}$ we use

$$\mathbf{g}_1 : \begin{pmatrix} u \\ v \end{pmatrix} \mapsto \begin{pmatrix} \sin uv \\ u + v \\ uv \end{pmatrix} , \text{ which gives the "surface" } X_1 \text{ shown in Figure 3.7.5,}$$

but this time the mapping $\mathbf{g}_1$ is emphatically not a parametrization.

FIGURE 3.7.5.

Left: The "surface" X_1 is the image of

$$\mathbf{g}_1 : \begin{pmatrix} u \\ v \end{pmatrix} \mapsto \begin{pmatrix} \sin uv \\ u + v \\ uv \end{pmatrix} ;$$

it is a subset of the surface Y of equation $x = \sin z$, which resembles a curved bench.

Right: The graph of $u \cos u^2 + u + 1 = 0$. The roots of that equation, marked with black dots, give values of the parameter u such that $\varphi(\mathbf{x}) = x + y + z$ restricted to X_1 has "critical points" at $\mathbf{g}_1 \begin{pmatrix} u \\ u \end{pmatrix}$. These are not true critical points, because we have no definition of a critical point of a function restricted to an object like X_1.

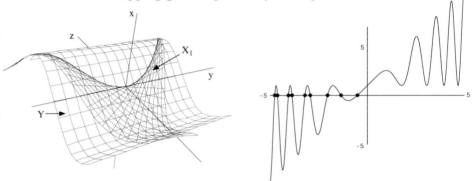

Since for any point $\begin{pmatrix} x \\ y \\ z \end{pmatrix} \in X_1$, $x = \sin z$, we see that X_1 is contained in the surface Y of equation $x = \sin z$, which is a graph of x as a function of z. But X_1 covers only part of Y, since $y^2 - 4z = (u+v)^2 - 4uv = (u-v)^2 \geq 0$; it only covers the part where $y^2 - 4z \geq 0$,[19] and it covers it twice, since $\mathbf{g}_1 \begin{pmatrix} u \\ v \end{pmatrix} = \mathbf{g}_1 \begin{pmatrix} v \\ u \end{pmatrix}$. The mapping $\mathbf{g}_1$ folds the (u, v)-plane over along the diagonal, and pastes the resulting half-plane onto the graph Y. Since $\mathbf{g}_1$ is not one to one, it does not qualify as parametrization (see Definition 3.1.21; it also fails to qualify because its derivative is not one to one. Can you justify this last statement?[20])

Exercise 3.7.2 asks you to show that the function φ has no critical points on Y: the plane of equation $x + y + z = c$ is never tangent to Y. But if you follow the same procedure as above, you will find that critical points of $\varphi \circ \mathbf{g}_1$ occur when $u = v$, and $u \cos u^2 + u + 1 = 0$. What has happened? The critical

---

[19]We realized that $y^2 - 4z = (u + v)^2 - 4uv = (u - v)^2 \geq 0$ while trying to understand the shape of X_1, which led to trying to understand the constraint imposed by the relationship of the second and third variables.

[20]The derivative of $\mathbf{g}_1$ is $\begin{bmatrix} \cos uv & \cos uv \\ 1 & 1 \\ v & u \end{bmatrix}$; at points where $u = v$, the columns of this matrix are not linearly independent.

points of $\varphi \circ \mathbf{g}_1$ now correspond to "fold points," where the plane $x + y + z = c_i$ is tangent not to the surface Y, nor to the "surface" X_1, whatever that would mean, but to the curve that is the image of $u = v$ by $\mathbf{g}_1$. $\triangle$

Lagrange multipliers

The proof of Theorem 3.7.1 relies on the parametrization $\mathbf{g}$ of X. What if you know a manifold only by equations? In this case, we can restate the theorem.

Suppose we are trying to maximize a function φ on a manifold X. Suppose further that we know X not by a parametrization but by a vector-valued equation, $\mathbf{F}(\mathbf{x}) = 0$, where $\mathbf{F} = \begin{bmatrix} F_1 \\ \vdots \\ F_m \end{bmatrix}$ goes from an open subset U of $\mathbb{R}^n$ to $\mathbb{R}^m$, and $[\mathbf{DF}(\mathbf{x})]$ is onto for every $\mathbf{x} \in X$.

Then, as stated by Theorem 3.2.7, for any $\mathbf{a} \in X$, the tangent space $T_{\mathbf{a}}X$ is the kernel of $[\mathbf{DF}(\mathbf{a})]$:

$$T_{\mathbf{a}}X = \ker[\mathbf{DF}(\mathbf{a})]. \qquad 3.7.3$$

So Theorem 3.7.1 asserts that for a mapping $\varphi : U \to \mathbb{R}$, at a critical point of φ on X, we have

$$\ker[\mathbf{DF}(\mathbf{a})] \subset \ker[\mathbf{D}\varphi(\mathbf{a})]. \qquad 3.7.4$$

This can be reinterpreted as follows.

Recall that the Greek λ is pronounced "lambda."

We call $F_1, \ldots, F_m$ constraint functions because they define the manifold to which φ is restricted.

Theorem 3.7.6 (Lagrange multipliers). *Let X be a manifold known by a vector-valued function $\mathbf{F}$. If φ restricted to X has a critical point at $\mathbf{a} \in X$, then there exist numbers $\lambda_1, \ldots, \lambda_m$ such that the derivative of φ at $\mathbf{a}$ is a linear combination of derivatives of the constraint functions:*

$$[\mathbf{D}\varphi(\mathbf{a})] = \lambda_1[\mathbf{D}F_1(\mathbf{a})] + \cdots + \lambda_m[\mathbf{D}F_m(\mathbf{a})]. \qquad 3.7.5$$

The numbers $\lambda_1, \ldots, \lambda_m$ are called *Lagrange multipliers*.
We will prove this theorem after giving some examples.

In our three examples of Lagrange multipliers, our constraint manifold is defined by a scalar-valued function F, not by a vector-valued function $\mathbf{F} = \begin{bmatrix} F_1 \\ \vdots \\ F_m \end{bmatrix}$. But the proof of the spectral theorem (Theorem 3.7.12) involves a vector-valued function.

Example 3.7.7 (Lagrange multipliers: a simple example). Suppose we want to maximize $\varphi \begin{pmatrix} x \\ y \end{pmatrix} = x + y$ on the ellipse $x^2 + 2y^2 = 1$. We have

$$F \begin{pmatrix} x \\ y \end{pmatrix} = x^2 + 2y^2 - 1, \quad \text{and} \quad \left[\mathbf{D}F \begin{pmatrix} x \\ y \end{pmatrix} \right] = [2x, 4y], \qquad 3.7.6$$

while $\left[\mathbf{D}\varphi \begin{pmatrix} x \\ y \end{pmatrix} \right] = [1, 1]$. So at a critical point, there will exist λ such that

$$[1, 1] = \lambda[2x, 4y]; \quad \text{i.e.,} \quad x = \frac{1}{2\lambda}; \quad y = \frac{1}{4\lambda}. \qquad 3.7.7$$

Inserting these values into the equation for the ellipse gives

$$\frac{1}{4\lambda^2} + 2\frac{1}{16\lambda^2} = 1; \quad \text{i.e.,} \quad \lambda = \pm\sqrt{\frac{3}{8}}.$$ 3.7.8

So the maximum of the function on the ellipse is

$$\underbrace{\frac{1}{2}\sqrt{\frac{8}{3}}}_{x} + \underbrace{\frac{1}{4}\sqrt{\frac{8}{3}}}_{y} = \frac{3}{4}\sqrt{\frac{8}{3}} = \sqrt{\frac{3}{2}}.$$ △ 3.7.9

Example 3.7.8 (Lagrange multipliers: a somewhat harder example).
What is the smallest number A such that any two squares S_1, S_2 of total area 1 can be put disjointly into a rectangle of area A?

Let us call a and b the lengths of the sides of S_1 and S_2, and we may assume that $a \geq b \geq 0$. Then the smallest rectangle that will contain the two squares disjointly has sides a and $a+b$, and area $a(a+b)$, as shown in Figure 3.7.6. The problem is to maximize the area $a^2 + ab$, subject to the constraints $a^2 + b^2 = 1$, and $a \geq b \geq 0$.

The Lagrange multiplier theorem tells us that at a critical point of the constrained function there exists a number λ such that

$$\underbrace{[2a + b, \ a]}_{\substack{\text{deriv. of} \\ \text{area function}}} = \lambda \underbrace{[2a, \ 2b]}_{\substack{\text{deriv. of} \\ \text{constraint func.}}}.$$ 3.7.10

So we need to solve the system of three simultaneous nonlinear equations

$$2a + b = 2a\lambda, \quad a = 2b\lambda, \quad a^2 + b^2 = 1.$$ 3.7.11

Substituting the value of a from the second equation into the first, we find

$$4b\lambda^2 - 4b\lambda - b = 0.$$ 3.7.12

This has one solution $b = 0$, but then we get $a = 0$, which is incompatible with $a^2 + b^2 = 1$. The other solution is

$$4\lambda^2 - 4\lambda - 1 = 0; \quad \text{i.e.,} \quad \lambda = \frac{1 \pm \sqrt{2}}{2}.$$ 3.7.13

Our remaining equations are now

$$\frac{a}{b} = 2\lambda = 1 \pm \sqrt{2} \quad \text{and} \quad a^2 + b^2 = 1,$$ 3.7.14

which, if we require a, $b \geq 0$, have the unique solution

$$\binom{a}{b} = \frac{1}{\sqrt{4 + 2\sqrt{2}}}\binom{1+\sqrt{2}}{1}.$$ 3.7.15

This satisfies the constraint $a \geq b \geq 0$, and leads to

$$A = a(a + b) = \frac{4 + 3\sqrt{2}}{4 + 2\sqrt{2}}.$$ 3.7.16

The value $\lambda = \sqrt{\frac{3}{8}}$ gives the maximum; $\lambda = -\sqrt{\frac{3}{8}}$ gives the minimum,

$$x + y = -\sqrt{3/2}.$$

Since the constraint manifold is an ellipse, there can be no saddle points.

Disjointly means having nothing in common.

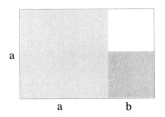

FIGURE 3.7.6.

The combined area of the two shaded squares is 1; we wish to find the smallest rectangle that will contain them both.

We don't use

$$\frac{a}{b} = 1 - \sqrt{2},$$

because since $1 - \sqrt{2} < 0$, that is incompatible with our requirement that a, $b \geq 0$.

The two endpoints correspond to the two extremes: all the area in one square and none in the other, or both squares with the same area: at $\begin{pmatrix} 1 \\ 0 \end{pmatrix}$, the larger square has area 1, and the smaller rectangle has area 0; at $\begin{pmatrix} \sqrt{2}/2 \\ \sqrt{2}/2 \end{pmatrix}$, the two squares are identical.

We must check (see Remark 3.6.5) that the maximum is not achieved at the endpoints of the constraint region, i.e., at the point with coordinates $a = 1, b = 0$ and the point with coordinates $a = b = \sqrt{2}/2$ It is easy to see that $(a + b)a = 1$ at both of these endpoints, and since $\frac{4+3\sqrt{2}}{4+2\sqrt{2}} > 1$, this is the unique maximum. $\triangle$

Example 3.7.9 (Lagrange multipliers: a third example). Find the critical points of the function xyz on the plane of equation

$$F\begin{pmatrix} x \\ y \\ z \end{pmatrix} = x + 2y + 3z - 1 = 0. \qquad 3.7.17$$

Theorem 3.7.6 asserts that a critical point is a solution to

(1) $\underbrace{[yz,\ xz,\ xy]}_{\text{deriv. of function } xyz} = \lambda \underbrace{[1,\ 2,\ 3]}_{\substack{\text{deriv. of } F \\ \text{(constraint)}}}$

(2) $\underbrace{x + 2y + 3z = 1}_{\text{constraint equation}}$

or

$$\begin{aligned} yz &= \lambda \\ xz &= 2\lambda \\ xy &= 3\lambda \\ 1 &= x + 2y + 3z. \end{aligned} \qquad 3.7.18$$

Note that Equation 3.7.18 is a system of four equations in four unknowns: x, y, z, λ. This is typical of what comes out of Lagrange multipliers except in the very simplest cases: you land on a system of nonlinear equations.

But this problem isn't quite typical, because there are tricks available for solving those equations. Often there are none, and the only thing to do is to use Newton's method.

In this case, there are tricks available. It is not hard to derive $xz = 2yz$ and $xy = 3yz$, so if $z \neq 0$ and $y \neq 0$, then $y = x/2$ and $z = x/3$. Substituting these values into the last equation gives $x = 1/3$, hence $y = 1/6$ and $z = 1/9$. At this point, the function has the value $1/162$.

Now we need to examine the cases where $z = 0$ or $y = 0$. If $z = 0$, then our Lagrange multiplier equation reads

$$[0,\ 0,\ xy] = \lambda[1,\ 2,\ 3] \qquad 3.7.19$$

which says that $\lambda = 0$, so one of x or y must also vanish. Suppose $y = 0$, then $x = 1$, and the value of the function is 0. There are two other similar points. Let us summarize: there are four critical points,

$$\begin{pmatrix} 1 \\ 0 \\ 0 \end{pmatrix}, \ \begin{pmatrix} 0 \\ 1/2 \\ 0 \end{pmatrix}, \ \begin{pmatrix} 0 \\ 0 \\ 1/3 \end{pmatrix}, \ \begin{pmatrix} 1/3 \\ 1/6 \\ 1/9 \end{pmatrix}; \qquad 3.7.20$$

at the first three our function is 0 and at the last it is $1/162$.

You are asked in Exercise 3.7.3 to show that the other critical points are saddles.

Is our last point a maximum? The answer is yes (at least, it is a local maximum), and you can see it as follows. The part of the plane of equation $x + 2y + 3z = 1$ that lies in the first octant, where $x, y, z \geq 0$, is compact, as $|x|, |y|, |z| \leq 1$ there; otherwise the equation of the plane cannot be satisfied. So our function does have a maximum in that octant. In order to be sure that this maximum is a critical point, we need to check that it isn't on the edge of the octant (see Remark 3.6.5). That is straightforward, since the function vanishes on the boundary, while it is positive at the fourth point. So this maximum is a critical point, hence it must be our fourth point.

Proof of Theorem 3.7.6.   Since $T_{\mathbf{a}}X = \ker[\mathbf{DF}(\mathbf{a})]$, the theorem follows from Theorem 3.7.1 and from the following lemma from linear algebra, using $A = [\mathbf{DF}(\mathbf{a})]$ and $\beta = [\mathbf{D}\varphi(\mathbf{a})]$.

A space with more constraints is smaller than a space with fewer constraints: more people belong to the set of musicians than belong to the set of red-headed, left-handed cello players with last names beginning with W. Here, $A\mathbf{x} = 0$ imposes m constraints, and $\beta\mathbf{x} = 0$ imposes only one.

Lemma 3.7.10. *Let* $A = \begin{bmatrix} \alpha_1 \\ \vdots \\ \alpha_m \end{bmatrix} : \mathbb{R}^n \to \mathbb{R}^m$ *be a linear transformation (i.e.,*

an $m \times n$ matrix), and $\beta : \mathbb{R}^n \to \mathbb{R}$ be a linear function (a row matrix n wide). Then

$$\ker A \subset \ker \beta \qquad\qquad 3.7.21$$

if and only if there exist numbers $\lambda_1, \ldots, \lambda_m$ such that

$$\beta = \lambda_1 \alpha_1 + \cdots + \lambda_m \alpha_m. \qquad\qquad 3.7.22$$

This is simply saying that the only linear consequences one can draw from a system of linear equations are the linear combinations of those equations.

Proof of Lemma 3.7.10.
    In one direction, if

$$\beta = \lambda_1 \alpha_1 + \cdots + \lambda_m \alpha_m, \qquad\qquad 3.7.23$$

and $\vec{\mathbf{v}} \in \ker A$, then $\vec{\mathbf{v}} \in \ker \alpha_i$ for $i = 1, \ldots, m$, so $\vec{\mathbf{v}} \in \ker \beta$.

We don't know anything about the relationship of n and m, but we know that $k \leq n$, since $n + 1$ vectors in $\mathbb{R}^n$ cannot be linearly independent.

Unfortunately, this isn't the important direction, and the other is a bit harder. Choose a maximal linearly independent subset of the α_i; by ordering we can suppose that these are $\alpha_1, \ldots, \alpha_k$. Denote the set A'. Then

$$\ker \underbrace{\begin{bmatrix} \alpha_1 \\ \vdots \\ \alpha_k \end{bmatrix}}_{A'} = \ker \underbrace{\begin{bmatrix} \alpha_1 \\ \vdots \\ \vdots \\ \alpha_m \end{bmatrix}}_{A}. \qquad\qquad 3.7.24$$

(Anything in the kernel of $\alpha_1, \ldots, \alpha_k$ is also in the kernel of their linear combinations.)

If β is not a linear combination of the $\alpha_1, \ldots, \alpha_m$, then it is not a linear combination of $\alpha_1, \ldots, \alpha_k$. This means that the $(k+1) \times n$ matrix

$$B = \begin{bmatrix} \alpha_{1,1} & \cdots & \alpha_{1,n} \\ \vdots & \cdots & \vdots \\ \alpha_{k,1} & \cdots & \alpha_{k,n} \\ \beta_1 & \cdots & \beta_n \end{bmatrix} \qquad\qquad 3.7.25$$

has $k + 1$ linearly independent rows, hence $k + 1$ linearly independent columns: the linear transformation $B : \mathbb{R}^n \to \mathbb{R}^{k+1}$ is onto. Then the set of equations

$$\begin{bmatrix} \alpha_{1,1} & \cdots & \alpha_{1,n} \\ \vdots & \cdots & \vdots \\ \alpha_{k,1} & \cdots & \alpha_{k,n} \\ \beta_1 & \cdots & \beta_n \end{bmatrix} \begin{bmatrix} v_1 \\ \vdots \\ \vdots \\ v_n \end{bmatrix} = \begin{bmatrix} 0 \\ \vdots \\ 0 \\ 1 \end{bmatrix} \qquad\qquad 3.7.26$$

has a nonzero solution. The first k lines say that $\vec{v}$ is in ker A', which is equal to ker A, but the last line says that it is not in ker β. $\square$

The spectral theorem for symmetric matrices

In this subsection we will prove what is probably the most important theorem of linear algebra. It goes under many names: the *spectral theorem*, the *principle axis theorem*, *Sylvester's principle of inertia*. The theorem is a statement about symmetric matrices; recall (Definition 1.2.18) that a symmetric matrix is a matrix that is equal to its transpose. For us, the importance of symmetric matrices is that they represent quadratic forms:

Proposition 3.7.11 (Quadratic forms and symmetric matrices). *For any symmetric matrix A, the function*

$$Q_A(\vec{x}) = \vec{x} \cdot A\vec{x} \qquad 3.7.27$$

is a quadratic form; conversely, every quadratic form

$$Q(\vec{x}) = \sum_{I \in \mathcal{I}_n^2} a_I \vec{x}^I \qquad 3.7.28$$

is of the form Q_A for a unique symmetric matrix A.

Actually, for any square matrix M the function $Q_M(\vec{x}) = \vec{x} \cdot A\vec{x}$ is a quadratic form, but there is a unique symmetric matrix A for which a quadratic form can be expressed as Q_A. This symmetric matrix is constructed as follows: each entry $A_{i,i}$ on the main diagonal is the coefficient of the corresponding variable squared in the quadratic form (i.e., the coefficient of x_i^2) while each entry $A_{i,j}$ is one-half the coefficient of the term $x_i x_j$. For example, for the matrix at left, $A_{1,1} = 1$ because in the corresponding quadratic form the coefficient of x_1^2 is 1, while $A_{2,1} = A_{1,2} = 0$ because the coefficient of $x_2 x_1 = x_1 x_2 = 0$. Exercise 3.7.4 asks you to turn this into a formal proof.

Theorem 3.7.12 (Spectral theorem). *Let A be a symmetric $n \times n$ matrix with real entries. Then there exists an orthonormal basis $\vec{v}_1, \ldots, \vec{v}_n$ of $\mathbb{R}^n$ and numbers $\lambda_1, \ldots, \lambda_n$ such that*

$$A\vec{v}_i = \lambda_i \vec{v}_i. \qquad 3.7.29$$

Definition 3.7.13 (Eigenvector, eigenvalue). For any square matrix A, a nonzero vector $\vec{v}$ such that $A\vec{v} = \lambda\vec{v}$ for some number λ is called an eigenvector of A. The number λ is the corresponding eigenvalue.

Generalizing the spectral theorem to infinitely many dimensions is one of the central problems of functional analysis.

For example,

$$\underbrace{\begin{bmatrix} x_1 \\ x_2 \\ x_3 \end{bmatrix}}_{\vec{x}} \cdot \underbrace{\begin{bmatrix} 1 & 0 & 1 \\ 0 & 1 & 2 \\ 1 & 2 & 9 \end{bmatrix} \begin{bmatrix} x_1 \\ x_2 \\ x_3 \end{bmatrix}}_{A\vec{x}}$$

$$= \underbrace{x_1^2 + x_2^2 + 2x_1 x_3 + 4x_2 x_3 + 9x_3^2}_{\text{quadratic form}}.$$

Square matrices exist that have no eigenvectors, or only one. Symmetric matrices are a very special class of square matrices, whose eigenvectors are guaranteed not only to exist but also to form an orthonormal basis.

The theory of eigenvalues and eigenvectors is the most exciting chapter in linear algebra, with close connections to differential equations, Fourier series, ...

"...when Werner Heisenberg discovered 'matrix' mechanics in 1925, he didn't know what a matrix was (Max Born had to tell him), and neither Heisenberg nor Born knew what to make of the appearance of matrices in the context of the atom. (David Hilbert is reported to have told them to go look for a differential equation with the same eigenvalues, if that would make them happier. They did not follow Hilbert's well-meant advice and thereby may have missed discovering the Schrödinger wave equation.)" —M. R. Schroeder, *Mathematical Intelligencer*, Vol. 7, No. 4

We use λ to denote both Lagrange multipliers and eigenvalues; we will see that eigenvalues are in fact Lagrange multipliers.

Example 3.7.14 (Eigenvectors). Let $A = \begin{bmatrix} 1 & 1 \\ 1 & 0 \end{bmatrix}$. You can easily check that

$$A \begin{bmatrix} \frac{1+\sqrt{5}}{2} \\ 1 \end{bmatrix} = \frac{1+\sqrt{5}}{2} \begin{bmatrix} \frac{1+\sqrt{5}}{2} \\ 1 \end{bmatrix} \quad \text{and} \quad A \begin{bmatrix} \frac{1-\sqrt{5}}{2} \\ 1 \end{bmatrix} = \frac{1-\sqrt{5}}{2} \begin{bmatrix} \frac{1-\sqrt{5}}{2} \\ 1 \end{bmatrix}, \quad 3.7.30$$

and that the two vectors are orthogonal since their dot product is 0:

$$\begin{bmatrix} \frac{1+\sqrt{5}}{2} \\ 1 \end{bmatrix} \cdot \begin{bmatrix} \frac{1-\sqrt{5}}{2} \\ 1 \end{bmatrix} = 0. \qquad 3.7.31$$

The matrix A is symmetric; why do the eigenvectors $\vec{\mathbf{v}}_1$ and $\vec{\mathbf{v}}_2$ not form the basis referred to in the spectral theorem ?[21]

Proof of Theorem 3.7.12 (Spectral theorem).   We will construct our basis one vector at a time. Consider the function $Q_A(\vec{\mathbf{x}}) : \mathbb{R}^n \to \mathbb{R} = \vec{\mathbf{x}} \cdot A\vec{\mathbf{x}}$. This function has a maximum (and a minimum) on the $(n-1)$-sphere S of equation $F_1(\vec{\mathbf{x}}) = |\vec{\mathbf{x}}|^2 = 1$. We know a maximum (and a minimum) exists, because a sphere is a compact subset of $\mathbb{R}^n$; see Theorem 1.6.7. We have

Exercise 3.7.5 asks you to justify the derivative in Equation 3.7.32, using the definition of a derivative as limit and the fact that A is symmetric.

$$[\mathbf{D}Q_A(\vec{\mathbf{a}})]\vec{\mathbf{h}} = \vec{\mathbf{a}} \cdot (A\vec{\mathbf{h}}) + \vec{\mathbf{h}} \cdot (A\vec{\mathbf{a}}) = \vec{\mathbf{a}}^\top A\vec{\mathbf{h}} + \vec{\mathbf{h}}^\top A\vec{\mathbf{a}} = 2\vec{\mathbf{a}}^\top A\vec{\mathbf{h}}, \qquad 3.7.32$$

whereas

$$[\mathbf{D}F_1(\vec{\mathbf{a}})]\vec{\mathbf{h}} = 2\vec{\mathbf{a}}^\top \vec{\mathbf{h}}. \qquad 3.7.33$$

So $2\vec{\mathbf{a}}^\top A$ is the derivative of the quadratic form Q_A, and $2\vec{\mathbf{a}}^\top$ is the derivative of the constraint function. Theorem 3.7.6 tells us that if the restriction of Q_A to the unit sphere has a maximum at $\vec{\mathbf{v}}_1$, then there exists λ_1 such that

In Equation 3.7.34 we take the transpose of both sides, remembering that

$$(AB)^\top = B^\top A^\top.$$

$$2\vec{\mathbf{v}}_1^\top A = \lambda_1 2\vec{\mathbf{v}}_1^\top, \quad \text{so} \quad A^\top \vec{\mathbf{v}}_1 = \lambda_1 \vec{\mathbf{v}}_1. \qquad 3.7.34$$

Since A is symmetric,

As often happens in the middle of an important proof, the point **a** at which we are evaluating the derivative has turned into a vector, so that we can perform vector operations on it.

$$A\vec{\mathbf{v}}_1 = \lambda_1 \vec{\mathbf{v}}_1. \qquad 3.7.35$$

This gives us our first eigenvector. Now let us continue by considering the maximum at $\vec{\mathbf{v}}_2$ of Q_A restricted to the space $S \cap (\vec{\mathbf{v}}_1)^\perp$ (where, as above, S is the unit sphere in $\mathbb{R}^n$, and $(\vec{\mathbf{v}}_1)^\perp$ is the space of vectors perpendicular to $\vec{\mathbf{v}}_1$).

---

[21]They don't have unit length; if we normalize them by dividing each vector by its length, we find that

$$\vec{\mathbf{v}}_1 = \sqrt{\frac{2}{5+\sqrt{5}}} \begin{bmatrix} \frac{1+\sqrt{5}}{2} \\ 1 \end{bmatrix} \quad \text{and} \quad \vec{\mathbf{v}}_2 = \sqrt{\frac{2}{5-\sqrt{5}}} \begin{bmatrix} \frac{1-\sqrt{5}}{2} \\ 1 \end{bmatrix} \quad \text{do the job.}$$

The first equality of Equation 3.7.39 uses the symmetry of A: if A is symmetric,

$$\vec{v} \cdot (A\vec{w}) = \vec{v}^\top (A\vec{w}) = (\vec{v}^\top A)\vec{w}$$
$$= (\vec{v}^\top A^\top)\vec{w} = (A\vec{v})^\top \vec{w}$$
$$= (A\vec{v}) \cdot \vec{w}.$$

The second uses Equation 3.7.35, and the third the fact that $\vec{v}_2 \in S \cap (\vec{v}_1)^\perp$.

If you've ever tried to find eigenvectors, you'll be impressed by how easily their existence dropped out of Lagrange multipliers. Of course we could not have done this without the existence of the maximum and minimum of the function Q_A, guaranteed by the non-constructive Theorem 1.6.7. In addition, we've only proved existence: there is no obvious way to find these constrained maxima of Q_A.

That is, we add a second constraint, F_2, maximizing Q_A subject to the two constraints

$$F_1(\vec{x}) = 1 \quad \text{and} \quad F_2(\vec{x}) = \vec{x} \cdot \vec{v}_1 = 0. \qquad 3.7.36$$

Since $[\mathbf{D}F_2(\vec{v}_2)] = \vec{v}_1$, Equations 3.7.32 and 3.7.33 and Theorem 3.7.6 tell us that there exist numbers λ_2 and $\mu_{2,1}$ such that

$$A\vec{v}_2 = \mu_{2,1}\vec{v}_1 + \lambda_2\vec{v}_2. \qquad 3.7.37$$

Take dot products of both sides of this equation with $\vec{v}_1$, to find

$$(A\vec{v}_2) \cdot \vec{v}_1 = \lambda_2\vec{v}_2 \cdot \vec{v}_1 + \mu_{2,1}\vec{v}_1 \cdot \vec{v}_1. \qquad 3.7.38$$

Using

$$(A\vec{v}_2) \cdot \vec{v}_1 = \vec{v}_2 \cdot (A\vec{v}_1) = \vec{v}_2 \cdot (\lambda_1\vec{v}_1) = 0, \qquad 3.7.39$$

Equation 3.7.38 becomes

$$0 = \mu_{2,1}|\vec{v}_1|^2 + \underbrace{\lambda_2\vec{v}_2 \cdot \vec{v}_1}_{= 0 \text{ since } \vec{v}_2 \perp \vec{v}1} = \mu_{2,1}, \qquad 3.7.40$$

so Equation 3.7.37 becomes

$$A\vec{v}_2 = \lambda_2\vec{v}_2. \qquad 3.7.41$$

We have found our second eigenvector.

It should be clear how to continue, but let us spell it out for one further step. Suppose that the restriction of Q_A to $S \cap \vec{v}_1^\perp \cap \vec{v}_2^\perp$ has a maximum at $\vec{v}_3$, i.e., maximize Q_A subject to the three constraints

$$F_1(\vec{x}) = 1, \qquad F_2(\vec{x}) = \vec{x} \cdot \vec{v}_1 = 0, \quad \text{and} \quad F_3(\vec{x}) = \vec{x} \cdot \vec{v}_2 = 0. \qquad 3.7.42$$

The same argument as above says that there then exist numbers $\lambda_3, \mu_{3,1}$ and $\mu_{3,2}$ such that

$$A\vec{v}_3 = \mu_{3,1}\vec{v}_1 + \mu_{3,2}\vec{v}_2 + \lambda_3\vec{v}_3. \qquad 3.7.43$$

Dot this entire equation with $\vec{v}_1$ (resp. $\vec{v}_2$); you will find $\mu_{3,1} = \mu_{3,2} = 0$, and we find $A\vec{v}_3 = \lambda_3\vec{v}_3$. $\square$

The spectral theorem gives us an alternative approach to quadratic forms, geometrically more appealing than the completing of squares used in Section 3.5.

Exercise 3.7.6 characterizes the norm in terms of eigenvalues.

Theorem 3.7.15. *If the quadratic form Q_A has signature (k, l), then A has k positive eigenvalues and l negative eigenvalues.*

3.8 GEOMETRY OF CURVES AND SURFACES

In which we return to curves and surfaces, applying what we have learned about Taylor polynomials, quadratic forms, and extrema to discussing their geometry: in particular, their curvature.

A curve acquires its geometry from the space in which it is embedded. Without that embedding, a curve is boring: geometrically it is a straight line. A one-dimensional worm living inside a smooth curve cannot tell whether the curve is straight or curvy; at most (if allowed to leave a trace behind him) it can tell whether the curve is closed or not.

Curvature in geometry manifests itself as gravitation.—C. Misner, K. S. Thorne, J. Wheeler, *Gravitation*

This is not true of surfaces and higher-dimensional manifolds. Given a long-enough tape measure you could prove that the earth is spherical without any recourse to ambient space; Exercise 3.8.1 asks you to compute how long a tape measure you would need.

The central notion used to explore these issues is *curvature*, which comes in many flavors. Its importance cannot be overstated: gravitation is the curvature of spacetime; the electromagnetic field is the curvature of the electromagnetic potential. Indeed, the geometry of curves and surfaces is an immense field, with many hundreds of books devoted to it; our treatment cannot be more than the barest overview.[22]

Recall (Remark 3.1.2) the fuzzy definition of "smooth" as meaning "as many times differentiable as is relevant to the problem at hand." In Sections 3.1 and 3.2, once continuously differentiable was sufficient; here it is not.

We will briefly discuss curvature as it applies to curves in the plane, curves in space, and surfaces in space. Our approach is the same in all cases: we write our curve or surface as the graph of a mapping in the coordinates best adapted to the situation, and read the curvature (and other quantities of interest) from quadratic terms of the Taylor polynomial for that mapping. Differential geometry only exists for functions that are twice continuously differentiable; without that hypothesis, everything becomes a million times harder. Thus the functions we discuss all have Taylor polynomials of degree at least 2. (For curves in space, we will need our functions to be three times continuously differentiable, with Taylor polynomials of degree 3.)

The geometry of plane curves

For a smooth curve in the plane, the "best coordinate system" X, Y at a point $\mathbf{a} = \begin{pmatrix} a \\ b \end{pmatrix}$ is the system centered at $\mathbf{a}$, with the X-axis in the direction of the tangent line, and the Y-axis orthogonal to the tangent at that point, as shown in Figure 3.8.1.

---

[22]For further reading, we recommend *Riemannian Geometry, A Beginner's Guide*, by Frank Morgan (A K Peters, Ltd., Wellesley, MA, second edition 1998) or *Differential Geometry of Curves and Surfaces*, by Manfredo P. do Carmo (Prentice-Hall, Inc., 1976).

In these X, Y coordinates, the curve is locally the graph of a function $Y = g(X)$, which can be approximated by its Taylor polynomial. This Taylor polynomial contains only quadratic and higher terms[23]:

$$Y = g(X) = \frac{A_2}{2}X^2 + \frac{A_3}{6}X^3 + \ldots, \qquad 3.8.1$$

where A_2 is the second derivative of g (see Equation 3.3.1). All the coefficients of this polynomial are *invariants* of the curve: numbers associated to a point of the curve that do not change if you translate or rotate the curve. (Of course they do depend on the point $\begin{pmatrix} X \\ Y \end{pmatrix}$ where you are on the curve.)

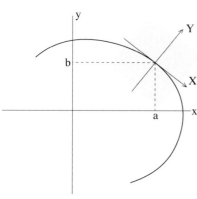

FIGURE 3.8.1.

To study a smooth curve at $\mathbf{a} = \begin{pmatrix} a \\ b \end{pmatrix}$, we make $\mathbf{a}$ the origin of our new coordinates, and place the X-axis in the direction of the tangent to the curve at $\mathbf{a}$. Within the shaded region, the curve is the graph of a function $Y = g(X)$ that starts with quadratic terms.

The Greek letter κ is "kappa." We could avoid the absolute value in Definition 3.8.1 by defining the *signed* curvature of an oriented curve, but we won't do so here, to avoid complications.

When $X = 0$, both $g(X)$ and $g'(0)$ vanish, while $g''(0) = -1$; the quadratic term for the Taylor polynomial is $\frac{1}{2}g''$.

The curvature of plane curves

The coefficient that will interest us is A_2, the second derivative of g.

Definition 3.8.1 (Curvature of a curve in $\mathbb{R}^2$). Let a curve in $\mathbb{R}^2$ be locally the graph of a function $g(X)$, with Taylor polynomial

$$g(X) = \frac{A_2}{2}X^2 + \frac{A_3}{6}X^3 + \ldots.$$

Then the curvature κ of the curve at $\mathbf{0}$ (i.e., at $\mathbf{a}$ in the original x, y coordinates) is $|A_2|$.

The curvature is normalized so that the unit circle has curvature 1. Indeed, near the point $\begin{pmatrix} 0 \\ 1 \end{pmatrix}$, the "best coordinates" for the unit circle are $X = x, Y = y - 1$, so the equation of the circle $y = \sqrt{1 - x^2}$ becomes

$$g(X) = Y = y - 1 = \sqrt{1 - X^2} - 1 \qquad 3.8.2$$

with the Taylor polynomial[24]

$$g(x) = -\frac{1}{2}X^2 + \ldots, \qquad 3.8.3$$

the dots representing higher degree terms. So the unit circle has curvature $|-1| = 1$.

Proposition 3.8.2 tells how to compute the curvature of a smooth plane curve that is locally the graph of a function $f(x)$. Note that when we use small letters, x and y, we are using the standard coordinate system.

---

[23]The point $\mathbf{a}$ has coordinates $X = 0$, $Y = 0$, so the constant term is 0; the linear term is 0 because the curve is tangent to the X-axis at $\mathbf{a}$.

[24]We avoided computing the derivatives for $g(X)$ by using the formula for the Taylor series of a binomial (Equation 3.4.7):

$$(1 + a)^m = 1 + ma + \frac{m(m-1)}{2!}a^2 + \frac{m(m-1)(m-2)}{3!}a^3 + \ldots.$$

In this case, m is $1/2$ and $a = -X^2$.

Proposition 3.8.2 (Computing the curvature of a plane curve known as a graph). *The curvature κ of the curve $y = f(x)$ at $\left(\begin{smallmatrix} a \\ f(a) \end{smallmatrix}\right)$ is*

$$\kappa = \frac{|f''(a)|}{\left(1 + f'(a)^2\right)^{3/2}}. \qquad 3.8.4$$

Proof. We express $f(x)$ as its Taylor polynomial, ignoring the constant term, since we can eliminate it by translating the coordinates, without changing any of the derivatives. This gives

$$f(x) = f'(a)x + \frac{f''(a)}{2}\,x^2 + \dots. \qquad 3.8.5$$

Now we rotate the coordinates by θ, using the rotation matrix

$$\begin{bmatrix} \cos\theta & \sin\theta \\ -\sin\theta & \cos\theta \end{bmatrix}. \qquad 3.8.6$$

The rotation matrix of Example 1.3.17:

$$\begin{bmatrix} \cos\theta & -\sin\theta \\ \sin\theta & \cos\theta \end{bmatrix},$$

is the inverse of the one we are using now; there we were rotating points, while here we are rotating coordinates.

Then

$$\begin{aligned} X &= x\cos\theta + y\sin\theta \\ Y &= -x\sin\theta + y\cos\theta \end{aligned} \quad \text{giving} \quad \begin{aligned} X\cos\theta - Y\sin\theta &= x(\cos^2\theta + \sin^2\theta) = x \\ X\sin\theta + Y\cos\theta &= y(\cos^2\theta + \sin^2\theta) = y. \end{aligned}$$

$$3.8.7$$

Substituting these into Equation 3.8.5 leads to

$$\underbrace{X\sin\theta + Y\cos\theta}_{y} = f'(a)\underbrace{(X\cos\theta - Y\sin\theta)}_{x} + \frac{f''(a)}{2}\underbrace{(X\cos\theta - Y\sin\theta)^2}_{x^2} + \dots.$$

$$3.8.8$$

Recall (Definition 3.8.1) that curvature is defined for a curve locally the graph of a function $g(X)$ whose Taylor polynomial starts with quadratic terms.

We want to choose θ so that this equation expresses Y as a function of X, with derivative 0, so that its Taylor polynomial starts with the quadratic term:

$$Y = g(X) = \frac{A_2}{2}X^2 + \dots. \qquad 3.8.9$$

If we subtract $X\sin\theta + Y\cos\theta$ from both sides of Equation 3.8.8, we can write the equation for the curve in terms of the X, Y coordinates:

Alternatively, we could say that X is a function of Y if D_1F is invertible.

Here,

$$D_2F = -f'(a)\sin\theta - \cos\theta$$

corresponds to Equation 2.9.21 in the implicit function theorem; it represents the "pivotal columns" of the derivative of F. Since that derivative is a line matrix, D_2F is a number, being nonzero and being invertible are the same.

$$F\left(\begin{smallmatrix} X \\ Y \end{smallmatrix}\right) = 0 = -X\sin\theta - Y\cos\theta + f'(a)(X\cos\theta - Y\sin\theta) + \dots, \qquad 3.8.10$$

with derivative

$$\left[\mathbf{D}F\left(\begin{smallmatrix} 0 \\ 0 \end{smallmatrix}\right)\right] = [\underbrace{f'(a)\cos\theta - \sin\theta}_{D_1F}, \underbrace{-f'(a)\sin\theta - \cos\theta}_{D_2F}]. \qquad 3.8.11$$

The implicit function theorem says that Y is a function $g(X)$ if D_2F is invertible, i.e., if $-f'(a)\sin\theta - \cos\theta \neq 0$. In that case, Equation 2.9.25 for the derivative of an implicit function tells us that in order to have $g'(0) = 0$ (so that $g(X)$ starts with quadratic terms) we must have $D_1F = f'(a)\cos\theta - \sin\theta = 0$, i.e., $\tan\theta = f'(a)$:

$$g'(0) = 0 = -[D_2F(0)]^{-1} \underbrace{[D_1F(0)]}_{\text{must be } 0} \qquad 3.8.12$$

$$\underset{\neq 0}{}$$

Setting $\tan\theta = f'(a)$ is simply saying that $f'(a)$ is the slope of the curve.

If we make this choice of θ, then indeed

$$-f'(a)\sin\theta - \cos\theta = \frac{-1}{\cos\theta} \neq 0, \qquad 3.8.13$$

so the implicit function theorem does apply. We can replace Y in Equation 3.8.10 by $g(X)$:

$$F\begin{pmatrix} X \\ g(X) \end{pmatrix} = 0 = -X\sin\theta - g(X)\cos\theta + f'(a)(X\cos\theta - g(X)\sin\theta)$$

$$+ \underbrace{\frac{f''(a)}{2}(X\cos\theta - g(X)\sin\theta)^2 + \dots,}_{\text{additional term; see Eq.3.8.8}} \qquad 3.8.14$$

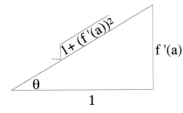

FIGURE 3.8.2.

This right triangle justifies Equation 3.8.18.

If we group the linear terms in $g(X)$ on the left, and put the linear terms in X on the right, we get

$$(f'(a)\sin\theta + \cos\theta)g(X) = \overbrace{(f'(a)\cos\theta - \sin\theta)}^{=0}X$$

$$+ \frac{f''(a)}{2}(\cos\theta X - \sin\theta g(X))^2 + \dots \qquad 3.8.15$$

$$= \frac{f''(a)}{2}(\cos\theta X - \sin\theta g(X))^2 + \dots.$$

We divide by $f'(a)\sin\theta + \cos\theta$ to obtain

Since $g'(0) = 0$, $g(X)$ starts with quadratic terms. Moreover, by Theorem 3.4.7, the function g is as differentiable as F, hence as f. So the term $Xg(X)$ is of degree 3, and the term $g(X)^2$ is of degree 4.

$$g(X) = \frac{1}{f'(a)\sin\theta + \cos\theta} \frac{f''(a)}{2}\Big(\cos^2\theta X^2$$

$$- \underbrace{2\cos\theta\sin\theta Xg(X) + \sin^2\theta(g(X)^2)}_{\text{these are of degree 3 or higher}}\Big) + \dots. \qquad 3.8.16$$

Now we express the coefficient of X^2 as $A_2/2$, getting

$$A_2 = \frac{f''(a)\cos^2\theta}{f'(a)\sin\theta + \cos\theta}. \qquad 3.8.17$$

Since $f'(a) = \tan\theta$, we have the right triangle of Figure 3.8.2, and

$$\sin\theta = \frac{f'(a)}{\sqrt{1 + (f'(a))^2}} \quad \text{and} \quad \cos\theta = \frac{1}{\sqrt{1 + (f'(a))^2}}. \qquad 3.8.18$$

Substituting these values in Equation 3.8.17 we have

$$A_2 = \frac{f''(a)}{\left(1 + (f'(a))^2\right)^{3/2}}, \quad \text{so that} \quad \kappa = |A_2| = \frac{|f''(a)|}{\left(1 + f'(a)^2\right)^{3/2}}. \quad \square \qquad 3.8.19$$

Geometry of curves parametrized by arc length

There is no reasonable generalization of this approach to surfaces, which do have intrinsic geometry.

There is an alternative approach to the geometry of curves, both in the plane and in space: parametrization by arc length. The existence of this method reflects the fact that curves have no interesting intrinsic geometry: if you were a one-dimensional bug living on a curve, you could not make any measurements that would tell whether your universe was a straight line, or all tangled up.

Recall (Definition 3.1.20) that a parametrized curve is a mapping $\gamma : I \to \mathbb{R}^n$, where I is an interval in $\mathbb{R}$. You can think of I as an interval of time; if you are traveling along the curve, the parametrization tells you where you are on the curve at a given time.

The vector $\vec{\gamma}'$ is the velocity vector of the parametrization γ.

Definition 3.8.3 (Arc length). The arc length of the segment $\gamma([a,b])$ of a curve parametrized by γ is given by the integral

$$\int_a^b |\vec{\gamma}'(t)|\, dt. \qquad 3.8.20$$

A more intuitive definition to consider is the lengths of straight line segments ("inscribed polygonal curves") joining points $\gamma(t_0), \gamma(t_1), \ldots, \gamma(t_m)$, where $t_0 = a$ and $t_m = b$, as shown in Figure 3.8.3. Then take the limit as the line segments become shorter and shorter.

If the odometer says you have traveled 50 miles, then you have traveled 50 miles on your curve.

In formulas, this means to consider

$$\sum_{i=0}^{m-1} |\gamma(t_{i+1}) - \gamma(t_i)|, \text{ which is almost } \sum_{i=0}^{m-1} |\vec{\gamma}'(t_i)|(t_{i+1} - t_i). \qquad 3.8.21$$

(If you have any doubts about the "which is almost," Exercise 3.8.2 should remove them when γ is twice continuously differentiable.) This last expression is a Riemann sum for $\int_a^b |\vec{\gamma}'(t)|\, dt$.

Computing the integral in Equation 3.8.22 is painful, and computing the inverse function $t(s)$ is even more so, so parametrization by arc length is more attractive in theory than in practice. Later we will see how to compute the curvature of curves known by arbitrary parametrizations.

If you select an origin $\gamma(t_0)$, then you can define $s(t)$ by the formula

$$\underbrace{s(t)}_{\substack{\text{odometer} \\ \text{reading} \\ \text{at time } t}} = \int_{t_0}^t \underbrace{|\vec{\gamma}'(u)|}_{\substack{\text{speedometer} \\ \text{reading} \\ \text{at time } u}} du; \qquad 3.8.22$$

Proposition 3.8.4 follows from Proposition 3.8.13.

$s(t)$ gives the odometer reading as a function of time: "how far have you gone since time t_0". It is a monotonically increasing function, so (Theorem 2.9.2) it has an inverse function $t(s)$ (at what time had you gone distance s on the curve?) Composing this function with $\gamma : I \to \mathbb{R}^2$ or $\gamma : I \to \mathbb{R}^3$ now says where you are in the plane, or in space, when you have gone a distance s along the curve (or, if $\gamma : I \to \mathbb{R}^n$, where you are in $\mathbb{R}^n$). The curve

$$\delta(s) = \gamma\big(t(s)\big) \qquad 3.8.23$$

is now parametrized by arc length: distances along the curve are exactly the same as they are in the parameter domain where s lives.

Proposition 3.8.4 (Curvature of a plane curve parametrized by arc length). *The curvature κ of a plane curve $\delta(s)$ parametrized by arc length is given by the formula*

$$\kappa\big(\delta(s)\big) = |\vec{\delta}''(s)|. \qquad\qquad 3.8.24$$

The best coordinates for surfaces

Let S be a surface in $\mathbb{R}^3$, and let $\mathbf{a}$ be a point in S. Then an adapted coordinate system for S at $\mathbf{a}$ is a system where X and Y are coordinates with respect to an orthonormal basis of the tangent plane, and the Z-axis is the normal direction, as shown in Figure 3.8.4. In such a coordinate system, the surface S is locally the graph of a function

$$Z = f\left(\begin{matrix}X\\Y\end{matrix}\right) = \underbrace{\frac{1}{2}(A_{2,0}X^2 + 2A_{1,1}XY + A_{0,2}Y^2)}_{\text{quadratic term of Taylor polynomial}} + \quad\text{higher degree terms.}$$

$$3.8.25$$

Many interesting things can be read off from the numbers $A_{2,0}$, $A_{1,1}$ and $A_{0,2}$: in particular, the *mean curvature* and the *Gaussian curvature*, both generalizations of the single curvature of smooth curves.

Definition 3.8.5 (Mean curvature of a surface). The mean curvature H of a surface at a point $\mathbf{a}$ is

$$H = \frac{1}{2}(A_{2,0} + A_{0,2}).$$

The mean curvature measures how far a surface is from being minimal. A *minimal surface* is one that locally minimizes surface area among surfaces with the same boundary.

Definition 3.8.6 (Gaussian curvature of a surface). The Gaussian curvature K of a surface at a point $\mathbf{a}$ is

$$K = A_{2,0}A_{0,2} - A_{1,1}^2. \qquad\qquad 3.8.26$$

The Gaussian curvature measures how big or small a surface is compared to a flat surface. The precise statement, which we will not prove in this book, is that the area of the disk $D_r(\mathbf{x})$ of radius r around a point $\mathbf{x}$ of a surface has the 4th degree Taylor polynomial

$$\underbrace{\text{Area}\big(D_r(\mathbf{x})\big)}_{\substack{\text{area of curved disk}}} \approx \underbrace{\pi r^2}_{\substack{\text{area of}\\\text{flat disk}}} - \frac{K(\mathbf{x})\pi}{12}\,r^4. \qquad\qquad 3.8.27$$

FIGURE 3.8.3.

A curve approximated by an inscribed polygon. While you may be more familiar with closed polygons, such as the hexagon and pentagon, a polygon does not need to be closed.

In Equation 3.8.25, the first index for the coefficient A refers to X and the second to Y, so $A_{1,1}$ is the coefficient for XY, and so on.

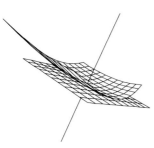

FIGURE 3.8.4.

In an adapted coordinate system, a surface is represented as the graph of a function from the tangent plane to the normal line. In those coordinates, the function starts with quadratic terms.

If the curvature is positive, the curved disk is smaller than a flat disk, and if the curvature is negative, it is larger. The disks have to be measured with a tape measure contained in the surface; in other words, $D_r(\mathbf{x})$ is the set of points which can be connected to $\mathbf{x}$ by a curve *contained in the surface* and of length at most r.

An obvious example of a surface with positive Gaussian curvature is the surface of a ball. Take a basketball and wrap a napkin around it; you will have extra fabric that won't lie smooth. This is why maps of the earth always distort areas: the extra "fabric" won't lie smooth otherwise.

An example of a surface with negative Gaussian curvature is a mountain pass. Another example is an armpit. If you have ever sewed a set-in sleeve on a shirt or dress, you know that when you pin the under part of the sleeve to the main part of the garment, you have extra fabric that doesn't lie flat; sewing the two parts together without puckers or gathers is tricky, and involves distorting the fabric.

Sewing is something of a dying art, but the mathematician Bill Thurston, whose geometric vision is legendary, maintains that it is an excellent way to acquire some feeling for the geometry of surfaces.

The *Gaussian curvature* is the prototype of all the really interesting things in differential geometry. It measures to what extent pieces of a surface can be made flat, without stretching or deformation—as is possible for a cone or cylinder but not for a sphere.

FIGURE 3.8.5. Did you ever wonder why the three Billy Goats Gruff were the sizes they were? The answer is Gaussian curvature. The first goat gets just the right amount of grass to eat; he lives on a flat surface, with Gaussian curvature zero. The second goat is thin. He lives on the top of a hill, with positive Gaussian curvature. Since the chain is heavy, and lies on the surface, he can reach less grass. The third goat is fat. His surface has negative Gaussian curvature; with the same length chain, he can get at more grass.

Computing curvature of surfaces

Proposition 3.8.2 tells us how to compute the curvature of a plane curve known as a graph. The analog for surfaces is a pretty frightful computation. Suppose we have a surface S, given as the graph of a function $f\begin{pmatrix} x \\ y \end{pmatrix}$, whose Taylor polynomial to degree 2 is

$$z = f\begin{pmatrix} x \\ y \end{pmatrix} = a_1 x + a_2 y + \frac{1}{2}(a_{2,0}x^2 + 2a_{1,1}xy + a_{0,2}y^2) + \dots . \qquad 3.8.28$$

(There is no constant term because we translate the surface so that the point we are interested in is the origin.)

A coordinate system adapted to S at the origin consists of coordinates X, Y, Z rotated with respect to the x, y, z axes so that the surface is tangent to the (X, Y)-plane. One coordinate system that satisfies this property is the following, where we set $c = \sqrt{a_1^2 + a_2^2}$ to lighten the notation:

$$x = -\frac{a_2}{c}X + \frac{a_1}{c\sqrt{1+c^2}}Y + \frac{a_1}{\sqrt{1+c^2}}Z$$

$$y = \frac{a_1}{c}X + \frac{a_2}{c\sqrt{1+c^2}}Y + \frac{a_2}{\sqrt{1+c^2}}Z \qquad 3.8.29$$

$$z = \frac{c}{\sqrt{1+c^2}}Y - \frac{1}{\sqrt{1+c^2}}Z.$$

That is, the new coordinates are taken with respect to the three basis vectors

$$\begin{bmatrix} -\dfrac{a_2}{c} \\ \dfrac{a_1}{c} \\ 0 \end{bmatrix}, \quad \begin{bmatrix} \dfrac{a_1}{c\sqrt{1+c^2}} \\ \dfrac{a_2}{c\sqrt{1+c^2}} \\ \dfrac{c}{\sqrt{1+c^2}} \end{bmatrix}, \quad \begin{bmatrix} \dfrac{a_1}{\sqrt{1+c^2}} \\ \dfrac{a_2}{\sqrt{1+c^2}} \\ \dfrac{-1}{\sqrt{1+c^2}} \end{bmatrix}. \qquad 3.8.30$$

The first vector is a horizontal unit vector in the tangent plane. The second is a unit vector orthogonal to the first, in the tangent plane. The third is a unit vector orthogonal to the previous two. It takes a bit of geometry to find them, but the proof of Proposition 3.8.7 will show that these coordinates are indeed adapted to the surface.

Remember that we set

$$c = \sqrt{a_1^2 + a_2^2}.$$

Note that Equations 3.8.33 and 3.8.34 are somehow related to Equation 3.8.4: in each case the numerator contains second derivatives ($a_{2,0}$, $a_{0,2}$, etc., are coefficients for the second degree terms of the Taylor polynomial) and the denominator contains something like $1 + |Df|^2$ (the a_1 and a_2 of $c = \sqrt{a_1^2 + a_2^2}$ are coefficients of the first degree term). A more precise relation can be seen if you consider the *surface* of equation $z = f(x)$, y arbitrary, and the plane *curve* $z = f(x)$. In that case the mean curvature of the surface is half the curvature of the plane curve. Exercise 3.8.3 asks you to check this.

We prove Proposition 3.8.7 after giving a few examples.

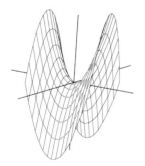

FIGURE 3.6.1, REPEATED.

The graph of $x^2 - y^2$, a typical saddle.

Proposition 3.8.7 (Computing curvature of surfaces). *(a) Let S be the surface of Equation 3.8.28, and X, Y, Z be the coordinates with respect to the orthonormal basis given by Equation 3.8.30. With respect to these coordinates, S is the graph of Z as a function F of X and Y:*

$$F\left(\begin{matrix} X \\ Y \end{matrix}\right) = \frac{1}{2}(A_{2,0}X^2 + 2A_{1,1}XY + A_{0,2}Y^2) + \dots, \qquad 3.8.31$$

which starts with quadratic terms.

(b) Setting $c = \sqrt{a_1^2 + a_2^2}$, the coefficients for the quadratic terms of F are

$$A_{2,0} = -\frac{1}{c^2\sqrt{1+c^2}}\left(a_{2,0}a_2^2 - 2a_{1,1}a_1a_2 + a_{0,2}a_1^2\right)$$

$$A_{1,1} = \frac{1}{c^2(1+c^2)}\left(a_1a_2(a_{2,0} - a_{0,2}) + a_{1,1}(a_2^2 - a_1^2)\right) \qquad 3.8.32$$

$$A_{0,2} = -\frac{1}{c^2(1+c^2)^{3/2}}\left(a_{2,0}a_1^2 + 2a_{1,1}a_1a_2 + a_{0,2}a_2^2\right).$$

(c) The Gaussian curvature of S is

$$K = \frac{a_{2,0}a_{0,2} - a_{1,1}^2}{(1+c^2)^2}, \qquad 3.8.33$$

and the mean curvature is

$$H = \frac{-1}{2(1+c^2)^{3/2}}\left(a_{2,0}(1+a_2^2) - 2a_1a_2a_{1,1} + a_{0,2}(1+a_1^2)\right). \qquad 3.8.34$$

Example 3.8.8 (Computing the Gaussian and mean curvature of a surface). Suppose we want to measure the Gaussian curvature at a point $\left(\begin{matrix} a \\ b \end{matrix}\right)$ of the surface given by the equation $z = x^2 - y^2$ (the saddle shown in Figure 3.6.1). We make that point our new origin; i.e., we use new translated coordinates, u, v, w, where

$$x = a + u$$
$$y = b + v \qquad 3.8.35$$
$$z = a^2 - b^2 + w.$$

(The u-axis replaces the original x-axis, the v-axis replaces the y-axis, and the w-axis replaces the z-axis.) Now we rewrite the equation $z = x^2 - y^2$ as

$$\underbrace{a^2 - b^2 + w}_{z} = \underbrace{(a+u)^2}_{x^2} - \underbrace{(b+v)^2}_{y^2} \qquad 3.8.36$$

$$= a^2 + 2au + u^2 - b^2 - 2bv - v^2,$$

which gives

$$w = 2au - 2bv + u^2 - v^2 = \underbrace{2a}_{a_1}u + \underbrace{-2b}_{a_2}v + \frac{1}{2}(\underbrace{2}_{a_{2,0}}u^2 + \underbrace{-2}_{a_{0,2}}v^2). \qquad 3.8.37$$

In Equation 3.8.38, remember that we set

$$c = \sqrt{a_1^2 + a_2^2}.$$

The mapping in Example 3.8.9 simultaneously rotates by a clockwise in the (x, y)-plane and lowers by a along the z-axis. We hope you recognized that the first two rows of the right-hand side of Equation 3.8.40 are the result of multiplying $\begin{bmatrix} x \\ y \end{bmatrix}$ by the rotation matrix we already saw in Equation 3.8.6, and whose inverse we saw in Example 1.3.17.

Now we have an equation of the form of Equation 3.8.28, and we can read off the Gaussian curvature, using Proposition 3.8.7, part (c):

$$K = \frac{\overbrace{(2 \cdot -2)}^{a_{2,0}a_{0,2}-a_{1,1}^2} - 0}{\underbrace{(1 + 4a^2 + 4b^2)^2}_{(1+c^2)^2}} = \frac{-4}{(1 + 4a^2 + 4b^2)^2}. \qquad 3.8.38$$

Looking at this formula for K, what can you say about this surface away from the origin?[25]

Similarly, we can compute the mean curvature:

$$H = \frac{4(b^2 - a^2)}{(1 + 4a^2 + 4b^2)^{3/2}}. \qquad \triangle \qquad 3.8.39$$

Example 3.8.9 (Computing the Gaussian and mean curvature of the helicoid). The helicoid is the surface of equation $y \cos z = x \sin z$. You can imagine it as swept out by a horizontal line going through the z-axis, and which turns steadily as the z-coordinate changes, making an angle z with the parallel to the x-axis through the same point, as shown in Figure 3.8.6.

A first thing to observe is that the mapping

$$\begin{pmatrix} x \\ y \\ z \end{pmatrix} \mapsto \begin{pmatrix} x \cos a + y \sin a \\ -x \sin a + y \cos a \\ z - a \end{pmatrix} \qquad 3.8.40$$

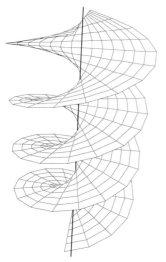

FIGURE 3.8.6.

The helicoid is swept out by a horizontal line, which rotates as it is lifted. Like the word "helix," "helicoid" comes from the Greek for "spiral."

is a rigid motion of $\mathbb{R}^3$ that sends the helicoid to itself. In particular, setting $a = z$, this rigid motion sends any point to a point of the form $\begin{pmatrix} r \\ 0 \\ 0 \end{pmatrix}$, and it is enough to compute the Gaussian curvature $K(r)$ at such a point.

We don't know the helicoid as a graph, but by the implicit function theorem, the equation of the helicoid determines z as a function $g_r\begin{pmatrix} x \\ y \end{pmatrix}$ near $\begin{pmatrix} r \\ 0 \\ 0 \end{pmatrix}$ when $r \neq 0$. What we need then is the Taylor polynomial of g_r. Introduce the new coordinate u such that $r + u = x$, and write

$$g_r\begin{pmatrix} u \\ y \end{pmatrix} = z = a_2 y + a_{1,1} uy + \frac{1}{2} a_{0,2} y^2 + \dots. \qquad 3.8.41$$

[25]The Gaussian curvature of this surface is always negative, but the further you go from the origin, the smaller it is, so the flatter the surface.

In rewriting

$$y \cos z = (r + u) \sin z$$

as Equation 3.8.42, we replace $\cos z$ by its Taylor polynomial,

$$1 - \frac{z^2}{2!} + \frac{z^4}{4!} \cdots,$$

keeping only the first term. (The term $z^2/2!$ is quadratic, but y times $z^2/2!$ is cubic.). We replace $\sin z$ by its Taylor polynomial,

$$\sin(z) = z - \frac{z^3}{3!} + \frac{z^5}{5!} \cdots,$$

keeping only the first term.

You should expect (Equation 3.8.43) that the coefficients a_2 and $a_{1,1}$ will blow up as $r \to 0$, since at the origin the helicoid does not represent z as a function of x and y. But the helicoid is a smooth surface at the origin.

We were glad to see that the linear terms in X and Y cancel, showing that we had indeed chosen adapted coordinates. Clearly, the linear terms in X do cancel. For the linear terms in Y, remember that $c = \sqrt{a_1^2 + a_2^2}$. So the linear terms on the right are

$$\frac{a_1^2 Y}{c\sqrt{1 + c^2}} + \frac{a_2^2 Y}{c\sqrt{1 + c^2}} = \frac{c^2 Y}{c\sqrt{1 + c^2}}$$

and on the left we have

$$\frac{cY}{\sqrt{1 + c^2}}.$$

Exercise 3.8.4 asks you to justify our omitting the terms $a_1 u$ and $a_{2,0} u^2$.

Introducing this into the equation $y \cos z = (r + u) \sin z$ and keeping only quadratic terms gives

$$\underbrace{y \cos z = y + \cdots}_{\text{see margin note}} = (r + u) \underbrace{\left(a_2 y + a_{1,1} u y + \frac{1}{2} a_{0,2} y^2 \right) + \ldots \ldots}_{z \text{ from Equation 3.8.41}}, \qquad 3.8.42$$

Identifying linear and quadratic terms gives

$$a_1 = 0 \ , \ a_2 = \frac{1}{r} \ , \ a_{1,1} = -\frac{1}{r^2} \ , \ a_{0,2} = 0 \ , \ a_{2,0} = 0. \qquad 3.8.43$$

We can now read off the Gaussian and mean curvatures:

$$\underbrace{K(r)}_{\substack{\text{Gaussian} \\ \text{curvature}}} = \frac{-1}{r^4(1 + 1/r^2)^2} = \frac{-1}{(1 + r^2)^2} \quad \text{and} \quad \underbrace{H(r)}_{\substack{\text{mean} \\ \text{curvature}}} = 0. \qquad 3.8.44$$

We see from the first equation that the Gaussian curvature is always negative and does not blow up as $r \to 0$: as $r \to 0$, $K(r) \to -1$. This is what we should expect, since the helicoid is a smooth surface. The second equation is more interesting yet. It says that the helicoid is a *minimal surface*: every patch of the helicoid minimizes area among surfaces with the same boundary.   △

Proof of Proposition 3.8.7.   In the coordinates X, Y, Z (i.e., using the values for x, y, and z given in Equation 3.8.29) the Equation 3.8.28 for S becomes

$$\overbrace{\frac{c}{\sqrt{1+c^2}} Y - \frac{1}{\sqrt{1+c^2}} Z}^{z \text{ from Equation 3.8.29}} = a_1 \Big(\overbrace{-\frac{a_2}{c} X + \frac{a_1}{c\sqrt{1+c^2}} Y + \frac{a_1}{\sqrt{1+c^2}} Z}^{x \text{ from Equation 3.8.29}} \Big)$$

$$+ a_2 \left(\frac{a_1}{c} X + \frac{a_2}{c\sqrt{1+c^2}} Y + \frac{a_2}{\sqrt{1+c^2}} Z \right)$$

$$+ \frac{1}{2} \Bigg(a_{2,0} \left(-\frac{a_2}{c} X + \frac{a_1}{c\sqrt{1+c^2}} Y + \frac{a_1}{\sqrt{1+c^2}} Z \right)^2$$

$$+ 2a_{1,1} \left(-\frac{a_2}{c} X + \frac{a_1}{c\sqrt{1+c^2}} Y + \frac{a_1}{\sqrt{1+c^2}} Z \right) \left(\frac{a_1}{c} X + \frac{a_2}{c\sqrt{1+c^2}} Y + \frac{a_2}{\sqrt{1+c^2}} Z \right)$$

$$+ a_{0,2} \left(\frac{a_1}{c} X + \frac{a_2}{c\sqrt{1+c^2}} Y + \frac{a_2}{\sqrt{1+c^2}} Z \right)^2 \Bigg) + \ldots . \qquad 3.8.45$$

We observe that all the linear terms in X and Y cancel, showing that this is an adapted system. The only remaining linear term is $-\sqrt{1 + c^2} Z$ and the coefficient of Z is not 0, so $D_3 F \neq 0$, so the implicit function theorem applies. Thus in these coordinates, Equation 3.8.45 expresses Z as a function of X and Y which starts with quadratic terms. This proves part (a).

To prove part (b), we need to multiply out the right-hand side. Remember that the linear terms in X and Y have canceled, and that we are interested

Since the expression of Z in terms of X and Y starts with quadratic terms, a term that is linear in Z is actually quadratic in X and Y.

We get $-\sqrt{1+c^2}\,Z$ on the left-hand side of Equation 3.8.46 by collecting all the linear terms in Z from Equation 3.8.45:

$$-\frac{1}{\sqrt{1+c^2}} - \frac{a_1^2}{\sqrt{1+c^2}} - \frac{a_2^2}{\sqrt{1+c^2}};$$

remember that $c = \sqrt{a_1^2 + a_2^2}$.

Equation 3.8.47 says

$$Z = \frac{1}{2}(A_{2,0}X^2 + 2A_{1,1}XY + A_{0,2}Y^2) + \dots .$$

This involves some quite miraculous cancellations. The mean curvature computation is similar, and left as Exercise 3.8.10; it also involves some miraculous cancellations.

Knot theory is a very active field of research today, with remarkable connections to physics (especially the latest darling of theoretical physicists: string theory).

only in terms up to degree 2; the terms in Z in the quadratic terms on the right now contribute terms of degree at least 3, so we can ignore them. We can thus rewrite Equation 3.8.45 as

$$-\sqrt{1+c^2}\,Z = \frac{1}{2}\left(a_{2,0}\left(-\frac{a_2}{c}X + \frac{a_1}{c\sqrt{1+c^2}}Y \right)^2 + \right.$$
$$2a_{1,1}\left(-\frac{a_2}{c}X + \frac{a_1}{c\sqrt{1+c^2}}Y \right)\left(\frac{a_1}{c}X + \frac{a_2}{c\sqrt{1+c^2}}Y \right) +$$
$$\left. a_{0,2}\left(\frac{a_1}{c}X + \frac{a_2}{c\sqrt{1+c^2}}Y \right)^2 \right) + \dots \qquad 3.8.46$$

If we multiply out, collect terms, and divide by $-\sqrt{1+c^2}$, this becomes

$$Z = \frac{1}{2}\left(-\overbrace{\frac{1}{c^2\sqrt{1+c^2}}\left(a_{2,0}a_2^2 - 2a_{1,1}a_1a_2 + a_{0,2}a_1^2 \right)}^{A_{2,0}}X^2 \right.$$
$$+ 2\frac{1}{c^2(1+c^2)}\left(-a_{2,0}a_1a_2 - a_{1,1}a_2^2 + a_{1,1}a_1^2 + a_{0,2}a_1a_2 \right)XY$$
$$\left. - \frac{1}{c^2(1+c^2)^{3/2}}\left(a_{2,0}a_1^2 + 2a_{1,1}a_1a_2 + a_{0,2}a_2^2 \right)Y^2 \right) + \dots \qquad 3.8.47$$

This proves part (b).

To see part (c), we just compute the Gaussian curvature, $K = A_{2,0}A_{0,2} - A_{1,1}^2$:

$$A_{2,0}A_{0,2} - A_{1,1}^2 =$$
$$\frac{1}{c^4(1+c^2)^2}\left(\left(a_{2,0}a_2^2 - 2a_{1,1}a_1a_2 + a_{0,2}a_1^2 \right)\left(a_{2,0}a_1^2 + 2a_{1,1}a_1a_2 + a_{0,2}a_2^2 \right) \right.$$
$$\left. - \left(a_{2,0}a_1a_2 + a_{1,1}a_2^2 - a_{1,1}a_1^2 - a_{0,2}a_1a_2 \right)^2 \right)$$
$$= \frac{a_{2,0}a_{0,2} - a_{1,1}^2}{(1+c^2)^2}. \qquad \square \qquad 3.8.48$$

Coordinates adapted to space curves

Curves in $\mathbb{R}^3$ have considerably simpler local geometry than do surfaces: essentially everything about them is in Propositions 3.8.12 and 3.8.13 below. Their global geometry is quite a different matter: they can tangle, knot, link, etc. in the most fantastic ways.

Suppose $C \subset \mathbb{R}^3$ is a smooth curve, and $\mathbf{a} \in C$ is a point. What new coordinate system X, Y, Z is well adapted to C at $\mathbf{a}$? Of course, we will take the origin of the new system at $\mathbf{a}$, and if we demand that the X-axis be tangent to C at $\mathbf{a}$, and call the other two coordinates U and V, then near $\mathbf{a}$ the curve C will have an equation of the form

$$U = f(X) = \frac{1}{2}a_2 X^2 + \frac{1}{6}a_3 X^3 + \dots$$

$$V = g(X) = \frac{1}{2}b_2 X^2 + \frac{1}{6}b_3 X^3 + \dots,.$$

3.8.49

where both coordinates start with quadratic terms. But it turns out that we can do better, at least when C is at least three times differentiable, and $a_2^2 + b_2^2 \neq 0$.

We again use the rotation matrix of Equation 3.8.6:

$$\begin{bmatrix} \cos\theta & \sin\theta \\ -\sin\theta & \cos\theta \end{bmatrix}.$$

Suppose we rotate the coordinate system around the X-axis by an angle θ, and call X, Y, Z the new (final) coordinates. Let $c = \cos\theta$ and $s = \sin\theta$; this means setting

$$U = cY + sZ \quad \text{and} \quad V = -sY + cZ.$$

3.8.50

Substituting these expressions into Equation 3.8.49 leads to

Remember that $s = \sin\theta$ and $c = \cos\theta$, so $c^2 + s^2 = 1$.

$$cY + sZ = \frac{1}{2}a_2 X^2 + \frac{1}{6}a_3 X^3 + \dots$$

$$-sY + cZ = \frac{1}{2}b_2 X^2 + \frac{1}{6}b_3 X^3 + \dots.$$

3.8.51

We solve these equations for Y by multiplying the first through by c and the second by $-s$ and adding the results:

$$Y(c^2 + s^2) = Y = \frac{1}{2}(ca_2 - sb_2)X^2 + \frac{1}{6}(ca_3 - sb_3)X^3.$$

3.8.52

A similar computation gives

$$Z = \frac{1}{2}(sa_2 + cb_2)X^2 + \frac{1}{6}(sa_3 + cb_3)X^3.$$

3.8.53

The point of all this is that we want to choose the angle θ (the angle by which we rotate the coordinate system around the X-axis) so that the Z-component of the curve begins with cubic terms. We achieve this by setting

$$A_2 = \sqrt{a_2^2 + b_2^2},$$

$$B_3 = \frac{-b_2 a_3 + a_2 b_3}{\sqrt{a_2^2 + b_2^2}}.$$

$$c = \cos\theta = \frac{a_2}{\sqrt{a_2^2 + b_2^2}} \quad \text{and} \quad s = \sin\theta = \frac{-b_2}{\sqrt{a_2^2 + b_2^2}}, \text{ so that } \tan\theta = \frac{-b_2}{a_2};$$

3.8.54

this gives

$$Y = \frac{1}{2}\overbrace{\sqrt{a_2^2 + b_2^2}}^{A_2}X^2 + \frac{1}{6}\frac{a_2 a_3 + b_2 b_3}{\sqrt{a_2^2 + b_2^2}}X^3 = \frac{A_2}{2}X^2 + \frac{A_3}{6}X^3 + \dots$$

3.8.55

$$Z = \frac{1}{6}\frac{-b_2 a_3 + a_2 b_3}{\sqrt{a_2^2 + b_2^2}}X^3 + \dots = \frac{B_3}{6}X^3 + \dots.$$

The word *osculating* comes from the Latin *osculari*, "to kiss."

We choose $\kappa = A_2$ to be the positive square root, $+\sqrt{a_2^2 + b_2^2}$; thus our definition of curvature of a space curve is compatible with Definition 3.8.1 of the curvature of a plane curve.

Note that the torsion is defined only when the curvature is not zero. The osculating plane is the plane that the curve is most nearly in, and the torsion measures how fast the curve pulls away from it. It measures the "non-planarity" of the curve.

A curve in $\mathbb{R}^n$ can be parametrized by arc length because curves have no intrinsic geometry; you could represent the Amazon River as a straight line without distorting its length. Surfaces and other manifolds of higher dimension cannot be parametrized by anything analogous to arc length; any attempt to represent the surface of the globe as a flat map necessarily distorts sizes and shapes of the continents. Gaussian curvature is the obstruction.

Imagine that you are driving in the dark, and that the first unit vector is the shaft of light produced by your headlights.

We know the acceleration must be orthogonal to the curve because your speed is constant; there is no component of acceleration in the direction you are going.

Alternatively, you can derive

$$2\vec{\delta}' \cdot \vec{\delta}'' = 0$$

from $|\vec{\delta}'|^2 = 1$.

The Z-component measures the distance of the curve from the (X, Y)-plane; since Z is small, then the curve stays mainly in that plane. The (X, Y)-plane is called the *osculating plane* to C at $\mathbf{a}$.

This is our best adapted coordinate system for the curve at $\mathbf{a}$, which exists and is unique unless $a_2 = b_2 = 0$. The number $\kappa = A_2 \geq 0$ is called the curvature of C at $\mathbf{a}$, and the number $\tau = B_3/A_2$ is called the *torsion* of C at $\mathbf{a}$.

Definition 3.8.10 (Curvature of a space curve). The curvature of a space curve C at $\mathbf{a}$ is

$$\kappa = A_2 \geq 0.$$

Definition 3.8.11 (Torsion of a space curve). The torsion of a space curve C at $\mathbf{a}$ is

$$\tau = B_3/A_2.$$

Parametrization of space curves by arc length: the Frenet frame

Usually, the geometry of space curves is developed using parametrizations by arc length rather than by adapted coordinates. Above, we emphasized adapted coordinates because they generalize to manifolds of higher dimension, while parametrizations by arc length do not.

The main ingredient of the approach using parametrization by arc length is the *Frenet frame*. Imagine driving at unit speed along the curve, perhaps by turning on cruise control. Then (at least if the curve is really curvy, not straight) at each instant you have a distinguished basis of $\mathbb{R}^3$. The first unit vector is the velocity vector, pointing in the direction of the curve. The second vector is the acceleration vector, normalized to have length 1. It is orthogonal to the curve, and points in the direction in which the force is being applied— i.e., in the opposite direction of the centrifugal force you feel. The third basis vector is the *binormal*, orthogonal to the other two vectors.

So, if $\delta : \mathbb{R} \to \mathbb{R}^3$ is the parametrization by arc length of the curve, the three vectors are:

$$\underbrace{\vec{\mathbf{t}}(s) = \vec{\delta}'(s)}_{\text{velocity vector}}, \quad \underbrace{\vec{\mathbf{n}}(s) = \frac{\vec{\mathbf{t}}'(s)}{|\vec{\mathbf{t}}'(s)|} = \frac{\vec{\delta}''(s)}{|\vec{\delta}''(s)|}}_{\substack{\text{normalized} \\ \text{acceleration vector}}}, \quad \underbrace{\vec{\mathbf{b}}(s) = \vec{\mathbf{t}}(s) \times \vec{\mathbf{n}}(s)}_{\text{binormal}}. \qquad 3.8.56$$

The propositions below relate the Frenet frame to the adapted coordinates; they provide another description of curvature and torsion, and show that the two approaches coincide. The same computations prove both; they are proved in Appendix A.11.

In particular, Equation 3.8.57 says that the point **a** in the old coordinates is the point $\begin{pmatrix} 0 \\ 0 \\ 0 \end{pmatrix}$ in the new coordinates, which is what we want.

Proposition 3.8.13: The derivatives $\vec{\mathbf{t}}'$, $\vec{\mathbf{n}}'$, and $\vec{\mathbf{b}}'$ are computed with respect to arc length. Think of the point **a** as the place where you set the odometer reading to 0, accounting for $\vec{\mathbf{t}}'(0)$, etc. The derivatives are easiest to compute at 0, but the equations $\vec{\mathbf{t}}'(0) = \kappa\vec{\mathbf{n}}(0)$ and so on are true at any point (any odometer reading).

Equation 3.8.58 corresponds to the antisymmetric matrix

$$\begin{bmatrix} 0 & \kappa & 0 \\ -\kappa & 0 & \tau \\ 0 & -\tau & 0 \end{bmatrix}.$$

Exercise 3.8.9 asks you to explain where this antisymmetry comes from.

Propositions 3.8.14 and 3.8.15 make the computation of curvature and torsion straightforward for any parametrized curve in $\mathbb{R}^3$.

Many texts refer to $\kappa(t)$ and $\tau(t)$, including this one in earlier printings. However, curvature and torsion are invariants of a curve: they depend only on the point at which they are evaluated. Strictly speaking, it does not make sense to think of them as functions of time.

Proposition 3.8.12 (Frenet frame).   *The vectors $\vec{\mathbf{t}}(0), \vec{\mathbf{n}}(0), \vec{\mathbf{b}}(0)$ form the orthonormal basis (Frenet frame) with respect to which our adapted coordinates are computed. Thus the point with coordinates X, Y, Z in the new, adapted coordinates is the point*

$$\mathbf{a} + X\vec{\mathbf{t}}(0) + Y\vec{\mathbf{n}}(0) + Z\vec{\mathbf{b}}(0) \qquad 3.8.57$$

in the old x, y, z coordinates.

Proposition 3.8.13 (Curvature and torsion of a space curve).   *The Frenet frame satisfies the following equations, where κ is the curvature of the curve at **a** and τ is its torsion:*

$$\begin{aligned} \vec{\mathbf{t}}'(0) &= \kappa\vec{\mathbf{n}}(0) \\ \vec{\mathbf{n}}'(0) &= -\kappa\vec{\mathbf{t}}(0) + \tau\vec{\mathbf{b}}(0) \qquad 3.8.58 \\ \vec{\mathbf{b}}'(0) &= -\tau\vec{\mathbf{n}}(0). \end{aligned}$$

Computing curvature and torsion of parametrized curves

We now have two equations that in principle should allow us to compute curvature and torsion of a space curve: Equations 3.8.55 and 3.8.58. Unfortunately, these equations are hard to use. Equation 3.8.55 requires knowing an adapted coordinate system, which leads to very cumbersome formulas, whereas Equation 3.8.58 requires a parametrization by arc length. Such a parametrization is only known as the inverse of a function which is itself an indefinite integral that can rarely be computed in closed form. However, the Frenet formulas can be adapted to any parametrized curve:

Proposition 3.8.14 (Curvature of a parametrized curve).   *The curvature κ of a curve parametrized by $\gamma : \mathbb{R} \to \mathbb{R}^3$ is*

$$\kappa\big(\gamma(t)\big) = \frac{|\vec{\gamma}'(t) \times \vec{\gamma}''(t)|}{|\vec{\gamma}'(t)|^3}. \qquad 3.8.59$$

Proposition 3.8.15 (Torsion of a parametrized curve).   *The torsion τ of a parametrized curve is*

$$\begin{aligned} \tau\big(\gamma(t)\big) &= \frac{(\vec{\gamma}'(t) \times \vec{\gamma}''(t)) \cdot \vec{\gamma}'''(t)}{(s'(t))^6} \frac{(s'(t))^6}{|\vec{\gamma}'(t) \times \vec{\gamma}''(t)|^2} \\ &= \frac{(\vec{\gamma}'(t) \times \vec{\gamma}''(t)) \cdot \vec{\gamma}'''(t)}{|\vec{\gamma}'(t) \times \vec{\gamma}''(t)|^2}. \end{aligned} \qquad 3.8.60$$

Example 3.8.16 (Computing curvature and torsion of a parametrized curve). Let $\gamma(t) = \begin{pmatrix} t \\ t^2 \\ t^3 \end{pmatrix}$. Then

$$\vec{\gamma}'(t) = \begin{pmatrix} 1 \\ 2t \\ 3t^2 \end{pmatrix}, \quad \vec{\gamma}''(t) = \begin{pmatrix} 0 \\ 2 \\ 6t \end{pmatrix}, \quad \vec{\gamma}'''(t) = \begin{pmatrix} 0 \\ 0 \\ 6 \end{pmatrix}. \qquad 3.8.61$$

So we find

$$\kappa(\gamma(t)) = \frac{1}{(1 + 4t^2 + 9t^4)^{3/2}} \left| \begin{pmatrix} 1 \\ 2t \\ 3t^2 \end{pmatrix} \times \begin{pmatrix} 0 \\ 2 \\ 6t \end{pmatrix} \right| = 2\frac{(1 + 9t^2 + 9t^4)^{1/2}}{(1 + 4t^2 + 9t^4)^{3/2}} \qquad 3.8.62$$

and

$$\tau(\gamma(t)) = \frac{1}{4(1 + 9t^2 + 9t^4)} \begin{pmatrix} 6t^2 \\ -6t \\ 2 \end{pmatrix} \cdot \begin{pmatrix} 0 \\ 0 \\ 6 \end{pmatrix} = \frac{3}{1 + 9t^2 + 9t^4}. \qquad 3.8.63$$

Since $\gamma(t) = \begin{pmatrix} t \\ t^2 \\ t^3 \end{pmatrix}$, $Y = X^2$ and $z = X^3$, so Equation 3.8.55 says that $Y = X^2 = \frac{2}{2}X^2$, so $A_2 = 2 \ldots$.

At the origin, the standard coordinates are adapted to the curve, so from Equation 3.8.55 we find $A_2 = 2$, $B_3 = 6$; hence $\kappa = A_2 = 2$ and $\tau = B_3/A_2 = 3$. This agrees with the formulas above when $t = 0$. $\triangle$

Proof of Proposition 3.8.14 (curvature of a parametrized curve). We will assume that we have a parametrized curve $\gamma : \mathbb{R} \to \mathbb{R}^3$; you should imagine that you are driving along some winding mountain road, and that $\gamma(t)$ is the position of your car at time t. Since our computation will use Equation 3.8.58, we will also use parametrization by arc length; we will denote by $\delta(s)$ the position of the car when the odometer is s, while γ denotes an arbitrary parametrization. These are related by the formula

Remark Strictly speaking, $\vec{t}$, $\vec{n}$, $\vec{b}$, κ, and τ should be considered as functions of $\delta(s(t))$: the point where the car is at a particular odometer reading, which itself depends on time. However, the notation is already fearsome, and we hesitate to make it any more difficult than it is.

To go from the first to the second line of Equation 3.8.66 we use Proposition 3.8.13, which says that $\vec{t}' = \kappa\vec{n}$.

Note in the second line of Equation 3.8.66 that we are adding vectors to get a vector:

$$\underbrace{\kappa(s(t))(s'(t))^2}_{\text{number}} \underbrace{\vec{n}(s(t))}_{\text{vec.}}$$
$$+ \underbrace{s''(t)}_{\text{no.}} \underbrace{\vec{t}(s(t))}_{\text{vec.}}$$

$$\gamma(t) = \delta(s(t)), \quad \text{where} \quad s(t) = \int_{t_0}^{t} |\vec{\gamma}'(u)|\, du, \qquad 3.8.64$$

and t_0 is the time when the odometer was set to 0. The function $s(t)$ gives you the odometer reading as a function of time. The unit vectors $\vec{t}$, $\vec{n}$ and $\vec{b}$ will be considered as functions of s, as will the curvature κ and the torsion τ.

We now use the chain rule to compute three successive derivatives of γ. In Equation 3.8.65, recall (Equation 3.8.56) that $\vec{\delta}' = \vec{t}$; in the second line of Equation 3.8.66, recall (Equation 3.8.58) that $\vec{t}'(0) = \kappa\vec{n}(0)$:

$$(1) \quad \vec{\gamma}'(t) = (\vec{\delta}'(s(t)))s'(t) = s'(t)\vec{t}(s(t)), \qquad 3.8.65$$

$$(2) \quad \vec{\gamma}''(t) = \vec{t}'(s(t))(s'(t))^2 + \vec{t}(s(t))s''(t)$$
$$= \kappa(s(t))(s'(t))^2\vec{n}(s(t)) + s''(t)\vec{t}(s(t)), \qquad 3.8.66$$

The derivative κ' is the derivative of κ with respect to arc length.

Equation 3.8.69: by Equations 3.8.65 and 3.8.66,

$$\vec{\gamma}'(t) \times \vec{\gamma}''(t)$$
$$= s'(t)\vec{t}(s(t))$$
$$\times \left[\kappa(s(t))(s'(t))^2 \vec{n}(s(t)) + s''(t)\vec{t}(s(t)) \right].$$

Since for any vector $\vec{t}$, $\vec{t} \times \vec{t} = 0$, and since (Equation 3.8.56)

$$\vec{t} \times \vec{n} = \vec{b}, \quad \text{this gives}$$

$$\vec{\gamma}'(t) \times \vec{\gamma}''(t)$$
$$= s'(t)\vec{t}(s(t))$$
$$\times \left[\kappa(s(t))(s'(t))^2 \vec{n}(s(t)) \right]$$
$$= \kappa(s(t))(s'(t))^3 \vec{b}(s(t)).$$

$$(3) \quad \vec{\gamma}'''(t) = \kappa'\big(s(t)\big)\vec{n}\big(s(t)\big)\big(s'(t)\big)^3 + 2\kappa\big(s(t)\big)\vec{n}'\big(s(t)\big)\big(s'(t)\big)^3 +$$
$$\kappa\big(s(t)\big)\vec{n}\big(s(t)\big)\big(s'(t)\big)\big(s''(t)\big)\vec{t}'\big(s(t)\big)\big(s'(t)\big)\big(s''(t)\big)$$
$$+ \vec{t}\big(s(t)\big)\big(s'''(t)\big)$$
$$= \Big(\big(\kappa(s(t))\big)^2 + s'''(t) \Big)\vec{t}(s(t)) + \tag{3.8.67}$$
$$\Big(\kappa'(s(t)(s'(t))^3 + 3\kappa(s(t))(s'(t))(s''(t)) \Big)\vec{n}\big(s(t)\big) +$$
$$\Big(\kappa\big(s(t)\big)\tau\big(s(t)\big) \Big)\vec{b}.$$

Since $\vec{t}$ has length 1, Equation 3.8.65 gives us

$$s'(t) = |\vec{\gamma}'(t)|, \tag{3.8.68}$$

which we already knew from the definition of s. Equations 3.8.65 and 3.8.66 give

$$\vec{\gamma}'(t) \times \vec{\gamma}''(t) = \kappa\big(s(t)\big)\big(s'(t)\big)^3 \vec{b}\big(s(t)\big), \tag{3.8.69}$$

since $\vec{t} \times \vec{n} = \vec{b}$. Since $\vec{b}$ has length 1,

$$|\vec{\gamma}'(t) \times \vec{\gamma}''(t)| = \kappa\big(s(t)\big)\big(s'(t)\big)^3, \tag{3.8.70}$$

Using Equation 3.8.68, this gives the formula for the curvature of Proposition 3.8.14. $\square$

Proof of Proposition 3.8.15 (Torsion of a parametrized curve). Since $\vec{\gamma}' \times \vec{\gamma}''$ points in the direction of $\vec{b}$, dotting it with $\vec{\gamma}'''$ picks out the coefficient of $\vec{b}$ for $\vec{\gamma}'''$. This leads to

$$\big(\vec{\gamma}'(t) \times \vec{\gamma}''(t)\big) \cdot \vec{\gamma}'''(t) = \tau\big(s(t)\big) \underbrace{\big(\kappa\big(s(t)\big)\big)^2 \big(s'(t)\big)^6}, \tag{3.8.71}$$
$$\text{square of Equation 3.8.70}$$

which gives us the formula for torsion found in Proposition 3.8.15. $\square$

3.9 Exercises for Chapter Three

Exercises for Section 3.1:

Curves and Surfaces

3.1.1   (a) For what values of the constant c is the locus of equation $\sin(x+y) = c$ a smooth curve?

(b) What is the equation for the tangent line to such a curve at a point $\begin{pmatrix} u \\ v \end{pmatrix}$?

3.1.2   (a) For what values of c is the set of equation $X_c = x^2 + y^3 = c$ a smooth curve?

(b) Give the equation of the tangent line at a point $\begin{pmatrix} u \\ v \end{pmatrix}$ of such a curve X_c.

(c) Sketch this curve for a representative sample of values of c.

3.1.3   (a) For what values of c is the set of equation $Y_c = x^2 + y^3 + z^4 = c$ a smooth surface?

We strongly advocate using *Matlab* or similar software.

(b) Give the equation of the tangent plane at a point $\begin{pmatrix} u \\ v \\ w \end{pmatrix}$ of the surface Y_c.

(c) Sketch this surface for a representative sample of values of c.

3.1.4   Show that every straight line in the plane is a smooth curve.

3.1.5   In Example 3.1.15, show that S^2 is a smooth surface, using $D_{x,y}$, $D_{x,z}$ and $D_{y,z}$; the half-axes $\mathbb{R}_z^+$, $\mathbb{R}_x^+$ and $\mathbb{R}_y^+$; and the mappings

$$\pm\sqrt{x^2 + y^2 - 1}, \quad \pm\sqrt{x^2 + z^2 - 1} \quad \text{and} \quad \pm\sqrt{y^2 + z^2 - 1}.$$

3.1.6   (a) Show that the set $\left\{ \begin{pmatrix} x \\ y \end{pmatrix} \in \mathbb{R}^2 \,\middle|\, x + x^2 + y^2 = 2 \right\}$ is a smooth curve.

(b) What is an equation for the tangent line to this curve at a point $\begin{pmatrix} 1 \\ 0 \end{pmatrix}$?

3.1.7   (a) Show that for all a and b, the sets X_a and Y_b of equation

Hint for Exercise 3.1.7 (a): This does *not* require the implicit function theorem.

$$x^2 + y^3 + z = a \quad \text{and} \quad x + y + z = b$$

respectively are smooth surfaces in $\mathbb{R}^3$.

(b) For what values of a and b is the intersection $X_a \cap Y_b$ a smooth curve? What geometric relation is there between X_a and Y_b for the other values of a and b?

3.1.8   (a) For what values of a and b are the sets X_a and Y_b of equation

$$x - y^2 = a \quad \text{and} \quad x^2 + y^2 + z^2 = b$$

respectively smooth surfaces in $\mathbb{R}^3$?

(b) For what values of a and b is the intersection $X_a \cap Y_b$ a smooth curve? What geometric relation is there between X_a and Y_b for the other values of a and b?

3.1.9   Show that if at a particular point $\mathbf{x}_0$ a surface is simultaneously the graph of z as a function of x and y, and y as a function of x and z, and x as a function of y and z (see Definition 3.1.13), then the corresponding equations for the tangent planes to the surface at $\mathbf{x}_0$ denote the same plane.

3.1.10   For each of the following functions f and points $\begin{pmatrix} a \\ b \end{pmatrix}$:

(a) State whether there is a tangent plane to the graph of f at the point
$$\begin{pmatrix} a \\ b \\ f\begin{pmatrix} a \\ b \end{pmatrix} \end{pmatrix}.$$

You are encouraged to use a computer, although it is not absolutely necessary.

(b) If there is, find its equation, and compute the intersection of the tangent plane with the graph.

(a) $f\begin{pmatrix} x \\ y \end{pmatrix} = x^2 - y^2$ at the point $\begin{pmatrix} 1 \\ 1 \end{pmatrix}$

(b) $f\begin{pmatrix} x \\ y \end{pmatrix} = \sqrt{x^2 + y^2}$ at the point $\begin{pmatrix} 0 \\ 0 \end{pmatrix}$

(c) $f\begin{pmatrix} x \\ y \end{pmatrix} = \sqrt{x^2 + y^2}$ at the point $\begin{pmatrix} 1 \\ -1 \end{pmatrix}$

(d) $f\begin{pmatrix} x \\ y \end{pmatrix} = \cos(x^2 + y)$ at the point $\begin{pmatrix} 0 \\ 0 \end{pmatrix}$

3.1.11   Find quadratic polynomials p and q for which the function

$$F\begin{pmatrix} x \\ y \end{pmatrix} = x^4 + y^4 + x^2 - y^2 \quad \text{of Example 3.1.11 can be written}$$

$$F\begin{pmatrix} x \\ y \end{pmatrix} = p(x)^2 + q(y)^2 - \frac{1}{2}.$$

Sketch the graphs of p, q, p^2, and q^2, and describe the connection between your graph and Figure 3.1.8.

Hint for Exercise 3.1.12, part (b): write that $\mathbf{x} - \gamma(t)$ is a multiple of $\gamma'(t)$, which leads to two equations in x, y, z, and t. Now eliminate t among these equations; it takes a bit of fiddling with the algebra.

3.1.12   Let $C \subset \mathbb{R}^3$ be the curve parametrized by $\gamma(t) = \begin{pmatrix} t \\ t^2 \\ t^3 \end{pmatrix}$. Let X be the union of all the lines tangent to C.

(a) Find a parametrization of X.

(b) Find an equation $f(\mathbf{x}) = 0$ for X.

Hint for part (c): show that the only common zeroes of f and $[Df]$ are the points of C; again this requires a bit of fiddling with the algebra.

*(c) Show that $X - C$ is a smooth surface. (Our solution uses material in Exercise 3.1.20.)

(d) Find the equation of the curve that is the intersection of X with the plane $x = 0$.

3.1.13   Let C be a helicoid parametrized by $\gamma(t) = \begin{pmatrix} \cos t \\ \sin t \\ t \end{pmatrix}$.

Part (b): A parametrization of this curve is not too hard to find, but a computer will certainly help in describing the curve.

(a) Find a parametrization for the union X of all the tangent lines to C. Use a computer program to visualize this surface.

(b) What is the intersection of X with the (x, z)-plane?

(c) Show that X contains infinitely many curves of double points, where X intersects itself; these curves are helicoids on cylinders $x^2 + y^2 = r_i^2$. Find an equation for the numbers r_i, and use Newton's method to compute r_1, r_2, r_3.

3.1.14   (a) What is the equation of the plane containing the point **a** perpendicular to the vector $\vec{v}$?

(b) Let $\gamma(t) = \begin{pmatrix} t \\ t^2 \\ t^3 \end{pmatrix}$, and let P_t be the plane through the point $\gamma(t)$ and perpendicular to $\vec{\gamma}'(t)$. What is the equation of P_t?

(c) Show that if $t_1 \neq t_2$, the planes P_{t_1} and P_{t_2} always intersect in a line. What are the equations of the line $P_1 \cap P_t$?

(d) What is the limiting position of the line $P_1 \cap P_{1+h}$ as h tends to 0?

3.1.15   In Example 3.1.17, what does the surface of equation

Hint: Think that $\sin \alpha = 0$ if and only if $\alpha = k\pi$ for some integer k.

$$f \begin{pmatrix} x \\ y \\ z \end{pmatrix} = \sin(x + yz) = 0 \qquad \text{look like?}$$

3.1.16   (a) Show that the set $X \subset \mathbb{R}^3$ of equation

$$x^3 + xy^2 + yz^2 + z^3 = 4 \quad \text{is a smooth surface.}$$

(b) What is the equation of the tangent plane to X at the point $\begin{pmatrix} 1 \\ 1 \\ 1 \end{pmatrix}$?

3.1.17   Let $f \begin{pmatrix} x \\ y \end{pmatrix} = 0$ be the equation of a curve $X \subset \mathbb{R}^2$, and suppose $\left[\mathbf{D}f \begin{pmatrix} x \\ y \end{pmatrix} \right] \neq 0$ for all $\begin{pmatrix} x \\ y \end{pmatrix} \in X$.

(a) Find an equation for the cone $CX \subset \mathbb{R}^3$ over X, i.e., the union of all the lines through the origin and a point $\begin{pmatrix} x \\ y \\ 1 \end{pmatrix}$ with $\begin{pmatrix} x \\ y \end{pmatrix} \in X$.

(b) If X has the equation $y = x^3$, what is the equation of CX?

(c) Show that $CX - \{\mathbf{0}\}$ is a smooth surface.

(d) What is the equation of the tangent plane to CX at any $\mathbf{x} \in CX$?

3.1.18   (a) Find a parametrization for the union X of the lines through the origin and a point of the parametrized curve $t \mapsto \begin{pmatrix} t \\ t^2 \\ t^3 \end{pmatrix}$.

(b) Find an equation for the closure $\overline{X}$ of X. Is $\overline{X}$ exactly X?

(c) Show that $\overline{X} - \{\mathbf{0}\}$ is a smooth surface.

(d) Show that

$$\begin{pmatrix} r \\ \theta \end{pmatrix} \mapsto \begin{pmatrix} r(1 + \sin\theta) \\ r\cos\theta \\ r(1 - \sin\theta) \end{pmatrix}$$

is another parametrization of $\overline{X}$. In this form you should have no trouble giving a name to the surface $\overline{X}$.

(e) Relate $\overline{X}$ to the set of non-invertible 2×2 matrices.

3.1.19   (a) What is the equation of the tangent plane to the surface S of equation $f \begin{pmatrix} x \\ y \\ z \end{pmatrix} = 0$ at the point $\begin{pmatrix} a \\ b \\ c \end{pmatrix} \in S$?

(b) Write the equations of the tangent planes P_1, P_2, P_3 to the surface of equation $z = Ax^2 + By^2$ at the points $\mathbf{p}_1, \mathbf{p}_2, \mathbf{p}_3$ with x, y-coordinates $\begin{pmatrix} 0 \\ 0 \end{pmatrix}$, $\begin{pmatrix} a \\ 0 \end{pmatrix}$, $\begin{pmatrix} 0 \\ b \end{pmatrix}$, and find the point $\mathbf{q} = P_1 \cap P_2 \cap P_3$.

(c) What is the volume of the tetrahedron with vertices at $\mathbf{p}_1, \mathbf{p}_2, \mathbf{p}_3$, and $\mathbf{q}$?

*3.1.20   Suppose $U \subset \mathbb{R}^2$ is open, $\mathbf{x}_0 \in U$ is a point, and $\mathbf{f} : U \to \mathbb{R}^3$ is a differentiable mapping with Lipschitz derivative. Suppose that $[\mathbf{Df}(\mathbf{x}_0)]$ is 1-1.

(a) Show that there are two basis vectors of $\mathbb{R}^3$ spanning a plane E_1 such that if $P : \mathbb{R}^3 \to E_1$ denotes the projection onto the plane spanned by these vectors, then $[\mathbf{D}(P \circ \mathbf{f})(\mathbf{x}_0)]$ is invertible.

(b) Show that there exists a neighborhood $V \subset E_1$ of $(P \circ \mathbf{f})(\mathbf{x}_0)$ and a mapping $\mathbf{g} : V \to \mathbb{R}^2$ such that $(P \circ \mathbf{f} \circ \mathbf{g})(\mathbf{y}) = \mathbf{y}$ for all $\mathbf{y} \in V$.

(c) Let $W = \mathbf{g}(V)$. Show that $\mathbf{f}(W)$ is the graph of $\mathbf{f} \circ \mathbf{g} : V \to E_2$, where E_2 is the line spanned by the third basis vector. Conclude that $\mathbf{f}(W)$ is a smooth surface.

Exercises for Section 3.2:

Manifolds

The "unit sphere" has radius 1; unless otherwise stated, it is always centered at the origin.

3.2.1   Consider the space X_l of positions of a rod of length l in $\mathbb{R}^3$, where one endpoint is constrained to be on the x-axis, and the other is constrained to be on the unit sphere centered at the origin.

(a) Give equations for X_l as a subset of $\mathbb{R}^4$, where the coordinates in $\mathbb{R}^4$ are the x-coordinate of the end of the rod on the x-axis (call it t), and the three coordinates of the other end of the rod.

(b) Show that near the point $\begin{pmatrix} 1+l \\ 1 \\ 0 \\ 0 \end{pmatrix}$, the set X_l is a manifold, and give the equation of its tangent space.

(c) Show that for $l \neq 1$, X_l is a manifold.

3.2.2   Consider the space X of positions of a rod of length 2 in $\mathbb{R}^3$, where one endpoint is constrained to be on the sphere of equation $(x - 1)^2 + y^2 + z^2 = 1$, and the other on the sphere of equation $(x + 1)^2 + y^2 + z^2 = 1$.

.tions for X as a subset of $\mathbb{R}^6$, where the coordinates in $\mathbb{R}^6$ are

.tes $\begin{pmatrix} x_1 \\ y_1 \\ z_1 \end{pmatrix}$ of the end of the rod on the first sphere, and the three

es $\begin{pmatrix} x_2 \\ y_2 \\ z_2 \end{pmatrix}$ of the other end of the rod.

Show that near the point in $\mathbb{R}^6$ shown in the margin, the set X is a ...ld, and give the equation of its tangent space. What is the dimension of ...ar this point?

3   In Example 3.2.1, show that knowing $\mathbf{x}_1$ and $\mathbf{x}_3$ determines exactly four ...sitions of the linkage if the distance from $\mathbf{x}_1$ to $\mathbf{x}_3$ is smaller than both $l_1 + l_2$ ...d $l_3 + l_4$ and greater than $|l_1 - l_2|$ and $|l_3 - l_4|$.

3.2.4   (a) Parametrize the positions of the linkage of Example 3.2.1 by the coordinates of $\mathbf{x}_1$, the polar angle θ_1 of the first rod with the horizontal line passing through $\mathbf{x}_1$, and the angle θ_2 between the first and the second: four numbers in all. For each value of θ_2 such that

$$(l_3 - l_4)^2 < l_1^2 + l_2^2 - 2l_1 l_2 \cos \theta_2 < (l_3 + l_4)^2,$$

...ow many positions of the linkage are there?

(b) What happens if either of the inequalities in Equation 3.2.4 above is an ...ality?

In Example 3.2.1, describe X_2 and X_3 when $l_1 = l_2 + l_3 + l_4$.

Let $M_k(n, m)$ be the space of $n \times m$ matrices of rank k.

...ow that the space $M_1(2, 2)$ of 2×2 matrices of rank 1 is a manifold ...d in $\text{Mat}\,(2, 2)$.

...how that the space $M_2(3, 3)$ of 3×3 matrices of rank 2 is a manifold ...d in $\text{Mat}\,(3, 3)$. Show (by explicit computation) that $[\mathbf{D}\det(A)] = 0$ if ... if A has rank < 2.

If $l_1 + l_2 = l_3 + l_4$, show that X_2 is not a manifold near the position ...ll four points are aligned with x_2 and x_4 between x_1 and x_3.

Let $O(n) \subset \text{Mat}\,(n, n)$ be the set of orthogonal matrices, i.e., matrices ...olumns form an orthonormal basis of $\mathbb{R}^n$. Let $S(n, n)$ be the space of ...ric $n \times n$ matrices, and $A(n, n)$ be the space of antisymmetric $n \times n$ ...s.

...w that $A \in O(n)$ if and only if $A^\top A = I$.

(b) Show that if A, $B \in O(n)$, then $AB \in O(n)$ and A^-

(c) Show that $A^\top A - I \in S(n,n)$.

(d) Define $F : \text{Mat}(n,n) \to S(n,n)$ to be $F(A) = A^\top A$ ⸱
$F^{-1}(0)$. Show that if A is invertible, then $\left[\mathbf{D}F\left(A\right) \right]$: Mat
onto.

(e) Show that $O(n)$ is a manifold embedded in Mat (n,n)
$A(n,n)$.

*3.2.9   Let $M_k(n,m)$ be the space of $n \times m$ matrices of
(a) Show that $M_1(n,m)$ is a manifold embedded in Mat (
1. Hint: It is rather difficult to write equations for $M_1(n,$
hard to show that $M_1(n,m)$ is locally the graph of a mapping
variables as functions of others. For instance, suppose

$$A = [\mathbf{a}_1, \ldots, \mathbf{a}_m] \in M_1(n,m),$$

and that $a_{1,1} \neq 0$. Show that all the entries of

$$\begin{bmatrix} a_{2,2} & \cdots & a_{2,m} \\ \vdots & \ddots & \vdots \\ a_{n,2} & \cdots & a_{n,m} \end{bmatrix}$$

are functions of the others, for instance $a_{2,2} = a_{1,2}a_{2,1}/a_{1,1}$.

(b) What is the dimension of $M_1(n,m)$?

*3.2.10   (a) Show that the mapping $\varphi_1 : (\mathbb{R}^m - \{0\}) \times \mathbb{R}$

$$\varphi_1 \left(\mathbf{a}, \begin{bmatrix} \lambda_2 \\ \vdots \\ \lambda_n \end{bmatrix} \right) \mapsto [\mathbf{a}, \lambda_2 \mathbf{a}, \ldots, \lambda_n \mathbf{a}]$$

is a parametrization of the subset $U_1 \subset M_1(m,n)$ of those m
column is not $\mathbf{0}$.

(b) Show that $M_1(m,n) - U_1$ is a manifold embedded in M
its dimension?

(c) How many parametrizations like φ_1 do you need to cc
$M_1(m,n)$?

Exercises for Section 3.3:   **3.3.1**   For the function f of Example 3.3.11, show that a⸱
Taylor Polynomials   partial derivatives exist everywhere, that the first partial de
tinuous, and that the second partial derivatives are not.

3.3.2   Compute

$$D_1(D_2f), \quad D_2(D_3f), \quad D_3(D_1f), \quad \text{and} \quad D_1(D_2(D_3f))$$

for the function $f\begin{pmatrix} x \\ y \\ z \end{pmatrix} = x^2y + xy^2 + yz^2.$

3.3.3   Consider the function

$$f\begin{pmatrix} x \\ y \end{pmatrix} = \begin{cases} \dfrac{x^2y(x-y)}{x^2+y^2} & \text{if } \begin{pmatrix} x \\ y \end{pmatrix} \neq \begin{pmatrix} 0 \\ 0 \end{pmatrix} \\ 0 & \text{if } \begin{pmatrix} x \\ y \end{pmatrix} = \begin{pmatrix} 0 \\ 0 \end{pmatrix}. \end{cases}$$

(a) Compute D_1f and D_2f. Is f of class C^1?

(b) Show that all second partial derivatives of f exist everywhere.

(c) Show that

$$D_1\left(D_2f\begin{pmatrix} 0 \\ 0 \end{pmatrix}\right) \neq D_2\left(D_1f\begin{pmatrix} 0 \\ 0 \end{pmatrix}\right).$$

(d) Why doesn't this contradict Theorem 3.3.9?

3.3.4   True or false? Suppose f is a function on $\mathbb{R}^2$ that satisfies Laplace's equation $D_1^2f + D_2^2f = 0$. Then the function

$$g\begin{pmatrix} x \\ y \end{pmatrix} = f\begin{pmatrix} x/(x^2+y^2) \\ y/(x^2+y^2) \end{pmatrix} \quad \text{also satisfies Laplace's equation.}$$

3.3.5   If $f\begin{pmatrix} x \\ y \end{pmatrix} = \varphi(x-y)$ for some twice continuously differentiable function $\varphi : \mathbb{R} \to \mathbb{R}$, show that $D_1^2f - D_2^2f = 0$.

3.3.6   (a) Write out the polynomial

$$\sum_{m=0}^{5} \sum_{I \in \mathcal{I}_3^m} a_I \mathbf{x}^I, \quad \text{where}$$

$$a_{(0,0,0)} = 4, \qquad a_{(0,1,0)} = 3, \qquad a_{(1,0,2)} = 4, \qquad a_{(1,1,2)} = 2,$$

$$a_{(2,2,0)} = 1, \qquad a_{(3,0,2)} = 2, \qquad a_{(5,0,0)} = 3,$$

and all other $a_I = 0$, for $I \in \mathcal{I}_3^m$ for $m \leq 5$.

(b) Use multi-exponent notation to write the polynomial

$$2x_2 + x_1x_2 - x_1x_2x_3 + x_1^2 + 5x_2^2x_3.$$

(c) Use multi-exponent notation to write the polynomial

$$3x_1x_2 - x_2x_3x_4 + 2x_2^2x_3 + x_2^2x_4^4 + x_2^5.$$

The object of this exercise is to illustrate how long successive derivatives become.

3.3.7   (a) Compute the derivatives of $(1 + f(x))^m$, up to and including the fourth derivative.

(b) Guess how many terms the fifth derivative will have.

**(c) Guess how many terms the nth derivative will have.

3.3.8   Prove Theorem 3.3.1. Hint: Compute

$$\lim_{h \to 0} \frac{f(a + h) - \left(f(a) + f'(a)h + \cdots + \frac{f^{(k)}(a)}{k!}h^k\right)}{h^k}$$

by differentiating, k times, the top and bottom with respect to h, and checking each time that the hypotheses of l'Hôpital's rule are satisfied.

3.3.9   (a) Redo Example 3.3.16, finding the Taylor polynomial of degree 3.
(b) Repeat, for degree 4.

3.3.10   Following the format of Example 3.3.16, write the terms of the Taylor polynomial of degree 2, of a function f with three variables, at $\mathbf{a}$.

3.3.11   Find the Taylor polynomial of degree 3 of the function

$$f \begin{pmatrix} x \\ y \\ z \end{pmatrix} = \sin(x + y + z) \quad \text{at the point} \quad \begin{pmatrix} \pi/6 \\ \pi/4 \\ \pi/3 \end{pmatrix}.$$

3.3.12   Find the Taylor polynomial of degree 2 of the function

$$f \begin{pmatrix} x \\ y \end{pmatrix} = \sqrt{x + y + xy} \quad \text{at the point} \quad \begin{pmatrix} -2 \\ -3 \end{pmatrix}.$$

3.3.13   Let $f(x) = e^x$, so that $f(1) = e$. Use Corollary A9.3 (a bound for the remainder of a Taylor polynomial in one dimension) to show that

Exercise 3.3.13 uses Taylor's theorem with remainder in one dimension, Theorem A9.1, stated and proved in Appendix A9.

$$e = \sum_{i=0}^{k} \frac{1}{i!} + r_{k+1}, \quad \text{where} \quad |r_{k+1}| \leq \frac{3}{(k + 1)!}.$$

(b) Prove that e is irrational: if $e = a/b$ for some integers a and b, deduce from part (a) that

$$|k!a - bm| \leq \frac{3b}{k + 1}, \quad \text{where } m \text{ is the integer } \frac{k!}{0!} + \frac{k!}{1!} + \frac{k!}{2!} + \cdots + \frac{k!}{k!}.$$

Conclude that if k is large enough, then $k!a - bm$ is an integer that is arbitrarily small, and therefore 0.

(c) Finally, observe that k does not divide m evenly, since it does divide every summand but the last one. Since k may be freely chosen, provided only that it is sufficiently large, take k to be a prime number larger than b. Then in $k!a = bm$ we have that k divides the left side, but does not divide m. What conclusion do you reach?

3.3.14 Let f be the function

$$f\left(\begin{matrix} x \\ y \end{matrix}\right) = \mathrm{sgn}(y)\sqrt{\frac{-x + \sqrt{x^2 + y^2}}{2}}$$

where $\mathrm{sgn}(y)$ is the sign of y, i.e., $+1$ when $y > 0$, 0 when $y = 0$ and -1 when $y < 0$.

(a) Show that f is continuously differentiable on the complement of the half-line $y = 0, x \leq 0$.

Note: Part (a) is almost obvious except when $y = 0, x > 0$, where y changes sign. It may help to show that this mapping can be written $(r, \theta) \mapsto \sqrt{r}\sin(\theta/2)$ in polar coordinates.

(b) Show that if $\mathbf{a} = \left(\begin{matrix} -1 \\ -\epsilon \end{matrix}\right)$ and $\vec{\mathbf{h}} = \begin{bmatrix} 0 \\ 2\epsilon \end{bmatrix}$, then although both $\mathbf{a}$ and $\mathbf{a} + \vec{\mathbf{h}}$ are in the domain of definition of f, Taylor's theorem with remainder (Theorem A9.5) is not true.

(c) What part of the statement is violated? Where does the proof fail?

3.3.15 Show that if $I \in \mathcal{I}_n^m$, then $(x\vec{\mathbf{h}})^I = x^m \vec{\mathbf{h}}^I$.

*3.3.16 A homogeneous polynomial in two variables of degree four is an expression of the form

A homogeneous polynomial is a polynomial in which all terms have the same degree.

$$p(x, y) = ax^4 + bx^3y + cx^2y^2 + dxy^3 + ey^4.$$

Consider the function

$$f\left(\begin{matrix} x \\ y \end{matrix}\right) = \begin{cases} \dfrac{p(x, y)}{x^2 + y^2} & \text{if } \left(\begin{matrix} x \\ y \end{matrix}\right) \neq \left(\begin{matrix} 0 \\ 0 \end{matrix}\right) \\ 0 & \text{if } \left(\begin{matrix} x \\ y \end{matrix}\right) = \left(\begin{matrix} 0 \\ 0 \end{matrix}\right), \end{cases}$$

where p is a homogeneous polynomial of degree 4. What condition must the coefficients of p satisfy in order for the crossed partials $D_1(D_2(f))$ and $D_2(D_1(f))$ to be equal at the origin?

Exercises for Section 3.4:

Rules for Computing Taylor Polynomials

Hint for Exercise 3.4.2, part (a): It is easier to substitute $x + y^2$ in the Taylor polynomial for $\sin u$ than to compute the partial derivatives. Hint for part (b): Same as above, except that you should use the Taylor polynomial of $1/(1 + u)$.

3.4.1 Prove the formulas of Proposition 3.4.2.

3.4.2 (a) What is the Taylor polynomial of degree 3 of $\sin(x + y^2)$ at the origin?

(b) What is the Taylor polynomial of degree 4 of $1/(1 + x^2 + y^2)$ at the origin?

3.4.3 Write, to degree 2, the Taylor polynomial of

$$f\left(\begin{matrix} x \\ y \end{matrix}\right) = \sqrt{1 + \sin(x + y)} \quad \text{at the origin.}$$

3.4.4   Write, to degree 3, the Taylor polynomial $P_{f,0}^3$ of

$$f \begin{pmatrix} x \\ y \end{pmatrix} = \cos\left(1 + \sin(x^2 + y)\right) \quad \text{at the origin.}$$

Hint for Exercise 3.4.5, part (a): this is easier if you use

$$\sin(\alpha + \beta)$$
$$= \sin\alpha\cos\beta + \cos\alpha\sin\beta.$$

Part (b) should be in with exercises for Section 3.6, where critical points are discussed.

3.4.5   (a) What is the Taylor polynomial of degree 2 of the function $f \begin{pmatrix} x \\ y \end{pmatrix} = \sin(2x + y)$ at the point $\begin{pmatrix} \pi/6 \\ \pi/3 \end{pmatrix}$?

(b) Show that $f \begin{pmatrix} x \\ y \end{pmatrix} + \dfrac{1}{2}\left(2x + y - \dfrac{2\pi}{3}\right) - \left(x - \dfrac{\pi}{6}\right)^2$ has a critical point at $\begin{pmatrix} \pi/6 \\ \pi/3 \end{pmatrix}$. What kind of critical point is it?

Exercises for Section 3.5:

Quadratic Forms

3.5.1   Let V be a vector space. A *symmetric bilinear function* on V is a mapping $B : V \times V \to \mathbb{R}$ such that

(1) $B(a\mathbf{v}_1 + b\mathbf{v}_2, \mathbf{w}) = aB(\mathbf{v}_1, \mathbf{w}) + bB(\mathbf{v}_2, \mathbf{w})$ for all $\mathbf{v}_1, \mathbf{v}_2, \mathbf{w} \in V$ and $a, b \in \mathbb{R}$;

(2) $B(\mathbf{v}, \mathbf{w}) = B(\mathbf{w}, \mathbf{v})$ for all $\mathbf{v}, \mathbf{w} \in V$.

(a) Show that if A is a symmetric $n \times n$ matrix, the mapping $B_A(\mathbf{v}, \mathbf{w}) = \mathbf{v}^\top A\mathbf{w}$ is a symmetric bilinear function.

(b) Show that every symmetric bilinear function on $\mathbb{R}^n$ is of the form B_A for a unique symmetric matrix A.

(c) Let P_k be the space of polynomials of degree at most k. Show that the function $B : P_k \times P_k \to \mathbb{R}$ given by $B(p, q) = \int_0^1 p(t)q(t)\, dt$ is a symmetric bilinear function.

(d) Denote by $p_1(t) = 1, p_2(t) = t, \ldots, p_{k+1}(t) = t^k$ the usual basis of P_k, and by Φ_p the corresponding "concrete to abstract" linear transformation. Show that $B\big(\Phi_p(\mathbf{a}), \Phi_p(\mathbf{b})\big)$ is a symmetric bilinear function on $\mathbb{R}^n$, and find its matrix.

3.5.2   If B is a symmetric bilinear function, denote by $Q_B : V \to \mathbb{R}$ the function $Q_B(\mathbf{v}) = B(\mathbf{v}, \mathbf{v})$. Show that every quadratic form on $\mathbb{R}^n$ is of the form Q_B for some bilinear function B.

3.5.3   Show that

$$Q(p) = \int_0^1 \big(p(t)\big)^2 \, dt$$

Exercise 3.5.4: by "represents the quadratic form" we mean that Q can be written as $\vec{\mathbf{x}} \cdot A\vec{\mathbf{x}}$ (see Proposition 3.7.11).

(see Example 3.5.2) is a quadratic form if p is a cubic polynomial, i.e., if $p(t) = a_0 + a_1 t + a_2 t^2 + a_3 t^3$.

3.5.4   Confirm that the symmetric matrix $A = \begin{bmatrix} 1 & 0 & 1/2 \\ 0 & 0 & -1/2 \\ 1/2 & -1/2 & -1 \end{bmatrix}$ represents the quadratic form $Q = x^2 + xz - yz - z^2$.

3.5.5   (a) Let P_k be the space of polynomials of degree at most k. Show that the function

$$Q(p) = \int_0^1 (p(t))^2 - (p'(t))^2\, dt \quad \text{is a quadratic form on } P_k.$$

(b) What is the signature of Q when $k = 2$?

3.5.6   Let P_k be the space of polynomials of degree at most k.

(a) Show that the function $\delta_a : P_k \to \mathbb{R}$ given by $\delta_a(p) = p(a)$ is a linear function.

(b) Show that $\delta_0, \ldots, \delta_k$ are linearly independent. First say what it means, being careful with the quantifiers. It may help to think of the polynomial

$$x(x-1)\ldots(x-(j-1))(x-(j+1))\ldots(x-k),$$

which vanishes at $0, 1, \ldots, j-1, j+1, \ldots, k$ but not at j.

(c) Show that the function

$$Q(p) = (p(0))^2 - (p(1))^2 + \cdots + (-1)^k (p(k))^2$$

is a quadratic form on P_k. When $k = 3$, write it in terms of the coefficients of $p(x) = ax^3 + bx^2 + cx + d$.

(d) What is the signature of Q when $k = 3$? There is the smart way, and then there is the plodding way $\ldots$.

3.5.7   For the quadratic form of Example 3.5.6,

$$Q(\vec{x}) = x^2 + 2xy - 4xz + 2yz - 4z^2,$$

(a) What decomposition into a sum of squares do you find if you start by eliminating the z terms, then the y terms, and finally the x terms?

(b) Complete the square starting with the x terms, then the y terms, and finally the z terms.

3.5.8   Consider the quadratic form of Example 3.5.7:

$$Q(\vec{x}) = xy - xz + yz.$$

(a) Verify that the decomposition

$$(x/2 + y/2)^2 - (x/2 - y/2 + z)^2 + z^2$$

is indeed composed of linearly independent functions.

(b) Decompose $Q(\vec{x})$ with a different choice of u, to support the statement that $u = x - y$ was not a magical choice.

3.5.9   Are the following quadratic forms degenerate or nondegenerate?

(a) $x^2 + 4xy + 4y^2$ on $\mathbb{R}^2$.

(b) $x^2 + 2xy + 2y^2 + 2yz + z^2$ on $\mathbb{R}^3$.

(c) $2x^2 + 2y^2 + z^2 + w^2 + 4xy + 2xz - 2xw - 2yw$ on $\mathbb{R}^4$.

3.5.10   Decompose each of the following quadratic forms by completing squares, and determine its signature.

(a) $x^2 + xy - 3y^2$ (b) $x^2 + 2xy - y^2$ (c) $x^2 + xy + yz$ (d) $xy + yz$

3.5.11   What is the signature of the following quadratic forms?

(a) $x^2 + xy$ on $\mathbb{R}^2$ (b) $xy + yz$ on $\mathbb{R}^3$

(c) $\det \begin{bmatrix} a & b \\ c & d \end{bmatrix}$ on $\mathbb{R}^4$ *(d) $x_1x_2 + x_2x_3 + \cdots + x_{n-1}x_n$ on $\mathbb{R}^n$

3.5.12   On $\mathbb{R}^4$ as described by $M = \begin{bmatrix} a & c \\ b & d \end{bmatrix}$, consider the quadratic form $Q(M) = \det M$. What is its signature?

3.5.13   Identify $\begin{pmatrix} a \\ b \\ d \end{pmatrix} \in \mathbb{R}^3$ with the upper triangular matrix $M = \begin{bmatrix} a & b \\ 0 & d \end{bmatrix}$.

(a) What kind of surface in $\mathbb{R}^3$ do you get by setting $\text{tr}(M^2) = 1$?

(b) What kind of surface in $\mathbb{R}^3$ do you get by setting $\text{tr}(MM^T) = 1$?

Hint for Exercise 3.5.14: The main point is to prove that if the quadratic form Q has signature $(k,0)$ with $k < n$, there is a vector $\vec{v} \neq \vec{0}$ such that $Q(\vec{v}) = 0$. You can find such a vector using the transformation T of Equation 3.5.26.

3.5.14   Show that a quadratic form on $\mathbb{R}^n$ is positive definite if and only if its signature is $(n,0)$.

3.5.15   Here is an alternative proof of Proposition 3.5.14. Let $Q : \mathbb{R}^n \to \mathbb{R}$ be a positive definite quadratic form. Show that there exists a constant $C > 0$ such that

$$Q(\vec{x}) \geq C|\vec{x}|^2 \qquad\qquad 3.5.30$$

for all $\vec{x} \in \mathbb{R}^n$, as follows.

(a) Let $S^{n-1} = \{\vec{x} \in \mathbb{R}^n \,|\, |\vec{x}| = 1\}$. Show that S^{n-1} is compact, so there exists $\vec{x}_0 \in S^{n-1}$ with $Q(\vec{x}_0) \leq Q(\vec{x})$ for all $\vec{x} \in S^{n-1}$.

(b) Show that $Q(\vec{x}_0) > 0$.

(c) Use the formula $Q(\vec{x}) = |\vec{x}|^2 Q(\vec{x}/|\vec{x}|)$ to prove Proposition 3.5.14.

Exercise 3.5.16:   See margin note for Exercise 3.5.4.

3.5.16   Show that a 2×2 symmetric matrix $G = \begin{bmatrix} a & b \\ b & d \end{bmatrix}$ represents a positive definite quadratic form if and only if $\det G > 0$, $a + d > 0$.

3.5.17   Consider the vector space of Hermitian 2×2 matrices:
$H = \begin{bmatrix} a & b+ic \\ b-ic & d \end{bmatrix}$, where a, b, c, d are real numbers. What is the signature of the quadratic form $Q(H) = \det H$?

3.5.18   Identify and sketch the conic sections and quadratic surfaces of equation $Q(\mathbf{x}) = 1$ when $Q(\mathbf{x})$ is a quadratic form defined by one of the following matrices:

(a) $\begin{bmatrix} 2 & 1 \\ 1 & 3 \end{bmatrix}$ (b) $\begin{bmatrix} 2 & 1 & 0 \\ 1 & 2 & 1 \\ 0 & 1 & 2 \end{bmatrix}$ (c) $\begin{bmatrix} 2 & 0 & 3 \\ 0 & 0 & 0 \\ 3 & 0 & -1 \end{bmatrix}$

(d) $\begin{bmatrix} -1 & 0 \\ 1 & 4 \end{bmatrix}$ (e) $\begin{bmatrix} 2 & 4 & -3 \\ 4 & 1 & 3 \\ -3 & 3 & -1 \end{bmatrix}$ (f) $\begin{bmatrix} 1 & 2 \\ 2 & 4 \end{bmatrix}$

3.5.19   For each of the following equations, determine the signature of the quadratic form represented by the left-hand side. Where possible, sketch the curve or surface represented by the equation.

(a) $x^2 + xy - y^2 = 1$ (b) $x^2 + 2xy - y^2 = 1$
(c) $x^2 + xy + yz = 1$ (d) $xy + yz = 1$

Exercises for Section 3.6: Classifying Critical Points

3.6.1   (a) Show that the function $f\begin{pmatrix} x \\ y \\ z \end{pmatrix} = x^2 + xy + z^2 - \cos y$ has a critical point at the origin.

(b) What kind of critical point does it have?

3.6.2   Find all the critical points of the following functions:
(a) $\sin x \cos y$ (b) $2x^3 - 24xy + 16y^3$
(c) $xy + \frac{8}{x} + \frac{1}{y}$ *(d) $\sin x + \sin y + \sin(x + y)$

For each function, find the second degree approximation at the critical points, and classify the critical point.

3.6.3   Complete the proof of Theorem 9.8.5 (behavior of functions near saddle points), showing that if f has a saddle at $\mathbf{a} \in U$, then in every neighborhood of $\mathbf{a}$ there are points $\mathbf{c}$ with $f(\mathbf{c}) < f(\mathbf{a})$.

3.6.4   (a) Find the critical points of the function $f\begin{pmatrix} x \\ y \end{pmatrix} = x^3 - 12xy + 8y^3$.

(b) Determine the nature of each of the critical points.

3.6.5   Use Newton's method (preferably by computer) to find the critical points of $-x^3 + y^3 + xy + 4x - 5y$. Classify them, still using the computer.

3.6.6   (a) Find the critical points of the function $f\begin{pmatrix} x \\ y \\ z \end{pmatrix} = xy + yz - xz + xyz$.

(b) Determine the nature of each of the critical points.

3.6.7   (a) Find the critical points of the function $f\begin{pmatrix} x \\ y \end{pmatrix} = 3x^2 - 6xy + 2y^3$.
(b) What kind of critical points are these?

3.7.1   Show that the mapping

$$\mathbf{g} : \begin{pmatrix} u \\ v \end{pmatrix} \mapsto \begin{pmatrix} \sin uv + u \\ u + v \\ uv \end{pmatrix}$$

is a parametrization of a smooth surface.

(a) Show that the image of $\mathbf{g}$ is contained in the locus S of equation

$$z = (x - \sin z)(\sin z - x + y).$$

(b) Show that S is a smooth surface.

(c) Show that g maps $\mathbb{R}^2$ onto S.

(d) Show that $\mathbf{g}$ is one to one, and that $[\mathbf{Dg}(\begin{smallmatrix} u \\ v \end{smallmatrix})]$ is one to one for every $\begin{pmatrix} u \\ v \end{pmatrix} \in \mathbb{R}^2$.

3.7.2   (a) Show that the function $\varphi \begin{pmatrix} x \\ y \\ z \end{pmatrix} = x + y + z$ constrained to the surface Y of equation $x = \sin z$ has no critical point.

(b) Explain geometrically why this is so.

Hint for Exercise 3.7.2, part (b): The tangent plane to Y at any point is always parallel to the y-axis.

3.7.3   (a) Show that the function xyz has four critical points on the plane of equation

$$\varphi \begin{pmatrix} x \\ y \\ z \end{pmatrix} = ax + by + cz - 1 = 0$$

when $a, b, c > 0$. (Use the equation of the plane to write z in terms of x and y; i.e., parametrize the plane by x and y.)

(b) Show that of these four critical points, three are saddles and one is a maximum.

3.7.4   Let $Q(\mathbf{x})$ be a quadratic form. Construct a symmetric matrix A as follows: each entry $A_{i,i}$ on the diagonal is the coefficient of x_i^2, while each entry $A_{i,j}$ is one-half the coefficient of the term $x_i x_j$.

(a) Show that $Q(\mathbf{x}) = \mathbf{x} \cdot A\mathbf{x}$.

(b) Show that A is the unique symmetric matrix with this property. Hint: consider $Q(\mathbf{e}_i)$, and $Q(a\mathbf{e}_i + b\mathbf{e}_j)$.

3.7.5   Justify Equation 3.7.32, using the definition of the derivative and the fact that A is symmetric.

3.7.6   Let A be any matrix (not necessarily square).

Part (c) of Exercise 3.7.6 uses the norm $\|A\|$ of a matrix A. The norm is defined (Definition 2.8.5) in an optional subsection of Section 2.8.

(a) Show that $A^\top A$ is symmetric.

(b) Show that all eigenvalues λ of $A^\top A$ are non-negative, and that they are all positive if and only if the kernel of A is $\{0\}$.

(c) Show that $\|A\| = \sup_{\lambda \text{ eigenvalue of } A^\top A} \sqrt{\lambda}$.

3.7.7   (a) Find the critical points of the function $x^3 + y^3 + z^3$ on the intersection of the planes of equation $x + y + z = 2$ and $x + y - z = 3$.

(b) Are the critical points maxima, minima, or neither?

3.7.8   Find all the critical points of the function

$$f\begin{pmatrix} x \\ y \\ z \end{pmatrix} = 2xy + 2yz - 2x^2 - 2y^2 - 2z^2 \quad \text{on the unit sphere of } \mathbb{R}^3.$$

3.7.9   What is the volume of the largest rectangular parallelepiped contained in the ellipsoid $x^2 + 4y^2 + 9z^2 \leq 9$?

3.7.10   Let A, B, C, D be a convex quadrilateral in the plane, with the vertices free to move but with a the length of AB, b the length of BC, c the length of CD and d the length of DA all assigned. Let φ be the angle at A and ψ be the angle at C.

(a) Show that the angles φ and ψ satisfy the constraint

$$a^2 + d^2 - 2d\cos\varphi = b^2 + c^2 - 2bc\cos\psi.$$

(b) Find a formula for the area of the quadrilateral in terms of φ, ψ and a, b, c, d.

(c) Show that the area is maximum if the quadrilateral can be inscribed in a circle. You may use the fact that a quadrilateral can be inscribed in a circle if the opposite angles add to π.

3.7.11   This exercise was a duplication of Exercise 3.7.7.

3.7.12   What is the maximum volume of a box of surface area 10, for which one side is exactly twice as long as another?

3.7.13   (a) Find the critical points of xyz, if x, y, z belong to the surface of equation $x + y + z^2 = 16$.

(b) Is there a maximum on the whole surface, and if so, which critical point is it?

Exercise 3.7.14 should be in Chapter 4, after multiple integrals are discussed in Section 4.5.

(c) Is there a maximum on the part of the surface where x, y, z are all positive?

3.7.14   (a) If $f\begin{pmatrix} x \\ y \end{pmatrix} = a + bx + cy$, what are

$$\int_0^1 \int_0^2 f\begin{pmatrix} x \\ y \end{pmatrix} |dx\,dy| \quad \text{and} \quad \int_0^1 \int_0^2 \left(f\begin{pmatrix} x \\ y \end{pmatrix} \right)^2 |dx\,dy|?$$

(b) Let f be as above. What is the minimum of $\int_0^1 \int_0^2 \left(f\left(\begin{smallmatrix} x \\ y \end{smallmatrix}\right)\right)^2 |dx\,dy|$ among all functions f such that

$$\int_0^1 \int_0^2 f\left(\begin{matrix} x \\ y \end{matrix}\right) |dx\,dy| = 1?$$

3.7.15   (a) Show that the set $X \subset \mathrm{Mat}\,(2,2)$ of matrices with determinant 1 is a smooth submanifold. What is its dimension?

(b) Find a matrix in X which is closest to the matrix $\begin{bmatrix} 0 & 1 \\ 1 & 0 \end{bmatrix}$.

**3.7.16   Let D be the closed domain bounded by the line of equation $x+y = 0$ and the circle of equation $x^2 + y^2 = 1$, whose points satisfy $x \geq -y$, as shaded in Figure 3.7.16.

(a) Find the maximum and minimum of the function $f\left(\begin{matrix} x \\ y \end{matrix}\right) = xy$ on D.

(b) Try it again with $f\left(\begin{matrix} x \\ y \end{matrix}\right) = x + 5xy$.
constraint

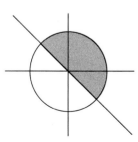

FIGURE 3.7.16.

Exercises for Section 3.8:

Geometry of Curves and Surfaces

Useful fact for Exercise 3.8.1

The arctic circle is those points that are $2\,607.5$ kilometers south of the north pole.

3.8.1   (a) How long is the arctic circle? How long would a circle of that radius be if the earth were flat?

(b) How big a circle around the pole would you need to measure in order for the difference of its length and the corresponding length in a plane to be one kilometer?

3.8.2   Suppose $\gamma(t) = \begin{pmatrix} \gamma_1(t) \\ \vdots \\ \gamma_n(t) \end{pmatrix}$ is twice continuously differentiable on a neighborhood of $[a, b]$.

(a) Use Taylor's theorem with remainder (or argue directly from the mean value theorem) to show that for any $s_1 < s_2$ in $[a, b]$, we have

$$|\gamma(s_2) - \gamma(s_1) - \gamma'(s_1)(s_2 - s_1)| \leq C|s_2 - s_1|^2, \qquad \text{where}$$

$$C = \sqrt{n} \sup_{j=1,\ldots,n} \sup_{t \in [a,b]} |\gamma_j''(t)|.$$

(b) Use this to show that

$$\lim \sum_{i=0}^{m-1} |\gamma(t_{i+1} - \gamma(t_i)| = \int_a^b |\gamma'(t)|\, dt,$$

where $a = t_0 < t_1 \cdots < t_m = b$, and we take the limit as the distances $t_{i+1} - t_i$ tend to 0.

3.8.3   Check that if you consider the *surface* of equation $z = f(x)$, y arbitrary, and the plane *curve* $z = f(x)$, the mean curvature of the surface is half the curvature of the plane curve.

3.8.4   (a) Show that the equation $y \cos z = x \sin z$ expresses z implicitly as a function $z = g_r \begin{pmatrix} x \\ y \end{pmatrix}$ near the point $\begin{pmatrix} r \\ 0 \\ 0 \end{pmatrix}$.

(b) Show that $D_1 g_r \begin{pmatrix} r \\ 0 \end{pmatrix} = D_1^2 g_r \begin{pmatrix} r \\ 0 \end{pmatrix} = 0$. Hint: The x-axis is contained in the surface.

3.8.5   Compute the Gaussian curvature and the mean curvature of the surface of equation $z = \sqrt{x^2 + y^2}$ at $\begin{pmatrix} x \\ y \\ z \end{pmatrix} = \begin{pmatrix} a \\ b \\ \sqrt{a^2 + b^2} \end{pmatrix}$. Explain your result.

3.8.6   (a) Draw the cycloid, given parametrically by

$$\begin{pmatrix} x \\ y \end{pmatrix} = \begin{pmatrix} a(t - \sin t) \\ a(1 - \cos t) \end{pmatrix}.$$

(b) Can you relate the name "cycloid" to "bicycle"?

(c) Find the length of one arc of the cycloid.

3.8.7   Do the same for the hypocycloid

$$\begin{pmatrix} x \\ y \end{pmatrix} = \begin{pmatrix} a \cos^3 t \\ a \sin^3 t \end{pmatrix}.$$

3.8.8   (a) Let $f : [a, b] \to \mathbb{R}$ be a smooth function satisfying $f(x) > 0$, and consider the surface obtained by rotating its graph around the x-axis. Show that the Gaussian curvature K and the mean curvature H of this surface depend only on the x-coordinate.

(b) Show that

$$K(x) = \frac{-f''(x)}{f(x)\left(1 + (f'(x))^2\right)^2}.$$

(c) Find a formula for the mean curvature in terms of f and its derivatives.

Hint for Exercise *3.8.9: The curve

$$F : t \mapsto \left[\vec{t}(t), \vec{n}(t), \vec{b}(t)\right] = T(t)$$

is a mapping $I \mapsto SO(3)$, so $t \mapsto T^{-1}(t_0)T(t)$ is a curve in $SO(3)$ passing through the identity at t_0.

*3.8.9   Use Exercise *3.2.8 to explain why the Frenet formulas give an antisymmetric matrix.

*3.8.10   Using the notation and the computations in the proof of Proposition 3.8.7, show that the mean curvature is given by the formula

$$H = \frac{1}{2(1 + c^2)^{3/2}}\left(a_{2,0}(1 + a_2^2) - 2a_1 a_2 a_{1,1} + a_{0,2}(1 + a_1^2)\right). \qquad 3.8.34$$

4

Integration

When you can measure what you are speaking about, and express it in numbers, you know something about it; but when you cannot measure it, when you cannot express it in numbers, your knowledge is of a meager and unsatisfactory kind: it may be the beginning of knowledge, but you have scarcely, in your thoughts, advanced to the stage of science.—
William Thomson, Lord Kelvin

4.0 INTRODUCTION

Chapters 1 and 2 began with algebra, then moved on to calculus. Here, as in Chapter 3, we dive right into calculus. We introduce the relevant linear algebra (determinants) later in the chapter, where we need it.

When students first meet integrals, integrals come in two very different flavors: Riemann sums (the idea) and anti-derivatives (the recipe), rather as derivatives arise as limits, and as something to be computed using Leibnitz's rule, the chain rule, etc.

Since integrals can be systematically computed (by hand) only as anti-derivatives, students often take this to be the definition. This is misleading: the definition of an integral is given by a Riemann sum (or by "area under the graph"; Riemann sums are just a way of making the notion of "area" precise). Section 4.1 is devoted to generalizing Riemann sums to functions of several variables. Rather than slice up the domain of a function $f : \mathbb{R} \to \mathbb{R}$ into little intervals and computing the "area under the graph" corresponding to each interval, we will slice up the "n-dimensional domain" of a function $f : \mathbb{R}^n \to \mathbb{R}$ into little n-dimensional cubes.

An actuary deciding what premium to charge for a life insurance policy needs integrals. So does a banker deciding what to charge for stock options. Black and Scholes received a Nobel prize for this work, which involves a very fancy stochastic integral.

Computing n-dimensional volume is an important application of multiple integrals. Another is probability theory; in fact probability has become such an important part of integration that integration has almost become a part of probability. Even such a mundane problem as quantifying how heavy a child is for his or her height requires multiple integrals. Fancier yet are the uses of probability that arise when physicists study turbulent flows, or engineers try to improve the internal combustion engine. They cannot hope to deal with one molecule at a time; any picture they get of reality at a macroscopic level is necessarily based on a probabilistic picture of what is going on at a microscopic level. We give a brief introduction to this important field in Section 4.2.

351

Section 4.3 discusses what functions are integrable; in the optional Section 4.4, we use the notion of *measure* to give a sharper criterion for integrability (a criterion that applies to more functions than the criteria of Section 4.3).

In Section 4.5 we discuss Fubini's theorem, which reduces computing the integral of a function of n variables to computing n ordinary integrals. This is an important theoretical tool. Moreover, whenever an integral can be computed in elementary terms, Fubini's theorem is the key tool. Unfortunately, it is usually impossible to compute anti-derivatives in elementary terms even for functions of one variable, and this tends to be truer yet of functions of several variables.

In practice, multiple integrals are most often computed using numerical methods, which we discuss in Section 4.6. We will see that although the theory is much the same in $\mathbb{R}^2$ or $\mathbb{R}^{10^{24}}$, the computational issues are quite different. We will encounter some entertaining uses of Newton's method when looking for optimal points at which to evaluate a function, and some fairly deep probability in understanding why Monte Carlo methods work in higher dimensions.

Defining volume using dyadic pavings, as we do in Section 4.1, makes most theorems easiest to prove, but such pavings are rigid; often we will want to have more "paving stones" where the function varies rapidly, and bigger ones elsewhere. Having some flexibility in choosing pavings is also important for the proof of the *change of variables formula*. Section 4.7 discusses more general pavings.

In Section 4.8 we return to linear algebra to discuss higher-dimensional determinants. In Section 4.9 we show that in all dimensions the determinant measures volumes; we use this fact in Section 4.10, where we discuss the change of variables formula.

Many of the most interesting integrals, such as those in Laplace and Fourier transforms, are not integrals of bounded functions over bounded domains. We will discuss these *improper integrals* in Section 4.11. Such integrals cannot be defined as Riemann sums, and require understanding the behavior of integrals under limits. The dominated convergence theorem is the key tool for this.

4.1 Defining the Integral

Integration is a summation procedure; it answers the question: how much is there in all? In one dimension, $\rho(x)$ might be the density at point x of a bar parametrized by $[a, b]$; in that case

The Greek letter ρ, or "rho," is pronounced "row."

$$\int_a^b \rho(x)\, dx \qquad\qquad 4.1.1$$

is the total mass of the bar.

If instead we have a rectangular plate parametrized by $a \le x \le b$, $c \le y \le d$, and with density $\rho\begin{pmatrix} x \\ y \end{pmatrix}$, then the total mass will be given by the *double integral*

$$\int\int_{[a,b]\times[c,d]} \rho\begin{pmatrix} x \\ y \end{pmatrix} dx\, dy, \qquad\qquad 4.1.2$$

where $[a, b] \times [c, d]$, i.e., the plate, is the domain of the entire double integral $\int \int$.

We will see in Section 4.5 that the double integral of Equation 4.1.2 can be written

$$\int_c^d \left(\int_a^b \rho \begin{pmatrix} x \\ y \end{pmatrix} dx \right) dy.$$

We are not presupposing this equivalence in this section. One difference worth noting is that $\int_a^b$ specifies a direction: from a to b. (You will recall that direction makes a difference: $\int_a^b = -\int_b^a$.) Equation 4.1.2 specifies a domain, but says nothing about direction.

We will define such multiple integrals in this chapter. But you should always remember that the example above is too simple. We might want to understand the total rainfall in Britain, whose coastline is a very complicated boundary. (A celebrated article analyzes that coastline as a fractal, with infinite length.) Or we might want to understand the total potential energy stored in the surface tension of a foam; physics tells us that a foam assumes the shape that minimizes this energy.

Thus we want to define integration for rather bizarre domains and functions. Our approach will not work for truly bizarre functions, such as the function that equals 1 at all rational numbers and 0 at all irrational numbers; for that one needs Lebesgue integration, not treated in this book. But we still have to specify carefully what domains and functions we want to allow.

Our task will be somewhat easier if we keep the domain of integration simple, putting all the complication into the function to be integrated. If we wanted to sum rainfall over Britain, we would use $\mathbb{R}^2$, *not* Britain (with its fractal coastline!) as the domain of integration; we would then define our function to be rainfall over Britain, and 0 elsewhere.

Thus, for a function $f : \mathbb{R}^n \to \mathbb{R}$, we will define the multiple integral

$$\int_{\mathbb{R}^n} f(\mathbf{x}) \, |d^n\mathbf{x}|, \qquad 4.1.3$$

with $\mathbb{R}^n$ the domain of integration.

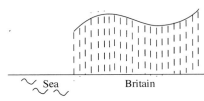

FIGURE 4.1.1.

The function that is rainfall over Britain and 0 elsewhere is discontinuous at the coast.

We emphatically do *not* want to assume that f is continuous, because most often it is not: if for example f is defined to be total rainfall for October over Britain, and 0 elsewhere, it will be discontinuous over most of the border of Britain, as shown in Figure 4.1.1. What we actually have is a function g (e.g., rainfall) defined on some subset of $\mathbb{R}^n$ larger than Britain. We then consider that function only over Britain, by setting

$$f(\mathbf{x}) = \begin{cases} g(\mathbf{x}) & \text{if } \mathbf{x} \in \text{Britain} \\ 0 & \text{otherwise.} \end{cases} \qquad 4.1.4$$

We can express this another way, using the *characteristic function* χ.

Definition 4.1.1 (Characteristic function). For any bounded subset $A \subset \mathbb{R}^n$, the characteristic function χ_A is:

The characteristic function χ_A is pronounced "kye sub A," the symbol χ being the Greek letter chi.

$$\chi_A(\mathbf{x}) = \begin{cases} 1 & \text{if } \mathbf{x} \in A \\ 0 & \text{if } \mathbf{x} \notin A. \end{cases} \qquad 4.1.5$$

Equation 4.1.4 can then be rewritten

$$f(\mathbf{x}) = g(\mathbf{x})\chi_{\text{Britain}}(\mathbf{x}). \qquad 4.1.6$$

We tried several notations before choosing $|d^n\mathbf{x}|$. First we used $dx_1 \ldots dx_n$. That seemed clumsy, so we switched to dV. But it failed to distinguish between $|d^2\mathbf{x}|$ and $|d^3\mathbf{x}|$, and when changing variables we had to tack on subscripts to keep the variables straight.

But dV had the advantage of suggesting, correctly, that we are not concerned with direction (unlike integration in first year calculus, where $\int_a^b dx \neq \int_b^a dx$). We hesitated at first to convey the same message with absolute value signs, for fear the notation would seem forbidding, but decided that the distinction between oriented and unoriented domains is so important (it is a central theme of Chapter 6) that our notation should reflect that distinction.

The notation Supp (support) should not be confused with sup (least upper bound).

Recall that "least upper bound" and "supremum" are synonymous, as are "greatest lower bound" and "infimum" (Definitions 1.6.4 and 1.6.6).

This doesn't get rid of difficulties like the coastline of Britain—indeed, such a function f will usually have discontinuities on the coastline—but putting all the difficulties on the side of the function will make our definitions easier (or at least shorter).

So while we really want to integrate g (i.e., rainfall) over Britain, we define that integral in terms of the integral of f over $\mathbb{R}^n$, setting

$$\int_{\text{Britain}} g\,|d^n\mathbf{x}| \;=\; \int_{\mathbb{R}^n} f\,|d^n\mathbf{x}|. \qquad\qquad 4.1.7$$

More generally, when integrating over a subset $A \subset \mathbb{R}^n$,

$$\int_A g(\mathbf{x})\,|d^n\mathbf{x}| = \int_{\mathbb{R}^n} g(\mathbf{x})\chi_A(\mathbf{x})\,|d^n\mathbf{x}|. \qquad\qquad 4.1.8$$

Some preliminary definitions and notation

Before defining the Riemann integral, we need a few definitions.

Definition 4.1.2 (Support of a function: $\mathrm{Supp}(f)$). The support of a function $f : \mathbb{R}^n \to \mathbb{R}$ is

$$\mathrm{Supp}(f) = \{\mathbf{x} \in \mathbb{R}^n \mid f(\mathbf{x}) \neq 0\}. \qquad\qquad 4.1.9$$

Definition 4.1.3 ($M_A(f)$ and $m_A(f)$). If $A \subset \mathbb{R}^n$ is an arbitrary subset, we will denote by

$$
\begin{aligned}
M_A(f) &= \sup_{\mathbf{x}\in A} f(\mathbf{x}), \text{ the supremum of } f(\mathbf{x}) \text{ for } \mathbf{x} \in A \\
m_A(f) &= \inf_{\mathbf{x}\in A} f(\mathbf{x}), \text{ the infimum of } f(\mathbf{x}) \text{ for } \mathbf{x} \in A.
\end{aligned}
\qquad 4.1.10
$$

Definition 4.1.4 (Oscillation). The oscillation of f over A, denoted $\mathrm{osc}_A(f)$, is the difference between its least upper bound and greatest lower bound:

$$\mathrm{osc}_A(f) = M_A(f) - m_A(f). \qquad\qquad 4.1.11$$

Definition of the Riemann integral: dyadic pavings

In Sections 4.1–4.9 we will discuss only integrals of functions f satisfying

(1) $|f|$ is bounded, and

(2) f has bounded support, i.e., there exits R such that $f(\mathbf{x}) = 0$ when $|\mathbf{x}| > R$.

With these restrictions on f, and for any subset $A \subset \mathbb{R}^n$, each quantity $M_A(f)$, $m_A(f)$, and $\mathrm{osc}_A(f)$, is a well-defined finite number. This is not true for a function like $f(x) = 1/x$, defined on the open interval $(0, 1)$. In that case $|f|$ is not bounded, and $\sup f(x) = \infty$.

There is quite a bit of choice as to how to define the integral; we will first use the most restrictive definition: *dyadic pavings* of $\mathbb{R}^n$.

In Section 4.7 we will see that much more general pavings can be used.

To compute an integral in one dimension, we decompose the domain into little intervals, and construct on each the tallest rectangle that fits under the graph and the shortest rectangle that contains it, as shown in Figure 4.1.2.

We call our pavings *dyadic* because each time we divide by a factor of 2; "dyadic" comes from the Greek *dyas*, meaning two. We could use decimal pavings instead, cutting each side into ten parts each time, but dyadic pavings are easier to draw.

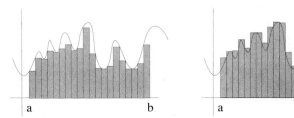

FIGURE 4.1.2. Left: Lower Riemann sum for $\int_a^b f(x)\,dx$. Right: Upper Riemann sum. If the two sums converge to a common limit, that limit is the integral of the function.

The dyadic upper and lower sums correspond to decomposing the domain first at the integers, then at the half-integers, then at the quarter-integers, etc.

If, as we make the rectangles skinnier and skinnier, the sum of the area of the upper rectangles approaches that of the lower rectangles, the function is integrable. We can then compute the integral by adding areas of rectangles— either the lower rectangles, the upper rectangles, or rectangles constructed some other way, for example by using the value of the function at the middle of each column as the height of the rectangle. The choice of the point at which to measure the height doesn't matter since the areas of the lower rectangles and the upper rectangles can be made arbitrarily close.

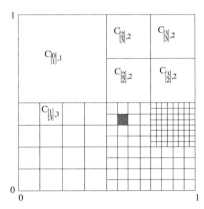

FIGURE 4.1.3.

A dyadic decomposition in $\mathbb{R}^2$. The entire figure is a "cube" in $\mathbb{R}^2$ at level $N = 0$, with side length $1/2^0 = 1$. At level 1 (upper left quadrant), cubes have side length $1/2^1 = 1/2$; at level 2 (upper right quadrant), they have side length $1/2^2 = 1/4$; and so on.

To use dyadic pavings in $\mathbb{R}^n$ we do essentially the same thing. We cut up $\mathbb{R}^n$ into cubes with sides 1 long, like the big square of Figure 4.1.3. (By "cube" we mean an interval in $\mathbb{R}$, a square in $\mathbb{R}^2$, a cube in $\mathbb{R}^3$, and analogs of cubes in higher dimensions.) Next we cut each side of a cube in half, cutting an interval in half, a square into four equal squares, a cube into eight equal cubes, ... At the next level we cut each side of those in half, and so on.

To define dyadic pavings in $\mathbb{R}^n$ precisely, we must first say what we mean by an n-dimensional "cube." For every

$$\mathbf{k} = \begin{bmatrix} k_1 \\ \vdots \\ k_n \end{bmatrix} \in \mathbb{Z}^n, \quad \text{where } \mathbb{Z} \text{ represents the integers,} \qquad 4.1.12$$

we define the cube

$$C_{\mathbf{k},N} = \left\{ \mathbf{x} \in \mathbb{R}^n \,\middle|\, \frac{k_i}{2^N} \leq x_i < \frac{k_i+1}{2^N} \text{ for } 1 \leq i \leq n \right\}. \qquad 4.1.13$$

Each cube C has two indices. The first index, $\mathbf{k}$, locates each cube: it gives the numerators of the coordinates of the cube's lower left-hand corner, when the denominator is 2^N. The second index, N, tells which "level" we are considering, starting with 0; you may think of N as the "fineness" of the cube. The length of a side of a cube is $1/2^N$, so when $N = 0$, each side of a cube is length 1; when $N = 1$, each side is length $1/2$; when $N = 2$, each side is length $1/4$. The bigger N is, the finer the decomposition and the smaller the cubes.

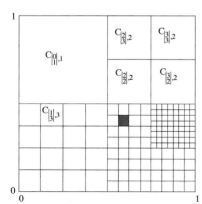

FIGURE 4.1.3.

Example 4.1.5 (Dyadic cubes). The small shaded cube in the lower right-hand quadrant of Figure 4.1.3 (repeated at left) is

$$C_{\left[\begin{smallmatrix}9\\6\end{smallmatrix}\right],4} = \left\{ \mathbf{x} \in \mathbb{R}^2 \,\middle|\, \underbrace{\frac{9}{16} \leq x < \frac{10}{16}}_{\text{width of cube}}; \underbrace{\frac{6}{16} \leq y < \frac{7}{16}}_{\text{height of cube}} \right\}. \qquad 4.1.14$$

For a three-dimensional cube, $\mathbf{k}$ has three entries, and each cube $C_{\mathbf{k},N}$ consists of the $\mathbf{x} = \begin{pmatrix} x \\ y \\ z \end{pmatrix} \in \mathbb{R}^3$ such that

In Equation 4.1.13, we chose the inequalities $\leq$ to the left of x_i and $<$ to the right so that at every level, every point of $\mathbb{R}^n$ is in exactly one cube. We could just as easily put them in the opposite order; allowing the edges to overlap wouldn't be a problem either.

$$\underbrace{\frac{k_1}{2^N} \leq x < \frac{k_1+1}{2^N}}_{\text{width of cube}}; \underbrace{\frac{k_2}{2^N} \leq y < \frac{k_2+1}{2^N}}_{\text{length of cube}}; \underbrace{\frac{k_3}{2^N} \leq z < \frac{k_3+1}{2^N}}_{\text{height of cube}}. \quad \triangle \qquad 4.1.15$$

The collection of all these cubes *paves* $\mathbb{R}^n$:

Definition 4.1.6 (Dyadic pavings). The collection of cubes $C_{\mathbf{k},N}$ at a single level N, denoted $\mathcal{D}_N(\mathbb{R}^n)$, is the Nth dyadic paving of $\mathbb{R}^n$.

We use vol_n to denote n-dimensional volume.

The n-dimensional volume of a cube C is the product of the lengths of its sides. Since the length of one side is $1/2^N$, the n-dimensional volume is

$$\mathrm{vol}_n C = \left(\frac{1}{2^N} \right)^n; \quad \text{i.e.,} \quad \mathrm{vol}_n C = \frac{1}{2^{Nn}}. \qquad 4.1.16$$

Note that all $C \in \mathcal{D}_N$ (all cubes at a given resolution) have the same n-dimensional volume.

You are asked to prove Equation 4.1.17 in Exercise 4.1.5.

The distance between two points $\mathbf{x}, \mathbf{y}$ in a cube $C \in \mathcal{D}_N$ is

$$|\mathbf{x} - \mathbf{y}| \leq \frac{\sqrt{n}}{2^N}. \qquad 4.1.17$$

Thus two points in the same cube C are close if N is large.

Upper and lower sums using dyadic pavings

With a Riemann sum in one dimension we sum the areas of the upper rectangles and the areas of the lower rectangles, and say that a function is integrable if the upper and lower sums approach a common limit as the decomposition becomes finer and finer. The common limit is the integral.

As in Definition 4.1.3, $M_C(f)$ denotes the least upper bound, and $m_C(f)$ denotes the greatest lower bound.

We will do the same thing here. We define the Nth upper and lower sums

$$\underset{\text{upper sum}}{U_N(f)} = \sum_{C \in \mathcal{D}_N} M_C(f) \operatorname{vol}_n C, \qquad \underset{\text{lower sum}}{L_N(f)} = \sum_{C \in \mathcal{D}_N} m_C(f) \operatorname{vol}_n C. \qquad 4.1.18$$

Since we are assuming that f has bounded support, these sums have only finitely many terms. Each term is finite, since f itself is bounded.

For the Nth upper sum we compute, for each cube C at level N, the product of the least upper bound of the function over the cube and the volume of the cube, and we add the products together. For the lower sum we do the same thing, using the greatest lower bound. Since for these pavings all the cubes have the same volume, it can be factored out:

$$U_N(f) = \underbrace{\frac{1}{2^{nN}}}_{\text{vol. of cube}} \sum_{C \in \mathcal{D}_N} M_C(f), \quad L_N(f) = \underbrace{\frac{1}{2^{nN}}}_{\text{vol. of cube}} \sum_{C \in \mathcal{D}_N} m_C(f). \qquad 4.1.19$$

Proposition 4.1.7. *As N increases, the sequence $U_N(f)$ decreases, and the sequence $L_N(f)$ increases.*

We invite you to turn this argument into a formal proof.

Think of a two-dimensional function, whose graph is a surface with mountains and valleys. At a coarse level, where each cube (i.e., square) covers a lot of area, a square containing both a mountain peak and a valley will contribute a lot to the upper sum; the mountain peak will be the least upper bound for the entire large square. As N increases, the peak is the least upper bound for a much smaller square; other small squares that were part of the original big square will have a much smaller least upper bound.

The same argument holds, in reverse, for the lower sum; if a large square contains a deep valley, the entire square will have a low greatest lower bound, contributing to a small lower sum. As N increases and the squares get smaller, the valley will have less of an impact, and the lower sum will increase.

We are now ready to define the multiple integral. First we will define upper and lower integrals.

Definition 4.1.8 (Upper and lower integrals). We call

$$U(f) = \lim_{N \to \infty} U_N(f) \quad \text{and} \quad L(f) = \lim_{N \to \infty} L_N(f) \qquad 4.1.20$$

the upper and lower integrals of f.

Definition 4.1.9 (Integrable function). A function $f : \mathbb{R}^n \to \mathbb{R}$ is integrable if its upper and lower integrals are equal; its integral is then denoted

$$\int_{\mathbb{R}^n} f|d^n\mathbf{x}| = U(f) = L(f). \qquad 4.1.21$$

It is rather hard to find integrals that can be computed directly from the definition; here is one.

Example 4.1.10 (Computing an integral). Let

$$f(x) = \begin{cases} x & \text{if } 0 \le x < 1 \\ 0 & \text{otherwise,} \end{cases} \qquad 4.1.22$$

which we could express (using the characteristic function) as the product

$$f(x) = x\chi_{[0,1)}(x). \qquad 4.1.23$$

First, note that f is bounded with bounded support. Unless $0 \le k/2^N < 1$, we have

$$m_{C_{k,N}}(f) = M_{C_{k,N}}(f) = 0. \qquad 4.1.24$$

Of course, we are simply computing

$$\int_0^1 f(x)|dx| = \left[\frac{x^2}{2}\right]_0^1 = \frac{1}{2}.$$

The point of this example is to show that this integral, almost the easiest that calculus provides, can be evaluated by dyadic sums.

Since we are in dimension 1, our cubes are intervals:

$$C_{k,N} = \left[\frac{k}{2^N}, \frac{k+1}{2^N}\right).$$

If $0 \le k/2^N < 1$, then

$$\underbrace{m_{C_{k,N}}(f) = \frac{k}{2^N}}_{\substack{\text{greatest lower bound of } f \text{ over } C_{k,N} \\ \text{is the beginning of the interval}}} \quad \text{and} \quad \underbrace{M_{C_{k,N}}(f) = \frac{k+1}{2^N}.}_{\substack{\text{lowest upper bound of } f \text{ over } C_{k,N} \\ \text{is the beginning of the } next \text{ interval}}} \qquad 4.1.25$$

Thus

$$L_N(f) = \frac{1}{2^N}\sum_{k=0}^{2^N-1}\frac{k}{2^N} \quad \text{and} \quad U_N(f) = \frac{1}{2^N}\sum_{k=1}^{2^N}\frac{k}{2^N}. \qquad 4.1.26$$

In particular, $U_N(f) - L_N(f) = 2^N/2^{2N} = 1/2^N$, which tends to 0 as N tends to ∞, so f is integrable. Evaluating the integral requires the formula $1 + 2 + \cdots + m = m(m+1)/2$. Using this formula, we find

$$L_N(f) = \frac{1}{2^N}\frac{(2^N-1)2^N}{2\cdot 2^N} = \frac{1}{2}\left(1 - \frac{1}{2^N}\right)$$

$$\text{and} \quad U_N(f) = \frac{1}{2^N}\frac{2^N(2^N+1)}{2\cdot 2^N} = \frac{1}{2}\left(1 + \frac{1}{2^N}\right). \qquad 4.1.27$$

Clearly both sums converge to $1/2$ as N tends to ∞. $\triangle$

Riemann sums

Computing the upper integral $U(f)$ and the lower integral $L(f)$ may be difficult. Suppose we know that f is integrable. Then, just as for Riemann sums in one dimension, we can choose any point $\mathbf{x}_{\mathbf{k},N} \in C_{\mathbf{k},N}$ we like, such as the center of each cube, or the lower left-hand corner, and consider the Riemann sum

$$R(f,N) = \sum_{\mathbf{k} \in \mathbb{Z}^n} \overbrace{\operatorname{vol}(C_{\mathbf{k},N})}^{\text{"width"}} \overbrace{f(\mathbf{x}_{\mathbf{k},N})}^{\text{"height"}}. \qquad 4.1.28$$

Then since the value of the function at some arbitrary point $\mathbf{x}_{\mathbf{k},N}$ is bounded above by the least upper bound, and below by the greatest lower bound,

$$m_{C_{\mathbf{k},N}} f \leq f(\mathbf{x}_{\mathbf{k},N}) \leq M_{C_{\mathbf{k},N}} f, \qquad 4.1.29$$

the Riemann sums $R(f,N)$ will converge to the integral.

Computing multiple integrals by Riemann sums is conceptually no harder than computing one-dimensional integrals; it simply takes longer. Even when the dimension is only moderately large (for instance 3 or 4) this is a serious problem. It becomes much more serious when the dimension is 9 or 10; even in those dimensions, getting a numerical integral correct to six significant digits may be unrealistic.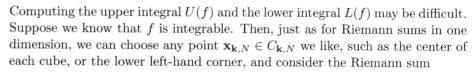

Some rules for computing multiple integrals

A certain number of results are more or less obvious:

> **Proposition 4.1.11 (Rules for computing multiple integrals).**
> (a) If two functions $f, g : \mathbb{R}^n \to \mathbb{R}$ are both integrable, then $f + g$ is also integrable, and the integral of $f + g$ equals the sum of the integral of f and the integral of g:
>
> $$\int_{\mathbb{R}^n} (f+g) |d^n\mathbf{x}| = \int_{\mathbb{R}^n} f |d^n\mathbf{x}| + \int_{\mathbb{R}^n} g |d^n\mathbf{x}|. \qquad 4.1.30$$
>
> (b) If f is an integrable function, and $a \in \mathbb{R}$, then the integral of af equals a times the integral of f:
>
> $$\int_{\mathbb{R}^n} (af) |d^n\mathbf{x}| = a \int_{\mathbb{R}^n} f |d^n\mathbf{x}|. \qquad 4.1.31$$
>
> (c) If f, g are integrable functions with $f \leq g$ (i.e., $f(\mathbf{x}) \leq g(\mathbf{x})$ for all $\mathbf{x}$), then the integral of f is less than or equal to the integral of g:
>
> $$\int_{\mathbb{R}^n} f |d^n\mathbf{x}| \leq \int_{\mathbb{R}^n} g |d^n\mathbf{x}|. \qquad 4.1.32$$

Warning! Before doing this you must know that your function is integrable: that the upper and lower sums converge to a *common limit*. It is perfectly possible for a Riemann sum to converge without the function being integrable (see Exercise 4.1.6). In that case, the limit doesn't mean much, and should be viewed with distrust.

In computing a Riemann sum, any point will do, but some are better than others. The sum will converge faster if you use the center point rather than a corner.

When the dimension gets really large, like 10^{24}, as happens in quantum field theory and statistical mechanics, even in straightforward cases no one knows how to evaluate such integrals, and their behavior is a central problem in the mathematics of the field. We give an introduction to Riemann sums as they are used in practice in Section 4.6.

An example of Equation 4.1.33: if f and g are functions of census tracts, f assigning to each per capita income for April through September, and g per capita income for October through March, then the sum of the maximum value for f and the maximum value for g must be at least the maximum value of $f+g$, and very likely more: a community dependent on the construction industry might have the highest per capita income in the summer months, while a ski resort might have the highest per capita income in the winter.

Proof. (a) For any subset $A \subset \mathbb{R}^n$, we have

$$M_A(f) + M_A(g) \geq M_A(f+g) \quad \text{and} \quad m_A(f) + m_A(g) \leq m_A(f+g). \quad 4.1.33$$

Applying this to each cube $C \in \mathcal{D}_N(\mathbb{R}^n)$ we get

$$U_N(f) + U_N(g) \geq U_N(f+g) \quad \geq \quad L_N(f+g) \geq L_N(f) + L_N(g). \quad 4.1.34$$

Since the outer terms have a common limit as $N \to \infty$, the inner ones have the same limit, giving

$$\underbrace{U_N(f) + U_N(g)}_{\int_{\mathbb{R}^n}(f)\,|d^n\mathbf{x}| + \int_{\mathbb{R}^n}(g)|d^n\mathbf{x}|} = \underbrace{U_N(f+g)}_{\int_{\mathbb{R}^n}(f+g)\,|d^n\mathbf{x}|} = L_N(f+g) = L_N(f) + L_N(g).$$

$$4.1.35$$

(b) If $a > 0$, then $U_N(af) = aU_N(f)$ and $L_N(af) = aL_N(f)$ for any N, so the integral of af is a times the integral of f.

If $a < 0$, then $U_N(af) = aL_N(f)$ and $L_N(af) = aU_N(f)$, so the result is also true: multiplying by a negative number turns the upper limit into a lower limit and vice versa.

(c) This is clear: $U_N(f) \leq U_N(g)$ for every N. □

The following statement follows immediately from Fubini's theorem, which is discussed in Section 4.5, but it fits in nicely here.

You can read Equation 4.1.37 to mean "the integral of the product $f_1(\mathbf{x})f_2(\mathbf{y})$ equals the product of the integrals," but please note that we are *not* saying, and it is *not* true, that for two functions with the same variable, the integral of the product is the product of the integrals. There is no formula for $\int f_1(\mathbf{x})f_2(\mathbf{x})$. The two functions of Proposition 4.1.37 have *different* variables.

Proposition 4.1.12. If $f_1(\mathbf{x})$ is integrable on $\mathbb{R}^n$ and $f_2(\mathbf{y})$ is integrable on $\mathbb{R}^m$, then the function

$$g(\mathbf{x}, \mathbf{y}) = f_1(\mathbf{x})f_2(\mathbf{y}) \quad 4.1.36$$

on $\mathbb{R}^{n+m}$ is integrable, and

$$\int_{\mathbb{R}^{n+m}} g\,|d^n\mathbf{x}||d^m\mathbf{y}| = \left(\int_{\mathbb{R}^n} f_1\,|d^n\mathbf{x}| \right) \left(\int_{\mathbb{R}^m} f_2\,|d^m\mathbf{y}| \right). \quad 4.1.37$$

The general case of Proposition 4.1.12 can be proved using $f_1 = f_1^+ + f_1^-$, $f_2 = f_2^+ + f_2^-$ (see Definition 4.3.4). This is worked out on the book web page.

Proof. We will prove this in the case where f_1 and f_2 are nonnegative. For any $A_1 \subset \mathbb{R}^n$, and $A_2 \subset \mathbb{R}^m$, we have

$$M_{A_1 \times A_2}(g) = M_{A_1}(f_1)M_{A_2}(f_2) \quad \text{and} \quad m_{A_1 \times A_2}(g) = m_{A_1}(f_1)m_{A_2}(f_2).$$

$$4.1.38$$

Since any $C \in \mathcal{D}_N(\mathbb{R}^{n+m})$ is of the form $C_1 \times C_2$ with $C_1 \in \mathcal{D}_N(\mathbb{R}^n)$ and $C_2 \in \mathcal{D}_N(\mathbb{R}^m)$, applying Equation 4.1.38 to each cube separately gives

$$U_N(g) = U_N(f_1)U_N(f_2) \quad \text{and} \quad L_N(g) = L_N(f_1)L_N(f_2). \quad 4.1.39$$

The result follows immediately. □

Volume defined more generally

The computation of volumes, historically the main motivation for integrals, remains an important application. We used the volume of cubes to define the integral; we now use integrals to define volume more generally.

Definition 4.1.13 (*n*-dimensional volume). When χ_A is integrable, the n-dimensional volume of A is

$$\text{vol}_n A = \int_{\mathbb{R}^n} \chi_A \, |d^n\mathbf{x}|. \qquad 4.1.40$$

Thus vol_1 is length of subsets of $\mathbb{R}$, vol_2 is area of subsets of $\mathbb{R}^2$, and so on.

We already defined the volume of dyadic cubes in Equation 4.1.16. In Proposition 4.1.16 we will see that these definitions are consistent.

Some texts refer to pavable sets as "contented" sets: sets with content.

Definition 4.1.14 (Pavable set: a set with well-defined volume). A set is pavable if it has a well-defined volume, i.e., if its characteristic function is integrable.

Lemma 4.1.15 (Length of interval). *An interval $I = [a, b]$ has volume (i.e., length) $|b - a|$.*

Proof. Of the cubes (i.e., intervals) $C \in \mathcal{D}_N(\mathbb{R})$, at most two contain one of the endpoints a or b. All the others are either entirely in I or entirely outside, so on those

The volume of a cube is $\frac{1}{2^{nN}}$, but here $n = 1$.

$$M_C(\chi_I) = m_C(\chi_I) = \begin{cases} 1 & \text{if } C \subset I \\ 0 & \text{if } C \cap I = \phi, \end{cases} \qquad 4.1.41$$

where ϕ denotes the empty set. Therefore the difference between upper and lower sums is at most two times the volume of a single cube:

Recall from Section 0.3 that
$$P = I_1 \times \cdots \times I_n \subset \mathbb{R}^n.$$
means
$$P = \{\mathbf{x} \in \mathbb{R}^n \mid x_i \in I_i\};$$
thus P is a rectangle if $n = 2$, a box if $n = 3$, and an interval if $n = 1$.

$$U_N(\chi_I) - L_N(\chi_I) \leq 2\frac{1}{2^N}, \qquad 4.1.42$$

which tends to 0 as $N \to \infty$, so the upper and lower sums converge to the same limit: χ_I is integrable, and I has volume. We leave its computation as Exercise 4.1.13. $\square$

Similarly, parallelepipeds with sides parallel to the axes have the volume one expects, namely, the product of the lengths of the sides. Consider

$$P = I_1 \times \cdots \times I_n \subset \mathbb{R}^n. \qquad 4.1.43$$

Proposition 4.1.16 (Volume of parallelepiped). *The parallelepiped*

$$P = I_1 \times \cdots \times I_n \subset \mathbb{R}^n \qquad\qquad 4.1.44$$

formed by the product of intervals $I_i = [a_i, b_i]$ *has volume*

$$\mathrm{vol}_n(P) = |b_1 - a_1|\,|b_2 - a_2|\,\ldots\,|b_n - a_n|. \qquad\qquad 4.1.45$$

In particular, the n-dimensional volume of a cube $C \in \mathcal{D}_N(\mathbb{R}^n)$ is

$$\mathrm{vol}_n C = \frac{1}{2^{nN}}. \qquad\qquad 4.1.46$$

Proof.   This follows immediately from Proposition 4.1.12, applied to

$$\chi_P(\mathbf{x}) = \chi_{I_1}(x_1)\chi_{I_2}(x_2)\ldots\chi_{I_n}(x_n). \quad\square \qquad\qquad 4.1.47$$

The following elementary result has powerful consequences (though these will only become clear later).

Disjoint means having no points in common.

Theorem 4.1.17 (Sum of volumes).   *If two disjoint sets A, B in $\mathbb{R}^n$ are pavable, then so is their union, and the volume of the union is the sum of the volumes:*

$$\mathrm{vol}_n(A \cup B) = \mathrm{vol}_n A + \mathrm{vol}_n B. \qquad\qquad 4.1.48$$

Proof.   Since $\chi_{A \cup B} = \chi_A + \chi_B$, this follows from Proposition 4.1.11, (a). $\square$

Proposition 4.1.18 (Set with volume 0).   *A set $X \subset \mathbb{R}^n$ has volume 0 if and only if for every $\epsilon > 0$ there exists N such that*

$$\sum_{\substack{C \in \mathcal{D}_N(\mathbb{R}^n) \\ C \cap X \neq \emptyset}} \mathrm{vol}_n(C) \leq \epsilon. \qquad\qquad 4.1.49$$

You are asked to prove Proposition 4.1.18 in Exercise 4.1.4.

Unfortunately, at the moment there are very few functions we can integrate; we will have to wait until Section 4.5 before we can compute any really interesting examples.

4.2 Probability and Integrals

Computing areas and volumes is one important application of multiple integrals. There are many others, coming from a wide range of different fields: geometry, mechanics, probability, ... Here we touch on a couple: computing centers of gravity and computing probabilities. They sound quite different, but the formulas are so similar that we think each helps in understanding the other.

Definition 4.2.1 (Center of gravity of a body). (a) If a body $A \subset \mathbb{R}^n$ (i.e., a pavable set) is made of some homogeneous material, then the center of gravity of A is the point $\overline{\mathbf{x}}$ whose ith coordinate is

$$\overline{x}_i = \frac{\displaystyle\int_A x_i |d^n\mathbf{x}|}{\displaystyle\int_A |d^n\mathbf{x}|}. \qquad 4.2.1$$

(b) More generally, if a body A (not necessarily made of a homogeneous material) has density μ, then the mass M of such a body is

$$M = \int_A \mu(\mathbf{x})|d^n\mathbf{x}|, \qquad 4.2.2$$

and the center of gravity $\overline{\mathbf{x}}$ is the point whose ith coordinate is

$$\overline{x}_i = \frac{\int_A x_i \mu(\mathbf{x})|d^n\mathbf{x}|}{M}. \qquad 4.2.3$$

Integrating density gives mass.

Here μ (mu) is a function from A to $\mathbb{R}$; to a point of A it associates a number giving the density of A at that point.

In physical situations μ will be non-negative.

We will see that in many problems in probability there is a similar function μ, giving the "density of probability."

A brief introduction to probability theory

In probability there is at the outset an experiment, which has a *sample space* S and a *probability measure* Prob. The sample space consists of all possible outcomes of the experiment. For example, if the experiment consists of throwing a six-sided die, then $S = \{1, 2, 3, 4, 5, 6\}$. The probability measure Prob takes a subset $A \subset S$, called an *event*, and returns a number $\text{Prob}(A) \in [0, 1]$, which corresponds to the probability of an outcome of the experiment being in A. Thus the probability can range from 0 (it is certain that the outcome will not be in A) to 1 (it is certain that it will be in A). We could restate the latter statement as $\text{Prob}(S) = 1$.

When the probability space S consists of a finite number of outcomes, then Prob is completely determined by knowing the probabilities of the individual outcomes. When the outcomes are all equally likely, the probability assigned any one outcome is 1 divided by the number of outcomes; 1/6 in the case of the die. But often the outcomes are not equally likely. If the die is loaded so that it lands on 4 half the time, while the other outcomes are equally likely, then the $\text{Prob}\{4\} = 1/2$, while the probability of each of the other five outcomes is 1/10.

When an event A consists of several outcomes, $\text{Prob}(A)$ is computed by adding together the weights corresponding to the elements of A. If the experiment consists of throwing the loaded die described above, and $A = \{3, 4\}$, then

$\mathrm{Prob}(A) = 1/10 + 1/2 = 3/5$. Since $\mathrm{Prob}(S) = 1$, the sum of all the weights for a given experiment always equals 1.

In the margin:

In the case of throwing a die, $\mathrm{Prob}\{3,4\}$ means the probability of landing on either 3 or 4.

Integrals come into play when a probability space is infinite. We might consider the experiment of measuring how late (or early) a train is; the sample space is then some interval of time. Or we might play "spin the wheel," in which case the sample space is the circle, and if the game is fair, the wheel has an equal probability of pointing in any direction.

A third example, of enormous theoretical interest, consists of choosing a number $x \in [0,1]$ by choosing its successive digits at random. For instance, you might write x in base 2, and choose the successive digits by tossing a fair coin infinitely many times, writing 1 if the toss comes up heads, and 0 if it comes up tails.

In the margin:

If you have a ten-sided die, with the sides marked $0 \ldots 9$ you could write your number in base 10 instead.

In these cases, the probability measure cannot be understood in terms of the probabilities of the individual outcomes, because each individual outcome has probability 0. Any particular infinite sequence of coin tosses is infinitely unlikely. Some other scheme is needed. Let us see how to understand probabilities in the last example above. It is true that the probability of any particular number, like $\{1/3\}$ or $\{\sqrt{2}/2\}$, is 0. But there are some subsets whose probabilities are easy to compute. For instance $\mathrm{Prob}([0,1/2)) = 1/2$. Why? Because $x \in [0,1/2)$, which in base 2 is written $x \in [0,.1)$, means exactly that the first digit of x is 0. More generally, any dyadic interval $I \in \mathcal{D}_N(\mathbb{R})$ has probability $1/2^N$, since it corresponds to x starting with a particular sequence of N digits, and then makes no further requirement about the others. (Again, remember that our numbers are in base 2.)

So for every dyadic interval, its probability is exactly its length. In fact, since length (i.e., vol_1) is defined in terms of dyadic intervals, we see that the probability of any pavable subset of $A \subset [0,1]$ is precisely

$$\mathrm{Prob}(A) = \int_{\mathbb{R}} \chi_A |dx|. \qquad 4.2.4$$

A similar description is probably possible in the case of late trains: there is likely a function $g(t)$ such that the probability of a train arriving in some time interval $[a,b]$ is given by $\int_{[a,b]} g(t)\,dt$. One might imagine that the function looks like a bell curve, perhaps centered at the scheduled time t_0, but perhaps several minutes later if the train is systematically late. It might also happen that the curve is not bell-shaped, but camel-backed, reflecting the fact that if the train misses a certain light then it will be set back by some definite amount of time.

In many cases where the sample space is $\mathbb{R}^k$, something of the same sort is true: there is a function $\mu(\mathbf{x})$ such that

$$\mathrm{Prob}(A) = \int_A \mu(\mathbf{x}) |d^k \mathbf{x}|. \qquad 4.2.5$$

In this case, μ is called a *probability density*; to be a probability density the function μ must satisfy

$$\mu(\mathbf{x}) \geq 0 \quad \text{and} \quad \int_{\mathbb{R}^k} \mu(\mathbf{x}) \, |d^k\mathbf{x}| = 1. \qquad 4.2.6$$

We will first look at an example in one variable; later we will build on this example to explore a use of multiple integrals (which are, after all, the reason we have written this section).

Example 4.2.2 (Height of 10-year-old girls). Consider the experiment consisting of choosing a 10-year-old girl at random in the U.S., and measuring her height. Our sample space is $\mathbb{R}$. As in the case of choosing a real number from 0 to 10, it makes no sense to talk about the probability of landing on any one particular point in $\mathbb{R}$. (No theoretical sense, at least; in practice, we are limited in our measurements, so this could be treated as a finite probability space.) What we can do is determine a "density of probability" function that will enable us to compute the probability of landing in some region of $\mathbb{R}$, for example, height between 54 and 55 inches.

Every pediatrician has growth charts furnished by the Department of Health, Education, and Welfare, which graph height and weight as a function of age, for girls from 2 to 18 years old; each consists of seven curves, representing the 5th, 10th, 25th, 50th, 75th, 90th, and 95th percentiles, as shown in Figure 4.2.1.

Of course some heights are impossible. Clearly, the height of such a girl will not fall in the range 10–12 inches, or 15–20 feet. Including such impossible outcomes in a sample space is standard practice. As William Feller points out in *An Introduction to Probability Theory and Its Applications*, vol. 1 (pp. 7–8), "According to formulas on which modern mortality tables are based, the proportion of men surviving 1 000 years is of the order of magnitude of one in $10^{10^{36}}$ .... This statement does not make sense from a biological or sociological point of view, but considered exclusively from a statistical standpoint it certainly does not contradict any experience .... Moreover, if we were seriously to discard the possibility of living 1 000 years, we should have to accept the existence of a maximum age, and the assumption that it should be possible to live x years and impossible to live x years and two seconds is as unappealing as the idea of unlimited life."

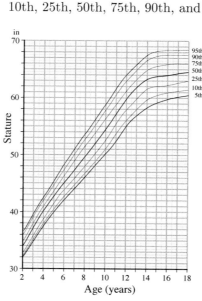
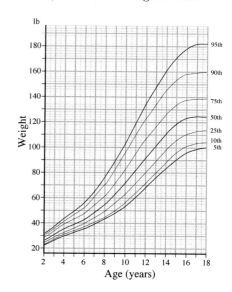

FIGURE 4.2.1. Charts graphing height and weight as a function of age, for girls from 2 to 18 years old.

Looking at the height chart and extrapolating (connecting the dots), we can construct the bell-shaped curve shown in Figure 4.2.2, with a maximum at $x =$

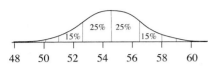

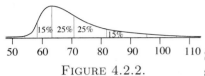

FIGURE 4.2.2.

Top: graph of μ_h, giving the "density of probability" for height for 10-year-old girls.

Bottom: graph of μ_w, giving the "density of probability" for weight.

It is not always possible to find a function μ that fits the available data; in that case there is still a probability measure, but it is not given by a density probability function.

The name *random function* is more accurate, but nonstandard.

The same experiment (i.e., same sample space and probability measure) can have more than one random variable.

The words *expectation, expected value, mean,* and *average* are all synonymous.

Since an expectation E corresponds not just to a random variable f but also to an experiment with density of probability μ, it would be more precise to denote it by something like $E_\mu(f)$.

54.5, since a 10-year-old girl who is 54.5 inches tall falls in the 50th percentile for height. This curve is the graph of a function that we will call μ_h; it gives the "density of probability" for height for 10-year-old girls. For each particular range of heights, it gives the probability that a 10-year-old girl chosen at random will be that tall:

$$\text{Prob}\{h_1 \leq h \leq h_2\} = \int_{h_1}^{h_2} \mu_h(h)\,|dh|. \qquad 4.2.7$$

Similarly, we can construct a "density of probability" function μ_w for weight, such that the probability of a child having weight w satisfying $w_1 \leq w \leq w_2$ is

$$\text{Prob}\{w_1 \leq w \leq w_2\} = \int_{w_1}^{w_2} \mu_w(w)\,|dw|. \qquad 4.2.8$$

The integrals $\int_{\mathbb{R}} \mu_h(h)|dh|$ and $\int_{\mathbb{R}} \mu_w(w)|dw|$ must of course equal 1.

Remark. Sometimes, as in the case of unloaded dice, we can figure out the appropriate probability measure on the basis of pure thought. More often, as in the "height experiment," it is constructed from real data. A major part of the work of statisticians is finding probability density functions that fit available data. $\triangle$

Once we know an experiment's probability measure, we can compute the expectation of a *random variable* associated with the experiment.

Definition 4.2.3 (Random variable). Let S be the sample space of outcomes of an experiment. A random variable is a function $f : S \to \mathbb{R}$.

If the experiment consists of throwing two dice, we might choose as our random variable the function that gives the total obtained. For the height experiment, we might choose the function f_H that gives the height; in that case, $f_H(x) = x$.

For each random variable, we can compute its expectation.

Definition 4.2.4 (Expectation). The expectation $E(f)$ of a random variable f is the value one would expect to get if one did the experiment a great many times and took the average of the results. If the sample space S is finite, $E(f)$ is computed by adding up all the outcomes s (elements of S), each weighted by its probability of occurrence. If S is continuous, and μ is the density of probability function, then

$$E(f) = \int_S f(s)\mu(s)\,|ds|.$$

Example 4.2.5 (Expectation). The experiment consisting of throwing two unloaded dice has 36 outcomes, each equally likely: for any $s \in S, \text{Prob}(s) = \frac{1}{36}$.

In the dice example, the weight for the total 3 is 2/36, since there are two ways of achieving the total 3: (2,1) and (1,2).

If you throw two dice 500 times and figure the average total, it should be close to 7; if not, you would be justified in suspecting that the dice are loaded.

Equation 4.2.9: As in the finite case, we compute the expectation by "adding up" the various possible outcomes, each weighted by its probability.

Let f be the random variable that gives the total obtained (i.e., the integers 2 through 12). To determine the expectation, we add up the possible totals, each weighted by its probability:

$$2\frac{1}{36} + 3\frac{2}{36} + 4\frac{3}{36} + 5\frac{4}{36} + 6\frac{5}{36} + 7\frac{6}{36} + 8\frac{5}{36} + 9\frac{4}{36} + 10\frac{3}{36} + 11\frac{2}{36} + 12\frac{1}{36} = 7.$$

For the "height experiment" and "weight experiment" the expectations are

$$E(f_H) = \int_{\mathbb{R}} h\mu_h(h)\,|dh| \quad \text{and} \quad E(f_W) = \int_{\mathbb{R}} w\mu_w(w)\,|dw|. \qquad \triangle \qquad 4.2.9$$

Note that an expectation does not need to be a realizable number; if our experiment consists of rolling a single die, and f consists of seeing what number we get, then $E(f) = \frac{1}{6} + \frac{2}{6} + \frac{3}{6} + \frac{4}{6} + \frac{5}{6} + \frac{6}{6} = 3.5$. Similarly, the average family may be said to have 2.2 children . . . .

Variance and standard deviation

The expectation of a random variable is useful, but it can be misleading. Suppose the random variable f assigns income to an element of the sample space S, and S consists of 1000 supermarket cashiers and Bill Gates (or, indeed, 1000 school teachers or university professors and Bill Gates); if all you knew was the average income, you might draw very erroneous conclusions. For a less extreme example, if a child's weight is different from average, her parents may well want to know whether it falls within "normal" limits. The *variance* and the *standard deviation* address the question of how spread out a function is from its mean.

Since Var (f) is defined in terms of the expectation of f, and computing the expectation requires knowing a probability measure, Var (f) is associated to a particular probability measure. The same is true of the definitions of standard deviation, covariance, and correlation coefficient.

Definition 4.2.6 (Variance). The variance of a random variable f, denoted Var (f), is given by the formula

$$\text{Var}\,(f) = E\left(\left(f - E(f)\right)^2\right) = \int_S \left(f(\mathbf{x}) - E(f)\right)^2 |d^k\mathbf{x}|. \qquad 4.2.10$$

Why the squared term in this formula? What we want to compute is how far f is, on average, from its average. But of course f will be less than the average just as much as it will be more than the average, so $E(f - E(f))$ is 0. We could solve this problem by computing the *mean absolute deviation*, $E|f - E(f)|$. But this quantity is difficult to compute. In addition, squaring $f - E(f)$ emphasizes the deviations that are far from the mean (the income of Bill Gates, for example), so in some sense it gives a better picture of the "spread" than does the absolute mean deviation.

But of course squaring $f - E(f)$ results in the variance having different units than f. The *standard deviation* corrects for this:

Definition 4.2.7 (Standard deviation). The standard deviation of a random variable f, denoted $\sigma(f)$, is given by

$$\sigma(f) = \sqrt{\operatorname{Var}(f)}. \qquad 4.2.11$$

The name for the Greek letter σ is "sigma."

Example 4.2.8 (Variance and standard deviation). If the experiment is throwing two dice, and the random variable gives the total obtained, then the variance is

$$\frac{1}{36}(2-7)^2 + \frac{2}{36}(3-7)^2 + \cdots + \frac{6}{36}(7-7)^2 + \cdots + \frac{2}{36}(11-7)^2 + \frac{1}{36}(12-7)^2 = 5.833\ldots, \qquad 4.2.12$$

and the standard deviation is $\sqrt{5.833\ldots} \approx 2.415$.

The mean absolute deviation for the "total obtained" random variable is approximately 1.94, significantly less than the standard deviation. Because of the square in the formula for the variance, the standard deviation weights more heavily values that are far from the expectation than those that are close, whereas the mean absolute deviation treats all deviations equally.

Probabilities and multiple integrals

Earlier we discussed the functions μ_h and μ_w, the first giving probabilities for height of a 10-year-old girl chosen at random, the second giving probabilities for weight. Can these functions answer the question: what is the probability that a 10-year-old girl chosen at random will have height between 54 and 55 inches, and weight between 70 and 71 pounds? The answer is no. Computing a "joint" probability as the product of "single" probabilities only works when the probabilities under study are *independent*. We certainly can't expect weight to be independent of height.

Indeed, a very important application of probability theory is to determine whether phenomena are related or not. Is a person subjected to second-hand smoke more likely to get lung cancer than someone who is not? Is total fat consumption related to the incidence of heart disease? Does participating in Head Start increase the chances that a child from a poor family will graduate from high school?

To construct a probability density function μ in two variables, height and weight, one needs more information than the information needed to construct μ_h and μ_w separately. One can imagine collecting thousands of file cards, each one giving the height and weight of a 10-year-old girl, and distributing them over a big grid; the region of the grid corresponding to 54-55 inches tall, 74-75 pounds, would have a very tall stack of cards, while the region corresponding to 50-51 inches and 100-101 pounds would have a much smaller stack; the region corresponding to 50-51 inches and 10-11 pounds would have none. The distribution of these cards corresponds to the density probability function μ. Its graph will probably look like a mountain, but with a ridge along some curve of the form $w = ch^3$, since roughly you would expect the weight to scale like the volume, which should be roughly proportional to the cube of the height.

We can compute μ_h and μ_w from μ:

$$\mu_h(h) = \int_{\mathbb{R}} \mu \begin{pmatrix} h \\ w \end{pmatrix} |dw|$$

$$\mu_w(w) = \int_{\mathbb{R}} \mu \begin{pmatrix} h \\ w \end{pmatrix} |dh|. \qquad 4.2.13$$

You might expect that

$$\int_{\mathbb{R}^2} h \mu \begin{pmatrix} h \\ w \end{pmatrix} |dh\,dw|$$
$$= \int_{\mathbb{R}} h \mu_h(h) \, |dh|.$$

This is true, and is a form of Fubini's theorem to be developed in Section 4.5. If we have thrown away information about height-weight distribution, we can still figure out height expectation from the cards on the height-axis (and weight expectation from the cards on the weight-axis). We've just lost all information about the correlation of height and weight.

But the converse is not true. If we have our file cards neatly distributed over the height-weight grid, we could cut each file card in half and put the half giving height on the corresponding interval of the h-axis and the half giving weight on the corresponding interval of the w-axis, which results in μ_h and μ_w. (This corresponds to Equation 4.2.13). In the process we throw away information: from our stacks on the h and w axes we would not know how to distribute the cards on the height-weight grid.[1]

Computing the expectation for a random variable associated to the height-weight experiment requires a double integral. If you were interested in the average weight for 10-year-old girls whose height is close to average, you might compute the expectation of the random variable f satisfying

$$f \begin{pmatrix} h \\ w \end{pmatrix} = \begin{cases} w & \text{if } \left(h - \frac{E(h)}{2} \right) \leq h \leq \left(h + \frac{E(h)}{2} \right) \\ 0 & \text{otherwise.} \end{cases} \qquad 4.2.14$$

The expectation of this function would be

$$E(f) = \int_{\mathbb{R}^2} f \begin{pmatrix} h \\ w \end{pmatrix} \mu \begin{pmatrix} h \\ w \end{pmatrix} |dh\,dw|. \qquad 4.2.15$$

A double integral would also be necessary to compute the *covariance* of the random variables f_H and f_W.

If two random variables are independent, their covariance is 0, as is their correlation coefficient. The converse is not true.

By "independent," we mean that the corresponding probability measures are independent: if f is associated with μ_h and g is associated with μ_w, then f and g are independent, and μ_h and μ_w are independent, if

$$\mu_h(x)\mu_w(y) = \mu \begin{pmatrix} x \\ y \end{pmatrix},$$

where μ is the density probability function corresponding to the variable $\begin{pmatrix} x \\ y \end{pmatrix}$.

Definition 4.2.9 (Covariance). Let S_1 be the sample space of one experiment, and S_2 be the sample space for another. If $f : S_1 \to \mathbb{R}$ and $g : S_2 \to \mathbb{R}$ are random variables, their covariance, denoted $\mathrm{Cov}\,(f, g)$, is:

$$\mathrm{Cov}\,(f, g) = E\Big(\big(f - E(f) \big)\big(g - E(g) \big) \Big). \qquad 4.2.16$$

The product $(f - E(f))(g - E(g))$ is positive when both f and g are on the same side of their mean (both less than average, or both more than average), and negative when they are on opposite sides, so the covariance is positive when f and g vary "together," and negative when they vary "opposite."

Finally, we have the *correlation coefficient* of f and g:

Definition 4.2.10 (Correlation coefficient). The correlation coefficient of two random variables f and g, denoted $\mathrm{corr}\,(f, g)$, is given by the formula

$$\mathrm{corr}\,(f, g) = \frac{\mathrm{Cov}\,(f, g)}{\sigma(f)\sigma(g)}. \qquad 4.2.17$$

The correlation is always a number between -1 and 1, and has no units.

---

[1]If μ_h and μ_w were independent, then we could compute μ from μ_h and μ_w; in that case, we would have $\mu = \mu_h\mu_w$.

You should notice the similarities between these definitions and

Exercise 4.2.2 explores these analogies.

the length squared of vectors,	analogous to the variance;
the length of vectors,	analogous to the standard deviation;
the dot product,	analogous to the covariance;
the cosine of the angle between two vectors,	analogous to the correlation.

In particular, "correlation 0" corresponds to "orthogonal."

Central limit theorem

One probability density is ubiquitous in probability theory: the *normal distribution* given by

The graph of the normal distribution is a bell curve.

$$\mu(x) = \frac{1}{\sqrt{2\pi}}\, e^{-t^2/2}.\qquad\qquad 4.2.18$$

The object of this subsection is to explain why.

The theorem that makes the normal distribution important is the *central limit theorem*. Suppose you have an experiment and a random variable, with expected value E and standard deviation σ. Suppose that you repeat the experiment n times, with results $x_1, \ldots, x_n$. Then the central limit theorem asserts that the average

$$\overline{x} = \frac{1}{n}(x_1 + \cdots + x_n)\qquad\qquad 4.2.19$$

As n grows, all the detail of the original experiment gets ironed out, leaving only the normal distribution.

is approximately distributed according to the normal distribution with mean E and standard deviation $\sigma/\sqrt{n}$, the approximation getting better and better as $n \to \infty$. Whatever experiment you perform, if you repeat it and average, the normal distribution will describe the results.

The standard deviation of the new experiment (the "repeat the experiment n times and take the average" experiment) is the standard deviation of the initial experiment divided by $\sqrt{n}$.

Below we will justify this statement in the case of coin tosses. First let us see how to translate the statement above into formulas. There are two ways of doing it. One is to say that the probability that $\overline{x}$ is between A and B is approximately

Equation 4.2.20 puts the complication in the exponent; Equation 4.2.21 puts it in the domain of integration.

$$\frac{\sqrt{n}}{\sqrt{2\pi}\,\sigma}\int_A^B e^{-\frac{n}{2}\left(\frac{x-E}{\sigma}\right)^2}\,dx.\qquad\qquad 4.2.20$$

We will use the other in our formal statement of the theorem. For this we make the change of variables $A = E + \sigma a/\sqrt{n}$, $B = E + \sigma b/\sqrt{n}$.

The exponent for e in Equation 4.2.20 is so small it may be hard to read; it is

$$-\frac{n}{2}\left(\frac{x-E}{\sigma}\right)^2.$$

There are a great many improvements on and extensions of the central limit theorem; we cannot hope to touch upon them here.

Theorem 4.2.11 (The central limit theorem). *If an experiment and a random variable have expectation E and standard deviation σ, then if the experiment is repeated n times, with average result $\overline{x}$, the probability that $\overline{x}$ is between $E + \frac{\sigma}{\sqrt{n}}a$ and $E + \frac{\sigma}{\sqrt{n}}b$ is approximately*

$$\frac{1}{\sqrt{2\pi}}\int_a^b e^{-y^2/2}\,dy.\qquad\qquad 4.2.21$$

We prove a special case of the central limit theorem in Appendix A.12. The proof uses *Stirling's formula*, a very useful result showing how the factorial $n!$ behaves as n becomes large. We recommend reading it if time permits, as it makes interesting use of some of the notions we have studied so far (Taylor polynomials and Riemann sums) as well as some you should remember from high school (logarithms and exponentials).

Example 4.2.12 (Coin toss). As a first example, let us see how the central limit theorem answers the question: what is the probability that a fair coin tossed 1000 times will come up heads between 510 and 520 times?

In principle, this is straightforward: just compute the sum

$$\frac{1}{2^{1000}} \sum_{k=510}^{520} \binom{1000}{k}. \qquad 4.2.22$$

Recall the binomial coefficient:
$$\binom{n}{k} = \frac{n!}{k!(n-k)!}.$$

In practice, computing these numbers would be extremely cumbersome; it is much easier to use the central limit theorem. Our individual experiment consists of throwing a coin, and our random variable returns 1 for "heads" and 0 for "tails." This random variable has expectation $E = .5$ and standard deviation $\sigma = .5$ also, and we are interested in the probability of the average being between .51 and .52. Using the version of the central limit theorem in Equation 4.2.20, we see that the probability is approximately

$$\frac{\sqrt{1000}}{\sqrt{2\pi}\,\frac{1}{2}} \int_{.51}^{.52} e^{-\frac{1000}{2}\left(\frac{x-.5}{.5}\right)^2} dx. \qquad 4.2.23$$

One way to get the answer in Equation 4.2.24 is to look up a table giving values for the "standard normal distribution function."

Another is to use some software. With MATLAB, we use .5 erf to get:
EDU> a= .5*erf(20/sqrt(2000))
EDU> a = 0.236455371567231
EDU> b= .5*erf(40/sqrt(2000))
EDU> b = 0.397048394633966
EDU> b-a
ans = 0.160593023066735

The "error function" erf is related to the "standard normal distribution function" as follows:
$$\frac{1}{2\pi} \int_0^a e^{-\frac{t^2}{2}} dt = \frac{1}{2}\,\mathrm{erf}\left(\frac{a}{\sqrt{2}}\right).$$

Computations like this are used everywhere: when drug companies figure out how large a population to try out a new drug on, when industries figure out how long a product can be expected to last, etc.

Now we set

$$1000\left(\frac{x-.5}{.5}\right)^2 = t^2, \quad \text{so that} \quad 2\sqrt{1000}\,dx = dt.$$

Substituting t^2 and dt in Equation 4.2.23 we get

$$\frac{1}{\sqrt{2\pi}} \int_{20/\sqrt{1000}}^{40/\sqrt{1000}} e^{-t^2/2} dt \approx 0.1606. \qquad 4.2.24$$

Does this seem large to you? It does to most people.   △

Example 4.2.13 (Political poll). How many people need to be polled to call an election, with a probability of 95% of being within 1% of the "true value"? A mathematical model of this is tossing a biased coin, which falls heads with unknown probability p and tails with probability $1-p$. If we toss this coin n times (i.e., sample n people) and return 1 for heads and 0 for tails (1 for candidate A and 0 for candidate B), the question is: how large does n need to be in order to achieve 95% probability that the average we get is within 1% of p?

We need to know something about the bell curve to answer this, namely that 95% of the mass is within two standard deviations of the mean (and it is a good idea to memorize Figure 4.2.3). That means that we want $\frac{1}{2}\%$ to be the standard deviation of the experiment of asking n people. The experiment of asking one person has standard deviation $\sigma = \sqrt{p(1-p)}$. Of course, p is what we don't know, but the maximum of $\sqrt{p(1-p)}$ is $1/2$ (which occurs for $p = 1/2$). So we will be safe if we choose n so that the standard deviation $\sigma/\sqrt{n}$ is

Figure 4.2.3 gives three typical values for the area under the bell curve; these values are useful to know. For other values, you need to use a table or software, as described in Example 4.2.12.

$$\frac{1}{2\sqrt{n}} = \frac{1}{200}; \quad \text{i.e.,} \quad n = 10\,000. \qquad 4.2.25$$

How many would you need to ask if you wanted to be 95% sure to be within 2% of the true value? Check below.[2]   △

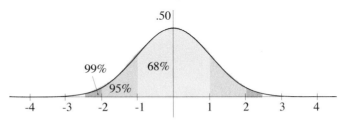

FIGURE 4.2.3. For the normal distribution, 68 percent of the probability is within one standard deviation; 95 percent is within two standard deviations; 99 percent is within 2.5 standard deviations.

4.3   WHAT FUNCTIONS CAN BE INTEGRATED?

What functions are integrable? It would be fairly easy to build up a fair collection by ad hoc arguments, but instead we prove in this section three theorems answering that question. They will tell us what functions are integrable, and in particular will guarantee that all usual functions are.

The first is based on our notion of dyadic pavings. The second states that any continuous function on $\mathbb{R}^n$ with bounded support is integrable. The third is stronger than the second; it tells us that a function with bounded support does not have to be continuous everywhere to be integrable; it is enough to require that it be continuous except on a set of volume 0.

This third criterion is adequate for most functions that you will meet. However, it is not the strongest possible statement. In the optional Section 4.4 we prove a harder result: a function $f : \mathbb{R}^n \to \mathbb{R}$, bounded and with bounded support, is integrable if and only if it is continuous except on a set of *measure* 0. The notion of measure 0 is rather subtle and surprising; with this notion,

---

[2]The number is 2500. Note that gaining 1% quadrupled the price of the poll.

we see that some very strange functions are integrable. Such functions actually arise in statistical mechanics.

First, the theorem based on dyadic pavings. Although the index under the sum sign may look unfriendly, the proof is reasonably easy, which doesn't mean that the criterion for integrability that it gives is easy to verify in practice. We don't want to suggest that this theorem is not useful; on the contrary, it is the foundation of the whole subject. But if you want to use it directly, proving that your function satisfies the hypotheses is usually a difficult theorem in its own right. The other theorems state that entire classes of functions satisfy the hypotheses, so that verifying integrability becomes a matter of seeing whether a function belongs to a particular class.

Theorem 4.3.1 (Criterion for integrability). *A function $f : \mathbb{R}^n \to \mathbb{R}$, bounded and with bounded support, is integrable if and only if for all $\epsilon > 0$, there exists N such that*

$$\overbrace{\sum_{\{C \in \mathcal{D}_N \mid \, \mathrm{osc}_C(f) > \epsilon\}} \mathrm{vol}_n\, C}^{\substack{\text{volume of all cubes for which} \\ \text{the oscillation of } f \text{ over the cube is } > \epsilon}} < \epsilon. \qquad 4.3.1$$

In Equation 4.3.1 we sum the volume of only those cubes for which the oscillation of the function is more than epsilon. If, by making the cubes very small (choosing N sufficiently large) the sum of their volumes is less than epsilon, then the function is integrable: we can make the difference between the upper sum and the lower sum arbitrarily small; the two have a common limit. (The other cubes, with small oscillation, contribute arbitrarily little to the difference between the upper and the lower sum.)

You may object that there will be a whole lot of cubes, so how can their volume be less than epsilon? The point is that as N gets bigger, there are more and more cubes, but they are smaller and smaller, and (if f is integrable) the total volume of those where $\mathrm{osc}_C > \epsilon$ tends to 0.

Example 4.3.2 (Integrable functions). Consider the characteristic function χ_D that is 1 on a disk and 0 outside, shown in Figure 4.3.1. Cubes C that are completely inside or completely outside the disk have $\mathrm{osc}_C(\chi_D) = 0$. Cubes straddling the border have oscillation equal to 1. (Actually, these cubes are squares, since $n = 2$.) By choosing N sufficiently large (i.e., by making the squares small enough), you can make the area of those that straddle the boundary arbitrarily small. Therefore χ_D is integrable.

Of course, when we make the squares small, we need more of them to cover the border, so that the sum of areas won't necessarily be less than ϵ. But as we divide the original border squares into smaller ones, some of them no

This follows the rule that there is no free lunch: we don't work very hard, so we don't get much for our work.

Recall that $\mathcal{D}_N$ denotes the collection of all cubes at a single level N, and that $\mathrm{osc}_C(f)$ denotes the oscillation of f over C: the difference between its least upper bound and greatest lower bound, over C.

Epsilon has the units of vol_n. If $n = 2$, epsilon is measured in centimeters (or meters ...) squared; if $n = 3$ it is measured in centimeters (or whatever) cubed.

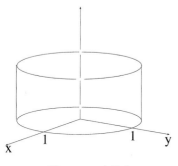

FIGURE 4.3.1.

The graph of the characteristic function of the unit disk, χ_D.

FIGURE 4.3.2.

The function $\sin \frac{1}{x}$ is integrable over any bounded interval. The dyadic intervals sufficiently near 0 will always have oscillation 2, but they have small length when the dyadic paving is fine.

The center region of Figure 4.3.2 is black because there are infinitely many oscillations in that region.

Note again the surprising but *absolutely standard way* in which we prove that something (here, the difference between upper and lower sums) is zero: we prove that it is smaller than an arbitrary $\epsilon > 0$. (Or equivalently, that it is smaller than $u(\epsilon)$, when u is a function such that $u(\epsilon) \to 0$ as $\epsilon \to 0$. Theorem 1.5.10 states that these conditions are equivalent.)

longer straddle the border. This is not quite a proof; it is intended to help you understand the meaning of the statement of Theorem 4.3.1.

Figure 4.3.2 shows another integrable function, $\sin \frac{1}{x}$. Near 0, we see that a small change in x produces a big change in $f(x)$, leading to a large oscillation. But we can still make the difference between upper and lower sums arbitrarily small by choosing N sufficiently large, and thus the intervals sufficiently small. Theorem 4.3.10 justifies our statement that this function is integrable.

Example 4.3.3 (A nonintegrable function). The function that is 1 at rational numbers in $[0,1]$ and 0 elsewhere is not integrable. No matter how small you make the cubes (intervals in this case), choosing N larger and larger, *each* cube will still contain both rational and irrational numbers, and will have osc $= 1$. $\triangle$

Proof of Theorem 4.3.1. First we will prove that the existence of such an N implies integrability: i.e., that the lower sum $U_N(f)$ and the upper sum $L_N(f)$ converge to a common limit. Choose any $\epsilon > 0$, and let N satisfy Equation 4.3.1. Then

$$U_N(f) - L_N(f) \quad \leq \quad \overbrace{\sum_{\{C \in \mathcal{D}_N \mid \operatorname{osc}_C(f) > \epsilon\}} 2 \sup |f| \operatorname{vol}_n C}^{\text{contribution from cubes with osc} > \epsilon} + \overbrace{\sum_{\substack{\{C \in \mathcal{D}_N \mid \operatorname{osc}_C(f) \leq \epsilon \\ \text{and } C \cap \operatorname{Supp}(f) \neq \emptyset\}}} \epsilon \operatorname{vol}_n C}^{\text{contribution from cubes with osc} \leq \epsilon}$$

$$\leq \epsilon \big(2 \sup |f| + \operatorname{vol}_n C_{\operatorname{Supp}} \big),$$

4.3.2

where $\sup |f|$ is the supremum of $|f|$, and C_{Supp} is a cube that contains the support of f (see Definition 4.1.2).

The first sum on the right-hand side of Equation 4.3.2 concerns only those cubes for which osc $> \epsilon$. Each such cube contributes at most $2 \sup |f| \operatorname{vol}_n C$ to the maximum difference between upper and lower sums. (It is $2 \sup |f|$ rather than $\sup |f|$ because the value of f over a single cube might swing from a positive number to a negative one. We could also express this difference as $\sup f - \inf f$.)

The second sum concerns the cubes for which osc $\leq \epsilon$. We must specify that we count only those cubes for which f has, at least somewhere in the cube, a nonzero value; that is why we say $\{C \mid C \cap \operatorname{Supp}(f) \neq \emptyset\}$. Since by definition the oscillation for each of those cubes is at most ϵ, each contributes at most $\epsilon \operatorname{vol}_n C$ to the difference between upper and lower sums.

We have assumed that it is possible to choose N such that the cubes for which osc $> \epsilon$ have total volume less than ϵ, so we replace the first sum by $2\epsilon \sup |f|$. Factoring out ϵ, we see that by choosing N sufficiently large, the upper and lower sums can be made arbitrarily close. Therefore, the function is integrable. This takes care of the "if" part of Theorem 4.3.1.

For the "only if" part we must prove that if the function is integrable, then there exists an appropriate N. Suppose not. Then there exists *one* epsilon, $\epsilon_0 > 0$, such that for *all* N we have

$$\sum_{\{C \in \mathcal{D}_N \mid \operatorname{osc}_C(f) > \epsilon_0\}} \operatorname{vol}_n C \geq \epsilon_0. \qquad 4.3.3$$

To review how to negate statements, see Section 0.2.

In this case, for any N we will have

$$U_N(f) - L_N(f) = \sum_{C \in \mathcal{D}_N} \operatorname{osc}_C(f) \operatorname{vol}_n C$$

$$\geq \sum_{\{C \in \mathcal{D}_N \mid \operatorname{osc}_C(f) > \epsilon_0\}} \overbrace{\operatorname{osc}_C(f)}^{> \epsilon_0} \overbrace{\operatorname{vol}_n C}^{\geq \epsilon_0} \geq \epsilon_0^2. \qquad 4.3.4$$

You might object that in Equation 4.3.2 we argued that by making the ϵ in the last line small, we could get the upper and lower sums to converge to a common limit. Now in Equation 4.3.4 we argue that the ϵ_0^2 in the last line means the sums *don't* converge; yet the square of a small number is smaller yet. The crucial difference is that Equation 4.3.4 concerns *one particular* $\epsilon_0 > 0$, which is fixed and won't get any smaller, while Equation 4.3.2 concerns *any* $\epsilon > 0$, which we can choose arbitrarily small.

The sum of $\operatorname{vol}_n C$ is at least ϵ_0, by Equation 4.3.3, so the upper and the lower integrals will differ by at least ϵ_0^2, and will not tend to a common limit. But we started with the assumption that the function is integrable. $\square$

Theorem 4.3.1 has several important corollaries. Sometimes it is easier to deal with non-negative functions than with functions that can take on both positive and negative values; Corollary 4.3.5 shows how to deal with this.

Definition 4.3.4 (f^+ and f^-). If $f : \mathbb{R}^n \to \mathbb{R}$ is any function, then set

$$f^+(\mathbf{x}) = \begin{cases} f(\mathbf{x}) & \text{if } f(\mathbf{x}) \geq 0 \\ 0 & \text{if } f(\mathbf{x}) < 0 \end{cases} \quad \text{and} \quad f^-(\mathbf{x}) = \begin{cases} -f(\mathbf{x}) & \text{if } f(\mathbf{x}) \leq 0 \\ 0 & \text{if } f(\mathbf{x}) > 0. \end{cases}$$

Clearly both f^+ and f^- are non-negative functions, and $f = f^+ - f^-$.

Corollary 4.3.5. *A bounded function f with bounded support is integrable if and only if both f^+ and f^- are integrable.*

Proof. If f^+ and f^- are both integrable, then so is f by Proposition 4.1.11. For the converse, suppose that f is integrable. Consider a dyadic cube $C \in \mathcal{D}_N(\mathbb{R}^n)$. If f is non-negative on C, then $\operatorname{osc}_C(f) = \operatorname{osc}_C(f^+)$ and $\operatorname{osc}_C(f^-) = 0$. Similarly, if f is non-positive on C, then $\operatorname{osc}_C(f) = \operatorname{osc}_C(f^-)$ and $\operatorname{osc}_C(f^+) = 0$. Finally, if f takes both positive and negative values, then $\operatorname{osc}_C(f^+) < \operatorname{osc}_C(f)$ and $\operatorname{osc}_C(f^-) < \operatorname{osc}_C(f)$. Then Theorem 4.3.1 says that both f^+ and f^- are integrable. $\square$

Proposition 4.3.6 tells us why the characteristic function of the disk discussed in Example 4.3.2 is integrable. We argued in that example that we can make the area of cubes straddling the boundary arbitrarily small. Now we justify that argument. The boundary of the disk is the union of two graphs of functions;

Proposition 4.3.6 says that any bounded part of the graph of an integrable function has volume 0.[3]

The graph of a function $f : \mathbb{R}^n \to \mathbb{R}$ is n-dimensional but it lives in $\mathbb{R}^{n+1}$, just as the graph of a function $f : \mathbb{R} \to \mathbb{R}$ is a curve drawn in the (x, y)-plane. The graph $\Gamma(f)$ can't intersect the cube C_0 because $\Gamma(f)$ is in $\mathbb{R}^{n+1}$ and C_0 is in $\mathbb{R}^n$. We have to add a dimension by using $C_0 \times \mathbb{R}$.

Proposition 4.3.6 (Bounded part of graph has volume 0). *Let* $f : \mathbb{R}^n \to \mathbb{R}$ *be an integrable function with graph* $\Gamma(f)$, *and let* $C_0 \subset \mathbb{R}^n$ *be any dyadic cube. Then*

$$\mathrm{vol}_{n+1}\big(\underbrace{\Gamma(f) \cap (C_0 \times \mathbb{R})}_{\text{bounded part of graph}} \big) = 0 \qquad\qquad 4.3.5$$

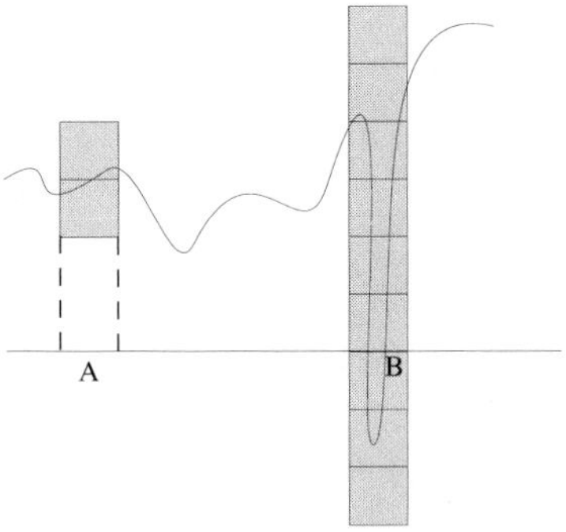

Proof. The proof is not so very hard, but we have two types of dyadic cubes that we need to keep straight: the $(n + 1)$-dimensional cubes that intersect the graph of the function, and the n-dimensional cubes over which the function itself is evaluated. Figure 4.3.3 illustrates the proof with the graph of a function from $\mathbb{R} \to \mathbb{R}$; in that figure, the x-axis plays the role of $\mathbb{R}^n$ in the theorem, and the (x, y)-plane plays the role of $\mathbb{R}^{n+1}$. In this case we have squares that intersect the graph, and intervals over which the function is evaluated. In keeping with that figure, let us denote the cubes in $\mathbb{R}^{n+1}$ by S (for squares) and the cubes in $\mathbb{R}^n$ by I (for intervals).

We need to show that the total volume of the cubes $S \in \mathcal{D}_N(\mathbb{R}^{n+1})$ that intersect $\Gamma(f) \cap (C_0 \times \mathbb{R})$ is small when N is large. Let us choose ϵ, and N satisfying the requirement of Equation 4.3.1 for that ϵ: we decompose C_0 into n-dimensional cubes I small enough so that the total n-dimensional volume of the cubes over which $\mathrm{osc}(f) > \epsilon$ is less than ϵ.

FIGURE 4.3.3.

The graph of a function from $\mathbb{R} \to \mathbb{R}$. Over the interval A, the function has $\mathrm{osc} < \epsilon$; over the interval B, it has $\mathrm{osc} > \epsilon$. Above A, we keep the two cubes that intersect the graph; above B, we keep the entire tower of cubes, including the basement.

Now we count the $(n+1)$-dimensional cubes S that intersect the graph. There are two kinds: those whose projection on $\mathbb{R}^n$ are cubes I with $\mathrm{osc}(f) > \epsilon$, and the others. In Figure 4.3.3, B is an example of an interval with $\mathrm{osc}(f) > \epsilon$, while A is an example of an interval with $\mathrm{osc}(f) < \epsilon$.

For the first sort (large oscillation), think of each n-dimensional cube I over which $\mathrm{osc}(f) > \epsilon$ as the ground floor of a tower of $(n + 1)$-dimensional cubes S that is at most $\sup|f|$ high and goes down (into the basement) at most $-\sup|f|$. To be sure we have enough, we add an extra cube S at top and bottom. Each

---

[3]It would be simpler if we could just write $\mathrm{vol}_{n+1}\big(\Gamma(f)\big) = 0$. The problem is that our definition for integrability requires that an integrable function have bounded support. Although the function is bounded with bounded support, it is *defined* on all of $\mathbb{R}^n$. So even though it has value 0 outside of some fixed big cube, its graph still exists outside the fixed cube, and the characteristic function of its graph does not have bounded support. We fix this problem by speaking of the volume of the intersection of the graph with the $(n + 1)$-dimensional bounded region $C_0 \times \mathbb{R}$. You should imagine that C_0 is big enough to contain the support of f, though the proof works in any case. In Section 4.11, where we define integrability of functions that are not bounded with bounded support, we will be able to say (Corollary 4.11.8) that a graph has volume 0.

tower then contains $2(\sup|f|+1)\cdot 2^N$ such cubes. (We multiply by 2^N because that is the inverse of the height of a cube S. At $N=0$, the height of a cube is 1; at $N=2$, the height is $1/2$, so we need twice as many cubes to make the same height tower.) You will see from Figure 4.3.3 that we are counting more squares than we actually need.

How many such towers of cubes will we need? We chose N large enough so that the total n-dimensional volume of all cubes I with osc $>\epsilon$ is less than ϵ. The inverse of the volume of a cube I is 2^{nN}, so there are $2^{nN}\epsilon$ intervals for which we need towers. So to cover the region of large oscillation, we need in all

$$\underbrace{\epsilon 2^{nN}}_{\substack{\text{no. of cubes }I\\\text{with osc}>\epsilon}}\qquad \underbrace{2(\sup|f|+1)2^N}_{\substack{\text{no. of cubes }S\\\text{for one }I\text{ with osc}(f)>\epsilon}} \qquad\qquad 4.3.6$$

In Equation 4.3.7 we are counting more cubes than necessary: we are using the entire n-dimensional volume of C_0, rather than subtracting the parts over which $\mathrm{osc}(f)>\epsilon$.

$(n+1)$-dimensional cubes S.

For the second sort (small oscillation), for each cube I we require at most $2^N\epsilon+2$ cubes S , giving in all

$$\underbrace{2^{nN}\,\mathrm{vol}_n(C_0)}_{\substack{\text{no. of cubes }I\\\text{to cover }C_0}}\qquad \underbrace{(2^N\epsilon+2)}_{\substack{\text{no. of cubes }S\\\text{for one }I\text{ with osc}(f)<\epsilon}}\qquad . \qquad\qquad 4.3.7$$

Adding these numbers (Equations 4.3.6 and 4.3.7), we find that the bounded part of the graph is covered by

$$2^{(n+1)N}\left(2\epsilon(\sup|f|+1)+\left(\epsilon+\frac{2}{2^N}\right)\mathrm{vol}_n(C_0)\right)\quad\text{cubes }S. \qquad 4.3.8$$

This is of course an enormous number, but recall that each cube has $(n+1)$-dimensional volume $1/2^{(n+1)N}$, so the total volume is

$$2\epsilon(\sup|f|+1)+\left(\epsilon+\frac{2}{2^N}\right)\mathrm{vol}_n(C_0), \qquad\qquad 4.3.9$$

which can be made arbitrarily small. $\square$

As you would expect, a curve in the plane has area 0, a surface in $\mathbb{R}^3$ has volume 0, and so on. Below we must stipulate that such manifolds be compact, since we have defined volume only for bounded subsets of $\mathbb{R}^n$.

Proposition 4.3.7. *If $M\subset\mathbb{R}^n$ is a manifold embedded in $\mathbb{R}^n$, of dimension $k<n$, then any compact subset $X\subset M$ satisfies $\mathrm{vol}_n(X)=0$. In particular, any bounded part of a subspace of dimension $k<n$ has n-dimensional volume 0.*

In Section 4.11 (Proposition 4.11.7) we will be able to drop the requirement that X be compact.

Proof.   We can choose for each $\mathbf{x}\in X$ a neighborhood $U\subset\mathbb{R}^n$ of $\mathbf{x}$ such that $M\cap U$ is a graph of a function expressing $n-k$ coordinates in terms of the other k. Since X is compact, a finite number of these neighborhoods cover

X, so it is enough to prove $\text{vol}_n(M \cap U) = 0$ for such a neighborhood. In the case $k = n - 1$, this follows from Proposition 4.3.6. Otherwise, there exists a

mapping $\mathbf{g} = \begin{bmatrix} g_1 \\ \vdots \\ g_{n-k} \end{bmatrix} : U \to \mathbb{R}^{n-k}$ such that $M \cap U$ is defined by the equation

$\mathbf{g}(\mathbf{x}) = \mathbf{0}$, and such that $[\mathbf{Dg}(\mathbf{x})]$ is onto $\mathbb{R}^{n-k}$ for every $\mathbf{x} \in M \cap U$. Then the locus M_1 given by just the first of these equations, $g_1(\mathbf{x}) = 0$, is a manifold of dimension $n - 1$ embedded in U, so it has n-dimensional volume 0, and since $M \cap U \subset M_1$, we also have $\text{vol}_n(M \cap U) = 0$. $\quad\square$

> The equation $\mathbf{g}(\mathbf{x}) = \mathbf{0}$ that defines $M \cap U$ is $n - k$ equations in n unknowns. Any point $\mathbf{x}$ satisfying $\mathbf{g}(\mathbf{x}) = \mathbf{0}$ necessarily satisfies $g_1(\mathbf{x}) = 0$, so $M \cap U \subset M_1$.

> The second part of the proof is just spelling out the obvious fact that since (for example) a surface in $\mathbb{R}^3$ has three-dimensional volume 0, so does a curve on that surface.

Integrability of continuous functions with bounded support

What functions satisfy the hypothesis of Theorem 4.3.1? One important class is the class of continuous functions with bounded support.

Theorem 4.3.8. *Any continuous function on $\mathbb{R}^n$ with bounded support is integrable.*

> The terms *compact support* and *bounded support* mean the same thing.

> Our proof of Theorem 4.3.8 actually proves a famous and much stronger theorem: every continuous function with bounded support is *uniformly continuous* (see Section 0.2 for a discussion of uniform continuity).

> This is stronger than Theorem 4.3.8 because it shows that the oscillation of a continuous function is small everywhere, whereas integrability requires only that it be small except on a small set. (For example, the characteristic function of the disk in Example 4.3.2 is integrable, although the oscillation is not small on the cubes that straddle the boundary.)

Our previous criterion for integrability, Theorem 4.3.1, defines integrability in terms of dyadic decompositions. It might appear that whether or not a function is integrable could depend on where the function fits on the grid of dyadic cubes; if you nudge the function a bit, might you get different results? Theorem 4.3.8 says nothing about dyadic decompositions, so we see that integrability does not depend on how the function is nudged; in mathematical language, integrability is *translation invariant*.

To prove Theorem 4.3.8 we will need a result from topology—Theorem 1.6.2 about convergent subsequences.

Proof. Suppose the theorem is false; then there certainly exists an $\epsilon_0 > 0$ such that for every N, the total volume of all cubes $C_N \in \mathcal{D}_N$ with $\text{osc} > \epsilon_0$ is at least ϵ_0. In particular, a cube $C_N \in \mathcal{D}_N$ must exist such that $\text{osc}_C(f) > \epsilon_0$. We can restate this in terms of distance between points: C_N contains two points $\mathbf{x}_N, \mathbf{y}_N$ such that

$$|f(\mathbf{x}_N) - f(\mathbf{y}_N)| > \epsilon_0. \qquad 4.3.10$$

These points are in the support of f, so they form two bounded sequences: the infinite sequence composed of the points $\mathbf{x}_N$ for all N, and the infinite sequence composed of the points $\mathbf{y}_N$ for all N. By Theorem 1.6.2 we can extract a convergent subsequence $\mathbf{x}_{N_i}$ that converges to some point $\mathbf{a}$. By Equation 4.1.17,

$$|\mathbf{x}_{N_i} - \mathbf{y}_{N_i}| \leq \frac{\sqrt{n}}{2^{N_i}}, \qquad 4.3.11$$

so we see that $\mathbf{y}_{N_i}$ also converges to $\mathbf{a}$.

Since f is continuous at $\mathbf{a}$, then for any ϵ there exists δ such that if $|\mathbf{x}-\mathbf{a}| < \delta$ then $|f(\mathbf{x}) - f(\mathbf{a})| < \epsilon$; in particular we can choose $\epsilon = \epsilon_0/4$, so $|f(\mathbf{x}) - f(\mathbf{a})| < \epsilon_0/4$.

For N sufficiently large, $|\mathbf{x}_{N_i} - \mathbf{a}| < \delta$ and $|\mathbf{y}_{N_i} - \mathbf{a}| < \delta$. Thus (using the triangle inequality, Theorem 1.4.9),

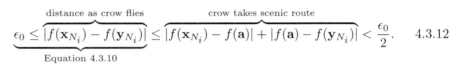

$$\epsilon_0 \leq \underbrace{\overbrace{|f(\mathbf{x}_{N_i}) - f(\mathbf{y}_{N_i})|}^{\text{distance as crow flies}}}_{\text{Equation 4.3.10}} \leq \overbrace{|f(\mathbf{x}_{N_i}) - f(\mathbf{a})| + |f(\mathbf{a}) - f(\mathbf{y}_{N_i})|}^{\text{crow takes scenic route}} < \frac{\epsilon_0}{2}. \qquad 4.3.12$$

But $\epsilon_0 < \epsilon_0/2$ is false, so our hypothesis is faulty: f is integrable. $\quad\square$

Corollary 4.3.9. *Any bounded part of the graph of a continuous function has volume 0.*

A function need not be continuous everywhere to be integrable, as our third theorem shows. This theorem is much harder to prove than the first two, but the criterion for integrability is much more useful.

Theorem 4.3.10. *A function $f : \mathbb{R}^n \to \mathbb{R}$, bounded with bounded support, is integrable if it is continuous except on a set of volume 0.*

Note that Theorem 4.3.10, like Theorem 4.3.8 but unlike Theorem 4.3.1, is not an "if and only if" statement. As will be seen in the optional Section 4.4, it is possible to find functions that are discontinuous at all the rationals, yet still are integrable.

Proof. Denote by Δ ("delta") the set of points where f is discontinuous:

$$\Delta = \{\mathbf{x} \in \mathbb{R}^n \mid f \text{ is not continuous at } \mathbf{x}\}. \qquad 4.3.13$$

Choose some $\epsilon > 0$. Since f is continuous except on a set of volume 0, we have $\text{vol}_n \Delta = 0$. So (by Definition 4.1.18) there exists N and some finite union of cubes $C_1, \ldots, C_k \in \mathcal{D}_N(\mathbb{R}^n)$ such that

$$\Delta \subset C_1 \cup \cdots \cup C_k \quad \text{and} \quad \sum_{i=1}^{k} \text{vol}_n C_i \leq \frac{\epsilon}{3^n}. \qquad 4.3.14$$

Now we create a "buffer zone" around the discontinuities: let L be the union of the C_i and all the surrounding cubes at level N, as shown in Figure 4.3.4. As illustrated by Figure 4.3.5, we can completely surround each C_i, using $3^n - 1$ cubes (3^n including itself). Since the total volume of all the C_i is less than $\epsilon/3^n$,

$$\text{vol}_n(L) \leq \epsilon. \qquad 4.3.15$$

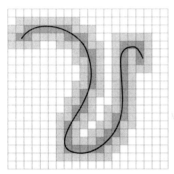

FIGURE 4.3.4.

The black curve represents Δ; the darkly shaded region consists of the cubes at some level that intersect Δ. The lightly shaded region consists of the cubes at the same depth that border at least one of the previous cubes.

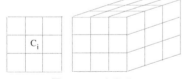

FIGURE 4.3.5.

In $\mathbb{R}^2, 3^2 - 1 = 8$ cubes are enough to completely surround a cube C_i. In $\mathbb{R}^3, 3^3 - 1 = 26$ cubes are enough to completely surround a cube C_i. If we include the cube C_i, then 3^2 cubes are enough in $\mathbb{R}^2$, and 3^3 in $\mathbb{R}^3$.

Moreover, since the length of a side of a cube is $1/2^N$, every point of $\mathbb{R}^n - L$ is at least $1/2^N$ away from Δ.

All that remains is to show that there exists $M \geq N$ such that if $C \in \mathcal{D}_M$ and $C \not\subset L$, then $\mathrm{osc}_C(f) \leq \epsilon$. If we can do that, we will have shown that a decomposition exists at which the total volume of all cubes over which $\mathrm{osc}(f) > \epsilon$ is less than ϵ, which is the criterion for integrability given by Theorem 4.3.1.

Suppose no such M exists. Then for every $M \geq N$, there is a cube $C \in \mathcal{D}_M$ and points $\mathbf{x}_M, \mathbf{y}_M \in C$ with $|f(\mathbf{x}_M) - f(\mathbf{y}_M)| > \epsilon$.

The $\mathbf{x}_M$ are a bounded sequence in $\mathbb{R}^n$, so we can extract a subsequence $\mathbf{x}_{M_i}$ that converges to some point $\mathbf{a}$. Since (again using the triangle inequality)

$$|f(\mathbf{x}_{M_i}) - f(\mathbf{a})| + |f(\mathbf{a}) - f(\mathbf{y}_{M_i})| \geq |f(\mathbf{x}_{M_i}) - f(\mathbf{y}_{M_i})| > \epsilon, \qquad 4.3.16$$

we see that at least one of $|f(\mathbf{x}_{M_i}) - f(\mathbf{a})|$ and $|f(\mathbf{y}_{M_i}) - f(\mathbf{a})|$ does not converge to 0, so f is not continuous at $\mathbf{a}$, i.e., $\mathbf{a} \in \Delta$. But this contradicts the fact that $\mathbf{a}$ is a limit of points outside of L. Since all $\mathbf{x}_{M_i}$ are at least $1/2^N$ away from points of Δ, $\mathbf{a}$ is also at least $1/2^N$ away from points of Δ. $\square$

Corollary 4.3.11. *If f is an integrable function on $\mathbb{R}^n$, and g is another bounded function such that $f = g$ except on a set of volume 0, then g is integrable, and*

$$\int_{\mathbb{R}^n} f \, |d^n \mathbf{x}| = \int_{\mathbb{R}^n} g \, |d^n \mathbf{x}|. \qquad 4.3.17$$

Corollary 4.3.12 says that virtually all examples that occur in "vector calculus" examples are integrable.

Corollary 4.3.12. *Let $A \subset \mathbb{R}^n$ be a region bounded by a finite union of graphs of continuous functions, and let $f : A \to \mathbb{R}$ be continuous. Then the function $\widetilde{f} : \mathbb{R}^n \to \mathbb{R}$ that is $f(\mathbf{x})$ for $\mathbf{x} \in A$ and 0 outside A is integrable.*

In particular, the characteristic function of the disk is integrable, since the disk is bounded by the graphs of the two functions

Exercise 4.3.1 asks you to give an explicit bound for the number of cubes of $\mathcal{D}_N(\mathbb{R}^2)$ needed to cover the unit circle.

$$y = +\sqrt{1 - x^2} \quad \text{and} \quad y = -\sqrt{1 - x^2}. \qquad 4.3.18$$

4.4 INTEGRATION AND MEASURE ZERO (OPTIONAL)

There is measure in all things.—Horace

We mentioned in Section 4.3 that the criterion for integrability given by Theorem 4.3.10 is not sharp. It is not necessary that a function (bounded and with bounded support) be continuous except on a set of volume 0 to be integrable: it is sufficient that it be continuous except on a set of *measure* 0.

The measure theory approach to integration, *Lebesgue integration*, is superior to Riemann integration from several points of view. It makes it possible to integrate otherwise unintegrable functions, and it is better behaved than Riemann integration with respect to limits $f = \lim f_n$. However, the theory takes much longer to develop and is poorly adapted to computation. For the kinds of problems treated in this book, Riemann integration is adequate.

Our boxes B_i are open cubes. But the theory applies to boxes B_i that have other shapes: Definition 4.4.1 works with the B_i as arbitrary sets with well-defined volume. Exercise 4.4.2 asks you to show that you can use balls, and Exercise 4.4.3 asks you to show that you can use arbitrary pavable sets.

FIGURE 4.4.1.

The set X, shown as a heavy line, is covered by boxes that overlap.

We say that the sum of the lengths is less than ϵ because some of the intervals overlap.

The set U_ϵ is interesting in its own right; Exercise A18.1 explores some of its bizarre properties.

Measure theory is a big topic, beyond the scope of this book. Fortunately, the notion of *measure* 0 is much more accessible. "Measure 0" is a subtle notion with some bizarre consequences; it gives us a way, for example, of saying that the rational numbers "don't count." Thus it allows us to use Riemann integration to integrate some quite interesting functions, including one we explore in Example 4.4.3 as a reasonable model for space averages in statistical mechanics.

In the definition below, a *box* B in $\mathbb{R}^n$ of side $\delta > 0$ will be a cube of the form

$$\{\mathbf{x} \in \mathbb{R}^n \mid a_i < x_i < a_i + \delta,\ i = 1, \dots, n\}. \qquad 4.4.1$$

There is no requirement that the a_i or δ be dyadic.

Definition 4.4.1 (Measure 0). A set $X \subset \mathbb{R}^n$ has measure 0 if and only if for every $\epsilon > 0$, there exists an infinite sequence of open boxes B_i such that

$$X \subset \cup B_i \quad \text{and} \quad \sum \text{vol}_n(B_i) \leq \epsilon. \qquad 4.4.2$$

That is, the set can be contained in a possibly infinite sequence of boxes (intervals in $\mathbb{R}$, squares in $\mathbb{R}^2$, ...) whose total volume is $\leq$ epsilon. The crucial difference between measure and volume is the word *infinite* in Definition 4.4.1. A set with volume 0 can be contained in a *finite* sequence of cubes whose total volume is arbitrary small. A set with volume 0 necessarily has measure 0, but it is possible for a set to have measure 0 but not to have a defined volume, as shown in Example 4.4.2.

We speak of boxes rather than cubes to avoid confusion with the cubes of our dyadic pavings. In dyadic pavings, we considered "families" of cubes all of the same size: the cubes at a particular resolution N, and fitting the dyadic grid. The boxes B_i of Definition 4.4.1 get small as i increases, since their total volume is less than ϵ, but it is not necessarily the case that any particular box is smaller than the one immediately preceding it. The boxes can overlap, as illustrated in Figure 4.4.1, and they are not required to square with any particular grid.

Finally, you may have noticed that the boxes in Definition 4.4.2 are open, while the dyadic cubes of our paving are semi-open. In both cases, this is just for convenience; the theory could be built just as well with closed cubes and boxes (see Exercise 4.4.1).

Example 4.4.2 (A set with measure 0, undefined volume). The set of rational numbers in the interval $[0,1]$ has measure 0. You can list them in order $1, 1/2, 1/3, 2/3, 1/4, 2/4, 3/4, 1/5 \dots$. (The list is infinite and includes some numbers more than once.) Center an open interval of length $\epsilon/2$ at 1, an open interval of length $\epsilon/4$ at $1/2$, an open interval of length $\epsilon/8$ at $1/3$, and so on. Call U_ϵ the union of these intervals. The sum of the lengths of these intervals (i.e., $\sum \text{vol}_1$) will be less than $\epsilon (1/2 + 1/4 + 1/8 + \dots) = \epsilon$.

The set of Example 4.4.2 is a good one to keep in mind while trying to picture the boxes B_i, because it helps us to see that while the sequence B_i is made of $B_1, B_2, \ldots$, in order, these boxes may skip around. The "boxes" here are the intervals; if B_1 is centered at $1/2$, then B_2 is centered at $1/3$, B_3 at $2/3$, B_4 at $1/4$, and so on. We also see that some boxes may be contained in others: for example, depending on the choice of ϵ, the interval centered at $17/32$ may be contained in the interval centered at $1/2$.

Statistical mechanics is an attempt to apply probability theory to large systems of particles, to estimate average quantities, like temperature, pressure, etc., from the laws of mechanics. Thermodynamics, on the other hand, is a completely macroscopic theory, trying to relate the same macroscopic quantities (temperature, pressure, etc.) on a phenomenological level. Clearly, one hopes to explain thermodynamics by statistical mechanics.

You can place all the rationals in $[0,1]$ in intervals that are infinite in number but whose total length is arbitrarily small! The set thus has measure 0. However it does not have a defined volume: if you were to try to measure the volume, you would fail because you could never divide the interval $[0,1]$ into intervals so small that they contain only rational numbers.

We already ran across this set in Example 4.3.3, when we found that we could not integrate the function that is 1 at rational numbers in the interval $[0,1]$ and 0 elsewhere. This function is discontinuous everywhere; in every interval, no matter how small, it jumps from 0 to 1 and from 1 to 0.   △

In Example 4.4.3 we see a function that looks similar but is very different. This function *is* continuous except over a set of measure 0, and thus *is* integrable. It arises in real life (statistical mechanics, at least).

Example 4.4.3 (An integrable function with discontinuities on a set of measure 0). The function

$$f(x) = \begin{cases} \frac{1}{q} & \text{if } x = \frac{p}{q} \text{ is rational, } |x| \le 1 \text{ and written in lowest terms} \\ 0 & \text{if } x \text{ is irrational, or } |x| > 1 \end{cases} \quad 4.4.3$$

is integrable. The function is discontinuous at values of x for which $f(x) \ne 0$. For instance, $f(3/4) = 1/4$, while arbitrarily close to $3/4$ we have irrational numbers such that $f(x) = 0$. But such values form a set of measure 0. The function is continuous at the irrationals: arbitrarily close to any irrational number x you will find rational numbers p/q, but you can choose a neighborhood of x that includes only rational numbers with arbitrarily large denominators q, so that $f(y)$ will be arbitrarily small.   △

The function of Example 4.4.3 is important because it is a model for functions that show up in an essential way in statistical mechanics (unlike the function of Example 4.4.2, which, as far as we know, is only a pathological example, devised to test the limits of mathematical statements).

In statistical mechanics, one tries to describe a system, typically a gas enclosed in a box, made up of perhaps 10^{25} molecules. Quantities of interest might be temperature, pressure, concentrations of various chemical compounds, etc.

A state of the system is a specification of the position and velocity of each molecule (and rotational velocity, vibrational energy, etc., if the molecules have inner structure); to encode this information one might use a point in some gadzillion dimensional space.

Mechanics tells us that at the beginning of our experiment, the system is in some state that evolves according to the laws of physics, "exploring" as time proceeds some part of the total state space (and exploring it quite fast relative to our time scale: particles in a gas at room temperature typically travel at several hundred meters per second, and undergo millions of collisions per second.)

The guess underlying thermodynamics is that the quantity one measures, which is really a time average of the quantity as measured along the trajectory of the system, should be nearly equal in the long run to the average over all possible states, called the *space average*. (Of course the "long run" is quite a short run by our clocks.)

This equality of time averages and space averages is called *Boltzmann's ergodic hypothesis*. There aren't many mechanical systems where it is mathematically proved to be true, but physicists believe that it holds in great generality, and it is the key hypothesis that connects statistical mechanics to thermodynamics.

We discussed Example 4.4.3 to show that such bizarre functions can have physical meaning. However, we do not mean to suggest that because the rational numbers have measure 0, trajectories with rational slopes are never important for understanding the evolution of dynamical systems. On the contrary: questions of rational vs. irrational numbers are central to understanding the intricate interplay of chaotic and stable behavior exhibited, for example, by the *lakes of Wada*. (For more on this topic, see J.H. Hubbard, *What it Means to Understand a Differential Equation*, The College Mathematics Journal, Vol. 25, (Nov. 5, 1994), 372-384.)

Now what does this have to do with our function f above? Even if you believe that a generic time evolution will explore state space fairly evenly, there will always be some trajectories that don't. Consider the (considerably simplified) model of a single particle, moving without friction on a square billiard table, with ordinary bouncing when it hits an edge (the angle of incidence equal to the angle of reflection). Then most trajectories will evenly fill up the table, in fact precisely those that start with irrational slope. But those with rational slopes emphatically will not: they will form closed trajectories, which will go over and over the same closed path. Still, as shown in Figure 4.4.2, these closed paths will visit more and more of the table as the denominator of the slope becomes large.

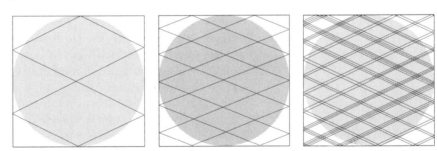

FIGURE 4.4.2. The trajectory with slope 2/5, at center, visits more of the square than the trajectory with slope 1/2, at left. The slope of the trajectory at right closely approximates an irrational number; if allowed to continue, this trajectory would visit nearly every part of the square. (In practice, the drawing would soon become black.)

Suppose further that the quantity to be observed is some function f on the table with average 0, which is positive near the center and very negative near the corners. Moreover, suppose we start our particle at the center of the table but don't specify its direction. This is some caricature of reality, where in the laboratory we set up the system in some macroscopic configuration, like having one gas in half a box and another in another half, and remove the partition. This corresponds to knowing something about the initial state, but is a very long way from knowing it exactly.

Trajectories through the center of the table, and with slope 0, will have positive time averages, as will trajectories with slope ∞. Similarly, we believe that the average, over time, of each trajectory with rational slope will also be positive; the trajectory will miss the corners. But trajectories with irrational slope will have 0 time averages: given enough time these trajectories will visit each part of the table equally. And trajectories with rational slopes with large denominators will have time averages close to 0.

Because the rational numbers have measure 0, their contribution to the average does not matter; in this case, at least, Boltzmann's ergodic hypothesis seems correct. $\triangle$

Integrability of "almost" continuous functions

We are now ready to prove Theorem 4.4.4:

Theorem 4.4.4 is stronger than Theorem 4.3.10, since any set of volume 0 also has measure 0, and not conversely. It is also stronger than Theorem 4.3.8.

But it is not stronger than Theorem 4.3.1. Theorems 4.4.4 and 4.3.1 both give an "if and only if" condition for integrability; they are exactly equivalent. But it is often easier to verify that a function is integrable using Theorem 4.4.4. It also makes it clear that whether or not a function is integrable does not depend on where a function is placed on some arbitrary grid.

We prune the list of boxes by throwing away any box that is contained in an earlier one. We could prove our result without pruning the list, but it would make the argument more cumbersome.

Recall (Definition 4.1.4) that $\mathrm{osc}_{B_i}(f)$ is the oscillation of f over B_i: the difference between the least upper bound of f over B_i and the greatest lower bound of f over B_i.

Theorem 4.4.4. *A function* $f : \mathbb{R}^n \to \mathbb{R}$, *bounded and with bounded support, is integrable if and only if it is continuous except on a set of measure 0.*

Proof.   Since this is an "if and only if" statement, we must prove both directions. We will start with the harder one: if a function $f : \mathbb{R}^n \to \mathbb{R}$, bounded and with bounded support, is continuous except on a set of measure 0, then it is integrable. We will use the criterion for integrability given by Theorem 4.3.1; thus we want to prove that for all $\epsilon > 0$ there exists N such that the cubes $C \in \mathcal{D}_N$ over which $\mathrm{osc}(f) > \epsilon$ have a combined volume less than ϵ.

We will denote by Δ the set of points where f is not continuous, and we will choose some $\epsilon > 0$ (which will remain fixed for the duration of the proof). By Definition 4.4.1 of measure 0, there exists a sequence of boxes B_i such that

$$\Delta \subset \cup B_i \quad \text{and} \quad \sum \mathrm{vol}_n B_i < \epsilon. \qquad 4.4.4$$

We will choose our boxes so that no box is contained in any other.

The proof is fairly involved. First, we want to get rid of infinity.

Lemma 4.4.5.   *There are only finitely many boxes* B_i *on which* $\mathrm{osc}_{B_i}(f) > \epsilon$.

We will denote such boxes B_{i_j}, and denote by L the union of the B_{i_j}.

Proof of Lemma 4.4.5.   We will prove Lemma 4.4.5 by contradiction. Assume it is false. Then there exist an infinite subsequence of boxes B_{i_j} and two infinite sequences of points, $\mathbf{x}_j, \mathbf{y}_j \in B_{i_j}$, such that $|f(\mathbf{x}_j) - f(\mathbf{y}_j)| > \epsilon$.

The sequence $\mathbf{x}_j$ is bounded, since the support of f is bounded and $\mathbf{x}_j$ is in the support of f. So (by Theorem 1.6.2) it has a convergent subsequence $\mathbf{x}_{j_k}$ converging to some point $\mathbf{p}$. Since $|f(\mathbf{x}_j) - f(\mathbf{y}_j)| \to 0$ as $j \to \infty$, the subsequence $\mathbf{y}_{j_k}$ also converges to $\mathbf{p}$.

The point $\mathbf{p}$ has to be in a particular box, which we will call B_p. (Since the boxes can overlap, it could be in more than one, but we just need one.) Since x_{j_k} and y_{j_k} converge to $\mathbf{p}$, and since the B_{i_j} get small as j gets big (their total volume being less than ϵ), then all $B_{i_{j_k}}$ after a certain point will be contained in B_p. But this contradicts our assumption that we had pruned our list of B_i so that no one box was contained in any other. Therefore Lemma 4.4.5 is correct: there are only finitely many B_i on which $\mathrm{osc}_{B_i} f > \epsilon$. (Our indices have proliferated in an unpleasant fashion. As illustrated in Figure 4.4.3, B_i are the sequences of boxes that cover Δ, i.e., the set of discontinuities; B_{i_j} are those B_i's where osc $> \epsilon$; and $B_{i_{j_k}}$ are those B_{i_j}'s that form a convergent subsequence.)

Now we assert that if we use dyadic pavings to pave the support of our function f, then:

Lemma 4.4.6. *There exists N such that if $C \in \mathcal{D}_N(\mathbb{R}^n)$ and $\mathrm{osc}_C f > \epsilon$, then $C \subset L$.*

That is, we assert that f can have osc $> \epsilon$ only over C's that are in L. If we prove this, we will be finished, because by Theorem 4.3.1, a bounded function with bounded support is integrable if there exists an N at which the total volume of cubes with osc $> \epsilon$ is less than ϵ. We know that L is a finite set of B_i, and (Equation 4.4.4) that the B_i have total volume $\le \epsilon$.

To prove Lemma 4.4.6, we will again argue by contradiction. Suppose the lemma is false. Then for every N, there exists a C_N not a subset of L such that $\mathrm{osc}_{C_N} f > \epsilon$. In other words,

$$\exists \text{ points } \mathbf{x}_N, \mathbf{y}_N, \mathbf{z}_N \text{ in } C_N, \text{ with } \mathbf{z}_N \notin L, \text{ and } |f(\mathbf{x}_N) - f(\mathbf{y}_N)| > \epsilon. \quad 4.4.5$$

Since $\mathbf{x}_N, \mathbf{y}_N$, and $\mathbf{z}_N$ are infinite sequences (for $N = 1, 2, \ldots$), then there exist convergent subsequences x_{N_i}, y_{N_i} and z_{N_i}, all converging to the same point, which we will call $\mathbf{q}$.

What do we know about $\mathbf{q}$?

- $\mathbf{q} \in \Delta$: i.e., it is a discontinuity of f. (No matter how close $\mathbf{x}_{N_i}$ and $\mathbf{y}_{N_i}$ get to $\mathbf{q}$, $|f(\mathbf{x}_{N_i}) - f(\mathbf{y}_{N_i})| > \epsilon$.) Therefore (since all the discontinuities of the function are contained in the B_i), it is in some box B_i, which we'll call B_q.

- $\mathbf{q} \notin L$. (The set L is open, so its complement is closed; since no point of the sequence $\mathbf{z}_{N_i}$ is in L, its limit, $\mathbf{q}$, is not in L either.)

Since $\mathbf{q} \in B_q$, and $\mathbf{q} \notin L$, we know that B_q is not one of the boxes with osc $> \epsilon$. But that isn't true, because x_{N_i} and y_{N_i} are in B_q for N_i large enough, so that $\mathrm{osc}_{B_q} f < \epsilon$ contradicts $|f(x_{N_i}) - f(y_{N_i})| > \epsilon$.

Therefore, we have proved Lemma 4.4.6, which, as we mentioned above, means that we have proved Theorem 4.4.4 in one direction: if a bounded function with bounded support is continuous except on a set of measure 0, then it is integrable.

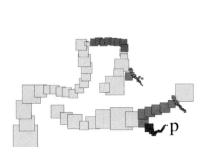

FIGURE 4.4.3.

The collection of boxes covering Δ is lightly shaded; those with osc $> \epsilon$ are shaded slightly darker. A convergent subsequence of those is shaded darker yet: the point p to which they converge must belong to some box.

You may ask, how do we know they converge to the same point? Because $\mathbf{x}_{N_i}, \mathbf{y}_{N_i}$, and $\mathbf{z}_{N_i}$ are all in the same cube, which is shrinking to a point as $N \to \infty$.

The boxes B_i making up L are in $\mathbb{R}^n$, so the complement of L is $\mathbb{R}^n - L$. Recall (Definition 1.5.4) that a closed set $C \subset \mathbb{R}^n$ is a set whose complement $\mathbb{R}^n - C$ is open.

Now we need to prove the other direction: if a function $f : \mathbb{R}^n \to \mathbb{R}$, bounded and with bounded support, is integrable, then it is continuous except on a set of measure 0. This is easier, but the fact that we chose our dyadic cubes half-open, and our boxes open, introduces a little complication.

Since f is integrable, we know (Theorem 4.3.1) that for any $\epsilon > 0$, there exists N such that the finite union of cubes

$$\{C \in \mathcal{D}_N(\mathbb{R}^n) \mid \operatorname{osc}_C(f) > \epsilon\} \qquad 4.4.6$$

has total volume less than ϵ.

Apply Equation 4.4.6, setting $\epsilon_1 = \delta/4$, with $\delta > 0$. Let $\mathcal{C}_{N_1}$ be the finite collection of cubes $C \in \mathcal{D}_{N_1}(\mathbb{R}^n)$ with $\operatorname{osc}_C f \geq \delta/4$. These cubes have total volume less than $\delta/4$. Now we set $\epsilon_2 = \delta/8$, and let $\mathcal{C}_{N_2}$ be the finite collection of cubes $C \in \mathcal{D}_{N_2}(\mathbb{R}^n)$ with $\operatorname{osc}_C f \geq \delta/8$; these cubes have total volume less than $\delta/8$. Continue with $\epsilon_3 = \delta/16, \dots$.

Finally, consider the infinite sequence of open boxes $B_1, B_2, \dots$ obtained by listing first the interiors of the elements of $\mathcal{C}_{N_1}$, then those of the elements of $\mathcal{C}_{N_2}$, etc.

This almost solves our problem: the total volume of our sequence of boxes is at most $\delta/4 + \delta/8 + \dots = \delta/2$. The problem is that discontinuities on the boundary of dyadic cubes may go undetected by oscillation on dyadic cubes: as shown in Figure 4.4.4, the value of the function over one cube could be 0, and the value over an adjacent cube could be 1; in each case the oscillation over the cube would be 0, but the function would be discontinuous at points on the border between the two cubes.

To deal with this, we simply shift our cubes by an irrational amount, as shown to the right of Figure 4.4.4, and repeat the above process.

To do this, we set

$$\widetilde{f}(\mathbf{x}) = f(\mathbf{x} - \mathbf{a}), \quad \text{where} \quad \mathbf{a} = \begin{bmatrix} \sqrt{2} \\ \vdots \\ \sqrt{2} \end{bmatrix}. \qquad 4.4.7$$

(We could translate $\mathbf{x}$ by any number with irrational entries, or indeed by a rational like $1/3$.) Repeat the argument, to find a sequence $\widetilde{B}_1, \widetilde{B}_2, \dots$.

Now translate these back: set

$$B'_i = \left\{ \mathbf{x} - \mathbf{a} \,\middle|\, \mathbf{x} \in \widetilde{B}_i \right\}. \qquad 4.4.8$$

Now we claim that the sequence $B_1, B'_1, B_2, B'_2, \dots$ solves our problem. We have

$$\operatorname{vol}_n(B'_1) + \operatorname{vol}_n(B'_2) + \dots < \frac{\delta}{4} + \frac{\delta}{8} + \dots = \frac{\delta}{2}, \qquad 4.4.9$$

so the total of volume of the sequence $B_1, B'_1, B_2, B'_2, \dots$ is less than δ.

Now we need to show that f is continuous on the complement of

$$B_1 \cup B'_1 \cup B_2 \cup B'_2, \cup \dots, \qquad 4.4.10$$

FIGURE 4.4.4.

The function that is identically 1 on the indicated dyadic cube and 0 elsewhere is discontinuous on the boundary of the dyadic cube. For instance, the function is 0 on one of the indicated sequences of points, but its value at the limit is 1. This point is in the interior of the shaded cube of the dotted grid.

Definition 4.4.1 specifies that the boxes B_i are open. Equation 4.1.13 defining dyadic cubes shows that they are half-open: x_i is greater than or equal to one amount, but strictly less than another:

$$\frac{k_i}{2^N} \leq x_i < \frac{k_i + 1}{2^N}.$$

i.e., on $\mathbb{R}^n$ minus the union of the B_i and B_i'. Indeed, if $\mathbf{x}$ is a point where f is not continuous, then at least one of $\mathbf{x}$ and $\widetilde{\mathbf{x}} = \mathbf{x} + \mathbf{a}$ is in the interior of a dyadic cube.

Suppose that the first is the case. Then there exists N_k and a sequence $\mathbf{x}_i$ converging to $\mathbf{x}$ so that $|f(\mathbf{x}_i) - f(\mathbf{x})| > \delta/2^{1+N_k}$ for all i; in particular, that cube will be in the set $\mathcal{C}_{N_k}$, and $\mathbf{x}$ will be in one of the B_i.

If $\mathbf{x}$ is not in the interior of a dyadic cube, then $\widetilde{\mathbf{x}}$ is a point of discontinuity of $\widetilde{f}$, and the same argument applies. $\square$

4.5 Fubini's Theorem and Iterated Integrals

We now know—in principle, at least—how to determine whether a function is integrable. Assuming it is, how do we integrate it? Fubini's theorem allows us to compute multiple integrals by hand, or at least reduce them to the computation of one-dimensional integrals. It asserts that if $f : \mathbb{R}^n \to \mathbb{R}$ is integrable, then

$$\int_{\mathbb{R}^n} f \, |d^n\mathbf{x}| = \int_{-\infty}^{\infty} \left(\dots \left(\int_{-\infty}^{\infty} f \begin{pmatrix} x_1 \\ \vdots \\ x_n \end{pmatrix} dx_1 \right) \dots \right) dx_n. \qquad 4.5.1$$

The expression on the left-hand side of Equation 4.5.1 doesn't specify the order in which the variables are taken, so the iterated integral on the right could be written in any order: we could integrate first with respect to x_n, or any other variable, rather than x_1. This is important for both theoretical and computational uses of Fubini's theorem.

That is, first we hold the variables $x_2, \dots, x_n$ constant and integrate with respect to x_1; then we integrate the resulting (no doubt complicated) function with respect to x_2, and so on.

Remark. The above statement is not quite correct, because some of the functions in parentheses on the right-hand side of Equation 4.5.1 may not be integrable; this problem is discussed (Example A13.1) in Appendix A.13. We state Fubini's theorem correctly at the end of this section. For now, just assume that we are in the (common) situation where the above statement works. $\triangle$

In practice, the main difficulty in setting up a multiple integral as an iterated one-dimensional integral is dealing with the "boundary" of the region over which we wish to integrate the function. We tried to sweep difficulties like the fractal coastline of Britain under the rug by choosing to integrate over all of $\mathbb{R}^n$, but of course those difficulties are still there. This is where we have to come to terms with them: we have to figure out the upper and lower limits of the integrals.

If the domain of integration looks like the coastline of Britain, it is not at all obvious how to go about this. For domains of integration bounded by smooth curves and surfaces, formulas exist in many cases that are of interest (particularly during calculus exams), but this is still the part that gives students the most trouble.

Before computing any multiple integrals, let's see how to set them up. While a multiple integral is *computed* from inside out— first with respect to the variable in the inner parentheses—we recommend *setting up* the problem from outside in, as shown in Examples 4.5.1 and 4.5.2.

Example 4.5.1 (Setting up multiple integrals: an easy example). Suppose we want to integrate a function $f \begin{pmatrix} x \\ y \end{pmatrix}$ over the triangle

By "integrate over the triangle" we mean that we imagine that the function f is defined by some formula inside the triangle, and outside the triangle $f = 0$.

$$T = \left\{ \begin{pmatrix} x \\ y \end{pmatrix} \in \mathbb{R}^2 \,\middle|\, 0 \le 2x \le y \le 2 \right\} \qquad 4.5.2$$

shown in Figure 4.5.1. This triangle is the intersection of the three regions (in this case, half-planes) defined by the three inequalities $0 \le x$, $2x \le y$, and $y \le 2$.

Say we want to integrate first with respect to y. We set up the integral as follows, temporarily omitting the limits of integration:

$$\iint_{\mathbb{R}^2} f \begin{pmatrix} x \\ y \end{pmatrix} \, dx \, dy = \int \left(\int f \, dy \right) dx. \qquad 4.5.3$$

(We just write f for the function, as we don't want to complicate issues by specifying a particular function.) Starting with the outer integral—thinking first about x—we hold a pencil parallel to the y-axis and roll it over the triangle from left to right. We see that the triangle (the domain of integration) starts at $x = 0$ and ends at $x = 1$, so we write in those limits:

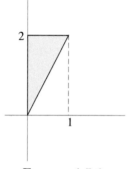

FIGURE 4.5.1.

The triangle defined by Equation 4.5.2.

$$\int_0^1 \left(\int f \, dy \right) dx. \qquad 4.5.4$$

Once more we roll the pencil from $x = 0$ to $x = 1$, this time asking ourselves what are the upper and lower values of y for each value of x? The upper value is always $y = 2$. The lower value is given by the intersection of the pencil with the hypotenuse of the triangle, which lies on the line $y = 2x$. Therefore the lower value is $y = 2x$, and we have

$$\int_0^1 \left(\int_{2x}^2 f \, dy \right) dx. \qquad 4.5.5$$

If we want to start by integrating f with respect to x, we write

$$\iint_{\mathbb{R}^2} f \begin{pmatrix} x \\ y \end{pmatrix} \, dx \, dy = \int \left(\int f \, dx \right) dy, \qquad 4.5.6$$

and, again starting with the outer integral, we hold our pencil parallel to the x-axis and roll it from the bottom of the triangle to the top, from $y = 0$ to $y = 2$. As we roll the pencil, we ask what are the lower and upper values of x for each value of y. The lower value is always $x = 0$, and the upper value is set by the hypotenuse, but we express it now in terms of x, getting $x = y/2$. This gives us

$$\int_0^2 \left(\int_0^{\frac{y}{2}} f \, dx \right) dy. \qquad 4.5.7$$

Now suppose we are integrating over only part of the triangle, as shown in Figure 4.5.2. What limits do we put in the expression $\int(\int f\,dy)\,dx$? Try it yourself before checking the answer in the footnote.[4] $\triangle$

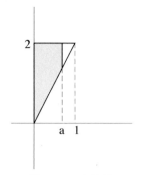

FIGURE 4.5.2.

The shaded area represents a truncated part of the triangle of Figure 4.5.1

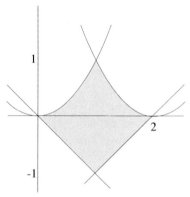

FIGURE 4.5.3.

The region of integration for Example 4.5.2.

Example 4.5.2 (Setting up multiple integrals: a somewhat harder example). Now let's integrate an unspecified function $f\begin{pmatrix} x \\ y \end{pmatrix}$ over the area bordered on the top by the parabolas $y = x^2$ and $y = (x-2)^2$ and on the bottom by the straight lines $y = -x$ and $y = x - 2$, as shown in Figure 4.5.3.

Let's start again by sweeping our pencil from left to right, which corresponds to the outer integral being with respect to x. The limits for the outer integral are clearly $x = 0$ and $x = 2$, giving

$$\int_0^2 \left(\int f\,dy \right) dx. \qquad 4.5.8$$

As we sweep our pencil from left to right, we see that the lower limit for y is set by the straight line $y = -x$, and the upper limit by the parabola $y = x^2$, so we are tempted to write

$$\int_0^2 \left(\int_{-x}^{x^2} f\,dy \right) dx. \qquad 4.5.9$$

But once our pencil arrives at $x = 1$, we have a problem. The lower limit is now set by the straight line $y = x - 2$, and the upper limit by the parabola $y = (x-2)^2$. How can we express this? Try it yourself before looking at the answer in the footnote below.[5] $\triangle$

Exercise 4.5.2 asks you to set up the multiple integral for Example 4.5.2 when the outer integral is with respect to y. Exercise 4.5.3 asks you to set up the multiple integral $\int(\int f\,dx)\,dy$ for the truncated triangle shown in Figure 4.5.2. In both cases the answer will be a sum of integrals.

Example 4.5.3 (Setting up a multiple integral in $\mathbb{R}^3$). As you might imagine, already in $\mathbb{R}^3$ this kind of visualization becomes much harder. Here

---

[4]When the domain of integration is the truncated triangle in Figure 4.5.2, the integral is written

$$\int_0^a \left(\int_{2x}^2 f\,dy \right) dx.$$

In the other direction writing the integral is harder; we will return to it in Exercise 4.5.3.

[5]We need to break up this integral into a sum of integrals:

$$\int_0^1 \left(\int_{-x}^{x^2} f\,dy \right) dx + \int_1^2 \left(\int_{x-2}^{(x-2)^2} f\,dy \right) dx.$$

Exercise 4.5.1 asks you to justify our ignoring that we have counted the line $x = 1$ twice.

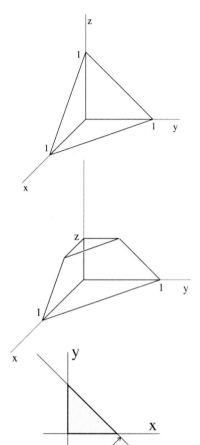

FIGURE 4.5.4.

Top: The pyramid over which we are integrating in Example 4.5.3. Middle: The same pyramid, truncated at height z. Bottom: The plane at height z shown in the middle figure, put flat.

is an unrealistically simple example. Suppose we want to integrate a function over the pyramid P shown in the top of Figure 4.5.4, and given by the formula

$$P = \left\{ \begin{pmatrix} x \\ y \\ z \end{pmatrix} \in \mathbb{R}^3 \mid 0 \le x; 0 \le y; 0 \le z; x + y + z \le 1 \right\}. \qquad 4.5.10$$

We want to figure out the limits of integration for the multiple integral

$$\iint_{\mathbb{R}^3} \int f \begin{pmatrix} x \\ y \\ z \end{pmatrix} \, dx \, dy \, dz = \int \left(\int \left(\int f \, dx \right) dy \right) dz. \qquad 4.5.11$$

There are six ways of applying Fubini's theorem, which in this case because of the symmetries will result in the same expressions with the variables permuted.

Let us think of varying z first, for instance by lifting a piece of paper and seeing how it intersects the pyramid at various heights. Clearly the paper will only intersect the pyramid when its height is between 0 and 1. This leads to writing

$$\int_0^1 (\qquad\qquad) dz \qquad 4.5.12$$

where the space needs be filled in by the double integral of f over the part of the pyramid P at height z, pictured in the middle of Figure 4.5.4, and again at the bottom, this time drawn flat.

This time we are integrating over a triangle (which depends on z), just as in Example 4.5.1. If we vary y next, rolling a horizontal pencil up, we see that clearly the relevant y-values are between 0 and $1 - z$. So we write

$$\int_0^1 \left(\int_0^{1-z} (\qquad\qquad) dy \right) dz, \qquad 4.5.13$$

where now the space represents the integral over part of the horizontal line segment at height z and "depth" y (if depth is the name of the y-coordinate). These x-values are between 0 and $1 - z - y$, so finally the integral is

$$\int_0^1 \left(\int_0^{1-z} \left(\int_0^{1-y-z} f \begin{pmatrix} x \\ y \\ z \end{pmatrix} \, dx \right) dy \right) dz \quad \triangle \qquad 4.5.14$$

Now let's actually compute a few multiple integrals.

Example 4.5.4 (Computing a multiple integral). Suppose we have a function $f \begin{pmatrix} x \\ y \end{pmatrix} = xy$ defined on the unit square, as shown in Figure 4.5.5. Then

In Equation 4.5.15, recall that to compute

$$\left[\frac{x^2 y}{2}\right]_{x=0}^{x=1}$$

we evaluate $x^2 y/2$ at both $x = 1$ and $x = 0$, subtracting the second from the first.

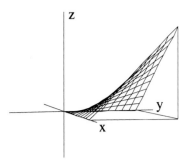

FIGURE 4.5.5.

The integral in Equation 4.5.15 is $1/4$: the volume under the surface defined by $f\begin{pmatrix} x \\ y \end{pmatrix} = xy$, and above the unit square, is $1/4$.

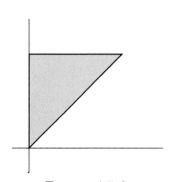

FIGURE 4.5.6.

The triangle of Example 4.5.5.

$$\iint_{\mathbb{R}^2} f\begin{pmatrix} x \\ y \end{pmatrix} \, dx\, dy = \int_0^1 \left(\int_0^1 xy \, dx \right) dy$$

$$= \int_0^1 \left[\frac{x^2 y}{2}\right]_{x=0}^{x=1} dy = \int_0^1 \frac{y}{2}\, dy = \frac{1}{4}. \quad \triangle$$

4.5.15

In Example 4.5.4 it is clear that we could have taken the integral in the opposite order and found the same result, since our function is $f\begin{pmatrix} x \\ y \end{pmatrix} = xy$, and $xy = yx$. Fubini's theorem says that this is always true as long as the functions involved are integrable. This fact can be useful; sometimes a multiple integral can be computed in elementary terms when written in one direction, but not in the other, as you will see in Example 4.5.5. It may also be easier to determine the limits of integration if the problem is set up in one direction rather than another, as we already saw in the case of the truncated triangle shown in Figure 4.5.2.

Example 4.5.5 (Choose the easy direction). Let us integrate the function e^{-y^2} over the triangle shown in Figure 4.5.6:

$$T = \left\{ \begin{pmatrix} x \\ y \end{pmatrix} \in \mathbb{R}^2 \mid 0 \le x \le y \le 1 \right\}. \qquad 4.5.16$$

Fubini's theorem gives us two ways of writing this integral as an iterated one-dimensional integral:

$$(1) \quad \int_0^1 \left(\int_x^1 e^{-y^2}\, dy \right) dx \quad \text{and} \quad (2) \quad \int_0^1 \left(\int_0^y e^{-y^2}\, dx \right) dy. \qquad 4.5.17$$

The first cannot be computed in elementary terms, since e^{-y^2} does not have an elementary anti-derivative.

But the second can:

$$\int_0^1 \left(\int_0^y e^{-y^2}\, dx \right) dy = \int_0^1 ye^{-y^2}\, dy = -\frac{1}{2}\left[e^{-y^2} \right]_0^1 = \frac{1}{2}\left(1 - \frac{1}{e} \right). \quad \triangle \quad 4.5.18$$

Older textbooks contain many examples of this sort of computational miracle. We are not sure the phenomenon was ever very important, but today it is sounder to take a serious interest in the numerical theory, and go lightly over computational tricks, which do not work in any great generality in any case.

Example 4.5.6 (Volume of a ball in $\mathbb{R}^n$). Let $B_R^n(0)$ be the ball of radius R in $\mathbb{R}^n$, centered at 0, and let $b_n(R)$ be its volume. Clearly $b_n(R) = R^n b_n(1)$. We will denote $b_n(1) = \beta_n$ the volume of the unit ball.

By Fubini's theorem,

$$
\underbrace{\beta_n}_{\substack{\text{vol. of}\\\text{unit ball in }\mathbb{R}^n}} = \int_{B_1^n(0)} |d^n\mathbf{x}| = \int_{-1}^{1} \left(\overbrace{\int_{B_{\sqrt{1-x_n^2}}^{n-1}(0)} |d^{n-1}\mathbf{x}|}^{\substack{(n-1)\text{-dimensional vol.}\\\text{of one slice of }B_1^n(0)}} \right) dx_n
$$

The ball $B_1^n(0)$ is the ball of radius 1 in $\mathbb{R}^n$, centered at the origin;

$$B_{\sqrt{1-x_n^2}}^{n-1}(0)$$

is the ball of radius $\sqrt{1-x_n^2}$ in $\mathbb{R}^{n-1}$, still centered at the origin.

In the first line of Equation 4.5.19 we imagine slicing the n-dimensional ball horizontally and computing the $n-1$)-dimensional volume of each slice.

$$
= \int_{-1}^{1} \underbrace{b_{n-1}\left(\sqrt{1-x_n^2}\right)}_{\substack{\text{vol. ball of radius}\\ r=\sqrt{1-x_n^2}\text{ in }\mathbb{R}^{n-1}}} dx_n = \int_{-1}^{1} \underbrace{(1-x_n^2)^{\frac{n-1}{2}}}_{r^{n-1}} \underbrace{\beta_{n-1}}_{\substack{\text{vol. ball of}\\\text{radius 1}\\\text{in }\mathbb{R}^{n-1}}} dx_n
$$

$$
= \underbrace{\beta_{n-1}}_{\substack{\text{vol. of unit}\\\text{ball in }\mathbb{R}^{n-1}}} \int_{-1}^{1} (1-x_n^2)^{\frac{n-1}{2}} dx_n. \qquad 4.5.19
$$

This reduces the computation of b_n to computing the integral

$$
c_n = \int_{-1}^{1} (1-t^2)^{\frac{n-1}{2}} dt. \qquad 4.5.20
$$

This is a standard tricky problem from one-variable calculus: Exercise 4.5.4, (a) asks you to show that

$$
c_n = \frac{n-1}{n} c_{n-2}, \quad \text{for } n \geq 2. \qquad 4.5.21
$$

You should learn how to handle simple examples using Fubini's theorem, and you should learn some of the standard tricks that work in some more complicated situations; these will be handy on exams, particularly in physics and engineering classes. But in real life you are likely to come across nastier problems, which even a professional mathematician would have trouble solving "by hand"; most often you will want to use a computer to compute integrals for you. We discuss numerical methods of computing integrals in Section 4.6.

So if we can compute c_0 and c_1, we can compute all the other c_n. Exercise 4.5.4, (b) asks you to show that $c_0 = \pi$ and $c_1 = 2$ (the second is pretty easy).

n	$c_n = \frac{n-1}{n} c_{n-2}$	Volume of ball $\beta_n = c_n \beta_{n-1}$
0	π	
1	2	2
2	$\frac{\pi}{2}$	π
3	$\frac{4}{3}$	$\frac{4\pi}{3}$
4	$\frac{3\pi}{8}$	$\frac{\pi^2}{2}$
5	$\frac{16}{15}$	$\frac{8\pi^2}{15}$

FIGURE 4.5.7. Computing the volume of a ball in $\mathbb{R}^1$ through $\mathbb{R}^5$.

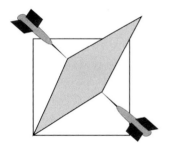

FIGURE 4.5.8.

Choosing a random parallelo-gram: one dart lands at $\begin{pmatrix} x_1 \\ y_1 \end{pmatrix}$, the other at $\begin{pmatrix} x_2 \\ y_2 \end{pmatrix}$.

Remember that in $\mathbb{R}^2$, the determinant is

$$\det \begin{bmatrix} x_1 & x_2 \\ y_1 & y_2 \end{bmatrix} = x_1 y_2 - x_2 y_1.$$

We have $d^2\mathbf{x}$ and $d^2\mathbf{y}$ because $\mathbf{x}$ and $\mathbf{y}$ have two coordinates:

$$d^2\mathbf{x} = dx_1 dx_2 \text{ and } d^2\mathbf{y} = dy_1 dy_2.$$

Saying that we are choosing our points at random in the square means that our density of probability for each dart is the characteristic function of the square.

An *integrand* is what comes after an integral sign: for $\int x\, dx$, the integrand is $x\, dx$. In Equation 4.5.24, the integrand for the innermost integral is $|x_1 y_2 - x_2 y_1|\, dy_2$; the integrand for the integral immediately to the left of the innermost integral is

$$\left(\int_0^1 |x_1 y_2 - x_2 y_1|\, dy_2 \right) dx_2.$$

This allows us to make the table of Figure 4.5.7. It is easy to continue the table (what is β_6? Check below.[6]) If you enjoy inductive proofs, you might try Exercise 4.5.5, which asks you to show that

$$\beta_{2k} = \frac{\pi^k}{k!} \quad \text{and} \quad \beta_{2k+1} = \frac{\pi^k\, k!\, 2^{2k+1}}{(2k+1)!}. \quad \triangle \qquad 4.5.22$$

Computing probabilities using integrals

As we mentioned in Section 4.1, an important use of integrals is in computing probabilities.

Example 4.5.7 (Using Fubini to compute a probability). Choose at random two pairs of positive numbers between 0 and 1 and use those numbers as the coordinates (x_1, y_1), (x_2, y_2) of two vectors anchored at the origin, as shown in Figure 4.5.8. (You might imagine throwing a dart at the unit square.) What is the expected (average) area of the parallelogram spanned by those vectors? In other words (recall Proposition 1.4.14), what is the expected value of the absolute value of the determinant?

This average is

$$\int_C |\underbrace{x_1 y_2 - y_1 x_2}_{\det}|\, |d^2\mathbf{x}||d^2\mathbf{y}|, \qquad 4.5.23$$

where C is the unit cube in $\mathbb{R}^4$. (Each possible parallelogram corresponds to two points in the unit square, each with two coordinates, so each point in $C \in \mathbb{R}^4$ corresponds to one parallelogram.) Our computation will be simpler if we consider only the cases $x_1 \geq y_1$; i.e., we assume that our first dart lands below the diagonal of the square. Since the diagonal divides the square symmetrically, the cases where the first dart lands below the diagonal and the cases where it lands above contribute the same amount to the integral. Thus we want to compute *twice* the quadruple integral

$$\int_0^1 \int_0^{x_1} \int_0^1 \int_0^1 |x_1 y_2 - x_2 y_1|\, dy_2\, dx_2\, dy_1\, dx_1. \qquad 4.5.24$$

(Note that the integral $\int_0^{x_1}$ goes with dy_1: the innermost integral goes with the innermost integrand, and so on. The second integral is $\int_0^{x_1}$ because $y_1 \leq x_1$.)

Now we would like to get rid of the absolute values, by considering separately the case where $\det = x_1 y_2 - x_2 y_1$ is negative, and the case where it is positive. Observe that when $y_2 < y_1 x_2 / x_1$, the determinant is negative, whereas when $y_2 > y_1 x_2 / x_1$ it is positive. Another way to say this is that on one side of the

---

[6] $c_6 = \frac{5}{6} c_4 = \frac{15\pi}{48}$, so $\beta_6 = c_6 \beta_5 = \frac{\pi^3}{6}$.

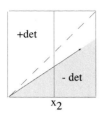

FIGURE 4.5.9.

The arrow represents the first dart. If the second dart (with coordinates x_2, y_2) lands in the shaded area, the determinant will be negative. Otherwise it will be positive.

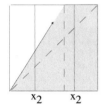

FIGURE 4.5.10.

The situation if the first dart lands above the diagonal. If the second dart lands in the shaded area, then the determinant is negative because the second vector is clockwise from the first (Proposition 1.4.14). For values of x_2 to the right of the vertical dotted line (which has x-coordinate x_1), y_2 goes from 0 to 1. For values of x_2 to the left of the vertical dotted line, we must consider separately values of y_2 from 0 to $\frac{y_1 x_2}{x_1}$ (the shaded region, with $-\det$) and values of y_2 from $\frac{y_1 x_2}{x_1}$ to 1 (the unshaded region, with $+\det$).

line $y_2 = \frac{y_1}{x_1} x_2$ (the shaded side in Figure 4.5.9), the determinant is negative, and on the other side it is positive.

Since we have assumed that the first dart lands below the diagonal of the square, then whatever the value of x_2, when we integrate with respect to y_2, we will have two choices: if y_2 is in the shaded part, the determinant will be negative; otherwise it will be positive. So we break up the innermost integral into two parts:

$$\int_0^1 \int_0^{x_1} \int_0^1 \left(\overbrace{\int_0^{(y_1 x_2)/x_1} (x_2 y_1 - x_1 y_2)\, dy_2}^{-\det} + \overbrace{\int_{(y_1 x_2)/x_1}^1 (x_1 y_2 - x_2 y_1)\, dy_2}^{+\det} \right) dx_2\, dy_1\, dx_1.$$
$$\text{4.5.25}$$

(If we had not restricted the first dart to below the diagonal, we would have the situation of Figure 4.5.10, and our integral would be more complicated.[7])

The rest of the computation is a matter of carefully computing four ordinary integrals, keeping straight what is constant and what is the variable of integration at each step. First we compute the inner integral, with respect to y_2. The first term gives

$$\left[x_2 y_1 y_2 - x_1 \frac{y_2^2}{2} \right]_0^{y_1 x_2/x_1} = \overbrace{\frac{x_2 y_1^2 x_2}{x_1} - \frac{1}{2} x_1 \frac{y_1^2 x_2^2}{x_1^2}}^{\text{eval. at } y_2 = y_1 x_2/x_1} - \overbrace{0}^{\text{eval. at } y_2 = 0} = \frac{1}{2} \frac{y_1^2 x_2^2}{x_1}.$$

The second gives

$$\left[\frac{x_1 y_2^2}{2} - x_2 y_1 y_2 \right]_{y_1 x_2/x_1}^1 = \overbrace{\left(\frac{x_1}{2} - x_2 y_1 \right)}^{\text{eval. at } y_2 = 1} - \overbrace{\left(\frac{x_1 y_1^2 x_2^2}{2 x_1^2} - \frac{x_2 y_1^2 x_2}{x_1} \right)}^{\text{eval. at } y_2 = y_1 x_2/x_1}$$
$$\text{4.5.26}$$

$$= \frac{x_1}{2} - x_2 y_1 + \frac{x_2^2 y_1^2}{2 x_1}.$$

Continuing with Equation 4.5.25, we get

---

[7]Exercise 4.5.7 asks you to compute the integral that way. The integral is then the sum of the integral for the half of the square below the diagonal (given by Equation 4.5.25), and the integral for the half above the diagonal. The latter is

$$\int_0^1 \int_{x_1}^1 \int_0^{y_1} \left(\overbrace{\int_0^{\frac{y_1 x_2}{x_1}} \underbrace{(x_2 y_1 - x_1 y_2)}_{-\det}\, dy_2 + \int_{\frac{y_1 x_2}{x_1}}^1 \underbrace{(x_1 y_2 - x_2 y_1)}_{+\det}\, dy_2}^{\text{for } x_2 \text{ to the left of the vertical line with } x\text{-coordinate } x_1} \right) dx_2\, dy_1\, dx_1$$

$$+ \int_0^1 \int_{x_1}^1 \int_{\frac{x_1}{y_1}}^1 \left(\int_0^1 (x_2 y_1 - x_1 y_2)\, dy_2 \right) dx_2\, dy_1\, dx_1.$$
$$\underbrace{\phantom{\int_0^1 \int_{x_1}^1 \int_{\frac{x_1}{y_1}}^1}}_{\substack{\text{for } x_2 \text{ to the right of the vertical line} \\ \text{with } x\text{-coordinate } x_1; \, -\det}}$$

$$\int_0^1 \int_0^{x_1} \int_0^1 \left(\int_0^{(y_1 x_2)/x_1} (x_2 y_1 - x_1 y_2) dy_2 + \int_{(y_1 x_2)/x_1}^1 (x_1 y_2 - x_2 y_1) dy_2 \right) dx_2 \, dy_1 \, dx_1,$$

$$= \int_0^1 \int_0^{x_1} \underbrace{\int_0^1 \left(\frac{x_1}{2} - x_2 y_1 + \frac{x_2^2 y_1^2}{x_1} \right) dx_2}_{\left[\frac{x_1 x_2}{2} - \frac{x_2^2 y_1}{2} + \frac{x_2^3 y_1^2}{3 x_1} \right]_{x_2=0}^{x_2=1}} \, dy_1 \, dx_1$$

$$= \int_0^1 \underbrace{\int_0^{x_1} \left(\frac{x_1}{2} - \frac{y_1}{2} + \frac{y_1^2}{3 x_1} \right) dy_1}_{\left[\frac{x_1 y_1}{2} - \frac{y_1^2}{4} + \frac{y_1^3}{9 x_1} \right]_{y_1=0}^{y_1=x_1}} \, dx_1 = \int_0^1 \left(\frac{x_1^2}{2} - \frac{x_1^2}{4} + \frac{x_1^2}{9} \right) dx_1$$

$$= \frac{13}{36} \int_0^1 x_1^2 \, dx_1 = \frac{13}{36} \left[\frac{x_1^3}{3} \right]_0^1 = \frac{13}{108}. \qquad 4.5.27$$

So the expected area is twice $13/108$, i.e., $13/54$, or slightly less than $1/4$.

Stating Fubini's theorem more precisely

We will now give a precise statement of Fubini's theorem. The statement is not as strong as what we prove in Appendix A.13, but it keeps the statement simpler.

Theorem 4.5.8 (Fubini's theorem). *Let f be an integrable function on $\mathbb{R}^n \times \mathbb{R}^m$, and suppose that for each $\mathbf{x} \in \mathbb{R}^n$, the function $\mathbf{y} \mapsto f(\mathbf{x}, \mathbf{y})$ is integrable. Then the function*

$$\mathbf{x} \mapsto \int_{\mathbb{R}^m} f(\mathbf{x}, \mathbf{y}) |d^m \mathbf{y}|$$

is integrable, and

$$\int_{\mathbb{R}^{n+m}} f(\mathbf{x}, \mathbf{y}) |d^n \mathbf{x}| |d^m \mathbf{y}| = \int_{\mathbb{R}^n} \left(\int_{\mathbb{R}^m} f(\mathbf{x}, \mathbf{y}) |d^m \mathbf{y}| \right) |d^n \mathbf{x}|.$$

What if we choose at random three vectors in the unit cube? Then we would be integrating over a nine-dimensional cube. Daunted by such an integral, we might be tempted to use a Riemann sum. But even if we used a rather coarse decomposition the computation would be forbidding. Say we divide each side into 10, choosing (for example) the midpoint of each mini-cube. We turn the nine coordinates of that point into a 3×3 matrix and compute its determinant. That gives 10^9 determinants to compute—a *billion* determinants, each requiring 18 multiplications and five additions.

Go up one more dimension and the computation is really out of hand. Yet physicists like Nobel laureate Kenneth Wilson routinely work with integrals in dimensions of thousands or more. Actually carrying out the computations is clearly impossible. The technique most often used is a sophisticated version of throwing dice, known as *Monte Carlo integration*. It is discussed in Section 4.6.

4.6 NUMERICAL METHODS OF INTEGRATION

In a great many cases, Fubini's theorem does not lead to expressions that can be calculated in closed form, and integrals must be computed numerically. In one dimension, this subject has been extensively investigated, and there is an enormous literature on the subject. In higher dimensions, the literature is still extensive but the field is not nearly so well known. We will begin with a reminder about the one-dimensional case.

One-dimensional integrals

In first year calculus you probably heard of the trapezoidal rule and of Simpson's rule for computing ordinary integrals (and quite likely you've forgotten them too). The trapezoidal rule is not of much practical interest, but Simpson's rule is probably good enough for anything you will need unless you become an engineer or physicist. In it, the function is sampled at regular intervals and different "weights" are assigned the samples.

Definition 4.6.1 (Simpson's rule). Let f be a function on $[a,b]$, choose an integer n, and sample f at $2n+1$ equally distributed points, $x_0, x_1, \ldots, x_{2n}$, where $x_0 = a$ and $x_{2n} = b$. Then Simpson's approximation to

$$\int_a^b f(x)\,dx \qquad \text{in } n \text{ steps is}$$

Here in speaking of weights starting and ending with 1, and so forth, we are omitting the factor of $(b-a)/6n$.

Why do we multiply by $(b-a)/6n$? Think of the integral as the sum of the area of n "rectangles," each with width $(b-a)/n$—i.e., the total length divided by the number of "rectangles." Multiplying by $(b-a)/n$ gives the width of each rectangle. The height of each "rectangle" should be some sort of average of the value of the function over the interval, but in fact we have weighted that value by 6. Dividing by 6 corrects for that weight.

$$S_{[a,b]}^n(f) = \frac{b-a}{6n}\Big(f(x_0)+4f(x_1)+2f(x_2)+4f(x_3)+\cdots+4f(x_{2n-1})+f(x_{2n})\Big).$$

For example, if $n=3$, $a=-1$ and $b=1$, then we divide the interval $[-1,1]$ into six equal parts and compute

$$\frac{1}{9}\Big(f(-1)+4f(-2/3)+2f(-1/3)+4f(0)+2f(1/3)+4f(2/3)+f(1)\Big). \quad 4.6.1$$

Why do the weights start and end with 1, and alternate between 4 and 2 for the intermediate samples? As shown in Figure 4.6.1, the pattern of weights is not $1,4,2,\ldots,4,1$ but $1,4,1$: each 1 that is not an endpoint is counted twice, so it becomes the number 2. We are actually breaking up the interval into n subintervals, and integrating the function over each subpiece

$$\int_a^b f(x)\,dx = \int_a^{x_2} f(x)\,dx + \int_{x_2}^{x_4} f(x)\,dx + \cdots. \qquad 4.6.2$$

Each of these n sub-integrals is computed by sampling the function at the beginning point and endpoint of the subpiece (with weight 1) and at the center of the subpiece (with weight 4), giving a total of 6.

Theorem 4.6.2 (Simpson's rule). *(a) If f is a piecewise cubic function, exactly equal to a cubic polynomial on the intervals $[x_{2i}, x_{2i+2}]$, then Simpson's rule computes the integral exactly.*

(b) If a function f is four times continuously differentiable, then there exists $c \in (a,b)$ such that

$$S_{[a,b]}^n(f) - \int_a^b f(x)\,dx = \frac{(b-a)^5}{2880n^4} f^{(4)}(c). \qquad 4.6.3$$

FIGURE 4.6.1. To compute the integral of a function f over $[a, b]$, Simpson's rule breaks the interval into n pieces. Within each piece, the function is evaluated at the beginning, the midpoint, and the endpoint, with weight 1 for the beginning and endpoint, and weight 4 for the midpoint. The endpoint of one interval is the beginning point of the next, so it is counted twice and gets weight 2. At the end, the result is multiplied by $(b - a)/6n$.

Theorem 4.6.2 tells when Simpson's rule computes the integral exactly, and when it gives an approximation.

Simpson's rule is a *fourth-order method*; the error (if f is sufficiently differentiable) is of order h^4, where h is the step size.

Proof. Figure 4.6.2 proves part (a); in it, we compute the integral for constant, linear, quadratic, and cubic functions, over the interval $[-1, 1]$, with $n = 1$. Simpson's rule gives the same result as computing the integral directly.

By cubic polynomial we mean polynomials of degree up to and including 3: constant functions, linear functions, quadratic polynomials and cubic polynomials.

If we can split the domain of integration into smaller intervals such that a function f is exactly equivalent to a cubic polynomial over each interval, then Simpson's rule will compute the integral of f exactly.

	Simpson's rule	Integration
Function	$1/3\left(f(-1) + 4(f(0)) + f(1)\right)$	$\int_{-1}^{1} f(x)\,dx$
$f(x) = 1$	$1/3\left(1 + 4 + 1\right) = 2$	2
$f(x) = x$	0	0
$f(x) = x^2$	$1/3\left(1 + 0 + 1\right) = 2/3$	$\int_{-1}^{1} x^2\,dx = 2/3$
$f(x) = x^3$	0	0

FIGURE 4.6.2. Using Simpson's rule to integrate a cubic function gives the exact answer.

A proof of part (b) is sketched in Exercise 4.6.8.

Of course, you don't often encounter in real life a piecewise cubic polynomial (the exception being computer graphics). Usually, Simpson's method is used to approximate integrals, not to compute them exactly.

Example 4.6.3 (Approximating integrals with Simpson's rule). Use Simpson's rule with $n = 100$ to compute

$$\int_{1}^{4} \frac{1}{x}\,dx = \ln 4 = 2\ln 2, \qquad 4.6.4$$

In the world of computer graphics, piecewise cubic polynomials are everywhere. When you construct a smooth curve using a drawing program, what the computer is actually making is a piecewise cubic curve, usually using the Bezier algorithm for cubic interpolation. The curves drawn this way are known as *cubic splines*.

When first drawing curves with such a program it comes as a surprise how few control points are needed.

By "normalize" we mean that we choose a definite domain of integration rather than an interval $[a, b]$; the domain $[-1, 1]$ allows us to take advantage of even and odd properties.

which is infinitely differentiable. Since $f^{(4)} = 24/x^5$, which is largest at $x = 1$, Theorem 4.6.2 asserts that the result will be correct to within

$$\frac{24 \cdot 3^5}{2880 \cdot 100^4} = 2.025 \cdot 10^{-8} \qquad 4.6.5$$

so at least seven decimals will be correct. △

The integral of Example 4.6.3 can be approximated to the same precision with far fewer evaluations, using *Gaussian rules*.

Gaussian rules

Simpson's rule integrates cubic polynomials exactly. Gaussian rules are designed to integrate higher degree polynomials exactly with the smallest number of function evaluations possible.   Let us normalize the problem as follows, integrating from -1 to 1:

Find points $x_1, \ldots, x_m$ and weights $w_1, \ldots, w_m$ with m as small as possible so that, for all polynomials p of degree $\leq d$,

$$\int_{-1}^{1} p(x) \, dx = \sum_{i=1}^{m} w_i p(x_i). \qquad 4.6.6$$

We will require that the points x_i satisfy $-1 \leq x_i \leq 1$ and that $w_i > 0$ for all i.

Think first of how many unknowns we have, and how many equations: the requirement of Equation 4.6.6 for each of the polynomials $1, x, \ldots, x^d$ give $d + 1$ equations for the $2m$ unknowns $x_1, \ldots, x_m, w_1, \ldots, w_m$, so we can reasonably hope that the equations might have a solution when $2m \geq d + 1$.

Example 4.6.4 (Gaussian rules). The simplest case (already interesting) is when $d = 3$ and $m = 2$. Showing that this integral is exact for polynomials of degree ≤ 3 amounts to the four equations

In Equation 4.6.7 we are integrating $\int_{-1}^{1} x^n \, dx$ for n from 0 to 3, using

$$\int_{-1}^{1} x^n \, dx = \begin{cases} 0 & \text{if } n \text{ is odd} \\ \frac{2}{n+1} & \text{if } n \text{ is even.} \end{cases}$$

$$\begin{array}{lll}
\text{for } f = 1 & w_1 \; + \; w_2 = 2 & \\
\text{for } f(x) = x & w_1 x_1 + w_2 x_2 = 0 & \\
\text{for } f(x) = x^2 & w_1 x_1^2 + w_2 x_2^2 = \frac{2}{3} & 4.6.7 \\
\text{for } f(x) = x^3 & w_1 x_1^3 + w_2 x_2^3 = 0 &
\end{array}$$

This is a system of four nonlinear equations in four unknowns, and it looks intractable, but in this case it is fairly easy to solve by hand: first, observe that if we set $x_1 = -x_2 = x > 0$ and $w_1 = w_2 = w$, making the formula symmetric around the origin, then the second and fourth equations are automatically satisfied, and the other two become

$$2w = 2 \quad \text{and} \quad 2wx^2 = 2/3, \qquad 4.6.8$$

i.e., $w = 1$ and $x = 1/\sqrt{3}$. △

Remark. This means that whenever we have a piecewise cubic polynomial function, we can integrate it exactly by sampling it at two points per piece. For a piece corresponding to the interval $[-1, 1]$, the samples should be taken at $-1/\sqrt{3}$ and $1/\sqrt{3}$, with equal weights. Exercise 4.6.3 asks you to say where the samples should be taken for a piece corresponding to an arbitrary interval $[a, b]$. $\triangle$

Whenever $m = 2k$ is even, we can do something similar, making the formula symmetric about the origin and considering only the integral from 0 to 1. This allows us to cut the number of variables in half; instead of $4k$ variables ($2k$ w's and $2k$ x's), we have $2k$ variables. We then consider the system of $2k$ equations

Exercises 4.6.4 and 4.6.5 invite you to explore how Gaussian integration can be adapted to such integrals.

$$w_1 + w_2 + \cdots + w_k = 1$$

$$w_1 x_1^2 + w_2 x_2^2 + \cdots + w_k x_k^2 = \frac{1}{3}$$

$$\cdots \qquad \cdots$$

$$w_1 x_1^{4k-2} + w_2 x_2^{4k-2} + \cdots + w_k x_k^{4k-2} = \frac{1}{4k - 1}.$$

4.6.9

If this system has a solution, then the corresponding integration rule gives the approximation

$$\int_{-1}^{1} f(x)\, dx \approx \sum_{i=-k}^{k} w_i \big(f(x_i) + f(-x_i)\big),$$

4.6.10

and this formula will be exact for all polynomials of degree $\leq 2k - 1$.

We say that Newton's method works "reasonably" well because you need to start with a fairly good initial guess in order for the procedure to converge; some experiments are suggested in Exercise 4.6.2.

A lot is known about solving the system of Equation 4.6.9. The principal theorem states that there is a unique solution to the equations with $0 \leq x_1 < \cdots < x_k < 1$, and that then all the w_i are positive. The main tool is the theory of orthogonal polynomials, which we don't discuss in this volume. Another approach is to use Newton's method, which works reasonably well for $k \leq 6$ (as far as we have looked).

Gaussian rules are well adapted to problems where we need to integrate functions with a particular weight, such as

In $\int_0^\infty f(x)e^{-x}\, dx$, the function $f(x)$ is weighted by e^{-x}; when we speak of "functions with a particular weight" we are not referring to the weights of Guassian rules or Simpson's method. The weight e^{-x} is particularly relevant when computing Laplace transforms, and the weight $\frac{1}{\sqrt{1-x^2}}$ comes up often in problems with spherical symmetry.

$$\int_0^\infty f(x)e^{-x}\, dx \quad \text{or} \quad \int_{-1}^{1} \frac{f(x)}{\sqrt{1 - x^2}}\, dx.$$

4.6.11

Exercises 4.6.4 and 4.6.5 explore how to choose the sampling points and the weights in such settings.

Product rules

Every one-dimensional integration rule has a higher-dimensional counterpart, called a *product rule*. If the rule in one dimension is

$$\int_a^b f(x)\, dx \approx \sum_{i=1}^{k} w_i f(p_i),$$

4.6.12

then the corresponding rule in n dimensions is

$$\int_{[a,b]^n} f \, |d^n \mathbf{x}| \approx \sum_{1 \le i_1, \ldots, i_n \le k} w_{i_1} \ldots w_{i_n} f \begin{pmatrix} p_{i_1} \\ \vdots \\ p_{i_n} \end{pmatrix}. \qquad 4.6.13$$

The following proposition shows why product rules are a useful way of adapting one-dimensional integration rules to several variables.

Proposition 4.6.5 (Product rules). *If $f_1, \ldots, f_n$ are functions that are integrated exactly by an integration rule:*

$$\int_a^b f_j(x) \, dx = \sum w_i f_j(x_i) \quad \text{for } j = 1, \ldots, n, \qquad 4.6.14$$

then the product

$$f(\mathbf{x}) \overset{\text{def}}{=} f_1(x_1) f_2(x_2) \ldots f_n(x_n) \qquad 4.6.15$$

is integrated exactly by the corresponding product rule over $[a,b]^n$.

$$
\begin{array}{ccccccc}
1 & 4 & 2 & 4 & \ldots & 4 & 1 \\
4 & 16 & 8 & 16 & \ldots & 16 & 4 \\
2 & 8 & 4 & 8 & \ldots & 8 & 2 \\
4 & 16 & 8 & 16 & \ldots & 16 & 4 \\
\vdots & \vdots & \vdots & \vdots & \ddots & \vdots & \vdots \\
4 & 16 & 8 & 16 & \ldots & 16 & 4 \\
1 & 4 & 2 & 4 & \ldots & 4 & 1
\end{array}
$$

FIGURE 4.6.3.

Weights for approximating the integral over a square, using the two-dimensional Simpson's rule. Each weight is multiplied by

$$(b-a)^2/(36n^2).$$

Proof. This follows immediately from Proposition 4.1.12. Indeed,

$$
\begin{aligned}
\int_{[a,b]^n} f(\mathbf{x}) \, |d^n \mathbf{x}| &= \left(\int_a^b f_1(x_1) \, dx_1 \right) \ldots \left(\int_a^b f_n(x_n) \, dx_n \right) \\
&= \left(\sum w_i f_1(p_i) \right) \ldots \left(\sum w_i f_n(p_i) \right) \qquad 4.6.16 \\
&= \sum_{i_1, \ldots i_n} w_{i_1} \ldots w_{i_n} f \begin{pmatrix} p_{i_1} \\ \vdots \\ p_{i_n} \end{pmatrix}. \quad \square
\end{aligned}
$$

Example 4.6.6 (Simpson's rule in two dimensions). The two-dimensional form of Simpson's rule will approximate the integral over a square, using the weights shown in Figure 4.6.3 (each multiplied by $(b-a)^2/(36n^2)$).

In the very simple case where we divide the square into only four subsquares, and sample the function at each vertex, we have nine samples in all, as shown in Figure 4.6.4. If we do this with the square of side length 2 centered at 0, Equation 4.6.13 then becomes

The integral

$$\int_{[-1,1]^n} f(\mathbf{x}) \, |d^n\mathbf{x}|$$

is a multiple integral, over the region in $\mathbb{R}^n$ where all the coordinates are in [-1,1] .

Here $b - a = 2$ (since we are integrating from -1 to 1), and $n = 1$ (this square corresponds, in the one-dimensional case, to the first piece out of n pieces, as shown in Figure 4.6.1). So

$$\left(\frac{b-a}{6n}\right)^2 = \left(\frac{1}{3}\right)^2 = \frac{1}{9}.$$

Each two-dimensional weight is the product of two of the one-dimensional Simpson weights, w_1, w_2, and w_3, where

$$w_1 = w_3 = \frac{b-a}{6n} \cdot 1 = \frac{1}{3}$$

and

$$w_2 = \frac{b-a}{6n} \cdot 4 = \frac{4}{3}.$$

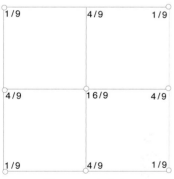

FIGURE 4.6.4.

If we divide a square into only four subsquares, Simpson's method in two dimensions gives the weights above.

$$\int_{[-1,1]^n} f(\mathbf{x}) \, |d^n\mathbf{x}| = \underbrace{w_1 w_1}_{1/9} f\begin{pmatrix} -1 \\ -1 \end{pmatrix} + \underbrace{w_1 w_2}_{4/9} f\begin{pmatrix} -1 \\ 0 \end{pmatrix} + \underbrace{w_1 w_3}_{1/9} f\begin{pmatrix} -1 \\ 1 \end{pmatrix}$$

$$+ \underbrace{w_2 w_1}_{4/9} f\begin{pmatrix} 0 \\ -1 \end{pmatrix} + \underbrace{w_2 w_2}_{16/9} f\begin{pmatrix} 0 \\ 0 \end{pmatrix} + \underbrace{w_2 w_3}_{4/9} f\begin{pmatrix} 0 \\ 1 \end{pmatrix} \qquad 4.6.17$$

$$+ \underbrace{w_3 w_1}_{1/9} f\begin{pmatrix} 1 \\ -1 \end{pmatrix} + \underbrace{w_3 w_2}_{4/9} f\begin{pmatrix} 1 \\ 0 \end{pmatrix} + \underbrace{w_3 w_3}_{1/9} f\begin{pmatrix} 1 \\ 1 \end{pmatrix};$$

Theorem 4.6.2 and Proposition 4.6.5 tell us that this two-dimensional Simpson's method will integrate exactly the polynomials

$$1, \ x, \ y, \ x^2, \ xy, \ y^2, x^3, \ x^2 y, \ xy^2, \ y^3, \qquad 4.6.18$$

and many others (for instance, $x^2 y^3$), but not x^4. They will also integrate functions which are piecewise polynomials of degree at most three on each of the unit squares, as in Figure 4.6.4. $\triangle$

Gaussian rules also lead to product rules for integrating functions in several variables, which will very effectively integrate polynomials in several variables of high degree.

Problems with higher dimensional Riemann sums

Both Simpson's rule and Gaussian rules are versions of Riemann sums. There are at least two serious difficulties with Riemann sums in higher dimensions. One is that the fancier the method, the smoother the function to be integrated needs to be in order for the method to work according to specs. In one dimension this usually isn't serious; if there are discontinuities, you break up the interval into several intervals at the points where the function has singularities. But in several dimensions, especially if you are trying to evaluate a volume by integrating a characteristic function, you will only be able to maneuver around the discontinuity if you already know the answer. For integrals of this sort, it isn't clear that delicate, high-order methods like Gaussians with many points are better than plain midpoint Riemann sums.

The other problem has to do with the magnitude of the computation. In one dimension, there is nothing unusual in using 100 or 1000 points for Simpson's method or Gaussian rules, in order to gain the desired accuracy (which might be 10 significant digits). As the dimension goes up, this sort of thing becomes first alarmingly expensive, and then utterly impossible. In dimension 4, a Simpson approximation using 100 points to a side involves 100 000 000 function evaluations, within reason for today's computers if you are willing to wait a while; with 1 000 points to a side it involves 10^{12} function evaluations, which would tie up the biggest computers for several days. By the time you get to dimension 9, this sort of thing becomes totally unreasonable unless you decrease your desired accuracy: $100^9 = 10^{18}$ function evaluations would take more than a billion seconds (about 32 years) even on the very fastest computers, but 10^9 is within reason, and should give a couple of significant digits.

When the dimension gets higher than 10, Simpson's method and all similar methods become totally impossible, even if you are satisfied with one significant digit, just to give an order of magnitude. These situations call for the probabilistic methods described below. They very quickly give a couple of significant digits (with high probability: you are never sure), but we will see that it is next to impossible to get really good accuracy (say six significant digits).

Monte Carlo methods

Suppose that we want to to find the average of $|\det A|$ for all $n \times n$ matrices with all entries chosen at random in the unit interval. We computed this integral in Example 4.5.7 when $n = 2$, and found $13/54$. The thought of computing the integral exactly for 3×3 matrices is awe-inspiring. How about numerical integration? If we want to use Simpson's rule, even with just 10 points on the side of the cube, we will need to evaluate 10^9 determinants, each a sum of six products of three numbers. This is not out of the question with today's computers, but a pretty massive computation. Even then, we still will probably know only two significant digits, because the integrand isn't differentiable.

In this situation, there is a much better approach. Simply pick points at random in the nine-dimensional cube, evaluate the determinant of the 3×3 matrix that you make from the coordinates of each point, and take the average. A similar method will allow you to evaluate (with some precision) integrals even of domains of dimension 20, or 100, or perhaps more.

The theorem that describes Monte Carlo methods is the *central limit theorem* from probability, stated (as Theorem 4.2.11) in Section 4.2 on probability.

When trying to approximate $\int_A f(\mathbf{x})|d^n\mathbf{x}|$, the individual experiment is to choose a point in A at random, and evaluate f there. This experiment has a certain expected value E, which is what we are trying to discover, and a certain standard deviation σ.

Unfortunately, both are unknown, but running the Monte Carlo algorithm gives you an approximation of both. It is wiser to compute both at once, as the approximation you get for the standard deviation gives an idea of how accurate the approximation to the expected value is.

A random number generator can be used to construct a code: you can add a random sequence of bits to your message, bit by bit (with no carries, so that $1+1 = 0$); to decode, subtract it again. If your message (encoded as bits) is the first line below, and the second line is generated by a random number generator, then the sum of the two will appear random as well, and thus undecipherable:

> 10 11 10 10 1111 01 01
>
> 01 01 10 10 0000 11 01
>
> 11 10 00 00 1111 10 00

The points in A referred to in Definition 4.6.7 will no doubt be chosen using some pseudo-random number generator. If this is biased, the bias will affect both the expected value and the expected variance, so the entire scheme becomes unreliable. On the other hand, off-the-shelf random number generators come with the guarantee that if you can detect a bias, you can use that information to factor large numbers and, in particular, crack most commercial encoding schemes. This could be a quick way of getting rich (or landing in jail).

The *Monte Carlo* program is found in Appendix B.2, and at the website given in the preface.

> **Definition 4.6.7 (Monte Carlo method).** The Monte Carlo algorithm for computing integrals consists of
>
> (1) Choosing points $\mathbf{x}_i, i = 1, \ldots, N$ in A at random, equidistributed in A.
>
> (2) Evaluating $a_i = f(\mathbf{x}_i)$ and $b_i = (f(\mathbf{x}_i))^2$.
>
> (3) Computing $\bar{a} = \frac{1}{N}\sum_{i=1}^{N} a_i$ and $\bar{s}^2 = \frac{1}{N}\sum_{i=1}^{N}(a_i - \bar{a})^2$.

Probabilistic methods of integration are like political polls. You don't pay much (if anything) for going to higher dimensions, just as you don't need to poll more people about a Presidential race than for a Senate race.

The real difficulty with Monte Carlo methods is making a good random number generator, just as in polling the real problem is making sure your sample is not biased. In the 1936 presidential election, the *Literary Digest* predicted that Alf Landon would beat Franklin D. Roosevelt, on the basis of two million mock ballots returned from a mass mailing. The mailing list was composed of people who owned cars or telephones, which during the Depression was hardly a random sampling.

Pollsters then began polling far fewer people (typically, about 10 thousand), paying more attention to getting representative samples. Still, in 1948 the *Tribune* in Chicago went to press with the headline, "Dewey Defeats Truman"; polls had unanimously predicted a crushing defeat for Truman. One problem was that some interviewers avoided low-income neighborhoods. Another was calling the election too early: Gallup stopped polling two weeks before the election.

Why $|d^9\mathbf{x}|$ in Equation 4.6.23? To each point $\mathbf{x} \in \mathbb{R}^9$, with coordinates $x_1, \ldots, x_9$, we can associate the determinant of the 3×3 matrix

$$A = \begin{bmatrix} x_1 & x_4 & x_7 \\ x_2 & x_5 & x_8 \\ x_3 & x_6 & x_9 \end{bmatrix}.$$

The number $\overline{a}$ is our approximation to the integral, and the number $\overline{s}$ is our approximation to the standard deviation σ.

The central limit theorem asserts that the probability that $\overline{a}$ is between

$$E + a\sigma/\sqrt{N} \quad \text{and} \quad E + b\sigma/\sqrt{N} \qquad 4.6.19$$

is approximately

$$\frac{1}{\sqrt{2\pi}} \int_a^b e^{-\frac{t^2}{2}} \, dt. \qquad 4.6.20$$

In principle, everything can be derived from this formula: let us see how this allows us to see how many times the experiment needs to be repeated in order to know an integral with a certain precision and a certain confidence.

For instance, suppose we want to compute an integral to within one part in a thousand. We can't do that by Monte Carlo: we can never be *sure* of anything. But we can say that with probability 98%, the estimate $\overline{a}$ is correct to one part in a thousand, i.e., that

$$\frac{E - \overline{a}}{E} \leq .001. \qquad 4.6.21$$

This requires knowing something about the bell curve: with probability 98% the result is within 2.36 standard deviations of the mean. So to arrange our desired relative error, we need

$$\frac{2.4\sigma}{\sqrt{N}E} \leq .001, \quad \text{i.e.,} \quad N \geq \frac{5.56 \cdot 10^6 \cdot \sigma^2}{E^2}. \qquad 4.6.22$$

Example 4.6.8 (Monte Carlo). In Example 4.5.7 we computed the expected value for the determinant of a 2×2 matrix. Now let us run the program *Monte Carlo* to approximate

$$\int_C |\det A| |d^9\mathbf{x}|, \qquad 4.6.23$$

i.e., to evaluate the average absolute value of the determinant of a 3×3 matrix with entries chosen at random in $[0, 1]$.

Several runs of length $10\,000$ (essentially instantaneous)[8] gave .127, .129, .129, .128 as values for $\overline{s}$ (guesses for the standard deviation σ). For these same runs, the computer the following estimates of the integral:

$$.13625, .133150, .135197, .13473. \qquad 4.6.24$$

It seems safe to guess that $\sigma < .13$, and also $E \approx .13$; this last guess is not as precise as we would like, neither do we have the confidence in it that is

---

[8]On a 1998 computer, a run of $5\,000\,000$ repetitions of the experiment took about 16 seconds. This involves about 3.5 billion arithmetic operations (additions, multiplications, divisions), about 3/4 of which are the calls to the random number generator.

required. Using these numbers to estimate how many times the experiment should be repeated so that with probability 98%, the result has a relative error at most .001, we use Equation 4.6.22, which says that we need about 5 000 000 repetitions to achieve this precision and confidence. This time the computation is not instantaneous, and yields $E = 0.134712$, with probability 98% that the absolute error is at most 0.000130. This is good enough: surely the digits 134 are right, but the fourth digit, 7, might be off by 1. $\triangle$

Note that when estimating how many times we need to repeat an experiment, we don't need several digits of σ; only the order of magnitude matters.

4.7 OTHER PAVINGS

The dyadic paving is the most rigid and restrictive we can think of, making most theorems easiest to prove. But in many settings the rigidity of the dyadic paving $\mathcal{D}_N$ is not necessary or best. Often we will want to have more "paving tiles" where the function varies rapidly, and bigger ones elsewhere, shaped to fit our domain of integration. In some situations, a particular paving is more or less imposed.

To measure the standard deviation of the income of Americans, you would want to subdivide the U.S. by census tracts, not by closely spaced latitudes and longitudes, because that is how the data is provided.

Example 4.7.1 (Measuring rainfall).   Imagine that you wish to measure rainfall in liters per square kilometer that fell over South America during October, 1999. One possibility would be to use dyadic cubes (squares in this case), measuring the rainfall at the center of each cube and seeing what happens as the decomposition gets finer and finer. One problem with this approach, which we discuss in Chapter 5, is that the dyadic squares lie in a plane, and the surface of South America does not.

Another problem is that using dyadic cubes would complicate the collection of data. In practice, you might break South America up into countries, and assign to each the product of its area and the rainfall that fell at a particular point in the country, perhaps its capital; you would then add these products together. To get a more accurate estimate of the integral you would use a finer decomposition, like provinces or counties. $\triangle$

Here we will show that very general pavings can be used to compute integrals.

The set of all $P \in \mathcal{P}$ completely paves $X \subset \mathbb{R}^n$, and two "tiles" can overlap only in a set of volume 0.

Definition 4.7.2 (A paving of $X \subset \mathbb{R}^n$).   A paving of a subset $X \subset \mathbb{R}^n$ is a collection $\mathcal{P}$ of subsets $P \subset X$ such that

$$\cup_{P \in \mathcal{P}} P = X, \text{ and } \operatorname{vol}_n(P_1 \cap P_2) = 0 \text{ (when } P_1, P_2 \in \mathcal{P} \text{ and } P_1 \neq P_2). \text{ 4.7.1}$$

Definition 4.7.3 (The boundary of a paving of $X \subset \mathbb{R}^n$).   The boundary $\partial \mathcal{P}$ of $\mathcal{P}$ is the union of the boundaries of the tiles:

$$\partial \mathcal{P} = \bigcup_{P \in \mathcal{P}} \partial P.$$

If you think of the $P \in \mathcal{P}$ as tiles, then the boundary $\partial\mathcal{P}$ is like the grout lines between the tiles—exceedingly thin grout lines, since we will usually be interested in pavings such that vol $\partial\mathcal{P} = 0$.

In contrast to the upper and lower sums of the dyadic decompositions (Equation 4.1.18), where $\mathrm{vol}_n C$ is the same for any cube C at a given resolution N, in Equation 4.7.3, $\mathrm{vol}_n P$ is not necessarily the same for all "paving tiles" $P \in \mathcal{P}_N$.

Recall that $M_P(f)$ is the maximum value of $f(\mathbf{x})$ for $\mathbf{x} \in P$; similarly, $m_P(f)$ is the minimum.

Definition 4.7.4 (Nested partition). A sequence $\mathcal{P}_N$ of pavings of $X \subset \mathbb{R}^n$ is called a nested partition of X if

(1) $\mathcal{P}_{N+1}$ refines $\mathcal{P}_N$: every piece of $\mathcal{P}_{N+1}$ is contained in a piece of $\mathcal{P}_N$.

(2) All the boundaries have volume 0: $\mathrm{vol}_n(\partial\mathcal{P}_N) = 0$ for every N.

(3) The pieces of $\mathcal{P}_N$ shrink to points as $N \to \infty$:

$$\lim_{N\to\infty} \sup_{P\in\mathcal{P}_N} \mathrm{diam}\, P = 0. \qquad 4.7.2$$

For example, paving the United States by counties refines the paving by states: no county lies partly in one state and partly in another. A further refinement is provided by census tracts. (But this is not a nested partition, because the third requirement isn't met.)

We can define an upper sum $U_{\mathcal{P}_N}(f)$ and a lower sum $L_{\mathcal{P}_N}(f)$ with respect to any paving:

$$U_{\mathcal{P}_N}(f) = \sum_{P\in\mathcal{P}_N} M_P(f)\,\mathrm{vol}_n P \quad \text{and} \quad L_{\mathcal{P}_N}(f) = \sum_{P\in\mathcal{P}_N} m_P(f)\,\mathrm{vol}_n P. \qquad 4.7.3$$

What we called $U_N(f)$ in Section 4.1 would be called $U_{\mathcal{D}_N}(f)$ using this notation. We will often omit the subscript $\mathcal{D}_N$ (which you will recall denotes the collection of cubes C at a single level N) when referring to the dyadic decompositions, both to lighten the notation and to avoid confusion between $\mathcal{D}$ and $\mathcal{P}$, which, set in small subscript type, can look similar.

Theorem 4.7.5 (Integrals using arbitrary pavings). Let $X \subset \mathbb{R}^n$ be a bounded subset, and $\mathcal{P}_N$ be a nested partition of X. If the boundary ∂X satisfies $\mathrm{vol}_n(\partial X) = 0$, and $f : \mathbb{R}^n \to \mathbb{R}$ is integrable, then the limits

$$\lim_{N\to\infty} U_{\mathcal{P}_N}(f) \quad \text{and} \quad \lim_{N\to\infty} L_{\mathcal{P}_N}(f) \qquad 4.7.4$$

both exist, and are equal to

$$\int_X f(\mathbf{x})\,|d^n\mathbf{x}|. \qquad 4.7.5$$

The theorem is proved in Appendix A.14.

4.8 DETERMINANTS

In higher dimensions the determinant is important because it has a geometric interpretation, as a signed volume.

The *determinant* is a function of square matrices. In Section 1.4 we introduced determinants of 2×2 and 3×3 matrices, and saw that they have a geometric interpretation: the first gives the area of the parallelogram spanned by two vectors; the second gives the volume of the parallelepiped spanned by three vectors. In higher dimensions the determinant also has a geometric interpretation, as a *signed volume*; it is this that makes the determinant important.

We will use determinants heavily throughout the remainder of the book: forms, to be discussed in Chapter 6, are built on the determinant.

As we did for the determinant of a 2×2 or 3×3 matrix, we will think of the determinant as a function of n vectors rather than as a function of a matrix. This is a minor point, since whenever you have n vectors in $\mathbb{R}^n$, you can always place them side by side to make an $n \times n$ matrix.

Once matrices are bigger than 3×3, the formulas for computing the determinant are far too messy for hand computation—and too time-consuming even for computers, once a matrix is even moderately large. We will see (Equation 4.8.21) that the determinant can be computed much more reasonably by row (or column) reduction.

In order to obtain the volume interpretation most readily, we shall define the determinant by the three properties that characterize it.

Definition 4.8.1 (The determinant). The determinant

$$\det A = \det \begin{bmatrix} | & | & & | \\ \vec{a}_1, & \vec{a}_2, & \dots, & \vec{a}_n \\ | & | & & | \end{bmatrix} = \det(\vec{a}_1, \vec{a}_2, \dots, \vec{a}_n) \qquad 4.8.1$$

is the unique real-valued function of n vectors in $\mathbb{R}^n$ with the following properties:

(1) *Multilinearity*: det A is linear with respect to each of its arguments. That is, if one of the arguments (one of the vectors) can be written

$$\vec{a}_i = \alpha \vec{u} + \beta \vec{w}, \qquad 4.8.2$$

then

$$\det \left(\vec{a}_1, \dots, \vec{a}_{i-1}, (\alpha \vec{u} + \beta \vec{w}), \vec{a}_{i+1}, \dots, \vec{a}_n \right)$$
$$= \alpha \det(\vec{a}_1, \dots, \vec{a}_{i-1}, \vec{u}, \vec{a}_{i+1}, \dots, \vec{a}_n) \qquad 4.8.3$$
$$+ \beta \det(\vec{a}_1, \dots, \vec{a}_{i-1}, \vec{w}, \vec{a}_{i+1}, \dots, \vec{a}_n).$$

The properties of multilinearity and antisymmetry will come up often in Chapter 6.

(2) *Antisymmetry*: det A is antisymmetric. Exchanging any two arguments changes its sign:

$$\det \left(\vec{a}_1, \dots, \vec{a}_i, \dots, \vec{a}_j, \dots, \vec{a}_n \right) = - \det \left(\vec{a}_1, \dots, \vec{a}_j, \dots, \vec{a}_i, \dots, \vec{a}_n \right). \quad 4.8.4$$

(3) *Normalization*: the determinant of the identity matrix is 1:

$$\det I = \det \left(\vec{e}_1, \vec{e}_2, \dots, \vec{e}_n \right) = 1, \qquad 4.8.5$$

where $\vec{e}_1, \dots, \vec{e}_n$ are the standard basis vectors.

More generally, normalization means "setting the scale." For example, physicists may normalize units to make the speed of light 1. Normalizing the determinant means setting the scale for n-dimensional volume: deciding that the unit "n-cube" has volume 1.

Example 4.8.2 (Properties of the determinant). (1) Multilinearity: if $\alpha = -1, \beta = 2$, and

$$\vec{u} = \begin{bmatrix} 1 \\ 0 \\ 1 \end{bmatrix}, \vec{w} = \begin{bmatrix} 2 \\ 2 \\ 3 \end{bmatrix}, \text{ so that } \alpha \vec{u} + \beta \vec{w} = \begin{bmatrix} -1 \\ 0 \\ -1 \end{bmatrix} + \begin{bmatrix} 4 \\ 4 \\ 6 \end{bmatrix} = \begin{bmatrix} 3 \\ 4 \\ 5 \end{bmatrix},$$

then

$$\det \begin{bmatrix} 1 & 3 & 3 \\ 2 & 4 & 1 \\ 0 & 5 & 1 \end{bmatrix} = -1 \det \begin{bmatrix} 1 & 1 & 3 \\ 2 & 0 & 1 \\ 0 & 1 & 1 \end{bmatrix} + 2 \det \begin{bmatrix} 1 & 2 & 3 \\ 2 & 2 & 1 \\ 0 & 3 & 1 \end{bmatrix}, \qquad 4.8.6$$

$$\underbrace{}_{23} \qquad \underbrace{}_{-1 \times 3 = -3} \qquad \underbrace{}_{2 \times 13 = 26}$$

as you can check using Definition 1.4.15.

(2) Antisymmetry:

$$\det \begin{bmatrix} 1 & 3 & 3 \\ 2 & 4 & 1 \\ 0 & 5 & 1 \end{bmatrix} = - \det \begin{bmatrix} 1 & 3 & 3 \\ 2 & 1 & 4 \\ 0 & 1 & 5 \end{bmatrix}. \qquad 4.8.7$$

$$\underbrace{}_{23} \qquad \underbrace{}_{-23}$$

(3) Normalization:

$$\det \begin{bmatrix} 1 & 0 & 0 \\ 0 & 1 & 0 \\ 0 & 0 & 1 \end{bmatrix} = 1\big((1 \times 1) - 0\big) = 1. \qquad 4.8.8$$

Remark 4.8.3. Exercise 4.8.4 explores some immediate consequences of Definition 4.8.1: if a matrix has a column of zeroes, or if it has two identical columns, its determinant is 0.

Our examples are limited to 3×3 matrices because we haven't shown yet how to compute larger ones. $\triangle$

In order to see that Definition 4.8.1 is reasonable, we will want the following theorem:

Theorem 4.8.4 (Existence and uniqueness of the determinant).
There exists a function $\det A$ *satisfying the three properties of the determinant, and it is unique.*

The proofs of existence and uniqueness are quite different, with a somewhat lengthy but necessary construction for each. The outline for the proof is as follows:

First we shall use a computer program to construct a function $D(A)$ by a process called "development according to the first column." Of course this could be developed differently, e.g., according to the first row, but you can show in Exercise 4.8.13 that the result is equivalent to this definition. Then (in Appendix A.15) we shall prove that $D(A)$ satisfies the properties of $\det A$, thus establishing *existence* of a function that satisfies the definition of determinant.

Finally we shall proceed by "column operations" to evaluate this function $D(A)$ and show that it is unique, which will prove *uniqueness* of the determinant. This will simultaneously give an effective algorithm for computing determinants.

Development according to the first column. Consider the function

$$D(A) = \sum_{i=1}^{n} (-1)^{1+i} a_{i,1} D(A_{i,1}), \qquad 4.8.9$$

In Equation 4.8.9 we have $A_{i,1}$, the $(n-1) \times (n-1)$ matrix obtained from A by erasing the ith row and the first column.

where A is an $n \times n$ matrix and $A_{i,j}$ is the $(n-1) \times (n-1)$ matrix obtained from A by erasing the ith row and the jth column, as illustrated by Example 4.8.5. The formula may look unfriendly, but it's not really complicated. As shown in Equation 4.8.10, each term of the sum is the product of the entry $a_{i,1}$ and D of the new, smaller matrix obtained from A by erasing the ith row and the first column; the $(-1)^{1+i}$ simply assigns a sign to the term.

$$D(A) = \sum_{i=1}^{n} \underbrace{(-1)^{1+i}}_{\substack{\text{tells whether} \\ + \text{ or } -}} \underbrace{a_{i,1} D(A_{i,1})}_{\substack{\text{product of } a_{i,1} \text{ and} \\ D \text{ of smaller matrix}}}. \qquad 4.8.10$$

Our candidate determinant D is thus recursive: D of an $n \times n$ matrix is the sum of n terms, each involving D's of $(n-1) \times (n-1)$ matrices; in turn, the D of each $(n-1) \times (n-1)$ matrix is the sum of $(n-1)$ terms, each involving D's of $(n-2) \times (n-2)$ matrices .... (Of course, when one deletes the *first* column of the $(n-1) \times (n-1)$ matrix, it is the *second* column of the original matrix, and so on.)

For this to work we must say that the D of a 1×1 "matrix," i.e., a number, is the number itself. For example, $D[7] = 7$.

Example 4.8.5 (The function $D(A)$). If

$$A = \begin{bmatrix} 1 & 3 & 4 \\ 0 & 1 & 1 \\ 1 & 2 & 0 \end{bmatrix}, \quad \text{then} \quad A_{2,1} = \begin{bmatrix} 1 & 3 & 4 \\ \cancel{0} & \cancel{1} & \cancel{1} \\ 1 & 2 & 0 \end{bmatrix} = \begin{bmatrix} 3 & 4 \\ 2 & 0 \end{bmatrix}, \qquad 4.8.11$$

and Equation 4.8.9 corresponds to

$$D(A) = \underbrace{1 \, D\left(\begin{bmatrix} 1 & 1 \\ 2 & 0 \end{bmatrix}\right)}_{i=1} \underbrace{- 0 \, D\left(\begin{bmatrix} 3 & 4 \\ 2 & 0 \end{bmatrix}\right)}_{i=2} \underbrace{+ 1 \, D\left(\begin{bmatrix} 3 & 4 \\ 1 & 1 \end{bmatrix}\right)}_{i=3}. \qquad 4.8.12$$

The first term is positive because when $i = 1$, then $1 + i = 2$ and we have $(-1)^2 = 1$; the second is negative, because $(-1)^3 = -1$, and so on.

Applying Equation 4.8.9 to each of these 2×2 matrices gives:

$$D\left(\begin{bmatrix} 1 & 1 \\ 2 & 0 \end{bmatrix}\right) = 1D(0) - 2D(1) = 0 - 2 = -2;$$

$$D\left(\begin{bmatrix} 3 & 4 \\ 2 & 0 \end{bmatrix}\right) = 3D(0) - 2D(4) = -8; \qquad 4.8.13$$

$$D\left(\begin{bmatrix} 3 & 4 \\ 1 & 1 \end{bmatrix}\right) = 3D(1) - 1D(4) = -1,$$

so that D of our original 3×3 matrix is $1(-2) - 0 + 1(-1) = -3$. $\triangle$

The Pascal program *Determinant*, in Appendix B.3, implements the development of the determinant according to the first column. It will compute $D(A)$ for any square matrix of side at most 10; it will run on a personal computer and in 1998 would compute the determinant of a 10×10 matrix in half a second.[9]

Please note that this program is *very time consuming*. Suppose that D takes time $T(k)$ to compute the determinant of a $k \times k$ matrix. Then, since it makes k "calls" of D for a $(k-1) \times (k-1)$ matrix, as well as k multiplications, $k-1$ additions, and k calls of the subroutine "erase," we see that

$$T(k) > kT(k-1), \qquad\qquad 4.8.14$$

so that $T(k) > k!\,T(1)$. In 1998, on a fast personal computer, one floating point operation took about 2×10^{-9} second. The time to compute determinants by this method is at least the factorial of the size of the matrix. For a 15×15 matrix, this means $15! \approx 1.3 \times 10^{12}$ calls or operations, which translates into roughly 45 minutes. And 15×15 is not a big matrix; engineers modeling bridges or airplanes and economists modeling a large company routinely use matrices that are more than $1\,000 \times 1\,000$. So if this program were the only way to compute determinants, they would be of theoretical interest only. But as we shall soon show, determinants can also be computed by row or column reduction, which is immensely more efficient when the matrix is even moderately large.

However, the construction of the function $D(A)$ is most convenient in proving existence in Theorem 4.8.4.

This program embodies the recursive nature of the determinant as defined above: the key point is that the function D calls itself. It would be quite a bit more difficult to write this program in Fortran or Basic, which do not allow that sort of thing.

The number of operations that would be needed to compute the determinant of a 40×40 matrix using development by the first column is *bigger than the number of seconds that have elapsed since the beginning of the universe*. In fact, bigger than the number of billionths of seconds that have elapsed: if you had set a computer computing the determinant back in the days of the dinosaurs, it would have barely begun

The effective way to compute determinants is by column operations.

Proving the existence and uniqueness of the determinant

We prove existence by verifying that the function $D(A)$ does indeed satisfy properties (1), (2), and (3) for the determinant det A. This is a messy and uninspiring exercise in the use of induction, and we have relegated it to Appendix A.15.

Of course, there might be other functions satisfying those properties, but we will now show that in the course of row reducing (or rather column reducing) a matrix, we simultaneously compute the determinant. Column reduction of an $n \times n$ matrix takes about n^3 operations. For a 40×40 matrix, this means $64\,000$ operations, which would take a reasonably fast computer much less than one second.

At the same time this algorithm proves uniqueness, since, by Theorem 2.1.8, given any matrix A, there exists a unique matrix $\widetilde{A}$ in echelon form that can be obtained from A by row operations. Our discussion will use only properties (1), (2), and (3), without the function $D(A)$.

We saw in Section 2.3 that a column operation is equivalent to multiplying a matrix on the right by an elementary matrix.

---

[9]In about 1990, the same computation took about an hour; in 1996, about a minute.

Let us check how each of the three column operations affect the determinant. It turns out that *each multiplies the determinant* by an appropriate factor μ:

(1) *Multiply a column through by a number $m \neq 0$* (multiplying by a type 1 elementary matrix). Clearly, by multilinearity (property (1) above), this has the effect of multiplying the determinant by the same number, so

$$\mu = m. \qquad\qquad 4.8.15$$

(2) *Add a multiple of one column onto another* (multiplying by a type 2 elementary matrix). By property (1), this does not change the determinant, because

$$\det\left(\vec{\mathbf{a}}_1, \ldots, \vec{\mathbf{a}}_i, \ldots, (\vec{\mathbf{a}}_j + \beta\vec{\mathbf{a}}_i), \ldots, \vec{\mathbf{a}}_n\right) \qquad\qquad 4.8.16$$

$$= \det(\vec{\mathbf{a}}_1, \ldots, \vec{\mathbf{a}}_i, \ldots, \vec{\mathbf{a}}_j, \ldots, \vec{\mathbf{a}}_n) \;\; + \underbrace{\beta \det(\vec{\mathbf{a}}_1, \ldots, \vec{\mathbf{a}}_i, \ldots, \vec{\mathbf{a}}_i, \ldots, \vec{\mathbf{a}}_n)}_{=\,0 \text{ because 2 identical terms } \vec{\mathbf{a}}_i}.$$

The second term on the right is zero, since two columns are equal (Exercise 4.8.4 b). Therefore

$$\mu = 1. \qquad\qquad 4.8.17$$

As mentioned earlier, we will use column operations (Definition 2.1.11) rather than row operations in our construction, because we defined the determinant as a function of the n column vectors. This convention makes the interpretation in terms of volumes simpler, and in any case you will be able to show in Exercise 4.8.13 that row operations could have been used just as well.

(3) *Exchange two columns* (multiplying by a type 3 elementary matrix). By antisymmetry, this changes the sign of the determinant, so

$$\mu = -1. \qquad\qquad 4.8.18$$

Any square matrix can be column reduced until at the end, you either get the identity, or a matrix with a column of zeroes. A sequence of matrices resulting from column operations can be denoted as follows, with the multipliers μ_i of the corresponding determinants on top of arrows for each operation:

$$A \xrightarrow{\;\mu_1\;} A_1 \xrightarrow{\;\mu_2\;} A_2 \xrightarrow{\;\mu_3\;} \cdots \xrightarrow{\;\mu_{n-1}\;} A_{n-1} \xrightarrow{\;\mu_n\;} A_n, \qquad\qquad 4.8.19$$

with A_n in column echelon form. Then, working backwards,

$$\det A_{n-1} = \frac{1}{\mu_n}\, \det A_n;$$

$$\det A_{n-2} = \frac{1}{\mu_{n-1}\,\mu_n}\, \det A_n;$$

$$\downarrow$$

$$\det A = \frac{1}{\mu_1 \mu_2 \cdots \mu_{n-1}\, \mu_n}\, \det A_n. \qquad\qquad 4.8.20$$

Therefore:

(1) If $A_n = I$, then by property (3) we have det $A_n = 1$, so by Equation 4.8.20,

$$\det A = \frac{1}{\mu_1 \mu_2 \ldots \mu_{n-1} \mu_n}. \qquad 4.8.21$$

Equation 4.8.21 is the formula that is really used to compute determinants.

(2) If $A_n \neq I$, then by property (1) we have det $A_n = 0$ (see Exercise 4.8.4), so

$$\det A = 0. \qquad 4.8.22$$

Proof of uniqueness of determinant. Suppose we have another function, $D_1(A)$, which obeys properties (1), (2), and (3). Then for any matrix A,

$$D_1(A) = \frac{1}{\mu_1 \mu_2 \ldots \mu_n} \det A_n = D(A); \qquad 4.8.23$$

You may object that a different sequence of column operations might lead to a different sequence of μ's, with a different product. If that were the case, it would show that the axioms for the determinant were inconsistent; we know they are consistent because of the existence part of Theorem 4.8.4, proved in Appendix A.15.

i.e., $D_1 = D$. $\square$

Theorems relating matrices and determinants

In this subsection we group several useful theorems that relate matrices and their determinants.

Theorem 4.8.6. *A matrix A is invertible if and only if its determinant is not zero.*

Proof. This follows immediately from the column-reduction algorithm and the uniqueness proof, since along the way we showed, in Equations 4.8.21 and 4.8.22, that a square matrix has a nonzero determinant if and only if it can be column-reduced to the identity. We know from Theorem 2.3.2 that a matrix is invertible if and only if it can be row reduced to the identity; the same argument applies to column reduction. $\square$

Now we come to a key property of the determinant, for which we will see a geometric interpretation later. It was in order to prove this theorem that we defined the determinant by its properties.

A definition that defines an object or operation by its properties is called an *axiomatic* definition. The proof of Theorem 4.8.7 should convince you that this can be a fruitful approach. Imagine trying to prove

$$D(A)D(B) = D(AB)$$

from the recursive definition.

Theorem 4.8.7. *If A and B are $n \times n$ matrices, then*

$$\det A \det B = \det(AB). \qquad 4.8.24$$

Proof. (a) The serious case is the one in which A is invertible. If A is invertible, consider the function

$$f(B) = \frac{\det(AB)}{\det A}. \qquad 4.8.25$$

As you can readily check (Exercise 4.8.5), it has the properties (1), (2), and (3), which characterize the determinant function. Since the determinant is uniquely characterized by those properties, then $f(B) = \det B$.

(b) The case where A is not invertible is easy, using what we know about images and dimensions of linear transformations. If A is not invertible, $\det A$ is zero (Theorem 4.8.6), so the left-hand side of the theorem is zero. The right-hand side must be zero also: since A is not invertible, rank $A < n$. Since $\text{Img}(AB) \subset \text{Img}\, A$, then $\text{rank}\,(AB) \leq \text{rank}\, A < n$, so AB is not invertible either, and $\det AB = 0$. $\square$

$$\begin{bmatrix} 1 & 0 & 0 & 0 \\ 0 & 1 & 0 & 0 \\ 0 & 0 & 2 & 0 \\ 0 & 0 & 0 & 1 \end{bmatrix}$$

The determinant of this type 1 elementary matrix is 2.

Theorem 4.8.7, combined with Equations 4.8.15, 4.8.18, and 4.8.17, give the following determinants for elementary matrices.

Theorem 4.8.8. *The determinant of an elementary matrix equals the determinant of its transpose:*

$$\det E = \det E^{\top}. \qquad 4.8.26$$

$$\begin{bmatrix} 1 & 0 & -3 \\ 0 & 1 & 0 \\ 0 & 0 & 1 \end{bmatrix}$$

The determinant of all type 2 elementary matrices is 1.

Corollary 4.8.9 (Determinants of elementary matrices). *The determinant of a type 1 elementary matrix is m, where $m \neq 0$ is the entry on the diagonal not required to be 1. The determinant of a type 2 elementary matrix is 1, and that of a type 3 elementary matrix is -1:*

$$\det E_1(i, m) = \quad m$$
$$\det E_2(i, j, x) = \quad 1$$
$$\det E_3(i, j) = -1.$$

$$\begin{bmatrix} 0 & 1 & 0 \\ 1 & 0 & 0 \\ 0 & 0 & 1 \end{bmatrix}$$

The determinant of all type 3 elementary matrices is -1.

Proof. The three types of elementary matrices are described in Definition 2.3.5. For the first type and the third types, $E = E^{\top}$, so there is nothing to prove. For the second type, all the entries on the main diagonal are 1, and all other entries are 0 except for one, which is nonzero. Call that nonzero entry, in the ith row and jth column, a. We can get rid of a by multiplying the ith column by $-a$ and adding the result to the jth column, creating a new matrix $E' = I$, as shown in the example below, where $i = 2$ and $j = 3$.

$$\text{If} \quad E = \begin{bmatrix} 1 & 0 & 0 & 0 \\ 0 & 1 & a & 0 \\ 0 & 0 & 1 & 0 \\ 0 & 0 & 0 & 1 \end{bmatrix}, \quad \text{then} \qquad 4.8.27$$

$$-a \times \underbrace{\begin{bmatrix} 0 \\ 1 \\ 0 \\ 0 \end{bmatrix}}_{\substack{i\text{th} \\ \text{column}}} = \begin{bmatrix} 0 \\ -a \\ 0 \\ 0 \end{bmatrix}; \text{ and } \underbrace{\begin{bmatrix} 0 \\ a \\ 1 \\ 0 \end{bmatrix}}_{\substack{j\text{th} \\ \text{column}}} + \begin{bmatrix} 0 \\ -a \\ 0 \\ 0 \end{bmatrix} = \begin{bmatrix} 0 \\ 0 \\ 1 \\ 0 \end{bmatrix}. \qquad 4.8.28$$

We know (Equation 4.8.16) that adding a multiple of one column onto another does not change the determinant, so $\det E = \det I$. By property (3) of the determinant (Equation 4.8.5), $\det I = 1$, so $\det E = 1$. The transpose $E^\top$ is identical to E except that instead of $a_{i,j}$ we have $a_{j,i}$; by the argument above, $\det E^\top = 1$. $\square$

We are finally in a position to prove the following result.

One easy consequence of Theorem 4.8.10 is that a matrix with a row of zeroes has determinant 0.

Theorem 4.8.10. *For any $n \times n$ matrix A,*

$$\det A = \det A^\top. \tag{4.8.29}$$

Proof. Column reducing a matrix A to echelon form $\widetilde{A}$ is the same as multiplying it on the right by a succession of elementary matrices $E_1 \ldots E_k$:

$$\widetilde{A} = A(E_1 \ldots E_k). \tag{4.8.30}$$

By Theorem 1.2.17, $(AB)^\top = B^\top A^\top$, so

$$\widetilde{A}^\top = (E_1 \ldots E_k)^\top A^\top. \tag{4.8.31}$$

We need to consider two cases.

First, suppose $\widetilde{A} = I$, the identity. Then $\widetilde{A}^\top = I$, and

$$A = E_k^{-1} \ldots E_1^{-1} \quad \text{and} \quad A^\top = \left(E_k^{-1} \ldots E_1^{-1}\right)^\top = (E_1^{-1})^\top \ldots (E_k^{-1})^\top, \tag{4.8.32}$$

so

The fact that determinants are numbers, and that therefore multiplication of determinants is commutative, is much of the point of determinants; essentially everything having to do with matrices that does not involve noncommutativity can be done using determinants.

$$\det A = \det\left(E_k^{-1} \ldots E_1^{-1}\right) = \det E_k^{-1} \ldots \det E_1^{-1};$$

$$\det A^\top = \det\left((E_1^{-1})^\top \ldots (E_k^{-1})^\top\right)$$
$$= \det(E^{-1})_1^\top \ldots \det(E^{-1})_k^\top \underbrace{=}_{\substack{\text{Theorem} \\ 4.8.8}} \det E_1^{-1} \ldots \det E_k^{-1}. \tag{4.8.33}$$

A determinant is a number, not a matrix, so multiplication of determinants is commutative: $\det E_1^{-1} \ldots \det E_k^{-1} = \det E_k^{-1} \ldots \det E_1^{-1}$. This gives us $\det A = \det A^\top$.

Recall Corollary 2.5.13: A matrix A and its transpose $A^\top$ have the same rank.

If $\widetilde{A} \neq I$, then rank $A < n$, so rank $A^\top < n$, so $\det A = \det A^\top = 0$. $\square$

One important consequence of Theorem 4.8.10 is that throughout this text, whenever we spoke of column operations, we could just as well have spoken of row operations.

Some matrices have a determinant that is easy to compute: the triangular matrices (See Definition 1.2.19).

Theorem 4.8.11. *If a matrix is triangular, then its determinant is the product of the entries along the diagonal.*

Proof. We will prove the result for upper triangular matrices; the result for lower triangular matrices then follows from Theorem 4.8.10. The proof is by induction. Theorem 4.8.11 is clearly true for a 1×1 triangular matrix (note that any 1×1 matrix is triangular). If A is triangular of size $n \times n$ with $n > 1$, the submatrix $A_{1,1}$ (A with its first row and first column removed) is also triangular, of size $(n-1) \times (n-1)$, so we may assume by induction that

An alternative proof is sketched in Exercise 4.8.6.

$$\det A_{1,1} = a_{2,2} \ldots a_{n,n}. \qquad 4.8.34$$

Since $a_{1,1}$ is the only nonzero entry in the first column, development according to the first column gives:

$$\det A = (-1)^2 a_{1,1} \det A_{1,1} = a_{1,1} a_{2,2} \ldots a_{n,n}. \quad \square \qquad 4.8.35$$

Here are some more characterizations of invertible matrices.

Theorem 4.8.12. *If a matrix A is invertible, then*

$$\det A^{-1} = \frac{1}{\det A}. \qquad 4.8.36$$

Proof. This is a simple consequence of Theorem 4.8.7:

$$\det A \det A^{-1} = \det (AA^{-1}) = \det I = 1. \quad \square \qquad 4.8.37$$

The following theorem acquires its real significance in the context of abstract vector spaces, where we will see that it means that the determinant function is basis independent. This is discussed in volume two; in this volume we will find the theorem useful in proving Corollary 4.8.22.

Theorem 4.8.13. *If P is invertible, then*

$$\det A = \det(P^{-1}AP). \qquad 4.8.38$$

Proof. This follows immediately from Theorems 4.8.7 and 4.8.12. $\square$

Theorem 4.8.14. *If A is an $n \times n$ matrix and B is an $m \times m$ matrix, then for the $(n+m) \times (n+m)$ matrix formed with these as diagonal elements,*

$$\det \begin{bmatrix} A & 0 \\ 0 & B \end{bmatrix} = \det A \det B. \qquad 4.8.39$$

The proof of Theorem 4.8.14 is left to the reader as Exercise 4.8.7.

The signature of a permutation

Some treatments of the determinant start out with the *signature* of a permutation, and proceed to define the determinant by Equation 4.8.46. We approached the problem differently because we wanted to emphasize the effect of row operations on the determinant, which is easier using our approach.

Recall that a *permutation* of $\{1,\dots,n\}$ is a one to one map $\sigma : \{1,\dots,n\} \to \{1,\dots,n\}$. One permutation of $\{1,2,3\}$ is $\{2,1,3\}$; another is $\{2,3,1\}$. There are several ways of denoting a permutation; the permutation that maps 1 to 2, 2 to 3, and 3 to 1 can be denoted

$$\sigma : \begin{bmatrix} 1 \\ 2 \\ 3 \end{bmatrix} \mapsto \begin{bmatrix} 2 \\ 3 \\ 1 \end{bmatrix} \quad \text{or}$$

$$\sigma = \begin{bmatrix} 1 & 2 & 3 \\ 2 & 3 & 1 \end{bmatrix}.$$

Permutations can be composed: if

$$\sigma : \begin{bmatrix} 1 \\ 2 \\ 3 \end{bmatrix} \mapsto \begin{bmatrix} 2 \\ 1 \\ 3 \end{bmatrix} \quad \text{and}$$

$$\tau : \begin{bmatrix} 1 \\ 2 \\ 3 \end{bmatrix} \mapsto \begin{bmatrix} 2 \\ 3 \\ 1 \end{bmatrix},$$

then we have

$$\tau \circ \sigma : \begin{bmatrix} 1 \\ 2 \\ 3 \end{bmatrix} \mapsto \begin{bmatrix} 3 \\ 2 \\ 1 \end{bmatrix} \quad \text{and}$$

$$\sigma \circ \tau : \begin{bmatrix} 1 \\ 2 \\ 3 \end{bmatrix} \mapsto \begin{bmatrix} 1 \\ 3 \\ 2 \end{bmatrix}.$$

We see that a permutation matrix acts on any element of $\mathbb{R}^n$ by permuting its coordinates.

In the language of group theory, the transformation that associates to a permutation its matrix is called a *group homomorphism*.

There are a great many possible definitions of the signature of a permutation, all a bit unsatisfactory.

One definition is to write the permutation as a product of *transpositions*, a transposition being a permutation in which exactly two elements are exchanged. Then the signature is $+1$ if the number of transpositions is even, and -1 if it is odd. The problem with this definition is that there are a great many different ways to write a permutation as a product of transpositions, and it isn't clear that they all give the same signature.

Indeed, showing that different ways of writing a permutation as a product of transpositions all give the same signature involves something like the existence part of Theorem 4.8.4; that proof, in Appendix A.15, is distinctly unpleasant. But armed with this result, we can get the signature almost for free.

First, observe that we can associate to any permutation σ of $\{1,\dots,n\}$ its permutation matrix M_σ, by the rule

$$(M_\sigma)\vec{e}_i = \vec{e}_{\sigma(i)}. \qquad 4.8.40$$

Example 4.8.15 (Permutation matrix). Suppose we have a permutation σ such that $\sigma(1) = 2, \sigma(2) = 3$, and $\sigma(3) = 1$, which we may write

$$\begin{bmatrix} 1 \\ 2 \\ 3 \end{bmatrix} \mapsto \begin{bmatrix} 2 \\ 3 \\ 1 \end{bmatrix}, \quad \text{or simply} \quad (2,3,1).$$

This permutation puts the first coordinate in second place, the second in third place, and the third in first place, *not* the first coordinate in third place, the second in first place, and the third in second place.

The first column of the permutation matrix is $M_\sigma \vec{e}_1 = \vec{e}_{\sigma(1)} = \vec{e}_2$. Similarly, the second column is $\vec{e}_3$ and the third column is $\vec{e}_1$:

$$M_\sigma = \begin{bmatrix} 0 & 0 & 1 \\ 1 & 0 & 0 \\ 0 & 1 & 0 \end{bmatrix}. \qquad 4.8.41$$

You can easily confirm that this matrix puts the first coordinate of a vector in $\mathbb{R}^3$ into second position, the second coordinate into third position, and the third coordinate into first position:

$$M_\sigma \begin{bmatrix} a \\ b \\ c \end{bmatrix} = \begin{bmatrix} c \\ a \\ b \end{bmatrix}. \quad \triangle \qquad 4.8.42$$

Exercise 4.8.9 asks you to check that the transformation $\sigma \mapsto M_\sigma$ that associates to a permutation its matrix satisfies $M_{\sigma \circ \tau} = M_\sigma M_\tau$.

The determinant of such a *permutation matrix* is obviously ± 1, since by exchanging rows repeatedly it can be turned into the identity matrix; each time two rows are exchanged, the sign of the determinant changes.

Definition 4.8.16 (Signature of a permutation). The signature of a permutation σ, denoted $\mathrm{sgn}(\sigma)$, is defined by

$$\mathrm{sgn}(\sigma) = \det M_\sigma. \qquad 4.8.43$$

Some authors denote the signature $(-1)^\sigma$.

Permutations of signature $+1$ are called *even permutations*, and permutations of signature -1 are called *odd permutations*. Almost all properties of the signature follow immediately from the properties of the determinant; we will explore them at some length in the exercises.

Example 4.8.17 (Signatures of permutations). There are six permutations of the numbers 1, 2, 3:

Remember that by

$$\sigma_3 = (3, 1, 2)$$

we mean the permutation such that $\begin{bmatrix} 1 \\ 2 \\ 3 \end{bmatrix} \mapsto \begin{bmatrix} 3 \\ 1 \\ 2 \end{bmatrix}$.

$$\sigma_1 = (1,2,3), \quad \sigma_2 = (2,3,1), \quad \sigma_3 = (3,1,2)$$
$$\sigma_4 = (1,3,2), \quad \sigma_5 = (2,1,3), \quad \sigma_6 = (3,2,1). \qquad 4.8.44$$

The first three permutations are even; the last three are odd. We gave the permutation matrix for σ_2 in Example 4.8.15; its determinant is $+1$. Here are three more:

$$\det \begin{bmatrix} 1 & 0 & 0 \\ 0 & 1 & 0 \\ 0 & 0 & 1 \end{bmatrix} = 1 \det \begin{bmatrix} 1 & 0 \\ 0 & 1 \end{bmatrix}$$
$$-0 \det \begin{bmatrix} 0 & 0 \\ 0 & 1 \end{bmatrix} + 0 \det \begin{bmatrix} 0 & 0 \\ 1 & 0 \end{bmatrix} = 1.$$

$$\det M_{\sigma_1} = \det \begin{bmatrix} 1 & 0 & 0 \\ 0 & 1 & 0 \\ 0 & 0 & 1 \end{bmatrix} = +1, \qquad \det M_{\sigma_3} = \det \begin{bmatrix} 0 & 1 & 0 \\ 0 & 0 & 1 \\ 1 & 0 & 0 \end{bmatrix} = +1,$$
$$4.8.45$$
$$\det M_{\sigma_4} = \det \begin{bmatrix} 1 & 0 & 0 \\ 0 & 0 & 1 \\ 0 & 1 & 0 \end{bmatrix} = -1.$$

Exercise 4.8.10 asks you to verify the signature of σ_5 and σ_6. △

Remark. In practice signatures aren't computed by computing the permutation matrix. If a signature is a composition of k transpositions, then the signature is positive if k is even and negative if k is odd, since each transposition corresponds to exchanging two columns of the permutation matrix, and hence changes the sign of the determinant. The second permutation of Example 4.8.17 has positive signature because two transpositions are required: exchanging 1 and 3, then exchanging 3 and 2 (or first exchanging 3 and 2, and then exchanging 1 and 3). △

The notation $\mathrm{Perm}(1, \ldots, n)$ denotes the set of permutations of $1, \ldots, n$.

We can now state one more formula for the determinant.

Theorem 4.8.18 (Determinant in terms of permutations). *Let A be an $n \times n$ matrix with entries denoted $(a_{i,j})$. Then*

$$\det A = \sum_{\sigma \in \mathrm{Perm}(1,\ldots,n)} \mathrm{sgn}(\sigma) a_{1,\sigma(1)} \ldots a_{n,\sigma(n)}. \qquad 4.8.46$$

Each term of the sum in Equation 4.8.46 is the product of n entries of the matrix A, chosen so that there is exactly one from each row and one from each column; no two are from the same column or the same row. These products are then added together, with an appropriate sign.

In Equation 4.8.46 we are summing over each permutation σ of the numbers $1, \ldots, n$. If $n = 3$, there will be six such permutations, as shown in Example 4.8.17. For each permutation σ, we see what σ does to the numbers $1, \ldots, n$, and use the result as the second index of the matrix entries. For instance, if $\sigma(1) = 2$, then $a_{1,\sigma(1)}$ is the entry $a_{1,2}$ of the matrix A.

Example 4.8.19 (Computing the determinant by permutations). Let $n = 3$, and let A be the matrix

$$A = \begin{bmatrix} 1 & 2 & 3 \\ 4 & 5 & 6 \\ 7 & 8 & 9 \end{bmatrix}. \qquad 4.8.47$$

Then we have

$\sigma_1 = (123)$	$+$	$a_{1,1}a_{2,2}a_{3,3} = 1 \cdot 5 \cdot 9 = 45$
$\sigma_2 = (231)$	$+$	$a_{1,2}a_{2,3}a_{3,1} = 2 \cdot 6 \cdot 7 = 84$
$\sigma_3 = (312)$	$+$	$a_{1,3}a_{2,1}a_{3,2} = 3 \cdot 4 \cdot 8 = 96$
$\sigma_4 = (132)$	$-$	$a_{1,1}a_{2,3}a_{3,2} = 1 \cdot 6 \cdot 8 = 48$
$\sigma_5 = (213)$	$-$	$a_{1,2}a_{2,1}a_{3,3} = 2 \cdot 4 \cdot 9 = 72$
$\sigma_6 = (321)$	$-$	$a_{1,3}a_{2,2}a_{3,1} = 3 \cdot 5 \cdot 7 = 105$

So $\det A = 45 + 84 + 96 - 48 - 72 - 105 = 0$. Can you see why this determinant had to be 0?[10] $\triangle$

In Example 4.8.19 it would be quicker to compute the determinant directly, using Definition 1.4.15. Theorem 4.8.18 does not provide an effective algorithm for computing determinants; for 2×2 and 3×3 matrices, which are standard in the classroom (but not anywhere else), we have explicit and manageable formulas. When they are large, column reduction (Equation 4.8.21) is immeasurably faster: for a 30×30 matrix, roughly the difference between one second and the age of the universe.

Proof of Theorem 4.8.18. So as not to prejudice the issue, let us temporarily call the function of Theorem 4.8.18 $D(A)$:

$$D(A) = \sum_{\sigma \in \mathrm{Perm}(1,\ldots,n)} \mathrm{sgn}(\sigma) a_{1,\sigma(1)} \cdots a_{n,\sigma(n)}. \qquad 4.8.48$$

We will show that the function D has the three properties that characterize the determinant. Normalization is satisfied: $D(I) = 1$, since if σ is not the identity, the corresponding product is 0, so the sum above amounts to multiplying

---

[10] Denote by $\vec{a}_1, \vec{a}_2, \vec{a}_3$ the columns of A. Then $\vec{a}_3 - \vec{a}_2 = \begin{bmatrix} 1 \\ 1 \\ 1 \end{bmatrix}$ and $\vec{a}_2 - \vec{a}_1 = \begin{bmatrix} 1 \\ 1 \\ 1 \end{bmatrix}$.

So $\vec{a}_1 - 2\vec{a}_2 + \vec{a}_3 = 0$; the columns are linearly dependent, so the matrix is not invertible, and its determinant is 0.

together the entries on the diagonal, which are all 1, and assigning the product the signature of the identity, which is +1. Multilinearity is straightforward: each term $a_{1,\sigma(1)}\cdots a_{n,\sigma(n)}$ is multilinear as a function of the columns, so any linear combination of such terms is also multilinear.

Now let's discuss antisymmetry. Let $i \neq j$ be the indices of two columns of an $n \times n$ matrix A, and let τ be the permutation of $\{1,\ldots,n\}$ that exchanges them and leaves all the others where they are. Further, denote by A' the matrix formed by exchanging the ith and jth columns of A. Then Equation 4.8.46, applied to the matrix A', gives

$$D(A') = \sum_{\sigma \in \mathrm{Perm}(1,\ldots,n)} \mathrm{sgn}(\sigma)a'_{1,\sigma(1)}\cdots a'_{n,\sigma(n)}$$

$$= \sum_{\sigma \in \mathrm{Perm}(1,\ldots,n)} \mathrm{sgn}(\sigma)a_{1,\tau\circ\sigma(1)}\cdots a_{n,\tau\circ\sigma(n)},$$

4.8.49

since the entry of A' in position (k,l) is the same as the entry of A in position $(k,\tau(l))$.

As σ runs through all permutations, $\sigma' = \tau \circ \sigma$ does too, so we might as well write

$$D(A') = \sum_{\sigma' \in \mathrm{Perm}(1,\ldots,n)} \mathrm{sgn}(\tau^{-1}\circ\sigma')a_{1,\sigma'(1)}\cdots a_{n,\sigma'(n)},$$

4.8.50

and the result follows from $\mathrm{sgn}(\tau^{-1}\circ\sigma') = \mathrm{sgn}(\tau^{-1})(\mathrm{sgn}(\sigma)) = -\mathrm{sgn}(\sigma)$, since $\mathrm{sgn}(\tau) = \mathrm{sgn}(\tau^{-1}) = -1$. $\square$

The trace of
$$\begin{bmatrix} 1 & 0 & 3 \\ 1 & 2 & 1 \\ 0 & 1 & -1 \end{bmatrix}$$
is $1 + 2 + (-1) = 2$.

Using sum notation, Equation 4.8.51 is
$$\mathrm{tr}A = \sum_{i=1}^{n} a_{i,i}.$$

The trace and the derivative of the determinant

Another interesting function of a square matrix is its *trace*, denoted tr.

Definition 4.8.20 (The trace of a matrix). The trace of a $n \times n$ matrix A is the sum of its diagonal elements:

$$\mathrm{tr}A = a_{1,1} + a_{2,2} + \cdots + a_{n,n}.$$

4.8.51

The trace is easy to compute, much easier than the determinant, and it is a linear function of A:

$$\mathrm{tr}(aA + bB) = a\,\mathrm{tr}\,A + b\,\mathrm{tr}\,B.$$

4.8.52

The trace doesn't look as if it has anything to do with the determinant, but Theorem 4.8.21 shows that they are closely related.

Of course A and B are both square matrices, elements of Mat(n,n).

Note that in Equation 4.8.53, $[\mathbf{D}\det(I)]$ is the derivative of the determinant function evaluated at I. (It should not be read as the derivative of $\det(I)$, which is 0, since $\det(I) = 1$.) In other words, $[\mathbf{D}\det(I)]$ is a linear transformation from Mat(n,n) to $\mathbb{R}$.

Part (b) is a special case of part (c), but it is interesting in its own right. We will prove it first, so we state it separately. Computing the derivative when A is not invertible is a bit trickier, and is explored in Exercise 4.8.14.

Theorem 4.8.21 (Derivative of the determinant). *(a) The determinant function* $\det : \text{Mat}\,(n,n) \to \mathbb{R}$ *is differentiable.*

(b) The derivative of the determinant at the identity is given by

$$[\mathbf{D}\det(I)]B = \operatorname{tr} B. \qquad 4.8.53$$

(c) If $\det A \neq 0$, *then* $[\mathbf{D}\det(A)]B = \det A \,\operatorname{tr}(A^{-1}B)$.

Proof. (a) By Theorem 4.8.18, the determinant is a polynomial in the entries of the matrix, hence certainly differentiable. (For instance, the formula $ad - bc$ is a polynomial in the variables a, b, c, d.)

(b) Since (Proposition 1.7.12) $[\mathbf{D}\det(I)]B$ is the directional derivative

$$\lim_{h \to 0} \frac{\det(I + hB) - \det I}{h}, \qquad 4.8.54$$

it is enough to evaluate that limit. Put another way, we want to find the terms that are linear in h of the expansion given by Equation 4.8.46, for

Try the 2×2 case of Equation 4.8.55:

$$\det\left(I + h\begin{bmatrix} a & b \\ c & d \end{bmatrix}\right)$$
$$= \det\begin{bmatrix} 1 + ha & hb \\ hc & 1 + hd \end{bmatrix}$$
$$= (1 + ha)(1 + hd) - h^2 bc$$
$$= 1 + h(a + d) + h^2(ad - bc).$$

$$\det(I + hB) = \det\begin{bmatrix} 1 + hb_{1,1} & hb_{1,2} & \cdots & hb_{1,n} \\ hb_{2,1} & 1 + hb_{2,2} & \cdots & hb_{2,n} \\ \vdots & \vdots & \ddots & \vdots \\ hb_{n,1} & hb_{n,2} & \cdots & 1 + hb_{n,n} \end{bmatrix}. \qquad 4.8.55$$

Equation 4.8.46 shows that If a term has one factor off the diagonal, then it must have at least two (as illustrated for the 2×2 case in the margin): a permutation that permutes all symbols but one to themselves must take the last symbol to itself also, as it has no other place to go. But all the off-diagonal terms contain a factor of h, so only the term corresponding to the identity permutation contributes any linear terms in h. That term, which has signature $+1$, is

$$(1 + hb_{1,1})(1 + hb_{2,2}) \dots (1 + hb_{n,n})$$
$$= 1 + h(b_{1,1} + b_{2,2} + \cdots + b_{n,n}) + \cdots + h^n b_{1,1}b_{2,2}\dots b_{n,n}, \qquad 4.8.56$$

and we see that the linear term is exactly $b_{1,1} + b_{2,2} + \cdots + b_{n,n} = \operatorname{tr} B$.

(c) Again, take directional derivatives:

The limit in the fourth line is the directional derivative of det at I in the direction $A^{-1}B$, which by Proposition 1.7.12 can be written $[\mathbf{D}\det(I)](A^{-1}B)$, which by part (b) of the theorem is $\operatorname{tr}(A^{-1}B)$.

$$\lim_{h \to 0} \frac{\det(A + hB) - \det A}{h} = \lim_{h \to 0} \frac{\det(A(I + hA^{-1}B)) - \det A}{h}$$
$$= \lim_{h \to 0} \frac{\det A \det(I + hA^{-1}B) - \det A}{h}$$
$$= \det A \lim_{h \to 0} \frac{\det(I + hA^{-1}B) - 1}{h} \qquad 4.8.57$$
$$= \det A \lim_{h \to 0} \frac{\det(I + hA^{-1}B) - \det I}{h}$$
$$= \det A \,\operatorname{tr}(A^{-1}B). \qquad \square$$

Theorem 4.8.21 allows easy proofs of many properties of the trace that are not at all obvious from the definition.

Equation 4.8.58 looks like Equation 4.8.38 from Theorem 4.8.13, but it is not true for the same reason. Theorem 4.8.13 follows immediately from Theorem 4.8.7:

$$\det(AB) = \det A \det B.$$

This is not true for the trace: the trace of a product is not the product of the traces. Corollary 4.8.22 is usually proved by showing first that $\operatorname{tr} AB = \operatorname{tr} BA$. Exercise 4.8.11 asks you to prove $\operatorname{tr} AB = \operatorname{tr} BA$ algebraically; Exercise 4.8.12 asks you to prove it using 4.8.22.

Corollary 4.8.22. *If P is invertible, then for any matrix A we have*

$$\operatorname{tr}(P^{-1}AP) = \operatorname{tr} A. \qquad 4.8.58$$

Proof. This follows from the corresponding result for the determinant (Theorem 4.8.13):

$$
\begin{aligned}
\operatorname{tr}(P^{-1}AP) &= \lim_{h \to 0} \frac{\det(I + hP^{-1}AP) - \det I}{h} \\
&= \lim_{h \to 0} \frac{\det(P^{-1}(I + hA)P) - \det I}{h} \\
&= \lim_{h \to 0} \frac{\det(P^{-1}) \det(I + hA) \det P - \det I}{h} \\
&= \lim_{h \to 0} \frac{\det(I + hA) - \det I}{h} = \operatorname{tr} A. \qquad \square
\end{aligned}
\qquad 4.8.59
$$

4.9 Volumes and Determinants

Recall that "pavable" means "having a well-defined volume," as stated in Definition 4.1.14.

In this section, we will show that *in all dimensions the determinant measures volumes*. This generalizes Propositions 1.4.14 and 1.4.20, which concern the determinant in $\mathbb{R}^2$ and $\mathbb{R}^3$.

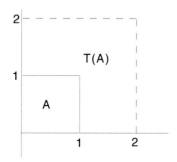

FIGURE 4.9.1.
The transformation given by

$\begin{bmatrix} 2 & 0 \\ 0 & 2 \end{bmatrix}$ turns the square with

side length 1 into the square with side length 2. The area of the first is 1; the area of the second is $|\det[T]|$ times 1; i.e., 4.

Theorem 4.9.1 (The determinant measures volume). *Let $T : \mathbb{R}^n \to \mathbb{R}^n$ be a linear transformation given by the matrix $[T]$. Then for any pavable set $A \subset \mathbb{R}^n$, its image $T(A)$ is pavable, and*

$$\operatorname{vol}_n T(A) = |\det[T]| \operatorname{vol}_n A. \qquad 4.9.1$$

The determinant $|\det[T]|$ scales the volume of A up or down to get the volume of $T(A)$; it *measures the ratio of the volume of $T(A)$ to the volume of A.*

Remark. A linear transformation T corresponds to multiplication by the matrix $[T]$. If A is a pavable set, then what does $T(A)$ correspond to in terms of matrix multiplication? It can't be $[T]A$; a matrix can only multiply a matrix or a vector. Applying T to A corresponds to multiplying *each point* of A by $[T]$. (To do this of course we write points as vectors.) If for example A is the unit square with lower left-hand corner at the origin and $T(A)$ is the square with same left-hand corner but side length 2, as shown in Figure 4.9.1, then $[T]$

is the matrix $\begin{bmatrix} 2 & 0 \\ 0 & 2 \end{bmatrix}$; multiplying $\begin{bmatrix} 1 \\ 0 \end{bmatrix}$ by $[T]$ gives $\begin{bmatrix} 2 \\ 0 \end{bmatrix}$, multiplying $\begin{bmatrix} 1/2 \\ 1/2 \end{bmatrix}$ by $[T]$ gives $\begin{bmatrix} 1 \\ 1 \end{bmatrix}$, and so on. $\triangle$

For this section and for Chapter 5 we need to define what we mean by a k-dimensional parallelogram, also called a k-parallelogram.

> **Definition 4.9.2 (k-parallelogram).** The k-parallelogram spanned by $\vec{v}_1, \dots, \vec{v}_k$ is the set of all $t_1\vec{v}_1 + \cdots + t_k\vec{v}_k$ with $0 \le t_i \le 1$ for i from 1 to k. It is denoted $P(\vec{v}_1, \dots, \vec{v}_k)$.

In Definition 4.9.2, the business with t_i is a precise way of saying that a k-parallelogram is the object spanned by $\vec{v}_1, \dots, \vec{v}_k$, including its boundary and its inside.

A k-dimensional parallelogram, or k-parallelogram, is an interval when $k = 1$, a parallelogram when $k = 2$, a parallelepiped when $k = 3$, and higher dimensional analogs when $k > 3$. (We first used the term k-parallelepiped; we dropped it when one of our daughters said "piped" made her think of a creature with $3.1415\dots$ legs.)

In the proof of Theorem 4.9.1 we will make use of a special case of the k-parallelogram: the n-dimensional unit cube. While the unit *disk* is traditionally centered at the origin, our unit cube has one corner anchored at the origin:

> **Definition 4.9.3 (Unit n-dimensional cube).** The unit n-dimensional cube is the n-dimensional parallelogram spanned by $\vec{e}_1, \dots, \vec{e}_n$. We will denote it Q_n, or, when there is no ambiguity, Q.

Anchoring Q at the origin is just a convenience; if we cut it from its moorings and let it float freely in n-dimensional space, it will still have n-dimensional volume 1, which is what we are interested in.

Note that if we apply a linear transformation T to Q, the resulting $T(Q)$ is the n-dimensional parallelogram spanned by the columns of $[T]$. This is nothing more than the fact, illustrated in Example 1.2.5, that the ith column of a matrix $[T]$ is $[T]\vec{e}_i$; if the vectors making up $[T]$ are $\vec{v}_1, \dots, \vec{v}_n$, this gives $\vec{v}_i = [T]\vec{e}_i$, and we can write $T(Q) = P(\vec{v}_1, \dots, \vec{v}_n)$.

Proof of Theorem 4.9.1 (The determinant measures volume). If $[T]$ is not invertible, the theorem is true because both sides of Equation 4.9.1 vanish:

$$\operatorname{vol}_n T(A) = |\det[T]| \operatorname{vol}_n A. \tag{4.9.1}$$

The right side vanishes because $\det[T] = 0$ when $[T]$ is not invertible (Theorem 4.8.6). The left side vanishes because if $[T]$ is not invertible, then $T(\mathbb{R}^n)$ is a subspace of $\mathbb{R}^n$ of dimension less than n, and $T(A)$ is a bounded subset of this subspace, so (by Proposition 4.3.7) it has n-dimensional volume 0.

This leaves the case where $[T]$ is invertible. This proof is much more involved. We will start by denoting by $T(\mathcal{D}_N)$ the paving of $\mathbb{R}^n$ whose blocks are all the $T(C)$ for $C \in \mathcal{D}_N(\mathbb{R}^n)$. We will need to prove the following statements:

Note that for $T(\mathcal{D}_N)$ to be a paving of $\mathbb{R}^n$ (Definition 4.7.2), T must be invertible. The first requirement for a paving of $\mathbb{R}^n$, that

$$\cup_{C \in \mathcal{D}_N} T(C) = \mathbb{R}^n,$$

is satisfied because T is onto, and the second, that no two tiles overlap, is satisfied because T is one to one.

(1) The sequence of pavings $T(\mathcal{D}_N)$ is a nested partition.

(2) If $C \in \mathcal{D}_N(\mathbb{R}^n)$, then

$$\operatorname{vol}_n T(C) = \operatorname{vol}_n T(Q) \operatorname{vol}_n C.$$

(3) If A is pavable, then its image $T(A)$ is pavable, and

$$\operatorname{vol}_n T(A) = \operatorname{vol}_n T(Q) \operatorname{vol}_n A.$$

$$(4) \ \operatorname{vol}_n T(Q) = |\det[T]|.$$

We will take them in order.

Lemma 4.9.4. *The sequence of pavings $T(\mathcal{D}_N)$ is a nested partition.*

Proof of Lemma 4.9.4.   We must check the three conditions of Definition 4.7.4 of a nested partition. The first condition is that small paving pieces must fit inside big paving pieces: if we pave $\mathbb{R}^n$ with blocks $T(C)$, then if

$$C_1 \in \mathcal{D}_{N_1}(\mathbb{R}^n), \ C_2 \in \mathcal{D}_{N_2}(\mathbb{R}^n), \ \text{and} \ C_1 \subset C_2, \qquad 4.9.2$$

we have

$$T(C_1) \subset T(C_2). \qquad 4.9.3$$

This is clearly met; for example, if you divide the square A of Figure 4.9.1 into four smaller squares, the image of each small square will fit inside $T(A)$.

We use the linearity of T in meeting the second and third conditions. The second condition is that the boundary of the sequence of pavings must have n-dimensional volume 0. The boundary $\partial \mathcal{D}_N(\mathbb{R}^n)$ is a union of subspaces of dimension $n - 1$, hence $\partial T\big(\mathcal{D}_N(\mathbb{R}^n)\big)$ is also. Moreover, only finitely many intersect any bounded subset of $\mathbb{R}^n$, so (by Corollary 4.3.7) the second condition is satisfied.

The third condition is that the pieces $T(C)$ shrink to points as $N \to \infty$. This is also met: since

$$\operatorname{diam}(C) = \frac{\sqrt{n}}{2^N} \ \text{when} \ C \in \mathcal{D}_N(\mathbb{R}^n), \ \text{we have} \ \operatorname{diam}\big(T(C)\big) \leq |T| \frac{\sqrt{n}}{2^N}. \quad 4.9.4$$

So $\operatorname{diam}\big(T(C)\big) \to 0$ as $N \to \infty$.[11]

Proof of Theorem 4.9.1: second statement.

Now for the second statement. Recall that Q is the unit (n-dimensional) cube, with n-dimensional volume 1. We will now show that $T(Q)$ is pavable, as are all $T(C)$ for $C \in \mathcal{D}_N$. Since C is Q scaled up or down by 2^N in all directions, and $T(C)$ is $T(Q)$ scaled by the same factor, we have

$$\frac{\operatorname{vol}_n T(C)}{\operatorname{vol}_n T(Q)} = \frac{\operatorname{vol}_n C}{\operatorname{vol}_n Q} = \frac{\operatorname{vol}_n C}{1}, \qquad 4.9.5$$

which we can write

$$\operatorname{vol}_n T(C) = \operatorname{vol}_n T(Q) \operatorname{vol}_n(C). \qquad 4.9.6$$

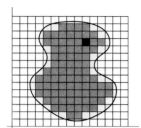

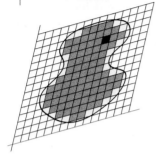

FIGURE 4.9.2.

The potato-shaped area at top is the set A; it is mapped by T to its image $T(A)$, at bottom. If C is the small black square in the top figure, $T(C)$ is the small black parallelogram in the bottom figure. The ensemble of all the $T(C)$ for C in $\mathcal{D}_N(\mathbb{R}^n)$ is denoted $T(\mathcal{D}_N)$. The volume of $T(A)$ is the limit of the sum of the volumes of the $T(C)$, where $C \in \mathcal{D}_N(\mathbb{R}^n)$ and $C \subset A$. Each of these has the same volume

$$\operatorname{vol}_n T(C) = \operatorname{vol}_n C \operatorname{vol}_n T(Q).$$

The diameter of X is the maximum distance between points $\mathbf{x}, \mathbf{y} \in X$; it is denoted $\operatorname{diam}(X)$.

---

[11] If this is not clear, consider that for any points $\mathbf{a}$ and $\mathbf{b}$ in C (which we can think of as joined by the vector $\vec{\mathbf{v}}$),

$$|T(\mathbf{a}) - T(\mathbf{b})| = |T(\mathbf{a} - \mathbf{b})| = \left|[T]\vec{\mathbf{v}}\right| \underbrace{\leq}_{\text{Prop. 1.4.11}} |[T]||\vec{\mathbf{v}}|.$$

So the diameter of $T(C)$ can be at most $|[T]|$ times the length of the longest vector joining two points of C: i.e. $\sqrt{n}/2^N$.

Proof of Theorem 4.9.1: third statement.   We know that A is pavable; as illustrated in Figure 4.9.2, we can compute its volume by taking the limit of the lower sum (the cubes $C \in \mathcal{D}$ that are entirely inside A) or the limit of the upper sum (the cubes either entirely inside A or straddling A).

Since $T(\mathcal{D}_N)$ is a nested partition, we can use it as a paving to measure the volume of $T(A)$, with upper and lower sums:

In reading Equation 4.9.7, it's important to pay attention to which C's one is summing over:

$$C \cap A \neq \phi = \quad C\text{'s in } A \text{ or straddling } A$$

$$C \subset A = \quad C\text{'s entirely in } A$$

Subtracting the second from the first gives C's straddling A.

$$\overbrace{\sum_{T(C) \cap T(A) \neq \phi} \mathrm{vol}_n T(C)}^{\text{upper sum for } \chi_{T(A)}} = \sum_{C \cap A \neq \phi} \overbrace{\mathrm{vol}_n(C) \, \mathrm{vol}_n T(Q)}^{= \, \mathrm{vol}_n T(C) \text{ by Eq. 4.9.6}} = \mathrm{vol}_n T(Q) \underbrace{\sum_{C \cap A \neq \phi} \mathrm{vol}_n C}_{\text{limit is } \mathrm{vol}_n(A)};$$

$$\underbrace{\phantom{\sum_{T(C) \cap T(A) \neq \phi}}}_{\text{limit is } \mathrm{vol}_n T(A)}$$

$$\overbrace{\sum_{T(C) \subset T(A)} \mathrm{vol}_n T(C)}^{\text{lower sum for } \chi_{T(A)}} = \sum_{C \subset A} \mathrm{vol}_n(C) \, \mathrm{vol}_n T(Q) = \mathrm{vol}_n T(Q) \overbrace{\sum_{C \subset A} \mathrm{vol}_n C}^{\text{limit is } \mathrm{vol}_n(A)}. \quad 4.9.7$$

Subtracting the lower sum from the upper sum, we get

You may recall that in $\mathbb{R}^2$ and especially $\mathbb{R}^3$ the proof that the determinant measures volume was a difficult computation. In $\mathbb{R}^n$, such a computational proof is out of the question.

Exercise 4.9.1 suggests a different proof: showing that $\mathrm{vol}_n T(Q)$ satisfies the axiomatic definition of the absolute value of determinant, Definition 4.8.1.

$$U_N(\chi_{T(A)}) - L_N(\chi_{T(A)}) = \sum_{\substack{C \text{ straddles} \\ \text{boundary of } A}} \mathrm{vol}_n T(C)$$

$$= \mathrm{vol}_n T(Q) \sum_{\substack{C \text{ straddles} \\ \text{boundary of } A}} \mathrm{vol}_n C.$$

Since A is pavable, the right-hand side can be made arbitrarily small, so $T(A)$ is also pavable, and

$$\mathrm{vol}_n T(A) = \mathrm{vol}_n T(Q) \, \mathrm{vol}_n A. \qquad 4.9.8$$

Proof of Theorem 4.9.1: fourth statement.   This leaves (4): why is $\mathrm{vol}_n T(Q)$ the same as $|\det[T]|$? There is no obvious relation between volumes and the immensely complicated formula for the determinant. Our strategy will be to reduce the theorem to the case where T is given by an elementary matrix, since the determinant of elementary matrices is straightforward.

What does $E(A)$ mean when the set A is defined in geometric terms, as above? If you find this puzzling, look again at Figure 4.9.1. We think of E as a transformation; applying that transformation to A means multiplying each point of A by E to obtain the corresponding point of $E(A)$.

The following lemma is the key to reducing the problem to the case of elementary matrices.

Lemma 4.9.5.   *If $S, T : \mathbb{R}^n \to \mathbb{R}^n$ are linear transformations, then*

$$\mathrm{vol}_n(S \circ T)(Q) = \mathrm{vol}_n S(Q) \, \mathrm{vol}_n T(Q). \qquad 4.9.9$$

Proof of Lemma 4.9.5.   This follows from Equation 4.9.8, substituting S for T and $T(Q)$ for A:

$$\mathrm{vol}_n(S \circ T)(Q) = \mathrm{vol}_n S(T(Q)) = \mathrm{vol}_n S(Q) \, \mathrm{vol}_n T(Q). \quad \square \qquad 4.9.10$$

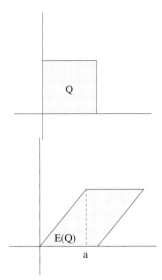

FIGURE 4.9.3.

The second type of elementary matrix, in $\mathbb{R}^2$, simply takes the unit square Q to a parallelogram with base 1 and height 1.

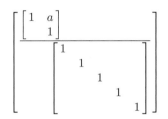

FIGURE 4.9.4.

Here $n = 7$; from the 7×7 matrix E we created the 2×2 matrix $E_1 = \begin{bmatrix} 1 & a \\ 0 & 1 \end{bmatrix}$ and the 5×5 identity matrix E_2.

Any invertible linear transformation T, identified to its matrix, can be written as the product of elementary matrices,

$$[T] = E_k E_{k-1} \ldots E_1, \qquad 4.9.11$$

since $[T]$ row reduces to the identity. So by Lemma 4.9.5, it is enough to prove (4) for elementary matrices: i.e., to prove

$$\mathrm{vol}_n\, E(Q) = |\det E|. \qquad 4.9.12$$

Elementary matrices come in three kinds, as described in Definition 2.3.5. (Here we discuss them in terms of columns, as we did in Section 4.8, not in terms of rows.)

(1) If E is a type 1 elementary matrix, multiplying a column by a nonzero number m, then $\det E = m$ (Corollary 4.8.9), and Equation 4.9.12 becomes $\mathrm{vol}_n\, E(Q) = |m|$. This result was proved in Proposition 4.1.16, because $E(Q)$ is then a parallelepiped all of whose sides are 1 except one side, whose length is $|m|$.

(2) The case where E is type 2, adding a multiple of one column onto another, is a bit more complicated. Without loss of generality, we may assume that a multiple of the first is being added to the second.

First let us verify it for the case $n = 2$, where E is the matrix

$$E = \begin{bmatrix} 1 & a \\ 0 & 1 \end{bmatrix}, \quad \text{with} \quad \det E = 1. \qquad 4.9.13$$

As shown in Figure 4.9.3, the image of the unit cube, $E(Q)$, is then a parallelogram still with base 1 and height 1, so $\mathrm{vol}(E(Q)) = |\det E| = 1$.[12]

If $n > 2$, write $\mathbb{R}^n = \mathbb{R}^2 \times \mathbb{R}^{n-2}$. Correspondingly, we can write $Q = Q_1 \times Q_2$, and $E = E_1 \times E_2$, where E_2 is the identity, as shown in Figure 4.9.4.

Then by Proposition 4.1.12,

$$\mathrm{vol}_n(E(Q)) = \mathrm{vol}_2(E_1(Q_1))\, \mathrm{vol}_{n-2}(Q_2) = 1 \cdot 1 = 1. \qquad 4.9.14$$

(3) If E is type 3, then $|\det E| = 1$, so that Equation 4.9.12 becomes $\mathrm{vol}_n\, E(Q) = 1$. Indeed, since $E(Q)$ is just Q with vertices relabeled, its volume is 1. $\square$

---

[12]But is this a proof? Are we using our definition of volume (area in this case) using pavings, or some "geometric intuition," which is right but difficult to justify precisely? One rigorous justification uses Fubini's theorem:

$$E(Q) = \int_0^1 \left(\int_{ay}^{ay+1} dx \right) dy = 1.$$

Another possibility is suggested in Exercise 4.9.2.

Note that since $\mathrm{vol}_n\, T(Q) = |\det[T]|$, Equation 4.9.9 can be rewritten

$$|\det[S]|\,|\det[T]| = |\det[ST]|. \qquad 4.9.15$$

Of course, this was clear from Theorem 4.8.7. But that result did not have a very transparent proof, whereas Equation 4.9.9 has a clear geometric meaning. Thus this interpretation of the determinant as a volume gives a reason why Theorem 4.8.7 should be true.

Linear change of variables

It is always more or less equivalent to speak about volumes or to speak about integrals; translating Theorem 4.9.1 ("the determinant measures volume") into the language of integrals gives the following theorem.

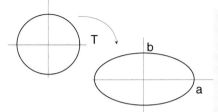

FIGURE 4.9.5.

The linear transformation

$$T = \begin{bmatrix} a & 0 \\ 0 & b \end{bmatrix}$$

takes the unit circle to the ellipse shown above.

Theorem 4.9.6 (Linear change of variables).   Let $T : \mathbb{R}^n \to \mathbb{R}^n$ be an invertible linear transformation, and $f : \mathbb{R}^n \to \mathbb{R}$ an integrable function. Then $f \circ T$ is integrable, and

$$\int_{\mathbb{R}^n} f(\mathbf{y})\,|d^n\mathbf{y}| = \underbrace{|\det T|}_{\substack{\text{corrects for} \\ \text{stretching by } T}} \int_{\mathbb{R}^n} \underbrace{f\big(T(\mathbf{x})\big)}_{f(\mathbf{y})}\,|d^n\mathbf{x}| \qquad 4.9.16$$

where $\mathbf{x}$ is the variable of the first $\mathbb{R}^n$ and $\mathbf{y}$ is the variable of the second $\mathbb{R}^n$.

In Equation 4.9.16, $|\det T|$ corrects for the distortion induced by T.

Example 4.9.7 (Linear change of variables). The linear transformation given by $T = \begin{bmatrix} a & 0 \\ 0 & b \end{bmatrix}$ transforms the unit circle into an ellipse, as shown in Figure 4.9.5. The area of the ellipse is then given by

$$\text{Area of ellipse } = \int_{\text{ellipse}} |d^2\mathbf{y}| = \underbrace{\left|\det \begin{bmatrix} a & 0 \\ 0 & b \end{bmatrix}\right|}_{ab} \underbrace{\int_{\text{circle}} |d^2\mathbf{x}|}_{\pi = \text{area of circle}} = |ab|\pi. \qquad 4.9.17$$

If we had integrated some function $f : \mathbb{R}^2 \to \mathbb{R}$ over the unit circle and wanted to know what the same function would give when integrated over the ellipse, we would use the formula

$$\int_{\text{ellipse}} f(\mathbf{y})|d^2\mathbf{y}| = |ab| \int_{\text{circle}} f \underbrace{\begin{pmatrix} ax_1 \\ bx_2 \end{pmatrix}}_{T(\mathbf{x})} |d^2\mathbf{x}| \qquad \triangle \quad 4.9.18$$

Proof of Theorem 4.9.6.

$$\int_{\mathbb{R}^n} f\big(T(\mathbf{x})\big)|\det T||d^n\mathbf{x}| = \lim_{N\to\infty}\sum_{C\in\mathcal{D}_N(\mathbb{R}^n)} M_C\Big((f\circ T)|\det T|\Big)\,\mathrm{vol}_n(C)$$

$$= \lim_{N\to\infty}\sum_{C\in\mathcal{D}_N(\mathbb{R}^n)} M_C(f\circ T)\,\underbrace{\mathrm{vol}_n\big(T(C)\big)}_{=|\det T|\,\mathrm{vol}_n(C)} \qquad 4.9.19$$

$$= \lim_{N\to\infty}\sum_{P\in(T(\mathcal{D}_N(\mathbb{R}^n)))} M_P(f)\,\mathrm{vol}_n(P) = \int_{\mathbb{R}^n} f(\mathbf{y})|d^n\mathbf{y}|. \quad\square$$

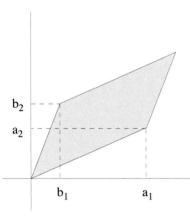

FIGURE 4.9.6.

The vectors $\vec{\mathbf{a}}, \vec{\mathbf{b}}$ span a parallelogram of positive area; the vectors $\vec{\mathbf{b}}, \vec{\mathbf{a}}$ span a parallelogram of negative area.

Alternatively, we can say that just as the volume of $T(Q)$ is $|\det T|$, the *signed volume* of $T(Q)$ is $\det T$.

Of course, "counterclockwise" is not a mathematical term; finding that the determinant of some 2×2 matrix is positive cannot tell you in which direction the arms of your clock move. What this really means is that the smallest angle from $\vec{\mathbf{v}}_1$ to $\vec{\mathbf{v}}_2$ should be in the same direction as the smallest angle from $\vec{\mathbf{e}}_1$ to $\vec{\mathbf{e}}_2$.

Signed volumes

The fact that the absolute value of the determinant is the volume of the image of the unit cube allows us to define the notion of *signed volume*.

Definition 4.9.8 (Signed volume).   The signed k-dimensional volume of the k-parallelogram spanned by $\vec{\mathbf{v}}_1,\ldots,\vec{\mathbf{v}}_k \in \mathbb{R}^k$ is the determinant

$$\det \begin{bmatrix} | & \cdots & | \\ \vec{\mathbf{v}}_1 & \cdots & \vec{\mathbf{v}}_k \\ | & \cdots & | \end{bmatrix}. \qquad 4.9.20$$

Thus the determinant not only measures volume; it also attributes a sign to the volume. In $\mathbb{R}^2$, two vectors $\vec{\mathbf{v}}_1$ and $\vec{\mathbf{v}}_2$, *in that order*, span a parallelogram of positive area if and only if the smallest angle from $\vec{\mathbf{v}}_1$ to $\vec{\mathbf{v}}_2$ is counterclockwise, as shown in Figure 4.9.6.

In $\mathbb{R}^3$, three vectors, $\vec{\mathbf{v}}_1, \vec{\mathbf{v}}_2$, and $\vec{\mathbf{v}}_3$, *in that order*, span a parallelepiped of positive signed volume if and only if they form a right-handed coordinate system. Again, what we really mean is that the same hand that fits $\vec{\mathbf{v}}_1, \vec{\mathbf{v}}_2$, and $\vec{\mathbf{v}}_3$ will fit $\vec{\mathbf{e}}_1, \vec{\mathbf{e}}_2$, and $\vec{\mathbf{e}}_3$; it is by convention that they are drawn counterclockwise, to accommodate the right hand.

4.10 THE CHANGE OF VARIABLES FORMULA

We discussed linear changes of variables in higher dimensions in Section 4.9. This section is devoted to nonlinear changes of variables in higher dimensions. You will no doubt have run into changes of variables in one-dimensional integrals, perhaps under the name of the *substitution method* in methods of integration theory.

Example 4.10.1 (Change of variables in one dimension: substitution method). To compute

$$\int_0^\pi \sin x e^{\cos x}\, dx. \tag{4.10.1}$$

Traditionally, one says: set

$$u = \cos x, \quad \text{so that} \quad du = -\sin x\, dx. \tag{4.10.2}$$

Then when $x = 0$, we have $u = \cos 0 = 1$, and when $x = \pi$, we have $u = \cos \pi = -1$, so

$$\int_0^\pi \sin x e^{\cos x}\, dx = \int_1^{-1} -e^u\, du = \int_{-1}^1 e^u\, du = e - \frac{1}{e}. \quad \triangle \tag{4.10.3}$$

In this section we want to generalize this sort of computation to several variables. There are two parts to this: transforming the integrand, and transforming the domain of integration. In Example 4.10.1 we transformed the integrand by setting $u = \cos x$, so that $du = -\sin x\, dx$ (whatever du means), and we transformed the domain of integration by noting that $x = 0$ corresponds to $u = \cos 0 = 1$, and $x = \pi$ corresponds to $u = \cos \pi = -1$.

Both parts are harder in several variables, especially the second. In one dimension, the domain of integration is usually an interval, and it is not too hard to see how intervals correspond. Domains of integration in $\mathbb{R}^n$, even in the traditional cases of disks, sectors, balls, cylinders, etc., are quite a bit harder to handle. Much of our treatment will be concerned with making precise the "correspondences of domains" under change of variables.

There is another difference between the way you probably learned the change of variables formula in one dimension, and the way we will present it now in higher dimensions. The way it is typically presented in one dimension makes the conceptual basis harder but the computations easier. In particular, you didn't have to make the domains correspond exactly; it was enough if the endpoints matched. Now we will have to make sure our domains correspond precisely, which will complicate our computations.

The meaning of expressions like du is explored in Chapter 6. In volume two we will see that we can use the change of variables formula in higher dimensions without requiring exact correspondence of domains, using the language of *forms*. This is what you were using (more or less blindly) in one dimension.

Three important changes of variables

Before stating the change of variables formula in general, we will first explore what it says for polar coordinates in the plane, and spherical and cylindrical coordinates in space. This will help you understand the general case. In addition, many real systems (encountered for instance in physics courses) have a central symmetry in the plane or in space, or an axis of symmetry in space, and in all those cases, these particular changes of variables are the useful ones. Finally, a great many of the standard multiple integrals are computed using these changes of variables.

Polar coordinates

Definition 4.10.2 (Polar coordinates map). The polar coordinate map P maps a point in the (r, θ)-plane to a point in the (x, y)-plane:

$$P : \begin{pmatrix} r \\ \theta \end{pmatrix} \mapsto \begin{pmatrix} x = r \cos \theta \\ y = r \sin \theta \end{pmatrix}, \qquad 4.10.4$$

where r measures distance from the origin along the spokes, and the *polar angle* θ measures the angle (in radians) formed by a spoke and the positive x axis.

Thus, as shown in Figure 4.10.1, a rectangle in the domain of P becomes a curvilinear "rectangle" in the image of P.

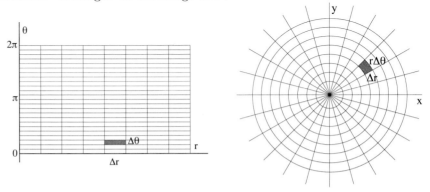

FIGURE 4.10.1. The polar coordinate map P maps the rectangle at left, with dimensions Δr and $\Delta \theta$, to the curvilinear box at right, with two straight sides of length Δr and two curved sides measuring $r \Delta \theta$ (for different values of r).

In Equation 4.10.5, the r in $r \, dr \, d\theta$ plays the role of $|\det T|$ in the linear change of variables formula (Theorem 4.9.6): it corrects for the distortion induced by the polar coordinate map P. We could put $|\det T|$ in front of the integral in the linear formula because it is a constant. Here, we cannot put r in front of the integral: since P is nonlinear, the amount of distortion is not constant but depends on the point at which P is applied.

In Equation 4.10.5, we could replace

$$\int_B f \begin{pmatrix} r \cos \theta \\ r \sin \theta \end{pmatrix}$$

by $\int_B f \left(P \begin{pmatrix} r \\ \theta \end{pmatrix} \right)$,

which is the format we used in Theorem 4.9.6 concerning the linear case.

Proposition 4.10.3 (Change of variables for polar coordinates). *Suppose f is an integrable function defined on $\mathbb{R}^2$, and suppose that the polar coordinate map P maps a region $B \subset (0, \infty) \times [0, 2\pi)$ of the (r, θ)-plane to a region A in the (x, y)-plane. Then*

$$\int_A f \begin{pmatrix} x \\ y \end{pmatrix} |dx \, dy| = \int_B f \begin{pmatrix} r \cos \theta \\ r \sin \theta \end{pmatrix} r \, |dr \, d\theta|. \qquad 4.10.5$$

Note that the mapping $P : B \to A$ is necessarily bijective (one to one and onto), since we required $\theta \in [0, 2\pi)$. Moreover, to every A there corresponds such a B, except that 0 should not belong to A (since there is no well-defined polar angle at the origin). This restriction does not matter: the behavior of an integrable function on a set of volume 0 does not affect integrals (Theorem 4.3.10). Requiring that θ belong to $[0, 2\pi)$ is essentially arbitrary; the interval

$[-\pi, \pi)$ would have done just as well. Moreover, there is no need to worry about what happens when $\theta = 0$ or $\theta = 2\pi$, since those are also sets of volume 0.

We will postpone the discussion of where Equation 4.10.5 comes from, and proceed to some examples.

Example 4.10.4 (Volume beneath a paraboloid of revolution). Consider the paraboloid of Figure 4.10.2, given by

$$z = f\begin{pmatrix} x \\ y \end{pmatrix} = \begin{cases} x^2 + y^2 & \text{if } x^2 + y^2 \leq R^2 \\ 0 & \text{if } x^2 + y^2 > R^2. \end{cases} \tag{4.10.6}$$

This was originally computed by Archimedes, who invented a lot of the integral calculus in the process. No one understood what he was doing for about 2 000 years.

Usually one would write the integral

$$\int_{\mathbb{R}^2} f\begin{pmatrix} x \\ y \end{pmatrix} |dx\, dy| \qquad \text{as} \qquad \int_{D_R} (x^2 + y^2)\, dx\, dy, \tag{4.10.7}$$

where

$$D_R = \left\{ \begin{pmatrix} x \\ y \end{pmatrix} \in \mathbb{R}^2 \,\middle|\, x^2 + y^2 \leq R^2 \right\} \tag{4.10.8}$$

is the disk of radius R centered at the origin.

This integral is fairly complicated to compute using Fubini's theorem; Exercise 4.10.1 asks you to do this. Using the change of variables formula 4.10.5, it is straightforward:

$$\int_{\mathbb{R}^2} f\begin{pmatrix} x \\ y \end{pmatrix} dx\, dy = \int_0^{2\pi} \int_0^R f\begin{pmatrix} r\cos\theta \\ r\sin\theta \end{pmatrix} r\, dr\, d\theta$$

$$= \int_0^{2\pi} \int_0^R (r^2)(\cos^2\theta + \sin^2\theta)\, r\, dr\, d\theta \tag{4.10.9}$$

$$= \int_0^{2\pi} \int_0^R (r^2)\, r\, dr\, d\theta = 2\pi \left[\frac{r^4}{4}\right]_0^R = \frac{\pi}{2} R^4. \quad \triangle$$

FIGURE 4.10.2.

In Example 4.10.4 we are measuring the region inside the cylinder and outside the paraboloid.

Most often, polar coordinates are used when the domain of integration is a disk or a sector of a disk, but they are also useful in many cases where the equation of the boundary is well suited to polar coordinates, as in Example 4.10.5.

Example 4.10.5 (Area of a lemniscate). The *lemniscate* looks like a figure eight; the name comes from the Latin word for ribbon. We will compute the area of the right-hand lobe A of the lemniscate given by the equation $r^2 = \cos 2\theta$, i.e., the area bounded by the right loop of the figure eight shown in Figure 4.10.3. (Exercise 4.10.2 asks you to write the equation of the lemniscate in complex notation.)

FIGURE 4.10.3.

The lemniscate of equation

$$r^2 = \cos 2\theta.$$

Of course this area can be written $\int_A dx\,dy$, which could be computed by Riemann sums, but the expressions you get applying Fubini's theorem are dismayingly complicated. Using polar coordinates simplifies the computations.

The region A (the right lobe) corresponds to the region B in the (r, θ)-plane where

$$B = \left\{ \begin{pmatrix} r \\ \theta \end{pmatrix} \,\middle|\, -\frac{\pi}{4} \le \theta \le \frac{\pi}{4},\ 0 < r < \sqrt{\cos 2\theta} \right\}. \qquad 4.10.10$$

Thus in polar coordinates, the integral becomes

$$\int_{-\pi/4}^{\pi/4} \left(\int_0^{\sqrt{\cos 2\theta}} r\,dr \right) d\theta = \int_{-\pi/4}^{\pi/4} \left[\frac{r^2}{2} \right]_0^{\sqrt{\cos 2\theta}} d\theta$$

$$= \int_{-\pi/4}^{\pi/4} \frac{\cos 2\theta}{2}\,d\theta \qquad 4.10.11$$

$$= \left[\frac{\sin 2\theta}{4} \right]_{-\pi/4}^{\pi/4} = \frac{1}{2}. \quad \triangle$$

The formula for change of variables for polar coordinates, Equation 4.10.5, has a function f on both sides of the equation. Since we are computing area here, the function is simply 1.

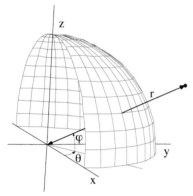

FIGURE 4.10.4.

In spherical coordinates, a point is specified by its distance from the origin (r), its longitude (θ), and its latitude (φ); longitude and latitude are measured in radians, not in degrees.

The $r^2 \cos \varphi$ corrects for distortion induced by the mapping.

Spherical coordinates

Spherical coordinates are important whenever you have a center of symmetry in $\mathbb{R}^3$.

Definition 4.10.6 (Spherical coordinates map). The spherical coordinate map S maps a point in space (e.g., a point inside the earth) known by its distance r from the center, its longitude θ, and its latitude φ, to a point in (x, y, z)-space:

$$S : \begin{pmatrix} r \\ \theta \\ \varphi \end{pmatrix} \mapsto \begin{pmatrix} x = r \cos \theta \cos \varphi \\ y = r \sin \theta \cos \varphi \\ z = r \sin \varphi \end{pmatrix}. \qquad 4.10.12$$

This is illustrated by Figure 4.10.4.

Proposition 4.10.7 (Change of variables for spherical coordinates). *Suppose f is an integrable function defined on $\mathbb{R}^3$, and suppose that the spherical coordinate map S maps a region B of (r, θ, φ)-space to a region A in (x, y, z)-space. Further, suppose that $B \subset (0, \infty) \times [0, 2\pi) \times (-\pi/2, \pi/2)$. Then*

$$\int_A f \begin{pmatrix} x \\ y \\ z \end{pmatrix} dx\,dy\,dz = \int_B f \begin{pmatrix} r \cos \theta \cos \varphi \\ r \sin \theta \cos \varphi \\ r \sin \varphi \end{pmatrix} r^2 \cos \varphi\,dr\,d\theta\,d\varphi. \qquad 4.10.13$$

Again, we will postpone the justification for this formula.

Example 4.10.8 (Spherical coordinates). Integrate the function z over the upper half of the unit ball:

$$\int_A z \, dx \, dy \, dz, \qquad 4.10.14$$

where A is the upper half of the unit ball, i.e., the region

$$A = \left\{ \begin{pmatrix} x \\ y \\ z \end{pmatrix} \in \mathbb{R}^3 \,\middle|\, x^2 + y^2 + z^2 \leq 1, \; z \geq 0 \right\}. \qquad 4.10.15$$

The region B corresponding to this region under S is

For spherical coordinates, many authors use the angle from the North Pole rather than latitude. Mainly because most people are comfortable with the standard latitude, we prefer this form. The formulas using the North Pole are given in Exercise 4.10.10.

$$B = \left\{ \begin{pmatrix} r \\ \theta \\ \varphi \end{pmatrix} \in \underset{r}{(0,\infty)} \times \underset{\theta}{[0, 2\pi)} \times \underset{\varphi}{(-\pi/2, \pi/2)} \,\middle|\, r \leq 1, \; \varphi \geq 0 \right\}. \qquad 4.10.16$$

As shown in Figure 4.10.4, r goes from 0 to 1, φ from 0 to $\pi/2$ (from the Equator to the North Pole), and θ from 0 to 2π.

At $\varphi = -\pi/2$ and $\varphi = \pi/2$, $r = 0$.

Thus our integral becomes

$$\int_B \underbrace{(r \sin \varphi)}_{z}(r^2 \cos \varphi) \, dr \, d\theta \, d\varphi = \int_0^1 \left(\int_0^{\pi/2} \left(\int_0^{2\pi} r^3 \sin \varphi \cos \varphi \, d\theta \right) d\varphi \right) dr$$

$$= 2\pi \int_0^1 r^3 \left[\frac{\sin^2 \varphi}{2} \right]_0^{\pi/2} dr$$

$$= 2\pi \int_0^1 \frac{r^3}{2} \, dr = 2\pi \left[\frac{r^4}{8} \right]_0^1 = \frac{\pi}{4}. \quad \triangle \quad 4.10.17$$

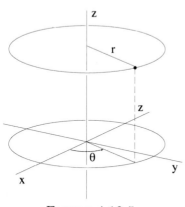

z

FIGURE 4.10.5.

In cylindrical coordinates, a point is specified by its distance r from the z-axis, the polar angle θ shown above, and the z coordinate.

Cylindrical coordinates

Cylindrical coordinates are important whenever you have an axis of symmetry. They correspond to describing a point in space by its altitude (i.e., its z-coordinate), and the polar coordinates r, θ of the projection onto the (x, y)-plane, as shown in Figure 4.10.5.

Definition 4.10.9 (Cylindrical coordinates map). The cylindrical coordinates map C maps a point in space known by its altitude z and by the polar coordinates r, θ of its projection in the (x, y)-plane, to a point in (x, y, z)-space:

$$C : \begin{pmatrix} r \\ \theta \\ z \end{pmatrix} \mapsto \begin{pmatrix} r \cos \theta \\ r \sin \theta \\ z \end{pmatrix}. \qquad 4.10.18$$

In Equation 4.10.19, the r in $r\,dr\,d\theta\,dz$ corrects for distortion induced by the cylindrical coordinate map C. Note that this is the same "distortion corrector" as for polar coordinates.

Exercise 4.10.3 asks you to derive the change of variable formula for cylindrical coordinates from the polar formula and Fubini's theorem.

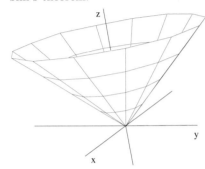

FIGURE 4.10.6.

The region we are integrating over is bounded by this cone, with a flat top on it.

Equation 4.10.20, second line: since the integrand r^3z doesn't depend on θ, the integral with respect to θ just multiplies the result by 2π, which we did at the end of the second line of Equation 4.10.20.

"Injective" and "one to one" are synonyms.

Proposition 4.10.10 (Change of variables for cylindrical coordinates). *Suppose f is an integrable function defined on $\mathbb{R}^3$, and suppose that the cylindrical coordinate map C maps a region $B \subset (0, \infty) \times [0, 2\pi) \times \mathbb{R}$ of (r, θ, z)-space to a region A in (x, y, z)-space. Then*

$$\int_A f\begin{pmatrix} x \\ y \\ z \end{pmatrix} dx\,dy\,dz = \int_B f\begin{pmatrix} r\cos\theta \\ r\sin\theta \\ z \end{pmatrix} r\,dr\,d\theta\,dz. \qquad 4.10.19$$

Example 4.10.11 (Integrating a function over a cone). Let us integrate $(x^2 + y^2)z$ over the region $A \subset \mathbb{R}^3$ that is the part of the inverted cone $z^2 \geq x^2 + y^2$ where $0 \leq z \leq 1$, as shown in Figure 4.10.6. This corresponds under the cylindrical coordinate map C to the region B where $r \leq z \leq 1$. Thus our integral becomes

$$\int_A (x^2 + y^2)z\,dx\,dy\,dz = \int_B r^2 z\overbrace{(\cos^2\theta + \sin^2\theta)}^{=1} r\,dr\,d\theta\,dz = \int_B (r^2 z)\,r\,dr\,d\theta\,dz$$

$$= \int_0^{2\pi}\left(\int_0^1 \left(\int_r^1 r^3 z\,dz \right) dr \right) d\theta = 2\pi \int_0^1 r^3 \left[\frac{z^2}{2} \right]_r^1 dr$$

$$= 2\pi \int_0^1 r^3 \left(\frac{1}{2} - \frac{r^2}{2} \right) dr = 2\pi \left[\frac{r^4}{8} - \frac{r^6}{12} \right]_0^1$$

$$= 2\pi \left(\frac{1}{8} - \frac{1}{12} \right) = \frac{\pi}{12}. \qquad 4.10.20$$

Note that it would have been unpleasant to express the flat top of the cone in spherical coordinates.   △

General change of variables formula

Now let's consider the general change of variables formula.

Theorem 4.10.12 (General change of variables formula). *Let X be a compact subset of $\mathbb{R}^n$, with boundary of volume 0, and U an open neighborhood of X. Let $\Phi : U \to \mathbb{R}^n$ be a C^1 mapping with Lipschitz derivative, that is injective on $(X - \partial X)$, and such that $[\mathbf{D}\Phi(\mathbf{x})]$ is invertible at every $\mathbf{x} \in (X - \partial X)$. Set $Y = \Phi(X)$.*

Then if $f : Y \to \mathbb{R}$ is integrable, $(f \circ \Phi)\left|\det[\mathbf{D}\Phi]\right|$ is integrable on X, and

$$\int_Y f(\mathbf{y})\,|d^n\mathbf{y}| = \int_X (f \circ \Phi)(\mathbf{x})\left|\det[\mathbf{D}\Phi(\mathbf{x})]\right| |d^n\mathbf{x}|. \qquad 4.10.21$$

Once we have introduced *improper integrals*, in Section 4.11, we will be able to give a cleaner version (Theorem 4.11.16) of the change of variables theorem.

Let us see how our examples are special cases of this formula. Consider polar coordinates

$$P\left(\begin{matrix} r \\ \theta \end{matrix}\right) = \left(\begin{matrix} r\cos\theta \\ r\sin\theta \end{matrix}\right),$$
4.10.22

and let $f : \mathbb{R}^2 \to \mathbb{R}$ be an integrable function. Suppose that the support of f is contained in the disk of radius R. Then set

$$X = \left\{ \left(\begin{matrix} r \\ \theta \end{matrix}\right) \mid 0 \le r \le R,\ 0 \le \theta \le 2\pi \right\},$$
4.10.23

and take U to be any bounded neighborhood of X, for instance the disk centered at the origin in the (r, θ)-plane of radius $R + 2\pi$. We claim all the requirements are satisfied: here P, which plays the role of Φ, is of class C^1 in U with Lipschitz derivative, and it is injective (one to one) on $X - \partial X$ (but not on the boundary). Moreover, $[\mathbf{D}P]$ is invertible in $X - \partial X$, since $\det[DP] = r$, which is only zero on the boundary of X.

The case of spherical coordinates

$$S : \left(\begin{matrix} r \\ \theta \\ \varphi \end{matrix}\right) = \left(\begin{matrix} r\cos\varphi\cos\theta \\ r\cos\varphi\sin\theta \\ r\sin\varphi \end{matrix}\right)$$
4.10.24

is very similar. If as before the function f to be integrated has its support in the ball of radius R around the origin, take

$$X = \left\{ \left(\begin{matrix} r \\ \theta \\ \varphi \end{matrix}\right) \mid 0 \le r \le R,\ -\frac{\pi}{2} \le \varphi \le \frac{\pi}{2}, 0 \le \theta \le 2\pi \right\},$$
4.10.25

and U any bounded open neighborhood of X. Then indeed S is C^1 on U with Lipschitz derivative; it is injective on $X - \partial X$, and its derivative is invertible there, since the determinant of the derivative is $r^2 \cos\varphi$, which only vanishes on the boundary.

Remark 4.10.13. The requirement that Φ be injective (one to one) often creates great difficulties. In first year calculus, you didn't have to worry about the mapping being injective. This was because the integrand dx of one-dimensional calculus is actually a *form field*, integrated over an oriented domain: $\int_a^b f\, dx = -\int_b^a f\, dx$.

For instance, consider $\int_1^4 dx$. If we set $x = u^2$, so that $dx = 2u\, du$, then $x = 4$ corresponds to $u = \pm 2$, while $x = 1$ corresponds to $u = \pm 1$. If we choose $u = -2$ for the first and $u = 1$ for the second, then the change of variable formula gives

$$3 = \int_1^4 dx = \int_1^{-2} 2u\, du = [u^2]_1^{-2} = 4 - 1 = 3,$$
4.10.26

even though the change of variables was *not* injective. We will discuss forms in Chapter 6. The best statement of the change of variables formula makes use of forms, but it is beyond the scope of this book.   △

Theorem 4.10.12 is proved in Appendix A.16; below we give an argument that is reasonably convincing without being rigorous.

A heuristic derivation of the change of variables formulas

It is not hard to see why the change of variables formulas above are correct, and even the general formula. For each of the coordinate systems above, the standard paving $\mathcal{D}_N$ in the new space induces a paving in the original space.

Actually, when using polar, spherical, or cylindrical coordinates, you will be better off if you use paving blocks with side length $\pi/2^N$ in the angular directions, rather than the $1/2^N$ of standard dyadic cubes. (Since π is irrational, dyadic fractions of radians do not fill up the circle exactly, but dyadic pieces of turns do.) We will call this paving $\mathcal{D}_N^{new}$, partly to specify these dimensions, but mainly to remember what space is being paved.

The paving of $\mathbb{R}^2$ corresponding to polar coordinates is shown in Figure 4.10.7; the paving of $\mathbb{R}^3$ corresponding to spherical coordinates is shown in Figure 4.10.8.

In the case of polar, spherical, and cylindrical coordinates, the paving $\mathcal{D}_N^{new}$ clearly forms a nested partition. (When we make more general changes of variables Φ, we will need to impose requirements that will make this true.) Thus given a change of variables mapping Φ with respect to the paving $\mathcal{D}_N^{new}$ we have

$$\int_V f\,|d^n\mathbf{v}| = \lim_{N\to\infty}\sum_{C\in\mathcal{D}_N^{new}} M_{\Phi(C)}(f)\,\mathrm{vol}_n\,\Phi(C)$$

$$= \lim_{N\to\infty}\sum_{C\in\mathcal{D}_N^{new}} M_C(f\circ\Phi)\,\frac{\mathrm{vol}_n\,\Phi(C)}{\mathrm{vol}_n\,C}\,\mathrm{vol}_n\,C. \qquad 4.10.27$$

This looks like the integral over U of the product of $f\circ\Phi$ and the limit of the ratio

$$\frac{\mathrm{vol}_n\,\Phi(C)}{\mathrm{vol}_n\,C} \qquad 4.10.28$$

as $N\to\infty$, so that C becomes small. This would give

$$\int_V f\,|d^n\mathbf{v}| \sim \int_U \left((f\circ\Phi)\lim_{N\to\infty}\frac{\mathrm{vol}_n\,\Phi(C)}{\mathrm{vol}_n\,C}\right)|d^n\mathbf{u}|. \qquad 4.10.29$$

This isn't meaningful because the product of $f\circ\Phi$ and the ratio of Equation 4.10.28 isn't a function, so it can't be integrated.

But recall (Equation 4.9.1) that the determinant is precisely designed to measure ratios of volumes under linear transformations. Of course our change

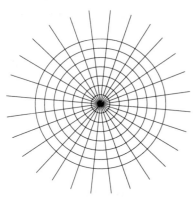

FIGURE 4.10.7.

The paving $P(\mathcal{D}_N^{new})$ of $\mathbb{R}^2$ corresponding to polar coordinates; the dimension of each block in the angular direction (the direction of the spokes) is $\pi/2^N$.

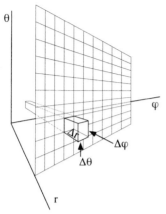

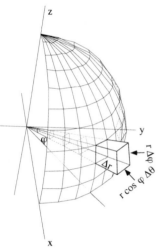

of variable map Φ isn't linear, but if it is differentiable, then it is almost linear on small cubes, so we would expect

$$\frac{\operatorname{vol}_n \Phi(C)}{\operatorname{vol}_n C} \sim \left|\det[\mathbf{D}\Phi(\mathbf{u})]\right| \qquad 4.10.30$$

when $C \in \mathcal{D}_N^{new}(\mathbb{R}^n)$ is small (i.e., N is large), and $\mathbf{u} \in C$. So we might expect our integral $\int_A f$ to be equal to

$$\int_V f(\mathbf{v})\,|d^n\mathbf{v}| = \int_U (f \circ \Phi)(\mathbf{u}) \big|\det[\mathbf{D}\Phi(\mathbf{u})]\big| |d^n\mathbf{u}|. \qquad 4.10.31$$

We find the above argument completely convincing; however, it is not a proof. Turning it into proof is an unpleasant but basically straightforward exercise, found in Appendix A.16.

Example 4.10.14 (Ratio of areas for polar coordinates). Consider the ratio of Equation 4.10.30 in the case of polar coordinates, when $\Phi = P$, the polar coordinates map. If a rectangle C in the (r, θ) plane, containing the point $\begin{pmatrix} r_0 \\ \theta_0 \end{pmatrix}$, has sides of length Δr and $\Delta \theta$, then the corresponding piece $P(C)$ of the (x, y) plane is *approximately* a rectangle with sides $r_0 \Delta \theta, \Delta r$, as shown by Figure 4.10.1. Thus its area is approximately $r_0 \Delta r \Delta \theta$, and the ratio of areas is approximately r_0. Thus we would expect that

$$\int_V f|d^n\mathbf{v}| = \int_U (f \circ P)\,r\,dr\,d\theta, \qquad 4.10.32$$

where the r on the right is the ratio of the volumes of infinitesimal paving blocks.

Indeed, for polar coordinates we find

$$\left[\mathbf{D}P\begin{pmatrix} r \\ \theta \end{pmatrix}\right] = \begin{bmatrix} \cos\theta & -r\sin\theta \\ \sin\theta & r\cos\theta \end{bmatrix}, \quad \text{so that} \quad \left|\det\left[\mathbf{D}P\begin{pmatrix} r \\ \theta \end{pmatrix}\right]\right| = r, \qquad 4.10.33$$

explaining the r in the change of variables formula, Equation 4.10.5. $\triangle$

FIGURE 4.10.8.

Under the spherical coordinate map S, a box with dimensions $\Delta r, \Delta \theta$, and $\Delta \varphi$, and anchored at $\begin{pmatrix} r \\ \theta \\ \varphi \end{pmatrix}$ (top) is mapped to a curvilinear "box" with dimensions

$$\Delta r, r\cos\varphi\,\Delta\theta, \text{ and } r\Delta\varphi.$$

Example 4.10.15 (Ratio of volumes for spherical coordinates). In the case of spherical coordinates, where $\Phi = S$, the image $S(C)$ of a box $C \in \mathcal{D}_N^{new}(\mathbb{R}^3)$ with sides $\Delta r, \Delta \theta, \Delta \varphi$ is approximately a box with sides Δr, $r\Delta\varphi$, and $r\cos\varphi\,\Delta\theta$, so the ratio of the volumes is approximately $r^2\cos\varphi$.

Indeed, for spherical coordinates, we have

$$\left[\mathbf{D}S\begin{pmatrix} r \\ \theta \\ \varphi \end{pmatrix}\right] = \begin{bmatrix} \cos\theta\cos\varphi & -r\sin\theta\cos\varphi & -r\cos\theta\sin\varphi \\ \sin\theta\cos\varphi & r\cos\theta\cos\varphi & -r\sin\theta\sin\varphi \\ \sin\varphi & 0 & r\cos\varphi \end{bmatrix}, \qquad 4.10.34$$

so that

$$\left| \det \left[\mathbf{D}S \begin{pmatrix} r \\ \theta \\ \varphi \end{pmatrix} \right] \right| = r^2 \cos \varphi. \quad \triangle \qquad 4.10.35$$

Example 4.10.16 (A less standard change of variables). The region T

$$\left(\frac{x}{1-z} \right)^2 + \left(\frac{y}{1+z} \right)^2 \leq 1, \quad -1 < z < 1 \qquad 4.10.36$$

looks like the curvy-sided tetrahedron pictured in Figure 4.10.9. We will compute its volume. The map $\gamma : [0, 2\pi] \times [0, 1] \times [-1, 1] \to \mathbb{R}^3$ given by

$$\gamma \begin{pmatrix} \theta \\ t \\ z \end{pmatrix} = \begin{pmatrix} t(1-z)\cos\theta \\ t(1+z)\sin\theta \\ z \end{pmatrix} \qquad 4.10.37$$

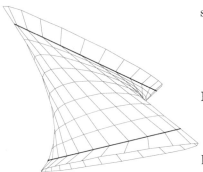

FIGURE 4.10.9.

The region T resembles a cylinder flattened at the ends. Horizontal sections of T are ellipses, which degenerate to lines when $z = \pm 1$.

parametrizes T. The determinant of $[\mathbf{D}\gamma]$ is

$$\det \begin{bmatrix} -t(1-z)\sin\theta & (1-z)\cos\theta & -t\cos\theta \\ t(1+z)\cos\theta & (1+z)\sin\theta & t\sin\theta \\ 0 & 0 & 1 \end{bmatrix} = -t(1-z^2). \qquad 4.10.38$$

Thus the volume is given by the integral

Exercise 4.5.18 asks you to solve a problem of the same sort.

$$\int_0^{2\pi} \int_0^1 \int_{-1}^1 \left| -t(1-z^2) \right| \, dz \, dt \, d\theta = \frac{4\pi}{3}. \quad \triangle \qquad 4.10.39$$

4.11   Improper Integrals

So far all our work has involved the integrals of bounded functions with bounded support. In this section we will relax both of these conditions, studying *improper integrals*: integrals of functions that are not bounded or do not have bounded support, or both.

There are many reasons to study improper integrals. An essential one is the Fourier transform, the fundamental tool of engineering and signal processing (not to mention harmonic analysis). Improper integrals are also ubiquitous in probability theory.

Improper integrals in one dimension

In one variable, you probably already encountered improper integrals: integrals like

$$\int_{-\infty}^{\infty} \frac{1}{1+x^2} \, dx = [\arctan x]_{-\infty}^{\infty} = \pi \qquad 4.11.1$$

$$\int_0^{\infty} x^n e^{-x} \, dx = n! \qquad 4.11.2$$

$$\int_0^1 \frac{1}{\sqrt{x}} \, dx = [2\sqrt{x}]_0^1 = 2. \qquad 4.11.3$$

In the cases above, even though the domain is unbounded, or the function is unbounded (or both), the function can be integrated, although you have to work a bit to define the integral: upper and lower sums do not exist. For the first two examples above, one can imagine writing upper and lower sums with respect to a dyadic partition; instead of being finite, these sums are infinite series whose convergence needs to be checked. For the third example, any upper sum will be infinite, since the maximum of the function over the cube containing 0 is infinity.

We will see below how to define such integrals, and will see that there are analogous multiple integrals, like

$$\int_{\mathbb{R}^n} \frac{|d^n\mathbf{x}|}{1+|\mathbf{x}|^{n+1}}. \tag{4.11.4}$$

There are other improper one-dimensional integrals, like

$$\int_0^\infty \frac{\sin x}{x}\,dx, \tag{4.11.5}$$

which are much more troublesome. You can define this integral as

$$\lim_{A\to\infty} \int_0^A \frac{\sin x}{x}\,dx \tag{4.11.6}$$

and show that the limit exists, for instance, by saying that the series

$$\sum_{k=0}^\infty \int_{k\pi}^{(k+1)\pi} \frac{\sin x}{x}\,dx \tag{4.11.7}$$

is a decreasing alternating series whose terms go to 0 as $k \to \infty$. But this works only because positive and negative terms cancel: the area between the graph of $\sin x/x$ and the x axis is infinite, and the limit

$$\lim_{A\to\infty} \int_0^A \left|\frac{\sin x}{x}\right|\,dx \tag{4.11.8}$$

does not exist. Improper integrals like this, whose existence depends on cancellations, do not generalize at all well to the framework of multiple integrals. In particular, no version of Fubini's theorem or the change of variables formula is true for such integrals, and we will carefully avoid them.

Defining improper integrals

It is harder to define improper integrals—integrals of functions that are unbounded, or have unbounded support, or both—than to define "proper" integrals. It is not enough to come up with a coherent definition: without Fubini's theorem and the change of variables formula, integrals aren't of much interest, so we need a definition for which these theorems are true, in appropriately modified form.

We will proceed in two steps: first we will define improper integrals of non-negative functions; then we will deal with the general case. Our basic approach will be to cut off a function so that it is bounded with bounded support, integrate the truncated function, and then let the cut-off go to infinity, and see what happens in the limit.

Let $f : \mathbb{R}^n \to \mathbb{R} \cup \{\infty\}$ be a function satisfying $f(\mathbf{x}) \geq 0$ everywhere. We allow the value $+\infty$, because we want to integrate functions like

$$f(x) = \frac{e^{-x^2}}{\sqrt{|x|}}; \qquad\qquad 4.11.9$$

Using $\overline{\mathbb{R}} = \mathbb{R} \cup \{-\infty, +\infty\}$ rather than $\mathbb{R}$ is purely a matter of convenience: it avoids speaking of functions defined except on a set of volume 0. Allowing infinite values does not affect our results in any substantial way; if a function were ever going to be infinite on a set that didn't have volume 0, none of our theorems would apply in any case.

setting this function equal to $+\infty$ at the origin avoids having to say that the function is undefined there. We will denote by $\overline{\mathbb{R}}$ the real numbers extended to include $+\infty$ and $-\infty$:

$$\overline{\mathbb{R}} = \mathbb{R} \cup \{+\infty, -\infty\}. \qquad\qquad 4.11.10$$

In order to define the improper integral, or *I-integral*, of a function f that is not bounded with bounded support, we will use truncated versions of f, which *are* bounded with bounded support, as shown in Figure 4.11.1.

For example, if we truncate the function $f(x) = x^2$ by $R = 3$, then

$$[f]_3(\mathbf{x}) = \begin{cases} f(\mathbf{x}) & \text{if } |\mathbf{x}| \leq 3, \, f(\mathbf{x}) \leq 3 \\ 3 & \text{if } |\mathbf{x}| \leq 3, f(\mathbf{x}) > 3 \\ 0 & \text{if } |\mathbf{x}| > 3, \end{cases}$$

so

$$f(1) = 1, \qquad [f]_3(1) = 1,$$
$$f(2) = 4, \qquad [f]_3(2) = 3,$$
$$f(3) = 9, \qquad [f]_3(3) = 3,$$
$$f(4) = 16, \quad [f]_3(4) = 0.$$

We will use the term *I-integral* to mean "improper integral," and *I-integrable* to mean "improperly integrable."

In Equation 4.11.13 we could write $\lim_{R \to \infty}$ rather than $\sup_R$. The condition for I-integrability says that the integral in Equation 4.11.13 must be finite, for any choice of R.

Definition 4.11.1 (R-truncation). The R-truncation $[f]_R$ is given by the formula

$$[f]_R(\mathbf{x}) = \begin{cases} f(\mathbf{x}) & \text{if } |\mathbf{x}| \leq R \text{ and } f(\mathbf{x}) \leq R; \\ R & \text{if } |\mathbf{x}| \leq R \text{ and } f(\mathbf{x}) > R; \qquad 4.11.11 \\ 0 & \text{if } |\mathbf{x}| > R. \end{cases}$$

Note that if $R_1 < R_2$, then $[f]_{R_1} \leq [f]_{R_2}$. In particular, if all $[f]_R$ are integrable, then

$$\int_{\mathbb{R}^n} [f]_{R_1}(\mathbf{x})|d^n\mathbf{x}| \leq \int_{\mathbb{R}^n} [f]_{R_2}(\mathbf{x})|d^n\mathbf{x}|. \qquad 4.11.12$$

Definition 4.11.2 (Improper integral). If the function $f : \mathbb{R}^n \to \overline{\mathbb{R}}$ is non-negative (i.e., satisfies $f(\mathbf{x}) \geq 0$), it is improperly integrable if all $[f]_R$ are integrable, and if

$$\sup_{R \to \infty} \int_{\mathbb{R}^n} [f]_R(\mathbf{x})|d^n\mathbf{x}| < \infty. \qquad\qquad 4.11.13$$

The supremum is then called the improper integral, or I-integral, of f.

If f has both positive and negative values, write $f = f^+ - f^-$, where both f^+ and f^- are non-negative (Definition 4.3.4). Then f is I-integrable if and only if both f^+ and f^- are I-integrable, and

$$\int_{\mathbb{R}^n} f(\mathbf{x})|d^n\mathbf{x}| = \int_{\mathbb{R}^n} f^+(\mathbf{x})|d^n\mathbf{x}| - \int_{\mathbb{R}^n} f^-(\mathbf{x})|d^n\mathbf{x}|. \qquad 4.11.14$$

Note that since f^+ and f^- are both I-integrable, the I-integrability of f does not depend on positive and negative terms canceling each other.

A function that is not I-integrable may qualify for a weaker form of integrability, *local integrability*:

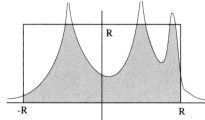

FIGURE 4.11.1.

Graph of a function f, truncated at R to form $[f]_R$; unlike f, the function $[f]_R$ is bounded with bounded support.

Definition 4.11.3 (Local integrability). A function $f : \mathbb{R}^n \to \overline{\mathbb{R}}$ is locally integrable if all the functions $[f]_R$ are integrable.

For example, the function 1 is locally integrable but not I-integrable.

Of course a function that is I-integrable is also locally integrable, but *improper integrability* and *local integrability* address two very different concerns. Local integrability, as its name suggests, concerns local behavior; the only way a bounded function with bounded support, like $[f]_R$, can fail to be integrable is if it has "local nonsense," like the function which is 1 on the rationals and 0 on the irrationals. This is usually not the question of interest when we are discussing improper integrals; there the real issue is how the function grows at infinity: knowing whether the integral is finite.

Generalities about improper integrals

Proposition 4.11.4 (Linearity of improper integrals). *If $f, g : \mathbb{R}^n \to \overline{\mathbb{R}}$ are I-integrable, and $a, b \in \mathbb{R}$, then $af + bg$ is I-integrable, and*

$$\int_{\mathbb{R}^n} (af(\mathbf{x}) + bg(\mathbf{x}))|d^n\mathbf{x}| = a\int_{\mathbb{R}^n} f(\mathbf{x})|d^n\mathbf{x}| + b\int_{\mathbb{R}^n} g(\mathbf{x})|d^n\mathbf{x}|. \qquad 4.11.15$$

Remember that $\overline{\mathbb{R}}$ denotes the real numbers extended to include $+\infty$ and $-\infty$.

Proof. It is enough to prove the result when f and g are non-negative. In that case, the proposition follows from the computation:

$$a\int_{\mathbb{R}^n} f(\mathbf{x})|d^n\mathbf{x}| + b\int_{\mathbb{R}^n} g(\mathbf{x})|d^n\mathbf{x}|$$

$$= a\sup_R \int_{\mathbb{R}^n} [f]_R(\mathbf{x})|d^n\mathbf{x}| + b\sup_R \int_{\mathbb{R}^n} [g]_R(\mathbf{x})|d^n\mathbf{x}|$$

$$= \sup_R \int_{\mathbb{R}^n} \Big(a[f]_R(\mathbf{x}) + b[g]_R(\mathbf{x})\Big) |d^n\mathbf{x}| \qquad 4.11.16$$

$$= \int_{\mathbb{R}^n} (af + bg)(\mathbf{x}) |d^n\mathbf{x}|. \qquad \square$$

Proposition 4.11.5 (Criterion for improper integrals). *A function* $f : \mathbb{R}^n \to \overline{\mathbb{R}}$ *is I-integrable if and only if it is locally integrable and* $|f|$ *is I-integrable.*

Proof. If f is locally integrable, so are f^+ and f^-, and since

$$\int_{\mathbb{R}^n} [f]_R^+(\mathbf{x}) |d^n\mathbf{x}| \leq \int_{\mathbb{R}^n} [|f|]_R(\mathbf{x}) |d^n\mathbf{x}| \leq \int_{\mathbb{R}^n} |f(\mathbf{x})| \, |d^n\mathbf{x}| \qquad 4.11.17$$

is bounded, we see that f^+ (and analogously f^-) are both I-integrable. Conversely, if f is I-integrable, then $|f| = f^+ + f^-$ is also. $\square$

Proposition 4.11.5 gives a criterion for integrability.

Volume of unbounded sets

In Section 4.1 we defined the n-dimensional volume of a *bounded* subset $A \subset \mathbb{R}^n$. Now we can define the volume of any subset.

Definition 4.11.6 (Volume of a subset of $\mathbb{R}^n$). The volume of any subset $A \subset \mathbb{R}^n$ is

$$\text{vol}_n A = \int_A |d^n\mathbf{x}| = \int_{\mathbb{R}^n} \chi_A(\mathbf{x}) \, |d^n\mathbf{x}| = \sup_R \int_{\mathbb{R}^n} [\chi_A]_R(\mathbf{x}) \, |d^n\mathbf{x}|.$$

When we spoke of the volume of graphs in Section 4.3, the best we could do (Corollary 4.3.6) was to say that any *bounded* part of the graph of an integrable function has volume 0. Now we can drop that annoying qualification.

A curve has length but no area (its two-dimensional volume is 0). A plane has area, but its three-dimensional volume is 0 . . . .

Thus a subset A has volume 0 if its characteristic function χ_A is I-integrable, with I-integral 0.

With this definition, several earlier statements where we had to insert "any bounded part of" become true without that restriction:

Proposition 4.11.7 (Manifold of lower dimension has volume 0). *(a) Any manifold $M \subset \mathbb{R}^n$ that is a closed subset of $\mathbb{R}^n$ and of dimension less than n has n-dimensional volume 0.*

(b) In particular, any subspace $E \subset \mathbb{R}^n$ with $\dim E < n$ has n-dimensional volume 0.

Corollary 4.11.8 (Graph has volume 0). *If $f : \mathbb{R}^n \to \mathbb{R}$ is an integrable function, then its graph $\Gamma(f) \subset \mathbb{R}^{n+1}$ has $(n+1)$-dimensional volume 0.*

Integrals and limits

The presence of sup in Definition 4.11.2 tells us that we are going to need to know something about how integrals of limits of functions behave if we are going to prove anything about improper integrals.

What we would like to be able to say is that if f_k is a convergent sequence of functions, then, as $k \to \infty$, the integral of the limit of the f_k is the same as the limit of the integral of f_k. There is one setting where this is true and straightforward: *uniformly convergent* sequences of integrable functions, all with support in the same bounded set.

The key condition is that given ϵ, the same K works for all $\mathbf{x}$.

Definition 4.11.9 (Uniform convergence). A sequence of functions f_k : $\mathbb{R}^k \to \mathbb{R}$ converges uniformly to a function f if for every $\epsilon > 0$, there exists K such that when $k \geq K$, then $|f_k(\mathbf{x}) - f(\mathbf{x})| < \epsilon$.

Exercise 4.11.1 asks you to verify these statements.

Instead of writing "the sequence $p_k \chi_A$" we could write "p_k restricted to A." We use $p_k \chi_A$ because we will use such restrictions in integration, and we use χ to define the integral over a subset A:

$$\int_A p(x) = \int_{\mathbb{R}^n} p(x) \chi_A(x)$$

(see Equation 4.1.6 concerning the coastline of Britain).

The three sequences of functions in Example 4.11.11 below provide typical examples of non-uniform convergence. Uniform convergence on all of $\mathbb{R}^n$ isn't a very common phenomenon, unless something is done to cut down the domain. For instance, suppose that

$$p_k(x) = a_{0,k} + a_{1,k} x + \cdots + a_{m,k} x^m \qquad 4.11.18$$

is a sequence of polynomials all of degree $\leq m$, and that this sequence "converges" in the "obvious" sense that for each degree i (i.e., each x^i), the sequence of coefficients $a_{i,0}, a_{i,1}, a_{i,2}, \ldots$ converges. Then p_k does not converge uniformly on $\mathbb{R}$. But for any bounded set A, the sequence $p_k \chi_A$ does converge uniformly.

Theorem 4.11.10 (When the limit of an integral equals the integral of the limit). *If f_k is a sequence of bounded integrable functions, all with support in a fixed ball $B_R \subset \mathbb{R}^n$, and converging uniformly to a function f, then f is integrable, and*

$$\lim_{k \to \infty} \int_{\mathbb{R}^n} f_k(\mathbf{x}) \, |d^n \mathbf{x}| = \int_{\mathbb{R}^n} f(\mathbf{x}) \, |d^n \mathbf{x}|. \qquad 4.11.19$$

Equation 4.11.20: if you picture this as Riemann sums in one variable, ϵ is the difference between the height of the lower rectangles for f, and the height of the lower rectangles for f_k, while the total width of all the rectangles is $\mathrm{vol}_n(B_R)$, since B_R is the support for f_k.

Proof. Choose $\epsilon > 0$ and K so large that $\sup_{\mathbf{x} \in \mathbb{R}^n} |f(\mathbf{x}) - f_k(\mathbf{x})| < \epsilon$ when $k > K$. Then

$$L_N(f) > L_N(f_k) - \epsilon \, \mathrm{vol}_n(B_R) \quad \text{and} \quad U_N(f) < U_N(f_k) + \epsilon \, \mathrm{vol}_n(B_R) \quad 4.11.20$$

when $k > K$. Now choose N so large that $U_N(f_k) - L_N(f_k) < \epsilon$; we get

$$U_N(f) - L_N(f) \leq \underbrace{U_N(f_k) - L_N(f_k)}_{< \epsilon} + 2\epsilon \, \mathrm{vol}_n(B_R), \qquad 4.11.21$$

The behavior of integrals under limits is a big topic; the main *raison d'être* for the Lebesgue integral is that it is better behaved under limits than the Riemann integral. We will not introduce the Lebesgue integral in this book, but in this subsection we will give the strongest statement that is possible using the Riemann integral.

yielding $U(f) - L(f) < \epsilon(1 + 2 \, \mathrm{vol}_n(B_R))$. Since ϵ is arbitrary, this gives the result. $\square$

In many cases Theorem 4.11.10 is good enough, but it cannot deal with unbounded functions, or functions with unbounded support. Example 4.11.11 shows some of the things that can go wrong.

Example 4.11.11 (Cases where the mass of an integral gets lost). Here are three sequences of functions where the limit of the integral is *not* the integral of the limit.

(1) When f_k is defined by

$$f_k(x) = \begin{cases} 1 & \text{if } k \leq x \leq k+1 \\ 0 & \text{otherwise,} \end{cases} \qquad 4.11.22$$

the mass of the integral is contained in a square 1 high and 1 wide; as $k \to \infty$ this mass drifts off to infinity and gets lost:

$$\lim_{k \to \infty} \int_0^\infty f_k(x)\, dx = 1, \quad \text{but } \int_0^\infty \lim_{k \to \infty} f_k(x)\, dx = \int_0^\infty 0\, dx = 0. \qquad 4.11.23$$

(2) For the function

$$f_k(x) = \begin{cases} k & \text{if } 0 \leq x \leq \frac{1}{k} \\ 0 & \text{otherwise,} \end{cases} \qquad 4.11.24$$

the mass is contained in a rectangle k high and $1/k$ wide; as $k \to \infty$, the height of the box approaches ∞ and its width approaches 0:

$$\lim_{k \to \infty} \int_0^1 f_k(x)\, dx = 1, \text{but } \int_0^1 \lim_{k \to \infty} f_k(x)\, dx = \int_0^1 0\, dx = 0. \qquad 4.11.25$$

(3) The third example is less serious, but still a nasty irritant. Let us make a list $a_1, a_2, \ldots$ of the rational numbers between 0 and 1. Now define

$$f_k(x) = \begin{cases} 1 & \text{if } x \in \{a_1, \ldots, a_k\} \\ 0 & \text{otherwise.} \end{cases} \qquad 4.11.26$$

$$\text{Then } \int_0^1 f_k(x)\, dx = 0 \quad \text{for all } k, \qquad 4.11.27$$

but $\lim_{k \to \infty} f_k$ is the function which is 1 on the rationals and 0 on the irrationals between 0 and 1, and hence not integrable. $\triangle$

The dominated convergence theorem avoids the problem of non-integrability illustrated by Equation 4.11.26, by making the local integrability of f part of the hypothesis.

The dominated convergence theorem is one of the fundamental results of Lebesgue integration theory. The difference between our presentation, which uses Riemann integration, and the Lebesgue version is that we have to assume that f is locally integrable, whereas this is part of the conclusion in the Lebesgue theory. It is hard to overstate the importance of this difference.

The f_k in Equation 4.11.26 provide an example of a sequence of I-integrable functions whose limit is not locally integrable.

The "dominated" in the title refers to the $|f_k|$ being dominated by g.

Our treatment of integrals and limits will be based on the *dominated convergence theorem*, which avoids the pitfalls of disappearing mass. This theorem is the strongest statement that can be made concerning integrals and limits if one is restricted to the Riemann integral.

Theorem 4.11.12 (Dominated convergence theorem for Riemann integrals). *Let $f_k : \mathbb{R}^n \to \overline{\mathbb{R}}$ be a sequence of I-integrable functions, let $f : \mathbb{R}^n \to \overline{\mathbb{R}}$ be a locally integrable function, and let $g : \mathbb{R}^n \to \overline{\mathbb{R}}$ be I-integrable. Suppose that all f_k satisfy $|f_k| \leq g$, and*

$$\lim_{k \to \infty} f_k(\mathbf{x}) = f(\mathbf{x})$$

except perhaps for $\mathbf{x}$ in a set B of volume 0. Then

$$\lim_{k \to \infty} \int_{\mathbb{R}^n} f_k(\mathbf{x})\, |d^n\mathbf{x}| = \int_{\mathbb{R}^n} \lim_{k \to \infty} f_k(\mathbf{x})\, |d^n\mathbf{x}| = \int_{\mathbb{R}^n} f(\mathbf{x})\, |d^n\mathbf{x}|. \qquad 4.11.28$$

The crucial condition above is that all the $|f_k|$ are bounded by the I-integrable function g; this prevents the mass of the integral of the f_k from escaping to infinity, as in the first two functions of Example 4.11.22. The requirement that f be locally integrable prevents the kind of "local nonsense" we saw in the third function of Example 4.11.22.

The proof, in Appendix A.18, is quite difficult and very tricky.

Before rolling out the consequences, let us state another result, which is often easier to use.

Saying $f_1 \le f_2$ means that for any $\mathbf{x}$, $f_1(\mathbf{x}) \le f_2(\mathbf{x})$.

The conclusions of the dominated convergence theorem and the monotone convergence theorem are not identical; the integrals in Equation 4.11.28 are finite while those in Equation 4.11.30 may be infinite.

Theorem 4.11.13 (Monotone convergence theorem). *Let $f_k : \mathbb{R}^n \to \overline{\mathbb{R}}$ be a sequence of I-integrable functions, and $f : \mathbb{R}^n \to \overline{\mathbb{R}}$ be a locally integrable function, such that*

$$0 \le f_1 \le f_2 \le \dots, \quad \text{and} \quad \sup_{k \to \infty} f_k(\mathbf{x}) = f(\mathbf{x}) \qquad 4.11.29$$

except perhaps for $\mathbf{x}$ in a set B of volume 0. Then

$$\sup_k \int_{\mathbb{R}^n} f_k(\mathbf{x}) |d^n\mathbf{x}| = \int_{\mathbb{R}^n} \underbrace{f(\mathbf{x})}_{\sup_k f_k(\mathbf{x})} |d^n\mathbf{x}|, \qquad 4.11.30$$

in the sense that they are either both infinite, or they are both finite and equal.

Note that the requirement in the dominated convergence theorem that $|f_k| \le g$ is replaced in the monotone convergence theorem by the requirement that the f_k be monotone increasing: $0 \le f_1 \le f_2 \le \dots$.

Proof. By the dominated convergence theorem,

$$\sup_k \int_{\mathbb{R}^n} [f_k]_R(\mathbf{x}) |d^n\mathbf{x}| = \int_{\mathbb{R}^n} [f]_R(\mathbf{x}) |d^n\mathbf{x}|, \qquad 4.11.31$$

since all the $[f_k]_R$ are bounded by $[f]_R$, which is I-integrable (i.e., $[f]_R$ plays the role of g in the dominated convergence theorem). Taking the sup as $R \to \infty$ of both sides gives

Unlike limits, sups can always be exchanged, so in Equation 4.11.32 we can rewrite $\sup_R \sup_k$ as $\sup_k \sup_R$.

$$\sup_R \int_{\mathbb{R}^n} [f]_R(\mathbf{x}) |d^n\mathbf{x}| = \sup_R \sup_k \int_{\mathbb{R}^n} [f_k]_R(\mathbf{x}) |d^n\mathbf{x}| = \sup_k \sup_R \int_{\mathbb{R}^n} [f_k]_R(\mathbf{x}) |d^n\mathbf{x}|, \qquad 4.11.32$$

and either both sides are infinite, or they are both finite and equal. But

Equation 4.11.33 is the definition of I-integrability (Definition 4.11.2), applied to f; Equation 4.11.34 is the same definition, applied to k_k.

$$\sup_R \int_{\mathbb{R}^n} [f]_R(\mathbf{x}) |d^n\mathbf{x}| = \int_{\mathbb{R}^n} f(\mathbf{x}) |d^n\mathbf{x}| \qquad 4.11.33$$

$$\text{and} \quad \sup_k \sup_R \int_{\mathbb{R}^n} [f_k]_R(\mathbf{x}) |d^n\mathbf{x}| = \sup_k \int_{\mathbb{R}^n} f_k(\mathbf{x}) |d^n\mathbf{x}|. \quad \square \qquad 4.11.34$$

Fubini's theorem and improper integrals

We will now show that if you state it carefully, Fubini's theorem is true for improper integrals.

> **Theorem 4.11.14 (Fubini's theorem for improper integrals).** *Let $f : \mathbb{R}^n \times \mathbb{R}^m \to \overline{\mathbb{R}}$ be a function such that*
>
> (1) *The functions $\mathbf{y} \mapsto [f]_R(\mathbf{x}, \mathbf{y})$ are integrable;*
> (2) *The function $h(\mathbf{x}) = \int_{\mathbb{R}^m} f(\mathbf{x}, \mathbf{y}) |d^m \mathbf{y}|$ is locally integrable as a function of $\mathbf{x}$;*
> (3) *The function f is locally integrable.*
>
> *Then f is I-integrable if and only if h is I-integrable, and if both are I-integrable, then*
> $$\int_{\mathbb{R}^n \times \mathbb{R}^m} f(\mathbf{x}, \mathbf{y}) |d^n \mathbf{x}| |d^n \mathbf{y}| = \int_{\mathbb{R}^n} \left(\int_{\mathbb{R}^m} f(\mathbf{x}, \mathbf{y}) |d^n \mathbf{y}| \right) |d^n \mathbf{x}|. \qquad 4.11.35$$

There is one function
$$\mathbf{y} \mapsto [f]_R(\mathbf{x}, \mathbf{y})$$
for each value of $\mathbf{x}$.

Here, $\mathbf{x}$ represents n entries of a point in $\mathbb{R}^{n+m}$, and $\mathbf{y}$ represents the remaining m entries.

Proof. To lighten notation, let us denote by h_R the function
$$h_R(\mathbf{x}) = \int_{\mathbb{R}^m} [f]_R(\mathbf{x}, \mathbf{y}) |d^m \mathbf{y}|; \quad \text{note that} \quad \lim_{R \to \infty} h_R(\mathbf{x}) = h(\mathbf{x}). \qquad 4.11.36$$

Applying Fubini's theorem (Theorem 4.5.8) gives
$$\int_{\mathbb{R}^n \times \mathbb{R}^m} [f]_R(\mathbf{x}, \mathbf{y}) |d^n \mathbf{x}| |d^m \mathbf{y}| = \int_{\mathbb{R}^n} \left(\int_{\mathbb{R}^m} [f]_R(\mathbf{x}, \mathbf{y}) |d^m \mathbf{y}| \right) |d^n \mathbf{x}|$$
$$= \int_{\mathbb{R}^n} h_R(\mathbf{x}) |d^n \mathbf{x}|. \qquad 4.11.37$$

Taking the sup of both sides as $R \to \infty$ and (for the second equality) applying the monotone convergence theorem to h_R, which we can do because h is locally integrable and the h_R are increasing as R increases, gives
$$\sup_R \int_{\mathbb{R}^n \times \mathbb{R}^m} [f]_R(\mathbf{x}, \mathbf{y}) |d^n \mathbf{x}| |d^m \mathbf{y}| = \sup_R \int_{\mathbb{R}^n} h_R(\mathbf{x}) |d^n \mathbf{x}|$$
$$= \int_{\mathbb{R}^n} \sup_R h_R(\mathbf{x}) |d^n \mathbf{x}|. \qquad 4.11.38$$

Thus we have

The terms connected by $\downarrow$ are equal. On the left-hand side of Equation 4.11.39, the first line is simply the definition of the improper integral of f on the third line. On the right-hand side, to go from the first to the second line we use the monotone convergence theorem, applied to $[f]_R$.

$$\sup \int_{\mathbb{R}^n \times \mathbb{R}^m} [f]_R(\mathbf{x},\mathbf{y}) |d^n\mathbf{x}| |d^m\mathbf{y}| = \int_{\mathbb{R}^n} \overbrace{\sup \int_{\mathbb{R}^m} [f]_R(\mathbf{x},\mathbf{y})|d^m\mathbf{y}|}^{h_R(\mathbf{x})} |d^n\mathbf{x}|$$

$$\downarrow \qquad\qquad \downarrow$$

$$= \int_{\mathbb{R}^n} \left(\int_{\mathbb{R}^m} \sup[f]_R(\mathbf{x},\mathbf{y})|d^m\mathbf{y}| \right) |d^n\mathbf{x}|$$

$$\downarrow \qquad\qquad \downarrow$$

$$\int_{\mathbb{R}^n \times \mathbb{R}^m} f(\mathbf{x},\mathbf{y}) |d^n\mathbf{x}| |d^m\mathbf{y}| = \int_{\mathbb{R}^n} \left(\int_{\mathbb{R}^m} f(\mathbf{x},\mathbf{y})|d^m\mathbf{y}| \right) |d^n\mathbf{x}|. \quad \square$$

$$4.11.39$$

Example 4.11.15 (Using Fubini to discover that a function is not I-integrable). Let us try to compute

$$\int_{\mathbb{R}^2} \frac{1}{1 + x^2 + y^2} |dx\,dy|. \qquad 4.11.40$$

It's not immediately apparent that $1/(1 + x^2 + y^2)$ is not integrable; it looks very similar to the function in one variable, $1/(1+x^2)$ of Equation 4.11.1, which is integrable.

According to Theorem 4.11.14, this integral will be finite (i.e., the function $\frac{1}{1+x^2+y^2}$ is I-integrable) if

$$\int_{\mathbb{R}} \left(\int_{\mathbb{R}} \frac{1}{1 + x^2 + y^2} |dy| \right) |dx| \qquad 4.11.41$$

is finite, and in that case they are equal. In this case the function

$$h(x) = \int_{\mathbb{R}} \frac{1}{1 + x^2 + y^2} |dy| \qquad 4.11.42$$

can be computed by setting $y^2 = (1 + x^2)u^2$, leading to

We get π in the numerator in Equation 4.11.43 because

$$\int \frac{1}{1 + u^2} \, du = \arctan u;$$

$$\lim_{u \to \pm\infty} \arctan u = \pm\frac{\pi}{2}.$$

$$h(x) = \int_{\mathbb{R}} \frac{1}{1 + x^2 + y^2} |dy| = \int_{\mathbb{R}} \frac{\sqrt{1 + x^2}}{(1 + x^2)(1 + u^2)} |du| = \frac{\pi}{\sqrt{1 + x^2}}. \qquad 4.11.43$$

But $h(x)$ is not integrable, since $1/\sqrt{1 + x^2} > \frac{1}{2x}$ when $x > 1$, and

$$\int_1^A \frac{1}{2x} \, dx = \frac{1}{2} \ln A, \qquad 4.11.44$$

which tends to infinity as $A \to \infty$. $\quad \triangle$

The change of variables formula for improper integrals

Note how much cleaner this statement is than our previous change of variables theorem, Theorem 4.10.12. In particular, it makes no reference to any particular behavior of Φ on the boundary of U. This will be a key to setting up surface integrals and similar things in Chapters 5 and 6.

A *diffeomorphism* is a differentiable mapping $\Phi : U \to V$ that is bijective (one to one and onto), and such that its inverse $\Phi^{-1} : V \to U$ is also differentiable.

Theorem 4.11.16 (Change of variables for improper integrals). *Let U and V be open subsets of $\mathbb{R}^n$ whose boundaries have volume 0, and $\Phi : U \to V$ be a C^1 diffeomorphism, with locally Lipschitz derivative. If $f : V \to \overline{\mathbb{R}}$ is an integrable function, then $(f \circ \Phi)|\det[\mathbf{D}\Phi]|$ is also integrable, and*

$$\int_V f(\mathbf{v})|d^n\mathbf{v}| = \int_U (f \circ \Phi)(\mathbf{u}) \left|\det[\mathbf{D}\Phi(\mathbf{u})]\right| |d^n\mathbf{u}|. \qquad 4.11.45$$

Proof. As usual, by considering $f = f^+ - f^-$, it is enough to prove the result if f is non-negative. Choose $R > 0$, and let U_R be the points $\mathbf{x} \in U$ such that $|\mathbf{x}| < R$. Choose N so that

$$\sum_{\substack{C \in \mathcal{D}_N(\mathbb{R}^n), \\ \overline{C} \cap \partial U_R \neq \emptyset}} \mathrm{vol}_n\, C < \frac{1}{R}, \qquad 4.11.46$$

which is possible since the boundary of U has volume 0. Set

$$X_R = \bigcup_{\substack{C \in \mathcal{D}_N(\mathbb{R}^n), \\ \overline{C} \subset U}} \overline{C}, \qquad 4.11.47$$

Recall (Definition 1.5.17) that $\overline{C}$ is the *closure* of C: the subset of $\mathbb{R}^n$ made up of the set of all limits of sequences in C that converge in $\mathbb{R}^n$.

and finally $Y_R = \Phi(X_R)$.

Note that X_R is compact, and has boundary of volume 0, since it is a union of finitely many cubes. The set Y_R is also compact, and its boundary also has volume 0. Moreover, if f is an I-integrable function on V, then in particular $[f]_R$ is integrable on Y_R. Thus Theorem 4.10.12 (the general change of variables) applies, and gives

$$\int_{X_R} [f]_R \circ \Phi(\mathbf{x})| \det[\mathbf{D}\Phi(\mathbf{x})]||d^n\mathbf{x}| = \int_{Y_R} [f]_R(\mathbf{y})|d^n\mathbf{y}|. \qquad 4.11.48$$

Now take the supremum of both sides as $R \to \infty$. By the monotone convergence theorem (Theorem 4.11.13), the left side and right side converge respectively to

$$\int_U f \circ \Phi(\mathbf{x})| \det[\mathbf{D}\Phi(\mathbf{x})]||d^n\mathbf{x}| \quad \text{and} \quad \int_V f(\mathbf{y})|d^n\mathbf{y}|, \qquad 4.11.49$$

in the sense that they are either both infinite, or both finite and equal. Since $\int_V f(\mathbf{y})|d^n\mathbf{y}| < \infty$, they are finite and equal. $\square$

The Gaussian integral

The integral of the Gaussian bell curve is one of the most important integrals in all of mathematics. The central limit theorem (see Section 4.6) asserts that if you repeat the same experiment over and over, independently each time, and make some measurement each time, then the probability that the average of the measurements will lie in an interval $[a, b]$ is

$$\int_a^b \frac{1}{\sqrt{2\pi\sigma}} e^{-\frac{(x-\bar{x})^2}{2\sigma}} \, dx, \qquad 4.11.50$$

where $\bar{x}$ is the expected value of x, and σ represents the standard deviation. Since most of probability is concerned with repeating experiments, the Gaussian integral is of the greatest importance.

Example 4.11.17 (Gaussian integral). An integral of immense importance, which underlies all of probability theory, is

$$\int_{-\infty}^{\infty} e^{-x^2} \, dx = \sqrt{\pi}. \qquad 4.11.51$$

But the function e^{-x^2} doesn't have an anti-derivative that can be computed in elementary terms.[13]

One way to compute the integral is to use improper integrals in two dimensions. Indeed, let us set

$$\int_{-\infty}^{\infty} e^{-x^2} \, dx = A. \qquad 4.11.52$$

Then

$$A^2 = \left(\int_{-\infty}^{\infty} e^{-x^2} \, dx \right) \left(\int_{-\infty}^{\infty} e^{-y^2} \, dy \right) = \int_{\mathbb{R}^2} e^{-(x^2+y^2)} \, |d^2\mathbf{x}|. \qquad 4.11.53$$

Note that we have used Fubini, and we now use the change of variables formula, passing to polar coordinates:

The polar coordinates map (Equation 4.10.4):

$$P : \begin{pmatrix} r \\ \theta \end{pmatrix} \mapsto \begin{pmatrix} r\cos\theta \\ r\sin\theta \end{pmatrix}.$$

Here,

$$x^2 + y^2 = r^2(\cos^2\theta + \sin^2\theta) = r^2.$$

$$\int_{\mathbb{R}^2} e^{-(x^2+y^2)} \, |d^2\mathbf{x}| = \int_0^{2\pi} \int_0^{\infty} e^{-r^2} r \, dr \, d\theta. \qquad 4.11.54$$

The factor of r that comes from the change of variables makes this straightforward to evaluate:

$$\int_0^{2\pi} \int_0^{\infty} re^{-r^2} \, dr \, d\theta = 2\pi \left[-\frac{e^{-r^2}}{2} \right]_0^{\infty} = \pi. \qquad 4.11.55$$

So $A = \sqrt{\pi}$. $\triangle$

---

[13]This is a fairly difficult result; see *Integration in Finite Terms* by R. Ritt, Columbia University Press, New York, 1948. Of course, it depends on your definition of elementary; the anti-derivative $\int_{-\infty}^{x} e^{-t^2} \, dt$ is a tabulated function, called the *error function*.

When does the integral of a derivative equal the derivative of an integral?

Very often we will need to differentiate a function that is itself an integral. This is particularly the case for Laplace transforms and Fourier transforms, as we will see below. Given a function that we will integrate with respect to one variable, and differentiate with respect to a different variable, under what circumstances does first integrating and then differentiating give the same result as first differentiating, then integrating? Using the dominated convergence theorem, we get the following very general result.

This theorem is a major result with far-reaching consequences.

Theorem 4.11.18 (Differentiating under the integral sign). *Let* $f(t, \mathbf{x}) : \mathbb{R}^{n+1} \to \mathbb{R}$ *be a function such that for each fixed* t, *the integral*

$$F(t) = \int_{\mathbb{R}^n} f(t, \mathbf{x}) \, |d^n\mathbf{x}| \qquad\qquad 4.11.56$$

exists. Suppose moreover that $D_t f$ *exists for all* $\mathbf{x}$ *except perhaps for* $\mathbf{x}$ *in a set of volume 0, and that there exists an I-integrable function* $g(\mathbf{x})$ *such that*

We need the condition of Equation 4.11.57 so that we can apply the dominated convergence theorem in the proof, moving the limit inside the integral sign.

$$\left| \frac{f(s, \mathbf{x}) - f(t, \mathbf{x})}{s - t} \right| \le g(\mathbf{x}) \qquad\qquad 4.11.57$$

for all $s \neq t$. *Then* $F(t)$ *is differentiable, and its derivative is*

$$DF(t) = \int_{\mathbb{R}^n} D_t f(t, \mathbf{x}) \, |d^n\mathbf{x}|. \qquad\qquad 4.11.58$$

Proof.   Just compute:

$$DF(t) = \lim_{h \to 0} \frac{F(t+h) - F(t)}{h} = \lim_{h \to 0} \int_{\mathbb{R}^n} \frac{f(t+h, \mathbf{x}) - f(t, \mathbf{x})}{h} |d^n\mathbf{x}|$$

$$\qquad\qquad\qquad 4.11.59$$

$$= \int_{\mathbb{R}^n} \lim_{h \to 0} \frac{f(t+h, \mathbf{x}) - f(t, \mathbf{x})}{h} |d^n\mathbf{x}| = \int_{\mathbb{R}^n} D_t f(t, \mathbf{x}) \, |d^n\mathbf{x}|;$$

moving the limit inside the integral sign is justified by the dominated convergence theorem (Theorem 4.11.12). $\square$

The Greek letter ξ is xi.

Applications to the Fourier and Laplace transforms

Equation 4.11.60: recall (Equation 0.6.7) that the length, or absolute value, of a complex number $a + ib$ is $\sqrt{a^2 + b^2}$. Since $e^{it} = \cos t + i \sin t$, we have $|e^{it}| = \sqrt{\cos^2 t + \sin^2 t} = 1$.

Fourier transforms and Laplace transforms give important example of differentiation under the integral sign. If f is an I-integrable function on $\mathbb{R}$, then so is $f(x)e^{i\xi x}$ for each $\xi \in \mathbb{R}$, since

$$|f(x)e^{i\xi x}| = |f(x)|. \qquad\qquad 4.11.60$$

So we can consider the function

$$\hat{f}(\xi) = \int_{\mathbb{R}} f(x)e^{i\xi x} \, dx. \qquad\qquad 4.11.61$$

This function $\hat{f}$ is called the *Fourier transform* of f. Passing from f to $\hat{f}$ is one of the central constructions of mathematical analysis; many books are written about it. We want to use it as an example of differentiation under the integral sign.

According to Theorem 4.11.18,

We could also write $D\hat{f}(\xi)$ as $(\hat{f})'$.

$$D\hat{f}(\xi) = \int_{\mathbb{R}} D_\xi\big(f(x)e^{ix\xi}\big)\,dx = \int_{\mathbb{R}} f(x)D_\xi(e^{ix\xi})\,dx = \int_{\mathbb{R}} ixf(x)e^{ix\xi}\,dx = \widehat{ixf}(\xi),$$

4.11.62

provided that the difference quotients

This condition is the condition given by Equation 4.11.57 of Theorem 4.11.18.

$$\left|\frac{e^{i(\xi+h)x} - e^{i\xi x}}{h}f(x)\right| = \left|\frac{e^{ihx} - 1}{h}\right||f(x)| \qquad 4.11.63$$

are all bounded by a single integrable function. Since

$$|e^{ia} - 1| = 2|\sin(a/2)| \le |a|$$

for any real number a, we see that this is satisfied if $xf(x)$ is an I-integrable function.

More generally, if $x^p f(x)$ is an I-integrable function, then $\hat{f}$ is p times differentiable, and

$$D^p\hat{f}(\xi) = \int_{\mathbb{R}} D_\xi^p\big(e^{ix\xi}f(x)\big)\,dx = \int_{\mathbb{R}} (ix)^p f(x)e^{ix\xi}\,dx = \widehat{(ix)^p f}(\xi). \qquad 4.11.64$$

In other words, the faster f decreases at infinity, the smoother the Fourier transform $\hat{f}$ is. This brings out one feature of the Fourier transform: growth conditions on the original function f get translated into smoothness conditions for the Fourier transform.

You might think that if f is I-integrable, then f must tend to 0 at infinity. This isn't true; for instance f could have spikes of height 1 and width $1/n^2$ at all the integers. But if f' is also I-integrable, then f does have to tend to 0 at infinity.

Rather than differentiating the Fourier transform, we might want to Fourier transform the derivative, which we can do if both f and f' are I-integrable. This is best done by integration by parts:

$$\widehat{(f')}(\xi) = \int_{\mathbb{R}} f'(x)e^{i\xi x}dx = \lim_{A\to\infty}\int_{-A}^{A} f'(x)e^{i\xi x}dx$$

4.11.65

$$= \lim_{A\to\infty}\big[f(x)e^{i\xi x}\big]_{-A}^{x=A} - \lim_{A\to\infty}\int_{-A}^{A} i\xi f(x)e^{i\xi x}dx.$$

Since f and f' are I-integrable, $\lim_{x\to\pm\infty} f(x) = 0$, so

$$\lim_{A\to\infty}\big[f(x)e^{i\xi x}\big]_{-A}^{A} = 0, \qquad 4.11.66$$

and

$$\widehat{(f')}(\xi) = -i\xi\int_{\mathbb{R}} f(x)e^{i\xi x}dx = -i\xi\hat{f}(\xi). \qquad 4.11.67$$

This shows another feature of the Fourier transform: it turns differentiation into multiplication. For instance, the Fourier transform of the differential equation

$$a_p D^p f + \cdots + a_0 f = g \quad \text{is} \quad \underbrace{\left(a_p(-i\xi)^p + a_{p-1}(-i\xi)^{p-1} + \cdots + a_0\right) \hat{f}}_{\text{product of } \hat{f} \text{ and a polynomial}} = \hat{g},$$

4.11.68

which gives

$$\hat{f} = \frac{\hat{g}}{(-i\xi)^p a_p + \cdots + a_0}. \qquad 4.11.69$$

This gives $\hat{f}$, and if you know how to undo the Fourier transform, it gives f. This ability to change the analytic operation of differentiation into the algebraic operation of multiplication is one important reason why Fourier series are essential in the theory of differential equations, especially partial differential equations.

The Laplace transform

The Laplace transform $\mathcal{L}(f)$ of f is defined by the formula

Depending on the range of values of s you are interested in, the Laplace transform $\mathcal{L}(f)$ exists for quite a broad range of functions f. For instance, it is a continuous function of $s \in [0, \infty)$ if f is I-integrable, and it is defined and continuous on $(0, \infty)$ if f grows more slowly than some polynomial.

$$\mathcal{L}(f)(s) = \int_0^\infty f(t) e^{-st} dt. \qquad 4.11.70$$

(Note that the integral is from 0 to ∞, not $-\infty$ to ∞.) Again, under appropriate circumstances, we can differentiate under the integral sign:

$$D(\mathcal{L}f)(s) = \int_0^\infty D_s(f(t)e^{-st}) dt = \int_0^\infty -tf(t)e^{-st} dt = \big(\mathcal{L}(-tf)\big)(s). \quad 4.11.71$$

If f grows more slowly than some polynomial, and $s \in (0, \infty)$, then Equation 4.11.57 tells us that this differentiation under the integral sign is justified. Indeed, the family of functions

As in the case of the Fourier transform, there is much more to do with the Laplace transform, but it is beyond the scope of this book.

$$\frac{e^{-(s+h)t} - e^{-st}}{h} f(t) = e^{-st} f(t) \frac{e^{-ht} - 1}{h} \qquad 4.11.72$$

is then bounded by $e^{-st}|tf(t)|$, which is I-integrable.

4.12 Exercises for Chapter Four

4.1.1   (a) What is the two-dimensional volume (i.e., area) of a dyadic cube $C \in \mathcal{D}_3(\mathbb{R}^2)$? of $C \in \mathcal{D}_4(\mathbb{R}^2)$? of $C \in \mathcal{D}_5(\mathbb{R}^2)$?

(b) What is the volume of a dyadic cube $C \in \mathcal{D}_3(\mathbb{R}^3)$? of $C \in \mathcal{D}_4(\mathbb{R}^3)$? of $C \in \mathcal{D}_5(\mathbb{R}^3)$?

4.1.2   In each group of dyadic cubes below, which has the smallest volume? the largest?

(a) $C_{\begin{bmatrix} 1 \\ 2 \end{bmatrix},4}$; $C_{\begin{bmatrix} 1 \\ 2 \end{bmatrix},2}$; $C_{\begin{bmatrix} 1 \\ 2 \end{bmatrix},6}$ (b) $C \in \mathcal{D}_2(\mathbb{R}^3)$; $C \in \mathcal{D}_1(\mathbb{R}^3)$; $C \in \mathcal{D}_8(\mathbb{R}^3)$

4.1.3   What is the volume of each of the following dyadic cubes? What dimension is the volume (i.e., are the cubes two-dimensional, three-dimensional or what)? What information is given below that you don't need to answer those two questions?

(a) $C_{\begin{bmatrix} 1 \\ 2 \end{bmatrix},3}$ (b) $C_{\begin{bmatrix} 0 \\ 1 \\ 3 \end{bmatrix},2}$ (c) $C_{\begin{bmatrix} 0 \\ 1 \\ 1 \\ 1 \end{bmatrix},3}$ (d) $C_{\begin{bmatrix} 0 \\ 1 \\ 4 \end{bmatrix},3}$

4.1.4   Prove Proposition 4.1.18.

4.1.5   Prove that the distance between two points $\mathbf{x}, \mathbf{y}$ in the same cube $C \in \mathcal{D}_N(\mathbb{R}^n)$ is

$$|\mathbf{x} - \mathbf{y}| \leq \frac{\sqrt{n}}{2^N}.$$

4.1.6   Consider the function

$$f(x) = \begin{cases} 0 & \text{if } |x| > 1, \text{ or } x \text{ is rational} \\ 1 & \text{if } |x| \leq 1, \text{ and } x \text{ is irrational.} \end{cases}$$

(a) What value do you get for the "left-hand Riemann sum," where for the interval

$$C_{k,N} = \left\{ x \,\middle|\, \frac{k}{2^N} \leq x < \frac{k+1}{2^N} \right\}$$

you choose the left endpoint $k/2^N$? The right-hand Riemann sum? The midpoint Riemann sum?

(b) What value do you get for the "geometric mean" Riemann sum, where the point you choose in each $C_{k,N}$ is the geometric mean of the two endpoints,

$$\sqrt{\left(\frac{k}{2^N}\right)\left(\frac{k+1}{2^N}\right)} = \frac{\sqrt{k(k+1)}}{2^N} ?$$

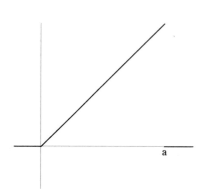

FIGURE 4.1.8.

The heavy line is the graph of the function $x\chi_{[0,a)}(x)$

4.1.7   (a) Calculate $\sum_{i=0}^n i$.

(b) Calculate directly from the definition the integrals

$$\int_{\mathbb{R}} x\chi_{[0,1)}(x)|dx|, \quad \int_{\mathbb{R}} x\chi_{[0,1]}(x)|dx|, \quad \int_{\mathbb{R}} x\chi_{(0,1]}(x)|dx|, \quad \int_{\mathbb{R}} x\chi_{(0,1)}(x)|dx|.$$

In particular show that they all exist, and that they are equal.

4.1.8   (a) If you did not do Exercise 4.1.7(a), calculate $\sum_{i=0}^n i$.

(b) Choose $a > 0$, and calculate directly from the definition the integrals

$$\int_{\mathbb{R}} x\chi_{[0,a)}(x)|dx|, \quad \int_{\mathbb{R}} x\chi_{[0,a]}(x)|dx|, \quad \int_{\mathbb{R}} x\chi_{(0,a]}(x)|dx|, \quad \int_{\mathbb{R}} x\chi_{(0,a)}(x)|dx|.$$

(The first is shown in Figure 4.1.8.) In particular show that they all exist, and that they are equal.

(c) If $a < b$, show that $x\chi_{[a,b]}$, $x\chi_{[a,b)}$, $x\chi_{(a,b]}$, $x\chi_{(a,b)}$ are all integrable and compute their integrals, which are all equal.

In Exercises 4.1.8 and 4.1.9, you need to distinguish between the cases where a and b are "dyadic," i.e., endpoints of dyadic intervals, and the cases where they are not.

4.1.9   (a) Calculate $\sum_{i=0}^{n} i^2$.

(b) Choose $a > 0$, and calculate directly from the definition the integrals

$$\int_{\mathbb{R}} x^2\chi_{[0,a)}(x)|dx|, \quad \int_{\mathbb{R}} x^2\chi_{[0,a]}(x)|dx|, \quad \int_{\mathbb{R}} x^2\chi_{(0,a]}(x)|dx|, \quad \int_{\mathbb{R}} x^2\chi_{(0,a)}(x)|dx.|$$

In particular show that they all exist, and that they are equal.

(c) If $a < b$, show that $x^2\chi_{[a,b]}$, $x^2\chi_{[a,b)}$, $x^2\chi_{(a,b]}$, $x^2\chi_{(a,b)}$ are all integrable and compute their integrals, which are all equal.

4.1.10   Let $Q \subset \mathbb{R}^2$ be the unit square $0 \le x, \ y \le 1$. Show that the function

$$f \begin{pmatrix} x \\ y \end{pmatrix} = \sin(x - y)\chi_Q \begin{pmatrix} x \\ y \end{pmatrix}$$

is integrable by providing an explicit bound for $U_N(f) - L_N(f)$ which tends to 0 as $N \to \infty$.

Exercise 4.1.11 should be with the exercises of Section 4.2.

4.1.11   (a) Let $A = [a_1, b_1] \times \cdots \times [a_n, b_n]$ be a box in $\mathbb{R}^n$, of constant density $\mu = 1$. Show that the center of gravity is the center of the box, i.e., the point $\mathbf{c}$ with coordinates $c_i = (a_i + b_i)/2$.

(b) Let A and B be two disjoint bodies, with densities μ_1 and μ_2, and set $C = A \cup B$. Show that

$$\overline{\mathbf{x}}(C) = \frac{M(A)\overline{\mathbf{x}}(A) + M(B)\overline{\mathbf{x}}(B)}{M(A) + M(B)}.$$

Part (b) of Exercise 4.1.12 is really a separate exercise; ignore the reference to part (a).

4.1.12   Define the dilation by a of a function $f : \mathbb{R}^n \to \mathbb{R}$ by the formula

$$D_a f(\mathbf{x}) = f\left(\frac{\mathbf{x}}{a}\right).$$ Show that if f is integrable, then so is $D_{2^N} f$, and

$$\int_{\mathbb{R}^n} D_{2^N} f(\mathbf{x})|d^n\mathbf{x}| = 2^{nN} \int_{\mathbb{R}^n} f(\mathbf{x}) |d^n\mathbf{x}|.$$

(b) Recall that the dyadic cubes are half open, half closed. (You should have used this in part (a)). Show that the closed cubes also have the same volume. (This is remarkably harder to carry out than you might expect.)

4.1.13   Complete the proof of Lemma 4.1.15.

4.1.14   Evaluate the limit

$$\lim_{N \to \infty} \frac{1}{N^2} \sum_{k=1}^{N} \sum_{l=1}^{2N} e^{\frac{k+l}{N}}.$$

4.1.15   (a) What are the upper and lower sums $U_1(f)$ and $L_1(f)$ for the function

$$f \begin{pmatrix} x \\ y \end{pmatrix} = \begin{cases} x^2 + y^2 & \text{if } 0 < x, y < 1 \\ 0 & \text{otherwise} \end{cases}$$

i.e., the upper and lower sums for the partition $\mathcal{D}_1(\mathbb{R}^2)$, shown in Figure 4.1.15?

(b) Compute the integral of the function f and show that it is between the upper and lower sum.

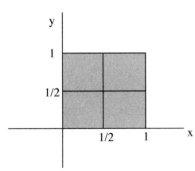

FIGURE 4.1.15.

4.1.16   (a) The question in previous printings was badly stated and should be ignored.

(b) Show that if X and Y have volume 0, then $X \cap Y$, $X \times Y$, and $X \cup Y$ have volume 0.

(c) Show that $\left\{ \begin{pmatrix} x_1 \\ 0 \end{pmatrix} \in \mathbb{R}^2 \middle| 0 \le x_1 \le 1 \right\}$ has volume 0 (i.e., $\text{vol}_2 = 0$).

(d) Show that $\left\{ \begin{pmatrix} x_1 \\ x_2 \\ 0 \end{pmatrix} \in \mathbb{R}^3 \middle| 0 \le x_1, x_2 \le 1 \right\}$ has volume 0 (i.e., $\text{vol}_3 = 0$).

Exercise 4.1.17 shows that the behavior of an integrable function $f : \mathbb{R}^n \to \mathbb{R}$ on the boundaries of the cubes of $\mathcal{D}_N$ does not affect the integral.

Starred exercises are difficult; exercises with two stars are more difficult yet.

*4.1.17   (a) Let S be the unit cube in $\mathbb{R}^n$, and choose $a \in [0, 1)$. Show that the subset $\{ \mathbf{x} \in S \mid x_i = a \}$ has n-dimensional volume 0.

(b) Let $X_N = \partial \mathcal{D}_N$ be the set made up of the boundaries of all the cubes $C \in \mathcal{D}_N$. Show that $\text{vol}_n(X_N \cap S) = 0$.

(c) For each

$$C = \left\{ \mathbf{x} \in \mathbb{R}^n \middle| \frac{k_i}{2^N} \le x_i < \frac{k_i + 1}{2^N} \right\}, \qquad \text{set}$$

$$\overset{\circ}{C} = \left\{ \mathbf{x} \in \mathbb{R}^n \middle| \frac{k_i}{2^N} < x_i < \frac{k_i + 1}{2^N} \right\} \text{ and } \overline{C} = \left\{ \mathbf{x} \in \mathbb{R}^n \middle| \frac{k_i}{2^N} \le x_i \le \frac{k_i + 1}{2^N} \right\}.$$

These are called the *interior* and the *closure* of C respectively.

Show that if $f : \mathbb{R}^n \to \mathbb{R}$ is integrable, then

$$\overset{\circ}{U}(f) = \lim_{N \to \infty} \sum_{C \in \mathcal{D}_N(\mathbb{R}^n)} M_{\overset{\circ}{C}}(f) \, \text{vol}_n(C)$$

$$\overset{\circ}{L}(f) = \lim_{N \to \infty} \sum_{C \in \mathcal{D}_N(\mathbb{R}^n)} m_{\overset{\circ}{C}}(f) \, \text{vol}_n(C)$$

Hint: You may assume that the support of f is contained in Q, and that $|f| \le 1$. Choose $\epsilon > 0$, then choose N_1 to make $U_N(f) - L_N(f) < \epsilon/2$, then choose $N_2 > N_1$ to make $\text{vol}(X_{\partial \mathcal{D}_{N_2}} < \epsilon/2$. Now show that for $N > N_2$,

$$\overset{\circ}{U}_N(f) - \overset{\circ}{L}_N(f) < \epsilon \quad \text{and}$$
$$\overline{U}_N(f) - \overline{L}_N(f) < \epsilon.$$

$$\overline{U}(f) = \lim_{N \to \infty} \sum_{C \in \mathcal{D}_N(\mathbb{R}^n)} M_{\overline{C}}(f) \operatorname{vol}_n(C)$$

$$\overline{L}(f) = \lim_{N \to \infty} \sum_{C \in \mathcal{D}_N(\mathbb{R}^n)} m_{\overline{C}}(f) \operatorname{vol}_n(C)$$

all exist, and are all equal to $\int_{\mathbb{R}^n} f(\mathbf{x})|d^n\mathbf{x}|$.

(d) Suppose $f : \mathbb{R}^n \to \mathbb{R}$ is integrable, and that $f(-\mathbf{x}) = -f(\mathbf{x})$. Show that $\int_{\mathbb{R}^n} f|d^n\mathbf{x}| = 0$.

Exercises for Section 4.2:

Probability

4.2.1   (a) Suppose an experiment consists of throwing two 6-sided dice, each loaded so that it lands on 4 half the time, while the other outcomes are equally likely. The random variable f gives the total obtained on each throw. What are the probability weights for each outcome?

(b) Repeat part (a), but this time one die is loaded as above, and the other falls on 3 half the time, with the other outcomes equally likely.

4.2.2   Suppose a probability space X consists of n outcomes, $\{1, 2, \ldots, n\}$, each with probability $1/n$. Then a random function f on X can be identified with an element $\vec{f} \in \mathbb{R}^n$.

(a) Show that $E(f) = 1/n(\vec{f} \cdot \vec{1})$, where $\vec{1} = \begin{bmatrix} 1 \\ 1 \\ \vdots \\ 1 \end{bmatrix}$.

(b) Show that

$$\operatorname{Var}(f) = \frac{1}{n}\left|\vec{f} - E(f)\vec{1}\right|^2 \quad \text{and} \quad \sigma(f) = \frac{1}{\sqrt{n}}\left|\vec{f} - E(f)\vec{1}\right|.$$

(c) Show that

$$\operatorname{Cov}(f, g) = \frac{1}{n}\left(\vec{f} - E(f)\vec{1}\right) \cdot \left(\vec{g} - E(g)\vec{1}\right);$$

$$\operatorname{corr}(f, g) = \cos\theta,$$

where θ is the angle between the vectors $\vec{f} - E(f)\vec{1}$ and $\vec{g} - E(g)\vec{1}$.

Hint for Exercise 4.3.1 (a): imitate the proof of Proposition 4.3.6, writing the unit circle as the union of four graphs of functions: $y = \sqrt{1 - x^2}$ for $|x| \leq \sqrt{2}/2$, and the three other curves obtained by rotating this curve around the origin by multiples of $\pi/2$.

Exercises for Section 4.3:

What Functions

Can Be Integrated

4.3.1   (a) Give an explicit upper bound (in terms of N) for the number of squares $C \in \mathcal{D}_N(\mathbb{R}^2)$ needed to cover the unit circle in $\mathbb{R}^2$ (the circle, not the disk). See the hint in the margin.

(b) Now try the same exercise for the unit sphere $S^2 \subset \mathbb{R}^3$.

4.3.2   For any real numbers $a < b$, let

$$Q_{a,b}^n = \{\mathbf{x} \in \mathbb{R}^n \mid a \leq x_i \leq b \text{ for all } 1 \leq i \leq n\},$$

and let $P_{a,b}^n \subset Q_{a,b}^n$ be the subset where $a \leq x_1 \leq x_2 \cdots \leq x_n \leq b$.

(a) Let $f : \mathbb{R}^n \to \mathbb{R}$ be an integrable function that is symmetric in the sense that

$$f\begin{pmatrix} x_1 \\ \vdots \\ x_n \end{pmatrix} = f\begin{pmatrix} x_{\sigma(1)} \\ \vdots \\ x_{\sigma(n)} \end{pmatrix} \qquad \text{for any permutation } \sigma \text{ of the symbols } 1, 2, \ldots, n.$$

Show that

$$\int_{Q_{a,b}^n} f(\mathbf{x})\,|d^n\mathbf{x}| = n! \int_{P_{a,b}^n} f(\mathbf{x})\,|d^n\mathbf{x}|.$$

(b) Let $f : [a, b] \to \mathbb{R}$ be an integrable function. Show that

We will give further applications of this result in Exercise 4.5.17.

$$\int_{P_{a,b}^n} f(x_1)f(x_2)\ldots f(x_n)|d^n\mathbf{x}| = \frac{1}{n!}\left(\int_a^b f(x)|dx|\right)^n.$$

4.3.3   Prove Corollary 4.3.11.

4.3.4   Let P be the region $x^2 < y < 1$. Prove that the integral

$$\int_P \sin y^2 |dx\, dy|$$

exists. You may either apply theorems or prove the result directly. If you use theorems, you must show that they actually apply.

Exercises for Section 4.4:
Integration and Measure Zero

4.4.1   Use Definition 4.1.13 of n-dimensional volume to show that the same sets have measure 0 whether you define measure 0 using open or closed boxes.

4.4.2   Show that $X \subset \mathbb{R}^n$ has measure 0 if and only if for any $\epsilon > 0$ there exists an infinite sequence of balls

$$B_i = \{\mathbf{x} \in \mathbb{R}^n \mid |\mathbf{x} - \mathbf{a}_i| < r_i\} \text{ with } \sum_{i=1}^{\infty} r_i^n < \epsilon$$

such that $X \subset \cup_{i=1}^{\infty} B_i$.

4.4.3   Show that if X is a subset of $\mathbb{R}^n$ such that for any $\epsilon > 0$, there exists a sequence of pavable sets B_i, $i = 1, 2, \ldots$ satisfying

$$X \subset \bigcup_{i=1}^{\infty} B_i \quad \text{and} \quad \sum_{i=1}^{\infty} \text{vol}_n(B_i) < \epsilon, \quad \text{then } X \text{ has measure 0.}$$

4.4.4   (a) Show that $\mathbb{Q} \subset \mathbb{R}$ has measure 0. More generally, show that any countable subset of $\mathbb{R}$ has measure 0.

(b) Show that a countable union of sets of measure 0 has measure 0.

**4.4.5   Consider the subset $U \subset [0, 1]$ that is the union of the open intervals

$$\left(\frac{p}{q} - \frac{C}{q^3}, \frac{p}{q} + \frac{C}{q^3} \right)$$

for all rational numbers $p/q \in [0, 1]$. Show that for $C > 0$ sufficiently small, U is not pavable. What would happen if the 3 were replaced by a 2? (This is really hard.)

Exercises for Section 4.5:

Fubini's Theorem

and Iterated Integrals

4.5.1   In Example 4.5.2, why can you ignore the fact that the line $x = 1$ is counted twice?

4.5.2   (a) Set up the multiple integral for Example 4.5.2, where the outer integral is with respect to y rather than x. Be careful about which square root you are using.

(b) If in (a) you replace $+\sqrt{y}$ by $-\sqrt{y}$ and vice versa, what would be the corresponding region of integration?

4.5.3   Set up the multiple integral $\int (\int f \, dx) \, dy$ for the truncated triangle shown in Figure 4.5.2.

4.5.4   (a) In the context of Example 4.5.6 (Equation 4.5.21), show that if

$$c_n = \int_{-1}^{1} (1 - t^2)^{(n-1)/2} \, dt, \quad \text{then} \quad c_n = \frac{n-1}{n} c_{n-2}, \quad \text{for } n \geq 2.$$

(b) Show that $c_0 = \pi$ and $c_1 = 2$.

4.5.5   Again for Example 4.5.6, show that

$$\beta_{2k} = \frac{\pi^k}{k!} \quad \text{and} \quad \beta_{2k+1} = \frac{\pi^k \, k! \, 2^{2k+1}}{(2k + 1)!}.$$

4.5.6   Write each of the following double integrals as iterated integrals in two ways, and compute them:

(a) The integral of $\sin(x + y)$ over the region $x^2 < y < 2$.

(b) The integral of $x^2 + y^2$ over the region $1 \leq |x|, |y| \leq 2$.

4.5.7   In Example 4.5.7, compute the integral without assuming that the first dart falls below the diagonal (see the footnote after Equation 4.5.25).

4.5.8   Write as an iterated integral, and in three different ways, the triple integral of xyz over the region $x, y, z \geq 0, \ x + 2y + 3z \leq 1$.

4.5.9   (a) Use Fubini's theorem to express

$$\int_0^{\pi} \left(\int_y^{\pi} \frac{\sin x}{x} \, dx \right) dy \qquad \text{as a double integral.}$$

(b) Write the integral as an iterated integral in the other order.

(c) Compute the integral.

4.5.10 (a) Represent the iterated integral

$$\int_0^a \left(\int_{x^2}^{a^2} \sqrt{y}e^{-y^2}\, dy \right) dx$$

as the integral of $\sqrt{y}e^{-y^2}$ over a region of the plane which you should sketch.

(b) Use Fubini's theorem to make this integral into an iterated integral, first with respect to x and then with respect to y.

(c) Evaluate the integral.

4.5.11 Theorem 3.3.9 asserts that if the second partials of a function are continuous, the crossed partials are equal:

$$D_1\big(D_2(f)\big) = D_2\big(D_1(f)\big).$$

The proof given in Appendix A.6 is surprisingly difficult. There is an easier proof using Fubini's theorem.

(a) Show that if $U \subset \mathbb{R}^2$ is an open set, and $f : U \to \mathbb{R}$ is a function such that

$$D_2(D_1(f)) \quad \text{and} \quad D_1(D_2(f))$$

both exist and are continuous, and if $D_1(D_2(f))\begin{pmatrix} a \\ b \end{pmatrix} \neq D_2(D_1(f))\begin{pmatrix} a \\ b \end{pmatrix}$ for some point $\begin{pmatrix} a \\ b \end{pmatrix}$, then there exists a square $S \subset U$ such that either

$$D_2(D_1(f)) > D_1(D_2(f)) \text{ on } S \text{ or } D_1(D_2(f)) > D_2(D_1(f)) \quad \text{on } S.$$

(b) Apply Fubini's theorem to the double integral

$$\int\int_S \Big(D_2\big(D_1(f)\big) - D_1\big(D_2(f)\big)\Big)\, dx\, dy$$

to derive a contradiction.

(c) The function

$$f\begin{pmatrix} x \\ y \end{pmatrix} = \begin{cases} xy\frac{x^2-y^2}{x^2+y^2} & \text{if } \begin{pmatrix} x \\ y \end{pmatrix} \neq \begin{pmatrix} 0 \\ 0 \end{pmatrix} \\ 0 & \text{otherwise,} \end{cases}$$

is the standard example of a function where $D_1(D_2 f)) \neq D_2(D_1(f))$. What happens to the proof above?

4.5.12 (a) Set up in two different ways the integral of $\sin y$ over the region $0 \leq x \leq \cos y$, $0 \leq y \leq \pi/6$ as an iterated integral.

(b) Write the integral

$$\int_1^2 \int_{y^3}^{3y^3} \frac{1}{x}\, dx\, dy$$

as an integral, first integrating with respect to y, then with respect to x.

4.5.13   Set up the iterated integral to find the volume of the slice of cylinder $x^2 + y^2 \leq 1$ between the planes

$$z = 0, \quad z = 2, \quad y = \frac{1}{2}, \quad y = -\frac{1}{2}.$$

4.5.14   Compute the integral of the function z over the region R described by the inequalities $x > 0$, $y > 0$, $z > 0$, $x + 2y + 3z < 1$.

4.5.15   Compute the integral of the function $|y - x^2|$ over the unit square $0 \leq x, y \leq 1$.

4.5.16   Find the volume of the region $z \geq x^2 + y^2$, $z \leq 10 - x^2 - y^2$.[14]

4.5.17   Recall from Exercise 4.3.2 the definitions of $P_{a,b}^n \subset Q_{a,b}^n$. Apply the result of Exercise 4.3.2 to compute the following integrals.
  (a) Let $M_r(\mathbf{x})$ be the rth largest of the coordinates $x_1, \ldots, x_n$ of $\mathbf{x}$. Then

$$\int_{Q_{0,1}^n} M_r(\mathbf{x})|d^n\mathbf{x}| = \frac{r}{n+1}.$$

  (b) Let $n \geq 2$ and $0 < b < 1$. Then

$$\int_{Q_{0,1}^n} \min\left(1, \frac{b}{x_1}, \ldots, \frac{b}{x_n}\right) |d^n\mathbf{x}| = \frac{nb - b^n}{n - 1}.$$

4.5.18   What is the volume of the region

$$\frac{x^2}{(z^3 - 1)^2} + \frac{y^2}{(z^3 + 1)^2} \leq 1, \quad -1 \leq z \leq 1,$$

shown in Figure 4.5.18?

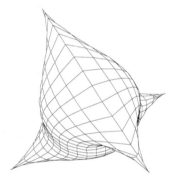

FIGURE 4.5.18.
The region

$$\frac{x^2}{(z^3 - 1)^2} + \frac{y^2}{(z^3 + 1)^2} \leq 1,$$

$$-1 \leq z \leq 1,$$

which looks like a peculiar pillow.

4.5.19   What is the z-coordinate of the center of gravity of the region

$$\frac{x^2}{(z^3 - 1)^2} + \frac{y^2}{(z^3 + 1)^2} \leq 1, \quad 0 \leq z \leq 1?$$

Exercises for Section 4.6:
Numerical Methods
of Integration

4.6.1   (a) Write out the sum given by Simpson's method with one step, for the integral

$$\int_Q f(\mathbf{x})|d^n\mathbf{x}|$$

---

[14]Exercise 4.5.16 is borrowed from Tiberiu Trif, "Multiple integrals of symmetric functions," *American Mathematical Monthly*, Vol. 104, No. 7 (1997), pp. 605–608.

when Q is the unit square in $\mathbb{R}^2$ and the unit cube in $\mathbb{R}^3$. There should be 9 and 27 terms respectively.

(b) Evaluate these sums when

$$f\begin{pmatrix} x \\ y \end{pmatrix} = \frac{1}{1+x+y},$$

and compare the approximation to the exact value of the integral.

4.6.2 Find the weights and control points for the Gaussian integration scheme by solving the system of equations 4.6.9, for $k = 2, 3, 4, 5$. Hint: Entering the equations is fairly easy. The hard part is finding good initial conditions. The following work:

$k = 1$	$w_1 = 17$	$x_1 = .57$	$k = 2$	$w_1 = .6$ $w_2 = .4$	$x_1 = .3$ $x_2 = .8$

$k = 3$	$w_1 = .5$ $w_2 = .3$ $w_3 = .2$	$x_1 = .2$ $x_2 = .7$ $x_3 = .9$	$k = 4$	$w_1 = .35$ $w_2 = .3$ $w_3 = .2$ $w_4 = .1$	$x_1 = .2$ $x_2 = .5$ $x_3 = .8$ $x_4 = .95$

The pattern should be fairly clear; experiment to find initial conditions when $k = 5$.

4.6.3 Find the formula relating the weights W_i and the sampling points X_i needed to compute $\int_a^b f(x)\, dx$ to the weights w_i and the points x_i appropriate for $\int_{-1}^1 f(x)dx$.

4.6.4 (a) Find the equations that must be satisfied by points $x_1 < \cdots < x_m$ and weights $w_1 < \cdots < w_m$ so that the equation

$$\int_0^\infty p(x)e^{-x}dx = \sum_{k=1}^m w_k f(x_k)$$

Part (d) of Exercise 4.6.4: this can be solved using Student MAT-LAB only for values of m up to 4; for $m = 5$, the professional version of MATLAB is needed.

is true for all polynomials p of degree $\le d$.

(b) For what number d does this lead to as many equations as unknowns?

(c) Solve the system of equations when $m = 1$.

(d) Use Newton's method to solve the system for $m = 2, \ldots, 5$.

(e) For each of the five values of m above, approximate

$$\int_0^\infty e^{-x}\sin x\, dx \quad \text{and} \quad \int_0^\infty e^{-x}\ln x\, dx.$$

and compare the approximations with the exact values.

4.6.5   Repeat Exercise 4.6.4, but this time for the weight e^{-x^2}, and with limits of integration $-\infty$ to ∞; i.e., find points x_i and w_i such that

$$\int_{-\infty}^{\infty} p(x)e^{-x^2}\, dx = \sum_{i=0}^{k} w_i p(x_i)$$

is true for all polynomials of degree $\leq 2k-1$.

4.6.6   (a) Show that if

$$\int_a^b f(x)\, dx = \sum_{i=1}^{n} c_i f(x_i) \quad \text{and} \quad \int_a^b g(x)\, dx = \sum_{i=1}^{n} c_i g(x_i),$$

then

$$\int_{[a,b]\times[a,b]} f(x)g(y)|dx\, dy| = \sum_{i=1}^{n}\sum_{j=1}^{n} c_i c_j f(x_i)g(x_j).$$

(b) What is the Simpson approximation with one step of the integral

$$\int_{[0,1]\times[0,1]} x^2 y^3 |dx\, dy|.$$

4.6.7   Show that there exist c and u such that

$$\int_{-1}^{1} f(x)\frac{1}{\sqrt{1-x^2}}\, dx = c\big(f(u) + f(-u)\big)$$

when f is a polynomial of degree $d \leq 3$.

*4.6.8   In this exercise we will sketch a proof of Equation 4.6.3. There are many parts to the proof, and many of the intermediate steps are of independent interest.

(a) Show that if the function f is continuous on $[a_0, a_n]$ and n times differentiable on (a_0, a_n), and f vanishes at the $n+1$ distinct points $a_0 < a_1 < \cdots < a_n$, then there exists $c \in (a_0, a_n)$ such that $f^{(n)}(c) = 0$.

Exercise 4.6.8 was largely inspired by a corresponding exercises in Michael Spivak's *Calculus*.

(b) Now prove the same thing if the function vanishes with *multiplicities*. The function f vanishes with multiplicity $k+1$ at a if $f(a) = f'(a) = \cdots = f^{(k)}(a) = 0$. Then if f vanishes with multiplicity $k_i + 1$ at a_i, and if f is $N = n + \sum_{i=0}^{n} k_i$ times differentiable, then there exists $c \in (a_0, a_n)$ such that $f^{(N)}(c) = 0$.

Hint for Exercise 4.6.8 (c): Show that the function $g(t) = q(x)(f(t-p(t)) - q(t)(f(x) - p(x))$ vanishes $n+2$ times; and recall that the $n+1$st derivative of a polynomial of degree n is zero.

(c) Let f be n times differentiable on $[a_0, a_n]$, and let p be a polynomial of degree n (in fact the unique one, by Exercise 2.5.16) such that $f(a_i) = p(a_i)$, and let

$$q(x) = \prod_{i=0}^{n}(x - a_i).$$

Show that there exists $c \in (a_0, a_n)$ such that

$$f(x) - p(x) = \frac{f^{(n+1)}(c)}{(n+1)!} q(x).$$

(d) Let f be 4 times continuously differentiable on $[a, b]$, and p be the polynomial of degree 3 such that

$$f(a) = p(a), \; f\left(\frac{a+b}{2}\right) = p\left(\frac{a+b}{2}\right), \; f'\left(\frac{a+b}{2}\right) = p'\left(\frac{a+b}{2}\right), \; f(b) = p(b).$$

Show that

$$\int_a^b f(x)\, dx = \frac{b-a}{6}\left(f(a) + 4f\left(\frac{a+b}{2}\right) + f(b)\right) - \frac{(b-a)^5}{2880} f^{(4)}(c)$$

for some $c \in [a, b]$.

(e) Prove part (b) of Simpson's rule (Theorem 4.6.2): If f is four times continuously differentiable, then there exists $c \in (a, b)$ such that

$$S_{[a,b]}^n(f) - \int_a^b f(x)\, dx = \frac{(b-a)^5}{2880n^4} f^{(4)}(c).$$

Exercises for Section 4.7: **Other Pavings**	**4.7.1** (a) Show that the limit

Exercises for Section 4.7:
Other Pavings

Hint for Exercise 4.7.1: This is a fairly obvious Riemann sum. You are allowed (and encouraged) to use all the theorems of Section 4.3.

4.7.1   (a) Show that the limit

$$\lim_{N \to \infty} \frac{1}{N^3} \sum_{0 \le n, m < N} m e^{-nm/N^2} \qquad \text{exists.}$$

(b) Compute the limit above.

4.7.2   (a) Let $A(R)$ be the number of points with integer entries in the disk $x^2 + y^2 \le R^2$. Show that the limit

$$\lim_{R \to \infty} \frac{A(R)}{R^2} \qquad \text{exists, and evaluate it.}$$

(b) Now do the same for the function $B(R)$ that counts how many points of the triangular grid

$$\left\{ n\begin{pmatrix}1\\0\end{pmatrix} + m\begin{pmatrix}1/2\\ \sqrt{3}/2\end{pmatrix} \; \middle| \; n, m \in \mathbb{Z} \right\} \qquad \text{are in the disk.}$$

Exercises for Section 4.8:
Determinants

4.8.1   Compute the determinants of the following matrices, using development by the first column:

(a) $\begin{bmatrix} 1 & -2 & 3 & 0 \\ 4 & 0 & 1 & 2 \\ 5 & -1 & 2 & 1 \\ 3 & 2 & 1 & 0 \end{bmatrix}$
(b) $\begin{bmatrix} 1 & 1 & 2 & 1 \\ 0 & 3 & 4 & 1 \\ 1 & 2 & 3 & 1 \\ 2 & 1 & 0 & 4 \end{bmatrix}$
(c) $\begin{bmatrix} 1 & 2 & 3 & 4 \\ 0 & 1 & -1 & 3 \\ 3 & 0 & 1 & 1 \\ 1 & 2 & -2 & 0 \end{bmatrix}$

4.8.2   (a) What is the determinant of the matrix

$$\begin{bmatrix} b & a & 0 & 0 \\ 0 & b & a & 0 \\ 0 & 0 & b & a \\ a & 0 & 0 & b \end{bmatrix}?$$

(b) What is the determinant of the corresponding $n \times n$ matrix, with b's on the diagonal and a's on the slanted line above the diagonal and in the lower left-hand corner?

(c) For each n, what are the values of a and b for which the matrix in (b) is not invertible? Hint: remember complex numbers.

4.8.3   Spell out exactly what the three conditions defining the determinant (Definition 4.8.1) mean for 2×2 matrices, and prove them.

Hint for Exercise 4.8.4: think of multiplying the column through by 2, or by -4.

4.8.4   (a) Show that if a square matrix has a column of zeroes, its determinant must be zero, using the multilinearity property (property (1)).

(b) Show that if two columns of a square matrix A are equal, the determinant must be 0.

4.8.5   If A and B are $n \times n$ matrices, and A is invertible, show that the function

$$f(B) = \frac{\det(AB)}{\det A}$$

has properties (1), (2), and (3) (multilinearity, antisymmetry, normalization) and that therefore $f(B) = \det B$.

4.8.6   Give an alternative proof of Theorem 4.8.11, by showing that

(a) If all the entries on the diagonal are nonzero, you can use column operations (of type 2) to make the matrix diagonal, without changing the entries on the main diagonal.

(b) If some entry on the main diagonal is zero, column operations can be used to get a column of zeroes.

4.8.7   Prove Theorem 4.8.14: If A is an $n \times n$ matrix and B is an $m \times m$ matrix, then for the $(n+m) \times (n+m)$ matrix formed with these as diagonal elements, $\det \begin{bmatrix} A & 0 \\ 0 & B \end{bmatrix} = \det A \det B$.

4.8.8   What elementary matrices are permutation matrices? Describe the corresponding permutation.

4.8.9   Given two permutations, σ and τ, show that the transformation that associates to each its matrix (M_σ and M_τ respectively) is a *group homomorphism*: it satisfies $M_{\sigma \circ \tau} = M_\sigma M_\tau$.

4.8.10   In Example 4.8.17, verify that the signature of σ_5 and σ_6 is -1.

4.8.11   Show by direct computation that if A and B are 2×2 matrices, then $\mathrm{tr}(AB) = \mathrm{tr}(BA)$.

4.8.12   Show that if A and B are $n \times n$ matrices, then $\mathrm{tr}(AB) = \mathrm{tr}(BA)$. Start with Corollary 4.8.22, and set $C = P$, $D = AP^{-1}$. This proves the formula when C is invertible; complete the proof by showing that if C_n is a sequence of matrices converging to C, and $\mathrm{tr}(C_n D) = \mathrm{tr}(DC_n)$ for all n, then $\mathrm{tr}(CD) = \mathrm{tr}(DC)$.

Hint for Exercise 4.8.13: think of using Theorem 4.8.10. It can also be proved, with more work, by induction on the size of the matrix.

*4.8.13   For a matrix A, we defined the determinant $D(A)$ recursively by development according to the first column. Show that it could have been defined, with the same result, as development according to the first row.

*4.8.14   (a) Show that if A is an $n \times n$ matrix of rank $n-1$, then $[\mathbf{D} \det(A)] :$ Mat $(n, n) \to \mathbb{R}$ is not the zero transformation.

(b) Show that if A is an $n \times n$ matrix with $\mathrm{rank}(A) \le n-2$, then $[\mathbf{D} \det(A)] :$ Mat $(n, n) \to \mathbb{R}$ is the zero transformation.

Exercises for Section 4.9:

Volumes and Determinants

If Δ is symmetric, switching two columns does not change the value.

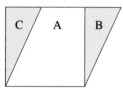

FIGURE 4.9.2.

Exercise 4.9.3, part (a): Yes, do use Fubini.

Exercise 4.9.3, part (b): No, do not use Fubini. Find a linear transformation S such that $S(T_1) = T_2$.

*4.9.1   (a) Show that $\Delta(T) = |\det T|$ is the unique mapping $\Delta :$ Mat $(n, n) \to \mathbb{R}$ that satisfies

(1) For all $T \in$ Mat (n, n), we have $\Delta(T) \ge 0$.

(2) The function Δ is a symmetric function of the columns.

(3) For all $T = [\vec{v}_1, \vec{v}_2, \ldots, \vec{v}_n] \in$ Mat (n, n), we have

$$\Delta[a\vec{v}_1, \vec{v}_2, \ldots, \vec{v}_n] = |a|\Delta[\vec{v}_1, \vec{v}_2, \ldots, \vec{v}_n].$$

(4) For all $T = [\vec{v}_1, \vec{v}_2, \ldots, \vec{v}_n] \in$ Mat (n, n), we have

$$\Delta[\vec{v}_1, \vec{v}_2, \ldots, \vec{v}_n] = \Delta[\vec{v}_1 + a\vec{v}_2, \vec{v}_2, \ldots, \vec{v}_n].$$

(5) $\Delta(I_n) = 1$.

(b) Show that $T \mapsto \mathrm{vol}_n(T(Q))$ satisfies the properties that characterize Δ.

4.9.2   Use "dissection" (as suggested in Figure 4.9.2) to prove Equation 4.9.12 when E is type 2.

4.9.3   (a) What is the volume of the tetrahedron T_1 with vertices

$$\begin{bmatrix} 0 \\ 0 \\ 0 \end{bmatrix}, \quad \begin{bmatrix} 1 \\ 0 \\ 0 \end{bmatrix}, \quad \begin{bmatrix} 0 \\ 1 \\ 0 \end{bmatrix}, \quad \begin{bmatrix} 0 \\ 0 \\ 1 \end{bmatrix}?$$

(b) What is the volume of the tetrahedron T_2 with vertices

$$\begin{bmatrix} 0 \\ 0 \\ 0 \end{bmatrix}, \quad \begin{bmatrix} 2 \\ 1 \\ 1 \end{bmatrix}, \quad \begin{bmatrix} -1 \\ 3 \\ 1 \end{bmatrix}, \quad \begin{bmatrix} -2 \\ -5 \\ 2 \end{bmatrix}?$$

4.9.4   What is the n-dimensional volume of the region

$$\{\mathbf{x} \in \mathbb{R}^n \mid x_i \geq 0 \text{ for all } i = 1, \ldots, n \text{ and } x_1 + \cdots + x_n \leq 1\}?$$

4.9.5   Let $T : \mathbb{R}^n \to \mathbb{R}^n$ be given by the matrix

$$\begin{bmatrix} 1 & 0 & 0 & \cdots & 0 \\ 2 & 2 & 0 & \cdots & 0 \\ 3 & 3 & 3 & \cdots & 0 \\ \vdots & \vdots & \vdots & \ddots & \vdots \\ n & n & n & \cdots & n \end{bmatrix},$$

and let $A \subset \mathbb{R}^n$ be given by the region given by

$$|x_1| + |x_2|^2 + |x_3|^3 + \cdots + |x_n|^n \leq 1.$$

What is $\text{vol}_n(T(A))/\text{vol}_n(A)$?

4.9.6   What is the n-dimensional volume of the region

$$\{\mathbf{x} \in \mathbb{R}^n \mid x_i \geq 0 \text{ for all } i = 1, \ldots, n \text{ and } x_1 + 2x_2 + \cdots + nx_n \leq n\}?$$

4.9.7   Let $q(x)$ be a continuous function on $\mathbb{R}$, and suppose that $f(x)$ and $g(x)$ satisfy the differential equations

$$f''(x) = q(x)f(x), \ \ g''(x) = q(x)g(x).$$

Express the area $A(x)$ of the parallelogram spanned by $\begin{bmatrix} f(x) \\ f'(x) \end{bmatrix}$, $\begin{bmatrix} g(x) \\ g'(x) \end{bmatrix}$ in terms of $A(0)$. Hint: you may want to differentiate $A(x)$.

4.9.8   (a) Find an expression for the area of the parallelogram spanned by $\vec{\mathbf{v}}_1$ and $\vec{\mathbf{v}}_2$, in terms of $|\vec{\mathbf{v}}_1|, |\vec{\mathbf{v}}_2|$, and $|\vec{\mathbf{v}}_1 - \vec{\mathbf{v}}_2|$.

(b) Prove Heron's formula: the area of a triangle with sides of length a, b, and c, is

$$\sqrt{p\,(p-a)(p-b)(p-c)}\,, \ \ \text{where} \ \ p = \frac{a+b+c}{2}.$$

4.9.9   Compute the area of the parallelograms spanned by the two vectors in (a) and (b), and the volume of the parallelepipeds spanned by the three vectors in (c) and (d).

(a) $\begin{bmatrix} 5 \\ 2 \end{bmatrix}, \begin{bmatrix} -2 \\ -1 \end{bmatrix}$ (c) $\begin{bmatrix} 1 \\ 3 \\ -1 \end{bmatrix}, \begin{bmatrix} 2 \\ -1 \\ -1 \end{bmatrix}, \begin{bmatrix} 3 \\ 6 \\ 0 \end{bmatrix}$

(b) $\begin{bmatrix} 6 \\ 4 \end{bmatrix}, \begin{bmatrix} 3 \\ 2 \end{bmatrix}$ (d) $\begin{bmatrix} 2 \\ 3 \\ 4 \end{bmatrix}, \begin{bmatrix} -1 \\ 2 \\ 3 \end{bmatrix}, \begin{bmatrix} 0 \\ 1 \\ 2 \end{bmatrix}$

Exercises for Section 4.10:

Change of Variables

4.10.1 Using Fubini, compute the integral of Example 4.10.4:

$$\int_{D_R} (x^2 + y^2)\, dx\, dy, \qquad \text{where}$$

$$D_R = \left\{ \begin{pmatrix} x \\ y \end{pmatrix} \in \mathbb{R}^2 \mid x^2 + y^2 \leq R^2 \right\}.$$

4.10.2 Show that in complex notation, with $z = x + iy$, the equation of the lemniscate can be written $|z^2 + 1| = 1$.

Hint for Exercise 4.10.4 (a): use the variables $u = x/a$, $v = y/b$.

4.10.3 Derive the change of variables formula for cylindrical coordinates from the polar formula and Fubini's theorem.

4.10.4 (a) What is the area of the ellipse $x^2/a^2 + y^2/b^2 \leq 1$?

(b) What is the volume of the ellipsoid $x^2/a^2 + y^2/b^2 + z^2/c^2 \leq 1$?

4.10.5 (a) Sketch the curve in the plane given in polar coordinates by the equation

$$r = 1 + \sin\theta, \quad 0 \leq \theta \leq 2\pi.$$

(b) Find the area that it encloses.

4.10.6 A semi-circle of radius R has density $\rho\begin{pmatrix} x \\ y \end{pmatrix} = m(x^2 + y^2)$ proportional to the square of the distance to the center. What is its mass?

Hint for Exercise 4.10.7: You may want to use Theorem 3.7.12.

4.10.7 Let A be an $n \times n$ symmetric matrix, such that the quadratic form $Q_A(\vec{x}) = \vec{x} \cdot A\vec{x}$ is positive definite. What is the volume of the region $Q(\vec{x}) \leq 1$?

4.10.8 Let

$$V = \left\{ \begin{pmatrix} x \\ y \\ z \end{pmatrix} \in \mathbb{R}^3 \;\middle|\; x > 0,\ y > 0,\ z > 0,\ \frac{x^2}{a^2} + \frac{y^2}{b^2} + \frac{z^2}{c^2} \leq 1 \right\}.$$

Compute $\int_V xyz\, |dx\, dy\, dz|$.

4.10.9 (a) What is the analog of spherical coordinates in four dimensions? What does the change of variables formula say in that case?

(b) What is the integral of $|\mathbf{x}|$ over the ball of radius R in $\mathbb{R}^4$?

Exercise 4.10.10 belongs with exercises for Section 5.2, as it requires the relaxed definition of a parametrization introduced there.

4.10.10 Show that the mapping

$$S : \begin{pmatrix} r \\ \theta \\ \varphi \end{pmatrix} \mapsto \begin{pmatrix} r\sin\varphi\cos\theta \\ r\sin\varphi\sin\theta \\ r\cos\varphi \end{pmatrix}$$

with $0 \leq r < \infty$, $0 \leq \theta \leq 2\pi$, and $0 \leq \varphi \leq \pi$, parametrizes space by the distance from the origin r, the polar angle θ, and the angle φ from the north pole.

4.10.11   Justify that the volume of the sphere of radius R is $\frac{4}{3}\pi R^3$.

4.10.12   Evaluate the iterated integral

$$\int_{-2}^{2} \int_{0}^{\sqrt{4-x^2}} \int_{0}^{\sqrt{4-x^2-y^2}} (x^2 + y^2 + z^2)^{3/2} \, dz \, dy \, dx.$$

Hint: First transform it into a multiple integral, then pass to spherical coordinates.

4.10.13   Find the volume of the region between the cone of equation $z^2 = x^2 + y^2$ and the paraboloid of equation $z = x^2 + y^2$.

4.10.14   (a) Let Q be the part of the unit ball $x^2+y^2+z^2 \leq 1$ where $x, y, z \geq 0$. Using spherical coordinates, set up the integral

$$\int_{Q} (x + y + z) \, |d^3\mathbf{x}| \quad \text{as an iterated integral.}$$

(b) Compute the integral.

4.10.15   (a) Sketch the curve given in polar coordinates by $r = \cos 2\theta$, $|\theta| < \pi/4$.

Hint for Exercise 4.10.15: You may wish to use the trigonometric formulas

$$4\cos^3\theta = \cos 3\theta + 3\cos\theta$$

$$2\cos\varphi\cos\theta = \cos(\theta + \varphi)$$

$$+ \cos(\theta - \varphi).$$

(b) Where is the center of gravity of the region $0 < r \leq \cos 2\theta$, $|\theta| < \pi/4$?

4.10.16   What is the center of gravity of the region A defined by the inequalities $x^2 + y^2 \leq z \leq 1$, $x \geq 0$, $y \geq 0$?

4.10.17   Let $Q_a = [0, a] \times [0, a] \subset \mathbb{R}^2$ be the square of side a in the first quadrant, with two sides on the axes, let $\Phi : \mathbb{R}^2 \to \mathbb{R}^2$ be given by

$$\Phi \begin{pmatrix} u \\ v \end{pmatrix} = \begin{pmatrix} u - v \\ e^u + e^v \end{pmatrix},$$

and $A = \Phi(Q_a)$.

(a) Sketch A, by computing the image of each of the sides of Q_a (it might help to begin by drawing carefully the curves of equation $y = e^x + 1$ and $y = e^{-x} + 1$).

(b) Show that $\Phi : Q_a \to A$ is one to one.

(c) What is $\int_A y \, |dx \, dy|$?

4.10.18   What is the volume of the part of the ball $x^2 + y^2 + z^2 \leq 4$ where $z^2 \geq x^2 + y^2$, $z > 0$?

4.10.19   Let $Q = [0, 1] \times [0, 1]$ be the unit square in $\mathbb{R}^2$, and let $\Phi : \mathbb{R}^2 \to \mathbb{R}^2$ be given by

$$\Phi \begin{pmatrix} u \\ v \end{pmatrix} = \begin{pmatrix} u - v^2 \\ u^2 + v \end{pmatrix} \quad \text{and} \quad A = \Phi(Q).$$

(a) Sketch A, by computing the image of each of the sides of Q (they are all arcs of parabolas).

(b) Show that $\Phi : Q \to A$ is 1-1.

(c) What is $\int_A x \, |dx \, dy|$?

4.10.20   The moment of inertia of a body $X \subset \mathbb{R}^3$ around an axis is the integral

$$\int_X (r(\mathbf{x}))^2 |d^3\mathbf{x}|,$$

where $r(\mathbf{x})$ is the distance from $\mathbf{x}$ to the axis.

(a) Let f be a positive continuous function of $x \in [a, b]$. What is the moment of inertia around the x-axis of the body obtained by rotating the region $0 \le y \le f(x)$, $a \le x \le b$ around the x-axis?

(b) What number does this give when

$$f(x) = \cos x, \quad a = -\frac{\pi}{2}, \; b = \frac{\pi}{2}?$$

Exercises for Section 4.11:
Improper Integrals

4.11.1   (a) Show that the three sequences of functions in Example 4.11.11 do not converge uniformly.

*(b) Show that the sequence of polynomials of degree $\le m$

$$p_k(x) = a_{0,k} + a_{1,k}x + \cdots + a_{m,k}x^m$$

does not converge uniformly on $\mathbb{R}$, unless the sequence $a_{i,k}$ is eventually constant for all $i > 0$ and the sequence $a_{0,k}$ converges.

(c) Show that if the sequences $a_{i,k}$ converge for each $i \le m$, and A is a bounded set, then the sequence $p_k \chi_A$ converges uniformly.

4.11.2   Let $a_n = \frac{(-1)^{n+1}}{n}$.

(a) Show that the series $\sum a_n$ is convergent.

*(b) Show that $\sum_{n=1}^\infty a_n = \ln 2$.

(c) Explain how to rearrange the terms of the series so it converges to 5.

(d) Explain how to rearrange the terms of the series so that it diverges.

4.11.3   For the first two sequences of functions in Example 4.11.11, show that

$$\lim_{k\to\infty} \lim_{R\to\infty} \int_{\mathbb{R}} [f_k(x)]_R \, dx \ne \lim_{R\to\infty} \lim_{k\to\infty} \int_{\mathbb{R}} [f_k(x)]_R \, dx.$$

4.11.4   In this exercise we will show that

$$\int_0^\infty \frac{\sin x}{x} dx = \frac{\pi}{2}. \qquad\qquad 4.11.4$$

This function is not integrable in the sense of Section 4.11, and the integral should be understood as

$$\int_0^\infty \frac{\sin x}{x}\,dx = \lim_{a\to\infty}\int_0^a \frac{\sin x}{x}\,dx.$$

(a) Show that

$$\int_a^b\left(\int_0^\infty e^{-px}\sin x\,dx\right)dp = \int_0^\infty\left(\int_a^b e^{-px}\sin x\,dp\right)dx.$$

for all $0 < a < b < \infty$.

(b) Use (a) to show

$$\arctan b - \arctan a = \int_a^b \frac{(e^{-ax}-e^{-bx})\sin x}{x}\,dx.$$

(c) Why does Theorem 4.11.12 not imply that

$$\lim_{a\to 0}\lim_{b\to\infty}\int_0^\infty \frac{((e^{-ax}-e^{-bx})\sin x}{x}\,dx = \int_0^\infty \frac{\sin x}{x}\,dx\,?$$

(d) As it turns out, the equation in part (c) is true anyway; prove it. The following lemma is the key: If $c_n(t) > 0$ are monotone increasing functions of t, with $\lim_{t\to\infty} c_n(t) = C_n$, and decreasing as a function of n for each fixed t, tending to 0, then

$$\lim_{t\to\infty}\sum_{n=1}^\infty (-1)^n c_n(t) = \sum_{n=1}^\infty (-1)^n C_n.$$

Remember that the next omitted term is a bound for the error for each partial sum.

(e) Write

$$\int_0^\infty \frac{((e^{-ax}-e^{-bx})\sin x}{x}\,dx = \sum_{n=0}^\infty \int_{k\pi}^{(k+1)\pi} (-1)^k \frac{((e^{-ax}-e^{-bx})|\sin x|}{x}\,dx,$$

and use (d) to prove Equation 4.11.4.

4.11.5   Show that the integral $\int_0^\infty \frac{\sin x}{x}\,dx$ of Equation 4.11.5 is equal to the sum of the series

$$\sum_{m=1}^\infty \left(\int_{m\pi}^{(m+1)\pi} \frac{\sin x}{x}\,dx\right).$$

Hint for Exercise 4.11.6: you will need the dominated convergence theorem (Theorem 4.11.12) to prove this.

4.11.6   Let P_k be the space of polynomials of degree at most k. Consider the function $F : P_k \to \mathbb{R}$ given by $p \mapsto \int_0^1 |p(x)|\,dx$.

(a) Show that F is differentiable except at 0, and compute the derivative.

*(b) Show that if p has only simple roots between 0 and 1, then F is twice differentiable at p.

5

Lengths of Curves, Areas of Surfaces, ...

5.0 INTRODUCTION

In Chapter 4 we saw how to integrate over subsets of $\mathbb{R}^n$, first using dyadic pavings, and then more general pavings. But these subsets are flat. What if we want to integrate over a (curvy) surface in $\mathbb{R}^3$, or more generally, a k-manifold in $\mathbb{R}^n$? There are many situations of obvious interest, like the area of a surface, or the total energy stored in the surface tension of a soap bubble, or the amount of fluid flowing through a pipe, which clearly are some sort of surface integral. In a physics course, for example, you may have learned that the *electric flux* through a closed surface is proportional to the electric charge inside that surface.

A first thing to realize is that you can't just consider a surface S as a subset of $\mathbb{R}^3$ and integrate a function in $\mathbb{R}^3$ over S. The surface S has three-dimensional volume 0, so such an integral will certainly vanish. Instead, we need to rethink the whole process of integration.

At heart, integration is always the same:

> *Break up the domain into little pieces, assign a little number to each little piece, and finally add together all the numbers.* Then break the domain into littler pieces and repeat, taking the limit as the decomposition becomes infinitely fine. The *integrand* is the thing that assigns the number to the little piece of the domain.

If this process is to be a useful way of measuring the total amount of something, the limit as the decomposition becomes infinitely fine must exist, and it should not depend on how the domain is broken up. It is not so very easy to come up with reasonable integrands. This issue is explored in Exercises 6.1.1 and 6.1.2.

In this chapter we will show how to compute things like arc length (already discussed in Section 3.8), surface area, and higher dimensional analogs, including fractals. We will be integrating expressions like $f(\mathbf{x})|d^k\mathbf{x}|$ over k-dimensional manifolds, where $|d^k\mathbf{x}|$ assigns to a k-dimensional manifold its area. Later, in Chapter 6, we will study a different kind of integrand, which assigns numbers to *oriented* manifolds.

What does "little piece" mean?

The words "little piece" in the heuristic description above needs to be pinned down to something more precise before we can do anything useful. There is quite a bit of leeway here; choosing a decomposition of a surface into little pieces is analogous to choosing a paving, and as we saw in Section 4.7, there were many possible choices besides the dyadic paving. We will choose to approximate curves, surfaces, and more general k-manifolds by k-parallelograms. These are described, and their volumes computed, in the next section.

We can only integrate over parametrized domains, and if we stick with the definition of parametrizations introduced in Chapter 3, we will not be able to parametrize even such simple objects as the circle and the sphere. Fortunately, for the purposes of integration, a looser definition of parametrization will suffice; we discuss this in Section 5.2.

5.1 Parallelograms and their Volumes

We specify a k-parallelogram in $\mathbb{R}^n$ by the point $\mathbf{x}$ where it is anchored, and the k vectors which span it. More precisely:

It follows from Definition 5.1.1 that the order in which we take the vectors doesn't matter:

$$P_{\mathbf{x}}(\vec{\mathbf{v}}_1, \vec{\mathbf{v}}_2) = P_{\mathbf{x}}(\vec{\mathbf{v}}_2, \vec{\mathbf{v}}_1),$$

$$P_{\mathbf{x}}(\vec{\mathbf{v}}_1, \vec{\mathbf{v}}_2, \vec{\mathbf{v}}_3) = P_{\mathbf{x}}(\vec{\mathbf{v}}_1, \vec{\mathbf{v}}_3, \vec{\mathbf{v}}_2),$$

and so on.

Definition 5.1.1 (k-parallelogram in $\mathbb{R}^n$). A k-parallelogram in $\mathbb{R}^n$ is the subset of $\mathbb{R}^n$

$$P_{\mathbf{x}}(\vec{\mathbf{v}}_1, \ldots, \vec{\mathbf{v}}_k) = \left\{ \mathbf{x} + t_1 \vec{\mathbf{v}}_1 + \cdots + t_k \vec{\mathbf{v}}_k \mid 0 \le t_1, \ldots, t_k \le 1 \right\},$$

where $\mathbf{x} \in \mathbb{R}^n$ is a point and $\vec{\mathbf{v}}_1, \ldots, \vec{\mathbf{v}}_k$ are k vectors. The corner $\mathbf{x}$ is part of the data, but the order in which the vectors are listed is not.

For example,

(1) $P_{\mathbf{x}}(\vec{\mathbf{v}})$ is the line segment joining $\mathbf{x}$ to $\mathbf{x} + \vec{\mathbf{v}}$.

(2) $P_{\mathbf{x}}(\vec{\mathbf{v}}_1, \vec{\mathbf{v}}_2)$ is the (ordinary) parallelogram with its four vertices at $\mathbf{x}$, $\mathbf{x} + \vec{\mathbf{v}}_1$, $\mathbf{x} + \vec{\mathbf{v}}_2$, and $\mathbf{x} + \vec{\mathbf{v}}_1 + \vec{\mathbf{v}}_2$.

(3) $P_{\mathbf{x}}(\vec{\mathbf{v}}_1, \vec{\mathbf{v}}_2, \vec{\mathbf{v}}_3)$ is the (ordinary) parallelepiped with its eight vertices at

Notice that a 3-parallelogram in $\mathbb{R}^2$ must be squashed flat, and it can perfectly well be squashed flat even if $n > 2$: this will happen if $\vec{\mathbf{v}}_1, \vec{\mathbf{v}}_2, \vec{\mathbf{v}}_3$ are linearly dependent.

$$\mathbf{x}, \quad \mathbf{x} + \vec{\mathbf{v}}_1, \quad \mathbf{x} + \vec{\mathbf{v}}_2, \quad \mathbf{x} + \vec{\mathbf{v}}_3, \quad \mathbf{x} + \vec{\mathbf{v}}_1 + \vec{\mathbf{v}}_2,$$

$$\mathbf{x} + \vec{\mathbf{v}}_1 + \vec{\mathbf{v}}_3, \quad \mathbf{x} + \vec{\mathbf{v}}_2 + \vec{\mathbf{v}}_3, \quad \mathbf{x} + \vec{\mathbf{v}}_1 + \vec{\mathbf{v}}_2 + \vec{\mathbf{v}}_3.$$

The volume of k-parallelograms

Clearly the k-dimensional volume of a k-parallelogram $P_{\mathbf{x}}(\vec{\mathbf{v}}_1, \ldots, \vec{\mathbf{v}}_k)$ does not depend on the position of $\mathbf{x}$ in $\mathbb{R}^n$. But it isn't obvious how to compute this volume. Already the area of a parallelogram in $\mathbb{R}^3$ is the length of the cross product of the two vectors spanning it (Proposition 1.4.19), and the formula is quite messy. How will we compute the area of a parallelogram in $\mathbb{R}^4$, where the cross product does not exist, never mind a 3-parallelogram in $\mathbb{R}^5$?

It comes as a nice surprise that there is a very pretty formula that covers all cases. The following proposition, which seems so innocent, is the key.

Proposition 5.1.2 (Volume of a k-parallelogram in $\mathbb{R}^k$). Let $\vec{\mathbf{v}}_1, \ldots, \vec{\mathbf{v}}_k$ be k vectors in $\mathbb{R}^k$, so that $T = [\vec{\mathbf{v}}_1, \ldots, \vec{\mathbf{v}}_k]$ is a square $k \times k$ matrix. Then

$$\mathrm{vol}_k\, P(\vec{\mathbf{v}}_1, \ldots, \vec{\mathbf{v}}_k) = \sqrt{\det(T^\top T)}. \qquad 5.1.1$$

In the proof of Proposition 5.1.2 we use Theorem 4.8.7:

$$\det A \det B = \det(AB)$$

and Theorem 4.8.10 (for square matrices):

$$\det A = \det A^\top.$$

We follow the common convention according to which the square root symbol of a positive number denotes the positive square root: $\sqrt{a} = +a$, not $\pm a$.

Proof. We have

$$\sqrt{\det(T^\top T)} = \sqrt{(\det T^\top)(\det T)} = \sqrt{(\det T)^2} = |\det T|. \quad \square \qquad 5.1.2$$

Proposition 5.1.2 is obviously true, but why is it interesting? The product $T^\top T$ works out to be

$$5.1.3$$

(We follow the format for matrix multiplication introduced in Example 1.2.3.) The point of this is that the entries of $T^\top T$ are all dot products of the vectors $\vec{\mathbf{v}}_i$. This has the consequence that they are computable from the lengths of the vectors $\vec{\mathbf{v}}_1, \ldots, \vec{\mathbf{v}}_k$ and angles between these vectors; no further information about the vectors is needed. Note that $T^\top T$ is symmetric; this makes computing it easier.

Recall (Definition 1.4.7) that

$$\vec{\mathbf{x}} \cdot \vec{\mathbf{y}} = |\vec{\mathbf{x}}|\,|\vec{\mathbf{y}}|\cos\alpha,$$

where α is the angle between $\vec{\mathbf{x}}$ and $\vec{\mathbf{y}}$.

Example 5.1.3 (Computing the volume of parallelograms in $\mathbb{R}^2$ and $\mathbb{R}^3$). When $k = 2$, we have

$$\det(T^\top T) = \det \begin{bmatrix} |\vec{\mathbf{v}}_1|^2 & \vec{\mathbf{v}}_1 \cdot \vec{\mathbf{v}}_2 \\ \vec{\mathbf{v}}_1 \cdot \vec{\mathbf{v}}_2 & |\vec{\mathbf{v}}_2|^2 \end{bmatrix} = |\vec{\mathbf{v}}_1|^2 |\vec{\mathbf{v}}_2|^2 - (\vec{\mathbf{v}}_1 \cdot \vec{\mathbf{v}}_2)^2. \qquad 5.1.4$$

If you write $\vec{\mathbf{v}}_1 \cdot \vec{\mathbf{v}}_2 = |\vec{\mathbf{v}}_1||\vec{\mathbf{v}}_2|\cos\theta$, this becomes

$$\det(T^\top T) = |\vec{\mathbf{v}}_1|^2 |\vec{\mathbf{v}}_2|^2 (1 - \cos^2\theta) = |\vec{\mathbf{v}}_1|^2 |\vec{\mathbf{v}}_2|^2 \sin^2\theta, \qquad 5.1.5$$

so that the area of the parallelogram spanned by $\vec{\mathbf{v}}_1$, $\vec{\mathbf{v}}_2$ is

$$\sqrt{\det(T^\top T)} = |\vec{\mathbf{v}}_1||\vec{\mathbf{v}}_2||\sin\theta|. \qquad 5.1.6$$

Of course, this should come as no surprise; we got the same thing in Equation 1.4.35. But exactly the same computation in the case $n = 3$ leads to a much less familiar formula. Suppose $T = [\vec{\mathbf{v}}_1, \vec{\mathbf{v}}_2, \vec{\mathbf{v}}_3]$, and let us call θ_1 the angle between $\vec{\mathbf{v}}_2$ and $\vec{\mathbf{v}}_3$, θ_2 the angle between $\vec{\mathbf{v}}_1$ and $\vec{\mathbf{v}}_3$, and θ_3 the angle between $\vec{\mathbf{v}}_1$ and $\vec{\mathbf{v}}_2$. Then

$$T^\top T = \begin{bmatrix} |\vec{\mathbf{v}}_1|^2 & \vec{\mathbf{v}}_1 \cdot \vec{\mathbf{v}}_2 & \vec{\mathbf{v}}_1 \cdot \vec{\mathbf{v}}_3 \\ \vec{\mathbf{v}}_1 \cdot \vec{\mathbf{v}}_2 & |\vec{\mathbf{v}}_2|^2 & \vec{\mathbf{v}}_2 \cdot \vec{\mathbf{v}}_3 \\ \vec{\mathbf{v}}_1 \cdot \vec{\mathbf{v}}_3 & \vec{\mathbf{v}}_2 \cdot \vec{\mathbf{v}}_3 & |\vec{\mathbf{v}}_3|^2 \end{bmatrix} \qquad 5.1.7$$

and $\det T^\top T$ is given by

$$|\vec{\mathbf{v}}_1|^2|\vec{\mathbf{v}}_2|^2|\vec{\mathbf{v}}_3|^2 + 2(\vec{\mathbf{v}}_1 \cdot \vec{\mathbf{v}}_2)(\vec{\mathbf{v}}_2 \cdot \vec{\mathbf{v}}_3)(\vec{\mathbf{v}}_1 \cdot \vec{\mathbf{v}}_3) \qquad 5.1.8$$
$$- |\vec{\mathbf{v}}_1|^2(\vec{\mathbf{v}}_2 \cdot \vec{\mathbf{v}}_3)^2 - |\vec{\mathbf{v}}_2|^2(\vec{\mathbf{v}}_1 \cdot \vec{\mathbf{v}}_3)^2 - |\vec{\mathbf{v}}_3|^2(\vec{\mathbf{v}}_1 \cdot \vec{\mathbf{v}}_2)^2$$
$$= |\vec{\mathbf{v}}_1|^2|\vec{\mathbf{v}}_2|^2|\vec{\mathbf{v}}_3|^2\big(1 + 2\cos\theta_1\cos\theta_2\cos\theta_3 - (\cos^2\theta_1 + \cos^2\theta_2 + \cos^2\theta_3)\big).$$

For instance, the volume of a parallelepiped spanned by three unit vectors, each making an angle of $\pi/4$ with the others, is

This would not be easy to compute directly from $\det T$, since we don't actually know the vectors.

$$\sqrt{1 + 2\cos^3\frac{\pi}{4} - 3\cos^2\frac{\pi}{4}} = \sqrt{\frac{\sqrt{2}-1}{2}}. \quad \triangle \qquad 5.1.9$$

Thus we have a formula for the volume of a parallelogram that depends only on the lengths and angles of the vectors that span it; we do not need to know what or where the vectors actually are. In particular, this formula is *just as good* for a k-parallelogram in any $\mathbb{R}^n$, even (and especially) if $n > k$.

This leads to the following theorem.

Note that one consequence of Theorem 5.1.4 is that if T is any matrix, $T^\top T$ always has a non-negative determinant:

$$\det(T^\top T) \geq 0.$$

This is not at all obvious from Equation 5.1.11.

Theorem 5.1.4 (Volume of a k-parallelogram in $\mathbb{R}^n$). Let $\vec{\mathbf{v}}_1, \dots, \vec{\mathbf{v}}_k$ be k vectors in $\mathbb{R}^n$, and T be the $n \times k$ matrix with these vectors as its columns: $T = [\vec{\mathbf{v}}_1, \dots, \vec{\mathbf{v}}_k]$. Then the k-dimensional volume of $P_{\mathbf{x}}(\vec{\mathbf{v}}_1, \dots, \vec{\mathbf{v}}_k)$ is

$$\mathrm{vol}_k P_{\mathbf{x}}(\vec{\mathbf{v}}_1, \dots, \vec{\mathbf{v}}_k) = \sqrt{\det(T^\top T)}. \qquad 5.1.10$$

Proof. If we compute the $n \times n$ matrix $T^\top T$, we find

$$T^\top T = \begin{bmatrix} |\vec{\mathbf{v}}_1|^2 & \vec{\mathbf{v}}_1 \cdot \vec{\mathbf{v}}_2 & \dots & \vec{\mathbf{v}}_1 \cdot \vec{\mathbf{v}}_k \\ \vec{\mathbf{v}}_2 \cdot \vec{\mathbf{v}}_1 & |\vec{\mathbf{v}}_2|^2 & \dots & \vec{\mathbf{v}}_2 \cdot \vec{\mathbf{v}}_k \\ \vdots & \vdots & \ddots & \dots \\ \vec{\mathbf{v}}_k \cdot \vec{\mathbf{v}}_1 & \vec{\mathbf{v}}_k \cdot \vec{\mathbf{v}}_2 & \dots & |\vec{\mathbf{v}}_k|^2 \end{bmatrix}, \qquad 5.1.11$$

Exercise 5.1.3 asks you to show that if $\vec{\mathbf{v}}_1, \dots, \vec{\mathbf{v}}_k$ are linearly dependent, $\mathrm{vol}_k(P(\vec{\mathbf{v}}_1, \dots, \vec{\mathbf{v}}_k)) = 0$. In particular, this shows that if $k > n$, $\mathrm{vol}_k(P(\vec{\mathbf{v}}_1, \dots, \vec{\mathbf{v}}_k)) = 0$

which is precisely our formula for the k-dimensional volume of a k-parallelogram in terms of lengths and angles. $\square$

Example 5.1.5 (Volume of a 3-parallelogram in $\mathbb{R}^4$). What is the volume of the 3-parallelogram in $\mathbb{R}^4$ spanned by $\vec{\mathbf{v}}_1 = \begin{bmatrix} 1 \\ 0 \\ 0 \\ 1 \end{bmatrix}, \vec{\mathbf{v}}_2 = \begin{bmatrix} 0 \\ 1 \\ 0 \\ 1 \end{bmatrix}, \vec{\mathbf{v}}_3 = \begin{bmatrix} 0 \\ 0 \\ 1 \\ 1 \end{bmatrix}$?

Set $T = [\vec{\mathbf{v}}_1, \vec{\mathbf{v}}_2, \vec{\mathbf{v}}_3]$; then

$$T^\top T = \begin{bmatrix} 2 & 1 & 1 \\ 1 & 2 & 1 \\ 1 & 1 & 2 \end{bmatrix} \quad \text{and} \quad \det(T^\top T) = 4,$$

so the volume is 2. $\triangle$

5.2 PARAMETRIZATIONS

In this section we are going to relax our definition of a parametrization. In Chapter 3 (Definition 3.2.5) we said that a parametrization of a manifold M is a C^1 mapping γ from an open subset $U \subset \mathbb{R}^n$ to M, which is one to one and onto, and whose derivative is also one to one.

The problem with this definition is that most manifolds do not admit a parametrization. Even the circle does not; neither does the sphere, nor the torus. On the other hand, our entire theory of integration over manifolds is going to depend on parametrizations, and we cannot simply give up on most examples.

Let us examine what goes wrong for the circle and the sphere. The most obvious parametrization of the circle is $\gamma : t \mapsto \begin{pmatrix} \cos t \\ \sin t \end{pmatrix}$. The problem is choosing a domain: If we choose $(0, 2\pi)$, then γ is not onto. If we choose $[0, 2\pi]$, the domain is not open, and γ is not one to one. If we choose $[0, 2\pi)$, the domain is not open. why is this important?

For the sphere, spherical coordinates

$$\gamma : \begin{pmatrix} \theta \\ \varphi \end{pmatrix} \mapsto \begin{pmatrix} \cos\varphi\cos\theta \\ \cos\varphi\sin\theta \\ \sin\varphi \end{pmatrix} \qquad 5.2.1$$

present the same sort of problem. If we use as domain $(-\pi/2, \pi/2) \times (0, 2\pi)$, then γ is not onto; if we use $[-\pi/2, \pi/2] \times [0, 2\pi]$, then the map is not one to one, and the derivative is not one to one at points where $\varphi = \pm\pi/2, \dots$.

The key point for both these examples is that *the trouble occurs on sets of volume* 0, and therefore it should not matter when we integrate. Our new definition of a parametrization will be exactly the old one, except that we allow things to go wrong on sets of k-dimensional volume 0 when parametrizing k-dimensional manifolds.

Sets of k-dimensional volume 0 in $\mathbb{R}^n$

Let X be a subset of $\mathbb{R}^n$. We need to know when X is negligible as far as k-dimensional integrals are concerned. Intuitively it should be fairly clear what this means: points are negligible for 1-dimensional integrals or higher, points and curves are negligible for 2-dimensional integrals, etc.

The cubes in Equation 5.2.2 have side length $1/2^N$. We are summing over cubes that intersect X.

Proposition A19.1 in the Appendix explains why Definition 5.2.1 is reasonable: if X has k-dimensional volume 0, then its projection onto k coordinates has k-dimensional volume 0.

It is possible to define the k-dimensional volume of an arbitrary subset $X \subset \mathbb{R}^n$, and we will touch on this in Section 5.6 on fractals. That definition is quite elaborate; it is considerably simpler to say when such a subset has k-dimensional volume 0.

Definition 5.2.1 (k-dimensional volume 0 of a subset of $\mathbb{R}^n$). A bounded subset $X \subset \mathbb{R}^n$ has k-dimensional volume 0 if

$$\lim_{N \to \infty} \sum_{\substack{C \in \mathcal{D}_N(\mathbb{R}^n) \\ C \cap X \neq \phi}} \left(\underbrace{\frac{1}{2^N}}_{\substack{\text{sidelength} \\ \text{of } C}} \right)^k = 0. \qquad 5.2.2$$

An arbitrary subset $X \subset \mathbb{R}^n$ has k-dimensional volume 0 if for all R, the bounded set $X \cap B_R(\mathbf{0})$ has k-dimensional volume 0.

New definition of parametrization

From now on when we use the word "parametrization" we will mean the following; if we wish to refer to the more demanding definition of Chapter 3, we will call it a "strict parametrization."

Definition 5.2.2: typically, U will be closed, and X will be its boundary. But there are many cases where this isn't quite the case, including many which come up in practice like 5.2.3 below, where it is desirable to allow X to be larger.

The mapping γ is a strict parametrization on $U - X$, i.e., on U minus the boundary and any other trouble spots of k-dimensional volume 0.

Definition 5.2.2 (Parametrization of a manifold). Let $M \subset \mathbb{R}^n$ be a k-dimensional manifold and U be a subset of $\mathbb{R}^k$ with boundary of k-dimensional volume 0; let $X \subset U$ have k-dimensional volume 0, and let $U - X$ be open. Then a continuous mapping $\gamma : U \to \mathbb{R}^n$ parametrizes M if

(1) $\gamma(U) \supset M$;

(2) $\gamma(U - X) \subset M$;

(3) $\gamma : (U - X) \to M$ is one to one, of class C^1, with locally Lipschitz derivative;

(4) the derivative $[D\gamma(\mathbf{u})]$ is one to one for all $\mathbf{u}$ in $U - X$;

(5) $\gamma(X)$ has k-dimensional volume 0.

Often condition (1) will be an equality; for example, if M is a sphere and U a closed rectangle mapped to M by spherical coordinates, then $\gamma(U) = M$. In that case, X is the boundary of U, and $\gamma(X)$ consists of the poles and half a great circle (the international date line, for example), giving $\gamma(U - X) \subset M$ for condition (2).

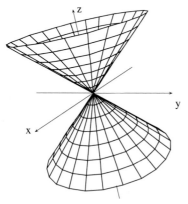

FIGURE 5.2.1.

The subset of $\mathbb{R}^3$ of equation $x^2 + y^2 = z^2$ is not a manifold at the vertex.

Example 5.2.3 (Parametrization of a cone). The subset of $\mathbb{R}^3$ of equation $x^2 + y^2 - z^2$, shown in Figure 5.2.1, is not a manifold in the neighborhood of the vertex, which is at the origin. However, the subset

$$M = \left\{ \begin{pmatrix} x \\ y \\ z \end{pmatrix} \,\middle|\, x^2 + y^2 - z^2 = 0, \quad 0 < z < 1 \right\} \qquad 5.2.3$$

is a manifold. Consider the map $\gamma : [0,1] \times [0,2\pi] \to \mathbb{R}^3$ given by

$$\gamma : \begin{pmatrix} r \\ \theta \end{pmatrix} \mapsto \begin{pmatrix} r\cos\theta \\ r\sin\theta \\ r \end{pmatrix}. \qquad 5.2.4$$

If we let $U = [0,1] \times [0,2\pi]$, and $X = \partial U$, then γ is a parametrization of M. Indeed, $\gamma([0,1] \times [0,2\pi]) \supset M$ (it contains the vertex and the circle of radius 1 in the plane $z = 1$, in addition to M), and γ does map $(0,1) \times (0,2\pi)$ into M (this time, it omits the line segment $x = z, y = 0$). The map is one to one on $(0,1) \times (0,2\pi)$, and so is its derivative. $\triangle$

A small catalog of parametrizations

We will see below (Theorem 5.2.6) that essentially all manifolds can be parametrized using the new definition. But it is one thing to construct a parametrization using the implicit function theorem, and another to write down a parametrization explicitly. Below we give a few examples, which frequently show up in applications and exam problems.

Graphs. If U is an open subset of $\mathbb{R}^k$ with boundary ∂U of k-dimensional volume 0, and $\mathbf{f} : U \to \mathbb{R}^{n-k}$ is a C^1 mapping, then the graph of $\mathbf{f}$ is a manifold in $\mathbb{R}^n$, and the map

$$\mathbf{x} \mapsto \begin{bmatrix} \mathbf{x} \\ \mathbf{f(x)} \end{bmatrix} \qquad 5.2.5$$

is a parametrization.

There are many cases where the idea of parametrizing as a graph still works, even though the conditions above are not satisfied: those where you can "solve" the defining equation for $n - k$ of the variables in terms of the other k.

Example 5.2.4 (Parametrizing as a graph). Consider the surface in $\mathbb{R}^3$ of equation $x^2 + y^3 + z^5 = 1$. In this case you can "solve" for x as a function of y and z:

$$x = \pm\sqrt{1 - y^3 - z^5}. \qquad 5.2.6$$

You could also solve for y or for z, as a function of the other variables, and the three approaches give different views of the surface, as shown in Figure 5.2.2. Of course, before you can call any of these a parametrization, you have to specify exactly what the domain is. When the equation is solved for x, the domain is the subset of the (y,z)-plane where $1 - y^3 - z^5 \geq 0$. When solving for y, remember that every number has a unique cube root, so the function $y = \left(1 - x^2 - z^5\right)^{1/3}$ is defined at every point, but it is not differentiable when

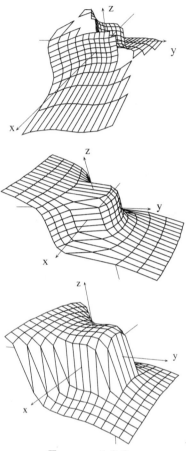

FIGURE 5.2.2.
The surface of equation $x^2 + y^3 + z^5 = 1$. Top: The surface seen as a graph of x as a function of y and z (i.e., parametrized by y and z). The graph consists of two pieces: the positive square root and the negative square root.
Middle: Parametrizing by x and z. Bottom: Parametrizing by x and y. Note in the bottom two graphs that the lines are drawn differently: different parametrizations give different resolutions to different areas.

$x^2 + z^5 = 1$, so this curve must be included in the set X of trouble points that can be ignored (using the notation of Definition 5.2.2.)   △

Surfaces of revolution.   The graph of a function $f(x)$ is the curve C of equation $y = f(x)$. Let us suppose that f takes only positive values, and rotate C around the x-axis, to get the surface of revolution of equation

$$y^2 + z^2 = \big(f(x)\big)^2. \tag{5.2.7}$$

This surface can be parametrized by

$$\gamma : \begin{pmatrix} x \\ \theta \end{pmatrix} \mapsto \begin{pmatrix} x \\ f(x)\cos\theta \\ f(x)\sin\theta \end{pmatrix}. \tag{5.2.8}$$

Again, to be precise one must specify the domain of γ. Suppose that $f : (a,b) \to \mathbb{R}$ is defined and continuously differentiable on (a,b). Then the domain of γ is $(a,b) \times [0, 2\pi]$, and γ is one to one, with derivative also one to one on $(a,b) \times (0, 2\pi)$.

If C is a parametrized curve, (not necessarily a graph), say parametrized by $t \mapsto \begin{pmatrix} u(t) \\ v(t) \end{pmatrix}$, the surface obtained by rotating C can still be parametrized by

$$\begin{pmatrix} t \\ \theta \end{pmatrix} \mapsto \begin{pmatrix} u(t) \\ v(t)\cos\theta \\ v(t)\sin\theta \end{pmatrix}. \tag{5.2.9}$$

Spherical coordinates on the sphere of radius R are a special case of this construction: If C is the semi-circle of radius R in the (x, z)-plane, parametrized by

$$\begin{pmatrix} x = R\cos\varphi \\ z = R\sin\varphi \end{pmatrix}, \quad -\pi/2 \le \phi \le \pi/2, \tag{5.2.10}$$

then the surface obtained by rotating this circle is precisely the sphere of radius R centered at the origin in $\mathbb{R}^3$, parametrized by

$$\begin{pmatrix} \phi \\ \theta \end{pmatrix} \mapsto \begin{pmatrix} R\cos\phi\cos\theta \\ R\cos\phi\sin\theta \\ R\sin\phi \end{pmatrix}, \tag{5.2.11}$$

the parametrization of the sphere by latitude and longitude.

Example 5.2.5 (Surface obtained by rotating a curve).   Consider the surface shown in Figure 5.2.3, which is obtained by rotating the curve of equation $(1 - x)^3 = z^2$ in the (x, z)-plane around the z-axis. This surface has the equation $\big(1 - \sqrt{x^2 + y^2}\big)^3 = z^2$. The curve can be parametrized by

$$t \mapsto \begin{pmatrix} x = 1 - t^2 \\ z = t^3 \end{pmatrix}, \tag{5.2.12}$$

so the surface can be parametrized by

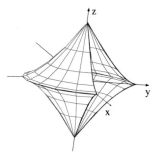

FIGURE 5.2.3.

The surface obtained by rotating the curve of equation

$$(1 - x)^3 = z^2$$

around the z-axis.   The surface drawn corresponds to the region where $|z| \le 1$, rotated only three quarters of a full turn.

If you take any equation representing a curve in the (x, z)-plane and replace x by $\sqrt{x^2 + y^2}$, you get the equation of the surface obtained by rotating the original curve around the x-axis. The surface is symmetric about the z-axis; if $y = 0$, the two equations are identical.

$$\begin{pmatrix} t \\ \theta \end{pmatrix} \mapsto \begin{pmatrix} (1 - t^2)\cos\theta \\ (1 - t^2)\sin\theta \\ t^3 \end{pmatrix}. \qquad\qquad 5.2.13$$

The exercises contain further examples of parametrizations.

Figure 5.2.3 represents the image of this parametrization for $|t| \leq 1$, $0 \leq \theta \leq 3\pi/2$. It can be guessed from the picture (and proved from the formula) that the subset of $[-1, 1] \times [0, 3\pi/2]$ where $t = \pm 1$ are trouble points (they correspond to the top and bottom "cone points"), and so is the subset $\{0\} \times [0, 3\pi/2]$, which corresponds to a "curve of cusps." $\triangle$

The existence of parametrizations

Since our entire theory of integrals over manifolds will be based on parametrizations, it would be nice to know that manifolds, or at least some fairly large class of manifolds, actually can be parametrized.

Remark. There is here some ambiguity as to what "actually" means. In the above examples, we came up with a formula for the parametrizing map, and that is what you would always like, especially if you want to evaluate an integral. Unfortunately, when a manifold is given by equations (the usual situation), it is usually impossible to find formulas for parametrizations; the parametrizing mappings only exist in the sense that the implicit function theorem guarantees their existence. If you really want to know the value of the mapping at a point, you will need to solve a system of nonlinear equations, presumably using Newton's method; you will not be able to find a formula. $\triangle$

Recall (Definition 1.5.17) that if $A \subset \mathbb{R}^n$ is a subset, the *closure* of A, denoted $\overline{A}$, is the set of all limits of sequences in A that converge in $\mathbb{R}^n$.

Theorem 5.2.6 (What manifolds can be parametrized). *Let $M \subset \mathbb{R}^n$ be a k-dimensional manifold, such that there are finitely many open subsets $U_i \subset M$ covering M, corresponding subsets $V_i \subset \mathbb{R}^k$ all with boundaries of k-dimensional volume 0, and continuous mappings $\gamma_i : \overline{V}_i \to \overline{M}$ which are one to one on V_i, with derivatives which are also one to one. Then M can be parametrized.*

It is rather hard to think of any manifold that does not satisfy the hypotheses of the theorem, hence be parametrized. Any compact manifold satisfies the hypotheses, as does any open subset of a manifold with compact closure. We will assume that our manifolds can all be parametrized. The proof of this theorem is technical and not very interesting; we do not give it in this book.

Change of parametrization

Our theory of integration over manifolds will be set up in terms of parametrizations, but of course we want the quantities computed (arc length, surface area, fluxes of vector fields, etc.), to depend only on the manifold and the integrand, not the chosen parametrization. In the next three sections we show that the length of a curve, the area of a surface, and, more generally, the volume of a manifold, are independent of the parametrization used in computing the length, area, or volume. In all three cases, the tool we use is the change of variables formula for improper integrals, Theorem 4.11.16: we set up a change of variables mapping and apply the change of variables formula to it. We need to justify this procedure, by showing that our change of variables mapping is something to which the change of variables formula can be applied: i.e., that satisfies the hypotheses of Theorem 4.11.16.

Recall that Theorem 4.11.16, the change of variables for improper integrals, says nothing about the behavior of the change of variables map on the boundary of its domain. This is important since often, as shown in Example 5.2.7, the mapping is not defined on the boundary. If we had only our earlier version of the change of variables formula (Theorem 4.10.12), we would not be able to use it to justify our claim that integration does not depend on the choice of parametrization.

Suppose we have a k-dimensional manifold M and two parametrizations

$$\gamma_1 : \overline{U}_1 \to M \quad \text{and} \quad \gamma_2 : \overline{U}_2 \to M, \qquad 5.2.14$$

where U_1 and U_2 are subsets of $\mathbb{R}^k$. Our candidate for the change of variables mapping is $\Phi = \gamma_2^{-1} \circ \gamma_1$, i.e.,

$$\overline{U}_1 \underset{\gamma_1}{\to} M \underset{\gamma_2^{-1}}{\to} \overline{U}_2. \qquad 5.2.15$$

But this mapping can have serious difficulties, as shown by the following example.

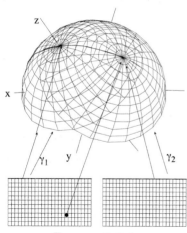

FIGURE 5.2.4.

The black dot in the left rectangle is mapped by γ_1 to a point in the hemisphere that is a pole for the parametrization γ_2; the inverse mapping γ_2^{-1} then maps that point to an entire line segment. To avoid this kind of problem, we must define our change of variables mapping carefully.

Example 5.2.7 (Problems when changing parametrizations). Let γ_1 and γ_2 be two parametrizations of S^2 by spherical coordinates, but with different poles. Call P_1, P_1' the poles for γ_1 and P_2, P_2' the poles for γ_2. Then $\gamma_2^{-1} \circ \gamma_1$ is not defined at $\gamma_1^{-1}(P_2)$ or at $\gamma_1^{-1}(P_2')$. Indeed, some single point in the domain of γ_1 maps to P_2.[1] But as shown in Figure 5.2.4, γ_2 maps a whole segment to P_2, so that $\gamma_2^{-1} \circ \gamma_1$ maps a point to a line segment, which is nonsense. The only way to deal with this is to remove $\gamma_1^{-1}(\{P_2, P_2'\})$ from the domain of $\Phi = \gamma_2^{-1} \circ \gamma_1$, and hope that the boundary still has k-volume 0. In this case this is no problem: we just removed two points from the domain, and two points certainly have area 0.

Definition 5.2.2 of a parametrization was carefully calculated to make the analogous statement true in general.

Let us set up our change of variables with a bit more precision. Let U_1 and U_2 be subsets of $\mathbb{R}^k$. Following the notation of Definition 5.2.2, denote by X_1 the negligible "trouble spots" of γ_1, and by X_2 the trouble spots of γ_2. In Example 5.2.7, X_1 and X_2 consist of the points that are mapped to the poles (i.e., the lines marked in bold in Figure 5.2.4).

_____

[1]If P_2 happens to be on the date line with respect to γ_1, two points map to P_2: in Figure 5.2.4, a point on the right-hand boundary of the rectangle, and the corresponding point on the left-hand boundary.

Theorem 4.11.16 (change of variables for improper integrals): Let U and V be open subsets of $\mathbb{R}^n$ whose boundaries have volume 0, and $\Phi : U \to V$ a C^1 diffeomorphism, with locally Lipschitz derivative. If $f : V \to \overline{\mathbb{R}}$ is an integrable function, then

$$(f \circ \Phi)|\det[\mathbf{D}\Phi]|$$

is also integrable, and

$$\int_V f(\mathbf{v})|d^n\mathbf{v}| =$$
$$\int_U (f \circ \Phi)(\mathbf{u})|\det[\mathbf{D}\Phi(\mathbf{u})]|\,|d^n\mathbf{u}|.$$

Call

$$Y_1 = (\gamma_2^{-1} \circ \gamma_1)(X_1), \quad \text{and} \quad Y_2 = (\gamma_1^{-1} \circ \gamma_2)(X_2). \qquad 5.2.16$$

In Figure 5.2.4, the dark dot in the rectangle at left is Y_2, which is mapped by γ_1 to a pole of γ_2 and then by γ_2^{-1} to the dark line at right; Y_1 is the (unmarked) dot in the right rectangle that maps to the pole of γ_1.

Set

$$U_1^{ok} = U_1 - (X_1 \cup Y_2) \quad \text{and} \quad U_2^{ok} = U_2 - (X_2 \cup Y_1); \qquad 5.2.17$$

i.e., we use the superscript "ok" to denote the domain or range of a change of mapping with any trouble spots of volume 0 removed.

Theorem 5.2.8. *Both U_1^{ok} and U_2^{ok} are open subsets of $\mathbb{R}^k$ with boundaries of k-dimensional volume 0, and*

$$\Phi : U_1^{ok} \to U_2^{ok} = \gamma_2^{-1} \circ \gamma_1$$

is a C^1 diffeomorphism with locally Lipschitz inverse.

Theorem 5.2.8 says that Φ is something to which the change of variables formula applies. It is proved in Appendix A.19.

5.3 Arc Length

Sometimes the element of arc length is denoted dl or ds.

Note that

$$|\vec{v}| = \sqrt{\det\left(\vec{v}^\top\vec{v}\right)},$$

so that Equation 5.3.1 is a special case of Equation 5.1.10.

Archimedes (287-212 BC) used this process to prove that

$$223/71 < \pi < 22/7.$$

In his famous paper *The Measurement of the Circle*, he approximated the circle by an inscribed and a circumscribed 96-sided regular polygon. That was the beginning of integral calculus.

The integrand $|d^1\mathbf{x}|$, called the *element of arc length*, is an integrand to be integrated over curves. As such, it should take a 1-parallelogram $P_\mathbf{x}(\vec{v})$ in $\mathbb{R}^n$ (i.e., a line segment) and return a number, and that is what it does:

$$|d^1\mathbf{x}|\big(P_\mathbf{x}(\vec{v})\big) = |\vec{v}|. \qquad 5.3.1$$

More generally, if f is a function on $\mathbb{R}^n$, then the integrand $f|d^1\mathbf{x}|$ is defined by the formula

$$f|d^1\mathbf{x}|\big(P_\mathbf{x}(\vec{v})\big) = f(\mathbf{x})|\vec{v}|. \qquad 5.3.2$$

If $C \subset \mathbb{R}^n$ is a smooth curve, the integral

$$\int_C |d^1\mathbf{x}| \qquad 5.3.3$$

is the number obtained by the following process: approximate C by little line segments as in Figure 5.3.1, apply $|d^1\mathbf{x}|$ to each to get its length, and add. Then let the approximation become infinitely fine; the limit is by definition the length of C.

In Section 3.8, we carried out this computation when I is the interval $[a,b]$, and $C \subset \mathbb{R}^3$ is a smooth curve parametrized by $\gamma : I \to \mathbb{R}^3$, and showed that the limit is given by

$$\int_I |d^1\mathbf{x}|P_{\gamma(t)}(\gamma'(t))|\,dt = \int_I |\vec{\gamma}'(t)|\,|dt|. \qquad 5.3.4$$

In particular, the arc length

$$\int_I |\vec{\gamma}'(t)|\,|dt| \qquad\qquad 5.3.5$$

depends only on C and not on the parametrization.

Example 5.3.1 (Graph of function). The graph of a C^1 function $f(x)$, for $a \leq x \leq b$, is parametrized by

$$x \mapsto \begin{pmatrix} x \\ f(x) \end{pmatrix}, \qquad\qquad 5.3.6$$

and hence its arc length is given by the integral

$$\int_{[a,b]} \left\| \begin{bmatrix} 1 \\ f'(x) \end{bmatrix} \right\| |dx| = \int_a^b \sqrt{1 + (f'(x))^2}\,dx. \qquad 5.3.7$$

FIGURE 5.3.1.

A curve approximated by an inscribed polygon, shown already in Section 3.8.

Because of the square root, these integrals tend to be unpleasant or impossible to calculate in elementary terms. The following example, already pretty hard, is still one of the simplest. The length of the arc of parabola $y = ax^2$ for $0 \leq x \leq A$ is given by

$$\int_0^A \left\| \begin{bmatrix} 1 \\ f'(x) \end{bmatrix} \right\| |dx| = \int_0^A \left\| \begin{bmatrix} 1 \\ 2ax \end{bmatrix} \right\| |dx| = \int_0^A \sqrt{1 + 4a^2x^2}\,dx. \qquad 5.3.8$$

A table of integrals will tell you that

$$\int \sqrt{1 + u^2}\,du = \frac{u}{2}\sqrt{u^2 + 1} + \frac{1}{2}\ln\left|u + \sqrt{1 + u^2}\right|. \qquad 5.3.9$$

Setting $2ax = u$, so that $dx = \frac{du}{2a}$, this leads to

$$\int_0^A \sqrt{1 + 4a^2x^2}\,dx = \frac{1}{4a}\left[2ax\sqrt{1 + 4a^2x^2} + \ln\left|2ax + \sqrt{1 + 4a^2x^2}\right|\right]_0^A$$

$$= \frac{1}{4a}\left(2aA\sqrt{1 + 4a^2A^2} + \ln\left|2aA + \sqrt{1 + 4a^2A^2}\right|\right). \qquad 5.3.10$$

Moral: if you want to compute arc length, brush up on your techniques of integration and dust off the table of integrals. $\triangle$

Curves in $\mathbb{R}^n$ have lengths even when $n > 3$, as Example 5.3.2 illustrates.

Example 5.3.2 (Length of a curve in $\mathbb{R}^4$). Let p, q be two integers, and consider the curve in $\mathbb{R}^4$ parametrized by

$$\gamma(t) = \begin{pmatrix} \cos pt \\ \sin pt \\ \cos qt \\ \sin qt \end{pmatrix}, \qquad 0 \leq t \leq 2\pi. \qquad 5.3.11$$

Its length is given by

The curve of Example 5.3.2 is important in mechanics, as well as in geometry. It is contained in the unit sphere $S^3 \subset \mathbb{R}^4$, and in this sphere is knotted when p and q are both greater than 1 and are relatively prime (share no common factor).

$$\int_0^{2\pi} \sqrt{(-p\sin pt)^2 + (p\cos pt)^2 + (-q\sin qt)^2 + (q\cos qt)^2}\, dt \qquad 5.3.12$$

$$= 2\pi\sqrt{p^2 + q^2}. \quad \triangle$$

We can also measure data other than pure arc length, using the integral

$$\int_C f(\mathbf{x})\,|d^1\mathbf{x}| \stackrel{\text{def}}{=} \int_I f(\gamma(t))\,|\vec{\gamma}'(t)|\,|dt|, \qquad 5.3.13$$

Integrating the length of the velocity vector, $|\vec{\gamma}'(t)|$, gives us distance along the curve. Proposition 5.3.3 says that if you take the same route (the same curve) from New York to Boston two times, making good time the first, and caught in a traffic jam the second, you will go the same distance both times.

for instance if $f(\mathbf{x})$ gives the density of a wire of variable density, the integral above would give the mass of the wire.

In other contexts (particularly surface area), it will be much harder to define the analogs of "arc length" independently of a parametrization. So here we give a direct proof that the arc length given by Equation 5.3.5 does not depend on the chosen parametrization; later we will adapt this proof to harder problems.

Proposition 5.3.3 (Arc length independent of parametrization).
Suppose $\gamma_1 : I_1 \to \mathbb{R}^3$ and $\gamma_2 : I_2 \to \mathbb{R}^3$ are two parametrizations of the same curve $C \subset \mathbb{R}^3$. Then

$$\int_{I_1} |\vec{\gamma_1}'(t_1)|\,|dt_1| = \int_{I_2} |\vec{\gamma_2}'(t_2)|\,|dt_2|. \qquad 5.3.14$$

Example 5.3.4 (Parametrizing a half-circle). We can parametrize the upper half of the unit circle by

In Equation 5.3.16 we write $\int_{[-1,1]}$ rather than $\int_{-1}^1$ because we are not concerned with orientation: it doesn't matter whether we go from -1 to 1, or from 1 to -1. For the same reason we write $|dx|$ not dx.

$$x \mapsto \begin{pmatrix} x \\ \sqrt{1-x^2} \end{pmatrix}, -1 \le x \le 1, \qquad \text{or by} \qquad t \mapsto \begin{pmatrix} \cos t \\ \sin t \end{pmatrix}, 0 \le t \le \pi. \quad 5.3.15$$

In both cases we get length π. With the first parametrization we get

$$\int_{[-1,1]} \left| \begin{bmatrix} 1 \\ \frac{-x}{\sqrt{1-x^2}} \end{bmatrix} \right| |dx| = \int_{[-1,1]} \sqrt{1 + \frac{x^2}{1-x^2}}\, |dx| \qquad 5.3.16$$

$$= \int_{[-1,1]} \frac{1}{\sqrt{1-x^2}}\, |dx| = [\arcsin x]_{-1}^1 = \frac{\pi}{2} - \left(-\frac{\pi}{2}\right) = \pi.$$

The second gives

$$\int_{[0,\pi]} \left| \begin{bmatrix} -\sin t \\ \cos t \end{bmatrix} \right| |dt| = \int_{[0,\pi]} \sqrt{\sin^2 t + \cos^2 t}\, |dt|$$

$$= \int_{[0,\pi]} |dt| = \pi. \quad \triangle \qquad 5.3.17$$

This map Φ goes from an open subset of I_1 to an open subset of I_2, with both boundaries of one-dimensional volume 0. In fact, the boundaries consist of finitely many points. This is an easy case of the harder general case discussed in Section 5.2: defining Φ so that the change of variables formula applies to it.

Proof of Proposition 5.3.3.   Denote by $\Phi = \gamma_2^{-1} \circ \gamma_1 : I_1^{ok} \to I_2^{ok}$ the "change of parameters map" such that $\Phi(t_1) = t_2$. This map Φ goes from an open subset of I_1 to an open subset of I_2 by way of the curve; γ_1 takes a point of I_1 to the curve, and then γ_2^{-1} takes it "backwards" from the curve to I_2.

Substituting γ_2 for $\mathbf{f}$, Φ for $\mathbf{g}$ and t_1 for $\mathbf{a}$ in Equation 1.8.12 of the chain rule gives

$$[\mathbf{D}(\gamma_2 \circ \Phi)(t_1)] = [\mathbf{D}\gamma_2(\Phi(t_1))][\mathbf{D}\Phi(t_1)]. \qquad 5.3.18$$

Since

$$\gamma_2 \circ \Phi = \gamma_2 \circ \overbrace{\gamma_2^{-1} \circ \gamma_1}^{\Phi} = \gamma_1, \qquad 5.3.19$$

we can substitute γ_1 for $(\gamma_2 \circ \Phi)$ in Equation 5.3.18 to get

$$\underbrace{[\mathbf{D}\gamma_1(t_1)]}_{\vec{\gamma_1}'(t_1)} = \underbrace{[\mathbf{D}\gamma_2(\Phi(t_1))]}_{\vec{\gamma_2}'(t_2)} \underbrace{[\mathbf{D}\Phi(t_1)]}_{(\gamma_2^{-1} \circ \gamma_1)'}. \qquad 5.3.20$$

Recall the change of variables formula:

$$\int_V f(\mathbf{v})|d^n\mathbf{v}| =$$

$$\int_U (f \circ \Phi)(\mathbf{u})|\det[\mathbf{D}\Phi(\mathbf{u})]\,|d^n\mathbf{u}|.$$

In Equation 5.3.23, $|\vec{\gamma_2}'|$ plays the role of f.

Recall that $\Phi(t_1) = t_2$.

Since $[\mathbf{D}\Phi(t_1)]$ is a number,

$$\det[\mathbf{D}\Phi(t_1)] = [\mathbf{D}\Phi(t_1)].$$

The second equality is Equation 5.3.22, and the third is Equation 5.3.21.

Note that the matrices

$$[\mathbf{D}\gamma_1(t_1)] = \vec{\gamma_1}'(t_1) \quad \text{and} \quad [\mathbf{D}\gamma_2(\Phi(t_1))] = \vec{\gamma_2}'(\Phi(t_1)) = \vec{\gamma_2}'(t_2) \qquad 5.3.21$$

are really column vectors (they go from $\mathbb{R}$ to $\mathbb{R}^3$) and that $[\mathbf{D}\Phi(t_1)]$, which goes from $\mathbb{R}$ to $\mathbb{R}$, is a 1×1 matrix, i.e., really a number. So when we take absolute values, Proposition 1.4.11 gives an equality, not an inequality:

$$\left|[\mathbf{D}\gamma_1(t_1)]\right| = \left|[\mathbf{D}\gamma_2(\Phi(t_1))]\right| \left|[\mathbf{D}\Phi(t_1)]\right|. \qquad 5.3.22$$

We now apply the change of variables formula (Theorem 4.10.12), to get

$$\int_{I_2} |\vec{\gamma_2}'(t_2)|\,|dt_2| = \int_{I_1} |[\mathbf{D}\gamma_2(\Phi(t_1))]|\,|\overbrace{[\mathbf{D}\Phi(t_1)]}^{\det[\mathbf{D}\Phi(t_1)]}|\,|dt_1| \qquad 5.3.23$$

$$= \int_{I_1} |[\mathbf{D}\gamma_1(t_1)]|\,|dt_1| = \int_{I_1} |\vec{\gamma_1}'(t_1)|\,|dt_1|. \quad \square$$

5.4 SURFACE AREA

Exercise 5.4.1 asks you to show that Equation 5.4.1 is another way of stating Equation 5.4.2.

The "element of area," which we denote $|d^2\mathbf{x}|$, is often denoted dA.

The integrand $|d^2\mathbf{x}|$ takes a parallelogram $P_{\mathbf{x}}(\vec{\mathbf{v}}_1, \vec{\mathbf{v}}_2)$ and returns its area. In $\mathbb{R}^3$, this means

$$|d^2\mathbf{x}|\big(P_{\mathbf{x}}(\vec{\mathbf{v}}_1, \vec{\mathbf{v}}_2)\big) = |\vec{\mathbf{v}}_1 \times \vec{\mathbf{v}}_2|; \qquad 5.4.1$$

the general formula, which works in all dimensions, and which is a special case of Theorem 5.1.4, is

$$|d^2\mathbf{x}|\big(P_{\mathbf{x}}(\vec{\mathbf{v}}_1, \vec{\mathbf{v}}_2)\big) = \sqrt{\det\big([\vec{\mathbf{v}}_1, \vec{\mathbf{v}}_2]^\top [\vec{\mathbf{v}}_1, \vec{\mathbf{v}}_2]\big)}. \qquad 5.4.2$$

We speak of triangles rather than parallelograms for the same reason that you would want a three-legged stool rather than a chair if your floor were uneven. You can make all three vertices of a triangle touch a curved surface, which you can't do for the four vertices of a parallelogram.

To integrate $|d^2\mathbf{x}|$ over a surface, we wish to break up the surface into little parallelograms, add their areas, and take the limit as the decomposition becomes fine. Similarly, to integrate $f|d^2\mathbf{x}|$ over a surface, we go through the same process, except that instead of adding areas we add $f(\mathbf{x}) \times$ area of parallelogram.

But while it is quite easy to define arc length as the limit of the length of inscribed polygons, and not much harder to prove that Equation 5.3.5 computes it, it is much harder to define surface area. In particular, the obvious idea of taking the limit of the area of inscribed triangles as the triangles become smaller and smaller only works if we are careful to prevent the triangles from becoming skinny as they get small, and then it isn't obvious that such inscribed polyhedra exist at all (see Exercise 5.4.13). The difficulties are not insurmountable, but they are daunting.

Instead, we will use Equation 5.4.3 as our definition of surface area. Since this depends on a parametrization, Proposition 5.4.4, the analog of Proposition 5.3.3, becomes not a luxury but an essential step in making surface area well defined.

Definition 5.4.1 (Surface area). Let $S \subset \mathbb{R}^3$ be a smooth surface parametrized by $\gamma : U \to S$, where U is an open subset of $\mathbb{R}^2$. Then the area of S is

$$\int_U |d^2\mathbf{x}| \Big(P_{\gamma(\mathbf{u})} \big(\vec{D_1\gamma}(\mathbf{u}), \vec{D_2\gamma}(\mathbf{u}) \big) \Big) |d^2\mathbf{u}| = \int_U \sqrt{\det\Big([\mathbf{D}\gamma(\mathbf{u})]^\top [\mathbf{D}\gamma(\mathbf{u})]\Big)} \, |d^2\mathbf{u}|.$$

5.4.3

Let us see why this ought to be right. The area should be

$$\lim_{N \to \infty} \sum_{C \in \mathcal{D}_N(\mathbb{R}^2)} \text{Area of } \gamma(C \cap U).$$

5.4.4

That is, we make a dyadic decomposition of $\mathbb{R}^2$ and see how γ maps to S the dyadic squares C that are in U or straddle it. We then sum the areas of $\gamma(C \cap U)$, which, for $C \subset U$, is the same as $\gamma(C)$; for C that straddle U, we add to the sum the area of the part of C that is in U.

The side length of a square C is $1/2^N$, so at least when $C \subset U$, the set $\gamma(C \cap U)$ is, as shown in Figure 5.4.1, approximately the parallelogram

$$P_{\gamma(\mathbf{u})} \Big(\frac{1}{2^N} \vec{D_1\gamma}(\mathbf{u}), \frac{1}{2^N} \vec{D_2\gamma}(\mathbf{u}) \Big),$$

5.4.5

where $\mathbf{u}$ is the lower left hand corner of C.

That parallelogram has area

$$\frac{1}{2^{2N}} \sqrt{\det[\mathbf{D}\gamma(\mathbf{u})]^\top [\mathbf{D}\gamma(\mathbf{u})]}.$$

5.4.6

So it seems reasonable to expect that the error we make by replacing

$$\text{Area of } \gamma(C \cap U) \quad \text{by} \quad \text{vol}_2(C) \sqrt{\det[\mathbf{D}\gamma(\mathbf{u})]^\top [\mathbf{D}\gamma(\mathbf{u})]}$$

5.4.7

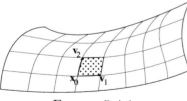

FIGURE 5.4.1.

A surface approximated by parallelograms. The point $\mathbf{x}_0$ corresponds to $\gamma(\mathbf{u})$, and the vectors $\vec{v}_1$ and $\vec{v}_2$ correspond to the vectors

$$\frac{1}{2^N} \vec{D_1\gamma}(\mathbf{u}) \quad \text{and} \quad \frac{1}{2^N} \vec{D_2\gamma}(\mathbf{u}).$$

will disappear in the limit as $N \to \infty$. And the area given by Equation 5.4.7 is precisely a Riemann sum for the integral giving surface area:

$$\lim_{N \to \infty} \sum_{C \in \mathcal{D}_N(\mathbb{R}^2)} \mathrm{vol}_2\, C \sqrt{\det[\mathbf{D}\gamma(\mathbf{u})]^\top [\mathbf{D}\gamma(\mathbf{u})]} = \underbrace{\int_U \sqrt{\det[\mathbf{D}\gamma(\mathbf{u})]^\top [\mathbf{D}\gamma(\mathbf{u})]}\, |d^2\mathbf{u}|}_{\text{area of surface by Eq. 5.4.3}}.$$

5.4.8

Unfortunately, this argument isn't entirely convincing. The parallelograms above can be imagined as some sort of tiling of the surface, gluing small flat tiles at the corners of a grid drawn on the surface, a bit like using ceramic tiles to cover a curved counter. It is true that we get a better and better fit by choosing smaller and smaller tiles, but is it good enough? Our definition involves a parametrization γ; only when we have shown that surface area is independent of parametrization can we be sure that Definition 5.4.1 is correct. We will verify this after computing a couple of examples of surface integrals.

Example 5.4.2 (Area of a torus). Choose $R > r > 0$. We obtain the torus shown in Figure 5.4.2 by taking the circle of radius r in the (x, z)-plane that is centered at $x = R$, $z = 0$, and rotating it around the z-axis.

This surface is parametrized by

$$\gamma\begin{pmatrix} u \\ v \end{pmatrix} = \begin{pmatrix} (R + r \cos u) \cos v \\ (R + r \cos u) \sin v \\ r \sin u \end{pmatrix},$$

5.4.9

as Exercise 5.4.2 asks you to verify. Then the surface area of the torus is given by the integral[2]

$$\int_{[0,2\pi] \times [0,2\pi]} \left| \overbrace{\begin{bmatrix} -r \sin u \cos v \\ -r \sin u \sin v \\ r \cos u \end{bmatrix}}^{\overrightarrow{D_1\gamma}} \times \overbrace{\begin{bmatrix} -(R + r \cos u) \sin v \\ (R + r \cos u) \cos v \\ 0 \end{bmatrix}}^{\overrightarrow{D_2\gamma}} \right| |du\, dv|$$

$$= \int_{[0,2\pi] \times [0,2\pi]} r(R + r \cos u) \overbrace{\sqrt{(\sin u)^2 + (\cos u \sin v)^2 + (\cos u \cos v)^2}}^{=1} |du\, dv|$$

$$= r \int_0^{2\pi} \int_0^{2\pi} (R + r \cos u)\, du\, dv = 4\pi^2 rR. \quad \triangle$$

5.4.10

Example 5.4.3 (Surface area: a harder problem). What is the area of the graph of the function $x^2 + y^3$ above the unit square $Q \subset \mathbb{R}^2$? Applying Equation 5.2.5, we parametrize the surface by

$$\gamma\begin{pmatrix} x \\ y \end{pmatrix} \mapsto \begin{pmatrix} x \\ y \\ x^2 + y^3 \end{pmatrix}, \quad \text{and apply Equation 5.4.3:}$$

---

[2]Here and in Example 5.4.3 we compute the area using the length of the cross-product, which for surface area is equivalent to the general formula $\sqrt{\det T^\top T}$ (see Proposition 1.4.19) and perhaps a little simpler to compute.

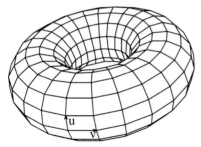

FIGURE 5.4.2.

The torus with the u and v coordinates drawn. You should imagine the straight lines as curved. By "torus" we mean the surface of the object. The solid object is called a solid torus.

We write $|du\, dv|$ in Equation 5.4.10 to avoid having to put subscripts on our variables; we could have used u_1 and u_2 rather than u and v, and then used the integrand $|d^2\mathbf{u}|$.

In the final, double integral, we are integrating over an *oriented* interval, from 0 to 2π, so we write $du\, dv$ rather than $|du\, dv|$.

Note that the answer has the right units: r and R have units of length so $4\pi^2 rR$ has units length squared.

In Example 5.4.2 the square root that inevitably shows up (since we are computing the length of a vector) was simply $\sqrt{1}$. It is exceptional that the square root of any function can be integrated in elementary terms. Example 5.4.3 is more typical.

$$\int_Q \left| \overbrace{\begin{bmatrix} 1 \\ 0 \\ 2x \end{bmatrix}}^{\overrightarrow{D_1\gamma}} \times \overbrace{\begin{bmatrix} 0 \\ 1 \\ 3y^2 \end{bmatrix}}^{\overrightarrow{D_2\gamma}} \right| |dx\,dy| = \int_0^1 \int_0^1 \sqrt{1 + 4x^2 + 9y^4}\,dx\,dy. \qquad 5.4.11$$

Again, on the right-hand side of Equation 5.4.11 we drop the absolute value signs, writing $dx\,dy$, because we now have an oriented interval, from 0 to 1: $\int_0^1$.

The integral with respect to x is one we can calculate (just barely in our case, checking our result with a table of integrals). First we get

$$\int \sqrt{u^2 + a^2}\,du = \frac{u\sqrt{u^2 + a^2}}{2} + \frac{a^2 \ln(u + \sqrt{u^2 + a^2})}{2}. \qquad 5.4.12$$

This leads to the integral

$$\int_0^1 \int_0^1 \sqrt{1 + 4x^2 + 9y^4}\,dx\,dy$$

$$= \int_0^1 \left[\frac{x\sqrt{4x^2 + 1 + 9y^4}}{2} + \frac{1 + 9y^4 \ln(2x + \sqrt{4x^2 + 1 + 9y^4})}{4} \right]_0^1 dy$$

$$= \int_0^1 \left(\frac{\sqrt{5 + 9y^4}}{2} + \frac{1 + 9y^4}{4} \ln \frac{2 + \sqrt{5 + 9y^4}}{\sqrt{1 + 9y^4}} \right) dy. \qquad 5.4.13$$

Even Example 5.4.3 is computationally nicer than is standard: we were able to integrate with respect to x. If we had asked for the area of the graph of $x^3 + y^4$, we couldn't have integrated in elementary terms with respect to either variable, and would have needed the computer to evaluate the double integral.

And what's wrong with that? Integrals exist whether or not they can be computed in elementary terms, and a fear of numerical integrals is inappropriate in this age of computers. If you restrict yourself to surfaces whose areas can be computed in elementary terms, you are restricting yourself to a minute class of surfaces.

It is hopeless to try to integrate this mess in elementary terms: the first term requires elliptic functions, and we don't know of any class of special functions in which the second term could be expressed. But numerically, this is no big problem; Simpson's method with 20 steps gives the approximation $1.93224957\ldots$. $\triangle$

Surface area is independent of the choice of parametrization

As shown by Exercise 5.4.13, it is quite difficult to give a rigorous definition of surface area that does not rely on a parametrization. In Definition 5.4.1 we defined surface area using a parametrization; now we need to show that two different parametrizations of the same surface give the same area.

Like Proposition 5.3.3 (the analogous statement for curves), this is an application of the change of variables formula.

Proposition 5.4.4 (Surface area independent of parametrization).
Let S be a smooth surface in $\mathbb{R}^3$ and $\gamma_1 : U \to \mathbb{R}^3$, $\gamma_2 : V \to \mathbb{R}^3$ be two parametrizations. Then

$$\int_U \sqrt{\det\big([\mathbf{D}\gamma_1(\mathbf{u})]^\top[\mathbf{D}\gamma_1(\mathbf{u})]\big)}\,|d^2\mathbf{u}| = \int_V \sqrt{\det\big([\mathbf{D}\gamma_2(\mathbf{v})]^\top[\mathbf{D}\gamma_2(\mathbf{v})]\big)}\,|d^2\mathbf{v}|.$$
$$5.4.14$$

Proof. We begin as we did with the proof of Proposition 5.3.3. Define $\Phi = \gamma_2^{-1} \circ \gamma_1 : U^{ok} \to V^{ok}$ to be the "change of parameters" map such that $\mathbf{v} = \Phi(\mathbf{u})$.

Notice that the chain rule applied to the equation

$$\gamma_1 = \gamma_2 \circ \underbrace{\gamma_2^{-1} \circ \gamma_1}_{\Phi} = \gamma_2 \circ \Phi$$

gives

$$[\mathbf{D}(\gamma_2 \circ \Phi)(\mathbf{u})] = [\mathbf{D}\gamma_1(\mathbf{u})] = [\mathbf{D}\gamma_2(\Phi(\mathbf{u}))][\mathbf{D}\Phi(\mathbf{u})]. \qquad 5.4.15$$

To go from line one to line two of Equation 5.4.16 we use the change of variables formula. To go from line two to line three we use the fact that $[\mathbf{D}\Phi(\mathbf{u})]$ is a square matrix, and that for a square matrix A, $\det A = \det A^\top$, so

$$[\mathbf{D}\Phi(\mathbf{u})] = \sqrt{\det[\mathbf{D}\Phi(\mathbf{u})]^\top[\mathbf{D}\Phi(\mathbf{u})]},$$

To go from line three to four first replace $\det AB$ by $\det B \det A$ and then remember that each of these dets is a number, so they can be multiplied in any order. To go from line four to line five we use the chain rule (Equation 5.4.15).

If we apply the change of variables formula, we find

$$\int_V \sqrt{\det[\mathbf{D}\gamma_2(\mathbf{v})]^\top[\mathbf{D}\gamma_2(\mathbf{v})])} \,|d^2\mathbf{v}|$$

$$= \int_U \sqrt{\det([\mathbf{D}\gamma_2(\Phi(\mathbf{u}))]^\top[\mathbf{D}\gamma_2(\Phi(\mathbf{u}))])} \,\left|\det[\mathbf{D}\Phi(\mathbf{u})]\right| \,|d^2\mathbf{u}|$$

$$= \int_U \sqrt{\det([\mathbf{D}\gamma_2(\Phi(\mathbf{u}))]^\top[\mathbf{D}\gamma_2(\Phi(\mathbf{u}))])} \,\sqrt{\det([\mathbf{D}\Phi(\mathbf{u})]^\top[\mathbf{D}\Phi(\mathbf{u})])} \,|d^2\mathbf{u}|$$

$$= \int_U \sqrt{\det([\mathbf{D}\Phi(\mathbf{u})]^\top[\mathbf{D}\gamma_2(\Phi(\mathbf{u}))]^\top[\mathbf{D}\gamma_2(\Phi(\mathbf{u}))][\mathbf{D}\Phi(\mathbf{u})])} \,|d^2\mathbf{u}|$$

$$= \int_U \sqrt{\det([\mathbf{D}(\gamma_2 \circ \Phi)(\mathbf{u})]^\top[\mathbf{D}(\gamma_2 \circ \Phi)(\mathbf{u})])} \,|d^2\mathbf{u}| \qquad 5.4.16$$

$$= \int_U \sqrt{\det[\mathbf{D}\gamma_1(\mathbf{u})])^\top[\mathbf{D}\gamma_1(\mathbf{u})])} \,|d^2\mathbf{u}|. \quad \square$$

Areas of surfaces in $\mathbb{R}^n$, $n > 3$

A surface (i.e., a two-dimensional manifold) embedded in $\mathbb{R}^n$ should have an area for any n, not just for $n = 3$.

A first difficulty is that it is hard to imagine such surfaces, and perhaps impossible to visualize them. But it isn't particularly hard to describe them mathematically.

For instance, the subset of $\mathbb{R}^4$ given by the two equations $x_1^2 + x_2^2 = r_1^2$, $x_3^2 + x_4^2 = r_2^2$, is a surface; it corresponds to two equations in four unknowns. This surface is discussed in Example 5.4.5. More generally, we saw in Section 3.2 that the set $X \subset \mathbb{R}^n$ defined by the $n - k$ equations in n variables

$$f_1\begin{pmatrix} x_1 \\ \vdots \\ x_n \end{pmatrix} = 0, \ \ldots, \ f_m\begin{pmatrix} x_1 \\ \vdots \\ x_n \end{pmatrix} = 0 \qquad 5.4.17$$

defines a k-dimensional manifold if $[\mathbf{Df}(\mathbf{x})] : \mathbb{R}^n \to \mathbb{R}^m$ is onto for each $\mathbf{x} \in X$.

Example 5.4.5 (Area of a surface in $\mathbb{R}^4$). The surface described above, the subset of $\mathbb{R}^4$ given by the two equations

$$x_1^2 + x_2^2 = r_1^2 \quad \text{and} \quad x_3^2 + x_4^2 = r_2^2, \qquad 5.4.18$$

is parametrized by

$$\gamma\left(\begin{matrix} u \\ v \end{matrix}\right) = \begin{pmatrix} r_1 \cos u \\ r_1 \sin u \\ r_2 \cos v \\ r_2 \sin v \end{pmatrix}, 0 \le u, v \le 2\pi, \qquad 5.4.19$$

and since

$$\left[\mathbf{D}\gamma\left(\left(\begin{matrix} u \\ v \end{matrix}\right)\right)\right]^{\top} \left[\mathbf{D}\gamma\left(\left(\begin{matrix} u \\ v \end{matrix}\right)\right)\right]$$

$$= \begin{bmatrix} -r_1 \sin u & r_1 \cos u & 0 & 0 \\ 0 & 0 & -r_2 \sin v & r_2 \cos v \end{bmatrix} \begin{bmatrix} -r_1 \sin u & 0 \\ r_1 \cos u & 0 \\ 0 & -r_2 \sin v \\ 0 & r_2 \cos v \end{bmatrix}$$

$$= \begin{bmatrix} r_1^2 & 0 \\ 0 & r_2^2 \end{bmatrix},$$

$$5.4.20$$

The statement that the area is given by Equation 5.4.21 is justified by Theorem 5.1.4 giving the volume of a k-parallelogram in $\mathbb{R}^n$, combined with Proposition 5.5.1, which says that the volume of a manifold is independent of the parametrization used. Thus Proposition 5.4.4 concerning surfaces in $\mathbb{R}^3$ is true for surfaces in $\mathbb{R}^n$.

the area of the surface is given by

$$\int_{[0,2\pi]\times[0,2\pi]} \sqrt{\det\left(\left[\mathbf{D}\gamma\left(\begin{matrix} u \\ v \end{matrix}\right)\right]^{\top} \left[\mathbf{D}\gamma\left(\begin{matrix} u \\ v \end{matrix}\right)\right]\right)} \, |du\, dv|$$

$$5.4.21$$

$$= \int_{[0,2\pi]\times[0,2\pi]} \sqrt{r_1^2 r_2^2} \, |du\, dv| = 4\pi^2 r_1 r_2. \quad \triangle$$

Another class of surfaces in $\mathbb{R}^4$ which is important in many applications, and which leads to remarkably simpler computations than one might expect, uses complex variables. Consider for instance the graph of the function $f(z) = z^2$, where z is complex. This graph has the equation $z_2 = z_1^2$ in $\mathbb{C}^2$, or

$$x_2 = x_1^2 - y_1^2, \quad y_2 = 2x_1 y_1, \quad \text{in } \mathbb{R}^4. \qquad 5.4.22$$

Equation 5.4.3 tells us how to compute the areas of such surfaces, if we manage to parametrize them. If $S \subset \mathbb{R}^n$ is a surface, $U \subset \mathbb{R}^2$ is an open subset, and $\gamma : U \to \mathbb{R}^n$ is a parametrization of S, then the area of S is given by

$$\int_S |d^2\mathbf{x}| = \int_U \sqrt{\det([\mathbf{D}\gamma(\mathbf{u})]^{\top} [\mathbf{D}\gamma(\mathbf{u})])} \, |d^2\mathbf{u}|. \qquad 5.4.23$$

Example 5.4.6 (Area of a surface in $\mathbb{C}^2$). Let us tackle the surface in $\mathbb{C}^2$ of Equation 5.4.22. More precisely, let us compute the area of the part of the surface of equation $z_2 = z_1^2$, where $|z_1| \le 1$. Polar coordinates for z_1 give a nice way to parametrize the surface:

$$\gamma\left(\begin{matrix} r \\ \theta \end{matrix}\right) = \begin{pmatrix} r \cos \theta \\ r \sin \theta \\ r^2 \cos 2\theta \\ r^2 \sin 2\theta \end{pmatrix}, \quad 0 \le r \le 1, \quad 0 \le \theta \le 2\pi. \qquad 5.4.24$$

Again we need to compute the area of the parallelogram spanned by the two partial derivatives. Since

$$\left[\mathbf{D}\gamma\begin{pmatrix}r\\\theta\end{pmatrix}\right]^{\top}\left[\mathbf{D}\gamma\begin{pmatrix}r\\\theta\end{pmatrix}\right]$$

Notice (last line of Equation 5.4.26) that the determinant ends up being a perfect square; the square root that created such trouble in Example 5.3.1, which deals with the real curve of equation $y = x^2$ causes none for the complex surface with the same equation $z_2 = z_1^2$.

$$= \begin{bmatrix} \cos\theta & \sin\theta & 2r\cos 2\theta & 2r\sin 2\theta \\ -r\sin\theta & r\cos\theta & -2r^2\sin 2\theta & 2r^2\cos 2\theta \end{bmatrix} \begin{bmatrix} \cos\theta & -r\sin\theta \\ \sin\theta & r\cos\theta \\ 2r\cos 2\theta & -2r^2\sin 2\theta \\ 2r\sin 2\theta & 2r^2\cos 2\theta \end{bmatrix}$$

$$= \begin{bmatrix} 1 + 4r^2 & 0 \\ 0 & r^2(1 + 4r^2) \end{bmatrix}, \qquad\qquad 5.4.25$$

Equation 5.4.3 says that the area is

This "miracle" happens for *all* manifolds in $\mathbb{C}^n$ given by "complex equations," such as polynomials in complex variables. Several examples are presented in the exercises.

$$\int_{[0,1]\times[0,2\pi]}\sqrt{\det\left(\left[\mathbf{D}\gamma\begin{pmatrix}r\\\theta\end{pmatrix}\right]^{\top}\left[\mathbf{D}\gamma\begin{pmatrix}r\\\theta\end{pmatrix}\right]\right)}\,|dr\,d\theta| \qquad 5.4.26$$

$$= \int_{[0,1]\times[0,2\pi]}\underbrace{\sqrt{r^2(1+4r^2)^2}}_{\substack{\text{square root}\\\text{of a perfect square}}}\,|dr\,d\theta| = 2\pi\left[\frac{r^2}{2} + r^4\right]_0^1 = 3\pi. \quad \triangle$$

5.5 Volume of Manifolds

Everything we have done so far in this chapter works for a manifold M of any dimension k, embedded in any $\mathbb{R}^n$. The k-dimensional volume of such a manifold is written

$$\int_M |d^k\mathbf{x}|,$$

where $|d^k\mathbf{x}|$ is the integrand that takes a k-parallelogram and returns its k-dimensional volume. Heuristically, this integral is defined by cutting up the manifold into little k-parallelograms, adding their k-dimensional volumes and taking the limits of the sums as the decomposition becomes infinitely fine.

The way to do this precisely is to parametrize M, using our relaxed definition of parametrizations (Definition 5.2.2).

Then the k-dimensional volume is defined to be

The argument why this should correspond to the heuristic description is exactly the same as the one in Section 5.4, and we won't repeat it.

$$\int_U \sqrt{\det([\mathbf{D}\gamma(\mathbf{u})]^{\top}[\mathbf{D}\gamma(\mathbf{u})])}\,|d^k\mathbf{u}|. \qquad 5.5.1$$

The independence of this integral from the chosen parametrization also goes through without any change at all.

The proof of Proposition 5.5.1 is identical to the case of surfaces, given in Equation 5.4.16. The difference is the width of the matrix $[\mathbf{D}\gamma_2(\mathbf{v})]$. In the case of surfaces, $[\mathbf{D}\gamma_2(\mathbf{v})]$ is $n \times 2$; here it is $n \times k$. (Of course the transpose is also different; now it is $k \times n$.)

Proposition 5.5.1 (Volume of manifold independent of parametrization). *Let M be a k-dimensional manifold in $\mathbb{R}^n$. If U and V are subsets of $\mathbb{R}^k$ and $\gamma_1 : U \to M$, $\gamma_2 : V \to M$ are two parametrizations of M, then*

$$\int_U \sqrt{\det([\mathbf{D}\gamma_1(\mathbf{u})]^\top[\mathbf{D}\gamma_1(\mathbf{u})])} \; |d^k\mathbf{u}| = \int_V \sqrt{\det([\mathbf{D}\gamma_2(\mathbf{v})]^\top[\mathbf{D}\gamma_2(\mathbf{v})])} \; |d^k\mathbf{v}|.$$

Example 5.5.2 (Volume of a three-dimensional manifold in $\mathbb{R}^4$). Let $U \subset \mathbb{R}^3$ be an open set, and $f : U \to \mathbb{R}$ be a C^1 function. Then the graph of f is a three-dimensional manifold in $\mathbb{R}^4$, and it comes with the natural parametrization

$$\gamma \begin{pmatrix} x \\ y \\ z \end{pmatrix} = \begin{pmatrix} x \\ y \\ z \\ f \begin{pmatrix} x \\ y \\ z \end{pmatrix} \end{pmatrix}. \tag{5.5.2}$$

We then have

$$\det \left(\left[\mathbf{D}\gamma \begin{pmatrix} x \\ y \\ z \end{pmatrix} \right]^\top \left[\mathbf{D}\gamma \begin{pmatrix} x \\ y \\ z \end{pmatrix} \right] \right)$$

$$= \det \left(\begin{bmatrix} 1 & 0 & 0 & D_1 f \\ 0 & 1 & 0 & D_2 f \\ 0 & 0 & 1 & D_3 f \end{bmatrix} \begin{bmatrix} 1 & 0 & 0 \\ 0 & 1 & 0 \\ 0 & 0 & 1 \\ D_1 f & D_2 f & D_3 f \end{bmatrix} \right) \tag{5.5.3}$$

$$= \det \begin{bmatrix} 1 + (D_1 f)^2 & (D_1 f)(D_2 f) & D_1 f)(D_3 f) \\ (D_1 f)(D_2 f) & 1 + (D_2 f)^2 & D_2 f)(D_3 f) \\ D_1 f)(D_3 f) & (D_2 f)(D_3 f) & 1 + (D_3 f)^2 \end{bmatrix}$$

$$= 1 + (D_1 f)^2 + (D_2 f)^2 + (D_2 f)^2.$$

So the three-dimensional volume of the graph of f is

$$\int_U \sqrt{1 + (D_1 f)^2 + (D_2 f)^2 + (D_2 f)^2} |d^3\mathbf{x}|. \tag{5.5.4}$$

It is a challenge to find any function for which this can be integrated in elementary terms. Let us try to find the area of the graph of

$$f \begin{pmatrix} x \\ y \\ z \end{pmatrix} = \frac{1}{2}(x^2 + y^2 + z^2) \tag{5.5.5}$$

above the ball of radius R centered at the origin, $B_R(\mathbf{0})$.

Using spherical coordinates, this leads to

$$\int_{B_0(R)} \sqrt{1+(D_1f)^2+(D_2f)^2+(D_2f)^2}|d^3\mathbf{x}| = \int_{B_0(R)} \sqrt{1+x^2+y^2+z^2}\,|d^3\mathbf{x}|$$

$$= \int_0^{2\pi}\int_{-\pi/2}^{\pi/2}\int_0^R \sqrt{1+(r\cos\theta\cos\varphi)^2+(r\sin\theta\cos\varphi)^2+r^2\sin^2\varphi}\,r^2\cos\varphi\,dr\,d\varphi\,d\theta$$

$$= \int_0^{2\pi}\int_{-\pi/2}^{\pi/2}\int_0^R \sqrt{1+r^2}\,r^2\cos\varphi\,dr\,d\varphi\,d\theta = 4\pi\int_0^R \sqrt{1+r^2}\,r^2\,dr \qquad 5.5.6$$

$$= \pi\left(R(1+R^2)^{3/2} - \frac{1}{2}\ln(R+\sqrt{1+R^2}) - \frac{1}{2}R\sqrt{1+R^2}\right). \quad \triangle$$

Again, this is an integral which will stretch you ability with techniques of integration. You are asked in Exercise 5.5.5 to justify the last step of this computation.

Example 5.5.3 (Volume of an n-dimensional sphere in $\mathbb{R}^{n+1}$). For a final example, let us compute $\mathrm{vol}_n S^n$, where $S^n \subset \mathbb{R}^{n+1}$ is the unit sphere. It would be possible to do this using some generalization of spherical coordinates, and you are asked to do so for the 3-sphere in Exercise 5.5.7. These computations become quite cumbersome, and there is an easier method. It relates $\mathrm{vol}_n S^n$ to the $(n+1)$-dimensional volume of an $(n+1)$-dimensional sphere, $\mathrm{vol}_{n+1} B^{n+1}$.

First, how might we relate the *length* of a circle (i.e., a one-dimensional sphere, S^1) to the *area* of a disk (i.e., a two-dimensional ball, B^2)? We could fill up the disk with concentric rings, and add together their areas, each approximately the length of the corresponding circle times some δr representing the spacing of the circles. The length of the circle of radius r is $r \times$ the length of the circle of radius 1. More generally, this approach gives

We can compute the area of the unit disk from the length of the unit circle:

$$\mathrm{vol}_2 B^2 = \int_0^1 r\,\mathrm{vol}_1 S^1\,dr$$

$$= \int_0^1 2\pi r\,dr = \left[\frac{2\pi r^2}{2}\right]_0^1 = \pi.$$

$$\mathrm{vol}_{n+1} B^{n+1} = \int_0^1 \mathrm{vol}_n S^n(r)\,dr = \int_0^1 r^n\,\mathrm{vol}_n S^n\,dr = \frac{1}{n+1}\,\mathrm{vol}_n(S^n). \quad 5.5.7$$

The part of B^{n+1} between r and $r+\Delta r$ should have volume $\Delta r\big(\mathrm{vol}_n(S^n(r))\big)$;
This allows us to add one more column to Table 4.5.7:

n	$c_n = \frac{n-1}{n}c_{n-2}$	Volume of ball $\beta_n = c_n\beta_{n-1}$	$\mathrm{vol}_n S^n = (n+1)\beta_{n+1}$
0	π		2
1	2	2	2π
2	$\frac{\pi}{2}$	π	4π
3	$\frac{4}{3}$	$\frac{4\pi}{3}$	$2\pi^2$
4	$\frac{3\pi}{8}$	$\frac{\pi^2}{2}$	$\frac{8\pi^2}{3}$
5	$\frac{16}{15}$	$\frac{8\pi^2}{15}$	π^3

5.6 FRACTALS AND FRACTIONAL DIMENSION

In 1919, Felix Hausdorff showed that dimensions are not limited to length, area, volume, ... : we can also speak of *fractional dimension*. This discovery acquired much greater significance with the work of Benoit Mandelbrot showing that many objects in nature (the lining of the lungs, the patterns of frost on windows, the patterns formed by a film of gasoline on water, for example) are fractals, with fractional dimension.

Example 5.6.1 (Koch snowflake). We construct the Koch snowflake curve K as follows. Start with a line segment, say $0 \leq x \leq 1$, $y = 0$ in $\mathbb{R}^2$. Replace its middle third by the top of an equilateral triangle, as shown in Figure 5.6.1. This gives four segments, each one-third the length of the original segment. Now replace the middle third of each by the top of an equilateral triangle, and so on.

What is the length of this "curve"? At resolution $N = 0$, we get length 1. At resolution $N = 1$, when the curve consists of four segments, we get length $4 \cdot 1/3$. At the next resolution, the length is $16 \cdot 1/9$. As our decomposition becomes infinitely fine, the length becomes infinitely long!

"Length" is the wrong word to apply to the Koch snowflake, which is neither a curve nor a surface. It is a fractal, with fractional dimension: the Koch snowflake has dimension $\ln 4 / \ln 3 \approx 1.26$.

Let us see why this might be the case. Call A the part of the curve constructed on $[0, 1/3]$, and B the whole curve, as in Figure 5.6.2. Then B consists of four copies of A. (This is true at any level, but it is easiest to see at the first level, the top graph in Figure 5.6.1.). Therefore, in any dimension d, it should be true that $\text{vol}_d(B) = 4 \text{vol}_d(A)$.

However, if you expand A by a factor of 3, you get B. (This is true in the limit, after the construction has been carried out infinitely many times.) According to the principle that area goes as the square of the length, volume goes as the cube of the length, etc., we would expect d-dimensional volume to go as the dth power of the length, which leads to

$$\text{vol}_d(B) = 3^d \text{vol}_d(A). \qquad 5.6.1$$

If you put this equation together with $\text{vol}_d(B) = 4 \text{vol}_d(A)$, you will see that the only dimension in which the volume of the Koch curve can be different from 0 or ∞ is the one for which $4 = 3^d$, i.e., $d = \ln 4 / \ln 3$.

If we break up the Koch curve into the pieces built on the sides constructed at the nth level (of which there are 4^n, each of length $1/3^n$), and raise their side-lengths to the dth power, we find

$$4^n \left(\frac{1}{3}\right)^{n \ln 4 / \ln 3} = 4^n e^{n \frac{\ln 4}{\ln 3}(\ln \frac{1}{3})} = 4^n e^{n \frac{\ln 4}{\ln 3}(-\ln 3)} = 4^n e^{-n \ln 4} = \frac{4^n}{4^n} = 1. \quad 5.6.2$$

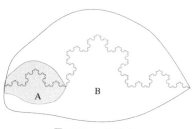

FIGURE 5.6.1.

The first five steps in constructing the Koch snowflake. Its length is infinite, but length is the wrong way to measure this fractal object.

FIGURE 5.6.2.

(In Equation 5.6.2 we use the fact that $a^x = e^{x \ln a}$.) Although the terms have not been defined precisely, you might expect the computation above to mean

$$\int_K |d\mathbf{x}^{\ln 4 / \ln 3}| = 1. \quad \triangle \qquad\qquad 5.6.3$$

Example 5.6.2 (Sierpinski gasket). While the Koch snowflake looks like a thick curve, the Sierpinski gasket looks more like a thin surface. This is the subset of the plane obtained by taking a filled triangle of side length l, removing the central inscribed subtriangle, then removing the central subtriangles from the three triangles that are left, then removing the central subtriangles from the nine triangles that are left, and so on; the process is sketched in Figure 5.6.3. We claim that this is a set of dimension $\ln 3 / \ln 2$: at the nth stage of the construction, sum, over all the little pieces, the side-length to the power p:

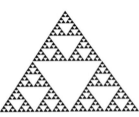

$$3^n \left(\frac{l}{2^n} \right)^p. \qquad\qquad 5.6.4$$

(If measuring length, $p = 1$; if measuring area, $p = 2$.) If the set really had a length, then the sum would converge when $p = 1$, as $n \to \infty$; in fact, the sum is infinite. If it really had an area, then the power $p = 2$ would lead to a finite limit; in fact, the sum is 0. But when $p = \ln 3 / \ln 2$, the sum converges to $l^{\ln 3 / \ln 2} \approx l^{1.58}$. This is the only dimension in which the Sierpinski gasket has finite, nonzero measure; in dimensions greater than $\ln 3 / \ln 2$, the measure is 0, and in dimensions less than $\ln 3 / \ln 2$ it is infinite. $\triangle$

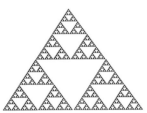

FIGURE 5.6.3.

The second, fourth, fifth and sixth steps of the Sierpinski gasket

5.7 EXERCISES FOR CHAPTER FIVE

Exercises for Section 5.1:

Parallelograms

5.1.1   What is vol_3 of the 3-parallelogram in $\mathbb{R}^4$ spanned by

$$\vec{\mathbf{v}}_1 = \begin{bmatrix} 1 \\ 0 \\ 1 \\ 1 \end{bmatrix}, \vec{\mathbf{v}}_2 = \begin{bmatrix} 0 \\ 2 \\ 1 \\ 1 \end{bmatrix}, \vec{\mathbf{v}}_3 = \begin{bmatrix} 1 \\ 1 \\ 0 \\ 2 \end{bmatrix}?$$

5.1.2   What is the volume of a parallelepiped with three sides emanating from the same vertex having lengths 1, 2, and 3, and with angles between them $\pi/3, \pi/4,$ and $\pi/6$?

Hint for Exercise 5.1.3: Show that $\mathrm{rank}(T^\top T) \leq \mathrm{rank}\, T < k$.

5.1.3   Show that if $\vec{\mathbf{v}}_1, \ldots, \vec{\mathbf{v}}_k$ are linearly dependent, $\mathrm{vol}_k(P(\vec{\mathbf{v}}_1, \ldots, \vec{\mathbf{v}}_k)) = 0$.

Exercises for Section 5.2:

Parametrizations

5.2.1   (a) Show that the segment of diagonal $\left\{ \begin{pmatrix} x \\ x \end{pmatrix} \in \mathbb{R}^2 \mid |x| \leq 1 \right\}$ does not have one-dimensional volume 0.

(b) Show that the curve in $\mathbb{R}^3$ parametrized by $t \mapsto \begin{pmatrix} \cos t \\ \cos t \\ \sin t \end{pmatrix}$ does have two-dimensional volume 0, but does not have one-dimensional volume 0.

5.2.2 Show that Proposition A19.1 is not true if X is unbounded. For instance, produce an unbounded subset of $\mathbb{R}^2$ of length 0, whose projection onto the x-axis does not have length 0.

5.2.3 Choose numbers $0 < r < R$, and consider the circle in the (x, z)-plane of radius r centered at $\left(\begin{smallmatrix} x = R \\ z = 0 \end{smallmatrix} \right)$. Let S be the surface obtained by rotating this circle.

(a) Write an equation for S, and check that it is a smooth surface.

(b) Write a parametrization for S, paying close attention to the sets U and X used.

(c) Now parametrize the part of S where
  i) $z > 0$;
  ii) $x > 0, y > 0$;
  iii) $z > x + y$. This is much harder, and even after finding an equation for the curve bounding the parametrizing region, you may need a computer to visualize it.

5.2.4 Let $f : [a, b] \to \mathbb{R}$ be a smooth positive function. Find a parametrization for the surface of equation

$$\frac{x^2}{A^2} + \frac{y^2}{B^2} = \left(f(z) \right)^2.$$

5.2.5 Show that if $M \subset \mathbb{R}^n$ has dimension less than k, then $\mathrm{vol}_k(M) = 0$.

*5.2.6 Consider the open subset of $\mathbb{R}$ constructed in Example 4.4.2: list the rationals between 0 and 1, say $a_1, a_2, a_3, \ldots$ and take the union

$$U = \bigcup_{i=10}^{\infty} \left(a_i - \frac{1}{2^{i+k}}, a_i + \frac{1}{2^{i+k}} \right)$$

for some integer $k > 1$. Show that U is a one-dimensional manifold, and that it cannot be parametrized according to Definition 5.2.2.

Exercises for Section 5.3:
Arc Length

5.3.1 (a) Let $\begin{pmatrix} r(t) \\ \theta(t) \end{pmatrix}$ be a parametrization of a curve in polar coordinates. Show that the length of the piece of curve between $t = a$ and $t = b$ is given by the integral

$$\int_a^b \sqrt{(r'(t))^2 + (r(t))^2(\theta'(t))^2} \, dt. \qquad 5.3.1$$

(b) What is the length of the spiral given in polar coordinates by

$$\begin{pmatrix} r(t) \\ \theta(t) \end{pmatrix} = \begin{pmatrix} e^{-\alpha t} \\ t \end{pmatrix}, \ \alpha > 0$$

between $t = 0$ and $t = a$? What is the limit of this length as $\alpha \to 0$?

(c) Show that the spiral turns infinitely many times around the origin as $t \to \infty$. Does the length tend to ∞ as $t \to \infty$?

5.3.2   Use Equation 5.3.1 of Exercise 5.3.1 to give the length of the curve

$$\begin{pmatrix} r(t) \\ \theta(t) \end{pmatrix} = \begin{pmatrix} 1/t \\ t \end{pmatrix},$$

between $t = 1$ and $t = a$. Is the limit of the length finite as $a \to \infty$?

5.3.3   For what values of α does the spiral

$$\begin{pmatrix} r(t) \\ \theta(t) \end{pmatrix} = \begin{pmatrix} 1/t^\alpha \\ t \end{pmatrix}, \alpha > 0$$

between $t = 1$ and $t = \infty$ have finite length?

5.3.4   (a) Suppose now that $\begin{pmatrix} r(t) \\ \varphi(t) \\ \theta(t) \end{pmatrix}$ is a parametrization of a curve in $\mathbb{R}^3$, written in spherical coordinates. Find the formula analogous to Equation 5.3.1 of Exercise 5.3.1 for the length of the arc between $t = a$ and $t = b$.

(b) What is the length of the curve parametrized by $r(t) = \cos t, \varphi(t) = t, \theta(t) = \tan t$ between $t = 0$ and $t = a$, where $0 < a < \pi/2$?

Exercises for Section 5.4:
Surface Area

5.4.1   Show that Equation 5.4.1 is another way of stating Equation 5.4.2, i.e., show that

$$|\vec{\mathbf{v}}_1 \times \vec{\mathbf{v}}_2| = \sqrt{\det\left([\vec{\mathbf{v}}_1, \vec{\mathbf{v}}_2]^\top [\vec{\mathbf{v}}_1, \vec{\mathbf{v}}_2]\right)}.$$

5.4.2   Verify that Equation 5.4.9 does indeed parametrize the torus obtained by taking the circle of radius r in the (x, z)-plane that is centered at $x = R$, $z = 0$, and rotating it around the z-axis.

5.4.3   Compute $\int_S (x^2 + y^2 + 3z^2) |d^2\mathbf{x}|$, where S is the part of the paraboloid of revolution $z = x^2 + y^2$ where $z \leq 9$.

5.4.4   What is the surface area of the part of the paraboloid of revolution $z = x^2 + y^2$ where $z \leq 1$?

In Exercise 5.4.5, part (b), a computer and appropriate software will help.

5.4.5   (a) Set up an integral to compute the integral $\int_S (x+y+z) |d^2\mathbf{x}|$, where S is the part of the graph of $x^3 + y^4$ above the unit circle.

(b) Can you evaluate it numerically?

For Exercise 5.4.6, part (b): the earth's diameter is 40 000 kilometers. The polar caps are the regions bounded by the Arctic and the Antarctic circles, which are 23°27′ from the North and South poles respectively. The tropics is the region between the tropic of Cancer (23°27′ north of the equator) and the tropic of Capricorn (23°27′ south of the equator).

An ellipsoid is a solid of which every plane section is an ellipse or a circle.

Exercise 5.4.11, part (a): since $z = x + iy$, using polar coordinates $x = r\cos\theta$, $y = r\sin\theta$ for z means setting $z = r(\cos\theta + i\sin\theta)$.

5.4.6   (a) Let S_1 be the part of the cylinder of equation $x^2 + y^2 = 1$ with $-1 \leq z \leq 1$, and let S_2 be the unit sphere. Show that the horizontal radial projection $S_1 \to S_2$ is area-preserving.

(b) What is the area of the polar cap on earth? The tropics?

(c) Find a formula for the area $A_R(r)$ of a disk of radius r on a sphere of radius R (the radius is measured on the sphere, not inside the ball). What is the Taylor polynomial of $A(r)$ to degree 4?

5.4.7   Compute the area of the graph of the function $f\begin{pmatrix} x \\ y \end{pmatrix} = \frac{2}{3}(x^{3/2} + y^{3/2})$ above the region $0 \leq x, y \leq 1$.

5.4.8   (a) Give a parametrization of the surface of the ellipsoid

$$\frac{x^2}{a^2} + \frac{y^2}{b^2} + \frac{z^2}{c^2} = 1 \quad \text{analogous to spherical coordinates.}$$

(b) Set up an integral to compute the surface area of the ellipsoid.

5.4.9   (a) Set up an integral to compute the surface area of the unit sphere.

(b) Compute the surface area (if you know the formula, as we certainly hope you do, simply giving it is not good enough).

5.4.10   Let $f(x)$ be a positive C^1 function of $x \in [a, b]$.
(a) Find a parametrization of the surface in $\mathbb{R}^3$ obtained by rotating the graph of f around the x-axis.

(b) What is the area of this surface? (The answer should be in the form of a one-dimensional integral.)

5.4.11   Let $X \subset \mathbb{C}^2$ be the graph of the function $w = z^k$, where both $z = x + iy = re^{i\theta}$ and $w = u + iv$ are complex variables.
(a) Parametrize X in terms of the polar coordinates r, θ for z.

(b) What is the area of the part of X where $|z| \leq R$?

5.4.12   The *total curvature* $K(S)$ of a surface $S \subset \mathbb{R}^3$ is given by

$$K(S) = \int_S |K(\mathbf{x})||d^2\mathbf{x}|.$$

(a) What is the total curvature of the sphere $S_R^2 \subset \mathbb{R}^3$ of radius R?

(b) What is the total curvature of the graph of the function $f\begin{pmatrix} x \\ y \end{pmatrix} = x^2 - y^2$? (See Example 3.8.8.)

(c) What is the total curvature of the part of the helicoid of equation $y\cos z = x\sin z$ (see Example 3.8.9) with $0 \leq z \leq a$?

*5.4.13   Let $\mathbf{f}\left(\begin{smallmatrix} u \\ v \end{smallmatrix}\right) = \begin{pmatrix} u \\ u^2 \\ v \end{pmatrix}$ be a parametrization of a parabolic cylinder.

If T is a triangle with vertices $\mathbf{a}, \mathbf{b}, \mathbf{c} \in \mathbb{R}^2$, the image triangle will be by definition the triangle in $\mathbb{R}^3$ with vertices $\mathbf{f}(\mathbf{a}), \mathbf{f}(\mathbf{b}), \mathbf{f}(\mathbf{c})$. Show that there exists a triangulation of the unit square in the (u, v)-plane such that the sum of the areas of the image triangles is arbitrarily large.

Triangulation of the unit square means decomposing the square into triangles.

Exercises for Section 5.5:

Volume of Manifolds

5.5.1   Let $M_1(n, m)$ be the space of $n \times m$ matrices of rank 1. What is the three-dimensional volume of the part of $M_1(2, 2)$ made up of matrices A with $|A| \leq 1$?

5.5.2   A gas has density C/r, where $r = \sqrt{x^2 + y^2 + z^2}$. If $0 < a < b$, what is the mass of the gas between the concentric spheres $r = a$ and $r = b$?

5.5.3   What is the center of gravity of a uniform wire, whose position is the parabola of equation $y = x^2$, where $0 \leq x \leq a$?

5.5.4   Let $X \subset \mathbb{C}^2$ be the graph of the function $w = e^z + e^{-z}$, where both $z = x + iy$ and $w = u + iv$ are complex variables. What is the area of the part of X where $-1 \leq x, y \leq 1$?

5.5.5   Justify the result in Equation 5.5.6 by computing the integral.

5.5.6   The function $\cos z$ of the complex variable z is by definition

$$\cos z = \frac{e^{iz} + e^{-iz}}{2}.$$

(a) If $z = x + iy$, write the real and imaginary parts of $\cos z$ in terms of x and y.

(b) What is the area of the part of the graph of $\cos z$ where $-\pi \leq x \leq \pi, -1 \leq y \leq 1$?

5.5.7   (a) Show that the mapping

$$\gamma \begin{pmatrix} \theta \\ \varphi \\ \psi \end{pmatrix} \mapsto \begin{pmatrix} \cos \psi \cos \varphi \cos \theta \\ \cos \psi \cos \varphi \sin \theta \\ \cos \psi \sin \varphi \\ \sin \psi \end{pmatrix}$$

parametrizes the unit sphere S^3 in $\mathbb{R}^4$ when $-\pi/2 \leq \varphi \leq \pi/2$ and $-\pi/2 \leq \psi \leq \pi/2, 0 \leq \theta < 2\pi$.

(b) Use this parametrization to compute $\mathrm{vol}_3(S^3)$.

5.5.8   What is the area of the surface in $\mathbb{C}^3$ parametrized by

$$\gamma(z) = \begin{pmatrix} z^p \\ z^q \\ z^r \end{pmatrix}, \quad z \in \mathbb{C}, \ |z| \leq 1?$$

Exercises for Section 5.6:

Fractals

FIGURE 5.6.1.

5.6.1   Consider the set C obtained by taking $[0, 1]$, and removing first the open middle third $(1/3, 2/3)$, then the open middle third of each of the segments left, then the open middle third of each of the segments left, etc., as shown in Figure 5.6.1.

(a) Show that an alternative description of C is that it is the set of points that can be written in base 3 without using the digit 1. Use this to show that C is an uncountable set. (Hint: For instance, the number written in base 3 as .02220000222002222... is in it.)

(b) Show that C is a pavable set, with one-dimensional volume 0.

(c) Show that the only dimension in which C can have volume different from 0 or infinity is $\ln 2/\ln 3$.

5.6.2   Now let the set C be obtained from the unit interval by omitting the open middle 1/5th, then the open middle fifth of each remaining interval, then the open middle fifth of each remaining interval, etc.

(a) Show that an alternative description of C is that it is the set of points which can be written in base 5 without using the digit 2. Use this to show that C is an uncountable set.

(b) Show that C is a pavable set, with one-dimensional volume 0.

(c) What is the only dimension in which C can have volume different from 0 or infinity?

5.6.3   This time let the set C be obtained from the unit interval by omitting the open middle $1/n$th, then the open middle $1/n$ of each of the remaining intervals, then the open middle $1/n$ of the remaining intervals, etc. (When n is even, this means omitting an open interval equivalent to $1/n$th of the unit interval, leaving equal amounts on both sides, and so on.)

(a) Show that C is a pavable set, with one-dimensional volume 0.

(b) What is the only dimension in which C can have volume different from 0 or infinity? What is this dimension when $n = 2$?

6

Forms and Vector Calculus

Gradient a 1-form? How so? Hasn't one always known the gradient as a vector? Yes, indeed, but only because one was not familiar with the more appropriate 1-form concept.—C. Misner, K. S. Thorne, J. Wheeler, *Gravitation*

6.0 INTRODUCTION

What really makes calculus work is the fundamental theorem of calculus: that differentiation, having to do with speeds, and integration, having to do with areas, are somehow inverse operations.

Obviously, we will want to generalize the fundamental theorem of calculus to higher dimensions. Unfortunately, we cannot do so using the techniques of Chapter 4 and Chapter 5, where we integrated using $|d^n\mathbf{x}|$. The reason is that $|d^n\mathbf{x}|$ always returns a positive number; it does not concern itself with the orientation of the subset over which it is integrating, unlike the dx of one-dimensional calculus, which does:

$$\int_a^b f(x)\,dx = -\int_b^a f(x)\,dx.$$

To get a fundamental theorem of calculus in higher dimensions, we need to introduce new tools. If we were willing to restrict ourselves to $\mathbb{R}^2$ and $\mathbb{R}^3$ we could use the techniques of *vector calculus*. We will use a different approach, *forms*, which work in any $\mathbb{R}^n$. Forms are integrands over *oriented* domains; they provide the theory of expressions containing dx or $dx\,dy\ldots$.

You have in fact been using forms without realizing it. When you write $d(t^2) = 2t\,dt$ you are saying something about forms, 1-forms to be precise.

Our treatment of forms, especially the exterior derivative, was influenced by the book *Mathematical Methods of Classical Mechanics*, by Vladimir Arnold (Springer-Verlag, 1978).

Because forms work in any dimension, they are the natural way to approach two towering subjects that are inherently four-dimensional: electromagnetism and the theory of relativity. They also provide a unified treatment of differentiation and of the fundamental theorem of calculus: one operator (the exterior derivative) works in all dimensions, and one short, elegant statement (the generalized Stokes's theorem) generalizes the fundamental theorem of calculus to all dimensions. In contrast, vector calculus requires special formulas, operators, and theorems for each dimension where it works.

On the other hand, the language of vector calculus is used in many science courses, particularly at the undergraduate level. So while in theory we could provide a unified treatment of higher dimensional calculus using only forms, this would probably not mesh well with other courses. If you are studying physics, for example, you definitely need to know vector calculus. In addition, the functions and vector fields of vector calculus are more intuitive than forms. A vector field is an object that one can picture, as in Figure 6.0.1. Coming to terms with forms requires more effort. We can't draw you a picture of a form. A k-form is, as we shall see, something like the determinant: it takes k vectors, fiddles with them until it has a square matrix, and then takes its determinant.

We said at the beginning of this book that the object of linear algebra "is at least in part to extend to higher dimensions the geometric language and intuition we have concerning the plane and space, familiar to us all from everyday experience." Here too we want to extend to higher dimensions the geometric language and intuition we have concerning the plane and space. We hope that translating forms into the language of vector calculus will help you do that.

Section 6.1 gives a brief discussion of integrands over oriented domains. In Section 6.2 we introduce k-forms: integrands that take a little piece of oriented domain and return a number. In Section 6.3 we define oriented k-parallelograms and show how to integrate *form fields*—functions that assign a form at each point—over parametrized domains. Section 6.4 translates the language of forms on $\mathbb{R}^3$ into the language of vector calculus. Section 6.5 gives the definitions of orientation necessary to integrate form fields over oriented domains, while Section 6.6 discusses boundary orientation. Section 6.7 introduces the *exterior derivative*, which Section 6.8 relates to vector calculus via the grad, div, and curl. Sections 6.9 and 6.10 discuss the generalized Stokes's theorem and its four different embodiments in the language of vector calculus. Section 6.11 addresses the question, important in both physics and geometry, of when a vector field is the gradient of a function.

6.1 FORMS AS INTEGRANDS OVER ORIENTED DOMAINS

In Chapter 4 we studied the integrand $|d^n\mathbf{x}|$, which takes a subset $A \subset \mathbb{R}^n$ and returns its n-dimensional volume, $\mathrm{vol}_n A$. In Chapter 5 we showed how to integrate the integrand $|d^1\mathbf{x}|$ (the element of arc length) over a curve, to determine its length, and how to integrate the integrand $|d^2\mathbf{x}|$ over a surface, to determine its area. More generally, we saw how to integrate $|d^k\mathbf{x}|$ over a k-dimensional manifold in $\mathbb{R}^n$, to determine its k-dimensional volume.

Such integrands take a little piece (of curve, surface, or higher-dimensional manifold) and return a number. They require no mention of the orientation of the piece; non-orientable surfaces like the Moebius strip shown in Figure 6.1.1 have a perfectly well-defined area, obtained by integrating $|d^2\mathbf{x}|$ over them.

The integrands above are thus fundamentally different from the integrand dx of one variable calculus, which requires *oriented* intervals. In one variable calculus, the standard integrand $f(x)\,dx$ takes a piece $[x_i, x_{i+1}]$ of the domain, and returns the number $f(x_i)(x_{i+1} - x_i)$: the area of a rectangle with height

FIGURE 6.0.1.

The radial vector field

$$\vec{F}\begin{pmatrix} x \\ y \end{pmatrix} = \begin{bmatrix} x \\ y \end{bmatrix}.$$

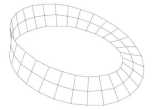

FIGURE 6.1.1.

The Moebius strip was discovered in 1858 by August Moebius, a German mathematician and astronomer. Game for a rainy day: Make a big Moebius strip out of paper. Give one young child a yellow crayon, another a blue crayon, and start them coloring on opposite sides of the strip.

$f(x_i)$ and width $x_{i+1} - x_i$. Note that dx returns $x_{i+1} - x_i$, not $|x_{i+1} - x_i|$; that is why

$$\int_{-1}^{1} f(x)\, dx = -\int_{1}^{-1} f(x)\, dx. \qquad \qquad 6.1.1$$

In order to generalize the fundamental theorem of calculus to higher dimensions, we need integrands over oriented objects. Forms are such integrands.

Example 6.1.1 (Flux form of a vector field: $\Phi_{\vec{F}}$). Suppose we are given a vector field $\vec{F}$ on some open subset U of $\mathbb{R}^3$. It may help to imagine this vector field as the velocity vector field of some fluid with a steady flow (not changing with time). Then the integrand $\Phi_{\vec{F}}$ associates to a little piece of surface the *flux* of $\vec{F}$ through that piece; if you imagine the vector field as the flow of a fluid, then $\Phi_{\vec{F}}$ associates to a little piece of surface the amount of fluid that flows through it in unit time.

But there's a catch: to define the flux of a vector field through a surface, you must orient the surface, for instance by coloring the sides yellow and blue, and counting how much flows from the blue side to the yellow side (counting the flow negative if the fluid flows in the opposite direction). It obviously does not make sense to calculate the flow of a vector field through a Moebius strip. $\triangle$

6.2 Forms on $\mathbb{R}^n$

You should think of this section as a continuation of Section 4.8. There we saw that there is a unique antisymmetric and multilinear function of n vectors in $\mathbb{R}^n$ that gives 1 if evaluated on the standard basis vectors: the determinant. Because of the connection between the determinant and volume described in Section 4.9, the determinant is fundamental to multiple integrals, as we saw in Section 4.10.

Here we will study the multilinear antisymmetric functions of k vectors in $\mathbb{R}^n$, where $k \geq 0$ may be any integer, though we will soon see that the only interesting case is when $k \leq n$. Again there is a close relation to volumes, and in fact these objects, called forms, are the right integrands for integrating over oriented domains.

Definition 6.2.1 (k-form on $\mathbb{R}^n$). A k-form on $\mathbb{R}^n$ is a function φ that takes k vectors in $\mathbb{R}^n$ and returns a number, such that $\varphi(\vec{v}_1, \ldots, \vec{v}_k)$ is multilinear and antisymmetric.

That is, a k-form φ is linear with respect to each of its arguments, and changes sign if two of the arguments are exchanged.

It is rather hard to imagine forms, so we start with an example, which will turn out to be the fundamental example.

While it is easy to show that a Moebius strip has only one side and therefore can't be oriented, it is less easy to define orientation with precision. How might a flat worm living *inside* a Moebius strip be able to tell that it was in a surface with only one side? We will discuss orientation further in Section 6.5.

Analogs to the Moebius strip exist in all higher dimensions, but all curves are orientable.

The important difference between determinants and k-forms is that a k-form on $\mathbb{R}^n$ is a function of k vectors, while the determinant on $\mathbb{R}^n$ is a function of n vectors; determinants are only defined for square matrices.

The words *antisymmetric* and *alternating* are synonymous.

Antisymmetry

If you exchange any two of the arguments of φ, you change the sign of φ:

$$\varphi(\vec{v}_1, \ldots, \vec{v}_i, \ldots, \vec{v}_j, \ldots, \vec{v}_k)$$
$$= -\varphi(\vec{v}_1, \ldots, \vec{v}_j, \ldots, \vec{v}_i, \ldots, \vec{v}_k).$$

Multilinearity

If φ is a k-form and $\vec{v}_i = a\vec{u} + b\vec{w}$, then

$$\varphi\left(\vec{v}_1, \ldots, (a\vec{u} + b\vec{w}), \ldots, \vec{v}_k\right) =$$
$$a\varphi(\vec{v}_1, \ldots, \vec{v}_{i-1}, \vec{u}, \vec{v}_{i+1}, \ldots, \vec{v}_k) +$$
$$b\varphi(\vec{v}_1, \ldots, \vec{v}_{i-1}, \vec{w}, \vec{v}_{i+1}, \ldots, \vec{v}_k).$$

The entire expression

$$dx_{i_1} \wedge \cdots \wedge dx_{i_k}$$

is just the name for this k-form; for now think of it as a single item without worrying about the component parts. The reason for the wedge $\wedge$ will be explained at the end of this section, where we discuss the *wedge product*; we will see that the use of $\wedge$ in the wedge product is consistent with its use here. In Section 6.8 we will see that the use of d in our notation here is consistent with its use to denote the *exterior derivative*.

Note (Equation 6.2.4) that to give an example of a 3-form we had to add a third vector. You cannot evaluate a 3-form on two vectors (or on four); a k-form is a function of k vectors. If you have more than k vectors, or fewer, then you will end up with a matrix that is not square, which will not have a determinant. But you *can* evaluate a 2-form on two vectors in $\mathbb{R}^4$, as we did above, or in $\mathbb{R}^{16}$. This is not the case for the determinant, which is a function of n vectors in $\mathbb{R}^n$.

In the top line of Equation 6.2.5 we could write

$$|d^1\mathbf{x}|(\vec{\mathbf{v}}) = \sqrt{\det\left(\vec{\mathbf{v}}^\top \vec{\mathbf{v}}\right)},$$

in keeping with the other formulas, since det of a number equals that number.

Example 6.2.2 (k-form). Let $i_1, \ldots, i_k$ be any k integers between 1 and n. Then $dx_{i_1} \wedge \cdots \wedge dx_{i_k}$ is that function of k vectors $\vec{\mathbf{v}}_1, \ldots, \vec{\mathbf{v}}_k$ in $\mathbb{R}^n$ that puts these vectors side by side, making the $n \times k$ matrix

$$\begin{bmatrix} v_{1,1} & \cdots & v_{1,k} \\ \vdots & \cdots & \vdots \\ v_{n,1} & \cdots & v_{n,k} \end{bmatrix} \qquad \begin{matrix} | & & | \\ \vec{\mathbf{v}}_1 & \cdots & \vec{\mathbf{v}}_k \\ | & & | \end{matrix} \qquad 6.2.1$$

and selects first the i_1th row, then the i_2 row, etc, and finally the i_kth row, making the square $k \times k$ matrix

$$\begin{bmatrix} v_{i_1,1} & \cdots & v_{i_1,k} \\ \vdots & \cdots & \vdots \\ v_{i_k,1} & \cdots & v_{i_k,k} \end{bmatrix}, \qquad 6.2.2$$

and finally takes its determinant. For instance,

$$\underbrace{dx_1 \wedge dx_2}_{\text{2-form}}\left(\begin{bmatrix} 1 \\ 2 \\ -1 \\ 1 \end{bmatrix}, \begin{bmatrix} 3 \\ -2 \\ 1 \\ 2 \end{bmatrix}\right) = \det \underbrace{\begin{bmatrix} 1 & 3 \\ 2 & -2 \end{bmatrix}}_{\substack{\text{1st and 2nd rows} \\ \text{of original matrix}}} = -8. \qquad 6.2.3$$

$$\underbrace{dx_1 \wedge dx_2 \wedge dx_4}_{\text{3-form}}\left(\begin{bmatrix} 1 \\ 2 \\ -1 \\ 1 \end{bmatrix}, \begin{bmatrix} 3 \\ -2 \\ 1 \\ 2 \end{bmatrix}, \begin{bmatrix} 0 \\ 1 \\ 2 \\ 1 \end{bmatrix}\right) = \det \begin{bmatrix} 1 & 3 & 0 \\ 2 & -2 & 1 \\ 1 & 2 & 1 \end{bmatrix} = -7 \quad \triangle \;\; 6.2.4$$

Remark. The integrand $|d^k\mathbf{x}|$ of Chapter 5 also takes k vectors in $\mathbb{R}^n$ and gives a number:

$$|d^1\mathbf{x}|(\vec{\mathbf{v}}) = |\vec{\mathbf{v}}| = \sqrt{\vec{\mathbf{v}}^\top \vec{\mathbf{v}}} = \sqrt{\det(\vec{\mathbf{v}}^\top \vec{\mathbf{v}})},$$

$$|d^2\mathbf{x}|(\vec{\mathbf{v}}_1, \vec{\mathbf{v}}_2) = \sqrt{\det\left([\vec{\mathbf{v}}_1, \vec{\mathbf{v}}_2]^\top [\vec{\mathbf{v}}_1, \vec{\mathbf{v}}_2]\right)}, \qquad 6.2.5$$

$$|d^k\mathbf{x}|(\vec{\mathbf{v}}_1, \ldots, \vec{\mathbf{v}}_k) = \sqrt{\det\left([\vec{\mathbf{v}}_1, \ldots, \vec{\mathbf{v}}_k]^\top [\vec{\mathbf{v}}_1, \ldots, \vec{\mathbf{v}}_k]\right)}.$$

Unlike forms, these are neither multilinear nor antisymmetric. $\triangle$

Geometric meaning of k-forms

Evaluating the 2-form $dx_1 \wedge dx_2$ on the vectors $\vec{\mathbf{a}}$ and $\vec{\mathbf{b}}$, we have:

$$dx_1 \wedge dx_2\left(\begin{bmatrix} a_1 \\ a_2 \\ a_3 \end{bmatrix}, \begin{bmatrix} b_1 \\ b_2 \\ b_3 \end{bmatrix}\right) = \det \begin{bmatrix} a_1 & b_1 \\ a_2 & b_2 \end{bmatrix} = a_1 b_2 - a_2 b_1, \qquad 6.2.6$$

which can be understood geometrically. If we project $\vec{a}$ and $\vec{b}$ onto the (x_1, x_2)-plane, we get the vectors

$$\begin{bmatrix} a_1 \\ a_2 \end{bmatrix} \quad \text{and} \quad \begin{bmatrix} b_1 \\ b_2 \end{bmatrix}; \qquad\qquad 6.2.7$$

the determinant in Equation 6.2.6 gives the *signed area of the parallelogram that they span*, as described in Proposition 1.4.14.

Rather than imagining projecting $\vec{a}$ and $\vec{b}$ onto the plane to get the vectors of Equation 6.2.7, we could imagine projecting the parallelogram spanned by $\vec{a}$ and $\vec{b}$ onto the plane to get the parallelogram spanned by the vectors of Equation 6.2.7.

Thus $dx_1 \wedge dx_2$ deserves to be called the (x_1, x_2) component of signed area. Similarly, $dx_2 \wedge dx_3$ and $dx_1 \wedge dx_3$ deserve to be called the (x_2, x_3) and the (x_1, x_3) components of signed area.

We can now interpret Equations 6.2.3 and 6.2.4 geometrically. The 2-form form $dx_1 \wedge dx_2$ tells us that the (x_1, x_2) component of signed area of the parallelogram spanned by the two vectors in Equation 6.2.3 is -8. The 3-form $dx_1 \wedge dx_2 \wedge dx_4$ tells us that the (dx_1, dx_2, dx_4) component of signed volume of the parallelepiped spanned by the three vectors in Equation 6.2.4 is -7.

Similarly, the 1-form dx gives the x component of signed length of a vector, while dy gives its y component:

$$dx \left(\begin{bmatrix} 2 \\ -3 \\ 1 \end{bmatrix} \right) = \det 2 = 2 \quad \text{and} \quad dy \left(\begin{bmatrix} 2 \\ -3 \\ 1 \end{bmatrix} \right) = \det(-3) = -3.$$

More generally (and an advantage of k-forms is that they generalize so easily to higher dimensions), we see that

$$dx_i \left(\begin{bmatrix} v_1 \\ \vdots \\ v_n \end{bmatrix} \right) = \det[v_i] = v_i \qquad\qquad 6.2.8$$

is the ith component of the signed length of $\vec{v}$, and that $dx_{i_1} \wedge \cdots \wedge dx_{i_k}$, evaluated on $(\vec{v}_1, \ldots, \vec{v}_k)$, gives the $(x_{i_1}, \ldots, x_{i_k})$ component of signed k-dimensional volume of the k-parallelogram spanned by $\vec{v}_1, \ldots, \vec{v}_k$.

Elementary forms

There is a great deal of redundancy in the expressions $dx_{i_1} \wedge \cdots \wedge dx_{i_k}$. Consider for instance $dx_1 \wedge dx_3 \wedge dx_1$. This 3-form takes three vectors in $\mathbb{R}^n$, stacks them side by side to make an $n \times 3$ matrix, selects the first row, then the third, then the first again, to make a 3×3 and takes its determinant. So far, so good; but observe that the determinant in question is always 0, independent of what the vectors were; we have taken the determinant of a 3×3 matrix for which the third row is the same as the first; such a determinant is always 0. (Do you see why?[1]) So

$$dx_1 \wedge dx_3 \wedge dx_1 = 0; \qquad\qquad 6.2.9$$

it takes three vectors and returns 0.

But that is of course not the only way to write the form that takes three vectors and returns the number 0; both $dx_1 \wedge dx_1 \wedge dx_3$ and $dx_2 \wedge dx_3 \wedge dx_3$ do so as well, and there are others. More generally, if any two of the indices $i_1, \ldots, i_k$ are equal, we have $dx_{i_1} \wedge \cdots \wedge dx_{i_k} = 0$: the k-form $dx_{i_1} \wedge \cdots \wedge dx_{i_k}$, where two indices are equal, is the k-form which takes k vectors and returns 0.

Next, consider $dx_1 \wedge dx_3$ and $dx_3 \wedge dx_1$. Evaluated on

$$\vec{\mathbf{a}} = \begin{bmatrix} a_1 \\ \vdots \\ a_n \end{bmatrix} \quad \text{and} \quad \vec{\mathbf{b}} = \begin{bmatrix} b_1 \\ \vdots \\ b_n \end{bmatrix}, \quad \text{we find}$$

$$dx_1 \wedge dx_3(\vec{\mathbf{a}}, \vec{\mathbf{b}}) = \det \begin{bmatrix} a_1 & b_1 \\ a_3 & b_3 \end{bmatrix} = a_1 b_3 - a_3 b_1$$

$$dx_3 \wedge dx_1(\vec{\mathbf{a}}, \vec{\mathbf{b}}) = \det \begin{bmatrix} a_3 & b_3 \\ a_1 & b_1 \end{bmatrix} = a_3 b_1 - a_1 b_3. \qquad 6.2.10$$

Clearly $dx_1 \wedge dx_3 = -dx_3 \wedge dx_1$; these two 2-forms, evaluated on the same two vectors, always return opposite numbers.

Recall (Definition 4.8.16) that the signature of a permutation σ, denoted $\mathrm{sgn}(\sigma)$, is $\mathrm{sgn}(\sigma) = \det M_\sigma$, where M_σ is the permutation matrix of σ. Theorem 4.8.18 gives a formula for the determinant using the signature.

More generally, if the integers $i_1, \ldots, i_k$ and $j_1, \ldots, j_k$ are the same integers, just taken in a different order, so that $j_1 = i_{\sigma(1)}, j_2 = i_{\sigma(2)}, \ldots, j_k = i_{\sigma(k)}$ for some permutation σ of $\{1, \ldots, k\}$, then

$$dx_{j_1} \wedge \cdots \wedge dx_{j_k} = \mathrm{sgn}(\sigma) dx_{i_1} \wedge \cdots \wedge dx_{i_k}. \qquad 6.2.11$$

Indeed, $dx_{j_1} \wedge \cdots \wedge dx_{j_k}$ computes the determinant of the same matrix as $dx_{i_1} \wedge \cdots \wedge dx_{i_k}$, only with the rows permuted by σ. For instance, $dx_1 \wedge dx_2 = -dx_2 \wedge dx_1$, and

$$dx_1 \wedge dx_2 \wedge dx_3 = dx_2 \wedge dx_3 \wedge dx_1 = dx_3 \wedge dx_1 \wedge dx_2. \qquad 6.2.12$$

To eliminate this redundancy, we make the following definition: an *elementary k-form* is of the form

$$dx_{i_1} \wedge \cdots \wedge dx_{i_k} \quad \text{with } 1 \le i_1 < i_2 \cdots < i_k \le n; \qquad 6.2.13$$

putting the indices in increasing order selects one particular permutation for any set of distinct integers $j_1, \ldots, j_k$.

---

[1]The determinant of a square matrix containing two identical columns is always 0, since exchanging them reverses the sign of the determinant, while keeping it the same. Since (Theorem 4.8.10), $\det A = \det A^\top$, the determinant of a matrix is also 0 if two of its rows are identical.

Definition 6.2.3 (Elementary k-forms on $\mathbb{R}^n$). A elementary k-form on $\mathbb{R}^n$ is an expression of the form

$$dx_{i_1} \wedge \cdots \wedge dx_{i_k}, \qquad\qquad 6.2.14$$

where $1 \leq i_1 < \cdots < i_k \leq n$ (and $0 \leq k \leq n$). Evaluated on the vectors $\vec{\mathbf{v}}_1, \ldots, \vec{\mathbf{v}}_k$, it gives the determinant of the $k \times k$ matrix obtained by selecting the $i_1, \ldots, i_k$ rows of the matrix whose columns are the vectors $\vec{\mathbf{v}}_1, \ldots, \vec{\mathbf{v}}_k$. The only elementary 0-form is the form, denoted 1, which evaluated on zero vectors returns 1.

What elementary k-forms exist on $\mathbb{R}^4$?[2]

Note that there are no elementary k forms on $\mathbb{R}^n$ when $k > n$; indeed, there are no nonzero forms at all when $k > n$: there is no function φ that takes $k > n$ vectors in $\mathbb{R}^n$ and returns a number, such that $\varphi(\vec{\mathbf{v}}_1, \ldots, \vec{\mathbf{v}}_k)$ is multilinear and antisymmetric. If $\vec{\mathbf{v}}_1, \ldots, \vec{\mathbf{v}}_k$ are vectors in $\mathbb{R}^n$ and $k > n$, then the vectors are not linearly independent, and at least one of them is a linear combination of the others, say

That there are no *elementary* k-forms when $k > n$ is an example of the "pigeon hole" principle: if you have more than n pigeons in n holes one hole must contain at least two pigeons. Here we would need to select more than n distinct integers between 1 and n.

$$\vec{\mathbf{v}}_k = \sum_{i=1}^{k-1} a_i \vec{\mathbf{v}}_i. \qquad\qquad 6.2.15$$

Then if φ is a k-form on $\mathbb{R}^n$, evaluation on the vectors $\vec{\mathbf{v}}_1, \ldots, \vec{\mathbf{v}}_k$ gives

$$\varphi(\vec{\mathbf{v}}_1, \ldots, \vec{\mathbf{v}}_k) = \varphi\Big(\vec{\mathbf{v}}_1, \ldots, \sum_{i=1}^{k-1} a_i \vec{\mathbf{v}}_i\Big)$$

$$= \sum_{i=1}^{k-1} a_i \varphi(\vec{\mathbf{v}}_1, \ldots, \vec{\mathbf{v}}_i, \ldots, \vec{\mathbf{v}}_i). \qquad 6.2.16$$

Each term in this last sum will evaluate φ on k vectors, two of which are equal, and so (by antisymmetry) will return 0.

In terms of the geometric description, this should come as no surprise: you would expect any kind of three-dimensional volume in $\mathbb{R}^2$ to be zero, and more generally any k-dimensional volume in $\mathbb{R}^n$ to be 0 when $k > n$.

---

[2]On $\mathbb{R}^4$ there exist

(1) one elementary 0-form, the number 1.
(2) four elementary 1-forms: dx_1, dx_2, dx_3 and dx_4.
(3) six elementary 2-forms: $dx_1 \wedge dx_2, dx_1 \wedge dx_3, dx_1 \wedge dx_4, dx_2 \wedge dx_3, dx_2 \wedge dx_4$, and $dx_3 \wedge dx_4$.
(4) four elementary 3-forms: $dx_1 \wedge dx_2 \wedge dx_3, dx_1 \wedge dx_2 \wedge dx_4, dx_1 \wedge dx_3 \wedge dx_4$, $dx_2 \wedge dx_3 \wedge dx_4$.
(5) one elementary 4-form: $dx_1 \wedge dx_2 \wedge dx_3 \wedge dx_4$.

All forms are linear combinations of elementary forms

We said above that $dx_{i_1} \wedge \cdots \wedge dx_{i_k}$ is the fundamental example of a k-form. Now we will justify this statement, by showing that any k-form is a linear combination of elementary k-forms.

The following definitions say that speaking of such linear combinations makes sense: we can add k-forms and multiply them by scalars in the obvious way.

The Greek letter ψ is pronounced "psi."

Definition 6.2.4 (Addition of k-forms). Let φ and ψ be two k-forms. Then

$$(\varphi + \psi)(\vec{\mathbf{v}}_1, \ldots, \vec{\mathbf{v}}_k) = \varphi(\vec{\mathbf{v}}_1, \ldots, \vec{\mathbf{v}}_k) + \psi(\vec{\mathbf{v}}_1, \ldots, \vec{\mathbf{v}}_k).$$

Adding k-forms

$3\,dx \wedge dy + 2\,dx \wedge dy = 5\,dx \wedge dy.$

Multiplying k-forms by scalars

$5\,(dx \wedge dy + 2\,dx \wedge dz)$

$= 5\,dx \wedge dy + 10\,dx \wedge dz.$

Saying that the space of k-forms in $\mathbb{R}^n$ is a vector space (Definition 2.6.1) means that k-forms can be manipulated in familiar ways, for example

$$\alpha(\beta\varphi) = (\alpha\beta)\varphi$$

(associativity for multiplication),

$$(\alpha + \beta)\varphi = \alpha\varphi + \beta\varphi$$

(distributive law for scalar addition), and

$$\alpha(\varphi + \psi) = \alpha\varphi + \alpha\psi$$

(distributive law for addition of forms).

Definition 6.2.5 (Multiplication of k-forms by scalars). If φ is a k-form and a is a scalar, then

$$(a\varphi)(\vec{\mathbf{v}}_1, \ldots, \vec{\mathbf{v}}_k) = a(\varphi(\vec{\mathbf{v}}_1, \ldots, \vec{\mathbf{v}}_k)).$$

Using these definitions of addition and multiplication by scalars, the space of k-forms in $\mathbb{R}^n$ is a vector space. We will now show that the elementary k-forms form a basis of this space.

Definition 6.2.6 ($A^k(\mathbb{R}^n)$). The space of k-forms in $\mathbb{R}^n$ is denoted $A^k(\mathbb{R}^n)$.

Theorem 6.2.7. *The elementary k-forms form a basis for $A^k(\mathbb{R}^n)$.*

In other words, every multilinear and antisymmetric function φ of k vectors in $\mathbb{R}^n$ can be uniquely written

$$\varphi = \sum_{1 \le i_1 < \cdots < i_k \le n} a_{i_1 \ldots i_k} \, dx_{i_1} \wedge \cdots \wedge dx_{i_k}, \qquad 6.2.17$$

and in fact the coefficients are given by

$$a_{i_1 \ldots i_k} = \varphi(\vec{\mathbf{e}}_{i_1}, \ldots, \vec{\mathbf{e}}_{i_k}). \qquad 6.2.18$$

Proof. Most of the work is already done, in the proof of Theorem 4.8.4, showing that the determinant exists and is uniquely characterized by its properties of multilinearity, antisymmetry, and normalization. (In fact, Theorem 6.2.7 is Theorem 4.8.4 when $k = n$.) We will illustrate it for the particular case of 2-forms on $\mathbb{R}^3$; this contains the idea of the proof while avoiding hopelessly complicated notation. Let φ be such a 2-form. Then, using multilinearity, we get the following computation. Forget about the coefficients, and notice that this equation says that φ *is completely determined by what it does to the standard basis vectors*:

$$\varphi\left(\underbrace{\begin{bmatrix} v_1 \\ v_2 \\ v_3 \end{bmatrix}}_{\vec{v}}, \underbrace{\begin{bmatrix} w_1 \\ w_2 \\ w_3 \end{bmatrix}}_{\vec{w}}\right) = \varphi(\underbrace{v_1\vec{e}_1 + v_2\vec{e}_2 + v_3\vec{e}_3}_{\vec{v}}, \underbrace{w_1\vec{e}_1 + w_2\vec{e}_2 + w_3\vec{e}_3}_{\vec{w}})$$

$$= \varphi(v_1\vec{e}_1, \underbrace{w_1\vec{e}_1 + w_2\vec{e}_2 + w_3\vec{e}_3}_{\vec{w}}) + \varphi(v_2\vec{e}_2, \underbrace{w_1\vec{e}_1 + w_2\vec{e}_2 + w_3\vec{e}_3}_{\vec{w}})$$

$$+ \varphi(v_3\vec{e}_3, \underbrace{w_1\vec{e}_1 + w_2\vec{e}_2 + w_3\vec{e}_3}_{\vec{w}}) \qquad 6.2.19$$

$$= \varphi(v_1\vec{e}_1, w_1\vec{e}_1) + \varphi(v_1\vec{e}_1, w_2\vec{e}_2) + \varphi(v_1\vec{e}_1, w_3\vec{e}_3) + \dots$$

$$= (v_1w_2 - v_2w_1)\varphi(\vec{e}_1, \vec{e}_2) + (v_1w_3 - v_3w_1)\varphi(\vec{e}_1, \vec{e}_3) + (v_2w_3 - v_3w_2)\varphi(\vec{e}_2, \vec{e}_3).$$

An analogous but messier computation will show the same for any k and n: φ is determined by its values on sequences $\vec{e}_{i_1}, \dots, \vec{e}_{i_k}$, with ascending indices. (The coefficients will be complicated expressions that give determinants, as in the case above, but you don't need to know that.) So any k-form that gives the same result when evaluated on every sequence $\vec{e}_{j_1}, \dots, \vec{e}_{j_k}$ with ascending indices coincides with φ. Thus it is enough to check that

$$\sum_{1 \leq i_1 < \dots < i_k \leq n} a_{i_1 \dots i_k} dx_{i_1} \wedge \dots \wedge dx_{i_k}(\vec{e}_{j_1}, \dots, \vec{e}_{j_k}) = a_{j_1 \dots j_k}. \qquad 6.2.20$$

This is fairly easy to see. If $i_1, \dots, i_k \neq j_1, \dots, j_k$, then there is at least one i that does not appear among the j's, so the corresponding dx_i, acting on the matrix $\vec{e}_{j_1}, \dots, \vec{e}_{j_k}$, selects a row of zeroes. Thus

$$dx_{i_1, \dots, i_k}(\vec{e}_{j_1}, \dots, \vec{e}_{j_k}) \qquad 6.2.21$$

is the determinant of a matrix with a row of zeroes, so it vanishes. But

$$dx_{j_1, \dots, j_k}(\vec{e}_{j_1}, \dots, \vec{e}_{j_k}) = 1, \qquad 6.2.22$$

since it is the determinant of the identity matrix. $\quad\square$

Theorem 6.2.8 (Dimension of $A^k(\mathbb{R}^n)$). *The space $A^k(\mathbb{R}^n)$ has dimension equal to the binomial coefficient*

$$\binom{n}{k} = \frac{n!}{k!(n-k)!}. \qquad 6.2.23$$

Proof. This is just a matter of counting the elements of the basis: i.e., the number of elementary k-forms on $\mathbb{R}^n$. Not for nothing is the binomial coefficient called "n choose k". $\quad\square$

When reading Theorem 6.2.8, remember that $0! = 1$.

Note that

$$\binom{n}{k} = \binom{n}{n-k},$$

since

$$\binom{n}{n-k}$$

$$= \frac{n!}{(n-k)!\left(n - (n-k)\right)!}$$

$$= \frac{n!}{k!(n-k)!}.$$

In particular, for a given n, there are equal numbers of elementary k-forms and elementary $(n-k)$-forms: $A^k(\mathbb{R}^n)$ and $A^{n-k}(\mathbb{R}^n)$ have the same dimension. Thus, in $\mathbb{R}^3$, there is one elementary 0-form and one elementary 3-form, so the spaces $A^0(\mathbb{R}^3)$ and $A^3(\mathbb{R}^3)$ can be identified with numbers. There are three elementary 1-forms and three elementary 2-forms, so the spaces $A^1(\mathbb{R}^3)$ and $A^2(\mathbb{R}^3)$ can be identified with vectors.

(handwritten margin notes)
$5v \cdot w = 5(v \cdot w)$
$5v \cdot 3w = 5 \cdot 3(v \cdot w)$

so $\varphi(v_1\vec{e}_1, w_2\vec{e}_2) = v_1 w_2 \varphi(\vec{e}_1, \vec{e}_2)$

Example 6.2.9 (Dimension of $A^k(\mathbb{R}^3)$). The dimension of $A^0(\mathbb{R}^3)$ and of $A^3(\mathbb{R}^3)$ is 1, and the dimension of $A^1(\mathbb{R}^3)$ and of $A^2(\mathbb{R}^3)$ is 3, because on $\mathbb{R}^3$ we have

$$\binom{3}{0} = \frac{3!}{0!(3)!} = 1 \text{ elementary 0-form}; \quad \binom{3}{1} = \frac{3!}{1!(2)!} = 3 \text{ elementary 1-forms};$$

$$\binom{3}{2} = \frac{3!}{2!(1)!} = 3 \text{ elementary 2-forms}; \quad \binom{3}{3} = \frac{3!}{3!(0)!} = 1 \text{ elementary 3-form}.$$

Forms on vector spaces

So far we have been studying k-forms on $\mathbb{R}^n$. When defining orientation, we will make vital use of k-forms on a subspace $E \subset \mathbb{R}^n$. It is no harder to write the definition when E is an abstract vector space.

Remember that although an element of an (abstract) vector space is called a vector, it need not be a column vector. But we will concentrate on subspaces $E \subset \mathbb{R}^n$ whose elements are column vectors.

Definition 6.2.10 (The space $A^k(E)$). Let E be a vector space. Then $A^k(E)$ is the set of k-forms on E, i.e., the set of functions that take k vectors in E and return a number, and which are multilinear and anti-symmetric.

Some texts use the notation $\Omega^k(E)$ rather than $A^k(E)$; yet others use $\Lambda^k(E^*)$.

The main result we will need is the following:

In Definition 6.2.10 we do not require that E be a subset of $\mathbb{R}^n$: a vector in E need not be something we can write as a column vector. But until we assign E a basis we don't know how to write down such a k-form.

Proposition 6.2.11 (Dimension of $A^k(E)$). If E has dimension m, then $A^k(E)$ has dimension $\binom{m}{k} = \frac{m!}{k!(m-k)!}$.

Proof. We already know the result when $E = \mathbb{R}^m$, and we will use a basis to translate from the concrete world of $\mathbb{R}^m$ to the abstract world of E. Let $\underline{\mathbf{b}}_1, \ldots, \underline{\mathbf{b}}_m$ be a basis of E.

Then the transformation $\Phi_{\{\underline{\mathbf{b}}\}} : \mathbb{R}^m \to E$ given by

Just as when $E = \mathbb{R}^n$, the space $A^k(E)$ is a vector space using the obvious addition and multiplication by scalars.

$$\begin{bmatrix} a_1 \\ \vdots \\ a_m \end{bmatrix} \mapsto a_1\underline{\mathbf{b}}_1 + \cdots + a_m\underline{\mathbf{b}}_m \qquad 6.2.24$$

is an invertible linear transformation, which performs the translation "concrete to abstract." (See Definition 2.6.12.) We will use the inverse dictionary $\Phi_{\{\underline{\mathbf{b}}\}}^{-1}$.

We claim that the forms $\varphi_{i_1,\ldots,i_k}$, $1 \le i_1, < \cdots < i_k \le m$, defined by

$$\varphi_{i_1,\ldots,i_k}(\underline{\mathbf{v}}_1, \ldots, \underline{\mathbf{v}}_k) = dx_{i_1} \wedge \cdots \wedge dx_{i_k}\left(\Phi_{\{\underline{\mathbf{b}}\}}^{-1}(\underline{\mathbf{v}}_1), \ldots, \Phi_{\{\underline{\mathbf{b}}\}}^{-1}(\underline{\mathbf{v}}_k)\right) \qquad 6.2.25$$

form a basis of $A^k(E)$. There is not much to prove: all the properties follow immediately from the corresponding properties in $\mathbb{R}^m$. We need to check that the $\varphi_{i_1,\ldots,i_k}$ are multilinear and antisymmetric, that they are linearly independent, and that they span $A^k(E)$.

Let us see for instance that the $\varphi_{i_1,\dots,i_k}$ are linearly independent. Suppose that

$$\sum_{1 \le i_1, < \cdots < i_k \le m} a_{i_1,\dots,i_k} \varphi_{i_1,\dots,i_k} = 0. \qquad 6.2.26$$

Then applied to the particular vectors

$$\underline{\mathbf{b}}_{j_1}, \dots, \underline{\mathbf{b}}_{j_k} = \Phi_{\{\underline{\mathbf{b}}\}}(\vec{\mathbf{e}}_{j_1}), \dots, \Phi_{\{\underline{\mathbf{b}}\}}(\vec{\mathbf{e}}_{j_k}) \qquad 6.2.27$$

we will still get 0. But

$$\sum_{1 \le i_1, < \cdots < i_k \le m} a_{i_1,\dots,i_k} \varphi_{i_1,\dots,i_k}\left(\Phi_{\{\underline{\mathbf{b}}\}}(\vec{\mathbf{e}}_{j_1}), \dots, \Phi_{\{\underline{\mathbf{b}}\}}(\vec{\mathbf{e}}_{j_k})\right)$$

Checking the other properties is similar, and is left as Exercise 6.2.1.

The first equality on the last line of Equation 6.2.28 is Equation 6.2.20.

$$= \sum_{1 \le i_1, < \cdots < i_k \le m} a_{i_1,\dots,i_k} dx_{i_1} \wedge \cdots \wedge dx_{i_k}(\vec{\mathbf{e}}_{j_1}, \dots, \vec{\mathbf{e}}_{j_k}) \qquad 6.2.28$$

$$= \quad a_{j_1,\dots,j_k} = 0.$$

So all the coefficients are 0, and the forms are linearly independent. $\quad\square$

The case of greatest interest to us is the case when $m = k$:

Corollary 6.2.12. *If E is a k-dimensional vector space, then $A^k(E)$ is a vector space of dimension 1.*

Regarding Corollary 6.2.12, there is one "trivial case" where the argument above doesn't quite apply: when $E = \{\vec{\mathbf{0}}\}$. But it is easy to see the result directly: an element of $A^0(\{\vec{\mathbf{0}}\})$ is simply a real number; it takes zero vectors and returns a number. So $A^0(\{\vec{\mathbf{0}}\})$ is not just one-dimensional; it *is* in fact $\mathbb{R}$.

The wedge product

We have used the wedge $\wedge$ to write down forms. Now we will see what it means: it denotes the wedge product, also known as the *exterior product*.

The wedge product is a messy thing: a complicated summation, over various shuffles of vectors, of the product of two k-forms, each given the sign $+$ or $-$ according to the rule for permutations.

Figure 6.2.1 explains the use of the word "shuffle" to describe the σ over which we are summing.

Definition 6.2.13 (Wedge product). Let φ be a k-form and ψ be a l-form, both on $\mathbb{R}^n$. Then their wedge product $\varphi \wedge \psi$ is a $(k + l)$-form that acts on $k + l$ vectors. It is defined by the following sum, where the summation is over all permutations σ of the numbers $1, 2, 3, \dots, k + l$ such that $\sigma(1) < \sigma(2) < \cdots < \sigma(k)$ and $\sigma(k+1) < \cdots < \sigma(k+l)$:

$$\overbrace{\varphi \wedge \psi(\vec{\mathbf{v}}_1, \vec{\mathbf{v}}_2, \dots, \vec{\mathbf{v}}_{k+l})}^{\substack{\text{wedge product evaluated} \\ \text{on } k+l \text{ vectors}}}$$

$$= \sum_{\substack{\text{shuffles} \\ \sigma \in \text{Perm}(k,l)}} \text{sgn}(\sigma) \varphi \underbrace{(\vec{\mathbf{v}}_{\sigma(1)}, \dots, \vec{\mathbf{v}}_{\sigma(k)})}_{k \text{ vectors}} \psi \underbrace{(\vec{\mathbf{v}}_{\sigma(k+1)}, \dots, \vec{\mathbf{v}}_{\sigma(k+l)})}_{l \text{ vectors}}.$$

We start on the left with a $(k + l)$-form evaluated on $k + l$ vectors. On the right we have a somewhat complicated expression involving a k-form φ acting on k vectors, and a l-form ψ acting on l vectors. To understand the right-hand side, first consider all possible permutations of the $k+l$ vectors $\vec{\mathbf{v}}_1, \vec{\mathbf{v}}_2, \dots, \vec{\mathbf{v}}_{k+l}$, dividing each permutation with a bar line | so that there are k vectors to the left and l vectors to the right, since φ acts on k vectors and ψ acts on l vectors. (For example, if $k = 2$ and $l = 1$, one permutation would be written $\vec{\mathbf{v}}_1\vec{\mathbf{v}}_2|\vec{\mathbf{v}}_3$, another would be written $\vec{\mathbf{v}}_2\vec{\mathbf{v}}_3|\vec{\mathbf{v}}_1$, and a third $\vec{\mathbf{v}}_3\vec{\mathbf{v}}_2|\vec{\mathbf{v}}_1$.)

Next, chose only those permutations where the indices for the k-form (to the left of the dividing bar) and the indices for the l-form (to the right of the bar) are each, separately and independently, in *ascending* order, as illustrated by Figure 6.2.1. (For $k = 2$ and $l = 1$, the only allowable choice is $\vec{\mathbf{v}}_1\vec{\mathbf{v}}_2|\vec{\mathbf{v}}_3$.) We assign each chosen permutation its sign, according to the rule given in Definition 4.8.16, and finally, take the sum.

FIGURE 6.2.1.

Take a pack of $k + 1$ cards, cut it to produce subpacks of k cards and l cards, and shuffle them. The permutation you obtain is one where the order of the cards in the subpacks remains unchanged.

More simply, we note that the first permutation involves an even number (0) of exchanges, or *transpositions*, so the signature is positive, while the second involves an odd number (1), so the signature is negative.

The wedge product of a 0-form α with a k-form ψ is a k-form, $\alpha \wedge \psi = \alpha\psi$. In this case, the wedge product coincides with multiplication by numbers.

Example 6.2.14 (The wedge product of two 1-forms). If φ and ψ are both 1-forms, we have two permutations, $\vec{\mathbf{v}}_1|\vec{\mathbf{v}}_2$ and $\vec{\mathbf{v}}_2|\vec{\mathbf{v}}_1$, both allowable under our "ascending order" rule. The sign for the first is positive, since

$$\begin{bmatrix} \vec{\mathbf{v}}_1 \\ \vec{\mathbf{v}}_2 \end{bmatrix} \rightarrow \begin{bmatrix} \vec{\mathbf{v}}_1 \\ \vec{\mathbf{v}}_2 \end{bmatrix} \quad \text{gives the permutation matrix} \quad \begin{bmatrix} 1 & 0 \\ 0 & 1 \end{bmatrix},$$

with determinant $+1$. The sign for the second is negative, since

$$\begin{bmatrix} \vec{\mathbf{v}}_1 \\ \vec{\mathbf{v}}_2 \end{bmatrix} \rightarrow \begin{bmatrix} \vec{\mathbf{v}}_2 \\ \vec{\mathbf{v}}_1 \end{bmatrix} \quad \text{gives the permutation matrix} \quad \begin{bmatrix} 0 & 1 \\ 1 & 0 \end{bmatrix},$$

with determinant -1. So in this case the equation of Definition 6.2.13 becomes

$$(\varphi \wedge \psi)(\vec{\mathbf{v}}_1, \vec{\mathbf{v}}_2) = \varphi(\vec{\mathbf{v}}_1)\psi(\vec{\mathbf{v}}_2) - \varphi(\vec{\mathbf{v}}_2)\psi(\vec{\mathbf{v}}_1). \qquad 6.2.29$$

We see that the 2-form $dx_1 \wedge dx_2$

$$dx_1 \wedge dx_2(\vec{\mathbf{a}}, \vec{\mathbf{b}}) = \det \begin{bmatrix} a_1 & b_1 \\ a_2 & b_2 \end{bmatrix} = a_1 b_2 - a_2 b_1, \qquad 6.2.30$$

is indeed equal to the wedge product of the 1-forms dx_1 and dx_2, which, evaluated on the same two vectors, gives

$$dx_1 \wedge dx_2(\vec{\mathbf{a}}, \vec{\mathbf{b}}) = dx_1(\vec{\mathbf{a}})dx_2(\vec{\mathbf{b}}) - dx_1(\vec{\mathbf{b}})dx_2(\vec{\mathbf{a}}) = a_1 b_2 - a_2 b_1. \qquad 6.2.31$$

So our use of the wedge in naming the elementary forms is coherent with its use to denote this special kind of multiplication.

Example 6.2.15 (The wedge product of a 2-form and a 1-form). If φ is a 2-form and ψ is a 1-form, then we have the six permutations

$$\vec{\mathbf{v}}_1\vec{\mathbf{v}}_2|\vec{\mathbf{v}}_3, \quad \vec{\mathbf{v}}_1\vec{\mathbf{v}}_3|\vec{\mathbf{v}}_2, \quad \vec{\mathbf{v}}_2\vec{\mathbf{v}}_3|\vec{\mathbf{v}}_1, \quad \vec{\mathbf{v}}_3\vec{\mathbf{v}}_1|\vec{\mathbf{v}}_2, \quad \vec{\mathbf{v}}_2\vec{\mathbf{v}}_1|\vec{\mathbf{v}}_3, \text{ and } \vec{\mathbf{v}}_3\vec{\mathbf{v}}_2|\vec{\mathbf{v}}_1. \qquad 6.2.32$$

The first three are in ascending order, so we have three permutations to sum,

$$+(\vec{\mathbf{v}}_1\vec{\mathbf{v}}_2|\vec{\mathbf{v}}_3), \quad -(\vec{\mathbf{v}}_1\vec{\mathbf{v}}_3|\vec{\mathbf{v}}_2), \quad +(\vec{\mathbf{v}}_2\vec{\mathbf{v}}_3|\vec{\mathbf{v}}_1), \qquad 6.2.33$$

giving the wedge product

$$\varphi \wedge \psi(\vec{\mathbf{v}}_1, \vec{\mathbf{v}}_2, \vec{\mathbf{v}}_3) = \varphi(\vec{\mathbf{v}}_1, \vec{\mathbf{v}}_2)\,\psi(\vec{\mathbf{v}}_3) - \varphi(\vec{\mathbf{v}}_1, \vec{\mathbf{v}}_3)\,\psi(\vec{\mathbf{v}}_2) + \varphi(\vec{\mathbf{v}}_2, \vec{\mathbf{v}}_3)\,\psi(\vec{\mathbf{v}}_1). \quad 6.2.34$$

Again, let's compare this result with what we get using Definition 6.2.3, setting $\varphi = dx_1 \wedge dx_2$ and $\psi = dx_3$; to avoid double indices we will rename the vectors $\vec{\mathbf{v}}_1, \vec{\mathbf{v}}_2, \vec{\mathbf{v}}_3$, calling them $\vec{\mathbf{u}}, \vec{\mathbf{v}}$, and $\vec{\mathbf{w}}$. Using Definition 6.2.3 we get

The wedge product $\varphi \wedge \psi$ satisfies a) and b) of Definition 6.2.1 for a form (multilinearity and antisymmetry). Multilinearity is not hard to see; antisymmetry is harder (as was the proof of antisymmetry for the determinant).

$$\underbrace{(dx_1 \wedge dx_2)}_{\varphi} \wedge \underbrace{dx_3}_{\psi}(\vec{\mathbf{u}}, \vec{\mathbf{v}}, \vec{\mathbf{w}}) = \det \begin{bmatrix} u_1 & v_1 & w_1 \\ u_2 & v_2 & w_2 \\ u_3 & v_3 & w_3 \end{bmatrix} \qquad 6.2.35$$

$$= u_1 v_2 w_3 - u_1 v_3 w_2 - u_2 v_1 w_3 + u_2 v_3 w_1 + u_3 v_1 w_2 - u_3 v_2 w_1.$$

If instead we use Equation 6.2.34 for the wedge product, we get

$$(dx_1 \wedge dx_2) \wedge dx_3(\vec{\mathbf{u}}, \vec{\mathbf{v}}, \vec{\mathbf{w}}) = (dx_1 \wedge dx_2)(\vec{\mathbf{u}}, \vec{\mathbf{v}})\, dx_3(\vec{\mathbf{w}})$$

$$- (dx_1 \wedge dx_2)(\vec{\mathbf{u}}, \vec{\mathbf{w}})\, dx_3(\vec{\mathbf{v}})$$

$$+ (dx_1 \wedge dx_2)(\vec{\mathbf{v}}, \vec{\mathbf{w}})\, dx_3(\vec{\mathbf{u}})$$

$$= \det \begin{bmatrix} u_1 & v_1 \\ u_2 & v_2 \end{bmatrix} w_3 - \det \begin{bmatrix} u_1 & w_1 \\ u_2 & w_2 \end{bmatrix} v_3 + \det \begin{bmatrix} v_1 & w_1 \\ v_2 & w_2 \end{bmatrix} u_3 \qquad 6.2.36$$

$$= u_1 v_2 w_3 - u_2 v_1 w_3 - u_1 v_3 w_2 + u_2 v_3 w_1 + u_3 v_1 w_2 - u_3 v_2 w_1. \quad \triangle$$

Properties of the wedge product

The wedge product behaves much like ordinary multiplication, except that one needs to be careful about the sign, because of skew commutativity:

You are asked to prove Proposition 6.2.16 in Exercise 6.2.4. Part (2) is quite a bit harder than the other two. Exercise 6.2.5 asks you to verify that Example 6.2.14 does not commute, and that Example 6.2.15 does.

Part (2) justifies the omission of parentheses in the k-form

$$dx_{i_1} \wedge dx_{i_2} \wedge \cdots \wedge dx_{i_k};$$

all the ways of putting parentheses in the expression give the same result.

Proposition 6.2.16 (Properties of the wedge product). *The wedge product has the following properties:*

(1) distributivity: $\qquad \varphi \wedge (\psi_1 + \psi_2) = \varphi \wedge \psi_1 + \varphi \wedge \psi_2.$ $\qquad$ 6.2.37

(2) associativity: $\qquad (\varphi_1 \wedge \varphi_2) \wedge \varphi_3 = \varphi_1 \wedge (\varphi_2 \wedge \varphi_3).$ $\qquad$ 6.2.38

(3) skew commutativity: $\qquad$ *If φ is a k-form and ψ is an l-form, then*

$$\varphi \wedge \psi = (-1)^{kl}\psi \wedge \varphi. \qquad 6.2.39$$

Note that in Equation 6.2.39 the φ and ψ change positions. For example, if $\varphi = dx_1 \wedge dx_2$ and $\psi = dx_3$, skew commutativity says that

$$(dx_1 \wedge dx_2) \wedge dx_3 = (-1)^2 dx_3 \wedge (dx_1 \wedge dx_2), \quad \text{i.e.,}$$

$$\det \begin{bmatrix} u_1 & v_1 & w_1 \\ u_2 & v_2 & w_2 \\ u_3 & v_3 & w_3 \end{bmatrix} = \det \begin{bmatrix} u_3 & v_3 & w_3 \\ u_1 & v_1 & w_1 \\ u_2 & v_2 & w_2 \end{bmatrix}, \qquad 6.2.40$$

which you can confirm either by observing that the two matrices differ by two exchanges of rows (changing the sign twice) or by carrying out the computation.

6.3 Integrating Form Fields over Parametrized Domains

The objective of this chapter is to define integration and differentiation over *oriented domains*. We now make our first stab at defining integration of forms; we will translate these results into the language of vector calculus in Section 6.4 and will return to orientation and integration of form fields in Section 6.5.

When $k = 2$, then if $\vec{e}_1$ and $\vec{e}_2$ are drawn in the standard way, a direct basis $\vec{v}_1, \vec{v}_2$ is one where $\vec{v}_2$ lies counterclockwise from $\vec{v}_1$. In $\mathbb{R}^3$, a basis $\vec{v}_1, \vec{v}_2, \vec{v}_3$ is direct if it is right-handed, again if $\vec{e}_1, \vec{e}_2$ and $\vec{e}_3$ are drawn in the standard right-handed way. (The right-hand rule is described in Section 1.4.)

We say that k linearly independent vectors $\vec{v}_1, \ldots, \vec{v}_k$ in $\mathbb{R}^k$ form a *direct basis* of $\mathbb{R}^k$ if $\det[\vec{v}_1, \ldots, \vec{v}_k] > 0$, otherwise an indirect basis. Of course, this depends on the order in which the vectors $\vec{v}_1, \ldots, \vec{v}_k$ are taken. We want to think of things like the k-parallelogram $P_{\mathbf{x}}(\vec{v}_1, \ldots, \vec{v}_k)$ in $\mathbb{R}^k$ (which is simply a subset of $\mathbb{R}^k$) *plus the information that the spanning vectors form a direct or an indirect basis.*

The situation when there are k vectors in $\mathbb{R}^n$ and $k \neq n$ is a little different. Consider a parallelogram in $\mathbb{R}^3$ spanned by two vectors, for instance

$$\vec{v}_1 = \begin{bmatrix} 1 \\ 1 \\ -1 \end{bmatrix} \quad \text{and} \quad \vec{v}_2 = \begin{bmatrix} 1 \\ -1 \\ 1 \end{bmatrix}. \tag{6.3.1}$$

This parallelogram has two orientations, but neither is more "direct" than the other. Below we define orientation for such objects.

An oriented k-parallelogram in $\mathbb{R}^n$, denoted $\pm P_{\mathbf{x}}^o(\vec{v}_1, \ldots, \vec{v}_k)$, is a k-parallelogram as defined in Definition 5.1.1, except that this time all the symbols written are part of the data: the anchor point, the vectors $\vec{v}_i$, and the sign. The sign is usually omitted when it is positive.

In $P_{\mathbf{x}}^o(\vec{v}_1, \ldots, \vec{v}_k)$, the little o is there to remind you that this is an oriented parallelogram: an *oriented* subset of $\mathbb{R}^n$.

Definition 6.3.1 (Oriented k-parallelogram). An oriented k-parallelogram $\pm P_{\mathbf{x}}^o(\vec{v}_1, \ldots, \vec{v}_k)$ is a k-parallelogram in which the sign and the order of the vectors are part of the data. The oriented k-parallelograms

$$P_{\mathbf{x}}^o(\vec{v}_1, \ldots, \vec{v}_k) \quad \text{and} \quad -P_{\mathbf{x}}^o(\vec{v}_1, \ldots, \vec{v}_k)$$

have opposite orientations, as do two oriented k-parallelograms

$$P_{\mathbf{x}}^o(\vec{v}_1, \ldots, \vec{v}_k) \quad \text{if two of the vectors are exchanged.}$$

Two oriented k-parallelograms are *opposite* if the data for the two is the same, except that either (1) the sign is changed, or (2) two of the vectors are exchanged (or, more generally, there is an odd number of transpositions of vectors). They are equal if the data is the same except that (1) the order of the vectors differs by an even number of transpositions, or (2) the order differs by an odd number of transpositions, and the sign is changed. For example,

Another way of saying this is that if a permutation σ is applied to the vectors, the parallelogram is multiplied by the signature of σ:

$$P_{\mathbf{x}}^{o}\left(\vec{\mathbf{v}}_{\sigma(1)}, \ldots, \vec{\mathbf{v}}_{\sigma(k)}\right)$$
$$= \operatorname{sgn}(\sigma) P_{\mathbf{x}}^{o}(\vec{\mathbf{v}}_1, \ldots, \vec{\mathbf{v}}_k).$$

$$
\begin{array}{llll}
P_{\mathbf{x}}^{o}(\vec{\mathbf{v}}_1, \vec{\mathbf{v}}_2) & \text{and} & P_{\mathbf{x}}^{o}(\vec{\mathbf{v}}_2, \vec{\mathbf{v}}_1) & \text{are opposite;} \\
P_{\mathbf{x}}^{o}(\vec{\mathbf{v}}_1, \vec{\mathbf{v}}_2) & \text{and} & -P_{\mathbf{x}}^{o}(\vec{\mathbf{v}}_2, \vec{\mathbf{v}}_1) & \text{are equal;} \\
P_{\mathbf{x}}^{o}(\vec{\mathbf{v}}_1, \vec{\mathbf{v}}_2, \vec{\mathbf{v}}_3) & \text{and} & P_{\mathbf{x}}^{o}(\vec{\mathbf{v}}_2, \vec{\mathbf{v}}_1, \vec{\mathbf{v}}_3) & \text{are opposite;} \\
P_{\mathbf{x}}^{o}(\vec{\mathbf{v}}_1, \vec{\mathbf{v}}_2, \vec{\mathbf{v}}_3) & \text{and} & -P_{\mathbf{x}}^{o}(\vec{\mathbf{v}}_2, \vec{\mathbf{v}}_3, \vec{\mathbf{v}}_1) & \text{are opposite;} \\
P_{\mathbf{x}}^{o}(\vec{\mathbf{v}}_1, \vec{\mathbf{v}}_2, \vec{\mathbf{v}}_3) & \text{and} & P_{\mathbf{x}}^{o}(\vec{\mathbf{v}}_2, \vec{\mathbf{v}}_3, \vec{\mathbf{v}}_1) & \text{are equal.}
\end{array}
$$

Are $P_{\mathbf{x}}^{o}(\vec{\mathbf{v}}_1, \vec{\mathbf{v}}_3, \vec{\mathbf{v}}_2), P_{\mathbf{x}}^{o}(\vec{\mathbf{v}}_3, \vec{\mathbf{v}}_1, \vec{\mathbf{v}}_2), -P_{\mathbf{x}}^{o}(\vec{\mathbf{v}}_2, \vec{\mathbf{v}}_1, \vec{\mathbf{v}}_3)$ equal or opposite?[3]

Form fields

The word "field" means data that varies from point to point. The number a form field gives depends on the point at which it is evaluated. A k-form field is also called a "differential form." We find "differential" a mystifying word; it is almost impossible to make sense of the word "differential" as it is used in first year calculus. We know a professor who claims that he has been teaching "differentials" for 20 years and still doesn't know what they are.

Most often, rather than integrate a k-form, we will integrate a k-*form field*. A k-form field φ on an open subset U of $\mathbb{R}^n$ assigns a k-form $\varphi(\mathbf{x})$ to every point $\mathbf{x} \in U$. While the number returned by a k-form depends only on k vectors, the number returned by a k-form field depends also on the point at which is evaluated: a k-form is a function of k vectors, but a k-form field is a function of an oriented k-parallelogram $P_{\mathbf{x}}^{o}(\vec{\mathbf{v}}_1, \ldots, \vec{\mathbf{v}}_k)$, which is anchored at $\mathbf{x}$.

Definition 6.3.2 (k-form field). A k-form field on an open subset $U \subset \mathbb{R}^n$ is a function that takes k vectors $\vec{\mathbf{v}}_1, \ldots, \vec{\mathbf{v}}_k$ anchored at a point $\mathbf{x} \in \mathbb{R}^n$, and which returns a number. It is multilinear and antisymmetric as a function of the $\vec{\mathbf{v}}$'s.

We already know how to write k-form fields: it is any expression of the form

$$\varphi = \sum_{1 \le i_1 < \cdots < i_k \le n} a_{i_1, \ldots, i_k}(\mathbf{x}) \, dx_{i_1} \wedge \cdots \wedge dx_{i_k}, \qquad 6.3.2$$

where the $a_{i_1, \ldots, i_k}$ are real-valued functions of $\mathbf{x} \in U$.

Example 6.3.3 (A 2-form field on $\mathbb{R}^3$). The form field $\cos(xz) \, dx \wedge dy$ is a 2-form field on $\mathbb{R}^3$. Below it is evaluated twice, each time on the same vectors, but at different points:

$$\cos(xz) \, dx \wedge dy \left(P^{o}_{\begin{pmatrix} 1 \\ 2 \\ \pi \end{pmatrix}} \left(\begin{bmatrix} 1 \\ 0 \\ 1 \end{bmatrix}, \begin{bmatrix} 2 \\ 2 \\ 3 \end{bmatrix} \right) \right) = \left(\cos(1 \cdot \pi) \right) \det \begin{bmatrix} 1 & 2 \\ 0 & 2 \end{bmatrix} = -2.$$

$$\cos(xz) \, dx \wedge dy \left(P^{o}_{\begin{pmatrix} 1/2 \\ 2 \\ \pi \end{pmatrix}} \left(\begin{bmatrix} 1 \\ 0 \\ 1 \end{bmatrix}, \begin{bmatrix} 2 \\ 2 \\ 3 \end{bmatrix} \right) \right) = \left(\cos(1/2 \cdot \pi) \right) \det \begin{bmatrix} 1 & 2 \\ 0 & 2 \end{bmatrix} = 0. \quad \triangle$$

From now on we will use the terms form and form field interchangeably; when we wish to speak of a form that returns the same number regardless of the point at which it is evaluated, we will call it a *constant form*.

---

[3] $P_{\mathbf{x}}^{o}(\vec{\mathbf{v}}_3, \vec{\mathbf{v}}_1, \vec{\mathbf{v}}_2) = -P_{\mathbf{x}}^{o}(\vec{\mathbf{v}}_2, \vec{\mathbf{v}}_1, \vec{\mathbf{v}}_3)$. Both are opposite to $P_{\mathbf{x}}^{o}(\vec{\mathbf{v}}_1, \vec{\mathbf{v}}_3, \vec{\mathbf{v}}_2)$.

Integrating form fields over parametrized domains

A *singularity* is a point where a subset fails to meet the criteria for a smooth manifold; for a curve, it could be a point where the curve intersects itself. But a singularity can be much worse than that; for example, the curve could go through itself infinitely many times at such a point.

Before we can integrate form fields over oriented domains, we must define the orientation of domains; we will do this in Section 6.5. Here, as an introduction, we will show how to integrate form fields over domains that come naturally equipped with orientation-preserving parametrizations: *parametrized domains*.

A parametrized k-dimensional domain in $\mathbb{R}^n$ is the *image* $\gamma(A)$ of a C^1 mapping γ that goes from a pavable subset A of $\mathbb{R}^k$ to $\mathbb{R}^n$. Such a domain $\gamma(A)$ may well not be a smooth manifold; a mapping γ always parametrizes something or other in $\mathbb{R}^n$, but $\gamma(A)$ may have horrible singularities (although it is more likely to be mainly a k-dimensional manifold with some bad points). If we had to assign orientation to $\gamma(A)$ this would be a problem; we will see in Section 6.5 how to assign orientation to a manifold, but we don't know how to assign orientation to something that is "mainly a k-dimensional manifold with some bad points."

Fortunately, for our purposes here it doesn't matter how nasty the image is. We don't need to know what $\gamma(A)$ looks like, and we don't have to determine its orientation. We are not thinking of $\gamma(A)$ in its own right, but as "the result of γ acting on A." A parametrization by a mapping γ automatically carries an orientation: γ maps an oriented k-parallelogram $P_{\mathbf{x}}^o(\vec{\mathbf{e}}_1, \dots, \vec{\mathbf{e}}_k)$ to a curvilinear parallelogram that can be approximated by $P_{\gamma(\mathbf{x})}^o(\overrightarrow{D_1\gamma}(\mathbf{x}), \dots, \overrightarrow{D_k\gamma}(\mathbf{x}))$; the order of the vectors in this k-parallelogram depends on the order of the variables in $\mathbb{R}^k$. To the extent that $\gamma(A)$ has an orientation, it is oriented by this order of vectors.

The k-parallelogram in Equation 6.3.3 is oriented: it comes with the spanning vectors in a particular order, which depends on the order in which we took our variables in $\mathbb{R}^k$.

The image $\gamma(A)$ comes with a natural decomposition into little pieces: take some N, and decompose $\gamma(A)$ into the little pieces $\gamma(C\cap A)$, where $C \in \mathcal{D}_N(\mathbb{R}^n)$. Such a piece $\gamma(C \cap A)$ is naturally well approximated by a k-parallelogram: if $\mathbf{u} \in \mathbb{R}^k$ is the lower left-hand corner of C, the parallelogram

$$P_{\gamma(\mathbf{u})}^o \left(\frac{1}{2^N} \overrightarrow{D_1\gamma}(\mathbf{u}), \dots, \frac{1}{2^N} \overrightarrow{D_k\gamma}(\mathbf{u}) \right) \qquad 6.3.3$$

is the image of C by the linear approximation to γ at $\mathbf{u}$.

$$\mathbf{w} \mapsto \gamma(\mathbf{u}) + [\mathbf{D}\gamma(\mathbf{u})](\overrightarrow{\mathbf{w} - \mathbf{u}}). \qquad 6.3.4$$

We hope that this discussion convinces you that Definition 6.3.4 corresponds to the heuristic description of how the integral of a form should work.

So if φ is a k-form field on $\mathbb{R}^n$ (or at least on a neighborhood of $\gamma(A)$), an approximation to

$$\int_{\gamma(A)} \varphi \qquad 6.3.5$$

should be

$$\sum_{\substack{C \in \mathcal{D}_N(\mathbb{R}^n) \\ A \cap C \neq \varnothing}} \varphi \left(P_{\gamma(\mathbf{u})}^o \left(\frac{1}{2^N} \overrightarrow{D_1\gamma}(\mathbf{u}), \dots, \frac{1}{2^N} \overrightarrow{D_k\gamma}(\mathbf{u}) \right) \right)$$

$$= \operatorname{vol}_k(C) \sum_{\substack{C \in \mathcal{D}_N(\mathbb{R}^n) \\ A \cap C \neq \varnothing}} \varphi \left(P_{\gamma(\mathbf{u})}^o \left(\overrightarrow{D_1\gamma}(\mathbf{u}), \dots, \overrightarrow{D_k\gamma}(\mathbf{u}) \right) \right). \qquad 6.3.6$$

But this last sum is a Riemann sum for the integral

$$\int_A \varphi\left(P^o_{\gamma(\mathbf{u})}\left(\overrightarrow{D_1\gamma}(\mathbf{u}),\ldots,\overrightarrow{D_k\gamma}(\mathbf{u})\right)\right)|d^k\mathbf{u}|. \qquad 6.3.7$$

To be rigorous, we *define* $\int_{\gamma(A)}\varphi$ to be the above integral:

Definition 6.3.4 (Integrating a k-form field over a parametrized domain). Let $A \subset \mathbb{R}^k$ be a pavable set and $\gamma : A \to \mathbb{R}^n$ be a C^1 mapping. Then the integral of the k-form field φ over $\gamma(A)$ is

$$\int_{\gamma(A)}\varphi = \int_A \underbrace{\varphi\left(P^o_{\gamma(\mathbf{u})}\left(\overrightarrow{D_1\gamma}(\mathbf{u}),\ldots,\overrightarrow{D_k\gamma}(\mathbf{u})\right)\right)}_{\text{This is a function of } \mathbf{u}.}|d^k\mathbf{u}|. \qquad 6.3.8$$

Example 6.3.5 (Integrating a 1-form field over a parametrized curve). Consider a case where $k = 1, n = 2$. We will use $\gamma(u) = \begin{pmatrix} R\cos u \\ R\sin u \end{pmatrix}$ and will take A to be the interval $[0,\alpha]$, for some $\alpha > 0$. If we integrate the 1-form field $x\,dy - y\,dx$ over $\gamma(A)$ using the above definition, we find

$$\int_{\gamma(A)} (x\,dy - y\,dx) = \int_{[0,\alpha]} (x\,dy - y\,dx)\left(P^o_{\begin{pmatrix} R\cos u \\ R\sin u \end{pmatrix}}\begin{bmatrix} -R\sin u \\ R\cos u \end{bmatrix}\right)|du|$$

$$= \int_{[0,\alpha]} \left(R\cos u R\cos u - (R\sin u)(-R\sin u)\right)|du| = \int_{[0,\alpha]} R^2\,|du| \qquad 6.3.9$$

$$= \int_0^\alpha R^2\,du = R^2\alpha.$$

What would we have gotten if $\alpha < 0$? Until the bottom line, everything is the same. But then we have to decide how to interpret $[0,\alpha]$. Should we write

$$\int_0^\alpha R^2 du \quad \text{or} \quad \int_\alpha^0 R^2\,du\,? \qquad 6.3.10$$

We have to choose the second, because we are now integrating over an oriented interval, and we must choose the positive orientation. So the answer is $-R^2\alpha$, which is still positive, since $\alpha < 0$ $\triangle$

Example 6.3.6 (Another parametrized curve). In Example 6.3.5, you probably saw that γ was parametrizing an arc of circle. To carry out the sort of computation we are discussing, the image need not be a smooth curve. For that matter, we don't need to have any idea what $\gamma(A)$ looks like.

Sidebar (left margin):

Since $k = 1$, in the first line of Equation 6.3.9, we have the single vector $\begin{bmatrix} -R\sin u \\ R\cos u \end{bmatrix}$ rather than the $\overrightarrow{D_1\gamma}(\mathbf{u}),\ldots,\overrightarrow{D_k\gamma}(\mathbf{u})$ of Definition 6.3.4.

In the second line of Equation 6.3.9, the first $R\cos u$ is x, while the second is the number given by dy evaluated on the parallelogram. Similarly, $R\sin u$ is y, and $(-R\sin u)$ is the number given by dx evaluated on the parallelogram.

Remember that φ is the integrand on the left side of Equation 6.3.8, just as $|d^k\mathbf{u}|$ is the integrand on the right. We use φ to avoid writing

$$\sum_{1\le i_1<\ldots<i_k\le n} a_{i_1,\ldots,i_k}(\mathbf{x})dx_{i_1}\wedge\ldots\wedge dx_{i_k}.$$

Why must we choose the positive orientation? The interval $[\alpha,0]$ is a subset of $\mathbb{R}$, so the orientation is determined by the orientation of $\mathbb{R}$. The standard orientation of $\mathbb{R}$ is from negative to positive.

Similarly, $\mathbb{R}^2$ and $\mathbb{R}^3$ also have a standard orientation: that in which the standard basis vectors are written in the normal way, giving $\det[\vec{e}_1, \vec{e}_2] = 1 > 0$ and $\det[\vec{e}_1, \vec{e}_2, \vec{e}_3] = 1 > 0$. For $\mathbb{R}^3$, this is equivalent to the right-hand rule (see Proposition 1.4.20). This will be discussed further in Section 6.5.

Take for instance $\gamma(t) = \begin{pmatrix} 1+t^2 \\ \arctan t \end{pmatrix}$, set $A = [0,a]$ for some $a > 0$ and $\varphi = x\,dy$. Then

$$\int_{\gamma(A)} \varphi = \int_{[0,a]} x\,dy \left(P_{\begin{pmatrix} 1+t^2 \\ \arctan t \end{pmatrix}} \begin{bmatrix} 2t \\ 1/(1+t^2) \end{bmatrix} \right) |dt| \qquad \text{6.3.11}$$

$$= \int_0^a \frac{1+t^2}{1+t^2}\,dt = a. \quad \triangle$$

In Equations 6.3.11 and 6.3.14 we drop the absolute value signs around dt and $ds\,dt$ when we are integrating over an oriented domain like $\int_0^a$ and $\int_0^1 \int_0^1$ as opposed to an unoriented one like $\gamma(C)$.

Example 6.3.7 (Integrating a 2-form field over a parametrized surface in $\mathbb{R}^3$). Let us compute

$$\int_{\gamma(C)} dx \wedge dy + y\,dx \wedge dz \qquad \text{6.3.12}$$

over the parametrized domain $\gamma(C)$ where

$$\gamma\begin{pmatrix} s \\ t \end{pmatrix} = \begin{pmatrix} s+t \\ s^2 \\ t^2 \end{pmatrix}, \quad C = \left\{ \begin{pmatrix} s \\ t \end{pmatrix} \;\middle|\; 0 \le s, t \le 1 \right\}. \qquad \text{6.3.13}$$

Applying Definition 6.3.4, we find
$$\frac{\partial \gamma_i}{\partial s} \qquad \frac{\partial \gamma_i}{\partial t} \qquad i=1,2,3$$

$$\int_{\gamma(C)} dx \wedge dy + y\,dx \wedge dz$$

$$= \int_0^1 \int_0^1 (dx \wedge dy + y\,dx \wedge dz) \left(P_{\begin{pmatrix} s+t \\ s^2 \\ t^2 \end{pmatrix}} \left(\begin{bmatrix} 1 \\ 2s \\ 0 \end{bmatrix}, \begin{bmatrix} 1 \\ 0 \\ 2t \end{bmatrix} \right) \right) ds\,dt$$

$$= \int_0^1 \int_0^1 \left(\det \begin{bmatrix} 1 & 1 \\ 2s & 0 \end{bmatrix} + s^2 \det \begin{bmatrix} 1 & 1 \\ 0 & 2t \end{bmatrix} \right) ds\,dt \quad \checkmark$$

$$= \int_0^1 \int_0^1 \left(-2s + s^2(2t) \right) ds\,dt$$

$$= \int_0^1 \left[s^2 + \frac{s^3}{3}2t \right]_{s=0}^{s=1} dt = \int_0^1 \left(-1 + \frac{2t}{3} \right) dt \qquad \text{6.3.14}$$

$$= \left[-t + \frac{t^2}{3} \right]_0^1 = -\frac{2}{3}. \quad \triangle$$

6.4 FORMS AND VECTOR CALCULUS

The real difficulty with forms is imagining what they are. What "is" $dx_1 \wedge dx_2 + dx_3 \wedge dx_4$? We have seen that it is the function that takes two vectors in $\mathbb{R}^4$, projects them first onto the (x_1, x_2)-plane and takes the signed area of the resulting parallelogram, then projects them onto the (x_3, x_4)-plane, takes the signed area of that parallelogram, and finally adds the two signed volumes.

But that description is extremely convoluted, and although it isn't too hard to use it in computations, it hardly expresses understanding.

However, in $\mathbb{R}^3$, it really is possible to visualize all forms and form fields, because they can be described in terms of functions and vector fields. There are four kinds of forms on $\mathbb{R}^3$: 0-forms, 1-forms, 2-forms, and 3-forms. Each has its own personality.

Acquiring an intuitive understanding of what sort of information a k-form encodes really is difficult. In some sense, a first course in electromagnetism is largely a matter of understanding what sort of beast the electromagnetic field is, namely a 2-form field on $\mathbb{R}^4$.

0-form fields. In $\mathbb{R}^3$ and in any $\mathbb{R}^n$, a 0-form is simply a number, and a 0-form field is simply a function. If $U \subset \mathbb{R}^n$ is open and $f : U \to \mathbb{R}$ is a function, then the rule $f(P_\mathbf{x}^o) = f(\mathbf{x})$ makes f into a 0-form field. The requirement of antisymmetry then says that $f(-P_\mathbf{x}^o) = -f(\mathbf{x})$.

We use f to denote both a function and a 0-form field; the function f is the 0-form field f.

1-form fields. Let $\vec{F}$ be a vector field on an open subset $U \subset \mathbb{R}^n$. We can then associate to $\vec{F}$ a 1-form field $W_{\vec{F}}$, which we call the *work form field*:

Definition 6.4.1 (Work form field). The work form field $W_{\vec{F}}$ of a vector field $\vec{F} = \begin{bmatrix} F_1 \\ \vdots \\ F_n \end{bmatrix}$ is the 1-form field defined by

$$W_{\vec{F}}\big(P_\mathbf{x}^o(\vec{v})\big) = \vec{F}(\mathbf{x}) \cdot \vec{v}. \qquad 6.4.1$$

The 1-form field $x\,dx + y\,dy + z\,dz$ is the work form field of the vector field $\vec{F}\begin{pmatrix} x \\ y \\ z \end{pmatrix} = \begin{bmatrix} x \\ y \\ z \end{bmatrix}$, the radial vector field shown (in $\mathbb{R}^2$) in Figure 1.1.6:

$$(x\,dx + y\,dy + z\,dz)\big(P_\mathbf{x}^o(\vec{v})\big) =$$

$$x \cdot \det v_1 + y \cdot \det v_2 + z \cdot \det v_3$$

$$= \begin{bmatrix} x \\ y \\ z \end{bmatrix} \cdot \begin{bmatrix} v_1 \\ v_2 \\ v_3 \end{bmatrix}.$$

This can also be written in coordinates: the work form field $W_{\vec{F}}$ of a vector field $\vec{F} = \begin{bmatrix} F_1 \\ \vdots \\ F_n \end{bmatrix}$ is the 1-form field $F_1 dx_1 + \cdots + F_n dx_n$. Indeed,

$$\big(F_1 dx_1 + \cdots + F_n dx_n\big)\big(P_\mathbf{x}^o(\vec{v})\big) = \big(F_1(\mathbf{x})dx_1 + \cdots + F_n(\mathbf{x})dx_n\big)\begin{bmatrix} v_1 \\ \vdots \\ v_n \end{bmatrix}$$

$$= F_1(\mathbf{x})v_1 + \cdots + F_n(\mathbf{x})v_n = \vec{F}(\mathbf{x}) \cdot \vec{v}.$$

In this form, it is clear from Theorem 6.2.7 that every 1-form on U is the work of some vector field.

What have we gained by saying that a 1-form field is the work form of a vector field? Mainly that it is quite easy to visualize $W_{\vec{F}}$ and to understand what it measures: if $\vec{F}$ is a force field, its work form field associates to a little line segment the work that the force field does along the line segment. To really understand this you need a little bit of physics, but even without it you can see what it means. Suppose for instance that $\vec{F}$ is the force field of gravity. In the absence of friction, it requires no work to push a wagon of mass m horizontally from $\mathbf{a}$ to $\mathbf{b}$; the vector $\overrightarrow{\mathbf{b} - \mathbf{a}}$ and the constant vector field representing gravity

In Equation 6.4.2, g represents the acceleration of gravity at the surface of the earth, and m is mass; $-gm$ is weight, a force; it is negative because it points down.

The unit of a force field such as the gravitation field is energy per length. It's the per length that tells us that the integrand to associate to a force field should be something to be integrated over curves. Since direction makes a difference—it takes work to push a wagon uphill but the wagon rolls down by itself—the appropriate integrand is a 1-form field, which is integrated over *oriented* curves.

The Φ in the flux form field $\Phi_{\vec{F}}$ is of course unrelated to the Φ of the "concrete to abstract" function $\Phi_{\{\underline{v}\}}$ introduced in Section 2.6.

It may be easier to remember the coordinate definition of $\Phi_{\vec{F}}$ if it is written

$$\Phi_{\vec{F}} =$$
$$F_1\, dy \wedge dz + F_2\, dz \wedge dx + F_3\, dx \wedge dy$$

(changing the order and sign of the middle term). Then (for $x = 1, y = 2, z = 3$) you can think that the first term goes $(1,2,3)$, the second $(2,3,1)$, and the third $(3,1,2)$. For instance,

$$\Phi \begin{pmatrix} x \\ y \\ z \end{pmatrix} =$$
$$x\, dy \wedge dz + y\, dz \wedge dx + z\, dx \wedge dy.$$

are orthogonal to each other, with dot product zero:

$$\begin{bmatrix} 0 \\ 0 \\ -gm \end{bmatrix} \cdot \begin{bmatrix} b_1 - a_1 \\ b_2 - a_2 \\ 0 \end{bmatrix} = 0. \qquad 6.4.2$$

But if the wagon rolls down an inclined plane, the force field of gravity does "work" on the wagon equal to the dot product of gravity and the displacement vector of the wagon:

$$\begin{bmatrix} 0 \\ 0 \\ -gm \end{bmatrix} \cdot \begin{bmatrix} b_1 - a_1 \\ b_2 - a_2 \\ b_3 - a_3 \end{bmatrix} = -gm(b_3 - a_3), \qquad 6.4.3$$

which is positive, since $b_3 - a_3$ is negative. If you want to push the wagon back up the inclined plane, you will need to furnish the work, and the force field of gravity will do negative work.

For what vector field $\vec{F}$ can the 1-form field $x_2\, dx_1 + x_2 x_4\, dx_2 + x_1^2\, dx_4$ be written as $W_{\vec{F}}$?[4]

2-forms. If $\vec{F}$ is a vector field on an open subset $U \subset \mathbb{R}^3$, then we can associate to it a 2-form field on U called its *flux form field* $\Phi_{\vec{F}}$, which we first saw in Example 6.1.1.

Definition 6.4.2 (Flux form field). The flux form field $\Phi_{\vec{F}}$ is the 2-form field defined by

$$\Phi_{\vec{F}}\big(P_{\mathbf{x}}^o(\vec{v}, \vec{w})\big) = \det[\vec{F}(\mathbf{x}),\ \vec{v},\ \vec{w}]. \qquad 6.4.4$$

In coordinates, this becomes $\Phi_{\vec{F}} = F_1\, dy \wedge dz - F_2\, dx \wedge dz + F_3\, dx \wedge dy$:

$$(F_1\, dy \wedge dz - F_2\, dx \wedge dz + F_3\, dx \wedge dy)\, P_{\mathbf{x}}^o \left(\begin{bmatrix} v_1 \\ v_2 \\ v_3 \end{bmatrix} \begin{bmatrix} w_1 \\ w_2 \\ w_3 \end{bmatrix} \right)$$
$$\qquad 6.4.5$$
$$= F_1(\mathbf{x})(v_2 w_3 - v_3 w_2) - F_2(\mathbf{x})(v_1 w_3 - v_3 w_1) + F_3(\mathbf{x})(v_1 w_2 - v_2 w_1)$$
$$= \det[\vec{F}(\mathbf{x}),\ \vec{v},\ \vec{w}].$$

In this form, it is clear, again from Theorem 6.2.7, that all 2-form fields on $\mathbb{R}^3$ are flux form fields of a vector field: the flux form field is a linear combination of all the elementary 2-forms on $\mathbb{R}^3$, so it is just a question of using the coefficients of the elementary forms to make a vector field.

---

[4] It is the work form field of the vector field $\vec{F} \begin{pmatrix} x_1 \\ x_2 \\ x_3 \\ x_4 \end{pmatrix} = \begin{bmatrix} x_2 \\ x_2 x_4 \\ 0 \\ x_1^2 \end{bmatrix}$.

Once more, what we have gained is an ability to visualize, as suggested by Figure 6.4.1: the flux form field of a vector field associates to a parallelogram the flow of the vector field through the parallelogram.

If a vector field represents the flow of a fluid, what units will it have? Clearly the vector field measures how much fluid flows through a unit surface perpendicular to the direction of flow in unit time: the units should be mass/(length^2). The length^2 in the denominator tips us off that the appropriate integrand to associate to this vector field is a 2-form, or at least an integrand to be integrated over a surface. You might go one step further, and say it is a 3-form on spacetime: the result of integrating it over a surface in space and an interval in time is the total mass flowing through that region of spacetime. In general, any $(n-1)$-form field in $\mathbb{R}^n$ can be considered a flux form field.

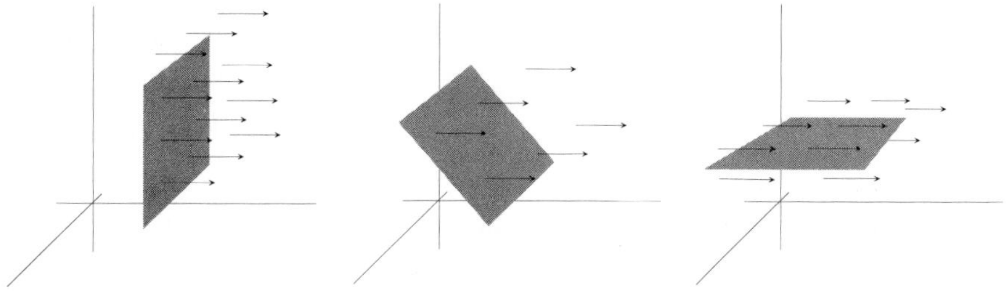

FIGURE 6.4.1. The flow of $\vec{F}$ through a surface depends on the angle between $\vec{F}$ and the surface. Left: $\vec{F}$ is orthogonal to the surface, providing maximum flow. This corresponds to $\vec{F}(\mathbf{x})$ being perpendicular to the parallelogram spanned by $\vec{v}, \vec{w}$. (The volume of the parallelepiped spanned by $\vec{F}, \vec{v}, \vec{w}$ is $\det[\vec{F}, \vec{v}, \vec{w}] = \vec{F} \cdot (\vec{v} \times \vec{w})$, which is greatest when the angle θ between $\vec{F}$ and $\vec{v} \times \vec{w}$ is 0, since $\vec{x} \cdot \vec{y} = |\vec{x}||\vec{y}| \cos\theta$.) Middle: $\vec{F}$ is not orthogonal to the surface, allowing less flow. Right: $\vec{F}$ is parallel to the surface; the flow is 0. In this case $P_{\mathbf{x}}^o(\vec{F}(\mathbf{x}), \vec{v}, \vec{w})$ is flat. This corresponds to $\vec{F} \cdot (\vec{v} \times \vec{w}) = 0$, i.e., $\vec{F}$ is perpendicular to $\vec{v} \times \vec{w}$ and therefore parallel to the parallelogram spanned by $\vec{v}$ and $\vec{w}$.

If $\vec{F}$ is the velocity vector field of a fluid, the integral of its flux form field over a surface measures the amount of fluid flowing through the surface. Indeed, the fluid that flows through the parallelogram $P_{\mathbf{x}}^o(\vec{v}, \vec{w})$ in unit time will fill the parallelepiped $P_{\mathbf{x}}^o(\vec{F}(\mathbf{x}), \vec{v}, \vec{w})$: the particle which at time 0 was at the corner $\mathbf{x}$ is now at $\mathbf{x} + \vec{F}(\mathbf{x})$. The sign is positive if $\vec{F}$ is on the same side of the parallelogram as $\vec{v} \times \vec{w}$, otherwise negative (and 0 if $\vec{F}$ is parallel to the parallelogram; indeed, nothing flows through it then).

Recall that ρ is the Greek letter "rho."

The 3-form $dx \wedge dy \wedge dz$ is another name for the determinant:

$$dx \wedge dy \wedge dz(\vec{v}_1, \vec{v}_2, \vec{v}_3)$$
$$= \det[\vec{v}_1, \vec{v}_2, \vec{v}_3].$$

The characteristic of functions which really should be considered as densities is that they have units *something/cubic length*, such as ordinary density (kg/m^3) or charge density (coulombs/m^3).

3-forms. Any 3-form on an open subset of $\mathbb{R}^3$ is the 3-form $dx \wedge dy \wedge dz$ (alias the determinant) multiplied by a function: we will denote by ρ_f the 3-form $f \, dx \wedge dy \wedge dz$, and call it the *density of f*.

Definition 6.4.3 (Density form of a function). Let U be a subset of $\mathbb{R}^3$. The density form ρ_f of a function $f : U \to \mathbb{R}$ is the 3-form defined by

$$\underbrace{\rho_f}_{\substack{\text{density} \\ \text{form} \\ \text{of } f}} \big(P_{\mathbf{x}}^o(\vec{v}_1, \vec{v}_2, \vec{v}_3) \big) = f(\mathbf{x}) \underbrace{\det[\vec{v}_1, \vec{v}_2, \vec{v}_3]}_{\text{signed volume of } P} . \qquad 6.4.6$$

Summary: work, flux, and density forms on $\mathbb{R}^3$

Let f be a function on $\mathbb{R}^3$ and $\vec{F} = \begin{bmatrix} F_1 \\ F_2 \\ F_3 \end{bmatrix}$ be a vector field. Then

$$W_{\vec{F}} = F_1 dx + F_2 dy + F_3 dz \tag{6.4.7}$$

$$\Phi_{\vec{F}} = F_1 dy \wedge dz - F_2 dx \wedge dz + F_3 dx \wedge dy \tag{6.4.8}$$

$$\rho_f = f\, dx \wedge dy \wedge dz. \tag{6.4.9}$$

To keep straight the order of the elementary 2-forms in Equation 6.4.8, think that the first one omits dx, the second omits dy and the third omits dz.

Recall that the work form field $W_{\vec{F}}$ of a vector field $\vec{F}$ is the 1-form field

$$W_{\vec{F}}\left(P_{\mathbf{x}}^o(\vec{\mathbf{v}})\right) = \vec{F}(\mathbf{x}) \cdot \vec{\mathbf{v}}.$$

Integrating work, flux and density form fields over parametrized domains

Now let us translate Definition 6.3.4 (integrating a k-form field over a parametrized domain) into the language of vector calculus.

Example 6.4.4 (Integrating a work form over a parametrized curve).
When integrating the work form field over a parametrized curve $\gamma(A) = C$, the equation of Definition 6.3.4:

$$\int_{\gamma(A)} \varphi = \int_A \varphi\left(P_{\gamma(\mathbf{u})}^o(\overrightarrow{D_1\gamma}(\mathbf{u}), \ldots, \overrightarrow{D_k\gamma}(\mathbf{u}))\right)|d^k\mathbf{u}| \tag{6.3.8}$$

becomes

$$\int_{\gamma(A)} W_{\vec{F}} = \int_A W_{\vec{F}}\left(P_{\gamma(u)}^o(\underbrace{\overrightarrow{D_1\gamma}(u)}_{\vec{\gamma}'(u)})\right)|du| = \int_A \vec{F}\big(\gamma(u)\big) \cdot \vec{\gamma}'(u)|du| \tag{6.4.10}$$

This integral measures the work of the force field $\vec{F}$ along the curve.   △

Example 6.4.5 (Integrating a work form over a helix). What is the work of the vector field

Note that the vectors $\vec{\mathbf{v}}, \vec{\mathbf{v}}_1, \vec{\mathbf{v}}_2$, and $\vec{\mathbf{v}}_3$ of the definitions for the work form, flux form, and density form, are replaced in the integration formulas by derivatives of the parametrizations: i.e., by $\vec{\gamma}'(t), \overrightarrow{D_1\gamma}, \overrightarrow{D_2\gamma}$, and $[\mathbf{D}\gamma(\mathbf{u})]$.

$$\vec{F}\begin{pmatrix} x \\ y \\ z \end{pmatrix} = \begin{bmatrix} y \\ -x \\ 0 \end{bmatrix} \quad \text{over the helix parametrized by} \tag{6.4.11}$$

$$\gamma(t) = \begin{bmatrix} \cos t \\ \sin t \\ t \end{bmatrix}, \quad 0 < t < 4\pi ? \tag{6.4.12}$$

By Equation 6.4.10 this is

$$\int_0^{4\pi} \begin{bmatrix} \sin t \\ -\cos t \\ 0 \end{bmatrix} \cdot \begin{bmatrix} -\sin t \\ \cos t \\ 1 \end{bmatrix} dt = \int_0^{4\pi} (-\sin^2 t - \cos^2 t)\, dt = -4\pi. \quad \triangle \tag{6.4.13}$$

Example 6.4.6 (Integrating a flux form over a 2-dimensional parametrized domain in $\mathbb{R}^3$). Let U be a subset of $\mathbb{R}^2$, $\gamma : U \to \mathbb{R}^3$ be a parametrized domain, and $\vec{F}$ be a vector field defined on a neighborhood of $\gamma(U)$. Then

The flux form field $\Phi_{\vec{F}}$ of a vector field $\vec{F}$ is the 2-form field

$$\Phi_{\vec{F}}\left(P_{\mathbf{x}}^o(\vec{\mathbf{v}}_1, \vec{\mathbf{v}}_2)\right) = \det[\vec{F}(\mathbf{x}), \vec{v}, \vec{w}].$$

If $\vec{F}$ is the velocity vector field of a fluid, the integral of a flux form field measures the amount of fluid flowing through a surface.

$$\int_{\gamma(U)} \Phi_{\vec{F}} = \int_U \Phi_{\vec{F}}\left(P_{\gamma(\mathbf{u})}^o\left(\overrightarrow{D_1\gamma}(\mathbf{u}), \overrightarrow{D_2\gamma}(\mathbf{u})\right)\right) |d^2\mathbf{u}|$$

$$= \int_U \det\left[\vec{F}\big(\gamma(\mathbf{u})\big), \overrightarrow{D_1\gamma}, \overrightarrow{D_2\gamma}\right] |d^2\mathbf{u}|. \qquad 6.4.14$$

If $\vec{F}$ is the velocity vector field of a fluid, this integral measures the amount of fluid flowing through $\gamma(U)$. $\triangle$

Example 6.4.7. The flux of the vector field $\vec{F}\begin{pmatrix} x \\ y \\ z \end{pmatrix} = \begin{bmatrix} x \\ y^2 \\ z \end{bmatrix}$ through the

parametrized domain

$$\begin{pmatrix} u \\ v \end{pmatrix} \mapsto \begin{pmatrix} u^2 \\ uv \\ v^2 \end{pmatrix}, 0 \le u, v \le 1 \qquad \text{is}$$

$$\int_0^1 \int_0^1 \det\begin{bmatrix} u^2 & 2u & 0 \\ u^2v^2 & v & u \\ v^2 & 0 & 2v \end{bmatrix} du\, dv = \int_0^1 \int_0^1 (2u^2v^2 - 4u^3v^3 + 2u^2v^2)\, du\, dv$$

$$= \int_0^1 \left[\frac{4}{3}u^3v^2 - u^4v^3\right]_{u=0}^1 dv = \int_0^1 \left(\frac{4}{3}v^2 - v^3\right) dv = \frac{7}{36}. \quad \triangle \qquad 6.4.15$$

The density form field ρ_f is the 3-form

$$\rho_f\left(P_{\mathbf{x}}^o(\mathbf{v}_1, \mathbf{v}_2, \mathbf{v}_3)\right)$$
$$= f(\mathbf{x})\det[\vec{\mathbf{v}}_1, \vec{\mathbf{v}}_2, \vec{\mathbf{v}}_3].$$

In coordinates, ρ_f is written

$$f(\mathbf{x})\, dx \wedge dy \wedge dz.$$

Example 6.4.8 (Integrating a density form over a parametrized piece of $\mathbb{R}^3$). Let $U, V \subset \mathbb{R}^3$ be open sets, and $\gamma : U \to V$ be a C^1 mapping. If $f : V \to \mathbb{R}$ is a function then

$$\int_{\gamma(U)} \rho_f = \int_U \rho_f\left(P_{\gamma(\mathbf{u})}^o\left(\overrightarrow{D_1\gamma}(\mathbf{u}), \overrightarrow{D_2\gamma}(\mathbf{u}), \overrightarrow{D_3\gamma}(\mathbf{u})\right)\right) |d^3\mathbf{u}|$$

$$= \int_U f\big(\gamma(\mathbf{u})\big)\det[\mathbf{D}\gamma(\mathbf{u})]\, |d^3\mathbf{u}|. \qquad 6.4.16$$

There is a particularly important special case of such a mapping $\gamma : U \to V$: the case where $V = U$ and $\gamma(\mathbf{x}) = \mathbf{x}$ is the identity. In that case, the formula for integrating a density form field becomes

$$\int_{\gamma(U)} \rho_f = \int_U f(\mathbf{u})\, |d^3\mathbf{u}|, \qquad 6.4.17$$

i.e., the integral of ρ_f is simply what we had called the integral of f in Section 4.2. If f is the density of some object, then this integral measures its mass. $\triangle$

Example 6.4.9 (Integrating a density form). Let f be the function

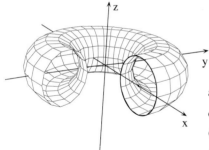

$$f \begin{pmatrix} x \\ y \\ z \end{pmatrix} = x^2 + y^2, \qquad 6.4.18$$

and for $r < R$, let $T_{r,R}$ be the torus obtained by rotating the circle of radius r centered at $\begin{pmatrix} R \\ 0 \end{pmatrix}$ in the (x, z)-plane around the z-axis, as shown in Figure 6.4.2. Compute the integral of ρ_f over the region bounded by $T_{r,R}$ (i.e., the *inside* of the torus). Here, using the identity parametrization would lead to quite a clumsy integral. The following parametrization, with $0 \le u \le r$, $0 \le v, w \le 2\pi$, is better adapted:

$$\gamma \begin{pmatrix} u \\ v \\ w \end{pmatrix} = \begin{pmatrix} (R + u\cos v)\cos w \\ (R + u\cos v)\sin w \\ u \sin v \end{pmatrix}. \qquad 6.4.19$$

The integral becomes

$$\int_0^{2\pi} \int_0^{2\pi} \int_0^r f\left(\gamma \begin{pmatrix} u \\ v \\ w \end{pmatrix}\right) \det\left[\mathbf{D}\gamma \begin{pmatrix} u \\ v \\ w \end{pmatrix}\right] \, du\, dv\, dw$$

$$= \int_0^{2\pi} \int_0^{2\pi} \int_0^r \overbrace{-(R + u\cos v)^2}^{f(\gamma(\mathbf{u}))} \overbrace{u(R + u\cos v)}^{\det[\mathbf{D}\gamma]} \, du\, dv\, dw$$

$$= 2\pi \int_0^{2\pi} \int_0^r (-R^3 u - 3R^2 u^2 \cos v - 3R u^3 \cos^2 v - u^4 \cos^3 v) \, du\, dv$$

$$= -2\pi \int_0^{2\pi} \left(\frac{R^3 r^2}{2} + R^2 r^3 \cos v + \frac{3Rr^4 \cos^2 v}{4} + \frac{r^5 \cos^3 v}{5} \right) dv$$

$$= -\pi^2 \left(2R^3 r^2 + \frac{3Rr^4}{2} \right) \quad \triangle \qquad 6.4.20$$

Work, flux and density in $\mathbb{R}^n$

In all dimensions,

(1) 0-form fields are functions.

(2) Every 1-form field is the work form field of a vector field.

(3) Every $(n-1)$-form field is the flux form field of a vector field

(4) Every n-form is the density form field of a function.

FIGURE 6.4.2.

This torus was discussed in Example 5.4.2. This time we are interested in the solid torus.

When computing integrals by hand, the choice of parametrization can make a big difference in how hard it is. It's always a good idea to choose a parametrization that reflects the symmetries of the problem. Here the torus is symmetrical around the z-axis; Equation 6.4.19 reflects that symmetry.

You might wonder whether Equation 6.4.20 has anything to do with the integral we would have obtained using the identity parametrization. Since it involves $\det[\mathbf{D}\gamma]$, the change of variables formula might well say that the integrals are equal. This is not true: the absolute value that appears in the change of variables formula isn't present here, which changes the sign of the integral. Really figuring out whether the absolute value is needed will be a lengthy story, involving a precise definition of orientation.

Every $(n-1)$-form field on $\mathbb{R}^n$ is the flux form field of a vector field $\vec{F}$: an $(n-1)$-form field acts on $n-1$ vectors, so those vectors, plus the vector $\vec{F}(\mathbf{x})$, can be used to form the $n \times n$ matrix

$$\left[\vec{F}(\mathbf{x}), \vec{v}_1, \ldots, \vec{v}_{n-1}\right],$$

which has a determinant. Fewer than $n-1$ vectors, or more, would not work, because the matrix would not be square.

Notice that a vector field in $\mathbb{R}^2$ defines two different 1-forms: the work and the flux.

Exercise 6.4.8 asks you to verify that $\det[\vec{F}(\mathbf{x}), \vec{v}_1, \ldots, \vec{v}_{n-1}]$ is an $(n-1)$-form field; they are related by the rule

$$W_{\vec{F}}(\vec{v}) = \Phi_{\vec{F}}\left(\begin{bmatrix} 0 & 1 \\ -1 & 0 \end{bmatrix} \vec{v}\right).$$

Equation 6.4.23: in the first term, where $i=1$, we omit dx_1; in the second term, where $i=2$, we omit dx_2, and so on.

In dimensions higher than $\mathbb{R}^3$, some form fields cannot be expressed in terms of vector fields and functions: in particular, 2-forms on $\mathbb{R}^4$, which are of great interest in physics, since the electromagnetic field is such a 2-form on spacetime. The language of vector calculus is not suited to describing integrands over surfaces in higher dimensions, while the language of forms is.

We've already seen this for 0-form fields and 1-form fields. In $\mathbb{R}^3$, the flux form field is of course a $2 = (n-1)$-form field; its definition can be generalized:

Definition 6.4.10 (Flux form field on $\mathbb{R}^n$). If $\vec{F}$ is a vector field on $U \subset \mathbb{R}^n$, then the flux form field $\Phi_{\vec{F}}$ is the $(n-1)$-form field defined by the formula

$$\Phi_{\vec{F}} P_{\mathbf{x}}^o(\vec{v}_1, \ldots, \vec{v}_{n-1}) = \det\left[\vec{F}(\mathbf{x}), \vec{v}_1, \ldots, \vec{v}_{n-1}\right]. \qquad 6.4.21$$

In coordinates, this becomes

$$\begin{aligned}
\Phi_{\vec{F}} &= \sum_{i=1}^{n} (-1)^{i-1} F_i \, dx_1 \wedge \cdots \wedge \widehat{dx_i} \wedge \cdots \wedge dx_n \\
&= F_1 \, dx_2 \wedge \cdots \wedge dx_n - F_2 \, dx_1 \wedge dx_3 \wedge \cdots \wedge dx_n + \ldots \\
&\quad + (-1)^{n-1} F_n \, dx_1 \wedge dx_2 \wedge \cdots \wedge dx_{n-1},
\end{aligned} \qquad 6.4.22$$

where the term under the hat is omitted.

For instance, the flux of the radial vector field $\vec{F}\begin{pmatrix} x_1 \\ \vdots \\ x_n \end{pmatrix} = \begin{bmatrix} x_1 \\ \vdots \\ x_n \end{bmatrix}$ is

$$\Phi_{\vec{F}} = (x_1 \, dx_2 \wedge \cdots \wedge dx_n) - (x_2 \, dx_1 \wedge dx_3 \wedge \cdots \wedge dx_n) + \cdots \pm x_n \, dx_1 \wedge \cdots \wedge dx_{n-1}, \qquad 6.4.23$$

where the last term is positive if n is odd, and negative if it is even.

In any dimension n, n-form fields are multiples of the determinant, so all n-form fields are densities of functions:

Definition 6.4.11 (Density form field on $\mathbb{R}^n$). Let U be a subset of $\mathbb{R}^n$. The density form field ρ_f of a function $f : U \to \mathbb{R}$ is given by

$$\rho_f = f \, dx_1 \wedge \cdots \wedge dx_n.$$

The correspondences between form fields, functions, and vectors, summarized in Table 6.4.3, explain why vector calculus works in $\mathbb{R}^3$—and why it doesn't work in dimensions higher than 3. For k-forms on $\mathbb{R}^n$, when k is anything other than $0, 1, n-1$, or n, there is no interpretation of form fields in terms of functions or vector fields.

A particularly important example is the electromagnetic field, which is a 6-component object, and thus cannot be represented either as a function (a 1-component object) or a vector field (in $\mathbb{R}^4$, a 4-component object).

The standard way of dealing with the problem is to choose coordinates x, y, z, t, in particular choosing a specific space-like subspace and a specific time-like subspace, quite likely those of your laboratory. Experiment indicates the following force law: there are two vector fields, $\vec{\mathbf{E}}$ (the electric field) and

$\vec{\mathbf{B}}$ (the magnetic field), with the property that a charge q at $(\mathbf{x}, t)$ and with velocity $\vec{\mathbf{v}}$ (in the laboratory coordinates) is subject to the force

$$q\left(\vec{\mathbf{E}}(\mathbf{x}) + \frac{\vec{\mathbf{v}}}{c} \times \vec{\mathbf{B}}(\mathbf{x})\right). \qquad 6.4.24$$

The "c" in Equations 6.4.25 and 6.4.25 represents the speed of light. It is necessary to put it in so that $W_{\vec{\mathbf{E}}} \wedge cdt$ and $\Phi_{\vec{\mathbf{B}}}$ have the same units: force/(charge $\times$ length^2). We are using the *cgs* system of units (centimeter, gram, second). In this system the unit of charge is the *statcoulomb*, which is designed to remove constants from Coulomb's law. (The *mks* system, based on meters, kilograms and seconds, uses a different unit of charge, the *coulomb*, which results in constants μ_0 and ϵ_0 that clutter up the equations). We could go one step further and use the mathematicians' privilege of choosing units arbitrarily, setting $c = 1$, but that offends intuition.

But $\vec{\mathbf{E}}$ and $\vec{\mathbf{B}}$ are not really vector fields. A true vector field keeps its individuality when you change coordinates. In particular, if a vector field is $\vec{\mathbf{0}}$ in one coordinate system, it will be $\vec{\mathbf{0}}$ in every coordinate system. This is not true of the electric and magnetic fields. If in one coordinate system the charge is at rest and the electric field is $\vec{\mathbf{0}}$, then the particle will not be accelerated in those coordinates. In another system moving at constant velocity with respect to the first (on a train rolling through the laboratory, for instance) it will still not be accelerated. But it now feels a force from the magnetic field, which must be compensated for by an electric field, which cannot now be zero.

Is there something natural that the electric field and the magnetic field together represent? The answer is yes: there is a 2-form field on $\mathbb{R}^4$, namely

$$E_x dx \wedge cdt + E_y dy \wedge cdt + E_3 dz \wedge cdt + B_x dy \wedge dz + B_y dz \wedge dx + B_z dx \wedge dy$$

$$= W_{\vec{\mathbf{E}}} \wedge cdt + \Phi_{\vec{\mathbf{B}}}. \qquad 6.4.25$$

Exercise 6.8.8 asks you to use form fields to write Maxwell's laws.

This 2-form field, which the distinguished physicists Charles Misner, Kip Thorne, and J. Archibald Wheeler call the *Faraday* (in their book *Gravitation*, the bible of general relativity), is really a natural object, the same in every inertial frame. Thus form fields are really the natural language in which to write Maxwell's equations.

Form Fields	Vector Calculus	
	$\mathbb{R}^3$	$\mathbb{R}^n$
0-form field	Function	Function
1-form field	Vector field (via work form field)	Vector field
$(n-2)$-form field	Same as 1-form	No Equivalent
$(n-1)$-form field	Vector field (via flux form field)	Vector field
n-form field	Function (via density form field)	Function

FIGURE 6.4.3. Correspondence between forms and vector calculus. In all dimensions, 0-form fields, 1-form fields, $(n-1)$-form fields, and n-form fields can be identified to a vector field or a function. Other form fields have no equivalence in vector calculus.

6.5 ORIENTATION AND INTEGRATION OF FORM FIELDS

" ... the great thing in this world is not so much where we stand, as in what direction we are moving."—Oliver Wendell Holmes

Compatible orientations of parametrized manifolds

We have discussed how to integrate k-form fields over k-dimensional parametrized domains. We have seen that where integrands like $|d^k\mathbf{x}|$ are concerned, the integral does not depend on the parametrization. Is this still true for form fields? The answer is "not quite": for two parametrizations to give the same result, they have to *induce the same orientation on the image.*

Let us see this by trying to prove the (false) statement that the integral does not depend on the parametrization, and discovering where we go wrong. Let $M \subset \mathbb{R}^n$ be a k-dimensional manifold, U, V be subsets of $\mathbb{R}^k$, and $\gamma_1 : U \to M$, $\gamma_2 : V \to M$ be two parametrizations, each inducing its own orientation. Let φ be a k-form on a neighborhood of M.

Define as in Theorem 5.2.8 the "change of parameters" map $\Phi = \gamma_2^{-1} \circ \gamma_1 : U^{ok} \to V^{ok}$.

Then Definition 6.3.4 (integrating a k-form field over a parametrized domain) and the change of variables formula, give

$$\int_{\gamma_2(V)} \varphi = \int_V \varphi\Big(P_{\gamma_2(\mathbf{v})}\big(\overrightarrow{D_1\gamma_2}(\mathbf{v}), \dots, \overrightarrow{D_k\gamma_2}(\mathbf{v})\big)\Big)|d^k\mathbf{v}| \qquad 6.5.1$$

$$= \int_U \varphi\Big(P_{\gamma_2\circ\Phi(\mathbf{u})}\big(\overrightarrow{D_1\gamma_2}(\Phi(\mathbf{u})), \dots, \overrightarrow{D_k\gamma_2}(\Phi(\mathbf{u}))\big)\Big)\Big|\det[\mathbf{D}\Phi(\mathbf{u})]\Big||d^k\mathbf{u}|.$$

We want to express everything in terms of γ_1. There is no trouble with the point $(\gamma_2 \circ \Phi)(\mathbf{u}) = \gamma_1(\mathbf{u})$ where the parallelogram is anchored, but the vectors that span it are more troublesome, and will require the following lemma.

Lemma 6.5.1. *If $\vec{\mathbf{w}}_1, \dots, \vec{\mathbf{w}}_k$ are any k vectors in $\mathbb{R}^k$, then*

$$\varphi\Big(P_{\gamma_2(\mathbf{v})}\big(\overrightarrow{D_1\gamma_2}(\mathbf{v}), \dots, \overrightarrow{D_k\gamma_2}(\mathbf{v})\big)\Big)\det[\vec{\mathbf{w}}_1, \dots, \vec{\mathbf{w}}_k]$$

$$= \varphi\Big(P_{\gamma_2(\mathbf{v})}\big([\mathbf{D}\gamma_2(\mathbf{v})]\vec{\mathbf{w}}_1, \dots, [\mathbf{D}\gamma_2(\mathbf{v})]\vec{\mathbf{w}}_k\big)\Big). \qquad 6.5.2$$

Proof. Since the vectors $[\mathbf{D}\gamma_2(\mathbf{v})]\vec{\mathbf{w}}_1, \dots, [\mathbf{D}\gamma_2(\mathbf{v})]\vec{\mathbf{w}}_k$ in the second line of Equation 6.5.2 depend on $\vec{\mathbf{w}}_1, \dots, \vec{\mathbf{w}}_k$, we can consider the entire right-hand side of that line as a function of $\mathbf{v}$ and $\vec{\mathbf{w}}_1, \dots, \vec{\mathbf{w}}_k$, multilinear and antisymmetric with respect to the $\vec{\mathbf{w}}$. The latter are k vectors in $\mathbb{R}^k$, so the right-hand side can be written as a multiple of the determinant: $a(\mathbf{v})\det[\vec{\mathbf{w}}_1, \dots, \vec{\mathbf{w}}_k]$ for some function $a(\mathbf{v})$.

We found this argument by working through the proof of the statement that integrals of manifolds with respect to $|d^k\mathbf{x}|$ are independent of parametrization (Proposition 5.5.1, proved for surfaces in Equation 5.4.16) and noting the differences. You may find it instructive to compare the two arguments. Superficially, the equations may seem very different, but note the similarities. The first line of Equation 5.4.16 corresponds to the right-hand side of the first line of Equation 6.5.1; in both we have

$$\int_V (--)|d^k\mathbf{v}|.$$

In the second lines of both equations, we have

$$\int_U (--)\ |\det[\mathbf{D}\Phi(\mathbf{u})]|.$$

To find $a(\mathbf{v})$, we set $\vec{\mathbf{w}}_1, \ldots, \vec{\mathbf{w}}_k = \vec{\mathbf{e}}_1, \ldots, \vec{\mathbf{e}}_k$. Since $[\mathbf{D}\gamma_2(\mathbf{v})]\vec{\mathbf{e}}_i = \overrightarrow{D_i\gamma_2}(\mathbf{v})$, substituting $\vec{\mathbf{e}}_1, \ldots, \vec{\mathbf{e}}_k$ for $\vec{\mathbf{w}}_1, \ldots, \vec{\mathbf{w}}_k$ in the second line of Equation 6.5.2 gives

$$\varphi\Big(P_{\gamma_2(\mathbf{v})}\big([\mathbf{D}\gamma_2(\mathbf{v})]\vec{\mathbf{e}}_1, \ldots, [\mathbf{D}\gamma_2(\mathbf{v})]\vec{\mathbf{e}}_k\big)\Big) = \varphi\Big(P_{\gamma_2(\mathbf{v})}\big(\overrightarrow{D_1\gamma_2}(\mathbf{v}), \ldots, \overrightarrow{D_k\gamma_2}(\mathbf{v})\big)\Big)$$

$$= a(\mathbf{v})\det[\vec{\mathbf{e}}_1, \ldots, \vec{\mathbf{e}}_k] = a(\mathbf{v}).   \quad 6.5.3$$

So

$$\varphi\Big(P_{\gamma_2(\mathbf{v})}\big([\mathbf{D}\gamma_2(\mathbf{v})]\vec{\mathbf{w}}_1, \ldots, [\mathbf{D}\gamma_2(\mathbf{v})]\vec{\mathbf{w}}_k\big)\Big) = a(\mathbf{v})\det[\vec{\mathbf{w}}_1, \ldots, \vec{\mathbf{w}}_k]$$

$$= \varphi\Big(P_{\gamma_2(\mathbf{v})}\big(\overrightarrow{D_1\gamma_2}(\mathbf{v}), \ldots, \overrightarrow{D_k\gamma_2}(\mathbf{v})\big)\Big)\det[\vec{\mathbf{w}}_1, \ldots, \vec{\mathbf{w}}_k]. \quad \square$$

$$6.5.4$$

Now we write down the function being integrated on the second line of Equation 6.5.1, *except* that we take $\det[\mathbf{D}\Phi(\mathbf{u})]$ out of absolute value signs, so that we will be able to apply Lemma 6.5.1 to go from the second to the third line:

$$\varphi\Big(P_{\gamma_2\circ\Phi(\mathbf{u})}\big(\overrightarrow{D_1\gamma_2}(\Phi(\mathbf{u})), \ldots, \overrightarrow{D_k\gamma_2}(\Phi(\mathbf{u}))\big)\Big)\det[\mathbf{D}\Phi(\mathbf{u})]$$

$$= \varphi\Big(P_{\gamma_2\circ\Phi(\mathbf{u})}\big(\overrightarrow{D_1\gamma_2}(\Phi(\mathbf{u})), \ldots, \overrightarrow{D_k\gamma_2}(\Phi(\mathbf{u}))\big)\Big)\det\big[\underbrace{\overrightarrow{D_1\Phi}(\mathbf{u}), \ldots, \overrightarrow{D_k\Phi}(\mathbf{u})}_{\vec{\mathbf{w}}_1, \ldots, \vec{\mathbf{w}}_k}\big]$$

$$= \varphi\Big(P_{\gamma_2\circ\Phi(\mathbf{u})}\big([\mathbf{D}\gamma_2(\underbrace{\Phi(\mathbf{u})}_{\mathbf{v}})](\underbrace{\overrightarrow{D_1\Phi}(\mathbf{u})}_{\vec{\mathbf{w}}_1}), \ldots, [\mathbf{D}\gamma_2(\Phi(\mathbf{u}))](\underbrace{\overrightarrow{D_k\Phi}(\mathbf{u})}_{\vec{\mathbf{w}}_k})\big)\Big)$$

$$= \varphi\Big(P_{\gamma_1(\mathbf{u})}\big(\overrightarrow{D_1\gamma_1}(\mathbf{u}), \ldots, \overrightarrow{D_k\gamma_1}(\mathbf{u})\big)\Big). \quad 6.5.5$$

The first line of Equation 6.5.5 is the function being integrated on the second line of Equation 6.5.1: everything between the $\int_U$ and the $|d^k\mathbf{u}|$, *with the important difference that here the* $\det[\mathbf{D}\Phi(\mathbf{u})]$ *is not between absolute value signs.*

The second line is identical to the first, except that $\det[\mathbf{D}\Phi(\mathbf{u})]$ has been rewritten in terms of the partial derivatives.

To pass from the second to the third line of Equation 6.5.5 we use Lemma 6.5.1, setting $\vec{\mathbf{w}}_j = D_j\Phi(\mathbf{u})$ and $\mathbf{v} = \Phi(\mathbf{u})$. (We have marked some of these correspondences with underbraces.) We use the chain rule to go from the third to the fourth line.

Now we come to the key point. The second line of Equation 6.5.1 has $|\det[\mathbf{D}\Phi(\mathbf{u})]|$, while the first line of Equation 6.5.5 has $\det[\mathbf{D}\Phi(\mathbf{u})]$. Therefore the integral

$$\int_U \varphi\big(P_{\gamma_1(\mathbf{u})}(\overrightarrow{D_1\gamma_1}(\mathbf{u}), \ldots, \overrightarrow{D_k\gamma_1}(\mathbf{u}))\big)\,|d^k\mathbf{u}| \quad 6.5.6$$

obtained using γ_1 and the integral

$$\int_V \varphi\big(P_{\gamma_2(\mathbf{v})}(\overrightarrow{D_1\gamma_2}(\mathbf{v}), \ldots, \overrightarrow{D_k\gamma_2}(\mathbf{v}))\big)\,|d^k\mathbf{v}| \quad 6.5.7$$

obtained using γ_2 will be the same *only if* $|\det[\mathbf{D}\Phi(\mathbf{u})]| = \det[\mathbf{D}\Phi(\mathbf{u})]$. That is, they will be identical if $\det[\mathbf{D}\Phi] > 0$ for all $\mathbf{u} \in U$, *and otherwise probably not.* If $\det[\mathbf{D}\Phi] < 0$ for all $\mathbf{u} \in U$ then

$$\int_{\gamma_1(U)} \varphi = -\int_{\gamma_2(V)} \varphi. \quad 6.5.8$$

If $\det[\mathbf{D}\Phi(\mathbf{u})]$ is positive in some regions of U and negative in others, then the integrals are probably unrelated.

If $\det[\mathbf{D}\Phi] > 0$, we say that the two parametrizations of M induce *compatible orientations* of M.

In Definition 6.5.2, recall that when γ_1 goes from U to M, and γ_2 from V to M, $\Phi = \gamma_2^{-1} \circ \gamma_1$ is only defined on U^{ok}.

Definition 6.5.2 (Compatible orientation). Let γ_1 and γ_2 be two parametrizations, with the "change of parameters" map $\Phi = \gamma_2^{-1} \circ \gamma_1$. The two parametrizations γ_1 and γ_2 are compatible if $\det[\mathbf{D}\Phi] > 0$.

This leads to the following theorem.

Theorem 6.5.3 (Integral independent of compatible parametrizations). *Let $M \subset \mathbb{R}^n$ be a k-dimensional oriented manifold, U, V open subsets of $\mathbb{R}^k$, and $\gamma_1 : U \to \mathbb{R}^n$ and $\gamma_2 : V \to \mathbb{R}^n$ be two parametrizations of M that induce compatible orientations of M. Then for any k-form φ defined on a neighborhood of M,*

$$\int_{\gamma_1(U)} \varphi = \int_{\gamma_2(V)} \varphi. \qquad 6.5.9$$

Orientation of manifolds

When using a parametrization to integrate a k-form field over an oriented domain, clearly we must take into account the orientation induced by the parametrization. We would like to be able to relate this to some characteristic of the domain of integration itself. What kind of structure can we bestow on an oriented curve, surface, or higher-dimensional manifold that would enable us to decide how to check whether a parametrization is appropriate?

There are two ways to approach the somewhat challenging topic of orientation. One is the *ad hoc approach*: to limit the discussion to points, curves, surfaces, and three-dimensional objects. This has the advantage of being more concrete, and the disadvantage that the various definitions appear to have nothing to do with each other. The other is the *unified approach*: to discuss orientation of k-dimensional manifolds, showing how orientation of points, curves, surfaces, etc., are embodiments of a general definition. This has the disadvantage of being abstract. We will present the ad hoc approach first, followed by the unified theory.

The ad hoc world: orienting the objects

We will treat orientations of the objects first, followed by orientation-preserving parametrizations.

FIGURE 6.5.1.

A curve is oriented by the choice of a unit tangent vector field that depends continuously on **x**. We could give this curve the opposite orientation by choosing tangent vectors pointing in the opposite direction.

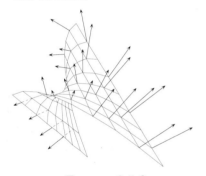

FIGURE 6.5.2.

To orient a surface, we choose a normal vector field that depends continuously on **x**. (Recall that "normal" means "orthogonal.")

Recall (part (b) of Proposition 1.4.20) that the determinant of three vectors is positive if they satisfy the right-hand rule, and negative otherwise. (This assumes that the standard basis vectors satisfy the right-hand rule.)

> **Definition 6.5.4 (Orientation of a point).** An orientation of a point is a choice of $\pm$: an oriented point is "plus the point" or "minus the point."

It is easy to understand orientations of curves (in any $\mathbb{R}^n$): give a direction to go along the curve. The following definition is a more formal way of saying the same thing; it is illustrated in Figure 6.5.1. By "unit tangent vector field" we mean a field of vectors tangent to the curve and of length 1.

> **Definition 6.5.5 (Orientation of a curve in $\mathbb{R}^n$).** An orientation of a curve $C \subset \mathbb{R}^3$ is the choice of a unit tangent vector field $\vec{T}$ that depends continuously on **x**.

We orient a surface $S \subset \mathbb{R}^3$ by choosing a normal vector at every point, as shown in Figure 6.5.2 and defined more formally below.

> **Definition 6.5.6 (Orientation of a surface in $\mathbb{R}^3$).** To orient a surface in $\mathbb{R}^3$, choose a unit vector field $\vec{N}$ orthogonal to the surface, and such that the vector field $\vec{N}$ depends continuously on the point.

At each point **x** there are two vectors $\vec{N}(\mathbf{x})$ orthogonal to the surface; Definition 6.5.6 says to choose one at each point, so that $\vec{N}$ varies continuously. This is possible for an orientable surface like a sphere or a torus: choose either the outer-pointing normal or the inward-pointing normal. But it is impossible on a Moebius strip. This definition does not extend at all easily to a surface in $\mathbb{R}^4$: at every point there is a whole normal plane, and choosing a normal vector field does not provide an orientation.

> **Definition 6.5.7 (Orientation of open subsets of $\mathbb{R}^3$).** One orientation of an open subset X of $\mathbb{R}^3$ is given by det; the opposite orientation is given by $-$ det. The standard orientation is by det.

We will use orientations to say whether three vectors $\vec{v}_1, \vec{v}_2, \vec{v}_3$ form a *direct basis* of $\mathbb{R}^3$; with the standard orientation, $\vec{v}_1, \vec{v}_2, \vec{v}_3$ being direct means that $\det[\vec{v}_1, \vec{v}_2, \vec{v}_3] > 0$. If we have drawn $\vec{e}_1, \vec{e}_2, \vec{e}_3$ in the standard way, so that they fit the right hand, then $\vec{v}_1, \vec{v}_2, \vec{v}_3$ will be direct precisely if those vectors also satisfy the right-hand rule.

The unified approach: orienting the objects

All three notions of orientation are reasonably intuitive, but they do not appear to have anything in common. Signs of points, directions on curves, normals to surfaces, right hands: how can we make all four be examples of a single construction?

We will see that orienting manifolds means orienting their tangent spaces, so before orienting manifolds we need to see how to orient vector spaces. We saw in Section 6.2 (Corollary 6.2.12) that for any k-dimensional vector space E, the space $A^k(E)$ of k-forms in E has dimension one. Now we will use this space to show that the different definitions of orientation we gave at the beginning of this section are all special cases of a general definition.

If E is k-dimensional, then $A^k(E)$ is a line, but it is not $\mathbb{R}$; it is a line with no built-in orientation. Every nonzero point on the line is a basis for $A^k(E)$, and choosing such a point ω chooses an orientation: the positive multiples of ω define the same orientation as ω, and negative multiples of ω define the opposite orientation.

Definition 6.5.8 (Orienting the space $A^k(E)$). If E is of dimension k, the one-dimensional space $A^k(E)$ is oriented by choosing a nonzero element ω of $A^k(E)$. An element $a\omega$, with $a > 0$, gives the same orientation as ω, while $b\omega$, with $b < 0$, gives the opposite orientation.

Definition 6.5.9 (Orienting a finite-dimensional vector space). An orientation of a k-dimensional vector space E is specified by a nonzero element of $A^k(E)$. Two nonzero elements specify the same orientation if one is a multiple of the other by a positive number.

Definition 6.5.9 makes it clear that every finite-dimensional vector space (in particular every subspace of $\mathbb{R}^n$) has two orientations.

Equivalence of the ad hoc and the unified approaches for subspaces of $\mathbb{R}^3$

Unlike the ad hoc definition of orientation, which does not work for a surface in $\mathbb{R}^4$, the unified definition applies in all dimensions.

Let $E \subset \mathbb{R}^k$ be a line, oriented in the ad hoc sense by a nonzero vector $\vec{v} \in E$, and oriented in the unified sense by a nonzero element $\omega \in A^1(E)$. Then these two orientations coincide precisely if $\omega(\vec{v}) > 0$.

For instance, if $E \subset \mathbb{R}^2$ is the line of equation $x + y = 0$, then the vector $\begin{bmatrix} 1 \\ -1 \end{bmatrix}$ defines an ad hoc orientation, whereas dx provides a unified orientation. They do coincide: $dx \begin{bmatrix} 1 \\ -1 \end{bmatrix} = 1 > 0$. The element of $A^1(E)$ corresponding to dy also defines an orientation of E, in fact the opposite orientation. Why does $dx + dy$ not define an orientation of this line?[5]

Since E is a plane, the normal vector field $\vec{N}$ orienting it "ad hoc" is constant.

Now suppose that $E \subset \mathbb{R}^3$ is a plane, oriented "ad hoc" by a normal $\vec{n}$, and oriented "unified" by $\omega \in A^2(E)$. Then the orientations coincide if for any two linearly independent vectors $\vec{v}_1, \vec{v}_2 \in E$, the number $\omega(\vec{v}_1, \vec{v}_2)$ is a positive multiple of $\det[\vec{n}, \vec{v}_1, \vec{v}_2]$.

For instance, suppose E is the plane of equation $x + y + z = 0$, oriented "ad hoc" by $\begin{bmatrix} 1 \\ 1 \\ 1 \end{bmatrix}$, and oriented "unified" by $dx \wedge dy$. Any two vectors in E can be

---

[5]Because any vector in E can be written $\begin{bmatrix} a \\ -a \end{bmatrix}$, and $(dx+dy)\begin{bmatrix} a \\ -a \end{bmatrix} = 0$, so $dx+dy$ corresponds to $0 \in A^1(E)$.

written

$$\begin{bmatrix} a \\ b \\ -a-b \end{bmatrix}, \quad \begin{bmatrix} c \\ d \\ -c-d \end{bmatrix}, \qquad \text{6.5.10}$$

so we have

$$\text{unified approach}: \quad dx \wedge dy \left(\begin{bmatrix} a \\ b \\ -a-b \end{bmatrix}, \begin{bmatrix} c \\ d \\ -c-d \end{bmatrix} \right) = ad - bc, \quad \text{6.5.11}$$

$$\text{ad hoc approach}: \quad \det \begin{bmatrix} 1 & a & c \\ 1 & b & d \\ 1 & -a-b & -c-d \end{bmatrix} = 3(ad - bc). \qquad \text{6.5.12}$$

These orientations coincide, since $3 > 0$. What if we had chosen $dy \wedge dz$ or $dx \wedge dz$ as our nonzero element of $A^2(E)$?[6]

We see that in most cases the choice of orientation is arbitrary: the choice of one nonzero element of $A^k(E)$ will give one orientation, while the choice of another may well give the opposite orientation. But $\mathbb{R}^n$ itself and $\{\mathbf{0}\}$ (the zero subspace of $\mathbb{R}^n$), are exceptions; these two trivial subspaces of $\mathbb{R}^n$ do have a standard orientation. For $\{\vec{\mathbf{0}}\}$, we have $A^0(\{\vec{\mathbf{0}}\}) = \mathbb{R}$, so one orientation is specified by $+1$, the other by -1; the positive orientation is standard. The trivial subspace $\mathbb{R}^n$ is oriented by $\omega = \det$; and $\det > 0$ is standard.

Orienting manifolds

Most often we will be integrating a form over a curve, surface, or higher-dimensional manifold, not simply over a line, plane, or $\mathbb{R}^3$. A k-manifold is oriented by orienting $T_{\mathbf{x}}M$, the tangent space to the manifold at $\mathbf{x}$, for each $\mathbf{x} \in M$: we orient the manifold M by choosing a nonzero element of $A^k(T_{\mathbf{x}}M)$.

> **Definition 6.5.10 (Orientation of a k-dimensional manifold).** An orientation of a k-dimensional manifold $M \subset \mathbb{R}^n$ is an orientation of the tangent space $T_{\mathbf{x}}M$ at every point $\mathbf{x} \in M$, so that the orientation varies continuously with $\mathbf{x}$. To orient the tangent space, we choose a nonzero element of $A^k(T_{\mathbf{x}}M)$.

---

[6]The first gives the same orientation as $dx \wedge dy$, and the second gives the opposite orientation: evaluated on the vectors of Equation 6.5.10, which we'll call $\vec{\mathbf{v}}_1$ and $\vec{\mathbf{v}}_2$, they give

$$dy \wedge dz(\vec{\mathbf{v}}_1, \vec{\mathbf{v}}_2) = \det \begin{bmatrix} b & d \\ -a-b & -c-d \end{bmatrix} = -bc - bd + ad + bd = ad - bc,$$

$$dx \wedge dz(\vec{\mathbf{v}}_1, \vec{\mathbf{v}}_2) = \det \begin{bmatrix} a & c \\ -a-b & -c-d \end{bmatrix} = -(ad - bc).$$

Once again, we use a linearization (the tangent space) in order to deal with nonlinear objects (curves, surfaces, and higher-dimensional manifolds).

What does it mean to say that the "orientation varies continuously with $\mathbf{x}$"? This is best understood by considering a case where you cannot choose such an orientation, a Moebius strip. If you imagine yourself walking along the surface of a Moebius strip, planting a forest of normal vectors, one at each point, all pointing "up" (in the direction of your head), then when you get back to where you started there will be vectors arbitrarily close to each other, pointing in opposite directions.

Recall (Section 3.1) that the *tangent space* to a smooth curve, surface or manifold is the set of vectors tangent to the curve, surface or manifold, at the point of tangency. The tangent space to a curve C at $\mathbf{x}$ is denoted $T_{\mathbf{x}}C$ and is one-dimensional; the tangent space to a surface S at $\mathbf{x}$ is denoted $T_{\mathbf{x}}S$ and is two-dimensional, and so on.

The ad hoc world: when does a parametrization preserve orientation?

We can now define what it means for a parametrization to preserve orientation. For a curve, this means that the parameter increases in the specified direction: a parametrization $\gamma : [a,b] \mapsto C$ preserves orientation if C is oriented from $\gamma(a)$ to $\gamma(b)$. The following definition spells this out; it is illustrated by Figure 6.5.3.

Definition 6.5.11 (Orientation-preserving parametrization of a curve). Let $C \subset \mathbb{R}^n$ be a curve oriented by the choice of unit tangent vector field $\vec{T}$. Then the parametrization $\gamma : (a,b) \to C$ is orientation preserving if at every $t \in (a,b)$, we have

$$\vec{\gamma}'(t) \cdot \vec{T}(\gamma(t)) > 0. \qquad 6.5.13$$

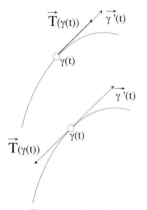

FIGURE 6.5.3.

Top: $\vec{\gamma}'(t)$ (the velocity vector of the parametrization) points in the same direction as the vector orienting the curve; the parametrization γ preserves orientation. Below: $\vec{\gamma}'(t)$ points in the opposite direction of the orientation; γ is orientation reversing.

Equation 6.5.13 says that the velocity vector of the parametrization points in the same direction as the vector orienting the curve. Remember that

$$\vec{v}_1 \cdot \vec{v}_2 = (\cos\theta)|\vec{v}_1|\,|\vec{v}_2|, \qquad 6.5.14$$

where θ is the angle between the two vectors. So the angle between $\vec{\gamma}'(t)$ and $\vec{T}(\gamma(t))$ is less than 90°. Since the angle must be either 0 or 180°, it is 0.

It is harder to understand what it means for a parametrization of an oriented surface to preserve orientation. In Definition 6.5.12, $\overrightarrow{D_1\gamma}(\mathbf{u})$ and $\overrightarrow{D_2\gamma}(\mathbf{u})$ are two vectors tangent to the surface at $\gamma(\mathbf{u})$.

Definition 6.5.12 (Orientation-preserving parametrization of a surface). Let $S \subset \mathbb{R}^3$ be a surface oriented by a choice of normal vector field $\vec{N}$. Let $U \subset \mathbb{R}^2$ be open and $\gamma : U \to S$ be a parametrization. Then γ is orientation preserving if at every $\mathbf{u} \in U$,

$$\det[\vec{N}(\gamma(\mathbf{u})), \overrightarrow{D_1\gamma}(\mathbf{u}), \overrightarrow{D_2\gamma}(\mathbf{u})] > 0. \qquad 6.5.15$$

Definition 6.5.13 (Orientation-preserving parametrization of an open subset of $\mathbb{R}^3$). An open subset U of $\mathbb{R}^3$ carries a standard orientation, defined by the determinant. If V is another open subset of $\mathbb{R}^3$, and $\gamma : V \to U$ is a parametrization (i.e., a change of variables), then γ is orientation preserving if $\det[\mathbf{D}\gamma(\mathbf{v})] > 0$ for all $\mathbf{v} \in V$.

The unified approach: when does a parametrization preserve orientation?

First let us define what it means for a linear transformation to be orientation preserving.

In Definition 6.5.14, $\mathbb{R}^k$ is oriented in the standard way, by det, and

$$\det[\vec{\mathbf{e}}_1, \dots, \vec{\mathbf{e}}_k] = 1 > 0.$$

If the orientation of V by ω also gives a positive number when applied to $T(\vec{\mathbf{e}}_1), \dots, T(\vec{\mathbf{e}}_k)$, then T is orientation preserving.

Exercise 6.5.2 asks you to prove that if a linear transformation T is not one to one, then it is not orientation preserving or reversing.

In Definition 6.5.15, the derivative $[\mathbf{D}\gamma(\mathbf{u})]$ is of course a linear transformation; we use Definition 6.5.14 to determine whether it preserves orientation.

Since $U \subset \mathbb{R}^k$ is open, it is necessarily k-dimensional. (We haven't actually defined dimension of a set; what we mean is that $U \subset \mathbb{R}^k$ is a k-dimensional manifold.) Think of a disk and a ball in $\mathbb{R}^3$; the ball can be open but the disk cannot. The "fence" of a disk in $\mathbb{R}^3$ is the entire disk, and it is impossible (Definition 1.5.3) to surround each point of the disk with a ball contained entirely in the disk.

Definition 6.5.14 (Orientation-preserving linear transformation). If $V \subset \mathbb{R}^m$ is a k-dimensional subspace oriented by $\omega \in A^k(V)$, and $T : \mathbb{R}^k \to V$ is a linear transformation, T is orientation preserving if

$$\omega\big(T(\vec{\mathbf{e}}_1), \dots, T(\vec{\mathbf{e}}_k)\big) > 0. \qquad 6.5.16$$

It is orientation reversing if

$$\omega\big(T(\vec{\mathbf{e}}_1), \dots, T(\vec{\mathbf{e}}_k)\big) < 0. \qquad 6.5.17$$

Note that for a linear transformation to preserve orientation, the domain and the range must have the same dimension, and they must be oriented.

As usual, faced with a nonlinear problem, we linearize it: a (nonlinear) parametrization of a manifold is orientation preserving if the *derivative* of the parametrization is orientation preserving.

Definition 6.5.15 (Orientation-preserving parametrization of a manifold). Let M be a k-dimensional manifold oriented by ω, $U \subset \mathbb{R}^k$ be an open set, and $\gamma : U \to M$ be a parametrization. Then γ is orientation preserving if $[\mathbf{D}\gamma(\mathbf{u})] : \mathbb{R}^k \to T_{\gamma(\mathbf{u})}M$ is orientation preserving for every $\mathbf{u} \in U$, i.e., if

$$\omega\big([\mathbf{D}\gamma(\mathbf{u})](\vec{\mathbf{e}}_1), \dots, [\mathbf{D}\gamma(\mathbf{u})](\vec{\mathbf{e}}_k)\big) = \omega\big(\overrightarrow{D_1\gamma}(\mathbf{u}), \dots, \overrightarrow{D_k\gamma}(\mathbf{u})\big) > 0.$$

Example 6.5.16 (Orientation-preserving parametrization). Consider the surface S in $\mathbb{C}^3$ parametrized by

$$z \mapsto \begin{pmatrix} z \\ z^2 \\ z^3 \end{pmatrix}, \ |z| < 1, \qquad 6.5.18$$

We will denote points in $\mathbb{C}^3$ by $\begin{pmatrix} z_1 \\ z_2 \\ z_3 \end{pmatrix} = \begin{pmatrix} x_1 + iy_1 \\ x_2 + iy_2 \\ x_3 + iy_3 \end{pmatrix}.$

Orient S, using $\omega = dx_1 \wedge dy_1$.

If we parametrize the surface by

In this parametrization we are writing the complex number z in terms of polar coordinates:

$$z = r(\cos\theta + i\sin\theta).$$

(See Equation 0.6.10).

$$\gamma : \begin{pmatrix} r \\ \theta \end{pmatrix} \mapsto \begin{pmatrix} x_1 = r\cos\theta \\ y_1 = r\sin\theta \\ x_2 = r^2\cos 2\theta \\ y_2 = r^2\sin 2\theta \\ x_3 = r^3\cos 3\theta \\ y_3 = r^3\sin 3\theta \end{pmatrix}, \qquad 6.5.19$$

does that parametrization preserve orientation? It does, since

$$dx_1 \wedge dy_1\big(D_1\gamma(\mathbf{u}), D_2\gamma(\mathbf{u})\big) = dx_1 \wedge dy_1 \begin{bmatrix} \cos\theta & -r\sin\theta \\ \sin\theta & r\cos\theta \\ \vdots & \vdots \\ \vdots & \vdots \end{bmatrix} \qquad 6.5.20$$

$$= \det \begin{bmatrix} \cos\theta & -r\sin\theta \\ \sin\theta & r\cos\theta \end{bmatrix} = r\cos^2\theta + r\sin^2\theta = r > 0. \quad \triangle$$

Exercise 6.5.4 asks you to show that our three ad hoc definitions of orientation-preserving parametrizations are special cases of Definition 6.5.15.

Compatibility of orientation-preserving parametrizations

Theorem 6.5.3 said the result of integrating a k-form over an oriented manifold does not depend on the choice of parametrization, as long as the parametrizations induce compatible orientations. Now we show that the integral is independent of parametrization if the parametrization is orientation preserving. Most of the work was done in proving Theorem 6.5.3. The only thing we need to show is that two orientation-preserving parametrizations define compatible orientations.

Recall (Definition 6.5.2) that two parametrizations γ_1 and γ_2 with the "change of parameters" map $\Phi = \gamma_2^{-1} \circ \gamma_1$ are compatible if $\det[\mathbf{D}\Phi] > 0$.

Theorem 6.5.17 (Orientation-preserving parametrizations define compatible orientations). *If M is an oriented k-manifold, U_1 and U_2 are open subsets of $\mathbb{R}^k$, and $\gamma_1 : U_1 \to M$, $\gamma_2 : U_2 \to M$ are orientation-preserving parametrizations, then they define compatible orientations.*

Proof. Consider two points $\mathbf{u}_1 \in U_1, \mathbf{u}_2 \in U_2$ such that $\gamma_1(\mathbf{u}_1) = \gamma_2(\mathbf{u}_2) = \mathbf{x} \in M$. The derivatives then give us maps

$$\mathbb{R}^k \xrightarrow{[\mathbf{D}\gamma_1(\mathbf{u}_1)]} T_\mathbf{x}M \xleftarrow{[\mathbf{D}\gamma_2(\mathbf{u}_2)]} \mathbb{R}^k, \qquad 6.5.21$$

where both derivatives are one to one linear transformations. Moreover, we have $\omega(\mathbf{x}) \neq 0$ in the one-dimensional vector space $A^k(T_\mathbf{x}M)$. What we must show is that if M is oriented by ω, and

$$\omega(\mathbf{x})\big(\overrightarrow{D_1\gamma_1}(\mathbf{u}_1),\ldots,\overrightarrow{D_k\gamma_1}(\mathbf{u}_1)\big) > 0 \quad \text{and} \quad \omega(\mathbf{x})\big(\overrightarrow{D_1\gamma_2}(\mathbf{u}_2),\ldots,\overrightarrow{D_k\gamma_2}(\mathbf{u}_2)\big) > 0,$$

then $\det([\mathbf{D}\gamma_2(\mathbf{u}_2)])^{-1}[\mathbf{D}\gamma_1(\mathbf{u}_1)] > 0$.

Note that

$$\omega(\mathbf{x})\big([\mathbf{D}\gamma_1(\mathbf{u}_1)](\vec{\mathbf{v}}_1),\ldots,[\mathbf{D}\gamma_1(\mathbf{u}_1)](\vec{\mathbf{v}}_k)\big) = \alpha \det[\vec{\mathbf{v}}_1,\ldots,\vec{\mathbf{v}}_k]$$
$$\omega(\mathbf{x})\big([\mathbf{D}\gamma_2(\mathbf{u}_2)](\vec{\mathbf{w}}_1),\ldots,[\mathbf{D}\gamma_2(\mathbf{u}_2)](\vec{\mathbf{w}}_k)\big) = \beta \det[\vec{\mathbf{w}}_1,\ldots,\vec{\mathbf{w}}_k]. \qquad 6.5.22$$

for some positive numbers α and β. Indeed, both left-hand sides are nonzero elements of the one-dimensional vector space $A^k(\mathbb{R}^k)$, hence nonzero multiples of the determinant, and they return positive values if evaluated on the standard basis vectors. Now write

$$\alpha = \omega(\mathbf{x})\big(\overrightarrow{D_1\gamma_1}(\mathbf{u}_1),\ldots,\overrightarrow{D_k\gamma_1}(\mathbf{u}_1)\big) = \omega(\mathbf{x})([\mathbf{D}\gamma_1(\mathbf{u}_1)]\vec{\mathbf{e}}_1,\ldots,[\mathbf{D}\gamma_1(\mathbf{u}_1)]\vec{\mathbf{e}}_k)$$
$$= \omega(\mathbf{x})([\mathbf{D}\gamma_2(\mathbf{u}_2)]\,([\mathbf{D}\gamma_2(\mathbf{u}_2)])^{-1}\,[\mathbf{D}\gamma_1(\mathbf{u}_1)]\vec{\mathbf{e}}_1,\ldots,$$
$$[\mathbf{D}\gamma_2(\mathbf{u}_2)]\,([\mathbf{D}\gamma_2(\mathbf{u}_2)])^{-1}\,[\mathbf{D}\gamma_1(\mathbf{u}_1)]\vec{\mathbf{e}}_k)$$
$$= \beta \det\Big[([\mathbf{D}\gamma_2(\mathbf{u}_2)])^{-1}\,[\mathbf{D}\gamma_1(\mathbf{u}_1)]\vec{\mathbf{e}}_1,\ldots,([\mathbf{D}\gamma_2(\mathbf{u}_2)])^{-1}\,[\mathbf{D}\gamma_1(\mathbf{u}_1)]\vec{\mathbf{e}}_k\Big]$$
$$= \beta \det\Big(([\mathbf{D}\gamma_2(\mathbf{u}_2)])^{-1}\,[\mathbf{D}\gamma_1(\mathbf{u}_1)]\Big). \qquad 6.5.23$$

So $\det\Big(([\mathbf{D}\gamma_2(\mathbf{u}_2)])^{-1}\,[\mathbf{D}\gamma_1(\mathbf{u}_1)]\Big) = \det[\mathbf{D}\Phi] = \alpha/\beta > 0$, which means (Definition 6.5.2) that γ_1 and γ_2 define compatible orientations. $\square$

Using the notation for oriented parallelograms, we could write

$$\omega(\mathbf{x})\big([\overrightarrow{D_1\gamma_1}(\mathbf{u}_1),\ldots,\overrightarrow{D_k\gamma_1}(\mathbf{u}_1)\big)$$

as

$$\omega\big(P_\mathbf{x}^o(\overrightarrow{D_1\gamma_1}(\mathbf{u}_1),\ldots,\overrightarrow{D_k\gamma_1}(\mathbf{u}_1)\big).$$

Remember that

$$\gamma_1(\mathbf{u}_1) = \gamma_2(\mathbf{u}_2) = \mathbf{x};$$

in the first line of Equation 6.5.22, $\mathbf{x} = \gamma_1(\mathbf{u}_1)$; in the second line, $\mathbf{x} = \gamma_2(\mathbf{u}_2)$.

To go from fourth to the last line of Equation 6.5.23, remember that $T\vec{\mathbf{e}}_i$ is the ith column of T so

$$[\mathbf{D}\gamma_2(\mathbf{u}_2)]^{-1}[\mathbf{D}\gamma_1(\mathbf{u}_1)]\vec{\mathbf{e}}_i$$

is the ith column of

$$[\mathbf{D}\gamma_2(\mathbf{u}_2)]^{-1}[\mathbf{D}\gamma_1(\mathbf{u}_1)];$$

this matrix is the "change of parameters" map $\Phi = \gamma_2^{-1} \circ \gamma_1$.

Corollary 6.5.18 (Integral independent of orientation-preserving parametrizations). *Let M be an oriented k-manifold, U and V be open subsets of $\mathbb{R}^k$, and $\gamma_1 : U \to M$, $\gamma_2 : V \to M$ be orientation-preserving parametrizations of M. Then for any k-form φ defined on a neighborhood of M, we have*

$$\int_{\gamma_1(U)} \varphi = \int_{\gamma_2(V)} \varphi. \qquad 6.5.24$$

Integrating form fields over oriented manifolds

Now we know everything we need to know in order to integrate form fields over oriented manifolds. We saw in Section 5.4 how to integrate form fields over parametrized domains. Corollary 6.5.18 says that we can use the same formula to integrate over oriented manifolds, as long as we use an orientation-preserving parametrization. This gives the following:

Definition 6.5.19 (Integral of a form field over an oriented manifold). Let M be a k-dimensional oriented manifold, φ be a k-form field on a neighborhood of M, and $\gamma : U \to M$ be any orientation-preserving parametrization of M. Then

$$\int_M \varphi = \int_{\gamma(U)} \varphi = \int_U \varphi\left(P^o_{\gamma(\mathbf{u})}\left(\overrightarrow{D_1\gamma}(\mathbf{u}), \ldots, \overrightarrow{D_k\gamma}(\mathbf{u})\right)\right)|d^k\mathbf{u}|.$$

Example 6.5.20 (Integrating a flux form over an oriented surface). What is the flux of the vector field $\vec{F}\begin{pmatrix} x \\ y \\ z \end{pmatrix} = \begin{bmatrix} y \\ -x \\ z \end{bmatrix}$ through the piece of the plane P defined by $x + y + z = 1$ where $x, y, z \geq 0$, and which is oriented by the normal $\begin{bmatrix} 1 \\ 1 \\ 1 \end{bmatrix}$?

This surface is the graph of $z = 1 - x - y$, so that

$$\gamma\begin{pmatrix} x \\ y \end{pmatrix} = \begin{pmatrix} x \\ y \\ 1 - x - y \end{pmatrix} \qquad 6.5.25$$

This is an example of the first class of parametrizations listed in Section 5.2, parametrizations as graphs; see Equation 5.2.5.

is a parametrization, if x and y are in the triangle $T \subset \mathbb{R}^2$ given by $x, y \geq 0$, $x + y \leq 1$. Moreover, this parametrization preserves orientation (see Definition 6.5.12), since $\det[\vec{N}(\gamma(\mathbf{u})), \overrightarrow{D_1\gamma}(\mathbf{u}), \overrightarrow{D_2\gamma}(\mathbf{u})]$ is

$$\det\left[\begin{bmatrix} 1 \\ 1 \\ 1 \end{bmatrix}, \begin{bmatrix} 1 \\ 0 \\ -1 \end{bmatrix}, \begin{bmatrix} 0 \\ 1 \\ -1 \end{bmatrix}\right] = 1 > 0. \qquad 6.5.26$$

At right we check that γ preserves orientation.

By Definition 6.4.6, the flux is

Now we compute the integral.

$$\int_P \Phi\begin{bmatrix} y \\ -x \\ z \end{bmatrix} = \int_T \det\left[\overbrace{\begin{bmatrix} y \\ -x \\ 1-x-y \end{bmatrix}}^{\vec{F}(\gamma(\vec{\mathbf{u}}))}, \overbrace{\begin{bmatrix} 1 \\ 0 \\ -1 \end{bmatrix}}^{\overrightarrow{D_1\gamma}}, \overbrace{\begin{bmatrix} 0 \\ 1 \\ -1 \end{bmatrix}}^{\overrightarrow{D_2\gamma}}\right] |dx\, dy|$$

Note that the formula for integrating a flux form over a surface in $\mathbb{R}^3$ enables us to transform an integral over a surface in $\mathbb{R}^3$ into a integral over a piece of $\mathbb{R}^2$, as studied in Chapter 4.

$$= \int_T (1 - 2x)\,|dx\,dy| = \int_0^1 \left(\int_0^{1-y} (1 - 2x)\,dx\right) dy \qquad 6.5.27$$

$$= \int_0^1 \left[x - x^2\right]_0^{1-y} dy = \int_0^1 (y - y^2)\,dy = \left[\frac{y^2}{2} - \frac{y^3}{3}\right]_0^1 = \frac{1}{6}. \quad \triangle$$

Example 6.5.21 (Integrating a 2-form field over a parametrized surface in $\mathbb{C}^3 = \mathbb{R}^6$). Consider again the surface S in $\mathbb{C}^3$ of Example 6.5.16.

The parametrization of Example 6.5.16 is given by Equation 6.5.19:

$$\gamma\colon \begin{pmatrix} r \\ \theta \end{pmatrix} \mapsto \begin{pmatrix} r\cos\theta \\ r\sin\theta \\ r^2\cos 2\theta \\ r^2\sin 2\theta \\ r^3\cos 3\theta \\ r^3\sin 3\theta \end{pmatrix}.$$

We know from that example that this parametrization preserves orientation, as long as we choose $r > 0$.

Note that in both cases in Example 6.5.23 we are integrating over the *same* oriented point, $\mathbf{x} = +2$. We use curly brackets to avoid confusion between integrating over the point $+2$ with negative orientation, and integrating over the point -2.

We need orientation of domains and their boundaries so that we can integrate and differentiate forms, but orientation is important for other reasons. Homology theory, one of the big branches of algebraic topology, is an enormous abstraction of the constructions in our discussion of the unified approach to orientation.

What is

$$\int_S dx_1 \wedge dy_1 + dx_2 \wedge dy_2 + dx_3 \wedge dy_3 \; ? \qquad 6.5.28$$

If we parametrize the surface using the parametrization (repeated in the margin) of Example 6.5.16, then

$$(dx_1 \wedge dy_1 + dx_2 \wedge dy_2 + dx_3 \wedge dy_3)\left(P^o_{\gamma\binom{r}{\theta}}\left(\overrightarrow{D_1\gamma}\begin{pmatrix} r \\ \theta \end{pmatrix}, \overrightarrow{D_2\gamma}\begin{pmatrix} r \\ \theta \end{pmatrix}\right)\right)$$

$$= \det\begin{bmatrix} \cos\theta & -r\sin\theta \\ \sin\theta & r\cos\theta \end{bmatrix} + \det\begin{bmatrix} 2r\cos 2\theta & -2r^2\sin 2\theta \\ 2r\sin 2\theta & 2r^2\cos 2\theta \end{bmatrix} + \det\begin{bmatrix} 3r^2\cos 3\theta & -3r^3\sin 3\theta \\ 3r^2\sin 3\theta & 3r^3\cos 3\theta \end{bmatrix}$$

$$= r + 4r^3 + 9r^5.$$

6.5.29

So we find the following for our integral. (Do you see why the upper limit of integration for r is 1?[7])

$$2\pi \int_0^1 (r + 4r^3 + 9r^5)\,dr = 6\pi. \quad \triangle \qquad 6.5.30$$

For completeness, we show the case where φ is a 0-form field.

Example 6.5.22 (Integrating a 0-form over an oriented point). Let $\mathbf{x}$ be an oriented point, and f a function (i.e., a 0-form field) defined in some neighborhood of $\mathbf{x}$. Then

$$\int_{+\mathbf{x}} f = +f(\mathbf{x}) \quad \text{and} \quad \int_{-\mathbf{x}} f = -f(\mathbf{x}). \quad \triangle \qquad 6.5.31$$

Example 6.5.23 (Integrating over an oriented point).

$$\int_{+\{+2\}} x^2 = 4 \quad \text{and} \quad \int_{-\{+2\}} x^2 = -4. \quad \triangle \qquad 6.5.32$$

6.6 BOUNDARY ORIENTATION

Stokes's theorem, the generalization of the fundamental theorem of calculus, is all about comparing integrals over manifolds and integrals over their boundaries. Here we will define exactly what a "manifold with boundary" is; we will see moreover that if a "manifold with boundary" is oriented, its boundary carries a natural orientation, called, naturally enough, the *boundary orientation*.

---

[7]We know from Example 6.5.16 that $|z| < 1$, and in our parametrization are writing $z = r(\cos\theta + i\sin\theta)$. So (see Equation 0.6.7) $|z| = \sqrt{r^2(\cos^2\theta + \sin^2\theta)} = r < 1$.

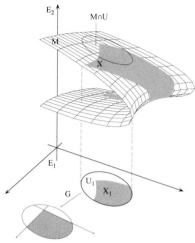

FIGURE 6.6.1.

Above, M is a manifold and X a piece-with-boundary of that manifold. Locally, the manifold M is a graph; every point $\mathbf{x} \in M$ has a neighborhood U such that $M \cap U$ is the graph of a mapping. Above, U is not shown; it is the ball that intersects M in the egg-shaped region labeled $M \cap U$. Such a ball U exists for each $\mathbf{x} \in M$.

The shaded part of $M \cap U$ is $X \cap U$. It is also the graph of a function: $\mathbf{f}(X_1) = X \cap U$. For X to satisfy our definition of a piece-with-boundary of a manifold, every point $\mathbf{a} \in X_1$ must satisfy $G_i(\mathbf{a}) > 0$: the diffeomorphism G takes X_1 to the intersection of $G(U_1)$ and the first "quadrant" of $\mathbb{R}^k$: the region where all variables are positive.

A diffeomorphism is a differentiable mapping with differentiable inverse. It takes smooth curves to smooth curves and angles to angles; any angle $0 < \theta < 180°$ can be deformed by a diffeomorphism into an angle of $90°$.

You may think of a "piece-with-boundary" of a k-dimensional manifold as a piece one can carve out of the manifold, such that the boundary of the piece is part of it (the piece is thus closed). However, the boundary can't have any arbitrary shape. In many treatments the boundaries are restricted to being smooth. In such a treatment, if the manifold is three-dimensional, the boundary of a piece of the manifold must be a smooth surface; if it is two-dimensional, the boundary must be a smooth curve.

We will be less restrictive, and will allow our boundaries to have corners. There are two reasons for this. First, in many cases, we wish to apply Stokes's theorem to things like the region in the sphere where in spherical coordinates, $0 \le \theta \le \pi/2$, and such a region has corners (at the poles). Second, we would like k-parallelograms to be manifolds with boundary, and they most definitely have corners. Fortunately, allowing our boundaries to have corners doesn't make any of the proofs more difficult.

However, we won't allow the boundaries to be just anything: the boundary can't be fractal, like the Koch snowflake we saw in Section 5.6; neither can it contain cusps. (Fractals would really cause problems; cusps would be acceptable, but would make our definitions too involved.) You should think that a region of the boundary either is smooth or contains a corner. Being smooth means being a manifold: locally the graph of a function of some variables in terms of others. What do we mean by corner? Roughly (we will be painfully rigorous below) if you should think of the kind of curvilinear "angles" you can get if you drew the (x, y)-axes on a piece of rubber and stretched it, or if you squashed a cube made of foam rubber.

Definition 6.6.1 is illustrated by Figure 6.6.1.

Definition 6.6.1 (Piece-with-boundary of a manifold). Let $M \subset \mathbb{R}^n$ be a k-dimensional manifold. A subset $X \subset M$ will be called a piece-with-boundary if for every $\mathbf{x} \in X$, there exist

(1) Open subsets $U_1 \subset E_1$ and $U \subset \mathbb{R}^n$ with $\mathbf{x} \in U$ and $\mathbf{f} : U_1 \to E_2$ a C^1 mapping such that $M \cap U$ is the graph of $\mathbf{f}$. (This is Definition 3.2.2 of a manifold.)

(2) A diffeomorphism $G = \begin{pmatrix} G_1 \\ \vdots \\ G_k \end{pmatrix} : U_1 \to \mathbb{R}^k$,

such that $X \cap U$ is $\mathbf{f}(X_1)$, where $X_1 \subset U_1$ is the subset where

$$G_1 \ge 0, \ldots, G_k \ge 0.$$

Example 6.6.2 (A k-parallelogram seen as a piece-with-boundary of a manifold). A k-parallelogram $P_{\mathbf{x}}^o(\vec{\mathbf{v}}_1, \ldots, \vec{\mathbf{v}}_k)$ in $\mathbb{R}^n$ is a piece-with-boundary of an oriented k-dimensional submanifold of $\mathbb{R}^n$ when the vectors $\vec{\mathbf{v}}_1, \ldots, \vec{\mathbf{v}}_k$ are linearly independent. Indeed, if $M \subset \mathbb{R}^n$ is the set parametrized by

$$\begin{pmatrix} t_1 \\ \vdots \\ t_k \end{pmatrix} \mapsto \mathbf{x} + t_1 \vec{\mathbf{v}}_1 + \cdots + t_k \vec{\mathbf{v}}_k, \qquad 6.6.1$$

Equation 6.6.1 looks similar to Definition 5.1.1, but here we are not restricting t.

then M is a k-dimensional manifold in $\mathbb{R}^n$. It is the translation by $\mathbf{x}$ of the subspace spanned by $\vec{\mathbf{v}}_1, \ldots, \vec{\mathbf{v}}_k$ (it is not itself a subspace because it doesn't contain the origin). For every $\mathbf{a} \in M$, the tangent space $T_{\mathbf{a}} M$ is the space spanned by $\vec{\mathbf{v}}_1, \ldots, \vec{\mathbf{v}}_k$. The manifold M is oriented by the choice of a nonzero element $\omega \in A^k(T_{\mathbf{a}} M)$, and ω gives the standard orientation if

Definition 6.6.3 distinguishes between the smooth boundary and the rest (with corners).

$$\omega(\vec{\mathbf{v}}_1, \ldots, \vec{\mathbf{v}}_k) > 0. \qquad 6.6.2$$

The k-parallelogram $P_{\mathbf{x}}^o(\vec{\mathbf{v}}_1, \ldots, \vec{\mathbf{v}}_k)$ is a piece-with-boundary of M, and thus it carries the orientation of M. $\triangle$

The G_i should be thought of as coordinate functions. Think of (x, y, z)-space, i.e., $k = 3$. The (x, z)-plane is the set where y vanishes, the (y, z)-plane is the set where x vanishes, and the (x, y)-plane is the set where z vanishes. This corresponds to the $m = 2$ dimensional stratum, where $k - m$ (i.e., $3 - 2 = 1$) of the G_i vanish. Similarly, the x axis is the set where the y and z vanish; this corresponds to part of the $m = 1$ dimensional stratum, where $k - m$ (i.e., $3 - 1 = 2$) of the G_i vanish.

Definition 6.6.3 (Boundary of a piece-with-boundary of a manifold). If X is a piece-with-boundary of a manifold M, its boundary ∂X is the set of points where at least one of the $G_i = 0$; the smooth boundary is the set where exactly one of the G_i vanishes.

Remark. We can think of a piece-with-boundary of a k-dimensional manifold as composed of strata of various dimensions: the interior of the piece and the various strata of the boundary, just as a cube is stratified into its interior and its two-dimensional faces, one-dimensional edges, and 0-dimensional vertices. When integrating a k-form over a piece-with-boundary of a k-dimensional manifold, we can disregard the boundary; similarly, when integrating a $(k-1)$-form over the boundary, we can ignore strata of dimension less than $k-1$. More precisely, the m-dimensional stratum of the boundary is the set where exactly $k - m$ of the G_i of Definitions 6.6.1 and 6.6.3 vanish, so the inside of the piece is the k-dimensional stratum, the smooth boundary is the $(k-1)$-dimensional stratum, etc. The m-dimensional stratum is an m-dimensional manifold in $\mathbb{R}^n$, hence has m'-dimensional volume 0 for any $m' > m$ (see Exercise 5.2.5); it can be ignored when integrating m'-forms. $\triangle$

We will be interested only in the inside (the k-dimensional stratum) and the smooth boundary (the $(k-1)$-dimensional stratum), since Stokes's theorem relates the integrals of k-forms and $(k-1)$-forms.

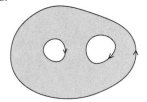

FIGURE 6.6.2.

The boundary of the shaded region of $\mathbb{R}^2$ consists of the three curves drawn, with the indicated orientations. If you walk along those curves, the region will always be to your left.

Boundary orientation: the ad hoc world

The faces of a cube are oriented by the outward-pointing normal, but the other strata of the boundary carry no distinguished orientation at all: there is no particularly natural way to draw an arrow on the edges. More generally, we will only be able to orient the smooth boundary of a piece-with-boundary.

The oriented boundary of a piece-with-boundary of an oriented curve is simply its endpoint minus its beginning point:

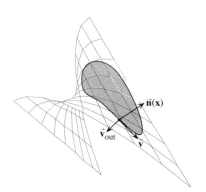

FIGURE 6.6.3.

The shaded area is the piece-with-boundary S_1 of the surface S. The vector $\vec{\mathbf{v}}_{\text{out}}$ is tangent to S at a point in the boundary C_i of S, and points out of S_1. The unit vector $\vec{\mathbf{v}}$ is tangent to C_i. Since

$$\det[\vec{\mathbf{n}}(\mathbf{x}), \vec{\mathbf{v}}_{\text{out}}, \vec{\mathbf{v}}] > 0,$$

the boundary orientation is defined by $\vec{\mathbf{v}}$. If we rotated $\vec{\mathbf{v}}$ by $180°$, then the vectors would obey the left-hand rule instead of the right-hand rule, and the orientation would be reversed.

In Definition 6.6.6, whether $\vec{N}$ is put first or last, or whether the vectors should point inward or outward, is entirely a matter of convention. The order we use is standard, but not universal.

The word "closed" is used here with its common meaning: like an unbroken rubber band, with no ends.

Definition 6.6.4 (Oriented boundary of a piece-with-boundary of an oriented curve). Let C be a curve oriented by the unit tangent vector field $\vec{T}$, and let $P \subset C$ be a piece-with-boundary of C. Then the oriented boundary of P consists of the two endpoints of P, taken with sign $+1$ if the tangent vector points out of P at that point, and with sign -1 if it points in.

If the piece-with-boundary consists of several such P_i, its oriented boundary is the sum of all the endpoints, each taken with the appropriate sign.

Definition 6.6.5 (Oriented boundary of a piece-with-boundary of $\mathbb{R}^2$). If $U \subset \mathbb{R}^2$ is a two-dimensional piece-with-boundary, then its boundary is a union of smooth curves C_i. We orient all the C_i so that if you walk along them in that direction, U will be to your left, as shown in Figure 6.6.2.

When $\mathbb{R}^2$ is given its standard orientation by $+\det$, Definition 6.6.5 says that when you walk on the curves, your head is pointing in the direction of the z-axis. With this definition, the boundary of the unit disk $\{x^2 + y^2 \leq 1\}$ is the unit circle oriented counterclockwise.

For a surface in $\mathbb{R}^3$ oriented by a unit normal, the normal vector field tells you on which side of the surface to walk. Let $S \subset \mathbb{R}^3$ be a surface oriented by a normal vector field $\vec{N}$, and let U be a piece-with-boundary of S, bounded by some union of curves C_i. An obvious example is the upper hemisphere bounded by the equator. If you walk eastwards along the boundary so that your head points in the direction of $\vec{N}$, and U is to your left, you are walking in the direction of the boundary orientation. Translating this into mathematically meaningful language gives the following, illustrated by Figure 6.6.3.

Definition 6.6.6 (Oriented boundary of a piece-with-boundary of an oriented surface). Let $S \subset \mathbb{R}^3$ be a surface oriented by a normal vector field $\vec{N}$, and let S_1 be a piece-with-boundary of S, bounded by some union of closed curves C_i. At a point $\mathbf{x} \in C_i$, let $\vec{\mathbf{v}}_{\text{out}}$ be a vector tangent to S and pointing out of S_1. Then the boundary orientation is defined by the unit vector $\vec{\mathbf{v}}$ tangent to C_i, chosen so that

$$\det\left[\vec{N}(\mathbf{x}), \vec{\mathbf{v}}_{\text{out}}, \vec{\mathbf{v}}\right] > 0. \qquad 6.6.3$$

Since the system composed of your head, your right arm, and your left arm also satisfies the right-hand rule, this means that to walk in the direction of ∂S_1, you should walk with your head in the direction of $\vec{N}$, and the surface to your left.

Finally let's consider the three-dimensional case:

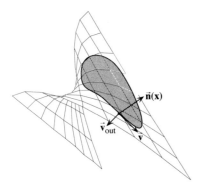

FIGURE 6.6.3.

The shaded area is the piece-with-boundary S_1 of the surface S. The vector $\vec{v}_{out}$ is tangent to S at a point in the boundary C_i of S, and points out of S_1. The unit vector $\vec{v}$ is tangent to C_i. Since

$$\det[\vec{n}(\mathbf{x}), \vec{v}_{out}, \vec{v}] > 0,$$

the boundary orientation is defined by $\vec{v}$. If we rotated $\vec{v}$ by $180°$, then the vectors would obey the left-hand rule instead of the right-hand rule, and the orientation would be reversed.

For a 2-manifold, i.e., a surface, the "outward-pointing vector tangent to M" is illustrated by Figure 6.6.3.

The point $\mathbf{x}_1$ in Definition 6.6.8 is the project of $x \in \partial X$ onto E_1.

In Definition 6.6.9, ω is a k-form and ω_∂ is a $(k-1)$-form, so the two can't be equal; Equation 6.6.5 says not that the forms are equal, but that evaluated on the appropriate vectors, they return the same number.

Definition 6.6.7 (Oriented boundary of a piece-with-boundary of $\mathbb{R}^3$). Let $U \subset \mathbb{R}^3$ be piece-with-boundary of $\mathbb{R}^3$, whose smooth boundary is a union of surfaces S_i. We will suppose that U is given the standard orientation of $\mathbb{R}^3$. Then the orientation of the boundary of U (i.e., the orientation of the surfaces) is specified by the outward-pointing normal.

Boundary orientation: the unified approach

Now we will see that our ad hoc definitions of oriented boundaries of curves, surfaces, and open subsets of $\mathbb{R}^3$ are all special cases of a general definition.

We need first to define outward-pointing vectors.

Let $M \subset \mathbb{R}^n$ be a manifold, $X \subset M$ a piece-with-boundary, and $\mathbf{x} \in \partial X$ a point of the smooth boundary of X. By Definitions 6.6.1 and 6.6.3, this means that there is a neighborhood $U \subset \mathbb{R}^n$ of $\mathbf{x}$ and a function $g : U \to \mathbb{R}$ (the G_i of the smooth boundary in Definition 6.6.3) such that $g(\mathbf{x}) = 0$, and that $X \cap U$ is the subset of $M \cap U$ where $g \geq 0$. At $\mathbf{x}$, the tangent space $T_{\mathbf{x}}(\partial X) = \ker[\mathbf{D}g(\mathbf{x})] \cap T_{\mathbf{x}}M$ has dimension one less than the dimension of $T_{\mathbf{x}}M$; it subdivides the tangent space $T_{\mathbf{x}}M$ into the outward-pointing vectors and the inward-pointing vectors.

Definition 6.6.8 (Outward-pointing and inward-pointing vectors). Let $\vec{v} \in T_{\mathbf{x}}(\partial X_1)$ and write

$$\vec{v} = \begin{bmatrix} \vec{v}_1 \\ \vec{v}_2 \end{bmatrix} \text{ with } \vec{v}_1 \in E_1, \vec{v}_2 \in E_2. \qquad 6.6.4$$

Then $\vec{v}$ is

$$\text{outward pointing if } \quad [\mathbf{D}g(\mathbf{x}_1)]\vec{v}_1 < 0, \quad \text{and}$$
$$\text{inward pointing if } \quad [\mathbf{D}g(\mathbf{x}_1)]\vec{v}_1 > 0.$$

If $\vec{v}$ is outward pointing, we denote it $\vec{v}_{out}$.

Definition 6.6.9 (Oriented boundary of piece-with-boundary of an oriented manifold). Let M be a k-dimensional manifold oriented by ω, and P be a piece-with-boundary of M. Let $\mathbf{x}$ be in ∂P, and $\vec{v}_{out} \in T_{\mathbf{x}}M$ be an outward-pointing vector tangent to M. Then, at $\mathbf{x}$, the boundary ∂P of P is oriented by ω_∂, where

$$\overbrace{\omega_\partial(\vec{v}_1, \ldots, \vec{v}_{k-1})}^{\text{orienting boundary}} = \overbrace{\omega(\vec{v}_{out}, \vec{v}_1, \ldots, \vec{v}_{k-1})}^{\text{orienting manifold}}. \qquad 6.6.5$$

Example 6.6.10 (Oriented boundary of a piece-with-boundary of an oriented curve). If C is a curve oriented by ω, and P is a piece-with-boundary of C, then at an endpoint $\mathbf{x}$ of P (i.e., a point in ∂P), with an outward-pointing vector $\vec{\mathbf{v}}_{\text{out}}$ anchored at $\mathbf{x}$, the boundary point $\mathbf{x}$ is oriented by the nonzero number $\omega_\partial = \omega(\vec{\mathbf{v}}_{\text{out}})$. Thus it has the sign $+1$ if ω_∂ is positive, and the sign -1 if ω_∂ is negative. (In this case, ω takes only one vector.)

This is consistent with the ad hoc definition (Definition 6.6.4). If $\omega(\vec{\mathbf{v}}) = \vec{\mathbf{t}} \cdot \vec{\mathbf{v}}$, then the condition $\omega_\partial > 0$ means exactly that $\vec{\mathbf{t}}(\mathbf{x})$ points out of P. $\triangle$

Example 6.6.11 (Oriented boundary of a piece-with-boundary of $\mathbb{R}^2$). Let the smooth curve C be the smooth boundary of a piece-with-boundary S of $\mathbb{R}^2$. If $\mathbb{R}^2$ is oriented in the standard way (i.e., by det), then at a point $\mathbf{x} \in C$, the boundary C is oriented by

$$\omega_\partial(\vec{\mathbf{v}}) = \det(\vec{\mathbf{v}}_{\text{out}}, \vec{\mathbf{v}}). \qquad 6.6.6$$

Suppose we have drawn the standard basis vectors in the plane in the standard way, with $\vec{\mathbf{e}}_2$ counterclockwise from $\vec{\mathbf{e}}_1$. Then

$$\det(\vec{\mathbf{v}}_{\text{out}}, \vec{\mathbf{v}}) > 0 \qquad 6.6.7$$

If we think of $\mathbb{R}^2$ as the horizontal plane in $\mathbb{R}^3$, then a piece-with-boundary of $\mathbb{R}^2$ is a special case of a piece-with-boundary of an oriented surface. Our definitions in the two cases coincide; in the ad hoc language, this means that the orientation of $\mathbb{R}^2$ by det is the orientation defined by the normal pointing upward.

if, when you look in the direction of $\vec{\mathbf{v}}$, the vector $\vec{\mathbf{v}}_{\text{out}}$ is on your right. In this case S is on your left, as was already shown in Figure 6.6.2. $\triangle$

Example 6.6.12 (Oriented boundary of a piece-with-boundary of an oriented surface in $\mathbb{R}^3$). Let $S_1 \subset S$ be a piece-with-boundary of an oriented surface S. Suppose that at $\mathbf{x} \in \partial S_1$, S is oriented by $\omega \in A^2(T_\mathbf{x}(S))$, and that $\vec{\mathbf{v}}_{\text{out}} \in T_\mathbf{x}S$ is tangent to S at $\mathbf{x}$ but points out of S_1. Then the curve ∂S_1 is oriented by

$$\omega_\partial(\vec{\mathbf{v}}) = \omega(\vec{\mathbf{v}}_{\text{out}}, \vec{\mathbf{v}}). \qquad 6.6.8$$

Equation 6.6.9 is justified in the subsection, "Equivalence of the ad hoc and the unified approaches for subspaces of $\mathbb{R}^3$."

This is consistent with the ad hoc definition, illustrated by Figure 6.6.3. In the ad hoc definition, where S is oriented by a normal vector field $\vec{N}$, the corresponding ω is

$$\omega(\vec{\mathbf{v}}_1, \vec{\mathbf{v}}_2) = \det(\vec{N}(\mathbf{x}), \vec{\mathbf{v}}_1, \vec{\mathbf{v}}_2), \qquad 6.6.9$$

so that

$$\omega_\partial(\vec{\mathbf{v}}) = \det(\vec{N}(\mathbf{x}), \vec{\mathbf{v}}_{\text{out}}, \vec{\mathbf{v}}). \qquad 6.6.10$$

Thus if the vectors $\vec{\mathbf{e}}_1, \vec{\mathbf{e}}_2, \vec{\mathbf{e}}_3$ are drawn in the standard way, satisfying the right-hand rule, then $\vec{\mathbf{v}}$ defines the orientation of ∂S_1 if $\vec{N}(\mathbf{x}), \vec{\mathbf{v}}_{\text{out}}, \vec{\mathbf{v}}$ satisfy the right-hand rule also. $\triangle$

Example 6.6.13 (Oriented boundary of a piece-with-boundary of $\mathbb{R}^3$). Suppose U is a piece-with-boundary of $\mathbb{R}^3$ with boundary $\partial U = S$, and U is oriented in the standard way, by det. Then S is oriented by

$$\omega_\partial(\vec{\mathbf{v}}_1, \vec{\mathbf{v}}_2) = \det(\vec{\mathbf{v}}_{\text{out}}, \vec{\mathbf{v}}_1, \vec{\mathbf{v}}_2). \qquad 6.6.11$$

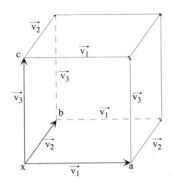

FIGURE 6.6.4.
The cube spanned by the vectors $\vec{v}_1, \vec{v}_2,$ and $\vec{v}_3$, anchored at $\mathbf{x}$. To lighten notation we set $\mathbf{a} = \mathbf{x} + \vec{v}_1$, $\mathbf{b} = \mathbf{x} + \vec{v}_2$, and $\mathbf{c} = \mathbf{x} + \vec{v}_3$. The original three vectors are drawn in dark lines; the translates in lighter or dotted lines. The cube's boundary is its six faces:

$P_{\mathbf{x}}^{o}(\vec{v}_1, \vec{v}_2)$	the bottom
$P_{\mathbf{x}}^{o}(\vec{v}_2, \vec{v}_3)$	the left side
$P_{\mathbf{x}}^{o}(\vec{v}_1, \vec{v}_3)$	the front
$P_{\mathbf{a}}^{o}(\vec{v}_2, \vec{v}_3)$	the right side
$P_{\mathbf{b}}^{o}(\vec{v}_1, \vec{v}_3)$	the back
$P_{\mathbf{c}}^{o}(\vec{v}_1, \vec{v}_2)$	the top

If i is odd, the expression

$$\left(P_{\mathbf{x}+\vec{v}_i}^{o}(\vec{v}_1, \ldots, \widehat{\vec{v}}_i, \ldots, \vec{v}_k) \right.$$
$$\left. - P_{\mathbf{x}}^{o}(\vec{v}_1, \ldots, \widehat{\vec{v}}_i, \ldots, \vec{v}_k) \right)$$

is preceded by a plus sign; if i is even, it is preceded by a minus sign.

Did you expect the right-hand side of Equation 6.6.14 to be

$$P_{\mathbf{x}+\vec{v}}^{o}(\vec{v}) - P_{\mathbf{x}}^{o}(\vec{v})?$$

Remember that $\vec{v}$ is precisely the $\vec{v}_i$ that is being omitted.

If we wish to think of orienting S in the ad hoc language, i.e., by a field of normals $\vec{N}$, this means exactly that for any $\mathbf{x} \in S$ and any two vectors $\vec{v}_1, \vec{v}_2 \in T_{\mathbf{x}}S$, the two numbers

$$\det\left(\vec{N}(\mathbf{x}), \vec{v}_1, \vec{v}_2\right) \quad \text{and} \quad \det(\vec{v}_{\text{out}}, \vec{v}_1, \vec{v}_2) \qquad 6.6.12$$

should have the same sign, i.e., $\vec{N}(\mathbf{x})$ should point out of U. $\triangle$

The oriented boundary of an oriented k-parallelogram

We saw above that an oriented k-parallelogram $P_{\mathbf{x}}^{o}(\vec{v}_1, \ldots, \vec{v}_k)$ is a piece-with-boundary of an oriented manifold if the vectors $\vec{v}_1, \ldots, \vec{v}_k$ are linearly independent (i.e., the parallelogram is not squished flat). As such its boundary carries an orientation.

Proposition 6.6.14 (Oriented boundary of an oriented k-parallelogram).

The oriented boundary of an oriented k-parallelogram $P_{\mathbf{x}}^{o}(\vec{v}_1, \ldots, \vec{v}_k)$ is given by

$$\partial P_{\mathbf{x}}^{o}(\vec{v}_1, \ldots, \vec{v}_k) =$$
$$\sum_{i=1}^{k} (-1)^{i-1} \left(P_{\mathbf{x}+\vec{v}_i}^{o}(\vec{v}_1, \ldots, \widehat{\vec{v}}_i, \ldots, \vec{v}_k) - P_{\mathbf{x}}^{o}(\vec{v}_1, \ldots, \widehat{\vec{v}}_i, \ldots, \vec{v}_k) \right), \qquad 6.6.13$$

where a hat over a term indicates that it is being omitted.

This business of hats indicating an omitted term may seem complicated. Recall that the boundary of an object always has one dimension less than the object itself: the boundary of a disk is a curve, the boundary of a box consists of the six rectangles making up its sides, and so on. The boundary of a k-dimensional parallelogram is made up of $(k-1)$-parallelograms, so omitting a vector gives the right number of vectors. For the faces of the form $P_{\mathbf{x}}(\vec{v}_1, \ldots, \widehat{\vec{v}}_i, \ldots, \vec{v}_k)$, each of the k vectors has a turn at being omitted. (In Figure 6.6.4, these faces are the three faces that include the point $\mathbf{x}$.) For the faces of the type $P_{\mathbf{x}+\vec{v}_i}^{o}(\vec{v}_1, \ldots, \widehat{\vec{v}}_i, \ldots, \vec{v}_k)$, the omitted vector is the vector added to the point $\mathbf{x}$.

Before the proof, let us give some examples, which should make the formula easier to read.

Example 6.6.15 (The boundary of an oriented 1-parallelogram). The boundary of $P_{\mathbf{x}}^{o}(\vec{v})$ is

$$\partial P_{\mathbf{x}}^{o}(\vec{v}) = P_{\mathbf{x}+\vec{v}}^{o} - P_{\mathbf{x}}^{o}. \qquad 6.6.14$$

So the boundary of an oriented line segment is its end minus its beginning, as you probably expect.

Example 6.6.16 (The boundary of an oriented 2-parallelogram). A look at Figure 6.6.5 will probably lead you to guess that the boundary of an oriented parallelogram is

$$\underbrace{\partial P_{\mathbf{x}}^o(\vec{\mathbf{v}}_1, \vec{\mathbf{v}}_2)}_{\text{boundary}} = \underbrace{P_{\mathbf{x}}^o(\vec{\mathbf{v}}_1)}_{\text{1st side}} + \underbrace{P_{\mathbf{x}+\vec{\mathbf{v}}_1}^o(\vec{\mathbf{v}}_2)}_{\text{2nd side}} - \underbrace{P_{\mathbf{x}+\vec{\mathbf{v}}_2}^o(\vec{\mathbf{v}}_1)}_{\text{3rd side}} - \underbrace{P_{\mathbf{x}}^o(\vec{\mathbf{v}}_2)}_{\text{4th side}}, \qquad 6.6.15$$

which agrees with Proposition 6.6.14.   △

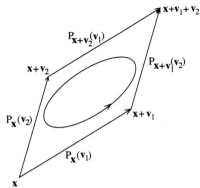

FIGURE 6.6.5.

If you start at $\mathbf{x}$ in the direction of $\vec{\mathbf{v}}_1$ and keep going around the boundary of the parallelogram, you will find the sum in Equation 6.6.15. The last two edges of that sum are negative because you are traveling against the direction of the vectors in question.

Example 6.6.17 (Boundary of a cube). For the faces of a cube shown in Figure 6.6.4 we have:

$$(i = 1 \quad \text{so} \quad (-1)^{i-1} = \ \ 1); \qquad + \left(\underbrace{P_{\mathbf{x}+\vec{\mathbf{v}}_1}^o(\vec{\mathbf{v}}_2, \vec{\mathbf{v}}_3)}_{\text{right side}} - \underbrace{P_{\mathbf{x}}^o(\vec{\mathbf{v}}_2, \vec{\mathbf{v}}_3)}_{\text{left side}} \right)$$

$$(i = 2 \quad \text{so} \quad (-1)^{i-1} = -1); \qquad - \left(\underbrace{P_{\mathbf{x}+\vec{\mathbf{v}}_2}^o(\vec{\mathbf{v}}_1, \vec{\mathbf{v}}_3)}_{\text{back}} - \underbrace{P_{\mathbf{x}}^o(\vec{\mathbf{v}}_1, \vec{\mathbf{v}}_3)}_{\text{front}} \right) \qquad 6.6.16$$

$$(i = 3 \quad \text{so} \quad (-1)^{i-1} = \ \ 1); \qquad + \left(\underbrace{P_{\mathbf{x}+\vec{\mathbf{v}}_3}^o(\vec{\mathbf{v}}_1, \vec{\mathbf{v}}_2)}_{\text{top}} - \underbrace{P_{\mathbf{x}}^o(\vec{\mathbf{v}}_1, \vec{\mathbf{v}}_2)}_{\text{bottom}} \right). \quad △$$

How many "faces" make up the boundary of a 4-parallelogram? What is each face? How would you describe the boundary following the format used for the cube in Figure 6.6.4? Check your answer below.[8]

The important consequence of preceding each term by $(-1)^{i-1}$ is that the *boundary of the boundary* is 0. The boundary of the boundary of a cube consists of the edges of each face, each edge appearing twice, once positively and once negatively, so that the two cancel.

Proof of Proposition 6.6.14. As in Example 6.6.2, denote by M the manifold of which $P_{\mathbf{x}}^o(\vec{\mathbf{v}}_1, \ldots, \vec{\mathbf{v}}_k)$ is a piece-with-boundary. The boundary $\partial P_{\mathbf{x}}^o(\vec{\mathbf{v}}_1, \ldots, \vec{\mathbf{v}}_k)$ is composed of its $2k$ faces (four for a parallelogram, six for a cube ...), each of the form

$$P_{\mathbf{x}+\vec{\mathbf{v}}_i}^o(\vec{\mathbf{v}}_1, \ldots, \widehat{\vec{\mathbf{v}}}_i, \ldots, \vec{\mathbf{v}}_k), \quad \text{or} \quad P_{\mathbf{x}}^o(\vec{\mathbf{v}}_1, \ldots, \widehat{\vec{\mathbf{v}}}_i, \ldots, \vec{\mathbf{v}}_k), \qquad 6.6.17$$

where a hat over a term indicates that it is being omitted. The problem is to show that the orientation of this boundary is consistent with Definition 6.6.9 of the oriented boundary of a piece-with-boundary.

---

[8]A 4-parallelogram has eight "faces," each of which is a 3-parallelogram (i.e., a parallelepiped, for example a cube). A 4-parallelogram spanned by the vectors $\vec{\mathbf{v}}_1, \vec{\mathbf{v}}_2, \vec{\mathbf{v}}_3$, and $\vec{\mathbf{v}}_4$, anchored at $\mathbf{x}$, is denoted $P_{\mathbf{x}}^o(\vec{\mathbf{v}}_1, \vec{\mathbf{v}}_2, \vec{\mathbf{v}}_3, \vec{\mathbf{v}}_4)$. The eight "faces" of its boundary are

$$P_{\mathbf{x}}^o(\vec{\mathbf{v}}_1, \vec{\mathbf{v}}_2, \vec{\mathbf{v}}_3), \quad P_{\mathbf{x}}^o(\vec{\mathbf{v}}_2, \vec{\mathbf{v}}_3, \vec{\mathbf{v}}_4), \quad P_{\mathbf{x}}^o(\vec{\mathbf{v}}_1, \vec{\mathbf{v}}_3, \vec{\mathbf{v}}_4), \quad P_{\mathbf{x}}(\vec{\mathbf{v}}_1, \vec{\mathbf{v}}_2, \vec{\mathbf{v}}_4),$$

$$P_{\mathbf{x}+\vec{\mathbf{v}}_1}^o(\vec{\mathbf{v}}_2, \vec{\mathbf{v}}_3, \vec{\mathbf{v}}_4), \qquad P_{\mathbf{x}+\vec{\mathbf{v}}_2}^o(\vec{\mathbf{v}}_1, \vec{\mathbf{v}}_3, \vec{\mathbf{v}}_4),$$

$$P_{\mathbf{x}+\vec{\mathbf{v}}_3}^o(\vec{\mathbf{v}}_1, \vec{\mathbf{v}}_2, \vec{\mathbf{v}}_4), \qquad P_{\mathbf{x}+\vec{\mathbf{v}}_4}^o(\vec{\mathbf{v}}_1, \vec{\mathbf{v}}_2, \vec{\mathbf{v}}_3).$$

Let $\omega \in A^k(M)$ define the orientation of M, so that $\omega(\vec{\mathbf{v}}_1, \ldots, \vec{\mathbf{v}}_k) > 0$. At a point of $P^o_{\mathbf{x}+\vec{\mathbf{v}}_i}(\vec{\mathbf{v}}_1, \ldots, \widehat{\vec{\mathbf{v}}}_i, \ldots, \vec{\mathbf{v}}_k)$, the vector $\vec{\mathbf{v}}_i$ is outward pointing, whereas at a point of $P^o_{\mathbf{x}}(\vec{\mathbf{v}}_1, \ldots, \widehat{\vec{\mathbf{v}}}_i, \ldots, \vec{\mathbf{v}}_k)$, the vector $-\vec{\mathbf{v}}_i$ is outward pointing, as shown in Figure 6.6.6. Thus the standard orientation of $P^o_{\mathbf{x}+\vec{\mathbf{v}}_i}(\vec{\mathbf{v}}_1, \ldots, \widehat{\vec{\mathbf{v}}}_i, \ldots, \vec{\mathbf{v}}_k)$ is consistent with the boundary orientation of $P^o_{\mathbf{x}}(\vec{\mathbf{v}}_1, \ldots, \widehat{\vec{\mathbf{v}}}_i, \ldots, \vec{\mathbf{v}}_k)$ precisely if

$$\omega(\vec{\mathbf{v}}_i, \vec{\mathbf{v}}_1, \ldots, \widehat{\vec{\mathbf{v}}}_i, \ldots, \vec{\mathbf{v}}_k) > 0,$$

i.e., precisely if the permutation σ_i on k symbols which consists of taking the ith element and putting it in first position is a positive permutation. But the signature of σ_i is $(-1)^{i-1}$, because you can obtain σ_i by switching the ith symbol first with the $(i-1)$th, then the $(i-2)$th, etc., and finally the first, doing $i-1$ transpositions. This explains why $P^o_{\mathbf{x}+\vec{\mathbf{v}}_i}(\vec{\mathbf{v}}_1, \ldots, \widehat{\vec{\mathbf{v}}}_i, \ldots, \vec{\mathbf{v}}_k)$ occurs with sign $(-1)^{i-1}$.

A similar argument holds for $P^o_{\mathbf{x}}(\vec{\mathbf{v}}_1, \ldots, \widehat{\vec{\mathbf{v}}}_i, \ldots, \vec{\mathbf{v}}_k)$. This oriented parallelogram has orientation compatible with the boundary orientation precisely if $\omega(-\vec{\mathbf{v}}_i, \vec{\mathbf{v}}_1, \ldots, \widehat{\vec{\mathbf{v}}}_i, \ldots, \vec{\mathbf{v}}_k) > 0$, which occurs if the permutation σ_i is odd. This explains why $P^o_{\mathbf{x}}(\vec{\mathbf{v}}_1, \ldots, \widehat{\vec{\mathbf{v}}}_i, \ldots, \vec{\mathbf{v}}_k)$ occurs in the sum with sign $(-1)^i$. $\square$

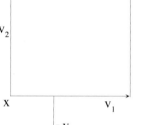

FIGURE 6.6.6.

The vector $\vec{\mathbf{v}}_1$ anchored at $\mathbf{x}$ is identical to $P^o_{\mathbf{x}}(\vec{\mathbf{v}}_1, \widehat{\vec{\mathbf{v}}}_2)$. On this segment of the boundary of the parallelogram $P^o_{\mathbf{x}}(\vec{\mathbf{v}}_1, \vec{\mathbf{v}}_2)$, the outward-pointing vector is $-\vec{\mathbf{v}}_2$. The top edge of the parallelogram is $P^0_{\mathbf{x}+\vec{\mathbf{v}}_2}(\vec{\mathbf{v}}_1, \widehat{\vec{\mathbf{v}}}_2)$; on this edge, the outward-pointing vector is $+\vec{\mathbf{v}}_2$. (We have shortened the outward and inward pointing vectors for the purpose of the drawing.)

6.7 The Exterior Derivative

In which we differentiate forms.

Now we come to the construction that gives the theory of forms its power, making possible a fundamental theorem of calculus in higher dimensions. We have already discussed integrals for forms. A derivative for forms also exists. This derivative, often called the *exterior derivative*, generalizes the derivative of ordinary functions. We will first discuss the exterior derivative in general; later we will see that the three differential operators of vector calculus (div, curl, and grad) are embodiments of the exterior derivative.

Reinterpreting the derivative

What is the ordinary derivative? Of course, you know that

$$f'(x) = \lim_{h \to 0} \frac{1}{h}\big(f(x+h) - f(x)\big), \qquad 6.7.1$$

but we will reinterpret this formula as

Equations 6.7.1 and 6.7.2 say the same thing in different words. In the first we are evaluating f at the two points $x+h$ and x. In the second we are integrating f over the boundary of the segment $P^o_x(h)$.

$$f'(x) = \lim_{h \to 0} \frac{1}{h}\int_{\partial P^o_x(h)} f. \qquad 6.7.2$$

What does this mean? We are just using different words and different notation to describe the same operation. Instead of saying that we are evaluating f at the two points $x+h$ and x, we say that we are integrating the 0-form f

over the boundary of the oriented segment $[x, x + h] = P_x^o(h)$. This boundary consists of the two oriented points $+P_{x+h}^o$ and $-P_x^o$. The first point is the endpoint of $P_x^o(h)$, and the second its beginning point; the beginning point is taken with a minus sign, to indicate the orientation of the segment. Integrating the 0-form f over these two oriented points means evaluating f on those points (Example 6.5.22). So Equations 6.7.1 and 6.7.2 say exactly the same thing.

It may seem absurd to take Equation 6.7.1, which everyone understands perfectly well, and turn it into Equation 6.7.2, which looks like just a more complicated way of saying exactly the same thing. But the language generalizes nicely to forms.

Defining the exterior derivative

The exterior derivative d is an operator that takes a k-form φ and gives a $(k + 1)$-form, $d\varphi$. Since a $(k + 1)$-form takes an oriented $(k + 1)$-dimensional parallelogram and gives a number, to define the exterior derivative of a k-form φ, we must say what number it gives when evaluated on an oriented $(k + 1)$-parallelogram.

Compare the definition of the exterior derivative and Equation 6.7.2 for the ordinary derivative:

$$f'(x) = \lim_{h \to 0} \frac{1}{h} \int_{\partial P_x^o(h)} f.$$

One thing that makes Equation 6.7.3 hard to read is that the expression for the boundary is so long that one might almost miss the φ at the end. We are integrating the k-form φ over the boundary, just as in Equation 6.7.2 we are integrating f over the boundary.

Definition 6.7.1 (Exterior derivative). The exterior derivative d of a k-form φ, denoted $d\varphi$, takes a $k + 1$-parallelogram and returns a number, as follows:

$$\underbrace{d\varphi}_{\substack{(k+1)- \\ \text{form}}} \overbrace{\left(P_{\mathbf{x}}^o(\vec{\mathbf{v}}_1, \dots, \vec{\mathbf{v}}_{k+1})\right)}^{(k+1)\text{-parallelogram}} = \lim_{h \to 0} \frac{1}{h^{k+1}} \overbrace{\int_{\underbrace{\partial P_{\mathbf{x}}^o(h\vec{\mathbf{v}}_1, \dots, h\vec{\mathbf{v}}_{k+1})}_{\substack{\text{boundary of } k+1\text{-parallelogram,} \\ \text{smaller and smaller as } h \to 0}}}^{\text{integrating } \varphi \text{ over boundary}} \varphi . \quad 6.7.3$$

This isn't a formula that you just look at and say—"got it." We will work quite hard to see what the exterior derivative gives in particular cases, and to see how to compute it. That the limit exists at all isn't obvious. Nor is it obvious that the exterior derivative is a $(k + 1)$-form: we can see that $d\varphi$ is a function of $k + 1$ vectors, but it's not obvious that it is multilinear and alternating. Two of Maxwell's equations say that a certain 2-form on $\mathbb{R}^4$ has exterior derivative zero; a course in electromagnetism might well spend six months trying to really understand what this means. But observe that the definition makes sense; since $P_{\mathbf{x}}^o(\vec{\mathbf{v}}_1, \dots, \vec{\mathbf{v}}_{k+1})$ is $(k + 1)$-dimensional, its boundary is k-dimensional, so the boundary is something over which we can integrate the k-form φ.

Notice also that when $k = 0$, this boils down to Equation 6.7.1, as restated in Equation 6.7.2.

Remark 6.7.2. Here we see why we had to define the boundary of a piece-with-boundary as we did in Definition 6.6.9. The faces of the $(k + 1)$-parallelogram

$P_{\mathbf{x}}^o(\vec{\mathbf{v}}_1, \ldots, \vec{\mathbf{v}}_{k+1})$ are k-dimensional. Multiplying the edges of these faces by h should multiply the integral over each face by h^k. So it may seem that the limit above should not exist, because the individual terms behave like $h^k/h^{k+1} = 1/h$. But the limit *does* exist, because the faces come in pairs with opposite orientation, according to Equation 6.6.13, and the terms in h^k from each pair cancel, leaving something of order h^{k+1}.

This cancellation is *absolutely essential for a derivative to exist*; that is why we have put so much emphasis on orientation.   △

We said earlier that to generalize the fundamental theorem of calculus to higher dimensions we needed a theory of integration over oriented domains. This is why.

Computing the exterior derivative

Theorem 6.7.3 shows how to take the exterior derivative of any k-form. This is a big theorem, one of the major results of the subject.

Theorem 6.7.3 (Computing the exterior derivative of a k-form).

(a) *If the coefficients a of the k-form*

$$\varphi = \sum_{1 \le i_1 < \cdots < i_k \le n} a_{i_1, \ldots, i_k} \, dx_{i_1} \wedge \cdots \wedge dx_{i_k} \qquad 6.7.4$$

are C^2 functions on an open subset $U \subset \mathbb{R}^n$, then the limit in Equation 6.7.3 exists, and defines a $(k+1)$-form.

In particular, part (b) says that if $a = b = 1$, then

$$d(\varphi + \psi) = d\varphi + d\psi.$$

Part (c): recall that a constant form returns the same number regardless of the point at which it is evaluated. Elementary forms are a special case of constant forms.

In part (e), note that the form $dx_{i_1} \wedge \cdots \wedge dx_{i_k}$ is an elementary form. The general case, taking the exterior derivative of the wedge product of a k-form and an l-form, is given by Theorem A20.2:

$$d(\varphi \wedge \psi) = d\varphi \wedge \psi + (-1)^k \varphi \wedge d\psi,$$

since in (e), ψ is a constant form, with exterior derivative 0 by part (c).

(b) *The exterior derivative is linear over $\mathbb{R}$: if φ and ψ are k-forms on $U \subset \mathbb{R}^n$, and a and b are numbers (not functions), then*

$$d(a\,\varphi + b\,\psi) = a\,d\varphi + b\,d\psi. \qquad 6.7.5$$

(c) *The exterior derivative of a constant form is 0.*

(d) *The exterior derivative of the 0-form (i.e., function) f is given by the formula*

$$df = [\mathbf{D}f] = \sum_{i=1}^n (D_i f)\,dx_i. \qquad 6.7.6$$

(e) *If f is a function, then*

$$d\left(f\,dx_{i_1} \wedge \cdots \wedge dx_{i_k}\right) = df \wedge dx_{i_1} \wedge \cdots \wedge dx_{i_k}. \qquad 6.7.7$$

Theorem 6.7.3 is proved in Appendix A.20.

These rules allow you to compute the exterior derivative of any k-form, as shown below for any k-form and as illustrated in the margin:

The first line of Equation 6.7.8 just says that $d\varphi = d\varphi$; for example, if $\varphi = f(dx \wedge dy) + g(dy \wedge dz)$, then

$$d\varphi = d\Big(f(dx \wedge dy) + g(dy \wedge dz)\Big).$$

The second line says that the exterior derivative of the sum is the sum of the exterior derivatives. For example:

$$d\Big(f(dx \wedge dy) + g(dy \wedge dz)\Big) =$$
$$d\Big(f(dx \wedge dy)\Big) + d\Big(g(dy \wedge dz)\Big).$$

The third line says that

$$d\Big(f(dx \wedge dy)\Big) = df \wedge dx \wedge dy$$

$$d\Big(g(dy \wedge dz)\Big) = dg \wedge dy \wedge dz.$$

The first term in the second line of Equation 6.7.11 is 0 because it contains two dx's; the second because it contains two dy's. Since exchanging two terms changes the sign of the wedge product, exchanging two identical terms changes the sign while leaving it unchanged, so the product must be 0. With a bit of practice computing exterior derivatives you will learn to ignore wedge products that contain two identical terms.

You will usually want to put the dx_i in ascending order, which may change the sign, as in the third line of Equation 6.7.12. The sign is not changed in the last line of Equation 6.7.11, because two exchanges are required.

$$
\begin{aligned}
d\varphi &= d\overbrace{\sum_{1 \le i_1 < \cdots < i_k \le n} a_{i_1,\ldots,i_k} dx_{i_1} \wedge \cdots \wedge dx_{i_k}}^{\text{writing } \varphi \text{ in full}} \\
&\underset{(b)}{=} \sum_{1 \le i_1 < \cdots < i_k \le n} d\big((a_{i_1,\ldots,i_k})(dx_{i_1} \wedge \cdots \wedge dx_{i_k})\big) \\
&\qquad\qquad\underbrace{}_{\text{exterior derivative of sum equals sum of exterior derivatives;}} \\
&\underset{(e)}{=} \sum_{1 \le i_1 < \cdots < i_k \le n} (d\underset{f}{\underbrace{a_{i_1,\ldots,i_k}}}) \wedge dx_{i_1} \wedge \cdots \wedge dx_{i_k} \\
&\qquad\underbrace{}_{\text{problem reduced to computing ext. deriv. of function}}
\end{aligned}
$$

6.7.8

Going from the first to the second line reduces the computation to computing exterior derivatives of elementary forms; going from the second to the third line reduces the computation to computing exterior derivatives of functions. In applying (e) we think of the coefficients $a_{i_1,\ldots,i_k}$ as the function f.

We compute the exterior derivative of the function $f = a_{i_1,\ldots,i_k}$ from part (d):

$$da_{i_1,\ldots,i_k} = \sum_{j=1}^{n} D_j a_{i_1,\ldots,i_k} dx_j. \qquad 6.7.9$$

For example, if f is a function in the three variables x, y, and z, then

$$df = D_1 f \, dx + D_2 f \, dy + D_3 f \, dz, \qquad 6.7.10$$

so

$$
\begin{aligned}
df \wedge dx \wedge dy &= (D_1 f \, dx + D_2 f \, dy + D_3 f \, dz) \wedge dx \wedge dy \\
&\underset{\substack{\text{distrib.} \\ \text{prop. } \wedge}}{=} D_1 f \underbrace{dx \wedge dx \wedge dy}_{0} + D_2 f \underbrace{dy \wedge dx \wedge dy}_{0} + D_3 f \, dz \wedge dx \wedge dy \\
&= D_3 f \, dz \wedge dx \wedge dy = D_3 f \, dx \wedge dy \wedge dz. \qquad 6.7.11
\end{aligned}
$$

Example 6.7.4 (Computing the exterior derivative of an elementary 2-form on $\mathbb{R}^4$). Computing the exterior derivative of $x_1 x_3 (dx_2 \wedge dx_4)$ gives

$$d(x_2 x_3) \wedge dx_2 \wedge dx_4$$

$$= \underbrace{(\overbrace{D_1(x_2 x_3)}^{0} dx_1 + D_2(x_2 x_3) \, dx_2 + D_3(x_2 x_3) \, dx_3 + \overbrace{D_4(x_2 x_3)}^{0} dx_4)}_{d(x_2 x_3)} \wedge dx_2 \wedge dx_4$$

$$= (x_3 \, dx_2 + x_2 \, dx_3) \wedge dx_2 \wedge dx_4 = (x_3 \, dx_2 \wedge dx_2 \wedge dx_4) + (x_2 \, dx_3 \wedge dx_2 \wedge dx_4)$$

$$= \underbrace{x_2 \, (dx_3 \wedge dx_2 \wedge dx_4)}_{dx\text{'s out of order}} = \underbrace{-x_2 \, (dx_2 \wedge dx_3 \wedge dx_4)}_{\substack{\text{sign changes as} \\ \text{order is corrected}}}. \quad \triangle \qquad 6.7.12$$

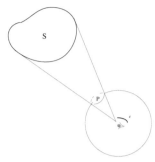

FIGURE 6.7.1.

Example 6.7.6: the origin is your eye; the "solid angle" with which you see the surface S is the cone shown above. The intersection of the cone and the sphere of radius 1 around your eye is the region P. The integral of $\Phi_{\vec{F}_3}$ over S equals its integral over P, as you are asked to prove in Exercise 6.9.8, when S is a parallelogram.

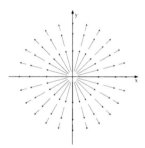

FIGURE 6.7.2.

The vector field $\vec{F}_2$. Note that because the vectors are scaled by $\frac{1}{|\mathbf{x}|}$, they get smaller the further they are from the origin. If $\vec{F}_2$ represents the flow of an incompressible fluid, then just as much flows through the unit circle as flows through a larger circle.

What is the exterior derivative of the 2-form on $\mathbb{R}^3$ $x_1 x_3^2\, dx_1 \wedge dx_2$? Check your answer below.[9]

Example 6.7.5 (Computing the exterior derivative of a 2-form). Compute the exterior derivative of the 2-form on $\mathbb{R}^4$,

$$\psi = x_1 x_2\, dx_2 \wedge dx_4 - x_2^2\, dx_3 \wedge dx_4, \qquad 6.7.13$$

which is the sum of two elementary 2-forms. We have

$$
\begin{aligned}
d\varphi &= d\big(x_1 x_2\, dx_2 \wedge dx_4\big) - d\big(x_2^2\, dx_3 \wedge dx_4\big)\\
&= \big(D_1(x_1 x_2)\, dx_1 + D_2(x_1 x_2)\, dx_2 + D_3(x_1 x_2)\, dx_3 + D_4(x_1 x_2) dx_4\big)\wedge dx_2 \wedge dx_4\\
&\qquad - \big(D_1(x_2^2)\, dx_1 + D_2(x_2^2)\, dx_2 + D_3(x_2^2)\, dx_3 + D_4(x_2^2)\, dx_4\big)\wedge dx_3 \wedge dx_4\\
&= \big(x_2\, dx_1 + x_1\, dx_2\big) \wedge dx_2 \wedge dx_4 - \big(2x_2\, dx_2 \wedge dx_3 \wedge dx_4\big)\\
&= x_2\, dx_1 \wedge dx_2 \wedge dx_4 + \underbrace{x_1\, dx_2 \wedge dx_2 \wedge dx_4}_{=0} - 2x_2\, dx_2 \wedge dx_3 \wedge dx_4 \qquad 6.7.14\\
&= x_2\, dx_1 \wedge dx_2 \wedge dx_4 - 2x_2\, dx_2 \wedge dx_3 \wedge dx_4. \quad \triangle
\end{aligned}
$$

Example 6.7.6 (Element of angle). The vector fields

$$
\vec{F}_2 = \frac{1}{x^2 + y^2}\begin{bmatrix} -y \\ x \end{bmatrix} \quad \text{and} \quad \vec{F}_3 = \frac{1}{(x^2 + y^2 + z^2)^{3/2}}\begin{bmatrix} x \\ y \\ z \end{bmatrix} \qquad 6.7.15
$$

satisfy the property that $dW_{\vec{F}_2} = 0$ and $d\Phi_{\vec{F}_3} = 0$. The forms $W_{\vec{F}_2}$ and $\Phi_{\vec{F}_3}$ can be called respectively the "element of polar angle" and the "element of solid angle"; the latter is depicted in Figure 6.7.1.

We will now find the analogs in any dimension. Using again a hat to denote a term that is omitted in the product, our candidate is the $(n-1)$-form on $\mathbb{R}^n$:

$$
\omega_n = \frac{1}{(x_1^2 + \cdots + x_n^2)^{n/2}}\sum_{i=1}^{n}(-1)^{i-1}x_i\, dx_1 \wedge \cdots \wedge \widehat{dx_i} \wedge \cdots \wedge dx_n, \qquad 6.7.16
$$

which can also be thought of as the flux of the vector field

$$
\vec{F}_n = \frac{1}{(x_1^2 + \cdots + x_n^2)^{n/2}}\begin{bmatrix} x_1 \\ \vdots \\ x_n \end{bmatrix}, \quad \text{which can be written} \quad \frac{\vec{\mathbf{x}}}{|\vec{\mathbf{x}}|^n}. \qquad 6.7.17
$$

9

$$
\begin{aligned}
d(x_1 x_3^2\, dx_1 \wedge dx_2) &= d(x_1 x_3^2) \wedge dx_1 \wedge dx_2\\
&= D_1(x_1 x_3^2)\, dx_1 \wedge dx_1 \wedge dx_2 + D_2(x_1 x_3^2)\, dx_2 \wedge dx_1 \wedge dx_2 + D_3(x_1 x_3^2)\, dx_3 \wedge dx_1 \wedge dx_2\\
&= 2x_1 x_3\, dx_3 \wedge dx_1 \wedge dx_2 = 2x_1 x_3\, dx_1 \wedge dx_2 \wedge dx_3.
\end{aligned}
$$

Equation 6.7.18: in going from the first to the second line we omitted the partial derivatives with respect to the variables that appear among the dx's to the right; i.e., we compute only the partial derivative with respect to x_i, since dx_i is the only dx that doesn't appear on the right.

Going from the second to the third line is just putting the dx_i in its proper position, which also gets rid of the $(-1)^{i-1}$; moving dx_i into its proper position requires $i-1$ transpositions.

The fourth equal sign is merely calculating the partial derivative and the fifth involves factoring out $(x_1^2 + \cdots + x_n^2)^{n/2-1}$ from the numerator and canceling with the same factor in the denominator.

The scaling $\frac{\mathbf{x}}{|\mathbf{x}|^n}$, which means that the vectors get smaller the further they are from the origin (as shown for $n=2$ in Figure 6.7.2), is chosen so that the flux outwards through a series of ever-larger spheres all centered at the origin remains constant. In fact, the flux is equal to $\mathrm{vol}_{(n-1)} S^{n-1}$.

The following shows that the exterior derivative of this flux form is 0:

$$
d\omega_n = d\Phi_{\vec{F}_n} = d\left(\frac{1}{(x_1^2 + \cdots + x_n^2)^{n/2}} \sum_{i=1}^{n} (-1)^{i-1} x_i \, dx_1 \wedge \cdots \wedge \widehat{dx_i} \wedge \cdots \wedge dx_n \right)
$$

$$
= \sum_{i=1}^{n} (-1)^{i-1} D_i \frac{x_i}{(x_1^2 + \cdots + x_n^2)^{n/2}} dx_i \wedge dx_1 \wedge \cdots \wedge \widehat{dx_i} \wedge \cdots \wedge dx_n
$$

$$
= \sum_{i=1}^{n} D_i \left(\frac{x_i}{(x_1^2 + \cdots + x_n^2)^{n/2}} \right) dx_1 \wedge \cdots \wedge dx_n
$$

$$
= \sum_{i=1}^{n} \left(\frac{(x_1^2 + \cdots + x_n^2)^{n/2} - n x_i^2 (x_1^2 + \cdots + x_n^2)^{n/2-1}}{(x_1^2 + \cdots + x_n^2)^n} \right) dx_1 \wedge \cdots \wedge dx_n
$$

$$
= \sum_{i=1}^{n} \left(\frac{x_1^2 + \cdots + x_n^2 - n x_i^2}{(x_1^2 + \cdots + x_n^2)^{n/2+1}} \right) dx_1 \wedge \cdots \wedge dx_n = 0. \qquad 6.7.18
$$

We get the last equality because the sum of the numerators cancel. For instance, when $n=2$ we have $x_1^2 + x_2^2 - 2x_1^2 + x_1^2 + x_2^2 - 2x_2^2 = 0$. $\triangle$

In the double sum in Equation 6.7.19, the terms corresponding to $i = j$ vanish, since they are followed by $dx_i \wedge dx_i$. If $i \neq j$, the pair of terms

$$
D_j D_i f dx_j \wedge dx_i
$$

and $\quad D_i D_j f dx_i \wedge dx_j$

cancel, since the crossed partials are equal, and $dx_j \wedge dx_i = -dx_i \wedge dx_j$.

The second equality in the second line of Equation 6.7.19 is part (d) of Theorem 6.7.3. Here $D_i f$ plays the role of f in part (d), giving

$$
d(D_i f) = \sum_{j=1}^{n} D_j D_i f dx_j;
$$

we have j in the subscript rather than i, since i is already taken.

Taking the exterior derivative twice

The exterior derivative of a k-form is a $(k+1)$-form; the exterior derivative of that $(k+1)$-form is a $(k+2)$-form. One remarkable property of the exterior derivative is that if you take it twice, you always get 0. (To be precise, we must specify that φ be twice continuously differentiable.)

Theorem 6.7.7. *For any k-form of class C^2 on an open subset $U \subset \mathbb{R}^n$, we have $d(d\varphi) = 0$.*

Proof. This can just be computed out. Let us see it first for 0-forms:

$$
ddf = d\left(\sum_{i=1}^{n} D_i f dx_i \right) = \sum_{i=1}^{n} d(D_i f dx_i)
$$

$$
= \sum_{i=1}^{n} dD_i f \wedge dx_i = \sum_{i=1}^{n} \sum_{j=1}^{n} D_j D_i f dx_j \wedge dx_i = 0. \qquad 6.7.19
$$

If $k > 0$, it is enough to make the following computation:

$$
d\big(d(f dx_{i_1} \wedge \cdots \wedge dx_{i_k})\big) = d(df \wedge dx_{i_1} \wedge \cdots \wedge dx_{i_k})
$$

$$
= (\underbrace{ddf}_{=0}) \wedge dx_{i_1} \wedge \cdots \wedge dx_{i_k} = 0. \quad \square \qquad 6.7.20
$$

There is also a conceptual proof of Theorem 6.7.7. Suppose φ is a k-form and you want to evaluate $d\,(d\varphi)$ on $k+2$ vectors. We get $d\,(d\varphi)$ by integrating $d\varphi$ over the boundary of the oriented $(k+2)$-parallelogram spanned by the vectors, i.e., by integrating φ over the boundary of the boundary. But what is the boundary of the boundary? It is empty! One way of saying this is that each face of the $(k+2)$-parallelogram is a $(k+1)$-dimensional parallelogram, and each edge of the $(k+1)$-parallelogram is also the edge of another $(k+1)$-parallelogram, but with opposite orientation, as Figure 6.7.3 suggests for $k=1$, and as Exercise 6.7.8 asks you to prove.

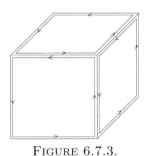

FIGURE 6.7.3.

Each edge of the cube is an edge of two faces of the cube, and is taken twice, with opposite orientations.

6.8 THE EXTERIOR DERIVATIVE IN THE LANGUAGE OF VECTOR CALCULUS

The operators *grad*, *div*, and *curl* are the workhorses of vector calculus. We will see that they are three different incarnations of the exterior derivative.

Definition 6.8.1 (Grad, curl and div). Let $f : U \to \mathbb{R}$ be a C^1 function on an open set $U \subset \mathbb{R}^n$, and let $\vec{F}$ be a C^1 vector field on U. Then the gradient of a function, the curl of a vector field, and the div of a vector field, are given by the formulas below:

$$\operatorname{grad} f = \begin{bmatrix} D_1 f \\ D_2 f \\ D_3 f \end{bmatrix} = \vec{\nabla} f$$

The gradient associates a vector field to a function. The curl associates a vector field to a vector field, and the divergence associates a function to a vector field.

We denote by the symbol $\vec{\nabla}$ ("nabla") the operator

$$\vec{\nabla} = \begin{bmatrix} D_1 \\ D_2 \\ D_3 \end{bmatrix}.$$

Some authors call $\vec{\nabla}$ "del."

$$\operatorname{curl} \vec{F} = \operatorname{curl} \begin{bmatrix} F_1 \\ F_2 \\ F_3 \end{bmatrix} = \vec{\nabla} \times \vec{F} = \begin{bmatrix} D_1 \\ D_2 \\ D_3 \end{bmatrix} \times \begin{bmatrix} F_1 \\ F_2 \\ F_3 \end{bmatrix} = \underbrace{\begin{bmatrix} D_2 F_3 - D_3 F_2 \\ D_3 F_1 - D_1 F_3 \\ D_1 F_2 - D_2 F_1 \end{bmatrix}}_{\text{cross product of } \vec{\nabla} \text{ and } \vec{F}}$$

Note that both the grad of a function, and the curl of a vector field, are vector fields, while the div of a vector field is a function.

Mnemonic: Both "curl" and "cross product" start with "c"; both "divergence" and "dot product" start with "d".

$$\operatorname{div} \vec{F} = \operatorname{div} \begin{bmatrix} F_1 \\ F_2 \\ F_3 \end{bmatrix} = \vec{\nabla} \cdot \vec{F} = \begin{bmatrix} D_1 \\ D_2 \\ D_3 \end{bmatrix} \cdot \begin{bmatrix} F_1 \\ F_2 \\ F_3 \end{bmatrix} = \underbrace{D_1 F_1 + D_2 F_2 + D_3 F_3}_{\text{dot product of } \vec{\nabla} \text{ and } \vec{F}}.$$

These operators all look kind of similar, some combination of partial derivatives. (Thus they are called *differential operators*.) We use the symbol $\vec{\nabla}$ to make it easier to remember the above formulas, which we can summarize:

$$\operatorname{grad} f = \vec{\nabla} f$$
$$\operatorname{curl} \vec{F} = \vec{\nabla} \times \vec{F} \qquad\qquad 6.8.1$$
$$\operatorname{div} \vec{F} = \vec{\nabla} \cdot \vec{F}.$$

We will use the words "grad," "curl," and "div" and the corresponding formulas of Equation 6.8.1 interchangeably, both in text and in equations. This may seem confusing at first, but it is important to learn both to see the name and think the formula, and to see the formula and think the name.

Example 6.8.2 (Curl and div). Let $\vec{F}$ be the vector field

$$\vec{F}\begin{pmatrix} x \\ y \\ z \end{pmatrix} = \begin{bmatrix} -z \\ xz^2 \\ x+y \end{bmatrix}; \quad \text{i.e.,} \quad F_1 = -z \ , \ F_2 = xz^2, \ F_3 = x+y.$$

The partial derivative D_2F_3 is the derivative with respect to the second variable of the function F_3, i.e., $D_2(x+y) = 1$. Continuing in this fashion we get

$$\mathrm{curl}\left(\begin{bmatrix} -z \\ xz^2 \\ x+y \end{bmatrix}\right) = \begin{bmatrix} D_2(x+y) - D_3(xz^2) \\ D_3(-z) - D_1(x+y) \\ D_1(xz^2) - D_2(-z) \end{bmatrix} = \begin{bmatrix} 1 - 2xz \\ -2 \\ z^2 \end{bmatrix}. \qquad 6.8.2$$

The divergence of the vector field $\vec{F}\begin{pmatrix} x \\ y \\ z \end{pmatrix} = \begin{bmatrix} x+y \\ x^2zy \\ yz \end{bmatrix}$ is $1 + x^2z + y$. $\quad \triangle$

What is the grad of the function $f = x^2y + z$? What are the curl and div of the vector field $\vec{F} = \begin{bmatrix} -y \\ x \\ xz \end{bmatrix}$? Check your answers below.[10]

The following theorem relates the exterior derivative to the *work*, *flux* and *density* form fields.

To compute the exterior derivative of a function f one can compute grad f. To compute the exterior derivative of the work form field of $\vec{F}$ one can compute curl $\vec{F}$. To compute the exterior derivative of the flux form field of $\vec{F}$ one can compute div $\vec{F}$.

Theorem 6.8.3 (Exterior derivative of form fields on $\mathbb{R}^3$). *Let f be a function on $\mathbb{R}^3$ and let $\vec{F}$ be a vector field. Then we have the following three formulas:*

(a) $\quad df \quad = \quad W_{\vec{\nabla}f}; \quad$ *i.e., df is the work form field of* grad f,

(b) $\quad dW_{\vec{F}} = \Phi_{\vec{\nabla}\times\vec{F}}; \quad$ *i.e., $dW_{\vec{F}}$ is the flux form field of* curl $\vec{F}$,

(c) $\quad d\Phi_{\vec{F}} \quad = \quad \rho_{\vec{\nabla}\cdot\vec{F}}; \quad$ *i.e., $d\Phi_{\vec{F}}$ is the density form field of* div $\vec{F}$.

Example 6.8.4 (Equivalence of df and $W_{\mathrm{grad}\,f}$). In the language of forms, to compute the exterior derivative of a function in $\mathbb{R}^3$, we can use part (d) of Theorem 6.7.3 to compute d of the 0-form f:

$$df = D_1f\,dx_1 + D_2f\,dx_2 + D_3f\,dx_3. \qquad 6.8.3$$

10

$$\mathrm{grad}\,f = \begin{bmatrix} 2xy \\ x^2 \\ 1 \end{bmatrix}; \ \mathrm{curl}\,\vec{F} = \begin{bmatrix} D_1 \\ D_2 \\ D_3 \end{bmatrix} \times \begin{bmatrix} -y \\ x \\ xz \end{bmatrix} = \begin{bmatrix} 0 \\ -z \\ 2 \end{bmatrix}; \ \mathrm{div}\,\vec{F} = \begin{bmatrix} D_1 \\ D_2 \\ D_3 \end{bmatrix} \cdot \begin{bmatrix} -y \\ x \\ xz \end{bmatrix} = x.$$

Note that writing

$$df = D_1 f dx_1 + D_2 f dx_2 + D_3 f dx_3$$

is exactly the same as writing

$$df = [\mathbf{D}f] = [D_1 f, D_2 f, D_3 f].$$

Both are linear transformations from $\mathbb{R}^3 \to \mathbb{R}$, and evaluated on a vector $\vec{v}$, they give the same result:

$$(D_1 f dx_1 + D_2 f dx_2 + D_3 f dx_3)(\vec{v})$$
$$= D_1 f v_1 + D_2 f v_2 + D_3 f v_3,$$

and

$$[D_1 f, D_2 f, D_3 f] \begin{bmatrix} v_1 \\ v_2 \\ v_3 \end{bmatrix}$$
$$= D_1 f v_1 + D_2 f v_2 + D_3 f v_3.$$

Evaluated on the vector $\vec{v} = \begin{bmatrix} v_1 \\ v_2 \\ v_3 \end{bmatrix}$, this 1-form gives

$$df(\vec{v}) = D_1 f v_1 + D_2 f v_2 + D_3 f v_3. \qquad 6.8.4$$

In the language of vector calculus, we can compute $W_{\text{grad } f} = W_{\vec{\nabla} f} = W_{\begin{bmatrix} D_1 f \\ D_2 f \\ D_3 f \end{bmatrix}}$,

which evaluated on $\vec{v}$ gives

$$W_{\vec{\nabla} f}(\vec{v}) = \begin{bmatrix} D_1 f \\ D_2 f \\ D_3 f \end{bmatrix} \cdot \begin{bmatrix} v_1 \\ v_2 \\ v_3 \end{bmatrix} = D_1 f v_1 + D_2 f v_2 + D_3 v_3. \quad \triangle \qquad 6.8.5$$

Example 6.8.5 (Equivalence of $dW_{\vec{F}}$ and $\Phi_{\vec{\nabla} \times \vec{F}}$). Let us compute the exterior derivative of the 1-form in $\mathbb{R}^3$

$$xy\, dx + z\, dy + yz\, dz, \quad \text{i.e.,} \quad W_{\vec{F}}, \quad \text{when} \quad \vec{F} = \begin{bmatrix} xy \\ z \\ yz \end{bmatrix}.$$

In the language of forms,

$$d(xy\, dx + z\, dy + yz\, dz) = d(xy) \wedge dx + d(z) \wedge dy + d(yz) \wedge dz$$
$$= (D_1 xy\, dx + D_2 xy\, dy + D_3 xy\, dz) \wedge dx + (D_1 z\, dx + D_2 z\, dy + D_3 z\, dz) \wedge dy$$
$$+ (D_1 yz\, dx + D_2 yz\, dy + D_3 yz\, dz) \wedge dz$$
$$= -x(dx \wedge dy) + (z - 1)(dy \wedge dz). \qquad 6.8.6$$

Since any 2-form in $\mathbb{R}^3$ can be written $\Phi_{\vec{G}} = G_1\, dy \wedge dz - G_2\, dx \wedge dz + G_3\, dx \wedge dy$,

Here we write $\Phi_{\vec{G}}$ to avoid confusion with our vector field

$$\vec{F} = \begin{bmatrix} xy \\ z \\ yz \end{bmatrix}.$$

the last line of Equation 6.8.6 can be written $\Phi_{\vec{G}}$ for $\vec{G} = \begin{bmatrix} z - 1 \\ 0 \\ -x \end{bmatrix}$.

This vector field is precisely the curl of $\vec{F}$:

$$\nabla \times \vec{F} = \begin{bmatrix} D_1 \\ D_2 \\ D_3 \end{bmatrix} \times \begin{bmatrix} xy \\ z \\ yz \end{bmatrix} = \begin{bmatrix} D_2 yz - D_3 z \\ -D_1 yz + D_3 xy \\ D_1 z - D_2 xy \end{bmatrix} = \begin{bmatrix} z - 1 \\ 0 \\ -x \end{bmatrix}. \quad \triangle$$

Exercise 6.8.3 asks you to work out a similar example showing the equivalence of $d\Phi_{\vec{F}}$ and $\rho_{\vec{\nabla} \cdot \vec{F}}$.

Proof of Theorem 6.8.3. The proof simply consists of using symbolic entries rather than the specific ones of Examples 6.8.4 and 6.8.5 and Exercise 6.8.3. For part (a), we find

$$df = D_1 f dx + D_2 f dy + D_3 f dz = W_{\begin{bmatrix} D_1 f \\ D_2 f \\ D_3 f \end{bmatrix}} = W_{\vec{\nabla} f}. \qquad 6.8.7$$

The *work form field* $W_{\vec{F}}$ of a vector field $\vec{F}$ is the 1-form field

$$W_{\vec{F}}\left(P_{\mathbf{x}}^o(\vec{\mathbf{v}})\right) = \vec{F}(\mathbf{x}) \cdot \vec{\mathbf{v}}.$$

So Theorem 6.8.3 part (a) says

$$df\left(P_{\mathbf{x}}^o(\vec{\mathbf{v}})\right) = W_{\vec{\nabla}f}\left(P_{\mathbf{x}}^o(\vec{\mathbf{v}})\right)$$
$$= \vec{\nabla}f(\mathbf{x}) \cdot \vec{\mathbf{v}}.$$

The *flux form field* $\Phi_{\vec{F}}$ of a vector field $\vec{F}$ is the 2-form field

$$\Phi_{\vec{F}}\left(P_{\mathbf{x}}^o(\vec{\mathbf{v}}_1, \vec{\mathbf{v}}_2)\right)$$
$$= \det[\vec{F}(\mathbf{x}), \vec{\mathbf{v}}_1, \vec{\mathbf{v}}_2].$$

So part (b) says

$$dW_{\vec{F}}\left(P_{\mathbf{x}}^o(\vec{\mathbf{v}}_1, \vec{\mathbf{v}}_2)\right)$$
$$= \det[(\vec{\nabla} \times \vec{F})(\mathbf{x}), \vec{\mathbf{v}}_1, \vec{\mathbf{v}}_2].$$

The *density form field* ρ_f of a function f is the 3-form field

$$\rho_f\left(P_{\mathbf{x}}^o(\vec{\mathbf{v}}_1, \vec{\mathbf{v}}_2, \vec{\mathbf{v}}_3)\right)$$
$$= f(\mathbf{x}) \det[\vec{\mathbf{v}}_1, \vec{\mathbf{v}}_2, \vec{\mathbf{v}}_3].$$

So part (c) of Theorem 6.8.3 says

$$d\Phi_{\vec{F}}\left(P_{\mathbf{x}}^o(\vec{\mathbf{v}}_1, \vec{\mathbf{v}}_2, \vec{\mathbf{v}}_3)\right)$$
$$= (\vec{\nabla} \cdot \vec{F})(\mathbf{x}) \det[\vec{\mathbf{v}}_1, \vec{\mathbf{v}}_2, \vec{\mathbf{v}}_3].$$

The diagram of Figure 6.8.1 commutes; if you start anywhere on the left, and go down and right, you will get the same answer as you get going first right and then down.

For part (b), a similar computation gives

$$\begin{aligned}
dW_{\vec{F}} &= d(F_1 dx + F_2 dy + F_3 dz) = dF_1 \wedge dx + dF_2 \wedge dy + dF_3 \wedge dz \\
&= (D_1 F_1 dx + D_2 F_1 dy + D_3 F_1 dz) \wedge dx \\
&\quad + (D_1 F_2 dx + D_2 F_2 dy + D_3 F_2 dz) \wedge dy \\
&\quad + (D_1 F_3 dx + D_2 F_3 dy + D_3 F_3 dz) \wedge dz \\
&= (D_1 F_2 - D_2 F_1) dx \wedge dy + (D_1 F_3 - D_3 F_1) dx \wedge dz \\
&\quad + (D_2 F_3 - D_3 F_2) dy \wedge dz \\
&= \Phi_{\left[\begin{smallmatrix} D_2 F_3 - D_3 F_2 \\ D_3 F_1 - D_1 F_3 \\ D_1 F_2 - D_2 F_1 \end{smallmatrix}\right]} = \Phi_{\vec{\nabla} \times \vec{F}}.
\end{aligned}$$

$$6.8.8$$

For part (c), the computation gives

$$\begin{aligned}
d\Phi_{\vec{F}} &= d(F_1 dy \wedge dz + F_2 dz \wedge dx + F_3 dx \wedge dy) \\
&= (D_1 F_1 dx + D_2 F_1 dy + D_3 F_1 dz) \wedge dy \wedge dz \\
&\quad + (D_1 F_2 dx + D_2 F_2 dy + D_3 F_2 dz) \wedge dz \wedge dx \\
&\quad + (D_1 F_3 dx + D_2 F_3 dy + D_3 F_3 dz) \wedge dx \wedge dy \\
&= (D_1 F_1 + D_2 F_2 + D_3 F_3) dx \wedge dy \wedge dz = \rho_{\vec{\nabla} \cdot \vec{F}}. \quad \square
\end{aligned}$$

$$6.8.9$$

Theorem 6.8.3 says that the three incarnations of the exterior derivative in $\mathbb{R}^3$ are precisely grad, curl, and div. Grad goes from 0-form fields to 1-form fields, curl goes from 1-form fields to 2-form fields, and div goes from 2-form fields to 3-form fields. This is summarized by the diagram in Figure 6.8.1, which you should learn.

Vector Calculus in $\mathbb{R}^3$		Form fields in $\mathbb{R}^3$
functions	$=$	0-form fields
$\downarrow$ gradient		$\downarrow$ d
vector fields	$\overset{\text{work } W}{\longrightarrow}$	1-form fields
$\downarrow$ curl		$\downarrow$ d
vector fields	$\overset{\text{flux } \Phi}{\longrightarrow}$	2-form fields
$\downarrow$ div		$\downarrow$ d
functions	$\overset{\text{density } \rho}{\longrightarrow}$	3-form fields

FIGURE 6.8.1. In $\mathbb{R}^3$, 0-form fields and 3-form fields can be identified with functions, and 1-form fields and 2-form fields can be identified with vector fields. The operators grad, curl, and div are three incarnations of the exterior derivative d, which takes a k-form field and gives a $(k+1)$-form field.

Geometric interpretation of the exterior derivative in $\mathbb{R}^3$

We already knew how to compute the exterior derivative of any k-form, and we had an interpretation of the exterior derivative of a k-form φ as integrating φ over the oriented boundary of a $(k+1)$-parallelogram. Why did we bring in grad, curl and div?

One reason is that being familiar with grad, curl, and div is essential in many physics and engineering courses. Another is that they give a different perspective on the exterior derivative in $\mathbb{R}^3$, with which many people are more comfortable.

Geometric interpretation of the gradient

The gradient of a function looks a lot like the Jacobian matrix. It is gotten simply by putting the entries of the line matrix $[\mathbf{D}f(\mathbf{x})]$ in a column instead of a row: $\operatorname{grad} f(\mathbf{x}) = [\mathbf{D}f(\mathbf{x})]^\top$. In particular,

$$\operatorname{grad} f(\mathbf{x}) \cdot \vec{\mathbf{v}} = [\mathbf{D}f(\mathbf{x})]\vec{\mathbf{v}}; \qquad 6.8.10$$

the dot product of $\vec{\mathbf{v}}$ with the gradient is the directional derivative in the direction $\vec{\mathbf{v}}$.

If θ is the angle between $\operatorname{grad} f(\mathbf{x})$ and $\vec{\mathbf{v}}$, we can write

$$\operatorname{grad} f(\mathbf{x}) \cdot \vec{\mathbf{v}} = |\operatorname{grad} f(\mathbf{x})|\, |\vec{\mathbf{v}}| \cos\theta, \qquad 6.8.11$$

which becomes $|\operatorname{grad} f(\mathbf{x})| \cos\theta$ if $\vec{\mathbf{v}}$ is constrained to have length 1. This is maximal when $\theta = 0$, giving $[\mathbf{D}f(\mathbf{x})]\vec{\mathbf{v}} = \operatorname{grad} f(\mathbf{x}) \cdot \vec{\mathbf{v}} = |\operatorname{grad} f(\mathbf{x})|$. So

> *the gradient of a function f at $\mathbf{x}$ points in the direction in which f increases the fastest, and has a length equal to its rate of increase in that direction.*

Remark. Some people find it easier to think of the gradient, which is a vector, and thus an element of $\mathbb{R}^n$, than to think of the derivative, which is a line matrix, and thus a linear *function* $\mathbb{R}^n \to \mathbb{R}$. They also find it easier to think that the gradient is orthogonal to the curve (or surface, or higher-dimensional manifold) of equation $f(\mathbf{x}) - c = 0$ than to think that $\ker[\mathbf{D}f(\mathbf{x})]$ is the tangent space to the curve (or surface or manifold).

Since the derivative is the transpose of the gradient, and vice versa, it may not seem to make any difference which perspective one chooses. But the derivative has an advantage that the gradient lacks: as Equation 6.8.10 makes clear, the derivative needs no extra geometric structure on $\mathbb{R}^n$, whereas the gradient requires the dot product. Sometimes (in fact usually) there is no natural dot product available. Thus the derivative of a function is the natural thing to consider.

But there is a place where gradients of functions really matter: in physics, gradients of potential energy functions are force fields, and we really want to

By *conservative* we mean that the integral on a closed path is zero: i.e., the total energy expended is zero. The gravity force field is conservative, but any force field involving friction is not: the potential energy you lose going down a hill on a bicycle is never quite recouped when you roll up the other side.

In French the curl is known as the *rotationnel*, and originally in English it was called the rotation of the vector field. It was to avoid the abbreviation *rot* that the word *curl* was substituted.

FIGURE 6.8.2.

Curl probe: put the paddle wheels at some spot of the fluid; the speed at which the paddle rotates will be proportional to the component of the curl in the direction of the axle.

think of force fields as vectors. For example, the gravitational force field is the vector $\begin{bmatrix} 0 \\ 0 \\ -gm \end{bmatrix}$, which we saw in Equation 6.4.2; this is the gradient of the height function (or rather, minus the gradient of the height function).

As it turns out, force fields are *conservative* exactly when they are gradients of functions, called potentials (discussed in Section 6.11). However, the potential is not observable, and discovering whether it exists from examining the force field is a big chapter in mathematical physics.   △

Geometric interpretation of the curl

The peculiar mixture of partials that go into the curl seems impenetrable. We aim to justify the following description.

The curl probe. Consider an axis, free to rotate in a bearing that you hold, and having paddles attached, as in Figure 6.8.2.

We will assume that the bearing is packed with a viscous fluid, so that its angular speed (not acceleration) is proportional to the torque exerted by the paddles. If a fluid is in constant motion with velocity vector field $\vec{F}$, then the curl of the velocity vector field at $\mathbf{x}$, $(\vec{\nabla} \times \vec{F})(\mathbf{x})$, is measured as follows:

The curl of a vector field at a point $\mathbf{x}$ points in the direction such that if you insert the paddle of the curl probe with its axis in that direction, it will spin the fastest. The speed at which it spins is proportional to the magnitude of the curl.

Why should this be the case? Using Theorem 6.8.3 (b) and Definition 6.7.1 of the exterior derivative, we see that

$$\underbrace{\Phi_{\vec{\nabla} \times \vec{F}}\big(P_{\mathbf{x}}^o(\vec{\mathbf{v}}_1, \vec{\mathbf{v}}_2)\big)}_{dW_{\vec{F}}} = \lim_{h \to 0} \frac{1}{h^2} \int_{\partial P_{\mathbf{x}}^o(h\vec{\mathbf{v}}_1, h\vec{\mathbf{v}}_2)} W_{\vec{F}} \qquad 6.8.12$$

measures the work of $\vec{F}$ around the parallelogram spanned by $\vec{\mathbf{v}}_1$ and $\vec{\mathbf{v}}_2$ (i.e., over its oriented boundary). If $\vec{\mathbf{v}}_1$ and $\vec{\mathbf{v}}_2$ are unit vectors orthogonal to the axis of the probe and to each other, this work is approximately proportional to the torque to which the probe will be subjected.

Theorems 6.7.7 and 6.8.3 have the following important consequence in $\mathbb{R}^3$:

If f is a C^2 function on an open subset $U \subset \mathbb{R}^3$, then curl grad $f = 0$.

Therefore, in order for a vector field to be the gradient of a function, its curl must be zero. This may seem obvious in terms of a falling apple; gravity does not exert any torque and cause the apple to spin. In more complicated settings, it is less obvious; if you observed the motions of stars in a galaxy, you might be tempted to think there was some curl, but there isn't. (We will see in Section

6.11 that having curl zero does not quite guarantee that a vector field is the gradient of a function.)

Geometric interpretation of the divergence

The divergence is easier to interpret than the curl. If we put together the formula of Theorem 6.8.3 (c) and Definition 6.7.1 of the exterior derivative, we see that the divergence of $\vec{F}$ at a point $\mathbf{x}$ is proportional to the *flux* of $\vec{F}$ through the boundary of a small box around $\mathbf{x}$, i.e., the *net flow out* of the box. In particular, if the fluid is incompressible, the divergence of its velocity vector field is 0: exactly as much must flow in as out. Thus, *the divergence measures the extent to which flow along the vector field changes the density.*

> The flow is out if the box has the standard orientation. If not, the flow is in.

Again, Theorems 6.7.7 and 6.8.3 have the following consequence:

> *If $\vec{F}$ is a C^2 vector field on an open subset $U \subset \mathbb{R}^3$, then* $\operatorname{div} \operatorname{curl} \vec{F} = 0$.

> In $\mathbb{R}^3$ the Laplacian is
> $$D_1^2 + D_2^2 + D_3^2;$$
> acting on a function f it gives $\operatorname{div} \operatorname{grad} f$. The Laplacian measures to what extent the graph of a function is "tight." It shows up in electromagnetism, relativity, elasticity, complex analysis . . . .

Remark. Theorem 6.7.7 says nothing about

$$\operatorname{div} \operatorname{grad} f, \quad \operatorname{grad} \operatorname{div} \vec{F}, \quad \text{or} \quad \operatorname{curl} \operatorname{curl} \vec{F},$$

which are not 0. The first, $\operatorname{div} \operatorname{grad} f$, is the *Laplacian*, arguably the most important differential operator in existence; the second minus the third gives the Laplacian acting on each coordinate of a vector field. $\triangle$

6.9   THE GENERALIZED STOKES'S THEOREM

We worked pretty hard to define the exterior derivative, and now we are going to reap some rewards for our labor: we are going to see that there is a higher-dimensional analog of the fundamental theorem of calculus, Stokes's theorem. It covers in one statement the four integral theorems of vector calculus, which are explored in detail in Section 6.10.

> This theorem is also known as the generalized Stokes's theorem, to distinguish it from the special case (surfaces in $\mathbb{R}^3$) discussed in Section 6.10.

> Names associated with the generalized Stokes's theorem include Poincaré (1895), Volterra (1889), Brouwer (1906), and Elie Cartan, who formalized the theory of differential forms in the early 20th century.

Recall the fundamental theorem of calculus:

Theorem 6.9.1 (Fundamental theorem of calculus). *If f is a C^1 function on a neighborhood of $[a, b]$, then*

$$\int_a^b f'(t)\, dt = f(b) - f(a). \tag{6.9.1}$$

Restate this as

$$\int_{[a,b]} df = \int_{\partial[a,b]} f, \tag{6.9.2}$$

i.e., the integral of df over an oriented interval is equal to the integral of f over the oriented boundary of the interval. In this form, the statement generalizes to higher dimensions:

Note that the dimensions in Equation 6.9.3 make sense: if X is $(k+1)$-dimensional, ∂X is k-dimensional, and if φ is a k form, $d\varphi$ is a $(k+1)$-form, so $d\varphi$ can be integrated over X, and φ can be integrated over ∂X.

This is a wonderful theorem; it is probably the best tool mathematicians have for deducing global properties from local properties.

Theorem 6.9.2 (Generalized Stokes's theorem). *Let X be a compact piece-with-boundary of a $(k+1)$-dimensional oriented manifold $M \subset \mathbb{R}^n$. Give the boundary ∂X of X the boundary orientation, and let φ be a k-form defined on a neighborhood of X. Then*

$$\int_{\partial X} \varphi = \int_X d\varphi. \qquad 6.9.3$$

This beautiful, short statement is the main result of the theory of forms.

Example 6.9.3 (Integrating over the boundary of a square). You apply Stokes's theorem every time you use anti-derivatives to compute an integral: to compute the integral of the 1-form $f\,dx$ over the oriented line segment $[a,b]$, you begin by finding a function g such that $dg = f\,dx$, and then say

$$\int_a^b f\,dx = \int_{[a,b]} dg = \int_{\partial[a,b]} g = g(b) - g(a). \qquad 6.9.4$$

This isn't quite the way it is usually used in higher dimensions, where "looking for anti-derivatives" has a different flavor.

For instance, to compute the integral $\int_C x\,dy - y\,dx$, where C is the boundary of the square S described by the inequalities $|x|, |y| \le 1$, with the boundary orientation, one possibility is to parametrize the four sides of the square (being careful to get the orientations right), then to integrate $x\,dy - y\,dx$ over all four sides and add. Another possibility is to apply Stokes's theorem:

The square S has side length 2, so its area is 4.

$$\int_C x\,dy - y\,dx = \int_S (dx \wedge dy - dy \wedge dx) = \int_S 2\,dx \wedge dy = 8. \quad \triangle \qquad 6.9.5$$

What is the integral over C of $x\,dy + y\,dx$? Check below.[11]

Example 6.9.4 (Integrating over the boundary of a cube). Let us integrate the 2-form

$$\varphi = (x - y^2 + z^3)\,(dy \wedge dz + dx \wedge dz + dx \wedge dy) \qquad 6.9.6$$

over the boundary of the cube C_a given by $0 \le x, y, z \le a$. It is quite possible to do this directly, parametrizing all six faces of the cube, but Stokes's theorem simplifies things substantially.

Computing the exterior derivative of φ gives

$$\begin{aligned} d\varphi &= dx \wedge dy \wedge dz - 2y\,dy \wedge dx \wedge dz + 3z^2\,dz \wedge dx \wedge dy \\ &= (1 + 2y + 3z^2)\,dx \wedge dy \wedge dz, \end{aligned} \qquad 6.9.7$$

---

[11]$d(x\,dy + y\,dx = dx \wedge dy + dy \wedge dx = 0$, so the integral is 0.

so

$$\int_{\partial C_a} \varphi = \int_{C_a} (1 + 2y + 3z^2)\, dx \wedge dy \wedge dz$$

$$= \int_0^a \int_0^a \int_0^a (1 + 2y + 3z^2)\, dx\, dy\, dz \qquad 6.9.8$$

$$= a^2\left([x]_0^a + [y^2]_0^a + [z^3]_0^a\right) = a^2\left(a + a^2 + a^3\right). \quad \triangle$$

Example 6.9.5 (Stokes's theorem: a harder example). Now let's try something similar to Example 6.9.4, but harder, integrating

$$\varphi = (x_1 - x_2^2 + x_3^3 - \cdots \pm x_n^n)\left(\sum_{i=1}^n dx_1 \wedge \cdots \wedge \widehat{dx_i} \wedge \cdots \wedge dx_n\right) \qquad 6.9.9$$

over the boundary of the cube C_a given by $0 \le x_j \le a,\, j = 1, \ldots, n$.

Computing this exterior derivative is less daunting if you are alert for terms that can be discarded. Denote $(x_1 - x_2^2 + x_3^3 - \cdots \pm x_n^n)$ by f. Then $D_1 f = dx_1$, $D_2 f = -2x_2\, dx_2$, $D_3 f = 3x_3^2\, dx_3$ and so on, ending with $\pm n x_n^{n-1}\, dx_n$. For $D_1 f$, the only term of

$$\sum_{i=1}^n dx_1 \wedge \cdots \wedge \widehat{dx_i} \wedge \cdots \wedge dx_n$$

that survives is that in which $i = 1$, giving

$$dx_1 \wedge dx_2 \wedge \cdots \wedge dx_n.$$

For $D_2 f$, the only term of the sum that survives is $dx_1 \wedge dx_3 \wedge \cdots \wedge dx_n$, giving $-2x_2\, dx_2 \wedge dx_1 \wedge dx_3 \wedge \cdots \wedge dx_n$; when the order is corrected this gives

$$2x_2\, dx_1 \wedge dx_2 \wedge \cdots \wedge dx_n.$$

In the end, all the terms are followed simply by $dx_1 \wedge \cdots \wedge dx_n$, and any minus signs have become plus.

This time, the idea of computing the integral directly is pretty awesome: parametrizing all $2n$ faces of the cube, etc. Doing it using Stokes's theorem is also pretty awesome, but much more manageable. We know how to compute $d\varphi$, and it comes out to

$$d\varphi = \underbrace{(1 + 2x_2 + 3x_3^2 + \cdots + n x_n^{n-1})}_{\sum_{j=1}^n j x_j^{j-1}}\, dx_1 \wedge \cdots \wedge dx_n, \qquad 6.9.10$$

The integral of $j x_j^{j-1}\, dx_1 \wedge \cdots \wedge dx_n$ over C_a is

$$\int_0^a \cdots \int_0^a j x_j^{j-1} |d^n \mathbf{x}| = a^{j+n-1}, \qquad 6.9.11$$

so the whole integral is $\sum_{j=1}^n a^{j+n-1} = a^n(1 + a + \cdots + a^{n-1}). \quad \triangle$

The examples above bring out one unpleasant feature of Stokes's theorem: it only relates the integral of a $k - 1$ form to the integral of a k-form if the former is integrated over a boundary. It is often possible to skirt this difficulty, as in the example below.

Example 6.9.6 (Integrating over faces of a cube). Let S be the union of the faces of the cube C given by $-1 \le x, y, z \le 1$ except the top face, oriented by the outward pointing normal. What is $\int_S \Phi_{\vec{F}}$, where $\vec{F} = \begin{bmatrix} x \\ y \\ z \end{bmatrix}$?

The integral of $\Phi_{\vec{F}}$ over the whole boundary ∂C is by Stokes's theorem the integral over C of $d\Phi_{\vec{F}} = \operatorname{div} \vec{F}\, dx \wedge dy \wedge dz = 3\, dx \wedge dy \wedge dz$, so

This parametrization is "obvious" because x and y parametrize the top of the cube, and at the top, $z = 1$.

$$\int_{\partial C} \Phi_{\vec{F}} = \int_C \operatorname{div} \vec{F}\, dx \wedge dy \wedge dz = 3 \int_C dx \wedge dy \wedge dz = 24. \qquad 6.9.12$$

Now we must subtract from that the integral over the top. Using the obvious parametrization $\begin{pmatrix} s \\ t \end{pmatrix} \mapsto \begin{pmatrix} s \\ t \\ 1 \end{pmatrix}$ gives

The matrix in Equation 6.9.13 is

$$\left[\vec{F}\!\left(\gamma\!\begin{pmatrix} s \\ t \end{pmatrix} \right), \overrightarrow{D_1\gamma}\begin{pmatrix} s \\ t \end{pmatrix}, \overrightarrow{D_2\gamma}\begin{pmatrix} s \\ t \end{pmatrix} \right].$$

You could also argue that all faces must contribute the same amount to the flux, so the top must contribute $24/6 = 4$.

$$\int_{-1}^{1} \int_{-1}^{1} \det \begin{bmatrix} s & 1 & 0 \\ t & 0 & 1 \\ 1 & 0 & 0 \end{bmatrix} |ds\,dt| = 4. \qquad 6.9.13$$

So the whole integral is $24 - 4 = 20$.   △

Proof of the generalized Stokes's theorem.

Before starting the proof of the generalized Stokes's theorem, we want to sketch two proofs of the fundamental theorem of calculus, Theorem 6.9.1. You probably saw the first in first-year calculus, but it is the other that will generalize to prove Stokes's theorem.

First proof of the fundamental theorem of calculus

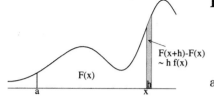

FIGURE 6.9.1.

Computing the derivative of F.

Set $F(x) = \int_a^x f(t)\,dt$. We will show that

$$F'(x) = f(x), \qquad 6.9.14$$

as Figure 6.9.1 suggests. Indeed,

$$F'(x) = \lim_{h \to 0} \frac{1}{h} \left(\int_a^{x+h} f(t)\,dt - \int_a^x f(t)\,dt \right)$$

$$= \lim_{h \to 0} \frac{1}{h} \underbrace{\int_x^{x+h} f(t)\,dt}_{\approx h f(x)} = f(x). \qquad 6.9.15$$

(The last integral is approximately $hf(x)$; the error disappears in the limit.) Now consider the function

$$f(x) - \underbrace{\int_a^x f'(t)\,dt}_{\text{with deriv. } f'(x)} . \qquad 6.9.16$$

The argument above shows that its derivative is zero, so it is constant; evaluating the function at $x = a$, we see that the constant is $f(a)$. Thus

$$f(b) - \int_a^b f'(t)\,dt = f(a). \quad \square \qquad 6.9.17$$

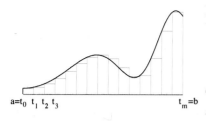

FIGURE 6.9.2.

A Riemann sum as an approximation to the integral in Equation 6.9.18.

Second proof of the fundamental theorem of calculus.

Here the appropriate drawing is the Riemann sum drawing of Figure 6.9.2.

By the very definition of the integral,

$$\int_a^b f(x)\,dx \approx \sum_i f(x_i)(x_{i+1} - x_i),$$

6.9.18

where $x_0 < x_1 < \cdots < x_m$ decompose $[a,b]$ into m little pieces, with $a = x_0$ and $b = x_m$.

You may take your pick as to which proof you prefer in the one-dimensional case but only the second proof generalizes well to a proof of the generalized Stokes's theorem. In fact, the proofs are almost identical.

By Taylor's theorem,

$$f(x_{i+1}) \approx f(x_i) + f'(x_i)(x_{i+1} - x_i).$$

6.9.19

These two statements together give

$$\int_a^b f'(x)\,dx \approx \sum_{i=0}^{m-1} f'(x_i)(x_{i+1} - x_i) \approx \sum_{i=0}^{m-1} \big(f(x_{i+1}) - f(x_i)\big).$$

6.9.20

In the far right-hand term all the interior x_i's cancel:

$$\sum_{i=0}^{m-1} \big(f(x_{i+1}) - f(x_i)\big) = f(x_1) - f(x_0) + f(x_2) - f(x_1) + \cdots + f(x_m) - f(x_{m-1}),$$

6.9.21

leaving $f(x_m) - f(x_0)$, i.e., $f(b) - f(a)$.

FIGURE 6.9.3.

Although the staircase is very close to the curve, its length is not close to the length of the curve, i.e., the curve does not fit well with a dyadic decomposition. In this case the informal proof of Stokes's theorem is not enough.

Let us analyze a little more closely the errors we are making at each step; we are adding more and more terms together as the partition becomes finer, so the errors had better be getting smaller faster, or they will not disappear in the limit. Suppose we have decomposed the interval into m pieces. Then when we replace the integral in Equation 6.9.20 by the first sum, we are making m errors, each bounded as follows. The first equality uses the fact that $A(b-a) = \int_a^b A$. Going from the first to the second line uses Corollary 1.9.2.

$$\left| \int_{x_i}^{x_{i+1}} f'(x)\,dx - \overbrace{f'(x_i)}^{A}\overbrace{(x_{i+1} - x_i)}^{b-a} \right| = \left| \int_{x_i}^{x_{i+1}} \big(f'(x) - \overbrace{f'(x_i)}^{A}\big)\,dx \right|$$

In the second line of Equation 6.9.22, we are multiplying $(x - x_i)$ by $\sup |f''|$.

We get the last equality in Equation 6.9.22 because the length of a little interval $x_{i+1} - x_i$ is precisely the original interval $b - a$ divided into m pieces:

$$x_{i+1} - x_i = \frac{b-a}{m}.$$

$$\leq \int_{x_i}^{x_{i+1}} (\sup |f''|)(x - x_i)\,dx$$

6.9.22

$$= \sup |f''| \int_{x_i}^{x_{i+1}} (x - x_i)\,dx$$

$$= \sup |f''| \frac{(x_{i+1} - x_i)^2}{2} = \sup |f''| \frac{(b-a)^2}{2m^2}.$$

We also need to remember the error term from Taylor's theorem, Equation 6.9.19, which turns out to be about the same. So all in all, we made m errors, each of which is $\leq C_1/m^2$, where C_1 is a constant that does not depend on m. Multiplying that maximal error for each piece by the number m of pieces leaves an m in the denominator, and a constant in the numerator, so the error tends to 0 as the decomposition becomes finer and finer. $\square$

An informal proof of Stokes's theorem

We find this argument convincing, but it is not quite rigorous. For a rigorous proof, see Appendix A.22. The problem with this informal argument is that the boundary of X does not necessarily fit well with the boundaries of the little cubes, as illustrated by Figure 6.9.3.

Suppose you decompose X into little pieces that are approximated by oriented $(k+1)$-parallelograms P_i^o:

$$P_i^o = P_{\mathbf{x}_i}^o(\vec{\mathbf{v}}_{1,i}, \vec{\mathbf{v}}_{2,i}, \ldots, \vec{\mathbf{v}}_{k+1,i}). \qquad 6.9.23$$

Then

$$\int_X d\varphi \approx \sum_i d\varphi(P_i^o) \approx \sum_i \int_{\partial P_i^o} \varphi \approx \int_{\partial X} \varphi. \qquad 6.9.24$$

The first approximate sign is just the definition of the integral; the $\approx$ becomes an equality in the limit as the decomposition becomes infinitely fine. The second approximate sign comes from our definition of the exterior derivative. When we add over all the P_i^o, all the internal boundaries cancel, leaving $\int_{\partial X} \varphi$.

As in the case of Riemann sums, we need to understand the errors that are signaled by our $\approx$ signs. If our parallelograms P_i^o have side ϵ, then there are on the order of $\epsilon^{-(k+1)}$ such parallelograms (precisely that many if X has volume 1). The errors in the first and second replacements are of order ϵ^{k+2}. For the first, it is our definition of the integral, and the error becomes small as the decomposition becomes infinitely fine. For the second, from the definition of the exterior derivative,

$$d\varphi(P_i^o) = \int_{\partial P_i^o} \varphi + \text{terms of order } (k+2), \qquad 6.9.25$$

so indeed the errors disappear in the limit. $\qquad \square$

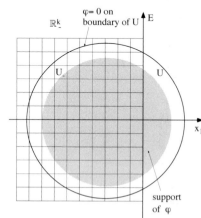

FIGURE 6.9.4.

The integral of $d\varphi$ over U_- equals the integral of φ over the boundary of U_-. Here E is the y-axis. The boundary of U_- consists of its eastern border (the intersection of E and U_-) and its curved border ($\partial U \cap U_-$), but only the eastern border contributes to the integral of φ over ∂U_-, since $\varphi = 0$ on the curved border.

Here the support of φ is shown smaller than U, but the support could be all of U; since U is open, and φ vanishes outside U, we have $\varphi = 0$ on the boundary of U.

The easy proof works in this case because the boundary of X (i.e., U_-) fits perfectly with the boundary of the dyadic cubes.

A situation where the easy proof works

We now describe a situation where this informal proof really does work. In this simple case, we have a $(k-1)$-form on $\mathbb{R}^k$, and the boundary of the piece over which we will integrate is the subspace $E \subset \mathbb{R}^k$ of equation $x_1 = 0$. There are no manifolds; nothing curvy. Figure 6.9.4 illustrates Proposition 6.9.7.

Proposition 6.9.7. *Let U be a bounded open subset of $\mathbb{R}^k$, and let U_- be the subset of U where the first coordinate is non-positive (i.e., $x_1 \leq 0$). Give U the standard orientation of $\mathbb{R}^k$ (by det), and give the boundary orientation to $\partial U_- = U \cap E$. Let φ be a $(k-1)$-form on $\mathbb{R}^k$ of class C^2, which vanishes identically outside U. Then*

$$\int_{\partial U_-} \varphi = \int_{U_-} d\varphi. \qquad 6.9.26$$

Proof. We will repeat the informal proof above, being a bit more careful about the bounds. Choose $\epsilon > 0$, and denote by $\mathbb{R}_-^k$ the subset of $\mathbb{R}^k$ where $x_1 \leq 0$.

Recall that when $d\varphi$ is a k-form (not a $(k+1)$-form as in Definition 6.7.1),

$$d\varphi(P^o_{\mathbf{x}}(\vec{\mathbf{e}}_1,\ldots,\vec{\mathbf{e}}_k))$$
$$= \frac{1}{h^k} \lim_{h\to 0} \int_{P^o_{\mathbf{x}}(h\vec{\mathbf{e}}_1,\ldots,h\vec{\mathbf{e}}_k)} \varphi.$$

So it is reasonable to expect that

$$d\varphi(P^o_{\mathbf{x}}(\vec{\mathbf{e}}_1,\ldots,\vec{\mathbf{e}}_k))$$
$$- \frac{1}{h^k} \int_{P^o_{\mathbf{x}}(h\vec{\mathbf{e}}_1,\ldots,h\vec{\mathbf{e}}_k)} \varphi$$

is of order h, and hence

$$d\varphi(P^o_{\mathbf{x}}(h\vec{\mathbf{e}}_1,\ldots,h\vec{\mathbf{e}}_k))$$
$$- \int_{P^o_{\mathbf{x}}(h\vec{\mathbf{e}}_1,\ldots,h\vec{\mathbf{e}}_k)} \varphi$$

is of order h^{k+1}.

When we evaluate $d\varphi$ on C in Equation 6.9.28, we are thinking of C as an oriented parallelogram, anchored at its lower left-hand corner.

One important motivation for allowing boundaries to have corners, rather than requiring that they be smooth, is that cubes have corners. Thus they are assumed under the general theory, and do not require separate treatment.

By Equation A20.15 (the proof of Theorem 6.7.3), there exists[12] a constant K and $\delta > 0$ such that when $|h| < \delta$,

$$\left| d\varphi\big(P^o_{\mathbf{x}}(h\vec{\mathbf{e}}_1,\ldots,h\vec{\mathbf{e}}_k)\big) - \int_{\partial P^o_{\mathbf{x}}(h\vec{\mathbf{e}}_1,\ldots,h\vec{\mathbf{e}}_k)} \varphi \right| \le Kh^{k+1}. \qquad 6.9.27$$

That is why we required φ to be of class C^2, so that the second derivatives of the coefficients of φ are bounded. Take the dyadic decomposition $\mathcal{D}_N(\mathbb{R}^k)$, where $h = 2^{-N}$. By taking N sufficiently large, we can guarantee that the difference between the integral of $d\varphi$ over U_- and the Riemann sum is less than $\epsilon/2$:

$$\left| \int_{U_-} d\varphi - \sum_{C\in\mathcal{D}_N(\mathbb{R}^k_-)} d\varphi(C) \right| < \frac{\epsilon}{2}. \qquad 6.9.28$$

Now we replace the k-parallelograms of Equation 6.9.27 by dyadic cubes, and evaluate the total difference between the exterior derivative of φ over the cubes C, and φ over the boundaries of the C. The number of cubes of $\mathcal{D}_N(\mathbb{R}^k_-)$ that intersect the support of φ is at most $L2^{kN}$ for some constant L, and since $h = 2^{-N}$, the bound for each error is now $K2^{-N(k+1)}$, so

$$\left| \sum_{C\in\mathcal{D}_N(\mathbb{R}^k_-)} d\varphi(C) - \sum_{C\in\mathcal{D}_N(\mathbb{R}^k_-)} \int_{\partial C} \varphi \right| \le \underbrace{L2^{kN}}_{\substack{\text{No. of cubes}}} \underbrace{K2^{-N(k+1)}}_{\substack{\text{bound for} \\ \text{each error}}} = LK2^{-N}.$$
$$6.9.29$$

This can also be made $< \epsilon/2$ by taking N sufficiently large—to be precise, by taking

$$N \ge \frac{\ln 2LK - \ln\epsilon}{\ln 2}. \qquad 6.9.30$$

Putting these inequalities together, we get

$$\overbrace{\left| \int_{U_-} d\varphi - \sum_{C\in\mathcal{D}_N(\mathbb{R}^k_-)} d\varphi(C) \right|}^{\le\epsilon/2} + \overbrace{\left| \sum_{C\in\mathcal{D}_N(\mathbb{R}^k_-)} d\varphi(C) - \sum_{C\in\mathcal{D}_N(\mathbb{R}^k_-)} \int_{\partial C} \varphi \right|}^{\le\epsilon/2} \le \epsilon,$$
$$6.9.31$$

so in particular, when N is sufficiently large we have

$$\left| \int_{U_-} d\varphi - \sum_{C\in\mathcal{D}_N(\mathbb{R}^k_-)} \int_{\partial C} \varphi \right| \le \epsilon, \qquad 6.9.32$$

Finally, all the internal boundaries in the sum

$$\sum_{C\in\mathcal{D}_N(\mathbb{R}^k_-)} \int_{\partial C} \varphi \qquad 6.9.33$$

---

[12]The constant in Equation A20.15 (there called C, not K), comes from Taylor's theorem with remainder, and involves the suprema of the second derivatives.

cancel, since each appears twice with opposite orientations. The only boundaries that count are those in $\mathbb{R}^{k-1}$. So (using C' to denote cubes of the dyadic composition of $\mathbb{R}^{k-1}$)

$$\sum_{C\in\mathcal{D}_N(\mathbb{R}^k_-)}\int_{\partial C}\varphi = \sum_{C'\in\mathcal{D}_N(E)}\int_{C'}\varphi = \int_E\varphi = \int_{\partial U_-}\varphi. \qquad 6.9.34$$

(We get the last equality because φ vanishes identically outside U, and therefore outside U_-.) So

Of course forms can be integrated only over oriented domains, so the E in the third term of Equation 6.9.34 must be oriented. But E is really $\mathbb{R}^{k-1}$, with coordinates $x_2,\dots,x_n$, and the boundary orientation of $\mathbb{R}^k_-$ is the standard orientation of $\mathbb{R}^{k-1}$. In Figure 6.9.4, it is shown as the line oriented from bottom to top.

$$\left|\int_{U_-} d\varphi - \int_{\partial U_-}\varphi\right| < \epsilon. \qquad 6.9.35$$

Since ϵ is arbitrary, the proposition follows. $\quad\square$

6.10 The Integral Theorems of Vector Calculus

The four forms of the generalized Stokes's theorem that make sense in $\mathbb{R}^2$ and $\mathbb{R}^3$ don't say anything that is not contained in that theorem, but each is of great importance in many applications; these theorems should all become personal friends, or at least acquaintances. They are used everywhere in electromagnetism, fluid mechanics, and many other fields.

Using a parametrization, Theorem 6.10.1 can easily be reduced to the ordinary fundamental theorem of calculus, Theorem 6.9.1, which it is if $n = 1$.

We could also call this the fundamental theorem for integrals over curves; "line integrals" is more traditional.

Theorem 6.10.1 (Fundamental theorem for line integrals). *Let C be an oriented curve in $\mathbb{R}^2$ or $\mathbb{R}^3$ (or for that matter any $\mathbb{R}^n$), with oriented boundary $(P^o_{\mathbf{b}} - P^o_{\mathbf{a}})$, and let f be a function defined on a neighborhood of C. Then*

$$\int_C df = f(\mathbf{b}) - f(\mathbf{a}). \qquad 6.10.1$$

Green's theorem and Stokes's theorem

Green's theorem is the special case of Stokes's theorem for surface integrals when the surface is flat.

Yes, we do need both *bounded*'s in Theorem 6.10.2. The exterior of the unit disk is bounded by the unit circle, but is not bounded.

Theorem 6.10.2 (Green's theorem). *Let S be a bounded region of $\mathbb{R}^2$, bounded by a curve C (or several curves C_i), carrying the boundary orientation as described in Definition 6.6.12. Let $\vec{F}$ be a vector field defined on a neighborhood of S. Then*

$$\int_S dW_{\vec{F}} = \int_C W_{\vec{F}}. \quad \text{or} \quad \int_S dW_{\vec{F}} = \sum_i \int_{C_i} W_{\vec{F}}. \qquad 6.10.2$$

This is traditionally written

$$\int_S (D_1 g - D_2 f)\, dx\, dy = \int_C f\, dx + g\, dy. \qquad 6.10.3$$

To see that the two versions are the same, write $W_{\vec{F}} = f\begin{pmatrix} x \\ y \end{pmatrix} dx + g\begin{pmatrix} x \\ y \end{pmatrix} dy$ and use Theorem 6.7.3 to compute its exterior derivative:

$$\begin{aligned} dW_{\vec{F}} &= d(f\, dx + g\, dy) = df \wedge dx + dg \wedge dy \\ &= (D_1 f\, dx + D_2 f\, dy) \wedge dx + (D_1 g\, dx + D_2 g\, dy) \wedge dy. \qquad 6.10.4 \\ &= D_2 f\, dy \wedge dx + D_1 g\, dx \wedge dy = (D_1 g - D_2 f)\, dx \wedge dy. \end{aligned}$$

There is a good deal of contention as to who should get credit for these important results. The Russians attribute them to Michael Ostrogradski, who presented them to the St. Petersburg Academy of Sciences in 1828. Green published his paper, privately, in 1828, but his result was largely overlooked until Lord Kelvin rediscovered it in 1846. Stokes proved Stokes's theorem, which he asked on an examination in Cambridge in 1854. Gauss proved the divergence theorem, also known as Gauss's theorem.

Example 6.10.3 (Green's theorem). What is the integral

$$\int_{\partial U} 2xy\, dy + x^2\, dx, \qquad 6.10.5$$

where U is the part of the disk of radius R centered at the origin where $y \geq 0$, with the standard orientation?

This corresponds to Green's theorem, with $f\begin{pmatrix} x \\ y \end{pmatrix} = x^2$ and $g\begin{pmatrix} x \\ y \end{pmatrix} = 2xy$, so that $D_1 g = 2y$ and $D_2 f = 0$. Green's theorem says

In the second line of Equation 6.10.6 we are switching to polar coordinates (Proposition 4.10.3).

$$\int_{\partial U} 2xy\, dy + x^2\, dx = \int_U (D_1 g - D_2 f)\, dx\, dy = \int_U 2y\, dx\, dy \qquad 6.10.6$$

$$= \int_0^\pi \int_0^R (2r\sin\theta)\, r\, dr\, d\theta = \frac{2R^3}{3} \int_0^\pi \sin\theta\, d\theta = \frac{4R^3}{3}.$$

What happens if we integrate over the boundary of the entire disk?[13]

The curve C in Theorem 6.10.4 may well consist of several pieces C_i.

Theorem 6.10.4 (Stokes's theorem). *Let S be an oriented surface in $\mathbb{R}^3$, bounded by a curve C that is given the boundary orientation. Let φ be a 1-form field defined on a neighborhood of S. Then*

$$\int_S d\varphi = \int_C \varphi. \qquad 6.10.7$$

Again, let's translate this into classical notation. First, and without loss of generality, we can write $\varphi = W_{\vec{F}}$, so that Theorem 6.10.4 becomes

$$\int_S dW_{\vec{F}} = \int_S \Phi_{\text{curl}\,\vec{F}} = \sum_i \int_{C_i} W_{\vec{F}}. \qquad 6.10.8$$

---

[13]It is 0, by symmetry: the integral of $2y$ over the top semi-disk cancels the integral over the bottom semi-disk.

The $\vec{N}\,|d^2\mathbf{x}|$ in the left-hand side of Equation 6.10.9 takes the parallelogram $P_\mathbf{x}(\vec{\mathbf{v}}, \vec{\mathbf{w}})$ and returns the vector

$$\vec{N}(\mathbf{x})\,|\vec{\mathbf{v}} \times \vec{\mathbf{w}}|,$$

since the integrand $|d^2\mathbf{x}|$ is the element of area; given a parallelogram, it returns its area, i.e., the length of the cross-product of its sides. When integrating over S, the only parallelograms $P_\mathbf{x}(\vec{\mathbf{v}}, \vec{\mathbf{w}})$ we will evaluate the integrand on are tangent to S at $\mathbf{x}$, and with compatible orientation. Since $\vec{\mathbf{v}}, \vec{\mathbf{w}}$ are tangent to S, $\vec{\mathbf{v}} \times \vec{\mathbf{w}}$ is perpendicular to S. So $\vec{\mathbf{v}} \times \vec{\mathbf{w}}$ is a multiple of $\vec{N}(\mathbf{x})$, which is also perpendicular to the surface. (It is a positive multiple because $\vec{\mathbf{v}} \times \vec{\mathbf{w}}$ and $\vec{N}(\mathbf{x})$ have the same orientation.) In fact

$$\vec{\mathbf{v}} \times \vec{\mathbf{w}} = |\vec{\mathbf{v}} \times \vec{\mathbf{w}}|\vec{N}(\mathbf{x}),$$

since $\vec{N}(\mathbf{x})$ has length 1. So

$$\left(\operatorname{curl}\vec{F}(\mathbf{x})\right) \cdot \left(\vec{N}(\mathbf{x})\right)\,|d^2\mathbf{x}|(\vec{\mathbf{v}}, \vec{\mathbf{w}})$$
$$= \left(\operatorname{curl}\vec{F}(\mathbf{x})\right) \cdot (\vec{\mathbf{v}} \times \vec{\mathbf{w}})$$
$$= \det[\operatorname{curl}\vec{F}(\mathbf{x}), \vec{\mathbf{v}}, \vec{\mathbf{w}}],$$

i.e., the flux of $\operatorname{curl}\vec{F}$ acting on $\vec{\mathbf{v}}$ and $\vec{\mathbf{w}}$.

Exercise 6.5.1 shows that for appropriate curves, orienting by decreasing polar angle means that the curve is oriented clockwise.

The cylinder is like an empty paper towel roll, not an empty tomato soup can. The surface S is obtained by cutting the top of the roll in some irregular fashion corresponding to its intersection with the surface $z = \sin sx + 2$; its boundary consists of the two curves: $\partial S = C + C_1$.

This still isn't the classical notation. Let $\vec{N}$ be the normal unit vector field on S defining the orientation, and $\vec{T}$ be the unit vector field on the C_i defining the orientation there. Then

$$\iint\limits_{S} \left(\operatorname{curl}\vec{F}(\mathbf{x})\right) \cdot \vec{N}(\mathbf{x})\,|d^2\mathbf{x}| = \sum_i \int_{C_i} \vec{F}(\mathbf{x}) \cdot \vec{T}(\mathbf{x})\,|d^1\mathbf{x}|. \qquad 6.10.9$$

The left-hand side of Equation 6.10.9 is discussed in the margin. Here let's compare the right-hand sides of Equations 6.10.8 and 6.10.9. Let us set $\vec{F} = \begin{bmatrix} F_1 \\ F_2 \\ F_3 \end{bmatrix}$. In the right-hand side of Equation 6.10.8, the integrand is $W_{\vec{F}} = F_1\,dx + F_2\,dy + F_3\,dz$; given a vector $\vec{\mathbf{v}}$, it returns the number $F_1v_1 + F_2v_2 + F_3v_3$.

In Equation 6.10.9, $\vec{T}(\mathbf{x})\,|d^1\mathbf{x}|$ is a complicated way of expressing the identity. Since the element of arc length $|d^1\mathbf{x}|$ takes a vector and returns its length, $\vec{T}(\mathbf{x})\,|d^1\mathbf{x}|$ takes a vector and returns $\vec{T}(\mathbf{x})$ times its length. Since $\vec{T}(\mathbf{x})$ has length 1, the result is a vector with length $|\vec{\mathbf{v}}|$, tangent to the curve. When integrating, we are only going to evaluate the integrand on vectors tangent to the curve and pointing in the direction of $\vec{T}$, so this process just takes such a vector and returns precisely the same vector. So $\vec{F}(\mathbf{x})\cdot\vec{T}(\mathbf{x})\,|d^1\mathbf{x}|$ takes a vector $\vec{\mathbf{v}}$ and returns the number

$$\left(\vec{F}(\mathbf{x})\cdot\underbrace{\vec{T}(\mathbf{x})\,|d^1\mathbf{x}|)(\vec{\mathbf{v}}}_{\vec{\mathbf{v}}}\right) = \begin{bmatrix} F_1 \\ F_2 \\ F_3 \end{bmatrix} \cdot \begin{bmatrix} v_1 \\ v_2 \\ v_3 \end{bmatrix} = F_1v_1 + F_2v_2 + F_3v_3 = W_{\vec{F}}(\vec{\mathbf{v}}). \quad 6.10.10$$

Example 6.10.5 (Stokes's theorem). Let C be the intersection of the cylinder of equation $x^2 + y^2 = 1$ with the surface of equation $z = \sin xy + 2$. Orient C so that the polar angle decreases along C. What is the work over C of the vector field

$$\mathbf{F}\begin{pmatrix} x \\ y \\ z \end{pmatrix} = \begin{bmatrix} y^3 \\ x \\ z \end{bmatrix}? \qquad 6.10.11$$

It's not so obvious how to visualize C, much less integrate over it. Stokes's theorem says there is an easier approach: compute the integral over the subsurface S consisting of the cylinder $x^2 + y^2 = 1$ bounded at the top by C and at the bottom by the unit circle C_1 in the (x, y)-plane, oriented counterclockwise.

By Stokes's theorem, $\int_C \varphi + \int_{C_1} \varphi = \int_S d\varphi$, so rather than integrate over the irregular curve C, we will integrate over S and then subtract the integral over C_1. First we integrate over S:

$$\int_C W_{\vec{F}} + \int_{C_1} W_{\vec{F}} = \int_S \Phi_{\operatorname{curl}\vec{F}} = \int_S \Phi_{\begin{bmatrix} 0 \\ 0 \\ 1-3y^2 \end{bmatrix}} = 0. \qquad 6.10.12$$

Since C is oriented clockwise, and C_1 is oriented counterclockwise, $C + C_1$ form the oriented boundary of S. If you walk on S along C, in the clockwise direction, with your head pointing away from the z-axis, the surface is to your left; if you do the same along C_1, counterclockwise, the surface is still to your left.

What if both curves were oriented clockwise? Denote by these curves by C^+ and C_1^+, and denote by C^- and C_1^- the curves oriented counterclockwise. Then (leaving out the integrands to simplify notation) we would have

$$\int_{C^+} + \left(-\int_{C_1^+} \right) = \int_S,$$

but

$$-\int_{C_1^+} = \int_{C_1^-},$$

so $\int_{C^+} W_{\vec{F}}$ remains unchanged.

If both were oriented counterclockwise, so that C did not have the boundary orientation of S, we would have

$$-\int_{C^-} + \int_{C_1^-} = \int_S = 0;$$

instead of

$$\int_{C^+} W_{\vec{F}} = -\int_{C_1^-} W_{\vec{F}} = -\frac{7}{4}\pi$$

we have

$$\int_{C^-} W_{\vec{F}} = \int_{C_1^-} W_{\vec{F}} = \frac{7}{4}\pi.$$

The integral is 0 because the vector field $\begin{bmatrix} 0 \\ 0 \\ 1-3y^2 \end{bmatrix}$ is vertical, and has no flow through the vertical cylinder. Finally parametrize C_1 in the obvious way:

$$t \mapsto \begin{bmatrix} \cos t \\ \sin t \\ 0 \end{bmatrix}, \qquad \text{6.10.13}$$

which is compatible with the counterclockwise orientation of C_1, and compute

$$\int_{C_1} W_{\vec{F}} = \int_0^{2\pi} \begin{bmatrix} (\sin t)^3 \\ \cos t \\ 0 \end{bmatrix} \cdot \begin{bmatrix} \cos t \\ \sin t \\ 0 \end{bmatrix} dt$$

$$= \int_0^{2\pi} \left((-\sin t)^4 + \cos^2 t \right) dt = \frac{3}{4}\pi + \pi = \frac{7}{4}\pi. \qquad \text{6.10.14}$$

So the work over C is

$$\int_C W_{\vec{F}} = -\frac{7}{4}\pi. \quad \triangle \qquad \text{6.10.15}$$

The divergence theorem

The divergence theorem is also known as *Gauss's theorem*.

> **Theorem 6.10.6 (The divergence theorem).** *Let M be a bounded domain in $\mathbb{R}^3$ with the standard orientation of space, and let its boundary ∂M be a union of surfaces S_i, each oriented by the outward normal. Let φ be a 2-form field defined on a neighborhood of M. Then*
>
> $$\int_M d\varphi = \sum_i \int_{S_i} \varphi. \qquad \text{6.10.16}$$

Again, let's make this look a bit more classical. Write $\varphi = \Phi_{\vec{F}}$, so that $d\varphi = d\Phi_{\vec{F}} = \rho_{\operatorname{div} \vec{F}}$, and let $\vec{N}$ be the unit outward-pointing vector field on the S_i; then Equation 6.10.16 can be rewritten

$$\int_M \rho_{\operatorname{div} \vec{F}} = \iiint_M \operatorname{div} \vec{F} \, dx \, dy \, dz = \sum_i \iint_{S_i} \vec{F} \cdot \vec{N} \, |d^2\mathbf{x}|. \qquad \text{6.10.17}$$

When we discussed Stokes's theorem, we saw that $\vec{F} \cdot \vec{N}$, evaluated on a parallelogram tangent to the surface, is the same thing as the flux of $\vec{F}$ evaluated on the same parallelogram. So indeed Equation 6.10.17 is the same as

$$\int_M d\Phi_{\vec{F}} = \int_M \rho_{\operatorname{div} \vec{F}} = \sum_i \int_{S_i} \Phi_{\vec{F}}. \qquad \text{6.10.18}$$

Remark. We think Equations 6.10.9 and 6.10.17 are a good reason to avoid the classical notation. They bring in $\vec{N}$, which usually involves dividing by the square root of the length; this is both messy and unnecessary, since the $|d^2\mathbf{x}|$ term cancels with the denominator. More seriously, the classical notation hides the resemblance of this special Stokes's theorem and the divergence theorem to the general one, Theorem 6.9.2. On the other hand, the classical notation has a geometric immediacy that speaks to people who are used to it. $\triangle$

The unit cube is as defined in Definition 4.9.3; in $\mathbb{R}^3$ it is the cube spanned by $\vec{\mathbf{e}}_1, \vec{\mathbf{e}}_2,$ and $\vec{\mathbf{e}}_1$. Thus it is not analogous to the unit circle and the unit disk, which are centered at the origin.

Example 6.10.7 (Divergence theorem). Let Q be the unit cube. What is the flux of the vector field $\begin{bmatrix} x^2y \\ -2yz \\ x^3y^2 \end{bmatrix}$ through the boundary of Q if Q carries the standard orientation of $\mathbb{R}^3$ and the boundary has the boundary orientation?

The divergence theorem asserts that

$$\int_{\partial Q} \Phi_{\begin{bmatrix} x^2y \\ -2yz \\ x^3y^2 \end{bmatrix}} = \int_Q \rho_{\,\mathrm{div}\begin{bmatrix} x^2y \\ -2yz \\ x^3y^2 \end{bmatrix}} = \int_Q (2xy - 2z)\,|d^3\mathbf{x}|. \qquad 6.10.19$$

Equation 6.10.20: Our general formula for integrating a density form over a parametrized piece of $\mathbb{R}^3$ is (Equation 6.4.16)

$$\int_{\gamma(U)} \rho_f =$$
$$\int_U f\big(\gamma(\mathbf{u})\big) \det[\mathbf{D}\gamma(\mathbf{u})]\,|d^3\mathbf{u}|.$$

Here we parametrize by the identity $\gamma(\mathbf{x}) = \mathbf{x}$, and use the special case (Equation 6.4.17)

$$\int_{\gamma(U)} \rho_f = \int_U f(\mathbf{u})\,|d^3\mathbf{u}|.$$

This can readily be computed by Fubini's theorem:

$$\int_0^1 \int_0^1 \int_0^1 (2xy - 2z)\,dx\,dy\,dz = \frac{1}{2} - 1 = -\frac{1}{2}. \quad \triangle \qquad 6.10.20$$

Example 6.10.8 (The principle of Archimedes). Archimedes is said to have been asked by Creon, the tyrant of Syracuse, to determine whether his crown was really made of gold. Archimedes discovered that by weighing the crown suspended in water, he could determine whether or not it was counterfeit. According to legend, he made the discovery in the bath, and proceeded to run naked through the streets, crying "Eureka" ("*I have found it*").

A crown made only partly of gold weighs less than a pure gold one of the same volume, so the assignment given Archimedes was to determine whether the crown's volume was greater than it should have been for its weight. According to some accounts, the solution found by Archimedes was to measure the water displaced by the crown and by an equal weight of gold.

The principle he claimed is the following: *A body immersed in a fluid receives a buoyant force equal to the weight of the displaced fluid.*

We do not understand how he came to this conclusion; the derivation we will give uses mathematics that was certainly not available to Archimedes.

The force the fluid exerts on the immersed body is due to the pressure p. Suppose that the body is M, with boundary ∂M made up of little oriented parallelograms P_i^o. The fluid exerts a force approximately

$$p(\mathbf{x}_i)\,\mathrm{Area}\,(P_i^o)\vec{\mathbf{n}}_i, \qquad 6.10.21$$

where $\vec{\mathbf{n}}$ is an inner pointing unit vector perpendicular to P_i^o and $\mathbf{x}_i$ is a point of P_i^o; this becomes a better and better approximation as P_i^o becomes small so that the pressure on it becomes approximately constant. The total force exerted by the fluid is the sum of the forces exerted on all the little pieces of the boundary.

Thus the force is naturally a surface integral, and in fact is really an integral of a 2-form field, since the orientation of ∂M matters. But we can't think of it

How did Archimedes find this result without the divergence theorem? He may have thought of the body as made up of little cubes, perhaps separated by little sheets of water. Then the force exerted on the body is the sum of the forces exerted on all the little cubes. Archimedes's law is easy to see for one cube of side s, where the vertical component of top of the cube is z, which is a negative number ($z = 0$ is the surface of the water).

The lateral forces obviously cancel, the force on the top is vertical, of magnitude $s^2 g\mu z$, and the force on the bottom is also vertical, of magnitude $-s^2 g\mu(z - s)$, so the total force is $s^3\mu g$, which is precisely the weight of a cube of the fluid of side s.

If a body is made of lots of little cubes separated by sheets of water, all the forces on the interior walls cancel, so it doesn't matter whether the sheets of water are there or not, and the total force on the body is buoyant, of magnitude equal to the weight of the displaced fluid. Note how similar this ad hoc argument is to the proof of Stokes's theorem.

as a single 2-form field: the force has three components, and we have to think of each of them as a 2-form field. In fact, the force is

$$
\begin{bmatrix}
\int_{\partial M} p\Phi_{\vec{\mathbf{e}}_1} \\
\int_{\partial M} p\Phi_{\vec{\mathbf{e}}_2} \\
\int_{\partial M} p\Phi_{\vec{\mathbf{e}}_3}
\end{bmatrix},
\qquad 6.10.22
$$

since

$$
\begin{bmatrix}
p\Phi_{\vec{\mathbf{e}}_1} \\
p\Phi_{\vec{\mathbf{e}}_2} \\
p\Phi_{\vec{\mathbf{e}}_3}
\end{bmatrix}
\left(P^o_{\mathbf{x}}(\vec{\mathbf{v}}_1, \vec{\mathbf{v}}_2)\right) = p(\mathbf{x})
\begin{bmatrix}
\det[\vec{\mathbf{e}}_1, \vec{\mathbf{v}}_1, \vec{\mathbf{v}}_2] \\
\det[\vec{\mathbf{e}}_2, \vec{\mathbf{v}}_1, \vec{\mathbf{v}}_2] \\
\det[\vec{\mathbf{e}}_3, \vec{\mathbf{v}}_1, \vec{\mathbf{v}}_2]
\end{bmatrix}
\qquad 6.10.23
$$

$$
= p(\mathbf{x})(\vec{\mathbf{v}}_1 \times \vec{\mathbf{v}}_2) = p(\mathbf{x})\text{Area}\left(P^o_{\mathbf{x}}(\vec{\mathbf{v}}_1, \vec{\mathbf{v}}_2)\right)\vec{\mathbf{n}}.
$$

In an incompressible fluid on the surface of the earth, the pressure is of the form $p(\mathbf{x}) = -\mu g z$, where μ is the density, and g is the gravitational constant. Thus the divergence theorem tells us that if ∂M is oriented in the standard way, i.e., by the outward normal, then

$$
\text{Total force} =
\begin{bmatrix}
\int_{\partial M} \mu g z \Phi_{\vec{\mathbf{e}}_1} \\
\int_{\partial M} \mu g z \Phi_{\vec{\mathbf{e}}_2} \\
\int_{\partial M} \mu g z \Phi_{\vec{\mathbf{e}}_3}
\end{bmatrix}
=
\begin{bmatrix}
\int_M \rho_{\vec{\nabla} \cdot (\mu g z \vec{\mathbf{e}}_1)} \\
\int_M \rho_{\vec{\nabla} \cdot (\mu g z \vec{\mathbf{e}}_2)} \\
\int_M \rho_{\vec{\nabla} \cdot (\mu g z \vec{\mathbf{e}}_3)}
\end{bmatrix}.
\qquad 6.10.24
$$

The divergences are:

$$
\vec{\nabla} \cdot (\mu g z \vec{\mathbf{e}}_1) = \vec{\nabla} \cdot (\mu g z \vec{\mathbf{e}}_2) = 0 \quad \text{and} \quad \vec{\nabla} \cdot (\mu g z \vec{\mathbf{e}}_3) = \mu g.
\qquad 6.10.25
$$

Thus the total force is

$$
\begin{bmatrix}
0 \\
0 \\
\int_M \rho_{\mu g}
\end{bmatrix},
\qquad 6.10.26
$$

and the third component is the weight of the displaced fluid; the force is oriented upwards.

This proves the Archimedes principle. $\triangle$

6.11 Potentials

A very important question that constantly comes up in physics is: when is a vector field conservative? The gravitational vector field is conservative: if you climb from sea level to an altitude of 500 meters by bicycle and then return to your starting point, the total work against gravity is zero, whatever your actual path. Friction is not conservative, which is why you actually get tired during such a trip.

A very important question that constantly comes up in geometry is: when does a space have a "hole" in it?

We will see in this section that these two questions are closely related.

Conservative vector fields and their potentials

Asking whether a vector field is *conservative* is equivalent to asking whether it is the gradient of a function.

Another way of stating independence of path is to require that the work around any closed path be zero; if γ_1 and γ_2 are two paths from $\mathbf{x}$ to $\mathbf{y}$, then $\gamma_1 - \gamma_2$ is a closed loop. Requiring that the integral around it be zero is the same as requiring that the works along γ_1 and γ_2 be equal. It should be clear why under these conditions the vector field is called *conservative*.

Theorem 6.11.1. *A vector field is the gradient of a function if and only if it is conservative: i.e., if and only if the work of the vector field along any path depends only on the endpoints, and not on the oriented path joining them.*

Proof. Suppose $\vec{F}$ is the gradient of a function f: $\vec{F} = \vec{\nabla} f$. Then by Theorem 6.9.2, for any parametrized path

$$\gamma : [a, b] \to \mathbb{R}^n \qquad \text{6.11.1}$$

we have (Theorem 6.10.1)

$$\int_{\gamma([a,b])} W_{\vec{\nabla} f} = f\big(\gamma(b)\big) - f\big(\gamma(a)\big). \qquad \text{6.11.2}$$

Clearly, the work of a vector field that is the gradient of a function depends only on the endpoints: the path taken between those points doesn't matter.

It is harder to show that path independence implies that the vector field is the gradient of a function. First we need a candidate for the function f, and there is an obvious choice: choose any point $\mathbf{x}_0$ in the domain of $\vec{F}$, and define

Why obvious? We are trying to undo a gradient, i.e., a derivative, so it is natural to integrate.

$$f(\mathbf{x}) = \int_{\gamma(\mathbf{x})} W_{\vec{F}}, \qquad \text{6.11.3}$$

where $\gamma(\mathbf{x})$ is an arbitrary path from $\mathbf{x}_0$ to $\mathbf{x}$: our independence of path condition guarantees that the choice does not matter.

Now we have to see that $\vec{F} = \vec{\nabla} f$, or alternatively that $W_{\vec{F}} = df$. We know from Definition 6.7.1 of the exterior derivative (here the ordinary directional derivative) that

Remember, if f is a function on $\mathbb{R}^3$ and $\vec{F}$ is a vector field, then $df = W_{\vec{\nabla} f}$. So if we show that $df = W_{\vec{F}}$, we will have shown that $\vec{F} = \vec{\nabla} f$, i.e., that $\vec{F}$ is the gradient of the function f.

Equation 6.11.5: the second equality is justified by Equation 6.4.10 for integrating work forms (or, alternatively, by Definition 6.5.19 for integrating a form field over an oriented manifold).

$$df\big(P_{\mathbf{x}}^o(\vec{v})\big) = \lim_{h \to 0} \frac{1}{h} \overbrace{\big(f(\mathbf{x} + h\vec{v}) - f(\mathbf{x})\big)}^{\int_{\partial f} f}, \qquad \text{6.11.4}$$

and, remembering the definition of f in Equation 6.11.3, we see that $-f(\mathbf{x}) + f(\mathbf{x} + h\vec{v})$ is the work of $\vec{F}$ first from $\mathbf{x}$ back to $\mathbf{x}_0$, then from $\mathbf{x}_0$ to $\mathbf{x} + h\vec{v}$. By independence of path, we may replace this by the work from $\mathbf{x}$ to $\mathbf{x} + h\vec{v}$ along the straight line. Parametrize the segment in the obvious way (by $\gamma : t \mapsto \mathbf{x} + t\vec{v}$, with $0 \le t \le h$) to get

$$df\big(P_{\mathbf{x}}^o(\vec{v})\big) = \lim_{h \to 0} \frac{1}{h} \int_{[\mathbf{x}, \mathbf{x} + h\vec{v}]} W_{\vec{F}} = \lim_{h \to 0} \frac{1}{h} \left(\int_0^h \underbrace{\vec{F}(\mathbf{x} + t\vec{v}) \cdot \vec{v}}_{\vec{F}(\gamma(t)) \cdot \gamma'(t)} \, dt \right)$$

$$= \vec{F}(\mathbf{x}) \cdot \vec{v}, \quad \text{i.e.,} \quad df = W_{\vec{F}}. \quad \square \qquad \text{6.11.5}$$

Definition 6.11.2 (Potential). A function f such that $\operatorname{grad} f = \vec{F}$ is called a potential of $\vec{F}$.

Actually, a function can have gradient 0 without being globally constant; for example, the function defined on the separate intervals $(0, 1$ and $(2, 3)$ and equal to 1 on the first and 2 on the other, has $\operatorname{grad} = 0$.

Theorem 6.11.1 provides one answer, but it isn't clear how to use it; it would mean checking the integral along all closed paths. (Of course you definitely can use it to show that a vector field is not a gradient: if you can find one closed path along which the work of a vector field is not 0, then the vector field is definitely not a gradient.)

A vector field has more than one potential, but pretty clearly, two such potentials f and g differ by a constant, since

$$\operatorname{grad}(f - g) = \operatorname{grad} f - \operatorname{grad} g = \vec{F} - \vec{F} = 0; \qquad 6.11.6$$

the only functions with gradient 0 are the constants.

So when does a vector field have a potential, and how do we find it? The first question turns out to be less straightforward than might appear. There is a *necessary* condition: in order for a vector field $\vec{F}$ to be the gradient of a function, it must satisfy

$$\operatorname{curl} \vec{F} = 0. \qquad 6.11.7$$

This follows immediately from Theorem 6.7.7: $d\,df = 0$. Since $df = W_{\vec{\nabla} f}$, then if $\vec{F} = \vec{\nabla} f$,

$$dW_{\vec{F}} = \Phi_{\operatorname{curl} \vec{F}} = d\,df = 0; \qquad 6.11.8$$

the flux of the curl of $\vec{F}$ can be 0 only if the curl is 0.

Some textbooks declare this condition to be sufficient also, but *this is not true*, as the following example shows.

Example 6.11.3 (Necessary but not sufficient). Consider the vector field

$$\vec{F} = \frac{1}{x^2 + y^2} \begin{bmatrix} -y \\ x \\ 0 \end{bmatrix} \qquad 6.11.9$$

on $\mathbb{R}^3$ with the z-axis removed. Then

$$\operatorname{curl} \vec{F} = \begin{bmatrix} 0 \\ 0 \\ D_1 \dfrac{x}{x^2 + y^2} - D_2 \dfrac{-y}{x^2 + y^2} \end{bmatrix}, \qquad 6.11.10$$

and the third entry gives

$$\frac{(x^2 + y^2) - 2x^2}{(x^2 + y^2)^2} + \frac{(x^2 + y^2) - 2y^2}{(x^2 + y^2)^2} = 0. \qquad 6.11.11$$

But $\vec{F}$ cannot be written $\vec{\nabla} f$ for any function $f : (\mathbb{R}^3 - z\text{-axis}) \to \mathbb{R}$. Indeed, using the standard parametrization

$$\gamma(t) = \begin{pmatrix} \cos t \\ \sin t \\ 0 \end{pmatrix} \qquad 6.11.12$$

Recall (Equation 5.6.1) that the formula for integrating a work form over an oriented surface is

$$\int_C W_{\vec{F}} = \int_a^b \vec{F}\big(\gamma(t)\big) \cdot \gamma'(t)\, dt.$$

The unit circle is often denoted S^1.

the work of $\vec{F}$ around the unit circle oriented counterclockwise gives

$$\int_{S^1} W_{\vec{F}} = \int_0^{2\pi} \underbrace{\frac{1}{\cos^2 t + \sin^2 t}\begin{bmatrix} -\sin t \\ \cos t \\ 0 \end{bmatrix}}_{\vec{F}(\gamma(t))} \cdot \underbrace{\begin{bmatrix} -\sin t \\ \cos t \\ 0 \end{bmatrix}}_{\gamma'(t)} dt = 2\pi. \qquad 6.11.13$$

This cannot occur for work of a conservative vector field: we started at one point and returned to the same point, so if the vector field were conservative, the work would be zero.

We will now play devil's advocate. We claim

$$\vec{F} = \vec{\nabla}\left(\arctan \frac{y}{x}\right); \qquad 6.11.14$$

and will leave the checking to you as Exercise 6.11.1. Why doesn't this contradict the statement above, that $\vec{F}$ cannot be written $\vec{\nabla}f$? The answer is that

$$\arctan \frac{y}{x} \qquad 6.11.15$$

is *not a function*, or at least, it cannot be defined as a continuous function on $\mathbb{R}^3$ minus the z-axis. Indeed, it really is the polar angle θ, and the polar angle cannot be defined on $\mathbb{R}^3$ minus the z-axis; if you take a walk counterclockwise on a closed path around the origin, taking your polar angle with you, when you get back where you started your angle will have increased by 2π. △

A pond is convex if you can swim in a straight line from any point of the pond to any other. A pond with an island (a pond with a "hole") is never convex.

Example 6.11.3 shows exactly what is going wrong. There isn't any problem with $\vec{F}$, the problem is with the domain. *We can expect trouble any time we have a domain with holes in it* (the hole in this case being the z-axis, since $\vec{F}$ is not defined there). The function f such that $\vec{\nabla}f = \vec{F}$ is determined only up to an additive constant, and if you go around the hole, there is no reason to think that you will not add on a constant in the process. So to get a converse to Equation 6.11.7, we need to restrict our domains to domains without holes. This is a bit complicated to define, so instead we will restrict them to *convex domains*.

Definition 6.11.4 (Convex domain). A domain $U \subset \mathbb{R}^n$ is convex if for any two points $\mathbf{x}$ and $\mathbf{y}$ of U, the straight line segment $[\mathbf{x}, \mathbf{y}]$ joining $\mathbf{x}$ to $\mathbf{y}$ lies entirely in U.

Theorem 6.11.5. If $U \subset \mathbb{R}^3$ is convex, and $\vec{F}$ is a vector field on U, then $\vec{F}$ is the gradient of a function f defined on U if and only if curl $\vec{F} = 0$.

The other direction of the theorem, that if $\vec{F} = \operatorname{grad} f$, then $\operatorname{curl} \vec{F} = 0$, is obvious, since $ddf = 0$.

A function f that is a potential can be thought of as an altitude function. Equation 6.11.16 says that we can determine the altitude of $\mathbf{x}$ (i.e., $f(\mathbf{x})$) by measuring how hard we have to work to climb to $\mathbf{x}$ (the work of $\vec{F}$ over $\gamma(\mathbf{x})$).

We have been considering the question, when is a 1-form (vector field) the exterior derivative (gradient) of a 0-form (function)? The *Poincaré lemma* addresses the general question, when is a k-form the exterior derivative of a $(k-1)$-form? In the case of a 2-form on $\mathbb{R}^4$, this question is of central importance for understanding electromagnetism. The 2-form

$$W_{\vec{\mathbf{E}}} \wedge c\,dt + \Phi_{\vec{\mathbf{B}}},$$

where $\vec{\mathbf{E}}$ is the electric field and $\vec{\mathbf{B}}$ is the magnetic field, is the force field of electromagnetism, known as the *Faraday*.

The statement that the Faraday is the exterior derivative of a 1-form ensures that the electromagnetic potential exists; it is the 1-form whose exterior derivative is the Faraday.

Unlike the gravitational potential, the electromagnetic potential is not unique up to an additive constant. Different 1-forms exist such that their exterior derivative is the Faraday. The choice of 1-form is called the choice of *gauge*; gauge theory is one of the dominant ideas of modern physics.

Proof. The proof is very similar to the proof of Theorem 6.11.1. Assume $\operatorname{curl} \vec{F} = 0$. We need a candidate function f, and again there is an "obvious" choice. Choose a point $\mathbf{x}_0 \in U$, and set

$$f(\mathbf{x}) = \int_{\gamma(\mathbf{x})} W_{\vec{F}}, \qquad 6.11.16$$

where this time $\gamma(\mathbf{x})$ is specifically the straight line joining $\mathbf{x}_0$ to $\mathbf{x}$. Note that this is possible because U is convex; if U were a pond with an island, the straight line might go through the island, where the vector field is undefined.

Now we need to show that $\vec{\nabla} f = \vec{F}$. Again,

$$\vec{\nabla} f(\mathbf{x}) \cdot \vec{\mathbf{v}} = \lim_{h \to 0} \frac{1}{h}\big(f(\mathbf{x} + h\vec{\mathbf{v}}) - f(\mathbf{x})\big), \qquad 6.11.17$$

and $f(\mathbf{x} + h\vec{\mathbf{v}}) - f(\mathbf{x})$ is the work of $\vec{F}$ along the path that goes straight from $\mathbf{x}$ to $\mathbf{x}_0$ and then straight on to $\mathbf{x} + h\vec{\mathbf{v}}$. We wish to replace this by the path that goes straight from $\mathbf{x}$ to $\mathbf{x} + h\vec{\mathbf{v}}$. We don't have path independence to allow this, but we can do it by Stokes's theorem. The three oriented segments $[\mathbf{x}, \mathbf{x}_0]$, $[\mathbf{x}_0, \mathbf{x} + h\vec{\mathbf{v}}]$, and $[\mathbf{x} + h\vec{\mathbf{v}}, \mathbf{x}]$ bound a triangle T, and the work of $\vec{F}$ around T equals the flux of $\operatorname{curl}\vec{F}$ through T, which is 0 since $\operatorname{curl}\vec{F} = 0$:

$$\int_{\partial T} W_{\vec{F}} = \int_T dW_{\vec{F}} = \int_T \Phi_{\operatorname{curl}\vec{F}} = 0. \qquad 6.11.18$$

We can now rewrite Equation 6.11.17:

$$\vec{\nabla} f(\mathbf{x}) \cdot \vec{\mathbf{v}} = \lim_{h \to 0} \frac{1}{h}\bigg(\underbrace{\int_{[\mathbf{x}, \mathbf{x}_o]} W_{\vec{F}}}_{-f(\mathbf{x})} + \underbrace{\int_{[\mathbf{x}_o, \mathbf{x} + h\vec{\mathbf{v}}]} W_{\vec{F}}}_{f(\mathbf{x} + h\vec{\mathbf{v}})}\bigg) = \lim_{h \to 0} \frac{1}{h} \int_{[\mathbf{x}, \mathbf{x} + h\vec{\mathbf{v}}]} W_{\vec{F}}.$$

$$6.11.19$$

The proof finishes as in Equation 6.11.5. $\square$

Example 6.11.6 (Finding the potential of a vector field). Let us carry out the computation in the proof above in one specific case. Consider the vector field

$$\vec{F}\begin{pmatrix} x \\ y \\ z \end{pmatrix} = \begin{bmatrix} \frac{y^2}{2} + yz \\ x(y + z) \\ xy \end{bmatrix} \qquad 6.11.20$$

whose curl is indeed 0:

$$\vec{\nabla} \times \vec{F} = \begin{bmatrix} D_1 \\ D_2 \\ D_3 \end{bmatrix} \times \begin{bmatrix} \frac{y^2}{2} + yz \\ x(y + z) \\ xy \end{bmatrix} = \begin{bmatrix} x - x \\ -y + y \\ y + z - (y + z) \end{bmatrix} = \begin{bmatrix} 0 \\ 0 \\ 0 \end{bmatrix}. \qquad 6.11.21$$

Since $\vec{F}$ is defined on all of $\mathbb{R}^3$, which is certainly convex, Theorem 6.11.5 asserts that $\vec{F} = \vec{\nabla} f$, and Equation 6.11.16 suggests how to find f; set

$$f(\mathbf{a}) = \int_{\gamma_{\mathbf{a}}} W_{\vec{F}}, \quad \text{for} \quad \gamma_{\mathbf{a}}(t) = t\mathbf{a}, \ 0 \le t \le 1, \qquad 6.11.22$$

i.e., $\gamma_{\mathbf{a}}$ is a parametrization of the segment joining $\mathbf{0}$ to $\mathbf{a}$. If we set $\mathbf{a} = \begin{pmatrix} a \\ b \\ c \end{pmatrix}$,

this leads to

$$
f\begin{pmatrix} a \\ b \\ c \end{pmatrix} = \int_0^1 \begin{bmatrix} \frac{(tb)^2}{2} + t^2 bc \\ (tb + tc)ta \\ t^2 ab \end{bmatrix} \cdot \begin{bmatrix} a \\ b \\ c \end{bmatrix} \, dt
$$

$$
= \left[\frac{t^3}{3} \right]_0^1 (3ab^2/2 + 3abc) = ab^2/2 + abc.
$$

6.11.23

This means that

$$
f\begin{pmatrix} x \\ y \\ z \end{pmatrix} = \frac{xy^2}{2} + xyz,
$$

6.11.24

and it is easy to check that $\vec{\nabla} f = \vec{F}$.

6.12 Exercises for Chapter Six

Exercises for Section 6.1:

**Forms as Integrands
Over Oriented Domains**

In Exercise 6.1.1, parts (f), (g), (h), f is a C^1 function on a neighborhood of $[0, 1]$.

6.1.1   An integrand should take a piece of the domain, and return a number, in such a way that if we decompose a domain into little pieces, evaluate the integrand on the pieces and add, the sums should have a limit as the decomposition becomes infinitely fine (and the limit should not depend on how the domain is decomposed). What will happen if we break up $[0, 1]$ into intervals $[x_i, x_{i+1}]$, for $i = 0, 1, \ldots, n - 1$, with $0 = x_0 < x_1 < \cdots < x_n = 1$, and assign one of the numbers below to each of the $[x_i, x_{i+1}]$?

(a) $|x_{i+1} - x_i|^2$ (b) $\sin |x_i - x_{i+1}|$ (c) $\sqrt{|x_i - x_{i+1}|}$

(d) $|(x_{i+1})^2 - (x_i)^2|$ (e) $|(x_{i+1})^3 - (x_i)^3|$ (f) $|f(x_{i+1}) - f(x_i)|$

(g) $|f\big((x_{i+1})^2\big) - f(x_i^2)|$ (h) $\big|\big(f(x_{i+1})\big)^2 - \big(f(x_i)\big)^2\big|$ (i) $|x_{i+1} - x_i| \ln |x_{i+1} - x_i|$

6.1.2   Same exercise as 6.1.1 but in $\mathbb{R}^2$, the integrand to be integrated over $[0, 1]^2$; again, the limit should not depend on how the domain is decomposed into rectangles. The integrand takes a rectangle $a < x < b, \quad c < y < d$ and returns the number

Exercises for Section 6.2:

Forms on $\mathbb{R}^n$

(a) $|b - a|^2 \sqrt{|c - d|}$ (b) $|ac - bd|$ (c) $(ad - bc)^2$

6.2.1   Complete the proof of Proposition 6.2.11.

6.2.2   Compute the following numbers:

(a) $dx_3 \wedge dx_2 \left(P_{\mathbf{0}}^o \left(\begin{bmatrix} 1 \\ 2 \\ 3 \\ 4 \end{bmatrix}, \begin{bmatrix} 0 \\ 1 \\ -1 \\ 1 \end{bmatrix} \right) \right)$ (b) $e^x dy \left(P_{\begin{pmatrix} 2 \\ 1 \end{pmatrix}}^o \begin{pmatrix} 3 \\ 2 \end{pmatrix} \right)$

(c) $x_1^2 \, dx_3 \wedge dx_2 \wedge dx_1 \left(P^o_{\begin{pmatrix} 2 \\ 0 \\ 0 \\ 0 \end{pmatrix}} \left(\begin{bmatrix} 1 \\ 2 \\ 3 \\ 4 \end{bmatrix}, \begin{bmatrix} 0 \\ 1 \\ -1 \\ 1 \end{bmatrix}, \begin{bmatrix} 1 \\ -1 \\ -1 \\ 0 \end{bmatrix} \right) \right)$

6.2.3   Compute the following functions:

(a) $\sin(x_4) \, dx_3 \wedge dx_2 \left(P^o_{\begin{pmatrix} x_1 \\ x_2 \\ x_3 \\ x_4 \end{pmatrix}} \left(\begin{bmatrix} 1 \\ 2 \\ 3 \\ 4 \end{bmatrix}, \begin{bmatrix} 0 \\ 1 \\ -1 \\ 1 \end{bmatrix} \right) \right)$ (b) $e^x \, dy \left(P^o_{\begin{pmatrix} x \\ y \end{pmatrix}} \begin{pmatrix} 3 \\ 2 \end{pmatrix} \right)$

(c) $x_1^2 e^{x_2} \, dx_3 \wedge dx_2 \wedge dx_1 \left(P^o_{\begin{pmatrix} -x_1 \\ x_2 \\ -x_3 \\ x_4 \end{pmatrix}} \left(\begin{bmatrix} 1 \\ 2 \\ 3 \\ 4 \end{bmatrix}, \begin{bmatrix} 0 \\ 1 \\ -1 \\ 1 \end{bmatrix}, \begin{bmatrix} 1 \\ -1 \\ -1 \\ 0 \end{bmatrix} \right) \right)$

6.2.4   Prove Proposition 6.2.16.

6.2.5   Verify that Example 6.2.14 does not commute, and that Example 6.2.15 does.

Exercises for Section 6.3:

Integration Over Parametrized Domains

6.3.1   Set up each of the following integrals of form fields over parametrized domains as an ordinary multiple integral, and compute it.

(a) $\int_{\gamma(I)} x \, dy + y \, dz$, where $I = [-1, 1]$, and $\gamma(t) = \begin{pmatrix} \sin t \\ \cos t \\ t \end{pmatrix}$.

(b) $\int_{\gamma(U)} x \, dy \wedge dz$, where $U = [-1, 1] \times [-1, 1]$, and $\gamma \begin{pmatrix} u \\ v \end{pmatrix} = \begin{pmatrix} u^2 \\ u + v \\ v^3 \end{pmatrix}$.

You are strongly encouraged to use MAPLE or the equivalent for parts (c) and (d) of Exercise 6.3.1.

(c) $\int_{\gamma(U)} x_1 \, dx_2 \wedge dx_3 + x_2 \, dx_3 \wedge dx_4$, where $U = \left\{ \begin{pmatrix} u \\ v \end{pmatrix} \,\middle|\, 0 \leq u, v; \ u + v \leq 2 \right\}$,

and $\gamma \begin{pmatrix} u \\ v \end{pmatrix} = \begin{pmatrix} uv \\ u^2 + v^2 \\ u - v \\ \ln(u + v + 1) \end{pmatrix}$.

(d) $\int_{\gamma(U)} x_2 \, dx_1 \wedge dx_3 \wedge dx_4$, where $U = \left\{ \begin{pmatrix} u \\ v \\ w \end{pmatrix} \,\middle|\, 0 \leq u, v, w; \ u + v + w \leq 3 \right\}$, and $\gamma \begin{pmatrix} u \\ v \\ w \end{pmatrix} = \begin{pmatrix} uv \\ u^2 + w^2 \\ u - v \\ w \end{pmatrix}$.

6.3.2   Set up each of the following integrals of form fields over parametrized domains as an ordinary multiple integral.

(a) $\int_{\gamma(I)} y^2 dy + x^2 dz$, where $I = [0, a]$ and $\gamma(t) = \begin{pmatrix} t^3 \\ t^2 + 1 \\ t^2 - 1 \end{pmatrix}$

(b) $\int_{\gamma(U)} \sin y^2 \, dx \wedge dz$, where $U = [0, a] \times [0, b]$, and $\gamma \begin{pmatrix} u \\ v \end{pmatrix} = \begin{pmatrix} u^2 - v \\ uv \\ v^4 \end{pmatrix}$.

(c) $\int_{\gamma(U)} (x_1 + x_4) \, dx_2 \wedge dx_3$, where $U = \left\{ \begin{pmatrix} u \\ v \end{pmatrix} \mid |v| \le u \le 1 \right\}$,

and $\gamma \begin{pmatrix} u \\ v \end{pmatrix} = \begin{pmatrix} e^u \\ e^{-v} \\ \cos u \\ \sin v \end{pmatrix}$.

(d) $\int_{\gamma(U)} x_2 x_4 \, dx_1 \wedge dx_3 \wedge dx_4$, where

$$U = \left\{ \begin{pmatrix} u \\ v \\ w \end{pmatrix} \mid (w-1)^2 \ge u^2 + v^2, \ 0 \le w \le 1 \right\}, \text{ and } \gamma \begin{pmatrix} u \\ v \\ w \end{pmatrix} = \begin{pmatrix} u + v \\ u - v \\ w + v \\ w - v \end{pmatrix}.$$

Exercises for Section 6.4:
Form Fields and Vector Calculus

6.4.1   (a) Write in coordinate form the work form field and the flux form field of the vector fields $\vec{F} = \begin{bmatrix} x^2 \\ xy \\ -z \end{bmatrix}$ and $\vec{F} = \begin{bmatrix} x^2 \\ xy \\ x \end{bmatrix}$.

(b) For what vector field $\vec{F}$ is each of the following 1-form fields in $\mathbb{R}^3$ the work form field $W_{\vec{F}}$?
(i) $xy \, dx - y^2 \, dz$;   (ii) $y \, dx + 2 \, dy - 3x \, dz$.

(c) For what vector field $\vec{F}$ is each of the following 2-form fields in $\mathbb{R}^3$ the flux form field $\Phi_{\vec{F}}$?

(i) $2z^4 \, dx \wedge dy + 3y \, dy \wedge dz - x^2 z \, dx \wedge dz$; (ii) $x_2 x_3 \, dx_1 \wedge dx_3 - x_1^2 x_3 \, dx_2 \wedge dx_3$.

6.4.2   What is the work form field $W_{\vec{F}}(P_{\mathbf{a}}^o(\vec{\mathbf{u}}))$ of the vector field

$$\vec{F} \begin{pmatrix} x \\ y \\ z \end{pmatrix} = \begin{bmatrix} x^2 y \\ x - y \\ -z \end{bmatrix}, \text{ at } \mathbf{a} = \begin{pmatrix} 0 \\ 1 \\ 2 \end{pmatrix}, \text{ evaluated on the vector } \vec{\mathbf{u}} = \begin{bmatrix} 1 \\ -1 \\ 1 \end{bmatrix}?$$

6.4.3   What is the flux form field $\Phi_{\vec{F}}$ of the vector field $\vec{F}\begin{pmatrix} x \\ y \\ z \end{pmatrix} = \begin{bmatrix} -x \\ y^2 \\ xy \end{bmatrix}$,

evaluated on $P^o\left(\begin{bmatrix} 1 \\ 0 \\ 1 \end{bmatrix}, \begin{bmatrix} 0 \\ 1 \\ 0 \end{bmatrix} \right)$ at the point $\mathbf{x} = \begin{pmatrix} 1 \\ 2 \\ -1 \end{pmatrix}$?

6.4.4   Evaluate the work of each the following vector fields $\vec{F}$ on the given 1-parallelograms:

(a) $\vec{F} = \begin{bmatrix} x \\ y \end{bmatrix}$ on $P^o_{\begin{pmatrix} 1 \\ 1 \end{pmatrix}} \begin{bmatrix} 2 \\ 3 \end{bmatrix}$ (b) $\vec{F} = \begin{bmatrix} x^2 \\ \sin xy \end{bmatrix}$ on $P^o_{\begin{pmatrix} -1 \\ -\pi \end{pmatrix}} \begin{bmatrix} e \\ \pi \end{bmatrix}$

(c) $\vec{F} = \begin{bmatrix} y \\ x \\ z \end{bmatrix}$ on $P^o_{\begin{pmatrix} 1 \\ 0 \\ 1 \end{pmatrix}} \begin{bmatrix} 2 \\ 3 \\ -1 \end{bmatrix}$ (d) $\vec{F} = \begin{bmatrix} \sin y \\ \cos(x+z) \\ e^x \end{bmatrix}$ on $P^o_{\begin{pmatrix} 0 \\ 1 \\ -1 \end{pmatrix}} \begin{bmatrix} 0 \\ 1 \\ 0 \end{bmatrix}$

6.4.5   What is the density form of the function $f\begin{pmatrix} x \\ y \\ z \end{pmatrix} = xy + z^2$, evaluated

at the point $\mathbf{x} = \begin{pmatrix} 1 \\ 2 \\ 1 \end{pmatrix}$ on the vectors $\begin{bmatrix} 1 \\ 0 \\ 1 \end{bmatrix}, \begin{bmatrix} 2 \\ 1 \\ 1 \end{bmatrix}$, and $\begin{bmatrix} 0 \\ 1 \\ 1 \end{bmatrix}$?

6.4.6   Given the vector field $\vec{F}\begin{pmatrix} x \\ y \\ z \end{pmatrix} = \begin{bmatrix} y^2 \\ x+z \\ xz \end{bmatrix}$, the function $f\begin{pmatrix} x \\ y \\ z \end{pmatrix} = xz+$

zy, the point $\mathbf{x} = \begin{pmatrix} 1 \\ 1 \\ -1 \end{pmatrix}$, and the vectors $\vec{v}_1 = \begin{bmatrix} 0 \\ 1 \\ 1 \end{bmatrix}, \vec{v}_2 = \begin{bmatrix} 1 \\ 1 \\ 0 \end{bmatrix}, \vec{v}_3 = \begin{bmatrix} -1 \\ 1 \\ 1 \end{bmatrix}$,

what is

(1) the work form $W_{\vec{F}}\left(P^o_{\mathbf{x}}(\vec{v}_1)\right)$?

(2) the flux form $\Phi_{\vec{F}}\left(P^o_{\mathbf{x}}(\vec{v}_1, \vec{v}_2)\right)$?

(3) the density form $\rho_f\left(P^o_{\mathbf{x}}(\vec{v}_1, \vec{v}_2, \vec{v}_3)\right)$?

6.4.7   Evaluate the flux of each the following vector fields $\vec{F}$ on the given 2-parallelograms:

(a) $\vec{F} = \begin{bmatrix} y \\ x \\ z \end{bmatrix}$ on $P^o_{\begin{pmatrix} 1 \\ 0 \\ 1 \end{pmatrix}} \left(\begin{bmatrix} 1 \\ 1 \\ 0 \end{bmatrix}, \begin{bmatrix} 0 \\ 0 \\ -1 \end{bmatrix} \right)$.

(b) $\vec{F} = \begin{bmatrix} \sin y \\ \cos(x + z) \\ e^x \end{bmatrix}$ on $P^o_{\begin{pmatrix} 0 \\ 1 \\ -1 \end{pmatrix}} \left(\begin{bmatrix} 0 \\ 1 \\ 0 \end{bmatrix}, \begin{bmatrix} 1 \\ 2 \\ 0 \end{bmatrix} \right)$.

6.4.8   Verify that $\det[\vec{F}(\mathbf{x}), \vec{\mathbf{v}}_1, \dots, \vec{\mathbf{v}}_{n-1}]$ is an $(n-1)$-form field, so that Definition 6.4.10 of the flux form on $\mathbb{R}^n$ makes sense.

Exercises for Section 6.5:
Orientation

6.5.1   (a) Let $C \subset \mathbb{R}^2$ be the circle of equation $x^2 + y^2 = 1$. Find the unit vector field $\vec{T}$ describing the orientation "increasing polar angle."

(b) Now do the same for the circle of equation $(x - 1)^2 + y^2 = 4$.

(c) Explain carefully why the phrase "increasing polar angle" does not describe an orientation of the circle of equation $(x - 2)^2 + y^2 = 1$.

6.5.2   Prove that if a linear transformation T is not one to one, then it is not orientation preserving or reversing.

6.5.3   (a) In Example 6.5.16, where does $dx_1 \wedge dy_2$ define an orientation of S?

(b) Where does $dx_1 \wedge dy_1$ define an orientation of S?

(c) Where do these orientations coincide?

6.5.4   Show that the ad hoc definitions of orientation-preserving parametrizations (Definitions 6.5.11 and 6.5.12) are special cases of Definition 6.5.15.

6.5.5   Let $z_1 = x_1 + iy_1, z_2 = x_2 + iy_2$ be coordinates in $\mathbb{C}^2$. Consider the surface S in $\mathbb{C}^2$ parametrized by

$$\gamma : z \mapsto \begin{pmatrix} e^z \\ e^{-z} \end{pmatrix}, \quad z = x + iy, |x| \le 1, |y| \le 1$$

which we will orient by requiring that $\mathbb{C}$ be given the standard orientation, and that γ be orientation preserving. What is

$$\int_S dx_1 \wedge dy_1 + dy_1 \wedge dx_2 + dx_2 \wedge dy_2 \, ?$$

6.5.6   Let $z_1 = x_1 + iy_1$, $z_2 = x_2 + iy_2$ be coordinates in $\mathbb{C}^2$.
Compute the integral of $dx_1 \wedge dy_1 + dx_2 \wedge dy_2$ over the part of the locus of equation $z_2 = z_1^k$ where $|z_1| < 1$.

6.5.7   Which of the surfaces in Figure 6.5.7 are orientable?

6.5.8   (a) Let $X \subset \mathbb{R}^n$ be a manifold of the form $X = f^{-1}(0)$ where $f : \mathbb{R}^n \to \mathbb{R}$ is a C^1 function and $[\mathbf{D}f(\mathbf{x})] \neq 0$ for all $x \in X$. Let $\vec{\mathbf{v}}_1, \dots, \vec{\mathbf{v}}_{n-1}$ be elements of $T_{\mathbf{x}}(X)$. Show that

$$\omega(\vec{\mathbf{v}}_1, \dots \vec{\mathbf{v}}_{n-1}) = \det\left(\vec{\nabla} f(x), \vec{\mathbf{v}}_1, \dots, \vec{\mathbf{v}}_{n-1}\right)$$

defines an orientation of X.

a)

b)

c)
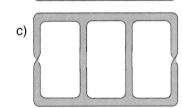

d)
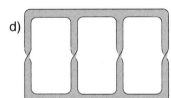

FIGURE 6.5.7.
Surfaces for Exercise 6.5.7:
which are orientable?

(b) What is the relation of this definition and the definition of the boundary orientation?

(c) Let $X \subset \mathbb{R}^n$ be an $(n-m)$-dimensional manifold of the form $X = \mathbf{f}^{-1}(\mathbf{0})$ where $\mathbf{f} : \mathbb{R}^n \to \mathbb{R}^m$ is a C^1 function and $[\mathbf{Df}(\mathbf{x})]$ is onto for all $x \in X$. Let $\vec{v}_1, \ldots, \vec{v}_{n-1}$ be elements of $T_{\mathbf{x}}(X)$. Show that

$$\omega(\vec{v}_1, \ldots, \vec{v}_{n-m}) = \det\left(\vec{\nabla} f_1(x), \ldots, \vec{\nabla} f_m(x), \vec{v}_1, \ldots, \vec{v}_{n-m}\right),$$

defines an orientation of X.

6.5.9   Consider the map $\mathbb{R}^2 \to \mathbb{R}^3$ given by spherical coordinates

$$\begin{pmatrix} \varphi \\ \theta \end{pmatrix} \mapsto \begin{pmatrix} \cos\varphi\cos\theta \\ \cos\varphi\sin\theta \\ \sin\varphi \end{pmatrix}.$$

The image of this mapping is the unit sphere, which we will orient by the outward-pointing normal. In what part of $\mathbb{R}^2$ is this mapping orientation preserving? In what part is it orientation reversing?

6.5.10   (a) Find a 2-form φ on the plane of equation $x + y + z = 0$ so that if the projection $\begin{pmatrix} x \\ y \\ z \end{pmatrix} \mapsto \begin{pmatrix} x \\ y \end{pmatrix}$ is oriented by φ, the projection is orientation-preserving.

(b) Repeat, but this time find a 2-form α so that if the projection is oriented by α, it is orientation reversing.

6.5.11   Let S be the part of the surface of equation $z = \sin xy + 2$ where $x^2 + y^2 \leq 1$ and $x \geq 0$, oriented by the upward-pointing normal. What is the flux of the vector field $\begin{bmatrix} 0 \\ 0 \\ x+y \end{bmatrix}$ through S?

6.5.12   Is the map

$$\begin{pmatrix} \varphi \\ \theta \end{pmatrix} \mapsto \begin{pmatrix} \cos\varphi\cos\theta \\ \cos\varphi\sin\theta \\ \sin\varphi \end{pmatrix}, \quad 0 \leq \theta, \varphi \leq \pi$$

an orientation-preserving parametrization of the unit sphere oriented by the outward-pointing normal? the inward-pointing normal?

6.5.13   What is the integral

$$\int_S x_3 \, dx_1 \wedge dx_2 \wedge dx_4$$

where S is the part of the three-dimensional manifold of equation $x_4 = x_1 x_2 x_3$ where $0 \leq x_1, x_2, x_3 \leq 1$, oriented by $dx_1 \wedge dx_2 \wedge dx_3$? Hint: this surface is a graph, so it is easy to parametrize it.

6.5.14   Find the work of the vector field $\vec{F}\begin{pmatrix} x \\ y \end{pmatrix} = \begin{pmatrix} xy \\ ye^x \end{pmatrix}$ around the boundary of the rectangle with vertices $\begin{pmatrix} 0 \\ 0 \end{pmatrix}, \begin{pmatrix} 0 \\ a \end{pmatrix}, \begin{pmatrix} b \\ a \end{pmatrix}, \begin{pmatrix} b \\ 0 \end{pmatrix}$, oriented so that these vertices appear in that order.

6.5.15   Find the work of the vector field

$$F\begin{pmatrix} x \\ y \\ z \end{pmatrix} = \begin{pmatrix} x^2 \\ y^2 \\ z^2 \end{pmatrix} \quad \text{over the arc of helix parametrized by } t \mapsto \begin{pmatrix} \cos t \\ \sin t \\ at \end{pmatrix},$$

with $0 \le t \le \alpha$, and oriented by increasing t.

In Exercises 6.5.16 and 6.5.17, part of the problem is finding parametrizations of S that preserve orientation.

6.5.16   Find the flux of the vector field $\vec{F}\begin{pmatrix} x \\ y \\ z \end{pmatrix} = r^a \begin{pmatrix} x \\ y \\ z \end{pmatrix}$, where $r = \sqrt{x^2 + z^2 + z^2}$, and a is a number, through the surface S, where S is the sphere of radius R oriented by the outward-pointing normal. The answer should be some function of a and R.

6.5.17   Find the flux of the vector field $\vec{F}\begin{pmatrix} x \\ y \\ z \end{pmatrix} = \begin{pmatrix} y \\ -z \\ yz \end{pmatrix}$, through S, where S is the part of the cone $z = \sqrt{x^2 + y^2}$ where $x, y \ge 0$, $x^2 + y^2 \le R$, and it is oriented by the upward pointing normal (i.e., the flux measures the amount flowing into the cone).

6.5.18   What is the flux of the vector field

$$F\begin{pmatrix} x \\ y \\ z \end{pmatrix} = \begin{bmatrix} x \\ -y \\ xy \end{bmatrix} \quad \text{through the surface } z = \sqrt{x^2 + y^2}, \ x^2 + y^2 \le 1,$$

oriented by the outward normal?

Hint for Exercise 6.5.19, part (b): Show that you cannot choose an orientation for $M_1(2,3)$ so that both φ_1 and φ_2, as defined in Exercise 3.2.10, are both orientation preserving.

Hint for Exercise 6.5.19, part (c): Use the same method as in (b); this time you can find an orientation of $M_1(3,3)$ such that all three of $\varphi_1, \varphi_2,$ and φ_3 are orientation preserving.

6.5.19   This exercise has Exercise 3.2.10 as a prerequisite. Let $M_1(n,m)$ be the space of $n \times m$ matrices of rank 1. (a) Show that $M_1(2,2)$ is orientable. (This follows from Exercise 3.2.6 (a) and 6.5.8(a).)

(*b) Show that $M_1(2,3)$ is not orientable.

(*c) Show that $M_1(3,3)$ is orientable.

6.5.20   Consider the surface S in $\mathbb{C}^3$ parametrized by

$$\gamma : z \mapsto \begin{pmatrix} z^p \\ z^q \\ z^s \end{pmatrix}, \ |z| < 1$$

which we will orient by requiring that $\mathbb{C}$ be given the standard orientation, and that γ be orientation-preserving. What is

$$\int_S dx_1 \wedge dy_1 + dx_2 \wedge dy_2 + dx_3 \wedge dy_3 ?$$

6.6.1   Consider the curve S of equation $x^2 + y^2 = 1$, oriented by the tangent vector $\begin{bmatrix} 0 \\ 1 \end{bmatrix}$ at the point $\begin{pmatrix} 1 \\ 0 \end{pmatrix}$.

(a) Show that the subset X where $x \geq 0$ is a piece-with-boundary of S. What is its oriented boundary?

(b) Show that the subset Y where $|x| \leq 1/2$ is a piece-with-boundary of S. What is its oriented boundary?

(c) Is the subset Z where $x > 0$ a piece-with-boundary? If so, what is its boundary?

6.6.2   Consider the region $X = P \cap B \subset \mathbb{R}^3$, where P is the plane of equation $x + y + z = 0$, and B is the ball $x^2 + y^2 + z^2 \leq 1$. We will orient P by the normal $\begin{pmatrix} 1 \\ 1 \\ 1 \end{pmatrix}$, and the sphere $x^2 + y^2 + z^2 = 1$ by the outward-pointing normal.

(a) Which of the forms $dx \wedge dy$, $dx \wedge dz$, $dy \wedge dz$ define the given orientation of P?

(b) Show that X is a piece-with-boundary of P, and that the mapping

$$t \mapsto \begin{pmatrix} \dfrac{\cos t}{\sqrt{2}} - \dfrac{\sin t}{\sqrt{6}} \\ -\dfrac{\cos t}{\sqrt{2}} - \dfrac{\sin t}{\sqrt{6}} \\ 2\dfrac{\sin t}{\sqrt{6}} \end{pmatrix}, \quad 0 \leq t \leq 2\pi$$

is a parametrization of ∂X.

(c) Is the parametrization compatible with the boundary orientation of ∂X.

(d) Do any of the 1-forms dx, dy, dz define its orientation at every point?

(e) Do any of the 1-forms $x\,dy - y\,dx$, $x\,dz - z\,dx$, $y\,dz - z\,dy$ define its orientation at every point?

6.7.1   What is the exterior derivative of

(a) $\sin(xyz)\,dx$ in $\mathbb{R}^3$;   (b) $x_1 x_3\,dx_2 \wedge dx_4$ in $\mathbb{R}^4$;

(c) $\sum_{i=1}^n x_i^2\,dx_1 \wedge \cdots \wedge \widehat{dx_i} \wedge \cdots \wedge dx_n$ in $\mathbb{R}^n$.

6.7.2   (a) Is there a function f on $\mathbb{R}^3$ such that

Exercise 6.7.2 really belongs in Section 6.11.

(1) $df = \cos(x + yz)\,dx + y\cos(x + yz)\,dy + z\cos(x + yz)\,dz$?

(2) $df = \cos(x + yz)\,dx + z\cos(x + yz)\,dy + y\cos(x + yz)\,dz$?

(b) Find the function when it exists.

6.7.3   Find all the 1-forms $\omega = p(y, z)\, dx + q(x, z)\, dy$ such that

$$d\omega = x\, dy \wedge dz + y\, dx \wedge dz.$$

6.7.4   (a) Let $\varphi = xyz\, dy$. Compute from the definition the number

$$d\varphi \left(P_{\begin{pmatrix} 1 \\ 2 \\ 3 \end{pmatrix}}(\vec{e}_2, \vec{e}_3) \right).$$

(b) What is $d\varphi$? Use your result to check the computation in (a).

6.7.5   (a) Let $\varphi = x_1 x_3\, dx_2 \wedge dx_4$. Compute from the definition the number

$$d\varphi \left(P^o_{\mathbf{e}_1}(\vec{e}_2, \vec{e}_3, \vec{e}_4) \right).$$

(b) What is $d\varphi$? Use your result to check the computation in (a).

6.7.6   (a) Let $\varphi = x_2^2\, dx_3$. Compute from the definition the number

$$d\varphi \left(P^o_{-\mathbf{e}_2}(\vec{e}_2, \vec{e}_3) \right).$$

(b) What is $d\varphi$? Use your result to check the computation in (b).

6.7.7   (a) There is an exponent m such that

$$\vec{\nabla} \cdot (x^2 + y^2 + z^2)^m \begin{bmatrix} x \\ y \\ z \end{bmatrix} = 0; \quad \text{find it.}$$

Exercise 6.7.7, part (b): The subscript in $\Phi_{r^{2m}\vec{r}}$ may be hard to read. It is $r^{2m}\vec{r}$.

(b*) More generally, there is an exponent m (depending on n) such that the $(n-1)$-form $\Phi_{r^{2m}\vec{r}}$ has exterior derivative 0, when $\vec{r}$ is the vector field $\begin{bmatrix} x_1 \\ \vdots \\ x_n \end{bmatrix}$, and $r = |\vec{r}|$. Can you find it? (Start with $n = 1, 2$.)

6.7.8   Show that each face of a $(k+2)$-parallelogram is a $(k+1)$-dimensional parallelogram, and that each edge of the $(k+1)$-parallelogram is also the edge of another $(k+1)$-parallelogram, but with opposite orientation.

Exercises for Section 6.8:
The Exterior Derivative in $\mathbb{R}^3$

6.8.1   Compute the gradients of the following functions:

(a) $f\begin{pmatrix} x \\ y \end{pmatrix} = x$ (b) $f\begin{pmatrix} x \\ y \end{pmatrix} = y^2$ (c) $f\begin{pmatrix} x \\ y \end{pmatrix} = x^2 + y^2$

(d) $f\begin{pmatrix} x \\ y \end{pmatrix} = x^2 - y^2$ (e) $f\begin{pmatrix} x \\ y \end{pmatrix} = \sin(x + y)$ (f) $f\begin{pmatrix} x \\ y \end{pmatrix} = \ln(x^2 + y^2)$

(g) $f\begin{pmatrix} x \\ y \\ z \end{pmatrix} = xyz$ (k) $f\begin{pmatrix} x \\ y \\ z \end{pmatrix} = \ln|x + y + z|$ (1) $f\begin{pmatrix} x \\ y \\ z \end{pmatrix} = \dfrac{xyz}{x^2 + y^2 + z^2}$

6.8.2   (a) For what vector field $\vec{F}$ is the 1-form on $\mathbb{R}^3$

$$x^2\, dx + y^2 z\, dy + xy\, dz \quad \text{the work form field } W_{\vec{F}}?$$

(b) Compute the exterior derivative of $x^2\, dx + y^2 z\, dy + xy\, dz$ using Theorem 6.7.3 (computing the exterior derivative of a k-form), and show that it is the same as $\Phi_{\vec{\nabla} \times \vec{F}}$.

6.8.3   (a) For what vector field $\vec{F}$ is the 2-form on $\mathbb{R}^3$
$(xy)\, dx \wedge dy + (x)\, dy \wedge dz + (xy)\, dx \wedge dz \quad$ the flux form field $\Phi_{\vec{F}}$?

(b) Compute the exterior derivative of $(xy)\, dx \wedge dy + (x)\, dy \wedge dz + (xy)\, dx \wedge dz$ using Theorem 6.7.3, and show that it is the same as the density form field of $\operatorname{div} \vec{F}$.

6.8.4   (a) Let $\vec{F} = \begin{bmatrix} F_1 \\ F_2 \end{bmatrix}$ be a vector field in the plane. Show that if $\vec{F}$ is the gradient of a C^2 function, then $D_2 F_1 = D_1 F_2$.

(b) Show that this is not necessarily true if f is twice differentiable but the second derivatives are not continuous.

Exercise 6.8.5 should be in Section 6.11.

6.8.5   Which of the vector fields of Exercise 1.1.5 are gradients of functions?

6.8.6   Prove the equations

$$\operatorname{curl}(\operatorname{grad} f) = 0 \quad \text{and} \quad \operatorname{div}(\operatorname{curl} \vec{F}) = 0$$

for any function f and any vector field $\vec{F}$ (at least of class C^2) using the formulas of Theorem 6.8.3.

6.8.7   (a) What is $dW_{\begin{bmatrix} 0 \\ 0 \\ x \end{bmatrix}}$? What is $dW_{\begin{bmatrix} 0 \\ 0 \\ x \end{bmatrix}}\left(P^o_{\begin{bmatrix} 0 \\ 0 \\ 0 \end{bmatrix}}(\vec{e}_1, \vec{e}_3)\right)$?

(b) Compute $dW_{\begin{bmatrix} 0 \\ 0 \\ x \end{bmatrix}}\left(P^o_{\begin{bmatrix} 0 \\ 0 \\ 0 \end{bmatrix}}(\vec{e}_1, \vec{e}_3)\right)$ directly from the definition.

6.8.8   (a) Find a book on electromagnetism (or a tee-shirt) and write Maxwell's laws.

Let $\vec{E}$ and $\vec{B}$ be two vector fields on $\mathbb{R}^4$, parametrized by x, y, z, t.

(b) Compute $d(W_{\vec{E}} \wedge c\, dt + \Phi_{\vec{B}})$.

(c) Compute $d(W_{\vec{B}} \wedge c\, dt - \Phi_{\vec{E}})$.

(d) Show that two of Maxwell's equations can be condensed into

$$d(W_{\vec{E}} \wedge c\, dt + \Phi_{\vec{B}}) = 0.$$

(e) How can you write the other two Maxwell's equations using forms?

6.8.9   (a) What is the exterior derivative of $W_{\begin{bmatrix} x \\ y \\ z \end{bmatrix}}$? (b) Of $\Phi_{\begin{bmatrix} x \\ y \\ z \end{bmatrix}}$?

6.8.10   Compute the divergence and curl of the vector fields

$$\text{(a)} \begin{bmatrix} x^2 y \\ -2yz \\ x^3 y^2 \end{bmatrix} \quad \text{and} \quad \text{(b)} \begin{bmatrix} \sin xz \\ \cos yz \\ xyz \end{bmatrix}.$$

6.8.11   (a) What is the divergence of $\vec{F}\begin{pmatrix} x \\ y \\ z \end{pmatrix} = \begin{bmatrix} x^2 \\ y^2 \\ yz \end{bmatrix}$?

(b) Use part (a) to compute

$$d\Phi_{\vec{F}} P^o_{\begin{pmatrix} 1 \\ 1 \\ 2 \end{pmatrix}} (\vec{e}_1, \vec{e}_2, \vec{e}_3).$$

(c) Compute it again, directly from the definition.

**Exercises for Section 6.9:
Stokes's Theorem in $\mathbb{R}^n$**

6.9.1   Let U be a compact piece-with-boundary of $\mathbb{R}^3$. Show that the volume of U is given by

$$\int_{\partial U} \frac{1}{3} \left(z\, dx \wedge dy + y\, dz \wedge dx + x\, dy \wedge dz \right).$$

6.9.2   (a) Find the unique polynomial p such that $p(1) = 1$ and such that if

$$\omega = x\, dy \wedge dz - 2zp(y)\, dx \wedge dy + yp(y)\, dz \wedge dx,$$

then $d\omega = dx \wedge dy \wedge dz$.

(b) For this polynomial p, find the integral $\int_S \omega$, where S is that part of the sphere $x^2 + y^2 + z^2 = 1$ where $z \geq \sqrt{2}/2$, oriented by the outward-pointing normal.

6.9.3   What is the integral

$$\int_C x\, dy \wedge dz + y\, dz \wedge dx + z\, dx \wedge dy$$

where C is that part of the cone of equation $z = a - \sqrt{x^2 + y^2}$ where $z \geq 0$, oriented by the upwards-pointing normal. (The volume of a full cone is $\frac{1}{3} \cdot$ height $\cdot$ area of base.)

6.9.4   Compute the integral of $x_1\, dx_2 \wedge dx_3 \wedge dx_4$ over the part of the three-dimensional manifold of equation

$$x_1 + x_2 + x_3 + x_4 = a, \ x_1, \ x_2, \ x_3, \ x_4 \geq 0,$$

oriented so that the projection to the (x_1, x_2, x_3)-coordinate 3-space is orientation preserving.

6.9.5   (a) Compute the exterior derivative of the 2-form
$$\varphi = \frac{x\,dy \wedge dz + y\,dz \wedge dx + z\,dx \wedge dy}{(x^2 + y^2 + z^2)^{3/2}}.$$

(b) Compute the integral of φ over the unit sphere $x^2 + y^2 + z^2 = 1$, oriented by the outward-pointing normal.

(c) Compute the integral of φ over the boundary of the cube of side 4, centered at the origin, and oriented by the outward-pointing normal.

(d) Can φ be written $d\psi$ for some 1-form ψ on $\mathbb{R}^3 - \{\mathbf{0}\}$?

6.9.6   What is the integral of
$$x\,dy \wedge dz + y\,dz \wedge dx + z\,dx \wedge dy$$

over the part S of the ellipsoid
$$\frac{x^2}{a^2} + \frac{y^2}{b^2} + \frac{z^2}{c^2} = 1,$$

where x, y, $z \geq 0$, oriented by the outward-pointing normal? (You may use Stokes's theorem, or parametrize the surface.)

6.9.7   (a) Parametrize the surface in 4-space given by the equations
$$x_1^2 + x_2^2 = a^2, \ x_3^2 + x_4^2 = b^2.$$

(b) Integrate the 2-form $x_1 x_2\,dx_2 \wedge dx_3$ over this surface.

(c) Compute $d(x_1 x_2\,dx_2 \wedge dx_3)$.

(d) Represent the surface as the boundary of a three-dimensional manifold in $\mathbb{R}^4$, and verify that Stokes's theorem is true in this case.

6.9.8   Use Stokes's theorem to prove the statement in the caption of Figure 6.7.1, in the special case where the surface S is a parallelogram: i.e., prove that the integral of the "element of solid angle" $\Phi_{\vec{F}_3}$ over a parallelogram S is the same as its integral over the corresponding P.

Exercises for Section 6.10:
The Integral Theorems
of Vector Calculus

6.10.1   Suppose $U \subset \mathbb{R}^3$ is open, $\vec{F}$ is a vector field on U, and $\mathbf{a}$ is a point of U. Let $S_r(\mathbf{a})$ be the sphere of radius r centered at $\mathbf{a}$, oriented by the outward pointing normal. Compute
$$\lim_{r \to 0} \frac{1}{r^3} \int_{S_r(\mathbf{a})} \Phi_{\vec{F}}.$$

6.10.2   (a) Let X be a bounded region in the (x, z)-plane where $x > 0$, and call Z_α the part of $\mathbb{R}^3$ swept out by rotating X around the z-axis by an angle α. Find a formula for the volume of Z_α, in terms of an integral over X.

Hint for Exercise 6.10.2:   use cylindrical coordinates.

(b) Let X be the circle of radius 1 in the (x, z)-plane, centered at the point $x = 2, z = 0$. What is the volume of the torus obtained by rotating it around the z-axis by a full circle?

(c) What is the flux of the vector field $\begin{bmatrix} x \\ y \\ z \end{bmatrix}$ through the part of the boundary of this torus where $y \geq 0$, oriented by the normal pointing out of the torus?

6.10.3   Let $\vec{F}$ be the vector field $\vec{F} = \vec{\nabla} \left(\frac{\sin xyz}{xy} \right)$. What is the work of $\vec{F}$ along the parametrized curve

$$\gamma(t) = \begin{bmatrix} t \cos \pi t \\ t \\ t \end{bmatrix}, \quad 0 \leq t \leq 1, \quad \text{oriented so that } \gamma \text{ is orientation preserving?}$$

In Exercise 6.9.6, S is a box without a top.

This is a "shaggy dog" exercise, with lots of irrelevant detail!

6.10.4   What is the integral of

$$W \begin{pmatrix} -y/(x^2 + y^2) \\ x/(x^2 + y^2) \end{pmatrix}$$

around the boundary of the 11-sided regular polygon inscribed in the unit circle, with a vertex at $\begin{pmatrix} 1 \\ 0 \end{pmatrix}$, oriented as the boundary of the polygon?

6.10.5   Let S be the surface of equation $z = 9 - y^2$, oriented by the upward-pointing normal.

(a) Sketch the piece $X \subset S$ where $x \geq 0$, $z \geq 0$ and $y \geq x$, indicating carefully the boundary orientation.

(b) Give a parametrization of X, being careful about the domain of the parametrizing map, and whether it is orientation preserving.

(c) Find the work of the vector field $\vec{F} \begin{pmatrix} x \\ y \\ z \end{pmatrix} = \begin{bmatrix} 0 \\ xz \\ 0 \end{bmatrix}$ around the boundary of X.

6.10.6   Let C be a closed curve in the plane. Show that the two vector fields $\begin{bmatrix} y \\ 0 \end{bmatrix}$ and $\begin{bmatrix} 0 \\ x \end{bmatrix}$ do opposite work around C.

6.10.7   Suppose $U \subset \mathbb{R}^3$ is open, $\vec{F}$ is a vector field on U, $\mathbf{a}$ is a point of U, and $\vec{v} \neq 0$ is a vector in $\mathbb{R}^3$. Let U_R be the disk of radius R in the plane of equation $(\mathbf{x} - \mathbf{a}) \cdot \vec{v} = 0$, centered at $\mathbf{a}$, oriented by the normal vector field $\vec{v}$, and let ∂U_R be its boundary, with the boundary orientation.

Compute

$$\lim_{R \to 0} \frac{1}{R^2} \int_{\partial U_R} W_{\vec{F}}.$$

6.10.8   Let $U \subset \mathbb{R}^3$ be a subset bounded by a surface S, which we will give the boundary orientation. What relation is there between the volume of U and the flux $\int_S \Phi_{\begin{bmatrix} x \\ y \\ z \end{bmatrix}}$?

Hint for Exercise 6.10.9: the first step is to find a closed curve of which C is a piece.

6.10.9   Compute the integral $\int_C W_{\vec{F}}$, where $\vec{F} \begin{pmatrix} x \\ y \end{pmatrix} = \begin{bmatrix} xy \\ \frac{\cos y}{y+1} + x \end{bmatrix}$, and C is the upper half-circle $x^2 + y^2 = 1$, $y \geq 0$, oriented clockwise.

6.10.10   Find the flux of the vector field $\begin{bmatrix} x^2 \\ y^2 \\ z^2 \end{bmatrix}$ through the surface of the unit sphere, oriented by the outward-pointing normal.

6.10.11   Use Green's theorem to calculate the area of the triangle with vertices

Hint: think of integrating $x \, dy$ around the triangle.

$$\begin{bmatrix} a_1 \\ b_1 \end{bmatrix}, \begin{bmatrix} a_2 \\ b_2 \end{bmatrix}, \begin{bmatrix} a_3 \\ b_3 \end{bmatrix}.$$

6.10.12   What is the work of the vector field $\begin{bmatrix} -3y \\ 3x \\ 1 \end{bmatrix}$ around the circle $x^2 + y^2 = 1, z = 3$ oriented by the tangent vector

$$\begin{bmatrix} 0 \\ -1 \\ 0 \end{bmatrix} \quad \text{at} \quad \begin{bmatrix} 1 \\ 0 \\ 3 \end{bmatrix}?$$

6.10.13   What is the flux of the vector field

$$\vec{F} \begin{pmatrix} x \\ y \\ z \end{pmatrix} = \begin{bmatrix} x + yz \\ y + xz \\ z + xy \end{bmatrix} \text{ through the boundary of the region in the first octant}$$

$x, y, z \geq 0$ where $z \leq 4$ and $x^2 + y^2 \leq 4$, oriented by the outward-pointing normal?

Exercises for Section 6.11:
Potentials

6.11.1   For the vector field of Example 6.11.3, show (Equation 6.11.14) that

$$\vec{F} = \vec{\nabla} \left(\arctan \frac{y}{x} \right).$$

6.11.2   A charge of c coulombs per meter on a vertical wire $x = a, y = b$ creates an electric potential

$$V\begin{pmatrix} x \\ y \\ z \end{pmatrix} = c\ln\big((x-a)^2 + (y-b)^2\big).$$

Several such wires produce a potential which is the sum of the potentials due to the individual wires.

(a) What is the electric field due to a single wire going through the point $\begin{pmatrix} 0 \\ 0 \end{pmatrix}$, with charge per length $c = 1\,\mathrm{coul}/m$, where coul is the unit for charge.

(b) Sketch the potential due to two wires, both charged with $1\,\mathrm{coul}/m$, one going through the point $\begin{pmatrix} 1 \\ 0 \end{pmatrix}$, and the other through $\begin{pmatrix} -1 \\ 0 \end{pmatrix}$.

(c) Do the same if the first wire is charged with $1\,\mathrm{coul}/m$ and the other with $-1\,\mathrm{coul}/m$.

6.11.3   (a) Is the vector field $\begin{bmatrix} \dfrac{x}{x^2+y^2} \\ \dfrac{y}{x^2+y^2} \end{bmatrix}$ the gradient of a function on $\mathbb{R}^2 - \{\mathbf{0}\}$?

(b) Is the vector field $\begin{bmatrix} x \\ y \\ z \end{bmatrix}$ on $\mathbb{R}^3$ the curl of another vector field?

6.11.4   Find a 1-form φ such that $d\varphi = y\,dz \wedge dx - x\,dy \wedge dz$.

6.11.5   Let $\vec{F}$ be the vector field on $\mathbb{R}^3$

$$\vec{F}\begin{pmatrix} x \\ y \\ z \end{pmatrix} = \begin{bmatrix} F_1(x,y) \\ F_2(x,y) \\ 0 \end{bmatrix}.$$

Suppose $D_2F_1 = D_1F_2$. Show that there exists a function $f : \mathbb{R}^3 \to \mathbb{R}$ such that $\vec{F} = \vec{\nabla}f$.

**6.11.6   (a) Show that a 1-form φ on $\mathbb{R}^2 - 0$ can be written df exactly when $d\varphi = 0$ and $\int_{S^1} \varphi = 0$, where S^1 is the unit circle, oriented counterclockwise.

(b) Show that a 1-form φ on $\mathbb{R}^2 - \left\{ \begin{pmatrix} 0 \\ 0 \end{pmatrix}, \begin{pmatrix} 1 \\ 0 \end{pmatrix} \right\}$ can be written df exactly when $d\varphi = 0$ and both $\int_{S_1} \varphi = 0$, $\int_{S_2} \varphi = 0$ where S_1 is the circle of radius $1/2$ centered at the origin, and S_2 is the circle of radius $1/2$ centered at $\begin{pmatrix} 1 \\ 0 \end{pmatrix}$, both oriented counterclockwise.

Appendix A: Some Harder Proofs

A.0 Introduction

When this book was first used in manuscript form as a textbook for the standard first course in multivariate calculus at Cornell University, all proofs were included in the main text and some students became anxious, feeling, despite assurances to the contrary, that because a proof was in the text, they were expected to understand it. We have thus moved to this appendix certain more difficult proofs. They are intended for students using this book for a class in analysis, and for the occasional student in a beginning course who has mastered the statement of the theorem and wishes to delve further.

In addition to proofs of theorems stated in the main text, the appendix includes material not covered in the main text, in particular rules for arithmetic involving o and O (Appendix A.8), Taylor's theorem with remainder (Appendix A.9), two theorems concerning compact sets (Appendix A.17), and a discussion of the pullback (Appendix A.21).

A.1 Proof of the Chain Rule

Theorem 1.8.2 (Chain rule). *Let $U \subset \mathbb{R}^n, V \subset \mathbb{R}^m$ be open sets, let $\mathbf{g} : U \to V$ and $\mathbf{f} : V \to \mathbb{R}^p$ be mappings, and let $\mathbf{a}$ be a point of U. If $\mathbf{g}$ is differentiable at $\mathbf{a}$ and $\mathbf{f}$ is differentiable at $\mathbf{g}(\mathbf{a})$, then the composition $\mathbf{f} \circ \mathbf{g}$ is differentiable at $\mathbf{a}$, and its derivative is given by*

$$[\mathbf{D}(\mathbf{f} \circ \mathbf{g})(\mathbf{a})] = [\mathbf{Df}(\mathbf{g}(\mathbf{a}))] \circ [\mathbf{Dg}(\mathbf{a})]. \qquad 1.8.12$$

Proof. To prove the chain rule, you must set about it the right way; this is already the case in one-variable calculus. The right approach (at least, one that works), is to define two "remainder" functions, $\mathbf{r}(\vec{\mathbf{h}})$ and $\mathbf{s}(\vec{\mathbf{k}})$. The function $\mathbf{r}(\vec{\mathbf{h}})$ gives the difference between the increment to the function $\mathbf{g}$ and its linear approximation at $\mathbf{a}$. The function $\mathbf{s}(\vec{\mathbf{k}})$ gives the difference between the increment to $\mathbf{f}$ and its linear approximation at $\mathbf{g}(\mathbf{a})$:

$$\underbrace{\mathbf{g}(\mathbf{a} + \vec{\mathbf{h}}) - \mathbf{g}(\mathbf{a})}_{\text{increment to function}} - \underbrace{[\mathbf{Dg}(\mathbf{a})]\vec{\mathbf{h}}}_{\text{linear approx.}} = \mathbf{r}(\vec{\mathbf{h}}) \qquad A1.1$$

$$\underbrace{\mathbf{f}(\mathbf{g}(\mathbf{a}) + \vec{\mathbf{k}}) - \mathbf{f}(\mathbf{g}(\mathbf{a}))}_{\text{increment to } \mathbf{f}} - \underbrace{[\mathbf{Df}(\mathbf{g}(\mathbf{a}))]\vec{\mathbf{k}}}_{\text{linear approx.}} = \mathbf{s}(\vec{\mathbf{k}}). \qquad A1.2$$

589

The hypotheses that $\mathbf{g}$ is differentiable at $\mathbf{a}$ and that $\mathbf{f}$ is differentiable at $\mathbf{g(a)}$ say exactly that

$$\lim_{\vec{\mathbf{h}} \to 0} \frac{\mathbf{r}(\vec{\mathbf{h}})}{|\vec{\mathbf{h}}|} = 0 \quad \text{and} \quad \lim_{\vec{\mathbf{k}} \to 0} \frac{\mathbf{s}(\vec{\mathbf{k}})}{|\vec{\mathbf{k}}|} = 0. \qquad A1.3$$

Now we rewrite Equations A1.1 and A1.2 in more convenient form:

$$\mathbf{g}(\mathbf{a} + \vec{\mathbf{h}}) = \mathbf{g(a)} + [\mathbf{Dg(a)}]\vec{\mathbf{h}} + \mathbf{r}(\vec{\mathbf{h}}) \qquad A1.4$$

$$\mathbf{f}\big(\mathbf{g(a)} + \vec{\mathbf{k}}\big) = \mathbf{f}\big(\mathbf{g(a)}\big) + [\mathbf{Df(g(a))}]\vec{\mathbf{k}} + \mathbf{s}(\vec{\mathbf{k}}), \qquad A1.5$$

and then write:

In the first line, we are just evaluating $\mathbf{f}$ at $\mathbf{g(a + h)}$, plugging in the value for $\mathbf{g(a + h)}$ given by the right-hand side of Equation A1.4. We then see that $[\mathbf{Dg(a)}]\mathbf{h} + \mathbf{r}(\vec{\mathbf{h}})$ plays the role of $\vec{\mathbf{k}}$ in the left side of Equation A1.5. In the second line we plug this value for $\vec{\mathbf{k}}$ into the right side of Equation A1.5.

To go from the second to the third line we use the linearity of $[\mathbf{Df(g(a))}]$:

$$[\mathbf{Df(g(a))}]\big([\mathbf{Dg(a)}]\vec{\mathbf{h}} + \mathbf{r}(\vec{\mathbf{h}})\big)$$
$$= [\mathbf{Df(g(a))}][\mathbf{Dg(a)}]\vec{\mathbf{h}}$$
$$+ [\mathbf{Df(g(a))}]\mathbf{r}(\vec{\mathbf{h}}).$$

$$\mathbf{f}\big(\mathbf{g}(\mathbf{a} + \vec{\mathbf{h}})\big) = \mathbf{f}\bigg(\overbrace{\mathbf{g(a)} + \underbrace{[\mathbf{Dg(a)}]\vec{\mathbf{h}} + \mathbf{r}(\vec{\mathbf{h}})}_{\vec{\mathbf{k}}, \text{ left-hand side Eq. A1.5}}}^{\text{from Equation A1.4}} \bigg) \qquad A1.6$$

$$= \mathbf{f}\big(\mathbf{g(a)}\big) + [\mathbf{Df(g(a))}] \underbrace{\big([\mathbf{Dg(a)}]\vec{\mathbf{h}} + \mathbf{r}(\vec{\mathbf{h}})\big)}_{\vec{\mathbf{k}}} + \mathbf{s} \underbrace{\big([\mathbf{Dg(a)}]\vec{\mathbf{h}} + \mathbf{r}(\vec{\mathbf{h}})\big)}_{\vec{\mathbf{k}}}$$

$$= \mathbf{f}\big(\mathbf{g(a)}\big) + [\mathbf{Df(g(a))}]\big([\mathbf{Dg(a)}]\vec{\mathbf{h}}\big) + \underbrace{[\mathbf{Df(g(a))}]\big(\mathbf{r}(\vec{\mathbf{h}})\big) + \mathbf{s}\big([\mathbf{Dg(a)}]\vec{\mathbf{h}} + \mathbf{r}(\vec{\mathbf{h}})\big)}_{\text{remainder}}.$$

We can subtract $\mathbf{f}\big(\mathbf{g(a)}\big)$ from both sides of Equation A1.6, to get

$$\underbrace{\mathbf{f}\big(\mathbf{g}(\mathbf{a} + \vec{\mathbf{h}})\big) - \mathbf{f}\big(\mathbf{g(a)}\big)}_{\text{increment to composition}} = \overbrace{[\mathbf{Df(g(a))}]}^{\substack{\text{linear approx.} \\ \text{to } \mathbf{f} \text{ at } \mathbf{g}(a)}} \underbrace{\overbrace{\big([\mathbf{Dg(a)}]}^{\substack{\text{linear approx.} \\ \text{to } \mathbf{g} \text{ at } \mathbf{a}}} \vec{\mathbf{h}}\big)}_{\text{composition of linear approximations}} + \quad \text{remainder.} \quad A1.7$$

The "composition of linear approximations" is the linear approximation of the increment to $\mathbf{f}$ at $\mathbf{g(a)}$ *as evaluated* on the linear approximation of the increment to $\mathbf{g}$ at $\vec{\mathbf{h}}$.

What we want to prove is that the linear approximation above is in fact the derivative of $\mathbf{f} \circ \mathbf{g}$ as evaluated on the increment $\vec{\mathbf{h}}$. To do this we need to prove that the limit of the remainder divided by $|\vec{\mathbf{h}}|$ is 0 as $\vec{\mathbf{h}} \to 0$:

$$\lim_{\vec{\mathbf{h}} \to 0} \frac{[\mathbf{Df(g(a))}]\big(\mathbf{r}(\vec{\mathbf{h}})\big) + \mathbf{s}\big([\mathbf{Dg(a)}]\vec{\mathbf{h}} + \mathbf{r}(\vec{\mathbf{h}})\big)}{|\vec{\mathbf{h}}|} = 0. \qquad A1.8$$

Let us look at the two terms in this limit separately. The first is straightforward. Since (Proposition 1.4.11)

$$\big|[\mathbf{Df(g(a))}]\mathbf{r}(\vec{\mathbf{h}})\big| \le \big|[\mathbf{Df(g(a))}]\big| \, |\mathbf{r}(\vec{\mathbf{h}})|, \qquad A1.9$$

we have

$$\lim_{\vec{\mathbf{h}} \to 0} \frac{\left| [\mathbf{Df}(\mathbf{g}(\mathbf{a}))] \mathbf{r}(\vec{\mathbf{h}}) \right|}{|\vec{\mathbf{h}}|} \leq \left| [\mathbf{Df}(\mathbf{g}(\mathbf{a}))] \right| \underbrace{\lim_{\vec{\mathbf{h}} \to 0} \frac{|\mathbf{r}(\vec{\mathbf{h}})|}{|\vec{\mathbf{h}}|}}_{= 0 \text{ by Eq. A1.3}} = 0. \qquad A1.10$$

The second term is harder. We want to show that

$$\lim_{\vec{\mathbf{h}} \to 0} \frac{\mathbf{s}\left([\mathbf{Dg}(\mathbf{a})] \vec{\mathbf{h}} + \mathbf{r}(\vec{\mathbf{h}}) \right)}{|\vec{\mathbf{h}}|} = 0. \qquad A1.11$$

First note that there exists $\delta > 0$ such that $|\mathbf{r}(\vec{\mathbf{h}})| \leq |\vec{\mathbf{h}}|$ when $|\vec{\mathbf{h}}| < \delta$ (by Equation A1.3).[1]

Thus, when $|\vec{\mathbf{h}}| < \delta$, we have

$$\left| [\mathbf{Dg}(\mathbf{a})] \vec{\mathbf{h}} + \underbrace{\mathbf{r}(\vec{\mathbf{h}})}_{\leq |\vec{\mathbf{h}}|} \right| \leq \left| [\mathbf{Dg}(\mathbf{a})] \vec{\mathbf{h}} \right| + |\vec{\mathbf{h}}| = \left(|[\mathbf{Dg}(\mathbf{a})]| + 1 \right) |\vec{\mathbf{h}}|. \qquad A1.12$$

Equation A1.3 also tells us that for any $\epsilon > 0$, there exists $0 < \delta' < \delta$ such that when $|\vec{\mathbf{k}}| \leq \delta'$, then $|\mathbf{s}(\vec{\mathbf{k}})| \leq \epsilon |\vec{\mathbf{k}}|$. If you don't see this right away, consider that for $|\vec{\mathbf{k}}|$ sufficiently small,

$$\frac{|\mathbf{s}(\vec{\mathbf{k}})|}{|\vec{\mathbf{k}}|} \leq \epsilon; \quad \text{i.e.,} \quad |\mathbf{s}(\vec{\mathbf{k}})| \leq \epsilon |\vec{\mathbf{k}}|. \qquad A1.13$$

Otherwise the limit as $\vec{\mathbf{k}} \to 0$ would not be 0. We specify "$|\vec{\mathbf{k}}|$ sufficiently small" by $|\vec{\mathbf{k}}| \leq \delta'$.

Now, when

$$|\vec{\mathbf{h}}| \leq \frac{\delta'}{|[\mathbf{Dg}(\mathbf{a})]| + 1}; \quad \text{i.e.,} \quad \left(|[\mathbf{Dg}(\mathbf{a})]| + 1 \right) |\vec{\mathbf{h}}| \leq \delta', \qquad A1.14$$

then Equation A1.12 gives

$$\left| [\mathbf{Dg}(\mathbf{a})] \vec{\mathbf{h}} + \mathbf{r}(\vec{\mathbf{h}}) \right| \leq \delta',$$

so we can substitute the expression $|[\mathbf{Dg}(\mathbf{a})] \vec{\mathbf{h}} + \mathbf{r}(\vec{\mathbf{h}})|$ for $|\vec{\mathbf{k}}|$ in the equation $|\mathbf{s}(\vec{\mathbf{k}})| \leq \epsilon |\vec{\mathbf{k}}|$, which is true when $|\vec{\mathbf{k}}| < \delta'$. This gives

$$\underbrace{\left| \mathbf{s}\left([\mathbf{Dg}(\mathbf{a})] \vec{\mathbf{h}} + \mathbf{r}(\vec{\mathbf{h}}) \right) \right|}_{|\mathbf{s}(\vec{\mathbf{k}})|} \leq \underbrace{\epsilon \left| [\mathbf{Dg}(\mathbf{a})] \vec{\mathbf{h}} + \mathbf{r}(\vec{\mathbf{h}}) \right|}_{\epsilon |\vec{\mathbf{k}}|} \underbrace{\leq}_{\text{Eq. A1.12}} \epsilon \left(|[\mathbf{Dg}(\mathbf{a})]| + 1 \right) |\vec{\mathbf{h}}|.$$

$$A1.15$$

Dividing by $|\vec{\mathbf{h}}|$ gives

$$\frac{\left| \mathbf{s}\left([\mathbf{Dg}(\mathbf{a})] \vec{\mathbf{h}} + \mathbf{r}(\vec{\mathbf{h}}) \right) \right|}{|\vec{\mathbf{h}}|} \leq \epsilon \left(|[\mathbf{Dg}(\mathbf{a})]| + 1 \right), \qquad A1.16$$

---

[1]In fact, by choosing a smaller δ, we could make $|\mathbf{r}(\vec{\mathbf{h}})|$ as small as we like, getting $|\mathbf{r}(\vec{\mathbf{h}})| \leq \epsilon |\vec{\mathbf{h}}|$ for any $\epsilon > 0$, but this will not be necessary; taking $\epsilon = 1$ is good enough (see Theorem 1.5.10).

and since this is true for every $\epsilon > 0$, we have proved that the limit in Equation A1.11 is 0:

$$\lim_{\vec{\mathbf{h}} \to 0} \frac{\mathbf{s}\big([\mathbf{Dg}(\mathbf{a})]\vec{\mathbf{h}} + \mathbf{r}(\mathbf{h})\big)}{|\vec{\mathbf{h}}|} = 0. \quad \square \qquad (A1.11)$$

A.2 Proof of Kantorovitch's Theorem

Theorem 2.7.11 (Kantorovitch's theorem). *Let* $\mathbf{a}_0$ *be a point in* $\mathbb{R}^n$, U *an open neighborhood of* $\mathbf{a}_0$ *in* $\mathbb{R}^n$ *and* $\mathbf{f} : U \to \mathbb{R}^n$ *a differentiable mapping, with its derivative* $[\mathbf{Df}(\mathbf{a}_0)]$ *invertible. Define*

$$\vec{\mathbf{h}}_0 = -[\mathbf{Df}(\mathbf{a}_0)]^{-1}\mathbf{f}(\mathbf{a}_0) \quad , \quad \mathbf{a}_1 = \mathbf{a}_0 + \vec{\mathbf{h}}_0 \quad , \quad U_0 = \Big\{ \mathbf{x} \,\Big|\, |\mathbf{x} - \mathbf{a}_1| \leq |\vec{\mathbf{h}}_0| \Big\}.$$
$$A2.1$$

If $U_0 \subset U$ *and the derivative* $[\mathbf{Df}(\mathbf{x})]$ *satisfies the Lipschitz condition*

$$\big|[\mathbf{Df}(\mathbf{u}_1)] - [\mathbf{Df}(\mathbf{u}_2)]\big| \leq M|\mathbf{u}_1 - \mathbf{u}_2| \quad \text{for all points } \mathbf{u}_1, \mathbf{u}_2 \in U_0, \qquad A2.2$$

and if the inequality

$$\big|\mathbf{f}(\mathbf{a}_0)\big| \, \big|[\mathbf{Df}(\mathbf{a}_0)]^{-1}\big|^2 M \leq \frac{1}{2} \qquad A2.3$$

is satisfied, the equation $\mathbf{f}(\mathbf{x}) = \mathbf{0}$ *has a unique solution in* U_0, *and Newton's method with initial guess* $\mathbf{a}_0$ *converges to it.*

Proof.   The proof is fairly involved, so we will first outline our approach. We prove existence by showing the following four facts:

Facts (1), (2) and (3) guarantee that the hypotheses about $\mathbf{a}_0$ of our theorem are also true of $\mathbf{a}_1$. We need (1) in order to define $\vec{\mathbf{h}}_1$, $\mathbf{a}_2$, and U_1. Statement (2) guarantees that $U_1 \subset U_0$, hence $[\mathbf{Df}(\mathbf{x})]$ satisfies the same Lipschitz condition on U_1 as on U_0. Statement (3) is needed to show that Inequality A2.3 is satisfied at $\mathbf{a}_1$. (Remember that the ratio M has not changed.)

(1) $[\mathbf{Df}(\mathbf{a}_1)]$ is invertible, allowing us to define $\vec{\mathbf{h}}_1 = -[\mathbf{Df}(\mathbf{a}_1)]^{-1}\mathbf{f}(\mathbf{a}_1)$;

(2) $|\vec{\mathbf{h}}_1| \leq \dfrac{|\vec{\mathbf{h}}_0|}{2}$;

(3) $\big|\mathbf{f}(\mathbf{a}_1)\big| \, \big|[\mathbf{Df}(\mathbf{a}_1)]^{-1}\big|^2 \leq \big|\mathbf{f}(\mathbf{a}_0)\big| \, \big|[\mathbf{Df}(\mathbf{a}_0)]^{-1}\big|^2$;

(4) $|\mathbf{f}(\mathbf{a}_1)| \leq \dfrac{M}{2}|\vec{\mathbf{h}}_0|^2$. $\qquad A2.4$

If (1), (2), (3) are true we can define sequences $\vec{\mathbf{h}}_i$, $\mathbf{a}_i$, U_i:

$$\vec{\mathbf{h}}_i = -[\mathbf{Df}(\mathbf{a}_i)]^{-1}\mathbf{f}(\mathbf{a}_i), \ \mathbf{a}_i = \mathbf{a}_{i-1} + \vec{\mathbf{h}}_{i-1}, \text{ and } U_i = \Big\{ \mathbf{x} \,\Big|\, |\mathbf{x} - \mathbf{a}_{i+1}| \leq |\vec{\mathbf{h}}_i| \Big\},$$

and at each stage all the hypotheses of Theorem 2.7.11 are true.

Statement (2), together with Proposition 1.5.30, also proves that the $\mathbf{a}_i$ converge; let us call the limit $\mathbf{a}$. Statement (4) will then say that $\mathbf{a}$ satisfies $\mathbf{f}(\mathbf{a}) = \mathbf{0}$. Indeed, by (2),

$$|\vec{\mathbf{h}}_i| \leq \frac{|\vec{\mathbf{h}}_0|}{2^i} \qquad A2.5$$

so by part (4),

$$|f(\mathbf{a}_i)| \leq \frac{M}{2}|\vec{\mathbf{h}}_{i-1}|^2 \leq \frac{M}{2^i}|\vec{\mathbf{h}}_0|^2, \qquad A2.6$$

and in the limit as $i \to \infty$, we have $|\mathbf{f}(\mathbf{a})| = 0$.

First we need to prove Proposition A2.1 and Lemma A2.3.

Proposition A2.1. *If $U \subset \mathbb{R}^n$ is an open ball and $\mathbf{f} : U \to \mathbb{R}^m$ is a differentiable mapping whose derivative satisfies the Lipschitz condition*

$$\big|[\mathbf{Df}(\mathbf{x})] - [\mathbf{Df}(\mathbf{y})]\big| \leq M|\mathbf{x} - \mathbf{y}| \text{ for all } \mathbf{x}, \mathbf{y} \in U, \qquad A2.7$$

then

$$\Big| \underbrace{\mathbf{f}(\mathbf{x} + \vec{\mathbf{h}}) - \mathbf{f}(\mathbf{x})}_{\text{increment to } \mathbf{f}} - \underbrace{[\mathbf{Df}(\mathbf{x})]\vec{\mathbf{h}}}_{\substack{\text{linear approx.} \\ \text{of increment to } \mathbf{f}}} \Big| \leq \frac{M}{2}|\vec{\mathbf{h}}|^2. \qquad A2.8$$

Before embarking on the proof, let us see why this statement is reasonable. The term $[\mathbf{Df}(\mathbf{x})]\vec{\mathbf{h}}$ is the linear approximation of the increment to the function in terms of the increment $\vec{\mathbf{h}}$ to the variable. You would expect the error term to be of second degree, i.e., some multiple of $|\vec{\mathbf{h}}|^2$, which thus gets very small as $\vec{\mathbf{h}} \to 0$. That is what Proposition A2.1 says, and it identifies the Lipschitz ratio M as the main ingredient of the coefficient of $|\vec{\mathbf{h}}|^2$.

The coefficient $M/2$ on the right is the *smallest* coefficient that will work for all functions $\mathbf{f} : U \to \mathbb{R}^m$, although it is possible to find functions where an inequality with a smaller coefficient is satisfied. Equality is achieved for the function $f(x) = x^2$: we have

$$[\mathbf{D}f(x)] = f'(x) = 2x, \quad \text{so} \quad |[\mathbf{D}f(x)] - [\mathbf{D}f(y)]| = 2|x - y|, \qquad A2.9$$

and the best Lipschitz ratio is $M = 2$:

$$|f(x + h) - f(x) - 2xh| = |(x + h)^2 - x^2 - 2xh| = h^2 = \frac{2}{2}h^2 = \frac{M}{2}h^2. \quad A2.10$$

If the derivative of f is *not* Lipschitz (as in Example A2.2) then it may be the case that there exists no C such that

$$|\mathbf{f}(\mathbf{x} + \vec{\mathbf{h}}) - \mathbf{f}(\mathbf{x}) - [\mathbf{Df}(\mathbf{x})]\vec{\mathbf{h}}| \leq C|\vec{\mathbf{h}}|^2. \qquad A2.11$$

In Example A2.2 we have no guarantee that the difference between the increment to f and its approximation by $f'(x)h$ behaves like h^2.

Example A2.2 (A derivative that is not Lipschitz). Let $f(x) = x^{4/3}$, so $[Df(x)] = f'(x) = \frac{4}{3}x^{1/3}$. In particular $f'(0) = 0$, so

$$|f(0 + h) - f(0) - f'(0)h| = h^{4/3}. \qquad A2.1?$$

But $h^{4/3}$ is not $\leq C|h|^2$ for any C, since $h^{4/3}/h^2 = 1/h^{2/3} \to \infty$ as $h \to 0$.

Proof of Proposition A2.1.   Consider the function $\mathbf{g}(t) = \mathbf{f}(\mathbf{x} + t\vec{\mathbf{h}})$. Each coordinate of $\mathbf{g}$ is a differentiable function of the single variable t, so the fundamental theorem of calculus says

$$\overbrace{\mathbf{f}(\mathbf{x} + \vec{\mathbf{h}})}^{\mathbf{g}(1)} - \overbrace{\mathbf{f}(\mathbf{x})}^{\mathbf{g}(0)} = \mathbf{g}(1) - \mathbf{g}(0) = \int_0^1 \mathbf{g}'(t)\, dt. \qquad A2.13$$

Using the chain rule, we see that

$$\mathbf{g}'(t) = [\mathbf{Df}(\mathbf{x} + t\vec{\mathbf{h}})]\vec{\mathbf{h}}, \qquad A2.14$$

which we will write as

$$\mathbf{g}'(t) = [\mathbf{Df}(\mathbf{x})]\vec{\mathbf{h}} + \Big([\mathbf{Df}(\mathbf{x} + t\vec{\mathbf{h}})]\vec{\mathbf{h}} - [\mathbf{Df}(\mathbf{x})]\vec{\mathbf{h}}\Big). \qquad A2.15$$

This leads to

$$\mathbf{f}(\mathbf{x} + \vec{\mathbf{h}}) - \mathbf{f}(\mathbf{x}) = \int_0^1 [\mathbf{Df}(\mathbf{x})]\vec{\mathbf{h}}\, dt + \int_0^1 \Big([\mathbf{Df}(\mathbf{x} + t\vec{\mathbf{h}})]\vec{\mathbf{h}} - [\mathbf{Df}(\mathbf{x})]\vec{\mathbf{h}}\Big)\, dt. \quad A2.16$$

The first term on the right is the integral from 0 to 1 of a constant, so it is simply that constant, so we can rewrite Equation A2.16 as

To go from the first to the second line of Equation A2.17 we use Equation A2.7, replacing $\mathbf{x}$ by $\mathbf{x} + t\vec{\mathbf{h}}$ and $\mathbf{y}$ by $\mathbf{x}$:

$$\Big|[\mathbf{Df}(\mathbf{x} + t\vec{\mathbf{h}})] - [\mathbf{Df}(\mathbf{x})]\Big|$$
$$\leq M|\mathbf{x} + t\vec{\mathbf{h}} - \mathbf{x}| = Mt|\vec{\mathbf{h}}|.$$

$$\left|\mathbf{f}(\mathbf{x} + \vec{\mathbf{h}}) - \mathbf{f}(\mathbf{x}) - [\mathbf{Df}(\mathbf{x})]\vec{\mathbf{h}}\right| = \left|\int_0^1 \Big([\mathbf{Df}(\mathbf{x} + t\vec{\mathbf{h}})]\vec{\mathbf{h}} - [\mathbf{Df}(\mathbf{x})]\vec{\mathbf{h}}\Big)\, dt\right| \quad A2.17$$

$$\leq \int_0^1 Mt|\vec{\mathbf{h}}||\vec{\mathbf{h}}|\, dt = \frac{M}{2}|\vec{\mathbf{h}}|^2. \qquad \square$$

Lemma A2.3.   *The matrix* $[\mathbf{Df}(\mathbf{a}_1)]$ *is invertible*, *and*

$$\left|[\mathbf{Df}(\mathbf{a}_1)]^{-1}\right| \leq 2\left|[\mathbf{Df}(\mathbf{a}_0)]^{-1}\right|. \qquad A2.18$$

Proving Lemma A2.3 is the hardest part of proving Theorem 2.7.11. At the level of this book, we don't know much about inverses of matrices, so we have to use "bare hands" techniques involving geometric series of matrices.

Proof.   We have required (Equation A2.2) that the derivative matrix not vary too fast, so it is reasonable to hope that

$$[\mathbf{Df}(\mathbf{a}_0)]^{-1}[\mathbf{Df}(\mathbf{a}_1)]$$

is not too far from the identity. Indeed, set

$$A = I - \big([\mathbf{Df}(\mathbf{a}_0)]^{-1}[\mathbf{Df}(\mathbf{a}_1)]\big) = \underbrace{[\mathbf{Df}(\mathbf{a}_0)]^{-1}[\mathbf{Df}(\mathbf{a}_0)]}_{I} - \big([\mathbf{Df}(\mathbf{a}_0)]^{-1}[\mathbf{Df}(\mathbf{a}_1)]\big)$$

$$= [\mathbf{Df}(\mathbf{a}_0)]^{-1}\Big([\mathbf{Df}(\mathbf{a}_0)] - [\mathbf{Df}(\mathbf{a}_1)]\Big). \qquad A2.19$$

By Equation A2.2 we know that $\big|[\mathbf{Df}(\mathbf{a}_0)] - [\mathbf{Df}(\mathbf{a}_1)]\big| \leq M|\mathbf{a}_0 - \mathbf{a}_1|$, and by definition (Equation A2.1) we know $|\vec{\mathbf{h}}_0| = |\mathbf{a}_1 - \mathbf{a}_0|$. So

$$|A| \leq \big|[\mathbf{Df}(\mathbf{a}_0)]^{-1}\big|\, |\vec{\mathbf{h}}_0|M. \qquad A2.20$$

By definition (Equation A2.1 again),

$$\vec{\mathbf{h}}_0 = -[\mathbf{Df}(\mathbf{a}_0)]^{-1}\mathbf{f}(\mathbf{a}_0), \qquad \text{so} \qquad |\vec{\mathbf{h}}_0| \leq \big|[\mathbf{Df}(\mathbf{a}_0)]^{-1}\big||\mathbf{f}(\mathbf{a}_0)| \qquad A2.21$$

(Proposition 1.4.11, once more). This gives us

$$|A| \leq \underbrace{\left|[\mathbf{Df}(\mathbf{a}_0)]^{-1}\right|\left|[\mathbf{Df}(\mathbf{a}_0)]^{-1}\right||\mathbf{f}(\mathbf{a}_0)|M}_{\text{left-hand side of Inequality A2.3}}. \qquad A2.22$$

Now Inequality A2.3 guarantees

$$|A| \leq \frac{1}{2}. \qquad A2.23$$

which we can use to show that $[\mathbf{Df}(\mathbf{a}_1)]$ is invertible, as follows. We know from Proposition 1.5.31 that if $|A| < 1$, then the geometric series

$$I + A + A^2 + A^3 + \ldots = B \qquad A2.24$$

converges, and that $B(I - A) = I$; i.e., that B and $(I - A)$ are inverses of each other. This tells us that $[\mathbf{Df}(\mathbf{a}_1)]$ is invertible: from Equation A2.19 we know that $I - A = [\mathbf{Df}(\mathbf{a}_0)]^{-1}[\mathbf{Df}(\mathbf{a}_1)]$; and if the product of two square matrices is invertible, then they both are invertible (see Exercise 2.5.14).

In fact, (by Proposition 1.2.15: $(AB)^{-1} = B^{-1}A^{-1}$) we have

$$B = (I - A)^{-1} = [\mathbf{Df}(\mathbf{a}_1)]^{-1}[\mathbf{Df}(\mathbf{a}_0)], \quad \text{so} \qquad A2.25$$

Note that in Equation A2.27 we use the number 1, not the identity matrix: $1 + |A| + |A|^2 + \ldots$, not $I + A + A^2 + \ldots$. This is crucial because since $|A| \leq 1/2$, we have

$$|A| + |A|^2 + |A|^3 + \cdots \leq 1,$$

since we have a geometric series as in Example 0.4.9.

When we first wrote this proof, adapting a proof using the norm of a matrix rather than the length, we factored before using the triangle inequality, and ended up with $I + A + A^2 + \ldots$. This was disastrous, because $|I| = \sqrt{n}$, *not* 1. The discovery that this could be fixed by factoring *after* using the triangle inequality was most welcome.

$$\begin{aligned}
[\mathbf{Df}(\mathbf{a}_1)]^{-1} &= B[\mathbf{Df}(\mathbf{a}_0)]^{-1} \\
&\overset{B \text{ by Eq. A2.24)}}{=} \overbrace{(I + A + A^2 + \cdots)}[\mathbf{Df}(\mathbf{a}_0)]^{-1} \\
&= [\mathbf{Df}(\mathbf{a}_0)]^{-1} + A[\mathbf{Df}(\mathbf{a}_0)]^{-1} + \cdots,
\end{aligned} \qquad A2.26$$

hence (by the triangle inequality and Proposition 1.4.11)

$$\begin{aligned}
\left|[\mathbf{Df}(\mathbf{a}_1)]^{-1}\right| &\leq \left|[\mathbf{Df}(\mathbf{a}_0)]^{-1}\right| + |A|\left|[\mathbf{Df}(\mathbf{a}_0)]^{-1}\right| + \cdots \\
&= \left|[\mathbf{Df}(\mathbf{a}_0)]^{-1}\right|\left(1 + |A| + |A|^2 + \cdots\right) \\
&\leq \left|[\mathbf{Df}(\mathbf{a}_0)]^{-1}\right|\underbrace{\left(1 + 1/2 + 1/4 + \cdots\right)}_{\text{since }|A| \leq 1/2,\ \text{Eq. A2.23}} = 2\left|[\mathbf{Df}(\mathbf{a}_0)]^{-1}\right|.
\end{aligned} \qquad A2.27$$

$\square$ Lemma A2.3

So far we have proved (1). This enables us to define the next step of Newton's method:

$$\vec{\mathbf{h}}_1 = -[\mathbf{Df}(\mathbf{a}_1)]^{-1}\mathbf{f}(\mathbf{a}_1), \quad \mathbf{a}_2 = \mathbf{a}_1 + \vec{\mathbf{h}}_1, \quad \text{and } U_1 = \left\{\mathbf{x} \,\middle|\, |\mathbf{x} - \mathbf{a}_2| \leq |\vec{\mathbf{h}}_1|\right\}. \qquad A2.28$$

Now we will prove (4), which we will call Lemma A2.4:

Lemma A2.4. *We have the inequality*

$$|\mathbf{f}(\mathbf{a}_1)| \leq \frac{M}{2}|\vec{\mathbf{h}}_0|^2. \qquad A2.29$$

Proof. This is a straightforward application of Proposition A2.1, which says that

$$\left|\mathbf{f}(\mathbf{a}_1) - \mathbf{f}(\mathbf{a}_0) - [\mathbf{Df}(\mathbf{a}_0)]\vec{\mathbf{h}}_0\right| \leq \frac{M}{2}|\vec{\mathbf{h}}_0|^2, \qquad A2.30$$

but a miracle happens during the computation: the third term in the sum on the left of Equation A2.30 cancels with the second term, since

$$-[\mathbf{Df}(\mathbf{a}_0)]\vec{\mathbf{h}}_0 = [\mathbf{Df}(\mathbf{a}_0)]\overbrace{[\mathbf{Df}(\mathbf{a}_0)]^{-1}\mathbf{f}(\mathbf{a}_0)}^{\vec{\mathbf{h}}_0} = \mathbf{f}(\mathbf{a}_0). \qquad A2.31$$

(Figure A2.1 explains why the cancellation occurs.) So we get

$$|\mathbf{f}(\mathbf{a}_1)| \leq \frac{M}{2}|\vec{\mathbf{h}}_0|^2 \text{ as required.} \quad \square \quad \text{Lemma A2.4} \qquad A2.32$$

Proof of Theorem 2.7.11 (Kantorovitch theorem) continued. Now we just string together the inequalities. We have proved (1) and (4). To prove statement (2), i.e., $|\vec{\mathbf{h}}_1| \leq |\vec{\mathbf{h}}_0|/2$, we consider

$$|\vec{\mathbf{h}}_1| \leq |\mathbf{f}(\mathbf{a}_1)|\,|[\mathbf{Df}(\mathbf{a}_1)]^{-1}| \leq \frac{M|\vec{\mathbf{h}}_0|^2}{2}2|[\mathbf{Df}(\mathbf{a}_0)]^{-1}|. \qquad A2.33$$

Cancel the 2's and write $|\vec{\mathbf{h}}_0|^2$ as two factors:

$$|\vec{\mathbf{h}}_1| \leq |\vec{\mathbf{h}}_0|M|[\mathbf{Df}(\mathbf{a}_0)]^{-1}|\,|\vec{\mathbf{h}}_0|. \qquad A2.34$$

Next, replace one of the $|\vec{\mathbf{h}}_0|$, using the definition $\vec{\mathbf{h}}_0 = -[\mathbf{Df}(\mathbf{a}_0)]^{-1}\mathbf{f}(\mathbf{a}_0)$, to get

$$|\vec{\mathbf{h}}_1| \leq |\vec{\mathbf{h}}_0|\Bigg(\underbrace{M|[\mathbf{Df}(\mathbf{a}_0)]^{-1}|\overbrace{|\mathbf{f}(\mathbf{a}_0)|\,|[\mathbf{Df}(\mathbf{a}_0)]^{-1}|}^{\geq|\vec{\mathbf{h}}_0|}}_{\leq 1/2 \text{ by Inequality A2.3}}\Bigg) \leq \frac{|\vec{\mathbf{h}}_0|}{2}. \qquad A2.35$$

Now to prove part (3), i.e.,

$$|\mathbf{f}(\mathbf{a}_1)|\,\big|[\mathbf{Df}(\mathbf{a}_1)]^{-1}\big|^2 \leq |\mathbf{f}(\mathbf{a}_0)|\,\big|[\mathbf{Df}(\mathbf{a}_0)]^{-1}\big|^2. \qquad A2.36$$

Using Lemma A2.4 to get a bound for $|\mathbf{f}(\mathbf{a}_1)|$, and Equation A2.18 to get a bound for $|[\mathbf{Df}(\mathbf{a}_1)]^{-1}|$, we write

$$|\mathbf{f}(\mathbf{a}_1)|\,|[\mathbf{Df}(\mathbf{a}_1)]^{-1}|^2 \leq \frac{M|\mathbf{h}_0|^2}{2}\left(4|[\mathbf{Df}(\mathbf{a}_0)]^{-1}|^2\right)$$

$$\leq 2|[\mathbf{Df}(\mathbf{a}_0)]^{-1}|^2M\overbrace{\left(|[\mathbf{Df}(\mathbf{a}_0)]^{-1}|\,|\mathbf{f}(\mathbf{a}_0)|\right)^2}^{\geq|\vec{\mathbf{h}}_0|^2} \qquad A2.37$$

$$\leq |[\mathbf{Df}(\mathbf{a}_0)]^{-1}|^2\,|\mathbf{f}(\mathbf{a}_0)|\,2\underbrace{|\mathbf{f}(\mathbf{a}_0)|\,|[\mathbf{Df}(\mathbf{a}_0)]^{-1}|^2M}_{\text{at most } 1/2 \text{ by A2.3}}$$

$$\leq |[\mathbf{Df}(\mathbf{a}_0)]^{-1}|^2|\mathbf{f}(\mathbf{a}_0)|. \quad \square \text{ existence, Thm. 2.7.11}$$

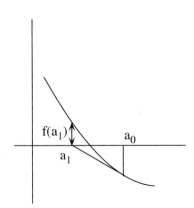

FIGURE A2.1.

The terms that cancel are exactly the value of the linearization to $\mathbf{f}$ at $\mathbf{a}_0$, evaluated at $\mathbf{a}_1$.

The first inequality in Equation A2.33 uses the definition of $\vec{\mathbf{h}}_1$ (Equation A2.28) and Proposition 1.4.11. The second inequality uses Lemmas A2.3 and A2.4.

Note that the middle term of Equation A2.33 has $\mathbf{a}_1$, while the right-hand term has $\mathbf{a}_0$.

Proof of uniqueness   To prove that the solution in U_0 is unique, we will prove that if $\mathbf{y} \in U_0$ and $\mathbf{f}(\mathbf{y}) = \mathbf{0}$, then

$$|\mathbf{y} - \mathbf{a}_{i+1}| \le \frac{1}{2}|\mathbf{y} - \mathbf{a}_i|. \qquad A2.38$$

First, set

$$\mathbf{0} = \mathbf{f}(\mathbf{y}) = \mathbf{f}(\mathbf{a}_i) + [\mathbf{Df}(\mathbf{a}_i)](\mathbf{y} - \mathbf{a}_i) + \vec{\mathbf{r}}_i, \qquad A2.39$$

where $\vec{\mathbf{r}}_i$ is the remainder necessary for the second equality to be true. This gives

$$\mathbf{y} - \mathbf{a}_i = \overbrace{-[\mathbf{Df}(\mathbf{a}_i)]^{-1}\mathbf{f}(\mathbf{a}_i)}^{\substack{+\vec{\mathbf{h}}_i \text{ by earlier part} \\ \text{of proof}} }-[\mathbf{Df}(\mathbf{a}_i)]^{-1}\vec{\mathbf{r}}_i, \qquad A2.40$$

which we can rewrite as

$$\mathbf{y} - \overbrace{(\mathbf{a}_i + \vec{\mathbf{h}}_i)}^{\mathbf{a}_{i+1}} = \mathbf{y} - \mathbf{a}_{i+1} = -[\mathbf{Df}(\mathbf{a}_i)]^{-1}\vec{\mathbf{r}}_i. \qquad A2.41$$

Now we return to Equation A2.39. By Proposition A2.1, since $\mathbf{y}, \mathbf{a}_i \in U_0$, we have

$$|\vec{\mathbf{r}}_i| = \Big|\underbrace{\mathbf{f}(\mathbf{y}) - \mathbf{f}(\mathbf{a}_i)}_{\substack{\text{increment to } \mathbf{f}}} - \underbrace{[\mathbf{Df}(\mathbf{a}_i)](\mathbf{y} - \mathbf{a}_i)}_{\substack{\text{linear approx.} \\ \text{of increment to } \mathbf{f}}}\Big| \le \frac{M}{2}|\mathbf{y} - \mathbf{a}_i|^2. \qquad A2.42$$

Next replace the $\vec{\mathbf{r}}_i$ in Equation A2.41 by the $\frac{M}{2}|\mathbf{y} - \mathbf{a}_i|^2$ of Equation A2.42, and take absolute values, to get

$$|\mathbf{y} - \mathbf{a}_{i+1}| \le \big|[\mathbf{Df}(\mathbf{a}_i)]^{-1}\big|\frac{M}{2}|\mathbf{y} - \mathbf{a}_i|^2. \qquad A2.43$$

Now we will prove Equation A2.38 by induction. To start the induction, note that

To get from the first to the second line of Equation A2.44, note that since $\mathbf{y} \in U_0$, it is in a ball of radius $\vec{\mathbf{h}}_0$, with center $\mathbf{a}_1$, and thus can be at most $2\vec{\mathbf{h}}_0$ away from $\mathbf{a}_0$. So we can replace one of the $|\mathbf{y} - \mathbf{a}_0|$ by $|2\vec{\mathbf{h}}_0|$. Going from the third to the fourth line uses Equation A2.3.

$$\begin{aligned} |\mathbf{y} - \mathbf{a}_1| &\le \big|[\mathbf{Df}(\mathbf{a}_0)]^{-1}\big|\frac{M}{2}|\mathbf{y} - \mathbf{a}_0|^2 \\ &\le \big|[\mathbf{Df}(\mathbf{a}_0)]^{-1}\big|\frac{M}{2}|2\vec{\mathbf{h}}_0||\mathbf{y} - \mathbf{a}_0| \\ &\le \big|[\mathbf{Df}(\mathbf{a}_0)]^{-1}\big| M \underbrace{|\mathbf{f}(\mathbf{a}_0)|\big|[\mathbf{Df}(\mathbf{a}_0)]^{-1}\big|}_{\ge |\vec{\mathbf{h}}_0| \text{ by Eq. A2.1}} |\mathbf{y} - \mathbf{a}_0| \\ &\le \frac{1}{2}|\mathbf{y} - \mathbf{a}_0|. \end{aligned} \qquad A2.44$$

Now assume by induction that

$$|\mathbf{y} - \mathbf{a}_j| \le \frac{1}{2}|\mathbf{y} - \mathbf{a}_{j-1}| \quad \text{for all } j \le i, \qquad A2.45$$

and rewrite Equation A2.43, dividing each side by $|\mathbf{y} - \mathbf{a}_i|$:

The inductive hypothesis is used to replace the $|\mathbf{y} - \mathbf{a}_i|$ in the first line of Equation A2.46 by $\frac{1}{2}|\mathbf{y} - \mathbf{a}_{i-1}|$ in the second line. Lemma A2.3 justifies replacing $|[\mathbf{Df}(\mathbf{a}_i)]^{-1}|$ in the first line by $2|[\mathbf{Df}(\mathbf{a}_{i-1})]^{-1}|$ in the second line.

$$
\begin{aligned}
\frac{|\mathbf{y} - \mathbf{a}_{i+1}|}{|\mathbf{y} - \mathbf{a}_i|} &\leq |[\mathbf{Df}(\mathbf{a}_i)]^{-1}| \frac{M}{2}|\mathbf{y} - \mathbf{a}_i| \\
&\leq 2|[\mathbf{Df}(\mathbf{a}_{i-1})]^{-1}| \frac{M}{2} \frac{|\mathbf{y} - \mathbf{a}_{i-1}|}{2} \\
&\leq \cdots \\
&\leq \frac{1}{2} M|\mathbf{y} - \mathbf{a}_0| \, |[\mathbf{Df}(\mathbf{a}_0)]^{-1}| \\
&\leq \frac{1}{2} M|2\vec{\mathbf{h}}_0| \, |[\mathbf{Df}(\mathbf{a}_0)]^{-1}| \, | \leq |\mathbf{f}(\mathbf{a}_0)||[\mathbf{Df}(\mathbf{a}_0)]^{-1}|^2 M \\
&\leq \frac{1}{2} \qquad \text{(We again use Equation A2.3.)}
\end{aligned}
$$
$$A2.46$$

Thus $|\mathbf{y} - \mathbf{a}_{i+1}| \leq \frac{1}{2}|\mathbf{y} - \mathbf{a}_i|$. This proves that $\mathbf{y} = \lim \mathbf{a}_i$, and that $\lim \mathbf{a}_i$ is the unique solution of $\mathbf{f}(\mathbf{x}) = \mathbf{0}$ in U_0. $\square$

Remark A2.5 : The smaller the set in which one can guarantee existence, the better: it is a stronger statement to say that there exists a William Ardvark in the town of Nowhere, NY, population 523, than to say there exists a William Ardvark in New York State.

The larger the set in which one can guarantee uniqueness, the better: it is a stronger statement to say there exists a unique John W. Smith in the state of California than to say there exists a unique John W. Smith in Tinytown, CA.

There are times, such as when proving uniqueness for the inverse function theorem, that one wants the Kantorovitch theorem stated for the larger ball U_{-1}. There are other times when the function is not Lipschitz on the larger space, or is not even defined on the larger space, and the original Kantorovitch theorem is the useful one.

Remark A2.5. If we change the hypotheses of Kantorovitch's theorem by defining a larger ball U_{-1}:

$$
U_{-1} = \{\mathbf{x} | |\mathbf{x} - \mathbf{x}_0| \leq 2|\vec{\mathbf{h}}_0|\},
$$

and require that U_{-1} be a subset of U, and that the Lipschitz condition A2.2 holds for all $\mathbf{u}_1, \mathbf{u}_2 \in U_{-1}$, then we can strengthen the conclusion to say that the equation $\mathbf{f}(\mathbf{x}) = \mathbf{0}$ has a unique solution in U_{-1}, and that Newton's method with initial guess $\mathbf{x}_0$ converges to it. The proof is exactly identical. $\triangle$

A.3 Proof of Lemma 2.8.4 (Superconvergence)

Here we prove Lemma 2.8.4, used in proving that Newton's method superconverges. Recall that

$$
c = \frac{1-k}{1-2k} |[\mathbf{Df}(\mathbf{a}_0)]^{-1}| \frac{M}{2}. \tag{2.8.3}
$$

Lemma 2.8.4. *If the conditions of Theorem 2.8.3 are satisfied, then for all i,*

$$
|\vec{\mathbf{h}}_{i+1}| \leq c|\vec{\mathbf{h}}_i|^2. \tag{A3.1}
$$

Proof. Look back at Lemma A2.4 (rewritten for $\mathbf{a}_i$):

$$
|\mathbf{f}(\mathbf{a}_i)| \leq \frac{M}{2} |\vec{\mathbf{h}}_{i-1}|^2. \tag{A3.2}
$$

The definition

$$
\vec{\mathbf{h}}_i = -[\mathbf{Df}(\mathbf{a}_i)]^{-1}\mathbf{f}(\mathbf{a}_i) \tag{A3.3}
$$

and Equation A3.2 give

$$|\vec{\mathbf{h}}_i| \le |[\mathbf{Df}(\mathbf{a}_i)]^{-1}| \, |\mathbf{f}(\mathbf{a}_i)| \le \frac{M}{2}|[\mathbf{Df}(\mathbf{a}_i)]^{-1}||\vec{\mathbf{h}}_{i-1}|^2. \qquad A3.4$$

This is an equation almost of the form $|\vec{\mathbf{h}}_i| \le c|\vec{\mathbf{h}}_{i-1}|^2$:

$$|\vec{\mathbf{h}}_i| \le \underbrace{\frac{M}{2}|[\mathbf{Df}(\mathbf{a}_i)]^{-1}|}_{c_i} \, |\vec{\mathbf{h}}_{i-1}|^2. \qquad A3.5$$

If we have such a bound, sooner or later superconvergence will occur.

The difference is that c_i is not a constant but depends on $\mathbf{a}_i$. So the $\vec{\mathbf{h}}_i$ will superconverge if we can find a bound on $|[\mathbf{Df}(\mathbf{a}_i)]|^{-1}$ valid for all i. (The term $M/2$ is not a problem because it is a constant.) We cannot find such a bound if the derivative $[\mathbf{Df}(\mathbf{a})]$ is not invertible at the limit point $\mathbf{a}$. (We saw this in one dimension in Example 2.8.1, where $f'(1) = 0$.) In such a case $|[\mathbf{Df}(\mathbf{a}_i)]^{-1}| \to \infty$ as $\mathbf{a}_i \to \mathbf{a}$. But Lemma A3.1 says that if the product of the Kantorovitch inequality is strictly less than $1/2$, we have such a bound.

Lemma A3.1 (A bound on $|[\mathbf{Df}(\mathbf{a}_n)]|^{-1}$). *If*

$$|\mathbf{f}(\mathbf{a}_0)|\,\big|[\mathbf{Df}(\mathbf{a}_0)]^{-1}\big|^2 M = k, \quad where \quad k < 1/2, \qquad A3.6$$

then all $[\mathbf{Df}(\mathbf{a}_n)]^{-1}$ exist and satisfy

$$\big|[\mathbf{Df}(\mathbf{a}_n)]^{-1}\big| \le \big|[\mathbf{Df}(\mathbf{a}_0)]^{-1}\big|\frac{1-k}{1-2k}. \qquad A3.7$$

You may find it helpful to refer to the proof of Lemma A2.3, as we are more concise here. Note that while Lemma A2.3 compares the derivative at $\mathbf{a}_1$ to the derivative at $\mathbf{a}_0$, here we compare the derivative at $\mathbf{a}_n$ to the derivative at $\mathbf{a}_0$.

Proof of Lemma A3.1.   The proof of this lemma is a rerun of the proof of Lemma A2.3. We will use Equation A2.35 in the form $|\vec{\mathbf{h}}_1| \le k|\vec{\mathbf{h}}_0|$, which by induction gives $|\vec{\mathbf{h}}_i| \le k|\vec{\mathbf{h}}_{i-1}|$, so that

$$|\mathbf{a}_n - \mathbf{a}_0| = \left|\sum_{i=0}^{n-1}\vec{\mathbf{h}}_i\right| \underbrace{\le}_{\substack{\text{triangle}\\\text{inequality}}} \sum_{i=0}^{n-1}|\vec{\mathbf{h}}_i| \le |\vec{\mathbf{h}}_0|\overbrace{\left(1 + k + \cdots + k^{n-1}\right)}^{\substack{<\sum_{n=0}^{\infty} k^n = 1/(1-k),\\ \text{by Eq. } 0.4.9}} \le |\vec{\mathbf{h}}_0|\frac{1}{1-k}.$$

$$A3.8$$

The A_n in Equation A3.9 (where we have $\mathbf{a}_n$) corresponds to the A in Equation A2.19 (where we had $\mathbf{a}_1$).

Next write

$$A_n = I - [\mathbf{Df}(\mathbf{a}_0)]^{-1}[\mathbf{Df}(\mathbf{a}_n)] = [\mathbf{Df}(\mathbf{a}_0)]^{-1}\underbrace{\left([\mathbf{Df}(\mathbf{a}_0)] - [\mathbf{Df}(\mathbf{a}_n)]\right)}_{\substack{\le M|\mathbf{a}_0-\mathbf{a}_n| \text{ by}\\ \text{Lipschitz condition}}}, \qquad A3.9$$

The second inequality of Equation A3.10 uses Equation A3.8. The third uses the inequality

$$\|\vec{\mathbf{h}}_0| \le |[\mathbf{Df}(\mathbf{a}_0)]^{-1}||\mathbf{f}(\mathbf{a}_0)|;$$

see Equation A2.21. The last equality uses the hypothesis of the lemma, Equation A3.6.

so that

$$|A_n| \le |[\mathbf{Df}(\mathbf{a}_0)]^{-1}|\, M\, |\mathbf{a}_0 - \mathbf{a}_n| \;\le\; |[\mathbf{Df}(\mathbf{a}_0)]^{-1}|\, M\frac{|\vec{\mathbf{h}}_0|}{1-k}$$

$$\le \frac{|[\mathbf{Df}(\mathbf{a}_0)]^{-1}|^2 M|\mathbf{f}(\mathbf{a}_0)|}{1-k} = \frac{k}{1-k}. \qquad A3.10$$

We are assuming $k < 1/2$, so $I - A_n$ is invertible (by Proposition 1.5.31), and the same argument that led to Equation A2.27 here gives

$$\left|[\mathbf{Df}(\mathbf{a}_n)]^{-1}\right| \leq \left|[\mathbf{Df}(\mathbf{a}_0)]^{-1}\right| \underbrace{(1 + |A_n| + |A_n|^2 + \dots)}_{= \dfrac{1}{1 - |A_n|}} \leq \frac{1 - k}{1 - 2k} \left|[\mathbf{Df}(\mathbf{a}_0)]^{-1}\right|$$

$$\square \quad A3.11$$

A.4 Proof of Differentiability of the Inverse Function

In Section 2.9 we proved the existence of an inverse function $\mathbf{g}$. As we mentioned there, a complete proof requires showing that $\mathbf{g}$ is continuously differentiable, and that $\mathbf{g}$ really is an inverse, not just a right inverse. We do this here.

Theorem 2.9.4 (The inverse function theorem). *Let $W \subset \mathbb{R}^m$ be an open neighborhood of $\mathbf{x}_0$, and $\mathbf{f} : W \to \mathbb{R}^m$ be a continuously differentiable function. Set $\mathbf{y}_0 = \mathbf{f}(\mathbf{x}_0)$, and suppose that the derivative $L = [\mathbf{Df}(\mathbf{x}_0)]$ is invertible.*

Let $R > 0$ be a number satisfying the following hypotheses:

(1) *The ball W_0 of radius $2R|L^{-1}|$ and centered at $\mathbf{x}_0$ is contained in W.*

(2) *In W_0, the derivative satisfies the Lipschitz condition*

$$\left|[\mathbf{Df}(\mathbf{u})] - [\mathbf{Df}(\mathbf{v})]\right| \leq \frac{1}{2R|L^{-1}|^2} |\mathbf{u} - \mathbf{v}|. \qquad 2.9.4$$

There then exists a unique continuously differentiable mapping $\mathbf{g}$ from the ball of radius R centered at $\mathbf{y}_0$ (which we will denote V) to the ball W_0:

$$\mathbf{g} : V \to W_0, \qquad 2.9.5$$

such that

Recall (Equation 2.9.8) that

$$f_{\mathbf{y}}(\mathbf{x}) \overset{\text{def}}{=} f(\mathbf{x}) - \mathbf{y} = 0.$$

$$\mathbf{f}\big(\mathbf{g}(\mathbf{y})\big) = \mathbf{y} \quad and \quad [\mathbf{Dg}(\mathbf{y})] = [\mathbf{Df}(\mathbf{g}(\mathbf{y}))]^{-1}. \qquad 2.9.6$$

Moreover, the image of $\mathbf{g}$ contains the ball of radius R_1 around $\mathbf{x}_0$, where

$$R_1 = 2R|L^{-1}|^2 \left(\sqrt{|L|^2 + \frac{1}{|L^{-1}|^2}} - |L| \right). \qquad 2.9.7$$

The first inequality on the second line of Equation A4.2 comes from the triangle inequality. We get the second inequality because at each step of Newton's method, $\vec{\mathbf{h}}_i$ is at most half of the previous. The last inequality comes from the fact (Equation 2.9.10 and Proposition 1.4.11) that $|\vec{\mathbf{h}}_0(\mathbf{y})| \leq |L^{-1}||\mathbf{y}_0 - \mathbf{y}|$.

(1) Proving that g is continuous at $\mathbf{y}_0$

Let us show first that $\mathbf{g}$ is continuous at $\mathbf{y}_0$: that for all $\epsilon > 0$, there exists $\delta > 0$ such that when $|\mathbf{y} - \mathbf{y}_0| < \delta$, then $|\mathbf{g}(\mathbf{y}) - \mathbf{g}(\mathbf{y}_0)| < \epsilon$. Since $\mathbf{g}(\mathbf{y})$ is the limit of Newton's method for the equation $f_{\mathbf{y}}(\mathbf{x}) = 0$, starting at $\mathbf{x}_0$, it can be expressed as $\mathbf{x}_0$ plus the sum of all the steps $(\vec{\mathbf{h}}_0(\mathbf{y}), \vec{\mathbf{h}}_1(\mathbf{y}), \dots)$:

$$\mathbf{g}(\mathbf{y}) = \mathbf{x}_0 + \sum_{i=0}^{\infty} \vec{\mathbf{h}}_i(\mathbf{y}). \qquad A4.1$$

So

$$|\mathbf{g}(\mathbf{y}) - \underbrace{\mathbf{g}(\mathbf{y}_0)}_{\mathbf{x}_0}| = \left|\mathbf{x}_0 + \sum_{i=0}^{\infty} \vec{\mathbf{h}}_i(\mathbf{y}) - \mathbf{x}_0\right| = \left|\sum_{i=0}^{\infty} \vec{\mathbf{h}}_i(\mathbf{y})\right| \qquad A4.2$$

$$\leq \sum_{i=0}^{\infty} |\vec{\mathbf{h}}_i(\mathbf{y})| \leq |\vec{\mathbf{h}}_0(\mathbf{y})| \underbrace{\left(1 + \frac{1}{2} + \dots\right)}_{=2} \leq 2|L^{-1}||\mathbf{y} - \mathbf{y}_0|.$$

If we set

$$\delta = \frac{\epsilon}{2|L^{-1}|}, \qquad A4.3$$

then when $|\mathbf{y} - \mathbf{y}_0| < \delta$, we have $|\mathbf{g}(\mathbf{y}) - \mathbf{g}(\mathbf{y}_0)| < \epsilon$.

(2) Proving that g is differentiable at $\mathbf{y}_0$

Next we must show that $\mathbf{g}$ is differentiable at $\mathbf{y}_0$, with derivative $[\mathbf{Dg}(\mathbf{y}_0)] = L^{-1}$; i.e., that

$$\lim_{\vec{\mathbf{k}} \to \mathbf{0}} \frac{\left(\mathbf{g}(\mathbf{y}_0 + \vec{\mathbf{k}}) - \mathbf{g}(\mathbf{y}_0)\right) - L^{-1}\vec{\mathbf{k}}}{|\vec{\mathbf{k}}|} = 0. \qquad A4.4$$

When $|\mathbf{y}_0 + \vec{\mathbf{k}}| \in V$, define $\vec{\mathbf{r}}(\vec{\mathbf{k}})$ to be the increment to $\mathbf{x}_0$ that under $\mathbf{f}$ gives the increment $\vec{\mathbf{k}}$ to $\mathbf{y}_0$:

$$\mathbf{f}\left(\mathbf{x}_0 + \vec{\mathbf{r}}(\vec{\mathbf{k}})\right) = \mathbf{y}_0 + \vec{\mathbf{k}}, \qquad A4.5$$

or, equivalently,

$$\mathbf{g}(\mathbf{y}_0 + \vec{\mathbf{k}}) = \mathbf{x}_0 + \vec{\mathbf{r}}(\vec{\mathbf{k}}), \qquad A4.6$$

Substituting the right-hand side of Equation A4.6 for $\mathbf{g}(\mathbf{y}_0 + \vec{\mathbf{k}})$ in the left-hand side of Equation A4.4, remembering that $\mathbf{g}(\mathbf{y}_0) = \mathbf{x}_0$, we find

To get the second line we just factor out L^{-1}.

$$\lim_{\vec{\mathbf{k}} \to \mathbf{0}} \frac{\mathbf{x}_0 + \vec{\mathbf{r}}(\vec{\mathbf{k}}) - \mathbf{x}_0 - L^{-1}\vec{\mathbf{k}}}{|\vec{\mathbf{k}}|} = \lim_{\vec{\mathbf{k}} \to \mathbf{0}} \frac{\vec{\mathbf{r}}(\vec{\mathbf{k}}) - L^{-1}\vec{\mathbf{k}}}{|\vec{\mathbf{k}}|} \frac{|\vec{\mathbf{r}}(\vec{\mathbf{k}})|}{|\vec{\mathbf{r}}(\vec{\mathbf{k}})|}$$

$$\qquad A4.7$$

$$= \lim_{\vec{\mathbf{k}} \to \mathbf{0}} \frac{L^{-1}\left(L\vec{\mathbf{r}}(\vec{\mathbf{k}}) - \overset{\vec{\mathbf{k}} \text{ by Eq. A4.5}}{\left(\mathbf{f}\left(\mathbf{x}_0 + \vec{\mathbf{r}}(\vec{\mathbf{k}})\right) - \mathbf{f}(\mathbf{x}_0)\right)}\right)}{|\vec{\mathbf{r}}(\vec{\mathbf{k}})|} \frac{|\vec{\mathbf{r}}(\vec{\mathbf{k}})|}{|\vec{\mathbf{k}}|}.$$

We know that $\mathbf{f}$ is differentiable at $\mathbf{x}_0$, so the term

$$\frac{L\vec{\mathbf{r}}(\vec{\mathbf{k}}) - \mathbf{f}(\mathbf{x}_0 + \vec{\mathbf{r}}(\vec{\mathbf{k}})) + \mathbf{f}(\mathbf{x}_0)}{|\vec{\mathbf{r}}(\vec{\mathbf{k}})|} \qquad A4.8$$

has limit 0 as $\vec{\mathbf{r}}(\vec{\mathbf{k}}) \to \mathbf{0}$. So we need to show that $\vec{\mathbf{r}}(\vec{\mathbf{k}}) \to \mathbf{0}$ when $\vec{\mathbf{k}} \to \mathbf{0}$. Using Equations A4.6 for the equality and A4.2 for the inequality, we have

$$\vec{\mathbf{r}}(\vec{\mathbf{k}}) = \mathbf{g}(\mathbf{y}_0 + \vec{\mathbf{k}}) - \mathbf{g}(\mathbf{y}_0) \leq 2|L^{-1}|(\mathbf{y}_0 + \vec{\mathbf{k}} - \mathbf{y}_0), \qquad A4.9$$

$$\text{i.e.,} \qquad \vec{\mathbf{r}}(\vec{\mathbf{k}}) \leq 2|L^{-1}|\vec{\mathbf{k}}. \qquad\qquad A4.10$$

So the limit is 0 as $\vec{\mathbf{k}} \to 0$. In addition, the term $|\vec{\mathbf{r}}(\vec{\mathbf{k}})|/|\vec{\mathbf{k}}|$ is bounded:

$$\frac{|\vec{\mathbf{r}}|}{|\vec{\mathbf{k}}|} \leq 2|L^{-1}|, \qquad\qquad A4.11$$

so Theorem 1.5.21, part (e) says that A4.4 is true.

(3) Proving that g is an inverse, not a just right inverse

We have already proved that $\mathbf{f}$ is onto the neighborhood V of $\mathbf{y}_0$; we want to show that $\mathbf{g}(\mathbf{y})$ is the *only* solution $\mathbf{x}$ of $\mathbf{f}_\mathbf{y}(\mathbf{x}) = 0$ with $\mathbf{x} \in W_0$. This is a stronger result than the original statement of Kantorovitch's theorem, but it is exactly the the same result as the modified statement of Kantorovitch's theorem discussed in Remark A2.5; the ball W_0 of the inverse function theorem is the ball U_{-1} discussed there.

A.5 Proof of the Implicit Function Theorem

Theorem 2.9.10 (The implicit function theorem). *Let W be an open neighborhood of $\mathbf{c} = \begin{pmatrix} \mathbf{a} \\ \mathbf{b} \end{pmatrix} \in \mathbb{R}^{n+m}$, and $\mathbf{F} : W \to \mathbb{R}^n$ be differentiable, with $\mathbf{F}(\mathbf{c}) = \mathbf{0}$. Suppose that the $n \times n$ matrix*

$$[D_1\mathbf{F}(\mathbf{c}), \dots, D_n\mathbf{F}(\mathbf{c})], \qquad\qquad A5.1$$

representing the first n columns of the derivative of $\mathbf{F}$, is invertible.

Then the following matrix, which we denote L, is invertible also:

$$L = \begin{bmatrix} [D_1\mathbf{F}(\mathbf{c}), \dots, D_n\mathbf{F}(\mathbf{c})] & [D_{n+1}\mathbf{F}(\mathbf{c}), \dots, D_m\mathbf{F}(\mathbf{c})] \\ \mathbf{0} & I_m \end{bmatrix}. \qquad A5.2$$

Let $W_0 = B_{2R|L^{-1}|}(\mathbf{c}) \subset \mathbb{R}^{n+m}$ be the ball of radius $2R|L^{-1}|$ centered at $\mathbf{c}$. Suppose that $R > 0$ satisfies the following hypotheses:

(1) It is small enough so that $W_0 \subset W$.

(2) In W_0, the derivative satisfies the Lipschitz condition

$$|[\mathbf{DF}(\mathbf{u})] - [\mathbf{DF}(\mathbf{v})]| \leq \frac{1}{2R|L^{-1}|^2}|\mathbf{u} - \mathbf{v}|. \qquad\qquad A5.3$$

Let $B_R(\mathbf{b}) \subset \mathbb{R}^m$ be the ball of radius R centered at $\mathbf{b}$.

There then exists a unique continuously differentiable mapping

$$\mathbf{g} : B_R(\mathbf{b}) \to B_{2R|L^{-1}|}(\mathbf{a}) \quad \text{such that} \quad \mathbf{F}\begin{pmatrix} \mathbf{g}(\mathbf{y}) \\ \mathbf{y} \end{pmatrix} = \mathbf{0} \quad \text{for all } \mathbf{y} \in B_R(\mathbf{b}),$$
$$A5.4$$

and the derivative of the implicit function $\mathbf{g}$ at $\mathbf{b}$ is

$$[\mathbf{Dg}(\mathbf{b})] = -[D_1\mathbf{F}(\mathbf{c}), \dots, D_n\mathbf{F}(\mathbf{c})]^{-1}[D_{n+1}\mathbf{F}(\mathbf{c}), \dots, D_{n+m}\mathbf{F}(\mathbf{c})]. \qquad A5.5$$

It would be possible, and in some sense more natural, to prove the theorem directly, using the Kantorovitch theorem. But this approach will avoid our having to go through all the work of proving that the implicit function is continuously differentiable.

When we add a tilde to $\mathbf{F}$, creating the function $\widetilde{\mathbf{F}}$ of Equation A5.6, we use $\mathbf{F}\begin{pmatrix} \mathbf{x} \\ \mathbf{y} \end{pmatrix}$ as the first n coordinates of $\widetilde{\mathbf{F}}$ and stick on $\mathbf{y}$ (m coordinates) at the bottom; $\mathbf{y}$ just goes along for the ride. We do this to fix the dimensions: $\widetilde{\mathbf{F}}$: $\mathbb{R}^{n+m} \to \mathbb{R}^{n+m}$ can have an inverse function, while $\mathbf{F}$ can't.

Proof. The inverse function theorem is obviously a special case of the implicit function theorem: the special case where $\mathbf{F}\begin{pmatrix} \mathbf{x} \\ \mathbf{y} \end{pmatrix} = \mathbf{f}(\mathbf{x}) - \mathbf{y}$; i.e., we can separate out the $\mathbf{y}$ from $\mathbf{F}\begin{pmatrix} \mathbf{x} \\ \mathbf{y} \end{pmatrix}$. There is a sneaky way of making the implicit function theorem be a special case of the inverse function theorem. We will create a new function $\widetilde{\mathbf{F}}$ to which we can apply the inverse function theorem. Then we will show how the inverse of $\widetilde{\mathbf{F}}$ will give us our implicit function $\mathbf{g}$.

Consider the function $\widetilde{\mathbf{F}} : W \to \mathbb{R}^n \times \mathbb{R}^m$, defined by

$$\widetilde{\mathbf{F}}\begin{pmatrix} \mathbf{x} \\ \mathbf{y} \end{pmatrix} = \begin{pmatrix} \mathbf{F}\begin{pmatrix} \mathbf{x} \\ \mathbf{y} \end{pmatrix} \\ \mathbf{y} \end{pmatrix}, \qquad A5.6$$

where $\mathbf{x}$ are n variables, which we have put as the first variables, and $\mathbf{y}$ the remaining m variables, which we have put last. Whereas $\mathbf{F}$ goes from the high-dimensional space $W \subset \mathbb{R}^{n+m}$ to the *lower-dimensional* space $\mathbb{R}^n$, and thus had no hope of having an inverse, the domain and range of $\widetilde{\mathbf{F}}$ have the same dimension: $n + m$, as illustrated by Figure A5.1.

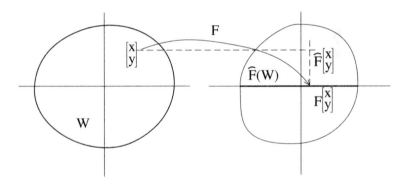

FIGURE A5.1. The mapping $\widetilde{\mathbf{F}}$ is designed to add dimensions to the image of $\mathbf{F}$ so that the image has the same dimension as the domain.

Exercise 2.3.6 addresses the question why L is invertible if $[D_1\mathbf{F}(\mathbf{c}), \ldots, D_n\mathbf{F}(\mathbf{c})]$ is invertible.

The derivative

$$[\mathbf{D}\widetilde{\mathbf{F}}(\mathbf{u})] = \begin{bmatrix} [\mathbf{DF}(\mathbf{u})] \\ 0 \mid I \end{bmatrix}$$

is an $(n + m) \times (n + m)$ matrix; the entry $[\mathbf{DF}(\mathbf{u})]$ is a matrix n tall and $n + m$ wide ; the 0 matrix is m high and n wide; the identity matrix is $m \times m$.

So now we will find an inverse of $\widetilde{\mathbf{F}}$, and we will show that the first coordinates of that inverse are precisely the implicit function $\mathbf{g}$.

The derivative of $\widetilde{\mathbf{F}}$ at $\mathbf{c}$ is

$$[\mathbf{D}\widetilde{\mathbf{F}}(\mathbf{c})] = \begin{bmatrix} [D_1\mathbf{F}(\mathbf{c}), \ldots, D_n\mathbf{F}(\mathbf{c})] & [D_{n+1}\mathbf{F}(\mathbf{c}), \ldots, D_{n+m}\mathbf{F}(\mathbf{c})] \\ \mathbf{0} & I \end{bmatrix} = L, \quad A5.7$$

showing that it is invertible at $\mathbf{c}$ precisely when $[D_1\mathbf{F}(\mathbf{c}), \ldots, D_n\mathbf{F}(\mathbf{c})]$ is invertible, i.e., the hypothesis of the inverse function theorem (Theorem 2.9.4).

Note that the conditions (1) and (2) above look the same as the conditions (1) and (2) of the inverse function theorem applied to $\widetilde{\mathbf{F}}$ (modulo a change of notation). Condition (1) is obviously met: $\widetilde{\mathbf{F}}$ is defined wherever $\mathbf{F}$ is. There is though a slight problem with condition (2): our hypothesis of Equation A5.3 refers to the derivative of $\mathbf{F}$ being Lipschitz; now we need the derivative of $\widetilde{\mathbf{F}}$ to be Lipschitz in order to show that it has an inverse. Since the derivative of $\widetilde{\mathbf{F}}$ is $[\mathbf{D}\widetilde{\mathbf{F}}(\mathbf{u})] = \begin{bmatrix} [\mathbf{DF}(\mathbf{u})] \\ 0 \mid I \end{bmatrix}$, when we compute $|[\mathbf{D}\widetilde{\mathbf{F}}(\mathbf{u})] - [\mathbf{D}\widetilde{\mathbf{F}}(\mathbf{v})]|$, the identity matrices cancel, giving

$$\left|[\mathbf{D}\widetilde{\mathbf{F}}(\mathbf{u})] - [\mathbf{D}\widetilde{\mathbf{F}}(\mathbf{v})]\right| = \left|[\mathbf{DF}(\mathbf{u})] - [\mathbf{DF}(\mathbf{v})]\right|. \qquad A5.8$$

We denote by $B_R \begin{pmatrix} 0 \\ \mathbf{b} \end{pmatrix}$ the ball of radius R centered at $\begin{pmatrix} 0 \\ \mathbf{b} \end{pmatrix}$.

While $\mathbf{G}$ is defined on all of $B_R \begin{pmatrix} 0 \\ \mathbf{b} \end{pmatrix}$, we will only be interested in points $\mathbf{G} \begin{pmatrix} 0 \\ \mathbf{y} \end{pmatrix}$.

Thus $\widetilde{\mathbf{F}}$ is locally invertible; there exists a unique inverse $\widetilde{\mathbf{G}} : B_R \begin{pmatrix} 0 \\ \mathbf{b} \end{pmatrix} \to W_0$. In particular, when $|\mathbf{y} - \mathbf{b}| < R$,

$$\widetilde{\mathbf{F}}\left(\widetilde{\mathbf{G}}\begin{pmatrix} \mathbf{x} \\ \mathbf{y} \end{pmatrix}\right) = \begin{pmatrix} \mathbf{x} \\ \mathbf{y} \end{pmatrix}. \qquad A5.9$$

Now let's denote by $\mathbf{G}$ the first n coordinates of $\widetilde{\mathbf{G}}$:

$$\widetilde{\mathbf{G}}\begin{pmatrix} \mathbf{x} \\ \mathbf{y} \end{pmatrix} \stackrel{\text{def}}{=} \begin{pmatrix} \mathbf{G}\begin{pmatrix} \mathbf{x} \\ \mathbf{y} \end{pmatrix} \\ \mathbf{y} \end{pmatrix}, \quad \text{so} \quad \widetilde{\mathbf{F}}\begin{pmatrix} \mathbf{G}\begin{pmatrix} \mathbf{x} \\ \mathbf{y} \end{pmatrix} \\ \mathbf{y} \end{pmatrix} = \begin{pmatrix} \mathbf{x} \\ \mathbf{y} \end{pmatrix}. \qquad A5.10$$

Equation A5.10: The function $\widetilde{\mathbf{G}}$ has exactly the same relationship to $\mathbf{G}$ as $\widetilde{\mathbf{F}}$ does to $\mathbf{F}$; to go from $\mathbf{G}$ to $\widetilde{\mathbf{G}}$ we stick on $\mathbf{y}$ at the bottom. Since $\widetilde{\mathbf{F}}$ does not change the second coordinate, its inverse cannot change it either.

Since $\widetilde{\mathbf{F}}$ is the inverse of $\widetilde{\mathbf{G}}$,

$$\widetilde{\mathbf{F}}\left(\widetilde{\mathbf{G}}\begin{pmatrix} 0 \\ \mathbf{y} \end{pmatrix}\right) = \begin{pmatrix} 0 \\ \mathbf{y} \end{pmatrix}. \qquad A5.11$$

By the definition of $\widetilde{\mathbf{G}}$ we have

$$\widetilde{\mathbf{F}}\left(\widetilde{\mathbf{G}}\begin{pmatrix} 0 \\ \mathbf{y} \end{pmatrix}\right) = \widetilde{\mathbf{F}}\begin{pmatrix} \mathbf{G}\begin{pmatrix} 0 \\ \mathbf{y} \end{pmatrix} \\ \mathbf{y} \end{pmatrix}, \quad \text{so} \quad \widetilde{\mathbf{F}}\begin{pmatrix} \mathbf{G}\begin{pmatrix} 0 \\ \mathbf{y} \end{pmatrix} \\ \mathbf{y} \end{pmatrix} = \begin{pmatrix} 0 \\ \mathbf{y} \end{pmatrix}. \qquad A5.12$$

Exercise A5.1 asks you to show that the implicit function found this way is unique.

Now set $\mathbf{g}(\mathbf{y}) = \mathbf{G}\begin{pmatrix} 0 \\ \mathbf{y} \end{pmatrix}$. This gives

$$\widetilde{\mathbf{F}}\begin{pmatrix} \mathbf{g}(\mathbf{y}) \\ \mathbf{y} \end{pmatrix} = \begin{pmatrix} \mathbf{F}\begin{pmatrix} \mathbf{g}(\mathbf{y}) \\ \mathbf{y} \end{pmatrix} \\ \mathbf{y} \end{pmatrix} = \begin{pmatrix} 0 \\ \mathbf{y} \end{pmatrix}, \quad \text{i.e.,} \quad \mathbf{F}\begin{pmatrix} \mathbf{g}(\mathbf{y}) \\ \mathbf{y} \end{pmatrix} = 0;$$

$\mathbf{g}$ is the required "implicit function": $\mathbf{F}\begin{pmatrix} \mathbf{x} \\ \mathbf{y} \end{pmatrix}$ implicitly defines $\mathbf{x}$ in terms of $\mathbf{y}$, and $\mathbf{g}$ makes this relationship explicit.

In Equation A5.13 $D\mathbf{g}(\mathbf{b})$ is an $n \times m$ matrix, I is $m \times m$, and 0 is the $n \times m$ zero matrix. In this equation we are using the fact that $\mathbf{g}$ is differentiable; otherwise we could not apply the chain rule.

Now we need to prove Equation A5.5 for the derivative of the implicit function $\mathbf{g}$. This follows from the chain rule. Since $\mathbf{F}\begin{pmatrix} \mathbf{g}(\mathbf{y}) \\ \mathbf{y} \end{pmatrix} = 0$, the derivative of the left side with respect to $\mathbf{y}$ is also 0, which gives (by the chain rule),

$$\left[\mathbf{DF}\begin{pmatrix} \mathbf{g}(\mathbf{b}) \\ \mathbf{b} \end{pmatrix}\right]\begin{bmatrix} D\mathbf{g}(\mathbf{b}) \\ I \end{bmatrix} = 0. \qquad A5.13$$

So

Remember that $\mathbf{c} = \begin{pmatrix} \mathbf{g(b)} \\ \mathbf{b} \end{pmatrix}$.

$$\underbrace{\left[D_1\mathbf{F(c)} \dots D_n\mathbf{F(c)}, \quad D_{n+1}\mathbf{F(c)}, \dots, D_{n+m}\mathbf{F(c)} \right]}_{\left[\mathbf{DF}\begin{pmatrix} \mathbf{g(b)} \\ \mathbf{b} \end{pmatrix} \right]} \begin{bmatrix} \mathbf{Dg(b)} \\ I \end{bmatrix} = \mathbf{0}, \quad A5.14$$

If A denotes the first n columns of $[\mathbf{DF(c)}]$ and B the last m columns, we have

$$A[\mathbf{Dg(b)}] + B = \mathbf{0}, \quad \text{so} \quad A[\mathbf{Dg(b)}] = -B, \quad \text{so} \quad [\mathbf{Dg(b)}] = -A^{-1}B. \quad A5.15$$

Substituting back, this is exactly what we wanted to prove:

$$[\mathbf{Dg(b)}] = \underbrace{-[D_1\mathbf{F(c)}, \dots, D_n\mathbf{F(c)}]^{-1}}_{-A^{-1}} \underbrace{[D_{n+1}\mathbf{F(c)}, \dots, D_{n+m}\mathbf{F(c)}]}_{B}. \quad \square$$

A.6   Proof of Theorem 3.3.9: Equality of Crossed Partials

As we saw in Equation 1.7.5, the partial derivative can be written in terms of the standard basis vectors:

$$D_i f(\mathbf{a}) = \lim_{h \to 0} \frac{f(\mathbf{a} + h\vec{e}_i) - f(\mathbf{a})}{h}.$$

Please observe that the part in parentheses in the last line of Equation A6.1 is completely symmetric with respect to $\vec{e}_i$ and $\vec{e}_j$; after all, $f(\mathbf{a} + h\vec{e}_i + k\vec{e}_j) = f(\mathbf{a} + k\vec{e}_j + h\vec{e}_i)$. So it may seem that the result is simply obvious. The problem is the order of the limits: you have to take them in the order in which they are written. For instance,

$$\lim_{x \to 0} \lim_{y \to 0} \frac{x^2 - y^2}{x^2 + y^2} = 1,$$

but

$$\lim_{y \to 0} \lim_{x \to 0} \frac{x^2 - y^2}{x^2 + y^2} = -1.$$

Theorem 3.3.9. *Let $f : U \to \mathbb{R}$ be a function such that all second partial derivatives exist and are continuous. Then for every pair of variables x_i, x_j, the crossed partials are equal:*

$$D_j(D_i f)(\mathbf{a}) = D_i(D_j f)(\mathbf{a}). \quad 3.3.20$$

Of course the second partials do not exist unless the first partials exist and are continuous, in fact, differentiable.

Proof. First, let us expand the definition of the second partial derivative. In the first line of Equation A6.1 we express the first partial derivative D_i as a limit, treating $D_j f$ as nothing more than the function to which D_i applies. In the second line we rewrite $D_j f$ as a limit:

$$D_i(D_j f)(\mathbf{a}) = \lim_{h \to 0} \frac{1}{h} \left(D_j f(\mathbf{a} + h\vec{e}_i) - D_j f(\mathbf{a}) \right)$$

$$= \lim_{h \to 0} \frac{1}{h} \left(\underbrace{\lim_{k \to 0} \frac{1}{k} \left(f(\mathbf{a} + h\vec{e}_i + k\vec{e}_j) - f(\mathbf{a} + h\vec{e}_i) \right)}_{D_j f(\mathbf{a} + h\vec{e}_i)} - \underbrace{\lim_{k \to 0} \frac{1}{k} \left(f(\mathbf{a} + k\vec{e}_j) - f(\mathbf{a}) \right)}_{D_j f(\mathbf{a})} \right)$$

$$= \lim_{h \to 0} \frac{1}{h} \left(\lim_{k \to 0} \frac{1}{k} \left(f(\mathbf{a} + h\vec{e}_i + k\vec{e}_j) - f(\mathbf{a} + h\vec{e}_i) \right) - \left(f(\mathbf{a} + k\vec{e}_j) - f(\mathbf{a}) \right) \right)$$

$$= \lim_{h \to 0} \lim_{k \to 0} \frac{1}{hk} \left(f(\mathbf{a} + h\vec{e}_i + k\vec{e}_j) - f(\mathbf{a} + h\vec{e}_i) - f(\mathbf{a} + k\vec{e}_j) + f(\mathbf{a}) \right). \quad A6.1$$

We now define the function

$$u(t) = f(\mathbf{a} + t\vec{e}_i + k\vec{e}_j) - f(\mathbf{a} + t\vec{e}_i), \quad \text{so that} \qquad A6.2$$

$$u(h) = f(\mathbf{a} + h\vec{e}_i + k\mathbf{e}_j) - f(\mathbf{a} + h\vec{e}_i) \text{ and } u(0) = f(\mathbf{a} + k\vec{e}_j) - f(\mathbf{a}). \quad A6.3$$

This allows us to rewrite Equation A6.1 as

$$D_i(D_j f)(\mathbf{a}) = \lim_{h \to 0} \lim_{k \to 0} \frac{1}{hk} \big(u(h) - u(0) \big). \qquad A6.4$$

Since u is a differentiable function, the mean value theorem (Theorem 1.6.9) asserts that for every $h > 0$, there exists h_1 between 0 and h satisfying

$$\frac{u(h) - u(0)}{h} = u'(h_1), \quad \text{so that} \quad u(h) - u(0) = hu'(h_1). \qquad A6.5$$

This allows us to rewrite Equation A6.4 as

$$D_i(D_j f)(\mathbf{a}) = \lim_{h \to 0} \lim_{k \to 0} \frac{1}{hk} h\, u'(h_1). \qquad A6.6$$

Since

$$u(h_1) = f(\mathbf{a} + h_1\vec{e}_i + k\vec{e}_j) - f(\mathbf{a} + h_1\vec{e}_i), \qquad A6.7$$

the derivative of $u(h_1)$ is the sum of the derivatives of its two terms:

$$u'(h_1) = D_i f(\mathbf{a} + h_1\vec{e}_i + k\vec{e}_j) - D_i f(\mathbf{a} + h_1\vec{e}_i), \quad \text{so} \qquad A6.8$$

This is a surprisingly difficult result. In Exercise 4.5.11 we give a very simple (but less obvious) proof using Fubini's theorem. Here, with fewer tools, we must work harder: we apply the mean value theorem twice, to carefully chosen functions. Even having said this, the proof isn't obvious.

$$D_i(D_j f)(\mathbf{a}) = \lim_{h \to 0} \lim_{k \to 0} \frac{1}{k} \underbrace{\Big(D_i f(\mathbf{a} + h_1\vec{e}_i + k\vec{e}_j) - D_i f(\mathbf{a} + h_1\vec{e}_i) \Big)}_{u'(h_1)}. \qquad A6.9$$

Now we create a new function so we can apply the mean value theorem again. We replace the part in brackets on the right-hand side of Equation A6.9 by the difference $v(k) - v(0)$, where v is the function defined by

$$v(k) = D_i f(\mathbf{a} + h_1\vec{e}_i + k\vec{e}_j). \qquad A6.10$$

This allows us to rewrite Equation A6.9 as

$$D_i(D_j f)(\mathbf{a}) = \lim_{h \to 0} \lim_{k \to 0} \frac{1}{k} \big(v(k) - v(0) \big). \qquad A6.11$$

Once more we use the mean value theorem.

Again v is differentiable, so there exists k_1 between 0 and k such that

$$v(k) - v(0) = kv'(k_1) = k\Big(D_j\big(D_i(f)\big)(\mathbf{a} + h_1\vec{e}_i + k_1\vec{e}_j)\Big). \qquad A6.12$$

Substituting this in Equation A6.11 gives

$$
\begin{aligned}
D_i(D_j f)(\mathbf{a}) &= \lim_{h \to 0} \lim_{k \to 0} \frac{1}{k} k v'(k_1) \\
&= \lim_{h \to 0} \lim_{k \to 0} \Big(D_j \big(D_i(f) \big)(\mathbf{a} + h_1 \vec{\mathbf{e}}_i + k_1 \vec{\mathbf{e}}_j) \Big).
\end{aligned}
\tag{A6.13}
$$

Now we use the hypothesis that the second partial derivatives are continuous. As h and k tend to 0, so do h_1 and k_1, so

$$
\begin{aligned}
D_i(D_j f)(\mathbf{a}) &= \lim_{h \to 0} \lim_{k \to 0} \big(D_j \big(D_i f(\mathbf{a} + h_1 \vec{\mathbf{e}}_i + k_1 \vec{\mathbf{e}}_j) \big) \big) \\
&= D_j(D_i f)(\mathbf{a}). \quad \square
\end{aligned}
\tag{A6.14}
$$

A.7 Proof of Proposition 3.3.19

Proposition 3.3.19 (Size of a function with many vanishing partial derivatives). *Let U be an open subset of $\mathbb{R}^n$ and $f : U \to \mathbb{R}$ be a C^k function. If at $\mathbf{a} \in U$ all partials up to order k vanish (including the 0th partial derivative, i.e., $f(\mathbf{a})$), then*

$$
\lim_{\vec{\mathbf{h}} \to 0} \frac{f(\mathbf{a} + \vec{\mathbf{h}})}{|\vec{\mathbf{h}}|^k} = 0.
\tag{3.3.39}
$$

The case $k = 1$ is the case where f is a C^1 function, once continuously differentiable.

Proof. The proof is by induction on k, starting with $k = 1$. The case $k = 1$ follows from Theorem 1.9.5: if f vanishes at $\mathbf{a}$, and its first partials are continuous, then f is differentiable at $\mathbf{a}$, and its derivative is given by the Jacobian matrix. So if the first partials vanish at $\mathbf{a}$, the derivative is 0:

$$
0 = \underbrace{\lim_{\vec{\mathbf{h}} \to 0} \frac{f(\mathbf{a} + \vec{\mathbf{h}}) - \overbrace{f(\mathbf{a})}^{0} - \overbrace{[\mathbf{D}f(\mathbf{a})]}^{0} \vec{\mathbf{h}}}{|\vec{\mathbf{h}}|}}_{= 0 \text{ since } f \text{ is differentiable}} = \lim_{\vec{\mathbf{h}} \to 0} \frac{f(\mathbf{a} + \vec{\mathbf{h}})}{|\vec{\mathbf{h}}|}.
\tag{A7.1}
$$

This proves the case $k = 1$.

Now we write $f(\mathbf{a} + \vec{\mathbf{h}})$ in a form that separates out the entries of the increment vector $\vec{\mathbf{h}}$, so that we can apply the mean value theorem.

Write

$$f(\mathbf{a} + \vec{\mathbf{h}}) = f(\mathbf{a} + \vec{\mathbf{h}}) - f(\mathbf{a}) =$$

$$\overbrace{f\begin{pmatrix} a_1 + h_1 \\ a_2 + h_2 \\ a_3 + h_3 \\ \vdots \\ a_{n-1} + h_{n-1} \\ a_n + h_n \end{pmatrix} \underbrace{-f\begin{pmatrix} a_1 \\ a_2 + h_2 \\ a_3 + h_3 \\ \vdots \\ a_{n-1} + h_{n-1} \\ a_n + h_n \end{pmatrix}}_{\text{minus}}}^{\text{changing only } h_1} \underbrace{\overbrace{+f\begin{pmatrix} a_1 \\ a_2 + h_2 \\ a_3 + h_3 \\ \vdots \\ a_{n-1} + h_{n-1} \\ a_n + h_n \end{pmatrix}}^{\text{changing only } h_2}}_{\text{plus}} \underbrace{-f\begin{pmatrix} a_1 \\ a_2 \\ a_3 + h_3 \\ \vdots \\ a_{n-1} + h_{n-1} \\ a_n + h_n \end{pmatrix}}_{\text{minus}}$$

This equation is simpler than it looks. At each step, we allow just one entry of the variable $\vec{\mathbf{h}}$ to vary. We first subtract, then add, identical terms, which cancel.

$$\underbrace{+ f\begin{pmatrix} a_1 \\ a_2 \\ a_3 + h_3 \\ \vdots \\ a_{n-1} + h_{n-1} \\ a_n + h_n \end{pmatrix}}_{\text{plus}}^{\text{changing only } h_3} \cdots \underbrace{\cdots + f\begin{pmatrix} a_1 \\ a_2 \\ a_3 \\ \vdots \\ a_{n-1} \\ a_n + h_n \end{pmatrix}}_{\text{plus}}^{\text{changing only } h_n} - f\begin{pmatrix} a_1 \\ a_2 \\ a_3 \\ \vdots \\ a_{n-1} \\ a_n \end{pmatrix}$$

$$= f\begin{pmatrix} a_1 + h_1 \\ a_2 + h_2 \\ a_3 + h_3 \\ \vdots \\ a_{n-1} + h_{n-1} \\ a_n + h_n \end{pmatrix} - f\begin{pmatrix} a_1 \\ a_2 \\ a_3 \\ \vdots \\ a_{n-1} \\ a_n \end{pmatrix}. \qquad A7.2$$

The mean value theorem: If $f : [a, a+h] \to \mathbb{R}$ is continuous, and f is differentiable on $(a, a+h)$, then there exists $b \in (a, a+h)$ such that

$$f'(b) = \frac{f(a+h) - f(a)}{h};$$

i.e.,

$$f(a+h) - f(a) = hf'(b).$$

By the mean value theorem, the ith term in Equation A7.2 is

$$\underbrace{f\begin{pmatrix} a_1 \\ \vdots \\ a_{i-1} \\ a_i + h_i \\ a_{i+1} + h_{i+1} \\ \vdots \\ a_n + h_n \end{pmatrix} - f\begin{pmatrix} a_1 \\ \vdots \\ a_{i-1} \\ a_i \\ a_{i+1} + h_{i+1} \\ \vdots \\ a_n + h_n \end{pmatrix}}_{f(a+h) - f(a)} = h_i \underbrace{D_i f}_{f'} \underbrace{\begin{pmatrix} a_1 \\ \vdots \\ a_{i-1} \\ \overbrace{b_i}^{\mathbf{b}_i} \\ a_{i+1} + h_{i+1} \\ \vdots \\ a_n + h_n \end{pmatrix}}_{hf'(\mathbf{b}_i)} \qquad A7.3$$

for some $b_i \in (a_i, a_i + h_i)$. Thus the ith term of $f(\mathbf{a} + \vec{\mathbf{h}})$ is $h_i D_i f(\mathbf{b}_i)$. This allows us to rewrite Equation A7.2 as

$$f(\mathbf{a} + \vec{\mathbf{h}}) = f(\mathbf{a} + \vec{\mathbf{h}}) - f(\mathbf{a}) = \sum_{i=1}^{n} h_i D_i f(\mathbf{b}_i). \qquad A7.4$$

Proposition 3.3.19 is a useful tool, but it does not provide an explicit bound on how much the function can change, given a specific change to the variable: a statement that allows us to say, for example, "All partial derivatives of f up to order $k = 3$ vanish, therefore, if we increase the variable by $h = 1/4$, the increment to f will be $\leq 1/64$, or $\leq c\,1/64$, where c is a constant which we can evaluate." Taylor's theorem with remainder (Theorem A9.5) will provide such a statement.

Now we can restate our problem; we want to prove that

$$\lim_{\vec{\mathbf{h}} \to 0} \frac{f(\mathbf{a} + \vec{\mathbf{h}})}{|\vec{\mathbf{h}}|^k} = \lim_{\vec{\mathbf{h}} \to 0} \frac{f(\mathbf{a} + \vec{\mathbf{h}})}{|\vec{\mathbf{h}}||\vec{\mathbf{h}}|^{k-1}} = \sum_{i=1}^{n} \lim_{\vec{\mathbf{h}} \to 0} \frac{h_i}{|\vec{\mathbf{h}}|} \frac{D_i f(\mathbf{b}_i)}{|\vec{\mathbf{h}}|^{k-1}} = 0. \qquad A7.5$$

Since $|h_i|/|\vec{\mathbf{h}}| \leq 1$, this comes down to proving that

$$\lim_{\vec{\mathbf{h}} \to 0} \frac{D_i f(\mathbf{b}_i)}{|\vec{\mathbf{h}}|^{k-1}} = 0. \qquad A7.6$$

Set $\mathbf{b}_i = \mathbf{a} + \vec{\mathbf{c}}_i$; i.e., $\vec{\mathbf{c}}_i$ is the increment to $\mathbf{a}$ that produces $\mathbf{b}_i$. If we substitute this value for $\mathbf{b}_i$ in Equation A7.6, we now need to prove

$$\lim_{\vec{\mathbf{h}} \to 0} \frac{D_i f(\mathbf{a} + \vec{\mathbf{c}}_i)}{|\vec{\mathbf{h}}|^{k-1}} = 0. \qquad A7.7$$

By definition, all partial derivatives of f to order k exist, are continuous on U and vanish at $\mathbf{a}$. By induction we may assume that Proposition 3.3.19 is true for $D_i f$, so that

$$\lim_{\vec{\mathbf{c}}_i \to 0} \frac{D_i f(\mathbf{a} + \vec{\mathbf{c}}_i)}{|\vec{\mathbf{c}}_i|^{k-1}} = 0; \qquad A7.8$$

In Equation A7.8 we are substituting $D_i f$ for f and $\vec{\mathbf{c}}_i$ for $\vec{\mathbf{h}}$ in Equation 3.3.39. You may object that in the denominator we now have $k-1$ instead of k. But Equation 3.3.39 is true when f is a C^k function, and if f is a C^k function, then $D_i f$ is a C^{k-1} function.

Thus we can assert that

$$\lim_{\vec{\mathbf{h}} \to 0} \frac{D_i f(\mathbf{a} + \vec{\mathbf{c}}_i)}{|\vec{\mathbf{h}}|^{k-1}} = 0. \qquad A7.9$$

You may object to switching the $\vec{\mathbf{c}}_i$ to $\vec{\mathbf{h}}$. But we know that $|\vec{\mathbf{c}}_i| \leq |\vec{\mathbf{h}}|$:

$$\vec{\mathbf{c}}_i = \begin{bmatrix} 0 \\ 0 \\ \vdots \\ c_i \\ h_{i+1} \\ \vdots \\ h_n \end{bmatrix}, \quad \text{and } c_i \text{ is between 0 and } h_i. \qquad A7.10$$

So Equation A7.8 is a stronger statement than Equation A7.9. Equation A7.9 tells us that for any ϵ, there exists a δ such that if $|\vec{\mathbf{h}}| < \delta$, then

$$\frac{D_i f(\mathbf{a} + \vec{\mathbf{h}})}{|\vec{\mathbf{h}}|^{k-1}} < \epsilon. \qquad A7.11$$

If $|\vec{\mathbf{h}}| < \delta$, then $|\vec{\mathbf{c}}_i| < \delta$. And putting the bigger number $|\vec{\mathbf{h}}|^{k-1}$ in the denominator just makes that quantity smaller. So we're done:

$$\lim_{\vec{\mathbf{h}} \to 0} \frac{f(\mathbf{a} + \vec{\mathbf{h}})}{|\vec{\mathbf{h}}|^k} = \sum_{i=1}^{n} \lim_{\vec{\mathbf{h}} \to 0} \frac{h_i}{|\vec{\mathbf{h}}|} \frac{D_i f(\mathbf{a} + \vec{\mathbf{c}}_i)}{|\vec{\mathbf{h}}|^{k-1}} = 0. \qquad \square \qquad A7.12$$

A.8 Proof of Rules for Taylor Polynomials

Proposition 3.4.3 (Sums and products of Taylor polynomials). *Let* $U \subset \mathbb{R}^n$ *be open, and* $f, g : U \to \mathbb{R}$ *be* C^k *functions. Then* $f + g$ *and* fg *are also of class* C^k, *and their Taylor polynomials are computed as follows.*

(a) The Taylor polynomial of the sum is the sum of the Taylor polynomials:

$$P^k_{f+g,\mathbf{a}}(\mathbf{a} + \vec{\mathbf{h}}) = P^k_{f,\mathbf{a}}(\mathbf{a} + \vec{\mathbf{h}}) + P^k_{g,\mathbf{a}}(\mathbf{a} + \vec{\mathbf{h}}). \qquad 3.4.8$$

(b) The Taylor polynomial of the product fg *is obtained by taking the product*

$$P^k_{f,\mathbf{a}}(\mathbf{a} + \vec{\mathbf{h}}) \cdot P^k_{g,\mathbf{a}}(\mathbf{a} + \vec{\mathbf{h}}) \qquad 3.4.9$$

and discarding the terms of degree $> k$.

Proposition 3.4.4 (Chain rule for Taylor polynomials). *Let* $U \subset \mathbb{R}^n$ *and* $V \subset \mathbb{R}$ *be open, and* $g : U \to V$, $f : V \to \mathbb{R}$ *be of class* C^k. *Then* $f \circ g : U \to \mathbb{R}$ *is of class* C^k, *and if* $g(\mathbf{a}) = b$, *then the Taylor polynomial* $P^k_{f \circ g, \mathbf{a}}(\mathbf{a} + \vec{\mathbf{h}})$ *is obtained by considering the polynomial*

$$P^k_{f,b}\left(P^k_{g,\mathbf{a}}(\mathbf{a} + \vec{\mathbf{h}})\right)$$

and discarding the terms of degree $> k$.

These results follow from some rules for doing arithmetic with little o and big O. Little o was defined in Definition 3.4.1.

Big O has an implied constant, while little o does not: big O provides more information.

Notation with big O "significantly simplifies calculations because it allows us to be sloppy—but in a satisfactorily controlled way."—Donald Knuth, Stanford University (*Notices of the AMS*, Vol. 45, No. 6, p. 688).

Definition A8.1 (Big O). If $h(x) > 0$ in some neighborhood of 0, then a function f is in $O(h)$ if there exist $\delta > 0$ and a constant C such that $|f(x)| \le Ch(x)$ when $0 < |x| < \delta$; this should be read "f is at most of order $h(x)$."

Below, to lighten the notation, we write $O(|\mathbf{x}|^k) + O(|\mathbf{x}|^l) = O(|\mathbf{x}|^k)$ to mean that if $f \in O(|\mathbf{x}|^k)$ and $g \in O(|\mathbf{x}|^l)$, then $f + g \in O(|\mathbf{x}|^k)$; we use similar notation for products and compositions.

Proposition A8.2 (Addition and multiplication rules for o and O).
Suppose that $0 \leq k \leq l$ are two integers. Then

For example, if $f \in O(|\mathbf{x}|^2)$ and $g \in O(|\mathbf{x}|^3)$, then $f + g$ is in $O(|\mathbf{x}|^2)$ (the least restrictive of the O, since big O is defined in a neighborhood of zero). However, the constants C for the two $O(|\mathbf{x}|^2)$ may differ.

1. $O(|\mathbf{x}|^k) + O(|\mathbf{x}|^l) = O(|\mathbf{x}|^k)$

2. $o(|\mathbf{x}|^k) + o(|\mathbf{x}|^l) = o(|\mathbf{x}|^k)$ *formulas for addition*

3. $o(|\mathbf{x}|^k) + O(|\mathbf{x}|^l) = o(|\mathbf{x}|^k)$ *if $k < l$*

4. $O(|\mathbf{x}|^k)\, O(|\mathbf{x}|^l) = O(|\mathbf{x}|^{k+l})$ *formulas for multiplication*

5. $O(|\mathbf{x}|^k)\, o(|\mathbf{x}|^l) = o(|\mathbf{x}|^{k+l})$

Similarly, if $f \in o(|\mathbf{x}|^2)$ and $g \in o(|\mathbf{x}|^3)$, then $f + g$ is in $o(|\mathbf{x}|^2)$, but for a given ϵ, the δ for $f \in o(|\mathbf{x}|^2)$ may not be the same as the δ for $f + g \in o(|\mathbf{x}|^2)$.

In Equation A8.2, note that the terms to the left and right of the second inequality are identical except that the $C_2|\mathbf{x}|^l$ on the left becomes $C_2|\mathbf{x}|^k$ on the right.

Proof.   The formulas for addition and multiplication are more or less obvious; half the work is figuring out exactly what they mean.

Addition formulas.   For the first of the addition formulas, the hypothesis is that we have functions $f(\mathbf{x})$ and $g(\mathbf{x})$, and that there exist $\delta > 0$ and constants C_1 and C_2 such that when $0 < |\mathbf{x}| < \delta$,

$$|f(\mathbf{x})| \leq C_1|\mathbf{x}|^k \quad \text{and} \quad |g(\mathbf{x})| \leq C_2|\mathbf{x}|^l. \qquad \text{A8.1}$$

If $\delta_1 = \inf\{\delta, 1\}$, $C = C_1 + C_2$, and $|\mathbf{x}| < \delta_1$, then

$$f(\mathbf{x}) + g(\mathbf{x}) \leq C_1|\mathbf{x}|^k + C_2|\mathbf{x}|^l \leq C_1|\mathbf{x}|^k + C_2|\mathbf{x}|^k = C|\mathbf{x}|^k. \qquad \text{A8.2}$$

For the second, the hypothesis is that

All these proofs are essentially identical; they are exercises in fine shades of meaning.

$$\lim_{|\mathbf{x}| \to 0} \frac{f(\mathbf{x})}{|\mathbf{x}|^k} = 0 \quad \text{and} \quad \lim_{|\mathbf{x}| \to 0} \frac{g(\mathbf{x})}{|\mathbf{x}|^l} = 0. \qquad \text{A8.3}$$

Since $l \geq k$, we have $\lim_{|\mathbf{x}| \to 0} \frac{g(\mathbf{x})}{|\mathbf{x}|^k} = 0$ also, so

$$\lim_{|\mathbf{x}| \to 0} \frac{f(\mathbf{x}) + g(\mathbf{x})}{|\mathbf{x}|^k} = 0. \qquad \text{A8.4}$$

The third follows from the second, since $g \in O(|\mathbf{x}|^l)$ implies that $g \in o(|\mathbf{x}|^k)$ when $l > k$. (Can you justify that statement?[2])

Multiplication formulas.   The multiplication formulas are similar. For the first, the hypothesis is again that we have functions $f(\mathbf{x})$ and $g(\mathbf{x})$, and that there exist $\delta > 0$ and constants C_1 and C_2 such that when $|\mathbf{x}| < \delta$,

$$|f(\mathbf{x})| \leq C_1|\mathbf{x}|^k, \quad |g(\mathbf{x})| \leq C_2|\mathbf{x}|^l. \qquad \text{A8.5}$$

Then $f(\mathbf{x})g(\mathbf{x}) \leq C_1C_2|\mathbf{x}|^{k+l}$.

For the second, the hypothesis is the same for f, and for g we know that for every ϵ, there exists η such that if $|\mathbf{x}| < \eta$, then $|g(\mathbf{x})| \leq \epsilon|\mathbf{x}|^l$. When $|\mathbf{x}| < \eta$,

---

[2]Let's set $l = 3$ and $k = 2$. Then in an appropriate neighborhood, $g(\mathbf{x}) \leq C|\mathbf{x}|^3 = C|\mathbf{x}||\mathbf{x}|^2$; by taking $|\mathbf{x}|$ sufficiently small, we can make $C|\mathbf{x}| < \epsilon$.

$$|f(\mathbf{x})g(\mathbf{x})| \leq C_1 \epsilon |\mathbf{x}|^{k+l}, \qquad A8.6$$

For the statements concerning composition, recall that f goes from U, a subset of $\mathbb{R}^n$, to V, a subset of $\mathbb{R}$, while g goes from V to $\mathbb{R}$, so $g \circ f$ goes from a subset of $\mathbb{R}^n$ to $\mathbb{R}$. Since g goes from a subset of $\mathbb{R}$ to $\mathbb{R}$, the variable for the first term is x, not $\mathbf{x}$.

so

$$\lim_{|\mathbf{x}| \to 0} \frac{|f(\mathbf{x})g(\mathbf{x})|}{|\mathbf{x}|^{k+l}} = 0. \quad \square \qquad A8.7$$

Composition rules

To speak of Taylor polynomials of compositions , we need to be sure that the compositions are defined. Let U be a neighborhood of $\mathbf{0}$ in $\mathbb{R}^n$, and V be a neighborhood of 0 in $\mathbb{R}$. We will write Taylor polynomials for compositions $g \circ f$, where $f : U - \{\mathbf{0}\} \to \mathbb{R}$ and $g : V \to \mathbb{R}$:

To prove the first and third statements about composition, the requirement that $l > 0$ is essential. When $l = 0$, saying that $f \in O(|x|^l) = O(1)$ is just saying that f is bounded in a neighborhood of 0; that does not guarantee that its values can be the input for g, or be in the region where we know anything about g.

$$\begin{array}{ccc} U - \{\mathbf{0}\} & \xrightarrow{\;f\;} & \mathbb{R} \\ & & \cup \\ V & \xrightarrow{\;g\;} & \mathbb{R} \end{array} \qquad A8.8$$

We must insist that g be defined at 0, since no reasonable condition will prevent 0 from being a value of f. In particular, when we require $g \in O(x^k)$, we need to specify $k \geq 0$. Moreover, $f(\mathbf{x})$ must be in V when $|\mathbf{x}|$ is sufficiently small; so if $f \in O(|\mathbf{x}|^l)$ we must have $l > 0$, and if $f \in o(|\mathbf{x}|^l)$ we must have $l \geq 0$. This explains the restrictions on the exponents in Proposition A8.3.

In the second statement about composition, saying $f \in o(1)$ precisely says that for all ϵ, there exists δ such that when $\mathbf{x} < \delta$, then $f(\mathbf{x}) \leq \epsilon$; i.e.,

$$\lim_{\mathbf{x} \to 0} f(\mathbf{x}) = 0.$$

So the values of f are in the domain of g for $|\mathbf{x}|$ sufficiently small.

Proposition A8.3 (Composition rules for o and O). *Let $f : U - \{\mathbf{0}\} \to \mathbb{R}$ and $g : V \to \mathbb{R}$ be functions, where U is a neighborhood of $\mathbf{0}$ in $\mathbb{R}^n$, and $V \subset \mathbb{R}$ is a neighborhood of 0. We will assume throughout that $k \geq 0$.*

1. *If $g \in O(|x|^k)$ and $f \in O(|\mathbf{x}|^l)$, then $g \circ f \in O(|\mathbf{x}|^{kl})$, if $l > 0$.*
2. *If $g \in O(|x|^k)$ and $f \in o(|\mathbf{x}|^l)$, then $g \circ f \in o(|\mathbf{x}|^{kl})$, if $l \geq 0$.*
3. *If $g \in o(|x|^k)$ and $f \in O(|\mathbf{x}|^l)$, then $g \circ f \in o(|\mathbf{x}|^{kl})$, if $l > 0$.*

Proof. For the formula 1, the hypothesis is that we have functions $f(\mathbf{x})$ and $g(x)$, and that there exist $\delta_1 > 0$, $\delta_2 > 0$, and constants C_1 and C_2 such that when $|x| < \delta_1$ and $|\mathbf{x}| < \delta_2$,

$$|g(x)| \leq C_1 |x|^k, \quad |f(\mathbf{x})| \leq C_2 |\mathbf{x}|^l. \qquad A8.9$$

We may have $f(\mathbf{x}) = 0$, but we have required that $g(0) = 0$ and that $k \geq 0$, so the composition is defined even at such values of $\mathbf{x}$.

Since $l > 0$, $f(\mathbf{x})$ is small when $|\mathbf{x}|$ is small, so the composition $g(f(\mathbf{x}))$ is defined for $|\mathbf{x}|$ sufficiently small: i.e., we may suppose that $\eta > 0$ is chosen so that $\eta < \delta_2$, and that $|f(\mathbf{x})| < \delta_1$ when $|\mathbf{x}| < \eta$. Then

$$\left| g\big(f(\mathbf{x})\big) \right| \leq C_1 |f(\mathbf{x})|^k \leq C_1 \left(C_2 |\mathbf{x}|^l \right)^k = C_1 C_2^k |\mathbf{x}|^{kl}. \qquad A8.10$$

For formula 2, we know as above that there exist C and $\delta_1 > 0$ such that $|g(x)| < C|x|^k$ when $|x| < \delta_1$. Choose $\epsilon > 0$; for f we know that there exists

$\delta_2 > 0$ such that $|f(\mathbf{x})| \le \epsilon|\mathbf{x}|^l$ when $|\mathbf{x}| < \delta_2$. Taking δ_2 smaller if necessary, we may also suppose $\epsilon|\delta_2|^l < \delta_1$. Then when $|\mathbf{x}| < \delta_2$, we have

$$\left|g\big(f(\mathbf{x})\big)\right| \le C|f(\mathbf{x})|^k \le C\left(\epsilon|\mathbf{x}|^l\right)^k = \underbrace{C\epsilon^k}_{\substack{\text{an arbitrarily} \\ \text{small } \epsilon}}|\mathbf{x}|^{kl}. \qquad A8.11$$

For formula 3, our hypothesis $g \in o(|x|^k)$ asserts that for any $\epsilon > 0$ there exists $\delta_1 > 0$ such that $|g(x)| < \epsilon|x|^k$ when $|x| < \delta_1$.

Now our hypothesis on f says that there exist C and $\delta_2 > 0$ such that $|f(\mathbf{x})| < C|\mathbf{x}|^l$ when $|\mathbf{x}| < \delta_2$; taking δ_2 smaller if necessary, we may further assume that $C|\delta_2|^l < \delta_1$. Then if $|\mathbf{x}| < \delta_2$,

This is where we are using the fact that $l > 0$. If $l = 0$, then making δ_2 small would not make $C|\delta_2|^l$ small.

$$\left|g\big(f(\mathbf{x})\big)\right| \le \epsilon|f(\mathbf{x})|^k \le \epsilon\big|C|\mathbf{x}|^l\big|^k = \epsilon C^k|\mathbf{x}|^{lk}. \quad \square \qquad A8.12$$

Proving Propositions 3.4.3 and 3.4.4

We are ready now to use Propositions A8.2 and A8.3 to prove Propositions 3.4.3 and 3.4.4. There are two parts to each of these propositions: one asserts that sums, products, and compositions of C^k functions are of class C^k; the other tells how to compute their Taylor polynomials.

The first part is proved by induction on k, using the second part. The rules for computing Taylor polynomials say that the $(k-1)$-partial derivatives of a sum, product, or composition are themselves complicated sums of products and compositions of derivatives, of order at most $k-1$, of the given C^k functions. As such, they are themselves continuously differentiable, by Theorems 1.8.1 and 1.8.2. So the sums, products, and compositions are of class C^k.

Computing sums and products of Taylor polynomials. The case of sums follows immediately from the second statement of Proposition A8.2. For products, suppose

$$f(\mathbf{x}) = p_k(\mathbf{x}) + r_k(\mathbf{x}) \quad \text{and} \quad g(\mathbf{x}) = q_k(\mathbf{x}) + s_k(\mathbf{x}), \qquad A8.13$$

with $r_k, s_k \in o(|\mathbf{x}|^k)$. Multiply

$$f(\mathbf{x})g(\mathbf{x}) = \big(p_k(\mathbf{x}) + r_k(\mathbf{x})\big)\big(q_k(\mathbf{x}) + s_k(\mathbf{x})\big) = P_k(\mathbf{x}) + R_k(\mathbf{x}), \qquad A8.14$$

where $P_k(\mathbf{x})$ is obtained by multiplying $p_k(\mathbf{x})q_k(\mathbf{x})$ and keeping the terms of degree between 1 and k. The remainder $R_k(\mathbf{x})$ contains the higher-degree terms of the product $p_k(\mathbf{x})q_k(\mathbf{x})$, which of course are in $o(|\mathbf{x}|^k)$. It also contains the products $r_k(\mathbf{x})s_k(\mathbf{x}), r_k(\mathbf{x})q_k(\mathbf{x})$, and $p_k(\mathbf{x})s_k(\mathbf{x})$, which are of the following forms:

$$O(1)s_k(\mathbf{x}) \in o(|\mathbf{x}|^k);$$

$$r_k(\mathbf{x})O(1) \in o(|\mathbf{x}|^k); \qquad A8.15$$

$$r_k(\mathbf{x})s_k(\mathbf{x}) \in o(|\mathbf{x}|^{2k}).$$

Computing compositions of Taylor polynomials. Finally we come to the compositions. Let us denote

$$f(\mathbf{a} + \vec{\mathbf{h}}) = \underbrace{b}_{\substack{\text{constant} \\ \text{term}}} + \underbrace{Q^k_{f,\mathbf{a}}(\vec{\mathbf{h}})}_{\substack{\text{polynomial terms} \\ 1 \leq \text{ degree } \leq k}} + \underbrace{r^k_{f,\mathbf{a}}(\vec{\mathbf{h}})}_{\text{remainder}}, \qquad A8.16$$

separating out the constant term; the polynomial terms of degree between 1 and k, so that $|Q^k_{f,\mathbf{a}}(\vec{\mathbf{h}})| \in O(|\vec{\mathbf{h}}|)$; and the remainder satisfying $r^k_{f,\mathbf{a}}(\vec{\mathbf{h}}) \in o(|\vec{\mathbf{h}}|^k)$.

Then

$$(g \circ f)(\mathbf{a} + \vec{\mathbf{h}}) = P^k_{g,b}\Big(b + Q^k_{f,\mathbf{a}}(\vec{\mathbf{h}}) + r^k_{f,\mathbf{a}}(\vec{\mathbf{h}})\Big) + r^k_{g,b}\Big(b + Q^k_{f,\mathbf{a}}(\vec{\mathbf{h}}) + r^k_{f,\mathbf{a}}(\vec{\mathbf{h}})\Big). \quad A8.17$$

Among the terms in the sum above, there are the terms of $P^k_{g,b}(b + Q^k_{f,\mathbf{a}}(\vec{\mathbf{h}}))$ of degree at most k in $\vec{\mathbf{h}}$; we must show that all the others are in $o(|\vec{\mathbf{h}}|^k)$.

Most prominent of these is

$$r^k_{g,b}\big(b + Q^k_{f,\mathbf{a}}(\vec{\mathbf{h}}) + r^k_{f,\mathbf{a}}(\vec{\mathbf{h}})\big) \in o\big(\big|O(|\vec{\mathbf{h}}|) + o(|\vec{\mathbf{h}}|^k)\big|^k\big) = o\big(\big|O(|\vec{\mathbf{h}}|)\big|^k\big) = o(|\vec{\mathbf{h}}|^k),$$
$$A8.18$$

using part (3) of Proposition A8.3.

The other terms are of the form

Note that m is an integer, not a multi-index, since g is a function of a single variable.

$$\frac{1}{m!} D_m g(b)\Big(b + Q^k_{f,\mathbf{a}}(\vec{\mathbf{h}}) + r^k_{f,\mathbf{a}}(\vec{\mathbf{h}})\Big)^m. \qquad A8.19$$

If we multiply out the power, we find some terms of degree at most k in the coordinates h_i of $\vec{\mathbf{h}}$, and no factors $r^k_{f,\mathbf{a}}(\vec{\mathbf{h}})$: these are precisely the terms we are keeping in our candidate Taylor polynomial for the composition. Then there are those of degree greater than k in the h_i and still have no factors $r^k_{f,\mathbf{a}}(\vec{\mathbf{h}})$, which are evidently in $o(|\vec{\mathbf{h}}|^k)$, and those which contain at least one factor $r^k_{f,\mathbf{a}}(\vec{\mathbf{h}})$. These last are in $O(1)o(|\vec{\mathbf{h}}|^k) = o(|\vec{\mathbf{h}}|^k)$. $\square$

In Landau's notation, Equation A9.1 says that if f is of class C^{k+1} near $\mathbf{a}$, then not only is

$$f(\mathbf{a} + \vec{\mathbf{h}}) - P^k_{f,\mathbf{a}}(\mathbf{a} + \vec{\mathbf{h}})$$

in $o(|\vec{\mathbf{h}}|^k)$; it is in fact in $O(|\vec{\mathbf{h}}|^{k+1})$; Theorem A9.7 gives a formula for the constant implicit in the O.

A.9 Taylor's Theorem with Remainder

It is all very well to claim (Theorem 3.3.18, part (b)) that

$$\lim_{\vec{\mathbf{h}} \to 0} \frac{f(\mathbf{a} + \vec{\mathbf{h}}) - P^k_{f,\mathbf{a}}(\mathbf{a} + \vec{\mathbf{h}})}{|\vec{\mathbf{h}}|^k} = 0; \qquad A9.1$$

that doesn't tell you how small the difference $f(\mathbf{a} + \vec{\mathbf{h}}) - P^k_{f,\mathbf{a}}(\mathbf{a} + \vec{\mathbf{h}})$ is for any particular $\vec{\mathbf{h}} \neq 0$.

Taylor's theorem with remainder gives such a bound, in the form of a multiple of $|\vec{\mathbf{h}}|^{k+1}$. You cannot get such a result without requiring a bit more about the function f; we will assume that all derivatives up to order $k + 1$ exist and are continuous.

Recall Taylor's theorem with remainder in one dimension:

Theorem A9.1 (Taylor's theorem with remainder in one dimension). *If g is $(k+1)$-times continuously differentiable on $(a - R, a + R)$, then, for $|h| < R$,*

When $k = 0$, Equation A9.2 is the fundamental theorem of calculus:

$$g(a + h) = g(a) + \int_0^h g'(a+t)\, dt\,.$$
$$\underbrace{}_{\text{remainder}}$$

$$\overbrace{g(a + h) =}^{} \quad \overbrace{g(a) + g'(a)h + \cdots + \frac{1}{k!}g^{(k)}(a)h^k}^{\substack{P_{g,a}^k(a+h) \\ \text{(Taylor polynomial of } g \text{ at } a, \text{ of degree } k)}}$$

$$+ \underbrace{\frac{1}{k!}\int_0^h (h - t)^k g^{(k+1)}(a+t)\, dt}_{\text{remainder}}\,. \qquad A9.2$$

Proof. The standard proof is by repeated integration by parts; you are asked to use that approach in Exercise A9.3. Here is an alternative proof (slicker and less natural). First, rewrite Equation A9.2 setting $x = a + h$:

We made the change of variables $s = a + t$, so that as t goes from 0 to h, s goes from a to x.

$$g(x) = g(a) + g'(a)(x - a) + \cdots + \frac{1}{k!}g^{(k)}(a)(x - a)^k$$
$$+ \frac{1}{k!}\int_a^x (x - s)^k g^{(k+1)}(s)\, dt. \qquad A9.3$$

Now think of both sides as functions of a, with x held constant. The two sides are equal when $a = x$: all the terms on the right-hand side vanish except the first, giving $g(x) = g(x)$. If we can show that as a varies and x stays fixed, the right-hand side stays constant, then we will know that the two sides are always equal. So we compute the derivative of the right-hand side:

$$g'(a) + \overbrace{\left(-g'(a) + (x-a)g''(a)\right)}^{=0} + \overbrace{\left(-(x-a)g''(a) + \frac{(x-a)^2 g'''(a)}{2!}\right)}^{=0} + \cdots$$

$$\cdots + \left(-\frac{(x-a)^{k-1}g^{(k)}(a)}{(k-1)!} + \overbrace{\frac{(x-a)^k g^{(k+1)}(a)}{k!}\right) \underbrace{-\frac{(x-a)^k g^{(k+1)}(a)}{k!}}_{\text{derivative of the remainder}}}^{=0}\,,$$
$$A9.4$$

where the last term is the derivative of the integral, computed by the fundamental theorem of calculus. A careful look shows that everything drops out. $\square$

Evaluating the remainder: in one dimension

To use Taylor's theorem with remainder, you must "evaluate" the remainder. It is *not* useful to compute the integral; if you do this, by repeated integrals by parts, you get exactly the other terms in the formula.

Another approach to Corollary A9.3 is to say that there exists c between a and x such that

$$\frac{1}{k!}\int_a^x (x-t)^k g^{(k+1)}(t)\, dt$$

$$= \frac{1}{k!}(x-a)(x-c)^k g^{(k+1)}(c).$$

A calculator that computes to eight places can store Equation A9.5, and spit it out when you evaluate sines; even hand calculation isn't out of the question. This is how the original trigonometric tables were computed.

Computing large factorials is quicker if you know that $6! = 720$.

It isn't often that high derivatives of functions can be so easily bounded; usually using Taylor's theorem with remainder is much messier.

Theorem A9.2. *There exists c between a and $a+h$ such that*

$$f(a+h) = P_{f,a}^k(a+h) + \frac{f^{(k+1)}(c)}{(k+1)!} h^{k+1}.$$

Corollary A9.3. *If $|f^{(k+1)}(a+t)| \le C$ for t between 0 and h, then*

$$|f(a+h) - P_{f,a}^k(a+h)| \le \frac{C}{(k+1)!} h^{k+1}$$

Example A9.4 (Finding a bound for the remainder in one dimension). A standard example of this sort of thing is to compute $\sin\theta$ to eight decimals when $|\theta| \le \pi/6$. Since the successive derivatives of $\sin\theta$ are all sines and cosines, they are all bounded by 1, so the remainder after taking k terms of the Taylor polynomial is at most

$$\frac{1}{(k+1)!}\left(\frac{\pi}{6}\right)^{k+1}. \qquad A9.5$$

for $|\theta| \le \pi/6$. Take $k = 8$ (found by trial and error); $1/9! = 3.2002048 \times 10^{-6}$ and $(\pi/6)^9 \approx 2.76349 \times 10^{-3}$; the error is then at most 8.8438×10^{-9}. Thus we can be sure that

$$\sin\theta = \theta - \frac{\theta^3}{3!} + \frac{\theta^5}{5!} - \frac{\theta^7}{7!} + \frac{\theta^9}{9!} \qquad A9.6$$

to eight decimals when $|\theta| \le \pi/6$. $\triangle$

Taylor's theorem with remainder in higher dimensions

Theorem A9.5 (Taylor's theorem with remainder in higher dimensions). *Let $U \subset \mathbb{R}^n$ be open, let $f : U \to \mathbb{R}$ a function of class C^{k+1}, and suppose that the interval $[\mathbf{a}, \mathbf{a}+\vec{\mathbf{h}}]$ is contained in U. Then there exists $\mathbf{c} \in [\mathbf{a}, \mathbf{a}+\vec{\mathbf{h}}]$ such that*

$$f(\mathbf{a}+\vec{\mathbf{h}}) = P_{f,\mathbf{a}}^k(\mathbf{a}+\vec{\mathbf{h}}) + \sum_{I \in \mathcal{I}_n^{k+1}} \frac{1}{I!} D_I f(\mathbf{c})\, \vec{\mathbf{h}}^I. \qquad A9.7$$

Proof. Define $\varphi(t) = \mathbf{a} + t\vec{\mathbf{h}}$, and consider the scalar-valued function of one variable $g(t) = f(\varphi(t))$. Theorem A9.2 applied to g when $h = 1$ and $a = 0$ says that there exists c with $0 < c < 1$ such that

$$g(1) = \underbrace{g(0) + \cdots + \frac{g^{(k)}(0)}{k!}}_{\text{Taylor polynomial}} + \underbrace{\frac{1}{k!} g^{(k+1)}(c)}_{\text{remainder}}. \qquad A9.8$$

We need to show that the various terms of Equation A9.8 are the same as the corresponding terms of Equation A9.7. That the two left-hand sides are equal is obvious; by definition, $g(1) = f(\mathbf{a} + \vec{\mathbf{h}})$. That the Taylor polynomials and the remainders are the same follows from the chain rule for Taylor polynomials.

To show that the Taylor polynomials are the same, we write

$$
P_{g,0}^k(t) = P_{f,\mathbf{a}}^k\big(P_{\varphi,0}^k(t)\big) = \sum_{m=0}^{k} \sum_{I \in \mathcal{I}_n^m} \frac{1}{I!} D_I f(\mathbf{a})(t\vec{\mathbf{h}})^I
$$

$$
= \sum_{m=0}^{k} \left(\sum_{I \in \mathcal{I}_n^m} \frac{1}{I!} D_I f(\mathbf{a})(\vec{\mathbf{h}})^I \right) t^m.
$$

A9.9

This shows that

$$
g(0) + \cdots + \frac{g^{(k)}(0)}{k!} = P_{f,\mathbf{a}}^k(\mathbf{a} + \vec{\mathbf{h}}).
$$

A9.10

For the remainder, set $\mathbf{c} = \varphi(c)$. Again the chain rule for Taylor polynomials gives

$$
P_{g,c}^{k+1}(t) = P_{f,\mathbf{c}}^{k+1}\big(P_{\varphi,c}^{k+1}(t)\big) = \sum_{m=0}^{k+1} \sum_{I \in \mathcal{I}_n^m} \frac{1}{I!} D_I f(\mathbf{c})(t\vec{\mathbf{h}})^I
$$

$$
= \sum_{m=0}^{k+1} \left(\sum_{I \in \mathcal{I}_n^m} \frac{1}{I!} D_I f(\mathbf{c})(\vec{\mathbf{h}})^I \right) t^m.
$$

A9.11

Looking at the terms of degree $k + 1$ on both sides gives the desired result:

$$
\frac{1}{k!} g^{(k+1)}(c) = \sum_{I \in \mathcal{I}_n^{k+1}} \frac{1}{I!} D_I f(\mathbf{c})(\vec{\mathbf{h}})^I. \quad \square
$$

A9.12

There are many different ways of turning this into a bound on the remainder; they yield somewhat different results. We will use the following lemma.

We call Lemma A9.6 the *polynomial formula* because it generalizes the binomial formula to polynomials. This result is rather nice in its own right, and shows how multi-index notation can simplify complicated formulas.

Lemma A9.6 (Polynomial formula).

$$
\sum_{I \in \mathcal{I}_n^k} \frac{1}{I!} \vec{\mathbf{h}}^I = \frac{1}{k!}(h_1 + \cdots + h_n)^k.
$$

A9.13

Proof. We will prove this by induction on n. When $n = 1$, there is nothing to prove: the lemma simply asserts $h^m = h^m$.

Suppose the formula is true for n, and let us prove it for $n + 1$. Now $\vec{\mathbf{h}} = \begin{bmatrix} h_1 \\ \vdots \\ h_{n+1} \end{bmatrix}$, and we will denote $\vec{\mathbf{h}}' = \begin{bmatrix} h_1 \\ \vdots \\ h_n \end{bmatrix}$. Let us simply compute:

$$\sum_{I \in \mathcal{I}_{n+1}^k} \frac{1}{I!} \mathbf{h}^I = \sum_{m=0}^{k} \sum_{J \in \mathcal{I}_n^m} \frac{1}{J!} (\vec{\mathbf{h}}')^J \frac{1}{(k-m)!} h_{n+1}^{k-m}$$

$$= \sum_{m=0}^{k} \underbrace{\frac{1}{m!} (h_1 + \cdots + h_n)^m}_{\text{by induction on } n} \frac{1}{(k-m)!} h_{n+1}^{k-m}$$

$$= \frac{1}{k!} \sum_{m=0}^{k} \frac{k!}{(k-m)!m!} (h_1 + \cdots + h_n)^m h_{n+1}^{k-m}$$

The last step is the binomial theorem.

$$= \frac{1}{k!} (h_1 + \cdots + h_n + h_{n+1})^m. \quad \square$$

A9.14

This, together with Theorem A9.5, immediately give the following result.

Theorem A9.7 (An explicit formula for the Taylor remainder). *Let $U \subset \mathbb{R}^n$ be open, and let $f : U \to \mathbb{R}$ be a function of class C^{k+1}. Suppose that the interval $[\mathbf{a}, \mathbf{a} + \vec{\mathbf{h}}]$ is contained in U. If*

$$\sup_{I \in \mathcal{I}_n^{k+1}} \sup_{\mathbf{c} \in [\mathbf{a}, \mathbf{a} + \vec{\mathbf{h}}]} |D_I f(\mathbf{c})| \leq C, \qquad A9.15$$

then

$$\left| f(\mathbf{a} + \vec{\mathbf{h}}) - P_{f,\mathbf{a}}^k(\mathbf{a} + \vec{\mathbf{h}}) \right| \leq C \left(\sum_{i=1}^{n} |h_i| \right)^{k+1}. \qquad A9.16$$

A.10 Proof of Theorem 3.5.3 (Procedure for Completing Squares)

Theorem 3.5.3 (Quadratic forms as sums of squares). *(a) For any quadratic form $Q(\vec{\mathbf{x}})$ on $\mathbb{R}^n$, there exist $m = k + l$ linearly independent linear functions $\alpha_1(\vec{\mathbf{x}}), \ldots, \alpha_m(\vec{\mathbf{x}})$ such that*

$$Q(\vec{\mathbf{x}}) = \left(\alpha_1(\vec{\mathbf{x}}) \right)^2 + \cdots + \left(\alpha_k(\vec{\mathbf{x}}) \right)^2 - \left(\alpha_{k+1}(\vec{\mathbf{x}}) \right)^2 - \cdots - \left(\alpha_{k+l}(\vec{\mathbf{x}}) \right)^2. \quad 3.5.3$$

(b) The number k of plus signs and the number l of minus signs in such a decomposition depends only on Q and not on the specific linear functions chosen.

Proof. Part (b) is proved in Section 3.5. To prove part (a) we need to formalize the completion of squares procedure; we will argue by induction on the number of variables appearing in Q.

Let $Q : \mathbb{R}^n \to \mathbb{R}$ be a quadratic form. Clearly, if only one variable x_i appears, then $Q(\vec{x}) = \pm a x_i^2$ with $a > 0$, so $Q(\vec{x}) = \pm(\sqrt{a}\, x_i)^2$, and the theorem is true. So suppose it is true for all quadratic forms in which at most $k - 1$ variables appear, and suppose k variables appear in the expression of Q. Let x_i be such a variable; there are then two possibilities: either (1), a term $\pm a x_i^2$ appears with $a > 0$, or (2), it doesn't.

(1)   If a term $\pm a x_i^2$ appears with $a > 0$, we can then write

$$Q(\vec{x}) = \pm\left(a x_i^2 + \beta(\vec{x})x_i + \frac{\big(\beta(\vec{x})\big)^2}{4a}\right) + Q_1(\vec{x}) = \pm\left(\sqrt{a}\, x_i + \frac{\beta(\vec{x})}{2\sqrt{a}}\right)^2 + Q_1(\vec{x})$$

A10.1

where β is a linear function of the $k - 1$ variables appearing in Q other than x_i, and Q_1 is a quadratic form in the same variables. By induction, we can write

$$Q_1(\vec{x}) = \pm\big(\alpha_1(\vec{x})\big)^2 \pm \cdots \pm \big(\alpha_m(\vec{x})\big)^2$$

A10.2

for some linearly independent linear functions $\alpha_i(\vec{x})$ of the $k - 1$ variables appearing in Q other than x_i.

We must check the linear independence of the linear functions $\alpha_0, \alpha_1, \dots, \alpha_m$, where by definition

$$\alpha_0(\vec{x}) = \sqrt{a}\, x_i + \frac{\beta(\vec{x})}{2\sqrt{a}}.$$

A10.3

Suppose

$$c_0\alpha_0 + \cdots + c_m\alpha_m = 0;$$

A10.4

then

$$(c_0\alpha_0 + \cdots + c_m\alpha_m)(\vec{x}) = 0$$

A10.5

for every $\vec{x}$, in particular, when $\vec{x} = \vec{e}_i$ (i.e., when $x_i = 1$ and all the other variables are 0). This leads to

Recall that β is a function of the variables other than x_i; thus when those variables are 0, so is $\beta(\vec{x})$ (as are $\alpha_1(\vec{e}_i), \dots, \alpha_m(\vec{e}_i)$).

$$c_0\sqrt{a} = 0, \quad \text{so} \quad c_0 = 0,$$

A10.6

so Equation A10.4 and the linear independence of $\alpha_1, \dots, \alpha_m$ imply $c_1 = \cdots = c_m = 0$.

(2)   If no term $\pm a x_i^2$ appears, then there must be a term of the form $\pm a x_i x_j$ with $a > 0$. Make the substitution $x_j = x_i + u$; we can now write

$$\begin{aligned} Q(\vec{x}) &= a x_i^2 + \beta(\vec{x}, u)x_i + \frac{\big(\beta(\vec{x}, u)\big)^2}{4a} + Q_1(\vec{x}, u) \\ &= \pm\left(\sqrt{a}\, x_i + \frac{\beta(\vec{x}, u)}{2\sqrt{a}}\right)^2 + Q_1(\vec{x}, u), \end{aligned}$$

A10.7

where β and Q_1 are functions (linear and quadratic respectively) of u and of the variables that appear in Q other than x_i and x_j. Now argue exactly as above; the only subtle point is that in order to prove $c_0 = 0$ you need to set $u = 0$, i.e., to set $x_i = x_j = 1$. $\square$

A.11   Proof of Propositions 3.8.12 and 3.8.13 (Frenet Formulas)

Proposition 3.8.12 (Frenet frame). *The vectors $\vec{\mathbf{t}}(0), \vec{\mathbf{n}}(0), \vec{\mathbf{b}}(0)$ form the orthonormal basis (Frenet frame) with respect to which our adapted coordinates are computed. Thus the point with coordinates X, Y, Z in the new, adapted coordinates is the point*

$$\mathbf{a} + X\vec{\mathbf{t}}(0) + Y\vec{\mathbf{n}}(0) + Z\vec{\mathbf{b}}(0) \qquad 3.8.57$$

in the old x, y, z coordinates.

Proposition 3.8.13 (Frenet frame related to curvature and torsion). *The Frenet frame satisfies the following equations, where κ is the curvature of the curve at a and τ is its torsion:*

$$\begin{aligned}
\vec{\mathbf{t}}'(0) &= & \kappa\vec{\mathbf{n}}(0) & \\
\vec{\mathbf{n}}'(0) &= -\kappa\vec{\mathbf{t}}(0) & & +\tau\vec{\mathbf{b}}(0) \\
\vec{\mathbf{b}}'(0) &= & -\tau\vec{\mathbf{n}}(0). &
\end{aligned}$$

Proof. We may assume that the curve C is written in its adapted coordinates, i.e., as in Equation 3.8.55, which we repeat here:

When Equation 3.8.55 first appeared we used dots (...) to denote the terms that can be ignored; here we denote these terms by $o(X^3)$.

$$Y = \frac{1}{2}\sqrt{a_2^2 + b_2^2}\,X^2 + \frac{a_2 a_3 + b_2 b_3}{6\sqrt{a_2^2 + b_2^2}}X^3 = \frac{A_2}{2}X^2 + \frac{A_3}{6}X^3 + o(X^3)$$

$$Z = \frac{-b_2 a_3 + a_2 b_3}{6\sqrt{a_2^2 + b_2^2}}X^3 + \cdots = \frac{B_3}{6}X^3 + o(X^3). \qquad A11.1$$

This means that we know (locally) the parametrization as a graph

$$\delta : X \mapsto \begin{pmatrix} X \\ \frac{A_2}{2}X^2 + \frac{A_3}{6}X^3 + o(X^3) \\ \frac{B_3}{6}X^3 + o(X^3) \end{pmatrix}, \qquad A11.2$$

whose derivative at X is

$$\delta'(X) = \begin{bmatrix} 1 \\ A_2 X + \frac{A_3 X^2}{2} + \cdots \\ \frac{B_3 X^2}{2} + \cdots \end{bmatrix}. \qquad A11.3$$

Equation 3.8.22:

$$s(t) = \int_{t_0}^{t} |\vec{\gamma}'(u)|\,du;$$

Parametrizing C by arc length means calculating X as a function of arc length s, or rather calculating the Taylor polynomial of $X(s)$ to degree 3. Equation 3.8.22 tells us how to compute $s(X)$; we will then need to invert this to find $X(s)$.

Lemma A11.1. *(a) The function*

$$s(X) = \int_0^X \underbrace{\sqrt{1 + \left(A_2 t + \frac{A_3}{2} t^2\right)^2 + \left(\frac{B_3}{2} t^2\right)^2 + o(t^2)}}_{\text{length of } \delta'(t)}\, dt \qquad A11.4$$

has the Taylor polynomial

$$s(X) = X + \frac{1}{6} A_2^2 X^3 + o(X^3). \qquad A11.5$$

(b) The inverse function $X(s)$ has the Taylor polynomial

$$X(s) = s - \frac{1}{6} A_2^2 s^3 + o(s^3) \qquad \text{to degree 3.} \qquad A11.6$$

Proof of Lemma A11.1.   (a) Using the binomial formula (Equation 3.4.7), we have

$$\sqrt{1 + \left(A_2 t + \frac{A_3}{2} t^2\right)^2 + \left(\frac{B_3}{2} t^2\right)^2 + o(t^2)} = 1 + \frac{1}{2} A_2^2 t^2 + o(t^2) \qquad A11.7$$

to degree 2, and integrating this gives

$$s(X) = \int_0^X \left(1 + \frac{1}{2} A_2^2 t^2 + o(t^2)\right) dt = X + \frac{1}{6} A_2^2 X^3 + o(X^3) \qquad A11.8$$

to degree 3. This proves part (a).

(b) The inverse function $X(s)$ has a Taylor polynomial; write it as $X(s) = \alpha s + \beta s^2 + \gamma s^3 + o(s^3)$, and use the equation $s(X(s)) = s$ and Equation A11.8 to write

$$
\begin{aligned}
s(X(s)) &= X(s) + \frac{1}{6} A_2^2 (X(s))^3 + o(s^3) \\
&= \left(\alpha s + \beta s^2 + \gamma s^3 + o(s^3)\right) + \frac{1}{6} A_2^2 \left(\alpha s + \beta s^2 + \gamma s^3 + o(s^3)\right)^3 + o(s^3) \\
&= s. \qquad\qquad A11.9
\end{aligned}
$$

Develop the cube and identify the coefficients of like powers to find

$$\alpha = 1 \quad , \quad \beta = 0 \quad , \quad \gamma = -\frac{A_2^2}{6}, \qquad A11.10$$

which is the desired result, proving part (b) of Lemma A11.1.

Proof of Propositions 3.8.12 and 3.8.13, continued.   Inserting the value of $X(s)$ given in Equation A11.6 into Equation A11.2 for the curve, we see that up to degree 3, the parametrization of our curve by arc length is given by

$$X(s) = s - \frac{1}{6}A_2^2 s^3 + o(s^3)$$

$$Y(s) = \frac{1}{2}A_2\left(s - \frac{1}{6}A_2^2 s^3\right)^2 + \frac{1}{6}A_3 s^3 + o(s^3) \;=\; \frac{1}{2}A_2 s^2 + \frac{1}{6}A_3 s^3 + o(s^3)$$

$$Z(s) = \frac{1}{6}B_3 s^3 + o(s^3). \tag{A11.11}$$

Differentiating these functions gives us the velocity vector

$$\vec{\mathbf{t}}(s) = \begin{bmatrix} 1 - \dfrac{A_2^2}{2}s^2 + o(s^2) \\[2mm] A_2 s + \dfrac{1}{2}A_3 s^2 + o(s^2) \\[2mm] \dfrac{B_3}{2}s^2 + o(s^2) \end{bmatrix} \quad \text{to degree 2, hence} \quad \vec{\mathbf{t}}(0) = \begin{bmatrix} 1 \\ 0 \\ 0 \end{bmatrix}. \tag{A11.12}$$

Now we want to compute $\vec{\mathbf{n}}(s)$. We have:

$$\vec{\mathbf{n}}(s) = \frac{\vec{\mathbf{t}}'(s)}{|\vec{\mathbf{t}}'(s)|} = \frac{1}{|\vec{\mathbf{t}}'(s)|}\begin{bmatrix} -A_2^2 s + o(s) \\ A_2 + A_3 s + o(s) \\ B_3 s + o(s) \end{bmatrix}, \tag{A11.13}$$

We need to evaluate $|\vec{\mathbf{t}}'(s)|$:

$$|\vec{\mathbf{t}}'(s)| = \sqrt{A_2^4 s^2 + A_2^2 + A_3^2 s^2 + 2A_2 A_3 s + B_3^2 s^2 + o(s^2)}$$

$$= \sqrt{A_2^2 + 2A_2 A_3 s + o(s)}. \tag{A11.14}$$

Therefore,

$$\frac{1}{|\vec{\mathbf{t}}'(s)|} \approx \left(A_2^2 + 2A_2 A_3 s + o(s)\right)^{-1/2} = \left(A_2^2\left(1 + \frac{2A_3}{A_2}s\right) + o(s)\right)^{-1/2}$$

$$= \frac{1}{A_2}\left(1 + \frac{2A_3 s}{A_2}\right)^{-1/2} + o(s). \tag{A11.15}$$

Again using the binomial theorem,

$$\frac{1}{|\vec{\mathbf{t}}'(s)|} = \frac{1}{A_2}\left(1 - \frac{1}{2}\left(\frac{2A_3}{A_2}s\right) + o(s)\right) = \frac{1}{A_2} - \frac{A_3}{A_2^2}s + o(s). \tag{A11.16}$$

So

$$\vec{\mathbf{n}}(s) = \begin{bmatrix} \left(\dfrac{1}{A_2} - \dfrac{A_3}{A_2^2}s\right)(-A_2^2 s) + o(s)) \\[2mm] \left(\dfrac{1}{A_2} - \dfrac{A_3}{A_2^2}s\right)(A_2 + A_3 s) + o(s)) \\[2mm] \left(\dfrac{1}{A_2} - \dfrac{A_3}{A_2^2}s\right)(B_3 s) + o(s)) \end{bmatrix} = \begin{bmatrix} -A_2 s + o(s) \\[2mm] 1 + o(s) \\[2mm] \dfrac{B_3}{A_2}s + o(s) \end{bmatrix}. \tag{A11.17}$$

Hence

$$\vec{\mathbf{n}}(0) = \begin{bmatrix} 0 \\ 1 \\ 0 \end{bmatrix} \quad \text{to degree 1, and} \quad \vec{\mathbf{b}}(0) = \vec{\mathbf{t}}(0) \times \vec{\mathbf{n}}(0) = \begin{bmatrix} 0 \\ 0 \\ 1 \end{bmatrix}. \qquad A11.18$$

Moreover, by Definitions 3.8.10 and 3.8.11 of κ and τ,

$$\vec{\mathbf{n}}'(0) = \begin{bmatrix} -A_2 \\ 0 \\ \frac{B_3}{A_2} \end{bmatrix} = -\kappa\vec{\mathbf{t}}(0) + \tau\vec{\mathbf{b}}(0), \quad \text{and} \quad \vec{\mathbf{t}}'(0) = \kappa\vec{\mathbf{n}}(0). \qquad A11.19$$

Now all that remains is to prove that $\vec{\mathbf{b}}'(0) = -\tau\vec{\mathbf{n}}(0)$, i.e., $\vec{\mathbf{b}}'(0) = -\dfrac{B_3}{A_2}\begin{bmatrix} 0 \\ 1 \\ 0 \end{bmatrix}$.

Ignoring higher degree terms,

$$\vec{\mathbf{b}}(s) = \vec{\mathbf{t}}(s) \times \vec{\mathbf{n}}(s) \approx \begin{bmatrix} 1 \\ A_2 s \\ 0 \end{bmatrix} \times \begin{bmatrix} -A_2 s \\ 1 \\ \frac{B_3}{A_2}s \end{bmatrix} = \begin{bmatrix} 0 \\ -\frac{B_3}{A_2}s \\ 1 \end{bmatrix}. \qquad A11.20$$

So

$$\vec{\mathbf{b}}'(0) = \begin{bmatrix} 0 \\ -\frac{B_3}{A_2} \\ 0 \end{bmatrix}. \quad \square \qquad A11.21$$

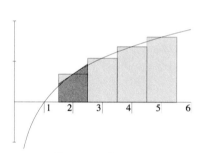

FIGURE A12.1.
The sum $\ln 1 + \ln 2 + \cdots + \ln n$ is a midpoint Riemann sum for the integral

$$\int_{1/2}^{n+1/2} \ln x \, dx.$$

The kth rectangle has the same area as the trapezoid whose top edge is tangent to the graph of $\ln x$ at $\ln n$, as illustrated when $k = 2$.

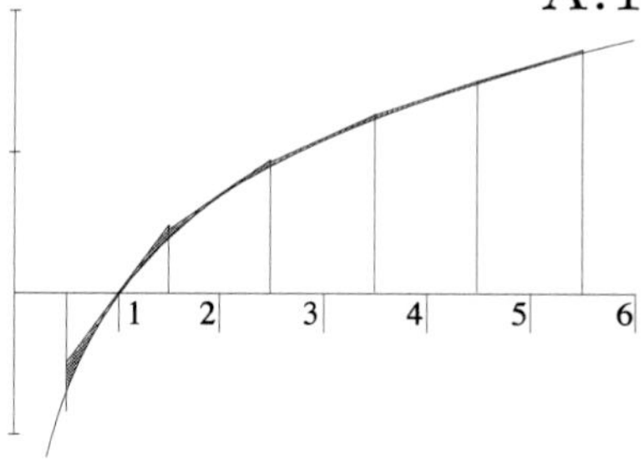

FIGURE A12.2.
The difference between the areas of the trapezoids and the area under the graph of the logarithm is the shaded region. It has finite total area, as shown in Equation A12.2.

A.12 PROOF OF THE CENTRAL LIMIT THEOREM

To explain why the central limit theorem is true, we will need to understand how the factorial $n!$ behaves as n becomes large. How big is 100!? How many digits does it have? *Stirling's formula* gives a very useful approximation.

Proposition A12.1 (Stirling's formula).   *The number $n!$ is approximately*

$$n! \approx \sqrt{2\pi}\left(\frac{n}{e}\right)^n \sqrt{n},$$

in the sense that the ratio of the two sides tends to 1 as n tends to ∞.

For instance,

$$\sqrt{2\pi}\,(100/e)^{100}\sqrt{100} \approx 9.3248 \cdot 10^{157} \quad \text{and} \quad 100! \approx 9.3326 \cdot 10^{157}, \qquad A12.1$$

for a ratio of about 1.0008.

Proof.   Define the number R_n by the formula

The second equality in Equation A12.3 comes from setting $x = n + t$ and writing

$$x = n + t = n\left(1 + \frac{t}{n}\right),$$

so

$$\ln x = \ln\left(n(1 + t/n)\right)$$

$$= \ln n + \ln(1 + t/n)$$

and

$$\int_{-1/2}^{1/2} \ln n\, dt = \ln n.$$

The next is justified by

$$\int_{-1/2}^{1/2} \frac{t}{n}\, dt = 0.$$

The last is Taylor's theorem with remainder:

$$\ln(1 + h) = h + \frac{1}{2}\left(-\frac{1}{(1+c)^2}\right)h^2$$

for some c with $|c| < |h|$; in our case, $h = t/n$ with $t \in [-1/2, 1/2]$ and $c = -1/2$ is the worst value.

Equation A12.5 comes from:

$$\ln(n + \frac{1}{2}) = \ln\left(n(1 + \frac{1}{2n})\right)$$

$$= \ln n + \ln(1 + \frac{1}{2n})$$

$$= \ln n + \frac{1}{2n} + O(\frac{1}{n^2}).$$

The epsilons $\epsilon_1(n)$ and $\epsilon_2(n)$ are unrelated, but both go to 0 as $n \to \infty$, as does $\epsilon(n) = \epsilon_1(n) + \epsilon_2(n)$.

$$\ln n! = \underbrace{\ln 1 + \ln 2 + \cdots + \ln n}_{\text{midpoint Riemann sum}} = \int_{1/2}^{n+1/2} \ln x\, dx + R_n. \qquad A12.2$$

(As illustrated by Figures 12.1 and 12.2, the left-hand side is a midpoint Riemann sum.) This formula is justified by the following computation, which shows that the R_n form a convergent sequence:

$$|R_n - R_{n-1}| = \left| \ln n - \int_{n-1/2}^{n+1/2} \ln x\, dx \right| = \left| \int_{-1/2}^{1/2} \ln\left(1 + \frac{t}{n}\right) dt \right|$$

$$= \left| \int_{-1/2}^{1/2} \left(\ln\left(1 + \frac{t}{n}\right) - \frac{t}{n}\right) dt \right| \qquad A12.3$$

$$\leq \left| -2 \int_{-1/2}^{1/2} \left(\frac{t}{n}\right)^2 dt \right| = \frac{1}{6n^2},$$

so the series formed by the $R_n - R_{n-1}$ is convergent, and the sequence converges to some limit R. Thus we can rewrite Equation A12.2 as follows:

$$\ln n! = \int_{1/2}^{n+1/2} \ln x\, dx + R + \epsilon_1(n) = [x \ln x - x]_{1/2}^{n+1/2} + R + \epsilon_1(n)$$

$$= \left(\left(n + \frac{1}{2}\right)\ln\left(n + \frac{1}{2}\right) - \left(n + \frac{1}{2}\right)\right) - \left(\frac{1}{2}\ln\frac{1}{2} - \frac{1}{2}\right) + R + \epsilon_1(n),$$

$$A12.4$$

where $\epsilon_1(n)$ tends to 0 as n tends to ∞. Now notice that

$$\left(n + \frac{1}{2}\right)\ln\left(n + \frac{1}{2}\right) = \left(n + \frac{1}{2}\right)\ln n + \frac{1}{2} + \epsilon_2(n), \qquad A12.5$$

where $\epsilon_2(n)$ includes all the terms that tend to 0 as $n \to \infty$. Putting all this together, we see that there is a constant

$$c = R - \left(\frac{1}{2}\ln\frac{1}{2} - \frac{1}{2}\right) \qquad A12.6$$

such that

$$\ln n! = n \ln n + \frac{1}{2}\ln n - n + c + \overbrace{\epsilon(n)}^{\epsilon_1(n) + \epsilon_2(n)}, \qquad A12.7$$

where $\epsilon(n) \to 0$ as $n \to \infty$. Exponentiating this gives exactly Stirling's formula, except for the determination of the constant C:

$$n! = Cn^n e^{-n} \sqrt{n} \underbrace{e^{\epsilon(n)}}_{\to 1 \text{ as } n \to \infty}, \qquad A12.8$$

where $C = e^c$. There isn't any obvious reason why it should be possible to evaluate C exactly, but it turns out that $C = \sqrt{2\pi}$; we will derive this at the end of this section, in Equation A12.20, using a result from Section 4.11 and a result developed below. Exercise A12.1 gives another way to derive it.   □

Proving the central limit theorem

We now prove the following version of the central limit theorem.

Theorem A12.2 (Central limit theorem). *If a fair coin is tossed $2n$ times, the probability that the number of heads is between $n + a\sqrt{n}$ and $n + b\sqrt{n}$ tends to*

$$\frac{1}{\sqrt{\pi}} \int_a^b e^{-t^2}\, dt \qquad\qquad A12.9$$

as n tends to ∞.

Proof. The probability of having between $n + a\sqrt{n}$ and $n + b\sqrt{n}$ heads is

$$\frac{1}{2^{2n}} \sum_{k=a\sqrt{n}}^{b\sqrt{n}} \binom{2n}{n+k} = \frac{1}{2^{2n}} \sum_{k=a\sqrt{n}}^{b\sqrt{n}} \frac{(2n)!}{(n+k)!(n-k)!}. \qquad A12.10$$

The idea is to use Stirling's formula to rewrite the sum on the right, cancel everything we can, and see that what is left is a Riemann sum for the integral in Equation A12.9 (more precisely, $1/\sqrt{\pi}$ times that Riemann sum).

In the margin: *In Equation A12.11 we use the version of Stirling's formula given in Equation A12.8, using C rather than $\sqrt{2\pi}$, as we have not yet proved that $C = \sqrt{2\pi}$.*

Let us begin by writing $k = t\sqrt{n}$, so that the sum is over those values of t between a and b such that $t\sqrt{n}$ is an integer; we will denote this set by $T_{[a,b]}$. These points are regularly spaced, $1/\sqrt{n}$ apart, between a and b, and hence are good candidates for the points at which to evaluate a function when forming a Riemann sum. With this notation, our sum becomes

$$\frac{1}{2^{2n}} \sum_{t \in T_{[a,b]}} \frac{(2n)!}{(n+t\sqrt{n})!(n-t\sqrt{n})!} \qquad\qquad A12.11$$

$$\approx \frac{1}{2^{2n}} \sum_{t \in T_{[a,b]}} \frac{C(2n)^{2n} e^{-2n} \sqrt{2n}}{\left(C(n+t\sqrt{n})^{(n+t\sqrt{n})} e^{-(n+t\sqrt{n})} \sqrt{n+t\sqrt{n}} \right) \left(C(n-t\sqrt{n})^{(n-t\sqrt{n})} e^{-(n-t\sqrt{n})} \sqrt{n-t\sqrt{n}} \right)}.$$

Now for some of the cancellations: $(2n)^{2n} = 2^{2n} n^{2n}$, and the powers of 2 cancel with the fraction in front of the sum. Also, all the exponential terms cancel, since $e^{-(n+t\sqrt{n})} e^{-(n-t\sqrt{n})} = e^{-2n}$. Also, one power of C cancels. This leaves

$$\cdots = \frac{1}{C} \sum_{t \in T_{[a,b]}} \frac{n^{2n} \sqrt{2n}}{\sqrt{n^2 - t^2 n}\,(n+t\sqrt{n})^{(n+t\sqrt{n})} (n-t\sqrt{n})^{(n-t\sqrt{n})}}. \qquad A12.12$$

Next, write $(n + t\sqrt{n})^{(n+t\sqrt{n})} = n^{(n+t\sqrt{n})} (1 + t/\sqrt{n})^{(n+t\sqrt{n})}$, and similarly for the term in $n - t\sqrt{n}$, note that the powers of n cancel with the n^{2n} in the numerator, to find

$$\cdots = \frac{1}{C} \sum_{t \in T_{[a,b]}} \underbrace{\sqrt{\frac{2n}{n^2 - t^2 n}}}_{\text{base}} \underbrace{\frac{1}{(1 + t/\sqrt{n})^{(n+t\sqrt{n})}(1 - t/\sqrt{n})^{(n-t\sqrt{n})}}}_{\text{height of rectangles for Riemann sum}} . \qquad A12.13$$

We denote by Δt the spacing of the points t, i.e., $1/\sqrt{n}$.

The term under the square root converges to $\sqrt{2/n} = \sqrt{2}\Delta t$, so it is the length of the base of the rectangles we need for our Riemann sum. For the other, remember that

$$\lim_{x \to \infty} \left(1 + \frac{a}{x}\right)^x = e^a. \qquad A12.14$$

We use Equation A12.14 repeatedly in the following calculation:

In the third line of Equation A12.15, the denominator of the first term tends to e^{-t^2} as $n \to \infty$, by Equation A12.14. By the same equation, the numerator of the second term tends to e^{-t^2}, and the denominator of the second term tends to e^{t^2}.

$$\frac{1}{(1 + t/\sqrt{n})^{(n+t\sqrt{n})}(1 - t/\sqrt{n})^{(n-t\sqrt{n})}}$$

$$= \frac{1}{(1 + t/\sqrt{n})^n (1 + t/\sqrt{n})^{t\sqrt{n}}(1 - t/\sqrt{n})^n (1 - t/\sqrt{n})^{-t\sqrt{n}}}$$

$$= \frac{1}{(1 - t^2/n)^n} \frac{(1 - \frac{t^2}{t\sqrt{n}})^{t\sqrt{n}}}{(1 + \frac{t^2}{t\sqrt{n}})^{t\sqrt{n}}} \qquad A12.15$$

$$\to \frac{1}{e^{-t^2}} \frac{e^{-t^2}}{e^{t^2}} = e^{-t^2}.$$

Putting this together, we see that

$$\frac{1}{C} \sum_{t \in T_{[a,b]}} \sqrt{\frac{2n}{n^2 - t^2 n}} \frac{1}{(1 + t/\sqrt{n})^{(n+t\sqrt{n})}(1 - t/\sqrt{n})^{(n-t\sqrt{n})}} \qquad A12.16$$

converges to

$$\frac{\sqrt{2}}{C} \frac{1}{\sqrt{n}} \sum_{t \in T_{[a,b]}} e^{-t^2}, \qquad A12.17$$

which is the desired Riemann sum. Thus as $n \to \infty$,

$$\frac{1}{2^{2n}} \sum_{k=n+a\sqrt{n}}^{n+b\sqrt{n}} \binom{2n}{k} \to \frac{\sqrt{2}}{C} \int_a^b e^{-t^2} \, dt. \qquad A12.18$$

We finally need to invoke a fact justified in Section 4.11 (Equation 4.11.51):

$$\int_{-\infty}^{\infty} e^{-t^2} \, dt = \sqrt{\pi}. \qquad A12.19$$

Now since when $a = -\infty$ and $b = +\infty$ we must have

$$\frac{\sqrt{2}}{C} \sqrt{\pi} = \frac{\sqrt{2}}{C} \int_{-\infty}^{\infty} e^{-t^2} \, dt = 1, \qquad A12.20$$

we see that $C = \sqrt{2\pi}$, and finally

$$\frac{1}{2^{2n}} \sum_{k=a\sqrt{n}}^{b\sqrt{n}} \binom{2n}{n+k} \quad \text{converges to} \quad \frac{1}{\sqrt{\pi}} \int_a^b e^{-t^2}\, dt. \quad \square \qquad A12.21$$

A.13 Proof of Fubini's Theorem

Theorem 4.5.8 (Fubini's theorem). *Let f be an integrable function on $\mathbb{R}^n \times \mathbb{R}^m$, and suppose that for each $\mathbf{x} \in \mathbb{R}^n$, the function $\mathbf{y} \mapsto f(\mathbf{x}, \mathbf{y})$ is integrable. Then the function*

$$\mathbf{x} \mapsto \int_{\mathbb{R}^m} f(\mathbf{x}, \mathbf{y})|d^m\mathbf{y}|$$

is integrable, and

$$\int_{\mathbb{R}^{n+m}} f(\mathbf{x}, \mathbf{y})|d^n\mathbf{x}||d^m\mathbf{y}| = \int_{\mathbb{R}^n} \left(\int_{\mathbb{R}^m} f(\mathbf{x}, \mathbf{y})|d^m\mathbf{y}| \right) |d^n\mathbf{x}|.$$

In fact, we will prove a stronger theorem: it turns out that the assumption "that for each $\mathbf{x} \in \mathbb{R}^n$, the function $\mathbf{y} \mapsto f(\mathbf{x}, \mathbf{y})$ is integrable" is not really necessary. But we need to be careful; it is not quite true that just because f is integrable, the function $\mathbf{y} \mapsto f(\mathbf{x}, \mathbf{y})$ is integrable, and we can't simply remove that hypothesis. The following example illustrates the difficulty.

By "rough statement" we mean Equation 4.5.1:

$$\int_{\mathbb{R}^n} f\,|d^n\mathbf{x}| =$$

$$\int_{-\infty}^{\infty}\left(\ldots \left(\int_{-\infty}^{\infty} f \begin{bmatrix} x_1 \\ \vdots \\ x_n \end{bmatrix} dx_1 \right) \ldots \right) dx_n.$$

Example A13.1 (A case where the rough statement of Fubini's theorem does not work). Consider the function $f\begin{pmatrix} x \\ y \end{pmatrix}$ that equals 0 outside the unit square, and 1 both inside the square and on its boundary, *except* for the boundary where $x = 1$. On that boundary, $f = 1$ when y is rational, and $f = 0$ when y is irrational:

$$f\begin{pmatrix} x \\ y \end{pmatrix} = \begin{cases} 1 & \text{if } 0 \le x < 1 \text{ and } 0 \le y \le 1 \\ 1 & \text{if } x = 1 \text{ and } y \text{ is rational} \\ 0 & \text{otherwise.} \end{cases} \qquad A13.1$$

Following the procedure we used in Section 4.5, we write the double integral

$$\iint_{\mathbb{R}^2} f\begin{pmatrix} x \\ y \end{pmatrix} dx\, dy = \int_0^1 \left(\int_0^1 f\begin{pmatrix} x \\ y \end{pmatrix} dy \right) dx. \qquad A13.2$$

However, the inner integral $F(x) = \int_0^1 f\begin{pmatrix} x \\ y \end{pmatrix} dy$ does not make sense when $x = 1$. Our function f is integrable on $\mathbb{R}^2$, but $f\begin{pmatrix} 1 \\ y \end{pmatrix}$ is not an integrable function of y. $\triangle$

In fact, the function F could be undefined on a much more complicated set than a single point, but this set will necessarily have volume 0, so it doesn't affect the integral $\int_{\mathbb{R}} F(x)\, dx$.

For example, if we have an integrable function $f\begin{pmatrix} x_1 \\ x_2 \\ y \end{pmatrix}$, we can think of it as a function on $\mathbb{R}^2 \times \mathbb{R}$, where we consider x_1 and x_2 as the horizontal variables and y as the vertical variable.

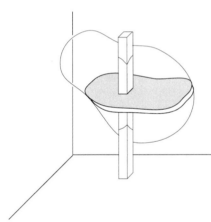

FIGURE A13.1.

Here we imagine that the x and y variables are horizontal and the z variable is vertical. Fixing a value of the horizontal variable picks out a French fry, and choosing a value of the vertical variable chooses a flat potato chip.

Fortunately, the fact that $F(1)$ is not defined is not a serious problem: since a point has one-dimensional volume 0, you could define $F(1)$ to be anything you want, without affecting the integral $\int_0^1 F(x)\, dx$. This always happens: if $f : \mathbb{R}^{n+m} \to \mathbb{R}$ is integrable, then $\mathbf{y} \mapsto f(\mathbf{x}, \mathbf{y})$ is always integrable except for a set of $\mathbf{x}$ of volume 0, which doesn't matter. We deal with this problem by using upper integrals and lower integrals for the inner integral.

Suppose we have a function $f : \mathbb{R}^{n+m} \to \mathbb{R}$, and that $\mathbf{x} \in \mathbb{R}^n$ denotes the first n variables of the domain and $\mathbf{y} \in \mathbb{R}^m$ denotes the last m variables. We will think of the $\mathbf{x}$ variables as "horizontal" and the $\mathbf{y}$ variables as "vertical." We denote by $f_{\mathbf{x}}$ the restriction of f to the vertical subset where the horizontal coordinate is fixed to be $\mathbf{x}$, and by $f^{\mathbf{y}}$ the restriction of the function to the horizontal subset where the vertical coordinate is fixed at $\mathbf{y}$.

With $f_{\mathbf{x}}(\mathbf{y})$ we hold the "horizontal" variables constant and look at the values of the vertical variables. You may imagine a bin filled with infinitely thin vertical sticks. At each point $\mathbf{x}$ there is a stick representing all the values of $\mathbf{y}$.

With $f^{\mathbf{y}}$ we hold the "vertical" variables constant, and look at the values of the horizontal variables. Here we imagine the bin filled with infinitely thin sheets of paper; for each value of $\mathbf{y}$ there is a single sheet, representing the values of $\mathbf{x}$. Either way, the entire bin is filled:

$$ f_{\mathbf{x}}(\mathbf{y}) = f^{\mathbf{y}}(\mathbf{x}) = f(\mathbf{x}, \mathbf{y}). \qquad \text{A13.3} $$

Alternatively, as shown in Figure A13.1, we can imagine slicing a potato vertically into French fries, or horizontally into potato chips.

As we saw in Example A13.1, it is unfortunately *not* true that if f is integrable, then $f_{\mathbf{x}}$ and $f^{\mathbf{y}}$ are also integrable for every $\mathbf{x}$ and $\mathbf{y}$. But the following ~~is~~ true:

Theorem A13.2 (Fubini's theorem). *Let f be an integrable function on $\mathbb{R}^n \times \mathbb{R}^m$. Then the four functions*

$$ U(f_{\mathbf{x}}), \qquad L(f_{\mathbf{x}}), \qquad U(f^{\mathbf{y}}), \qquad L(f^{\mathbf{y}}) \qquad \text{A13.4} $$

are all integrable, and

$$ \overbrace{\int_{\mathbb{R}^n} U(f_{\mathbf{x}})\, |d^n\mathbf{x}|}^{\text{adding upper sums for all columns}} = \overbrace{\int_{\mathbb{R}^n} L(f_{\mathbf{x}})\, |d^n\mathbf{x}|}^{\text{adding lower sums for all columns}} $$

$$ = \overbrace{\int_{\mathbb{R}^m} U(f^{\mathbf{y}})\, |d^n\mathbf{y}|}^{\text{adding upper sums for all rows}} = \overbrace{\int_{\mathbb{R}^m} L(f^{\mathbf{y}})\, |d^n\mathbf{y}|}^{\text{adding lower sums for all rows}} \qquad \text{A13.5} $$

$$ = \overbrace{\int_{\mathbb{R}^n \times \mathbb{R}^m} f\, |d^n\mathbf{x}|\, |d^n\mathbf{y}|}^{\text{integral of } f}. $$

Corollary A13.3. *The set of $\mathbf{x}$ such that $U(f_{\mathbf{x}}) \neq L(f_{\mathbf{x}})$ has volume 0. The set of $\mathbf{y}$ such that $U(f^{\mathbf{y}}) \neq L(f^{\mathbf{y}})$ has volume 0.*

In particular, the set of $\mathbf{x}$ such that $f_{\mathbf{x}}$ is not integrable has n-dimensional volume 0, and similarly, the set of $\mathbf{y}$ where $f^{\mathbf{y}}$ is not integrable has m-dimensional volume 0.

Proof of Corollary A13.3. If these volumes were not 0, the first and third equalities of Equation A13.5 would not be true. $\square$

Proof of Theorem A13.2. The underlying idea is straightforward. Consider a double integral over some bounded domain in $\mathbb{R}^2$. For every N, we have to sum over all the squares of some dyadic decomposition of the plane. These squares can be taken in any order, since only finitely many contribute a nonzero term (because the domain is bounded). Adding together the entries of each column and then adding the totals is like integrating $f_{\mathbf{x}}$; adding together the entries of each row and then adding the totals together is like integrating $f^{\mathbf{y}}$, as illustrated in Figure A13.2.

Of course, the same idea holds in $\mathbb{R}^3$: integrating over all the French fries and adding them up gives the same result as integrating over all the potato chips and adding them.

Equation A13.7: The first line is just the definition of an upper sum.

To go from the first to the second line, note that the decomposition of $\mathbb{R}^n \times \mathbb{R}^m$ into $C_1 \times C_2$ with $C_1 \in \mathcal{D}_N(\mathbb{R}^n)$ and $C_2 \in \mathcal{D}_{N'}(\mathbb{R}^m)$ is finer than $\mathcal{D}_N(\mathbb{R}^{n+m})$.

The third line: for each $C_1 \in \mathcal{D}_N(\mathbb{R}^n)$ we choose a point $\mathbf{x} \in C_1$, and for each $C_2 \in \mathcal{D}_{N'}(\mathbb{R}^m)$ we find the $\mathbf{y} \in C_2$ such that $f(\mathbf{x}, \mathbf{y})$ is maximal, and add these maxima. These maxima are restricted to all have the same $\mathbf{x}$-coordinate, so they are at most $M_{C_1 \times C_2} f$, and even if we now maximize over all $\mathbf{x} \in C_1$, we will still find less than if we had added the maxima independently; equality will occur only if all the maxima are above each other (i.e., all have the same $\mathbf{x}$-coordinate).

$$
\begin{array}{ccc}
1 & 5 \\
2 & 6 \\
3 & 7 \\
\underline{+4} & \underline{+8} \\
10 & + \quad 26 & = 36
\end{array}
\quad \text{gives the same result as} \quad
\begin{array}{rcr}
1+5= & & 6 \\
2+6= & & 8 \\
3+7= & & 10 \\
4+8= & & \underline{12} \\
& & 36
\end{array}
$$

FIGURE A13.2. To the left, we sum entries of each column and add the totals; this is like integrating $f_{\mathbf{x}}$. To the right, we sum entries of each row and add the totals; this is like integrating $f^{\mathbf{y}}$.

Putting this in practice requires a little attention to limits. The inequality that makes things work is that for any $N' \geq N$, we have (Lemma 4.1.7)

$$U_N(f) \geq U_N\big(U_{N'}(f_{\mathbf{x}})\big). \qquad \text{A13.6}$$

Indeed,

$$
U_N(f) = \sum_{C \in \mathcal{D}_N(\mathbb{R}^n \times \mathbb{R}^m)} M_C(f)\, \text{vol}_{n+m}\, C
$$

$$
\geq \sum_{C_1 \in \mathcal{D}_N(\mathbb{R}^n)} \sum_{C_2 \in \mathcal{D}_{N'}(\mathbb{R}^m)} M_{C_1 \times C_2}(f)\, \text{vol}_n\, C_1\, \text{vol}_m\, C_2 \qquad \text{A13.7}
$$

$$
\geq \sum_{C_1 \in \mathcal{D}_N(\mathbb{R}^n)} M_{C_1} \left(\sum_{C_2 \in \mathcal{D}_{N'}(\mathbb{R}^m)} M_{C_2}(f_{\mathbf{x}})\, \text{vol}_m\, C_2 \right) \text{vol}_n\, C_1.
$$

An analogous argument about lower sums gives

$$U_N(f) \geq U_N\big(U_{N'}(f_{\mathbf{x}})\big) \geq L_N\big(L_{N'}(f_{\mathbf{x}})\big) \geq L_N(f). \qquad A13.8$$

In Equations A13.8 and A13.9, expressions like $U_N(U_{N'}(f_{\mathbf{x}}))$ and $L_N(L(f_{\mathbf{x}}))$ may seem strange, but note that $U_{N'}(f_{\mathbf{x}})$ and $L(f_{\mathbf{x}})$ are just functions of $\mathbf{x}$, bounded with bounded support, so we can take Nth upper or lower sums of them.

Since f is integrable, we can make $U_N(f)$ and $L_N(f)$ arbitrarily close, by choosing N sufficiently large; we can squeeze the two ends of Equation A13.8 together, squeezing everything inside in the process. This is what we are going to do.

The limits as $N' \to \infty$ of $U_{N'}(f_{\mathbf{x}})$ and $L_{N'}(f_{\mathbf{x}})$ are the upper and lower integrals $U(f_{\mathbf{x}})$ and $L(f_{\mathbf{x}})$ (by Definition 4.1.9), so we can rewrite Equation A13.8:

$$U_N(f) \geq U_N\big(U(f_{\mathbf{x}})\big) \geq L_N\big(L(f_{\mathbf{x}})\big) \geq L_N(f). \qquad A13.9$$

Given a function f, $U(f) \geq L(f)$; in addition, if $f \geq g$, then $U_N(f) \geq U_N(g)$. So we see that $U_N\big(L(f_{\mathbf{x}})\big)$ and $L_N\big(U(f_{\mathbf{x}})\big)$ are between the inner values of Equation A13.9:

We don't know which is bigger, $U_N(L(f_{\mathbf{x}}))$ or $L_N(U(f_{\mathbf{x}}))$, but that doesn't matter. We know they are between the first and last terms of Equation A13.9, which themselves have a common limit as $N \to \infty$.

$$U_N\big(U(f_{\mathbf{x}})\big) \geq U_N\big(L(f_{\mathbf{x}})\big) \geq L_N\big(L(f_{\mathbf{x}})\big)$$
$$U_N\big(U(f_{\mathbf{x}})\big) \geq L_N\big(U(f_{\mathbf{x}})\big) \geq L_N\big(L(f_{\mathbf{x}})\big). \qquad A13.10$$

So $U_N\big(L(f_{\mathbf{x}})\big)$ and $L_N\big(L(f_{\mathbf{x}})\big)$ have a common limit, as do $U_N\big(U(f_{\mathbf{x}})\big)$ and $L_N\big(U(f_{\mathbf{x}})\big)$, showing that both $L(f_{\mathbf{x}})$ and $U(f_{\mathbf{x}})$ are integrable, and their integrals are equal, since they are both equal to

$$\int_{\mathbb{R}^n \times \mathbb{R}^m} f. \qquad A13.11$$

The argument about the functions $f^{\mathbf{y}}$ is similar. $\square$

A.14 JUSTIFYING THE USE OF OTHER PAVINGS

Here we prove Theorem 4.7.5, which says that we are not restricted to dyadic pavings when computing integrals.

Theorem 4.7.5 (Integrals using arbitrary pavings). *Let $X \subset \mathbb{R}^n$ be a bounded subset, and $\mathcal{P}_N$ be a nested partition of X. If the boundary ∂X satisfies $\mathrm{vol}_n(\partial X) = 0$, and $f : \mathbb{R}^n \to \mathbb{R}$ is integrable, then the limits*

$$\lim_{N \to \infty} U_{\mathcal{P}_N}(f) \quad and \quad \lim_{N \to \infty} L_{\mathcal{P}_N}(f) \qquad 4.7.4$$

both exist, and are equal to

$$\int_X f(\mathbf{x}) \, |d^n\mathbf{x}|. \qquad 4.7.5$$

Proof. Since the boundary of X has volume 0, the characteristic function χ_X is integrable (why?[3]), and we may replace f by $\chi_X f$, and suppose that the support of f is in X. We need to prove that for any ϵ, we can find M such that

$$U_{\mathcal{P}_M}(f) - L_{\mathcal{P}_M}(f) < \epsilon. \qquad A14.1$$

Since we know that the analogous statement for dyadic pavings is true, the idea of the proof is to use "other pavings" small enough so that each paving piece P will either be entirely inside a dyadic cube, or (if it touches or intersects a boundary between dyadic cubes) will contribute a negligible amount to the upper and lower sums.

First, using the fact that f is integrable, find N such that the difference between upper and lower sums of dyadic decompositions is less than $\epsilon/2$:

$$U_N(f) - L_N(f) < \frac{\epsilon}{2}. \qquad A14.2$$

Next, find $N' > N$ such that if B is the union of the cubes $C \in \mathcal{D}_{N'}$ whose closures intersect $\partial\mathcal{D}_N$, then the contribution of $\mathrm{vol}_n B$ to the integral of f is negligible. We do this by finding N' such that

Why the 8 in the denominator? Because it will give us the result we want; the ends justify the means.

$$\mathrm{vol}_n B \le \frac{\epsilon}{8 \sup |f|}. \qquad A14.3$$

Now, find N'' such that every $P \in \mathcal{P}_{N''}$ either is entirely contained in B, or is entirely contained in some $C \in \mathcal{D}_N$, or both.

We claim that this N'' works, in the sense that

$$U_{\mathcal{P}_{N''}}(f) - L_{\mathcal{P}_{N''}}(f) < \epsilon, \qquad A14.4$$

This is where we use the fact that the diameters of the tiles go to 0.

but it takes a bit of doing to prove it.

Every $\mathbf{x}$ is contained in some dyadic cube C. Let $C_N(\mathbf{x})$ be the cube at level N that contains $\mathbf{x}$. Now define the function $\overline{f}$ that assigns to each $\mathbf{x}$ the maximum of f over its cube:

$$\overline{f}(\mathbf{x}) = M_{C_N(\mathbf{x})}(f). \qquad A14.5$$

Similarly, every $\mathbf{x}$ is in some paving tile P. Let $P_M(\mathbf{x})$ be the paving tile at level M that contains $\mathbf{x}$, and define the function $\overline{g}$ that assigns to each $\mathbf{x}$ the maximum of f over its paving tile P if P is entirely within a dyadic cube at level N, and minus the sup of f if P intersects the boundary of a dyadic cube:

$$\overline{g}(\mathbf{x}) = \begin{cases} M_{P_{N''}(\mathbf{x})}(f) & \text{if } P_{N''}(\mathbf{x}) \cap \partial\mathcal{D}_N = \phi \\ -\sup|f| & \text{otherwise.} \end{cases} \qquad A14.6$$

Then $\overline{g} \le \overline{f}$; hence

$$\int_{\mathbb{R}^n} \overline{g}|d^n\mathbf{x}| \le \int_{\mathbb{R}^n} \overline{f}|d^n\mathbf{x}| = U_N(f). \qquad A14.7$$

---

[3]Theorem 4.3.10: A function $f : \mathbb{R}^n \to \mathbb{R}$, bounded with bounded support, is integrable if it is continuous except on a set of volume 0.

Now we compute the upper sum $U_{\mathcal{P}_{N''}}(f)$, as follows:

$$U_{\mathcal{P}_{N''}}(f) = \sum_{P \in \mathcal{P}_{N''}} M_P(f) \operatorname{vol}_n P \qquad A14.8$$

On the far right of the second line of Equation A14.8 we add $-\sup|f| + \sup|f| = 0$ to $M_P(f)$.

$$= \underbrace{\sum_{\substack{P \in \mathcal{P}_{N''}, \\ P \cap \partial \mathcal{D}_N = \phi}} M_P(f) \operatorname{vol}_n P}_{\substack{\text{contribution from } P \\ \text{entirely in dyadic cubes}}} + \underbrace{\sum_{\substack{P \in \mathcal{P}_{N''}, \\ P \cap \partial \mathcal{D}_N \neq \phi}} \big(M_P(f) \overbrace{- \sup|f| + \sup|f|}^{\text{cancels out}}\big) \operatorname{vol}_n P.}_{\substack{\text{contribution from } P \text{ that intersect} \\ \text{the boundary of dyadic cubes}}}$$

Now we make two sums out of the single sum on the far right:

$$\sum_{\substack{P \in \mathcal{P}_{N''}, \\ P \cap \partial \mathcal{D}_N \neq \phi}} \big(- \sup|f|\big) \operatorname{vol}_n P + \sum_{\substack{P \in \mathcal{P}_{N''}, \\ P \cap \partial \mathcal{D}_N \neq \phi}} \big(M_P(f) + \sup|f|\big) \operatorname{vol}_n P, \qquad A14.9$$

and add the first sum to the sum giving the contribution from P entirely in dyadic cubes, to get the integral of $\overline{g}$:

$$\sum_{\substack{P \in \mathcal{P}_{N''}, \\ P \cap \partial \mathcal{D}_N = \phi}} M_P(f) \operatorname{vol}_n P + \sum_{\substack{P \in \mathcal{P}_{N''}, \\ P \cap \partial \mathcal{D}_N \neq \phi}} \big(- \sup|f|\big) \operatorname{vol}_n P = \int_{\mathbb{R}^n} \overline{g} |d^n \mathbf{x}|. \quad A14.10$$

We can rewrite Equation A14.8 as:

Since $M_P(f)$ is the least upper bound just over P while $\sup|f|$ is the least upper bound over all of $\mathbb{R}^n$, we have $M_P(f) + \sup|f| \leq 2\sup|f|$.

$$U_{\mathcal{P}_{N''}}(f) = \int_{\mathbb{R}^n} \overline{g} |d^n \mathbf{x}| + \sum_{\substack{P \in \mathcal{P}_{N''}, \\ P \cap \partial \mathcal{D}_N \neq \phi}} \overbrace{\big(M_P(f) + \sup|f|\big)}^{\leq 2\sup|f| \text{ (see note in margin)}} \operatorname{vol}_n P. \quad A14.11$$

Using Equation A14.3 to give an upper bound on the volume of the paving pieces P that intersect the boundary, we get

$$\left| U_{\mathcal{P}_{N''}}(f) - \int_{\mathbb{R}^n} \overline{g} |d^n \mathbf{x}| \right| \leq 2\sup|f| \operatorname{vol}_n B \leq 2\sup|f| \frac{\epsilon}{8\sup|f|} = \frac{\epsilon}{4}. \quad A14.12$$

Substituting $U_N(f)$ for the integral of $\overline{g}$ (justified by Equation A14.7) gives us

$$\left| U_{\mathcal{P}_{N''}}(f) - \overbrace{U_N(f)}^{\geq \int_{\mathbb{R}^n} \overline{g} |d^n \mathbf{x}|} \right| \leq \frac{\epsilon}{4}, \quad \text{so} \quad U_{\mathcal{P}_{N''}}(f) \leq U_N(f) + \frac{\epsilon}{4}. \qquad A14.13$$

An exactly analogous argument leads to

$$L_{\mathcal{P}_{N''}}(f) \geq L_N(f) - \frac{\epsilon}{4}, \quad \text{i.e.,} \quad -L_{\mathcal{P}_{N''}}(f) \leq -L_N(f) + \frac{\epsilon}{4}. \qquad A14.14$$

Adding these together and using Equation A14.2, we get

$$U_{\mathcal{P}_{N''}}(f) - L_{\mathcal{P}_{N''}}(f) \leq U_N(f) - L_N(f) + \frac{\epsilon}{2} < \epsilon. \quad \square \qquad A14.15$$

A.15 EXISTENCE AND UNIQUENESS OF THE DETERMINANT

This is a messy and uninspiring exercise in the use of induction; students willing to accept the theorem on faith may wish to skip the proof, or save it for a rainy day.

The three $n \times n$ matrices of Equation A15.1 are identical except for the kth column, which are respectively $\mathbf{a}_k, \mathbf{b}$ and $\mathbf{c}$.

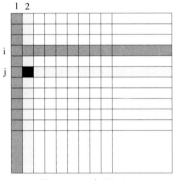

FIGURE A15.1.

Part (2) of the proof: Antisymmetry. The black square is in the ($j = 6$)th row of the matrix. But after removing the first column and ith row, it is in the 5th row of the new matrix. So when we exchange the first and second columns, the determinant of the unshaded matrix multiplies $a_{i,1}a_{j,2}$ in both cases, but contributes to the determinant with opposite sign.

In the second line of Equation A15.4, $A_{i,1} = B_{i,1} = C_{i,1}$ because A, B, and C are identical except for the first column, which is erased to produce $A_{i,1}, B_{i,1}$, and $C_{i,1}$.

Theorem 4.8.4 (Existence and uniqueness of determinants). *There exists a function* $\det A$ *satisfying the three properties of the determinant, and it is unique.*

Uniqueness is proved in Section 4.8; here we prove existence. We will verify that the function $D(A)$, the development along the first column, does indeed satisfy properties (1), (2), and (3) for the determinant $\det A$.

(1) *Multilinearity*

Let $\mathbf{b}, \mathbf{c} \in \mathbb{R}^n$, and suppose $\mathbf{a}_k = \beta\mathbf{b} + \gamma\mathbf{c}$. Set

$$A = [\mathbf{a}_1, \dots, \mathbf{a}_k, \dots, \mathbf{a}_n],$$

$$B = [\mathbf{a}_1, \dots, \mathbf{b}, \dots, \mathbf{a}_n], \qquad \text{A15.1}$$

$$C = [\mathbf{a}_1, \dots, \mathbf{c}, \dots, \mathbf{a}_n].$$

The object is to show that

$$D(A) = \beta D(B) + \gamma D(C), \qquad \text{A15.2}$$

We need to distinguish two cases: $k = 1$ (i.e., k is the first column) and $k > 1$.

The case $k > 1$ is proved by induction. Clearly multilinearity is true for D's of 1×1 matrices, which are just numbers. We will suppose multilinearity is true for D's of $(n-1) \times (n-1)$ matrices, such as $A_{i,1}$. Just write:

$$D(A) = \sum_{i=1}^{n} (-1)^{1+i} a_{i,1} D(A_{i,1}) \qquad \text{(Equation 4.8.9)}$$

$$= \sum_{i=1}^{n} (-1)^{1+i} a_{i,1} \big(\beta D(B_{i,1}) + \gamma D(C_{i,1})\big) \quad \text{(Inductive assumption)}$$

$$= \beta \sum_{i=1}^{n} (-1)^{1+i} a_{i,1} D(B_{i,1}) + \gamma \sum_{i=1}^{n} (-1)^{1+i} a_{i,1} D(C_{i,1})$$

$$= \beta D(B) + \gamma D(C). \qquad \text{A15.3}$$

This proves the case $k > 1$. Now for the case $k = 1$:

$$D(A) = \sum_{i=1}^{n} (-1)^{1+i} a_{i,1} D(A_{i,1}) = \sum_{i=1}^{n} (-1)^{1+i} \underbrace{(\beta b_{i,1} + \gamma c_{i,1})}_{= \, a_{i,1} \text{ by definition}} D(A_{i,1})$$

$$= \beta \sum_{i=1}^{n} (-1)^{1+i} b_{i,1} D \underbrace{(A_{i,1})}_{= (B_{i,1})} + \gamma \sum_{i=1}^{n} (-1)^{1+i} c_{i,1} D \underbrace{(A_{i,1})}_{= (C_{i,1})}$$

$$= \beta D(B) + \gamma D(C). \qquad \text{A15.4}$$

This proves multilinearity of our function D.

(2) *Antisymmetry*

The matrix $\widetilde{A}$ is the same as A with the jth and kth columns exchanged.

We want to prove $D(A) = -D(\widetilde{A})$, where $\widetilde{A}$ is formed by exchanging the jth and kth columns of A.

Again, we have two cases to consider. The first, where both j and k are greater than 1, is proved by induction: we assume the function D is antisymmetric for $(n-1) \times (n-1)$ matrices, so that in particular $D(A_{1,i}) = -D(\widetilde{A}_{1,i})$ for each i, and we will show that if so, it is true for $n \times n$ matrices.

We can see from Equation 4.8.9 that exchanging the columns of a 2×2 matrix changes the sign of D, so we can start the induction at $n = 2$.

$$D(A) = \sum_{i=1}^{n} (-1)^{i+1} a_{i,1} D(A_{i,1}) \underbrace{=}_{\substack{\text{by} \\ \text{induction}}} -\sum_{i=1}^{n} (-1)^{i+1} a_{i,1} D(\widetilde{A}_{1,i})$$

$$= -D(\widetilde{A}).$$

A15.5

We can restrict ourselves to $k = 2$ because if $k > 2$, say, $k = 5$, then we could switch $j = 1$ and k with a total of three exchanges: one to exchange the kth and the second position, one to exchange positions 1 and 2, and a third to exchange position 2 and the fifth position again. By our argument above we know that the first and third exchanges would each change the sign of the determinant, resulting in no net change; the only exchange that "counts" is the change of the first and second positions.

The case where either j or k equals 1 is more unpleasant. Let's assume $j = 1, k = 2$. Our approach will be to go one level deeper into our recursive formula, expressing $D(A)$ not just in terms of $(n-1) \times (n-1)$ matrices, but in terms of $(n-2) \times (n-2)$ matrices: the matrix $(A_{i,m;1,2})$, formed by removing the ith and mth rows and the first and second columns of A.

In the second line of Equation A15.6 below, the entire expression within big parentheses gives $D(A_{i,1})$, in terms of $D(A_{i,m;1,2})$:

$$D(A) = \sum_{i}^{n} (-1)^{i+1} a_{i,1} D(A_{i,1})$$

A15.6

$$= \underbrace{\sum_{i}^{n} (-1)^{i+1} a_{i,1} \left(\underbrace{\sum_{m=1}^{i-1} (-1)^{m+1} a_{m,2} D(A_{i,m;1,2})}_{\text{terms where } m<i} + \underbrace{\sum_{m=i+1}^{n} (-1)^{m} a_{m,2} D(A_{i,m;1,2})}_{\text{terms where } m>i} \right)}_{=D(A_{i,1}), \text{ in terms of } D \text{ of } (A_{i,m;1,2}), \text{ considered in two parts}}.$$

There are two sums within the term in parentheses, because in going from the matrix A to the matrix $A_{i,1}$, the ith row was removed, as shown in Figure A15.1. Then, in creating $A_{i,m;1,2}$ from $A_{i,1}$, we remove the mth row (and the second column) of A. When we write $D(A_{i,1})$, *we must thus remember that the ith row is missing*, and hence $a_{m,2}$ is in the $(m-1)$ row of $A_{i,1}$ when $m > i$. We do that by summing separately, for each value of i, the terms with m from 1 to $i-1$ and those with m from $i+1$ to n, carefully using the sign $(-1)^{m-1+1} = (-1)^{m}$ for the second batch. (For $i = 4$, how many terms are there for the sum with m from 1 to $i-1$? For the sum with m from $i+1$ to n?[4])

---

[4] As shown in Figure A15.1, there are three terms in the first sum, $m = 1, 2, 3$, and $n - 4$ terms in the second.

Exactly the same computation for $\widetilde{A}$ leads to

$$D(\widetilde{A}) = \sum_{j=1}^{n} (-1)^{j+1} \tilde{a}_{j,1} D(\widetilde{A}_{j,1}) \qquad A15.7$$

$$= \sum_{j=1}^{n} (-1)^{j+1} \tilde{a}_{j,1} \left(\sum_{p=1}^{j-1} (-1)^{p+1} \tilde{a}_{p,2} D(\widetilde{A}_{j,p;1,2}) + \sum_{p=j+1}^{n} (-1)^{p} \tilde{a}_{p,2} D(\widetilde{A}_{j,p;1,2}) \right).$$

Let us look at one particular term of the double sum of Equation A15.7, corresponding to some j and p:

$$\begin{cases} (-1)^{j+p} & \tilde{a}_{j,1}\,\tilde{a}_{p,2}\ D(\widetilde{A}_{j,p;1,2}) & \text{if } p < j \\ (-1)^{j+p+1} & \tilde{a}_{j,1}\,\tilde{a}_{p,2}\ D(\widetilde{A}_{j,p;1,2}) & \text{if } p > j. \end{cases} \qquad A15.8$$

Remember that $\tilde{a}_{j,1} = a_{j,2}$, $\tilde{a}_{p,2} = a_{p,1}$, and $\widetilde{A}_{j,p;1,2} = A_{p,j;1,2}$. Thus we can rewrite Equation A15.8 as

$$\begin{cases} (-1)^{j+p} & a_{j,2}\,a_{p,1}\ D(A_{p,j;1,2}) & \text{if } p < j \\ (-1)^{j+p+1} & a_{j,2}\,a_{p,1}\ D(A_{p,j;1,2}) & \text{if } p > j. \end{cases} \qquad A15.9$$

This term corresponds to $i = p$ and $m = j$ in Equation A15.6, but with the opposite sign. $\square$

Let us compare A and $\widetilde{A}$ in a particular case. Focus on the 2 and the 8:

What about the other terms in Equations A15.12 and A15.13? Each term from the expansion of A corresponds to a term of the expansion of $\widetilde{A}$, identical but with opposite sign. For example, the term $-2(5\,D\begin{bmatrix} - & - \\ - & - \end{bmatrix})$ of Equation A15.12 corresponds to the first term gotten by expanding the first term on the right-hand side of Equation A15.11.

In the matrix $A_{2,4;1,2}$ of Equation A15.12, the 4 in the subscript corresponds to the $(m = 4)$th row of the original matrix, i.e., the row containing $\underline{8}$. But that is the third row of the matrix $A_{2,1}$ in Equation A15.10.

$$D\begin{bmatrix} 1 & 5 & - & - \\ 2 & 6 & - & - \\ 3 & 7 & - & - \\ 4 & \underline{8} & - & - \end{bmatrix} = 1\,D\begin{bmatrix} 6 & - & - \\ 7 & - & - \\ 8 & - & - \end{bmatrix} - \underline{2}\,D\overbrace{\begin{bmatrix} 5 & - & - \\ 7 & - & - \\ \underline{8} & - & - \end{bmatrix}}^{A_{2,1}}$$

$$+ 3\,D\begin{bmatrix} 5 & - & - \\ 6 & - & - \\ 8 & - & - \end{bmatrix} - 4\,D\begin{bmatrix} 5 & - & - \\ 6 & - & - \\ 7 & - & - \end{bmatrix} \qquad A15.10$$

$$D\underbrace{\begin{bmatrix} 5 & 1 & - & - \\ 6 & 2 & - & - \\ 7 & 3 & - & - \\ \underline{8} & 4 & - & - \end{bmatrix}}_{\widetilde{A}} = 5\,D\begin{bmatrix} 2 & - & - \\ 3 & - & - \\ 4 & - & - \end{bmatrix} - 6\,D\begin{bmatrix} 1 & - & - \\ 3 & - & - \\ 4 & - & - \end{bmatrix}$$

$$+ 7\,D\begin{bmatrix} 1 & - & - \\ 2 & - & - \\ 4 & - & - \end{bmatrix} - \underline{8}\,D\underbrace{\begin{bmatrix} 1 & - & - \\ 2 & - & - \\ 3 & - & - \end{bmatrix}}_{A_{4,1}}. \qquad A15.11$$

Expanding the second term on the right-hand side of Equation A15.10 gives

$$-\underline{2}\left(5\,D\begin{bmatrix} - & - \\ - & - \end{bmatrix} - 7\,D\begin{bmatrix} - & - \\ - & - \end{bmatrix} + \underline{8}\,D\underbrace{\begin{bmatrix} - & - \\ - & - \end{bmatrix}}_{A_{2,4;1,2}} \right); \qquad A15.12$$

expanding the fourth term on the right-hand side of of Equation A15.11 gives

$$-\underline{8}\left(1\,D\begin{bmatrix} - & - \\ - & - \end{bmatrix} - \underline{2}\,D\underbrace{\begin{bmatrix} - & - \\ - & - \end{bmatrix}}_{A_{4,2;1,2}} + 3\,D\begin{bmatrix} - & - \\ - & - \end{bmatrix}\right). \qquad A15.13$$

The first gives $-16\,D\begin{bmatrix} - & - \\ - & - \end{bmatrix}$, the second $+16\,D\begin{bmatrix} - & - \\ - & - \end{bmatrix}$. The two blank matrices here are identical, so the terms are identical, with opposite signs.

Why does this happen? In the matrix A, the 8 in the second column is below the 2 in the first column, so when the second row (with the 2) is removed, the 8 is in the third row, not the fourth. Therefore, $8\,D\begin{bmatrix} - & - \\ - & - \end{bmatrix}$ comes with positive sign: $(-1)^{j+1} = (-1)^4 = +1$. In the matrix $\widetilde{A}$, the 2 in the second column is above the 8 in the first column, so when the fourth row (with the 8) is removed, the 2 is still in the second row. Therefore, $2\,D\begin{bmatrix} - & - \\ - & - \end{bmatrix}$ comes with negative sign: $(-1)^{i+1} = (-1)^3 = -1$.

We chose our 2 and 8 arbitrarily, so the same argument is true for any pair consisting of one entry from the first column and one from the second. (What would happen if we chose two entries from the same row, e.g., the 2 and 6 above?[5] What happens if the first two columns are identical?[6])

(3) *Normalization*

The normalization condition is much simpler. If $A = [\vec{\mathbf{e}}_1, \ldots, \vec{\mathbf{e}}_n]$, then in the first column, only the first entry $a_{1,1} = 1$ is nonzero, and $A_{1,1}$ is the identity matrix one size smaller, so that D of it is 1 by induction. So

$$D(A) = a_{1,1} D(A_{1,1}) = 1, \qquad A15.14$$

and we have also proved property (3). This completes the proof of existence; uniqueness is proved in Section 4.8. $\square$

A.16 RIGOROUS PROOF OF THE CHANGE OF VARIABLES FORMULA

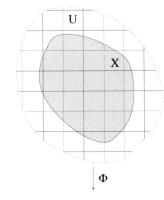

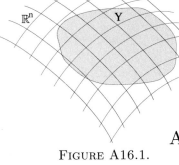

FIGURE A16.1.

The C^1 mapping Φ maps X to Y. We will use in the proof of the change of variables formula the fact that Φ is defined on U, not just on X.

Theorem 4.10.12 (Change of variables formula). *Let X be a compact subset of $\mathbb{R}^n$ with boundary ∂X of volume 0, and U an open neighborhood of X. Let $\Phi : U \to \mathbb{R}^n$ be a C^1 mapping with Lipschitz derivative, that is one to one on $(X - \partial X)$, and such that $[\mathbf{D}\Phi(\mathbf{x})]$ is invertible at every $\mathbf{x} \in (X - \partial X)$. Set $Y = \Phi(X)$.*

---

[5]This is impossible, since when we go one level deeper, that row is erased.

[6]The determinant is 0, since each term has a term that is identical to it but with opposite sign.

Then if $f : Y \to \mathbb{R}$ is integrable, then $(f \circ \Phi)|\det[\mathbf{D}\Phi]|$ is integrable on X, and

$$\int_Y f(\mathbf{y}) \, |d^n\mathbf{y}| = \int_X (f \circ \Phi)(\mathbf{x}) \, |\det[\mathbf{D}\Phi(\mathbf{x})]| \, |d^n\mathbf{x}|.$$

Proof. The proof is just a (lengthy) matter of dotting the i's of the sketch in Section 4.10. As shown in Figure A16.1, we will use the dyadic decomposition of X, and the image decomposition for Y, whose paving blocks are the $\Phi(C \cap X)$, $C \in \mathcal{D}_N(\mathbb{R}^n)$. We will call this partition $\Phi(\mathcal{D}_N(X))$. The outline of the proof is as follows:

The second line of Equation A16.1 is a Riemann sum; $\mathbf{x}_C$ is the point in C where Φ is evaluated: midpoint, lower left-hand corner, or some other choice.

The $\approx$ become equalities in the limit.

$$
\begin{aligned}
\int_Y f|d^n\mathbf{y}| &\approx \sum_{C \in \mathcal{D}_N(\mathbb{R}^n)} \overbrace{M_{\Phi(C)} f \, \mathrm{vol}_n \, \Phi(C)}^{\substack{\sup f \text{ over curvy cube} \\ \text{times vol. of curvy cube}}} \\
&\approx \sum_{C \in \mathcal{D}_N(\mathbb{R}^n)} M_C(f \circ \Phi)(\mathrm{vol}_n \, C|\det[\mathbf{D}\Phi(\mathbf{x}_C)]|) \\
&\approx \int_X (f \circ \Phi)\mathbf{x}|\det[\mathbf{D}\Phi(\mathbf{x})]|) \, |d^n\mathbf{x}|,
\end{aligned}
$$

A16.1

where the $\mathbf{x}_C$ in the second line is some $\mathbf{x}$ in C.

(1) To justify the first $\approx$, we need to show that the image decomposition of Y, $\Phi(\mathcal{D}_N(X))$, is a nested partition.

(2) To justify the second (this is the hard part) we need to show that as $N \to \infty$, the volume of a curvy cube of the image decomposition equals the volume of a cube of the original dyadic decomposition times $|\det[\mathbf{D}\Phi(\mathbf{x}_C)]|$.

(3) The third is simply the definition of the integral as the limit of a Riemann sum.

We need Proposition A16.1 (which is of interest in its own right) to prove (1): i.e., to show that $\Phi(\mathcal{D}_N(X))$ is a nested partition. It will also be used at the end of the proof of the change of variables formula.

Proposition A16.1 (Volume of the image by a C^1 map). *Let $Z \subset \mathbb{R}^n$ be a compact pavable subset of $\mathbb{R}^n$, U an open neighborhood of Z and $\Phi : U \to \mathbb{R}^n$ a C^1 mapping with bounded derivative. Set $K = \sup_{\mathbf{x} \in U} \|[\mathbf{D}\Phi(\mathbf{x})]\|$. Then*

$$\mathrm{vol}_n \, \Phi(Z) \; \leq \; (K\sqrt{n})^n \, \mathrm{vol}_n \, Z.$$

In particular, if $\mathrm{vol}_n \, Z = 0$, then $\mathrm{vol}_n \, \Phi(Z) = 0$.

The subset A consists of the closure of all the C's either entirely within Z or straddling the boundary of Z.

A cube $C \in \mathcal{D}_N(\mathbb{R}^n)$ has side-length $1/2^N$, and (see Exercise 4.1.5) the distance between two points $\mathbf{x}, \mathbf{y}$ in the same cube C is

$$|\mathbf{x} - \mathbf{y}| \le \frac{\sqrt{n}}{2^N}.$$

So the maximum distance between two points of $\Phi(C)$ is $\frac{K\sqrt{n}}{2^N}$, i.e., $\Phi(C)$ is contained in the box C' centered at $\Phi(\mathbf{z}_C)$ with side-length $K\sqrt{n}/2^N$.

Proof. Choose $\epsilon > 0$ and $N \ge 0$ so large that

$$A = \bigcup_{\substack{C \in \mathcal{D}_N(\mathbb{R}^n), \\ C \cap Z \ne \phi}} \overline{C} \subset U \quad \text{and} \quad \text{vol}_n A \le \text{vol}_n Z + \epsilon. \qquad A16.2$$

(Recall that $\overline{C}$ denotes the closure of C.) Let $\mathbf{z}_C$ be the center of one of the cubes C above. Then by Corollary 1.9.2, when $\mathbf{z} \in C$ we have

$$|\Phi(\mathbf{z}_C) - \Phi(\mathbf{z})| \le K|\mathbf{z}_C - \mathbf{z}|. \qquad A16.3$$

(The distance between the two points in the image is at most K times the distance between the corresponding points of the domain.) Therefore $\Phi(C)$ is contained in the box C' centered at $\Phi(\mathbf{z}_C)$ with side-length $K\sqrt{n}/2^N$.

Finally,

$$\Phi(Z) \subset \bigcup_{\substack{C \in \mathcal{D}_N(\mathbb{R}^n), \\ C \cap Z \ne \phi}} C', \quad \text{so} \qquad A16.4$$

To go from the third to the fourth term of Equation A16.5, we simultaneously multiply and divide by

$$\text{vol}_n C = \left(\frac{1}{2^N}\right)^n.$$

$$\text{vol}_n \Phi(Z) \le \sum_{\substack{C \in \mathcal{D}_N(\mathbb{R}^n), \\ C \cap Z \ne \phi}} \text{vol}_n C' \le \sum_{\substack{C \in \mathcal{D}_N(\mathbb{R}^n), \\ C \cap Z \ne \phi}} \left(\frac{K\sqrt{n}}{2^N}\right)^n = \underbrace{\frac{\left(\frac{K\sqrt{n}}{2^N}\right)^n}{\left(\frac{1}{2^N}\right)^n}}_{\substack{\text{ratio vol}_n C' \\ \text{to vol}_n C}} \sum_{\substack{C \in \mathcal{D}_N(\mathbb{R}^n), \\ C \cap Z \ne \phi}} \text{vol}_n C$$

$$= (K\sqrt{n})^n \, \text{vol}_n A \le (K\sqrt{n})^n (\text{vol}_n Z + \epsilon). \quad \square \qquad A16.5$$

Corollary A16.2. *The partition $\Phi(\mathcal{D}_N(X))$ is a nested partition of Y.*

Nested partitions are defined in Definition 4.7.4.

Proof. The three conditions to be verified are that the pieces are nested, that the diameters tend to 0 as N tends to infinity, and that the boundaries of the pieces have volume 0. The first is clear: if $C_1 \subset C_2$, then $\Phi(C_1) \subset \Phi(C_2)$. The second is the same as Equation A16.3, and the third follows from the second part of Proposition A16.1. $\square$

Our next proposition contains the real substance of the change of variables theorem. It says exactly why we can replace the volume of the little curvy parallelogram $\Phi(C)$ by its approximate volume $|\det[\mathbf{D}\Phi(\mathbf{x})]| \text{vol}_n C$.

Every time you want to compare balls and cubes in $\mathbb{R}^n$, there is a pesky $\sqrt{n}$ which complicates the formulas. We will need to do this several times in the proof of Proposition A16.4; the following lemma isolates what we need.

Lemma A16.3. *Choose $0 < a < b$, and let C_a and C_b be the cubes centered at the origin of side length $2a$ and $2b$ respectively, i.e., the cubes defined by $|x_i| < a$ (respectively $|x_i| < b$), $i = 1, \dots, n$. Then the ball of radius*

$$\frac{(b-a)|\mathbf{x}|}{a\sqrt{n}} \qquad A16.6$$

around any point $\mathbf{x} \in C_a$ is contained in C_b.

Proposition A16.4 is the main tool for proving Theorem 4.10.12. It says that for a change of variables mapping Φ such that $\Phi(0) = 0$, the image $\Phi(C)$ of a cube C centered at 0 is arbitrarily close to the image of C by the derivative of Φ at 0, as shown in Figures A16.2 and A16.3.

Why does Equation A16.9 prove the right-hand inclusion? We want to know that if $\mathbf{x} \in C$, then

$$\Phi(\mathbf{x}) \in (1 + \epsilon)[\mathbf{D}\Phi(0)](C),$$

or equivalently,

$$[\mathbf{D}\Phi(0)]^{-1}\Phi(\mathbf{x}) \in (1 + \epsilon)C.$$

(As noted in the text, $\Phi(\mathbf{x})$ is really the vector $\Phi(\mathbf{x}) - \Phi(0)$.) Since

$$[\mathbf{D}\Phi(0)]^{-1}\Phi(\mathbf{x}) = \mathbf{x} +$$
$$[\mathbf{D}\Phi(0)]^{-1}\Big(\Phi(\mathbf{x}) - [\mathbf{D}\Phi(0)](\mathbf{x})\Big),$$

$[\mathbf{D}\Phi(0)]^{-1}\Phi(\mathbf{x})$ is distance

$$\left|[\mathbf{D}\Phi(0)]^{-1}\Big(\Phi(\mathbf{x}) - [\mathbf{D}\Phi(0)](\mathbf{x})\Big)\right|$$

from $\mathbf{x}$. But the ball of radius $\epsilon|\mathbf{x}|/\sqrt{n}$ around any point $\mathbf{x} \in C$ is completely contained in $(1 + \epsilon)C$, by Lemma A16.3.

Proposition A2.1: If the derivative of $\mathbf{f}$ is Lipschitz with Lipschitz ratio M, then

$$\left|\mathbf{f}(\mathbf{x} + \vec{\mathbf{h}}) - \mathbf{f}(\mathbf{x}) - [\mathbf{Df}(\mathbf{x})]\vec{\mathbf{h}}\right|$$
$$\leq \frac{M}{2}|\vec{\mathbf{h}}|^2.$$

In the expression in brackets under Equation A16.9, Φ plays the role of $\mathbf{f}$, $\mathbf{x} - 0$ is $\vec{\mathbf{h}}$, and 0 is $\mathbf{x}$.

Proof. First note that if $\mathbf{x} \in C_a$, then $|\mathbf{x}| < a\sqrt{n}$. Let $\mathbf{x} + \vec{\mathbf{h}}$ be a point of the ball. Then

$$|h_i| \leq |\vec{\mathbf{h}}| \leq \frac{(b-a)|\mathbf{x}|}{a\sqrt{n}} \leq \frac{(b-a)a\sqrt{n}}{a\sqrt{n}} = b - a. \qquad A16.7$$

Thus $|x_i + h_i| \leq |x_i| + |h_i| < a + b - a = b.$ $\square$

Proposition A16.4. *Let U, V be open subsets in $\mathbb{R}^n$ with $0 \in U$ and $0 \in V$. Let $\Phi : U \to V$ be a differentiable mapping with $\Phi(0) = 0$. Suppose that Φ is bijective, $[\mathbf{D}\Phi]$ is Lipschitz, and that $\Phi^{-1} : V \to U$ is also differentiable with Lipschitz derivative. Let M be a Lipschitz constant for both $[\mathbf{D}\Phi]$ and $[\mathbf{D}\Phi^{-1}]$. Then*

(a) For any $\epsilon > 0$, there exists $\delta > 0$ such that if C is a cube centered at 0 of side $< 2\delta$, then

$$(1 - \epsilon)\,[\mathbf{D}\Phi(0)]C \subset \underbrace{\Phi(C)}_{\substack{\text{squeezed between} \\ \text{right and left sides} \\ \text{as } \epsilon \to 0}} \subset (1 + \epsilon)\,[\mathbf{D}\Phi(0)]C. \qquad A16.8$$

(b) We can choose δ to depend only on ϵ, $\big|[\mathbf{D}\Phi(0)]\big|$, $\big|[\mathbf{D}\Phi(0)]^{-1}\big|$, and the Lipschitz constant M, but no other information about Φ.

Proof. The right-hand and the left-hand inclusions of Equation A16.8 require slightly different treatments. They are both consequences of Proposition A2.1, and you should remember that the largest n-dimensional cube contained in a ball of radius r has side-length $2r/\sqrt{n}$.

The right-hand inclusion, illustrated by Figure A16.2, is gotten by finding a δ such that if the side-length of C is less than 2δ, and $\mathbf{x} \in C$, then

$$\left|[\mathbf{D}\Phi(0)]^{-1}\underbrace{\big(\Phi(\mathbf{x}) - [\mathbf{D}\Phi(0)](\mathbf{x})\big)}_{\underbrace{\Phi(\mathbf{x}) - \Phi(0)}_{\text{vector}} - [\mathbf{D}\Phi(0)]\underbrace{(\mathbf{x} - 0)}_{\text{vector}}}\right| \leq \frac{\epsilon|\mathbf{x}|}{\sqrt{n}}. \qquad A16.9$$

Note that although it looks as though $[\mathbf{D}\Phi(0)](\mathbf{x})$ is a matrix times a point, which doesn't make sense, $\mathbf{x}$ is really the vector $(\mathbf{x} - 0)$. Similarly, $\Phi(\mathbf{x})$ is the vector $\Phi(\mathbf{x}) - \Phi(0)$.

According to Proposition A2.1,

$$\left|[\mathbf{D}\Phi(0)]^{-1}\big(\Phi(\mathbf{x}) - [\mathbf{D}\Phi(0)](\mathbf{x})\big)\right| \leq \big|[\mathbf{D}\Phi(0)]^{-1}\big|\frac{M|\mathbf{x}|^2}{2}, \qquad A16.10$$

so it is enough to require that when $\mathbf{x} \in C$,

$$\frac{\big|[\mathbf{D}\Phi(0)]^{-1}\big|M|\mathbf{x}|^2}{2} \leq \frac{\epsilon|\mathbf{x}|}{\sqrt{n}}, \quad \text{i.e.,} \quad |\mathbf{x}| \leq \frac{2\epsilon}{M\sqrt{n}\,\big|[\mathbf{D}\Phi(0)]^{-1}\big|}. \qquad A16.11$$

Again, it isn't immediately obvious why the left-hand inclusion of Equation A16.8 follows from the inequality A16.13. We need to show that if $\mathbf{x} \in (1 - \epsilon)C$, then $[\mathbf{D}\Phi(0)]\mathbf{x} \in \Phi(C)$.

Apply Φ^{-1} to both sides to get $\Phi^{-1}([\mathbf{D}\Phi(0)]\mathbf{x}) \in C$. Inequality A16.13 asserts that

$$\Phi^{-1}([\mathbf{D}\Phi(0)]\mathbf{x}) \quad \text{is within}$$

$$\frac{\epsilon}{\sqrt{n}(1 - \epsilon)} |\mathbf{x}| \quad \text{of } \mathbf{x},$$

but the ball of that radius around any point of $(1 - \epsilon)C$ is contained in C, again by Lemma A16.3.

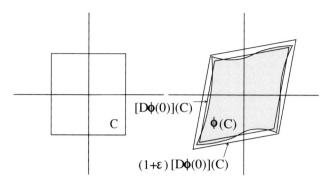

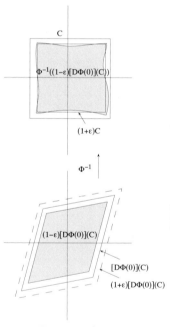

FIGURE A16.3.

The parallelepiped

$$(1 - \epsilon)[\mathbf{D}\Phi(0)](C)$$

is mapped by Φ^{-1} almost to $(1 - \epsilon)C$, and definitely inside C. Therefore, the image of C covers $(1 - \epsilon)[\mathbf{D}\Phi(0)](C)$.

FIGURE A16.2. The cube C is mapped to $\Phi(C)$, which is almost $[\mathbf{D}\Phi(0)](C)$, and definitely inside $(1 + \epsilon)[\mathbf{D}\Phi(0)](C)$. As $\epsilon \to 0$, the image $\Phi(C)$ becomes more and more exactly the parallelepiped $[\mathbf{D}\Phi(0)]C$.

Since $\mathbf{x} \in C$ and C has side-length 2δ, we have $|\mathbf{x}| \le \delta\sqrt{n}$, so the right-hand inclusion will be satisfied if

$$\delta = \frac{2\epsilon}{Mn\big|[\mathbf{D}\Phi(0)]^{-1}\big|}. \qquad A16.12$$

For the left-hand inclusion, illustrated by Figure A16.3, we need to find δ such that when C has side-length $< 2\delta$, then[7]

$$\left| \Phi^{-1}([\mathbf{D}\Phi(0)]\mathbf{x}) - \mathbf{x} \right| \le \frac{\epsilon}{\sqrt{n}(1 - \epsilon)} |\mathbf{x}| \qquad A16.13$$

when $\mathbf{x} \in (1 - \epsilon)C$. Again this follows from Proposition A2.1. Set $\mathbf{y} = [\mathbf{D}\Phi(0)]\mathbf{x}$. This time, Φ^{-1} plays the role of $\mathbf{f}$ in Proposition A2.1, 0 is $\mathbf{x}$, and $\mathbf{y}$ is $\vec{\mathbf{h}}$. We find

$$\left| \Phi^{-1}([\mathbf{D}\Phi(0)]\mathbf{x}) - \mathbf{x} \right| = \left| \Phi^{-1}\mathbf{y} - [\mathbf{D}\Phi^{-1}(0)]\mathbf{y} \right| = \left| \Phi^{-1}(0 + \mathbf{y}) - \Phi^{-1}(0) - [\mathbf{D}\Phi^{-1}(0)]\mathbf{y} \right|$$

$$\le \frac{M}{2} |\mathbf{y}|^2 \le \frac{M}{2} \big| [\mathbf{D}\Phi(0)]\mathbf{x} \big|^2 \le \frac{M}{2} \big| [\mathbf{D}\Phi(0)] \big|^2 |\mathbf{x}|^2. \qquad A16.14$$

Our inequality will be satisfied if

$$\frac{M}{2} \big| [\mathbf{D}\Phi(0)] \big|^2 |\mathbf{x}|^2 \le \frac{\epsilon}{\sqrt{n}(1 - \epsilon)} |\mathbf{x}|, \text{ i.e., } |\mathbf{x}| \le \frac{2\epsilon}{M(1 - \epsilon)\sqrt{n}\big| [\mathbf{D}\Phi(0)] \big|^2}. \qquad A16.15$$

Remember that $\mathbf{x} \in (1 - \epsilon)C$, so $|\mathbf{x}| \le (1 - \epsilon)\delta\sqrt{n}$, and the left-hand inclusion is satisfied if we take

$$\delta = \frac{2\epsilon}{(1 - \epsilon)^2 Mn\big| [\mathbf{D}\Phi(0)] \big|^2}. \qquad A16.16$$

Choose the smaller of the two deltas. $\quad\square$

---

[7] As before, $[\mathbf{D}\Phi(0)]\mathbf{x}$ is really $[\mathbf{D}\Phi(0)](\mathbf{x} - 0)$. This vector is treated as a point, and therefore something that can be the input for Φ^{-1}.

Proof of the change of variables formula, continued

Proposition A16.4 goes a long way to towards proving the change of variables formula; still, the integral is defined in terms of upper and lower sums, and we must translate the statement into that language.

Proposition A16.5. *Let U and V be bounded subsets in $\mathbb{R}^n$ and let $\Phi : U \to V$ be a differentiable mapping with Lipschitz derivative, that is bijective, and such that $\Phi^{-1} : V \to U$ is also differentiable with Lipschitz derivative.*

Then for any $\eta > 0$, there exists N such that if $C \in \mathcal{D}_N(\mathbb{R}^n)$ and $C \subset U$, then,

$$(1 - \eta)\, M_C\big(|\det[\mathbf{D}\Phi]|\big)\, \mathrm{vol}_n\, C \leq \mathrm{vol}_n\, \Phi(C)$$

$$\leq (1 + \eta)\, m_C\big(|\det[\mathbf{D}\Phi]|\big)\, \mathrm{vol}_n\, C. \qquad A16.17$$

Proof of Proposition A16.5. Choose $\eta > 0$, and find $\epsilon > 0$ so that

$$(1 + \epsilon)^{n+1} < 1 + \eta \quad \text{and} \quad (1 - \epsilon)^{n+1} > 1 - \eta.$$

For this ϵ, find N_1 such that Proposition A16.4 is true for every cube $C \in \mathcal{D}_{N_1}(\mathbb{R}^n)$ such that $C \subset U$.

Next find N_2 such that for every cube $C \in \mathcal{D}_{N_2}(\mathbb{R}^n)$ with $C \subset U$, we have

In Equation A16.18, $M_C|\det[\mathbf{D}\Phi]|$ is not M_C times $[\mathbf{D}\Phi]$; it is the last upper bound of $|\det[\mathbf{D}\Phi]|$. Similarly, $m_C|\det[\mathbf{D}\Phi]|$ is the greatest lower bound of $|\det[\mathbf{D}\Phi]|$.

$$\frac{M_C|\det[\mathbf{D}\Phi]|}{m_C|\det[\mathbf{D}\Phi]|} < 1 + \epsilon \quad \text{and} \quad \frac{m_C|\det[\mathbf{D}\Phi]|}{M_C|\det[\mathbf{D}\Phi]|} > 1 - \epsilon. \qquad A16.18$$

Actually the second inequality follows from the first, since $1/(1 + \epsilon) > 1 - \epsilon$.

If N is the larger of N_1 and N_2, together these give

For an appropriate N_1, Proposition A16.4 and Theorem 4.9.1 tell us that

$$\mathrm{vol}_n\, C$$

$$\leq \mathrm{vol}_n\Big((1 + \epsilon)[\mathbf{D}\Phi(0)]C\Big)$$

$$= (1 + \epsilon)^n|\det[\mathbf{D}\Phi(0)]C|\, \mathrm{vol}_n\, C;$$

$[\mathbf{D}\Phi(0)]$ is the T of Theorem 4.9.1, which says that

$$\mathrm{vol}_n\, T(A) = |\det[T]|\, \mathrm{vol}_n\, A.$$

$$\mathrm{vol}_n\, \Phi(C) < (1 + \epsilon)^n \big|\det[\mathbf{D}\Phi(0)]\big|\, \mathrm{vol}_n\, C, \qquad A16.19$$

and since $[\mathbf{D}\Phi(0)] \leq M_C|\det[\mathbf{D}\Phi]| < (1 + \epsilon)m_C|\det[\mathbf{D}\Phi]|$, we have

$$\mathrm{vol}_n\, \Phi(C) < (1 + \epsilon)^{n+1}m_C|\det[\mathbf{D}\Phi|\, \mathrm{vol}_n\, C. \qquad A16.20$$

An exactly similar argument leads to

$$\mathrm{vol}_n\, \Phi(C) > (1 - \epsilon)^{n+1}M_C|\det[\mathbf{D}\Phi]|\, \mathrm{vol}_n\, C. \quad \square \qquad A16.21$$

We can now prove the change of variables theorem. First, we may assume that the function f to be integrated is positive. Call M the Lipschitz constant of $[\mathbf{D}\Phi]$, and set

$$K = \sup_{\mathbf{x} \in X} |[\mathbf{D}\Phi(\mathbf{x})]| \quad \text{and} \quad L = \sup_{\mathbf{x} \in X} |f(\mathbf{x})|. \qquad A16.22$$

Choose $\eta > 0$. First choose N_1 sufficiently large that the union of the cubes $C \in \mathcal{D}_{N_1}(X)$ whose closures intersect the boundary of X have total volume $< \eta$. We will denote by Z the union of these cubes; it is a thickening of ∂X, the boundary of X.

Since $K' = \sup |[\mathbf{D}\Phi(\mathbf{x})]|^{-1}$, it is also $\sup |[\mathbf{D}\Phi(\mathbf{y})]|^{-1}$, accounting for the K^2 in the second line of Equation A16.23

Equation A16.24:

$$M_C\Big((f \circ \Phi)|\det[\mathbf{D}\Phi]|\Big)$$

is the least upper bound of the product of two functions,

$$g = (f \circ \Phi), \quad h = |\det[\mathbf{D}\Phi]|.$$

To go from line 4 to line 5, we use

$$M_C(gh) \le (M_C f)(M_C g);$$

we have

$$M_C(f \circ \Phi) = M_{\Phi(C)}(f)$$

and (using Proposition A16.5)

$$M_C|\det[\mathbf{D}\Phi]| \operatorname{vol}_n C$$
$$\le \frac{1}{1-\eta} \operatorname{vol}_n \Phi(C).$$

To go from the boundary cubes in line 3 to the last term of line 4, note that $M_C(f \circ \Phi) \le L$. The sum over boundary cubes C_∂ is then less than or equal to

$$\sum_{C_\partial} L M_C |\det[\mathbf{D}\Phi]| \operatorname{vol}_n C$$
$$\le L \sum_{C_\partial} \frac{\operatorname{vol}_n \Phi(C)}{1-\eta} = L \frac{\operatorname{vol}_n \Phi(Z)}{1-\eta}$$
$$\le \frac{L(K\sqrt{n})^n \operatorname{vol}_n Z}{1-\eta}$$
$$\le \frac{\eta L(K\sqrt{n})^n}{1-\eta};$$

Proposition A16.1 says that

$$\operatorname{vol}_n \Phi(Z) \le (K\sqrt{n})^n \operatorname{vol}_n Z,$$

and we know that $\eta = \operatorname{vol}_n Z$.

Lemma A16.6. *The closure of $X - Z$ is compact, and contains no point of ∂X.*

Proof. For the first part, X is bounded, so $X - Z$ is bounded, so its closure is closed and bounded. For the second, notice that for every point $\mathbf{a} \in \mathbb{R}^n$, there is an $r > 0$ such that the ball $B_r(\mathbf{a})$ is contained in the union of the cubes of $\mathcal{D}_{N_1}(\mathbb{R}^n)$ with $\mathbf{a}$ in their closure. So no sequence in $X - Z$ can converge to a point $\mathbf{a} \in \partial X$. $\square$

Proof of Proposition A16.5, cont. In particular, $[\mathbf{D}\Phi]^{-1}$ is bounded on $X - Z$, say by K', and it is also Lipschitz:

$$|[\mathbf{D}\Phi(\mathbf{x})]^{-1} - [\mathbf{D}\Phi(\mathbf{y})]^{-1}| = |[\mathbf{D}\Phi(\mathbf{x})]^{-1} \overbrace{([\mathbf{D}\Phi(\mathbf{y})] - [\mathbf{D}\Phi(\mathbf{x})])}^{\le M|\mathbf{y}-\mathbf{x}|}[\mathbf{D}\Phi(\mathbf{y})]^{-1}|$$
$$\le (K')^2 M|\mathbf{x} - \mathbf{y}|. \qquad A16.23$$

So we can choose $N_2 > N_1$ so that Proposition A16.5 is true for all cubes in $\mathcal{D}_{N_2}$ contained in $X - Z$. We will call the cubes of $\mathcal{D}_{N_2}$ in Z *boundary cubes*, and the others *interior cubes*.

Then we have

$$U_{N_2}\big((f \circ \Phi)|\det[\mathbf{D}\Phi]|\big) = \sum_{C \in \mathcal{D}_{N_2}(\mathbb{R}^n)} M_C\big((f \circ \Phi)|\det[\mathbf{D}\Phi]|\big) \operatorname{vol}_n C$$

$$= \sum_{\text{interior cubes } C} M_C\big((f \circ \Phi)|\det[\mathbf{D}\Phi]|\big) \operatorname{vol}_n C$$
$$+ \sum_{\text{boundary cubes } C} M_C\big((f \circ \Phi)|\det[\mathbf{D}\Phi]|\big) \operatorname{vol}_n C$$

$$\le \sum_{\text{interior cubes } C} M_C\big((f \circ \Phi)|\det[\mathbf{D}\Phi]|\big) \operatorname{vol}_n C + \frac{\eta L(K\sqrt{n})^n}{1-\eta} \qquad A16.24$$

$$\le \frac{1}{1-\eta} \sum_{C \in \mathcal{D}_{N_2}(\mathbb{R}^n)} M_{\Phi(C)}(f) \operatorname{vol}_n \Phi(C) + \frac{\eta L(K\sqrt{n})^n}{1-\eta}$$

$$= \frac{1}{1-\eta} U_{\Phi(\mathcal{D}_{N_2}(\mathbb{R}^n))}(f) + \frac{\eta L(K\sqrt{n})^n}{1-\eta}.$$

A similar argument about lower sums, using N_3, leads to

$$L_{N_3}\big((f \circ \Phi)|\det[\mathbf{D}\Phi]|\big) \ge \frac{1}{1+\eta} L_{\Phi(\mathcal{D}_{N_3}(\mathbb{R}^n))}(f) - \frac{\eta L(K\sqrt{n})^n}{1-\eta}. \qquad A16.25$$

Denoting by N the larger of N_2 and N_3, we get

$$\frac{1}{1+\eta} L_{\Phi(\mathcal{D}_N(\mathbb{R}^n))}(f) - \frac{\eta L(K\sqrt{n})^n}{1-\eta} \le L_N\big((f \circ \Phi)|\det[\mathbf{D}\Phi]|\big)$$

$$\le U_N\big((f \circ \Phi)|\det[\mathbf{D}\Phi]|\big) \le \frac{1}{1-\eta} U_{\Phi(\mathcal{D}_N(\mathbb{R}^n))}(f) + \frac{\eta L(K\sqrt{n})^n}{1-\eta}. \qquad A16.26$$

We can choose N larger yet so that the difference between upper and lower sums

$$U_{\Phi(\mathcal{D}_N(\mathbb{R}^n))}(f) - L_{\Phi(\mathcal{D}_N(\mathbb{R}^n))}(f), \qquad A16.27$$

is less than η, since f is integrable and $\Phi(\mathcal{D}(\mathbb{R}^n))$ is a nested partition. Denote by a the upper sum in Equation A16.27, by b the lower sum, and by c the term $L(K\sqrt{n})^n$. Then $|a - b| < \eta$, and

$$
\begin{aligned}
\frac{a+\eta c}{1-\eta} - \frac{b-\eta c}{1+\eta} &= \frac{(1+\eta)(a+\eta c) - (1-\eta)(b-\eta c)}{1-\eta^2} \\
&= \frac{a-b+\eta a+\eta b+2\eta c}{1-\eta^2} < \frac{\eta}{1-\eta^2}(1+a+b+2c),
\end{aligned}
\qquad A16.28
$$

which will be arbitrarily small when η is arbitrarily small, so

$$U_N\big((f\circ\Phi)|\det[\mathbf{D}\Phi]|\big) - L_N\big((f\circ\Phi)|\det[\mathbf{D}\Phi]|\big) \qquad A16.29$$

can be made arbitrarily small by choosing η sufficiently small (and the corresponding N sufficiently large). This proves that $(f\circ\Phi)|\det[\mathbf{D}\Phi]|$ is integrable, and that the integral is equal to the integral of f. $\square$

A.17 Two Extra Results in Topology

In this section, we give two more properties of compact subsets of $\mathbb{R}^n$, which we will need for proofs in Appendices A.18 and A.22. They are not particularly harder than the ones in Section 1.6, but it seemed a bad idea to load down that section with results which we did not need immediately.

Note that the hypothesis that the X_k are compact is essential. For instance, the intervals $(0, 1/n)$ form a decreasing intersection of non-empty sets, but their intersection is empty; similarly, the sequence of unbounded intervals $[k, \infty)$ is a decreasing sequence of non-empty closed subsets, but its intersection is also empty.

Theorem A17.1 (Decreasing intersection of nested compact sets). *If $X_k \subset \mathbb{R}^n$ is a sequence of non-empty compact sets, such that $X_1 \supset X_2 \supset \dots$, then*

$$\bigcap_{k=1}^{\infty} X_k \neq \emptyset. \qquad A17.1$$

Proof. For each k, choose $\mathbf{x}_k \in X_k$ (using the hypothesis that $X_k \neq \emptyset$). Since this is in particular a sequence in X_1, choose a convergent subsequence $\mathbf{x}_{k_i}$. The limit of this sequence is a point of the intersection $\cap_{k=1}^{\infty} X_k$, since the sequence beyond $\mathbf{x}_k, k \geq m$ is contained in X_m, hence the limit also since each X_m is closed. $\square$

The next proposition constitutes the definition of "compact" in general topology; all other properties of compact sets can be derived from it. It will not play such a central role for us, but we will need it in the proof of the general Stokes's theorem in Appendix A.22.

Theorem A17.2 (Heine-Borel theorem). *If $X \subset \mathbb{R}^n$ is compact, and $U_i \subset \mathbb{R}^n$ is a family of open subsets such that $X \subset \cup U_i$, then there exist finitely many of the open sets, say $U_1, \dots, U_N$, such that*

$$X \subset U_1 \cup \dots \cup U_N. \qquad A17.2$$

Proof. This is very similar to Theorem 1.6.2. We argue by contradiction: suppose it requires infinitely many of the U_i to cover X.

The set X is contained in a box $-10^N \leq x_i < 10^N$ for some N. Decompose this box into finitely many closed boxes of side 1 in the obvious way. If each of these boxes is covered by finitely many of the U_i, then all of X is also, so at least one of the boxes B_0 requires infinitely many of the U_i to cover it.

Now cut up B_0 into 10^n closed boxes of side $1/10$ (in the plane, 100 boxes; in $\mathbb{R}^3$, 1,000 boxes). At least one of these smaller boxes must again require infinitely many of the U_i to cover it. Call such a box B_1, and keep going: cut up B_1 into 10^n boxes of side $1/10^2$; again, at least one of these boxes must require infinitely many U_i to cover it; call one such box B_2, etc.

The boxes B_i form a decreasing sequence of compact sets, so there exists a point $\mathbf{x} \in \cap B_i$. This point is in X, so it is in one of the U_i. That U_i contains the ball of radius r around $\mathbf{x}$ for some $r > 0$, and hence around all the boxes B_j for j sufficiently large (to be precise, as soon as $\sqrt{n}/10^j < r$).

This is a contradiction. $\square$

A.18 PROOF OF THE DOMINATED CONVERGENCE THEOREM

The Italian mathematician Cesare Arzela proved the dominated convergence theorem in 1885.

Many famous mathematicians (Banach, Riesz, Landau, Hausdorff) have contributed proofs of their own. But the main contribution is certainly Lebesgue's; the result (in fact, a stronger result) is quite straightforward when Lebesgue integrals are used. The usual attitude of mathematicians today is that it is perverse to prove this result for the Riemann integral, as we do here; they feel that one should put it off until the Lebesgue integral is available, where it is easy and natural. We will follow the proof of a closely related result due to Eberlein, *Comm Pure App. Math.*, 10 (1957), pp. 357-360; the trick of using the parallelogram law is due to Marcel Riesz.

Theorem 4.11.12 (The dominated convergence theorem for Riemann integrals). *Let $f_k : \mathbb{R}^n \to \overline{\mathbb{R}}$ be a sequence of I-integrable functions, and let $f, g : \mathbb{R}^n \to \overline{\mathbb{R}}$ be two I-integrable functions such that*

(1) *$|f_k| \leq g$ for all k;*
(2) *the set of $\mathbf{x}$ where $\lim_{k \to \infty} f_k(\mathbf{x}) \neq f(\mathbf{x})$ has volume 0.*

Then

$$\lim_{k \to \infty} \int_{\mathbb{R}^n} f_k |d^n \mathbf{x}| = \int_{\mathbb{R}^n} f |d^n \mathbf{x}|.$$

Note that the term I-integrable refers to a form of the *Riemann* integral; see Definition 4.11.2.

Monotone convergence

We will first prove an innocent-looking result about interchanging limits and integrals. Actually, much of the difficulty is concentrated in this proposition, which could be used as the basis of the entire theory.

Equation A18.1: since $f_k \leq 1$, and A_k is a subset of the unit cube, which has volume 1, if $\mathrm{vol}_n A_k < K$, then the first term on the right-hand side of Equation A18.1 is less than K. The second term contributes at most K because $f_k \leq 1$, and $Q - A_k$ is a subset of the unit cube.

Recall (Definition 4.1.8) that we denote by $L(f)$ the lower integral of f:

$$L(f) = \lim_{N \to \infty} L_N(f).$$

Equation A18.2: the last inequality isn't quite obvious. It is enough to show that

$$L_N(\sup(f_k(\mathbf{x}), K)) \leq K + L_N(\chi_{A_k})$$

for any N. Take any cube $C \in \mathcal{D}_N(\mathbb{R}^n)$. Then either $m_C(f_k) \leq K$, in which case,

$$m_C(f_k)\, \mathrm{vol}_n C \leq K\, \mathrm{vol}_n C,$$

or $m_C(f_k) > K$. In the latter case, since $f_k \leq 1$,

$$m_C(f_k)\, \mathrm{vol}_n C \leq \mathrm{vol}_n C.$$

The first case contributes at most $K\,\mathrm{vol}_n Q = K$ to the lower integral. The second contributes at most $L_N(\chi_{A_k}) = \underline{\mathrm{vol}}_n(A_k)$, since any cube for which $m_C(f_k) > K$ is entirely within A_k, and thus contributes to the lower sum. This is why the possible non-pavability of A_k is just an irritant. For typical non-pavable sets, like the rationals or the irrationals, the lower volume is 0. Here there definitely are whole dyadic cubes completely contained in A_k.

Proposition A18.1 (Monotone convergence). *Let f_k be a sequence of integrable functions, all with support in the unit cube $Q \subset \mathbb{R}^n$, and satisfying $1 \geq f_1 \geq f_2 \geq \cdots \geq 0$. Let $B \subset Q$ be a pavable subset with $\mathrm{vol}_n(B) = 0$, and suppose that for all $\mathbf{x} \in Q$,*

$$\lim_{k \to \infty} f_k(\mathbf{x}) = 0 \quad \text{if } \mathbf{x} \notin B.$$

Then

$$\lim_{k \to \infty} \int_{\mathbb{R}^n} f_k |d^n\mathbf{x}| = \int_{\mathbb{R}^n} \lim_{k \to \infty} f_k |d^n\mathbf{x}| = 0.$$

Proof. The sequence $\int_{\mathbb{R}^n} f_k |d^n\mathbf{x}|$ is non-increasing and non-negative, so it has a limit, which we call $2K$. We will suppose that $K > 0$, and derive a contradiction.

The object is to find a set of positive volume of points $\mathbf{x} \notin B$ such that $\lim_{k \to \infty} f_k(\mathbf{x}) \geq K$, which contradicts the hypothesis. Let $A_k \subset Q$ be the set $A_k = \{\mathbf{x} \in Q \mid f_k(\mathbf{x}) \geq K\}$, so that since the sequence f_k is non-increasing, the sets A_k are nested: $A_1 \supset A_2 \supset \ldots$ What we want to show is that the intersection of the A_k is bigger than B.

It is tempting to rephrase this, and say that we need to show that $\mathrm{vol}_n \cap_k(A_k) \neq 0$, and that this is true, since the A_k are nested, and $\mathrm{vol}_n A_k \geq K$ for each k. Indeed, if $\mathrm{vol}_n A_k < K$ then

$$\underbrace{\int_{\mathbb{R}^n} f_k |d^n\mathbf{x}|}_{\geq 2K} = \int_Q f_k |d^n\mathbf{x}| = \int_{A_k} f_k |d^n\mathbf{x}| + \int_{Q-A_k} f_k |d^n\mathbf{x}| < K + K, \quad A18.1$$

which contradicts the assumption that $\int_Q f_k |d^n\mathbf{x}| \geq 2K$. Thus the intersection should have volume at least K, and since B has volume 0, there should be points in the intersection that are not in B, i.e., points $\mathbf{x} \notin B$ such that $\lim_{k \to \infty} f_k(\mathbf{x}) \geq K$.

The problem with this argument is that A_k might fail to be pavable (see Exercise A18.1), so we cannot blithely speak of its volume. In addition, even if the A_k are pavable, their intersection might not be pavable (see Exercise A18.2). In this particular case this is just an irritant, not a fatal flaw; we need to doctor the A_k's a bit. We can replace $\mathrm{vol}_n(A_k)$ by the *lower volume*, $\underline{\mathrm{vol}}_n(A_k)$, which can be thought of as the lower integral: $\underline{\mathrm{vol}}_n(A_k) = L(\chi_{A_k})$, or as the sum of the volumes of all the disjoint dyadic cubes of all sizes contained in A_k. Even this lower volume is at least K since $f_k(\mathbf{x}) = \inf(f_k(\mathbf{x}), K) + \sup(f_k(\mathbf{x}), K) - K$:

$$2K \leq \int_Q f_k |d^n\mathbf{x}| = \int_Q \inf(f_k(\mathbf{x}), K)|d^n\mathbf{x}| + \int_Q \sup(f_k(\mathbf{x}), K)|d^n\mathbf{x}| - K$$

$$\leq \int_Q \sup(f_k(\mathbf{x}), K)|d^n\mathbf{x}| = L(\sup(f_k(\mathbf{x}), K)) \leq K + \underline{\mathrm{vol}}_n(A_k). \quad A18.2$$

We want to see that if we remove all of B from the A_k, what is left still has positive volume. We start with these:

$$\underline{\mathrm{vol}}_n(A_k) \geq K$$

$$\mathrm{vol}_n B' < \frac{K}{3}$$

$$\underline{\mathrm{vol}}_n A'_k = \underline{\mathrm{vol}}_n(A_k - B') \geq \frac{2K}{3},$$

$$\underline{\mathrm{vol}}_n(A'_k - A''_k) < \epsilon^k.$$

Since the difference $(A'_k - A''_k)$ is very small, the A''_k almost fill up the A'_k, so their volume is also non-negligible. They are not nested, but the A'''_k are:

$$A'''_1 = A''_1,$$
$$A'''_2 = A''_1 \cap A''_2,$$
$$A'''_3 = A''_1 \cap A''_2 \cap A''_3,$$

and so on.

Because the A_k are nested,

$$\underline{\mathrm{vol}} A_k = \underline{\mathrm{vol}}(\cap_k A_k);$$

similarly,

$$\underline{\mathrm{vol}} A'_k = \underline{\mathrm{vol}}(\cap_k A'_k).$$

So $\underline{\mathrm{vol}}_n(A_k) \geq K$. Now we want to find points in the intersection of the A_k that are not in B. To do this we adjust our A_k's. First, choose a number N such that the union of all the dyadic cubes in $\mathcal{D}_N(\mathbb{R}^n)$ whose closures intersect B have total volume $< K/3$. Let B' be the union of all these cubes, and let $A'_k = A_k - B'$. Note that the A'_k are still nested, and $\underline{\mathrm{vol}}_n(A'_k) \geq 2K/3$. Next choose ϵ so small that $\epsilon/(1-\epsilon) < 2K/3$, and for each k let $A''_k \subset A'_k$ be a finite union of closed dyadic cubes, such that $\underline{\mathrm{vol}}_n(A'_k - A''_k) < \epsilon^k$. Unfortunately, now the A''_k are no longer nested, so define

$$A'''_k = A''_1 \cap A''_2 \cap \cdots \cap A''_k. \qquad A18.3$$

We need to show that the A'''_k are non-empty; this is true, since

$$\underbrace{\mathrm{vol}\, A'''_k)}_{\mathrm{vol}_n(A''_1 \cap A''_2 \cap \cdots \cap A''_k)} > \underbrace{\mathrm{vol}\, A'_k}_{\underline{\mathrm{vol}}_n(A'_1 \cap A'_2 \cap \cdots \cap A'_k)} - (\epsilon + \epsilon^2 + \cdots + \epsilon^k) \geq \frac{2K}{3} - \frac{\epsilon}{1-\epsilon} > 0.$$

$$A18.4$$

Now the punchline: The A'''_k form a decreasing intersection of compact sets, so their intersection is non-empty (see Theorem A17.1). Let $\mathbf{x} \in \cap_k A'''_k$, then all $f_k(\mathbf{x}) \geq K$, but $\mathbf{x} \notin B$. This is the contradiction we were after. $\square$

We use Proposition A18.1 below.

Lemma A18.2. *Let h_k be a sequence of integrable non-negative functions on Q, and h an integrable function on Q, satisfying $0 \leq h(\mathbf{x}) \leq 1$. If $B \subset Q$ is a pavable set of volume 0, and if $\sum_{k=1}^{\infty} h_k(\mathbf{x}) \geq h(\mathbf{x})$ when $\mathbf{x} \notin B$, then*

$$\sum_{k=1}^{\infty} \int_Q h_k(\mathbf{x}) |d^n \mathbf{x}| \geq \int_Q h(\mathbf{x}) |d^n \mathbf{x}|. \qquad A18.5$$

Proof. Set $g_k = \sum_{i=1}^{k} h_k$, which is a non-decreasing sequence of non-negative integrable functions, and $g'_k = \inf(g_k, h)$, which is still a non-decreasing sequence of non-negative integrable functions. Finally, set $f_k = h - g'_k$; these functions satisfy the hypotheses of Proposition A18.1. So

$$0 = \lim_{k \to \infty} \int f_k |d^n \mathbf{x}| = \int h |d^n \mathbf{x}| - \lim_{k \to \infty} \int g'_k |d^n \mathbf{x}| \geq \int h |d^n \mathbf{x}| - \lim_{k \to \infty} \int g_k |d^n \mathbf{x}|$$

$$= \int h |d^n \mathbf{x}| - \sum_{k=1}^{\infty} \int h_k |d^n \mathbf{x}|. \quad \square \qquad A18.6$$

Simplifications to the dominated convergence theorem

Let us simplify the statement of Theorem 4.11.12. First, by subtracting f from all the f_k, and replacing g by $g + |f|$, we may assume $f = 0$.

Second, by writing the $f_k = f_k^+ - f_k^-$, we see that it is enough to prove the result when all f_k satisfy $f_k \geq 0$.

Recall that f_k^+ and f_k^- are defined in Definition 4.3.4.

Since f is the limit of the f_k, and we have assumed $f = 0$ (and therefore $\int f = 0$), we need to show that L, the limit of the integrals, is also 0.

The point of this argument about "if $L \neq 0$" is to show that if there is a counterexample to Theorem 4.11.12, there is a counterexample when the functions are bounded by a single constant and have support in a single bounded set. So it is sufficient to prove the statement for such functions.

The main simplification is that the functions f_k all have their support in a single bounded set, the unit cube. Notice that the conclusion is identical to the conclusion of Proposition A18.1; the difference is that we are no longer requiring the f_k to form a nonincreasing sequence.

This makes the proof a great deal harder; when we call Proposition A18.3 a "simplified" version of the dominated convergence theorem, we don't mean that its proof is simple. It is among the harder proofs in this book, and certainly it is the trickiest.

Third, since when $f_k \geq 0$,

$$0 \leq \int_{\mathbb{R}^n} f_k |d^n\mathbf{x}| \leq \int_{\mathbb{R}^n} g |d^n\mathbf{x}|, \qquad A18.7$$

by passing to a subsequence we may assume that $\lim_{k\to\infty} \int_{\mathbb{R}^n} f_k(\mathbf{x})|d^n\mathbf{x}|$ exists. Call that limit L. If the theorem is true, then $L = 0$. Assume the theorem is false; i.e., that $L > 0$. We will show that if it is false for f_k, it is also false for truncated functions $[f_k]_R$.

So assume $L > 0$. Then there exists R such that

$$\left| \int_{\mathbb{R}^n} g|d^n\mathbf{x}| - \int_{\mathbb{R}^n} [g]_R|d^n\mathbf{x}| \right| < L/2. \qquad A18.8$$

It is then also true that

$$\left| \overbrace{\int_{\mathbb{R}^n} f_k|d^n\mathbf{x}|}^{\text{limit of this is } L} - \int_{\mathbb{R}^n} [f_k]_R|d^n\mathbf{x}| \right| < L/2. \qquad A18.9$$

Thus passing to a further subsequence if necessary, we may assume that

$$\lim_{k\to\infty} \int_{\mathbb{R}^n} [f_k]_R|d^n\mathbf{x}| \geq L/2. \qquad A18.10$$

Thus if the theorem is false, it will also be false for the functions $[f_k]_R$, so if it is true for the functions $[f_k]_R$, it is true in general; i.e., it is enough to prove the theorem for f_k satisfying $0 \leq f_k \leq R$, with support in the ball of radius R. By replacing these f_k by $\frac{[f_k]_R}{R}$, we may assume that our functions are bounded by 1, and by covering the ball of radius R by dyadic cubes of side 1 and making the argument for each separately, we may assume that all functions have support in one such cube.

To lighten notation, let us restate our theorem after all these simplifications.

Proposition A18.3 (Simplified dominated convergence theorem).

Suppose f_k is a sequence of integrable functions all satisfying $0 \leq f_k \leq 1$, and all having their support in the unit cube Q. If there exists a pavable subset $B \subset Q$ with $\mathrm{vol}_n(B) = 0$ such that $f_k(\mathbf{x}) \to 0$ when $\mathbf{x} \notin B$, then

$$\lim_{k\to\infty} \int_{\mathbb{R}^n} f_k|d^n\mathbf{x}| = \int_{\mathbb{R}^n} \lim_{k\to\infty} f_k|d^n\mathbf{x}| = 0.$$

Proof of the dominated convergence theorem

We will prove the dominated convergence theorem by proving Proposition

A18.3. By passing to a subsequence, we may assume that $\lim_{k\to\infty}\int_{\mathbb{R}^n} f_k|d^n\mathbf{x}| = C$; we will assume that $C > 0$ and derive a contradiction. Let us consider the set K_p of linear combinations

$$\sum_{m=p}^{\infty} a_m f_m \qquad\qquad A18.11$$

Note that the functions in K_p are all integrable, since they are *finite* linear combinations of integrable functions, all bounded by 1, and all have support in Q.

with all $a_m \geq 0$, only finitely many of the terms nonzero (so that the sum is actually finite), and $\sum_{m=p}^{\infty} a_m = 1$.

We will need two properties of the functions $g \in K_p$. First, for any point $\mathbf{x} \in (Q - B)$, and any sequence $g_p \in K_p$, we will have $\lim_{p\to\infty} g_p(\mathbf{x}) = 0$. Indeed, for any $\epsilon > 0$ we can find N such that all $f_m(\mathbf{x})$ satisfy $0 \leq f_m(\mathbf{x}) \leq \epsilon$ when $m \geq N$, so that when $p \geq N$ we have

$$g_p(\mathbf{x}) = \sum_{m=p}^{\infty} a_m f_m(\mathbf{x}) \leq \sum_{m=p}^{\infty} (a_m \epsilon) = \epsilon. \qquad\qquad A18.12$$

Second, again if $g_p \in K_p$, we have $\lim_{p\to\infty}\int_Q g_p|d^n\mathbf{x}| = C$. Indeed, choose $\epsilon > 0$, and N so large that $|\int_Q f_m|d^n\mathbf{x}| - C| < \epsilon$ when $m \geq N$. Then, when $p > N$ we have

$$\left|\int_Q g_p(\mathbf{x})|d^n\mathbf{x}| - C\right| = \left|\left(\sum_{m=p}^{\infty} a_m \int_Q f_m(\mathbf{x})|d^n\mathbf{x}|\right) - C\right| \qquad\qquad A18.13$$

$$= \left|\sum_{m=p}^{\infty} a_m \underbrace{\left(\int_Q f_m(\mathbf{x})|d^n\mathbf{x}|\right) - C}_{<\epsilon}\right| \leq \sum_{m=p}^{\infty} (a_m\epsilon) = \epsilon.$$

Let $d_p = \inf_{g\in K_p}\int_Q g^2(\mathbf{x})|d^n\mathbf{x}|$. Clearly the d_p form a non-decreasing sequence bounded by 1, hence convergent. Choose $g_p \in K_p$ so that $\int_Q g_p^2 < d_p + 1/p$.

The appearance of integrals of squares of functions in this argument appears to be quite unnatural. The reason they are used is that it is possible to express $(g_p - g_q)^2$ algebraically in terms of $(g_p + g_q)^2$, g_p^2, and g_q^2. We could write

$$|g_p - g_q| = 2\sup(g_p, g_q) - g_p - g_q,$$

but we don't know much about $\sup(g_p, g_q)$.

Lemma A18.4. *For all $\epsilon > 0$, there exists N such that when $p, q \geq N$,*

$$\int_Q (g_p - g_q)^2|d^n\mathbf{x}| < \epsilon. \qquad\qquad A18.14$$

Proof of Lemma A18.4. Algebra says that

$$\int_Q \left(\frac{1}{2}(g_p - g_q)\right)^2 |d^n\mathbf{x}| + \int_Q \left(\frac{1}{2}(g_p + g_q)\right)^2 |d^n\mathbf{x}| = \frac{1}{2}\int_Q g_p^2|d^n\mathbf{x}| + \frac{1}{2}\int_Q g_q^2|d^n\mathbf{x}|. \qquad A18.15$$

But $\frac{1}{2}(g_p + g_q)$ is itself in K_N, so $\int_Q(\frac{1}{2}(g_p + g_q))^2|d^n\mathbf{x}| \geq d_N$, so

$$\int_Q \left(\frac{1}{2}(g_p - g_q)\right)^2 \leq \frac{1}{2}\left(d_p + \frac{1}{p}\right) + \frac{1}{2}\left(d_q + \frac{1}{q}\right) - d_N. \qquad\qquad A18.16$$

Since the d_p converge, we see that this can be made arbitrarily small. $\square$

Using this lemma, we can choose a further subsequence h_q of the g_p so that

$$\sum_{q=1}^{\infty} \left(\int_Q (h_q - h_{q+1})^2 |d^n\mathbf{x}| \right)^{1/2} \qquad A18.17$$

converges. Notice that

$$h_q(\mathbf{x}) = (h_q - h_{q+1})(\mathbf{x}) + (h_{q+1} - h_{q+2})(\mathbf{x}) + \dots \quad \text{when } \mathbf{x} \notin B, \qquad A18.18$$

since

$$h_q(\mathbf{x}) - \sum_{i=q}^{m} (h_{i+1} - h_i)(\mathbf{x}) = h_{m+1}(\mathbf{x}), \qquad A18.19$$

which tends to 0 when $m \to \infty$ and $\mathbf{x} \notin B$ by Equation A18.12.

In particular, $h_q \leq \sum_{m=q}^{\infty} |h_{m+1} - h_m|$, and we can apply Lemma A18.2 to get the first inequality below; the second follows from Schwarz's lemma for integrals:

The second inequality follows from Schwarz's lemma for integrals (Exercise A18.3). Write

$$\left(\int_Q |h_m - h_{m+1}| \cdot 1 |d^n\mathbf{x}| \right)^2$$

$$\leq \left(\int_Q |h_m - h_{m+1}|^2 |d^n\mathbf{x}| \right)$$

$$\left(\int_Q 1^2 |d^n\mathbf{x}| \right)$$

$$= \left(\int_Q |h_m - h_{m+1}|^2 |d^n\mathbf{x}| \right).$$

$$\int_Q h_q |d^n\mathbf{x}| \leq \sum_{m=q}^{\infty} \int_Q |h_m - h_{m+1}| |d^n\mathbf{x}| \leq \sum_{m=q}^{\infty} \left(\int_Q (h_m - h_{m+1})^2 |d^n\mathbf{x}| \right)^{1/2}.$$
$$A18.20$$

The sum on the right can be made arbitrarily small by taking q sufficiently large. This contradicts Equation A18.13, and the assumption $C > 0$. This proves Proposition A18.3, hence also Theorem 4.11.12. $\quad\square$

A.19 Justifying the Change of Parametrization

Before restating and proving Theorem 5.2.8, we will prove the following proposition, which we will need in our proof. The proposition also explains why Definition 5.2.1 of k-dimensional volume 0 of a subset of $\mathbb{R}^n$ is reasonable.

We could state Proposition A19.1 projecting onto any k coordinates; they don't have to be the first.

Proposition A19.1. *If $X \subset \mathbb{R}^n$ is a bounded subset of k-dimensional volume 0, then its projection onto the first k coordinates also has k-dimensional volume 0.*

Proof. Let $\pi : \mathbb{R}^n \to \mathbb{R}^k$ denote the projection of $\mathbb{R}^n$ onto the first k coordinates. Choose $\epsilon > 0$, and N so large that

$$\sum_{\substack{C \in \mathcal{D}_N(\mathbb{R}^n) \\ C \cap X \neq \phi}} \left(\frac{1}{2^N} \right)^k < \epsilon. \qquad A19.1$$

Then

$$\epsilon > \sum_{\substack{C \in \mathcal{D}_N(\mathbb{R}^n) \\ C \cap X \neq \phi}} \left(\frac{1}{2^N} \right)^k \geq \sum_{\substack{C_1 \in \mathcal{D}_N(\mathbb{R}^k) \\ C_1 \cap \pi(X) \neq \phi}} \left(\frac{1}{2^N} \right)^k, \qquad A19.2$$

since for every k-dimensional cube $C_1 \in \mathcal{D}_N(\mathbb{R}^k)$ such that $C_1 \cap \pi(X) \neq \phi$, there is at least one n-dimensional cube $C \in \mathcal{D}_N(\mathbb{R}^n)$ with $C \subset \pi^{-1}(C_1)$ such that $C \cap X \neq \phi$. Thus $\mathrm{vol}_k(\pi(X)) < \epsilon$ for any $\epsilon > 0$. $\square$

Remark. The sum to the far right of Equation A19.2 is precisely our old definition of volume, vol_k in this case; we are summing over cubes C_1 that are in $\mathbb{R}^k$. In the sum to its left, we have the side length to the kth power for cubes in $\mathbb{R}^n$; it's less clear what that is measuring. $\triangle$

Justifying the change of parametrization

Now we will restate and prove Theorem 5.2.8, which explains why we can apply the change of variables formula to Φ, the function giving change of parametrization.

Let U_1 and U_2 be subsets of $\mathbb{R}^k$, and let γ_1 and γ_2 be two parametrizations of a k-dimensional manifold M:

$$\gamma_1 : \overline{U}_1 \to M \quad \text{and} \quad \gamma_2 : \overline{U}_2 \to M. \qquad A19.3$$

Following the notation of Definition 5.2.2, denote by X_1 the negligible "trouble spots" of γ_1, and by X_2 the trouble spots of γ_2 (illustrated by Figure A19.1, which we already saw in Section 5.2). Call

$$Y_1 = (\gamma_2^{-1} \circ \gamma_1)(X_1), \quad \text{and} \quad Y_2 = (\gamma_1^{-1} \circ \gamma_2)(X_2). \qquad A19.4$$

Theorem 5.2.8. *Both $U_1^{ok} = U_1 - (X_1 \cup Y_2)$ and $U_2^{ok} = U_2 - (X_2 \cup Y_1)$ are open subsets of $\mathbb{R}^k$ with boundaries of k-dimensional volume 0, and*

$$\Phi : U_1^{ok} \to U_2^{ok} = \gamma_2^{-1} \circ \gamma_1$$

is a C^1 diffeomorphism with locally Lipschitz inverse.

Proof. The mapping Φ is well defined and injective on U_1^{ok}. It is well defined because its domain excludes Y_2; it is injective because its domain excludes X_1.

We need to check two different kinds of things: that $\Phi : U_1^{ok} \to U_2^{ok}$ is a diffeomorphism with locally Lipschitz derivative, and that the boundaries of U_1^{ok} and U_2^{ok} have volume 0.

For the first part, it is enough to show that Φ is of class C^1 with locally Lipschitz derivative, since the same proof applied to

$$\Psi = \gamma_1^{-1} \circ \gamma_2 : U_2^{ok} \to U_1^{ok} \qquad A19.5$$

will show that the inverse is also of class C^1 with locally Lipschitz derivative.

Everything about the differentiability stems from the following lemma.

Lemma A19.2. *Let $M \subset \mathbb{R}^n$ be a k-dimensional manifold, U_1, $U_2 \subset \mathbb{R}^k$, and $\gamma_1 : U_1 \to M$, $\gamma_2 : U_2 \to M$ be two maps of class C^1, each with derivative*

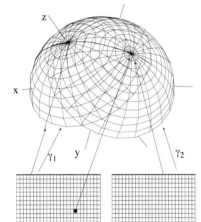

FIGURE A19.1.

Here X_1 consists of the dark line at the top of the rectangle at left, which is mapped by γ_1 to a pole and then by γ_2^{-1} to a point in the rectangle at right. The dark box in the rectangle at left is Y_2, which is mapped to a pole of γ_2 and then to the dark line at right. Excluding X_1 from the domain of Φ ensures that it is injective (one to one); excluding Y_2 ensures that it is well defined. Excluding X_2 and Y_1 from the range ensures that when X_1 and Y_2 are removed from the domain, Φ is still surjective (onto).

that is both Lipschitz and injective. Suppose that $\gamma_1(\mathbf{x}_1) = \gamma_2(\mathbf{x}_2) = \mathbf{x}$. Then there exist neighborhoods V_1 of $\mathbf{x}_1$ and V_2 of $\mathbf{x}_2$ such that $\gamma_2^{-1} \circ \gamma_1$ is defined on V_1 and is a diffeomorphism of V_1 onto V_2.

This looks quite a lot like the chain rule, which asserts that a composition of C^1 mappings is C^1, and that the derivative of the composition is the composition of the derivatives. The difficulty in simply applying the chain rule is that we have not defined what it means for γ_2^{-1} to be differentiable, since it is only defined on a subset of M, not on an open subset of $\mathbb{R}^n$. It is quite possible (and quite important) to define what it means for a function defined on a manifold (or on a subset of a manifold) to be differentiable, and to state an appropriate chain rule, etc., but we decided not to do it in this book, and here we pay for that decision.

Proof. By our definition of a manifold, there exist subspaces E_1, E_2 of $\mathbb{R}^n$, an open subset $W \subset E_1$, and a mapping $\mathbf{f} : W \to E_2$ such that near $\mathbf{x}$, M is the graph of $\mathbf{f}$. Let $\pi_1 : \mathbb{R}^n \to E_1$ denote the projection of $\mathbb{R}^n$ onto E_1, and denote by $F : W \to \mathbb{R}^n$ the mapping

$$F(\mathbf{y}) = \mathbf{y} + \mathbf{f}(\mathbf{y}) \qquad\qquad A19.6$$

so that $\pi_1(F(\mathbf{y})) = \mathbf{y}$.

Consider the mapping $\pi_1 \circ \gamma_2$ defined on some neighborhood of $\mathbf{x}_2$, and with values in some neighborhood of $\pi_1(\mathbf{x})$. Both domain and range are open subsets of $\mathbb{R}^k$, and $\pi_1 \circ \gamma_2$ is of class C^1. Moreover, $[\mathbf{D}(\pi_1 \circ \gamma_2)(\mathbf{x}_2)]$ is invertible, for the following reason. The derivative

$$[\mathbf{D}\gamma_2(\mathbf{x}_2)]$$

is injective, and its image is contained in (in fact is exactly) the tangent space $T_{\mathbf{x}}M$. The mapping π_1 has as its kernel E_2, which intersects $T_{\mathbf{x}}M$ only at the origin. Thus the kernel of $[\mathbf{D}(\pi_1 \circ \gamma_2)(\mathbf{x}_2)]$ is $\{0\}$, which means that $[\mathbf{D}(\pi_1 \circ \gamma_2)(\mathbf{x}_2)]$ is injective. But the domain and range are of the same dimension k, so $[\mathbf{D}(\pi_1 \circ \gamma_2)(\mathbf{x}_2)]$ is invertible.

We can thus apply the inverse function theorem, to assert that there exists a neighborhood W_1 of $\pi_1(\mathbf{x})$ in which $\pi_1 \circ \gamma_2$ has a C^1 inverse. In fact, the inverse is precisely $\gamma_2^{-1} \circ F$, which is therefore of class C^1 on W_1. Furthermore, on the graph, i.e., on M, $F \circ \pi_1$ is the identity.

Why not four? Because $\gamma_2^{-1} \circ F$ should be viewed as a single mapping, which we just saw is differentiable. We don't have a definition of what it would mean for γ_2^{-1} by itself to be differentiable.

Now write

$$\gamma_2^{-1} \circ \gamma_1 = \gamma_2^{-1} \circ F \circ \pi_1 \circ \gamma_1. \qquad\qquad A19.7$$

This represents $\gamma_2^{-1} \circ \gamma_1$ as a composition of three (not four) C^1 mappings, defined on the neighborhood $\gamma_1^{-1}(F(W_1))$ of $\mathbf{x}_1$, so the composition is of class C^1 by the chain rule. We leave it to you to check that the derivative is locally Lipschitz. To see that $\gamma_2^{-1} \circ \gamma_1$ is locally invertible, with invertible derivative, notice that we could make the argument exchanging γ_1 and γ_2, which would construct the inverse map. $\square$ Lemma A19.2

We now know that $\Phi : U_1^{ok} \to U_2^{ok}$ is a diffeomorphism.

The only thing left to prove is that the boundaries of U_1^{ok} and U_2^{ok} have volume 0. It is enough to show it for U_1^{ok}. The boundary of U_1^{ok} is contained in the union of

(1) the boundary of U_1, which has volume 0 by hypothesis;

(2) X_1, which has volume 0 by hypothesis; and

(3) Y_2, which also has volume 0, although this is not obvious.

First, it is clearly enough to show that $Y_2 - X_1$ has volume 0; the part of Y_2 contained in X_1 (if any) is taken care of since X_1 has volume 0. Next, it is enough to prove that every point $\mathbf{y} \in (Y_2 - X_1)$ has a neighborhood W_1 such that $Y_2 \cap W_1$ has volume 0; we will choose a neighborhood on which $\gamma_1^{-1} \circ F$ is a diffeomorphism. We can write

$$Y_2 = \gamma_1^{-1}\big(\gamma_2(X_2)\big) = \gamma_1^{-1} \circ F \circ \pi_1 \circ \gamma_2(X_2). \qquad A19.8$$

By hypothesis, $\gamma_2(X_2)$ has k-dimensional volume 0, so by Proposition A19.1, $\pi_1 \circ \gamma_2(X_2)$ also has volume 0. Therefore, the result follows from Proposition A16.1, as applied to $\gamma_1^{-1} \circ F$. $\square$

A.20 Computing the Exterior Derivative

Theorem 6.7.3 (Computing the exterior derivative of a k-form).

(a) If the coefficients a of the k-form

$$\varphi = \sum_{1 \le i_1 < \cdots < i_k \le n} a_{i_1,\dots,i_k} \, dx_{i_1} \wedge \cdots \wedge dx_{i_k} \qquad 6.7.4$$

are C^2 functions on $U \subset \mathbb{R}^n$, then the limit in Equation 6.7.3 exists, and defines a $(k+1)$-form.

(b) The exterior derivative is linear over $\mathbb{R}$: if φ and ψ are k-forms on $U \subset \mathbb{R}^n$, and a and b are numbers (not functions), then

$$d(a\,\varphi + b\,\psi) = a\,d\varphi + b\,d\psi. \qquad 6.7.5$$

(c) The exterior derivative of a constant form is 0.

(d) The exterior derivative of the 0-form (i.e., function) f is given by the formula

$$df = [\mathbf{D}f] = \sum_{i=1}^{n} (D_i f)\,dx_i. \qquad 6.7.6$$

(e) If f is a function, then

$$d\left(f\,dx_{i_1} \wedge \cdots \wedge dx_{i_k}\right) = df \wedge dx_{i_1} \wedge \cdots \wedge dx_{i_k}. \qquad 6.7.7$$

Proof. First, let us prove part (d): the exterior derivative of a 0-form field, i.e., of a function, is just its derivative. This is a restatement of Theorem 1.7.12:

$$df\left(P_x^o(\vec{\mathbf{v}})\right) \overset{\text{Def. 6.7.1}}{=} \lim_{h\to 0} \frac{f(\mathbf{x}+h\vec{\mathbf{v}})-f(\mathbf{x})}{h} = [\mathbf{D}f(\mathbf{x})]\vec{\mathbf{v}}$$

$$= [\mathbf{D}f(\mathbf{x})]^\top \cdot \vec{\mathbf{v}}.$$

A20.1

Now let us prove part (e), that

$$d(f\,dx_{i_1} \wedge \cdots \wedge dx_{i_k}) = df \wedge dx_{i_1} \wedge \cdots \wedge dx_{i_k}.$$

A20.2

It is enough to prove the result at the origin; this amounts to translating φ, and it simplifies the notation. The idea is to write $f = T^0(f) + T^1(f) + R(f)$ as a Taylor polynomial with remainder at the origin, where

the constant term is $T^0(f)(\vec{\mathbf{x}}) = f(0),$

the linear term is $T^1(f)(\mathbf{x}) = D_1 f(0)x_1 + \cdots + D_n f(0)x_n = [\mathbf{D}f(0)]\vec{\mathbf{x}},$

the remainder is $|R(\vec{\mathbf{x}})| \le C|\vec{\mathbf{x}}|^2,$ for some constant $C.$

We will then see that only the linear terms contribute to the limit.

Since φ is a k-form, the exterior derivative $d\varphi$ is a $(k+1)$-form; evaluating it on $k+1$ vectors involves integrating φ over the boundary (i.e., the faces) of $P_0^o(h\vec{\mathbf{v}}_1,\ldots,h\vec{\mathbf{v}}_{k+1})$. We can parametrize those faces by the $2(k+1)$ mappings

We need to parametrize the faces of

$$P_0^o(h\vec{\mathbf{v}}_1,\ldots,h\vec{\mathbf{v}}_{k+1})$$

because we only know how to integrate over parametrized domains. There are $k+1$ mappings $\gamma_{1,i}(\mathbf{t})$, one for each i from 1 to $k+1$: for each mapping, a different $\vec{\mathbf{v}}_i$ is omitted. The same is true for $\gamma_{0,i}(\mathbf{t})$.

$$\gamma_{1,i}\begin{pmatrix} t_1 \\ \vdots \\ t_k \end{pmatrix} = \gamma_{1,i}(\mathbf{t}) = h\vec{\mathbf{v}}_i + t_1\vec{\mathbf{v}}_1 + \cdots + t_{i-1}\vec{\mathbf{v}}_{i-1} + t_i\vec{\mathbf{v}}_{i+1} + \cdots + t_k\vec{\mathbf{v}}_{k+1},$$

$$\gamma_{0,i}\begin{pmatrix} t_1 \\ \vdots \\ t_k \end{pmatrix} = \gamma_{0,i}(\mathbf{t}) = \quad t_1\vec{\mathbf{v}}_1 + \cdots + t_{i-1}\vec{\mathbf{v}}_{i-1} + t_i\vec{\mathbf{v}}_{i+1} + \cdots + t_k\vec{\mathbf{v}}_{k+1},$$

for i from 1 to $k+1$, and where $0 \le t_j \le h$ for each $j = 1,\ldots,k$. We will denote by Q_h the domain of this parametrization.

Notice that $\gamma_{1,i}$ and $\gamma_{0,i}$ have the same partial derivatives, the k vectors $\vec{\mathbf{v}}_1,\ldots,\vec{\mathbf{v}}_{k+1}$, *excluding* the vector $\vec{\mathbf{v}}_i$; we will write the integrals over these faces under the same integral sign. So we can write the exterior derivative as the limit as $h \to 0$ of the sum

$$\sum_{i=1}^{k+1}(-1)^{i-1}\frac{1}{h^{k+1}} \int_{Q_h} \underbrace{\left(f\big(\gamma_{1,i}(\mathbf{t})\big)-f\big(\gamma_{0,i}(\mathbf{t})\big)\right)}_{\text{coefficient (function of }\mathbf{t})} \underbrace{dx_{i_1}\wedge\cdots\wedge dx_{i_k}}_{k\text{-form}} \overbrace{\underbrace{(\vec{\mathbf{v}}_1,\ldots,\widehat{\vec{\mathbf{v}}}_i,\ldots,\vec{\mathbf{v}}_{k+1})}_{k\text{ vectors}}}^{\substack{\text{partial derivatives}\\ \text{of }\gamma_{1,i}\text{ and }\gamma_{0,i}}} |d^k\mathbf{t}|,$$

A20.3

where each term

$$\int_{Q_h} \left(f\big(\gamma_{1,i}(\mathbf{t})\big)-f\big(\gamma_{0,i}(\mathbf{t})\big)\right) dx_{i_1}\wedge\cdots\wedge dx_{i_k}(\vec{\mathbf{v}}_1,\ldots,\widehat{\vec{\mathbf{v}}}_i,\ldots,\vec{\mathbf{v}}_{k+1})|d^k\mathbf{t}|$$

A20.4

is the sum of three terms, of which the second is the only one that counts (most of the work is in proving that the third one doesn't count):

$$\text{constant term:} \quad \int_{Q_h}\Big(\overbrace{T^0(f)\big(\gamma_{1,i}(\mathbf{t})\big)-T^0(f)\big(\gamma_{0,i}(\mathbf{t})\big)}^{\text{coefficient of }k\text{-form}}\Big)\overbrace{dx_{i_1}\wedge\dots\wedge dx_{i_k}}^{k\text{-form}}(\vec{\mathbf{v}}_1,\dots,\widehat{\vec{\mathbf{v}}}_i,\dots,\vec{\mathbf{v}}_{k+1})|d^k\mathbf{t}|+$$

$$\text{linear term:} \quad \int_{Q_h}\Big(T^1(f)\big(\gamma_{1,i}(\mathbf{t})\big)-T^1(f)\big(\gamma_{0,i}(\mathbf{t})\big)\Big)dx_{i_1}\wedge\dots\wedge dx_{i_k}(\vec{\mathbf{v}}_1,\dots,\widehat{\vec{\mathbf{v}}}_i,\dots,\vec{\mathbf{v}}_{k+1})|d^k\mathbf{t}|+$$

$$\text{remainder:} \quad \int_{Q_h}\Big(R(f)\big(\gamma_{1,i}(\mathbf{t})\big)-R(f)\big(\gamma_{0,i}(\mathbf{t})\big)\Big)dx_{i_1}\wedge\dots\wedge dx_{i_k}(\vec{\mathbf{v}}_1,\dots,\widehat{\vec{\mathbf{v}}}_i,\dots,\vec{\mathbf{v}}_{k+1})|d^k\mathbf{t}|.$$

The constant term cancels, since

$$\underbrace{T^0(f)(\text{anything})}_{\text{constant}}-\underbrace{T^0(f)(\text{anything})}_{\text{same constant}}=0. \qquad A20.5$$

In Equation A20.6 the derivatives are evaluated at 0 because that is where the Taylor polynomial is being computed.

The second equality in Equation A20.6 comes from linearity.

For the second term, note that

$$T^1(f)\big(\gamma_{1,i}(\mathbf{t})\big)-T^1(f)\big(\gamma_{0,i}(\mathbf{t})\big)=[\mathbf{D}f(0)]\big(\overbrace{h\vec{\mathbf{v}}_i+\gamma_{0,i}(\mathbf{t})}^{\gamma_{1,i}(\mathbf{t})}\big)-[\mathbf{D}f(0)]\big(\gamma_{0,i}(\mathbf{t})\big)$$
$$=h[\mathbf{D}f(0)]\vec{\mathbf{v}}_i, \qquad A20.6$$

which is a constant with respect to $\mathbf{t}$, so the entire sum for the linear terms becomes

$$\sum_{i=1}^{k+1}(-1)^{i-1}\frac{1}{h^{k+1}}\int_{Q_h}\Big(T^1(f)\big(\gamma_{1,i}(\mathbf{t})\big)-T^1(f)\big(\gamma_{0,i}(\mathbf{t})\big)\Big)dx_{i_1}\wedge\dots\wedge dx_{i_k}$$
$$(\vec{\mathbf{v}}_1,\dots,\widehat{\vec{\mathbf{v}}}_i,\dots,\vec{\mathbf{v}}_{k+1})\,|d^k\mathbf{t}|$$

In the third line of Equation A20.7, where does the h^{k+1} in the numerator come from? One h comes from the $h[\mathbf{D}f(0)]\vec{\mathbf{v}}_i$ in Equation A20.6. The other h^k come from the fact that we are integrating over Q_h, a cube of side length h in $\mathbb{R}^k$.

$$=\sum_{i=1}^{k+1}(-1)^{i-1}\frac{h^{k+1}}{h^{k+1}}\big([\mathbf{D}f(0)]\vec{\mathbf{v}}_i\big)dx_{i_1}\wedge\dots\wedge dx_{i_k}(\vec{\mathbf{v}}_1,\dots,\widehat{\vec{\mathbf{v}}}_i,\dots,\vec{\mathbf{v}}_{k+1})$$
$$=(df\wedge dx_{i_1}\wedge\dots\wedge dx_{i_k})\big(P_0^o(\vec{\mathbf{v}}_1,\dots,\vec{\mathbf{v}}_{k+1})\big) \qquad A20.7$$

by the definition of the wedge product.

The last equality in Equation A20.7 explains why we defined the oriented boundary as we did, each part of the boundary being given the sign $(-1)^{i-1}$: it was to make it compatible with the wedge product. Exercise A20.1 asks you to elaborate on our statement that this equality is "by the definition of the wedge product."

Now for the remainder. Since we have taken into account the constant and linear terms, we can expect that the remainder will be at most of order h^2, and Theorem A9.7, the version of Taylor's theorem that gives an explicit bound for the remainder, is the tool we need. We will use the following version of that theorem, for Taylor polynomials of degree 1:

$$\Big|f(\mathbf{a}+\vec{\mathbf{h}})-P^1_{f,\mathbf{a}}(\mathbf{a}+\vec{\mathbf{h}})\Big|\le C\left(\sum_{i=1}^n|h_i|\right)^2, \qquad A20.8$$

where

$$C=\sup_{I\in\mathcal{I}_n^{k+2}}\sup_{\mathbf{c}\in[\mathbf{a},\mathbf{a}+\vec{\mathbf{h}}]}|D_If(\mathbf{c})|. \qquad A20.9$$

This more or less obviously gives

$$|R(f)\big(\gamma_{0,i}(\mathbf{t})\big)|\le Kh^2 \quad\text{and}\quad |R(f)\big(\gamma_{1,i}(\mathbf{t})\big)|\le Kh^2, \qquad A20.10$$

where K is some number concocted out of the second derivatives of f and the lengths of the $\vec{\mathbf{v}}_j$. The following lemma gives a proof and a formula for K.

The *1-norm* $\|\vec{v}\|_1$ is not to be confused with the norm $\|A\|$ of a matrix A, discussed in Section 2.8 (Definition 2.8.5).

We see that the $\sum_{i=1}^{n} |h_i|$ in Equation A20.8 can be written $\|\vec{h}\|_1$.

Lemma A20.1. *Suppose that all second partials of f are bounded by C at all points $\gamma_{0,i}(\mathbf{t})$ and $\gamma_{1,i}(\mathbf{t})$ when $\mathbf{t} \in Q_h$. Then*

$$|R(f)(\gamma_{0,i}(\mathbf{t}))| \le Kh^2 \quad and \quad |R(f)(\gamma_{1,i}(\mathbf{t}))| \le Kh^2, \qquad \text{A20.11}$$

where $K = Cn(k+1)^2(\sup_j |\vec{v}_j|)^2$.

Proof. Let us denote $\|\vec{v}\|_1 = |v_1| + \cdots + |v_n|$ (this actually is the correct mathematical name). An easy computation shows that $\|\vec{v}\|_1 \le \sqrt{n}|\vec{v}|$ for any vector $\vec{v}$.

A bit of fiddling should convince you that

$$\|\gamma_{0,i}(\mathbf{t})\|_1 \le |h|\big(\|\vec{v}_1\|_1 + \cdots + \|\vec{v}_{k+1}\|_1\big)$$
$$\le |h|(k+1) \sup_j \|\vec{v}_j\|_1 \le |h|(k+1)\sqrt{n} \sup_j |\vec{v}_j|. \qquad \text{A20.12}$$

Now Taylor's theorem, Theorem A9.7, says that

$$|R(f)(\gamma_{0,i}(\mathbf{t})| \le C\|\gamma_{0,i}(\mathbf{t})\|_1^2 \le h^2 Cn(k+1)^2(\sup_j |\vec{v}_j|)^2 = h^2 K. \qquad \text{A20.13}$$

The same calculation applies to $\gamma_{1,i}(\mathbf{t})$. $\square$

Using Lemma A20.1, we can see that the remainder disappears in the limit, using

$$\left| R(f)\big(\gamma_{1,i}(\mathbf{t})\big) - R(f)\big(\gamma_{0,i}(\mathbf{t})\big) \right| \le \left| R(f)(\gamma_{1,i}(\mathbf{t})\right| + \left| R(f)(\gamma_{0,i}(\mathbf{t})) \right| \le 2h^2 K.$$
$$\text{A20.14}$$

Inserting this into the integral leads to

$$\left| \int_{Q_h} \underbrace{R(f)\big(\gamma_{1,i}(\mathbf{t})\big) - R(f)\big(\gamma_{0,i}(\mathbf{t})\big)}_{\le 2Kh^2}\, dx_{i_1} \wedge \ldots \wedge dx_{i_k}(\vec{v}_1, \ldots, \widehat{\vec{v}}_i, \ldots, \vec{v}_{k+1})\, |d^k\mathbf{t}| \right|$$

$$\le \int_{Q_h} \Big(2h^2 K \left| dx_{i_1} \wedge \cdots \wedge dx_{i_k} (\vec{v}_1, \ldots, \widehat{\vec{v}}_i, \ldots, \vec{v}_{k+1}) \right| |d^k\mathbf{t}| \Big) \qquad \text{A20.15}$$

$$\le h^{k+2} K (\sup_j |\vec{v}_j|)^k,$$

which still disappears in the limit after dividing by h^{k+1}.

This proves part (e). Now let us prove part (a):

$$d\Big(\sum a_{i_1 \ldots i_k} dx_{i_1} \wedge \cdots \wedge dx_{i_k}\Big) P_{\mathbf{x}}^o(\vec{v}_1, \ldots, \vec{v}_{k+1})$$

$$= \lim_{h \to 0} \frac{1}{h^{k+1}} \int_{\partial P_{\mathbf{x}}^o(\vec{v}_1, \ldots, \vec{v}_{k+1})} \Big(\sum a_{i_1 \ldots i_k} dx_{i_1} \wedge \cdots \wedge dx_{i_k}\Big) \qquad \text{A20.16}$$

$$= \sum_{1 \le i_1 < \cdots < i_k \le n} \lim_{h \to 0} \left(\frac{1}{h^{k+1}} \int_{\partial P_{\mathbf{x}}^o(\vec{v}_1, \ldots, \vec{v}_{k+1})} a_{i_1 \ldots i_k} dx_{i_1} \wedge \cdots \wedge dx_{i_k} \right)$$

$$\overset{\text{part e}}{=} \sum_{1 \le i_1 < \cdots < i_k \le n} (da_{i_1 \ldots i_k} \wedge dx_{i_1} \wedge \cdots \wedge dk_{i_k}) \Big(P_{\mathbf{x}}^o(\vec{v}_1, \ldots, \vec{v}_{k+1}) \Big).$$

This proves part (a); in particular, the limit in the second line exists because the limit in the third line exists, by part (e). Part (b) is now clear, and (c) follows immediately from (e) and (a). $\square$

Exercise A20.2 asks you to prove this. It is an application of Theorem 6.7.3, and part (e) of Theorem 6.7.3 is a special case of Theorem A20.2, the case where φ is a function and ψ is elementary.

The following result is one more basic building stone in the theory of the exterior derivative. It says that the exterior derivative of a wedge product satisfies an analog of Leibnitz's rule for differentiating products. There is a sign that comes in to complicate matters.

Theorem A20.2 (Exterior derivative of wedge product). *If φ is a k-form and ψ is an l-form, then*

$$d(\varphi \wedge \psi) = d\varphi \wedge \psi + (-1)^k \varphi \wedge d\psi.$$

A.21   THE PULLBACK

To prove Stokes's theorem, we will need a new notion: the *pullback* of form fields.

Pullbacks and the Exterior Derivative

The pullback describes how integrands transform under changes of variables. It has been used implicitly throughout Chapter 6, and indeed underlies the change of variables formula for integrals both in elementary calculus, and as developed in Section 4.10. When you write: "let $x = f(u)$, so that $dx = f'(u)\, du$," you are computing a pullback, $f^* dx = f'(u)\, du$. Forms were largely invented to keep track of such changes of variables in multiple integrals, so the pullback plays a central role in the subject. In this appendix we will give a bare bones treatment of the pullback; the central result is Theorem A21.8.

The pullback by a linear transformation

We will begin by the simplest case, pullbacks of forms by linear transformations.

$T^*\varphi$ is pronounced "T upper star phi."

Note that V and W do not need to have the same dimension.

Definition A21.1 (Pullback by a linear transformation). Let V, W be vector spaces, and $T : V \to W$ be a linear transformation. Then T^* is a linear transformation $A^k(W) \to A^k(V)$, defined as follows: if φ is a k-form on $\mathbb{R}^m$, then

$$T^*\varphi(\vec{v}_1, \ldots, \vec{v}_k) = \varphi\big(T(\vec{v}_1), \ldots, T(\vec{v}_k)\big). \qquad \text{A21.1}$$

The pullback of φ, $T^*\varphi$, acting on k vectors $\vec{v}_1, \ldots, \vec{v}_k$ in the domain of T, gives the same result as φ, acting on the vectors $T(\vec{v}_1), \ldots, T(\vec{v}_k)$ in the range.

Note that the domain and range can be of different dimensions: $T^*\varphi$ is a k-form on V, while φ is on W. But both forms must have the same degree: they both act on the same number of vectors.

It is an immediate consequence of Definition A21.1 that $T^* : A^k(W) \to A^k(V)$ is linear:

$$T^*(\varphi_1 + \varphi_2) = T^*\varphi_1 + T^*\varphi_2 \quad \text{and} \quad T^*(a\varphi) = aT^*\varphi, \qquad A21.2$$

as you are asked to show in Exercise A21.1.

The following proposition and the linearity of T^* give a cumbersome but straightforward way of computing the pullback of any form by a linear transformation $T : \mathbb{R}^n \to \mathbb{R}^m$.

Determinants of *minors*, i.e., of square submatrices of matrices, occur in many settings. The real meaning of this construction is given by Proposition A21.2.

Proposition A21.2 (Computing the pullback by a linear transformation). *Let $T : \mathbb{R}^n \to \mathbb{R}^m$ be a linear transformation, and denote by $x_1, \ldots, x_n$ the coordinates in $\mathbb{R}^n$ and by $y_1, \ldots, y_m$ the coordinates in $\mathbb{R}^m$. Then*

$$T^* dy_{i_1} \wedge \cdots \wedge dy_{i_k} = \sum_{1 \leq j_1 < \cdots < j_k \leq n} b_{j_1,\ldots,j_k} dx_{j_1} \wedge \cdots \wedge dx_{j_k}, \qquad A21.3$$

where $b_{j_1,\ldots,j_k}$ is the number obtained by taking the matrix of T, selecting its rows $i_1, \ldots, i_k$ in that order, and its columns $j_1, \ldots, j_k$, and taking the determinant of the resulting matrix.

Example A21.3 (Computing the pullback). Let $T : \mathbb{R}^4 \to \mathbb{R}^3$ be the linear transformation given by the matrix

$$[T] = \begin{bmatrix} 1 & 0 & 0 & 1 \\ 0 & 1 & 0 & 1 \\ 0 & 0 & 1 & 1 \end{bmatrix}. \qquad A21.4$$

then

$$T^* dy_2 \wedge dy_3 = b_{1,2} dx_1 \wedge dx_2 + b_{1,3} dx_1 \wedge dx_3 + b_{1,4} dx_1 \wedge dx_4 + b_{2,3} dx_2 \wedge dx_3$$
$$+ b_{2,4} dx_2 \wedge dx_4 + b_{3,4} dx_3 \wedge dx_4, \qquad A21.5$$

Since we are computing the pullback $T^* dy_2 \wedge dy_3$, we take the second and third rows of T, and then select out columns 1 and 2 for $b_{1,2}$, columns 1 and 3 for $b_{1,3}$, and so on.

where

$$b_{1,2} = \det \begin{bmatrix} 0 & 1 \\ 0 & 0 \end{bmatrix} = 0, \quad b_{1,3} = \det \begin{bmatrix} 0 & 0 \\ 0 & 1 \end{bmatrix} = 0, \quad b_{1,4} = \det \begin{bmatrix} 0 & 1 \\ 0 & 1 \end{bmatrix} = 0,$$

$$b_{2,3} = \det \begin{bmatrix} 1 & 0 \\ 0 & 1 \end{bmatrix} = 1, \quad b_{2,4} = \det \begin{bmatrix} 1 & 1 \\ 0 & 1 \end{bmatrix} = 1, \quad b_{3,4} = \det \begin{bmatrix} 0 & 1 \\ 1 & 1 \end{bmatrix} = -1.$$
$$A21.6$$

So

$$T^* dy_2 \wedge dy_3 = dx_2 \wedge dx_3 + dx_2 \wedge dx_4 - dx_3 \wedge dx_4. \qquad A21.7$$

Proof. Since any k-form on $\mathbb{R}^n$ is of the form

$$\sum_{1 \le j_1 < \cdots < j_k \le n} b_{j_1,\ldots,j_k} dx_{j_1} \wedge \cdots \wedge dx_{j_k}, \qquad A21.8$$

the only problem is to compute the coefficients. This is very analogous to Equation 6.2.20 in the proof of Theorem 6.2.7:

$$\begin{aligned} b_{j_1,\ldots,j_k} &= (T^* dy_{i_1} \wedge \cdots \wedge dy_{i_k})(\vec{\mathbf{e}}_{j_1}, \ldots, \vec{\mathbf{e}}_{j_k}) \\ &= (dy_{i_1} \wedge \cdots \wedge dy_{i_k})\big(T(\vec{\mathbf{e}}_{j_1}), \ldots, T(\vec{\mathbf{e}}_{j_k})\big). \end{aligned} \qquad A21.9$$

This is what we needed: $dy_{i_1} \wedge \cdots \wedge dy_{i_k}$ selects the corresponding lines from the matrix $[(T(\vec{\mathbf{e}}_{j_1}), \ldots, T(\vec{\mathbf{e}}_{j_k})]$, but this is precisely the matrix made up of the columns $j_1, \ldots, j_k$ of $[T]$. $\square$

Pullback of a k-form field by a C^1 mapping

If $U \subset \mathbb{R}^n$, $V \subset \mathbb{R}^m$ are open subsets, and $\mathbf{f} : U \to V$ is a C^1 mapping, then we can use $\mathbf{f}$ to pull back k-form fields on V to k-form fields on U. The definition is similar to Definition A21.1, except that we must replace $\mathbf{f}$ by its derivative.

In the case where φ is a a 0-form, i.e., a function g, then

$$\mathbf{f}^* g = g \circ \mathbf{f},$$

since

$$\begin{aligned} \mathbf{f}^* g(P_{\mathbf{x}}) &= g(P_{\mathbf{f}(\mathbf{x})} = g(\mathbf{f}(\mathbf{x})) \\ &= g \circ \mathbf{f}(P_{\mathbf{x}}). \end{aligned}$$

Definition A21.4 (Pullback by a C^1 mapping). If φ is a k-form field on V, and $\mathbf{f} : U \to V$ is a C^1 mapping, then $\mathbf{f}^*\varphi$ is the k-form field on U defined by

$$(\mathbf{f}^*\varphi)\Big(P_{\mathbf{x}}(\vec{\mathbf{v}}_1, \ldots, \vec{\mathbf{v}}_k)\Big) = \varphi\Big(P_{\mathbf{f}(\mathbf{x})}([\mathbf{Df}(\mathbf{x})]\vec{\mathbf{v}}_1, \ldots, [\mathbf{Df}(\mathbf{x})]\vec{\mathbf{v}}_k)\Big). \quad A21.10$$

If $k = n$, so that $\mathbf{f}(U)$ can be viewed as a parametrized domain, then our definition of the integral over a parametrized domain, Equation 6.3.8, is

$$\int_{\mathbf{f}(U)} \varphi = \int_U \mathbf{f}^*\varphi. \qquad A21.11$$

Thus we have been using pullbacks throughout Chapter 6.

Example A21.5 (Pullback by a C^1 mapping). Let $\mathbf{f} : \mathbb{R}^2 \to \mathbb{R}^3$ be given by

$$\mathbf{f}\begin{pmatrix} x_1 \\ x_2 \end{pmatrix} = \begin{pmatrix} x_1^2 \\ x_1 x_2 \\ x_2^2 \end{pmatrix}. \qquad A21.12$$

We will compute $\mathbf{f}^*(y_2\, dy_1 \wedge dy_3)$. Certainly

$$\mathbf{f}^*(y_2\ dy_1 \wedge dy_3) = b\ dx_1 \wedge dx_2 \qquad A21.13$$

for some function b, and the object is to compute that function:

Note that if U were a bounded subset of $\mathbb{R}^2$, then Equation 6.3.8 says exactly, and by the same computation, that

$$\int_{\mathbf{f}(U)} y_2\, dy_1 \wedge dy_3$$
$$= \int_U 4x_1^2 x_2^2 \, |dx_1 \, dx_2|.$$

$$b\begin{pmatrix} x_1 \\ x_2 \end{pmatrix} = \mathbf{f}^*(y_2\, dy_1 \wedge dy_3)\left(P_{\begin{pmatrix} x_1 \\ x_2 \end{pmatrix}}\left(\begin{bmatrix} 1 \\ 0 \end{bmatrix}, \begin{bmatrix} 0 \\ 1 \end{bmatrix}\right)\right)$$

$$= (y_2\, dy_1 \wedge dy_3)\left(P_{\begin{pmatrix} x_1^2 \\ x_1 x_2 \\ x_2^2 \end{pmatrix}}\left(\begin{bmatrix} 2x_1 \\ x_2 \\ 0 \end{bmatrix}, \begin{bmatrix} 0 \\ x_1 \\ 2x_2 \end{bmatrix}\right)\right) \qquad A21.14$$

$$= x_1 x_2 \det\begin{bmatrix} 2x_1 & 0 \\ 0 & 2x_2 \end{bmatrix} = 4x_1^2 x_2^2.$$

So
$$\mathbf{f}^*(y_2\, dy_1 \wedge dy_3) = 4x_1^2 x_2^2 \, dx_1 \wedge dx_2. \quad \triangle \qquad A21.15$$

Pullbacks and compositions

To prove Stokes's theorem, we will need to compute with pullbacks. One thing we will need to know is how pullbacks behave under composition. If S and T are linear transformations, then

The pullback behaves nicely under composition: $(S \circ T)^* = T^* S^*$.

$$(S \circ T)^* \varphi(\vec{\mathbf{v}}_1, \ldots, \vec{\mathbf{v}}_k) = \varphi\big((S \circ T)(\vec{\mathbf{v}}_1), \ldots, (S \circ T)(\vec{\mathbf{v}}_k)\big)$$
$$= S^* \varphi\big(T(\vec{\mathbf{v}}_1), \ldots, T(\vec{\mathbf{v}}_k)\big) \qquad A21.16$$
$$= T^* S^* \varphi(\vec{\mathbf{v}}_1, \ldots, \vec{\mathbf{v}}_k).$$

Thus $(S \circ T)^* = T^* S^*$. The same formula holds for pullbacks of form fields by C^1 mappings, which should not be surprising in view of the chain rule.

> **Proposition A21.6 (Compositions and pullbacks by nonlinear maps).** If $U \subset \mathbb{R}^n, V \subset \mathbb{R}^m$, and $W \subset \mathbb{R}^p$ are open, $\mathbf{f}: U \to V$, $\mathbf{g}: V \to W$ are C^1 mappings, and φ is a k-form on W, then
>
> $$(\mathbf{g} \circ \mathbf{f})^* \varphi = \mathbf{f}^* \mathbf{g}^* \varphi. \qquad A21.17$$

The first, third, and fourth equalities in Equation A21.18 are the definition of the pullback for $\mathbf{g} \circ \mathbf{f}$, $\mathbf{g}$ and $\mathbf{f}$ respectively; the second equality is the chain rule.

Proof. This follows from the chain rule:

$$(\mathbf{g} \circ \mathbf{f})^* \varphi\big(P_{\mathbf{x}}^o(\vec{\mathbf{v}}_1, \ldots, \vec{\mathbf{v}}_k)\big) = \varphi\Big(P_{(\mathbf{g}(\mathbf{f}(\mathbf{x})))}^o\big([\mathbf{D}(\mathbf{g} \circ \mathbf{f})(\mathbf{x})]\vec{\mathbf{v}}_1, \ldots, [\mathbf{D}(\mathbf{g} \circ \mathbf{f})(\mathbf{x})]\vec{\mathbf{v}}_k\big)\Big)$$

$$= \varphi\Big(P_{(\mathbf{g}(\mathbf{f}(\mathbf{x})))}^o\big([\mathbf{D}\mathbf{g}(\mathbf{f}(\mathbf{x}))][\mathbf{D}\mathbf{f}(\mathbf{x})]\vec{\mathbf{v}}_1, \ldots, [\mathbf{D}\mathbf{g}(\mathbf{f}(\mathbf{x}))][\mathbf{D}\mathbf{f}(\mathbf{x})]\vec{\mathbf{v}}_k\big)\Big)$$

$$= \mathbf{g}^* \varphi\Big(P_{\mathbf{f}(\mathbf{x})}^o\big([\mathbf{D}\mathbf{f}(\mathbf{x})]\vec{\mathbf{v}}_1, \ldots, [\mathbf{D}\mathbf{f}(\mathbf{x})]\vec{\mathbf{v}}_k\big)\Big)$$

$$= \mathbf{g}^* \mathbf{f}^* \varphi\big(P_{\mathbf{x}}^o(\vec{\mathbf{v}}_1, \ldots, \vec{\mathbf{v}}_k)\big). \quad \square \qquad A21.18$$

The pullback and wedge products

We will need to know how pullbacks are related to wedge products. The formula one might hope for is true.

> **Proposition A21.7 (Pullback and wedge products).** *If $U \subset \mathbb{R}^n$ and $V \subset \mathbb{R}^m$ are open subsets, $\mathbf{f} : U \to V$ is a C^1 mapping, and φ and ψ are respectively a k-form and an l-form on V, then*
>
> $$\mathbf{f}^*\varphi \wedge \mathbf{f}^*\psi = \mathbf{f}^*(\varphi \wedge \psi). \qquad \text{A21.19}$$

Proof. This is one of those proofs where you write down the definitions and follow your nose. Let us spell it out when $\mathbf{f} = T$ is linear; we will leave the general case as Exercise A21.2. Recall that the wedge product is a certain sum over all permutations σ of $\{1, \ldots, k+l\}$ such that $\sigma(1) < \cdots < \sigma(k)$ and $\sigma(k+1) < \cdots < \sigma(k+l)$; as in Definition 6.2.13, these permutations are denoted $\text{Perm}(k, l)$. We find

$$T^*(\varphi \wedge \psi)(\vec{\mathbf{v}}_1, \ldots, \vec{\mathbf{v}}_{k+l}) = (\varphi \wedge \psi)\big(T(\vec{\mathbf{v}}_1), \ldots, T(\vec{\mathbf{v}}_{k+l})\big)$$

$$= \sum_{\sigma \in \text{Perm}(k,l)} \text{sgn}(\sigma)\varphi\Big(T(\vec{\mathbf{v}}_{\sigma(1)}), \ldots, T(\vec{\mathbf{v}}_{\sigma(k)})\Big)\psi\Big(T(\vec{\mathbf{v}}_{\sigma(k+1)}), \ldots, T(\vec{\mathbf{v}}_{\sigma(k+l)})\Big)$$

$$= \sum_{\sigma \in \text{Perm}(k,l)} \text{sgn}(\sigma)T^*\varphi(\vec{\mathbf{v}}_{\sigma(1)}, \ldots, \vec{\mathbf{v}}_{\sigma(k)}) \, T^*\psi(\vec{\mathbf{v}}_{\sigma(k+1)}, \ldots, \vec{\mathbf{v}}_{\sigma(k+l)})$$

$$= (T^*\varphi \wedge T^*\psi)(\vec{\mathbf{v}}_1, \ldots, \vec{\mathbf{v}}_{k+l}). \quad \square \qquad \text{A21.20}$$

The exterior derivative is intrinsic.

The next theorem has the innocent appearance $d\mathbf{f}^* = \mathbf{f}^*d$. But this formula says something quite deep, and although we could have written the proof of Stokes's theorem without mentioning the pullback, the step which uses this result is very awkward.

Let us try say why this result matters. To define the exterior derivative, we used the parallelograms $P_{\mathbf{x}}^o(\vec{\mathbf{v}}_1, \ldots, \vec{\mathbf{v}}_k)$. For these parallelograms to exist requires the linear structure of $\mathbb{R}^n$: we have to know how to draw straight lines from one point to another.

It turns out that this isn't necessary; if we had used "curved parallelograms" it would have worked as well. This is the real content of Theorem A21.8.

> **Theorem A21.8 (Exterior derivative is intrinsic).** *Let $U \subset \mathbb{R}^n, V \subset \mathbb{R}^m$ be open sets, and $\mathbf{f} : U \to V$ be a C^1 mapping. If φ is a k-form field on V, then the exterior derivative of φ pulled back by $\mathbf{f}$ is the same as the pullback by $\mathbf{f}$ of the exterior derivative of φ:*
>
> $$d\mathbf{f}^*\varphi = \mathbf{f}^*d\varphi.$$

This proof is thoroughly unsatisfactory: it doesn't explain why the result is true. It is quite possible to give a conceptual proof, but this proof is as hard as (and largely a repetition of) the proof of Theorem 6.7.3. That proof is quite difficult, and the present proof really builds on the work we did there.

The $d(\mathbf{f}^* dx_i)$ in the first line of Equation A21.23 becomes the $dd\mathbf{f}^* x_i$ in the second line. This substitution is allowed by induction (it is the case $k = 0$) because x_i is a function. In fact $\mathbf{f}^* x_i = f_i$, the ith component of $\mathbf{f}$. Of course $dd\mathbf{f}^* x_i = 0$ since it is the exterior derivative taken twice.

Proof. We will prove this theorem by induction on k. The case $k = 0$, where $\varphi = g$ is a function, is an application of the chain rule:

$$\mathbf{f}^* dg\big(P_{\mathbf{x}}^o(\vec{\mathbf{v}})\big) = dg\big(P_{\mathbf{f}(\mathbf{x})}^o[\mathbf{Df}(\mathbf{x})]\vec{\mathbf{v}}\big) = [\mathbf{D}g(\mathbf{f}(\mathbf{x}))][\mathbf{D}g(\mathbf{x})]\vec{\mathbf{v}}$$

$$= [\mathbf{D}g \circ \mathbf{f}(\mathbf{x})]\vec{\mathbf{v}} = d(g \circ \mathbf{f})\big(P_{\mathbf{x}}^o(\vec{\mathbf{v}})\big) \qquad \text{A21.21}$$

$$= d(\mathbf{f}^* g)\big(P_{\mathbf{x}}^o(\vec{\mathbf{v}})\big).$$

If $k > 0$, it is enough to prove the result when we can write $\varphi = \psi \wedge dx_i$, where ψ is a $(k-1)$-form. Then

$$\mathbf{f}^* d(\psi \wedge dx_i) \overset{\substack{\text{Theorem} \\ \text{A20.2}}}{=} \mathbf{f}^*\Big(d\psi \wedge dx_i + \overset{\substack{0 \text{ since } ddx_1=0}}{(-1)^{k-1}\psi \wedge ddx_i}\Big)$$

$$\underset{\substack{\text{Prop.} \\ \text{A21.7}}}{=} \mathbf{f}^*(d\psi) \wedge \mathbf{f}^* dx_i \underset{\substack{\text{inductive} \\ \text{hypothesis}}}{=} d(\mathbf{f}^*\psi) \wedge \mathbf{f}^* dx_i, \qquad \text{A21.22}$$

whereas

$$d\mathbf{f}^*(\psi \wedge dx_i) = d\big(\mathbf{f}^*\psi \wedge \mathbf{f}^* dx_i\big)$$

$$= \big(d(\mathbf{f}^*\psi)\big) \wedge \mathbf{f}^* dx_i + (-1)^{k-1}\mathbf{f}^*\psi \wedge d(\mathbf{f}^* dx_i) \qquad \text{A21.23}$$

$$= \big(d(\mathbf{f}^*\psi)\big) \wedge \mathbf{f}^* dx_i + (-1)^{k-1}\mathbf{f}^*\psi \wedge dd\mathbf{f}^* x_i = \big(d(\mathbf{f}^*\psi)\big) \wedge \mathbf{f}^* dx_i. \quad \square$$

A.22   Proof of Stokes's Theorem

The proof of this theorem uses virtually every major theorem contained in this book. Exercise A22.1 asks you to find as many as you can, and explain where they are used.

Proposition A22.1 is a generalization of Proposition 6.9.7; here we allow for corners.

We repeat some of the discussion from Section 6.9, to make this proof self-contained.

Theorem 6.9.2 (Generalized Stokes's theorem). *Let X be a compact piece-with-boundary of a k-dimensional oriented manifold $M \subset \mathbb{R}^n$. Give the boundary ∂X of X the boundary orientation, and let φ be a $(k-1)$-form defined on a neighborhood of X. Then*

$$\int_{\partial X} \varphi = \int_X d\varphi. \qquad 6.9.3$$

A situation where the easy proof works

We will now describe a situation where the "proof" in Section 6.9 really does work. In this simple case, we have a $(k-1)$-form in $\mathbb{R}^k$, and the piece we will integrate over is the first "quadrant." There are no manifolds; nothing curvy.

Proposition A22.1 (Easy case of Stokes's theorem). *Let U be a bounded open subset of $\mathbb{R}^k$, and let U_+ be the part of U in the first quadrant, where $x_1 \geq 0, \ldots, x_k \geq 0$. Orient U by $\det$ on $\mathbb{R}^k$, and give ∂U_+ the boundary orientation. Let φ be a $(k-1)$-form on $\mathbb{R}^k$ of class C^2, which vanishes identically outside U. Then*

$$\int_{\partial U_+} \varphi = \int_{U_+} d\varphi. \qquad A22.1$$

Proof.   Choose $\epsilon > 0$. Recall from Equation A20.15 (in the proof of Theorem 6.7.3 on computing the exterior derivative of a k-form) that there exist a constant K and $\delta > 0$ such that when $|h| < \delta$,

It was in order to get Equation A22.2 that we required φ to be of class C^2, so that the second derivatives of the coefficients of φ have finite maxima (by Theorem 1.6.7).

$$\left| d\varphi\big(P_{\mathbf{x}}(h\vec{e}_1, \ldots, h\vec{e}_k)\big) - \int_{\partial P_{\mathbf{x}}(h\vec{e}_1, \ldots, h\vec{e}_k)} \varphi \right| \leq Kh^{k+1}. \qquad A22.2$$

The constant in Equation A20.15 (there called C, not K), comes from Theorem A9.7 (Taylor's theorem with remainder with explicit bound), and involves the suprema of the second derivatives.

In Equation A20.15 we have $\leq h^{k+2}K$ because there we are computing the exterior derivative of a k-form; here we are computing the exterior derivative of a $(k-1)$-form.

Denote by $\mathbb{R}_+^k$ the "first quadrant," i.e., the subset where all $x_i \geq 0$. Take the dyadic decomposition $\mathcal{D}_N(\mathbb{R}^k)$, where $h = 2^{-N}$. By taking N sufficiently large, we can guarantee that the difference between the integral of $d\varphi$ over U_+ and the Riemann sum is less than $\epsilon/2$:

$$\left| \int_{U_+} d\varphi - \sum_{C \in \mathcal{D}_N(\mathbb{R}_+^k)} d\varphi(C) \right| < \frac{\epsilon}{2}. \qquad A22.3$$

Now we replace the k-parallelograms of Equation A22.2 by dyadic cubes, and evaluate the total difference between the exterior derivative of φ over the cubes C, and φ over the boundaries of the C. The number of cubes of $\mathcal{D}_N(\mathbb{R}^k)$ that intersect the support of φ is at most $L2^{kN}$ for some constant L, and since $h = 2^{-N}$, the bound for each error is now $K2^{-N(k+1)}$, so

The constant L depends on the size of the support of φ. More precisely, it is the side-length of the support, to the kth power.

$$\left| \sum_{C \in \mathcal{D}_N(\mathbb{R}_+^k)} d\varphi(C) - \sum_{C \in \mathcal{D}_N(\mathbb{R}_+^k)} \int_{\partial C} \varphi \right| \leq \underbrace{L2^{kN}}_{\text{No. of cubes}} \underbrace{K2^{-N(k+1)}}_{\substack{\text{bound for} \\ \text{each error}}} = LK2^{-N}. \qquad A22.4$$

This can also be made $< \epsilon/2$ by taking N sufficiently large—to be precise, by taking

Equation A22.5 is unnecessary; arguing that "this can also be made $< \epsilon/2$ by taking N sufficiently large" is adequate.

$$N \geq \frac{\ln 2LK - \ln \epsilon}{\ln 2}. \qquad A22.5$$

Finally, all the internal boundaries in the sum

$$\sum_{C \in \mathcal{D}_N(\mathbb{R}_+^k)} \int_{\partial C} \varphi \qquad A22.6$$

cancel, since each appears twice with opposite orientations. So (using C' to denote cubes of the dyadic composition of $\partial\mathbb{R}_+^k$) we have

$$\sum_{C\in\mathcal{D}_N(\mathbb{R}_+^k)}\int_{\partial C}\varphi = \sum_{C'\in\mathcal{D}_N(\partial\mathbb{R}_+^k)}\int_{C'}\varphi = \int_{\partial\mathbb{R}_+^k}\varphi. \qquad A22.7$$

Putting these inequalities together, we get

The second term of Equation A22.8 is the left-hand side of Equation A22.4, except that now we have $\mathbb{R}_+^k$ rather than all of $\mathbb{R}^k$. By Equation A22.7,

$$\sum_{C\in\mathcal{D}_N(\mathbb{R}_+^k)}\int_{\partial C}\varphi = \int_{\partial\mathbb{R}_+^k}\varphi.$$

Since φ vanishes outside U,

$$\int_{\partial\mathbb{R}_+^k}\varphi = \int_{\partial U_+}\varphi.$$

justifying Equation A22.9.

$$\overbrace{\left|\int_{U_+}d\varphi - \sum_{C\in\mathcal{D}_N(\mathbb{R}_+^k)}d\varphi(C)\right|}^{\leq\epsilon/2 \text{ by Equation A22.3}} + \overbrace{\left|\sum_{C\in\mathcal{D}_N(\mathbb{R}_+^k)}d\varphi(C) - \sum_{C\in\mathcal{D}_N(\mathbb{R}_+^k)}\int_{\partial C}\varphi\right|}^{\leq\epsilon/2} \leq \epsilon,$$

$$A22.8$$

i.e.,

$$\left|\int_{U_+}d\varphi - \int_{\partial U_+}\varphi\right| < \epsilon. \qquad A22.9$$

Since ϵ is arbitrary, the proposition follows. $\square$

Partitions of unity

To prove Stokes's theorem, our tactic will be to reduce it to Proposition A22.1, by covering X with parametrizations that satisfy the requirements. Of course, this will mean cutting up X into pieces that are separately parametrized. This can be done as suggested above, but it is difficult. Rather than hacking X apart, we will use a softer technique: fading one parametrization out as we bring another in. The following lemma allows us to do this.

The sum $\sum_{m=1}^N \alpha_m = 1$ of Equation A22.10 is the sum such that for all $\mathbf{x}\in X$,

$$\sum_{m=1}^N \alpha_m(\mathbf{x}) = 1.$$

It is called a *partition of unity*, because it breaks up 1 into a sum of functions. These functions have the interesting property that they can have small support, which makes it possible to piece together global functions, forms, etc., from local ones. As far as we know, they are exclusively of theoretical use, never used in practice.

Lemma A22.2 (Partitions of unity). *Let X be a piece-with-boundary of k-manifold, and let φ be a $(k-1)$-form on X. If α_m, for $m = 1,\dots,N$, are smooth functions on X such that*

$$\sum_{m=1}^N \alpha_m = 1 \qquad \text{then} \qquad d\varphi = \sum_{m=1}^N d(\alpha_m\varphi). \qquad A22.10$$

Proof. This is an easy but non-obvious computation. The thing *not* to do is to use Theorem A20.2 to write $\sum d(\alpha_m\varphi) = \sum d\alpha_m \wedge \varphi + (-1)^0\sum\alpha_m\,d\varphi$; this leads to an awful mess. Instead take advantage of Equation A22.10 to write

$$\sum_{m=1}^N d(\alpha_m\varphi) = d\left(\left(\sum_{m=1}^N \alpha_m\right)\varphi\right) = d\varphi. \quad\square \qquad A22.11$$

This means that if we can prove Stokes's theorem for the forms $\alpha_m\varphi$, then it will be proved, since if

$$\int_X d(\alpha_m\varphi) = \int_{\partial X}\alpha_m\varphi \qquad A22.12$$

for each $m = 1, \ldots, N$, then

$$\int_X d\varphi = \int_X \Big(\sum_{m=1}^{N} d(\alpha_m \varphi) \Big) = \sum_{m=1}^{N} \int_{\partial X} \alpha_m \varphi = \int_{\partial X} \Big(\sum_{m=1}^{N} \alpha_m \Big) \varphi = \int_{\partial X} \varphi.$$

A22.13

We will choose our α_m so that in addition to the conditions of Equation A22.10, they have their supports in little subsets in which M has the standard form of Definition 6.6.1 (piece-with-boundary of a manifold): i.e., in little subsets where M is the graph of a function. It will be fairly easy to put these individual pieces into a form where Proposition A22.1 can be applied.

Choosing good parametrizations

Below we will need the "bump" function $\beta_R : \mathbb{R}^k \to \mathbb{R}$ given by

$$\beta_R(\mathbf{x}) = \begin{cases} 4 \left(\dfrac{|\mathbf{x}|^2}{R^2} - 1 \right)^4 & \text{if } |\mathbf{x}|^2 \le R^2 \\ 0 & \text{if } |\mathbf{x}|^2 > R^2, \end{cases}$$

A22.14

and shown in Figure A22.1.

Now go back to Definition 6.6.1 of a piece-with-boundary X of a manifold $M \subset \mathbb{R}^n$:

For every $\mathbf{x} \in X$, there exists a ball $U_\mathbf{x}$ of radius $R_\mathbf{x}$ around $\mathbf{x}$ in $\mathbb{R}^n$ such that

- $U_\mathbf{x} \cap M$ is the graph of a mapping $\mathbf{f} : U_1 \to E_2$, where U_1 is an open subset of the subspace E_1 spanned by the k standard basis vectors that, near $\mathbf{x}$, determine the values of the other variables, and E_2 is the subspace spanned by the other $n - k$. Then if we set $\tilde{\mathbf{f}} = \begin{pmatrix} \mathbf{u} \\ \mathbf{f(u)} \end{pmatrix}$, we have $\tilde{\mathbf{f}}(U_1) = M \cap U$.

- There is a diffeomorphism $G : U_1 \to V \subset \mathbb{R}^k$ such that

$$G_i(\mathbf{u}) \ge 0, \ i = 1, \ldots, k \quad \text{if and only if} \quad \tilde{\mathbf{f}}(\mathbf{u}) \in X \cap U. \qquad A22.15$$

In other words, a point in $X \cap U$ is taken by $G_i \circ \tilde{\mathbf{f}}^{-1}$ to a point in the first quadrant of $\mathbb{R}^k$, where all the coordinates are positive.

Since X is compact, the Heine-Borel theorem (Theorem A17.2) says that we can cover it by *finitely* many $U_{\mathbf{x}_1}, \ldots, U_{\mathbf{x}_N}$ satisfying the properties above. We will label the corresponding sets and functions $U^m = U_{\mathbf{x}_m}$, U_1^m, $\mathbf{f}^m$, G^m, E_1^m, V^m, and R_m.

To recapitulate: think of X as having a complicated shape, very likely considerably worse than the example we show in Figure A22.2. We cover this awkward shape with N little balls U^m. We choose these U^m so that for each one, $M \cap U^m$ is the graph of a single function $\mathbf{f}^m$. We project each $M \cap U^m$ onto E_1^m to get U_1^m. Thus $\tilde{\mathbf{f}}^m(U_1^m) = M \cap U^m$, and $\tilde{\mathbf{f}}^m(X_1^m) = X \cap U^m$, as

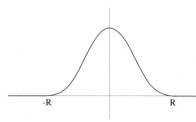

FIGURE A22.1.

Graph of the bump function β_R of Equation A22.14. The power 4 is used in Equation A22.14 to make sure that β_R is of class C^2; in Exercise A22.2 you are asked to show that it is of class C^3 on all of $\mathbb{R}^k$. It evidently vanishes off the ball of radius R and, since $4((1/2)^2 - 1)^4 = 324/256 > 1$, we have $\beta_R(\mathbf{x}) > 1$ when $|\mathbf{x}| \le \frac{R}{2}$. It is not hard to manufacture something analogous of class C^m for any m, and rather harder but still possible to manufacture something analogous of class C^∞. But it is absolutely impossible to do so using functions that are sums of their Taylor series.

This is where the assumption that X is compact is used, and it is absolutely essential.

Since (Definition 3.2.2) E_1 is the span of the k standard basis vectors corresponding to the k variables that, near $\mathbf{x}$, will determine the values of the other variables, different U^m will require different E_1.

shown in Figure A22.2. For each U_1^m there is a function $G^m : U_1^m \to V^m \subset \mathbb{R}^k$, which takes X_1^m to the first quadrant of V^m, denoted V_+^m in the figure. So we have gone from a big, awkward shape to a collection of N little, straight shapes.

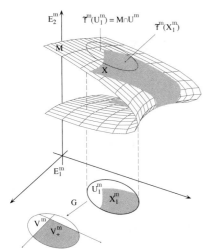

FIGURE A22.2.

Here we have relabeled Figure 6.6.1. Every point in X can be surrounded by a ball that is first projected onto an appropriate E_1^m and then taken by G^m to V^m; points in X are taken to V_+^m, where all coordinates are positive.

We also divide φ up into a sum of forms φ_m with very small support, which we construct with the help of the bump functions: Let $\beta_m : \mathbb{R}^n \to \mathbb{R}$ be the function

$$\beta_m(\mathbf{x}) = \beta_{R_m}(\mathbf{x} - \mathbf{x}_m), \qquad A22.16$$

so that β_m is a C^2 function on $\mathbb{R}^n$ with support in the ball U^m of radius R_m around $\mathbf{x}_m$. Set $\beta(\mathbf{x}) = \sum_{m=1}^N \beta_m$; this corresponds to a finite set of overlapping bump functions, so that we have $\beta(\mathbf{x}) > 0$ on a neighborhood of X. Then the functions

$$\alpha_m(\mathbf{x}) = \frac{\beta_m(\mathbf{x})}{\beta(\mathbf{x})} \qquad A22.17$$

are C^2 on some neighborhood of X. Clearly $\sum_{m=1}^N \alpha_m(\mathbf{x}) = 1$ for all $\mathbf{x} \in X$, so that if we set $\varphi_m = \alpha_m \varphi$, we can write

$$\varphi = \sum_{m=1}^N \alpha_m \varphi = \sum_{m=1}^N \varphi_m. \qquad A22.18$$

Now we can parametrize X using the functions

$$\mathbf{h}_m = \mathbf{f}^m \circ (G^m)^{-1} : V^m \to M,$$

Instead of integrating $d\varphi$ over X, we will integrate each $d\varphi_m$ over the corresponding $X \cap U^m$ and add the results.

and we can pull the forms $\alpha_m \varphi$ back to $\mathbb{R}^k$ using the pullbacks $\mathbf{h}_m^*(\alpha_m \varphi)$. We have satisfied the conditions of Proposition A22.1: each V_+^m satisfies the conditions for U_+ in that proposition, and using all N of the V_+^m we can account for all of X.

Completing the proof of Stokes's theorem

The 1st equality is Equation A22.18. The 2nd says that φ_m has its support in U^m, the 3rd that $\mathbf{h}_m$ parametrizes $U^m \cap X$.

The 4th is the first crucial step, using $d\mathbf{h}_m^* = \mathbf{h}_m^* d$, i.e., Theorem A21.8.

The 5th, which is also a crucial step, is Proposition A22.1.

Like the 3th, the 6th is that $\mathbf{h}_m$ parametrizes U^m, and for the 7th we once more use $\sum \alpha_m = 1$.

The proof of Stokes's theorem now consists of the following sequence of equalities:

$$\int_X d\varphi \overset{1}{=} \int_X \sum_{m=1}^N d\varphi_m \overset{2}{=} \sum_{m=1}^N \int_{X \cap U^m} d\varphi_m \overset{3}{=} \sum_{m=1}^N \int_{V_+^m} \mathbf{h}_m^*(d\varphi_m)$$

$$\overset{4}{=} \sum_{m=1}^N \int_{V_+^m} d(\mathbf{h}_m^* \varphi_m) \overset{5}{=} \sum_{m=1}^N \int_{\partial V_+^m} \mathbf{h}_m^*(\varphi_m)$$

$$\overset{6}{=} \sum_{m=1}^N \int_{\partial X} \varphi_m \overset{7}{=} \int_{\partial X} \varphi. \qquad A22.19$$

A.23 Exercises for the Appendix

A5.1   Using the notation of Theorem 2.9.10, show that the implicit function found by setting $\mathbf{g}(\mathbf{y}) = \mathbf{G}\begin{pmatrix} \mathbf{0} \\ \mathbf{y} \end{pmatrix}$ is the *unique* continuous function defined on $B_R(\mathbf{b})$ satisfying

$$\mathbf{F}\begin{pmatrix} \mathbf{g}(\mathbf{y}) \\ \mathbf{y} \end{pmatrix} = 0 \quad \text{and} \quad \mathbf{g}(\mathbf{b}) = \mathbf{a}.$$

A7.1   In the proof of Proposition 3.3.19, we start the induction at $k = 1$. Show that you could start the induction at $k = 0$ and that if you do so, Proposition 3.3.19 contains Theorem 1.9.5 as a special case.

A8.1   (a) Show that Proposition 3.4.4 (chain rule for Taylor polynomials) contains the chain rule as a special case, as long as the mappings are continuously differentiable.

(b) Go back to Appendix A1 (proof of the chain rule) and show how o and O notation can be used to shorten the proof.

A9.1   Let $f\begin{pmatrix} x \\ y \end{pmatrix} = e^{\sin(x+y^2)}$. Use Maple, Mathematica, or similar software.

(a) Calculate the Taylor polynomial $P^k_{f,\mathbf{a}}$ of degree $k = 1, 2$ and 4 at $\mathbf{a} = \begin{pmatrix} 1 \\ 1 \end{pmatrix}$.

(b) Estimate the maximum error $|P^k_{f,\mathbf{a}} - f|$ on the region $|x - 1| < .5$ and $|y - 1| < .5$, for $k = 1$ and $k = 2$.

(c) Similarly, estimate the maximum error in the region $|x - 1| < .25$ and $|y - 1| < .25$, for $k = 1$ and $k = 2$.

*A9.2   (a) Write the integral form of the remainder when $\sin(xy)$ is approximated by its Taylor polynomial of degree 2 at the origin.

(b) Give an upper bound for the remainder when $x^2 + y^2 \le 1/4$.

A9.3   Prove Equation A9.2 by induction, by first checking that when $k = 0$, it is the fundamental theorem of calculus, and using integration by parts to prove

$$\frac{1}{k!} \int_0^h (h - t)^k g^{(k+1)}(a + t)\, dt = \frac{1}{(k+1)!} g^{(k+1)}(a) h^{k+1} +$$

$$\frac{1}{(k+1)!} \int_0^h (h - t)^{k+1} g^{(k+2)}(a + t)\, dt.$$

A12.1   This exercise sketches another way to find the constant in Stirling's formula. We will show that if there is a constant C such that

$$n! = C\sqrt{n} \left(\frac{n}{e}\right)^n (1 + o(1)),$$

as is proved in Theorem A12.1, then $C = \sqrt{2\pi}$. The argument is fairly elementary, but not at all obvious. Let $c_n = \int_0^\pi \sin^n x\, dx$.

(a) Show that $c_n < c_{n-1}$ for all $n = 1, 2, \dots$.

(b) Show that for $n \geq 2$, we have $c_n = \frac{n-1}{n} c_{n-2}$. Hint: write $\sin^n x = \sin x \sin^{n-1} x$ and integrate by parts.

(c) Show that $c_0 = \pi$ and $c_1 = 2$, and use this and part (b) to show that

$$c_{2n} = \frac{2n-1}{2n} \cdot \frac{2n-3}{2n-2} \cdots \frac{1}{2} \pi = \frac{(2n)!\pi}{2^{2n}(n!)^2}$$

$$c_{2n+1} = \frac{2n}{2n+1} \cdot \frac{2n-2}{2n-1} \cdots \frac{2}{3} 2 = \frac{2^{2n}(n!)^2 2}{(2n+1)!}.$$

(d) Use Stirling's formula with constant C to show that

$$c_{2n} = \frac{1}{C} \sqrt{\frac{2}{n}}\, \pi \big(1 + o(1)\big)$$

$$c_{2n+1} = \frac{C}{\sqrt{2n+1}} \big(1 + o(1)\big).$$

Now use part (a) to show that $C^2 \leq 2\pi + o(1)$ and $C^2 \geq 2\pi + o(1)$.

A18.1   Show that there exists a continuous function $f : \mathbb{R} \to \mathbb{R}$, bounded with bounded support (and in particular integrable), such that the set

$$\{x \in \mathbb{R} \mid f(x) \geq 0\}$$

is not pavable. For instance, follow the following steps.

(a) Show that if $X \subset \mathbb{R}^n$ is any non-empty subset, then the function

$$f_X(\mathbf{x}) = \inf_{\mathbf{y} \in X} |\mathbf{x} - \mathbf{y}|$$

is continuous. Show that $f_X(\mathbf{x}) = 0$ if and only if $\mathbf{x} \in \overline{X}$.

(b) Take any non-pavable closed subset $X \subset [0, 1]$, such as the complement of the set U_ϵ that is constructed in Example 4.4.2, and let $X' = X \cup \{0, 1\}$. Set

$$f(x) = -\chi_{[0,1]}(x) f_{X'}(x).$$

Show that this function f satisfies our requirements.

A18.2   Make a list $a_1, a_2, \dots$ of the rationals in $[0, 1]$. Consider the function f_k such that

$f_k(x) = 0$ if $x \notin [0, 1]$, or if $x \in \{a_1, \dots, a_k\}$;
$f_k(x) = 1$ if $x \in [0, 1]$ and $x \notin \{a_1, \dots, a_k\}$.

Show that all the f_k are integrable, and that $f(x) = \lim_{k \to \infty} f_k(x)$ exists for every x, but that f is not integrable.

A18.3   Show that if f and g are any integrable functions on $\mathbb{R}^n$, then

$$\left(\int_{\mathbb{R}^n} f(\mathbf{x}) g(\mathbf{x}) |d^n \mathbf{x}| \right)^2 \leq \left(\int_{\mathbb{R}^n} (f(\mathbf{x}))^2 |d^n \mathbf{x}| \right) \left(\int_{\mathbb{R}^n} (g(\mathbf{x}))^2 |d^n \mathbf{x}| \right).$$

Hint: follow the proof of Schwarz's inequality (Theorem 1.4.6). Consider the quadratic polynomial

$$\int_{\mathbb{R}^n} ((f+tg)(\mathbf{x}))^2 |d^n\mathbf{x}| = \int_{\mathbb{R}^n} (f(\mathbf{x}))^2 |d^n\mathbf{x}| + 2t \int_{\mathbb{R}^n} f(\mathbf{x})g(\mathbf{x})|d^n\mathbf{x}| + t^2 \int_{\mathbb{R}^n} (g(\mathbf{x}))^2 |d^n\mathbf{x}|.$$

Since the polynomial is ≥ 0, its discriminant is non-positive.

A20.1   Show that the last equality of Equation A20.7 is "by the definition of the wedge product."

A20.2   Prove Theorem A20.2 concerning the derivative of wedge product:
(a) Show it for 0-forms, i.e.,

$$d(fg) = f\, dg + g\, df.$$

(b) Show that it is enough to prove the theorem when

$$\varphi = a(\mathbf{x})\, dx_{i_1} \wedge \cdots \wedge dx_{i_k};$$
$$\psi = b(\mathbf{x})\, dx_{j_1} \wedge \cdots \wedge dx_{j_l}.$$

(c) Prove the case in (b), using that

$$\varphi \wedge \psi = a(\mathbf{x})b(\mathbf{x})\, dx_{i_1} \wedge \cdots \wedge dx_{i_k} \wedge dx_{j_1} \wedge \cdots \wedge dx_{j_l}.$$

A21.1   (a) Show that if $T : V \to W$ is a linear transformation, the pullback $T^* : A^k(W) \to A^k(V)$ is linear.
(b) Now show that the pullback by a C^1 mapping is linear.

A21.2   In the text we proved Proposition A21.7 in the special case where the mapping is linear. Prove the general statement, where the mapping $\mathbf{f}$ is only assumed to be of class C^1.

A22.1   Identify the theorems used to prove Theorem 6.9.2, and show how they are used.

A22.2   Show (proof of Lemma A22.2) that β_R is of class C^3 on all of $\mathbb{R}^k$.

Appendix B: Programs

The programs given in this appendix can also be found at the web page
http://www.math.cornell.edu/~hubbard/vectorcalculus.html

B.1 MATLAB NEWTON PROGRAM

This program can be typed into the MATLAB window, and saved as an m-file called "Newton.m". It was created by a Cornell undergraduate, Jon Rosenberger. For explanations as to how to use it, see below.

The program evaluates the Jacobian matrix (derivative of the function) symbolically, using the link of MATLAB to MAPLE.

```
function[x] = newton(F, x0, iterations)
 vars = '[';
for i = 1:length(F)
  iS = num2str(i);
  vars = [vars 'x' iS ' '];
  eval(['x' iS ' = sym(''x' iS ''');']); % declare xn to be symbolic
end
 vars = [vars ']'];
 eval(['vars = ' vars ';']);
 J = jacobian(F, vars);
x = x0;
for i = 1:iterations
  JJ = double(subs(J, vars, x.'));
  FF = double(subs(F, vars, x.'));
  x = x - inv(JJ) * FF
 end
```

The following two lines give an example of how to use this program.

```
EDU>syms x1 x2
EDU>newton([cos(x1)-x1; sin(x2)], [.1; 3.0], 3)
```

The semicolons separating the entries in the first square brackets means that they are column vectors; this is MATLAB's convention for writing column vectors.

Use * to indicate multiplication, and ^ for power ; if $f_1 = x_1 x_2^2 - 1$, and $f_2 = x_2 - \cos x_1$, the first entry would be
[x1 * x2^2 -1; x2-cos(x1)].

The first lists the variables; they must be called $x1, x2, \ldots, xn$; n may be whatever you like. Do not separate by commas; if $n = 3$ write x1 x2 x3.

The second line contains the word newton and then various terms within parentheses. These are the arguments of the function Newton. The first argument, within the first square brackets, is the list of the functions f_1 up to f_n that you are trying to set to 0. Of necessity this n is the same n as for line one. Each f is a function of the n variables, or some subset of the n variables. The second entry, in the second square brackets, is the point at which to start Newton's method; in this example, $\begin{pmatrix} .0 \\ 3 \end{pmatrix}$. The third entry is the number of times to iterate. It is not in brackets. The three entries are separated by commas.

B.2 Monte Carlo Program

Like the determinant program in Appendix B.3, this program requires a Pascal compiler.

```
program montecarlo;

const lengthofrun = 100000;

var S,V,x,intguess,varguess,stddev,squarerootlerun,
    errfifty,errninety,errninetyfive:longreal; i,seed,answer:longint;
```

The nine lines beginning **function Rand** and ending **end** are a random number generator.

```
function Rand(var Seed:  longint):  real;
{Generate pseudo random number between 0 and 1}
const Modulus = 65536;
   Multiplier = 25173;
   Increment = 13849;
begin
   Seed := ((Multiplier * Seed) + Increment) mod Modulus;
   Rand := Seed / Modulus
end;
```

The six lines beginning **function randomfunction** and ending **end** define a random function that gives the absolute value of the determinant of a 2×2 matrix $\begin{bmatrix} a & b \\ c & d \end{bmatrix}$. For an $n \times n$ matrix you would enter n^2 "seeds." You can name them what you like; if $n = 3$, you could call them $x1, x2, \ldots, x9$, instead of $a, b, \ldots, i$. In that case you would write `x1:=rand(seed); x2:=rand(seed)` and so on. To define the random function you would use the formula

$$\det \begin{bmatrix} a_1 & b_1 & c_1 \\ a_2 & b_2 & c_2 \\ a_3 & b_3 & c_3 \end{bmatrix} =$$

$$a_1(b_2 c_3 - b_3 c_2) - a_2(b_1 c_3 - b_3 c_1)$$
$$+ a_3(b_1 c_2 - b_2 c_1).$$

```
function randomfunction:real;
var a,b,c,d:real;
begin
   a:=rand(seed);b:=rand(seed);c:=rand(seed);d:=rand(seed);
   randomfunction:=abs(a*d-b*c);
end;

begin
Seed := 21877;
repeat
   S:=0;V:=0;
   for i:=1 to lengthofrun do
   begin
      x:=randomfunction;
      S:=S+x;V:=V+sqr(x);
   end;
   intguess:=S/lengthofrun; varguess:= V/lengthofrun-sqr(intguess);
   stddev:= sqrt(varguess);
   squarerootlerun:=sqrt(lengthofrun);
   errfifty:= 0.6745*stddev/squarerootlerun;
   errninety:= 1.645*stddev/squarerootlerun;
   errninetyfive:= 1.960*stddev/squarerootlerun;
```

```
writeln('average for this run = ',intguess);
writeln('estimated standard deviation = ',stddev);
writeln('with probability 50',errfifty);
writeln('with probability 90',errninety);
writeln('with probability 95',errninetyfive);
writeln('another run? 1 with new seed, 2 without');
readln(answer);
if (answer=1) then
begin
   writeln('enter a new seed, which should be an integer');
   readln(seed);
end;
until (answer=0);

end.
```

Another example of using the Monte Carlo program:

In Pascal, x^2 is written `sqr(x)`.

To compute the area inside the unit square and above the parabola $y = x^2$, you would type

```
function randomfunction:real;
var x,y:real;
begin
  x:=rand(seed);y:=rand(seed);
  if (y-sqr(x) <0 ) then randomfunction:=0
  else randomfunction:=1;
end
```

B.3 Determinant Program

Program determinant;

Const maxsize = 10;

Type matrix = **record**

        size:integer;

        coeffs: **array**[1..maxsize, 1..maxsize] of real;

    end;

  submatrix =**record**

        size:**integer**;

        rows,cols:**array**[1..maxsize] of **integer**;

    end;

Var   M: Matrix;

       S: submatrix;

       d: real;

Function det(S:submatrix):real;

  Var   tempdet: real;

          i,sign: **integer**;

          S1: submatrix;

   Procedure erase(S:submatrix; i,j: integer; var S1:submatrix);

         Var k:integer;

         begin {erase}

            S1.size:=S.size-1;

            for k:=S.size-1 **downto** i do S1.cols[k]:=S.cols[k+1];

            for k:=i-1 **downto** 1 **do** S1.cols[k]:=S.cols[k];

            for k:=S.size-1 **downto** j **do** S1.rows[k]:=S.rows[k+1];

            for k:=j-1 **downto** 1 **do** S1.rows[k]:=S.rows[k];

         end;

begin{function det}

    If S.size = 1 **then** det := M.coeffs[S.rows[1],S.col[1]]

    else begin

      tempdet := 0; sign := 1;

      for i := 1 to S.size **do**

      begin

        erase(S,i,1,S1);

        tempdet := tempdet + sign*M.coeffs[S.rows[1],S.cols[i]]*det(S1);

        sign := -sign;

      end;

      det := tempdet;

    end;

end;

```
begin{function det}
      If S.size = 1 then det := M.coeffs[S.rows[1],S.col[1]]
      else begin
         tempdet := 0; sign := 1;
         for i := 1 to S.size do
         begin
            erase(S,i,1,S1);
            tempdet := tempdet + sign*M.coeffs[S.rows[1],S.cols[i]]*det(S1);
            sign := -sign;
         end;
         det := tempdet;
      end;
end;
Procedure   InitSubmatrix (Var S:submatrix);
   Var      k:integer;
   begin
            S.size := M.size;
            for k := 1 to S.size do begin S.rows[k] := k; S.cols[k] := k end;
   end;
Procedure   InitMatrix;
   begin    {define M.size and M.coeffs any way you like} end;

Begin       {main program}
            InitMatrix;
            InitSubmatrix(S);
            d := det(S);
            writeln('determinant = ',d);
end.
```

Bibliography

Calculus and Forms

Henri Cartan, *Differential Forms*, Hermann, Paris; H. Mifflin Co., Boston, 1970.

Jean Dieudonné, *Infinitesimal Calculus*, Houghton Mifflin Company, Boston, 1971. Aimed at a junior-level advanced calculus course.

Stanley Grossman, *Multivariable Calculus, Linear Algebra and Differential Equations*, Harcourt Brace College Publishers, Fort Worth, 1995.

Lynn Loomis and Shlomo Sternberg, *Advanced Calculus*, Addison-Wesley Publishing Company, Reading, MA, 1968.

Jerrold Marsden and Anthony Tromba, *Vector Calculus*, fourth edition, W. H. Freeman and Company, New York, 1996.

Michael Spivak, *Calculus*, second edition, Publish or Perish, Inc., Wilmington, DE, 1980 (one variable calculus).

Michael Spivak, *Calculus on Manifolds*, W. A. Benjamin, Inc., New York, 1965; now available from Publish or Perish, Inc., Wilmington, DE.

John Milnor, *Morse Theory*, Princeton University Press, 1963. (Lemma 2.2, p. 6, is referred to in Section 3.6.)

Differential Equations

What it Means to Understand a Differential Equation, The College Mathematics Journal, Vol. 25, Nov. 5 (Nov. 1994), 372-384.

John Hubbard and Beverly West, *Differential Equations, A Dynamical Systems Approach, Part I*, Texts in Applied Mathematics No. 5, Springer-Verlag, N.Y., 1991.

Differential Geometry

Frank Morgan, *Riemannian Geometry, A Beginner's Guide*, second edition, AK Peters, Ltd., Wellesley, MA, 1998.

Manfredo P. do Carmo, *Differential Geometry of Curves and Surfaces*, Prentice-Hall, Inc., 1976.

Fractals and Chaos

John Hubbard, *The Beauty and Complexity of the Mandelbrot Set*, Science TV, N.Y., Aug. 1989.

History

John Stillwell, *Mathematics and Its History*, Springer-Verlag, New York, 1989.

Linear Algebra

Jean-Luc Dorier, ed., *L'enseignement de l'Algèbre Linéaire en Question*, La Pensée Sauvage, Editions, 1997.

Index